HARVESTING THE SEA

UNVEILING MARINE BIORESOURCES FOR FUNCTIONAL FOODS AND HUMAN HEALTH ENHANCEMENT

HARVESTING THE SEA

UNVEILING MARINE BIORESOURCES FOR FUNCTIONAL FOODS AND HUMAN HEALTH ENHANCEMENT

Edited by

THILINA JAYAWARDENA
Department of Medicine, University of Alberta, Edmonton, AB, Canada

ASANKA SANJEEWA
Department of Biosystems Technology, Faculty of Technology, University of Sri Jayewardenepura, Homagama, Sri Lanka

K.H.I.N.M. HERATH
Department of Biosystems Engineering, Faculty of Agriculture and Plantation Management, Wayamba University of Sri Lanka, Makandura, Gonawila, Sri Lanka

ELSEVIER

ACADEMIC PRESS
An imprint of Elsevier

Academic Press is an imprint of Elsevier
125 London Wall, London EC2Y 5AS, United Kingdom
50 Hampshire Street, 5th Floor, Cambridge, MA 02139, United States

ISBN: 978-0-443-29190-6

For information on all Academic Press publications visit our website at https://www.elsevier.com/books-and-journals

Publisher: Mara Conner
Acquisitions Editor: Vaishali Singh
Editorial Project Manager: Kritika Sharma
Production Project Manager: Aishwarya R
Cover Designer: Arumugam Kothandan

Typeset by Integra Software Services

Dedication

To the mentors who inspire us to question, explore, and discover, and to future generations of scientists, may you continue pushing the boundaries of knowledge to better our world.

Contents

III

Marine-derived nutritional and bioactive compounds

V

Therapeutic properties of functional foods incorporated with marine ingredients

VI

Emerging prospects of functional foods from marine bioresources

Contributors

Takagi Anna School of Science, Monash University Malaysia, Subang Jaya, Selangor, Malaysia

Laura Arellano-García Nutrition and Obesity Group, Department of Pharmacy and Food Science, Faculty of Pharmacy and Lucio Lascaray Research Centre, University of the Basque Country [UPV/EHU], Vitoria-Gasteiz; CIBERobn Physiopathology of Obesity and Nutrition, National Institute of Health Carlos III, Madrid, Spain

Minthari Sakethanika Bandara Jeju Bio Research Center, Korea Institute of Ocean Science and Technology (KIOST), Jeju-si; Department of Marine Technology & Convergence Engineering (Marine Biotechnology), KIOST School, University of Science and Technology, Daejeon, Republic of Korea

Inoka Batugedara Department of Zoology, University of Sri Jayewardenepura, Nugegoda, Sri Lanka

Jayani Chandrapala School of Science, RMIT University, Bundoora, VIC, Australia

Chi Yie Choong School of Science, Monash University Malaysia, Subang Jaya, Selangor, Malaysia

D.M.C.B. Dissanayake Department of Biological Sciences, Faculty of Applied Sciences, Rajarata University of Sri Lanka, Mihintale, Sri Lanka

Brenda Juana Flores Agricultural and Environmental Biotechnology Club; School of Production, Technology, and Environment, National University of Río Negro, Villa Regina; Wikimedians of Jujuy, San Salvador de Jujuy, Jujuy, Argentina

Luis Fernando Flores Agricultural and Environmental Biotechnology Club, Villa Regina; Wikimedians of Jujuy, San Salvador de Jujuy, Jujuy, Argentina

Navindu Dinara Gajanayaka Jeju Bio Research Center, Korea Institute of Ocean Science and Technology (KIOST), Jeju-si; Department of Marine Technology & Convergence Engineering (Marine Biotechnology), KIOST School, University of Science and Technology, Daejeon, Republic of Korea

Devargya Ganguly Department of Life Sciences, GITAM (Deemed to be University), Visakhapatnam, Andhra Pradesh, India

Soumajit Ghosh Department of Life Sciences, GITAM (Deemed to be University), Visakhapatnam, Andhra Pradesh, India

Saioa Gómez-Zorita Nutrition and Obesity Group, Department of Pharmacy and Food Science, Faculty of Pharmacy and Lucio Lascaray Research Centre, University of the Basque Country [UPV/EHU]; BIOARABA Health Research Institute, Vitoria-Gasteiz; CIBERobn Physiopathology of Obesity and Nutrition, National Institute of Health Carlos III, Madrid, Spain

Maitane González-Arceo Nutrition and Obesity Group, Department of Pharmacy and Food Science, Faculty of Pharmacy and Lucio Lascaray Research Centre, University of the Basque Country [UPV/EHU], Vitoria-Gasteiz, Spain

Gargi Goswami Department of Life Sciences, GITAM (Deemed to be University), Visakhapatnam, Andhra Pradesh, India

D.M.N.M. Gunasekara Department of Biosystems Technology, Faculty of Technology, University of Sri Jayewardenepura, Homagama, Sri Lanka

Thilina L. Gunathilaka Department of Basic Science and Social Science for Nursing, Faculty of Nursing, University of Colombo, Colombo, Sri Lanka

K.H.I.N.M. Herath Department of Biosystems Engineering, Faculty of Agriculture and Plantation Management, Wayamba University of Sri Lanka, Makandura, Gonawila, Sri Lanka

Sachithra Amarin Hettiarachchi Department of Marine Biotechnology, Ocean University of Sri Lanka, Colombo, Sri Lanka

Susara Madduma Hewage St. Boniface Hospital Research Centre; Department of Animal Science, University of Manitoba, Winnipeg, MB, Canada

Aixin Hou Department of Environmental Sciences, College of the Coast & Environment, Louisiana State University, Baton Rouge, LA, United States

Sung-Pyo Hur Department of Marine Life Science, Jeju National University, Jeju, Republic of Korea

Kim Hyun-Soo Department of Seafood Science and Technology, Institute of Marine Industry, Gyongsang National University, South Korea

Jasintha Jayasanka Department of Biosystems Technology, Faculty of Technology, University of Sri Jayewardenepura, Homagama, Sri Lanka

Arachchige Maheshika Kumari Jayasinghe Department of Food Technology and Nutrition, Chonnam National University, Republic of Korea

Malith Chalitha Lalaksha Jayasinghe Department of Botany, University of Sri Jayewardenepura, Nugegoda, Sri Lanka

Thilina U Jayawardena Department of Medicine, University of Alberta, Edmonton, AB, Canada

Chathuranga Kiramage Faculty of Health, School of Medicine, Deakin University, Waurn Ponds; Institute for Mental and Physical Health and Clinical Translation (IMPACT), Deakin University, Geelonng, VIC, Australia

Kirinde Gedara Isuru Sandanuwan Kirindage Department of Food Technology and Nutrition, Chonnam National University, Republic of Korea

Hiruni S. Kumarasinghe Department of Interdisciplinary Graduate Program in Advanced Convergence Technology and Science, Jeju National University, Jeju, Republic of Korea

Yee-Ying Lee School of Science; Monash Industry Research Laboratory, Monash University Malaysia, Subang Jaya, Selangor, Malaysia

Nisansala Madushani Liyanage Department of Medicine, University of Alberta, Edmonton, AB, Canada

Gabriela Fabiana Malanga Faculty of Pharmacy and Biochemistry, University of Buenos Aires; Institute of Biochemistry and Molecular Medicine, National Scientific and Technical Research Council of Argentina - University of Buenos Aires, Ciudad Autónoma de Buenos Aires, Argentina

Svini Dileepa Marasinghe Korea Institute of Ocean Science and Technology, Jeju, Republic of Korea

Marcia Mazzuca Institute of Biological Sciences of Patagonia, National Scientific and Technical Research Council of Argentina; Department of Chemistry, Faculty of Natural Sciences and Health Sciences, National University of Patagonia San Juan Bosco, Comodoro Rivadavia, Chubut, Argentina

Nalin Medagoda Department of Marine Life Science, Jeju National University, Jeju, South Korea

Dineth Nagahwatta Department of Medicine, University of Alberta, Edmonton, AB, Canada

Chulhong Oh Jeju Bio Research Center, Korea Institute of Ocean Science and Technology (KIOST), Jeju-si; Department of Marine Technology & Convergence Engineering (Marine Biotechnology), KIOST School, University of Science and Technology, Daejeon, Republic of Korea

Thiruni Perera Department of Interdisciplinary Graduate Program in Advanced Convergence Technology and Science, Jeju National University, Jeju, Republic of Korea

María P. Portillo Nutrition and Obesity Group, Department of Pharmacy and Food Science, Faculty of Pharmacy and Lucio Lascaray Research Centre, University of the Basque Country [UPV/EHU]; BIOARABA Health Research Institute,

Vitoria-Gasteiz; CIBERobn Physiopathology of Obesity and Nutrition, National Institute of Health Carlos III, Madrid, Spain

Kalpa W. Samarakoon Institute for Combinatorial Advanced Research and Education (KDU-CARE), General Sir John Kotelawala Defence University, Ratmalana, Sri Lanka

K.K. Asanka Sanjeewa Department of Biosystems Technology, Faculty of Technology, University of Sri Jayewardenepura, Pitipana, Homagama, Sri Lanka

Uvin Eksith Senadheera Department of Biosystems Technology, Faculty of Technology, University of Sri Jayewardenepura, Homagama; Faculty of Graduate Studies, University of Sri Jayewardenepura, Gangodawila, Nugegoda, Sri Lanka

Y.T. Senaweera Faculty of Graduate Studies, General Sir John Kotelawala Defense University, Rathmalana, Sri Lanka

Sneha Sengupta Department of Life Sciences, GITAM (Deemed to be University), Visakhapatnam, Andhra Pradesh, India

Ishara De Silva School of Science, RMIT University, Bundoora, VIC, Australia

Mayumi Silva School of Science, RMIT University, Bundoora, VIC, Australia

R.C.N. Thilakarathna School of Science, Monash University Malaysia, Subang Jaya, Selangor, Malaysia

Sanduni Umayangani Department of Biosystems Technology, Faculty of Technology, University of Sri Jayewardenepura, Homagama, Sri Lanka

Lei Wang State Key Laboratory of Marine Food Processing & Safety Control, College of Food Science and Engineering, Ocean University of China, Qingdao; Sanya Oceanographic Institution, Ocean University of China, Sanya, China

Sashika Dilshan Warnasuriya Department of Environmental Sciences, College of the Coast & Environment, Louisiana State University, Baton Rouge, LA, United States

Bopitige Udith Ravindu Wijemanna Department of Animal Science, Iowa State University, Ames, IA, United States

H.M. Sarath M. Wijerathna Department of Aquaculture and Seafood Technology, Faculty of Fisheries and Ocean Sciences, Ocean University of Sri Lanka, Colombo, Sri Lanka

H.D. Tharushi Wijerathne Department of Seafood Science and Technology, Institute of Marine Industry, Gyongsang National University, South Korea

J.A.A.C. Wijesinghe Department of Biosystems Engineering, Faculty of Agriculture and Plantation Management, Wayamba University of Sri Lanka, Makandura, Gonawila, Sri Lanka

P. Wijethunga Department of Biological Sciences, Faculty of Applied Sciences, Rajarata University of Sri Lanka, Mihintale, Sri Lanka

About the editors

Dr. Thilina Jayawardena is a Research Scientist in the Department of Medicine at the University of Alberta with an interdisciplinary background spanning chemistry, marine natural products, and molecular cell biology. His research has included the discovery and characterization of bioactive compounds from marine organisms, contributing foundational insights to the themes of this book. He now applies his expertise in biochemical and cellular mechanisms to translational cardiovascular research, employing multiomics, advanced cell-based models, and in vivo approaches. He has an extensive publication record, including peer-reviewed articles and book chapters, and has previously served in editorial roles for academic volumes.

Prof. Asanka Sanjeewa earned his MSc and PhD in Marine Life Sciences from Jeju National University, South Korea. His research focuses on marine bioprospecting, particularly the discovery of novel bioactive secondary metabolites with applications in nutraceuticals, cosmeceuticals, and functional foods, alongside advancing knowledge of the antiinflammatory mechanisms of natural compounds related to fine dust-induced inflammation. He has authored over 100 peer-reviewed articles and book chapters and has contributed to and edited several academic books in marine biosciences. His work has received over 4000 citations and has an H-index of 37 on Google Scholar (as of May 2026). He is also an active member of several editorial and reviewer boards for academic journals. His accolades include the Elsevier-Stanford Top 2% Highly Cited Researcher Recognition (2022–2025) and the Best Researcher Award from the Korean Society of Phycology in 2019.

Dr. K.H.I.N.M. Herath is a Senior Lecturer in the Department of Biosystems Engineering, Faculty of Agriculture and Plantation Management, Wayamba University of Sri Lanka. She earned her BSc (Special) in Agriculture from Rajarata University, graduating with first-class honors and a gold medal, and subsequently completed an MSc and a PhD in Advanced Convergence Technology and Science at Jeju National University, South Korea. Her research focuses on functional foods, bioactive compounds, immunology, histology, and cell biology. She has authored and coauthored numerous peer-reviewed research articles and has also served as an editor and reviewer for several scientific journals. Her work has been widely recognized and cited, highlighting her significant contributions to both fundamental and applied marine biosystems research.

Preface

Harvesting the Sea: Unveiling Marine Bioresources for Functional Foods and Human Health Enhancement explores the remarkable potential hidden within the depths of our oceans. Covering more than two-thirds of the Earth's surface, the marine ecosystem harbors an astonishing diversity of life. Beneath its vast waters lies a treasure trove of bioactive compounds that have the power to revolutionize human health and well-being.

In an era where functional foods are increasingly valued for their ability to provide health benefits beyond basic nutrition, this book delves into the untapped potential of marine bioresources. Focusing on bioactive molecules derived from a wide range of marine organisms, from microscopic microorganisms to complex marine macroorganisms, this book traces the captivating journey of discovery, sustainable utilization, and the prospects of these valuable compounds.

The book emphasizes the urgent need for innovative solutions in nutrition and healthcare. In the face of modern lifestyles, rising stress, and the growing prevalence of chronic diseases, there is an increasing demand for functional foods that promote well-being and aid in disease prevention. *Harvesting the Sea* highlights the potential of marine-derived bioactive compounds and explores how they can be integrated into functional foods to improve health. Through careful and thorough research, this book offers a deeper understanding of how these compounds may contribute to a more vibrant, fulfilling life.

With its interdisciplinary approach, this book serves as an essential resource for nutritionists, food scientists, biotechnologists, researchers, and anyone interested in the profound relationship between our oceans and human health. As a bridge between conventional foods and dietary supplements, *Harvesting the Sea* illuminates the path toward a healthier, more functional future through the wonders of marine bioresources.

PART I

General overview of marine bioresources

CHAPTER

1

Major types of marine bioresources and their industrial applications

K.H.I.N.M. Herath[a], *Thilina U. Jayawardena*[b], *and K.K. Asanka Sanjeewa*[c]

[a]**Department of Biosystems Engineering, Faculty of Agriculture and Plantation Management, Wayamba University of Sri Lanka, Makandura, Gonawila, Sri Lanka**
[b]**Department of Medicine, University of Alberta, Edmonton, AB, Canada**
[c]**Department of Biosystems Technology, Faculty of Technology, University of Sri Jayewardenepura, Pitipana, Homagama, Sri Lanka**

1.1 Introduction

Marine bioresources encompass a diverse range of living organisms and the ecosystems found in the world's oceans, seas, and coastal regions. These resources, such as fish, algae (both micro and macro), mollusks, crustaceans, marine higher plant species, and microorganisms (such as bacteria, archaea, microalgae, and fungi), are essential for supporting biodiversity, human livelihoods, and global ecological balance. As the largest and most biodiverse ecosystem on Earth, marine environments cover over 70% of the planet's surface and host approximately 80% of all life forms (Cavicchioli et al., 2019; Urbanek et al., 2018). In general, marine environments are diverse, with wide environmental gradients in the physical, chemical, and hydrological parameters such as temperature, light intensity, salinity, and pressure (Min et al., 2021). Thus, marine organisms that thrive in these different ecosystems have adapted to these diverse environments by developing a broad spectrum of forms, functions, and strategies that play a crucial role in survival, adaptation, and living in these different marine ecosystems (Rotter et al., 2021). Beyond their ecological significance, marine bioresources serve as a foundation/raw material for global industries such as food, pharmaceuticals, cosmetics, energy, and environmental management (Fig. 1.1) (Karunarathne et al., 2020; Sanjeewa, 2024a, 2024b). The dynamic nature of marine ecosystems also makes them a critical component of Earth's climate regulation and carbon sequestration processes (Yuan et al., 2023). Despite the vast potential of marine bioresources in different industries, the sustainable management of these resources remains a complex challenge due to the pressures

Harvesting the Sea
https://doi.org/10.1016/B978-0-443-29190-6.00005-2

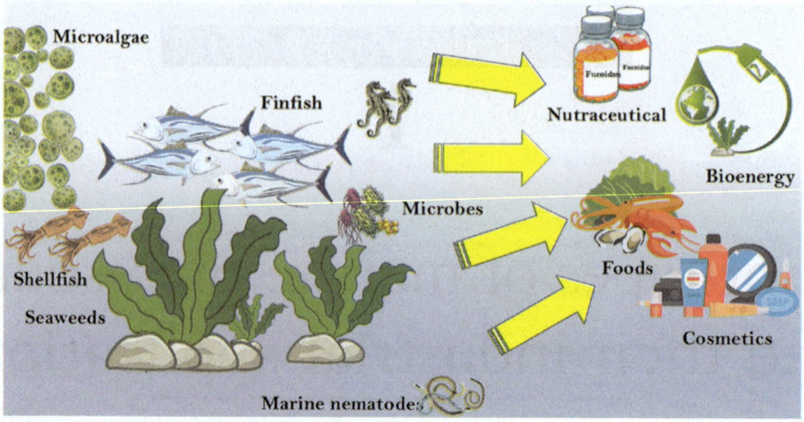

FIG. 1.1 Major marine bioresources and their commercial applications.

of overexploitation, pollution, and climate change (Fu et al., 2018). This chapter explains the key segments, applications, challenges, and future potential of marine bioresources in the context of environmental sustainability and human development.

1.2 Major segments of marine bioresources

Marine bioresources can be broadly classified into two primary categories: biological and genetic resources. Biological resources include commercially harvested or cultured marine organisms such as fish, shellfish, seaweeds, microalgae, and other marine organisms used for food, cosmeceutical, nutraceutical, pharmaceutical, animal feed production, and bioenergy applications. Genetic resources refer to the genetic material derived from marine organisms, which holds immense potential for biotechnology, medicine, and agriculture.

1.2.1 Marine fish and shellfish species

The marine ecosystem is home to a vast diversity of fish and shellfish species that play critical roles in maintaining ecological balance and supporting human livelihoods. Specifically, finfish and shellfish represent the largest category of marine bioresources utilized for human consumption. Marine fin fish are categorized broadly into pelagic (pelagic fish can be categorized as coastal and oceanic fish) and demersal species. Pelagic fish, such as tuna, mackerel, salmon, and sardines, inhabit the open ocean. Pelagic fish species play an important ecological role as they are in an intermediate trophic position, linking plankton to higher trophic levels and apex predators (such as marine mammals and seabirds) (Moyano et al., 2023). Moreover, pelagic fish species are prized for their high protein content, omega-3 fatty acids, docosahexaenoic acid (DHA), and eicosapentaenoic acid (EPA) (Majluf et al., 2024). Demersal fish, such as cod, catfish, croakers, soles, groupers, snappers, mullets, megalops, haddock, flatfish, dab, and flounder, reside closer to the seabed (Khedkar et al., 2003;

Rummel et al., 2016). Demersal fish are highly valued for their protein content and delicate flavors and play a major role in global fisheries (Khedkar et al., 2003).

Shellfish have economic and ecological importance and have been consumed as food for centuries. After finfish, shellfish form a major component of global seafood production (Sao Mai, 2014). Shellfish is a broad term that covers various aquatic mollusks, crustaceans, and echinoderms that are used as food (Gökoğlu, 2021). Categories of shellfish include shrimp, lobsters, oysters, squid, mussels, scallops, clams, cuttlefish, snails, crabs, crayfish, krill, abalone, prawns, and more (Sao Mai, 2014; Venugopal and Gopakumar, 2017; Gökoğlu, 2021; Wan Mahari et al., 2022; Filippini et al., 2023).

Other than their ecological importance, marine fish and shellfish species are indispensable to food security, providing millions of tons of seafood annually. According to The State of World Fisheries and Aquaculture 2024 report, in 2022, fisheries and aquaculture production reached an all-time high of 223.2 million tons, worth a record USD 472 billion and contributing an estimated 20.7 kg of aquatic animal foods per capita. This constituted about 15% of the animal protein supply, reaching over 50% in some countries located in Asia and on the African continent (The State of World Fisheries and Aquaculture, 2024). Despite these significant achievements, the marine fin fish and shellfish sector still faces major challenges from climate change and disasters, water scarcity, pollution, biodiversity loss, and other anthropogenic impacts. Environmental conditions determine the yield of marine fish landings (Baag and Mandal, 2022). Therefore, sustainable approaches are essential to further expand the industry and meet future demands driven by population growth.

1.2.2 Edible seaweeds

In general, seaweeds are divided into three groups: brown, red, and green, according to their appearance. Edible seaweeds are an important side dish and are used as an ingredient in traditional medicine to treat various diseases in Southeast Asian countries. According to the literature, seaweed has been consumed as food in Korea, Japan, and China for over 2000 years, while ancient Greeks used seaweed as animal feed as early as 100 BC (Sanjeewa et al., 2023). Beyond the historical value, edible marine macroalgae (seaweeds) are economically important bioresources with diverse industrial applications (Katiyar, 2024; Sanjeewa and Jeon, 2018). Most seaweed species are edible and rich in nutrients such as polysaccharides (fucoidan, agar, algin, and carrageenan), essential amino acids, polyunsaturated fatty acids, phytochemicals, dietary fibers, vitamins (A, B1, B2, B9, B12, C, D, E, and K), and minerals (calcium, iron, iodine, magnesium, phosphorus, potassium, zinc, copper, manganese, selenium, and fluoride) (Cotas et al., 2020; El-Beltagi et al., 2022; Dissanayake et al., 2024). Seaweed species such as Kombu (species of *Laminaria/Saccharina*), Nori (*Porphyra/Pyropia/Neoporphyra/Neopyropia* genera), and Wakame (*Undaria pinnatifida*) are examples of popular edible seaweed types (Kumar et al., 2021). Moreover, various food items, such as seaweed wraps, soups, pickles, halwa, spices, wafers, porridge, jellies, jams, cookies, and sauces, can be prepared from the aforementioned edible seaweeds (Kumar et al., 2021). Other than food and traditional medicinal applications, edible seaweeds can be used in numerous industries, such as functional foods, nutraceuticals, cosmeceuticals, the animal feed industry, fertilizers, biostimulants, and many other biotechnological applications.

1.2.3 Marine microalgae

Microalgae are microscopic organisms found in both marine and freshwater ecosystems. Microalgae include both prokaryotes and eukaryotes. Prokaryotes consist of cyanobacteria (blue-green algae), while eukaryotes include diatoms and green algae (Khavari et al., 2021). These unicellular species grow either individually or in clusters and efficiently perform photosynthesis by utilizing CO_2, contributing significantly to the reduction of greenhouse gases (Mondal and Khan, 2021; Vieira de Mendonça et al., 2021). Marine microalgae represent the largest primary biomass producers in ocean ecosystems, with growth rates 5–10 times faster than those of conventional food crops and land plants (Krzemińska et al., 2014; Loke Show, 2022). Most microalgae form the base of marine food webs and habitats, thriving in diverse environments, including extreme conditions such as cold regions, varying pH levels, and even hydrothermal vents (Moreno-Garrido, 2008; Martínez Andrade et al., 2018; Loke Show, 2022; Zhuang et al., 2022).

Recently, microalgae have gained considerable attention as a valuable feedstock for many industrial applications due to their diverse potential for different applications (Matsunaga et al., 2005). Microalgae are rich in nutrients, such as proteins, minerals, and beneficial compounds like pigments, astaxanthin, lipids, omega-3 and omega-6 long-chain polyunsaturated fatty acids, sterols, and various essential vitamins (including β-carotene, thiamin, riboflavin, niacin, pantothenic acid, pyridoxine, biotin, and folic acid) (Nazih and Bard, 2018; Kusmayadi et al., 2021; Martínez-Ruiz et al., 2022; Koller et al., 2014; Liu et al., 2022). Specifically, recent studies have highlighted the potential applications of microalgae, such as biofuel, animal feed, and bioactive metabolites for various industries, including cosmetics, pharmaceuticals, and functional foods (Martínez-Ruiz et al., 2022; Zhuang et al., 2022). Furthermore, microalgae are a well-known source of novel bioactive substances with notable health benefits, such as immunomodulatory, anticancer, antiinflammatory, antioxidant, antiviral, and antiallergy properties (Nazih and Bard, 2018; Sanjeewa et al., 2016a, 2016b).

1.2.4 Marine bacteria and fungi

Among the microorganisms, marine bacteria and fungi represent a vast and largely unexplored reservoir of bioactive compounds with diverse industrial applications (Zhivkoplias et al., 2024). Marine bacteria, fungi, and their isolates represent a major source of undiscovered scientific potential (Soliev et al., 2011). Specifically, products derived from marine bacteria and fungi have been reported to demonstrate enhanced stability and unique structural features, such as high levels of macrolides, alkaloids, peptides, terpenoids, meroterpenoids, uronic acid, and sulfate groups in exopolysaccharides (Soliev et al., 2011). The biosynthesis of these specific microbial metabolites depends on a number of biotic and abiotic conditions in the marine environment, such as temperature, nutrient availability, pressure of the water column, light availability, salinity, and interaction with other microorganisms (Gozari et al., 2021; Macedo et al., 2021). Moreover, these specific properties make marine microorganism-derived products ideal for use in various industries, such as food, pharmaceuticals, cosmetics, and biotechnology (Maldonado-Ruiz et al., 2024). Marine microbes have even been found to produce crude oil-degrading metabolites (*Bacillus licheniformis, Ochrobactrum anthropi IITR07, Microbacterium esteraromaticum IITR47, Pseudomonas aeruginosa*

IITR48, and *Stenotrophomonas maltophilia IITR87*) and have applications in bioremediation (*Novosphingobium*, *Gordonia*, *Stappia*, *Mesobacillus*, *Alcanivorax*, *Flexivirga*, *Cytobacillus*, *Thioclava*, and *Thalassospira*) (Gao and Sun, 2021; Ganesan et al., 2022; Srimathi et al., 2024; Liu et al., 2023). According to various sources, it is estimated that around 10%–20% of bacteria isolated from marine ecosystems have applications in the pharmaceutical industry, therapeutics, and biotechnological fields (Stincone and Brandelli, 2020). Specifically, marine bacterial isolates have been found to possess diverse biological properties, including cytotoxic, antiinflammatory, anticancer, antimicrobial, antioxidant, antiangiogenesis, and antidiabetic activities (Stincone and Brandelli, 2020; Mayer et al., 2021). However, to date, knowledge remains limited on marine microbes, their functions, and their influence on marine ecosystems. Thus, future studies are required to explore the applications of marine bacteria and fungi.

1.2.5 Marine nematodes

The phylum Nematoda is best known for its pathogenic species, which cause some of the most infectious diseases in humans and animals and contribute to significant crop losses (Moens et al., 2013). Nematodes are nonsegmented roundworms that are reported in most ecosystems on Earth (Bhat et al., 2022). Approximately 50% of humans are infected with gastrointestinal nematodes, with children being the most affected in severe cases. In addition, nematodes are responsible for an estimated 8%–15% of global crop losses (Kiontke and Fitch, 2013). The majority of nematode species and individuals are free-living, inhabiting terrestrial soils and aquatic sediments (Moens et al., 2013; Sanjeewa, 2024). Marine nematodes are most abundant in marine sediments, and compared to terrestrial nematodes, marine nematodes are ubiquitous across depths, latitudes, and biogeographic regions. Moreover, these marine nematodes exhibit different feeding strategies, tail shapes, body sizes, life histories, or a combination of several of these parameters (Moens et al., 2013). Marine nematodes can be used as bioindicators to determine the health and pollution levels of marine ecosystems (Ridall and Ingels, 2021). Beyond their ecological role, marine nematodes offer potential applications in economically important industries, including pathogen management, aquaculture, fisheries, nutrient cycling, and heavy metal bioremediation. However, further research is needed to fully explore their potential, particularly in areas such as nematode enzymes and heavy metal absorption (Sanjeewa, 2024). Integrating marine nematode-derived enzymes and their bioremediation properties into various industries could provide eco-friendly and efficient solutions.

1.3 Common applications of marine bioresources

Marine bioresources are indispensable across multiple sectors, ranging from being a nutrient-rich food source to complex biotechnological applications. These applications include food and nutrition, pharmaceuticals, cosmetics and personal care, bioenergy, bioremediation, and carbon sequestration. This section of the chapter briefly discusses the major applications of marine bioresources.

1.3.1 Food and nutritional applications of marine bioresources

Seafood is a primary protein source for billions of people around the globe. The consumption and demand for marine-based food products have increased over the last few decades, driven by increased awareness of their health benefits and the positive image of seafood among consumers (Zhang et al., 2012). Most marine bioresources are rich in essential nutrients such as polysaccharides, essential amino acids, omega-3 (ω3)-polyunsaturated fatty acids, phytochemicals, peptides, EPA, DHA, dietary fibers, vitamins, and minerals (Larsen et al., 2011). Moreover, marine organisms such as sea cucumbers, edible seaweeds, and mollusks are rich sources of bioactive secondary metabolites with therapeutic potential to treat numerous diseases (Larsen et al., 2011; Lobine et al., 2022; Sanjeewa and Herath, 2023). Specifically, functional foods (Box 1.1) with antioxidant, anticancer, antiinflammatory, antidiabetic, cardiovascular protection, and antiobesity properties, developed from marine organisms, represent a growing segment in the functional food industry (Freitas et al., 2015).

1.3.2 Pharmaceutical applications of marine organisms

Besides their nutritional value, marine organisms are a rich source of bioactive compounds with therapeutic potential (Karthikeyan et al., 2022). Many marine organisms, including seaweeds, have been recognized for their therapeutic potential in traditional folk medicine for thousands of years (Sanjeewa et al., 2019). Novel bioactive secondary metabolites isolated from marine organisms have been found to possess numerous therapeutic applications, including anticancer (Santaniello et al., 2023), antiobesity (Tammam et al., 2024), antidiabetic (Mayer et al., 2022), antihypertensive (Lee and Hur, 2017), antihyperlipidemic (Popov and Krivoshapko, 2013), antioxidant (Kim et al., 2021), anticoagulant (Chandika et al., 2022), liver protection (Hira et al., 2021), neuroprotective (Šimat et al., 2024), antiinflammatory (Herath et al., 2021), immunomodulatory (Cutolo et al., 2024), antiviral (Mayer et al., 2022), antifungal (Mayer et al., 2022), antibacterial (Mayer et al., 2022), and wound healing (Cruz et al., 2021).

1.3.3 Bioenergy-related applications of marine organisms

With the beginning of the 21st century, most countries around the world have started to adopt biofuels as alternatives to fossil fuels. Countries such as the United States, Brazil, Canada, Japan, India, China, and those located in the European Union have shown interest in the more eco-friendly biofuel and have implemented strategies to support its use (Manivasagan et al., 2015). This global interest in biofuel is driven by several key factors, such as rising oil prices and depleting reserves, environmental concerns, international agreements, and support for agriculture (Box 1.2). Among the potential biofuel sources, algae-based

Box 1.1 Definition of functional food

Functional foods are foods that provide health benefits beyond basic nutrition, such as improving heart health, enhancing immunity, or aiding in weight management.

Box 1.2 Why global interest is toward biofuels over fossil fuels

Rising oil prices and depleting reserves: The volatility in oil prices and the rapid depletion of global oil reserves have highlighted the need for sustainable energy sources. Biofuels, derived from renewable biomass, offer a promising solution to reduce dependence on finite fossil fuel resources.

Environmental concerns: The growing awareness of the environmental impact of fossil fuels, such as their contribution to greenhouse gas emissions, has spurred interest in cleaner energy alternatives.

International agreements: Commitments under global frameworks, such as the Kyoto Protocol and the Bali Action Plan, have emphasized reducing carbon emissions, further encouraging the adoption of renewable energy sources.

Support for agriculture: Biofuel production provides alternative markets for agricultural producers by utilizing crops like corn, sugarcane, and soybeans. This not only diversifies income streams for farmers but also strengthens rural economies.

biofuels offer a sustainable alternative to fossil fuels (Moshood et al., 2021). Specifically, macro- and microalgae can be cultivated in non-arable areas, reducing competition with food crops and minimizing environmental impact (Occhipinti et al., 2024). According to previous studies, microalgae such as *Dunaliella, Spirulina, Nannochloris,* and *Chlorella species* accumulate high levels of lipids (triacylglycerides) and are reported to have >40 times the productivity compared to land crops (Sathya et al., 2023).

1.3.4 Cosmetics and personal care

As the market size of the cosmetics industry increases, the safety and effectiveness of new products face stricter requirements. The marine environment offers a unique selection of bioactive compounds for the cosmetics and personal care industry (Ding et al., 2022). These compounds are increasingly valued for their potential to meet the demand for safe and effective cosmetic ingredients (Resende et al., 2021). Marine-derived ingredients, such as fucoidan, phlorotannins, alginates, carrageenan, and collagen, are widely used in skincare, haircare, and cosmetic formulations (Sanjeewa et al., 2016; Kang et al., 2024). The bioactive properties, such as antioxidant activity, tyrosinase inhibition, hyaluronidase inhibition, matrix metalloproteinase inhibitory activities, controlled histamine release, and enhanced IGF-1 mRNA expression, make them ideal active ingredients to use in the cosmetics and personal care industries (Ding et al., 2022).

1.4 Challenges in marine bioresource management

Despite the immense potential to be used in different industries, even as a food source, marine bioresources face significant threats from overexploitation, habitat destruction, pollution, and climate change. These challenges require sustainable management practices to ensure the long-term availability and health of marine ecosystems. The sustainable utilization of marine bioresources is essential to achieving global goals such as food security, climate resilience, and economic development. Innovations in biotechnology, conservation efforts, and policy frameworks are shaping the future of marine bioresource management.

1.5 Conclusion

Marine bioresources play an important role in sustaining life on Earth and driving economic and technological advancements. However, understanding the full potential of these marine bioresources requires a balance between utilization and conservation to prevent the overexploitation of these resources. By addressing existing challenges and adopting innovative strategies, humans can ensure the sustainable development of marine resources for future generations. Expanding global collaborations, investing in scientific research, and promoting awareness are essential steps in safeguarding these valuable resources and unlocking their transformative potential for society.

References

Baag, S., Mandal, S., 2022. Combined effects of ocean warming and acidification on marine fish and shellfish: a molecule to ecosystem perspective. Sci. Total Environ. 0048-9697. 802, 149807. doi:10.1016/j.scitotenv.2021.149807.

Bhat, K.A., Mir, R.A., Farooq, A., Manzoor, M., Hami, A., Allie, K.A., Wani, S.M., Khan, M.N., Sayyed, R.Z., Poczai, P., Almalki, W.H., Zargar, S.M., Shah, A.A., 2022. Advances in nematode identification: a journey from fundamentals to evolutionary aspects. Diversity. 1424-2818. 14 (7). Available from: https://www.mdpi.com/1424-2818/14/7/536/pdf?version=1656645372. doi:10.3390/d14070536.

Cavicchioli, R., Ripple, W.J., Timmis, K.N., Azam, F., Bakken, L.R., Baylis, M., Behrenfeld, M.J., Boetius, A., Boyd, P.W., Classen, A.T., Crowther, T.W., Danovaro, R., Foreman, C.M., Huisman, J., Hutchins, D.A., Jansson, J.K., Karl, D.M., Koskella, B., Welch, D.B.M., Martiny, J.B.H., Moran, M.A., Orphan, V.J., Reay, D.S., Remais, J.V., Rich, V.I., Singh, B.K., Stein, L.Y., Stewart, F.J., Sullivan, M.B., van Oppen, M.J.H., Weaver, S.C., Webb, E.A., Webster, N.S., 2019. Scientists' warning to humanity: microorganisms and climate change. Nat. Rev. Microbiol. 1740-1534. 17 (9), 569–586. Available from: http://www.nature.com/nrmicro/index.html. doi:10.1038/s41579-019-0222-5.

Chandika, P., Tennakoon, P., Kim, T.H., Kim, S.C., Je, J.Y., Kim, J.I., Lee, B., Ryu, B.M., Kang, H.W., Kim, H.W., Kim, Y.M., Kim, C.S., Choi, I.W., Park, W.S., Yi, M., Jung, W.K., 2022. Marine biological macromolecules and chemically modified macromolecules; potential anticoagulants. Mar. Drugs. 1660-3397. 20 (10). Available from: http://www.mdpi.com/journal/marinedrugs. doi:10.3390/md20100654.

Cotas, J., Leandro, A., Pacheco, D., Gonçalves, A.M.M., Pereira, L., 2020. A comprehensive review of the nutraceutical and therapeutic applications of red seaweeds (Rhodophyta). Life. 2075-1729. 10 (3), 19. doi:10.3390/life10030019.

Cruz, M.A., Araujo, T.A., Avanzi, I.R., Parisi, J.R., de Andrade, A.L.M., Rennó, A.C.M., 2021. Collagen from marine sources and skin wound healing in animal experimental studies: a systematic review. Mar. Biotechnol. 1436-2236. 23 (1). Available from: http://link.springer.com/journal/volumesAndIssues/10126. doi:10.1007/s10126-020-10011-6.

Cutolo, E.A., Campitiello, R., Caferri, R., Pagliuca, V.F., Li, J., Agathos, S.N., Cutolo, M., 2024. Immunomodulatory compounds from the sea: from the origins to a modern marine pharmacopoeia. Mar. Drugs. 1660-3397. 22 (7). Available from: http://www.mdpi.com/journal/marinedrugs. doi:10.3390/md22070304.

Ding, J., Wu, B., Chen, L., 2022. Application of marine microbial natural products in cosmetics. Front. Microbiol. 1664-302X. 13. doi:10.3389/fmicb.2022.892505.

Dissanayake, D.S., Sanjeewa, K.K.A., Jayawardena, T.U., 2024. What is seaweed? General facts about seaweeds. In: The Role of Seaweeds in Blue Bioeconomy. Bentham Science Publishers, Sri Lanka, pp. 13–24. Available from: https://benthambooks.com/book/9789815223644/. doi:10.2174/9789815223644124010004.

El-Beltagi, H.S., Mohamed, A.A., Mohamed, H.I., Ramadan, K.M.A., Barqawi, A.A., Mansour, A.T., 2022. Phytochemical and potential properties of seaweeds and their recent applications: a review. Mar. Drugs. 1660-3397. 20 (6). Available from: https://www.mdpi.com/1660-3397/20/6/342/pdf?version=1653475780. doi:10.3390/md20060342.

Filippini, G., Dafforn, K.A., Bugnot, A.B., 2023. Shellfish as a bioremediation tool: a review and meta-analysis. Environ. Pollut. 1873-6424. 316. Available from: https://www.journals.elsevier.com/environmental-pollution. doi:10.1016/j.envpol.2022.120614.

Freitas, A.C., Rodrigues, D., Carvalho, A.P., Pereira, L., Panteleitchouk, T., Gomes, A.M., Duarte, A.C., 2015. Marine functional foods. In: Springer Handbook of Marine Biotechnology. Springer, Berlin Heidelberg Portugal, pp. 969–994. Available from: http://dx.doi.org/10.1007/978-3-642-53971-8. doi:10.1007/978-3-642-53971-8_42.

Fu, X.M., Zhang, M.Q., Liu, Y., Shao, C.L., Hu, Y., Wang, X.Y., Su, L.R., Wang, N., Wang, C.Y., 2018. Protective exploitation of marine bioresources in China. Ocean Coast. Manag. 0964-5691. 163, 192–204. Available from: www.elsevier.com/inca/publications/store/4/0/5/8/8/9. doi:10.1016/j.ocecoaman.2018.06.018.

Ganesan, M., Mani, R., Sai, S., Kasivelu, G., Awasthi, M.K., Rajagopal, R., Wan Azelee, N.I., Selvi, P.K., Chang, S.W., Ravindran, B., 2022. Bioremediation by oil degrading marine bacteria: an overview of supplements and pathways in key processes. Chemosphere. 1879-1298. 303. Available from: www.elsevier.com/locate/chemosphere. doi:10.1016/j.chemosphere.2022.134956.

Gao, R., Sun, C., 2021. A marine bacterial community capable of degrading poly (ethylene terephthalate) and polyethylene. J. Hazard. Mater. 0304-3894. 416, 125928. doi:10.1016/j.jhazmat.2021.125928.

Gökoğlu, N., 2021. Shellfish Processing and Preservation. Springer International Publishing. doi:10.1007/978-3-030-60303-8.

Gozari, M., Alborz, M., El-Seedi, H.R., Jassbi, A.R., 2021. Chemistry, biosynthesis and biological activity of terpenoids and meroterpenoids in bacteria and fungi isolated from different marine habitats. Eur. J. Med. Chem. 1768-3254. 210. Available from: http://www.journals.elsevier.com/european-journal-of-medicinal-chemistry/. doi:10.1016/j.ejmech.2020.112957.

Herath, K.H.I.N.M., Kim, H.J., Lee, J.H., Je, J.G., Yu, H.S., Jeon, Y.J., Kim, H.J., Jee, Y., 2021. *Sargassum horneri* (Turner) C. Agardh containing polyphenols attenuates particulate matter-induced inflammatory response by blocking TLR-mediated MYD88-dependent MAPK signaling pathway in MLE-12 cells. J. Ethnopharmacol. 1872-7573. 265. Available from: www.elsevier.com/locate/jethpharm. doi:10.1016/j.jep.2020.113340.

Hira, K., Farhat, H., Sohail, N., Ansari, M., Ara, J., Ehteshamul-Haque, S., 2021. Hepatoprotective activity against acetaminophen-induced liver dysfunction and GC-MS profiling of a brown algae *Sargassum ilicifolium*. Clin. Phytosci. 2199-1197. 7 (1). doi:10.1186/s40816-021-00274-4.

Kang, J.W., Hyun, S.H., Kim, H.M., Park, S.Y., Lee, J.A., Lee, I.C., Bae, J.S., 2024. The effects of fucoidan-rich polysaccharides extracted from *Sargassum horneri* on enhancing collagen-related skin barrier function as a potential cosmetic product. J. Cosmet. Dermatol. 1473-2165. 23 (4), 1365–1373. Available from: http://onlinelibrary.wiley.com/journal/10.1111/(ISSN)1473-2165. doi:10.1111/jocd.16108.

Karthikeyan, A., Joseph, A., Nair, B.G., 2022. Promising bioactive compounds from the marine environment and their potential effects on various diseases. J. Genet. Eng. Biotechnol. 1687-157X. 20 (1), 14. doi:10.1186/s43141-021-00290-4.

Karunarathne, S., Mendis, E., Kim, S.K., 2020. The potential of developing additives from marine bioresources for the food industry. In: Encyclopedia of Marine Biotechnology. Wiley, Sri Lanka, pp. 3085–3107. Available from: https://onlinelibrary.wiley.com/doi/book/10.1002/9781119143802. doi:10.1002/9781119143802.ch137.

Katiyar, K., 2024. Marine resources: plethora of opportunities for sustainable future. In: Marine Biomass: Biorefinery, Bioproducts and Environmental Bioremediation. De Gruyter, India, pp. 367–388. Available from: https://www.degruyter.com/document/doi/10.1515/9783111353951/html. doi:10.1515/9783111353951-016.

Khavari, F., Saidijam, M., Taheri, M., Nouri, F., 2021. Microalgae: therapeutic potentials and applications. Mol. Biol. Rep. 0301-4851. 48 (5), 4757–4765. doi:10.1007/s11033-021-06422-w.

Khedkar, G.D., Jadhao, B.V., Chavan, N.V., Khedkar, C.D., 2003. FISH | demersal species of tropical climates. In: Encyclopedia of Food Sciences and Nutrition. Elsevier BV, pp. 2438–2442. doi:10.1016/b0-12-227055-x/00472-7.

Kim, H.J., Herath, K.H.I.N.M., Dinh, D.T.T., Kim, H.S., Jeon, Y.J., Kim, H.J., Jee, Y., 2021. Sargassum horneri ethanol extract containing polyphenols attenuates PM-induced oxidative stress via ROS scavenging and transition metal chelation. J. Funct. Foods. 1756-4646. 79. Available from: http://www.elsevier.com/wps/find/journaldescription.cws_home/717426/description#description. doi:10.1016/j.jff.2021.104401.

Kiontke, K., Fitch, D.H.A., 2013. Nematodes. Curr. Biol. 0960-9822. 23 (19), R862. doi:10.1016/j.cub.2013.08.009.

Koller, M., Muhr, A., Braunegg, G., 2014. Microalgae as versatile cellular factories for valued products. Algal Res. 2211-9264. 6, 52–63. doi:10.1016/j.algal.2014.09.002.

Krzemińska, I., Pawlik-Skowrońska, B., Trzcińska, M., Tys, J., 2014. Influence of photoperiods on the growth rate and biomass productivity of green microalgae. Bioprocess Biosyst. Eng. 1615-7591. 37 (4), 735–741. doi:10.1007/s00449-013-1044-x.

Kumar, Y., Tarafdar, A., Badgujar, P.C., Sobhy El-Sohaimy, 2021. Seaweed as a source of natural antioxidants: therapeutic activity and food applications. J. Food Qual. 1745-4557. 2021, 1–17. doi:10.1155/2021/5753391.

Kusmayadi, A., Leong, Y.K., Yen, H.W., Huang, C.Y., Chang, J.S., 2021. Microalgae as sustainable food and feed sources for animals and humans—biotechnological and environmental aspects. Chemosphere. 1879-1298. 271. Available from: www.elsevier.com/locate/chemosphere. doi:10.1016/j.chemosphere.2021.129800.

Larsen, R., Eilertsen, K.E., Elvevoll, E.O., 2011. Health benefits of marine foods and ingredients. Biotechnol. Adv. 0734-9750. 29 (5), 508–518. doi:10.1016/j.biotechadv.2011.05.017.

Lee, S.Y., Hur, S.J., 2017. Antihypertensive peptides from animal products, marine organisms, and plants. Food Chem. 1873-7072. 228, 506–517. Available from: www.elsevier.com/locate/foodchem. doi:10.1016/j.foodchem.2017.02.039.

Liu, R., Li, S., Tu, Y., Hao, X., Qiu, F., 2022. Recovery of value-added products by mining microalgae. J. Environ. Manage. 0301-4797. 307, 114512. doi:10.1016/j.jenvman.2022.114512.

Liu, R., Zhao, S., Zhang, B., Li, G., Fu, X., Yan, P., Shao, Z., 2023. Biodegradation of polystyrene (PS) by marine bacteria in mangrove ecosystem. J. Hazard. Mater. 0304-3894. 442, 130056. doi:10.1016/j.jhazmat.2022.130056.

Lobine, D., Rengasamy, K.R.R., Mahomoodally, M.F., 2022. Functional foods and bioactive ingredients harnessed from the ocean: current status and future perspectives. Crit. Rev. Food Sci. Nutr. 1549-7852. 62 (21), 5794–5823. Available from: https://www.tandfonline.com/loi/bfsn20. doi:10.1080/10408398.2021.1893643.

Loke Show, P., 2022. Global market and economic analysis of microalgae technology: status and perspectives. Bioresour. Technol. 0960-8524. 357, 127329. doi:10.1016/j.biortech.2022.127329.

Macedo, M.W.F.S., Cunha, N.B.D., Carneiro, J.A., Costa, R.A.D., Alencar, S.A.D., Cardoso, M.H., Franco, O.L., Dias, S.C., 2021. Marine organisms as a rich source of biologically active peptides. Front. Mar. Sci. 2296-7745. 8. Available from: https://www.frontiersin.org/journals/marine-science#. doi:10.3389/fmars.2021.667764.

Majluf, P., Matthews, K., Pauly, D., Skerritt, D.J., Palomares, M.L.D., 2024. A review of the global use of fishmeal and fish oil and the Fish In:Fish Out metric. Sci. Adv. 2375-2548. 10 (42). Available from: https://www.science.org/doi/epdf/10.1126/sciadv.adn5650. doi:10.1126/sciadv.adn5650.

Maldonado-Ruiz, K., Pedroza-Islas, R., Pedraza-Segura, L., 2024. Blue biotechnology: marine bacteria bioproducts. Microorganisms. 2076-2607. 12 (4), 697. doi:10.3390/microorganisms12040697.

Manivasagan, P., Venkatesan, J., Kim, S.K., 2015. Marine algae: an important source of bioenergy production. In: Marine Bioenergy: Trends and Developments. CRC Press, South Korea, pp. 45–70. Available from: http://www.tandfebooks.com/doi/book/10.1201/b18494. doi:10.1201/b18494-8.

Martínez Andrade, K.A., Lauritano, C., Romano, G., Ianora, A., 2018. Marine microalgae with anti-cancer properties. Mar. Drugs. 1660-3397. 16 (5). Available from: http://www.mdpi.com/1660-3397/16/5/165/pdf. doi:10.3390/md16050165.

Martínez-Ruiz, M., Martínez-González, C.A., Kim, D.H., Santiesteban-Romero, B., Reyes-Pardo, H., Villaseñor-Zepeda, K.R., Meléndez-Sánchez, E.R., Ramírez-Gamboa, D., Díaz-Zamorano, A.L., Sosa-Hernández, J.E., Coronado-Apodaca, K.G., Gámez-Méndez, A.M., Iqbal, H.M.N., Parra-Saldivar, R., 2022. Microalgae bioactive compounds to topical applications products—a review. Molecules. 1420-3049. 27 (11). Available from: https://www.mdpi.com/1420-3049/27/11/3512/pdf?version=1653909367. doi:10.3390/molecules27113512.

Matsunaga, T., Takeyama, H., Miyashita, H., Yokouchi, H., 2005. Marine microalgae. In: Advances in Biochemical Engineering/Biotechnology. Springer Science and Business Media LLC, Vol. 96, pp. 165–188. doi:10.1007/b135784.

Mayer, A.M.S., Guerrero, A.J., Rodríguez, A.D., Taglialatela-Scafati, O., Nakamura, F., Fusetani, N., 2021. Marine pharmacology in 2016–2017: marine compounds with antibacterial, antidiabetic, antifungal, anti-inflammatory, antiprotozoal, antituberculosis and antiviral activities; affecting the immune and nervous systems, and other miscellaneous mechanisms of action. Mar. Drugs. 1660-3397. 19 (2), 49. doi:10.3390/md19020049.

Mayer, A.M.S., Pierce, M.L., Howe, K., Rodríguez, A.D., Taglialatela-Scafati, O., Nakamura, F., Fusetani, N., 2022. Marine pharmacology in 2018: marine compounds with antibacterial, antidiabetic, antifungal, anti-inflammatory, antiprotozoal, antituberculosis and antiviral activities; affecting the immune and nervous systems, and other miscellaneous mechanisms of action. Pharmacol. Res. 1096-1186. 183. Available from: http://www.elsevier.com/inca/publications/store/6/2/2/9/3/1/index.htt. doi:10.1016/j.phrs.2022.106391.

Min, R., Liu, Z., Pereira, L., Yang, C., Sui, Q., Marques, C., 2021. Optical fiber sensing for marine environment and marine structural health monitoring: a review. Opt. Laser Technol. 0030-3992. 140, 107082. doi:10.1016/j.optlastec.2021.107082.

Moens, T., Braeckman, U., Derycke, S., Fonseca, G., Gallucci, F., Gingold, R., Guilini, K., Ingels, J., Leduc, D., Vanaverbeke, J., Colen, C.V., Vanreusel, A., Vincx, M., 2013. Ecology of free-living marine nematodes. In: Nematoda, Vol. 2. De Gruyter, Belgium, pp. 109–152. Available from: https://www.degruyter.com/document/doi/10.1515/9783110274257/html.

Mondal, M., Khan, A.A., 2021. Immobilized microalgae for removing industrial pollutants: a greener technique. In: Wastewater Treatment: Cutting-Edge Molecular Tools, Techniques and Applied Aspects. Elsevier, India, pp. 367–384. Available from: https://www.sciencedirect.com/book/9780128218815. doi:10.1016/B978-0-12-821881-5.00018-0.

Moreno-Garrido, I., 2008. Microalgae immobilization: current techniques and uses. Bioresour. Technol. 0960-8524. 99 (10), 3949–3964. doi:10.1016/j.biortech.2007.05.040.

Moshood, T.D., Nawanir, G., Mahmud, F., 2021. Microalgae biofuels production: a systematic review on socioeconomic prospects of microalgae biofuels and policy implications. Environ. Chall. 2667-0100. 5, 100207. doi:10.1016/j.envc.2021.100207.

Moyano, M., Illing, B., Akimova, A., Alter, K., Bartolino, V., Börner, G., Clemmesen, C., Finke, A., Gröhsler, T., Kotterba, P., Livdane, L., Mittermayer, F., Moll, D., von Nordheim, L., Peck, M.A., Schaber, M., Polte, P., 2023. Caught in the middle: bottom-up and top-down processes impacting recruitment in a small pelagic fish. Rev. Fish Biol. Fish. 1573-5184. 33 (1), 55–84. Available from: https://www.springer.com/journal/11160. doi:10.1007/s11160-022-09739-2.

Nazih, H., Bard, J.-M., 2018. Microalgae in Human Health. Elsevier BV, pp. 211–226. doi:10.1016/b978-0-12-811405-6.00010-4.

Occhipinti, P.S., Russo, N., Foti, P., Zingale, I.M., Pino, A., Romeo, F.V., Randazzo, C.L., Caggia, C., 2024. Current challenges of microalgae applications: exploiting the potential of non-conventional microalgae species. J. Sci. Food Agric. 1097-0010. 104 (7), 3823–3833. Available from: http://onlinelibrary.wiley.com/journal/10.1002/(ISSN)1097-0010. doi:10.1002/jsfa.13136.

Popov, A.M., Krivoshapko, O.N., 2013. Protective effects of polar lipids and redox-active compounds from marine organisms at modeling of hyperlipidemia and diabetes. J. Biomed. Sci. Eng. 1937-6871. 06 (05), 543–550. doi:10.4236/jbise.2013.65069.

Resende, D.I.S.P., Ferreira, M., Magalhães, C., Lobo, J.M.S., Sousa, E., Almeida, I.F., 2021. Trends in the use of marine ingredients in anti-aging cosmetics. Algal Res. 2211-9264. 55. Available from: http://www.sciencedirect.com/science/journal/aip/22119264. doi:10.1016/j.algal.2021.102273.

Ridall, A., Ingels, J., 2021. Suitability of free-living marine nematodes as bioindicators: status and future considerations. Front. Mar. Sci. 2296-7745. 8. doi:10.3389/fmars.2021.685327.

Rotter, A., Barbier, M., Bertoni, F., Bones, A.M., Cancela, M.L., Carlsson, J., Carvalho, M.F., Cegłowska, M., Chirivella-Martorell, J., Dalay, M.C., Cueto, M., Dailianis, T., Deniz, I., Díaz-Marrero, A.R., Drakulovic, D., Dubnika, A., Edwards, C., Einarsson, H., Erdoğan, A., Eroldoğan, O.T., Ezra, D., Fazi, S., FitzGerald, R.J., Gargan, L.M., Gaudêncio, S.P., Udovič, M.G., Ivošević Denardis, N., Jónsdóttir, R., Kataržytė, M., Klun, K.,

Kotta, J., Ktari, L., Ljubešić, Z., Lukić Bilela, L., Mandalakis, M., Massa-Gallucci, A., Matijošytė, I., Mazur-Marzec, H., Mehiri, M., Nielsen, S.L., Novoveská, L., Overlingė, D., Perale, G., Ramasamy, P., Rebours, C., Reinsch, T., Reyes, F., Rinkevich, B., Robbens, J., Röttinger, E., Rudovica, V., Sabotič, J., Safarik, I., Talve, S., Tasdemir, D., Schneider, X.T., Thomas, O.P., Toruńska-Sitarz, A., Varese, G.C., Vasquez, M.I., 2021. The essentials of marine biotechnology. Front. Mar. Sci. 2296-7745. 8. Available from: https://www.frontiersin.org/journals/marine-science#. doi:10.3389/fmars.2021.629629.

Rummel, C.D., Löder, M.G.J., Fricke, N.F., Lang, T., Griebeler, E.M., Janke, M., Gerdts, G., 2016. Plastic ingestion by pelagic and demersal fish from the North Sea and Baltic Sea. Mar. Pollut. Bull. 1879-3363. 102 (1), 134–141. Available from: www.elsevier.com/locate/marpolbul. doi:10.1016/j.marpolbul.2015.11.043.

Sanjeewa, K.K.A., 2024b. Exploring the Blue Bioeconomy: Marine Bioresources and Sustainable Applications. CRC Press, Sri Lanka, pp. 1–144. Available from: https://www.taylorfrancis.com/books/mono/10.1201/9781003477365/exploring-blue-bioeconomy-sanjeewa-kka. doi:10.1201/9781003477365.

Sanjeewa, K.K.A., Fernando, I.P.S., Samarakoon, K.W., Lakmal, H.H.C., Kim, E.-A., Kwon, O.N., Dilshara, M.G., Lee, J.-B., Jeon, Y.-J., 2016a. Anti-inflammatory and anti-cancer activities of sterol rich fraction of cultured marine microalga *Nannochloropsis oculata*. Algae 31 (3), 277–287. doi:10.4490/algae.2016.31.6.29.

Sanjeewa, K.K.A., Herath, K.H.I.N.M., 2023. Bioactive secondary metabolites in sea cucumbers and their potential to use in the functional food industry. Fish. Aquat. Sci. 2234-1757. 26 (2), 69–86. Available from: https://www.e-fas.org/info/about. doi:10.47853/FAS.2023.e6.

Sanjeewa, K.K.A., Herath, K.H.I.N.M., Kim, Y.S., Jeon, Y.J., Kim, S.K., 2023. Enzyme-assisted extraction of bioactive compounds from seaweeds and microalgae. TrAC Trends Anal. Chem. 1879-3142. 167. Available from: www.elsevier.com/locate/trac. doi:10.1016/j.trac.2023.117266.

Sanjeewa, K.K.A., Jayawardena, T.U., Lee, H.G., Herath, K.H.I.N.M., Jee, Y., Jeon, Y.J., 2019. The protective effect of *Sargassum horneri* against particulate matter-induced inflammation in lung tissues of an in vivo mouse asthma model. Food Funct. 2042-650X. 10 (12), 7995–8004. Available from: http://pubs.rsc.org/en/journals/journal/fo. doi:10.1039/c9fo02068c.

Sanjeewa, K.K.A., Jeon, Y.-J., 2018. Edible brown seaweeds: a review. J. Food Bioact. 2 (2), 37–50. doi:10.31665/jfb.2018.2139.

Sanjeewa, K.K.A., Kim, E.A., Son, K.T., Jeon, Y.J., 2016b. Bioactive properties and potentials cosmeceutical applications of phlorotannins isolated from brown seaweeds: a review. J. Photochem. Photobiol. B. 1873-2682. 162, 100–105. Available from: www.elsevier.com/locate/jphotobiol. doi:10.1016/j.jphotobiol.2016.06.027.

Sanjeewa, K., 2024a. Chapter 6. marine nematodes. In: Sanjeewa, K. (Ed.), Exploring the Blue Bioeconomy: Marine Bioresources and Sustainable Applications. CRC Press.

Santaniello, G., Nebbioso, A., Altucci, L., Conte, M., 2023. Recent advancement in anticancer compounds from marine organisms: approval, use and bioinformatic approaches to predict new targets. Mar. Drugs. 1660-3397. 21 (1), 24. doi:10.3390/md21010024.

Sao Mai, D., 2014. SHELLFISH (MOLLUSCS AND CRUSTACEA) | characteristics of the groups. In: Encyclopedia of Food Microbiology. Elsevier BV, pp. 376–388. doi:10.1016/b978-0-12-384730-0.00305-0.

Sathya, A.B., Thirunavukkarasu, A., Nithya, R., Nandan, A., Sakthishobana, K., Kola, A.K., Sivashankar, R., Tuan, H. A., Deepanraj, B., 2023. Microalgal biofuel production: potential challenges and prospective research. Fuel. 0016-2361. 332. Available from: http://www.journals.elsevier.com/fuel/. doi:10.1016/j.fuel.2022.126199.

Šimat, V., Čagalj, M., Phadke, G.G., Rathod, N.B., Hassoun, A., 2024. The sea as a source of neuroprotective and other health-protective molecules. In: Natural Molecules in Neuroprotection and Neurotoxicity. Elsevier, Croatia, pp. 29–60. Available from: https://www.sciencedirect.com/book/9780443237638. doi:10.1016/B978-0-443-23763-8.00002-6.

Soliev, A.B., Hosokawa, K., Enomoto, K., 2011. Bioactive pigments from marine bacteria: applications and physiological roles. Evid. Based Complement. Alternat. Med. 1741-4288. 2011. doi:10.1155/2011/670349.

Srimathi, M., Suganthi, M., Sugitha, S., Kumar, K.A., Ramprasath, C., Abirami, G., 2024. Characterization of crude oil degrading marine bacterium *Bacillus licheniformis*. Indian J. Microbiol. 0973-7715. 64 (4), 1697–1709. Available from: https://www.springer.com/journal/12088. doi:10.1007/s12088-024-01222-9.

Stincone, P., Brandelli, A., 2020. Marine bacteria as source of antimicrobial compounds. Crit. Rev. Biotechnol. 0738-8551. 40 (3), 306–319. doi:10.1080/07388551.2019.1710457.

Tammam, M.A., Aly, O., Pereira, F., Mahdy, A., El-Demerdash, A., 2024. Unveiling the potential of marine-derived diterpenes from the order *Alcyonacea* as promising anti-obesity agents. Curr. Res. Biotechnol. 2590-2628. 7, 100175. doi:10.1016/j.crbiot.2024.100175.

The State of World Fisheries and Aquaculture 2024. FAO, (2024), doi: 10.4060/cd0683en.

Urbanek, A.K., Rymowicz, W., Mirończuk, A.M., 2018. Degradation of plastics and plastic-degrading bacteria in cold marine habitats. Appl. Microbiol. Biotechnol. 1432-0614. 102 (18), 7669–7678. Available from: link.springer.de/link/service/journals/00253/index.htm. doi:10.1007/s00253-018-9195-y.

Venugopal, V., Gopakumar, K., 2017. Shellfish: nutritive value, health benefits, and consumer safety. Compr. Rev. Food Sci. Food Saf. 1541-4337. 16 (6), 1219–1242. doi:10.1111/1541-4337.12312.

Vieira de Mendonça, H., Assemany, P., Abreu, M., Couto, E., Maciel, A.M., Duarte, R.L., Barbosa Dos Santos, M.G., Reis, A., 2021. Microalgae in a global world: new solutions for old problems? Renew. Energy. 1879-0682. 165, 842–862. Available from: http://www.journals.elsevier.com/renewable-and-sustainable-energy-reviews/. doi:10.1016/j.renene.2020.11.014.

Wan Mahari, W.A., Waiho, K., Azwar, E., Fazhan, H., Peng, W., Ishak, S.D., Tabatabaei, M., Yek, P.N.Y., Almomani, F., Aghbashlo, M., Lam, S.S., 2022. A state-of-the-art review on producing engineered biochar from shellfish waste and its application in aquaculture wastewater treatment. Chemosphere. 1879-1298. 288. Available from: www.elsevier.com/locate/chemosphere. doi:10.1016/j.chemosphere.2021.132559.

Yuan, S., Li, Y., Bao, F., Xu, H., Yang, Y., Yan, Q., Zhong, S., Yin, H., Xu, J., Huang, Z., Lin, J., 2023. Marine environmental monitoring with unmanned vehicle platforms: present applications and future prospects. Sci. Total Environ. 0048-9697. 858 (Pt 1), 159741. doi:10.1016/j.scitotenv.2022.159741.

Zhang, C., Li, X., Kim, S.-K., 2012. Application of marine biomaterials for nutraceuticals and functional foods. Food Sci. Biotechnol. 21 (3), 625–631. doi:10.1007/s10068-012-0081-6.

Zhivkoplias, E., Jouffray, J.B., Dunshirn, P., Pranindita, A., Blasiak, R., 2024. Growing prominence of deep-sea life in marine bioprospecting. Nat. Sustain. 2398-9629. 7 (8), 1027–1037. Available from: https://www.nature.com/natsustain/. doi:10.1038/s41893-024-01392-w.

Zhuang, D., He, N., Khoo, K.S., Ng, E.P., Chew, K.W., Ling, T.C., 2022. Application progress of bioactive compounds in microalgae on pharmaceutical and cosmetics. Chemosphere. 1879-1298. 291. Available from: www.elsevier.com/locate/chemosphere. doi:10.1016/j.chemosphere.2021.132932.

PART II

Different categories of marine bioresources

CHAPTER

2

Marine microalgae

Sashika Dilshan Warnasuriya[a], *Inoka Batugedara*[b], *and Aixin Hou*[a]

[a]Department of Environmental Sciences, College of the Coast & Environment, Louisiana State University, Baton Rouge, LA, United States

[b]Department of Zoology, University of Sri Jayewardenepura, Nugegoda, Sri Lanka

2.1 Introduction

Algae are a diverse group of organisms that include both photoautotrophic eukaryotic protists and prokaryotic cyanobacteria (commonly known as blue–green algae), despite their unrelated phylogenetic origins (Burgunter-Delamare et al., 2024; Stengel and Connan, 2015; Thoré et al., 2023). Any organism that contains chlorophyll "a" can be considered algae, except for members of the kingdom Plantae, which possess stems, roots, leaves, and flowers (Burgunter-Delamare et al., 2024; Hachicha et al., 2022). They are ubiquitous and can be found in the air as vegetative cells or their propagules, in terrestrial environments associated with water, and in aquatic ecosystems, including freshwater and various saline conditions, ranging from oligohaline to hypersaline (Hachicha et al., 2022; Singh and Saxena, 2015). Algae are taxonomically classified into the major phyla Rhodophyta, Phaeophyta, and Chlorophyta. Based on size, they are categorized into macroalgae and microalgae. Macroalgae are multicellular organisms visible to the naked eye, while microalgae are microscopic (Burgunter-Delamare et al., 2024; Khan et al., 2018).

Microalgae are classified as microscopic algae, ranging in size from 0.5 μm to 1 mm. They are primarily single-celled, although some exceptions exist as simple multicellular organisms (Silva et al., 2019). They can exist in unicellular, colonial, and filamentous forms (Rooij et al., 2021). Most of them become visible when they produce large numbers of cells from the same species, appearing as green, red, or brown patches on the water. Microalgae are predominantly photoautotrophic, whereas some are mixotrophic, meaning they can be both autotrophic and heterotrophic in exceptional cases (Sutherland and Ralph, 2021). According to the available literature, the microalga *Prochlorococcus* sp. is the smallest known photosynthetic organism, with an average cell size of 0.5 μm (Partensky et al., 1999), and the microalga *Ostreococcus tauri* is the smallest known free-living eukaryote, with an average cell size of

Harvesting the Sea
https://doi.org/10.1016/B978-0-443-29190-6.00008-8

0.8 μm (Chrétiennot-Dinet et al., 2019). Microalgae account for about half of the global photosynthetic activity while serving as the major food source for >70% of the total biomass of organisms (Singh and Saxena, 2015; Thoré et al., 2023). To date, around 30,000 species of microalgae have been documented, and it is estimated that there are more than 300,000 species (Hemantkumar and Rahimbhai, 2019). Among the described microalgae, about 90% are eukaryotic, and diatoms account for more than 50% (Hopes and Mock, 2015).

As the main primary producers of virtually all aquatic systems, microalgae play crucial ecological roles in biogeochemical cycles. They contribute to carbon sequestration by facilitating the transfer of atmospheric carbon to ocean sediments (Thoré et al., 2023). Specific groups perform distinct functions. For example, coccolithophores drive the formation of calcium carbonate sediments in the ocean (Krumhardt et al., 2019); cyanobacteria play a vital role in nitrogen fixation, converting atmospheric nitrogen into bioavailable forms in aquatic systems (Li et al., 2018); and diatoms are critical in the global silica cycle (Taucher et al., 2022). In addition, microalgae engage in essential symbiotic relationships, such as mutualistic associations with coral polyps in reef ecosystems and with fungi in lichen formation (Thoré et al., 2023).

Microalgae are taxonomically diverse (Stengel and Connan, 2015) and are categorized into the prokaryotic phylum Cyanophyta and the eukaryotic phyla Euglenozoa, Loukozoa, Metamonada, Choanozoa, Picozoa, Cryptophyta, Haptophyta, Ciliophora, Miozoa, and Heterochontophyta (Hachicha et al., 2022). Due to their complexity in diversity, taxonomy, and phylogeny, various criteria are used for their classification. These include cell shape, cellular organization, the chemical composition of the cell wall, the organization of photosynthetic membranes, storage profile, and pigment composition, which serves as a key identifying factor (Hachicha et al., 2022; Singh and Saxena, 2015). In addition to morphological characteristics, molecular biology techniques play a crucial role in microalgae classification. Genes encoding 16S and 23S rRNA are used for cyanobacteria, while 18S and 28S rRNA serve as molecular markers for eukaryotic microalgae (Hachicha et al., 2022). However, phylogenetic analyses reveal that algae, including microalgae, are heterogeneous at the molecular level (Singh and Saxena, 2015). Hence, a polyphasic approach that integrates both morphological and molecular methods is widely used for their identification and classification. With advancements in molecular technologies, this integrative method is becoming increasingly popular in microalgae classification, similar to its use in classifying other organisms. Microalgae demonstrate a relatively rapid growth rate compared to other photosynthetic organisms (Hachicha et al., 2022). Their photosynthetic efficiency is typically around 3% under average conditions but can reach up to 10% under optimal growth conditions, which is 2–10 times higher than that of higher plants (Bayu et al., 2023). Microalgae can also synthesize complex metabolites using minimal resources while achieving high biomass (Rozenberg et al., 2024). They are regarded as a sustainable solution for renewable energy, bioactive compound synthesis, and food supply, as well as for the mitigation of elevated atmospheric carbon dioxide, wastewater treatment, and the removal of other pollutants (Khan et al., 2018; Rozenberg et al., 2024; Hachicha et al., 2022).

Marine microalgae are those inhabiting marine ecosystems such as open oceans, coastal waters, and brackish waters connected to the ocean. They are a promising solution for meeting nutritional demands and supporting various biotechnological applications, especially in response to the growing global population. Their cultivation requires significantly less land compared to conventional crops or livestock and can be effectively conducted in offshore environments, reducing competition for terrestrial resources, including freshwater (Bayu et al., 2023; Hachicha et al., 2022; Rozenberg et al., 2024). The cultivation of marine microalgae is particularly suitable for regions with extended periods of solar irradiance and proximity to coastal areas (Bayu et al., 2023). Although access to sufficient mono-specific biomass of microalgae in natural systems is limited and they require specific cultivation conditions (Stengel and Connan, 2015), global algae production has increased dramatically, from 0.56 million tons in 1950 to 35.8 million tons in 2019 (Cai and Galli, 2021). China is the world's largest producer of algae, accounting for 57% of global production. The remaining major producers include Indonesia, South Korea, and the Philippines (Kuech et al., 2023). Green microalgae are the most commonly produced, but *Spirulina* dominates production in Europe and globally when cyanobacteria are included (Cai and Galli, 2021). Despite their high diversity, algal cultivation is currently limited to very few species, including macroalgae and microalgae from both freshwater and marine water. Microalgae have been used for aquaculture feed production and are now also utilized as a human nutrient supplement. With the development of new technologies, other biotechnological applications have been increasing as well. To date, approximately 10 genera of microalgae, including *Spirulina, Chlorella, Nannochloropsis, Isochrysis, Muriellopsis, Odontella, Dunaliella, Haematococcus, Rhodella*, and *Porphyridium*, have been employed for industrial purposes, reflecting the economic challenges faced by microalgae-based industries (Bayu et al., 2023).

This chapter focuses on marine microalgae as bioresources, examining their biological features that qualify them as bioresources, their environmental adaptations, symbiotic relationships, and potential applications in nutraceuticals, pharmaceuticals, cosmetics, industry, and environmental sustainability. Furthermore, novel insights into marine microalgae isolation, cultivation, harvesting, and storage techniques are summarized, highlighting methods that enable their use as potential bioresources. Advancements in marine microalgae-related research and limitations are also discussed through a comprehensive analysis of the available literature, while also proposing possible solutions.

2.2 Cell biology of microalgae: A sustainable bioresource

The cellular biology of microalgae reflects their structural and functional characteristics. Each structural and functional component contributes to their value as a bioresource, as illustrated in Fig. 2.1.

Cell coverings: Almost every microalgal cell is covered by an extracellular matrix that provides structural and biochemical support for the cell. Such extracellular matrices include cell walls, mucilage, and scales. The matrix plays an important role in protecting the cell from environmental stressors, providing structural support, and facilitating adhesion, nutrient uptake, morphogenesis, reproduction, and cell-to-cell communication (Ghassemi et al., 2022).

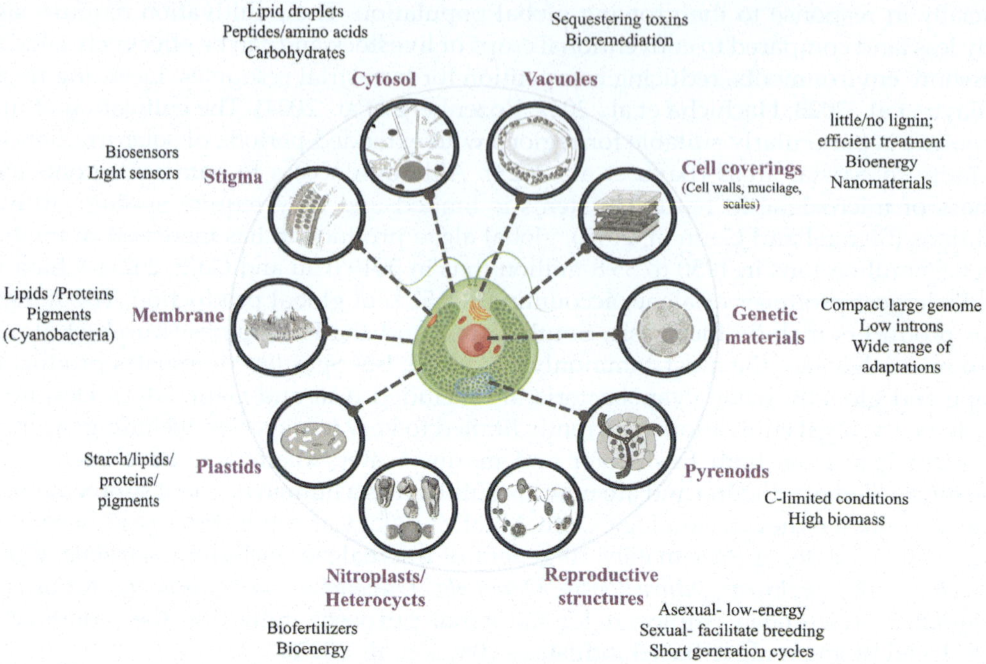

FIG. 2.1 Marine microalgal structural and functional components as bioresource.

Usually, microalgae have rigid, thick, protective cell walls that act as a barrier between the cytoplasm and the extracellular environment (Spain et al., 2021). The composition, complexity, and arrangement of cell walls vary between genera and even among strains of the same species, depending on life stage and growth conditions (Spain and Funk, 2022; Singh and Saxena, 2015). Mehariya et al. (2021) have described three different types of microalgae cell walls: Type I, consisting of two layers that include the innermost microfibrillar layer and an outer layer; Type II, with two microfibrillar layers without an outer layer; and Type III, with one microfibrillar layer. In some cases, such as in *Spirulina* sp., four layers have been observed in the cell wall: an innermost polysaccharide layer, a second peptidoglycan layer, a third polysaccharide layer, and an outermost layer with an acidic polysaccharide covering that is connected to the second layer. In some exceptional cases, microalgae lack a cell wall. For example, *Dunaliella salina*, a biflagellate microalga, is characterized by a mucilaginous coat or a plasticity membrane, which regulates intercellular glycerol levels and enzymatic activities to maintain osmotic balance in saline environments (Okoro et al., 2019). Similarly, *Euglena* sp., belonging to the euglenophytes, lack a cell wall; instead, they possess a pellicle composed of proteinaceous strips supported by microtubules (Borowitzka, 2018; Pierre et al., 2019). Chlorarachniophytes also lack rigid cell walls and feature amoeboid cells with multiple pseudopodia (Pierre et al., 2019). Cyanobacterial cell walls generally consist of a peptidoglycan layer and an outer membrane composed of various components, including proteins, lipids, carotenoids, and, in some cases, lipopolysaccharides (Borowitzka, 2018). Many cyanobacteria have additional layers, such as a polysaccharide sheath surrounding the cell filaments and a more diffuse layer of polysaccharides (Rossi and De Philippis, 2016). Some microalgae, such as diatoms and coccolithophores, possess extracellular scales made of silica and calcium, respectively, and, in some cases, organic scales (Gonçalves and Figueredo, 2020; Borowitzka, 2018). In contrast to other photosynthetic organisms, microalgae lack or contain little lignin material in their cell walls, which facilitates an efficient pretreatment process for microalgae-based industries (Khan et al., 2018).

Membrane: Microalgal cell membranes are a vital bioresource because of their unique characteristics (Gonçalves and Figueredo, 2020). The composition and functionality of microalgal membranes are closely related to their ecological roles, environmental adaptability, and metabolic processes. Microalgal cell membranes are highly enriched with lipids, pigments, and specialized membrane proteins. Structural lipids, such as glycolipids and phospholipids, contribute not only to membrane fluidity and permeability but also to photosynthesis and stress responses (Gonçalves and Figueredo, 2020). The production of high levels of polyunsaturated fatty acids (PUFAs), such as omega-3 and omega-6, including arachidonic acid, γ-linolenic acid, linolenic acid, eicosapentaenoic acid (EPA), docosahexaenoic acid (DHA), and alpha-linolenic acid, also helps maintain cell membrane fluidity while being excellent sources for human use (Kiran and Venkata Mohan, 2021). In addition, betaines and sulfolipids, like lipids, are unique to some microalgae and are responsible for salinity and temperature tolerance (Aveiro et al., 2020). Lipid derivatives, such as oxylipins, contribute to signaling and defense mechanisms in microalgae and are used in humans for inflammation and immune-related diseases (Gabbs et al., 2015).

Depending on their ecological and metabolic requirements, microalgae possess characteristic membrane proteins. Iron-transporting proteins are crucial for ionic balance in nutrient-limited or saline conditions (Sutak et al., 2020). Some unique membrane proteins, such as iron, nitrogen, and phosphorus transporters, are specialized for nutrient uptake (Brawley et al., 2017). These adaptations can be applied in bioengineering and environmental remediation applications. In addition, light-harvesting protein complexes exhibit different arrangements among microalgae species under varying light conditions (Larkum et al., 2020). Moreover, membranes are enriched with photosynthetic pigments, such as chlorophylls, carotenoids, and phycobiliproteins, which have various applications in health and other industries. These applications are discussed in the following sections.

Cytosol: The cytosol of microalgae stores energy in the form of carbohydrate granules (starch, floridean, laminarin) or lipid droplets, which are utilized in several biotechnological applications due to their massive production and storage capabilities. Microalgae can accumulate high lipid content, storing up to 50%–70% of their biomass as oil (Khan et al., 2018).

When considering cytoplasmic microalgal polysaccharides, the production of sulfated exopolysaccharides, known for their antimicrobial, anticoagulant, and antioxidant properties, highlights the potential of microalgae as valuable bioresources (Mousavian et al., 2022). Moreover, the greater diversity and extraction feasibility of microalgal polysaccharides compared to other sources enhance their suitability as bioresources. Meanwhile, microalgal proteins are characterized by their higher amino acid content and biochemical activity compared to other sources of proteins, positioning them as viable alternatives to meet the high demand for human proteins.

Plastids: Microalgal cells contain membrane-bound organelles, whereas prokaryotic cyanobacterial microalgae lack them (Singh and Saxena, 2015). Chloroplasts, chromoplasts, and leucoplasts are the prominent cell organelles found in eukaryotic microalgae. They have several biotechnological applications, including their use in the production of biofuel, biomaterials, bioplastics, and biopolymers, as well as in other environmental applications, such as the development of biohybrid solar cells and the rapid removal of waste carbon dioxide from emissions (Shlosberg et al., 2022). Chloroplasts contain chlorophyll, while chromoplasts primarily consist of carotenoids. Photosynthetic pigments, such as chlorophyll, carotenoids, and phycocyanin (found in cyanobacteria), in microalgae protect them from various stresses (Mehariya et al., 2021). Leucoplasts are colorless and primarily store starch, lipids, and proteins. In general, these plastids are responsible for photosynthesis and carbon fixation, biosynthesis of metabolic compounds, lipid biosynthesis, and the production of many other cellular compounds. Microalgal plastids can be considered biochemical factories. Chlorophyll, phycobiliproteins, and carotenoid pigments have antioxidant properties, which are utilized in nutraceutical, pharmaceutical, and cosmetic applications (Jacob-Lopes et al., 2020; Ren et al., 2021).

Vacuoles: Vacuoles in microalgae contribute to storing ions and metabolites and sequestering toxins, which help them withstand fluctuating aquatic environmental conditions. Recent research has emphasized the role of peroxisomes in the oxidation of lipids and detoxification of reactive oxygen species, which also protect cells against environmental fluctuations (Lacroux et al., 2022).

Nucleus: Generally, the genetic material or the nucleus of microalgae controls the cell by regulating gene expressions. However, the genetic material of microalgae exhibits characteristic features that enable them to serve as a more valuable bioresource than other organisms. Eukaryotic microalgal nuclei have compact genomes with relatively low noncoding regions, which are feasible for genetic manipulation and reduce the complexity of identifying functional genes (Stavridou et al., 2024). Certain microalgae have exceptionally large genomes that contain twice the number of base pairs compared to humans, revealing a wider range of adaptations (Stavridou et al., 2024). Unlike plants, these photosynthetic organisms have short growth cycles, enabling rapid adaptation to genetic changes and reducing experimental duration. Furthermore, microalgal genomes show remarkable flexibility, allowing the integration of exogenous genes from different species, which is typically a limitation in other organisms (Goold et al., 2024; Stavridou et al., 2024). The advancement of new technologies, such as imaging and omics, has focused on elucidating the interactions and dynamics of cell organelles.

Nitrogen-fixing organelles: Marine microalgae can play an important role in meeting nitrogen demand in agriculture, increasing bioenergy production, and enhancing ecological restoration. The nitrogen fixation ability of microalgae, particularly cyanobacterial microalgae, is one of the characteristic features that is crucial in oceanic nutrient cycling. In general, nitrogen fixation is mediated by nitrogenase enzymes, which reduce N_2 into NH_4^+ under anaerobic conditions. It occurs in specialized cellular compartments. For instance, some filamentous cyanobacteria from the orders Nostocales and Stigonematales have heterocysts specialized as anaerobic microenvironments. These morphologically and metabolically differentiated cells have thick cell walls to prevent oxygen diffusion and lack oxygenic photosynthesis (Zeng and Zhang, 2022). Other filamentous cyanobacteria, such as *Trichodesmium* sp., perform nitrogen fixation during the daytime through physiological adaptations such as localized suppression or rapid scavenging of oxygen (Luo et al., 2022). Nitrogen fixation is also observed in diazoplasts, specialized intracellular structures in which cyanobacteria engage in symbiosis with eukaryotic diatoms. These specialized compartments restrict oxygen exposure to the host diatoms, supporting diazotrophy, the biological process of nitrogen fixation (Llamas et al., 2023). Alternatively, some microalgae, such as *Synechococcus* sp. and *Crocosphaera* sp., fix nitrogen during nighttime to prevent oxygen interference by photosynthesis (Masuda et al., 2024; Zeng and Zhang, 2022).

Recently, researchers have discovered a cellular organelle known as the nitroplast, which can fix nitrogen in eukaryotic cells. This cellular structure was observed in *Braarudosphaera bigelowii*, a species of microalgae that is known as the "first nitrogen-fixing eukaryote" (Coale et al., 2024). Genetic analysis revealed that this organelle evolved approximately 100 million years ago from a symbiotic interaction between the algal ancestor and the bacterial species *Atelocyanobacterium thalassa* (Fletcher-Hoppe et al., 2023). Recent advancements in microscopy and molecular techniques, such as metagenomics and transcriptomics, have led to the investigation of the genetic basis, regulatory mechanisms, and structural features of nitrogen fixation organelles in marine microalgae. The findings not only elucidate their ecological role but also highlight their potential in biotechnological applications, opening new paths for the development of novel biofertilizers, bioremediation, and many other uses.

Reproductive structures: Cyanobacterial microalgae reproduce asexually, most commonly through binary fission, sometimes through fragmentation, and in rare cases, through baeocyte formation (Springstein et al., 2020). Under unfavorable conditions, they produce dormant structures called akinetes. Eukaryotic microalgae have both sexual and asexual forms of reproductive mechanisms, although asexual reproduction is the most prominent method. Therefore they do not require much energy to produce reproductive structures and can allocate more energy to trapping light for high biomass generation (Laurens, 2017; Singh and Saxena, 2015), making them potential bioresources for large-scale industrial applications. On the contrary, their sexual reproduction capabilities allow genetic manipulation and breeding to enhance their applications in bioresource technologies.

Physiology: The physiology of microalgae is remarkably different from that of other organisms, particularly with respect to their incredible adaptability, efficient metabolism, and diverse biochemical pathways. They have physiologically adapted to performing photosynthesis with a broad spectrum of light waves through the use of chlorophylls and other accessory pigments. This adaptation enables them to survive in light-limited environments, such as deep or turbid water. Pyrenoids, specialized organelles found in algae, including most microalgae, enhance photosynthetic efficiency in carbon dioxide-limited conditions. This carbon-concentrating mechanism increases carbon dioxide concentration around the enzyme RuBisCO through the involvement of pyrenoids (Barrett et al., 2021), which enables the use of microalgae as bioresources for carbon-neutral applications. Interestingly, depending on environmental conditions, microalgae exhibit remarkable metabolic flexibility. They can switch between photoautotrophic, heterotrophic, and mixotrophic modes of nutrient acquisition, enhancing survivability (Diankristanti and Ng, 2024). Under controlled growth conditions, this adaptability in nutrient acquisition can be applied to optimize industrial processes. Furthermore, microalgae can produce osmoprotectants, such as glycerol and betaines, which allow their survival in high-salinity conditions. These osmoprotectants can be effectively used in wastewater treatment, bioenergy production, and bioremediation. Under natural environmental conditions, microalgae can change their cell and colony size. During nutrient stress, cell diameters are largest in winter and smallest in summer (Yan et al., 2021). Small microalgae cells have a higher surface area-to-volume ratio, facilitating efficient nutrient uptake. They also exhibit a lower half-saturation constant for nutrient uptake and a higher growth rate (Yan et al., 2021), allowing them to thrive in low-nutrient environments.

2.3 Marine microalgal adaptations toward bioresource development

Due to their high adaptability, microalgae are ubiquitously distributed in the biosphere and can survive in varying environmental conditions, such as the oceanic surface layer, subsurface water column, the limits of the photic zone, and coastal regions where temperature, salinity, light intensity, pH, and nutrient availability greatly fluctuate (Gatamaneni et al., 2018; Hachicha et al., 2022). Microalgae are dominant in marine ecosystems (Hopes and Mock, 2015). Microalgae in coastal zones are adapted to turbulence, high nutrients, and low light intensities, while in the open ocean, they are adapted to high radiation and low nutrient levels (Hopes and Mock, 2015). In marine ecosystems, the distribution of microalgae depends on tidal movements, turbulence, and sea depth and is influenced by their chromatic adaptations (Hachicha et al., 2022). An overview of the adaptability and applicability of marine microalgae is illustrated in Fig. 2.2.

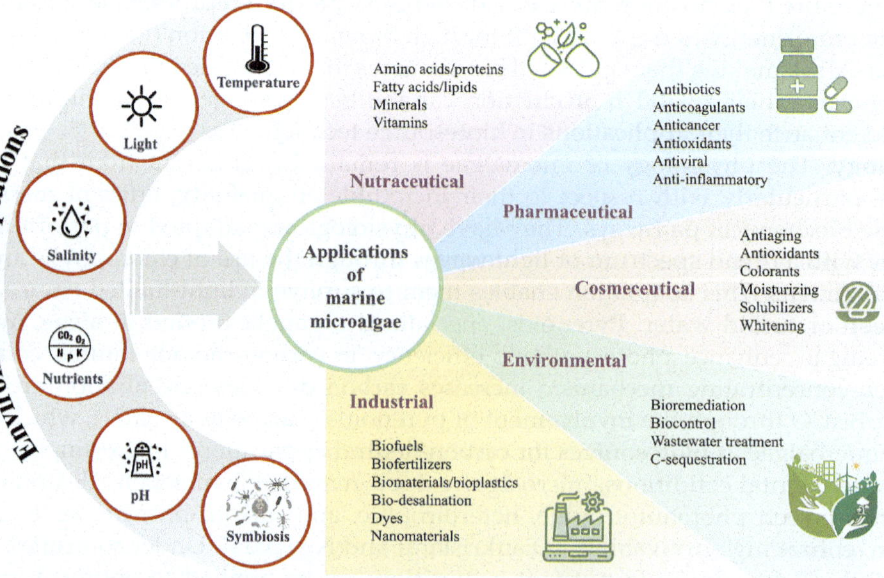

FIG. 2.2 Overview of the adaptability and applicability of marine microalgae.

Marine microalgae have been widely used in food, nutraceutical, pharmaceutical, cosmeceutical, industrial, and environmental applications. Some of their uses are summarized here.

Temperature: Temperature is a critical factor in microalgal growth and other metabolisms (Rezaei et al., 2024). Under cold conditions, microalgae change their cell wall permeability and strength, typically through their biopolymer algaenan or sporopollenin (González-Hourcade et al., 2023). Generally, lower temperatures increase the amount of PUFAs in the cytoplasmic membrane to maintain fluidity (González-Hourcade et al., 2023; Rezaei et al., 2024). Cold-adapted microalgae tolerate dense water while maintaining enzymatic activities in the cells (Kumari et al., 2022). The ability to cultivate these cold-adapted microalgae without controlled heat or light systems reduces costs, rendering them crucial as bioresources (Rezaei et al., 2024; Kumari et al., 2022). The optimal temperature range for most algae is 20°C–30°C (Khan et al., 2018), and no photosynthetic microalgae have been reported to grow beyond 75°C (Kumari et al., 2022). Carbon assimilation is reduced at lower temperatures. At high temperatures, photosynthetic activity is reduced primarily due to the degradation of enzymes or proteins and changes in energy balance in the cells (Khan et al., 2018). The effect of high temperatures is more deleterious than that of low temperatures on microalgal growth (Kumari et al., 2022).

Light: Effects of temperature and light intensity on microalgae are correlated (Hachicha et al., 2022). Microalgae in high-latitude regions are exposed to extremely low temperatures and darkness in winter, while being subjected to continuous light and high UV during summers (Kumari et al., 2022). Microalgae growth efficiency is low at both very low and very high light intensities. Optimum growth occurs at a specific light intensity that enhances carbon assimilation through photosynthesis while minimizing photorespiration and photoinhibition (Khan et al., 2018). On the contrary, both light and dark cycles of photosynthesis are important, as ATP and NADPH are produced during the light reaction, and carbon assimilation occurs during the dark reaction of photosynthesis (Khan et al., 2018). Interestingly, the composition of cellular lipids plays an important role in adapting to varying light intensities. Membrane lipids increase under low light intensities, and lipids are stored under high light conditions (Kumari et al., 2022). Accessory pigments, including carotenoids and phycobilins, enhance the ability to capture light in environments with limited light availability as an adaptation. In addition, in response to excess light energy, mechanisms like nonphotochemical quenching prevent photoinhibition by redirecting absorbed energy into heat dissipation (Ruban, 2016).

pH: Marine microalgae have adapted to survive in typical oceanic conditions, which are slightly alkaline with a pH of approximately 8.1. The alkalinity is regulated by the carbonate buffering system, and these conditions are favorable for photosynthesis, biochemical and physiochemical processes, and the growth of most microalgae. Microalgae possess carbon-concentrating mechanisms that help ribulose-1,5-bisphosphate carboxylase/oxygenase (RuBisCO) capture CO_2 efficiently (Mondal et al., 2016). In the carbonate buffering system, the carbonic anhydrase enzyme, which is a zinc-containing metalloenzyme, is usually overexpressed in microalgae (Li et al., 2018; Mondal et al., 2016). This enzyme catalyzes the interconversion of CO_2 and bicarbonate, ensuring a sufficient carbon supply for photosynthesis (Mondal et al., 2016). Bicarbonate is the dominant inorganic carbon source at a pH of around 8.1, which enhances carbon assimilation in microalgae (Zhu et al., 2020).

Marine pH levels are compatible with microalgal metabolic requirements. Essential nutrients, such as iron and organic acids, are stable at pH 8.1, maintaining effective nutrient uptake (Pérez-Rojas et al., 2022). Moreover, microalgal cellular components, including membrane phospholipids, remain stable under slightly alkaline conditions (Kumari et al., 2022). Nevertheless, certain microalgae have been observed to exhibit optimum growth at pH levels ranging from acidic to alkaline (Dolganyuk et al., 2020; Peng et al., 2020). Therefore they are adaptable to varying pH levels and are resilient to ocean acidification, although some of them, such as coccolithophorids, are very sensitive to small changes in pH (Kholssi et al., 2023). The adaptations of marine microalgae to varying pH conditions indicate their evolutionary relationship to changing aquatic environments, such as coastal regions and estuaries.

Salinity: Microalgae have various adaptations to salinity changes, which significantly influence their lipid content, production, and overall growth. High salinity levels increase lipid production and accumulation, while low salinity levels reduce cell growth (Shetty et al., 2019). The high lipid levels serve as an energy reserve and stabilize the membrane under high salinity conditions. Microalgae adapt to salinity level fluctuations through homeostasis and acclimatization. The presence of Na^+-transporting ATPase in the plasma membrane of marine microalgae is one example of salinity adaptation. From an evolutionary perspective, this enzyme cannot be observed in higher vascular plants in low-salinity environments (Matalin et al., 2021; Popova and Balnokin, 2013). This enzyme actively pumps sodium ions (Na^+) out of the cells while maintaining the intracellular ion concentration. However, unlike in other organisms, it does not transport potassium-like ions alongside sodium, as an adaptation to high-salinity environments (Matalin et al., 2021). Furthermore, the synthesis of glycerol, proline, and betaine-like compatible solutes helps with osmoregulation in microalgal cells (Lauritano et al., 2020). The use of salt-tolerant microalgae as bioresources can reduce competition for freshwater resources and increase lipid production under highly saline conditions in biotechnological applications.

Nutrient: Marine microalgae are adapted to nutrient-depleted conditions. Many of them have been shown to increase their lipid content in the form of triglycerides under nitrogen and phosphorus-limited conditions (Zhu et al., 2016). Nitrogen limitation triggers lipid synthesis, while phosphorus limitation enhances lipid accumulation (Gatamaneni et al., 2018). This is because, under limited essential nutrients, algae shift metabolic pathways from growth and division to an energy storage stage as a survival strategy (Ran et al., 2019). Similarly, they also accumulate starch under nutrient-limited conditions (Ran et al., 2019). Researchers have found that during nitrogen limitation, microalgal metabolism shifts from protein synthesis to lipid storage, a process influenced by the intercellular C/N ratio (Gatamaneni et al., 2018). Typically, marine environments are nitrogen and phosphorus-limited, and microalgae obtain these nutrients from synergistic interactions with other microorganisms, such as nitrogen-fixing, nitrifying, and phosphorus-solubilizing bacteria (Rozenberg et al., 2024). For example, the cocultivation of marine microalgae *Nannochloropsis salina* with heterotrophic bacteria has supported its growth under nutrient-limited conditions (Poddar et al., 2018). Optimizing environmental conditions such as pH, aeration, light exposure, and nutrient content further enhances productivity. In addition, strategies such as genetic engineering for pigment and nutrient synthesis have been shown to boost microalgal efficiency beyond bacterial interactions or CO_2 fixation alone (Rozenberg et al., 2024).

2.4 Harnessing microalgal symbiotic relationships

Associations between microalgae and other organisms enhance their potential as valuable bioresources. The symbiotic relationships can improve microalgal growth and their resistance to different environmental conditions (Hasnain et al., 2023; Rasheed et al., 2023). Most of the time, interactions between microalgae and other organisms exhibit synergistic relationships. Microalgae can provide carbon and energy sources for associated organisms, while host organisms provide growth stimulatory compounds such as antioxidants, siderophores, indole acetic acids, and vitamins (Rozenberg et al., 2024; Sandhya and Vijayan, 2019). Hosts also benefit from being protected from extreme environmental conditions and predation or grazing (Mutalipassi et al., 2021). In addition, hosts facilitate favorable environmental conditions by reducing oxygen tension to support microalgal growth (Sandhya and Vijayan, 2019). Symbiotic interactions have also shown high bioremediation potential in polluted aquatic environments while converting waste into valuable resources (Hasnain et al., 2023). In general, the use of microalgal symbiosis is gaining significant interest as a potential alternative for reducing production costs, increasing biomass, and producing derived metabolic products (Llamas et al., 2023).

Cyanobacteria interact with fungi, bacteria, diatoms, macroalgae, seagrasses, sponges, and other invertebrates (Mutalipassi et al., 2021). Usually, most cyanobacterial microalgae fix nitrogen, which helps host organisms thrive in nitrogen-limited conditions (Llamas et al., 2023; Mutalipassi et al., 2021). Symbiosis often activates silenced biosynthetic pathways that lead to the production of specific bioactive compounds (Mutalipassi et al., 2021), presenting opportunities for the food, cosmeceutical, nutraceutical, and pharmaceutical industries.

Interactions between microalgae and fungi play important roles in microbial food webs and biogeochemical cycles (Lauritano and Galasso, 2023). Microalgal symbiosis transfers photosynthetically fixed carbon to parasitic fungi, which subsequently affects bacterial colonization on microalgae (Tirichine and Piganeau, 2023). The coexistence of these organisms in the same habitats, including water columns and benthic substrates, facilitates biological, physical, and chemical interactions such as predation, nutrient search, and proliferation. Recent studies suggest that metabolites such as octadecadienoic acids and β-carbolines produced by fungi exposed to microalgae exhibit anticancer properties, highlighting the ecological and industrial importance of these metabolites (Lauritano and Galasso, 2023). Microalgae and fungi interactions further facilitate bioflocculation, in which fungi are involved in aggregating microalgal cells to form larger flocs (Ray et al., 2021). This significantly reduces time, cost, and energy requirements for harvesting microalgae, as well as avoiding the use of chemical flocculation. In addition, recent studies demonstrate that microalgal–fungal consortia have been effectively used to enhance wastewater treatment and overcome the limitations of single-strain decontamination systems. Synergistic interactions of these organisms have shown improvement in decontamination rates, cell aggregation, and harvesting efficiency (Lauritano and Galasso, 2023).

Marine microalgae exhibit complex interactions with bacteria in the photic zones of oceans, and chemical mediators, especially sulfur metabolites, are responsible for most of these interactions (Burgunter-Delamare et al., 2024). In these interactions, bacteria usually play an important role in enhancing nutrient uptake and vitamin production (Wang et al., 2024).

In addition, bacteria can reduce iron deficiency in microalgae through siderophore formation, whereas microalgae provide attachment sites, an oxygen-rich environment, and organic sources for bacterial growth (Chen et al., 2023; Wang et al., 2024). Model organisms such as the diatoms *Phaeodactylum tricornutum* and *Thalassiosira pseudonana* are used to study these symbioses. Research has shown that in the interaction between the coccolithophore *Emiliania huxleyi* and the bacterium *Phaeobacter gallaeciensis*, the bacteria produce growth-promoting phenylacetic acid and antibiotics to protect algae from other pathogenic bacteria, while the algae provide carbon and sulfur sources and adhesion sites (Burgunter-Delamare et al., 2024). This microalgae–bacteria interaction is practically applied in wastewater treatment, where bacteria create favorable conditions for microalgae to increase their biomass and lipid productivity. In the meantime, these coculture systems reduce operational costs by harvesting biomass through bioflocculation (Chen et al., 2023).

2.5 Applications of marine microalgae

2.5.1 Overview and history

Microalgae have been used since ancient civilizations in human history. About 1500 years ago, the Chinese used the cyanobacterium *Nostoc* sp. during famines, while the Aztecs and Chad people used *Spirulina* as a protein source (Mendes et al., 2022). Besides that, traditionally, microalgae such as *Nostoc* sp. were consumed in regions of China, Mongolia, Tartaria, and South America. In Japan, "suizenji-nori," derived from the cyanobacterium *Aphanotheca sacrum*, is used in their food culture. Southeast Asians in Burma, Thailand, Vietnam, and India have also historically used *Spirogyra* sp. and *Oedogonium* sp. in their diets. Furthermore, Spanish fishermen have prepared microalgae-based dry cakes, demonstrating early attempts at preservation and processing techniques in human history (García et al., 2017). In the 20th century, there was a significant shift toward exploring microalgae as a potential bioresource beyond nutrients. In the early stages of microalgal technology development, scientists developed microalgae as a protein supplement. Starting from that point, researchers explored the potential of microalgae, such as *Chlorella* and *Spirulina*, even for space missions (Karkos et al., 2011; Mapstone et al., 2022). Marine microalgae have been widely used in food and nutraceutical, pharmaceutical, cosmeceutical, industrial, and environmental applications. Some of their uses are summarized in Table 2.1 and Fig. 2.2.

2.5.2 Food and nutraceutical applications

In recent years, consumer preferences have shifted toward natural and sustainable sources of health-enhancing compounds. Nutraceuticals are commonly recognized for their ability to confer health advantages. The marine nutraceuticals and functional food sectors have demonstrated remarkable expansion over the past decades. Among them, microalgae are promising sources for nutraceutical applications, owing to their diverse bioactive compounds, including proteins, peptides, enzymes, minerals, PUFAs, pigments, vitamins, and phenolic compounds. A variety of microalgae species, including *Chlorella* sp., *Dunaliella* sp., *Arthrospira platensis*, *Scenedesmus obliquus*, *Nannochloropsis* sp., *Isochrysis galbana*, and *Haematococcus*

TABLE 2.1 A summary of applications of marine microalgae.

Application	Compound group/application subsector	Microalgae genus/species	Reference (s)
Food and nutraceuticals	Amino acids and proteins	*Tetraselmis* sp.	Schwenzfeier et al. (2011)
		Dunaliella salina	Tibbetts et al. (2015)
		Arthrospira sp.	Kiran and Venkata Mohan (2021)
		Chlorella sp.	
		Dunaliella sp.	
		Phaeodactylum tricornutum	Villar-Navarro et al. (2021)
		Isochrysis galbana	
	Fatty acids and lipids	*Schizochytrium* sp.	Guo et al. (2018)
		Chaetoceros calcitrans	Sahin et al. (2018)
		Chaetoceros muelleri	
		Crypthecodinium cohnii	
		Monodus subterraneus	
		Nannochloropsis	
		Nitzchia sp.	
		Pavlova salina	
		Phaeodactylum cruentum	
		Phaeodactylum tricornutum	
	Vitamins and minerals	*Dunaliella tertiolecta*	Koyande et al. (2019)
		Tetraselmis suecica	
	Human food	*Chlorella vulgaris*	Kusmayadi et al. (2021)
		Dunaliella sp.	
		Spirulina sp.	
		Crypthecodinium cohnii	Wu et al. (2021)
		Schizochytrium sp.	
		Tetraselmis chuii	Qazi et al. (2021)
		Chlorella minutissima	Ben Atitallah et al. (2019)
		Isochrysis galbana	
		Picochlorum sp.	
	Animal feed	*Arthrospira maxima*	Samarakoon and Jeon (2012)
		Schizochytrium sp.	Urrutia et al. (2016)
		Spirulina platensis	Zahroojian et al. (2011); Alagawany et al. (2021)
		Haematococcus pluvialis	Lorenz and Cysewski (2000)

Continued

TABLE 2.1 A summary of applications of marine microalgae—cont'd

Application	Compound group/application subsector	Microalgae genus/species	Reference (s)
		Chlorella stigmatophora	Guedes et al. (2015)
		Dunaliella tertiolecta	
		Isochrysis galbana	
		Tetraselmis suecica	
		Nannochloropsis sp.	Brown et al. (1997), Mamdouh et al. (2021)
		Chaetoceros sp.	Brown et al. (1997)
		Chlorella sp.	
		Pavlova sp.	
		Skeletonema sp.	
		Thalassiosira sp.	
		Tetraselmis sp.	
		Staurosira sp.	Pereira et al. (2020)
		Tetraselmis sp.	
Pharmaceuticals	Polysaccharides	*Cochlodinium polykrikoides*	Matsui et al. (2003)
		Porphyridium sp.	
		Gyrodinium impudium	Guo et al. (2017)
		Tribonema sp.	Li et al. (2016)
		Chlorella sp.	Song et al. (2018), Vo et al. (2011)
		Isochrysis sp.	Raposo et al. (2013)
		Porphyridium cruentum	
		Porphyridium purpureum	
		Tetraselmis sp.	
		Gelidiella acerosa	da Silva Chagas et al. (2020)
		Porphyridium sp.	Hans et al. (2021), Sami et al. (2021)
		Gyrodinium impudicum	Amaro et al. (2011)
		Spirulina sp.	Claus-Desbonnet et al. (2022)
		Arthrospira platensis	Radonic et al. (2010)
		Porphyridium purpureum	
	Lipids	*Tetraselmis* sp.	Talero et al. (2015)
		Nannochloropsis oculate	
		Isochrysis sp.	Anusree et al. (2023)
		Chaetoceros calcitrans	

Application	Compound group/application subsector	Microalgae genus/species	Reference (s)
		Nannochloropsis sp.	
		Schizochytrium sp.	
		Thalassiosira sp.	
	Peptides and proteins	*Navicula* sp.	Hempel et al. (2017)
		Phaeodactylum tricornutum	
		Tetraselmis sp.	Guzman et al. (2019)
		Scenedesmus subspicatus	Dantas et al. (2019)
	Carotenoids	*Chlorella vulgaris*	Bule et al. (2018), Sathisvam and Ki (2018)
		Chlorella zofigiensis	
		Chlorococcum sp.	
		Haematococcus pluvialis	
		Chlorella sp.	Perry et al. (2009), Saha et al. (2020)
		Dunaliella salina	
		Spirulina sp.	
		Chlorella ellipsoidea	Kim et al. (2020)
		Dunaliella tertiolecta	
		Phaeodactylum tricornutum	Bulet et al. (2018), Talero et al. (2015), Zhao et al. (2015)
		Porphyridium cruentum	Bermejo Román et al. (2002), Gamal and Shreadah (2024)
		Chlorella vulgaris	López and Soto (2020)
		Dunaliella salina	
		Amphora subtropica	Sasirekha et al. (2019)
Cosmeceuticals	–	*Nostoc flagelliform*	Hamed (2016)
		Porphyridium sp.	
		Skeletonema sp.	
		Dunaliella salina	Hamed (2016), Morocho-Jacome et al. (2020)
		Haematococcus pluvialis	
		Porphyridium	
		Spirulina sp.	
		Nannochloropsis salina	Koller et al. (2014)
		Nannochloropsis gaditana	Koller et al. (2014), Yarkent et al. (2020)
		Nannochloropsis oculata	
		Dunaliella tertiolecta	Brown et al. (1999)

Continued

TABLE 2.1 A summary of applications of marine microalgae—cont'd

Application	Compound group/application subsector	Microalgae genus/species	Reference (s)
		Tetraselmis suecica	
		Chlorella vulgaris	Gamal and Shreadah (2024), Stoyneva-Gärtner et al. (2020)
		Chondrus crispus	Resende et al. (2021)
		Kappaphycus alvarezii	
		Laminaria digitata	
Industrial applications	Biofuels	*Dunaliella* sp.	Gamal and Shreadah (2024)
		Scenedesmus obliquus	
		Botryococcus braunii	Khan et al. (2018)
		Chlorella sp.	Ievina and Romagnoli (2024)
		Chlamydomonas sp.	Iqbal et al. (2022), Fakhimi et al. (2024)
		Nannochloropsis sp.	
	Biofertilizers	*Chlorella vulgaris*	Dineshkumar et al. (2019)
		Spirulina platensis	
		Tetraselmis	Jochum et al. (2018), Saadaoui et al. (2019)
		Nannochloropsis	
	Biomaterials	Diatoms	Appiah et al. (2024)
	Bioplastics	*Chlamydomonas* sp.	Madadi et al. (2021), Tennakoon et al. (2023)
		Chlorella sp.	
		Klebsormidium flaccidum	
		Micractinium sp.	
		Nannochloropsis salina	
	Bio desalination	*Chlorella vulgaris*	Patel et al. (2021), Sahle-Demessie et al. (2019), Moayedi et al. (2021)
		Dunaliella salina	
		Nannochloropsis oculate	
		Scenedesmus quadricauda	
	Dyes	–	https://www.algaeing.com
	Nanomaterials	*Anabaena flosaquae*	Sidorowicz et al. (2023), Rosyidah et al. (2024)
		Calothrix pulvinata	
		Leptolyngbya foveolarum	
		Synechococcus moorigangae	
		Arthrospira platensis	Saravanakumar et al. (2024)
		Chlorella sp.	

Application	Compound group/application subsector	Microalgae genus/species	Reference (s)
		Dunaliella salina	
		Trichodesmium erythraeum	
Environmental applications	Bioremediation and wastewater treatment	*Synechococcus*	Srimongkol et al. (2019)
		Dunaliella salina	Liu and Yildiz (2018)
	Carbon sequestration	*Dunaliella tertiolect*	Chow et al. (2013)
		Scenedesmus obliquus	Alcántara et al. (2015)
		Dunaliella tertiolect	Chow et al. (2013)
		Botryococcus braunii	Ighalo et al. (2022)
		Chlorella kessler	
		Chlorella sp.	
		Cholorococcum littorale	
		Dunaliella salina	
		Isochrysis galbana	
		Nannochloropsis	
		Spirulina platensis	
	Biocontrol	*Arthrospira platensis*	Mishra et al. (2022)
		Chlorella minutissima	
		Chlorella vulgaris	
		Dunaliella salina	
		Haematococcus pluvialis	
		Haslea karadagensis	
		Isochrysis sp	
		Nannochloropsis sp	
		Skeletonema	
		Synechococcus elongates	
		Synechocystis sp.	
		Tetraselmis chui	
		Tetraselmis suecia	
		Thalassiosira	

pluvialis, are being extensively cultivated in the industrial sector for the production of food and dietary supplements.

Amino acids and proteins: Microalgae can be utilized as a viable protein source in food industries due to their high protein content, ranging from 40% to 60% (w/w) (Bleakley and Hayes, 2017). As reported by Schwenzfeier et al. (2011), *Tetraselmis* sp. contains 65% w/w protein in total, while *D. salina* has a protein content of 57% (w/w) (Tibbetts et al., 2016). The entire cellular biomass of *Arthrospira* sp., *Dunaliella* sp., and *Chlorella* sp. is commercially accessible as a protein supplement compliant with FAO/WHO standards (Kiran and Venkata Mohan, 2021). Microalgae possess the capability to generate nearly all essential and nonessential amino acid compounds. Therefore they are an ideal vegan protein alternative as they meet the essential amino acid profile (Kusmayadi et al., 2021). *P. tricornutum* and *I. galbana* have essential amino acid profile percentages of 57.07% and 48.7%, respectively (Villar-Navarro et al., 2021).

Fatty acids and lipids: Microalgal lipids are abundant in vital long-chain PUFAs, such as DHA and EPA, which are necessary components of a regular dietary regimen, as they cannot be synthesized naturally by humans or many animals. DHA has important functions in the human body. Owing to their richness in unsaturated fatty acids, microalgae have been proposed as a viable and sustainable source of PUFAs for incorporation into nutraceuticals. *Schizochytrium* sp. is a viable source of DHA, capable of accumulating up to 30% DHA of its total fatty acids (Guo et al., 2018). *P. tricornutum, Monodus subterraneus, Phaeodactylum cruentum, Chaetoceros calcitrans, Nannochloropsis, Crypthecodinium cohnii, I. galbana, Pavlova salina, Chaetoceros muelleri,* and *Nitzchia* sp. are some microalgal strains that are rich in PUFAs (Sahin et al., 2019).

Vitamins and minerals: Similar to macronutrients, vitamins and minerals are essential for growth and maintenance. Marine microalgae are considered a superior source of vitamins (Mahata et al., 2022). Microalgae contain vitamins B1, B2, B6, B12, C, E, niacin, biotin, and folic acid (Raposo and De Morais, 2015). *Spirulina* sp. is rich in vitamins A and B complex, which directly influence cognitive functions, cellular metabolism, and infection prevention. *Dunaliella tertiolecta* is an excellent source of vitamin B2, vitamin B12, and vitamin E, while *Tetraselmis suecica* is capable of synthesizing vitamin C and vitamin B complex (Koyande et al., 2019). Microalgae have a rich and diverse mineral makeup, with calcium, potassium, phosphorus, magnesium, iron, sodium, sulfur, zinc, and copper being the most prevalent (Fox and Zimba, 2018). The mineral composition of microalgae varies from 2.2% to 4.8% of total dry weight (Guedes et al., 2015).

Marine microalgae as human food: Algae-based compounds are more acceptable in food industries due to their natural intrinsic traits. *Chlorella vulgaris* is a popular edible microalga, and its safety is well-established. The global market is dominated by *Spirulina* sp., followed by *Chlorella* sp. (Levasseur et al., 2020). *Spirulina* sp. and *Dunaliella* sp., like other common marine microalgae, are consumed as sources of functional food in different forms, including dried powder, tablets, and capsules. They are used in various food items, including noodles, snacks, sweets, cookies, and beverages (Kusmayadi et al., 2021). The marine diatom *Odontella aurita* is approved as a functional food and has been commercialized (Fu et al., 2015). The US Food and Drug Administration approved the use of DHA from *C. cohnii* in 2001 and from *Schizochytrium* sp. in 2015 as dietary supplements for humans (Wu et al., 2021). Qazi et al. (2021) increased the protein content in bread with the use of *Tetraselmis chui*. The incorporation of *Spirulina* sp.

biomass into extruded snacks yielded a nutritional enhancement of 22.6% in proteins, 28.1% in lipids, and 46.4% in minerals without significantly affecting the product's physical qualities (Lucas et al., 2018). The addition of *C. minutissima*, *I. galbana*, and *Picochlorum* sp. powder into canned fish burgers increased sensory acceptability and texture (Ben Atitallah et al., 2019).

β-Carotene is used as a food colorant in various foods, beverages, and additives (Bule et al., 2018). Zeaxanthin is used as a yellow colorant in the food sector. Phycoerythrin, isolated from *Porphyridium cruentum*, is utilized as a food marker (Roman et al., 2002). Another nutritional consideration is the possible utilization of microalgae polysaccharides as a source of fiber and prebiotics (Raposo et al., 2013). The application of microalgae is important in the preservation, quality improvement, and shelf-life extension of seafood (Tavakoli et al., 2022).

Marine microalgae as animal feeds: Due to their capability to produce bioactive compounds, marine microalgae are a promising source of feed and supplements, either as whole biomass or specific metabolites. The integration of microalgae into the feedstock has yielded numerous physiological benefits for animals, including an improved feed conversion ratio and digestibility, enhanced immune response, increased resistance to disease and stress, greater weight gain and productivity, elevated survival rates, superior sensory and nutritional quality of end products (such as meat, eggs, and milk), and enhanced reproductive performance (Saadaoui et al., 2021).

Microalgal nutritional components are optimal for the diets of cattle, chickens, goats, and pigs, enhancing meat quality through increased PUFAs and amino acids (Wells et al., 2017). Meat quality has been improved with the addition of *Arthrospira maxima* (Samarakoon and Jeon, 2012) and *Schizochytrium* sp. (Urrutia et al., 2016) in swine and lamb feed, respectively.

Microalgal pigments hold significant importance in animal feed preparations. β-carotene has been employed in animal feed and has been shown to decrease the risks of age-related macular degeneration (Bule et al., 2018). Moreover, marine *Spirulina platensis* can be incorporated into hens' meals to improve egg yolk pigmentation by substituting synthetic pigments (Zahroojian et al., 2011). Fucoxanthin is utilized in the poultry industry to enhance the meat coloration of broiler chickens (Sasaki et al., 2010). Lutein, zeaxanthin, and canthaxanthin can also be employed in poultry to enhance chicken skin color, serving as an alternative to synthetic coloring additives (Esatbeyoglu et al., 2016). In aquaculture, astaxanthin derived from *H. pluvialis* is used as a pigmentation agent for salmon flesh (Lorenz and Cysewski, 2000).

Marine microalgae can be utilized as mineral sources for animal diets. Marine *S. platensis* is a model mineral source in feed formulation (Alagawany et al., 2021). For fish, *Chlorella stigmatophora*, *T. suecica*, *I. galbana*, and *D. tertiolecta* can be incorporated as mineral sources. The incorporation of mineral-rich microalgae into salmon diets has improved the flavor and texture of the fish (Guedes et al., 2015).

Microalgal biomass can be included in the feeds of domestic animals such as dogs, cats, as well as aquarium fish (Rizwan et al., 2018). *Chlorella*, *Tetraselmis*, *Isochrysis*, *Pavlova*, *Phaeodactylum*, *Chaetoceros*, *Nannochloropsis*, *Skeletonema*, and *Thalassiosira* species are utilized as larval food for mollusks and penaeid shrimp cultivations (Brown et al., 1997). Integrating *Nannochloropsis oculata* into tilapia diets has been shown to diminish mercury bioaccumulation while enhancing hematological health and alleviating organ damage (Mamdouh et al., 2021).

Defatted microalgal biomass, leftovers from lipid extraction for biofuel production, is used as an ingredient in animal feed. Soybeans in aquafeed and broiler chicken feed can be replaced by deoiled biomass of *Tetraselmis* sp. and defatted diatom *Staurosira* sp., respectively (Pereira et al., 2020). However, strain selection for animal feed must prioritize high digestibility and minimal toxin accumulation.

2.5.3 Pharmaceutical applications

Microalgae are ideal platforms for meeting the substantial demand for biopharmaceutical production. In addition, these factories offer a diverse array of pharmaceutical goods and medications that are either inaccessible or too costly to create using human cell lines or plant sources (Sharma et al., 2017). The pharmaceutical applications of microalgae rely on their antioxidant characteristics, anticancer activity, antiinflammatory effects, antibacterial capabilities, etc. (Anusree et al., 2023). High-value products generated from microalgae can be advantageous in health applications, serving as antioxidants, immunological boosters, neuroprotective agents, vitamin precursors, and many more. Potential therapeutic applications of some selected metabolites are briefly described in the following.

Marine microalgae-derived polysaccharides are prospective agents for antioxidant, anticancer, immunomodulatory, antiviral, and anticoagulant applications (Hwang et al., 2022). Red microalgae, *Porphyridium* sp. and *Cochlodinium polykrikoides,* produce sulfated polysaccharides with antiinflammatory properties (Matsui et al., 2003). The polysaccharide p-KG103, produced from the marine microalga *Gyrodinium impudicum,* inhibits tumor cell proliferation by enhancing cytokine production (Guo et al., 2017). The sulfated polysaccharide from *Tribonema* sp. has anticancer properties (Li et al., 2016). *C. vulgaris* polysaccharides have immunomodulatory activities and are important for chronic airway inflammation prevention (Song et al., 2018). β-1,3-glucan is harvested from *Chlorella* sp. and is an active immune stimulator that neutralizes free radicals and reduces blood lipids (Vo et al., 2011). Furthermore, *Chlorella* sp. contains polysaccharides, which have immunostimulating, antiinflammatory, and analgesic properties. *Tetraselmis* sp., *Isochrysis* sp., *P. cruentum*, and *Porphyridium purpureum* have polysaccharides with antiinflammatory, immunomodulating, and antiviral properties, as well as being used as joint lubricants (Raposo et al., 2013). Sulfated polysaccharides derived from the red algae *Gelidiella acerosa* exhibit anticoagulant, antiplatelet, and antithrombotic properties (Chagas et al., 2020). Polysaccharides obtained from *Porphyridium* sp., in combination with carrageenan, demonstrate potential effectiveness against the COVID-19 virus (Sami et al., 2021; Hans et al., 2021). Sulfated polysaccharides constitute a class of bioactive metabolites with significant antiviral activity. Microalgae *G. impudicum* possess a homopolysaccharide with sulfate groups, demonstrating significant efficacy against the encephalomyocarditis virus (Amaro et al., 2011). Sulfur polysaccharides derived from spirulina are commonly utilized as antiviral medications (Claus-Desbonnet et al., 2022). *A. platensis* and *P. purpureum* release polysaccharides, which show an antiviral effect on the Vaccinia virus and Ectromelia virus (Radonic et al., 2010).

Microalgae fatty acids, such as PUFAs, are the most studied microalgal lipid compounds. PUFAs, specifically EPA and DHA, are two main omega-3 fatty acids, influencing blood pressure, coagulation, platelet and endothelial function, and lipoprotein metabolism (Sarandria and Sarandria, 2022). Derivatives of PUFAs, specifically DHA, EPA, alpha-linolenic acid, and docosapentaenoic acid, have been utilized in the treatment of type 2 diabetes, inflammatory bowel disorders, dermatological conditions, and asthma (Khan et al., 2016). *Tetraselmis* sp. and *N. oculata* are two specific microalgae strains abundant in fatty acids. Both EPA and DHA play important roles in the treatment of diseases such as Parkinson's, Alzheimer's, psoriasis, cancer, atherosclerosis, rheumatoid arthritis, and other inflammatory conditions (Talero et al., 2015). The supplementation of algal DHA oils for breastfeeding moms also aids in the brain development of infants (Jensen et al., 2005). High PUFA levels are found in many microalgal species, including *Isochrysis* sp., *Pavlova* sp., *Nannochloropsis* sp., *C. calcitrans*, *Thalassiosira* sp., and *Schizochytrium* sp. (Anusree et al., 2023).

Microalgal peptides exhibit antiinflammatory and anticancer properties. Recently identified macromolecules, known as mycosporine-like amino acids, have demonstrated antiphotoaging properties in various skin types (Talero et al., 2015). *Navicula* sp. contains bioactive peptides with potent anticancer and antioxidant properties. Microalgae *P. tricornutum* is also used to produce therapeutic proteins (Hempel et al., 2017). *Tetraselmis* sp. exhibits antibacterial and antifungal properties. The peptide AQ-1766, extracted from *T. suecica*, has been reported to show broad-spectrum antibacterial activity (Guzmán et al., 2019). Extracts from *Scenedesmus subspicatus* have been found to inhibit the proliferation of microorganisms, including *Bacillus subtilis, Klebsiella pneumoniae,* and *Escherichia coli* (Dantas et al., 2019).

Carotenoids produced by microalgae include β-carotene, astaxanthin, lycopene, lutein, and canthaxanthin, etc. β-carotene exhibits significant provitamin A activity, essential for optimal vision and the maintenance of the immune system. Therefore it can be incorporated into multivitamin preparations. β-carotene may also be utilized in the therapy of conditions such as asthma and cardiovascular ailments (Suleria et al., 2016). Astaxanthin (red-orange color) is synthesized by algae such as *H. pluvialis, Chlorella zofingiensis, C. vulgaris,* and *Chlorococcum* sp. Its antioxidant properties significantly contribute to protection against UV radiation, photo-oxidation, inflammation, cancer, ulcers, bacterial infections, age-related conditions, and the enhancement of immune response, cardiovascular health, ocular health, hepatic function, joint health, and prostate health (Bule et al., 2018; Sathasivam and Ki, 2018). Zeaxanthin and lutein are produced by *D. salina, Spirulina,* and *Chlorella* sp. and are used as supplements for age-related macular degeneration. In addition to its antioxidant, antiinflammatory, and anticancer properties, lutein also offers other therapeutic benefits, including lowering the risk of heart stroke and heart attack, mitigating the effects of other severe metabolic disorders, and providing eye protection (Saha et al., 2020). *D. tertiolecta* and *Chlorella ellipsoidea* are other microalgae widely cultivated in saline environments to harvest violaxanthin pigment, which has anticarcinogenic and antiinflammatory properties (Kim et al., 2020). The diatom *P. tricornutum* is a commercial producer of fucoxanthin (golden-brown color), which exhibits anticancer, antiobesity, antiinflammatory, and antioxidant therapeutic properties with the potential to prevent cerebrovascular disorders (Bule et al., 2018; Talero et al., 2015; Zhao et al., 2015). Phycoerythrin, isolated from *P. cruentum,* is utilized as a reagent for fluorescence immunoassays (Roman et al., 2002).

The first antibacterial substance identified from the microalga *Chlorella* is chlorellin, which exhibits inhibitory effects on both gram-negative and gram-positive bacteria (Pratt et al., 1944). *Streptococcus mutans* and *Lactobacillus* sp. are two bacterial species that create biofilm on teeth, leading to the development of dental plaque. Extracts of *C. vulgaris* and *D. salina* may suppress biofilm formation and prevent tooth caries (Lourenço-Lopes et al., 2020). The red microalga

P. cruentum, a potential source of vitamin E, exerts a preventive effect on the body against various ailments, including atherosclerosis, coronary heart disease, and neurological conditions (Gamal and Shreadah, 2024). Silica nanoparticles are a promising drug carrier in nanomedicine, and borosilica derived from diatoms is a good alternative to synthetic silica. Natural silica from *Amphora subtropica* can be used in drug delivery applications (Sasirekha et al., 2019).

2.5.4 Cosmeceutical applications

The use of natural compounds has progressively increased in the cosmetics industry over synthetic chemicals. Cosmeceuticals are products containing physiologically active substances that offer medical or pharmaceutical benefits intended to enhance the structure, morphology, and appearance of the skin (Mourelle et al., 2017). This includes haircare, personal hygiene products, makeup, and perfumes (Dussably et al., 2022). The secondary metabolites produced by microalgae during stress conditions are useful as cosmeceuticals. Microalgae-derived cosmetics available on the market include extracts containing various bioactive components or their refined forms. The active ingredients facilitating the use of microalgae in cosmetics are highly diverse, including polysaccharides, proteins, peptides, amino acids, fatty acids, sterols, glycolipids, phospholipids, pigments, phenolic compounds, and vitamins (Morocho-Jacome et al., 2020).

Various microalgae, primarily from the genera *Aphanizomenon, Arthrospira, Chlorella, Desmodesmus, Dunaliella, Haematococcus, Nannochloropsis, Scenedesmus*, and *Spirulina*, are extensively utilized in cosmetic and cosmeceutical applications (Stoyneva-Gärtner et al., 2020). Phospholipids may constitute 10%–50% of the total lipid content in marine microalgae. Therefore they are widely employed in the cosmetic industry as solubilizers, wetting agents, liposomes, and emulsifiers (Mahata et al., 2022). Microalgae are rich in PUFAs, which have antioxidant and antiinflammatory properties, making them useful in different cosmetic formulations (Plaza et al., 2009). Polysaccharides in microalgae have the potential for applications in damaged skin repair, blemish prevention, moisturizing, and inhibition of inflammation; they are used as thickeners, moisturizers, and gelling agents (Kim et al., 2008). β-1,3-glucan, a microalgae-derived polysaccharide, is important in external aging prevention cosmetics targeting sensitive skin. The Chlorella genus, diatom *Skeletonema* sp., as well as *Porphyridium* sp. and *Nostoc flagelliforme*, are rich in β-glucans (Hamed, 2016).

It has been proposed to incorporate algal pigments into skin care lotions as antiaging agents (Favas et al., 2021). Pigments with antioxidant properties produced by marine microalgae are used as cosmetic antioxidants and natural colorants. This includes carotenoids such as β-carotenes from *D. salina*, astaxanthin from *H. pluvialis*, phycocyanobilins, and phycoerythrobilins from *Spirulina* and *Porphyridium*, which can be utilized in the decorative cosmetics sector, including eyeliners and lipsticks (Hamed, 2016; Morocho-Jacome et al., 2020). For make-up formulations, pink and purple colors could be formulated using natural colorants extracted from red microalgae. Astaxanthin, working against UV-induced photooxidation, can be used in natural sunscreen cosmetics and is a red-pink pigment useful for makeup (Marino et al., 2020). Canthaxanthin, mainly obtained from *Nannochloropsis* sp. like *N. salina, N. oculata*, or *Nannochloropsis gaditana*, is commercialized in tanning pills (Koller et al., 2014). Zeaxanthin, obtained from *N. oculata*, demonstrates notable antityrosinase activity, which can cause a reduction of hyperpigmentation and has the potential for skin whitening (Yarkent et al., 2020). The incorporation of microalgal metabolites, including mycosporine, scytonemin, and sporopollenin, as active ingredients in sunscreen formulations, has been proven to mitigate skin damage caused by UV radiation (Sharma et al., 2017).

Vitamin E is an effective antioxidant, widely used in cosmetic formulations. *D. tertiolecta* and *T. suecica* are two proven microalgae for the production of vitamin E (Brown et al., 1999). *C. vulgaris* has been reported to have collagen-repairing mechanisms incorporated into skin-aging cosmeceuticals. Furthermore, *C. vulgaris* extract is believed to clear skin pores and reduce wrinkles (Gamal and Shreadah, 2024) and is employed in shampoos to treat dandruff (Stoyneva-Gärtner et al., 2020).

Resende et al. (2021) reported that *Laminaria digitata, Kappaphycus alvarezii*, and *Chondrus crispus* are among the major marine microalgae extracts used in antiaging cosmetic formulations. Given that certain microalgae extracts are applicable in skincare formulations (such as antiaging creams, emollients, protective, refreshing, or regenerative products, antioxidant and antiirritant formulations, and deodorants), cosmetic business companies have engaged in proprietary microalgal production (Mourelle et al., 2017).

2.5.5 Industrial applications

Marine microalgae are prominent bioresources used in various industrial applications, including biofuels, biofertilizers, bioplastics, biodesalination, the dye industry, and nanotechnology.

Biofuels: The production of biofuels from marine microalgae is a sustainable alternative to fossil fuels. Due to their high lipid content, rapid biomass production, independence from arable lands and freshwater, tendency to fix carbon dioxide, nontoxicity, and wastewater treatment capabilities, marine microalgae are considered a valuable bioresource for biofuel production (Rai and Vasistha, 2023; Khan et al., 2018). Various types of biofuels are produced from microalgae, such as biodiesel, bioethanol, and biohydrogen. Biodiesel is produced through the transesterification of triacylglycerols with alcohol to produce long-chain fatty acid esters (Ng et al., 2015; Rai and Vasistha, 2023). Commonly studied marine microalgae with high biodiesel production potential include *Dunaliella* sp. and *S. obliquus* (Gamal and Shreadah, 2024), which have lipid contents ranging from 11% to 67% (Rai and Vasistha, 2023). Moreover, *Botryococcus braunii,* a potential microalga for biodiesel production, has been shown to accumulate up to 80% of its biomass as lipid (Khan et al., 2018), while most higher crops exhibit lipid yields of less than 5% of their biomass (Ng et al., 2015). Microalgae produce high amounts of fermentable carbohydrates such as glycogen, agar, cellulose, and starch, which can be converted to ethanol (Khan et al., 2018). Species such as *Chlorella* sp. have been reported to be high-potential sources for bioethanol production (Ievina and Romagnoli, 2024). *Chlamydomonas* sp. and *Nannochloropsis* sp. have been used in biohydrogen production in several studies (Iqbal et al., 2022; Fakhimi et al., 2024). Biohydrogen is also produced autotrophically by microalgae or through bacterial conversion of microalgal carbohydrates (Rai and Vasistha, 2023). However, the use of microalgae for biofuel production has some drawbacks, including high operational costs for harvesting, conversion, and maintenance (Khan et al., 2018; Rai and Vasistha, 2023; Rooij et al., 2021).

Biofertilizers: The application of marine microalgae as biofertilizers has gained attention in modern agriculture due to their environmental and economic benefits. They efficiently absorb nutrients from aquatic systems, including wastewater (Çakirsoy et al., 2022), where they produce and accumulate plant-available nitrogen and phosphorus, amino acids, vitamins, plant hormones, and minerals (Çakirsoy et al., 2022; Rai and Vasistha, 2023; Dineshkumar et al., 2019). The use of microalgal biomass-based biofertilizers ensures the slow release of nutrients, maintaining long-term nutrient availability in the soil (Çakirsoy et al., 2022; Rai and Vasistha, 2023), while extracellular polymeric substances in the biomass enhance the soil by forming biofilms, reducing erosion, and enhancing water retention (Gamal and Shreadah, 2024; Rai and Vasistha, 2023). In addition to providing plant nutrients and maintaining soil quality, microalgal biofertilizers also have antimicrobial properties, which act against plant pathogens (Rai and Vasistha, 2023; Dineshkumar et al., 2019). Several studies have reported the biofertilizer potential of marine microalgae such as *C. vulgaris, S. platensis* (Dineshkumar et al., 2019), *Tetraselmis,* and *Nannochloropsis* (Saadaoui et al., 2019; Jochum et al., 2018).

Bioplastics: Microalgal biomass can be used to produce derivatives of biopolymers, including bioplastics (Dino and Kishore, 2024). Bioplastics are a sustainable alternative to conventional plastics derived from fossil fuels. Due to their nontoxic nature, bioplastics can be used for food packaging, biomedical applications, pharmaceuticals, and cosmetics (Tennakoon et al., 2023; Onen Cinar et al., 2020). Conventional biopolymers are derived from food sources. Microalgae are not primary food sources, making them an excellent resource for sustainable bioplastic production without compromising food security (Dino and Kishore, 2024; Onen Cinar et al., 2020). They also exhibit rapid growth, lower water requirements compared to land crops, and the ability to mitigate CO_2 emissions (Roy Chong et al., 2022; Park and Lee, 2023). The absence of lignin in microalgal cell walls makes them a more efficient and sustainable bioresource for bioplastics and other biopolymer production (Roy Chong et al., 2022). In addition, the residual microalgal biomass, particularly the cell walls left after lipid extraction for biodiesel production, can be repurposed for bioplastic manufacturing. This drastically reduces the production costs associated with biodiesel production (Park and Lee, 2023). Marine microalgae, such as *Klebsormidium flaccidum, Nannochloropsis salina, Chlorella* sp., *Micractinium* sp., and *Chlamydomonas* sp., have been used to produce bioplastics (Madadi et al., 2021; Tennakoon et al., 2023).

Biodesalination: Biodesalination is a method that uses biological organisms or their products to purify saline water (Kumar Patel et al., 2021). Due to their adaptations to high-saline environments, marine microalgae are suitable for this process (Esmaeili et al., 2023). The removal of salts from saline water is facilitated by their metabolic activities and physiological mechanisms, including the regulation of ion transporters, accumulation of osmolytes, biosorption, adjustments in photosynthetic pathways, and mitigation of oxidative stress through reactive oxygen species scavenging (Esmaeili et al., 2023; Moayedi et al., 2021). Marine microalgae such as *D. salina, C. vulgaris, N. oculata,* and *Scenedesmus quadricauda* have demonstrated effective desalination abilities (Kumar Patel et al., 2021; Sahle-Demessie et al., 2019; Moayedi et al., 2021).

Dyes: Dyes derived from marine microalgae are sustainable alternatives to conventional textile dyes because of their eco-friendly properties. Algae-derived dyes exhibit color stability due to their natural antioxidants and a lasting dye-textile connection through affinity bonds such as hydrogen or covalent bonds. They are nontoxic and biodegradable, significantly reducing the ecological footprint of the textile industry. Companies such as Mounid and Wargon Innovation are developing dyes using microalgae (https://wargoninnovation.se/en) (https://www.mounid.se/). Colors in some cosmetics and skincare products are also derived from microalgae, which not only enhance skin properties but also provide health benefits, such as antioxidant activity (Thiyagarasaiyar et al., 2020). Synthetic colors can be replaced in food and beverages, providing a healthier alternative to artificial dyes (Rai and Vasistha, 2023; Thiyagarasaiyar et al., 2020). Furthermore, microalgal dyes are used in art and design fields to produce eco-friendly art materials, inks, and paints. Innovations like Algadye 3.0 have led to the development of algae-derived dyes for art supplies, also helping to reduce plastic waste in the art and design sector (https://www.algaeing.com/).

Nanotechnology: Nanotechnology is an advanced, emerging technology that has gained attention for its ability to reduce the use of hazardous chemicals and lower production costs. Recently, microalgae have been explored for the biosynthesis of nanoparticles (Rai and Vasistha, 2023). Their biological characteristics contribute to unique properties that make them applicable in sensor production, medicine, drug delivery, and as catalysts in chemical reactions (Rai and Vasistha, 2023). Marine microalgae have demonstrated significant potential as biocatalysts in the biosynthesis of various nanoparticles, including oxides, chalcogenides, and metallic forms (Saravanakumar et al., 2024). Cyanobacterial strains, such as *Anabaena flosaquae, Calothrix pulvinata, Synechococcus moorigangae,* and *Leptolyngbya foveolarum,* have been used for the green synthesis of gold (Au), silver (Ag), platinum (Pt), and palladium (Pd) nanoparticles (Rosyidah et al., 2024; Saravanakumar et al., 2024). In addition, eukaryotic microalgae *Chlorella* sp. have been effectively utilized for nanoparticle synthesis due to their resilience to high concentrations of heavy metals (Rai and Vasistha, 2023). Other species that have been used for the synthesis of nanoparticles include *A. platensis, D. salina,* and *Trichodesmium erythraeum* (Sidorowicz et al., 2023).

Biomaterials: Microalgae, such as diatoms, are excellent materials due to their silicon content and other constituents, including aluminum and iron oxides. These compounds have high absorptive capacity, large surface area, and low bulk density (Rotter et al., 2021), rendering them useful as filtering materials in the beverage industry, the oil industry, other food industries, and water treatment facilities. Furthermore, significant advancements in studies on diatom-based materials have demonstrated their potential for the synthesis of electrode materials for supercapacitors and batteries (Appiah et al., 2024). Their porous structure and high surface area make them ideal for energy storage, as well as efficient charge storage and transfer (Appiah et al., 2024; Chen et al., 2024).

2.5.6 Environmental applications

The use of marine microalgae supports sustainable environmental management strategies. Their rapid growth rates, high photosynthetic efficiency, and adaptability to diverse environmental conditions make them valuable bioresources for various applications, including bioremediation, phycoremediation, wastewater treatment, biocontrol, and climate change mitigation through carbon sequestration.

Bioremediation and wastewater treatment: The use of microalgae for bioremediation and wastewater treatment reduces energy input. Microalgae use solar energy, absorb carbon dioxide along with nutrients from wastewater, and produce oxygen. Synergistic microorganisms can use oxygen released from microalgae for aerobic degradation. Environmental decontamination mechanisms of marine microalgae can be broadly classified as biosorption, bioaccumulation, and biodegradation (Diankristanti and Ng, 2024). Marine microalgae are effective in treating highly polluted water. They are well adapted to osmotic pressure and can thrive in photoautotrophic, heterotrophic, and mixotrophic conditions, synergistically interacting with other microorganisms (Alcántara et al., 2015; Diankristanti and Ng, 2024). These metabolic capabilities and interactions help accumulate contaminants such as heavy metals, assimilate nutrients, and degrade organic pollutants from various types of wastewater (Srimongkol et al., 2022; Mohd Udaiyappan et al., 2017; Srinuanpan et al., 2019). Wastewater is categorized into municipal, agricultural, and industrial types based on its source (Srimongkol et al., 2022). Wastewater treatment depends on microalgal growth, the removal rate, and the development of intracellular compounds in the wastewater (Srimongkol et al., 2022). For nutrient removal, microalgae exhibit various mechanisms, such as photosynthesis-induced ammonia volatilization, where the increased pH during photosynthesis shifts ammonium into the gaseous ammonia phase (Alcántara et al., 2015). Similarly, increased pH influences phosphate precipitation with cations like calcium (Alcántara et al., 2015). Thus microalgae play a dual role in nutrient assimilation and abiotic transformation in wastewater treatment. Microalgae are capable of removing heavy metals and pathogens in wastewater. Mechanisms such as sorption, electrostatic interactions, complexation, chelation, and ion exchange facilitate metal ion removal, followed by irreversible covalent bonding or surface precipitation (Alcántara et al., 2015; Bulgariu and Gavrilescu, 2015). Furthermore, the excretion of metal-chelating exopolysaccharides by microalgae helps remove toxic heavy metals such as cadmium and chromium (Alcántara et al., 2015). In addition, increased pH influences heavy metal precipitation (Alcántara et al., 2015). The combined effects of photosynthesis, such as increased pH, dissolved oxygen levels, and temperature changes, significantly deactivate microbial contaminants in wastewater (Alcántara et al., 2015). Interestingly, the efficiency of sewage disinfection is generally estimated by the extent of total coliform removal, while environmental factors favorable for algal growth are unfavorable for the survival of coliforms (Amaro et al., 2023). Marine cyanobacterium *Synechococcus* sp. has been reported for the removal of ammonium from brackish aquaculture wastewater (Srimongkol et al., 2019), and *D. salina* for municipal wastewater treatment (Liu and Yildiz, 2018).

Nonliving microalgal biomass has been used for wastewater treatment. Characteristic features of nonliving biomass, such as stability at room temperature, long-term biosorption properties, and resistance to the toxicity of heavy metals, make it a remarkable agent for wastewater treatment (Bulgariu and Gavrilescu, 2015). In addition, reactive functional groups in the polysaccharide-rich cell walls of dead microalgae enhance biosorption efficiency (Bulgariu and Gavrilescu, 2015).

Furthermore, the integration of wastewater treatment with other biotechnological applications, such as biofuels, bioactive compounds, and biomaterials production, maximizes energy recovery, which aligns with circular economy principles and highlights the importance of microalgae as a bioresource (Oruganti et al., 2023).

Carbon sequestration: Marine microalgae are highly efficient for carbon sequestration due to their highly efficient photosynthetic capabilities (Udaypal et al., 2024; Min et al., 2024). The photosynthetic efficiency of microalgae is 10–50 times higher than that of terrestrial plants (Ng et al., 2015). This adaptability enables the development of large-scale industrial systems without requiring extensive gas pretreatment facilities. Remarkably, some robust microalgal species are capable of tolerating high concentrations of nitrogen, sulfur oxides, and CO_2 as well as high temperatures, making them suitable for mitigating industrial emissions (Ng et al., 2015). For example, *S. obliquus* exhibits resilience to CO_2 concentrations as high as 50%, making it suitable for direct treatment of flue gas (Alcántara et al., 2015). Furthermore, extracellular products synthesized by microalgae can increase carbon fixation by several times compared to cellular biomass alone (Ng et al., 2015). This enhanced CO_2 removal efficiency has been discovered in the microalgal species *D. tertiolecta,* which produces extracellular glycerol, acting as a sink for carbon sequestration and increasing carbon transformation by five times (Chow et al., 2013). In addition, species such as *B. braunii, Chlorella kessleri, I. galbana, Chlorococcum littorale, S. platensis, Nannochloropsis, D. salina, S. obliquus,* and *Chlorella* sp. have shown high biomolecule production, including lipids, at high carbon dioxide levels (Ighalo et al., 2022). Recent advancements in biotechnological and molecular techniques have been developed to enhance the carbon sequestration potential. Modifications in the ribulose-1,5-bisphosphate carboxylase-oxygenase (RuBisCO) enzyme and molecular interventions that redirect carbon flux toward lipid synthesis for biofuel production increase carbon removal from the atmosphere (Saravanakumar et al., 2024).

Biocontrol: Microalgae are capable of producing hydrolytic enzymes and biocidal compounds, rendering them potential biocontrol agents against pathogens (Berthon et al., 2021; Renuka et al., 2018; Rooij et al., 2021). Enzymes such as glucanases, chitinases, catalase, peroxidase, polyphenol oxidase, and phenylalanine ammonia-lyase have been recorded in cyanobacteria (Rooij et al., 2021). The use of marine microalgae as a biocontrol agent, while also serving as a food source in aquaculture facilities, helps reduce disease control costs. In addition, it offers a solution for antibiotic-resistant pathogens and serves as an alternative to antiviral vaccines and chemical fungicides (Berthon et al., 2021; Mishra et al., 2022). Marine microalgae, such as *A. platensis, C. minutissima, C. vulgaris, P. tricornutum, Tetraselmis suecia, Skeletonema, Thalassiosira, Synechococcus elongatus, Synechocystis* sp., *T. chui, Nannochloropsis* sp., and *Isochrysis* sp., are potential biocontrol agents (Mishra et al., 2022). Rani et al. (2021) have listed additional microalgal species with antimicrobial properties against bacterial pathogens, including *Vibrio, Aeromonas, Pseudomonas, Bacillus subtilis, Staphylococcus aureus, Micrococcus luteus,* and *Klebsiella* sp. Marine microalgae also exhibit antiviral properties against fish diseases (Falaise et al., 2016). *D. salina* and *H. pluvialis* have demonstrated enhanced antiviral potential, while cyanobacteria-derived sulfated polysaccharides have exhibited significant antiviral properties (Mishra et al., 2022). Moreover, marine microalgae can target fish disease-causing fungal pathogens in aquaculture, including *Aspergillus, Corollospora maritima, Dendryphiella salina, Lulworthia* sp., and *Saprolegnia* (Falaise et al., 2016). For example, *Haslea karadagensis* produces pigments with broad-spectrum antimicrobial activity (Mishra et al., 2022).

2.6 Marine microalgae: isolation, cultivation, harvesting, preservation, and application

2.6.1 Isolation techniques

Isolation of marine microalgae from the vast marine environment is critical. Marine ecosystems are highly diverse, with fluctuating environmental conditions depending on latitude, seasonal changes, and the position of the moon (Alam et al., 2019). The isolation methodologies for marine microalgae depend on research objectives and target species. These techniques employ various strategies to acquire pure and viable cultures. The ability to obtain pure cultures is essential for scientific research and microalgae-derived production at a commercial scale. Isolation techniques can be broadly divided into three steps: preisolation, isolation, and purification (Fernandez-Valenzuela et al., 2021). Before isolation, the sampling location and collection method must be determined. This is also referred to as preisolation techniques, which include collection, enrichment, and contamination removal steps. The sample collection step is highly important because damage to cells during collection can lead to isolation failures. Common sampling techniques include water collection, syringe sampling, and the use of Niskin bottles or rosette samplers for deep-water sampling (Fernandez-Valenzuela et al., 2021). During sample collection, environmental parameters should be recorded, which will help enhance the isolation process under laboratory conditions. Replicates from the sample collection may ensure successful isolation (Alam et al., 2019). Sometimes, the enrichment of samples during collection enhances the growth of target microalgae while suppressing contaminants (Fernandez-Valenzuela et al., 2021; Vu et al., 2018). However, excessive organic nutrients may suppress microalgal growth due to bacterial overgrowth.

In the initial stage, samples can undergo a contamination removal step using vortexing, sonication, or surfactant treatment, which removes extracellular polysaccharides and other adhesive compounds associated with microalgae (Vu et al., 2018). The second isolation step aims to separate individual cells from a mixed population. Commonly used isolation methods include single-cell isolation, agar-based isolation, and dilution techniques (Fernandez-Valenzuela et al., 2021; Vu et al., 2018). The single-cell isolation method can be conducted with micropipettes, micromanipulators, or optical tweezers. These methods are highly accurate, and the optical tweezer method is especially suitable for fragile species but requires specialized equipment and expertise (Favre-Bulle et al., 2019). Coccoids and diatoms are commonly isolated using agar-based isolation methods because of their ability to grow on solid media. Further cultivation can be carried out in liquid media (Fernandez-Valenzuela et al., 2021). The serial dilution method allows the isolation of dominant species in a sample (Alam et al., 2019). A primary challenge is the risk of contamination during the isolation process. Due to their autotrophic nature, microalgal cultures act as potential nutrient sources and may become contaminated with heterotrophic bacteria and fungi. This can complicate the acquisition and maintenance of axenic cultures. Once a single microalgal cell culture is obtained, purification methods, such as antibiotic treatment, enzymatic treatment, or antimicrobial agents, can be used to remove bacterial or fungal contaminants in algal samples (Fernandez-Valenzuela et al., 2021; Vu et al., 2018). Antibiotics such as penicillin, streptomycin, and gentamycin are usually used to treat bacterial contamination (Vu et al., 2018).

However, some microalgae may be sensitive to specific antibiotics, and careful selection and optimization are needed. Generally, cyanobacterial cultures are treated with lysozyme enzymes to eliminate bacterial contamination. In most cases, a combination of antibiotics and enzymes has demonstrated enhanced purification success (Fernandez-Valenzuela et al., 2021; Vu et al., 2018). In addition, chemicals such as sodium hypochlorite or triiodide ions have been used as antimicrobial decontaminants to reduce bacterial contamination (Fernandez-Valenzuela et al., 2021). Apart from the isolation techniques mentioned above, automated methods such as flow cytometry have been utilized with advancements in this field (Alam et al., 2019). The automation methods reduce time and enable large-scale studies (Fernandez-Valenzuela et al., 2021).

2.6.2 Cultivation systems

Marine microalgae cultivation is a promising field with considerable ecological and industrial significance, utilizing microscopic, photosynthetic organisms to produce biomass for various applications. The growth rate and maximal biomass yield of microalgal strains are influenced by abiotic factors (light, temperature, salinity, pH, CO_2, nitrate, phosphate concentrations, etc.), biotic factors (microalgae strain, cell density, cell history, and presence of other strains), and operational parameters such as fluid and hydrodynamic stress, culture depth, mixing, dilution rate, as well as harvesting frequency (Okoro et al., 2019; Zuccaro et al., 2019). Therefore the establishment of effective culture systems is essential for the mass production and industrial utilization of microalgae. They can be cultured via various metabolic pathways: photoautotrophic, heterotrophic, mixotrophic, and photoheterotrophic, by using different cultivation systems broadly categorized as open and closed systems (Zuccaro et al., 2019).

Open culture systems: Open culture systems represent the most straightforward approach to microalgae cultivation, classified into two main categories: natural open ponds and manmade ponds (raceway ponds). Raceway ponds are shallow concrete channels coated with thick plastic sheeting, with variations in depth and length. To guarantee the uniform distribution of nutrients and facilitate the circulation of microalgae culture, paddle wheels are incorporated into raceways (Zhang et al., 2024). The primary benefits of cultivating microalgae in raceway ponds include the simple layout, scalability, elevated production capacities, operational ease, and reduced operational and maintenance costs (Lee et al., 2024). Nonetheless, microbial contamination (bacteria, parasites, phycophages, and/or zooplankton), the proliferation of undesirable algal species, inadequate diffusion of atmospheric CO_2, evaporative losses, extensive land area requirements, suboptimal productivity, restriction to a limited number of strains, and adverse weather conditions (precipitation, temperature, and light intensity) constitute the primary drawbacks of open pond systems (Zhang et al., 2024; Ouada et al., 2024). (Hoeniges et al., 2022) suggested techniques such as using external mirrors to reflect additional sunlight onto the culture surface to increase production. *Chlorella* sp., *Dunaliella* sp., *Haematococcus* sp., and *Spirulina* sp. are commercially cultivated algae species in raceway systems (Suparmaniam et al., 2019).

Closed culture systems: Photobioreactors (PBRs) or lumostatic systems have been engineered as closed-culture systems to mitigate the drawbacks of open-culture systems

(Abdur Razzak et al., 2024). PBRs can be made of transparent polyvinyl chloride (PVC) materials, glass, plexiglass, silicone, and stainless steel, with internal or external illumination and equipped with controlled gas exchange and circulation media. Optical fibers have also been utilized as interior light sources. PBRs are less prone to contamination by other microorganisms, and the primary parameters that limit algae development, including temperature, light (type, quality, intensity, and duration), CO_2 levels, mixing, gas exchange, and evaporation loss, can be easily regulated and managed in such reactors. The design of PBRs is adaptable, suitable for indoor and outdoor use, and may be tailored to meet individual requirements and growth conditions (Manu et al., 2024). The primary constraints of these systems are the economic factors and substantial capital expenditures associated with installation and maintenance. Various types of closed PBRs include flat plate, helical (coil), airlift, horizontal tubular airlift, vertical tubular airlift, bubble column, membrane, plastic bag, solid-state, and hybrid systems (Penloglou et al., 2024; Uyar et al., 2024). The size and type of the PBR can be customized based on the strain type, end use of the biomass, and scalability. The proliferation of PBRs leads to an increased biomass-to-substrate conversion ratio and enhanced economic efficiency of production compared to open systems (Rasel and Karim, 2021).

Microalgae cocultivation strategy: Most studies on microalgae cultivation have concentrated on monoculture systems, which are susceptible to contamination, resulting in diminished productivity, culture failure, inferior biomass quality, and nutrient depletion. Consequently, considerable focus is now directed toward coculturing microalgae with other microorganisms in consortia to overcome these constraints (Kumsiri et al., 2021; Naseema Rasheed et al., 2023). Within cocultivation systems, whether open or closed, two types of cultures are encountered: axenic cultures, which contain a single species, and nonaxenic cultures, which consist of consortia. Coculturing of microalgae and bacteria is established and employed in biomanufacturing, with potential applications in the pharmaceutical, nutraceutical, agricultural, food, chemical, and biofuel sectors, as well as other industries related to bioremediation and methods for nutrient recycling (Pathom-aree et al., 2024). Cocultivating microalgae and fungi has been demonstrated in wastewater treatment, biodiesel production, chemical synthesis, and aquaculture feed applications (Padmaperuma et al., 2018; Rasheed et al., 2023; Yao et al., 2019). Coproduction of microalgae alleviates the issues associated with monoculture systems, providing benefits for large-scale microalgae cultivation (Tong et al., 2023).

2.6.3 Harvesting techniques

Algal cells can be harvested using various techniques, including mechanical, chemical, and electric operations in various combinations (Esteves et al., 2020; Singh and Patidar, 2018; Deepa et al., 2023), and no harvesting method is deemed superior, nor is any approach appropriate for all algae species (Uduman et al., 2010). The choice of harvesting technique is contingent upon the characteristics of microalgae, including density, size, and the value of the intended products (Vasistha et al., 2021). The challenges associated with algae harvesting arise from the diminutive size of micro-algal cells (predominantly under 30 μm), the similar density of algae to the growth medium, the negative surface charge on algae that leads to stability in suspension, and the rapid growth rates of algae necessitating increased harvesting frequency (Liu et al., 2017). The harvested microalgal biomass can degrade within hours if the moisture concentration surpasses 85%. Therefore the eventual moisture content of the harvested algal biomass is a critical factor in the choice of harvesting technique. It has been proposed that 20%–30% of the expenses associated with micro-algal biomass arise from harvesting expenditures (Verma et al., 2010). It is anticipated that 90% of the equipment expenses for algal biomass production in open systems may arise from harvesting and dewatering. Therefore the harvesting cost of microalgae needs to be reduced, and the most suitable harvesting technique should be followed. Microalgae harvesting can often be divided into a two-step process (Uduman et al., 2010; Ananthi et al., 2021). (1) Bulk harvesting: The purpose is to remove microalgal biomass from the bulk solution. Total solid matter can achieve 2%–7% of total suspended solids via sedimentation, flocculation, or flotation. (2) Thickening: The slurry is concentrated through filtering and centrifugation at 15%–25% of total suspended solids. This step demands more energy than bulk harvesting.

Gravity sedimentation: Gravity sedimentation is a low-cost microalga harvesting method that can typically yield concentrations of 1.5% solids (Muylaert et al., 2017; Uduman et al., 2010). The cytoplasm of marine microalgae has a density between 1030 and 1100 kgm^{-3}, while the density of marine diatoms and dinoflagellates varies between 1030 and 1230 kgm^{-3} and is close to the saltwater density of 1024.7 kgm^{-3}. Therefore there is only a small density difference driving micro-algal settlement (Milledge and Heaven, 2013). Because of the fluctuating density of algal cells, the reliability of sedimentation is also low and time-consuming (Gupta et al., 2017). Enhanced gravity sedimentation in microalgal harvesting can be achieved with flocculation (Gupta et al., 2017; Deepa et al., 2023).

Centrifugation: Centrifugation is perhaps the most rapid and reliable technique for recovering suspended algae (Vasistha et al., 2021). It is a great harvesting technique due to its rapidity, vast applicability, and high yield without using any flocculants or chemicals (Pragya et al., 2013). Therefore this approach is suitable for the production of food, beverages, and pharmaceuticals. Stack centrifuges, disc-bowl centrifuges, and solid-bowl centrifuges are some types. Centrifugation with a continuous-flow centrifuge is identified as a potential and cost-effective primary harvesting strategy for *Nannochloris* sp. (Dassey and Theegala, 2013). Hydrocyclones, a relatively low-energy centrifugation method, can only process a limited number of micro-algal strains, and efficiency is highly dependent on solids concentration. Hydro-cyclones have been shown to interfere with the natural flocculation of the marine microalgae *Phaeocystis* sp. and may also disintegrate microalgal flocs, hence complicating later harvesting efforts (Veldhuis et al., 2006).

Flotation: Flotation is a gravity-based separation technique in which air or gas bubbles adhere to solid particles, subsequently transporting them to the liquid surface. It is a rapid method compared to sedimentation. Based on the sizes of bubbles utilized in the flotation process, the applications can be categorized as dissolved air flotation, dispersed flotation, and electrolytic flotation (Deepa et al., 2023).

Filtration and screening: Various filter types have been employed to harvest algae, and filtering has proven effective in recovering rather large algal cells but inadequate for recovering microalgal species with diameters of <10 μm (Uduman et al., 2010). The suspension is introduced through a screen with a particular pore size. Several filtering techniques exist, including vacuum filtration, pressure filtration, microfiltration, ultrafiltration, dead-end filtration, and tangential flow filtration (Deepa et al., 2023). They can be categorized based on pore size: macrofiltration (>10 μm), microfiltration (between 0.1 and 10 μm), ultrafiltration (0.02–0.2 μm), and reverse osmosis (<0.001 μm) (Milledge and Heaven, 2013). A hybrid integrated dewatering system, combining tubular membranes and polymeric hollow fiber membranes via a cross-filtering system, has been utilized to dewater the marine microalgae *Nannochloropsis* sp. (Bhave et al., 2012). The marine microalgae *P. tricornutum* exhibited superior filtration efficacy compared to other species, attributed to their nonspherical morphology and larger cell size, resulting in less resistance to filtration (Marbelia et al., 2016).

Flocculation: Flocculation is a process whereby scattered particles combine to produce larger particles for settling, and it can accommodate substantial volumes of microalgal suspension and a diverse array of microalgae (Uduman et al., 2010; Deepa et al., 2023). It is frequently employed as a pretreatment to augment the particle size prior to using an alternative approach for algal harvesting. Flocculants can be initiated by both inorganic and organic compounds and microbes; nevertheless, flocculants may be specific to certain algal species, and the recovery and recycling of these flocculants present challenges (Ananthi et al., 2021).

Chemical flocculation induces flocculation by adding electrolytes and synthetic polymers (organic/inorganic) as coagulants to the microalgal culture. These flocculants neutralize the negative charge on microalgal cells. Inorganic flocculants effectively flocculate microalgae utilizing multivalent metal salts, including aluminum sulfate, aluminum chloride, ferric chloride, and ferric sulfate (Mubarak et al., 2019). However, the end product is contaminated with added salts. Therefore the metal component of these inorganic flocculants may be toxic if the harvested biomass is intended for food and nutraceutical uses (Wan et al., 2015). For example, in *N. salina,* high residual aluminum was reported in lipids after flocculation with aluminum nitrate sulfate (Rwehumbiza et al., 2012). Cationic starch, tannins, and chitosan have also been identified as natural polymeric coagulants, and a low dosage of tannin has been required with almost 100% recovery to harvest marine microalgae *N. oculata* (Roselet et al., 2017).

Autoflocculation occurs when the pH level increases in the medium due to a restricted supply of carbon dioxide resulting from extended cultivation (Şirin et al., 2012). An increase in pH is posited to induce flocculation through the development of magnesium and calcium precipitates in the culture medium. This method has been proven to be cost-effective in harvesting marine species. A flocculation efficiency above 90% at pH 9 and pH 9.3 was achieved in pH-induced experiments using *P. tricornutum* and *N. oculata.* Although this approach is a low-cost, low-energy-consuming, nontoxic method, it will yield a high concentration of minerals in the collected biomass due to elevated pH levels, which will result in the production of inorganic precipitates (Wu et al., 2012). Bioflocculation using biopolymers of algae, bacteria, or fungi is an economical method. The flocculating marine microalgae species *T. suecica* were used to harvest the nonflocculating microalgae *N. oleoabundans*, resulting in high recovery efficiency with faster sedimentation (Salim et al., 2011). Electrolytic coagulation and flocculation occur when the coagulation or flocculation process is initiated by the current produced by electrophoresis. The combination of electro-coagulation and flocculation techniques has proven effective in the harvesting of the marine microalga *P. tricornutum* (Vandamme et al., 2011).

Most studies on marine microalgae harvesting report the use of flocculation. In marine systems, the application of polyelectrolytes alongside inorganic flocculants, such as ferric salts, alum, and lime, has proven effective. However, the required dosage of flocculants is observed to be 5–10 times greater than that needed for freshwater microalgae (Uduman et al., 2010). The combined flocculation process can be used in a multistep manner by employing more than one type of flocculant. For marine water, ferric chloride/alum can be added to combine with polyelectrolytes, followed by ozone oxidation. Pérez et al., (2016) discovered that effective biomass recovery was attained with minimal dosage by utilizing a combination of ferric chloride and chitosan to harvest the marine microalgae *Chaetoceros gracilis*.

Magnetic nanomaterial-mediated separation: Microalgae's magnetophoretic capabilities are utilized by introducing an external magnetic field, such as ferric magnetic nanoparticles (Fe_2O_3) (Ananthi et al., 2021). These magnetic nanoparticles function as flocculants and facilitate the adhesion of microalgae cells to them. The resultant flocs are then concentrated and separated from the suspension using magnetic aid. More than 90% of the culture of *Nannochloropsis oceanica* has been harvested using iron oxide magnetic microparticles under different pH conditions (Boli et al., 2017). An efficiency of 95%–100% was successfully achieved when harvesting marine *Chlorella* sp. MTF-7 by magnetic ferrofluid in conjunction with flocculation (Ho et al., 2017). Magnetic nanoparticles (MNPs) demonstrated the ability to capture *Nannochloropsis maritima* with a harvesting efficiency of 99.5% (Fu et al., 2021). Although simplicity, low nutrient and energy costs, and reusable media are key benefits, the high cost of magnetic nanoparticles and the need for compressor equipment to remove particles from flocs are notable drawbacks.

2.6.4 Preservation techniques

After sufficient growth, marine microalgae must be harvested effectively to prepare stocks for further applications. The effective storage of microalgae stocks is crucial for both research and industrial applications. Proper storage techniques ensure the viability and quality of microalgae strains over extended periods, allowing for consistent use. For shorter-term storage, refrigerated conditions (typically at 4°C) are often employed. This method can preserve microalgae for weeks to months, depending on the species and initial biomass concentration. Refrigerated storage is generally more accessible and cost-effective than cryopreservation or freeze drying but may result in a gradual decline in viability and physiological quality over time. Several methods are utilized for the storage of marine microalgae, each with its advantages and limitations. The most common approaches include periodic transfer (routine subculture), freeze drying, lyophilization, cryopreservation, and refrigerated storage (Arguelles et al., 2020).

Periodic transfer: The most primitive method of microalgae preservation involves continuously transferring pure inoculum into fresh media from depleted media within a fixed time under aseptic conditions. Although this method is low cost and requires minimal skills, it is highly prone to contamination and genetic drift, resulting in a loss of desired traits due to phenotypical changes (Foo et al., 2023). This method is recommended only for short-term preservation.

Freeze drying: Freeze drying is an effective method for the long-term storage of microalgae. This technique involves removing moisture from algal biomass while it is frozen, resulting in a stable, dry powder that can be easily rehydrated for use (Foo et al., 2023). Freeze-dried microalgae maintain their biochemical properties and can be stored at ambient temperatures, thus simplifying logistics and reducing storage costs. However, the freeze-drying process can be costly and may lead to some loss of viability if not optimized properly (Arguelles et al., 2020).

Lyophilization (liquid drying or L-drying): A preservation technique in which microalgae are maintained in a liquid form at temperatures just above freezing, while moisture is rapidly extracted under vacuum conditions. It is a modification of freeze drying; the freezing stage is bypassed. It minimizes genetic drift contamination while providing an advantage in longevity (Tan Gana et al., 2014).

Cryopreservation: Cryopreservation is the preservation of living organisms (or their components) at ultra-low temperatures (usually below −130°C) to ensure their viability postthawing. Although using liquid nitrogen (−196°C) is the most common way of obtaining ultra-low temperatures (Prieto-Guevara et al., 2023), freezers capable of generating temperatures of −150°C can also be used. Some species, such as *Chlorella protothecoides*, may endure direct exposure to liquid nitrogen, tolerating an extraordinary chilling rate of 2000°C min^{-1}, but the overwhelming majority of organisms do not possess this remarkable ability. Small volumes of cultures (1.0–2.0 mL) are placed in properly labeled cryovials. Cultures should be in optimal condition, healthy, and free from any stress. Cryoprotective agents are employed to mitigate cellular damage during the freezing and thawing process. Intracellular ice development is potentially detrimental and must be mitigated for effective cryopreservation. Methanol (MeOH), dimethyl sulfoxide (DMSO; Me_2SO), and glycerol are the three predominant cryoprotectants utilized in algal protocols (Lourenço, 2020; Foo et al., 2023). Success of the cryopreservation varies among taxonomic groups. The cryopreservation of the majority of dinoflagellates, cryptophytes, synurophytes, and raphidophytes presents significant challenges. Notably, marine diatoms have been cryopreserved with more success than freshwater cultures. Nakanishi et al. (2012) reported a 50% survival rate during a 15-year cryostorage duration for four commercial species, namely *C. vulgaris* C-27 and M-207A7, *N. oculata* ST-4, and *Tetraselmis tetrathele* T-501. Methods such as encapsulation dehydration vitrification have been examined to preserve microalgae strains resistant to traditional cryopreservation, such as *D. tertiolecta* (Hirata et al., 1996). Guermazi et al. (2010) reported the possibility of cryopreserving marine microalgae, including *C. vulgaris, I. galbana,* and *D. salina.*

2.6.5 Application techniques

Marine microalgae are versatile organisms with significant applications in industries such as aquaculture, biofuels, pharmaceuticals, and environmental management. The effectiveness of these applications often depends on the methods of application and the carrier materials used to deliver the microalgae. Here, this text briefly explains various methods of application, categorizing them based on the type of carrier materials employed and their relevance to specific applications.

Liquid suspension carriers: Liquid suspensions are one of the most commonly used carriers for marine microalgae. In this method, microalgae are suspended in aqueous solutions, enabling easy handling, transport, and application. Liquid suspensions can be prepared using water, nutrient solutions, or culture media that maintain the viability and activity of microalgal cells (Miranda et al., 2024). The preparation involves concentrating micro-algal cultures to the desired density and ensuring optimal pH, salinity, and temperature conditions to sustain their physiological activity during storage and application. These suspensions can be supplemented with preservatives or stabilizers to enhance shelf life. Liquid suspensions are sprayed on crops or added to irrigation systems to promote plant growth as biofertilizers (Akhter, 2021; Lu and Xiao, 2022; Butkauskite et al., 2020).

Encapsulation techniques: Encapsulation involves embedding marine microalgae within a protective polymeric shell or matrix, such as alginate beads, chitosan capsules, or gelatin shells. This technique enhances stability and shelf-life, provides protection from external factors, improves bioavailability, and enables controlled release and targeted delivery. Encapsulated microalgae formulations are easier to incorporate into various products, such as cosmetics (Vieira et al., 2020; Ye et al., 2018). Spray drying, ionic gelation, and emulsion-based techniques are common approaches (da Silva et al., 2024). For wound healing, encapsulated *A.* platensis with silk serin (Bari et al., 2017) and encapsulated *S. platensis* with a chitosan complex are used (Hu et al., 2022). Encapsulated *D. salina* is used in the food industry (Konar et al., 2023).

Dry powders and granules: Microalgae can be processed into dry powders or granules by drying techniques such as spray drying or freeze drying. These forms are often mixed with inert carriers like silica, starch, or clay. Dry powders are used as dietary supplements due to their rich nutrient profile (Chen et al., 2022). Granules are applied to soil as slow-release biofertilizers (Akhter, 2021). *Spirulina* sp. and *Dunaliella* sp., like common marine algae, are available as dry powders and used in nutraceuticals (Kusmayadi et al., 2021). They are compatible with various application methods and easy to store. However, there is a potential loss of bioactivity during drying.

Biofilms and coatings: Marine microalgae can be integrated into biofilms or coated onto substrates such as biochar, ceramics, or biodegradable polymers. This method is particularly suitable for specific industrial or environmental applications (Moreno Osorio et al., 2021). These microalgae biofilms are incorporated into a wide range of treatment processes, including industrial wastewater treatment (Mohsenpour et al., 2021). Although there is an initial setup cost for coated materials, they are reusable and exhibit enhanced stability under various environmental conditions.

Immobilized systems: In immobilization systems, the independent movement of living free microalgal cells is prevented through the use of artificial, physical, or chemical means, and the cells are retained in a specific location to perform a desired biological activity (Han et al., 2023; Mallick, 2020). Microalgae are either attached to a matrix (passive immobilization) or entrapped within a porous matrix (active immobilization). Immobilized microalgae systems are highly efficient in wastewater treatment applications (Cao et al., 2022). Based on immobilized microalgae, companies like Gross-Wen Technologies, Onewater, and HydroMentia have commercialized their wastewater treatment solutions (Wollmann et al., 2019).

In addition to the above-mentioned systems, hybrid systems combine multiple carrier materials or application methods to leverage their individual advantages. For instance, encapsulated microalgae within biofilm-coated substrates can provide both stability and controlled release.

2.7 Advancements in marine microalgae-related research

Recent advancements in various technological fields are driving valuable research on marine microalgae and leading to practical applications. These recent developments include new isolation, cultivation, harvesting, and processing techniques, innovations in genetic engineering, advanced computational systems, and many other novel technologies.

Biomimicry: In marine microalgae research, biomimicry can be used to enhance productivity and optimize resource utilization, among many other sustainable biotechnological approaches. Biomimicry principles focus on developing efficient artificial photosynthesis systems and designing bioinspired reactors, biosensors, light sensors, and others. These strategies can help improve light penetration in dense microalgal cultures. For example, the stigma or eyespot in some marine microalgae, which has a phototaxis nature, can be used in biotechnological applications. Its photoreceptor proteins can be utilized for biobased light sensors and environmental biosensors. Optimizing phototaxis can enhance light capture and thereby improve microalgal growth (Böhm and Kreimer, 2020). The ability of microalgae to absorb carbon dioxide has inspired the development of carbon capture technology, aiming to reduce greenhouse gas emissions (Mohapatra et al., 2022). Their ability to withstand high pressure in the ocean can be utilized to develop pressure-resistant robots and new underwater transportation systems. Advanced biohybrid microalgae robots can facilitate drug delivery, imaging, and water decontamination (Zhang et al., 2024).

Genetic engineering and synthetic biology: The natural metabolic productivity of marine microalgae is sometimes insufficient to meet industrial demand. The development of novel genetic engineering techniques, such as CRISPR–Cas9, Transcription Activator-Like Effector Nucleases (TALENs), and Zinc Finger Nucleases (ZFNs), enables targeting specific genetic modifications that enhance microalgal capabilities in various biotechnological applications (Dhokane et al., 2023). For example, lipid biosynthesis in *P. tricornutum* has been enhanced using the CRISPR–Cas9 system (Hao et al., 2022). Moreover, genetic engineering and synthetic biology techniques benefit from advancements in artificial intelligence (AI) algorithms, improving the identification of target genes and editing efficiency in microalgae (Diankristanti and Ng, 2024; Teng et al., 2020). Synthetic biology has been applied to optimize the metabolic pathways of marine microalgae (Sun et al., 2024). Research focuses on cell-free systems and automated genome assembly, reflecting current advancements in synthetic biology. In addition, transcriptomic, proteomic, and metabolomic omics technologies are applied to identify and manipulate key genes in marine microalgae. Recently, the integration of genetic manipulation of microalgae with the use of multiomics data and AI algorithms has received growing attention from scientists (Helmy et al., 2023).

AI in marine microalgae research: Integrating AI and machine learning into marine microalgae research is experiencing exponential growth. Challenges in microalgae identification, cultivation, monitoring, processing, genetic engineering, and other environmental and industrial applications can be addressed by cutting-edge, emerging tools like AI (Alzahmi et al., 2024; Imamoglu, 2024; Yu et al., 2024). Conventional microalgae identification methods are labor-intensive and require expert knowledge. Even using molecular techniques for identification is complex (Ning et al., 2022). Meanwhile, recent advancements in AI-based approaches have demonstrated high accuracy and speed in microalgae identification, classification, and quantification (Alzahmi et al., 2024; Imamoglu, 2024; Otálora et al., 2023; Teng et al., 2020). Examples of success include the microscopic image analysis-based automated detection of *Scenedesmus coenobia* (Giraldo-Zuluaga et al., 2018), classification of Cyanobacteria and Chlorophyta groups using images (Sonmez et al., 2022), and identification and classification of microalgae such as *Cosmarium*, *Closterium*, *Micrasterias*, and *H. pluvialis* (Zheng et al., 2023). These new technologies enable data acquisition and processing while allowing real-time monitoring of species dynamics in complex marine environmental conditions (Otálora et al., 2023).

AI algorithms have also been used to optimize cultivation conditions, such as temperature, light, pH, and nutrients, to achieve high biomass yield and, ultimately, various biological compounds (Imamoglu, 2024), improving resource efficiency and reducing operational costs (Chong et al., 2023). Furthermore, ongoing AI-based research explores downstream processes such as harvesting, extraction, and purification of microalgae (Chong et al., 2023; Imamoglu, 2024). Several studies have emerged utilizing AI models in biodiesel production (Ning et al., 2022; Mayol et al., 2023; Srivastava et al., 2018; Sultana et al., 2022). AI models can improve the design of various environmental applications (Yu et al., 2024). They have been reported to help design microalgae and bacteria consortia for bioremediation purposes (Diankristanti and Ng, 2024). Large-scale bioremediation processes become more effective when integrating AI into remote sensing. Yet, AI poses some challenges, such as the high processing power and energy demands of computers, as well as concerns about the transparency and trustworthiness of AI algorithms (Imamoglu, 2024).

2.8 Research gaps, limitations, and future direction

Despite advancements in marine microalgae research, some gaps and limitations remain. Major challenges include difficulties in large-scale cultivation optimizations, limited understanding of community dynamics, and especially the underexplored biodiversity. The application of the biodiversity and taxonomy of marine microalgae is limited, as much of the research has only focused on a few species. Moreover, marine systems such as deep-sea hydrothermal vents, open ocean areas far away from land, and polar regions, which are home to diverse microalgae, have yet to be fully explored. Undiscovered taxa may have more suitable metabolic properties and tolerance for extreme environmental conditions. Robotic sampling or remote sensing technologies may expand the bioprospecting of marine microalgae in extreme conditions.

Studies on interactions between marine microalgae and other microorganisms, such as bacteria and fungi in natural environments, remain limited. Even though research on genetic manipulation in marine microalgae is ongoing, many species still lack standardized protocols, and the impact of genetic engineering on the surrounding environment remains debated. Furthermore, there is a considerable deficit in studies that validate the efficacy and safety of modified organisms as they enter the ocean, where ecosystems are interconnected. Another limitation is editing efficiency, which directly impacts the reliability of the modified organisms. Challenges include low transformation rates, off-target effects, and unstable integration of desired traits. In addition, self-DNA repair mechanisms in some marine microalgae reduce editing efficiency. Addressing these challenges requires optimization of transformation protocols, reliable selection markers, and a better understanding of species-specific genomic barriers.

The operation and maintenance of large-scale cultivation systems also face significant difficulties. Hybrid systems that combine AI-related automation may help to address these challenges. A major limitation is the lack of integrated datasets on biology, physiology, symbiosis, genetics, and environmental variables related to marine microalgae. Such data on marine microalgae at a global scale are needed. Moreover, the absence of policy and economic frameworks limits their utilization as a bioresource. Natural marine microalgae are facing numerous environmental challenges, including ocean acidification, rising temperatures, eutrophication, and marine pollution such as microplastics. These factors can alter their ecological functions. Future research must address how marine microalgae respond to these stressors in changing environments.

Algae have been commercially cultivated for over 70 years. Innovative advancements in microalgae-based industries are necessary to meet the growing demand for their products and applications. The discovery of new, robust strains that grow fast, naturally produce high levels of the desired compound, maintain high productivity across various environmental conditions, adapt well to planned culture systems, and enable cost-efficient large-scale production is essential for advancing microalgae-based industries and applications.

There is a significant shortage of financial resources and technical support in developing countries, which curtails comprehensive studies and the utilization of marine microalgae as a bioresource. Collaborative, interdisciplinary research initiatives are highly needed to prioritize the environment and establish standardized protocols. It is especially important to promote collaboration between scientific communities and industries to achieve innovative solutions for resource recovery, while also addressing climate change and biodiversity loss for a sustainable environment.

2.9 Conclusion

Marine microalgae represent one of the ocean's most remarkable biological groups and are increasingly recognized as sustainable bioresources. Their unique cellular structures, metabolic diversity, ability to form symbiotic relationships, and adaptability to various marine environments highlight their potential for diverse biotechnological applications. With high photosynthetic efficiency and the ability to synthesize complex metabolites using minimal resources, they offer promising solutions to many of the global challenges we face today, such as food security, climate change, and sustainable resource management.

The use of microalgae has expanded beyond traditional applications in food and nutrition into a wide range of biotechnological applications since the 20th century, and it continues to evolve. Despite their high diversity, commercial cultivation is currently limited to very few species. Advancements in marine microalgae isolation, cultivation, harvesting, and preservation techniques have significantly enhanced their feasibility for real-world applications. However, considering the use of marine microalgae as bioresources to their full potential requires overcoming challenges related to isolation and identification, large-scale production, product consistency, a deeper understanding of community interactions, the integration of comprehensive datasets, regulatory frameworks, and several aspects of technology transfer. Innovations in bioinstruments, genetic and molecular engineering, and bioprocess optimization will be important in addressing most of these issues. Advanced tools such as AI will serve as cutting-edge solutions for tackling many challenges. Moreover, future interdisciplinary research and supportive policy frameworks are necessary to incorporate marine microalgae into a broad range of applications.

The sea remains one of the least explored ecosystems on our planet, with an expansive, living frontier filled with unknown and potentially highly valuable life forms. Discovering these biological mysteries not only holds immense scientific interest but also opens new avenues for existing knowledge. When we harvest the sea, we should do so with care, knowledge, and respect for its intricate balance, and marine microalgae may become key drivers in the global transition toward environmental sustainability.

References

Abdur Razzak, S., Bahar, K., Islam, K.M.O., Haniffa, A.K., Faruque, M.O., Hossain, S.M.Z., Hossain, M.M., 2024. Microalgae cultivation in photobioreactors: sustainable solutions for a greener future. Green Chem. Eng. 26669528. 5 (4), 418–439. Available from: https://www.sciencedirect.com/science/journal/26669528. doi:10.1016/j.gce.2023.10.004.

Akhter, N., 2021. Microalgae-based biofertilizers: applications and potential. Agric. Sci. 12 (4), 345–356.

Alagawany, M., Taha, A.E., Noreldin, A., El-Tarabily, K.A., Abd El-Hack, M.E., 2021. Nutritional applications of species of Spirulina and Chlorella in farmed fish: a review. Aquaculture. 00448486. 542. Available from: http://www.journals.elsevier.com/aquaculture/. doi:10.1016/j.aquaculture.2021.736841.

Alam, M.A., Muhammad, G., Rehman, A., Russel, M., Shah, M., Wang, Z., 2019. Standard techniques and methods for isolating, selecting and monitoring the growth of microalgal strain. In: Microalgae Biotechnology for Development of Biofuel and Wastewater Treatment. Springer Singapore, China, pp. 75–93. Available from: http://www.springer.com/in/book/9789811322631. doi:10.1007/978-981-13-2264-8_4.

Alcántara, C., Posadas, E., Guieysse, B., Muñoz, R., 2015. Microalgae-based wastewater treatment. In: Handbook of Marine Microalgae. Elsevier BV, pp. 439–455. doi:10.1016/b978-0-12-800776-1.00029-7.

Alzahmi, A.S., Daakour, S., Nelson, D., Al-Khairy, D., Twizere, J.C., Salehi-Ashtiani, K., 2024. Enhancing algal production strategies: strain selection, AI-informed cultivation, and mutagenesis. Front. Sustain. Food Syst. 2571581X. 8. Available from: https://www.frontiersin.org/journals/sustainable-food-systems#. doi:10.3389/fsufs.2024.1331251.

Amaro, H., Guedes, A., Malcata, F., 2011. Antimicrobial activities of microalgae: an invited review. Sci. Against Microb. Pathog. Commun. Curr. Res. Technol. Adv. 2, 1272-1284.

Amaro, H.M., Salgado, E.M., Nunes, O.C., Pires, J.C., Esteves, A.F., 2023. Microalgae systems-environmental agents for wastewater treatment and further potential biomass valorisation. J. Environ. Manag. 337, 117678.

Ananthi, V., Balaji, P., Sindhu, R., Kim, S.-H., Arivalagan Pugazhendhi, A.A., 2021. A critical review on different harvesting techniques for algal based biodiesel production. Sci. Total Environ. 00489697. 780, 146467. doi:10.1016/j.scitotenv.2021.146467.

Anusree, M.K., Leela, K.M., Sreehari, M., Raj, S., Sreenikethanam, A., Bajhaiya, A.K., 2023. Marine microalgae: an emerging source of pharmaceuticals and bioactive compounds. In: New Horizons in Natural Compound Research. Elsevier, India, pp. 251–265. Available from: https://www.sciencedirect.com/book/9780443152320. doi:10.1016/B978-0-443-15232-0.00025-4.

Appiah, E.S., Dzikunu, P., Akinwamide, S.O., Fangnon, E.A.K., Mensah-Darkwa, K., Andrews, A., Agyemang, F.O., Nartey, M.A., Makgopa, K., Bossuyt, S., 2024. A review on progress and prospects of diatomaceous earth as a bio-template material for electrochemical energy storage: synthesis, characterization, and applications. Ionics.

18620760. 30 (12), 7809–7860. Available from: https://www.springer.com/journal/11581. doi:10.1007/s11581-024-05825-6.

Arguelles, E.D.L.R., NH, T.G., Monsalud, R.G., 2020. Maintenance and preservation of microalgal cultures. Methods Microalgal Stud. 1, 53–60.

Aveiro, S.S., Melo, T., Figueiredo, A., Domingues, P., Pereira, H., Maia, I.B., Joana Silva, M.R.D., Nunes, C., Moreira, A.S.P., 2020. The polar lipidome of cultured emiliania huxleyi: a source of bioactive lipids with relevance for biotechnological applications. Biomolecules. 2218-273X. 10 (10), 1434. doi:10.3390/biom10101434.

Bari, E., Arciola, C., Vigani, B., Crivelli, B., Moro, P., Marrubini, G., Sorrenti, M., Catenacci, L., Bruni, G., Chlapanidas, T., Lucarelli, E., Perteghella, S., Torre, M., 2017. In vitro effectiveness of microspheres based on silk sericin and *Chlorella vulgaris* or *Arthrospira platensis* for wound healing applications. Materials. 1996-1944. 10 (9), 983. doi:10.3390/ma10090983.

Barrett, J., Girr, P., Mackinder, L.C.M., 2021. Pyrenoids: CO2-fixing phase separated liquid organelles. Biochim. Biophys. Acta Mol. Cell Res. 01674889. 1868 (5), 118949. doi:10.1016/j.bbamcr.2021.118949.

Bayu, A., Noerdjito, D.R., Rahmawati, S.I., Putra, M.Y., Karnjanakom, S., 2023. Biological and technical aspects on valorization of red microalgae genera Porphyridium. Biomass Convers. Biorefin. 21906823. 13 (14), 12395–12411. Available from: https://www.springer.com/journal/13399. doi:10.1007/s13399-021-02167-5.

Ben Atitallah, A., Barkallah, M., Hentati, F., Dammak, M., Ben Hlima, H., Fendri, I., Attia, H., Michaud, P., Abdelkafi, S., 2019. Physicochemical, textural, antioxidant and sensory characteristics of microalgae-fortified canned fish burgers prepared from minced flesh of common barbel (*Barbus barbus*). Food Biosci. 22124306. 30. Available from: http://www.journals.elsevier.com/food-bioscience/. doi:10.1016/j.fbio.2019.100417.

Bermejo Román, R., Alvárez-Pez, J., Acién Fernández, F. and Molina Grima, E., (2002). Recovery of pure B-phycoerythrin from the microalga Porphyridium cruentum. J. Biotechnol. 93 (1), 73–85. doi:10.1016/S0168-1656(01)00385-6.

Berthon, J.Y., Michel, T., Wauquier, A., Joly, P., Gerbore, J., Filaire, E., 2021. Seaweed and microalgae as major actors of blue biotechnology to achieve plant stimulation and pest and pathogen biocontrol - a review of the latest advances and future prospects. J. Agric. Sci. 14695146. 159 (7-8), 523–534. Available from: http://journals.cambridge.org/action/displayJournal?jid=AGS. doi:10.1017/S0021859621000885.

Bhave, R., Kuritz, T., Powell, L., Adcock, D., 2012. Membrane-based energy efficient dewatering of microalgae in biofuels production and recovery of value added co-products. Environ. Sci. Technol. 0013-936X. 46 (10), 5599–5606. doi:10.1021/es204107d.

Bleakley, S., Hayes, M., 2017. Algal proteins: extraction, application, and challenges concerning production. Foods. 2304-8158. 6 (5), 33. doi:10.3390/foods6050033.

Böhm, M., Kreimer, G., 2020. Orient in the world with a single eye: the green algal eyespot and phototaxis. In: Progress in Botany. Springer Science and Business Media LLC, pp. 259–304. doi:10.1007/124_2020_38.

Boli, E., Savvidou, M., Logothetis, D., Louli, V., Pappa, G., Voutsas, E., Kolisis, F., Magoulas, K., 2017. Magnetic harvesting of marine algae *Nannochloropsis oceanica*. Sep. Sci. Technol. 56 (4), 730–737. doi:10.1080/01496395.2017.1296463.

Borowitzka, M.A., 2018. Biology of microalgae. In: Microalgae in Health and Disease Prevention. Elsevier, Australia, pp. 23–72. Available from: http://www.sciencedirect.com/science/book/9780128114056. doi:10.1016/B978-0-12-811405-6.00003-7.

Brawley, S.H., Blouin, N.A., Ficko-Blean, E., Wheeler, G.L., Lohr, M., Goodson, H.V., Jenkins, J.W., Blaby-Haas, C.E., Helliwell, K.E., Chan, C.X., Marriage, T.N., Bhattacharya, D., Klein, A.S., Badis, Y., Brodie, J., Cao, Y., Collén, J., Dittami, S.M., Gachon, C.M.M., Green, B.R., Karpowicz, S.J., Kim, J.W., Kudahl, U.J., Lin, S., Michel, G., Mittag, M., Olson, B.J.S.C., Pangilinan, J.L., Peng, Y., Qiu, H., Shu, S., Singer, J.T., Smith, A.G., Sprecher, B.N., Wagner, V., Wang, W., Wang, Z.Y., Yan, J., Yarish, C., Zäuner-Riek, S., Zhuang, Y., Zou, Y., Lindquist, E.A., Grimwood, J., Barry, K.W., Rokhsar, D.S., Schmutz, J., Stiller, J.W., Grossman, A.R., Prochnik, S.E., 2017. Insights into the red algae and eukaryotic evolution from the genome of *Porphyra umbilicalis* (*Bangiophyceae*, *Rhodophyta*). Proc. Natl. Acad. Sci. U. S. A. 10916490. 114 (31), E6361. Available from: http://www.pnas.org/content/114/31/E6361full.pdf. doi:10.1073/pnas.1703088114.

Brown, M.R., Jeffrey, S.W., Volkman, J.K., Dunstan, G.A., 1997. Nutritional properties of microalgae for mariculture. Aquaculture. 00448486. 151 (1-4), 315–331. doi:10.1016/s0044-8486(96)01501-3.

Brown, M.R., Mular, M., Miller, I., Farmer, C., Trenerry, C., 1999. The vitamin content of microalgae used in aquaculture. J. Appl. Phycol. 11 (3), 247–255. doi:10.1023/A:1008075903578.

Bule, M.H., Ahmed, I., Maqbool, F., Bilal, M., Iqbal, H.M.N., 2018. Microalgae as a source of high-value bioactive compounds. Frontiers in Bioscience, Ethiopia Frontiers in Bioscience - Scholar. 19450524. 10 (2), 197–216. Available from: https://www.bioscience.org/2018/v10s/af/509/list.htm. doi:10.2741/s509.

Bulgariu, L., Gavrilescu, M., 2015. Bioremediation of heavy metals by microalgae. In: Handbook of Marine Microalgae. Elsevier BV, pp. 457–469. doi:10.1016/b978-0-12-800776-1.00030-3.

Burgunter-Delamare, B., Shetty, P., Vuong, T., Mittag, M., 2024. Exchange or eliminate: the secrets of algal-bacterial relationships. Plants. 2223-7747. 13 (6), 829. doi:10.3390/plants13060829.

Butkauskite, A., Mikolaitiene, A., Slinksiene, R., 2020. The influence of algae as a bioactive additive on the properties of suspension fertilizers. Mater. Sci. 2029-7289. 26 (4), 482–488. doi:10.5755/j01.ms.26.4.23457.

Cai, J., Galli, G., 2021. Status and trends of world marine fish stocks. https://openknowledge.fao.org/server/api/core/bitstreams/97409d09-2f8e-4712-b11e-60105d89959b/content#:~:text=Status%20and%20trends%20of%20world,software/fishstatj/en.

Çakirsoy, I., Miyamoto, T., Ohtake, N., 2022. Physiology of microalgae and their application to sustainable agriculture: a mini-review. Front. Plant Sci. 1664462X. 13. Available from: https://www.frontiersin.org/journals/plant-science. doi:10.3389/fpls.2022.1005991.

Cao, S., Teng, F., Jiahui, L., Zhang, Q., Wang, T., Zhu, C., Li, X., Cai, Z., Xie, L., Tao, Y., 2022. Performance of an immobilized microalgae-based process for wastewater treatment and biomass production: nutrients removal, lipid induction, microalgae harvesting and dewatering. Bioresour. Technol. 09608524. 356, 127298. doi:10.1016/j.biortech.2022.127298.

Chen, C., Tang, T., Shi, Q., Zhou, Z., Fan, J., 2022. The potential and challenge of microalgae as promising future food sources. Trends Food Sci. Technol. 09242244. 126, 99–112. Available from: http://www.elsevier.com/wps/find/journaldescription.cws_home/601278/description#description. doi:10.1016/j.tifs.2022.06.016.

Chen, J., Ren, Z., Li, Z., Wang, B., Qi, Y., Yan, W., Liu, Q., Song, H., Han, Q., Zhang, L., 2023. Interaction of Scenedesmus quadricauda and native bacteria in marine biopharmaceutical wastewater for desirable lipid production and wastewater treatment. Chemosphere. 00456535. 313, 137473. doi:10.1016/j.chemosphere.2022.137473.

Chen, Y., Liu, H., Shen, Z., Yang, K., Xia, J., Yuan, P., Nie, Z., Liu, H., Xie, J., 2024. Developing a novel lithium-ion battery anode material via thiol functionalization of diatom frustules plus Ag modification. iScience. 25890042. 27 (2). Available from: https://www.sciencedirect.com/science/journal/25890042. doi:10.1016/j.isci.2024.108850.

Chong, J.W.R., Tang, D.Y.Y., Leong, H.Y., Khoo, K.S., Show, P.L., Chew, K.W., 2023. Bridging artificial intelligence and fucoxanthin for the recovery and quantification from microalgae. Bioengineered. 21655987. 14 (1), 2244232. doi:10.1080/21655979.2023.2244232.

Chow, Y.Y.S., Goh, S.J.M., Su, Z., Ng, D.H.P., Lim, C.Y., Lim, N.Y.N., Lin, H., Fang, L., Lee, Y.K., 2013. Continual production of glycerol from carbon dioxide by *Dunaliella tertiolecta*. Bioresour. Technol. 18732976. 136, 550–555. Available from: https://www.elsevier.com/locate/biortech. doi:10.1016/j.biortech.2013.03.040.

Chrétiennot-Dinet, M.-J., Courties, C., Vaquer, A., Neveux, J., Claustre, H., Lautier, J., Machado, M.C., 2019. A new marine picoeucaryote: *ostreococcus tauri* gen. et sp. nov. (*Chlorophyta, Prasinophyceae*). Phycologia. 0031-8884. 34 (4), 285–292. doi:10.2216/i0031-8884-34-4-285.1.

Claus-Desbonnet, H., Nikly, E., Nalbantova, V., Karcheva-Bahchevanska, D., Ivanova, S., Pierre, G., Benbassat, N., Katsarov, P., Michaud, P., Lukova, P., Delattre, C., 2022. Polysaccharides and their derivatives as potential antiviral molecules. Viruses. 1999-4915. 14 (2), 426. doi:10.3390/v14020426.

Coale, T.H., Loconte, V., Turk-Kubo, K.A., Vanslembrouck, B., Mak, W.K.E., Cheung, S., Ekman, A., Chen, J.H., Hagino, K., Takano, Y., Nishimura, T., Adachi, M., Gros, M.L., Larabell, C., Zehr, J.P., 2024. Nitrogen-fixing organelle in a marine alga. Science. 10959203. 384 (6692), 217–222. Available from: https://www.science.org/doi/epdf/10.1126/science.adk1075. doi:10.1126/science.adk1075.

da Silva Chagas, F.D., Lima, G.C., Dos Santos, V.I.N., Eduardo Castanheira Costa, L., de Sousa, W.M., Gonçalves Sombra, V., de Araújo, D.F., Barros, F.C.N., Marinho-Soriano, E., de Andrade Feitosa, J.P., de Paula, R.C.M., Gonçalves Pereira, M., Ponte Freitas, A.L., 2020. Sulfated polysaccharide from the red algae *Gelidiella acerosa*:

anticoagulant, antiplatelet and antithrombotic effects. Int. J. Biol. Macromol. 01418130. 159, 415–421. doi:10.1016/j.ijbiomac.2020.05.012.

da Silva, A.F., Moreira, A.F., Miguel, S.P., Coutinho, P., 2024. Recent advances in microalgae encapsulation techniques for biomedical applications. Adv. Colloid Interface Sci. 00018686. 333. Available from: https://www.sciencedirect.com/science/journal/00018686. doi:10.1016/j.cis.2024.103297.

Dantas, D.M.D.M., Oliveira, C.Y.B.D., Costa, R.M.P.B., Carneiro-da-Cunha, M.D.G., Gálvez, A.O., Bezerra, R.D.S., 2019. Evaluation of antioxidant and antibacterial capacity of green microalgae Scenedesmus subspicatus. Food Sci. Technol. Int. 15321738. 25 (4), 318–326. Available from: http://www.sagepub.com. doi:10.1177/1082013218825024.

Dassey, A.J., Theegala, C.S., 2013. Harvesting economics and strategies using centrifugation for cost effective separation of microalgae cells for biodiesel applications. Bioresour. Technol. 18732976. 128, 241–245. Available from: https://www.elsevier.com/locate/biortech. doi:10.1016/j.biortech.2012.10.061.

de Almeida Gonçalves, C., Figueredo, C.C., 2020. What we really know about the composition and function of microalgae cell coverings? – an overview. Acta Bot. Bras. 1677-941X. 34 (4), 599–614. doi:10.1590/0102-33062020abb0309.

de Oliveira Lourenço, S., 2020. Microalgae culture collections, strain maintenance, and propagation. In: Handbook of Microalgae-Based Processes and Products. Elsevier BV, pp. 49–84. doi:10.1016/b978-0-12-818536-0.00003-8.

Deepa, P., Sowndhararajan, K., Kim, S., 2023. A review of the harvesting techniques of microalgae. Water. 2073-4441. 15 (17), 3074. doi:10.3390/w15173074.

Dhokane, D., Shaikh, A., Yadav, A., Giri, N., Bandyopadhyay, A., Dasgupta, S., Bhadra, B., 2023. CRISPR-based bioengineering in microalgae for production of industrially important biomolecules. Front. Bioeng. Biotechnol. 22964185. 11. Available from: http://journal.frontiersin.org/journal/bioengineering-and-biotechnology#archive. doi:10.3389/fbioe.2023.1267826.

Diankristanti, P.A., Ng, I.-S., 2024. Marine microalgae for bioremediation and waste-to-worth valorization: recent progress and future prospects. Blue Biotechnol. 2948-2364. 1 (1). doi:10.1186/s44315-024-00010-w.

Dineshkumar, R., Rasheeq, A.A., Arumugam, A., Nambi, K.S., Sampathkumar, P., 2019. Marine microalgal extracts on cultivable crops as a considerable bio-fertilizer: A Review Indian J. Tradit. Knowl. 18 (4), 849–854.

Dineshkumar, R., Subramanian, J., Gopalsamy, J., Jayasingam, P., Arumugam, A., Kannadasan, S., Sampathkumar, P., 2019. The impact of using microalgae as biofertilizer in maize (*Zea mays* L.). Waste Biomass Valorization. 1877-2641. 10 (5), 1101–1110. doi:10.1007/s12649-017-0123-7.

Dino, Kishore, G., 2024. Microalgae: an emerging source of bioplastics production. Discov. Environ. 2 (1), 10. doi:10.1007/s44274-024-00038-0.

Dolganyuk, V., Belova, D., Babich, O., Prosekov, A., Ivanova, S., Katserov, D., Patyukov, N., Sukhikh, S., 2020. Microalgae: a promising source of valuable bioproducts. Biomolecules. 2218-273X. 10 (8), 1153. doi:10.3390/biom10081153.

Dussably, J., Mshvildadze, V., Pichette, A., Ripoll, L., 2022. Microalgae and diatom-potential pharmaceutical and cosmetic resources–review. J. Biomed. Res. Environ. Sci. 3 (9), 1082–1092. doi:10.37871/jbres1559.

Esatbeyoglu, T., Ewald, P., Yasui, Y., Yokokawa, H., Wagner, A.E., Matsugo, S., Winterhalter, P., Rimbach, G., 2016. Chemical characterization, free radical scavenging, and cellular antioxidant and anti-inflammatory properties of a stilbenoid-rich root extract of vitis vinifera. Oxid. Med. Cell. Longev. 19420994. 2016. Available from: http://www.hindawi.com/journals/oximed/. doi:10.1155/2016/8591286.

Esmaeili, A., Moghadam, H.A., Golzary, A., 2023. Application of marine microalgae in biodesalination and CO_2 biofixation: a review. Desalination. 00119164. 567, 116958. doi:10.1016/j.desal.2023.116958.

Esteves, A.F., Jesus Almeida, C., Luísa Gonçalves, A., Carlos Pires, J., 2020. Microalgae harvesting techniques. In: Handbook of Microalgae-Based Processes and Products. Elsevier BV, pp. 225–281. doi:10.1016/b978-0-12-818536-0.00010-5.

Fakhimi, N., Jesus Torres, M., Fernández, E., Galván, A., Dubini, A., González-Ballester, D., 2024. *Chlamydomonas reinhardtii* and *Microbacterium forte* sp. nov., a mutualistic association that favors sustainable hydrogen production. Sci. Total Environ. 00489697. 913, 169559. doi:10.1016/j.scitotenv.2023.169559.

Falaise, C., François, C., Travers, M.-A., Morga, B., Haure, J., Tremblay, R., Turcotte, F., Pasetto, P., Gastineau, R., Hardivillier, Y., Leignel, V., Mouget, J.-L., 2016. Antimicrobial compounds from eukaryotic microalgae against human pathogens and diseases in aquaculture. Mar. Drugs. 1660-3397. 14 (9), 159. doi:10.3390/md14090159.

Favas, R., Morone, J., Martins, R., Vasconcelos, V., Lopes, G., 2021. Cyanobacteria and microalgae bioactive compounds in skin-ageing: potential to restore extracellular matrix filling and overcome hyperpigmentation. J. Enzyme Inhib. Med. Chem. 1475-6366. 36 (1), 1829–1838. doi:10.1080/14756366.2021.1960830.

Favre-Bulle, I.A., Stilgoe, A.B., Scott, E.K., Rubinsztein-Dunlop, H., 2019. Optical trapping in vivo: theory, practice, and applications. Nanophotonics. 21928614. 8 (6), 1023–1040. Available from: http://www.degruyter.com/view/j/nanoph?rskey=PZEKdQ&result=1&q=Nanophotonics. doi:10.1515/nanoph-2019-0055.

Fernandez-Valenzuela, S., Chávez-Ruvalcaba, F., Beltran-Rocha, J.C., Claudio, P.M.S., Reyna-Martínez, R., 2021. Isolation and culturing axenic microalgae: mini–review. Open Microbiol. J. 1874-2858. 15 (1), 111–119. doi:10.2174/1874285802115010111.

Fletcher-Hoppe, C., Yeh, Y.-C., Raut, Y., Weissman, J.L., Fuhrman, J.A., 2023. Symbiotic UCYN-A strains co-occurred with El Niño, relaxed upwelling, and varied eukaryotes over 10 years off Southern California. ISME Commun. 2730-6151. 3 (1). doi:10.1038/s43705-023-00268-y.

Foo, S.C., Mok, C.Y., Ho, S.Y., Khong, N.M.H., 2023. Microalgal culture preservation: progress, trends and future developments. Algal Res. 22119264. 71. Available from: http://www.sciencedirect.com/science/journal/aip/22119264. doi:10.1016/j.algal.2023.103007.

Fox, J.M., Zimba, P.V., 2018. Minerals and trace elements in microalgae. In: Microalgae in Health and Disease Prevention. Elsevier, United States, pp. 177–193. Available from: http://www.sciencedirect.com/science/book/9780128114056. doi:10.1016/B978-0-12-811405-6.00008-6.

Fu, W., Wichuk, K., Brynjólfsson, S., 2015. Developing diatoms for value-added products: challenges and opportunities. New Biotechnol. 18764347. 32 (6), 547–551. Available from: http://www.elsevier.com. doi:10.1016/j.nbt.2015.03.016.

Fu, Y., Hu, F., Li, H., Cui, L., Qian, G., Zhang, D., Xu, Y., 2021. Application and mechanisms of microalgae harvesting by magnetic nanoparticles (MNPs). Sep. Purif. Technol. 13835866. 265, 118519. doi:10.1016/j.seppur.2021.118519.

Gabbs, M., Leng, S., Devassy, J.G., Monirujjaman, M., Aukema, H.M., 2015. Advances in our understanding of oxylipins derived from dietary PUFAs. Adv. Nutr. 21618313. 6 (5), 513–540. doi:10.3945/an.114.007732.

Gamal, R., Shreadah, M.A., 2024. Marine microalgae and their industrial biotechnological applications: a review. J. Genet. Eng. Biotechnol. 1687157X. 22 (4), 100407. doi:10.1016/j.jgeb.2024.100407.

García, J.L., de Vicente, M., Galán, B., 2017. Microalgae, old sustainable food and fashion nutraceuticals. Microb. Biotechnol. 17517915. 10 (5), 1017–1024. Available from: http://onlinelibrary.wiley.com/journal/10.1111/(ISSN)1751-7915. doi:10.1111/1751-7915.12800.

Gatamaneni, B.L., Orsat, V., Lefsrud, M., 2018. Factors affecting growth of various microalgal species. Environ. Eng. Sci. 15579018. 35 (10), 1037–1048. Available from: https://www.liebertonline.com/ees. doi:10.1089/ees.2017.0521.

Ghassemi, N., Poulhazan, A., Deligey, F., Mentink-Vigier, F., Marcotte, I., Wang, T., 2022. Solid-state NMR investigations of extracellular matrixes and cell walls of algae, bacteria, fungi, and plants. Chem. Rev. 15206890. 122 (10), 10036–10086. Available from: http://pubs.acs.org/journal/chreay. doi:10.1021/acs.chemrev.1c00669.

Giraldo-Zuluaga, J.H., Salazar, A., Diez, G., Gomez, A., Martínez, T., Vargas, J.F., Peñuela, M., 2018. Automatic identification of *Scenedesmus polymorphic* microalgae from microscopic images. Pattern Anal. Appl. 14337541. 21 (2), 601–612. Available from: http://link.springer-ny.com/link/service/journals/10044/index.htm. doi:10.1007/s10044-017-0662-3.

González-Hourcade, M., Fernando, D., Gentili, F.G., 2023. Morphological and cellular organization of green microalgae to cope with cold stress in subarctic environment. Algal Res. 22119264. 75, 103254. doi:10.1016/j.algal.2023.103254.

Goold, H.D., Moseley, J.L., Lauersen, K.J., 2024. The synthetic future of algal genomes. Cell Genom. 2666979X. 4 (3). Available from: https://www.sciencedirect.com/science/journal/2666979X. doi:10.1016/j.xgen.2024.100505.

Guedes, A.C., Sousa-Pinto, I., Malcata, F.X., 2015. Application of microalgae protein to aquafeed. In: Handbook of Marine Microalgae: Biotechnology Advances. Elsevier Inc., Portugal, pp. 93–125. Available from: http://www.sciencedirect.com/science/book/9780128007761. doi:10.1016/B978-0-12-800776-1.00008-X.

Guermazi, W., Sellami-Kammoun, A., Elloumi, J., Drira, Z., Aleya, L., Marangoni, R., Ayadi, H., Maalej, S., 2010. Microalgal cryo-preservation using dimethyl sulfoxide (Me_2SO) coupled with two freezing protocols: influence on the fatty acid profile. J. Therm. Biol. 03064565. 35 (4), 175–181. doi:10.1016/j.jtherbio.2010.03.001.

Guo, D.S., Ji, X.J., Ren, L.J., Li, G.L., Sun, X.M., Chen, K.Q., Gao, S., Huang, H., 2018. Development of a scale-up strategy for fermentative production of docosahexaenoic acid by *Schizochytrium* sp. Chem. Eng. Sci. 00092509.

176, 600–608. Available from: http://www.journals.elsevier.com/chemical-engineering-science/. doi:10.1016/j.ces.2017.11.021.

Guo, Q., Shao, Q., Xu, W., Rui, L., Sumi, R., Eguchi, F., Li, Z., 2017. Immunomodulatory and anti-IBDV activities of the polysaccharide AEX from Coccomyxa gloeobotrydiformis. Mar. Drugs. 1660-3397. 15 (2), 36. doi:10.3390/md15020036.

Gupta, Ansari, A., Bauddh, Singh, et al, 2017. Harvesting of microalgae for biofules: compreshensive performance evaluation of natural, inorganic and synthetic flocculants. Springer.

Guzmán, F., Wong, G., Román, T., Cárdenas, C., Alvárez, C., Schmitt, P., Albericio, F., Rojas, V., 2019. Identification of antimicrobial peptides from the microalgae *Tetraselmis suecica* (Kylin) butcher and bactericidal activity improvement. Mar. Drugs. 1660-3397. 17 (8), 453. doi:10.3390/md17080453.

Hachicha, R., Elleuch, F., Hlima, H.B., Dubessay, P., de Baynast, H., Delattre, C., Pierre, G., Hachicha, R., Abdelkafi, S., Michaud, P., Fendri, I., 2022. Biomolecules from microalgae and cyanobacteria: applications and market survey. Appl. Sci. 2076-3417. 12 (4), 1924. doi:10.3390/app12041924.

Hamed, I., 2016. The evolution and versatility of microalgal biotechnology: a review. Compr. Rev. Food Sci. Food Saf. 1541-4337. 15 (6), 1104–1123. doi:10.1111/1541-4337.12227.

Han, M., Zhang, C., Shih-Hsin, H., 2023. Immobilized microalgal system: an achievable idea for upgrading current microalgal wastewater treatment. Environ. Sci. Ecotechnol. 26664984. 14, 100227. doi:10.1016/j.ese.2022.100227.

Hans, N., Malik, A., Naik, S., 2021. Antiviral activity of sulfated polysaccharides from marine algae and its application in combating COVID-19: mini review. Bioresour. Technol. Rep. 2589014X. 13, 100623. doi:10.1016/j.biteb.2020.100623.

Hao, X., Chen, W., Amato, A., Jouhet, J., Maréchal, E., Moog, D., Hu, H., Jin, H., You, L., Huang, F., Moosburner, M., Allen, A.E., Gong, Y., 2022. Multiplexed CRISPR/Cas9 editing of the long-chain acyl-CoA synthetase family in the diatom *Phaeodactylum tricornutum* reveals that mitochondrial ptACSL3 is involved in the synthesis of storage lipids. New Phytol. 0028-646X. 233 (4), 1797–1812. doi:10.1111/nph.17911.

Hasnain, M., Zainab, R., Ali, F., Abideen, Z., Yong, J.W.H., El-Keblawy, A., Hashmi, S., Radicetti, E., 2023. Utilization of microalgal-bacterial energy nexus improves CO_2 sequestration and remediation of wastewater pollutants for beneficial environmental services. Ecotoxicol. Environ. Saf. 10902414. 267. Available from: http://www.elsevier.com/inca/publications/store/6/2/2/8/1/9/index.htt. doi:10.1016/j.ecoenv.2023.115646.

Helmy, M., Elhalis, H., Liu, Y., Chow, Y., Selvarajoo, K., 2023. Perspective: multiomics and machine learning help unleash the alternative food potential of microalgae. Adv. Nutr. 21618313. 14 (1), 1–11. doi:10.1016/j.advnut.2022.11.002.

Hemantkumar, J.N., Rahimbhai, M.I., 2019. Microalgae and its use in nutraceuticals and food supplements. In: Microalgae-from Physiology to Application (London, United Kingdom: IntechOpen). doi:10.5772/intechopen.90143.

Hempel, F., Maurer, M., Brockmann, B., Mayer, C., Biedenkopf, N., Kelterbaum, A., Becker, S., Maier, U.G., 2017. From hybridomas to a robust microalgal-based production platform: molecular design of a diatom secreting monoclonal antibodies directed against the Marburg virus nucleoprotein. Microb. Cell Fact. 14752859. 16 (1). Available from: http://www.microbialcellfactories.com/home/. doi:10.1186/s12934-017-0745-2.

Hirata, K., Phunchindawan, M., Tukamoto, J., Goda, S., Miyamoto, K., 1996. Cryopreservation of microalgae using encapsulation-dehydration. Cryo-Letters 17 (5), 321–328.

Ho, S.H., Chiu, S.Y., Kao, C.Y., Chen, T.Y., Chang, Y.B., Chang, J.S., Lin, C.S., 2017. Ferrofluid-assisted rapid and directional harvesting of marine microalgal *Chlorella* sp. used for biodiesel production. Bioresour. Technol. 18732976. 244, 1337–1340. Available from: https://www.elsevier.com/locate/biortech. doi:10.1016/j.biortech.2017.05.110.

Hoeniges, J., Welch, W., Pruvost, J., Pilon, L., 2022. A novel external reflecting raceway pond design for improved biomass productivity. Algal Res. 22119264. 65, 102742. doi:10.1016/j.algal.2022.102742.

Hopes, A., Mock, T., 2015. Evolution of microalgae and their adaptations in different marine ecosystems. In: Encyclopedia of Life Sciences. Wiley, pp. 1–9. doi:10.1002/9780470015902.a0023744.

Hu, H., Zhong, D., Li, W., Lin, X., He, J., Sun, Y., Wu, Y., Shi, M., Chen, X., Xu, F., Zhou, M., 2022. Microalgae-based bioactive hydrogel loaded with quorum sensing inhibitor promotes infected wound healing. Nano Today. 17480132. 42, 101368. doi:10.1016/j.nantod.2021.101368.

Hwang, J., Yadav, D., Lee, P.C.W., Jin, J., 2022. Immunomodulatory effects of polysaccharides from marine algae for treating cancer, infectious disease, and inflammation. Phytother. Res. 0951-418X. 36 (2), 761–777. doi:10.1002/ptr.7348.

Ievina, B., Romagnoli, F., 2024. Efficient low-temperature nutrient removal from agricultural digestate using microalgae. Environ. Clim. Technol. 22558837. 28 (1), 812–826. Available from: https://sciendo.com/journal/RTUECT. doi:10.2478/rtuect-2024-0062.

Ighalo, J.O., Dulta, K., Kurniawan, S.B., Omoarukhe, F.O., Ewuzie, U., Eshiemogie, S.O., Ojo, A.U., Abdullah, S.R.S., 2022. Progress in microalgae application for CO_2 sequestration. Clean. Chem. Eng. 27727823. 3, 100044. doi:10.1016/j.clce.2022.100044.

Imamoglu, E., 2024. Artificial intelligence and/or machine learning algorithms in microalgae bioprocesses. Bioengineering. 2306-5354. 11 (11), 1143. doi:10.3390/bioengineering11111143.

Iqbal, K., Saxena, A., Pande, P., Tiwari, A., Joshi, N.C., Varma, A., Mishra, A., 2022. Microalgae-bacterial granular consortium: striding towards sustainable production of biohydrogen coupled with wastewater treatment. Bioresour. Technol. 09608524. 354, 127203. doi:10.1016/j.biortech.2022.127203.

Jacob-Lopes, E., Queiroz, M.I., Zepka, L.Q., 2020. Pigments from Microalgae Handbook. Springer International Publishing, Brazil, pp. 1–653. Available from: https://link.springer.com/book/10.1007/978-3-030-50971-2. doi:10.1007/978-3-030-50971-2.

Jensen, C.L., Voigt, R.G., Prager, T.C., Zou, Y.L., Fraley, J.K., Rozelle, J.C., Turcich, M.R., Llorente, A.M., Anderson, R. E., Heird, W.C., 2005. Effects of maternal docosahexaenoic acid intake on visual function and neurodevelopment in breastfed term infants. Am. J. Clin. Nutr. 00029165. 82 (1), 125–132. Available from: http://www.ajcn.org/contents-by-date.2005.shtml. doi:10.1093/ajcn/82.1.125.

Jochum, M., Moncayo, L.P., Jo, Y.-K., Tsang, Y.F., 2018. Microalgal cultivation for biofertilization in rice plants using a vertical semi-closed airlift photobioreactor. PLOS ONE. 1932-6203. 13 (9), e0203456. doi:10.1371/journal.pone.0203456.

Karkos, P.D., Leong, S.C., Karkos, C.D., Sivaji, N., Assimakopoulos, D.A., 2011. Spirulina in clinical practice: evidence-based human applications. Evid. Based Complement. Alternat. Med. 1741-427X. 2011 (1). doi:10.1093/ecam/nen058.

Khan, M.I., Shin, J.H., Kim, J.D., 2018. The promising future of microalgae: current status, challenges, and optimization of a sustainable and renewable industry for biofuels, feed, and other products. Microb. Cell Fact. 14752859. 17 (1). Available from: http://www.microbialcellfactories.com/home/. doi:10.1186/s12934-018-0879-x.

Khan, M., Rahman, M.M., Zaman, S., Jahangir, T.A., Razu, M.H., 2016. Omega-3 polyunsaturated fatty acids from algae. In: Recent Advances in Microalgal Biotechnology (OMICS Group) Oster City. doi:10.4172/978-1-63278-066-9-67.

Kholssi, R., Lougraimzi, H., Moreno-Garrido, I., 2023. Effects of global environmental change on microalgal photosynthesis, growth and their distribution. Mar. Environ. Res. 01411136. 184, 105877. doi:10.1016/j.marenvres.2023.105877.

Kim, J., Kim, M., Lee, S., Jin, E., 2020. Development of a *Chlorella vulgaris* mutant by chemical mutagenesis as a producer for natural violaxanthin. Algal Res. 22119264. 46, 101790. doi:10.1016/j.algal.2020.101790.

Kim, S.K., Ravichandran, Y.D., Khan, S.B., Kim, Y.T., 2008. Prospective of the cosmeceuticals derived from marine organisms. Biotechnol. Bioprocess Eng. 12268372. 13 (5), 511–523. doi:10.1007/s12257-008-0113-5.

Kiran, Venkata Mohan, 2021. Microalgal cell biofactory—therapeutic, nutraceutical and functional food applications. Plants (Basel, Switzerland) 10 (5) 836.

Koller, M., Muhr, A., Braunegg, G., 2014. Microalgae as versatile cellular factories for valued products. Algal Res. 22119264. 6, 52–63. doi:10.1016/j.algal.2014.09.002.

Konar, N., Durmaz, Y., Gurbuz, B., Genc Polat, D., Mert, B., 2023. Spray-drying optimization for *Dunaliella salina* and *Porphyridium cruentum* biomass. Dry. Technol. 15322300. 41(15), 2371–2384. Available from: https://www.tandf.co.uk/journals/titles/07373937.asp. doi:10.1080/07373937.2023.2244058.

Koyande, A.K., Chew, K.W., Rambabu, K., Tao, Y., Chu, D.T., Show, P.L., 2019. Microalgae: a potential alternative to health supplementation for humans. Food Sci. Hum. Wellness. 22134530. 8 (1), 16–24. Available from: https://www.keaipublishing.com/en/journals/food-science-and-human-wellness/. doi:10.1016/j.fshw.2019.03.001.

Krumhardt, K.M., Lovenduski, N.S., Long, M.C., Levy, M., Lindsay, K., Moore, J.K., Nissen, C., 2019. Coccolithophore growth and calcification in an acidified ocean: insights from community earth system model simulations. J. Adv. Model. Earth Syst. 1942-2466. 11 (5), 1418–1437. doi:10.1029/2018ms001483.

Kuech A. Breuer M. Popescu I. 2023. The future of the EU algae sector https://www.europarl.europa.eu/thinktank/en/document/IPOL_STU(2023)733114

Kuech, A., Breuer, M., Popescu, I., 2023. Research for PECH Committee – The future of the EU algae sector. https://www.europarl.europa.eu/thinktank/en/document/IPOL_STU(2023)733114.

Kumari, S., Satapathy, S., Datta, M., Kumar, S., 2022. Adaptation of microalgae to temperature and light stress. In: Advances in Science, Technology and Innovation. Springer Nature, India, pp. 123–134. Available from: https://www.springer.com/series/15883. doi:10.1007/978-3-030-95365-2_8.

Kumsiri, B., Pekkoh, J., Pathom-aree, W., Lumyong, S., Phinyo, K., Pumas, C., Srinuanpan, S., 2021. Enhanced production of microalgal biomass and lipid as an environmentally friendly biodiesel feedstock through actinomycete co-culture in biogas digestate effluent. Bioresour. Technol. 09608524. 337, 125446. doi:10.1016/j.biortech.2021.125446.

Kusmayadi, A., Leong, Y.K., Yen, H.W., Huang, C.Y., Chang, J.S., 2021. Microalgae as sustainable food and feed sources for animals and humans – biotechnological and environmental aspects. Chemosphere. 18791298. 271. Available from: https://www.elsevier.com/locate/chemosphere. doi:10.1016/j.chemosphere.2021.129800.

Lacroux, J., Atteia, A., Brugière, S., Couté, Y., Vallon, O., Steyer, J.-P., van Lis, R., 2022. Proteomics unveil a central role for peroxisomes in butyrate assimilation of the heterotrophic Chlorophyte alga Polytomella sp. Front. Microbiol. 1664-302X. 13. doi:10.3389/fmicb.2022.1029828.

Larkum, A.W.D., Grossman, A.R., Raven, J.A., 2020. Photosynthesis in algae: biochemical and physiological mechanisms. In: Advances in Photosynthesis and Respiration, Vol. 45. Springer International Publishing. doi:10.1007/978-3-030-33397-3.

Laurens, L.M.L., 2017. State of Technology Review – Algae Bioenergy. IEA Bioenergy, https://www.ieabioenergy.com/wp-content/uploads/2017/01/IEA-Bioenergy-Algae-report-update-20170114.pdf.

Lauritano, C., Galasso, C., 2023. Microbial interactions between marine microalgae and fungi: from chemical ecology to biotechnological possible applications. Mar. Drugs. 1660-3397. 21 (5), 310. doi:10.3390/md21050310.

Lauritano, C., Rizzo, C., Giudice, A.L., Saggiomo, M., 2020. Physiological and molecular responses to main environmental stressors of microalgae and bacteria in polar marine environments. Microorganisms. 2076-2607. 8 (12), 1957. doi:10.3390/microorganisms8121957.

Lee, T.H.-W., Show, P.-L., Ong, H.C., Ling, T.C., Lan, J.C.-W., Chang, J.-S., 2024. Microalgae isolation and cultivation technology for mass production. In: Handbook of Biorefinery Research and Technology. Springer Science and Business Media LLC, pp. 651–678. doi:10.1007/978-981-97-7586-6_10.

Levasseur, W., Perré, P., Pozzobon, V., 2020. A review of high value-added molecules production by microalgae in light of the classification. Biotechnol. Adv. 07349750. 41, 107545. doi:10.1016/j.biotechadv.2020.107545.

Li, L., Li, H., Qian, J., He, Y., Zheng, J., Lu, Z., Xu, Z., Shi, J., 2016. Structural and immunological activity characterization of a polysaccharide isolated from *Meretrix* meretrix Linnaeus. Mar. Drugs. 1660-3397. 14 (1), 6. doi:10.3390/md14010006.

Li, X., Fonseca-Batista, D., Roevros, N., Dehairs, F., Chou, L., 2018. Environmental and nutrient controls of marine nitrogen fixation. Prog. Oceanogr. 00796611. 167, 125–137. Available from: https://www.elsevier.com/inca/publications/store/4/2/2. doi:10.1016/j.pocean.2018.08.001.

Liu, J., Song, Y., Qiu, W., 2017. Oleaginous microalgae Nannochloropsis as a new model for biofuel production: review & analysis. Renew. Sustain. Energy Rev. 13640321. 72, 154–162. doi:10.1016/j.rser.2016.12.120.

Liu, Y., Yildiz, I., 2018. The effect of salinity concentration on algal biomass production and nutrient removal from municipal wastewater by *Dunaliella salina*. Int. J. Energy Res. 0363907X. 42 (9), 2997–3006. doi:10.1002/er.3967.

Llamas, A., Leon-Miranda, E., Tejada-Jimenez, M., 2023. Microalgal and nitrogen-fixing bacterial consortia: from interaction to biotechnological potential. Plants. 2223-7747. 12 (13), 2476. doi:10.3390/plants12132476.

López, Y., and Soto, S.M., (2020). The Usefulness of Microalgae Compounds for Preventing Biofilm Infections. Antibiotics, 9 (1), 9. doi:10.3390/antibiotics9010009.

Lorenz, R.T., Cysewski, G.R., 2000. Commercial potential for *Haematococcus microalgae* as a natural source of astaxanthin. Trends Biotechnol. 01677799. 18 (4), 160–167. doi:10.1016/S0167-7799(00)01433-5.

Lourenço-Lopes, C., Fraga-Corral, M., Jimenez-Lopez, C., Pereira, A.G., Garcia-Oliveira, P., Carpena, M., Prieto, M. A., Gandara, 2020. Metabolites from macroalgae and its applications in the cosmetic industry: a circular economy approach. Resources 9. 10.3390/resources9090101

Lu, Q., Xiao, Y., 2022. From manure to high-value fertilizer: the employment of microalgae as a nutrient carrier for sustainable agriculture. Algal Res. 22119264. 67, 102855. doi:10.1016/j.algal.2022.102855.

Lucas, B.F., Morais, M.G.D., Santos, T.D., Costa, J.A.V., 2018. Spirulina for snack enrichment: nutritional, physical and sensory evaluations. LWT. 00236438. 90, 270–276. Available from: https://www.journals.elsevier.com/lwt. doi:10.1016/j.lwt.2017.12.032.

Luo, W., Inomura, K., Zhang, H., Luo, Y.W., 2022. N2 Fixation in trichodesmium does not require spatial segregation from photosynthesis. mSystems. 23795077. 7 (4). Available from: https://journals.asm.org/doi/10.1128/msystems.00538-22. doi:10.1128/msystems.00538-22.

Madadi, R., Maljaee, H., Serafim, L.S., Ventura, S.P.M., 2021. Microalgae as contributors to produce biopolymers. Mar. Drugs. 1660-3397. 19 (8), 466. doi:10.3390/md19080466.

Mahata, C., Das, P., Khan, S., Thaher, M.I.A., Abdul Quadir, M., Annamalai, S.N., Al Jabri, H., 2022. The potential of marine microalgae for the production of food, feed, and fuel (3F). Fermentation. 23115637. 8 (7). Available from: https://www.mdpi.com/2311-5637/8/7/316/pdf?version=1657022370. doi:10.3390/fermentation8070316.

Mallick, N., 2020. Immobilization of Microalgae. Immobilization of Enzymes and Cells, Methods in Molecular Biology, 2100 (Springer / Humana Press), pp. 453–471. doi: 10.1007/978-1-0716-0215-7_31.

Mamdouh, A.Z., Zahran, E., Mohamed, F., Zaki, V., 2021. Nannochloropsis oculate feed additive alleviates mercuric chloride-induced toxicity in Nile tilapia (*Oreochromis niloticus*). Aquat. Toxicol. 238, 105936.

Manu, L., Mokolensang, J.F., Ben Gunawan, W., Setyawardani, A., Salindeho, N., Syahputra, R.A., Iqhrammullah, M., Nurkolis, F., 2024. Photobioreactors are beneficial for mass cultivation of microalgae in terms of areal efficiency, climate implications, and metabolites content. J. Agric. Food Res. 26661543. 18. Available from: https://www.sciencedirect.com/science/journal/26661543. doi:10.1016/j.jafr.2024.101282.

Mapstone, L.J., Leite, M.N., Purton, S., Crawford, I.A., Dartnell, L., 2022. Cyanobacteria and microalgae in supporting human habitation on Mars. Biotechnol. Adv. 07349750. 59. Available from: https://www.elsevier.com/inca/publications/store/5/2/5/4/5/5/index.htt. doi:10.1016/j.biotechadv.2022.107946.

Marbelia, L., Mulier, M., Vandamme, D., Muylaert, K., Szymczyk, A., Vankelecom, I.F.J., 2016. Polyacrylonitrile membranes for microalgae filtration: influence of porosity, surface charge and microalgae species on membrane fouling. Algal Res. 22119264. 19, 128–137. Available from: http://www.sciencedirect.com/science/journal/aip/22119264. doi:10.1016/j.algal.2016.08.004.

Marino, T., Iovine, A., Casella, P., Martino, M., Chianese, S., Larocca, V., Musmarra, D., Molino, A., 2020. From haematococcus pluvialis microalgae a powerful antioxidant for cosmetic applications. Chem. Eng. Trans. 22839216. 79, 271–276. Available from: https://www.aidic.it/cet/20/79/046.pdf. doi:10.3303/CET2079046.

Masuda, T., Mareš, J., Shiozaki, T., Inomura, K., Fujiwara, A., Prášil, O., 2024. Crocosphaera watsonii–a widespread nitrogen-fixing unicellular marine cyanobacterium. J. Phycol. 60 (3), 604–620. doi:10.1111/jpy.13450.

Matalin, D.A., Khramov, D.E., Shuvalov, A.V., Volkov, V.S., Balnokin, Y.V., Popova, L.G., 2021. Cloning and characterization of two putative p-type atpases from the marine microalga dunaliella maritima similar to plant h+-atpases and their gene expression analysis under conditions of hyperosmotic salt shock. Plants. 22237747. 10 (12). Available from: https://www.mdpi.com/2223-7747/10/12/2667/pdf. doi:10.3390/plants10122667.

Matsui, M.S., Muizzuddin, N., Arad, S., Marenus, K., 2003. Sulfated polysaccharides from red microalgae have antiinflammatory properties in vitro and in vivo. Appl. Biochem. Biotechnol. 104 (1), 13–22. doi:10.1385/ABAB:104:1:13.

Mayol, A.P., Juan, J.L.S., Concepcion, R.S., Ubando, A.T., Culaba, A.B., 2023. Environmental impact optimization of microalgae biorefinery using life cycle assessment and evolutionary computing. Chem. Eng. Trans. 22839216. 103, 379–384. Available from: https://www.cetjournal.it/index.php/cet/article/view/CET23103064. doi:10.3303/CET23103064.

Mehariya, S., Goswami, R.K., Karthikeysan, O.P., Verma, P., 2021. Microalgae for high-value products: a way towards green nutraceutical and pharmaceutical compounds. Chemosphere. 18791298. 280. Available from: https://www.elsevier.com/locate/chemosphere. doi:10.1016/j.chemosphere.2021.130553.

Mendes, M., Navalho, S., Ferreira, A., Paulino, C., Figueiredo, D., Silva, D., Gao, F., Gama, F., Bombo, G., Jacinto, R., Aveiro, S., Schulze, P., Gonçalves, A.T., Pereira, H., Gouveia, L., Patarra, R., Abreu, M.H., Silva, J., Navalho, J.,

Varela, J., Speranza, L., 2022. Algae as food in Europe: an overview of species diversity and their application. Foods. 2304-8158. 11 (13), 1871. doi:10.3390/foods11131871.

Milledge, J.J., Heaven, S., 2013. A review of the harvesting of micro-algae for biofuel production. Rev. Environ. Sci. Biotechnol. 15691705. 12 (2), 165–178. doi:10.1007/s11157-012-9301-z.

Min, Z., Wang, K., Wang, H., Fu, W., Wu, B., 2024. Advances in carbon sequestration technology using marine microalgae. Mar. Biol. Res. 17451019. 20 (9-10), 401–417. Available from: http://www.tandf.co.uk/journals/titles/17451000.asp. doi:10.1080/17451000.2024.2371317.

Miranda, A.M., Hernandez-Tenorio, F., Villalta, F., Vargas, G.J., Sáez, A.A., 2024. Advances in the development of biofertilizers and biostimulants from microalgae. Biology. 20797737. 13 (3). Available from: https://www.mdpi.com/journal/biology. doi:10.3390/biology13030199.

Mishra, B., Tiwari, A., Mahmoud, A.E.D., 2022. Microalgal potential for sustainable aquaculture applications: bioremediation, biocontrol, aquafeed. Clean Technol. Environ. Policy. 1618-954X. doi:10.1007/s10098-021-02254-1.

Moayedi, A., Yargholi, B., Pazira, E., Babazadeh, H., 2021. Investigation of bio-desalination potential algae and their effect on water quality. Desalin. Water Treat. 19443986. 212, 78–86. doi:10.5004/dwt.2021.26638.

Mohapatra, R.K., Padhi, D., Sen, R., Nayak, M., 2022. Bio-inspired CO_2 capture and utilization by microalgae for bioenergy feedstock production: a greener approach for environmental protection. Bioresour. Technol. Rep. 2589014X. 19, 101116. doi:10.1016/j.biteb.2022.101116.

Mohd Udaiyappan, A.F., Abu Hasan, H., Takriff, M.S., Sheikh Abdullah, S.R., 2017. A review of the potentials, challenges and current status of microalgae biomass applications in industrial wastewater treatment. J. Water Process Eng. 22147144. 20, 8–21. Available from: http://www.journals.elsevier.com/journal-of-water-process-engineering/. doi:10.1016/j.jwpe.2017.09.006.

Mohsenpour, S.F., Hennige, S., Willoughby, N., Adeloye, A., Gutierrez, T., 2021. Integrating micro-algae into wastewater treatment: a review. Sci. Total Environ. 18791026. 752. Available from: https://www.elsevier.com/locate/scitotenv. doi:10.1016/j.scitotenv.2020.142168.

Mondal, M., Khanra, S., Tiwari, O.N., Gayen, K., Halder, G.N., 2016. Role of carbonic anhydrase on the way to biological carbon capture through microalgae—a mini review. Environ. Prog. Sustain. Energy. 1944-7442. 35 (6), 1605–1615. doi:10.1002/ep.12394.

Moreno Osorio, J.H., Pollio, A., Frunzo, L., Lens, P.N.L., Esposito, G., 2021. A review of microalgal biofilm technologies: definition, applications, settings and analysis. Front. Chem. Eng. 26732718. 3. Available from: https://www.frontiersin.org/journals/chemical-engineering. doi:10.3389/fceng.2021.737710.

Morocho-Jacome, A.L., Ruscinc, N., Martinez, R.M., Carvalho, Santos de Almeida, T., Rosado, C., Costa, J.G., Velasco, M.V.R., Baby, A.R., 2020. Technological aspects of microalgae pigments for cosmetics. Appl. Microbiol. Biotechnol. 104 (22), 9513–9522. doi:10.1007/s00253–020-10936-x.

Mourelle, M.L., Gómez, C.P., Legido, J.L., 2017. The potential use of marine microalgae and cyanobacteria in cosmetics and thalassotherapy. Cosmetics. 20799284. 4 (4). Available from: https://www.mdpi.com/2079-9284/4/4/46/pdf. doi:10.3390/cosmetics4040046.

Mousavian, Z., Safavi, M., Azizmohseni, F., Hadizadeh, M., Mirdamadi, S., 2022. Characterization, antioxidant and anticoagulant properties of exopolysaccharide from marine microalgae. AMB Express. 21910855. 12 (1). Available from: http://www.amb-express.com/. doi:10.1186/s13568-022-01365-2.

Mubarak, M., Shaija, A., Suchithra, T.V., 2019. Flocculation: an effective way to harvest microalgae for biodiesel production. J. Environ. Chem. Eng. 22133437. 7 (4), 103221. doi:10.1016/j.jece.2019.103221.

Mutalipassi, M., Riccio, G., Mazzella, V., Galasso, C., Somma, E., Chiarore, A., de Pascale, D., Zupo, V., 2021. Symbioses of cyanobacteria in marine environments: ecological insights and biotechnological perspectives. Mar. Drugs. 1660-3397. 19 (4), 227. doi:10.3390/md19040227.

Muylaert, K., Bastiaens, L., Vandamme, D., Gouveia, L., 2017. Harvesting of microalgae: overview of process options and their strengths and drawbacks. In: Microalgae-Based Biofuels and Bioproducts. Elsevier BV, pp. 113–132. doi:10.1016/b978-0-08-101023-5.00005-4.

Nakanishi, K., Deuchi, K., Kuwano, K., 2012. Cryopreservation of four valuable strains of microalgae, including viability and characteristics during 15 years of cryostorage. J. Appl. Phycol. 09218971. 24 (6), 1381–1385. Available from: https://www.wkap.nl/journalhome.htm/0921-8971. doi:10.1007/s10811-012-9790-8.

Naseema Rasheed, R., Pourbakhtiar, A., Mehdizadeh Allaf, M., Baharlooeian, M., Rafiei, N., Alishah Aratboni, H., Morones-Ramirez, J.R., Winck, F.V., 2023. Microalgal co-cultivation -recent methods, trends in omic-studies, applications, and future challenges. Front. Bioeng. Biotechnol. 22964185. 11. Available from: http://journal.frontiersin.org/journal/bioengineering-and-biotechnology#archive. doi:10.3389/fbioe.2023.1193424.

Ng, D.H.P., Ng, Y.K., Shen, H., Lee, Y.K., 2015. Microalgal biotechnology. In: Handbook of Marine Microalgae. Elsevier BV, pp. 69–80. doi:10.1016/b978-0-12-800776-1.00006-6.

Ning, H., Li, R., Zhou, T., 2022. Machine learning for microalgae detection and utilization. Front. Mar. Sci. 22967745. 9. Available from: https://www.frontiersin.org/journals/marine-science#. doi:10.3389/fmars.2022.947394.

Okoro, V., Azimov, U., Munoz, J., Hernandez, H.H., Phan, A.N., 2019. Microalgae cultivation and harvesting: growth performance and use of flocculants - a review. Renew. Sustain. Energy Rev. 18790690. 115. Available from: https://www.journals.elsevier.com/renewable-and-sustainable-energy-reviews. doi:10.1016/j.rser.2019.109364.

Onen Cinar, S., Chong, Z.K., Kucuker, M.A., Wieczorek, N., Cengiz, U., Kuchta, K., 2020. Bioplastic production from microalgae: a review. Int. J. Environ. Res. Public Health. 1660-4601. 17 (11), 3842. doi:10.3390/ijerph17113842.

Oruganti, R.K., Biji, A.P., Lanuyanger, T., Show, P.L., Sriariyanun, M., Upadhyayula, V.K.K., Gadhamshetty, V., Bhattacharyya, D., 2023. Artificial intelligence and machine learning tools for high-performance microalgal wastewater treatment and algal biorefinery: a critical review. Sci. Total Environ. 18791026. 876. Available from: https://www.elsevier.com/locate/scitotenv. doi:10.1016/j.scitotenv.2023.162797.

Otálora, P., Guzmán, J.L., Acién, F.G., Berenguel, M., Reul, A., 2023. An artificial intelligence approach for identification of microalgae cultures. New Biotechnol. 18716784. 77, 58–67. doi:10.1016/j.nbt.2023.07.003.

Ouada, H.B., Ammar, J., Ghoul, M., 2024. Photobioreactor engineering. In: Bioreactor Implementation in the Agro-Food Industries: Technology, Kinetics and Modelization (Wiley-ISTE).

Padmaperuma, G., Kapoore, R.V., Gilmour, D.J., Vaidyanathan, S., 2018. Microbial consortia: a critical look at microalgae co-cultures for enhanced biomanufacturing. Crit. Rev. Biotechnol. 15497801. 38 (5), 690–703. doi:10.1080/07388551.2017.1390728.

Park, Y. and Lee, J., (2023). Achievements in the production of bioplastics from microalgae. Phytochem. Rev. 22 (4), 1147–1165. doi:10.1007/s11101-021-09788-8.

Partensky, F., Hess, W.R., Vaulot, D., 1999. Prochlorococcus, a marine photosynthetic prokaryote of global significance. Microbiol. Mol. Biol. Rev. 1092-2172. 63 (1), 106–127. doi:10.1128/mmbr.63.1.106-127.1999.

Patel, A.K., Tseng, Y.S., Rani Singhania, R., Chen, C.W., Chang, J.S., Di Dong, C., 2021. Novel application of microalgae platform for biodesalination process: a review. Bioresour. Technol. 18732976. 337. Available from: https://www.elsevier.com/locate/biortech. doi:10.1016/j.biortech.2021.125343.

Pathom-aree, W., Sattayawat, P., Inwongwan, S., Cheirsilp, B., Liewtrakula, N., Maneechote, W., Rangseekaew, P., Ahmad, F., Mehmood, M.A., Gao, F., Srinuanpan, S., 2024. Microalgae growth-promoting bacteria for cultivation strategies: recent updates and progress. Microbiol. Res. 09445013. 286, 127813. doi:10.1016/j.micres.2024.127813.

Peng, X., Meng, F., Wang, Y., Yi, X., Cui, H., 2020. Effect of pH, temperature, and CO_2 concentration on growth and lipid accumulation of Nannochloropsis sp. MASCC 11. J. Ocean Univ. China. 19935021. 19 (5), 1183–1192. Available from: http://www.springerlink.com/content/1672-5182. doi:10.1007/s11802-020-4302-y.

Penloglou, G., Pavlou, A., Kiparissides, C., 2024. Recent advancements in photo-bioreactors for microalgae cultivation: a brief overview. Processes. 2227-9717. 12 (6), 1104. doi:10.3390/pr12061104.

Pereira, H., Sardinha, M., Santos, T., Gouveia, L., Barreira, L., Dias, J., Varela, J., 2020. Incorporation of defatted microalgal biomass (*Tetraselmis* sp. CTP4) at the expense of soybean meal as a feed ingredient for juvenile gilthead seabream (*Sparus aurata*). Algal Res. 22119264. 47, 101869. doi:10.1016/j.algal.2020.101869.

Pérez-Rojas, C.A., Martínez-Martínez, C.A., Palacios-Mechetnov, E., Lora-Vilchis, M.C., 2022. Design of a low-cost pH-Stat to study effects of ocean acidification on growth and nutrient consumption of diatoms. Aquac. Eng. 01448609. 99. Available from: https://www.elsevier.com/inca/publications/store/4/0/5/8/5/2. doi:10.1016/j.aquaeng.2022.102300.

Pérez, L., Salgueiro, J.L., Maceiras, R., Cancela, Á., Sánchez, Á., 2016. Influence of a combination of flocculants on harvesting of chaetoceros gracilis marine microalgae. Chem. Eng. Technol. 15214125. 39 (9), 1685–1692. Available from: http://www3.interscience.wiley.com/journal/10008333/home. doi:10.1002/ceat.201500564.

Perry, A., Rasmussen, H. and Johnson, E.J., (2009). Xanthophyll (lutein, zeaxanthin) content in fruits, vegetables and corn and egg products. J. Food Compos. Anal. 22 (1), 9–15. doi:10.1016/j.jfca.2008.07.006.

Pierre, G., Delattre, C., Dubessay, P., Jubeau, S., Vialleix, C., Cadoret, J.-P., Probert, I., Michaud, P., 2019. What is in store for EPS microalgae in the next decade? Molecules. 1420-3049. 24 (23), 4296. doi:10.3390/molecules24234296.

Plaza, M., Herrero, M., Cifuentes, A., Ibáñez, E., 2009. Innovative natural functional ingredients from microalgae. J. Agric. Food Chem. 0021-8561. 57 (16), 7159–7170. doi:10.1021/jf901070g.

Poddar, N., Sen, R., Martin, G.J.O., 2018. Glycerol and nitrate utilisation by marine microalgae *Nannochloropsis salina* and *Chlorella* sp. and associated bacteria during mixotrophic and heterotrophic growth. Algal Res. 22119264. 33, 298–309. doi:10.1016/j.algal.2018.06.002.

Popova, L.G., Balnokin, Y.V., 2013. Na+-ATPases of halotolerant microalgae. Russ. J. Plant Physiol. 10214437. 60 (4), 472–482. doi:10.1134/S1021443713040122.

Pragya, N., Pandey, K.K., Sahoo, P.K., 2013. A review on harvesting, oil extraction and biofuels production technologies from microalgae. Renew. Sustain. Energy Rev. 13640321. 24, 159–171. doi:10.1016/j.rser.2013.03.034.

Pratt, R., Daniels, T.C., Eiler, J.J., Gunnison, J.B., Kumler, W.D., Oneto, J.F., Strait, L.A., Spoehr, H.A., Hardin, G.J., Milner, H.W., Smith, J.H.C., Strain, H.H., 1944. Chlorellin, an antibacterial substance from Chlorella. Science. 00368075. 99 (2574), 351–352. doi:10.1126/science.99.2574.351.

Prieto-Guevara, M., Alarcón-Furnieles, J., Jiménez-Velásquez, C., Hernández-Julio, Y., Espinosa-Araujo, J., Atencio-García, V., 2023. Cryopreservation of the microalgae *Scenedesmus* sp. Cells. 2073-4409. 12 (4), 562. doi:10.3390/cells12040562.

Qazi, W.M., Ballance, S., Uhlen, A.K., Kousoulaki, K., Haugen, J. and Rieder, A., (2021). Protein enrichment of wheat bread with the marine green microalgae Tetraselmis chuii – Impact on dough rheology and bread quality. LWT, 143, 111115. doi:10.1016/j.lwt.2021.111115.

Radonic, A., Thulke, S., Achenbach, J., Kurth, A., Vreemann, A., 2010. Anionic polysaccharides from phototrophic microorganisms exhibit antiviral activities to vaccinia virus. J. Antivir. Antiretrovir. 19485964. 02 (04). doi:10.4172/jaa.1000023.

Rai, M.P., Vasistha, S., 2023. Biorefinery concept and value-added products. In: Microalgae Biotechnology for Wastewater Treatment, Resource Recovery and Biofuels. Springer Science and Business Media LLC, pp. 95–110. doi:10.1007/978-3-031-31674-6_8.

Ran, W., Wang, H., Liu, Y., Qi, M., Xiang, Q., Yao, C., Zhang, Y., Lan, X., 2019. Storage of starch and lipids in microalgae: biosynthesis and manipulation by nutrients. Bioresour. Technol. 09608524. 291, 121894. doi:10.1016/j.biortech.2019.121894.

Rani, A., Saini, K.C., Bast, F., Varjani, S., Mehariya, S., Bhatia, S.K., Sharma, N., Funk, C., 2021. A review on microbial products and their perspective application as antimicrobial agents. Biomolecules. 2218273X. 11 (12). Available from: https://www.mdpi.com/2218-273X/11/12/1860/pdf. doi:10.3390/biom11121860.

Raposo, M., De Morais, R., de Morais, A.B., 2013. Bioactivity and applications of sulphated polysaccharides from marine microalgae. Mar. Drugs. 1660-3397. 11 (1), 233–252. doi:10.3390/md11010233.

Raposo, M.F.D.J., De Morais, A.M.M.B., 2015. Microalgae for the prevention of cardiovascular disease and stroke. Life Sci. 18790631. 125, 32–41. Available from: https://www.elsevier.com/locate/lifescie. doi:10.1016/j.lfs.2014.09.018.

Rasel, I., Karim, S., 2021. A mini review on microalgae biomass production: recent progress in cultivation systems. Plants Ecosyst. 2789-7370. 1 (01), 15–20. doi:10.54479/pae.v1i01.6666.

Rasheed, N., Pourbakhtiar, A., Allaf, Baharlooeian, M., Rafiei, N., Aratboni, Morones-Ramirez, J.R., Winck, F.V., 2023. Microalgal co-cultivation—recent methods, trends in omic-studies, applications, and future challenges. Front. Bioeng. Biotechnol. 11, 1193424.

Ray, A., Banerjee, S., Das, D., 2021. Microalgal bio-flocculation: present scenario and prospects for commercialization. Environ. Sci. Pollut. Res. 16147499. 28 (21), 26294–26312. Available from: https://link.springer.com/journal/11356. doi:10.1007/s11356-021-13437-0.

Ren, Y., Sun, H., Deng, J., Huang, J., Chen, F., 2021. Carotenoid production from microalgae: biosynthesis, salinity responses and novel biotechnologies. Mar. Drugs. 1660-3397. 19 (12), 713. doi:10.3390/md19120713.

Renuka, N., Guldhe, A., Prasanna, R., Singh, P., Bux, F., 2018. Microalgae as multi-functional options in modern agriculture: current trends, prospects and challenges. Biotechnol. Adv. 07349750. 36 (4), 1255–1273. Available from: https://www.elsevier.com/inca/publications/store/5/2/5/4/5/5/index.htt. doi:10.1016/j.biotechadv.2018.04.004.

Resende, D.I.S.P., Ferreira, M., Magalhães, C., Lobo, J.M.S., Sousa, E., Almeida, I.F., 2021. Trends in the use of marine ingredients in anti-aging cosmetics. Algal Res. 22119264. 55. Available from: http://www.sciencedirect.com/science/journal/aip/22119264. doi:10.1016/j.algal.2021.102273.

Rezaei, A., Cheniany, M., Ahmadzadeh, H., Vaezi, J., 2024. Evaluation of lipid composition and growth parameters of cold-adapted microalgae under different conditions. Bioenergy Res. 19391242. 17 (1), 557–569. Available from: https://www.springer.com/journal/12155. doi:10.1007/s12155-023-10626-2.

Rizwan, M., Mujtaba, G., Memon, S.A., Lee, K., Rashid, N., 2018. Exploring the potential of microalgae for new biotechnology applications and beyond: a review. Renew. Sustain. Energy Rev. 18790690. 92, 394–404. Available from: https://www.journals.elsevier.com/renewable-and-sustainable-energy-reviews. doi:10.1016/j.rser.2018.04.034.

Roman, Alvarez-Pez, J.M., Fernandez, A., Grima, M., 2002. Recovery of pure Bphyco-erythrin from the microalga *Porphyridium cruentum*. J. Biotechnol. 93 (1), 73–85. doi:10.1016/S0168-1656(01)00385-6.

Rooij, V., Smets, G., Rüdelsheim, P.L., 2021. Taxonomy and risk classification of algae. https://cogem.net/app/uploads/2021/10/CGM-2021-01-Taxonomy-and-risk-classification-of-algae-met-kaft.pdf. (Netherlands: Netherlands Commission on Genetic Modification (COGEM)).

Roselet, F., Vandamme, D., Roselet, M., Muylaert, K., Abreu, P.C., 2017. Effects of pH, salinity, biomass concentration, and algal organic matter on flocculant efficiency of synthetic versus natural polymers for harvesting microalgae biomass. Bioenergy Res. 19391242. 10 (2), 427–437. Available from: http://www.springer.com/life+sci/plant+sciences/journal/12155. doi:10.1007/s12155-016-9806-3.

Rossi, F., De Philippis, R., 2016. Exocellular polysaccharides in microalgae and cyanobacteria: chemical features, role and enzymes and genes involved in their biosynthesis. In: The Physiology of Microalgae. Springer Science and Business Media LLC, pp. 565–590. doi:10.1007/978-3-319-24945-2_21.

Rosyidah, A., Purbani, D.C., Pratiwi, R.D., Muttaqien, S.E., Nantapong, N., Warsito, M.F., Fikri, M.N., Ruth, F., Gustini, N., Syahputra, G., Padri, M., Noerdjito, D.R., Nurkanto, A., Afani, H., 2024. Eco-friendly synthesis of gold nanoparticles by marine microalgae *Synechococcus moorigangae*: characterization, antimicrobial, and antioxidant properties. Kuwait J. Sci. 23074116. 51 (2). Available from: https://www.sciencedirect.com/science/journal/23074108. doi:10.1016/j.kjs.2024.100194.

Rotter, A., Barbier, M., Bertoni, F., Bones, A.M., Cancela, M.L., Carlsson, J., Carvalho, M.F., Cegłowska, M., Chirivella-Martorell, J., Dalay, M.C., Cueto, M., Dailianis, T., Deniz, I., Díaz-Marrero, A.R., Drakulovic, D., Dubnika, A., Edwards, C., Einarsson, H., Erdoğan, A., Eroldoğan, O.T., Ezra, D., Fazi, S., FitzGerald, R.J., Gargan, L.M., Gaudêncio, S.P., Udovič, M.G., Ivošević Denardis, N., Jónsdóttir, R., Kataržytė, M., Klun, K., Kotta, J., Ktari, L., Ljubešić, Z., Lukić Bilela, L., Mandalakis, M., Massa-Gallucci, A., Matijošytė, I., Mazur-Marzec, H., Mehiri, M., Nielsen, S.L., Novoveská, L., Overlingė, D., Perale, G., Ramasamy, P., Rebours, C., Reinsch, T., Reyes, F., Rinkevich, B., Robbens, J., Röttinger, E., Rudovica, V., Sabotič, J., Safarik, I., Talve, S., Tasdemir, D., Schneider, X.T., Thomas, O.P., Toruńska-Sitarz, A., Varese, G.C., Vasquez, M.I., 2021. The essentials of marine biotechnology. Front. Mar. Sci. 22967745. 8. Available from: https://www.frontiersin.org/journals/marine-science#. doi:10.3389/fmars.2021.629629.

Roy Chong, J.W., Tan, X., Khoo, K.S., Ng, H.S., Jonglertjunya, W., Yew, G.Y., Show, P.L., 2022. Microalgae-based bioplastics: future solution towards mitigation of plastic wastes. Environ. Res. 10960953. 206. Available from: http://www.elsevier.com/inca/publications/store/6/2/2/8/2/1/index.htt. doi:10.1016/j.envres.2021.112620.

Rozenberg, J.M., Sorokin, B.A., Mukhambetova, A.N., Emelianova, A.A., Kuzmin, V.V., Belogurova-Ovchinnikova, O.Y., Kuzmin, D.V., 2024. Recent advances and fundamentals of microalgae cultivation technology. Biotechnol. J. 18607314. 19 (3). Available from: http://onlinelibrary.wiley.com/journal/10.1002/(ISSN)1860-7314. doi:10.1002/biot.202300725.

Ruban, A.V., 2016. Nonphotochemical chlorophyll fluorescence quenching: mechanism and effectiveness in protecting plants from photodamage. Plant Physiol. 15322548. 170 (4), 1903–1916. Available from: http://www.plantphysiol.org/content/170/4/1903.full.pdf. doi:10.1104/pp.15.01935.

Rwehumbiza, V.M., Harrison, R., Thomsen, L., 2012. Alum-induced flocculation of preconcentrated Nannochloropsis salina: residual aluminium in the biomass, FAMEs and its effects on microalgae growth upon media recycling. Chem. Eng. J. 13858947. 200-202, 168–175. doi:10.1016/j.cej.2012.06.008.

Saadaoui, I., Rasheed, R., Aguilar, A., Cherif, M., Jabri, H.A., Sayadi, S., Manning, S.R., 2021. Microalgal-based feed: promising alternative feedstocks for livestock and poultry production. J. Anim. Sci. Biotechnol. 20491891. 12 (1). Available from: https://www.jasbsci.com/. doi:10.1186/s40104-021-00593-z.

Saadaoui, I., Sedky, R., Rasheed, R., Bounnit, T., Almahmoud, A., Elshekh, A., Dalgamouni, T., Al Jmal, K., Das, P., Al Jabri, H., 2019. Assessment of the algae-based biofertilizer influence on date palm (*Phoenix dactylifera* L.) cultivation. J. Appl. Phycol. 0921-8971. 31 (1), 457–463. doi:10.1007/s10811-018-1539-6.

Saha, S.K., Ermis, H., Murray, P., 2020. Marine microalgae for potential lutein production. Appl. Sci. 2076-3417. 10 (18), 6457. doi:10.3390/app10186457.

Sahin, M.S., Khazi, M.I., Demirel, Z., Dalay, M.C., 2019. Variation in growth, fucoxanthin, fatty acids profile and lipid content of marine diatoms Nitzschia sp. and Nanofrustulum shiloi in response to nitrogen and iron. Biocatal. Agric. Biotechnol. 18788181. 17, 390–398. Available from: http://www.journals.elsevier.com/biocatalysis-and-agricultural-biotechnology/. doi:10.1016/j.bcab.2018.12.023.

Sahle-Demessie, E., Hassan, A.A., El Badawy, A., 2019. Bio-desalination of brackish and seawater using halophytic algae. Desalination. 00119164. 465, 104–113. Available from: https://www.journals.elsevier.com/desalination. doi:10.1016/j.desal.2019.05.002.

Salim, S., Bosma, R., Vermuë, M.H., Wijffels, R.H., 2011. Harvesting of microalgae by bio-flocculation. J. Appl. Phycol. 09218971. 23 (5), 849–855. doi:10.1007/s10811-010-9591-x.

Samarakoon, K., Jeon, Y.J., 2012. Bio-functionalities of proteins derived from marine algae - a review. Food Res. Int. 09639969. 48 (2), 948–960. doi:10.1016/j.foodres.2012.03.013.

Sami, N., Ahmad, R., Fatma, T., 2021. Exploring algae and cyanobacteria as a promising natural source of antiviral drug against SARS-CoV-2. Biomed. J. 23194170. 44 (1), 54–62. Available from: https://www.journals.elsevier.com/biomedical-journal. doi:10.1016/j.bj.2020.11.014.

Sandhya, S.V., Vijayan, K.K., 2019. Symbiotic association among marine microalgae and bacterial flora: a study with special reference to commercially important Isochrysis galbana culture. J. Appl. Phycol. 0921-8971. 31 (4), 2259–2266. doi:10.1007/s10811-019-01772-2.

Sarandria, D., Sarandria, N., 2022. Omega 3 fatty acids, inflammation, and cardiovascular diseases prevention—an essay. World J. Cardiovasc. Dis. 2164-5329. 12 (03), 152–159. doi:10.4236/wjcd.2022.123016.

Saravanakumar, K., Sathyanarayanan, K., Khan, M.S., Park, S.J., Yoo, H.M., Cho, S.H., Ahn, G., Ahamed, M.A.A., Padmanabhan, S., MubarakAli, D., Cho, N., 2024. Recent progress in biotechnological approaches for diverse applications of algae: an overview. Int. J. Environ. Sci. Technol. 1735-1472. 21 (3), 3453–3474. doi:10.1007/s13762-023-05321-w.

Sasaki, K., Ishihara, K., Yamazaki, M., Nakashima, K., Abe, H., Oyamada, C., Motoyama, M., Mitsumoto, M., 2010. Oral administration of fucoxanthin increases plasma fucoxanthinol concentration and antioxidative ability and improves meat color in broiler chicks. J. Poult. Sci. 1346-7395. 47 (4), 316–320. doi:10.2141/jpsa.010019.

Sasirekha, R., Sheena, T.S., Sathiya Deepika, M., Santhanam, P., Townley, H.E., Jeganathan, K., Dinesh Kumar, S., Premkumar, K., 2019. Surface engineered *Amphora subtropica* frustules using chitosan as a drug delivery platform for anticancer therapy. Mater. Sci. Eng. C. 18730191. 94, 56–64. doi:10.1016/j.msec.2018.09.009.

Sathasivam, R., Ki, J.S., 2018. A review of the biological activities of microalgal carotenoids and their potential use in healthcare and cosmetic industries. Mar. Drugs. 16603397. 16 (1). Available from: http://www.mdpi.com/1660-3397/16/1/26. doi:10.3390/md16010026.

Schwenzfeier, A., Wierenga, P.A. and Gruppen, H., (2011). Isolation and characterization of soluble protein from the green microalgae Tetraselmis sp. Bioresour. Technol. 102 (19), 9121–9127. doi:10.1016/j.biortech.2011.07.046.

Sharma, P., Sharma, N., Keskin, C., 2017. Industrial and biotechnological applications of algae: a review. J. Adv. Plant Biol. 2638-4469. 1 (1), 1–25. doi:10.14302/issn.2638-4469.japb-17-1534.

Shetty, P., Gitau, M.M., Maróti, G., 2019. Salinity stress responses and adaptation mechanisms in eukaryotic green microalgae. Cells. 20734409. 8 (12). Available from: https://www.mdpi.com/2073-4409/8/12/1657/pdf. doi:10.3390/cells8121657.

Shlosberg, Y., Schuster, G., Adir, N., 2022. Harnessing photosynthesis to produce electricity using cyanobacteria, green algae, seaweeds and plants. Front. Plant Sci. 1664462X. 13. Available from: https://www.frontiersin.org/journals/plant-science. doi:10.3389/fpls.2022.955843.

Sidorowicz, A., Fais, G., Casula, M., Borselli, M., Giannaccare, G., Locci, A.M., Lai, N., Orrù, R., Cao, G., Concas, A., 2023. Nanoparticles from microalgae and their biomedical applications. Mar. Drugs. 1660-3397. 21 (6), 352. doi:10.3390/md21060352.

Silva, J., Alves, C., Pinteus, S., Reboleira, J., Pedrosa, R., Bernardino, S., 2019. Chlorella. In: Nonvitamin and Nonmineral Nutritional Supplements. Elsevier BV, pp. 187–193. doi:10.1016/b978-0-12-812491-8.00026-6.

Singh, G., Patidar, S.K., 2018. Microalgae harvesting techniques: a review. J. Environ. Manag. 03014797. 217, 499–508. doi:10.1016/j.jenvman.2018.04.010.

Singh, J., Saxena, R.C., 2015. An introduction to microalgae. In: Handbook of Marine Microalgae. Elsevier BV, pp. 11–24. doi:10.1016/b978-0-12-800776-1.00002-9.

Şirin, S., Trobajo, R., Ibanez, C., Salvadó, J., 2012. Harvesting the microalgae *Phaeodactylum tricornutum* with polyaluminum chloride, aluminium sulphate, chitosan and alkalinity-induced flocculation. J. Appl. Phycol. 0921-8971. 24 (5), 1067–1080. doi:10.1007/s10811-011-9736-6.

Song, H., Meilin, H., Chuankun, G., Wei, D., Liang, Y., Yan, J., Wang, C., 2018. Extraction optimization, purification, antioxidant activity, and preliminary structural characterization of crude polysaccharide from an Arctic Chlorella sp. Polymers. 2073-4360. 10 (3), 292. doi:10.3390/polym10030292.

Sonmez, M.E., Eczacıoglu, N., Gumuş, N.E., Aslan, M.F., Sabanci, K., Aşikkutlu, B., 2022. Convolutional neural network-support vector machine based approach for classification of cyanobacteria and chlorophyta microalgae groups. Algal Res. 61, 102568. doi:10.1016/j.algal.2021.102568.

Spain, O., Funk, C., 2022. Detailed characterization of the cell wall structure and composition of nordic green microalgae. J. Agric. Food Chem. 15205118. 70 (31), 9711–9721. Available from: http://pubs.acs.org/journal/jafcau. doi:10.1021/acs.jafc.2c02783.

Spain, O., Plöhn, M., Funk, C., 2021. The cell wall of green microalgae and its role in heavy metal removal. Physiol. Plant. 13993054. 173 (2), 526–535. Available from: http://onlinelibrary.wiley.com/journal/10.1111/(ISSN)1399-3054. doi:10.1111/ppl.13405.

Springstein, B.L., Nürnberg, D.J., Weiss, G.L., Pilhofer, M., Stucken, K., 2020. Structural determinants and their role in cyanobacterial morphogenesis. Life. 20751729. 10 (12), 1–33. Available from: https://www.mdpi.com/2075-1729/10/12/355/pdf. doi:10.3390/life10120355.

Srimongkol, P., Sangtanoo, P., Songserm, P., Watsuntorn, W., Karnchanatat, A., 2022. Microalgae-based wastewater treatment for developing economic and environmental sustainability: current status and future prospects. Front. Bioeng. Biotechnol. 22964185. 10. Available from: http://journal.frontiersin.org/journal/bioengineering-and-biotechnology#archive. doi:10.3389/fbioe.2022.904046.

Srimongkol, P., Thongchul, N., Phunpruch, S., Karnchanatat, A., 2019. Ability of marine cyanobacterium *Synechococcus* sp. VDW to remove ammonium from brackish aquaculture wastewater. Agric. Water Manag. 03783774. 212, 155–161. doi:10.1016/j.agwat.2018.09.006.

Srinuanpan, S., Cheirsilp, B., Kassim, M.A., 2019. Oleaginous microalgae cultivation for biogas upgrading and phytoremediation of wastewater. In: Microalgae Cultivation for Biofuels Production. Elsevier, Thailand, pp. 69–82. Available from: http://www.sciencedirect.com/science/book/9780128175361. doi:10.1016/B978-0-12-817536-1.00005-9.

Srivastava, G., Paul, A.K., Goud, V.V., 2018. Optimization of non-catalytic transesterification of microalgae oil to biodiesel under supercritical methanol condition. Energy Convers. Manag. 01968904. 156, 269–278. doi:10.1016/j.enconman.2017.10.093.

Stavridou, E., Karapetsi, L., Nteve, G.M., Tsintzou, G., Chatzikonstantinou, M., Tsaousi, M., Martinez, A., Flores, P., Merino, M., Dobrovic, L., Mullor, J.L., Martens, S., Cerasino, L., Salmaso, N., Osathanunkul, M., Labrou, N.E., Madesis, P., 2024. Landscape of microalgae omics and metabolic engineering research for strain improvement: an overview. Aquaculture. 00448486. 587. Available from: https://www.sciencedirect.com/science/journal/00448486. doi:10.1016/j.aquaculture.2024.740803.

Stengel, D.B., Connan, S., 2015. Marine algae: a source of biomass for biotechnological applications. Methods Mol. Biol. 10643745. 1308, 1–37. Available from: http://www.springer.com/series/7651. doi:10.1007/978-1-4939-2684-8_1.

Stoyneva-Gärtner, M., Uzunov, B., Gärtner, G., 2020. Enigmatic microalgae from aeroterrestrial and extreme habitats in cosmetics: the potential of the untapped natural sources. Cosmetics. 2079-9284. 7 (2), 27. doi:10.3390/cosmetics7020027.

Suleria, H.A.R., Gobe, G., Masci, P., Osborne, S.A., 2016. Marine bioactive compounds and health promoting perspectives; innovation pathways for drug discovery. Trends Food Sci. Technol. 09242244. 50, 44–55. Available from: http://www.elsevier.com/wps/find/journaldescription.cws_home/601278/description#description. doi:10.1016/j.tifs.2016.01.019.

Sultana, N., Zakir Hossain, S.M., Abusaad, M., Alanbar, N., Senan, Y., Razzak, S.A., 2022. Prediction of biodiesel production from microalgal oil using Bayesian optimization algorithm-based machine learning approaches. Fuel. 00162361. 309, 122184. doi:10.1016/j.fuel.2021.122184.

Sun, H., Wang, J., Li, Y., Yang, S., Di Chen, D., Tu, Y., Liu, J., Sun, Z., 2024. Synthetic biology in microalgae towards fucoxanthin production for pharmacy and nutraceuticals. Biochem. Pharmacol. 00062952. 220, 115958. doi:10.1016/j.bcp.2023.115958.

Suparmaniam, U., Lam, M.K., Uemura, Y., Lim, J.W., Lee, K.T., Shuit, S.H., 2019. Insights into the microalgae cultivation technology and harvesting process for biofuel production: a review. Renew. Sustain. Energy Rev. 18790690. 115. Available from: https://www.journals.elsevier.com/renewable-and-sustainable-energy-reviews. doi:10.1016/j.rser.2019.109361.

Sutak, R., Camadro, J.M., Lesuisse, E., 2020. Iron uptake mechanisms in marine phytoplankton. Front. Microbiol. 1664302X. 11. Available from: https://www.frontiersin.org/journals/microbiology#. doi:10.3389/fmicb.2020.566691.

Sutherland, D.L., Ralph, P.J., 2021. Differing growth responses in four related microalgal genera grown under autotrophic, mixotrophic and heterotrophic conditions. Aust. J. Appl. Phycol. 15735176. 33 (6), 3539–3553. Available from: https://www.wkap.nl/journalhome.htm/0921-8971. doi:10.1007/s10811-021-02593-y.

Talero, E., García-Mauriño, S., Ávila-Román, J., Rodríguez-Luna, A., Alcaide, A., Motilva, V., 2015. Bioactive compounds isolated from microalgae in chronic inflammation and cancer. Mar. Drugs. 1660-3397. 13 (10), 6152–6209. doi:10.3390/md13106152.

Tan Gana, N.H.A., Monsalud, E.D.L.R., G, R., 2014. Applicability of liquid-drying for the preservation of microalgal cultures. Philipp. Sci. 51, 128–138.

Taucher, J., Bach, L.T., Prowe, A.E.F., Boxhammer, T., Kvale, K., Riebesell, U., 2022. Enhanced silica export in a future ocean triggers global diatom decline. Nature. 14764687. 605 (7911), 696–700. Available from: http://www.nature.com/nature/index.html. doi:10.1038/s41586-022-04687-0.

Tavakoli, S., Regenstein, J.M., Daneshvar, E., Bhatnagar, A., Luo, Y., Hong, H., 2022. Recent advances in the application of microalgae and its derivatives for preservation, quality improvement, and shelf-life extension of seafood. Crit. Rev. Food Sci. Nutr. 1040-8398. 62 (22), 6055–6068. doi:10.1080/10408398.2021.1895065.

Teng, S.Y., Yew, G.Y., Sukačová, K., Show, P.L., Máša, V., Chang, J.S., 2020. Microalgae with artificial intelligence: a digitalized perspective on genetics, systems and products. Biotechnol. Adv. 07349750. 44. Available from: https://www.elsevier.com/inca/publications/store/5/2/5/4/5/5/index.htt. doi:10.1016/j.biotechadv.2020.107631.

Tennakoon, P., Chandika, P., Yi, M., Jung, W.K., 2023. Marine-derived biopolymers as potential bioplastics, an eco-friendly alternative. iScience. 25890042. 26 (4). Available from: https://www.cell.com/iscience. doi:10.1016/j.isci.2023.106404.

Thiyagarasaiyar, K., Goh, B.H., Jeon, Y.J., Yow, Y.Y., 2020. Algae metabolites in cosmeceutical: an overview of current applications and challenges. Mar. Drugs. 16603397. 18 (6). Available from: https://www.mdpi.com/1660-3397/18/6/323/pdf. doi:10.3390/md18060323.

Thoré, E.S.J., Muylaert, K., Bertram, M.G., Brodin, T., 2023. Microalgae. Curr. Biol. 18790445. 33 (3), R91. Available from: http://www.elsevier.com/journals/current-biology/0960-9822. doi:10.1016/j.cub.2022.12.032.

Tibbetts, S.M., Melanson, R.J., Park, K.C., Banskota, A.H., Stefanova, R., McGinn, P.J., 2016. Nutritional evaluation of whole and lipid-extracted biomass of the *Microalga Scenedesmus* sp. AMDD isolated in Saskatchewan, Canada for animal feeds: proximate, amino acid, fatty acid, carotenoid and elemental composition. Curr. Biotechnol. 22115501. 4 (4), 530–546. doi:10.2174/2211550104666150827201854.

Tirichine, L., Piganeau, G., 2023. Algal symbiotic relationships in freshwater and marine environments. Front. Plant Sci. 14, 1155759. doi:10.3389/fpls.2023.1155759.

Tong, C.Y., Honda, K., Derek, C.J.C., 2023. A review on microalgal-bacterial co-culture: the multifaceted role of beneficial bacteria towards enhancement of microalgal metabolite production. Environ. Res. 00139351. 228, 115872. doi:10.1016/j.envres.2023.115872.

Udaypal, Goswami, R.K., Mehariya, S., Verma, P., 2024. Advances in microalgae-based carbon sequestration: current status and future perspectives. Environ. Res. 00139351. 249, 118397. doi:10.1016/j.envres.2024.118397.

Uduman, N., Qi, Y., Danquah, M.K., Forde, G.M., Hoadley, A., 2010. Dewatering of microalgal cultures: a major bottleneck to algae-based fuels. J. Renew. Sustain. Energy. 19417012. 2 (1). doi:10.1063/1.3294480.

Urrutia, O., Antonio Mendizabal, J., Insausti, K., Soret, B., Purroy, A., Arana, A., te Pas, M.F.W., 2016. Effects of addition of linseed and marine algae to the diet on adipose tissue development, fatty acid profile, lipogenic gene expression, and meat quality in lambs. PLOS ONE. 1932-6203. 11 (6), e0156765. doi:10.1371/journal.pone.0156765.

Uyar, B., Ali, M.D., Uyar, G.E.O., 2024. Design parameters comparison of bubble column, airlift and stirred tank photobioreactors for microalgae production. Bioprocess Biosyst. Eng. 16157605. 47 (2), 195–209. Available from: https://www.springer.com/journal/449. doi:10.1007/s00449-023-02952-8.

Vandamme, D., Pontes, S.C.V., Goiris, K., Foubert, I., Pinoy, L.J.J., Muylaert, K., 2011. Evaluation of electro-coagulation-flocculation for harvesting marine and freshwater microalgae. Biotechnol. Bioeng. 10970290. 108 (10), 2320–2329. doi:10.1002/bit.23199.

Vasistha, S., Khanra, A., Clifford, M., Rai, M.P., 2021. Current advances in microalgae harvesting and lipid extraction processes for improved biodiesel production: a review. Renew. Sustain. Energy Rev. 13640321. 137, 110498. doi:10.1016/j.rser.2020.110498.

Veldhuis, M.J.W., Fuhr, F., Boon, J.P., Hallers-Tjabbers, C.C.T., 2006. Treatment of ballast water; how to test a system with a modular concept? Environ. Technol. 0959-3330. 27 (8), 909–921. doi:10.1080/09593332708618701.

Verma, N.M., Mehrotra, S., Shukla, A., Mishra, B.N., 2010. Prospective of biodiesel production utilizing microalgae as the cell factories: a comprehensive discussion. Afr. J. Biotechnol. 16845315. 9 (10), 1402–1411. Available from: http://www.academicjournals.org/AJB/PDF/pdf2010/8Mar/Verma%20et%20al.pdf. doi:10.5897/ajbx09.071.

Vieira, M.V., Pastrana, L.M., Fuciños, P., 2020. Microalgae encapsulation systems for food, pharmaceutical and cosmetics applications. Mar. Drugs. 16603397. 18 (12). Available from: https://www.mdpi.com/1660-3397/18/12/644. doi:10.3390/md18120644.

Villar-Navarro, E., Garrido-Pérez, C., Perales, J.A., 2021. The potential of different marine microalgae species to recycle nutrients from recirculating aquaculture systems (RAS) fish farms and produce feed additives. Algal Res. 22119264. 58, 102389. doi:10.1016/j.algal.2021.102389.

Vo, T.S., Ngo, D.H., Ta, Q.V., Kim, S.K., 2011. Marine organisms as a therapeutic source against herpes simplex virus infection. Eur. J. Pharm. Sci. 18790720. 44(1-2), 11–20. doi:10.1016/j.ejps.2011.07.005.

Vu, C.H.T., Lee, H.G., Chang, Y.K., Oh, H.M., 2018. Axenic cultures for microalgal biotechnology: establishment, assessment, maintenance, and applications. Biotechnol. Adv. 07349750. 36 (2), 380–396. Available from: https://www.elsevier.com/inca/publications/store/5/2/5/4/5/5/index.htt. doi:10.1016/j.biotechadv.2017.12.018.

Wan, C., Alam, M.A., Zhao, X.Q., Zhang, X.Y., Guo, S.L., Ho, S.H., Chang, J.S., Bai, F.W., 2015. Current progress and future prospect of microalgal biomass harvest using various flocculation technologies. Bioresour. Technol. 18732976. 184, 251–257. Available from: https://www.elsevier.com/locate/biortech. doi:10.1016/j.biortech.2014.11.081.

Wang, Z., Yu, S., Nie, Y., Liu, R., Zhu, W., Zhou, Z., Ma, Y., Diao, J., 2024. Effect of acetochlor on the symbiotic relationship between microalgae and bacteria. J. Hazard. Mater. 03043894. 463, 132848. doi:10.1016/j.jhazmat.2023.132848.

Wells, M.L., Potin, P., Craigie, J.S., Raven, J.A., Merchant, S.S., Helliwell, K.E., Smith, A.G., Camire, M.E., Brawley, S. H., 2017. Algae as nutritional and functional food sources: revisiting our understanding. J. Appl. Phycol. 15735176. 29 (2), 949–982. Available from: https://www.wkap.nl/journalhome.htm/0921-8971. doi:10.1007/s10811-016-0974-5.

Wollmann, F., Dietze, S., Ackermann, J., Bley, T., Walther, T., Steingroewer, J., Krujatz, F., 2019. Microalgae wastewater treatment: biological and technological approaches. Eng. Life Sci. 1618-0240. 19 (12), 860–871. doi:10.1002/elsc.201900071.

Wu, J., Gu, X., Yang, D., Xu, S., Wang, S., Chen, X., Wang, Z., 2021. Bioactive substances and potentiality of marine microalgae. Food Sci. Nutr. 2048-7177. 9 (9), 5279–5292. doi:10.1002/fsn3.2471.

Wu, Z., Zhu, Y., Huang, W., Zhang, C., Li, T., Zhang, Y., Li, A., 2012. Evaluation of flocculation induced by pH increase for harvesting microalgae and reuse of flocculated medium. Bioresour. Technol. 09608524. 110, 496–502. doi:10.1016/j.biortech.2012.01.101.

Yan, P., Guo, J.S., Zhang, P., Xiao, Y., Li, Z., Zhang, S.Q., Zhang, Y.X., He, S.X., 2021. The role of morphological changes in algae adaptation to nutrient stress at the single-cell level. Sci. Total Environ. 18791026. 754. Available from: https://www.elsevier.com/locate/scitotenv. doi:10.1016/j.scitotenv.2020.142076.

Yao, S., Lyu, S., An, Y., Lu, J., Gjermansen, C., Schramm, A., 2019. Microalgae-bacteria symbiosis in microalgal growth and biofuel production: a review. J. Appl. Microbiol. 13645072. 126 (2), 359–368. doi:10.1111/jam.14095.

Yarkent, Ç., Gürlek, C., Oncel, S.S., 2020. Potential of microalgal compounds in trending natural cosmetics: a review. Sustain. Chem. Pharm. 23525541. 17, 100304. doi:10.1016/j.scp.2020.100304.

Ye, Q., Georges, N., Selomulya, C., 2018. Microencapsulation of active ingredients in functional foods: from research stage to commercial food products. Trends Food Sci. Technol. 09242244. 78, 167–179. Available from: http://www.elsevier.com/wps/find/journaldescription.cws_home/601278/description#description. doi:10.1016/j.tifs.2018.05.025.

Yu, K.L., Ong, H.C., Zaman, H.B., 2024. Integrated energy informatics technology on microalgae-based wastewater treatment to bioenergy production: a review. J. Environ. Manag. 10958630. 368. Available from: https://www.sciencedirect.com/science/journal/03014797. doi:10.1016/j.jenvman.2024.122085.

Zahroojian, N., Moravej, H., Shivazad, M., 2011. Comparison of marine algae (*Spirulina platensis*) and synthetic pigment in enhancing egg yolk colour of laying hens. Br. Poult. Sci. 0007-1668. 52 (5), 584–588. doi:10.1080/00071668.2011.610779.

Zeng, X., Zhang, C.C., 2022. The making of a heterocyst in cyanobacteria. Annu. Rev. Microbiol. 15453251. 76, 597–618. Available from: http://www.annualreviews.org/journal/micro. doi:10.1146/annurev-micro-041320-093442.

Zhang, F., Li, Z., Chen, C., Luan, H., Fang, R.H., Zhang, L., Wang, J., 2024. Biohybrid microalgae robots: design, fabrication, materials, and applications. Adv. Mater. 15214095. 36 (3). Available from: http://onlinelibrary.wiley.com/journal/10.1002/(ISSN)1521-4095. doi:10.1002/adma.202303714.

Zhao, C., Wu, Y., Yang, C., Liu, B., Huang, Y., 2015. Hypotensive, hypoglycaemic and hypolipidaemic effects of bioactive compounds from microalgae and marine micro-organisms. Int. J. Food Sci. Technol. 0950-5423. 50 (8), 1705–1717. doi:10.1111/ijfs.12860.

Zheng, J., Cole, T., Zhang, Y., Bayinqiaoge, N., Yuan, D., Tang, S.Y., 2023. An automated and intelligent microfluidic platform for microalgae detection and monitoring. Lab Chip. 14730189. 24 (2), 244–253. Available from: http://pubs.rsc.org/en/journals/journal/lc. doi:10.1039/d3lc00851g.

Zhu, C., Zhai, X., Xi, Y., Wang, J., Kong, F., Zhao, Y., Chi, Z., 2020. Efficient CO_2 capture from the air for high microalgal biomass production by a bicarbonate Pool. J. CO2 Util. 22129820. 37, 320–327. Available from: http://www.journals.elsevier.com/journal-of-co2-utilization/. doi:10.1016/j.jcou.2019.12.023.

Zhu, L.D., Li, Z.H., Hiltunen, E., 2016. Strategies for lipid production improvement in microalgae as a biodiesel feedstock. Biomed. Res. Int. 2314-6133. 2016, 1–8. doi:10.1155/2016/8792548.

Zuccaro, G., Yousuf, A., Pollio, A., Steyer, J.P., 2019. Microalgae cultivation systems. In: Microalgae Cultivation For Biofuels Production. Elsevier, France, pp. 11–29. Available from: http://www.sciencedirect.com/science/book/9780128175361. doi:10.1016/B978-0-12-817536-1.00002-3.

CHAPTER

3

Marine fungi and bacteria

Sachithra Amarin Hettiarachchi[a], *Svini Dileepa Marasinghe*[b], *and Sung-Pyo Hur*[c]

[a]Department of Marine Biotechnology, Ocean University of Sri Lanka, Colombo, Sri Lanka
[b]Korea Institute of Ocean Science and Technology, Jeju, Republic of Korea
[c]Department of Marine Life Science, Jeju National University, Jeju, Republic of Korea

3.1 Introduction

Marine fungi and bacteria represent fundamental components of ocean ecosystems, playing pivotal roles in maintaining marine environmental health and driving essential biogeochemical processes. These microorganisms pervade every marine habitat, from sun-lit surface waters to the darkest depths of ocean trenches, demonstrating remarkable adaptability and functional diversity. Current research indicates that a single milliliter of seawater harbors approximately one million bacteria and thousands of fungal species, highlighting the remarkable density and diversity of marine microbial life (Djahnit et al., 2019; Bing et al., 2019). The total microbial biomass in the oceans is estimated to equal or exceed that of all marine plants and animals combined, emphasizing their ecological dominance in marine systems (Amend et al., 2019; Bar-On and Milo, 2019).

The significance of marine microorganisms extends far beyond their numerical abundance. In surface waters, photosynthetic bacteria like *Prochlorococcus* and *Synechococcus* contribute substantially to global primary production, generating an estimated 50% of the Earth's oxygen (Dwivedi and Ahmad, 2023). These organisms have evolved sophisticated light-harvesting mechanisms, including unique pigment complexes that allow them to utilize different wavelengths of light at various ocean depths. Deep-sea bacteria and fungi play crucial roles in organic matter decomposition and nutrient recycling, maintaining the ocean's biogeochemical balance (Nawaz et al., 2022). For instance, the recently discovered *Marinispora* bacteria, found in deep-sea sediments, have shown exceptional abilities in breaking down complex organic compounds under high-pressure and low-temperature conditions (Subramani and Sipkema, 2019).

Marine microorganisms demonstrate remarkable metabolic versatility that has significant implications for biotechnology and environmental applications. Recent studies have revealed their capacity to produce novel antibiotics, anticancer compounds, and industrial enzymes

https://doi.org/10.1016/B978-0-443-29190-6.00007-6

(Baindara et al., 2018; Srinivasan et al., 2021; Cheng et al., 2020). The marine environment, particularly extreme habitats like hydrothermal vents and deep-sea trenches, has proven to be a rich source of unique biochemical pathways and adaptive mechanisms (Sánchez and Stefanni, 2024). These discoveries have led to breakthrough applications in pharmaceutical development, industrial processing, and environmental remediation technologies.

The adaptation strategies developed by marine microorganisms represent some of the most sophisticated survival mechanisms known in biology. In the hadal zones, organisms have evolved to withstand pressures exceeding 1000 atmospheres through modifications in membrane composition and protein structure (Wenhao et al., 2022; Yinnan et al., 2021). Psychrophilic bacteria and fungi in polar regions produce antifreeze proteins and maintain enzyme activity at temperatures below 0°C, while thermophilic species near hydrothermal vents function optimally at temperatures above 100°C (Furhan, 2020; Yusof et al., 2021; Holden and Sistu, 2023; Zeng et al., 2021). These adaptations have significant implications for industrial applications, particularly in the development of pressure-resistant and temperature-stable enzymes for biotechnological processes.

The role of marine microorganisms in global climate regulation has gained increasing attention in recent years (Cavicchioli et al., 2019). Through their participation in the biological carbon pump, these organisms influence atmospheric carbon dioxide levels and ocean acidification patterns (Jiao et al., 2024). Marine bacteria and fungi are also crucial in the nitrogen cycle, with some species capable of fixing atmospheric nitrogen in nutrient-poor waters, while others facilitate the removal of excess nitrogen compounds from marine ecosystems. Understanding these processes has become critical for predicting and potentially mitigating the effects of climate change on marine ecosystems.

Technological advances in recent decades have revolutionized our understanding of marine microbial communities. The application of next-generation sequencing technologies has revealed previously unknown diversity, such as the discovery of the SAR11 clade, now recognized as one of the most abundant bacterial groups in the ocean (López-Pérez et al., 2020). Single-cell genomics has enabled the study of unculturable marine microorganisms, revealing novel metabolic capabilities and evolutionary adaptations (Pachiadaki et al., 2019). Metagenomic studies have uncovered new metabolic pathways and functional capabilities, particularly in extreme marine environments. For example, deep-sea hydrothermal vent bacteria have been found to possess unique enzymes capable of functioning at temperatures exceeding 100°C, presenting exciting possibilities for industrial applications (Fongaro et al., 2020).

The emerging field of marine microbial biotechnology has seen rapid growth, driven by the discovery of novel compounds and processes. Marine-derived pharmaceuticals, including antibiotics, antiviral agents, and anticancer compounds, are increasingly entering clinical trials (Brown and Wobst, 2021; Pandey et al., 2020). The marine actinobacterium *Salinispora tropica*, for instance, produces salinosporamide A, a promising compound for cancer treatment (Chhun et al., 2021). In addition, marine microorganisms have shown potential in addressing environmental challenges, with certain species demonstrating capabilities in plastic degradation, oil spill remediation, and heavy metal sequestration (Khandare et al., 2022; Luo et al., 2021; Manobala et al., 2019).

The study of marine microbial ecology has revealed complex interactions and symbiotic relationships that maintain ecosystem stability. Recent research has highlighted the importance

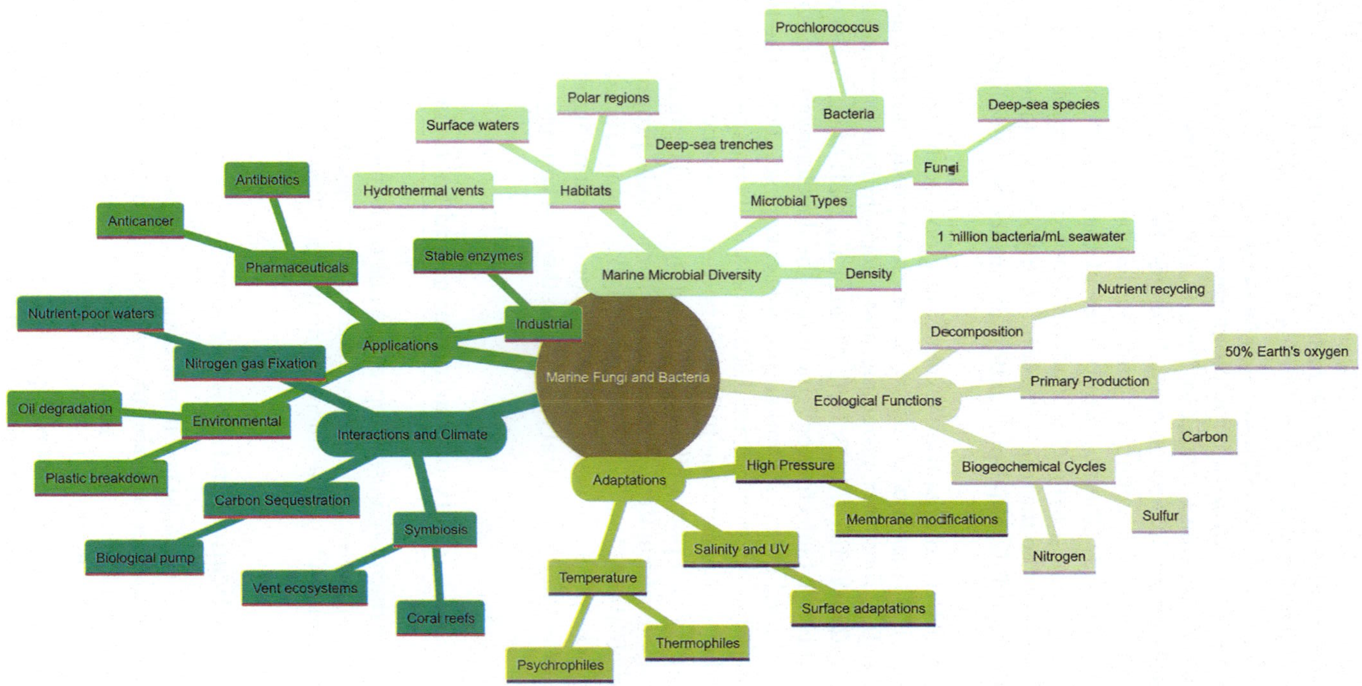

FIG. 3.1 **Mindmap summarizing the key aspects of marine fungi and bacteria**, including their diversity, ecological functions, environmental adaptations, applications, and climate interactions. The hierarchical structure illustrates the relationships between major themes and their specific components in marine microbial ecology.

of microbial communities in coral reef health, deep-sea hydrothermal vent ecosystems, and marine food webs (Glasl et al., 2019; Castelán-Sánchez et al., 2019; Potapov et al., 2019). These microorganisms form intricate networks of interactions, including nutrient exchange, chemical signaling, and defensive partnerships. Understanding these relationships has become crucial for marine conservation efforts and ecosystem management strategies.

Historical perspectives on marine microbiology reveal a field that has evolved rapidly since the first isolation of marine bacteria in the late 19th century. Early studies focused primarily on culturable organisms from coastal waters, but technological advances have progressively revealed the vast diversity of marine microbes across all ocean habitats. The discovery of deep-sea chemosynthetic communities in the 1970s fundamentally changed our understanding of life's requirements and distribution (Shimabukuro et al., 2020). Similarly, the recognition of the marine viral community's importance in the 1980s introduced new perspectives on microbial population dynamics and evolution (Ambalavanan et al., 2021).

The influence of marine microorganisms extends to global-scale processes and industrial applications. These organisms play crucial roles in marine food production through aquaculture, contribute to pharmaceutical discovery programs, and offer solutions for environmental remediation (Ahmed et al., 2022; Wichard, 2023; Dell' Anno et al., 2021; Mingxing et al., 2022) Their study has led to innovations in biotechnology, including the development of novel enzymes for industrial processes, biobased materials, and sustainable energy solutions. The potential applications continue to expand as new species are discovered and their capabilities are explored.

Looking forward, the study of marine fungi and bacteria continues to expand our understanding of life's diversity and potential applications (Fig. 3.1). The integration of advanced molecular techniques, bioinformatics, and cultivation methods promises to unlock more secrets of marine microbial life. These discoveries will likely play crucial roles in addressing global challenges, from healthcare to environmental conservation, highlighting the continuing importance of marine microbiological research in the 21st century.

3.2 Diversity of marine fungi and bacteria

The diversity of marine microorganisms represents one of the most complex and fascinating aspects of marine biology. This section explores taxonomic richness, unique adaptations, and distinctive characteristics that set marine microbes apart from their terrestrial counterparts.

3.2.1 Taxonomic classification and major groups

Marine bacteria exhibit remarkable taxonomic diversity across numerous phyla (Table 3.1). The Proteobacteria phylum, particularly abundant in marine environments, comprises several classes with distinct ecological roles. The Alphaproteobacteria class includes the metabolically versatile *Roseobacter* clade, which can contribute up to 20% of coastal bacterioplankton communities (Zoccarato and Grossart, 2019). These organisms possess diverse metabolic capabilities, including aerobic anoxygenic photosynthesis, carbon monoxide oxidation, and aromatic compound degradation (Lemfack et al., 2020). The Gammaproteobacteria class includes the genus

TABLE 3.1 Major marine microbial groups and their ecological roles.

Taxonomic group	Representative genera	Primary functions	Habitat distribution	References
Proteobacteria	*Alteromonas, Vibrio*	Organic matter degradation, Symbiosis	Surface waters, deep sea	(Gutierrez, 2019)
Cyanobacteria	*Prochlorococcus, Synechococcus*	Primary production, N_2 fixation	Photic zone	(Wang et al., 2023)
Bacteroidetes	*Flavobacterium, Polaribacter*	Polymer degradation	Coastal waters	(McKee et al., 2021)
Actinobacteria	*Salinispora, Marinispora*	Secondary metabolite production	Marine sediments	(Nair and Abraham, 2020)
Ascomycete fungi	*Corollospora, Lindra*	Wood degradation, decomposition	Coastal zones	(Bunbury-Blanchette and Walker, 2019)
Basidiomycete fungi	*Digitatispora, Halocyphina*	Symbiotic relationships	Coral reefs, Mangroves	(Heimland, 2022)

Alteromonas, capable of utilizing a wide range of organic compounds, and *Vibrio* sp., which form important symbiotic relationships with marine animals (Gutierrez, 2019; Destoumieux-Garzón et al., 2020; Essock-Burns et al., 2020). For example, *Vibrio fischeri* establishes bioluminescent symbioses with squid species through sophisticated molecular signaling mechanisms (Essock-Burns et al., 2021).

Cyanobacteria, represented by genera like *Prochlorococcus* and *Synechococcus*, dominate the photic zones of oceans (Wang et al., 2023). *Prochlorococcus*, discovered in 1988, is now recognized as the most abundant photosynthetic organism on Earth (Muñoz-Marín et al., 2020). This genus has evolved into multiple ecotypes adapted to different light intensities and nutrient conditions. High-light-adapted strains dominate surface waters, while low-light-adapted strains thrive at depths up to 200 m (Genge, 2021). Each ecotype possesses specific genomic adaptations, including modified photosynthetic apparatus and nutrient acquisition systems (Becker et al., 2024).

The Bacteroidetes phylum includes important polymer degraders in marine environments (McKee et al., 2021). Members of this phylum, particularly the Flavobacteria class, specialize in breaking down complex organic matter such as algal polysaccharides (Gavriilidou et al., 2020; Xinyang et al., 2024). These organisms possess extensive arrays of carbohydrate-active enzymes, enabling them to access carbon sources unavailable to other marine bacteria. The genus *Polaribacter*, for example, has evolved specialized mechanisms for adhering to and degrading algal cell walls (Pontiller et al., 2022).

Marine Actinobacteria, though less abundant than other bacterial groups, play crucial roles in secondary metabolite production (Girão et al., 2022). The genus *Salinispora*, the first obligately marine actinobacterium discovered, produces numerous bioactive compounds with pharmaceutical potential (Nair and Abraham, 2020; Subramani and Sipkema, 2019). These organisms possess unique biosynthetic gene clusters that reflect adaptation to marine conditions and contribute to their chemical diversity (Creamer, 2022).

Marine fungi, though less studied than bacteria, display significant diversity across both Ascomycota and Basidiomycota phyla. The Ascomycete genus *Corollospora* includes species specialized in degrading wooden substrates in marine environments (Bunbury-Blanchette and Walker, 2019). These fungi possess unique enzymes adapted to saline conditions and play essential roles in marine carbon cycling (Giddings and Newman, 2019). Marine Basidiomycetes, such as *Digitatispora marina*, have evolved sophisticated mechanisms for colonizing and breaking down marine plant material (Heimland, 2022).

Recent molecular studies have revealed previously unknown diversity among marine microorganisms. Culture-independent techniques have identified numerous candidate phyla that lack cultivated representatives but appear to play important ecological roles. These include the SAR324 cluster of Deltaproteobacteria, which contributes to sulfur oxidation in oxygen-minimum zones, and the Marine Group II Euryarchaeota, which participate in organic matter turnover in surface waters (Ding et al., 2024; Suominen et al., 2021).

The taxonomic classification of marine microorganisms continues to evolve as new molecular methods reveal previously unknown diversity and evolutionary relationships. This improved understanding of marine microbial taxonomy provides crucial insights into ecosystem function and potential biotechnological applications.

3.2.2 Adaptations to marine environments

Marine microorganisms have developed sophisticated mechanisms to thrive in oceanic conditions, demonstrating remarkable adaptability across various environmental challenges. These adaptations span multiple cellular and molecular levels, enabling survival and proliferation in some of Earth's most extreme environments.

Salinity adaptation represents one of the most fundamental challenges in marine environments. Marine bacteria, like *Halomonas* sp., produce specialized osmolytes such as ectoine and glycine betaine to maintain cellular osmotic balance (Hobmeier et al., 2022; Imhoff et al., 2021). These compounds not only protect against salt stress but also stabilize proteins and cellular structures. For example, *Halomonas elongata* can accumulate ectoine up to 20% of its cell dry weight when exposed to high-salinity conditions (Junxiong et al., 2024). The molecular mechanisms behind salt tolerance include modified ion transport systems, specialized membrane lipid compositions, and enhanced cell wall structures. Marine fungi, such as *Aspergillus sydowii*, have evolved unique Na^+/H^+ antiporters that efficiently regulate internal sodium concentrations even in hypersaline conditions exceeding 2 M NaCl (Rodríguez-Pupo et al., 2021).

Pressure adaptation is particularly evident in deep-sea microorganisms, which have evolved remarkable mechanisms to maintain cellular function under extreme hydrostatic pressure. The bacterium *Shewanella psychropiezotolerans*, isolated from the Mariana Trench, exhibits optimal growth at pressures of 0.1 MPa but can grow across a range of 0.1–40 MPa (Libo et al., 2021). This organism has evolved modified membrane lipids with increased proportions of unsaturated fatty acids, maintaining membrane fluidity under extreme pressure (Tamby et al., 2023). The adaptation includes pressure-sensitive gene regulation systems that modify protein expression patterns according to depth (Wang et al., 2021). Moreover, various deep-sea bacteria produce pressure-resistant enzymes that remain active at high hydrostatic pressures, featuring structural adaptations that prevent pressure-induced denaturation.

Temperature adaptation varies dramatically across marine environments, requiring specialized cellular mechanisms at both extreme cold and heat. Psychrophilic bacteria, such as *Pseudoalteromonas haloplanktis*, found in polar waters, produce cold-active enzymes that function efficiently at temperatures near 0°C (Ghosh et al., 2025a). These enzymes demonstrate higher catalytic efficiency at low temperatures compared to their mesophilic counterparts, achieved through structural modifications that increase flexibility around the active site. The adaptation includes the production of antifreeze proteins, cold-shock proteins, and modified membrane lipids that maintain fluidity at low temperatures (Jin et al., 2022). Conversely, thermophilic bacteria from hydrothermal vents, such as *Thermococcus barophilus*, produce heat-stable proteins that maintain functionality at temperatures exceeding 100°C (Crosby et al., 2019). These organisms possess enhanced DNA repair mechanisms and specialized chaperone proteins that prevent the thermal denaturation of cellular components (Ranawat and Rawat, 2020).

Light adaptation in marine microorganisms demonstrates remarkable sophistication, particularly in photosynthetic species. Marine cyanobacteria have evolved specialized pigment–protein complexes that optimize light harvesting at different ocean depths (Kolodny et al., 2022). For example, *Prochlorococcus* strains show distinct ecological differentiation based on their light-harvesting capability, with high-light- and low-light-adapted ecotypes possessing different versions of the photosynthetic apparatus (Soulier, 2021; Yan et al., 2020). Some deep-sea bacteria have evolved bioluminescence capabilities, using light production for communication, predation, or defense, demonstrating the diverse roles of light-based adaptations in marine environments (Bao et al., 2023; Vannier et al., 2020).

3.3 Ecology and functional roles

3.3.1 Nutrient cycling

Marine microorganisms serve as primary drivers of global biogeochemical cycles, facilitating essential nutrient transformations that sustain marine ecosystems. Their activities influence carbon, nitrogen, sulfur, and phosphorus cycling through complex, interconnected processes that maintain ocean productivity and global elemental balance.

The marine carbon cycle heavily depends on microbial activities across various oceanic zones. Photosynthetic bacteria, particularly *Prochlorococcus* and *Synechococcus*, annually fix approximately 4 gigatons of carbon dioxide into organic matter in surface waters (Flombaum et al., 2013; Veaudor et al., 2020). These organisms have evolved highly efficient carbon-concentrating mechanisms that enable photosynthesis even in low-CO_2 conditions. The microbial carbon pump, operated by heterotrophic bacteria like *Pelagibacter ubique* (SAR11), transforms labile organic carbon into recalcitrant forms, contributing to long-term carbon sequestration in the oceans (Bayer et al., 2019). In deeper waters, chemolithoautotrophic bacteria fix inorganic carbon using energy from reduced compounds (Kieft, 2021; Le Bris et al., 2019), contributing significantly to deep-sea carbon fixation. Deep-sea bacteria and fungi play crucial roles in carbon remineralization, with organisms like *Pseudoalteromonas* sp. efficiently breaking down complex organic compounds in the hadal zones (Peoples et al., 2019).

Nitrogen cycling in marine environments involves multiple microbial-mediated processes that maintain ecosystem productivity. Diazotrophic cyanobacteria, such as *Trichodesmium erythraeum*, can fix atmospheric nitrogen in nutrient-poor tropical waters, contributing an estimated 100–200 Tg of fixed nitrogen annually to marine ecosystems (Shijie et al., 2024). These organisms possess specialized cells called heterocysts that provide oxygen-free environments for nitrogen fixation (Liu et al., 2022b; Upadhyay et al., 2021). Nitrifying bacteria, including *Nitrosopumilus maritimus*, perform ammonia oxidation in the mesopelagic zone, producing nitrate that supports primary production. This organism represents a unique group of archaea that dominates ammonia oxidation in the open ocean (Bayer et al., 2019; Semedo et al., 2021). In oxygen-minimum zones, bacteria like *Scalindua brodae* conduct anaerobic ammonium oxidation (anammox), significantly influencing marine nitrogen budgets by converting fixed nitrogen back to N_2 gas (Russ et al., 2019; Vipindas et al., 2020).

The marine sulfur cycle features complex microbial interactions that influence both ocean chemistry and atmospheric processes. Sulfate-reducing bacteria, like *Desulfovibrio salexigens*, thrive in marine sediments, producing hydrogen sulfide that supports chemosynthetic communities (Barton and Fauque, 2022; Marietou et al., 2022). These organisms can couple organic matter oxidation to sulfate reduction, making them crucial players in carbon mineralization in anoxic environments (Jabir et al., 2021). Sulfur-oxidizing bacteria, including *Thiomicrospira crunogena* found near hydrothermal vents, convert reduced sulfur compounds back to sulfate, completing the sulfur cycle (Böhnke et al., 2019). Marine bacteria also mediate the production of dimethylsulfide from algal-produced dimethylsulfoniopropionate (DMSP), influencing global climate through cloud formation processes. The bacterial transformation of DMSP involves multiple biochemical pathways that have evolved specifically in marine environments (Lipsman et al., 2024; Young and Schmidt, 2020).

Phosphorus cycling, though less complex, remains crucial for marine productivity. Marine bacteria like *Ruegeria pomeroyi* efficiently mineralize organic compounds through the production of alkaline phosphatases (Wang et al., 2022), making this essential nutrient available for primary producers. Certain marine bacteria, such as *Cobetia marina*, can accumulate polyphosphate under favorable conditions, serving as phosphorus reservoirs in marine ecosystems (Balabanova et al., 2024; Fernández-Juárez et al., 2022). These organisms play vital roles in maintaining phosphorus availability in oligotrophic waters through sophisticated uptake and storage mechanisms.

3.3.2 Biodegradation and bioremediation

Marine microorganisms demonstrate remarkable capabilities in degrading various compounds and pollutants, making them crucial for ecosystem maintenance and environmental remediation. Their diverse metabolic capabilities enable the breakdown of both natural and anthropogenic materials, offering sustainable solutions for marine pollution management.

Natural organic matter decomposition involves complex microbial communities working synergistically. Marine bacteria like *Cellulophaga lytica* specialize in degrading complex polysaccharides such as cellulose and chitin (Baghel et al., 2024; Delbarre-Ladrat et al., 2019), producing multiple carbohydrate-active enzymes adapted to marine conditions. These organisms secrete specialized enzyme complexes called cellulosomes, which efficiently break down plant material in high-salt environments (Lisov et al., 2022). Marine fungi, particularly members of the genus *Corollospora*, excel at breaking down lignin-rich materials in coastal environments, utilizing unique oxidative enzymes that function in saline conditions (Poli et al., 2021). The degradation process involves the production of lignin peroxidases and manganese-dependent peroxidases, which remain active even at high salt concentrations (Vala and Dave, 2018).

Hydrocarbon degradation represents a crucial function of marine microorganisms in maintaining ocean health. The hydrocarbon-degrading bacterium *Alcanivorax borkumensis* has become particularly important in oil spill remediation (Kadri et al., 2022). Following the Deepwater Horizon oil spill in 2010, populations of this bacterium increased significantly in affected areas, demonstrating natural bioremediation capabilities (Joseph, 2019). *A. borkumensis* produces specialized alkane hydroxylases that initiate hydrocarbon breakdown, along with biosurfactants that increase oil bioavailability (Ibrahim et al., 2023). Similarly, *Cycloclasticus pugetii* specializes in degrading aromatic hydrocarbons, including the highly toxic polycyclic aromatic hydrocarbons, through a unique ring-cleaving dioxygenase system (Denaro et al., 2020; Vogel, 2024).

Plastic pollution degradation has emerged as a critical area of research in marine microbiology. The bacterium *Ideonella sakaiensis*, discovered in 2016, produces two unique enzymes capable of breaking down polyethylene terephthalate (PET) plastic (Duru and Duru, 2021). The first enzyme, PETase, converts PET to mono (2-hydroxyethyl) terephthalic acid (MHET), while the second enzyme, MHETase, further breaks down MHET into simpler compounds (Anuar et al., 2022). Marine fungi, such as *Zalerion maritimum*, have demonstrated the ability to degrade microplastics, reducing both the size and mass of plastic particles in seawater (Oliveira et al., 2020). Recent research has identified consortia of marine microorganisms capable of degrading multiple types of plastics, including polyethylene and polypropylene (Atanasova et al., 2021).

Heavy metal transformation by marine microorganisms involves various sophisticated mechanisms. The marine bacterium *Pseudomonas stutzeri* strain DN2 can effectively transform toxic chromium (VI) to less toxic chromium (III) through enzymatic reduction (Badar et al., 2000; Marzuki et al., 2023). This process involves a complex electron transport chain coupled with specialized chromate reductases. *Marinobacter* spp. have shown remarkable abilities in mercury detoxification, converting methylmercury to less toxic forms through the mer operon system (Rincón-Tomás et al., 2024; Sanz-Sáez, 2021). Some marine fungi, like *Aspergillus niger* isolated from marine environments with various marine organisms, create intricate networks that support ecosystem function and organism health. These relationships range from obligate symbioses crucial for survival to facultative associations that enhance host fitness.

Coral–microbe symbiosis represents one of the most studied and ecologically significant marine symbiotic relationships. The primary symbiosis involves zooxanthellae (*Symbiodinium* spp.) within coral tissues, providing up to 90% of the coral's energy requirements through photosynthesis (Hoadley et al., 2019). Different *Symbiodinium* clades show distinct temperature tolerances and photosynthetic efficiencies, influencing coral resilience to environmental stress (Hoadley et al., 2019; Turnham et al., 2003). Beyond this classical example, corals host diverse bacterial communities. The coral *Pocillopora damicornis* harbors specific strains of *Endozoicomonas* bacteria that aid in sulfur and nitrogen cycling within the coral holobiont (Maire et al., 2024; Ostria-Hernández et al., 2022). These bacteria also produce antimicrobial compounds that protect the coral from pathogens (Dungan et al., 2023). During thermal stress events, shifts in these microbial communities can serve as early warning indicators of coral health decline. Recent research has revealed that coral-associated bacteria also contribute to skeletal formation through calcium carbonate precipitation and pH regulation (Liang et al., 2021).

Marine sponges demonstrate particularly complex microbiomes, often hosting thousands of microbial species that can constitute up to 40% of the sponge's biomass (Orlić, 2019). These relationships show remarkable host specificity and metabolic integration. The deep-sea sponge *Geodia barretti* maintains a stable association with specific *Chloroflexi* and *Proteobacteria* species that produce secondary metabolites, providing chemical defense against predators (Cleary et al., 2024; Mazzella et al., 2024). These microorganisms also participate in nitrogen cycling, with evidence of both nitrification and denitrification occurring within the sponge tissue. The sponge species *Theonella swinhoei* harbors the bacterium *Entotheonella*, which produces multiple bioactive compounds, including the anticancer compound theopalauamide (Kuo et al., 2019; Uria and Wakimoto, 2021; Wakimoto, 2023). Recent metagenomic studies have revealed that sponge-associated bacteria possess unique biosynthetic gene clusters, indicating their potential for producing novel pharmaceutical compounds (Amelia et al., 2022; Gavriilidou et al., 2020).

Algal–bacterial relationships showcase diverse beneficial interactions that influence marine primary production. The green alga *Ulva mutabilis* requires specific bacterial strains for normal morphological development, demonstrating the complexity of these associations (Alsufyani et al., 2020; Ghaderiardakani et al., 2024). The bacterium *Roseovarius* sp. promotes algal cell division through the production of specific morphogens, while *Maribacter* sp. induces proper cell wall formation (Ulrich et al., 2022; Alsufyani et al., 2020). These bacteria also enhance nutrient acquisition through siderophore production and vitamin synthesis (Gonçalves et al., 2024). In the open ocean, heterotrophic bacteria associated with diatoms facilitate iron acquisition by producing siderophores, enabling diatom growth in iron-limited waters (Manck et al., 2022). The bacterium *Marinobacter algicola*, frequently associated with dinoflagellates, provides vitamin B12 essential for algal growth, while benefiting from organic carbon released by the algae (Zhu et al., 2023; Kim et al., 2024).

Deep-sea symbiosis often involves chemosynthetic bacteria, representing some of the most extreme examples of metabolic cooperation. The giant tubeworm *Riftia pachyptila*, found near hydrothermal vents, hosts sulfur-oxidizing bacteria within specialized organs called trophosomes (Hinzke et al., 2019). These bacteria convert hydrogen sulfide from vent fluids into organic carbon, providing all the tubeworm's nutritional needs. The complexity of this symbiosis is reflected in the extreme reduction of the host's digestive system and the sophisticated molecular mechanisms that regulate sulfide transport and carbon transfer (Lan, 2020; Lo Giudice and Rizzo, 2022). Similar relationships exist in cold seep environments, where methane-oxidizing bacteria form symbioses with mussels of the genus *Bathymodiolus*, enabling these organisms to thrive in methane-rich environments (McCowin et al., 2020; Rubin-Blum et al., 2019). Recent studies have revealed that these bacterial symbionts possess unique adaptations to high-pressure environments and specialized pathways for carbon fixation under extreme conditions.

3.4 Potential applications in biotechnology, medicine, and aquaculture

3.4.1 Biotechnological applications

Marine microorganisms offer unprecedented opportunities for biotechnological advancement, particularly in the development of novel enzymes, biofuels, and biomaterials (Fig. 3.2). Their unique adaptations to extreme marine conditions make them valuable sources of industrially relevant compounds and processes, with applications spanning multiple industrial sectors.

3.4.1.1 Industrial enzyme applications

The enzyme industry has benefited significantly from marine microbial diversity, with marine-derived enzymes offering unique properties that address specific industrial challenges. These enzymes demonstrate remarkable stability and activity under conditions that typically inhibit their terrestrial counterparts. Cold-adapted enzymes from psychrophilic bacteria, such as *Colwellia psychrerythraea*, have revolutionized various industrial processes by functioning effectively at temperatures as low as 4°C (Gupta et al., 2023), enabling energy-efficient cold-water washing processes (Chapadgaonkar et al., 2024). These enzymes show catalytic efficiency 2–10 times higher than mesophilic counterparts at low temperatures, reducing energy requirements, improving preservation of temperature-sensitive substrates, and demonstrating enhanced stability in the presence of chemical denaturants (Liu et al., 2023a; Gupta et al., 2023). Modern laundry detergent formulations incorporating these cold-active enzymes achieve energy savings of up to 30%, improve fabric preservation through lower washing temperatures, enhance removal of protein-based stains at low temperatures, and reduce environmental impact through lower energy consumption (Fernandes, 2023; Poonam Vashist et al., 2019).

Deep-sea extremophiles like *Thermococcus gammatolerans* produce hyperthermostable enzymes, including amylases that remain active at temperatures exceeding 100°C (Pedraza-Segura et al., 2023) and maintain 50% activity even after 48 h at 95°C, significantly outperforming traditional alternatives in applications such as starch processing at elevated temperatures, paper pulp treatment, biofuel production, and food processing under sterile conditions (Debnath et al., 2019; Ariaeenejad et al., 2023; Kanekar and Kanekar, 2022). Halophilic enzymes from *Marinobacter hydrocarbonoclasticus* maintain activity in solutions containing up to 4 M NaCl, show enhanced stability in organic solvents, facilitate biodiesel production in high-salt conditions, and enable efficient wastewater treatment in saline conditions (Daoud and Ali, 2020; Wang et al., 2023). Pressure-adapted enzymes from piezophilic bacteria like *Shewanella piezotolerans* function optimally under high pressure (up to 70 MPa), enabling high-pressure food processing, facilitating deep-sea oil degradation, and improving extraction processes in the food industry (Scoma et al., 2019; Ghosh et al., 2025a, 2025b). Marine fungi, particularly species from the genus *Aspergillus* isolated from marine sediments, produce unique enzyme variants, including cellulases active in high-salt conditions, xylanases for paper pulp processing, chitinases for chitin waste treatment, and proteases for marine feed processing (Samantaray and Sahu, 2021; Yadav et al., 2020; Das et al., 2019; Jannathulla and Dayal, 2023). Recent developments in enzyme engineering have further

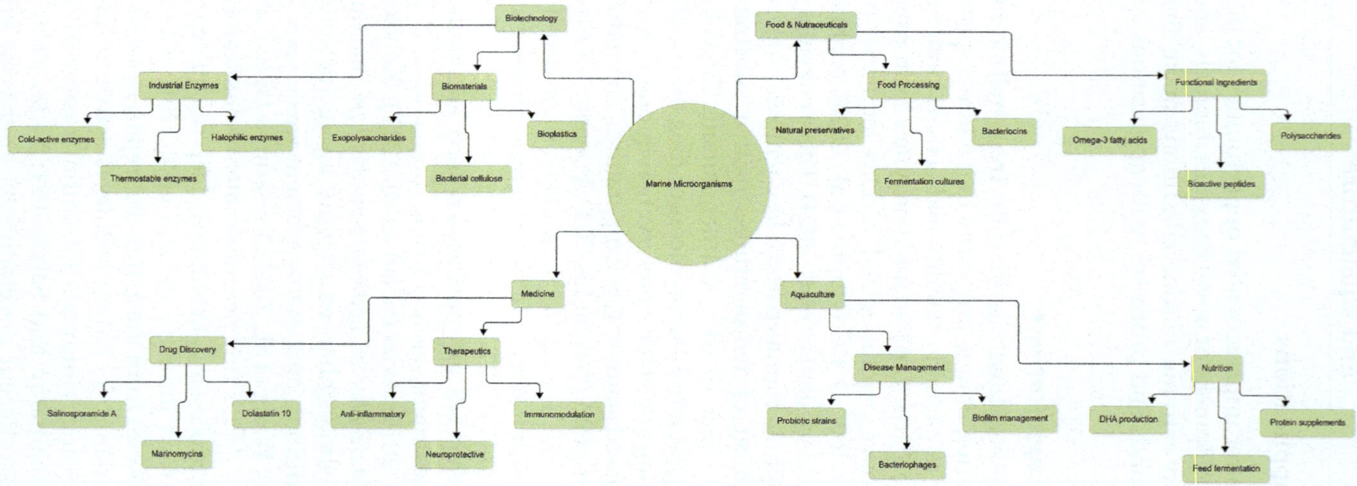

FIG. 3.2 **Applications of marine microorganisms across industrial sectors**. The hierarchical diagram illustrates four major application domains with their respective sub-categories and specific products, showcasing the diverse industrial and biomedical applications of marine microbial resources.

enhanced the industrial applicability of marine-derived enzymes through protein engineering for improved stability, enhancement of catalytic efficiency, and modification of substrate specificity.

3.4.1.2 *Novel biomaterial development*

Marine microorganisms are revolutionizing biomaterial development through their unique synthetic capabilities and environmental adaptations. The deep-sea bacterium *Alteromonas macleodii* produces exceptional exopolysaccharides with remarkable stability across temperatures (4–90°C) and pH values (2–12), making them valuable in food processing, cosmetics, and drug delivery systems (Concórdio-Reis et al., 2023; Ibrahim et al., 2022; Benhadda et al., 2023; Shukla et al., 2022). Their structural complexity offers enhanced functionality compared to terrestrial alternatives. Marine *Gluconacetobacter* strains have been engineered to produce modified bacterial cellulose under saline conditions, yielding materials with increased crystallinity, superior tensile strength, enhanced water retention capacity exceeding 200 times their dry weight, and improved biocompatibility (Beliah et al., 2020; Ibrahim et al., 2020; Singhania et al., 2021; Liu et al., 2020a). *Pseudoalteromonas* spp. generate unique adhesive proteins that maintain functionality in wet conditions, demonstrating strong aqueous adhesion, biodegradation resistance, and nontoxic properties across diverse surfaces (Nwuzor et al., 2021; Bayer, 2022). *Vibrio diabolicus* produces polysaccharides particularly effective for bone tissue regeneration, showing enhanced osteoblast activity (Norris et al., 2019), controlled biodegradation, superior mechanical properties, and natural antiinflammatory characteristics (Nisha et al., 2023; Faÿ et al., 2023; Ullah et al., 2019; Jeewon et al., 2023). The halophilic bacterium *Halomonas boliviensis* produces polyhydroxyalkanoates using seawater-based media, creating bioplastics with enhanced thermal stability, improved mechanical properties, and reduced production costs (Woo et al., 2024). *Pseudoalteromonas tunicata* offers environmentally friendly antifouling compounds with broad-spectrum activity and self-renewing capabilities (Ali, 2025). Deep-sea bacteria produce sophisticated hydrogel networks characterized by responsive behavior to environmental stimuli, excellent biocompatibility, unique mechanical properties suitable for soft tissue engineering, and controlled therapeutic compound release capabilities (Cuenot et al., 2022; Gélébart et al., 2022). These marine-derived biomaterials continue expanding possibilities across medical, industrial, and environmental applications, offering sustainable and high-performance alternatives to traditional materials.

3.4.2 Marine microbes in medicine

The pharmaceutical potential of marine microorganisms has generated intense research interest, yielding numerous promising compounds for medical applications (Table 3.2). This field has expanded rapidly with advances in cultivation techniques and molecular screening methods, leading to several breakthrough discoveries in drug development.

3.4.2.1 *Novel drug discovery and development*

Marine microorganisms have emerged as a transformative source of novel pharmaceutical compounds, particularly excelling in cancer treatment, antimicrobial therapy, and

TABLE 3.2 Marine microbial compounds and their medical applications.

Compound class	Source organism	Medical application	Development stage	Clinical significance
Salinosporamide A	*Salinispora tropica*	Multiple myeloma	Phase III trials	Proteasome inhibitor with a unique mechanism (Fenical et al., 2009)
Cephalosporin C	*Acremonium chrysogenum*	Bacterial infections	Approved drug	Broad-spectrum antibiotic (Lin et al., 2022)
Marinomycins	*Marinispora* sp.	MRSA infections	Preclinical	Active against resistant bacteria (Kwon et al., 2006)
Dolastatin 10	Symbiotic bacteria	Breast cancer	Clinical use	Microtubule inhibitor (Maderna et al., 2014)
Thiocoraline	Marine actinomycete	Solid tumors	Phase II trials	DNA polymerase α inhibitor (Jin et al., 2019)
Abyssomicin C	Deep-sea bacteria	Tuberculosis	Research phase	Novel antibiotic mechanism (Song et al., 2017)

immunomodulation (Stincone and Brandelli, 2020; Cyske et al., 2021; Riera-Romo et al., 2020). The unique environmental conditions of marine ecosystems have driven the evolution of distinctive biosynthetic pathways, resulting in compounds with unprecedented structural features and mechanisms of action. A significant breakthrough in this field is the discovery of salinosporamide A from the marine actinobacterium *S. tropica* (Jensen, 2022). This compound demonstrates potent proteasome inhibition through irreversible binding to the 20S proteasome β subunit, forming crucial tetrahydrofuran ring structures and selectively targeting cancer cells (Sweeney et al., 2024; Chhun et al., 2021; Bahrami et al., 2022).

The field has greatly benefited from novel, simple techniques (Hettiarachchi et al., 2017) to advanced screening techniques, including genome mining for cryptic biosynthetic gene clusters, high-throughput screening against multiple disease targets, and metabolomic profiling of marine microbial communities. Technological advances have enhanced the development process through improved cultivation methods for rare marine microorganisms, optimized fermentation techniques, and advanced structural elucidation methods. Recent discoveries have unveiled new compound classes, including cyclic peptides with enhanced membrane permeability, modified polyketides with improved pharmacological properties, and novel alkaloids targeting unexplored pathways (Ribeiro et al., 2022; Sukmarini et al., 2024; Mayer et al., 2024; De Rop et al., 2022).

In the field of antimicrobial compounds, marine-derived antibiotics have gained significant importance amid growing resistance concerns. Marinomycins, isolated from *Marinispora* strains, exhibit powerful activity against methicillin-resistant *Staphylococcus aureus* (MRSA) through novel membrane disruption mechanisms. These compounds demonstrate significant antimicrobial properties without developing resistance after multiple passages (Bailey et al., 2019). This resistance-evading characteristic suggests their potential as sustainable therapeutic agents for addressing persistent infections that typically develop resistance to conventional antibiotics. Deep-sea fungi, particularly *Aspergillus versicolor*, produce cyclic peptides demonstrating broad-spectrum activity against various pathogens, including influenza viruses, multidrug-resistant bacteria, and pathogenic yeasts (Saide et al., 2021; Shi et al., 2024; Minakshi et al., 2024).

Marine actinobacteria contribute significantly through compounds like anthracimycin, which effectively targets MRSA through DNA replication inhibition, *Bacillus anthracis* with low MIC (Minimum Inhibitory Concentration) values (Molina, 2024), vancomycin-resistant enterococci, and multiple drug-resistant tuberculosis strains (Liu et al., 2019; Stonik et al., 2020; Fenical, 2020). The cyanobacterium *Lyngbya majuscula* produces dolastatin 10, showing promising results in enhancing immune responses through dual mechanisms of direct cytotoxicity and immunomodulation (Verma et al., 2023; Saad et al., 2022). Clinical studies have demonstrated increased natural killer cell activity and T-cell proliferation, leading to the development of antibody-drug conjugates like brentuximab vedotin for lymphoma treatment (Srivastava et al., 2024; Vujović et al., 2025; Prabhu and Patravale, 2020; Robles-Bañuelos et al., 2022).

Marine-derived compounds often exhibit unique protein-targeting capabilities, exemplified by the deep-sea fungus *Aspergillus* sp., which produces gliotoxin-like compounds that selectively inhibit NF-κB signaling through novel cysteine-dependent mechanisms (Wei et al., 2021). Similarly, compounds from marine *Streptomyces* species modulate protein-protein interactions through allosteric mechanisms, offering new approaches to targeting previously "undruggable" proteins (Silva et al., 2021; Sánchez-Suárez et al., 2022). These discoveries have particularly impacted cancer therapy development, with compounds like salinosporamide A inducing apoptosis through nonclassical pathways (Fricker, 2020).

In the antiinflammatory domain, *Pseudoalteromonas phenolica* produces pseudopterosins that effectively treat chronic skin inflammation through novel pathways distinct from traditional drugs. Marine *Penicillium* fungi generate antiinflammatory molecules targeting the NFκB pathway, showing promising results in early clinical studies of inflammatory bowel disease (Chakraborty, 2023; Yuru et al., 2025; Pal et al., 2023; Jianzhou et al., 2019; Almasi et al., 2021; Qiu et al., 2024). Recent research has expanded into neurodegenerative disorders, where deep-sea *Penicillium* fungi produce neuroprotective compounds that reduce oxidative stress, prevent protein aggregation associated with Alzheimer's disease, enhance mitochondrial function, and promote neurite outgrowth (Verma et al., 2023; Anushree Pant, 2023; Semwal et al., 2022; Tilki et al., 2021).

Marine actinobacteria, particularly *Streptomyces* sp. CNQ-617, have shown promise in treating metabolic diseases by producing compounds that modulate glucose metabolism through AMPK activation, enhance insulin sensitivity, regulate hepatic glucose production, and protect pancreatic β-cells (Dinda et al., 2022; Liu et al., 2024; Ghareeb et al., 2024; Sun et al., 2024; Kumar et al., 2019). In addition, marine bacterial extracts demonstrate significant potential in regenerative medicine by stimulating stem cell proliferation, enhancing tissue regeneration, promoting wound healing through novel mechanisms, and regulating inflammatory responses during tissue repair (Shinde et al., 2019; Sahana and Rekha, 2020; Kim et al., 2023).

These diverse therapeutic applications and continued discoveries underscore the vast potential of marine microorganisms in pharmaceutical development, offering new hope for treating various diseases through novel mechanisms and approaches. The unique environmental adaptations of these organisms continue to yield compounds with unprecedented structural features and therapeutic properties, making them invaluable resources in modern drug discovery and development.

3.4.3 Role in sustainable aquaculture

Marine microorganisms have become increasingly crucial in sustainable aquaculture development, contributing to disease prevention, nutrition, and environmental management. Their applications have become particularly important as aquaculture intensifies to meet the growing global seafood demand while maintaining environmental sustainability. These microorganisms serve multiple functions across different aspects of aquaculture operations, from water quality management to feed production and disease control.

The implementation of probiotic strategies in aquaculture has evolved significantly through research on marine microorganisms. The bacterium *Pseudomonas fluorescens* strain AH2, isolated from healthy shrimp, provides protection against vibriosis in cultured shrimp through multiple mechanisms, including competitive exclusion of pathogens, enhancement of host immune responses, production of antimicrobial compounds, and improvement of water quality through nitrogen cycling (Alagawany et al., 2020; Rahayu et al., 2024). Probiotic formulations incorporating multiple complementary strains have demonstrated enhanced effectiveness in aquaculture applications (Niu et al., 2019). The combination of *Bacillus subtilis* and *Lactobacillus plantarum* enhances growth performance in tilapia by improving digestive enzyme activity, increasing nutrient absorption efficiency, strengthening the intestinal barrier, and modulating the gut microbiome composition (Guimarães et al., 2021, 2022; Wei et al., 2022b).

Marine microbial applications in feed production have advanced significantly, offering sustainable alternatives to traditional ingredients. The thraustochytrid *Schizochytrium* sp. has been optimized for large-scale production, achieving biomass yields of 80–100 g/L, DHA content exceeding 40% of total fatty acids, protein content comparable to fishmeal, and essential amino acid profiles matching fish requirements (Patel et al., 2021; Du et al., 2021; Trevi et al., 2023; Eroldoğan et al., 2023). Marine yeasts have revolutionized feed fermentation processes through the enhancement of plant-based protein digestibility, production of essential vitamins and minerals, reduction of antinutritional factors, and improvement of feed palatability (Madhulika et al., 2025; Meena et al., 2020; Chen et al., 2024; Siddik et al., 2024).

Advanced disease control methods incorporating marine microorganisms include specialized approaches for different aquaculture systems. Bacteriophages isolated from marine environments show remarkable specificity in controlling bacterial pathogens, with success rates of 70%–80% reduction in mortality during disease outbreaks, selective targeting of specific pathogens with minimal impact on beneficial bacteria, and no development of bacterial resistance (Dubey et al., 2021; Xionglin et al., 2024; Naureen et al., 2020; Schulz et al., 2022; Nachimuthu et al., 2021). Biofilm management using beneficial bacteria has emerged as an effective disease prevention strategy through the formation of protective biofilms on tank surfaces, prevention of pathogen colonization, improvement of water quality, and reduction in the need for chemical treatments (Rahman et al., 2024; Liu et al., 2023a; Schoina et al., 2022).

Marine microorganisms play a crucial role in maintaining water quality and reducing environmental impact through various mechanisms. Nitrifying bacteria facilitate ammonia removal and nitrogen cycle maintenance, while photosynthetic microalgae contribute to oxygen production and CO_2 fixation (Tadda et al., 2021). Sulfur-oxidizing bacteria aid in hydrogen sulfide elimination, and heterotrophic bacteria assist in organic matter decomposition (Patil et al., 2023; Wang et al., 2022). This comprehensive approach to environmental management ensures optimal water quality conditions for aquaculture species.

Specialized microbial consortia facilitate efficient waste management in aquaculture systems through the conversion of organic waste into valuable nutrients, reduction of suspended solids, prevention of harmful algal blooms, and enhancement of water reuse efficiency (Suresh Kunjiraman et al., 2024). These sustainable approaches to aquaculture management demonstrate the essential role of marine microorganisms in developing environmentally responsible food production systems. The integration of these various applications creates a holistic approach to sustainable aquaculture, ensuring both productivity and environmental stewardship. As research continues, new applications and refinements of existing techniques promise to further enhance the sustainability and efficiency of aquaculture operations, making marine microorganisms indispensable tools in modern aquaculture practices.

3.4.4 Marine microbes in food and nutraceuticals

Marine microorganisms represent a transformative frontier in food and nutritional science, offering sustainable and innovative solutions that are revolutionizing our approach to food production and functional nutrition. These microscopic marine inhabitants, including bacteria and fungi, are emerging as powerful sources of essential nutrients, bioactive compounds, and cutting-edge food technologies that aim to address critical challenges in global nutrition and sustainable food systems (Garcia-Vaquero and Rajauria, 2021). The realm of marine microbial applications spans multiple critical domains, beginning with bioactive compound production. Pioneering research has demonstrated the remarkable potential of marine bacteria like *Bacillus licheniformis* strains, which can produce polyunsaturated fatty acids—particularly eicosapentaenoic acid—through controlled fermentation processes (Speckmann et al., 2022). This breakthrough approach offers unique advantages, including sustainable production independent of marine ecosystem disruption, consistent product quality, year-round availability, and significantly reduced environmental impact (Ashaari et al., 2024; Zhu et al., 2024).

In the domain of fermentation applications, marine microorganisms are revolutionizing traditional food processing techniques. The halophilic bacterium *Tetragenococcus halophilus* has emerged as a critical player in fish sauce fermentation, delivering multiple benefits, including the production of unique flavor compounds (Jinjin et al., 2024), enhancement of umami taste, ensuring product safety through acid production, and degrading antinutritional factors (Zhang et al., 2024; Wang et al., 2024; Han et al., 2023a; Elhalis et al., 2024). Marine yeasts have demonstrated equally impressive capabilities, showcasing superior salt tolerance for fermented seafood products, generating novel flavor compounds, improving protein hydrolysis, and increasing overall nutritional value (Han et al., 2023a; Yao et al., 2023; Maya et al., 2023).

Marine microbes show remarkable potential in creating novel food preservatives (Ghosh et al., 2022). Unlike traditional synthetic additives, marine-derived natural preservatives offer a more sustainable and potentially healthier alternative (Bharathi and Lee, 2024). Bacteriocins from marine lactic acid bacteria, antifungal compounds from marine fungi, natural antioxidants, and biopreservative peptides are opening new frontiers in food preservation (Anumudu et al., 2022). The bacterium *Lactobacillus plantarum*, isolated from fermented fish, produces specialized bacteriocins that can inhibit food-borne pathogens, remain stable in high-salt conditions, demonstrate heat stability during processing, and exhibit broad-spectrum antimicrobial activity (Barbieri, 2023; Mgomi et al., 2023; Han et al., 2023a; Almeida-Santos et al., 2021).

Functional food ingredients derived from marine microorganisms present another exciting avenue of research and development (Arokiarajan et al., 2022). Polysaccharides and dietary fibers from marine microorganisms offer novel prebiotics, bioactive sulfated polysaccharides, and enhanced dietary fibers (Gurpilhares et al., 2019; Arokiarajan et al., 2022). Protein and peptide products include single-cell proteins for food fortification, bioactive peptides with potential health benefits, novel protein hydrolyzates, and innovative functional protein ingredients that could transform nutritional supplementation (Lobine et al., 2022).

The nutraceutical applications of marine microbes are equally promising, presenting a diverse range of health-promoting compounds (Quitério et al., 2021). Antioxidant compounds, such as carotenoids from marine bacteria, phenolic compounds from marine fungi, novel flavonoids from marine actinobacteria, and unique polyphenol derivatives, offer significant potential for health-focused nutritional interventions (Liu et al., 2020a; Vitale et al., 2020; Hozzein et al., 2021; Mendes et al., 2024). Immune system modulators, including beta-glucans from marine yeasts, immunostimulatory peptides, and polysaccharide–protein complexes, represent another critical area of research with profound implications for human health (Cutolo et al., 2024).

Advanced processing technologies have been crucial in unlocking the full potential of marine microbial products. Extraction technologies, such as supercritical fluid extraction, enzyme-assisted extraction, green solvent systems, and novel membrane technologies, have dramatically improved the efficiency and quality of marine microbial compound extraction (Bhadange et al., 2024). Stabilization methods, including sophisticated microencapsulation techniques, novel drying technologies, and advanced preservation methods, have further enhanced the practical application of these marine-derived ingredients (Rezagholizade-shirvan et al., 2024).

Quality control and safety remain vital in the development of marine microbial food and nutraceutical products. Researchers have developed comprehensive analytical methods, including advanced chromatographic techniques, spectroscopic analysis, biological activity assays, and rigorous safety assessment protocols (Valdés et al., 2022). Stringent quality parameters focus on purity standards, stability testing, bioactivity preservation, and contamination prevention strategies to ensure the highest standards of product safety and efficacy (Edo et al., 2024).

Looking toward future perspectives, the field of marine microbial food and nutraceuticals continues to evolve with exciting emerging trends. Innovations such as personalized nutrition products, smart food packaging materials, bioactive food coatings, and functional food additives are pushing the boundaries of what is possible. Sustainable production methods remain a key focus, with ongoing research dedicated to improved cultivation systems, enhanced productivity methods, reduced environmental impact, and increasingly cost-effective processing techniques.

As global challenges in nutrition, sustainability, and food security become increasingly complex, marine microorganisms offer a beacon of hope and innovation. These microscopic marine organisms are not just alternative sources of nutrients; they represent a fundamental reimagining of food production, nutritional science, and sustainable agricultural practices. By harnessing the extraordinary metabolic capabilities of marine bacteria and fungi, researchers are developing solutions that are both scientifically groundbreaking and environmentally responsible, promising to transform our approach to nutrition in the next century.

3.4.5 Climate change and environmental sustainability

Marine microorganisms are critical actors in global climate regulation and environmental sustainability, offering unique solutions to complex ecological challenges. Their role extends far beyond simple biological processes, encompassing sophisticated mechanisms of climate interaction and environmental adaptation. Carbon sequestration represents a primary mechanism of their environmental significance (Mistry et al., 2019). The coccolithophore *Emiliania huxleyi* demonstrates this through its calcium carbonate production, which contributes substantially to the biological carbon pump (Dean et al., 2024). This process not only sequesters carbon but also significantly influences ocean alkalinity by creating complex chemical interactions within marine ecosystems (Han et al., 2024).

Deep-sea bacteria, particularly those in the SAR202 clade, further expand our understanding of marine carbon cycling. These specialized organisms excel at degrading recalcitrant dissolved organic carbon, directly impacting long-term carbon storage in oceanic systems (Quero et al., 2020). Their metabolic processes facilitate more efficient carbon cycling, potentially mitigating carbon accumulation in marine and atmospheric environments (Changfei et al., 2024).

Methane regulation emerges as another critical domain of marine microbial activity. Methanotrophic bacteria, like *Methylococcus capsulatus*, play a crucial role in oxidizing methane in marine sediments (Kumar et al., 2019b), preventing its release into the atmosphere. In cold seep environments, these bacteria form complex microbial consortia with sulfate-reducing bacteria (Guerrero-Cruz et al., 2021; Li et al., 2020). The archaeon *Methanoperedens nitroreducens* represents an advanced example of this process, simultaneously oxidizing methane and reducing nitrate through a unique mechanism (Cai et al., 2021; Nie et al., 2021).

Environmental adaptation strategies provide insights into climate change mitigation. *Prochlorococcus marinus* demonstrates remarkable genetic diversity across ocean temperature gradients, suggesting potential for developing climate-resilient primary producers (Kujawinski et al., 2023). Marine fungi from mangrove ecosystems, particularly *Pestalotiopsis* sp., show enhanced capabilities in degrading complex organic matter under varying salinity conditions, indicating potential approaches to coastal ecosystem restoration (Salazar-Alekseyeva et al., 2024).

Addressing ocean acidification represents another critical area of marine microbial research. The bacterium *Pseudomonas* sp. strain ED321 produces calcium carbonate-precipitating enzymes that could potentially regulate pH in sensitive marine ecosystems (Seifan and Berenjian, 2019). Marine cyanobacteria enhance carbonate precipitation through photosynthetic activities, offering potential strategies for protecting vulnerable marine environments, such as coral reef systems, from acidification stress (Xu et al., 2019).

Marine microorganisms serve as sensitive environmental indicators, providing early detection of ecological transformations (Fangzhou et al., 2022). Their rapid genetic adaptability allows for quick responses to environmental changes, making them dynamic agents of ecological interaction (Cavicchioli et al., 2019). Biotechnological applications are increasingly exploring ways to harness their unique metabolic pathways for carbon capture, pollution remediation, and ecosystem restoration.

The interactions between marine microorganisms and their environment reveal the complex interconnectedness of global ecological systems (Heinrichs et al., 2019). Their activities influence marine environments, global carbon and nutrient cycles, atmospheric composition, and planetary climate regulation (Abirami et al., 2021). Each microorganism contributes to a complex ecological network that maintains the planetary environmental balance.

As climate change presents unique challenges, marine microorganisms offer scientifically robust approaches to environmental management (Hutchins et al., 2019). Their metabolic capabilities and adaptive potential suggest innovative strategies for addressing complex environmental issues (Lauritano et al., 2020). Ongoing research in this field promises to develop targeted interventions for environmental sustainability and climate resilience.

The potential of marine microorganisms extends beyond current understanding. Their ability to adapt, interact, and transform environmental conditions positions them as critical components in developing comprehensive approaches to global environmental challenges. Researchers continue to uncover the intricate ways these organisms contribute to ecological systems, offering hope for more effective environmental management strategies.

By investigating the metabolic processes, genetic adaptations, and ecological interactions of marine microorganisms, scientists are developing a more nuanced understanding of global environmental dynamics. These organisms represent more than simple biological entities; they are key participants in the complex system of global environmental regulation.

3.5 Challenges and future directions in marine microbiology research

The study and application of marine fungi and bacteria face several significant challenges, while simultaneously offering exciting opportunities for future research and development.

3.5.1 Cultivation and maintenance challenges

Marine microbial cultivation represents one of the most complex and challenging frontiers in microbiological research, with profound implications for scientific understanding and potential biotechnological applications (Alain and Querellou, 2009). The fundamental limitation of current cultivation techniques is starkly illustrated by the "great plate count anomaly," where traditional methods capture a mere 0.1%–1% of marine microbial diversity (Rodrigues and de Carvalho, 2022). This critical bottleneck severely constrains researchers' ability to explore and utilize the vast majority of marine microbial species, representing a significant gap in our understanding of marine ecosystems and their potential applications.

The case of the SAR11 clade bacteria epitomizes the extraordinary challenges in marine microbial cultivation (Rappé et al., 2002). For over a decade, these highly abundant marine microorganisms remained unculturable, defying conventional cultivation techniques. A breakthrough came only through the development of specialized media that meticulously replicated the oligotrophic marine conditions, a testament to the intricate and specific environmental requirements of these microorganisms (Vartoukian et al., 2010). This achievement underscores the necessity of advanced, highly specialized approaches in marine microbial research.

Nutritional complexity emerges as a primary obstacle in marine microbial cultivation efforts. Symbiotic marine fungi, particularly those associated with deep-sea corals, present particularly challenging cultivation scenarios (Schultz et al., 2022). These organisms rely on intricate chemical signaling networks and metabolic exchanges that are nearly impossible to reproduce in laboratory settings. The growth requirements often involve specific host-derived factors that cannot be easily synthesized or replicated, creating a fundamental barrier to their study and potential utilization (Amend et al., 2019).

Oligotrophic bacteria represent another critical challenge in marine microbial cultivation (Wei et al., 2022a). Adapted to nutrient-poor marine environments, these microorganisms have evolved highly specialized metabolic strategies that make them virtually impossible to cultivate using standard nutrient-rich media. Researchers must develop sophisticated low-nutrient cultivation approaches that carefully mimic the extreme nutritional limitations of deep-sea and open-ocean environments. This requires a significant understanding of microbial metabolism and environmental adaptation, which goes far beyond traditional microbiological techniques.

Environmental parameter simulation introduces additional layers of technical complexity. Deep-sea organisms demand extraordinarily sophisticated cultivation conditions, including pressure vessels capable of maintaining pressures up to 100 MPa. The deep-sea bacterium *Photobacterium profundum* exemplifies these challenges, requiring a constant pressure of 30 MPa for optimal growth (Campanaro et al., 2005; Garel et al., 2019). Such requirements necessitate specialized equipment that is not only technically complex but also prohibitively expensive, limiting large-scale research efforts. Moreover, temperature control represents another critical dimension of marine microbial cultivation. Thermal vent species, in particular, require incredibly precise temperature gradient management. Research systems must be capable of maintaining stable temperatures ranging from near-freezing 4°C to extreme temperatures exceeding 120°C (Baross, 2018). Custom-built incubation systems with unprecedented thermal stability and control become essential, pushing the boundaries of current technological capabilities.

Molecular techniques have begun to supplement traditional cultivation methods, offering new insights into marine microbial diversity (Mohamed et al., 2023). Metagenomics and single-cell genomics provide alternative approaches to studying unculturable microorganisms, allowing researchers to extract genetic information without relying on traditional cultivation techniques. These methods have revealed the genetic potential of marine microbes that remain impossible to grow under laboratory conditions, opening new avenues for understanding marine microbial ecosystems.

Symbiotic relationships further complicate cultivation efforts. Many marine microorganisms exist in intricate symbiotic networks that are nearly impossible to replicate in isolation (Liu et al.,

2022b; Kaboré et al., 2020). Coral-associated microbes, for instance, rely on complex interactions that involve multiple species, chemical signaling, and specific environmental conditions (Mohamed et al., 2023). Researchers must develop sophisticated cocultivation techniques that can maintain these delicate ecological relationships, requiring interdisciplinary approaches that combine microbiology, ecology, and advanced biotechnological methods.

Technological innovations are gradually addressing these cultivation challenges. Advanced microfluidic devices, specialized cultivation chambers, and high-pressure bioreactors are expanding the possibilities for marine microbial research (Cario et al., 2019; Wessel et al., 2013). These technologies allow for more precise control of environmental parameters, enabling researchers to create increasingly sophisticated simulation environments that more closely approximate natural marine conditions. The economic and scientific implications of marine microbial cultivation are profound. These microorganisms hold potential for numerous applications, including novel pharmaceutical compounds, biotechnological innovations, environmental remediation, and insights into extreme environment adaptation. Each breakthrough in cultivation techniques potentially unlocks new scientific understanding and technological possibilities.

Emerging approaches combine multiple methodological innovations. Artificial intelligence and machine learning are being applied to predict optimal cultivation conditions, while advanced molecular techniques provide unprecedented insights into microbial metabolism and growth requirements (Rabiya and Sen, 2022; Bemani et al., 2023; Kassem et al., 2023). Researchers are developing several approaches that integrate computational modeling, molecular analysis, and sophisticated cultivation technologies (Fang et al., 2020).

The future of marine microbial research depends on continued technological innovation and interdisciplinary collaboration. Overcoming cultivation challenges requires a holistic approach that combines expertise from microbiology, marine ecology, biotechnology, and advanced engineering (Rotter et al., 2020). Each incremental improvement in cultivation techniques promises to reveal new insights into the extraordinary world of marine microorganisms. As research progresses, the goal extends beyond mere cultivation. Scientists seek to understand the complex ecological roles, metabolic capabilities, and evolutionary adaptations of these remarkable microorganisms. The challenges of marine microbial cultivation represent not just a technical hurdle but a gateway to understanding the fundamental processes that sustain marine ecosystems and potentially address global environmental challenges.

3.5.2 Technological and methodological limitations

Current technological constraints significantly impact both research capabilities and practical applications in marine microbiology. These limitations affect various aspects of research, from sample collection to data analysis, creating substantial challenges to advancing our understanding of marine microorganisms.

Sample collection and preservation present significant technical challenges. Deep-sea sampling poses particular difficulties due to pressure changes during specimen retrieval. Many pressure-sensitive organisms lose viability during sample collection from deep ocean environments (Macdonald, 2021), with survival rates often dropping below 1% for organisms from depths exceeding 6000 m. The deep-sea bacterium *Thermococcus barophilus*, for example, loses viability rapidly when brought to atmospheric pressure (Scheffer and Gieg, 2023), necessitating specialized collection and preservation protocols. Temperature fluctuations during sampling can irreversibly damage psychrophilic enzymes, leading to the loss of biological activity and structural information (Wang et al., 2025). Current preservation methods often fail to maintain the original community structure, particularly for complex microbial consortia involved in symbiotic relationships.

Analytical challenges persist throughout the research pipeline. Metagenomic studies generate massive datasets with high proportions of unknown sequences, often exceeding 50% of the total sequence data in samples from extreme marine environments (Dávila-Ramos et al., 2019). This high percentage of uncharacterized genetic material, known as "biological dark matter" (Jiao et al., 2021) significantly masks our understanding of marine microbial diversity and function. Protein structure determination under high-pressure conditions requires specialized equipment that is both expensive and technically challenging to operate (Liu et al., 2020b). Current technologies for studying protein-protein interactions under extreme conditions are limited, restricting our understanding of molecular adaptations to deep-sea environments.

Real-time monitoring of microbial activities in situ remains technically challenging. Available sensors lack the sensitivity to detect many microbial processes at environmentally relevant scales (Valle et al., 2021). Current methodologies for measuring enzyme activities in extreme marine environments often provide incomplete or inaccurate results due to technical limitations in maintaining appropriate pressure and temperature conditions during analysis (Mesbah, 2022). The development of pressure-resistant monitoring devices faces significant engineering challenges, particularly in terms of maintaining long-term stability and accuracy under extreme conditions.

Scale-up difficulties present major obstacles for industrial applications. The transition from laboratory to industrial scale faces several challenges:

- Maintaining optimal conditions in large-scale cultivation systems requires sophisticated control systems.
- Product recovery from high-salt conditions presents unique downstream processing challenges.
- Sterilization of large volumes of seawater-based media is both technically challenging and expensive.
- Monitoring and controlling contamination in marine-based fermentation systems require specialized approaches.

In addition, current bioinformatics tools often struggle with the complexity of marine microbial datasets (Steen et al., 2020). Existing algorithms may fail to accurately predict gene functions in marine organisms due to their unique evolutionary adaptations and novel protein families. The integration of multiple data types (genomics, transcriptomics, metabolomics) remains computationally challenging, particularly for complex marine microbial communities.

3.5.3 Future research directions and opportunities

Despite current challenges, several promising research directions are emerging in marine microbiology, offering new opportunities for both fundamental research and practical applications. These developments span multiple technological and methodological frontiers, promising to revolutionize our understanding and utilization of marine microorganisms.

Advanced single-cell technologies are transforming our ability to study unculturable marine microorganisms (Grujcic et al., 2022). Single-cell genomics and metabolomics techniques now enable detailed analysis of previously inaccessible species (Hatzenpichler et al., 2020). Recent applications of single-cell Raman spectroscopy combined with stable isotope probing have revealed previously unknown metabolic capabilities in deep-sea bacteria (Wanying et al., 2023). These approaches have successfully characterized unculturable symbionts of marine sponges, leading to the discovery of novel biosynthetic pathways for pharmaceutical compounds. The integration of microfluidic devices with single-cell analysis has enabled real-time observation of cellular responses to environmental changes, providing unprecedented insights into microbial adaptation mechanisms (Ou et al., 2019).

Synthetic biology applications in marine microorganisms are expanding rapidly. CRISPR–Cas9 technology is being adapted for marine microorganisms, allowing genetic modification of previously intractable species (Hassanien et al., 2023). Recent success in engineering the marine bacterium *Vibrio natriegens* for rapid protein production demonstrates the potential for marine synthetic biology (Hoff et al., 2020). Advanced genome editing techniques have enabled the creation of enhanced biocatalysts for industrial applications, the development of optimized strains for biofuel production, the engineering of improved stress tolerance mechanisms, and the design of novel biosynthetic pathways (Shanmugam et al., 2020; Huang, 2024).

The integration of multiomics approaches is providing deeper insights into marine microbial functions (Satya et al., 2024). The combined application of genomics, transcriptomics, and metabolomics reveals complex metabolic networks in marine bacterial communities (Manochkumar et al., 2023). This integration has demonstrated synchronized nutrient cycling between different bacterial species in coral microbiomes, novel secondary metabolite production pathways in marine fungi, previously unknown mechanisms of environmental adaptation, and unexpected metabolic cooperation in microbial communities.

Emerging technologies in environmental monitoring and sampling are opening new research frontiers. Advanced in situ sampling devices equipped with real-time molecular analysis capabilities enable continuous monitoring of marine microbial communities (Huang, 2024). Novel cultivation approaches, including pressure-retaining sampling systems, advanced bioreactor designs mimicking natural environments, microfluidic cultivation platforms, and artificial marine matrices for biofilm studies, are revolutionizing our ability to study marine microorganisms in their natural environments. Machine learning and artificial intelligence applications are accelerating marine microbial research. These computational approaches enable the prediction of novel bioactive compounds, the optimization of cultivation conditions, the analysis of complex community interactions, and the identification of new biosynthetic gene clusters. The integration of these technologies with traditional approaches is creating unprecedented opportunities for understanding and harnessing marine microbial diversity.

3.6 Conclusion

Marine fungi and bacteria represent a vast and largely untapped resource for biological discovery and biotechnological innovation. Their diverse adaptations to marine environments, ranging from pressure-resistant enzymes to unique secondary metabolites, offer solutions to numerous industrial and environmental challenges.

The ecological significance of these microorganisms cannot be overstated. Their roles in global nutrient cycling, particularly in carbon sequestration and nitrogen fixation, make them crucial players in planetary health. The discovery of novel metabolic pathways, such as archaeal ammonia oxidation in deep ocean waters, continues to reshape our understanding of global biogeochemical cycles.

The biotechnological applications of marine microorganisms span an impressive range. From the development of new pharmaceuticals, exemplified by compounds like salinosporamide A, to the production of industrial enzymes and biopolymers, marine microbes offer sustainable solutions to various technological challenges. Their applications in aquaculture and food production demonstrate practical contributions to global food security.

Future prospects in marine microbiology are particularly promising in several areas:

1. Development of novel antimicrobial compounds to address antibiotic resistance
2. Discovery of enzymes for industrial applications under extreme conditions
3. Enhancement of aquaculture productivity through microbial management
4. Application of marine microbes in environmental remediation
5. Utilization of marine microbial processes for carbon sequestration

However, realizing these prospects requires addressing significant challenges in the cultivation, preservation, and scaling of marine microbial processes. Continued technological advancement, particularly in cultivation techniques and genetic tools, will be crucial for unlocking the full potential of marine microorganisms.

The sustainable utilization of marine microbial resources presents both opportunities and responsibilities. As we continue to discover and harness these microscopic organisms, maintaining balance in marine ecosystems becomes increasingly important. The future of marine microbiology lies not only in scientific discovery but also in the responsible application of these findings for the benefit of both humanity and the marine environment.

AI disclosure

During the preparation of this work, the author(s) used ChatGPT to assist in enhancing readability through grammar checks and sentence structure improvements. AI tools were also employed to organize the chapter's structural framework and create clear visual representations of marine microbial applications. All scientific content, data, and technical information were independently verified through primary literature sources and expert knowledge to ensure accuracy and reliability. After using this tool/service, the author(s) reviewed and edited the content as needed and take(s) full responsibility for the content of the publication.

References

Abirami, B., Radhakrishnan, M., Kumaran, S., Wilson, A., 2021. Impacts of global warming on marine microbial communities. Sci. Total Environ. 00489697. 791, 147905. doi:10.1016/j.scitotenv.2021.147905.

Ahmed, I., Asgher, M., Sher, F., Hussain, S.M., Nazish, N., Joshi, N., Sharma, A., Parra-Saldívar, R., Bilal, M., Iqbal, H. M.N., 2022. Exploring marine as a rich source of bioactive peptides: challenges and opportunities from marine pharmacology. Mar. Drugs 20 (3). doi:10.3390/md20030208.

Alagawany, M., El-Saadony, M.T., Farag, M.R., 2020. Probiotics in aquaculture and their applications against pathogenic bacteria. Indian J. Anim. Health. 0019-5057. 59 (2-Spl), 138–145. doi:10.36062/ijah.59.2spl.2020.138-145.

Alain, K., Querellou, J., 2009. Cultivating the uncultured: limits, advances and future challenges. Extremophiles. 1431-0651. 13 (4), 583–594. doi:10.1007/s00792-009-0261-3.

Ali, S., Discovery and characterization of novel biofilm. (2025), Available from: https://hdl.handle.net/10012/21376.

Almasi, F., Salimi, F., Mohammadipanah, F., 2021. Microbial metabolites as promising anti-inflammatory resources in biomedicine. Microbiology, Metabolites and Biotechnology 4 (1), 67–81. doi:10.22104/armmt.2023.5892.1078.

Almeida-Santos, A.C., Novais, C., Peixe, L., Freitas, A.R., 2021. Enterococcus spp. As a producer and target of bacteriocins: a double-edged sword in the antimicrobial resistance crisis context. Antibiotics. 20796382. 10 (10). Available from: https://www.mdpi.com/2079-6382/10/10/1215/pdf. doi:10.3390/antibiotics10101215.

Alsufyani, T., Alsufyani, T., Califano, G., Deicke, M., Grueneberg, J., Grueneberg, J., Weiss, A., Weiss, A., Engelen, A. H., Kwantes, M., Mohr, J.F., Mohr, J.F., Ulrich, J.F., Wichard, T., Wichard, T., 2020. Macroalgal-bacterial interactions: identification and role of thallusin in morphogenesis of the seaweed Ulva (Chlorophyta). J. Exp. Bot. 14602431. 71 (11), 3340–3349. Available from: http://jxb.oxfordjournals.org/. doi:10.1093/jxb/eraa066.

Ambalavanan, L., Iehata, S., Fletcher, R., Stevens, E.H., Zainathan, S.C., 2021. A review of marine viruses in coral ecosystem. J. Mar. Sci. Eng. 20771312. 9 (7). Available from: https://www.mdpi.com/2077-1312/9/7/711/pdf. doi:10.3390/jmse9070711.

Amelia, T.S.M., Suaberon, F.A.C., Vad, J., Fahmi, A.D.M., Saludes, J.P., Bhubalan, K., 2022. Recent advances of marine sponge-associated microorganisms as a source of commercially viable natural products. Mar. Biotechnol. 14362236. 24 (3), 492–512. Available from: http://link.springer.com/journal/volumesAndIssues/10126. doi:10.1007/s10126-022-10130-2.

Amend, A., Burgaud, G., Cunliffe, M., Edgcomb, V.P., Ettinger, C.L., Gutiérrez, M.H., Heitman, J., Hom, E.F.Y., Ianiri, G., Jones, A.C., Kagami, M., Picard, K.T., Quandt, C.A., Raghukumar, S., Riquelme, M., Stajich, J., Vargas-Muñiz, J., Walker, A.K., Yarden, O., Gladfelter, A.S., 2019. Fungi in the marine environment: open questions and unsolved problems. mBio. 21507511. 10 (2). Available from: https://mbio.asm.org/content/mbio/10/2/e01189-18.full.pdf. doi:10.1128/mBio.01189-18.

Anuar, N.F.S.K., Huyop, F., Ur-Rehman, G., Abdullah, F., Normi, Y.M., Sabullah, M.K., Wahab, R.A., 2022. An overview into polyethylene terephthalate (PET) hydrolases and efforts in tailoring enzymes for improved plastic degradation. Int. J. Mol. Sci. 14220067. 23 (20). Available from: http://www.mdpi.com/journal/ijms. doi:10.3390/ijms232012644.

Anumudu, C.K., Omoregbe, O., Hart, A., Miri, T., Eze, U.A., Onyeaka, H., 2022. Applications of bacteriocins of lactic acid bacteria in biotechnology and food preservation: a bibliometric review. Open Microbiol. J. 18742858. 16 (1). Available from: http://www.bentham.org/open/tomicroj/index.htm. doi:10.2174/18742858-v16-e2206300.

Anushree Pant, M.V., 2023. Endophytic fungi: a potential source for drugs against central nervous system disorders. Braz. J. Microbiol. 1517-8382. 54 (3), 1479–1499. doi:10.1007/s42770-023-00997-1.

Ariaeenejad, S., Kavousi, K., Zolfaghari, B., Roy, S., Koshiba, T., Salekdeh, G.H., 2023. Efficient bioconversion of lignocellulosic waste by a novel computationally screened hyperthermostable enzyme from a specialized microbiota. Ecotoxicol. Environ. Saf. 01476513. 252, 114587. doi:10.1016/j.ecoenv.2023.114587.

Arokiarajan, M.S., Thirunavukkarasu, R., Joseph, J., Ekaterina, O., Aruni, W., 2022. Advance research in biomedical applications on marine sulfated polysaccharide. Int. J. Biol. Macromol. 18790003. 194, 870–881. Available from: www.elsevier.com/locate/ijbiomac. doi:10.1016/j.ijbiomac.2021.11.142.

Ashaari, A., Iehata, S., Kim, H.J., Rasdi, N.W., 2024. Recent advancement of zooplankton enriched with nutrients and probiotic isolates from aquaculture systems: a review. J. Appl. Anim. Res. 09741844. 52 (1). Available from: http://www.tandfonline.com/loi/taar. doi:10.1080/09712119.2024.2417052.

Atanasova, N., Stoitsova, S., Paunova-Krasteva, T., Kambourova, M., 2021. Plastic degradation by extremophilic bacteria. Int. J. Mol. Sci. 1422-0067. 22 (11), 5610. doi:10.3390/ijms22115610.

Badar, U., Ahmed, N., Beswick, A.J., Pattanapipitpaisal, P., Macaskie, L.E., 2000. Reduction of chromate by microorganisms isolated from metal contaminated sites of Karachi, Pakistan. Biotechnol. Lett. 22 (10), 829–836. doi:10.1023/A:1005649113190.

Baghel, B., Singhania, R., Patel, A., Kurrey, N., Chen, C.-W., Dong, C.-D., 2024. Marine polysaccharides and potential enzymes for its degradation: marine polysaccharide degrading enzymes. Indian J. Exp. Biol. 62 (07), 447–462. doi:10.56042/ijeb.v62i07.12273.

Bahrami, Y., Bouk, S., Kakaei, E., Taheri, M., 2022. Natural products from actinobacteria as a potential source of new therapies against colorectal cancer: a review. Front. Pharmacol. 13. 10.3389/fphar.2022.929161

Bailey, C.S., Zarins-Tutt, J.S., Agbo, M., Gao, H., Diego-Taboada, A., Gan, M., Hamed, R.B., Abraham, E.R., Mackenzie, G., Evans, P.A., Goss, R.J.M., 2019. A natural solution to photoprotection and isolation of the potent polyene antibiotic, marinomycin A. Chem. Sci. 20416539. 10 (32), 7549–7553. Available from: http://pubs.rsc.org/en/journals/journal/sc. doi:10.1039/c9sc01375j.

Baindara, P., Korpole, S., Grover, V., 2018. Bacteriocins: perspective for the development of novel anticancer drugs. Appl. Microbiol. Biotechnol. 0175-7598. 102 (24), 10393–10408. doi:10.1007/s00253-018-9420-8.

Balabanova, L., Bakholdina, S., Buinovskaya, N., Noskova, Y., Kolpakova, O., Vlasova, V., Bondarev, G., Seitkalieva, A., Son, O., Tekutyeva, L., 2024. LPS-Dephosphorylating cobetia amphilecti alkaline phosphatase of PhoA family divergent from the multiple homologues of cobetia spp. Microorganisms. 2076-2607. 12 (3), 631. doi:10.3390/microorganisms12030631.

Bao, X.C., Tang, H.Z., Li, X.G., Li, A.Q., Qi, X.Q., Li, D.H., Liu, S.S., Wu, L.F., Zhang, W.J., 2023. Bioluminescence contributes to the adaptation of deep-sea bacterium photobacterium phosphoreum ANT-2200 to high hydrostatic pressure. Microorganisms. 20762607. 11 (6). Available from: www.mdpi.com/journal/microorganisms. doi:10.3390/microorganisms11061362.

Bar-On, Y.M., Milo, R., 2019. The biomass composition of the oceans: a blueprint of our blue planet. Cell. 10974172. 179 (7), 1451–1454. Available from: https://www.sciencedirect.com/journal/cell. doi:10.1016/j.cell.2019.11.018.

Barbieri, F., Bio-preservation and bio-active compounds as sustainable strategies to improve quality and safety of fresh and fermented products. (2023), Available from: https://hdl.handle.net/10481/84463.

Baross, J.A., 2018. Isolation and Cultivation of Hyperthermophilic Bacteria from Marine and Freshwater Habitats. Informa UK Limited, pp. 21–30. doi:10.1201/9780203752746-3.

Barton, L.L., Fauque, G.D., 2022. Ecology of Dissimilatory Sulfate Reducers: Life in Extreme Conditions and Activities of SRB. Springer Science and Business Media LLC, pp. 463–527. doi:10.1007/978-3-030-96703-1_9.

Bayer, B., Hansman, R.L., Bittner, M.J., Noriega-Ortega, B.E., Niggemann, J., Dittmar, T., Herndl, G.J., 2019. Ammonia-oxidizing archaea release a suite of organic compounds potentially fueling prokaryotic heterotrophy in the ocean. Environ. Microbiol. 21 (11), 4062–4075. doi:10.1111/1462-2920.14755.

Bayer, I.S., 2022. Recent advances in mucoadhesive interface materials, mucoadhesion characterization, and technologies. Adv. Mater. Interfaces. 21967350. 9 (18). Available from: http://onlinelibrary.wiley.com/journal/10.1002/(ISSN)2196-7350. doi:10.1002/admi.202200211.

Becker, J.W., Pollak, S., Berta-Thompson, J.W., Becker, K.W., Braakman, R., Dooley, K.D., Hackl, T., Coe, A., Arellano, A., LeGault, K.N., Berube, P.M., Biller, S.J., Cubillos-Ruiz, A., Van Mooy, B.A.S., Chisholm, S.W., 2024. Novel isolates expand the physiological diversity of Prochlorococcus and illuminate its macroevolution. mBio. 21507511. 15 (11). Available from: https://journals.asm.org/doi/10.1128/mbio.03497-23. doi:10.1128/mbio.03497-23.

Beliah, M., Ibrahim, H., Farag, A.M., El-Ahwany, A., Sabry, S., 2020. Utilization of marine algae as a carbon source for bacterial cellulose production by Gluconacetobacter xylinus. Egypt. J. Aquat. Biol. Fish. 2536-9814. 24 (6), 497–518. doi:10.21608/ejabf.2020.117323.

Bemani, A., Kazemi, A., Ahmadi, M., 2023. An insight into the microorganism growth prediction by means of machine learning approaches. J. Pet. Sci. Eng. 09204105. 220, 111162. doi:10.1016/j.petrol.2022.111162.

Benhadda, F., Zykwinska, A., Colliec-Jouault, S., Sinquin, C., Thollas, B., Courtois, A., Fuzzati, N., Toribio, A., Delbarre-Ladrat, C., 2023. Marine versus non-marine bacterial exopolysaccharides and their skincare applications. Mar. Drugs. 1660-3397. 21 (11), 582. doi:10.3390/md21110582.

Bhadange, Y.A., Carpenter, J., Saharan, V.K., 2024. A comprehensive review on advanced extraction techniques for retrieving bioactive components from natural sources. ACS Omega. 24701343. 9 (29), 31274–31297. Available from: https://pubs.acs.org/journal/acsodf. doi:10.1021/acsomega.4c02718.

Bharathi, D., Lee, J., 2024. Recent advances in marine-derived compounds as potent antibacterial and antifungal agents: a comprehensive review. Mar. Drugs. 1660-3397. 22 (8), 348. doi:10.3390/md22080348.

Bing, W., Hussain, M., Zhang, W., Stadler, M., Liu, X., Xiang, M., 2019. Current insights into fungal species diversity and perspective on naming the environmental DNA sequences of fungi. Mycology 10 (3), 127–140. doi:10.1080/21501203.2019.1614106.

Böhnke, S., Sass, K., Gonnella, G., Diehl, A., Kleint, C., Bach, W., Zitoun, R., Koschinsky, A., Indenbirken, D., Sander, S. G., Kurtz, S., Perner, M., 2019. Parameters governing the community structure and element turnover in kermadec volcanic ash and hydrothermal fluids as monitored by inorganic electron donor consumption, autotrophic CO2 fixation and 16S tags of the transcriptome in incubation experiments. Front. Microbiol. 1664302X. 10. Available from: https://www.frontiersin.org/journals/microbiology#. doi:10.3389/fmicb.2019.02296.

Brown, D.G., Wobst, H.J., 2021. A Decade of FDA-Approved Drugs (2010-2019): trends and Future Directions. J. Med. Chem. 64 (5), 2312–2338. doi:10.1021/acs.jmedchem.0c01516.

Bunbury-Blanchette, A.L., Walker, A.K., 2019. Occurrence and Distribution of Fungi in Saline Environments. Springer Science and Business Media LLC, pp. 19–38. doi:10.1007/978-3-030-18975-4_2.

Cai, C., Zhang, X., Wu, M., Liu, T., Lai, C.Y., Frank, J., He, B., Marcellin, E., Guo, J., Hu, S., Yuan, Z., 2021. Roles and opportunities for microbial anaerobic oxidation of methane in natural and engineered systems. Energy Environ. Sci. 17545706. 14 (9), 4803–4830. Available from: http://pubs.rsc.org/en/journals/journal/ee. doi:10.1039/d1ee00708d.

Campanaro, S., Vezzi, A., Vitulo, N., Lauro, F.M., D'Angelo, M., Simonato, F., Cestaro, A., Malacrida, G., Bertoloni, G., Valle, G., Bartlett, D.H., 2005. Laterally transferred elements and high pressure adaptation in Photobacterium profundum strains. BMC Genomics. 14712164. 6. Available from: http://www.biomedcentral.com/1471-2164/6/122. doi:10.1186/1471-2164-6-122.

Cario, A., Oliver, G.C., Rogers, K.L., 2019. Exploring the Deep Marine Biosphere: challenges, Innovations, and Opportunities. Front. Earth Sci. 22966463. 7. Available from: https://www.frontiersin.org/journals/earth-science. doi:10.3389/feart.2019.00225.

Castelán-Sánchez, H.G., Lopéz-Rosas, I., García-Suastegui, W.A., Peralta, R., Dobson, A.D.W., Alberto Batista-Garc ía, R., Dávila-Ramos, S., 2019. Extremophile deep-sea viral communities from hydrothermal vents: structural and functional analysis. Mar. Genomics 46, 16–28. doi:10.1016/j.margen.2019.03.001.

Cavicchioli, R., Ripple, W.J., Timmis, K.N., Azam, F., Bakken, L.R., Baylis, M., Behrenfeld, M.J., Boetius, A., Boyd, P. W., Classen, A.T., Crowther, T.W., Danovaro, R., Foreman, C.M., Huisman, J., Hutchins, D.A., Jansson, J.K., Karl, D.M., Koskella, B., Welch, D.B.M., Martiny, J.B.H., Moran, M.A., Orphan, V.J., Reay, D.S., Remais, J.V., Rich, V.I., Singh, B.K., Stein, L.Y., Stewart, F.J., Sullivan, M.B., van Oppen, M.J.H., Weaver, S.C., Webb, E.A., Webster, N.S., 2019. Scientists' warning to humanity: microorganisms and climate change. Nat. Rev. Microbiol. 17 (9), 569–586. doi:10.1038/s41579-019-0222-5.

Chakraborty, K., 2023. Recent Advances in Marine Biotechnology Frontiers in Aquaculture Biotechnology. Elsevier, India, pp. 187–217. Available from: https://www.sciencedirect.com/book/9780323912402. doi:10.1016/B978-0-323-91240-2.00002-6.

Changfei, H., Fucich, D., Sosa, A., Wang, H., Kan, J., Liu, J., Yongle, X., Jiao, N., Gonsior, M., Chen, F., 2024. Deep metagenomic sequencing unveils novel SAR202 lineages and their vertical adaptation in the ocean. Commun. Biol. 2399-3642. 7 (1). doi:10.1038/s42003-024-06535-5.

Chapadgaonkar, S.S., Das, B.B., Shourie, A., 2024. Harnessing the Untapped Potential of Cold-Adapted Enzymes. Ind. Biotechnol. 15509087. 20 (6), 257–267. Available from: www.liebertonline.com/ind. doi:10.1089/ind.2024.0017.

Chen, Y., Mitra, A., Rahimnejad, S., Chi, S., Kumar, V., Tan, B., Niu, J., Xie, S., 2024. Retrospect of fish meal substitution in Pacific white shrimp (Litopenaeus vannamei) feed: alternatives, limitations and future prospects. Rev. Aquac. 1753-5123. 16 (1), 382–409. doi:10.1111/raq.12843.

Cheng, T.H., Ismail, N., Kamaruding, N., Saidin, J., Danish-Daniel, M., 2020. Industrial enzymes-producing marine bacteria from marine resources. Biotechnol. Rep. 2215017X. 27, e00482. doi:10.1016/j.btre.2020.e00482.

Chhun, A., Sousoni, D., Aguiló-Ferretjans, M.D.M., Song, L., Corre, C., Christie-Oleza, J.A., 2021. Phytoplankton trigger the production of cryptic metabolites in the marine actinobacterium Salinispora tropica. Microb. Biotechnol. 17517915. 14 (1), 291–306. Available from: http://onlinelibrary.wiley.com/journal/10.1111/(ISSN)1751-7915. doi:10.1111/1751-7915.13722.

Cleary, D.F.R., van Bendegom, D.J., Gomes, N.C.M., de Voogd, N.J., 2024. Sponge-associated microbes in the twilight zone of Curaçao. Symbiosis. 0334-5114. 93 (1), 81–97. doi:10.1007/s13199-024-00992-6.

Concórdio-Reis, P., Ferreira, S.S., Alves, V.D., Moppert, X., Guézennec, J., Coimbra, M.A., Reis, M.A.M., Freitas, F., 2023. Rheological characterization of the exopolysaccharide produced by Alteromonas macleodii Mo 169. Int. J. Biol. Macromol. 18790003. 227, 619–629. Available from: www.elsevier.com/locate/ijbiomac. doi:10.1016/j.ijbiomac.2022.12.117.

Creamer, K.E., Experimental and phylogenetic approaches to understand natural product biosynthetic gene cluster evolution in the marine actinomycete genus salinispora. (2022).

Crosby, J.R., Laemthong, T., Lewis, A.M., Straub, C.T., Adams, M.W., Kelly, R.M., 2019. Extreme thermophiles as emerging metabolic engineering platforms. Curr. Opin. Biotechnol. 18790429. 59, 55–64. Available from: http://www.elsevier.com/locate/copbio. doi:10.1016/j.copbio.2019.02.006.

Cuenot, S., Gélébart, P., Sinquin, C., Colliec-Jouault, S., Zykwinska, A., 2022. Mechanical relaxations of hydrogels governed by their physical or chemical crosslinks. J. Mech. Behav. Biomed. Mater. 17516161. 133, 105343. doi:10.1016/j.jmbbm.2022.105343.

Cutolo, E.A., Campitiello, R., Caferri, R., Pagliuca, V.F., Li, J., Agathos, S.N., Cutolo, M., 2024. Immunomodulatory Compounds from the Sea: from the Origins to a Modern Marine Pharmacopoeia. Mar. Drugs. 16603397. 22 (7). Available from: http://www.mdpi.com/journal/marinedrugs. doi:10.3390/md22070304.

Cyske, Z., Jaroszewicz, W., Żabińska, M., Lorenc, P., Sochocka, M., Bielańska, P., Grabowski, Ł., Gaffke, L., Pierzynowska, K., Węgrzyn, G., 2021. Unexplored potential: biologically active compounds produced by microorganisms from hard-to-reach environments and their applications. Acta Biochim. Pol. 1734154X. 68 (4), 564–574. Available from: http://www.actabp.pl/. doi:10.18388/abp.2020_5887.

Daoud, L., Ali, M.B., 2020. Halophilic Microorganisms: Interesting Group of Extremophiles with Important Applications in Biotechnology and Environment Physiological and Biotechnological Aspects of Extremophiles. Elsevier, Tunisia, pp. 51–64. Available from https://www.sciencedirect.com/book/9780128183229. doi:10.1016/B978-0-12-818322-9.00005-8.

Das, S., Dey, P., Roy, D., Maiti, M.K., Sen, R., 2019. N-Acetyl-d-glucosamine production by a chitinase of marine fungal origin: a case study of potential industrial significance for valorization of waste chitins. Appl. Biochem. Biotechnol. 15590291. 187 (1), 407–423. Available from: http://www.springer.com/humana+press/journal/12010. doi:10.1007/s12010-018-2822-3.

Dávila-Ramos, S., Castelán-Sánchez, H.G., Martínez-ávila, L., Sánchez-Carbente, M.D.R., Peralta, R., Hernández-Mendoza, A., Dobson, A.D.W., Gonzalez, R.A., Pastor, N., Batista-García, R.A., 2019. A review on viral metagenomics in extreme environments. Front. Microbiol. 1664302X. 10. Available from: https://www.frontiersin.org/journals/microbiology#. doi:10.3389/fmicb.2019.02403.

De Rop, A.S., Rombaut, J., Willems, T., De Graeve, M., Vanhaecke, L., Hulpiau, P., De Maeseneire, S.L., De Mol, M. L., Soetaert, W.K., 2022. Novel alkaloids from marine actinobacteria: discovery and characterization. Mar. Drugs. 16603397. 20 (1). Available from: https://www.mdpi.com/1660-3397/20/1/6/pdf. doi:10.3390/md20010006.

Dean, C.L., Harvey, E.L., Johnson, M.D., Subhas, A.V., 2024. Microzooplankton grazing on the coccolithophore Emiliania huxleyi and its role in the global calcium carbonate cycle. Sci. Adv. 23752548. 10 (45). Available from: https://www.science.org/doi/epdf/10.1126/sciadv.adr5453. doi:10.1126/sciadv.adr5453.

Debnath, T., Kujur, R.R.A., Mitra, R., Das, S.K., 2019. Diversity of Microbes in Hot Springs and Their Sustainable Use. Springer Science and Business Media LLC, pp. 159–186. doi:10.1007/978-981-13-8315-1_6.

Delbarre-Ladrat, C., Verrez-Bagnis, V., Colliec-Jouault, S., Zykwinska, A., 2019. Enzymes Used to Produce Glycosaminoglycan Mimetics from Marine Polysaccharides. Informa UK Limited, pp. 467–494. doi:10.1201/9780429058653-21.

Dell' Anno, F., Rastelli, E., Sansone, C., Dell' Anno, A., Brunet, C., Ianora, A., 2021. Bacteria, fungi and microalgae for the bioremediation of marine sediments contaminated by petroleum hydrocarbons in the omics era. Microorganisms 9 (8). doi:10.3390/microorganisms9081695.

Denaro, R., Aulenta, F., Crisafi, F., Di Pippo, F., Viggi, C.C., Matturro, B., Tomei, P., Smedile, F., Martinelli, A., Di Lisio, V., Venezia, C., Rossetti, S., 2020. Marine hydrocarbon-degrading bacteria breakdown poly(ethylene terephthalate) (PET). Sci. Total Environ. 00489697. 749, 141608. doi:10.1016/j.scitotenv.2020.141608.

Destoumieux-Garzón, D., Canesi, L., Oyanedel, D., Travers, M.A., Charrière, G.M., Pruzzo, C., Vezzulli, L., 2020. Vibrio–bivalve interactions in health and disease. Environ. Microbiol. 14622920. 22 (10), 4323–4341. Available from: http://onlinelibrary.wiley.com/journal/10.1111/(ISSN)1462-2920. doi:10.1111/1462-2920.15055.

Dinda, B., Dinda, S., Chakraborty, M., 2022. Pharmacology of Anti-obesity and Antidiabetic Phytochemicals Isolated from Various Natural Sources (Plants, Seaweeds, Mushrooms, Marine Animals, and Microorganisms) Natural Products in Obesity and Diabetes: Therapeutic Potential and Role in Prevention and Treatment. Springer International Publishing, India, pp. 277–467. Available from https://link.springer.com/book/10.1007/978-3-030-92196-5. doi:10.1007/978-3-030-92196-5_5.

Ding, L., Wang, L., Jiang, F., Zeng, X., Qinzeng, X., Zhang, X., Zheng, Q., Shao, Z., 2024. Prokaryotic community structure and key taxa in the Arabian Sea's oxygen minimum zone. Front. Mar. Sci. 2296-7745. 11. doi:10.3389/fmars.2024.1380819.

Djahnit, N., Chernai, S., Catania, V., Hamdi, B., China, B., Cappello, S., Quatrini, P., 2019. Isolation, characterization and determination of biotechnological potential of oil-degrading bacteria from Algerian centre coast. J. Appl. Microbiol. 13645072. 126 (3), 780–795. doi:10.1111/jam.14185.

Du, F., Wang, Y.Z., Xu, Y.S., Shi, T.Q., Liu, W.Z., Sun, X.M., Huang, H., 2021. Biotechnological production of lipid and terpenoid from thraustochytrids. Biotechnol. Adv. 07349750. 48. Available from: www.elsevier.com/inca/publications/store/5/2/5/4/5/5/index.htt. doi:10.1016/j.biotechadv.2021.107725.

Dubey, S., Singh, A., Kumar, B.T.N., Singh, N.K., Tyagi, A., 2021. Isolation and Characterization of Bacteriophages from Inland Saline Aquaculture Environments to Control Vibrio parahaemolyticus Contamination in Shrimp. Indian J. Microbiol. 09737715. 61 (2), 212–217. Available from: http://www.springerlink.com/content/0046-8991/. doi:10.1007/s12088-021-00934-6.

Dungan, A.M., Geissler, L., Williams, A.S., Gotze, C.R., Flynn, E.C., Blackall, L.L., van Oppen, M.J.H., 2023. DNA from non-viable bacteria biases diversity estimates in the corals Acropora loripes and Pocillopora acuta. Environ. Microbiome. 25246372. 18 (1). Available from: https://environmentalmicrobiome.biomedcentral.com. doi:10.1186/s40793-023-00541-6.

Duru, C.E., Duru, I.A., 2021. Christian Ebere Enyoh, In silico binding affinity analysis of microplastic compounds on PET hydrolase enzyme target of Ideonella sakaiensis. Bull. Natl. Res. Cent. 2522-8307. 45 (1). doi:10.1186/s42269-021-00563-5.

Dwivedi, S., Ahmad, I.Z., 2023. Cyanobacteria in Ocean Current Status of Marine Water Microbiology. CRC Press, India, pp. 47–66. Available from https://doi.org/10.1007/978-981-99-5022-5. doi:10.1007/978-981-99-5022-5_4.

Edo, G.I., Obasohan, P., Makia, R.S., Tubi Abiola, O., Umelo, E.C., Jikah, A.N., Yousif, E., Isoje, E.F., Igbuku, U.A., Opiti, R.A., Essaghah, A.E.A., Ahmed, D.S., Umar, H., 2024. The use of quality control parameters in the evaluation of herbal drugs. A review. Discov. Med. 3004-8885. 1 (1). doi:10.1007/s44337-024-00177-6.

Elhalis, H., Chin, X.H., Chow, Y., 2024. Soybean fermentation: microbial ecology and starter culture technology. Crit. Rev. Food Sci. Nutr. 15497852. 64 (21), 7648–7670. Available from: https://www.tandfonline.com/loi/bfsn20. doi:10.1080/10408398.2023.2188951.

Eroldoğan, O.T., Glencross, B., Novoveska, L., Gaudêncio, S.P., Rinkevich, B., Varese, G.C., de Fátima Carvalho, M., Tasdemir, D., Safarik, I., Nielsen, S.L., Rebours, C., Lada, L.B., Robbens, J., Strode, E., Haznedaroğlu, B.Z., Kotta, J., Evliyaoğlu, E., Oliveira, J., Girão, M., Vasquez, M.I., Čabarkapa, I., Rakita, S., Klun, K., Rotter, A., 2023. From the sea to aquafeed: a perspective overview. Rev. Aquac. 17535131. 15 (3), 1028–1057. Available from: http://onlinelibrary.wiley.com/journal/10.1111/(ISSN)1753-5131. doi:10.1111/raq.12740.

Essock-Burns, T., Bennett, B.D., Arencibia, D., Moriano-Gutierrez, S., Medeiros, M., Mcfall-Ngai, M.J., Ruby, E.G., 2021. Bacterial quorum-sensing regulation induces morphological change in a key host tissue during the euprymna scolopes-vibrio fischeri symbiosis. mBio. 21507511. 12 (5). Available from: https://journals.asm.org/doi/10.1128/mBio.02402-21. doi:10.1128/mBio.02402-21.

Essock-Burns, T., Bongrand, C., Goldman, W.E., Ruby, E.G., McFall-Ngai, M.J., Graf, J., 2020. Interactions of symbiotic partners drive the development of a complex biogeography in the squid-vibrio symbiosis. mBio. 21507511. 11 (3). Available from: https://mbio.asm.org/content/mbio/11/3/e00853-20.full.pdf. doi:10.1128/mBio.00853-20.

Fang, X., Lloyd, C.J., Palsson, B.O., 2020. Reconstructing organisms in silico: genome-scale models and their emerging applications. Nat. Rev. Microbiol. 17401534. 18 (12), 731–743. Available from: http://www.nature.com/nrmicro/index.html. doi:10.1038/s41579-020-00440-4.

Fangzhou, M., Wang, C., Zhang, Y., Chen, J., Xie, R., Sun, Z., 2022. Development of microbial indicators in ecological systems. Int. J. Environ. Res. Public Health. 1660-4601. 19 (21), 13888. doi:10.3390/ijerph192113888.

Faÿ, F., Champion, M., Guennec, A., Moppert, X., Simon-Colin, C., Elie, M., 2023. Biobased anti-adhesive marine coatings from polyhydroxyalkanoates and polysaccharides. Coatings. 2079-6412. 13 (4), 766. doi:10.3390/coatings13040766.

Fenical, W., 2020. Marine microbial natural products: the evolution of a new field of science. J. Antibiot. 0021-8820. 73 (8), 481–487. doi:10.1038/s41429-020-0331-4.

Fenical, W., Jensen, P.R., Palladino, M.A., Lam, K.S., Lloyd, G.K., Potts, B.C., 2009. Discovery and development of the anticancer agent salinosporamide A (NPI-0052). Bioorg. Med. Chem. 09680896. 17 (6), 2175–2180. doi:10.1016/j.bmc.2008.10.075.

Fernandes, P., 2023. 9 Role of microbial oxidative enzymes. Microb. Oxid. Enzymes Biotechnol. Appl. 183. Available from. doi:10.1515/9783111062235-009.

Fernández-Juárez, V., Jaén-Luchoro, D., Brito-Echeverría, J., Agawin, N.S.R., Bennasar-Figueras, A., Echeveste, P., 2022. Everything is everywhere: physiological responses of the mediterranean sea and eastern pacific ocean epiphyte cobetia sp. to varying nutrient concentration. Microb. Ecol. 0095-3628. 83 (2), 296–313. doi:10.1007/s00248-021-01766-z.

Flombaum, P., Gallegos, J.L., Gordillo, R.A., Rincón, J., Zabala, L.L., Jiao, N., Karl, D.M., Li, W.K.W., Lomas, M.W., Veneziano, D., Vera, C.S., Vrugt, J.A., Martiny, A.C., 2013. Present and future global distributions of the marine cyanobacteria prochlorococcus and synechococcus. Proc. Natl. Acad. Sci. U. S. A. 110 (24), 9824–9829. doi:10.1073/pnas.1307701110.

Fongaro, G., Maia, G.A., Rogovski, P., Cadamuro, R.D., Lopes, J.C., Simões Moreira, R., Camargo, A.F., Scapini, T., Spitza Stefanski, F., Bonatto, C., Souza, D.S.M., Hermes Stoco, P., Duarte, R.T.D., 2020. Ariadne cristiane cabral da cruz, glauber wagner, helen treichel, extremophile microbial communities and enzymes for bioenergetic application based on multi-omics tools. Curr. Genomics 21 (4), 240–252. doi:10.2174/1389202921999200601144137.

Fricker, L.D., 2020. Proteasome Inhibitor Drugs. Annu. Rev. Pharmacol. Toxicol. 15454304. 60, 457–476. Available from: http://www.annualreviews.org/journal/pharmtox. doi:10.1146/annurev-pharmtox-010919-023603.

Furhan, J., 2020. Adaptation, production, and biotechnological potential of cold-adapted proteases from psychrophiles and psychrotrophs: recent overview. J. Genet. Eng. Biotechnol. 1687157X. 18 (1), 36. doi:10.1186/s43141-020-00053-7.

Garcia-Vaquero, M., Rajauria, G., 2021. Innovative and Emerging Technologies in the Bio-marine Food Sector: Applications, Regulations, and Prospects Innovative and Emerging Technologies in the Bio-marine Food Sector: Applications, Regulations, and Prospects. Elsevier, Ireland, pp. 1–529. Available from https://www.sciencedirect.com/book/9780128200964. doi:10.1016/C2019-0-01113-2.

Garel, M., Bonin, P., Martini, S., Guasco, S., Roumagnac, M., Bhairy, N., 2019. Fabrice armougom, christian tamburini, pressure-retaining sampler and high-pressure systems to study deep-sea microbes under in situ conditions. Front. Microbiol. 1664-302X. 10. doi:10.3389/fmicb.2019.00453.

Gavriilidou, A., Gutleben, J., Versluis, D., Forgiarini, F., Van Passel, M.W.J., Ingham, C.J., Smidt, H., Sipkema, D., 2020. Comparative genomic analysis of Flavobacteriaceae: insights into carbohydrate metabolism, gliding motility and secondary metabolite biosynthesis. BMC Genomics. 14712164. 21 (1). Available from: http://www.biomedcentral.com/bmcgenomics. doi:10.1186/s12864-020-06971-7.

Gélébart, P., Cuenot, S., Sinquin, C., Halgand, B., Sourice, S., Visage, C.L., Guicheux, J., Colliec-Jouault, S., Zykwinska, A., 2022. Microgels based on Infernan, a glycosaminoglycan-mimetic bacterial exopolysaccharide, as BMP-2 delivery systems. Carbohydr. Polym. 01448617. 284, 119191. doi:10.1016/j.carbpol.2022.119191.

Genge, L., Prochlorococcus growth responses to light and oxygen regimes mount allison university. (2021).

Ghaderiardakani, F., Ulrich, J.F., Barth, E., Quartino, M.L., Wichard, T., 2024. Algal Growth and Morphogenesis-Promoting Factors Released by Cold-Adapted Bacteria Contribute to the Resilience and Morphogenesis of the Seaweed Ulva (Chlorophyta) in Antarctica (Potter Cove). J. Plant Growth Regul. 0721–7595. doi:10.1007/s00344-024-11507-4.

Ghareeb, A., Fouda, A., Kishk, R.M., Kazzaz, W.M.E., 2024. Unlocking the therapeutic potential of bioactive exopolysaccharide produced by marine actinobacterium Streptomyces vinaceusdrappus AMG31: a novel approach to drug development. Int. J. Biol. Macromol. 18790003. 276. Available from: https://www.sciencedirect.com/science/journal/01418130. doi:10.1016/j.ijbiomac.2024.133861.

Ghosh, M., Heo, Y., Pulicherla, K.K., Ha, M.W., Do, K., Son, Y.O., 2025a. Cold-active enzymes from deep marine psychrophiles: harnessing their potential in enhanced food production and sustainability. Crit. Rev. Biotechnol. 15497801. Available from: http://www.tandfonline.com/loi/ibty20. doi:10.1080/07388551.2024.2435974.

Ghosh, M., Heo, Y., Pulicherla, K.K., Min Woo, H., Kyoungtag, D., Son, Y.-O., 2025b. Cold-active enzymes from deep marine psychrophiles: harnessing their potential in enhanced food production and sustainability. Crit. Rev. Biotechnol. doi:10.1080/07388551.2024.2435974.

Ghosh, S., Sarkar, T., Pati, S., Kari, Z.A., Edinur, H.A., Chakraborty, R., 2022. Novel Bioactive Compounds From Marine Sources as a Tool for Functional Food Development. Front. Mar. Sci. 22967745. 9. Available from: https://www.frontiersin.org/journals/marine-science#. doi:10.3389/fmars.2022.832957.

Giddings, L.-A., Newman, D.J., 2019. Bioactive Compounds from Extremophilic Marine Fungi. Springer Science and Business Media LLC, pp. 349–382. doi:10.1007/978-3-030-19030-9_18.

Girão, M., Ribeiro, I., de Fátima Carvalho, M., 2022. Actinobacteria from Marine Environments: A Unique Source of Natural Products Natural Products from Actinomycetes: Diversity, Ecology and Drug Discovery. Springer Nature, Portugal, pp. 1–45. Available from https://link.springer.com/book/10.1007/978-981-16-6132-7. doi:10.1007/978-981-16-6132-7_1.

Glasl, B., Bourne, D.G., Frade, P.R., Thomas, T., Schaffelke, B., Webster, N.S., 2019. Microbial indicators of environmental perturbations in coral reef ecosystems. Microbiome 7 (1). doi:10.1186/s40168-019-0705-7.

Gonçalves, M.A.L., López-Vela, M., Palacios-Espinosa, A., Romero-Bastidas, M., Rojas-Contreras, M., Magallón-Servín, P., 2024. Impact of Phycosphere-Isolated Marine Bacteria on Nutritional Value, Growth, and Nutrient Uptake of Co-Cultured Chaetoceros calcitrans. Resources. 20799276. 13 (9). Available from: www.mdpi.com/journal/resources. doi:10.3390/resources13090116.

Grujcic, V., Taylor, G.T., Foster, R.A., 2022. One Cell at a Time: advances in Single-Cell Methods and Instrumentation for Discovery in Aquatic Microbiology. Front. Microbiol. 1664302X. 13. Available from: https://www.frontiersin.org/journals/microbiology#. doi:10.3389/fmicb.2022.881018.

Guerrero-Cruz, S., Vaksmaa, A., Horn, M.A., Niemann, H., Pijuan, M., Ho, A., 2021. Methanotrophs: discoveries, Environmental Relevance, and a Perspective on Current and Future Applications. Front. Microbiol. 1664302X. 12. Available from: https://www.frontiersin.org/journals/microbiology#. doi:10.3389/fmicb.2021.678057.

Guimarães, M.C., Cerezo, I.M., Fernandez-Alarcon, M.F., Natori, M.M., Sato, L.Y., Kato, C.A.T., Moriñigo, M.A., Tapia-Paniagua, S., Dias, D.D.C., Ishikawa, C.M., Ranzani-Paiva, M.J.T., Cassiano, L.L., Bach, E.E., Clissa, P.B., Orefice, D.P., Tachibana, L., 2022. Oral Administration of Probiotics (Bacillus subtilis and Lactobacillus plantarum) in Nile Tilapia (Oreochromis niloticus) Vaccinated and Challenged with Streptococcus agalactiae. Fishes. 24103888. 7 (4). Available from: www.mdpi.com/journal/fishes/. doi:10.3390/fishes7040211.

Guimarães, M.C., da Silva Guimarães, A.I.C., Natori, M.M., Alarcon, M.F.F., Dias, D.D.C., Ishikawa, C.M., Tapia-Paniagua, S., Moriñigo, M.Á., Moyano, F.J., Tachibana, L., 2021. Oral administration of Bacillus subtilis and Lactobacillus plantarum modulates the gut microbiota and increases the amylase activity of Nile tilapia (Oreochromis niloticus). Aquac. Int. 1573143X. 29 (1), 91–104. Available from: www.wkap.nl/journalhome.htm/0967-6120. doi:10.1007/s10499-020-00612-2.

Gupta, V., Bhaskar, P., Thoudam, J., Bisht, S., Sharma, A., 2023. Rashmi Tripathi, Psychrophiles as a novel and promising source of cold-adapted industrial enzymes. Appl. Biol. Chem. J. 2582-8789, 54–68. doi:10.52679/tabcj.2023.0006.

Gurpilhares, D.D.B., Cinelli, L.P., Simas, N.K., Pessoa, A., Sette, L.D., 2019. Marine prebiotics: polysaccharides and oligosaccharides obtained by using microbial enzymes. Food Chem. 18737072. 280, 175–186. Available from: www.elsevier.com/locate/foodchem. doi:10.1016/j.foodchem.2018.12.023.

Gutierrez, T., 2019. Aerobic Hydrocarbon-Degrading Gammaproteobacteria: Porticoccus. Springer Science and Business Media LLC. Springer Science and Business Media LLC, pp. 181–189. doi:10.1007/978-3-030-14796-9_32.

Han, J., Jiang, J., Kong, T., Zhao, X., Zhao, X., Ping, L., 2023a. Qing Gu, Comparative analysis of volatile compounds and microbial community succession in Chinese fish sauce: three typical types of fast-fermented fish sauces. Food Biosci. 22124292. 54, 102918. doi:10.1016/j.fbio.2023.102918.

Han, Y., Steiner, Z., Cao, Z., Fan, D., Chen, J., Yu, J., Dai, M., 2024. Pelagic Coccolithophore Production and Dissolution and Their Impacts on Particulate Inorganic Carbon Cycling in the Western North Pacific, EGUsphere.

Hassanien, A., Saadaoui, I., Schipper, K., Al-Marri, S., Dalgamouni, T., Aouida, M., Saeed, S., Al-Jabri, H.M., 2023. Genetic engineering to enhance microalgal-based produced water treatment with emphasis on CRISPR/Cas9:

a review. Front. Bioeng. Biotechnol. 22964185. 10. Available from: http://journal.frontiersin.org/journal/bioengineering-and-biotechnology#archive. doi:10.3389/fbioe.2022.1104914.

Hatzenpichler, R., Krukenberg, V., Spietz, R.L., Jay, Z.J., 2020. Next-generation physiology approaches to study microbiome function at single cell level. Nat. Rev. Microbiol. 17401534. 18 (4), 241–256. Available from: http://www.nature.com/nrmicro/index.html. doi:10.1038/s41579-020-0323-1.

Heimland, E.N., Exploring the Biotechnological Potential of the Secretome from the Marine Fungus Digitatispora marina Proteomics, Bioinformatics and Heterologous Expression of Novel Proteases. (2022), Available from: https://hdl.handle.net/10037/31377.

Heinrichs, M.E., Mori, C., Dlugosch, L., 2019. Complex Interactions between Aquatic Organisms and Their Chemical Environment Elucidated from Different Perspectives. Springer Science and Business Media LLC, pp. 279–297. doi:10.1007/978-3-030-20389-4_15.

Hettiarachchi, S.A., Lee, S.-J., Lee, Y., Kwon, Y.-K., De Zoysa, M., Moon, S., Eunyoung, J., Kim, T., Kang, D.-H., Heo, S.-J., Chulhong, O., 2017. A rapid and efficient screening method for antibacterial compound-producing bacteria. J. Microbiol. Biotechnol. 27 (8), 1441–1448. doi:10.4014/jmb.1703.03012.

Hinzke, T., Kleiner, M., Breusing, C., Felbeck, H., Häsler, R., Sievert, S.M., Schlüter, R., Rosenstiel, P., Reusch, T.B.H., Schweder, T., Markert, S., 2019. Host-microbe interactions in the chemosynthetic riftia pachyptila symbiosis. mBio. 21507511. 10 (6). Available from: https://mbio.asm.org/content/mbio/10/6/e02243-19.full.pdf. doi:10.1128/mBio.02243-19.

Hoadley, K.D., Lewis, A.M., Wham, D.C., Pettay, D.T., Grasso, C., Smith, R., Kemp, D.W., LaJeunesse, T.C., Warner, M.E., 2019. Host–symbiont combinations dictate the photo-physiological response of reef-building corals to thermal stress. Sci. Rep. 9 (1). doi:10.1038/s41598-019-46412-4.

Hobmeier, K., Cantone, M., Nguyen, Q.A., Pflüger-Grau, K., Kremling, A., Kunte, H.J., Pfeiffer, F., Marin-Sanguino, A., 2022. Adaptation to Varying Salinity in Halomonas elongata: much More Than Ectoine Accumulation. Front. Microbiol. 1664302X. 13. Available from: https://www.frontiersin.org/journals/microbiology#. doi:10.3389/fmicb.2022.846677.

Hoff, J., Daniel, B., Daniel Stukenberg, B.W.T., Waldminghaus, T., Fritz, G., 2020. Vibrio natriegens: an ultrafast-growing marine bacterium as emerging synthetic biology chassis. Environ. Microbiol. 1462-2912. 22 (10), 4394–4408. doi:10.1111/1462-2920.15128.

Holden, J.F., Sistu, H., 2023. Formate and hydrogen in hydrothermal vents and their use by extremely thermophilic methanogens and heterotrophs. Front. Microbiol. 14. 10.3389/fmicb.2023.1093018

Hozzein, W.N., Mohany, M., Alhawsawi, S.M.M., Zaky, M.Y., Al-Rejaie, S.S., Alkhalifah, D.H.M., 2021. Flavonoids from marine-derived actinobacteria as anticancer drugs. Curr. Pharm. Des. 18734286. 27 (4), 505–512. Available from: https://www.eurekaselect.com/189185/article. doi:10.2174/1381612826666201216160154.

Huang, W., 2024. Application of Synthetic Biology in Directed Evolution to Enhance Enzyme Catalytic Efficiency. Biol. Evid. 1927–6478. doi:10.5376/be.2024.14.0015.

Hutchins, D.A., Jansson, J.K., Remais, J.V., Rich, V.I., Singh, B.K., Trivedi, P., 2019. Climate change microbiology — problems and perspectives. Nat. Rev. Microbiol. 17401534. 17 (6), 391–396. Available from: http://www.nature.com/nrmicro/index.html. doi:10.1038/s41579-019-0178-5.

Ibrahim, H.A.H., Beliah, M.M.B., Farag, A.M., ElAhwany, A.M.D., Sabry, S.A., 2020. Properties and aerogel applications of a marine algal origin biocellulose produced by the immobilized Gluconacetobacter xylinus ATCC 10245. Egypt. J. Aquat. Biol. Fish. 2536-9814. 24 (7), 51–71. doi:10.21608/ejabf.2020.117931.

Ibrahim, H.A.H., Elhassayeb, H.E.A., El-Sayed, W.M.M., 2022. Potential functions and applications of diverse microbial exopolysaccharides in marine environments. J. Genet. Eng. Biotechnol. 20905920. 20 (1). Available from: https://www.springer.com/journal/43141. doi:10.1186/s43141-022-00432-2.

Ibrahim, U.B., Muhammad, K., Nata'ala, A.S., 2023. Bacterial Alkane Hydroxylase; Diversity, Distribution and Influence in Degradation of HydrocarbonsBacterial Alkane Hydroxylase; Diversity, Distribution and Influence in Degradation of Hydrocarbons. Recent Adv. Mol. Biol. Biochem. 57, 57–78.

Imhoff, J.F., Rahn, T., Künzel, S., Keller, A., Neulinger, S.C., 2021. Osmotic adaptation and compatible solute biosynthesis of phototrophic bacteria as revealed from genome analyses. Microorganisms. 20762607. 9 (1), 1–32. Available from: https://www.mdpi.com/2076-2607/9/1/46/pdf. doi:10.3390/microorganisms9010046.

Jabir, T., Vipindas, P.V., Jesmi, Y., Divya, P.S., Adarsh, B.M., Miziriya, H.S.N., Hatha, A.A.M., 2021. Influence of environmental factors on benthic nitrogen fixation and role of sulfur reducing diazotrophs in a eutrophic tropical estuary. Mar. Pollut. Bull. 0025326X (165), 112126. doi:10.1016/j.marpolbul.2021.112126.

Jannathulla, R., Dayal, J.S., 2023. Beneficial effects, challenges and opportunities of the filamentous fungus, Aspergillus Niger with special reference to the shrimp feed industry—A review. Rev. Aquac. 17535131. 15 (4), 1311–1334. Available from: http://onlinelibrary.wiley.com/journal/10.1111/(ISSN)1753-5131. doi:10.1111/raq.12775.

Jeewon, R., Aullybux, A.A., Puchooa, D., Nazurally, N., Alrefaei, A.F., Zhang, Y., 2023. Marine Microbial Polysaccharides: an Untapped Resource for Biotechnological Applications. Mar. Drugs. 16603397. 21 (7). Available from: http://www.mdpi.com/journal/marinedrugs. doi:10.3390/md21070420.

Jensen, P.R., 2022. Microbe Profile: salinispora tropica: natural products and the evolution of a unique marine bacterium. Microbiology. 1350-0872. 168 (4). doi:10.1099/mic.0.001163.

Jianzhou, X., Mengqi, Y., Ding, L., Shan, H., 2019. A Review of Anti-Inflammatory Compounds from Marine Fungi, 2000–2018. Mar. Drugs. 1660-3397. 17 (11), 636. doi:10.3390/md17110636.

Jiao, J.Y., Liu, L., Hua, Z.S., Fang, B.Z., Zhou, E.M., Salam, N., Hedlund, B.P., Li, W.J., 2021. Microbial dark matter coming to light: challenges and opportunities. Natl. Sci. Rev. 2053714X. 8 (3). Available from: http://nsr.oxfordjournals.org/. doi:10.1093/nsr/nwaa280.

Jiao, N., Luo, T., Chen, Q., Zhao, Z., Xiao, X., Liu, J., Jian, Z., Xie, S., Thomas, H., Herndl, G.J., Benner, R., Gonsior, M., Chen, F., Cai, W.-J., Robinson, C., 2024. The microbial carbon pump and climate change. Nat. Rev. Microbiol. 22 (7), 408–419. doi:10.1038/s41579-024-01018-0.

Jin, J., Zhao, Y., Guo, W., Wang, B., Wang, Y., Liu, X., Xu, C., 2019. Thiocoraline mediates drug resistance in MCF-7 cells via PI3K/Akt/BCRP signaling pathway. China Cytotechnology. 15730778. 71 (1), 401–409. Available from: http://www.springer.com/chemistry/biotech/journal/10616. doi:10.1007/s10616-019-00301-w.

Jin, S., Wang, Y., Zhao, X., 2022. Cold-adaptive mechanism of psychrophilic bacteria in food and its application. Microb. Pathog. 08824010. 169 (105652). doi:10.1016/j.micpath.2022.105652.

Jinjin, M., Nie, Y., Zhang, L., Yan, X., 2024. The evolutionary mechanism and function analysis of two subgroups of histamine-producing and non-histamine-producing Tetragenococcus halophilus. Food Res. Int. 09639969. 176, 113744. doi:10.1016/j.foodres.2023.113744.

Joseph, R. 2019 The Deepwater Horizon disaster in 2010 led to the release of 200 million gallons of crude oil into the ocean, and really put into perspective the amount of damage that man can inflict on the environment. One way that first responders attempted to deal with the spill was by dispersing the oil into the water column with the use of a chemical mixture known as Corexit 9500. Being sprayed onto the slick from above, and down near the wellhead below, this mixture of surfactants and solvents was aimed at stopping the slick from reaching the coast and also speeding up the bioremediation process by increasing the specific surface area of the slick. And while Corexit has been shown in prior studies to be effective at performing this task, certain components that make up Corexit, mainly the surfactant dioctyl sodium sulfosuccinate (DOSS), have environmental concerns surrounding them including effects to human health, persistence in the ocean and possible impairment of the bacterial oil degradation. Therefore, there is a need for a more environmentally friendly dispersant. Recently, an alternative dispersant mixture being composed of the food grade amphiphiles lecithin and Tween 80 has shown much promise in exhibiting similar dispersant effectiveness as Corexit 9500. The studies performed in this thesis investigate these mixtures further, while also trying to understand the dispersion and degradation processes. The first manuscript investigates a solvent free mixture of lecithin and Tween 80, being composed of lecithin and Tween 80 assemblies formed in water. Investigation through Cryo TEM revealed that the combination of these two amphiphiles forms bicelles in water. Dispersant effectiveness testing showed the lecithin/Tween 80 solvent free system to be effective at forming a stable emulsion. The second manuscript investigates the oil biodegradation process of oil with Alcanivorax borkumensis when supplemented with a variety of dispersant mixtures. Emphasis was given to the newly developed lecithin-Tween 80 mixture, which was compared to the controls of no dispersant and Corexit 9500. Lecithin alone and Tween 80 alone dispersant mixtures were also tested in order to elucidate whether it is the rise in bioavailability or the presence of one of the two surfactants that gives the lecithin-Tween 80 mixture its degradation behavior. Microscopic analysis of the systems showed the presence of bacteria-oil agglomerates for all but Tween 80. And oil degradation quantification results after 144 hours showed lecithin/Tween 80 to enhance degradation the most, at 52%; compared to no dispersant at 25%, Corexit 9500 at 35%, Tween 80 at 35% and lecithin at 38%. ProQuest Dissertations and Theses 978-1-392-07195-3 Applied sciences Hydrocarbonoclastic bacteria Lipid vesicles Marine snow Oil degradation Oil dispersants Oil-in-water emulsion Chemical engineering Environmental engineering 0542:Chemical engineering 0775:Environmental engineering 9191 Surfactant Enhanced Bioremediation of Oil by Alcanivorax Borkumensis Using Food Grade Amphiphiles University of Rhode Island United States – rhode Island https://www.proquest.com/dissertations-theses/surfactant-enhanced-bioremediation-oil/docview/2213657877/se-2?accountid=10066

Junxiong, Y., Zhang, Y., Liu, H., Liu, Y., Mohsin, A., Liu, Z., Zheng, Y., Xing, J., Han, J., Zhuang, Y., Guo, M., Wang, Z., 2024. Temporal dynamics of stress response in Halomonas elongata to NaCl shock: physiological, metabolomic, and transcriptomic insights. Microb. Cell Fact. 1475-2859. 23 (1). doi:10.1186/s12934-024-02358-5.

Kaboré, O.D., Godreuil, S., Drancourt, M., 2020. Planctomycetes as Host-Associated Bacteria: a Perspective That Holds Promise for Their Future Isolations, by Mimicking Their Native Environmental Niches in Clinical Microbiology Laboratories. Front. Cell. Infect. Microbiol. 22352988. 10. Available from: http://www.frontiersin.org/Cellular_and_Infection_Microbiology/archive. doi:10.3389/fcimb.2020.519301.

Kadri, T., Miri, S., Robert, T., Brar, S.K., Rouissi, T., LaxmanPachapur, V., Lauzon, J.M., 2022. Pilot-scale production and in-situ application of petroleum-degrading enzyme cocktail from Alcanivorax borkumensis. Chemosphere. 18791298. 295. Available from: www.elsevier.com/locate/chemosphere. doi:10.1016/j.chemosphere.2022.133840.

Kanekar, P.P., Kanekar, S.P., 2022. Thermophilic, Thermotolerant Microorganisms. Springer Science and Business Media LLC, pp. 117–153. doi:10.1007/978-981-19-1573-4_4.

Kassem, A., Abbas, L., Coutinho, O., Opara, S., Najaf, H., Kasperek, D., Pokhrel, K., Xiaohua, L., 2023. Sonia Tiquia-Arashiro, Applications of Fourier Transform-Infrared spectroscopy in microbial cell biology and environmental microbiology: advances, challenges, and future perspectives. Front. Microbiol. 1664-302X. 14. doi:10.3389/fmicb.2023.1304081.

Khandare, S.D., Agrawal, D., Mehru, N., Chaudhary, D.R., 2022. Marine bacterial based enzymatic degradation of low-density polyethylene (LDPE) plastic. J. Environ. Chem. Eng. 22133437. 10 (3). Available from: http://www.journals.elsevier.com/journal-of-environmental-chemical-engineering/. doi:10.1016/j.jece.2022.107437.

Kieft, T.L., 2021. Lithotrophic ("stone-eating") Microbes Provide the Foundation for Deep Subsurface Ecosystems. Springer International Publishing, Cham, pp. 131–139. Available from https://doi.org/10.1007/978-3-030-63512-1_7. doi:10.1007/978-3-030-63512-1_7.

Kim, K.H., Kim, J.M., Baek, J.H., Jeong, S.E., Kim, H., Yoon, H.S., Jeon, C.O., 2024. Metabolic relationships between marine red algae and algae-associated bacteria. Mar. Life Sci. Technol. 26621746. 6 (2), 298–314. Available from: https://www.springer.com/journal/42995. doi:10.1007/s42995-024-00227-z.

Kim, T.H., Kim, S.C., Jung, W.K., 2023. Therapeutic effect of marine bioactive substances against periodontitis based on in vitro, in vivo, and clinical studies. Fish. Aquat. Sci. 22341757. 26 (1), 1–23. Available from: https://www.e-fas.org/info/about. doi:10.47853/FAS.2023.e1.

Kolodny, Y., Avrahami, Y., Zer, H., Frada, M.J., Paltiel, Y., Keren, N., 2022. Phycobilisome light-harvesting efficiency in natural populations of the marine cyanobacteria Synechococcus increases with depth. Commun. Biol. 23993642. 5 (1). Available from: https://www.springer.com/journal/11852. doi:10.1038/s42003-022-03677-2.

Kujawinski, E.B., Braakman, R., Longnecker, K., Becker, J.W., Chisholm, S.W., Dooley, K., Kido Soule, M.C., Swarr, G.J., Halloran, K., 2023. Metabolite diversity among representatives of divergent Prochlorococcus ecotypes. mSystems, 8 (5). doi:10.1128/msystems.01261-22.

Kumar, A., Aswal, S., Semwal, R.B., Chauhan, A., Joshi, S.K., Semwal, D.K., 2019. Role of plant-derived alkaloids against diabetes and diabetes-related complications: a mechanism-based approach. Phytochem. Rev. 1572980X. 18 (5), 1277–1298. Available from: www.springer.com/journal/11101. doi:10.1007/s11101-019-09648-6.

Kuo, J., Yang, Y.T., Lu, M.C., Wong, T.Y., Sung, P.J., Huang, Y.S., 2019. Antimicrobial activity and diversity of bacteria associated with Taiwanese marine sponge Theonella swinhoei. Ann. Microbiol. 18692044. 69 (3), 253–265. Available from: http://springerlink.com/content/121707/. doi:10.1007/s13213-018-1414-3.

Kwon, H.C., Kauffman, C.A., Jensen, P.R., Fenical, W., Marinomycins, A.-D., 2006. antitumor-antibiotics of a new structure class from a marine actinomycete of the recently discovered genus "Marinispora". J. Am. Chem. Soc. 128 (5), 1622–1632. doi:10.1021/ja0558948.

Lan, Y. 2020 Deep sea is a harsh environment characterized by high hydrostatic pressure and low temperatures. For instance, high hydrostatic pressure can disrupt actin organization and microtubules assembly, which have adverse influences on intracellular movements and cell motility. Nucleic acids structures can be influenced by not only high hydrostatic pressure but also cold temperatures. These unfavourable effects caused by the high hydrostatic pressure and coldness are challenging for deep-sea animals. In addition, there are chemosynthetic ecosystems, such as hydrothermal vents and hydrocarbon seeps, in the deep sea. Dense communities of megafauna rely on the nutrients produced by chemotrophic bacteria that are capable of utilizing chemical substances as energy resources. To study the potential mechanisms of biological adaptation to deep-sea environment, we performed genomic analysis of representative species from both deep-sea non-vent environment and deep-sea chemosynthetic environment, including Hirondellea gigas amphipod from Challenger Deep in Mariana Trench, Aldrovandia affinis fish from the Pacific Ocean, Phreagena okutanii clam from a deep-sea hydrothermal vent, Archivesica marissinica clam from a deep-sea methane seep, and Gigantopelta aegisgastropod from a hydrothermal vent. Our results suggest that genetic adaptation to the deep-sea environment are mediated by both gene family expansion and amino acid substitutions. Furthermore, the vent-endemic megafauna is highly dependent on the mutualistic relationships with their symbionts. The host provide intermediates to fulfill the metabolic needs of their endosymbionts, and in return, the symbionts are active to generate nutrients and provide to the hosts through lysis processes. Overall, my thesis study not only provides high-quality transcriptomes and genomes of a few representative deep-sea animals for future deep-sea studies but also facilitates a better understanding of the molecular adaptation to the deep sea. PQDT - Global 9798342782944 Cold Symbiosis Cytoskeleton Mollusks Biofilms Molecular biology Biosynthesis Bacteria Adaptation Amino acids Sulfur Heat Genomes Metabolism Genomics Energy resources Ecosystems Signal transduction Chromosomes Information processing Nutrients Biology Energy Genetics Microbiology 0307:Molecular biology 0306:Biology 0791:Energy 0410:Microbiology 0369:Genetics 230230 Molecular Adaptation of Deep-Sea Fauna Inferred from Omics Data Hong Kong University of Science and Technology (Hong Kong) Hong Kong https://www.proquest.com/dissertations-theses/molecular-adaptation-deep-sea-fauna-inferred/docview/3132860922/se-2?accountid=10066

Lauritano, C., Rizzo, C., Giudice, A.L., Saggiomo, M., 2020. Physiological and Molecular Responses to Main Environmental Stressors of Microalgae and Bacteria in Polar Marine Environments. Microorganisms. 2076-2607. 8 (12), 1957. doi:10.3390/microorganisms8121957.

Le Bris, N., Yücel, M., Das, A., Sievert, S.M., LokaBharathi, P.P., Girguis, P.R., 2019. Hydrothermal energy transfer and organic carbon production at the deep seafloor. Front. Mar. Sci. 22967745. 5. Available from: https://www.frontiersin.org/articles/10.3389/fmars.2018.00531/full. doi:10.3389/fmars.2018.00531.

Lemfack, M.-C., Bahl, H., Piechulla, B., Magnus, N., 2020. The Domain of Bacteria and Their Volatile Metabolic Potential. Springer Science and Business Media LLC. Springer Science and Business Media LLC, pp. 1–38. doi:10.1007/978-981-15-7293-7_1.

Li, H., Yang, Q., Zhou, H., 2020. Niche differentiation of sulfate-and iron-dependent anaerobic methane oxidation and methylotrophic methanogenesis in deep sea methane seeps. Front. Microbiol. 11. 10.3389/fmicb.2020.01409

Liang, J., Deng, C., Kefu, Y., Ruiqi, G., Yongqian, X., Qin, Z., Chen, B., Wang, Y., Hongfei, S., Huang, X., Huang, W., Wang, G., Gong, S., 2021. Cross-linked regulation of coral-associated dinoflagellates and bacteria in pocillopora sp. During high-temperature stress and recovery. Microorganisms 9 (9). doi:10.3390/microorganisms9091972.

Libo, Y., Jian, H., Gai, Y., Zhiwei, Y., Feng, Y., Qiu, X., Shao, Z., Tang, X., 2021. Characterization of two novel psychrophilic and piezotolerant strains, Shewanella psychropiezotolerans sp. nov. and Shewanella eurypsychrophilus sp. nov, adapted to an extreme deep-sea environment. Syst. Appl. Microbiol. 07232020. 44 (6), 126266. doi:10.1016/j.syapm.2021.126266.

Lin, X., Lambertz, J., Dahlmann, T.A., Nowaczyk, M.M., König, B., Kück, U., 2022. A Straightforward Approach to Synthesize 7-Aminocephalosporanic Acid In Vivo in the Cephalosporin C Producer Acremonium chrysogenum. J. Fungi. 2309608X. 8 (5). Available from: https://www.mdpi.com/2309-608X/8/5/450/pdf. doi:10.3390/jof8050450.

Lipsman, V., Shlakhter, O., Rocha, J., Segev, E., 2024. Bacteria contribute exopolysaccharides to an algal-bacterial joint extracellular matrix. Npj Biofilms Microbiomes. 2055-5008. 10 (1). doi:10.1038/s41522-024-00510-y.

Lisov, A.V., Kiselev, S.S., Trubitsina, L.I., Belova, O.V., Andreeva-Kovalevskaya, Z.I., Trubitsin, I.V., Shushkova, T. V., Leontievsky, A.A., 2022. Multifunctional Enzyme with Endoglucanase and Alginase/Glucuronan Lyase Activities from Bacterium Cellulophaga lytica. Biochemistry (Mosc. 16083040. 87 (7), 617–627. Available from: https://link.springer.com/journal/10541. doi:10.1134/S0006297922070045.

Liu, H., Zhang, C., Zhang, X., Tan, K., Zhang, H., Cheng, D., Ting, Y., Shengkang, L., Hongyu, M., Zheng, H., 2020a. A novel carotenoids-producing marine bacterium from noble scallop Chlamys nobilis and antioxidant activities of its carotenoid compositions. Food Chem. 03088146. 320, 126629. doi:10.1016/j.foodchem.2020.126629.

Liu, M., El-Hossary, E.M., Oelschlaeger, T.A., Donia, M.S., Quinn, R.J., Abdelmohsen, U.R., 2019. Potential of marine natural products against drug-resistant bacterial infections. Lancet Infect. Dis. 14744457. 19 (7), e237. Available from: http://www.journals.elsevier.com/the-lancet-infectious-diseases. doi:10.1016/S1473-3099(18)30711-4.

Liu, S., Moon, C.D., Zheng, N., Huws, S., Zhao, S., Wang, J., 2022b. Opportunities and challenges of using metagenomic data to bring uncultured microbes into cultivation. Microbiome. 20492618. 10 (1). Available from: http://www.microbiomejournal.com/. doi:10.1186/s40168-022-01272-5.

Liu, X.R., Zhang, M.M., Gross, M.L., 2020b. Mass Spectrometry-based protein footprinting for higher-order structure analysis: fundamentals and applications. Chem. Rev. 15206890. 120 (10), 4355–4454. Available from: http://pubs.acs.org/journal/chreay. doi:10.1021/acs.chemrev.9b00815.

Liu, Y., Zhang, N., Jie, M., Zhou, Y., Wei, Q., Tian, C., Fang, Y., Zhong, R., Chen, G., Zhang, S., 2023a. Advances in cold-adapted enzymes derived from microorganisms. Front. Microbiol. 1664-302X. 14. doi:10.3389/fmicb.2023.1152847.

Liu, Y., Zhou, Z., Sun, S., 2024. Prospects of marine-derived compounds as potential therapeutic agents for glioma. Pharm. Biol. 1388-0209. 62 (1), 513–526. doi:10.1080/13880209.2024.2359659.

Lo Giudice, A., Rizzo, C., 2022. Bacteria Associated with Benthic Invertebrates from Extreme Marine Environments: promising but Underexplored Sources of Biotechnologically Relevant Molecules. Mar. Drugs. 1660-3397. 20 (10), 617. doi:10.3390/md20100617.

Lobine, D., Rengasamy, K.R.R., Mahomoodally, M.F., 2022. Functional foods and bioactive ingredients harnessed from the ocean: current status and future perspectives. Crit. Rev. Food Sci. Nutr. 15497852. 62 (21), 5794–5823. Available from: https://www.tandfonline.com/loi/bfsn20. doi:10.1080/10408398.2021.1893643.

López-Pérez, M., Haro-Moreno, J.M., Coutinho, F.H., Martinez-Garcia, M., Rodriguez-Valera, F., 2020. The evolutionary success of the marine bacterium SAR11 analyzed through a metagenomic perspective. mSystems 5 (5). doi:10.1128/mSystems.00605-20.

Luo, Q., Hou, D., Jiang, D., Chen, W., 2021. Bioremediation of marine oil spills by immobilized oil-degrading bacteria and nutrition emulsion. Biodegradation 32 (2), 165–177. doi:10.1007/s10532-021-09930-5.

Macdonald, A., 2021. High-Pressure Equipment for Use in the Laboratory, at Sea and at Depth. Springer Science and Business Media LLC, pp. 353–417. doi:10.1007/978-3-030-67587-5_13.

Maderna, A., Doroski, M., Subramanyam, C., Porte, A., Leverett, C.A., Vetelino, B.C., Chen, Z., Risley, H., Parris, K., Pandit, J., Varghese, A.H., Shanker, S., Song, C., Sukuru, S.C.K., Farley, K.A., Wagenaar, M.M., Shapiro, M.J., Musto, S., Lam, M.H., Loganzo, F., O'Donnell, C.J., 2014. Discovery of cytotoxic dolastatin 10 analogues with N-terminal modifications. J. Med. Chem. 15204804. 57 (24), 10527–10543. Available from: http://pubs.acs.org/jmc. doi:10.1021/jm501649k.

Madhulika, M.M., Deepti, M., Soibam Ngasotter, S.S.G., Varghese, T., 2025. Functional Feed Additives in Aquaculture to Improve Food Security. Springer Science and Business Media LLC, pp. 375–396. doi:10.1007/978-3-031-75830-0_21.

Maire, J., Ching, S.J.T.M., Damjanovic, K., Epstein, H.E., Judd, L.M., Blackall, L.L., van Oppen, M.J.H., 2024. Tissue-associated and vertically transmitted bacterial symbiont in the coral Pocillopora acuta. ISME J. 18 (1). doi:10.1093/ismejo/wrad027.

Manck, L.E., Park, J., Tully, B.J., Poire, A.M., Bundy, R.M., Dupont, C.L., Barbeau, K.A., 2022. Petrobactin, a siderophore produced by Alteromonas, mediates community iron acquisition in the global ocean. ISME J. 17517370. 16 (2), 358–369. Available from: http://www.nature.com/ismej/marketing/aims_scope.html. doi:10.1038/s41396-021-01065-y.

Manobala, T., Shukla, S.K., Rao, T.S., Kumar, M.D., 2019. Uranium sequestration by biofilm-forming bacteria isolated from marine sediment collected from Southern coastal region of India. Int. Biodeterior. Biodegrad. 09648305. 145. Available from: www.elsevier.com/inca/publications/store/4/0/5/8/9/9. doi:10.1016/j.ibiod.2019.104809.

Manochkumar, J., Cherukuri, A.K., Kumar, R.S., Almansour, A.I., Ramamoorthy, S., Efferth, T., 2023. A critical review of machine-learning for "multi-omics" marine metabolite datasets. Comput. Biol. Med. 18790534. 165. Available from: www.elsevier.com/locate/compbiomed. doi:10.1016/j.compbiomed.2023.107425.

Marietou, A., Kjeldsen, K.U., Glombitza, C., Jørgensen, B.B., 2022. Response to substrate limitation by a marine sulfate-reducing bacterium. ISME J. 17517370. 16 (1), 200–210. Available from: http://www.nature.com/ismej/marketing/aims_scope.html. doi:10.1038/s41396-021-01061-2.

Marzuki, I., Rosmiati, R., Mustafa, A., Sahabuddin, S., Tarunamulia, T., Susianingsih, E., Hendrajat, E.A., Sahrijanna, A., Muslimin, M., Ratnawati, E., Kamariah, K., Nisaa, K., Herlambang, S., Gunawan, S., Santi, I.S., Isnawan, B.H., Kaseng, E.S., Septiningsih, E., Asaf, R., Athirah, A., Basri, B., 2023. Potential Utilization of Bacterial Consortium of Symbionts Marine Sponges in Removing Polyaromatic Hydrocarbons and Heavy Metals, Review. Biology. 20797737. 12 (1). Available from: www.mdpi.com/journal/biology. doi:10.3390/biology12010086.

Maya, C., Shertukde, S., Liu, C., 2023. Flavour of Novel Food Proteins. Royal Society of Chemistry (RSC), pp. 234–274. doi:10.1039/9781839165047-00234.

Mayer, M., Swanson-Mungerson, P., Rodríguez, N., 2024. Taglialatela-Scafati, Marine Pharmacology in 2019–2021: marine Compounds with Antibacterial, Antidiabetic, Antifungal, Anti-Inflammatory, Antiprotozoal, Antituberculosis and Antiviral Activities; Affecting the Immune and Nervous Systems, and Other Miscellaneous Mechanisms of Action. Mar. Drugs 22 (7). doi:10.3390/md22070309.

Mazzella, V., Dell'Anno, A., Etxebarría, N., González-Gaya, B., Nuzzo, G., Fontana, A., Núñez-Pons, L., 2024. High microbiome and metabolome diversification in coexisting sponges with different bio-ecological traits. Commun. Biol. 2399-3642. 7 (1). doi:10.1038/s42003-024-06109-5.

McCowin, M.F., Feehery, C., Rouse, G.W., 2020. Spanning the depths or depth-restricted: three new species of Bathymodiolus (Bivalvia, Mytilidae) and a new record for the hydrothermal vent Bathymodiolus thermophilus at methane seeps along the Costa Rica margin. Deep-Sea Res. Part I Oceanogr. Res. Pap. 09670637. 164. Available from: www.elsevier.com/inca/publications/store/2/1/6. doi:10.1016/j.dsr.2020.103322.

McKee, L.S., Rosa, S.L.L., Westereng, B., Eijsink, V.G., Pope, P.B., Larsbrink, J., 2021. Polysaccharide degradation by the Bacteroidetes: mechanisms and nomenclature. Environ. Microbiol. Rep. 17582229. 13 (5), 559–581. Available from: http://onlinelibrary.wiley.com/journal/10.1111/(ISSN)1758-2229. doi:10.1111/1758-2229.12980.

Meena, K., Sharma, V., Manzoor, M., Aseri, G.K., Sohal, J.S., Singh, D., Khare, N., Sharma, D., 2020. Mineral-enriched yeast biomass: a promising mineral food and feed supplement New and Future Developments in Microbial Biotechnology and Bioengineering. In: Panwar, J.S., Gehlot, P., (Eds.), Recent Advances in Application of Fungi and Fungal Metabolites: Environmental and Industrial Aspects. Elsevier, India, 155–170. Available from https://www.elsevier.com/books/new-and-future-developments-in-microbial-biotechnology-and-bioengineering/singh/978-0-12-821007-9. doi:10.1016/B978-0-12-821007-9.00013-9.

Mendes, A.R., Spínola, M.P., Lordelo, M., Prates, J.A.M., 2024. Advances in Bioprocess Engineering for Optimising Chlorella vulgaris Fermentation: biotechnological Innovations and Applications. Foods. 23048158. 13 (24). Available from: www.mdpi.com/journal/foods. doi:10.3390/foods13244154.

Mesbah, N.M., 2022. Industrial Biotechnology Based on Enzymes From Extreme Environments. Front. Bioeng. Biotechnol. 22964185. 10. Available from: http://journal.frontiersin.org/journal/bioengineering-and-biotechnology#archive. doi:10.3389/fbioe.2022.870083.

Mgomi, F.C., Yang, Y.R., Cheng, G., Yang, Z.Q., 2023. Lactic acid bacteria biofilms and their antimicrobial potential against pathogenic microorganisms. Biofilm. 25902075. 5. Available from: www.journals.elsevier.com/biofilm. doi:10.1016/j.bioflm.2023.100118.

Minakshi, S.P., Kumari, H., Kumar, A., 2024. Natural Bioactive Products from Marine Fungi against Bacterial Infection Fungi Bioactive Metabolites: Integration of Pharmaceutical Applications. Springer Nature, India, pp. 241–259. Available from https://link.springer.com/book/10.1007/978-981-99-5696-8. doi:10.1007/978-981-99-5696-8_8.

Mingxing, Q., Zheng, C., Wenhui, W., Guangli, Y., Wang, P., 2022. Exopolysaccharides from Marine Microbes: source, Structure and Application. Mar. Drugs 20 (8). doi:10.3390/md20080512.

Mistry, A.N., Ganta, U., Chakrabarty, J., Dutta, S., 2019. A review on biological systems for CO2 sequestration: organisms and their pathways. Environ. Prog. Sustain. Energy. 19447450. 38 (1), 127–136. Available from: http://onlinelibrary.wiley.com/journal/10.1002/(ISSN)1944-7450. doi:10.1002/ep.12946.

Mohamed, A.R., Ochsenkühn, M.A., Kazlak, A.M., Moustafa, A., Amin, S.A., 2023. The coral microbiome: towards an understanding of the molecular mechanisms of coral-microbiota interactions. FEMS Microbiol. Rev. 15746976. 47 (2). Available from: https://academic.oup.com/femsre/pages/About. doi:10.1093/femsre/fuad005.

Molina, C., Expanding the chemical space of microbial specialized metabolites: structure elucidation and biosynthesis of novel bioactive natural products from actinomycetes. (2024), Available from: https://hdl.handle.net/10481/89254.

Muñoz-Marín, M.C., Gómez-Baena, G., López-Lozano, A., Moreno-Cabezuelo, J.A., Díez, J., García-Fernández, J.M., 2020. Mixotrophy in marine picocyanobacteria: use of organic compounds by Prochlorococcus and Synechococcus. ISME J. 1751-7362. 14 (5), 1065–1073. doi:10.1038/s41396-020-0603-9.

Nachimuthu, R., Royam, M.M., Manohar, P., Leptihn, S., 2021. Application of bacteriophages and endolysins in aquaculture as a biocontrol measure. Biol. Control. 10499644. 160 (104678). doi:10.1016/j.biocontrol.2021.104678.

Nair, S., Abraham, J., 2020. Natural Products from Actinobacteria for Drug Discovery Advances in Pharmaceutical Biotechnology: Recent Progress and Future Applications. Springer Singapore, India, pp. 333–363. Available from http://dx.doi.org/10.1007/978-981-15-2195-9. doi:10.1007/978-981-15-2195-9_23.

Naureen, Z., Dautaj, A., Anpilogov, K., Camilleri, G., Dhuli, K., Tanzi, B., Maltese, P.E., Cristofoli, F., Antoni, L.D., Beccari, T., Dundar, M., Bertelli, M., 2020. Bacteriophages presence in nature and their role in the natural selection of bacterial populations. Acta Biomed. 25316745. 91, 1–13. Available from: https://www.mattioli1885journals.com/index.php/actabiomedica/article/download/10819/9319. doi:10.23750/abm.v91i13-S.10819.

Nawaz, M.Z., Sasidharan, R.S., Alghamdi, H.A., Dang, H., 2022. Understanding Interaction Patterns within Deep-Sea Microbial Communities and Their Potential Applications. Mar. Drugs 20 (2). doi:10.3390/md20020108.

Nie, W.B., Ding, J., Xie, G.J., Tan, X., Lu, Y., Peng, L., Liu, B.F., Xing, D.F., Yuan, Z., Ren, N., 2021. Simultaneous nitrate and sulfate dependent anaerobic oxidation of methane linking carbon, nitrogen and sulfur cycles. Water Res. 18792448. 194, 116928. doi:10.1016/j.watres.2021.116928.

Nisha, P., Thangavel, M., Mohan, A., Jyothilekshmi, C., Raslana, T.A., Razeena, K.A., 2023. Biopolymers from Marine Prokaryotes Handbook of Biopolymers. Springer Nature, India, pp. 135–166. Available from https://doi.org/10.1007/978-981-19-0710-4. doi:10.1007/978-981-19-0710-4_5.

Niu, K.M., Khosravi, S., Kothari, D., Lee, W.D., Lim, J.M., Lee, B.J., Kim, K.W., Lim, S.G., Lee, S.M., Kim, S.K., 2019. Effects of dietary multi-strain probiotics supplementation in a low fishmeal diet on growth performance, nutrient utilization, proximate composition, immune parameters, and gut microbiota of juvenile olive flounder (Paralichthys olivaceus). Fish Shellfish Immunol. 10959947. 93, 258–268. Available from: http://www.elsevier.com/inca/publications/store/6/2/2/8/3/2/index.htt. doi:10.1016/j.fsi.2019.07.056.

Norris, K., Mishukova, O.I., Zykwinska, A., Colliec-Jouault, S., Sinquin, C., Koptioug, A., Cuenot, S., Kerns, J.G., Surmeneva, M.A., Surmenev, R.A., Douglas, T.E.L., 2019. Marine polysaccharide-collagen coatings on Ti6Al4V alloy formed by self-assembly. Micromachines. 2072666X. 10 (1). Available from: https://www.mdpi.com/2072-666X/10/1/68/pdf. doi:10.3390/mi10010068.

Nwuzor, I.C., Idumah, C.I., Nwanonenyi, S.C., Ezeani, O.E., 2021. Emerging trends in self-polishing anti-fouling coatings for marine environment. Saf. Extrem. Environ. 25248189. 3 (1), 9–25. Available from: https://www.springer.com/journal/42797. doi:10.1007/s42797-021-00031-3.

Oliveira, J., Belchior, A., da Silva, V.D., Rotter, A., Petrovski, Ž., Almeida, P.L., Lourenço, N.D., Gaudêncio, S.P., 2020. Marine Environmental Plastic Pollution: mitigation by Microorganism Degradation and Recycling Valorization. Front. Mar. Sci. 22967745. 7. Available from: https://www.frontiersin.org/journals/marine-science#. doi:10.3389/fmars.2020.567126.

Orlić, S., 2019. Microbial Diversity of Sponge/Coral Microbiome. Springer Science and Business Media LLC, pp. 29–41. doi:10.1007/978-94-024-1612-1_3.

Ostria-Hernández, M.L., Hernández-Zulueta, J., Vargas-Ponce, O., Díaz-Pérez, L., Araya, R., Rodríguez-Troncoso, A. P., Ríos-Jara, E., Rodríguez-Zaragoza, F.A., 2022. Core microbiome of corals Pocillopora damicornis and Pocillopora verrucosa in the northeastern tropical Pacific. Mar. Ecol. 14390485. 43 (6). Available from: http://onlinelibrary.wiley.com/journal/10.1111/(ISSN)1439-0485. doi:10.1111/maec.12729.

Ou, X., Chen, P., Liu, B.F., 2019. Single-cell analysis with microfluidic devices. Anal. Sci. 13482246. 35 (6), 609–618. Available from: https://www.jstage.jst.go.jp/article/analsci/35/6/35_19R001/_pdf/-char/en. doi:10.2116/analsci.19R001.

Pachiadaki, M.G., Brown, J.M., Brown, J., Bezuidt, O., Berube, P.M., Biller, S.J., Poulton, N.J., Burkart, M.D., Clair, J.J. L., Chisholm, S.W., Stepanauskas, R., 2019. Charting the Complexity of the Marine Microbiome through Single-Cell Genomics. Cell 179 (7). doi:10.1016/j.cell.2019.11.017.

Pal, P.P., Begum, A.S., Basha, S.A., Araya, H., Fujimoto, Y., 2023. New natural pro-inflammatory cytokines (TNF-α, IL-6 and IL-1β) and iNOS inhibitors identified from Penicillium polonicum through in vitro and in vivo studies. Int. Immunopharmacol. 117. 10.1016/j.intimp.2023.109940

Pandey, A., Nikam, A.N., Shreya, A.B., Mutalik, S.P., Gopalan, D., Kulkarni, S., Padya, B.S., Fernandes, G., Mutalik, S., Prassl, R., 2020. Potential therapeutic targets for combating SARS-CoV-2: drug repurposing, clinical trials and recent advancements. Life Sci. 256. 10.1016/j.lfs.2020.117883

Patel, A., Karageorgou, D., Katapodis, P., Sharma, A., Rova, U., Christakopoulos, P., Matsakas, L., 2021. Bioprospecting of thraustochytrids for omega-3 fatty acids: a sustainable approach to reduce dependency on animal sources. Trends Food Sci. Technol. 09242244. 115, 433–444. doi:10.1016/j.tifs.2021.06.044.

Patil, P.K., Nagaraju, V.T., Baskaran, V., Avunje, S., Rameshbabu, R., Ghate, S.D., Solanki, H.G., 2023. Development of microbial enrichments for simultaneous removal of sulfur and nitrogenous metabolites in saline water aquaculture. J. Appl. Microbiol. 13652672. 134 (8). Available from: https://academic.oup.com/jambio. doi:10.1093/jambio/lxad173.

Pedraza-Segura, L., Esparza, K.M.R., Pedroza-Islas, R., 2023. Microbial Bioresource for the Production of Marine Enzymes Microbial Bioreactors for Industrial Molecules. wiley, Mexico, pp. 17–37. Available from:

https://www.wiley.com/en-us/Microbial+Bioreactors+for+Industrial+Molecules-p-9781119874065. doi:10.1002/9781119874096.ch2.

Peoples, L.M., Grammatopoulou, E., Pombrol, M., Xu, X., Osuntokun, O., Blanton, J., Allen, E.E., Nunnally, C.C., Drazen, J.C., Mayor, D.J., Bartlett, D.H., 2019. Microbial community diversity within sediments from two geographically separated hadal trenches. Front. Microbiol. 1664302X. 10. Available from: https://www.frontiersin.org/journals/microbiology#. doi:10.3389/fmicb.2019.00347.

Poli, A., Bovio, E., Perugini, I., Varese, G.C., Prigione, V., 2021. Corollospora mediterranea: a Novel Species Complex in the Mediterranean Sea. Appl. Sci. 2076-3417. 11 (12), 5452. doi:10.3390/app11125452.

Pontiller, B., Martínez-García, S., Joglar, V., Amnebrink, D., Pérez-Martínez, C., González, J.M., Lundin, D., Fernández, E., Teira, E., Pinhassi, J., 2022. Rapid bacterioplankton transcription cascades regulate organic matter utilization during phytoplankton bloom progression in a coastal upwelling system. ISME J. 17517370. 16 (10), 2360–2372. Available from: http://www.nature.com/ismej/marketing/aims_scope.html. doi:10.1038/s41396-022-01273-0.

Poonam Vashist, R.K., Devasia, V.L.A., Shirodkar, P.V., Muraleedharan, U.D., Meena, S.N., Naik, M.M., 2019. Chapter 7 - Biotechnological Implications of Hydrolytic Enzymes from Marine Microbes. Academic Press, pp. 103–118. Available from https://www.sciencedirect.com/science/article/pii/B9780128174975000070. doi:10.1016/B978-0-12-817497-5.00007-0.

Potapov, A.M., Brose, U., Scheu, S., Tiunov, A.V., 2019. Trophic position of consumers and size structure of food webs across aquatic and terrestrial ecosystems. Am. Nat. 194 (6), 823–839. doi:10.1086/705811.

Prabhu, R.H., Patravale, V.B., 2020. Marine-derived Pharmaceuticals for Oncotherapy: Clinical Trial and Fda-approved Compounds Encyclopedia of Marine Biotechnology. wiley, India, pp. 2607–2618. Available from https://onlinelibrary.wiley.com/doi/book/10.1002/9781119143802. doi:10.1002/9781119143802.ch117.

Qiu, Y., Chen, S., Miao, Y., Shi, J., Liu, J., Xiaoyang, L., Chen, J., Sun, X., Huang, G., Zheng, C., 2024. Natural Products from Marine-Derived Fungi with Anti-Inflammatory Activity. Mar. Drugs. 1660-3397. 22 (10), 433. doi:10.3390/md22100433.

Quero, G.M., Celussi, M., Relitti, F., Kovačević, V., Negro, P.D., Luna, G.M., 2020. Inorganic and Organic Carbon Uptake Processes and Their Connection to Microbial Diversity in Meso- and Bathypelagic Arctic Waters (Eastern Fram Strait). Microb. Ecol. 1432184X. 79 (4), 823–839. Available from: http://link.springer.de/link/service/journals/00248/index.htm. doi:10.1007/s00248-019-01451-2.

Quitério, E., Soares, C., Ferraz, R., Delerue-Matos, C., Grosso, C., 2021. Marine Health-Promoting Compounds: recent Trends for Their Characterization and Human Applications. Foods. 2304-8158. 10 (12), 3100. doi:10.3390/foods10123100.

Rabiya, R., Sen, R., 2022. Artificial intelligence driven advanced optimization strategy vis-à-vis response surface optimization of production medium: bacterial exopolysaccharide production as a case-study. Biochem. Eng. J. 1873295X (178) Available from www.elsevier.com/locate/bej. doi:10.1016/j.bej.2021.108271.

Rahayu, S., Amoah, K., Huang, Y., Cai, J., Wang, B., Shija, V.M., Jin, X., Anokyewaa, M.A., Jiang, M., 2024. Probiotics application in aquaculture: its potential effects, current status in China and future prospects. Front. Mar. Sci. 22967745. 11. Available from: https://www.frontiersin.org/journals/marine-science#. doi:10.3389/fmars.2024.1455905.

Rahman, M.A., Ashrafudoulla, M., Akter, S., Park, S.H., Ha, S.D., 2024. Probiotics and biofilm interaction in aquaculture for sustainable food security: a review and bibliometric analysis. Crit. Rev. Food Sci. Nutr. 15497852. 64 (33), 12319–12335. Available from: https://www.tandfonline.com/loi/bfsn20. doi:10.1080/10408398.2023.2249114.

Ranawat, P., Rawat, S., 2020. Radiation-Resistant Thermophiles: From High Temperature and Radiation to Engineered Bioremediation. Informa UK Limited, pp. 287–318. doi:10.1201/9780429284137-10.

Rappé, M.S., Connon, S.A., Vergin, K.L., Giovannoni, S.J., 2002. Cultivation of the ubiquitous SAR11 marine bacterioplankton clade. Nature. 1476-4687. 418 (6898), 630–633. Available from: https://doi.org/10.1038/nature00917. doi:10.1038/nature00917.

Rezagholizade-shirvan, A., Soltani, M., Shokri, S., Radfar, R., Arab, M., Shamloo, E., 2024. Bioactive compound encapsulation: characteristics, applications in food systems, and implications for human health. Food Chem. X 25901575 (24), 101953. doi:10.1016/j.fochx.2024.101953.

Ribeiro, R., Pinto, E., Fernandes, C., Sousa, E., 2022. Marine Cyclic Peptides: antimicrobial Activity and Synthetic Strategies. Mar. Drugs. 1660-3397. 20 (6), 397. doi:10.3390/md20060397.

Riera-Romo, M., Wilson-Savón, L., Hernandez-Balmaseda, I., 2020. Metabolites from marine microorganisms in cancer, immunity, and inflammation: a critical review. J. Pharm. Pharmacogn. Res. 07194250. 8 (1), 368–391. doi:10.56499/jppres19.739_8.5.368.

Rincón-Tomás, B., Lanzén, A., Sánchez, P., Estupiñán, M., Isabel Sanz-Sáez, M.E.B., Rojo, D., Mendibil, I., Pérez-Cruz, C., Ferri, M., Capo, E., Abad-Recio, I.L., Amouroux, D., Bertilsson, S., Sánchez, O., Acinas, S.G., Alonso-Sáez, L., 2024. Revisiting the mercury cycle in marine sediments: a potential multifaceted role for Desulfobacterota. J. Hazard. Mater. 03043894. 465, 133120. doi:10.1016/j.jhazmat.2023.133120.

Robles-Bañuelos, B., Durán-Riveroll, L.M., Rangel-López, E., Pérez-López, H.I., González-Maya, L., 2022. Marine Cyanobacteria as Sources of Lead Anticancer Compounds: a Review of Families of Metabolites with Cytotoxic, Antiproliferative, and Antineoplastic Effects. Molecules. 14203049. 27 (15). Available from: http://www.mdpi.com/journal/molecules. doi:10.3390/molecules27154814.

Rodrigues, C.J.C., de Carvalho, C.C.C.R., 2022. Cultivating marine bacteria under laboratory conditions: overcoming the "unculturable" dogma. Front. Bioeng. Biotechnol. 2296-4185. 10. doi:10.3389/fbioe.2022.964589.

Rodríguez-Pupo, E.C., Pérez-Llano, Y., Tinoco-Valencia, J.R., Sánchez, N.S., Padilla-Garfias, F., Calahorra, M., Sánchez, N.D.C., Sánchez-Reyes, A., Rodríguez-Hernández, M.D.R., Peña, A., Sánchez, O., Aguirre, J., Batista-García, R.A., Folch-Mallol, J.L., Sánchez-Carbente, M.D.R., 2021. Osmolyte signatures for the protection of aspergillus sydowii cells under halophilic conditions and osmotic shock. J. Fungi. 2309608X. 7 (6). Available from: https://www.mdpi.com/2309-608X/7/6/414/pdf. doi:10.3390/jof7060414.

Rotter, A., Bacu, A., Barbier, M., Bertoni, F., Bones, A.M., Cancela, M.L., Carlsson, J., Carvalho, M.F., Cegłowska, M., Dalay, M.C., Dailianis, T., Deniz, I., Drakulovic, D., Dubnika, A., Einarsson, H., Erdoğan, A., Eroldoğan, O.T., Ezra, D., Fazi, S., FitzGerald, R.J., Gargan, L.M., Gaudêncio, S.P., Ivošević Denardis, N., Joksimovic, D., Kataržytė, M., Kotta, J., Mandalakis, M., Matijošytė, I., Mazur-Marzec, H., Massa-Gallucci, A., Mehiri, M., Nielsen, S.L., Novoveská, L., Overlingė, D., Portman, M.E., Pyrc, K., Rebours, C., Reinsch, T., Reyes, F., Rinkevich, B., Robbens, J., Rudovica, V., Sabotič, J., Safarik, I., Talve, S., Tasdemir, D., Schneider, X.T., Thomas, O.P., Toruńska-Sitarz, A., Varese, G.C., Vasquez, M.I., 2020. A New Network for the Advancement of Marine Biotechnology in Europe and Beyond. Front. Mar. Sci. 22967745. 7. Available from: https://www.frontiersin.org/journals/marine-science#. doi:10.3389/fmars.2020.00278.

Rubin-Blum, M., Antony, C.P., Sayavedra, L., Martínez-Pérez, C., Birgel, D., Peckmann, J., Wu, Y.C., Cardenas, P., MacDonald, I., Marcon, Y., Sahling, H., Hentschel, U., Dubilier, N., 2019. Fueled by methane: deep-sea sponges from asphalt seeps gain their nutrition from methane-oxidizing symbionts. ISME J. 17517370. 13 (5), 1209–1225. Available from: http://www.nature.com/ismej/marketing/aims_scope.html. doi:10.1038/s41396-019-0346-7.

Russ, L., Van Alen, T.A., Jetten, M.S.M., Camp, H.J.M.O.D., Kartal, B., 2019. Interactions of anaerobic ammonium oxidizers and sulfide-oxidizing bacteria in a substrate-limited model system mimicking the marine environment. FEMS Microbiol. Ecol. 15746941. 95 (9). Available from: http://femsec.oxfordjournals.org/. doi:10.1093/femsec/fiz137.

Saad, M.H., El-Fakharany, E.M., Salem, M.S., Sidkey, N.M., 2022. The use of cyanobacterial metabolites as natural medical and biotechnological tools: review article. J. Biomol. Struct. Dyn. 15380254. 40 (6), 2828–2850. Available from: http://www.tandfonline.com/loi/tbsd20. doi:10.1080/07391102.2020.1838948.

Sahana, T.G., Rekha, P.D., 2020. A novel exopolysaccharide from marine bacterium Pantoea sp. YU16-S3 accelerates cutaneous wound healing through Wnt/β-catenin pathway. Carbohydr. Polym. 01448617. 238, 116191. doi:10.1016/j.carbpol.2020.116191.

Saide, A., Lauritano, C., Ianora, A., 2021. A Treasure of Bioactive Compounds from the Deep Sea. Biomedicines. 2227-9059. 9 (11), 1556. doi:10.3390/biomedicines9111556.

Salazar-Alekseyeva, K., Herndl, G.J., Baltar, F., 2024. Influence of Salinity on the Extracellular Enzymatic Activities of Marine Pelagic Fungi. J. Fungi. 2309608X. 10 (2). Available from: www.mdpi.com/journal/jof. doi:10.3390/jof10020152.

Samantaray, M.U.S., Sahu, M.S., 2021. Characterization and Cellulase Production Activity of different Aspergillus Sp. Isolated from "Puri, Odisha" Marine Fungal Strains. Int. J. Res. Appl. Sci. Biotechnol. 2349-8889. 8 (3). doi:10.31033/ijrasb.8.3.11.

Sánchez-Suárez, J., Villamil, L., Díaz, L., Coy-Barrera, E., 2022. Uncovering Streptomyces-Derived Compounds as Cosmeceuticals for the Development of Improved Skin Photoprotection Products: an In Silico Approach to Explore Multi-Targeted Agents. Sci. Pharm. 2218-0532. 90 (3), 48. doi:10.3390/scipharm90030048.

Sánchez, O., Stefanni, S., 2024. Punyasloke Bhadury, The deep sea biodiversity and conservation collection. Sci. Rep. 14 (1). doi:10.1038/s41598-024-77742-7.

Sanz-Sáez, I., Contribution of marine heterotrophic cultured bacteria to microbial diversity and mercury detoxification. (2021), Available from: http://hdl.handle.net/10803/671617.

Satya, S., Sharma, S., Choudhary, G., Kaushik, G., 2024. Advances in Environmental Microbiology: A Multi-omic Perspective Microbial Omics in Environment and Health. Springer Nature, India, pp. 175–204. Available from https://link.springer.com/book/10.1007/978-981-97-1769-9. doi:10.1007/978-981-97-1769-9_7.

Scheffer, G., Gieg, L.M., 2023. The Mystery of Piezophiles: understudied Microorganisms from the Deep, Dark Subsurface. Microorganisms. 2076-2607. 11 (7), 1629. doi:10.3390/microorganisms11071629.

Schoina, E., Doulgeraki, A.I., Miliou, H., Nychas, G.-J.E., 2022. Dynamics of water and biofilm bacterial community composition in a mediterranean recirculation aquaculture system. Aquac. J. 2673-9496. 2 (2), 164–179. doi:10.3390/aquacj2020008.

Schultz, J., Modolon, F., Rosado, A.S., Voolstra, C.R., Sweet, M., Peixoto, R.S., 2022. Methods and strategies to uncover coral-associated microbial dark matter. mSystems. 23795077. 7 (4). Available from: https://journals.asm.org/doi/10.1128/msystems.00367-22. doi:10.1128/msystems.00367-22.

Schulz, P., Pajdak-Czaus, J., Siwicki, A.K., 2022. In vivo bacteriophages' application for the prevention and therapy of aquaculture animals—chosen aspects. Animals. 20762615. 12 (10). Available from: https://www.mdpi.com/2076-2615/12/10/1233/pdf?version=1652189256. doi:10.3390/ani12101233.

Scoma, A., Heyer, R., Rifai, R., Dandyk, C., Marshall, I., Kerckhof, F.M., Marietou, A., Boshker, H.T.S., Meysman, F.J. R., Malmos, K.G., Vosegaard, T., Vermeir, P., Banat, I.M., Benndorf, D., Boon, N., 2019. Reduced TCA cycle rates at high hydrostatic pressure hinder hydrocarbon degradation and obligate oil degraders in natural, deep-sea microbial communities. ISME J. 17517370. 13 (4), 1004–1018. Available from: http://www.nature.com/ismej/marketing/aims_scope.html. doi:10.1038/s41396-018-0324-5.

Seifan, M., Berenjian, A., 2019. Microbially induced calcium carbonate precipitation: a widespread phenomenon in the biological world. Appl. Microbiol. Biotechnol. 0175-7598. 103 (12), 4693–4708. doi:10.1007/s00253-019-09861-5.

Semedo, M., Lopes, E., Baptista, M.S., Oller-Ruiz, A., Gilabert, J., Tomasino, M.P., Magalhães, C., 2021. Depth profile of nitrifying archaeal and bacterial communities in the remote oligotrophic waters of the north pacific. Front. Microbiol. 1664302X. 12. Available from: https://www.frontiersin.org/journals/microbiology#. doi:10.3389/fmicb.2021.624071.

Semwal, P., Painuli, S., Anand, J., Martins, N.C., Machado, M., Sharma, R., Batiha, G.E.S., Yaro, C.A., Lorenzo, J.M., Rahman, M.M., 2022. The Neuroprotective Potential of Endophytic Fungi and Proposed Molecular Mechanism: a Current Update. Evid. Based Complement. Alternat. Med. 17414288. 2022. Available from: http://www.hindawi.com/journals/ecam/contents.html. doi:10.1155/2022/6214264.

Shanmugam, S., Ngo, H.H., Wu, Y.R., 2020. Advanced CRISPR/Cas-based genome editing tools for microbial biofuels production: a review. Renew. Energy. 18790682. 149, 1107–1119. Available from: http://www.journals.elsevier.com/renewable-and-sustainable-energy-reviews/. doi:10.1016/j.renene.2019.10.107.

Shi, J., Miao, Y., Chen, W., Chen, S., Qiu, Y., Zhenyang, X., Wang, Y., Huang, G., Zheng, C., 2024. Recent Discovery of Nitrogen Heterocycles from Marine-Derived Aspergillus Species. Mar. Drugs. 1660-3397. 22 (7), 321. doi:10.3390/md22070321.

Shijie, W., Kaiqiang, Y., Wang, F., Tiezhu, M., Zhen, Y., 2024. Impact of low pH/high pCO2 on the physiological response and exopolysaccharide content in cyanobacteria Trichodesmium erythraeum. Aquat. Ecol. 1386-2588. 58 (4), 1225–1236. doi:10.1007/s10452-024-10137-y.

Shimabukuro, M., Alfaro-Lucas, J.M., Bernardino, A.F., Ramos, R.B., de Mahiques, M.M., Sumida, P.Y.G., 2020. Chemosynthetic Ecosystems on the Brazilian Deep-Sea Margin. Springer Science and Business Media LLC, pp. 109–132. doi:10.1007/978-3-030-53222-2_5.

Shinde, P., Banerjee, P., Mandhare, A., 2019. Marine natural products as source of new drugs: a patent review (2015–2018). Expert Opin. Ther. Pat. 1354-3776. 29 (4), 283–309. doi:10.1080/13543776.2019.1598972.

Shukla, P.J., Vhora, S.B., Murnal, A.G., Yagnik, U.B., Patadiya, M., 2022. Exopolysaccharide Production from Marine Bacteria and Its Applications. Informa UK Limited, pp. 337–368. doi:10.1201/9781003303909-18.

Siddik, M.A.B., Julien, B.B., Islam, S.M.M., Francis, D.S., 2024. Fermentation in aquafeed processing: achieving sustainability in feeds for global aquaculture production. Rev. Aquac. 17535131. 16 (3), 1244–1265. Available from: http://onlinelibrary.wiley.com/journal/10.1111/(ISSN)1753-5131. doi:10.1111/raq.12894.

Silva, M., Seijas, P., Otero, P., 2021. Exploitation of marine molecules to manage alzheimer's disease. Mar. Drugs. 1660-3397. 19 (7), 373. doi:10.3390/md19070373.

Singhania, R.R., Patel, A.K., Tsai, M.L., Chen, C.W., Di Dong, C., 2021. Genetic modification for enhancing bacterial cellulose production and its applications. Bioengineered. 21655987. 12 (1), 6793–6807. Available from: http://www.tandfonline.com/loi/kbie20. doi:10.1080/21655979.2021.1968989.

Song, Y., Li, Q., Qin, F., Sun, C., Liang, H., Wei, X., Wong, N.K., Ye, L., Zhang, Y., Shao, M., Ju, J., 2017. Neoabyssomicins A–C, polycyclic macrolactones from the deep-sea derived Streptomyces koyangensis SCSIO 5802. Tetrahedron. 14645416. 73 (36), 5366–5372. Available from: http://www.journals.elsevier.com/tetrahedron/. doi:10.1016/j.tet.2017.07.034.

Soulier, N.T., Light-harvesting proteins and complexes in cyanobacteria acclimated to far-red and low-light conditions. (2021).

Speckmann, B., Kleinbölting, J., Börner, F., Jordan, P.M., Werz, O., Pelzer, S., Tom Dieck, H., Wagner, T., Schön, C., 2022. Synbiotic Compositions of Bacillus megaterium and Polyunsaturated Fatty Acid Salt Enable Self-Sufficient Production of Specialized Pro-Resolving Mediators. Nutrients. 2072-6643. 14 (11), 2265. doi:10.3390/nu14112265.

Srinivasan, R., Kannappan, A., Shi, C., Lin, X., 2021. Marine bacterial secondary metabolites: a treasure house for structurally unique and effective antimicrobial compounds. Mar. Drugs 19 (10). doi:10.3390/md19100530.

Srivastava, R., Singh, N., Kanda, T., Yadav, S., Yadav, S., Atri, N., 2024. Antimicrobial Activity of Plant and Cyanobacterial Secondary Metabolites for Therapeutic Advancements in Modern Medicine. Springer Science and Business Media LLC, pp. 607–620. doi:10.1007/978-981-97-5272-0_28.

Steen, A.D., Kusch, S., Abdulla, H.A., Cakić, N., Coffinet, S., Dittmar, T., Fulton, J.M., Galy, V., Hinrichs, K.U., Ingalls, A.E., Koch, B.P., Kujawinski, E., Liu, Z., Osterholz, H., Rush, D., Seidel, M., Sepúlveda, J., Wakeham, S.G., 2020. Analytical and computational advances, opportunities, and challenges in marine organic biogeochemistry in an era of "Omics". Front. Mar. Sci. 22967745. 7. Available from: https://www.frontiersin.org/journals/marine-science#. doi:10.3389/fmars.2020.00718.

Stincone, P., Brandelli, A., 2020. Marine bacteria as source of antimicrobial compounds. Crit. Rev. Biotechnol. 0738-8551. 40 (3), 306–319. doi:10.1080/07388551.2019.1710457.

Stonik, V.A., Makarieva, T.N., Shubina, L.K., 2020. Antibiotics from Marine Bacteria. Biochemistry (Mosc. 0006-2979. 85 (11), 1362–1373. doi:10.1134/s0006297920110073.

Subramani, R., Sipkema, D., 2019. Marine rare actinomycetes: a promising source of structurally diverse and unique novel natural products. Mar. Drugs 17 (5). doi:10.3390/md17050249.

Sukmarini, L., Atikana, A., Hertiani, T., 2024. Antibiofilm activity of marine microbial natural products: potential peptide- and polyketide-derived molecules from marine microbes toward targeting biofilm-forming pathogens. J. Nat. Med. 1340-3443. 78 (1), 1–20. doi:10.1007/s11418-023-01754-2.

Sun, Y., Dong, Y., Cui, X., Guo, X., Zhang, J., Chong, Y., Zhang, M., Wang, H., 2024. Effects of marine natural products on liver diseases. Mar. Drugs. 1660-3397. 22 (7), 288. doi:10.3390/md22070288.

Suominen, S., van Vliet, D.M., Sánchez-Andrea, I., van der Meer, M.T.J., Damsté, J.S.S., Villanueva, L., 2021. Organic matter type defines the composition of active microbial communities originating from anoxic baltic sea sediments. Front. Microbiol. 1664302X. 12. Available from: https://www.frontiersin.org/journals/microbiology#. doi:10.3389/fmicb.2021.628301.

Suresh Kunjiraman, I.S.B.S., Sarasan, M., Puthumana, J., 2024. Immobilized microbial consortia: an eco-friendly and sustainable solution for aquaculture waste management. The Microbe. 29501946. 4, 100100. doi:10.1016/j.microb.2024.100100.

Sweeney, D., Chase, A.B., Bogdanov, A., Jensen, P.R., 2024. MAR4 Streptomyces: a Unique Resource for Natural Product Discovery. J. Nat. Prod. 15206025. 87 (2), 439–452. Available from: http://pubs.acs.org/journal/jnprdf. doi:10.1021/acs.jnatprod.3c01007.

Tadda, M.A., Li, C., Gouda, M., Abomohra, A.E.F., Shitu, A., Ahsan, A., Zhu, S., Liu, D., 2021. Enhancement of nitrite/ammonia removal from saline recirculating aquaculture wastewater system using moving bed bioreactor. J. Environ. Chem. Eng. 22133437. 9 (5). Available from: http://www.journals.elsevier.com/journal-of-environmental-chemical-engineering/. doi:10.1016/j.jece.2021.105947.

Tamby, A., Damsté, J.S.S., Villanueva, L., 2023. Microbial membrane lipid adaptations to high hydrostatic pressure in the marine environment. Front. Mol. Biosci. 2296889X. 9. Available from: http://journal.frontiersin.org/journal/molecular-biosciences. doi:10.3389/fmolb.2022.1058381.

Tilki, E.K., Engür Öztürk, S., Özarda, M.G., Cantürk, Z., Dikmen, M., 2021. Investigation of the neuroprotective and neuritogenic effects of halotolerant Penicillium flavigenum-derived sorbicillin-like compounds on PC-12 Adh cells. Cytotechnology. 15730778. 73 (6), 801–813. Available from: http://www.springer.com/chemistry/biotech/journal/10616. doi:10.1007/s10616-021-00498-9.

Trevi, S., Webster, T.U., Consuegra, S., de Leaniz, C.G., 2023. Benefits of the microalgae Spirulina and Schizochytrium in fish nutrition: a meta-analysis. Sci. Rep. 2045-2322. 13 (1). doi:10.1038/s41598-023-29183-x.

Turnham, K.E., Aschaffenburg, M.D., Pettay, D.T., Paz-García, D.A., Reyes-Bonilla, H., Pinzón, J., Timmins, E., Smith, R.T., McGinley, M.P., Warner, M.E., Lajeunesse, T.C., 2003. High physiological function for corals with thermally tolerant, host-adapted symbionts. Proc. R. Soc. B Biol. Sci. 290. 10.1098/rspb.2023.1021

Ullah, S., Khalil, A.A., Shaukat, F., Song, Y., 2019. Sources, extraction and biomedical properties of polysaccharides. Foods. 2304-8158. 8 (8), 304. doi:10.3390/foods8080304.

Ulrich, J.F., Gräfe, M.S., Dhiman, S., Wienecke, P., Arndt, H.D., Wichard, T., 2022. Thallusin quantification in marine bacteria and algae cultures. Mar. Drugs. 16603397. 20 (11). Available from: http://www.mdpi.com/journal/marinedrugs. doi:10.3390/md20110690.

Upadhyay, D., Shukla, K., Mishra, A., Jindal, T., Sharma, S., Shukla, S., 2021. Molecular Aspects and Oxygen Relations of Nitrogen Fixation in Cyanobacteria. Springer International Publishing, Cham, pp. 521–568. Available from https://doi.org/10.1007/978-3-030-71206-8_26. doi:10.1007/978-3-030-71206-8_26.

Uria, A.R., Wakimoto, T., 2021. Polyketides Biosynthesis in Marine Sponges of the Family Theonellidae. Springer Science and Business Media LLC, pp. 389–413. doi:10.1007/7081_2020_48.

Vala, A.K., Dave, B.P., 2018. Marine-Derived Fungi: Prospective Candidates for Bioremediation. Springer Science and Business Media LLC, pp. 17–37. doi:10.1007/978-3-319-68957-9_2.

Valdés, A., Álvarez-Rivera, G., Socas-Rodríguez, B., Herrero, M., Ibáñez, E., Cifuentes, A., 2022. Foodomics: analytical Opportunities and Challenges. Anal. Chem. 15206882. 94 (1), 366–381. Available from: http://pubs.acs.org/journal/ancham. doi:10.1021/acs.analchem.1c04678.

Valle, I.D., Fulk, E.M., Kalvapalle, P., Silberg, J.J., Masiello, C.A., Stadler, L.B., 2021. Translating New Synthetic Biology Advances for Biosensing Into the Earth and Environmental Sciences. Front. Microbiol. 1664302X. 11. Available from: https://www.frontiersin.org/journals/microbiology#. doi:10.3389/fmicb.2020.618373.

Vannier, T., Hingamp, P., Turrel, F., Tanet, L., Lescot, M., Timsit, Y., 2020. Diversity and evolution of bacterial bioluminescence genes in the global ocean. NAR Genom. Bioinform. 2631-9268. 2 (2). doi:10.1093/nargab/lqaa018.

Vartoukian, S.R., Palmer, R.M., Wade, W.G., 2010. Strategies for culture of 'unculturable' bacteria. FEMS Microbiol. Lett. 15746968. 309 (1), 1–7. Available from: http://www3.interscience.wiley.com/journal/118512081/home. doi:10.1111/j.1574-6968.2010.02000.x.

Veaudor, T., Blanc-Garin, V., Chenebault, C., Diaz-Santos, E., Sassi, J.-F., Cassier-Chauvat, C., Chauvat, F., 2020. Recent advances in the photoautotrophic metabolism of cyanobacteria: biotechnological implications. Life 10 (5). doi:10.3390/life10050071.

Verma, A., Rai, N., Kamble, S.C., Mishra, P., Barik, S., Kumar, R., Singh, S.K., Salvi, P., Gautam, V., 2023. Beyond the Synthetic Drugs: Fungal Endophytes Derived Bioactive Compounds in the Management of Neurodegenerative Disorders Traditional Medicine for Neuronal Health. Bentham Science Publishers, India, pp. 82–102. Available from https://benthambooks.com/book/9789815040197/. doi:10.2174/9789815040197123010007.

Vipindas, P.V., Krishnan, K.P., Rehitha, T.V., Jabir, T., Dinesh, S.L., 2020. Diversity of sediment associated Planctomycetes and its related phyla with special reference to anammox bacterial community in a high Arctic fjord. World J. Microbiol. Biotechnol. 0959-3993. 36 (7). doi:10.1007/s11274-020-02886-3.

Vitale, G.A., Coppola, D., Esposito, F.P., Buonocore, C., Ausuri, J., Tortorella, E., de Pascale, D., 2020. Antioxidant molecules from marine fungi: methodologies and perspectives. Antioxidants. 2076-3921. 9 (12), 1183. doi:10.3390/antiox9121183.

Vogel, A.L., Concentration-dependent PAH degradation and gene expression in marine Cycloclasticus spp Universität Tübingen]. (2024), Available from: 10.15496/publikation-97616.

Vujović, T., Paradžik, T., Babić Brčić, S., Piva, R., 2025. Unlocking the therapeutic potential of algae-derived compounds in hematological malignancies. Cancers. 2072-6694. 17 (2), 318. doi:10.3390/cancers17020318.

Wakimoto, T., 2023. Biosynthesis of Bioactive Natural Products Derived from Theonellidae Family Marine Sponges. Chem. Pharm. Bull. 0009-2363. 71 (1), 1–8. doi:10.1248/cpb.c22-00715.

Wang, C., Jiang, C., Gao, T., Peng, X., Shuanglong, M., Sun, Q., Xia, B., Xie, X., Bai, Z., Shengjun, X., Zhuang, X., 2022. Improvement of fish production and water quality in a recirculating aquaculture pond enhanced with bacteria-microalgae association. Aquaculture. 00448486. 547, 737420. doi:10.1016/j.aquaculture.2021.737420.

Wang, F., Guo, S., Liang, J., Sun, X., 2023. Water column stratification governs picophytoplankton community structure in the oligotrophic eastern Indian ocean. Mar. Environ. Res. 01411136. 189, 106074. doi:10.1016/j.marenvres.2023.106074.

Wang, H., Zhang, Y., Bartlett, D.H., Xiao, X., 2021. Transcriptomic analysis reveals common adaptation mechanisms under different stresses for moderately piezophilic bacteria. Microb. Ecol. 1432184X. 81 (3), 617–629. Available from: http://link.springer.de/link/service/journals/00248/index.htm. doi:10.1007/s00248-020-01609-3.

Wang, J., Xie, J., Mei, J., 2025. Research Progress Regarding Psychrotrophic Pseudomonas in Aquatic Products: psychrophilic Characteristics, Spoilage Mechanisms, Detection Methods, and Control Strategies. Foods. 2304-8158. 14 (3), 363. doi:10.3390/foods14030363.

Wang, Q., Cui, R., Liu, X., Zheng, X., Yao, Y., Zhao, G., 2024. Examining the impact of Tetragenococcus halophilus, Zygosaccharomyces rouxii, and Starmerella etchellsii on the quality of soy sauce: a comprehensive review of microbial population dynamics in fermentation. Crit. Rev. Food Sci. Nutr. 1040-8398. 64 (29), 10873–10884. doi:10.1080/10408398.2023.2230285.

Wanying, H., Cai, R., Shichuan, X., Yin, Z., Zengfeng, D., Luan, Z., Sun, C., Zhang, X., Garg, N., 2023. Study of Microbial Sulfur Metabolism in a Near Real-Time Pathway through Confocal Raman Quantitative 3D Imaging. Microbiol. Spectr. 2165-0497. 11 (2). doi:10.1128/spectrum.03678-22.

Wei, H., Zhang, H., Lin, X., Liu, R., Bartlam, M., 2022a. Yingying Wang, Characteristics, Biodiversity, and Cultivation Strategy of Low Nucleic Acid Content Bacteria. Front. Microbiol. 1664-302X. 13. doi:10.3389/fmicb.2022.900669.

Wei, L., Huang, X., Xunzheng, L., Jiang, B., Liu, C., Huang, Y., Youlu, S., 2022b. Effects of dietary Lactobacillus reuteri on growth performance, nutrient retention, gut health and microbiota of the Nile tilapia (Oreochromis niloticus). Aquac. Rep. 23525134. 26, 101275. doi:10.1016/j.aqrep.2022.101275.

Wei, Y., Liu, T., Zhang, W., Zhang, W., 2021. The toxic mechanism of gliotoxins and biosynthetic strategies for toxicity prevention. Int. J. Mol. Sci. 1422-0067. 22 (24), 13510. doi:10.3390/ijms222413510.

Wenhao, L., Wang, F., Jiang, S., Pan, B., Liu, Q., Qianghua, X., 2022. Morphological and molecular evolution of hadal amphipod's eggs provides insights into embryogenesis under high hydrostatic pressure. Front. Cell Dev. Biol. 2296-634X. 10. doi:10.3389/fcell.2022.987409.

Wessel, A.K., Hmelo, L., Parsek, M.R., Whiteley, M., 2013. Going local: technologies for exploring bacterial microenvironments. Nat. Rev. Microbiol. 17401534. 11 (5), 337–348. doi:10.1038/nrmicro3010.

Wichard, T., 2023. From model organism to application: bacteria-induced growth and development of the green seaweed Ulva and the potential of microbe leveraging in algal aquaculture. Semin. Cell Dev. Biol. 134, 69–78. doi:10.1016/j.semcdb.2022.04.007.

Woo, S.-G., Averesch, N.J.H., Berliner, A.J., Deutzmann, J.S., Pane, V.E., Chatterjee, S., Criddle, C.S., Spear, J.R., 2024. Isolation and characterization of a Halomonas species for non-axenic growth-associated production of biopolyesters from sustainable feedstocks. Appl. Environ. Microbiol. 0099-2240. 90 (8). doi:10.1128/aem.00603-24.

Xinyang, L., Cheng, X., Junjie, X., Jiajun, W., Chan, L.L., Cai, Z., Zhou, J., 2024. Dynamic patterns of carbohydrate metabolism genes in bacterioplankton during marine algal blooms. Microbiol. Res. 09445013. 286, 127785. doi:10.1016/j.micres.2024.127785.

Xionglin, L., Yuan, L., Huang, K., Chen, S., Zhou, S., Rongrong, M., Jiang, J., Xie, J., 2024. Isolation and characterization of a bacteriophage for biological control of Photobacterium damselae subsp. damselae in silver pomfret mariculture. Aquaculture. 00448486. 582, 740539. doi:10.1016/j.aquaculture.2023.740539.

Xu, H., Peng, X., Bai, S., Ta, K., Yang, S., Liu, S., Jang, H.B., Guo, Z., 2019. Precipitation of calcium carbonate mineral induced by viral lysis of cyanobacteria: evidence from laboratory experiments. Biogeosciences. 17264189. 16 (4), 949–960. Available from: http://www.biogeosciences.net/volumes_and_issues.html. doi:10.5194/bg-16-949-2019.

Yadav, K.K., Patil, P.B., Kumaraswamy, H.H., Kashyap, B.K., 2020. Ligninolytic Microbes and Their Role in Effluent Management of Pulp and Paper Industry. Springer Singapore, Singapore, pp. 309–350. Available from https://doi.org/10.1007/978-981-33-4347-4_13. doi:10.1007/978-981-33-4347-4_13.

Yan, W., Feng, X., Zhang, W., Zhang, R., Jiao, N., 2020. Research advances on ecotype and sub-ecotype differentiation of Prochlorococcus and its environmental adaptability. Sci. China Earth Sci. 1674-7313. 63 (11), 1691–1700. doi:10.1007/s11430-020-9651-0.

Yao, H., Liu, S., Liu, T., Ren, D., Zhou, Z., Yang, Q., Mao, J., 2023. Microbial-derived salt-tolerant proteases and their applications in high-salt traditional soybean fermented foods: a review. Bioresour. Bioprocess. 2197-4365. 10 (1). doi:10.1186/s40643-023-00704-w.

Yinnan, M., Bian, C., Liu, R., Wang, Y., Shao, G., Jia, L., Qiu, Y., Tianliang, H., Wanru, L., Jingqun, A., Shi, Q., Chen, X., 2021. Whole genome sequencing of a snailfish from the Yap Trench (~7,000 m) clarifies the molecular mechanisms underlying adaptation to the deep sea. PLoS Genet. 17 (5). doi:10.1371/journal.pgen.1009530.

Young, J.N., Schmidt, K., 2020. It's what's inside that matters: physiological adaptations of high-latitude marine microalgae to environmental change. New Phytol. 14698137. 227 (5), 1307–1318. Available from: http://onlinelibrary.wiley.com/journal/10.1111/(ISSN)1469-8137. doi:10.1111/nph.16648.

Yuru, L., Guangjie, L., Feng, J., Songlin, L., Liu, N., Advances in Research on Marine Natural Products for Modulating the Inflammatory Microenvironment. Phytotherapy Research. 0951–418X. (2025), doi: 10.1002/ptr.8418.

Yusof, N.A., Hashim, N.H.F., Bharudin, I., 2021. Cold adaptation strategies and the potential of psychrophilic enzymes from the antarctic yeast, glaciozyma Antarctica pi12. J. Fungi 7 (7). doi:10.3390/jof7070528.

Zeng, X., Alain, K., Shao, Z., 2021. Microorganisms from deep-sea hydrothermal vents. Mar. Life Sci. Technol. 3 (2), 204–230. doi:10.1007/s42995-020-00086-4.

Zhang, L., Zhang, Z., Huang, J., Zhou, R., Chongde, W., 2024. Co-culture of Tetragenococcus halophilus and Zygosaccharomyces rouxii to improve microbiota and metabolites in secondary fortified fermented soy sauce. Food Biosci. 22124292. 61, 104850. doi:10.1016/j.fbio.2024.104850.

Zhu, J., Tang, S., Cheng, K., Cai, Z., Chen, G., Zhou, J., 2023. Microbial community composition and metabolic potential during a succession of algal blooms from Skeletonema sp. to Phaeocystis sp. Front. Microbiol. 1664-302X. 14. doi:10.3389/fmicb.2023.1147187.

Zhu, J., Wakisaka, M., Omura, T., Yang, Z., Yin, Y., Fang, W., 2024. Advances in industrial harvesting techniques for edible microalgae: recent insights into sustainable, efficient methods and future directions. J. Clean. Prod. 09596526. 436, 140626. doi:10.1016/j.jclepro.2024.140626.

Zoccarato, L., Grossart, H.P., 2019. Relationship between Lifestyle and Structure of Bacterial Communities and Their Functionality in Aquatic Systems. Springer International Publishing, Cham, pp. 13–52. Available from https://doi.org/10.1007/978-3-030-16775-2_2. doi:10.1007/978-3-030-16775-2_2.

CHAPTER

4

Applications of marine mollusks and echinoderms

H.M. Sarath M. Wijerathna[a] and Chathuranga Kiramage[b,c]

[a]Department of Aquaculture and Seafood Technology, Faculty of Fisheries and Ocean Sciences, Ocean University of Sri Lanka, Colombo, Sri Lanka

[b]Faculty of Health, School of Medicine, Deakin University, Waurn Ponds, VIC, Australia

[c]Institute for Mental and Physical Health and Clinical Translation (IMPACT), Deakin University, Geelonng, VIC, Australia

4.1 Introduction to marine mollusks and echinoderms

Marine mollusks belong to the phylum Mollusca, an invertebrate group, and this phylum is considered the second-largest animal phylum. There are approximately 100,000–200,000 species within this phylum, with over 52,000 species having been investigated and classified (Cuezzo et al., 2020). The term *mollusks* is derived from the Latin word meaning "soft" (Ehrlich, 2019). Although some mollusks possess a shell, they are fundamentally soft-bodied animals (Salvador, 2022). The phylum Mollusca includes seven classes: Aplacophora, Monoplacophora, Polyplacophora, Bivalvia, Gastropoda, Scaphopoda, and Cephalopoda (Hodgson, 2018) (Fig. 4.1). Common, well-known species, such as clams, mussels, snails, squids, and octopuses, belong to these classes. The body of mollusks can be divided into three distinct regions: the head, which contains various sensory structures; the muscular foot; and the visceral mass, which is covered by the mantle or pallium, a thick epidermal layer. The mantle is responsible for secreting the shell (calcareous skeleton) in shelled mollusks (Cuezzo et al., 2020). Mollusks are commonly found in aquatic habitats, with some species also present in terrestrial environments (Prié, 2019). Marine mollusks are distributed from shallow waters to the deep sea. Most species inhabit either hard substrates or sediments in the marine environment (Guerra-García et al., 2021). Animals in the phylum Mollusca exhibit various feeding patterns, including filter feeding (Bivalvia), grazing and herbivory (Gastropoda), carnivory (Cephalopoda and some Gastropoda), and scavenging (some Gastropoda and Bivalvia) (Zhong and Wang, 2024; Gündoğdu et al., 2024). Due to the diversity and unique biological functions of mollusks, scientists have investigated their potential uses as food, for ecological

Harvesting the Sea
https://doi.org/10.1016/B978-0-443-29190-6.00006-4

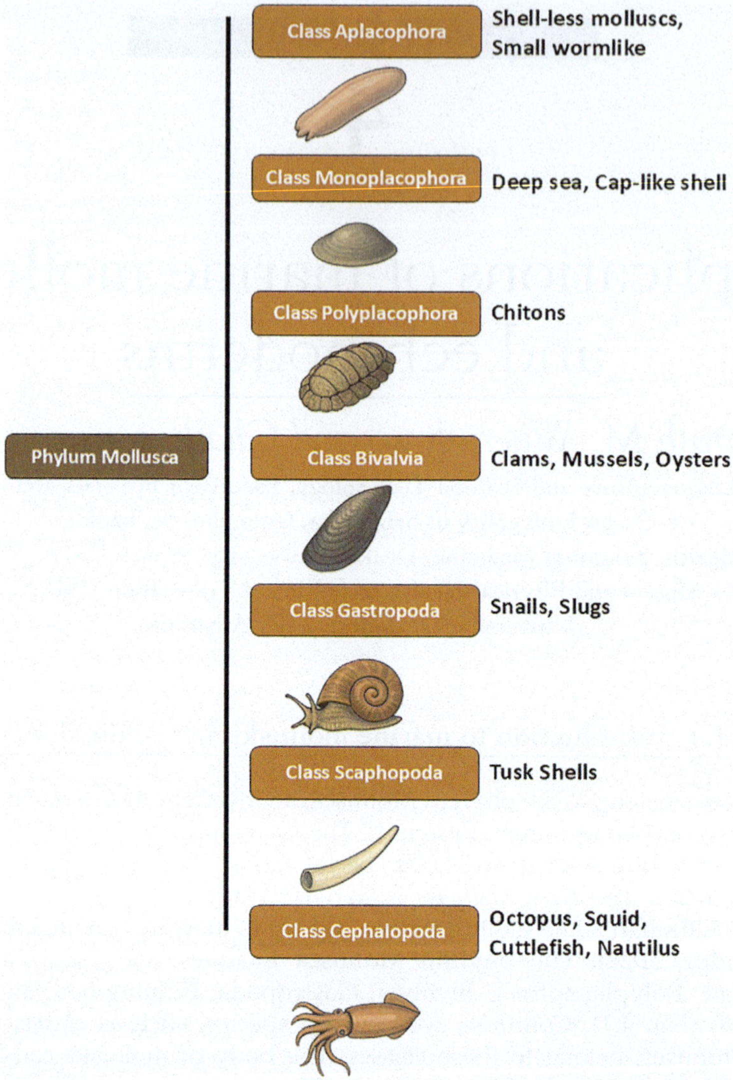

FIG. 4.1 Classification of phylum Mollusca.

functions, and as a source of various bioactive compounds (Ngandjui et al., 2024; Eghianruwa et al., 2019; Chen et al., 2022).

Echinoderms, which belong to the phylum Echinodermata, are widely distributed across all benthic marine habitats. Approximately 7000 species have been identified, belonging to five classes, including Asteroidea (starfish, seastars, and sea daisies), Ophiuroidea (brittle stars, basket stars, and snake stars), Holothuroidea (sea cucumbers), Crinoidea (feather stars and sea lilies), and Echinoidea (sea urchins, sand dollars, and heart urchins) (Doll et al., 2022)

(Fig. 4.2). Sea cucumbers (class Holothuroidea) and sea urchins (class Echinoidea) are examples of echinoderms that are economically important in commercial fisheries around the world (Baker-Médard and Ohl, 2019). Moreover, due to their unique biology, echinoderms are valuable creatures in the scientific field, including marine ecology, biology, and biotechnology. They support regional and international economies by being gathered for food, traditional medicine, and other purposes (Rahman et al., 2022). For instance, the Japanese harvest sea urchins for their roe to prepare their traditional cuisine, "Uni." In addition, echinoderms' shells are used in the production of jewelry and other decorative items.

Echinoderms are being investigated in biotechnology research due to their unique regenerative ability. For example, sea cucumbers and starfish have the ability to regenerate lost limbs or even entire bodies (Saleuddin et al., 2024). Investigating the mechanisms behind this regeneration could open up significant opportunities for utilizing these processes in human tissue regeneration and wound healing.

Investigation of marine mollusks and echinoderms is essential for understanding the functioning of marine ecosystems, as well as their economic applications, ecological significance, and scientific value, highlighting their importance in both natural and human contexts.

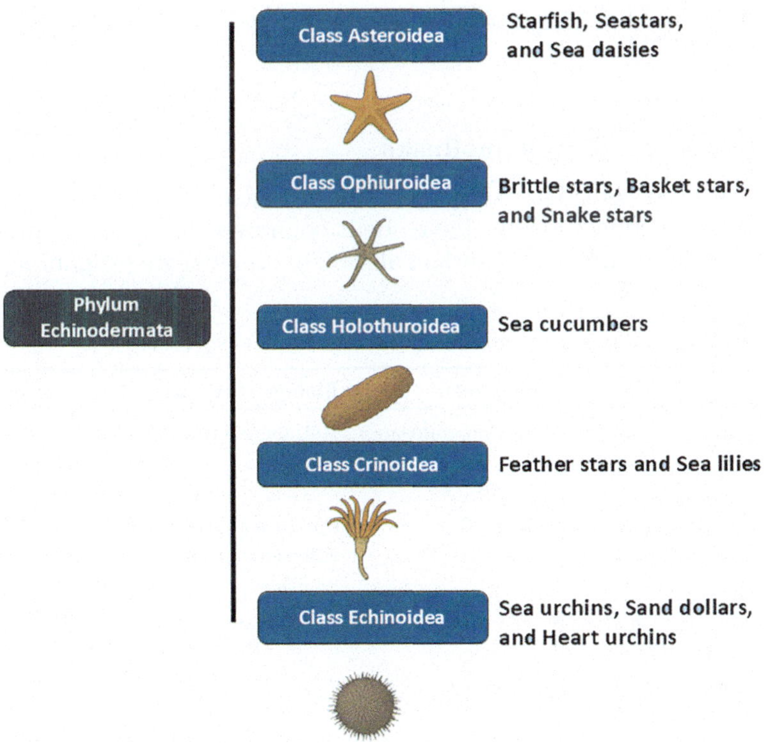

FIG. 4.2 Classification of phylum Echinodermata.

4.2 Bioactive compounds from marine mollusks and their applications

Mollusks lack adaptive immunity, but they possess a complex and effective innate immune system with humoral and cellular defenses similar to those in vertebrates. Although mollusks do not produce antibodies or exhibit antigen specificity, they can produce thousands of proteins through a limited number of genes, acting as effector molecules with high diversity to recognize and eliminate a variety of pathogens and respond to environmental stressors. Furthermore, mollusks produce bioactive substances with antimicrobial, antifungal, antiviral, antiprotozoal, anticancer, and antihelmintic activities (Canesi et al., 2022; Guo and Ford, 2016). In addition, some studies have demonstrated that mollusks produce secondary metabolites that can be applied in various pharmaceutical fields (Ramasamy et al., 2019). Some investigations have shown that lipid extractions from mollusks contain sterols, vitamin E, fatty acids (FAs), and polyunsaturated fatty acids (PUFAs), which exhibit effective wound healing properties and are rich in antioxidant biomolecules (Ngandjui et al., 2024; Aouji et al., 2023). Bivalve shell waste accumulation causes marine pollution due to inappropriate disposal and recycling issues. However, using shell waste as a raw material for the extraction of bioactive compounds would benefit both the pharmaceutical industry and waste management.

Several findings highlight the potential use of mollusks in pharmaceutical applications and functional foods (Table 4.1).

4.2.1 Calcium carbonate from mollusks

Mollusks have been considered organisms with beneficial therapeutic and nutritional potential. Mollusks, such as mussels, have shells composed of approximately 94% calcium carbonate as the major component (Clark et al., 2020). Previous investigations have aimed to

TABLE 4.1 Bioactive compounds in marine mollusks and their biological activities.

Taxa	Species	Tissue	Compound	Biological activity	References
Bivalve	*Modiolus modiolus*	Shell	Chitin and chitosan	Antimicrobial, antioxidant	Varma and Vasudevan (2020)
Gastropod	*Rapana venosa*	Soft tissue	Gelatin	Cell proliferation, cell-induced adhesion, antioxidant	Gaspar-Pintiliescu et al. (2019)
Bivalve	*Tergillarca granosa*	Muscle	Protein hydrolysates	Antioxidant	Yang et al. (2019)
Bivalve	*Pernacanaliculus*	Body	Glycosaminoglycans/Biolane	Antiinflammatory	Mutalipassi et al. (2021)
Bivalve	*Mytilus coruscus*	Mantle tissue	Myticusin-beta	Antimicrobial activity	Oh et al. (2020)

TABLE 4.1 Bioactive compounds in marine mollusks and their biological activities—cont'd

Taxa	Species	Tissue	Compound	Biological activity	References
Bivalve	*Mytilus edulis*	Muscle	Hydrolysate	Antiinflammatory effects	Starikova et al. (2021)
Cuttlefish	*Sepia pharaonis*	Bone	Chitin and chitosan	Antioxidation and oil-absorbing abilities	Hazeena et al. (2022)
Bivalve	*Anadara senilis*, *Crassostrea gasar*, and *Mytilus edulis*	Shells	Calcium, magnesium, and calcium carbonate	Food supplements, animal feeds, dental products, plant nutrient supplements, ornamental purposes, construction, and agricultural industries	Elegbede et al. (2023)

produce pure calcium salts, such as calcium levulinate, from *Mytilus galloprovincialis* for pharmaceutical applications (Mititelu et al., 2022). The American Pharmacopoeia initially used calcium levulinate as an injectable solution for calcium supplementation and allergy treatment (Mititelu et al., 2022). Furthermore, calcium levulinate is now used as a nutritional supplement (Sharath et al., 2019).

4.2.2 Protein hydrolysate from mollusks

One of the notable medicinal and pharmaceutical applications of mollusks is the use of protein hydrolysates. Protein hydrolysate consists of peptides of various sizes and free amino acids, produced by hydrolyzing the peptide bonds in proteins. The hydrolysis process alters the protein, affecting its properties, such as its bioactive, functional, and nutritional characteristics (Liceaga and Hall, 2018). These peptides may be inactive in their precursor form but become biologically active after conversion into peptides that interact with various receptors. These interactions can influence different body functions, imparting properties such as antibacterial, antiviral, antifungal, immunomodulatory, antioxidant, antihypertensive, antiproliferative, anticoagulant, hemolytic, calcium-binding, and opioid activities (Tkaczewska, 2020). These findings are important for the potential use of protein hydrolysate in treating chronic diseases, such as diabetes, cardiovascular disorders, and neurodegenerative conditions.

Previous studies have investigated the antibacterial activity of protein hydrolysate from cottonseed, finding that it positively correlates with the content of basic amino acids (Arg and Lys) and negatively correlates with the content of acidic amino acids (Glu and Asp) (Song et al., 2020). Another study demonstrated that protein hydrolysates from yellowfin tuna with a low molecular weight (< 3 kDa) exhibit highly significant antibacterial activity against both Gram-positive bacteria (*Listeria* and *Staphylococcus*) and Gram-negative bacteria (*Escherichia coli* and *Pseudomonas*), as well as strong antioxidant activity (Pezeshk et al., 2019). The extract from Gonggong sea snail (*Laevistrombus turturella*), with a molecular weight of 1.4 kDa, consists of 12 amino acid residues (RHPDYSVALLLR) and shows antimicrobial activity against *E. coli* and *Staphylococcus aureus* (Viruly et al., 2023).

In addition, several studies have investigated marine-derived compounds with antiviral activities against a plethora of viruses. Moreover, some of these compounds have been tested in clinical trials and approved for drug marketing. RvH and the functional units RvH1-a/RvH2-e, extracted from *Rapana venosa*, show antiviral activity against Epstein–Barr virus. Previous investigations on the antiviral activity of hemocyanin fractions from *Haliotis rubra* reveal antiviral activity against Herpes simplex virus (HSV-1) through binding to the gD, gB, and gC glycoproteins on the surface of the virus. Furthermore, modified and nanoencapsulated Myticin C from *M. galloprovincialis* shows antiviral activity against Ostreid herpesvirus 1 (OsHV-1), HSV-1, and HSV-2.

In addition to their medicinal effects, peptides extracted from mollusks have provided clues about their potential use in improving the absorption and bioavailability of drugs when used as pharmaceutical compounds. The YPRKDETGAERT peptide (PME-1), extracted from *Mytilus edulis,* has been investigated for its ability to enhance the proliferation and differentiation of osteoblasts, as well as its rapid absorption through the gastrointestinal tract (Xu et al., 2021). However, the use of mollusk peptides as drug delivery facilitators is still an open area for further investigation.

4.2.3 Collagen and gelatin from mollusks

Collagen and gelatin, derived from mollusks, have important applications in the pharmaceutical, medicinal, and cosmetic industries (Gaspar-Pintiliescu et al., 2019). Collagen is a crucial structural protein found in bones, tendons, cartilage, skin, and connective tissues in the human body and all animal tissues, accounting for 30% of its dry weight, while gelatin, a collagen-derived product, is widely used in these fields due to its gelling properties (Owczarzy et al., 2020; Cai et al., 2021). Marine mollusks, such as squids, octopuses, and certain bivalves, are considered valuable sources of collagen (Grabacki et al., 2022). Their specific properties, such as bioactivity, biocompatibility, and the ability to enhance tissue regeneration, make collagen an invaluable compound derived from marine mollusks (Rahman, 2019).

Collagen and gelatin are potential candidates in regenerative medicine to accelerate tissue repair, cell growth, and the healing of chronic wounds, surgical incisions, and burns (Fani et al., 2024). The use of jellyfish as the most common source of collagen is mentioned in about 91 patent documents, while mollusks, including cuttlefish, squid, and shellfish, are cited in 46 patents (Almeida et al., 2025). Investigations have revealed the potential application of gelatin extracted from the marine snail *R. venosa* in pharmaceutical and skincare products (Gaspar-Pintiliescu et al., 2024). Collagen derived from mollusks, such as mussels (*Geukensia demissa*), hard clams (*Meretrix lusoria*), and pearl oysters (*Pinctada fucata*), has been characterized for its biochemical and physicochemical properties (Wu et al., 2019). In addition, collagen has been identified as a carrier in controlled-release drug delivery systems, enhancing the bioavailability of certain drugs (Kim and Khora, 2020). Moreover, the application of other marine collagens and gelatin in skin rejuvenation and care in the cosmetic industry reveals the potential use of mollusk-derived collagen and gelatin in skincare products (Makgobole et al., 2024).

Taken together, collagen and gelatin extracts derived from marine mollusks exhibit significant potential for use in the medicine and cosmetics industries. Their applications are diverse, enhancing tissue regeneration, drug delivery, and skincare products. However, future investigations are needed to introduce more advanced products incorporating marine mollusk-derived collagen and gelatin for medicinal and cosmetic uses.

4.2.4 Lipids from mollusks

Marine mollusks are critically important as a valuable source of a variety of lipids, which exhibit unique composition, nutritional value, and bioactive properties. The lipids in the cell membranes of marine invertebrates serve as energy storage, and the commercial value of these invertebrates as food is directly determined by their lipid composition (Imbs et al., 2021). PUFAs from the n-3 and n-6 families are essential in physiological processes and are mainly derived from marine ecosystems. FAs such as PUFAs, docosahexaenoic acid (22:6n-3), and eicosapentaenoic acid (20:5n-3) are vital components for human health (Zhukova, 2019).

Lipid-based drug delivery systems are a promising method to enhance the solubility and bioavailability of poorly water-soluble drugs (Shrestha et al., 2014). Research has indicated that marine-derived lipid nanoparticles are potential candidates for use as vehicles for intranasal administration to manage neurodegenerative diseases, as they facilitate the penetration of drugs across the blood–brain barrier (Torres et al., 2023).

Moreover, lipids are crucial components that should be incorporated into cosmetic product formulations to maintain the structure and performance of the product (de Clermont-gallerande, 2020). Marine organisms exhibit a variety of lipids with the potential to be utilized in the cosmetic industry. Lipid extracts derived from *M. galloprovincialis* have antioxidant properties, making them a promising candidate for the healing of skin burns (Badiu et al., 2008).

In conclusion, marine mollusks represent a rich and diverse source of lipids with unique bioactive properties that hold significant promise across various industries. Their lipid composition not only contributes to their nutritional and commercial value as a food source but also plays a crucial role in the development of advanced drug delivery systems, particularly for neurodegenerative diseases. In addition, the antioxidant properties of lipid extracts from marine mollusks, such as *M. galloprovincialis*, offer exciting possibilities for the cosmetic industry, especially in skin care and wound healing applications. As research continues, the potential applications of marine-derived lipids in medicine, pharmaceuticals, and cosmetics are likely to expand, further highlighting the importance of marine mollusks in both health and industry.

4.2.5 Polysaccharides from marine mollusks

Polysaccharide extracts from marine mollusks have gained considerable attention due to their bioactive properties, unique structure, and potential applications in various industries worldwide. Among these, chitin, chitosan, and glycosaminoglycans play a major role in the biotechnology field. Chitin, the second most abundant polysaccharide in nature after cellulose, exhibits unique properties that are utilized in fields such as wound dressing, controlled drug release, anticancer drug delivery, enzyme fixation, and cosmetics (Dananjaya et al., 2020). Chitosan, a deacetylated form of chitin, is a nontoxic, stable, biodegradable, biocompatible, and sterilizable compound with medicinal properties, including slow drug release, the ability to cross-link with other polymers, gel formation, immunostimulation, macrophage activation, gas permeability, easy modification, and antimicrobial properties (Parhi, 2020). Modified chitosan, such as N-succinyl chitosan/gold nanocomposites, demonstrates excellent biocompatibility and antifungal activity (Dananjaya et al., 2020). In addition, N-succinyl chitosan nanoparticles show great potential for use in wound dressings due to their wound healing, antimicrobial, and low cytotoxicity properties (Thao et al., 2021). Several studies have been conducted to isolate and characterize chitosan from marine mollusks, such as the marine gastropod (*Telescopium telescopium*) (Thangaraj Jebarani et al., 2020), Horse Mussel (*Modiolus modiolus*) (Varma and Vasudevan, 2020), squid (*Dosidicus gigas*) (Moreno-Tovar et al., 2021), sea snail (*Conus inscriptus*) (Mohan et al., 2019), and cuttlefish (*Sepia brevimana*) (Narasimman and Ramachandran, 2024).

Glycosaminoglycans (GalAGs), such as chondroitin sulfate (CS), dermatan sulfate (DS), keratan sulfate, heparan sulfate, and hyaluronic acid, have been utilized in the medicinal field to treat joint pain, inflammation, and osteoarthritis (Sodhi and Panitch, 2021; Pomin et al., 2019). GalAGs are key functional molecules in vascular physiology and the pathogenesis of atherosclerosis, making them potential candidates for vascular regeneration (Lepedda et al., 2021). In addition, several studies have highlighted the application of GalAGs in tissue engineering (Sodhi and Panitch, 2021). Previous research has extracted GalAGs from various marine invertebrates, such as mollusks, starfish, and jellyfish. One investigation revealed the antioxidant, physicochemical, and antiproliferative properties of GalAGs extracted from the marine snail *R. venosa* (Gaspar-Pintiliescu et al., 2024).

In conclusion, polysaccharides from marine mollusks, such as chitin, chitosan, and glycosaminoglycans, hold significant promise in biotechnological applications. Chitin and chitosan are widely used for wound healing, drug delivery, and antimicrobial treatments, with modified chitosan showing enhanced biocompatibility and antifungal properties. Glycosaminoglycans, like CS and hyaluronic acid, are valuable in treating joint pain and supporting tissue regeneration. These bioactive polysaccharides offer exciting potential for medical and industrial applications, with ongoing research further expanding their uses.

4.3 Bioactive compounds from marine echinoderms and their applications

Marine echinoderms are among the marine organisms that possess a diverse range of bioactive compounds, holding significant promise for pharmaceutical, industrial, and agricultural applications. Echinoderms, including sea stars, sea cucumbers, sea urchins, brittle stars, and feather stars, have unique biochemical defense mechanisms and metabolic pathways that enable the synthesis of these diverse bioactive compounds (Rahman et al., 2022). The following section briefly explains the key bioactive compounds and their potential applications.

4.3.1 Saponins

Saponins, a type of glycoside, are secondary metabolites found in sea cucumbers (holothurians) and sea stars (asteroids), which they use for chemical defense. Sea cucumbers produce various types of saponins, such as frondosides, holothurins, and cucumariosides (Kamyab et al., 2019). Saponins are toxic to other organisms, including their predators, due to their ability to bind to cholesterol molecules on cell membranes, disrupting membrane integrity and leading to cell death (Thimmappa et al., 2022). However, investigations have revealed that saponins extracted from marine organisms, such as sea cucumbers, possess various biological properties, including antihyperuricemic (Amini et al., 2017), antiobesity (Seo et al., 2024), antifungal (Zhao et al., 2018), antibacterial (Darya et al., 2020), antitumor (Mohamed et al., 2021), and immunomodulatory activities (Nursid et al., 2021). In addition, studies have shown antidiabetic activity in saponins extracted from *Holothuria thomasi* (Barky et al., 2016). Several studies have highlighted the biological activities of different saponins, including frondosides, holothurins, and cucumariosides (Table 4.2).

TABLE 4.2 Biological activities and potential applications of saponins.

Saponins	Biological activity	References
Frondosides A	Anticancer activity • Inhibition of cancer cell growth (pancreatic, breast, leukemia, lung, and prostate cells) • Induction of apoptosis • Synergistic effects with chemotherapy • Inhibition of angiogenesis (inhibit tumor growth) • Inhibition of metastasis (prevent cancer cells from spreading) • Inhibition of p21-activated kinase 1 (PAK1) Antiinflammatory activity • Antagonist of prostaglanding E receptors Immunomodulatory activity Modulation of immune response	Ma et al. (2012), Adrian and Collin (2018), Ru et al. (2023), Ma et al. (2012), Ma et al. (2012), Adrian and Collin (2018)
Holothurins	Anticancer properties Antimicrobial activity Neurological effects	Wargasetia et al. (2020), Han et al. (2010) Kamyab et al. (2019), Darya et al. (2020) Thron et al. (1964)
Cucumariosides	Immunomodulatory effects • Macrophage activation • Cytotoxic activity • Intraperitoneal injection Antitumor activity • Cell cycle arrest • Apoptosis induction • Inhibition of tumor growth and metastasis Inhibit cancer cell migration and colony formation	Aminin (2016), Aminin et al. (2006), Aminin et al. (2001), Menchinskaya et al. (2014), Wargasetia and Widodo (2017), Menchinskaya et al. (2024)

4.3.2 Glycosaminoglycans

Glycosaminoglycans (GAGs) are linear, negatively charged polysaccharides found on cell surfaces and in the extracellular matrix of prokaryotes and invertebrates. Historically, GAGs were identified as compounds consisting of long, unbranched chains of repeating disaccharide units, which are present in all mammalian tissues for structural scaffolding and cell hydration (Afratis et al., 2012). However, advanced investigations have revealed that GAGs exhibit various biological functions, such as cell signaling, regulation of cell growth and proliferation, promotion of cell adhesion, wound repair, and anticoagulation (Casale et al., 2025). GAGs can be classified into four groups based on their core disaccharide units: keratan sulfate, hyaluronic acid, CS/DS, and heparin/heparan sulfate (Prydz, 2015).

Studies have found that sea cucumbers and starfish are rich sources of GAGs, which exhibit antiviral and antiinflammatory activities (Valcarcel et al., 2017). The major GAG molecule present in sea cucumbers is fucosylated CS (Vasconcelos and Pomin, 2017), while in starfish, it is DS and heparinoids (Bilan et al., 2023). In addition, other GAGs, such as CS found in squid and DS found in sea urchins, squirts, and sharks, have been identified as animals with CS/DS hybrids that exhibit nerve regeneration abilities (Valcarcel et al., 2017). Commercial GAGs, such as hyaluronic acid, heparin, and CS, are used in cosmetics and pharmaceuticals for various applications, including joint pain relief, antiaging skincare, and wound healing (Jin et al., 2021). Therefore marine-derived GAGs are potential candidates for use in the medicinal, pharmaceutical, and cosmetic fields.

4.3.3 Peptides and proteins

Echinoderms, including sea stars and sea urchins, contain various peptides and proteins, such as strongylocins and centrocins, which have antimicrobial properties. Some of these peptides, such as neuropeptides and collagen-related proteins, are involved in connective tissue function and muscle contraction (Elphick and Taghert, 2012; Smith et al., 2018).

Strongylocins are cysteine-rich antimicrobial peptides found in sea urchins, including *Strongylocentrotus droebachiensis* and *Strongylocentrotus purpuratus*, and are active against both Gram-positive and Gram-negative bacteria (Li et al., 2010). A study found that strongylocin-1 and 2 may respond to viral infections in sea urchins (*Loxechinus albus*) (Dettleff et al., 2020). Another study discovered a novel cysteine-rich antimicrobial peptide in *Patiria pectinifera*, named PpCrAMP, which plays a role in the innate immunity of echinoderms (Kim et al., 2018). In addition, a peptide derived from sea urchins, centrocin 1, extracted from the green sea urchin (*S. droebachiensis*), exhibits antiinflammatory activity by reducing the expression of inflammatory cytokines, including TLR2, TNF-α, IL-1β, IL-6, and IL-12p40 (Han et al., 2018). These investigations reveal the potential therapeutic applications of antimicrobial peptides extracted from echinoderms.

4.3.4 Polyhydroxylated sterols

Marine echinoderms are a valuable source of polyhydroxylated sterols (PHS) with bioactive properties, which can be used in pharmacological applications. Their biological activities arise from the steroid nucleus with multiple hydroxyl groups and unique structural modifications. Recent investigations have reported the therapeutic potential of echinoderm-derived PHS (Kalinin and Silchenko, 2022). PHS and their glycosylated derivatives are secondary metabolites found in echinoderms, such as starfish and sea cucumbers, and exhibit cytotoxic and antiinflammatory effects (Vien et al., 2022; Minale et al., 1994). Moreover, a study revealed the anticancer activity of sulfated polyhydroxysteroid glycosides from the Okhotsk starfish (*Henricia leviuscula spiculifera*) against human cancer cells (Kicha et al., 2024). In addition, polyhydroxylated steroids extracted from the starfish *Culcita novaeguineae* show cytotoxicity against human glioblastoma cell lines (Lu et al., 2022).

These findings highlight the significant therapeutic potential of PHS derived from marine echinoderms, emphasizing their promise as valuable compounds for future pharmacological research and development.

4.3.5 Gangliosides

Gangliosides are widely distributed components in the cell membranes of vertebrates' central nervous systems and have been tested in various clinical applications from different sources. Recent studies have highlighted the key role of gangliosides in cancer, infection, and diabetes (Magistretti et al., 2019). Furthermore, investigations are being carried out to explore the neuroprotective and restorative functions of bovine-free gangliosides, such as those derived from echinoderms, as bovine-derived gangliosides were associated with adverse effects in Europe during the 1990s (Magistretti et al., 2019). Gangliosides have the potential to regulate tumor growth, invasion, and metastasis in cancer cells, and antiganglioside antibodies targeting gangliosides have shown cytotoxicity in ganglioside-expressing cancer cells, indicating their potential therapeutic application in cancer treatment (Sasaki et al., 2021). In addition, gangliosides help prevent infection by blocking pathogenic binding in the intestine and are used as a supplement in infant formula to increase Bifidobacteria content while decreasing *E. coli* levels (Sasaki et al., 2021). A study also revealed that different forms of gangliosides in pancreatic beta cells promote glucose transporter 2 (Glut2) function, allowing for an adequate response to high glucose loads (Jennemann et al., 2020). However, some studies have suggested that gangliosides can induce insulin resistance in blood vessels by enhancing aging and inflammation in endothelial cells, emphasizing that targeting gangliosides could be a potential treatment for managing insulin resistance and other complications in the vascular system (Sasaki et al., 2019). Several reports have mentioned the availability of gangliosides in echinoderms (Kalinin and Silchenko, 2022; Ma et al., 2021). In conclusion, gangliosides, especially those derived from alternative sources like echinoderms, offer significant therapeutic potential for various diseases, but their complex roles in insulin resistance and vascular health require further exploration to fully realize their benefits.

4.4 Conclusion

Marine mollusks and echinoderms are biologically rich and chemically diverse invertebrates that play a vital role in marine ecosystems while offering great potential for pharmaceutical, nutraceutical, and biotechnological applications. Mollusks, through their unique innate immune systems, produce a multitude of bioactive compounds, including proteins, peptides, lipids, collagen, gelatin, and polysaccharides such as chitin, chitosan, and glycosaminoglycans. These compounds have shown a wide range of therapeutic properties, such as antimicrobial, antioxidant, antiviral, antiinflammatory, antihypertensive, immunomodulatory, tissue regenerative, and wound healing, making them highly valuable for applications in drug development, functional foods, cosmetics, and advanced biomedical materials. Echinoderms, such as sea cucumbers and sea urchins, bring additional health-promoting benefits to what the ocean already offers. In addition to their regenerative abilities, echinoderms are known to produce bioactive compounds such as saponins, glycosaminoglycans, peptides and proteins, PHS, and gangliosides. These molecules exhibit cytotoxic, antifungal, antiviral, antiinflammatory, anticoagulant, and anticancer activities and have also been identified for their use in immunotherapy, regenerative medicine, and chronic disease management. For instance, the use of sea cucumber-derived saponins in apoptosis induction and immunomodulation, or sea urchin-derived compounds in antiproliferative treatments, underscores the pharmacological relevance of these organisms. Both mollusks and echinoderms contribute significantly to global fisheries and traditional medicine and are increasingly recognized for their roles in sustainable industry and marine biotechnology. Their by-products, such as shells and internal organs, once considered waste, are now being utilized to extract high-value compounds, contributing to environmental sustainability and waste utilization in value addition. In conclusion, the bioactive compounds derived from marine mollusks and echinoderms represent a powerful resource for innovation across medical, pharmaceutical, and cosmetic industries. With their diverse biochemical profiles, ecological importance, and economic potential, these marine invertebrates stand at the forefront of sustainable biotechnological development. Continued research and responsible harvesting are essential to unlocking their full potential while ensuring the conservation of marine biodiversity.

AI disclosure

During the preparation of this work, the author(s) used ChatGPT to improve the readability and language of the work only. After using this tool/service, the author(s) reviewed and edited the content as needed and take(s) full responsibility for the content of the publication.

References

Adrian, T.E., Collin, P., 2018. The anti-cancer effects of frondoside A. Mar. Drugs. 1660-3397. 16 (2). Available from: http://www.mdpi.com/1660-3397/16/2/64/pdf. doi:10.3390/md16020064.

Afratis, N., Gialeli, C., Nikitovic, D., Tsegenidis, T., Karousou, E., Theocharis, A.D., Pavão, M.S., Tzanakakis, G.N., Karamanos, N.K., 2012. Glycosaminoglycans: key players in cancer cell biology and treatment. FEBS J. 1742-4658. 279 (7), 1177–1197. doi:10.1111/j.1742-4658.2012.08529.x.

Almeida, M., Silva, T., Solstad, R.G., Lillebø, A.I., Calado, R., Vieira, H., 2025. How significant are marine invertebrate collagens? Exploring trends in research and innovation. Mar. Drugs. 1660-3397. 23 (1). Available from: http://www.mdpi.com/journal/marinedrugs. doi:10.3390/md23010002.

Amini, E., Nabiuni, M., Baharara, J., Parivar, K., Asili, J., 2017. In-vitro pro apoptotic effect of crude Saponin from *Ophiocoma erinaceus* against cervical cancer. Iran. J. Pharm. Res. 1726-6890. 16 (1), 266–276. Available from: http://ijpr.sbmu.ac.ir/article_1972_d642abd4dd7e3f28c20c4ae82fe607d1.pdf.

Aminin, D., 2016. Immunomodulatory properties of sea cucumber triterpene glycosides. In: Marine and Freshwater Toxins. Springer Science and Business Media LLC, pp. 381–401. doi:10.1007/978-94-007-6419-4_3.

Aminin, D.L., Agafonova, I.G., Berdyshev, E.V., Isachenko, E.G., Avilov, S.A., Stonik, V.A., 2001. Immunomodulatory properties of cucumariosides from the edible far-Eastern holothurian *Cucumaria japonica*. J. Med. Food. 1096-620X. 4 (3), 127–135. doi:10.1089/109662001753165701.

Aminin, D.L., Pinegin, B.V., Pichugina, L.V., Zaporozhets, T.S., Agafonova, I.G., Boguslavski, V.M., Silchenko, A.S., Avilov, S.A., Stonik, V.A., 2006. Immunomodulatory properties of cumaside. Int. Immunopharmacol. 1567-5769. 6 (7), 1070–1082. doi:10.1016/j.intimp.2006.01.017.

Aouji, M., Imtara, H., Rkhaila, A., Bouhaddioui, B., Alahdab, A., Parvez, M., Alzahrani, M.S., Lrhorfi, L.A., Bengueddour, R., 2023. Nutritional composition, fatty acids profile, mineral content, antioxidant activity and acute toxicity of the flesh of Helix aspersa Müller. Molecules. 1420-3049. 28 (17), 6323. doi:10.3390/molecules28176323.

Badiu, D.L., Balu, A.M., Barbes, L., Luque, R., Nita, R., Radu, M., Tanase, E., Rosoiu, N., 2008. Physico-chemical characterisation of lipids from *Mytilus galloprovincialis* (L.) and *Rapana venosa* and their healing properties on skin burns. Lipids. 1558-9307. 43 (9), 829–841. doi:10.1007/s11745-008-3205-2.

Baker-Médard, M., Ohl, K.N., 2019. Sea cucumber management strategies: challenges and opportunities in a developing country context. Environ. Conserv. 1469-4387. 46 (4), 267–277. Available from: https://www.cambridge.org/core/journals/environmental-conservation/all-issues. doi:10.1017/S0376892919000183.

Barky, A.R.E., Hussein, S.A., Alm-Eldeen, A.A., Hafez, Y.A., Mohamed, T.M., 2016. Anti-diabetic activity of *Holothuria thomasi* saponin. Biomed. Pharmacother. 1950-6007. 84, 1472–1487. Available from: https://www.elsevier.com/locate/biomedpharm. doi:10.1016/j.biopha.2016.10.002.

Bilan, M.I., Anisimova, N.Y., Tokatly, A.I., Nikogosova, S.P., Vinnitskiy, D.Z., Ustyuzhanina, N.E., Dmitrenok, A.S., Tsvetkova, E.A., Kiselevskiy, M.V., Nifantiev, N.E., Usov, A.I., 2023. Glycosaminoglycans from the starfish *Lethasterias fusca*: structures and influence on hematopoiesis. Mar. Drugs. 1660-3397. 21 (4). Available from: http://www.mdpi.com/journal/marinedrugs. doi:10.3390/md21040205.

Cai, S., Zhao, K.-X., Jiang, M.-T., Han, S.-Y., Zheng, Y.-F., Liu, X., Zhao, M., Duan, J.-A., Liu, R., 2021. Collagen derived species-specific peptides for distinguishing donkey-hide gelatin (Asini Corii Colla). Chin. Herb. Med. 1674-6384. 13 (2), 261–266. doi:10.1016/j.chmed.2020.12.006.

Canesi, L., Auguste, M., Balbi, T., Prochazkova, P., 2022. Soluble mediators of innate immunity in annelids and bivalve mollusks: a mini-review. Front. Immunol. 1664-3224. 13. doi:10.3389/fimmu.2022.1051155.

Casale, J., Crane, J.S., (2025). Biochemistry, glycosaminoglycans. In: StatPearls, StatPearls Publishing Treasure Island (FL). http://www.ncbi.nlm.nih.gov/books/NBK544295/. (Accessed 27 April 2025).

Chen, Z.H., Guo, Y.W., Li, X.W., 2022. Recent advances on marine mollusk-derived natural products: chemistry, chemical ecology and therapeutical potential. Nat. Prod. Rep. 1460-4752. 40 (3), 509–556. Available from: http://pubs.rsc.org/en/journals/journal/np. doi:10.1039/d2np00021k.

Clark, M.S., Peck, L.S., Arivalagan, J., Backeljau, T., Berland, S., Cardoso, J.C.R., Caurcel, C., Chapelle, G., De Noia, M., Dupont, S., Gharbi, K., Hoffman, J.I., Last, K.S., Marie, A., Melzner, F., Michalek, K., Morris, J., Power, D.M., Ramesh, K., Sanders, T., Sillanpää, K., Sleight, V.A., Stewart-Sinclair, P.J., Sundell, K., Telesca, L., Vendrami, D.L.J., Ventura, A., Wilding, T.A., Yarra, T., Harper, E.M., 2020. Deciphering mollusc shell production: the roles of genetic mechanisms through to ecology, aquaculture and biomimetics. Biol. Rev. 1469-185X. 95 (6), 1812–1837. Available from: http://onlinelibrary.wiley.com/journal/10.1111/(ISSN)1469-185X. doi:10.1111/brv.12640.

Cuezzo, M. Gabriela, Eduardo Gutiérrez Gregoric, D., Pointier, J.-P., Vázquez, A.A., Ituarte, C., Mansur, M.C.D., Arruda, J.O., Barker, G.M., Dos Santos, S.B., Ovando, X.M.C., de Lacerda, L.E.M., Fernandez, M.A., Thiengo, S.C., de Mattos, A.C., da Silva, E.F., Berning, M.I., Collado, G.A., Miyahira, I.C., Antoniazzi, T.N., Pimpão, D.M., Damborenea, C., 2020. Phylum Mollusca. In: Thorp and Covich's Freshwater Invertebrates: Volume 5: Keys to Neotropical and Antarctic Fauna, pp. 261–430. doi:10.1016/B978-0-12-804225-0.00011-3.

Dananjaya, S.H.S., Edirisinghe, S.L., Thao, N.T.T., Kumar, R.S., Wijerathna, H.M.S.M., Mudiyanselage, A.Y., De Zoysa, M., Choi, D., 2020. Succinyl chitosan gold nanocomposite: preparation, characterization, in vitro and in vivo anticandidal activity. Int. J. Biol. Macromol. 1879-0003. 165, 63–70. Available from: https://www.elsevier.com/locate/ijbiomac. doi:10.1016/j.ijbiomac.2020.09.126.

Darya, M., Sajjadi, M.M., Yousefzadi, M., Sourinejad, I., Zarei, M., 2020. Antifouling and antibacterial activities of bioactive extracts from different organs of the sea cucumber *Holothuria leucospilota*. Helgol. Mar. Res. 1438-3888. 74 (1). Available from: https://link.springer.com/journal/10152. doi:10.1186/s10152-020-0536-8.

de Clermont-gallerande, H., 2020. Functional roles of lipids in make-up products. OCL. 2272-6977. 27, 33. doi:10.1051/ocl/2020026.

Dettleff, P., Villagra, M., González, J., Fuentes, M., Estrada, J.M., Valenzuela, C., Molina, A., Valdés, J.A., 2020. Effect of bacterial LPS, poly I:C and temperature on the immune response of coelomocytes in short term cultures of red sea urchin (*Loxechinus albus*). Fish Shellfish Immunol. 1050-4648. 107 (Pt A), 187–193. doi:10.1016/j.fsi.2020.09.028.

Doll, P.C., Caballes, C.F., Hoey, A.S., Uthicke, S., Ling, S.D., Pratchett, M.S., 2022. Larval settlement in echinoderms: a review of processes and patterns. In: Oceanography and Marine Biology: An Annual Review, Volume 60. CRC Press, Australia, pp. 433–494. Available from: https://www.routledge.com/Oceanography-and-Marine-Biology-An-Annual-Review-Volume-60/Hawkins/p/book/9781032265056. doi:10.1201/9781003288602-9.

Eghianruwa, Q.A., Osoniyi, O.R., Maina, N., Wachira, S., 2019. Bioactive peptides from marine molluscs—a review. Int. J. Biochem. Res. Rev. 2231-086X. 1–12. doi:10.9734/ijbcrr/2019/v27i430130.

Ehrlich, H., 2019. Spicular Structures in Molluscs. Springer Science and Business Media LLC, pp. 133–157. doi:10.1007/978-3-319-92483-0_10.

Elegbede, I., Lawal-Are, A., Favour, O., Jolaosho, T., 2023. Appolinaire Goussanou, chemical compositions of bivalves shells: *Anadara senilis*, *Crassostrea gasar*, and *Mytilus edulis* and their potential for a sustainable circular economy. SN Appl. Sci. 2523-3963. 5 (1). doi:10.1007/s42452-022-05267-7.

Elphick, M.R., Taghert, P.H., 2012. The protein precursors of peptides that affect the mechanics of connective tissue and/or muscle in the echinoderm *Apostichopus japonicus*. PLoS ONE. 1932-6203. 7 (8), e44492. doi:10.1371/journal.pone.0044492.

Fani, N., Moradi, M., Zavari, R., Parvizpour, F., Soltani, A., Arabpour, Z., Jafarian, A., 2024. Current advances in wound healing and regenerative medicine. Curr. Stem Cell Res. Ther. 2212-3946. 19 (3), 277–291. Available from: https://www.eurekaselect.com/article/129936. doi:10.2174/1574888X18666230301140659.

Gaspar-Pintiliescu, A., Stefan, L.M., Anton, E.D., Berger, D., Matei, C., Negreanu-Pirjol, T., Moldovan, L., 2019. Physicochemical and biological properties of gelatin extracted from marine snail *Rapana venosa*. Mar. Drugs 17 (10). doi:10.3390/md17100589.

Gaspar-Pintiliescu, A., Stefan, L.M., Mihai, E., Sanda, C., Manoiu, V.S., Berger, D., Craciunescu, O., 2024. Antioxidant and antiproliferative effect of a glycosaminoglycan extract from *Rapana venosa* marine snail. PLoS ONE. 1932-6203. 19 (2). Available from: https://journals.plos.org//plosone/article?id=10.1371/journal.pone.0297803. doi:10.1371/journal.pone.0297803.

Grabacki, S.T., Pathak, Y.V., Joshi, N.H., 2022. Nutraceuticals from Mollusks. In: Rádis-Baptista, G., (Ed.), Marine-based Bioactive Compounds: Applications in Nutraceuticals, CRC Press 6. (Accessed 20 April 2025). doi:10.1201/9781003128175-6.

Guerra-García, J.M., Martínez-Pita, I., García-García, F.J., Moreira, J., 2021. Diversity, community structure and habitat use of molluscs in marinas from the Iberian Peninsula and Northern Africa. Ocean Coast. Manag. 0964-5691. 212, 105795. doi:10.1016/j.ocecoaman.2021.105795.

Gündoğdu, S., Yeşilyurt, İ.N., Abbas, Z., Baylan, M., 2024. Effects of microplastics on aquatic organisms: a comprehensive review. Turkish J. Zool. 1303-6114. 48 (5), 248–285. Available from: https://journals.tubitak.gov.tr/cgi/viewcontent.cgi?article=3182&context=zoology. doi:10.55730/1300-0179.3182.

Guo, X., Ford, S.E., 2016. Infectious diseases of marine mollusks and host responses as revealed by genomic tools. Philos. Trans. R. Soc. Lond. B Biol. Sci. 1471-2970. 371, 1689. Available from: http://rstb.royalsocietypublishing.org/content/371/1689/20150206.full.pdf. doi:10.1098/rstb.2015.0206.

Han, H., Zhang, W., Yi, Y., Liu, B., Pan, M., Wang, X., 2010. A novel sulfated holostane glycoside from sea cucumber *Holothuria leucospilota*. Chem. Biodivers. 1612-1872. 7 (7), 1764–1769. doi:10.1002/cbdv.200900094.

Han, R., Blencke, H.M., Cheng, H., Li, C., 2018. The antimicrobial effect of CEN1HC-Br against propionibacterium acnes and its therapeutic and anti-inflammatory effects on acne vulgaris. Peptides. 1873-5169. 99, 36–43. Available from: https://www.elsevier.com/locate/peptides. doi:10.1016/j.peptides.2017.11.001.

Hazeena, S.H., Hou, C.Y., Zeng, J.H., Li, B.H., Lin, T.C., Liu, C.S., Chang, C.I., Hsieh, S.L., Shih, M.K., 2022. Extraction optimization and structural characteristics of chitosan from cuttlefish (*S. pharaonis* sp.) Bone. Materials. 1996-1944. 15 (22). Available from: http://www.mdpi.com/journal/materials. doi:10.3390/ma15227969.

Hodgson, A.N., 2018. Spermatogenesis and spermiogenesis in the mollusca. In: Encyclopedia of Reproduction. Elsevier, South Africa, pp. 336–343. doi:10.1016/B978-0-12-809633-8.20658-4.

Imbs, A.B., Ermolenko, E.V., Grigorchuk, V.P., Sikorskaya, T.V., Velansky, P.V., 2021. Current progress in lipidomics of marine invertebrates. Mar. Drugs. 1660-3397. 19 (12). Available from: https://www.mdpi.com/1660-3397/19/12/660/pdf. doi:10.3390/md19120660.

Jennemann, R., Kaden, S., Volz, M., Nordström, V., Herzer, S., Sandhoff, R., Gröne, H.J., 2020. Gangliosides modulate insulin secretion by pancreatic beta cells under glucose stress. Glycobiology. 1460-2423. 30 (9), 722–734. Available from: http://glycob.oxfordjournals.org/. doi:10.1093/glycob/cwaa022.

Jin, W., Zhang, F., Linhardt, R.J., 2021. Bioengineered production of glycosaminoglycans and their analogues. Syst. Microbiol. Biomanuf. 2662-7663. 1 (2), 123–130. Available from: https://www.springer.com/journal/43393. doi:10.1007/s43393-020-00011-x.

Kalinin, V.I., Silchenko, A.S., 2022. Echinoderms metabolites: structure, functions and biomedical perspectives II. Mar. Drugs. 1660-3397. 20 (8). Available from: http://www.mdpi.com/journal/marinedrugs. doi:10.3390/md20080492.

Kamyab, E., Kellermann, M.Y., Kunzmann, A., Schupp, P.J., 2019. Chemical Biodiversity and Bioactivities of Saponins in *Echinodermata* with an Emphasis on Sea Cucumbers (*Holothuroidea*). Springer Science and Business Media LLC, Springer Science and Business Media LLC, pp. 121–157. doi:10.1007/978-3-030-20389-4_7.

Kicha, A.A., Tolkanov, D.K., Malyarenko, T.V., Malyarenko, O.S., Kuzmich, A.S., Kalinovsky, A.I., Popov, R.S., Stonik, V.A., Ivanchina, N.V., Dmitrenok, P.S., 2024. Sulfated polyhydroxysteroid glycosides from the Sea of Okhotsk Starfish *Henricia leviuscula spiculifera* and potential mechanisms for their observed anti-cancer activity against several types of human cancer cells. Mar. Drugs. 1660-3397. 22 (7). Available from: http://www.mdpi.com/journal/marinedrugs. doi:10.3390/md22070294.

Kim, C.H., Go, H.J., Oh, H.Y., Park, J.B., Lee, T.K., Seo, J.K., Elphick, M.R., Park, N.G., 2018. Identification of a novel antimicrobial peptide from the sea star *Patiria pectinifera*. Dev. Comp. Immunol. 1879-0089. 86, 203–213. Available from: https://www.sciencedirect.com/science/journal/0145305X. doi:10.1016/j.dci.2018.05.002.

Kim, S., Khora, S.S., 2020. Marine collagen as a source of biomaterial. In: Encyclopedia of Marine Biotechnology. Wiley, pp. 1175–1193. doi:10.1002/9781119143802.ch48.

Lepedda, A.J., Nieddu, G., Formato, M., Baker, M.B., Fernández-Pérez, J., Moroni, L., 2021. Glycosaminoglycans: from vascular physiology to tissue engineering applications. Front. Chem. 2296-2646. 9. Available from: http://journal.frontiersin.org/journal/chemistry. doi:10.3389/fchem.2021.680836.

Li, C., Haug, T., Moe, M.K., Styrvold, O.B., Stensvåg, K., 2010. Centrocins: isolation and characterization of novel dimeric antimicrobial peptides from the green sea urchin, *Strongylocentrotus droebachiensis*. Dev. Comp. Immunol. 0145-305X. 34 (9), 959–968. Available from: https://www.sciencedirect.com/science/journal/0145305X. doi:10.1016/j.dci.2010.04.004.

Liceaga, A.M., Hall, F., 2018. Nutritional, functional and bioactive protein hydrolysates. In: Encyclopedia of Food Chemistry. Elsevier, United States, pp. 456–464. doi:10.1016/B978-0-08-100596-5.21776-9.

Lu, Y.Y., Wang, M.C., Liu, K., Zhang, W., Liu, Y., Tang, H.F., 2022. Two new polyhydroxylated steroidal glycosides from the starfish *Culcita novaeguineae*. Nat. Prod. Res. 1478-6427. 36(8), 2118–2124. Available from: https://www.tandf.co.uk/journals/titles/14786419.asp. doi:10.1080/14786419.2020.1845676.

Ma, X., Kundu, N., Collin, P.D., Goloubeva, O., Fulton, A.M., 2012. Frondoside A inhibits breast cancer metastasis and antagonizes prostaglandin E receptors EP4 and EP2. Breast Cancer Res. Treat. 1573-7217. 132 (3), 1001–1008. doi:10.1007/s10549-011-1675-z.

Ma, Y., Wang, X., Wang, Z., Cong, P., Xu, J., Xue, C., 2021. Characterization of gangliosides in three sea urchin species by HILIC-ESI-MS/MS. J. Agric. Food Chem. 1520-5118. 69 (27), 7641–7651. Available from: http://pubs.acs.org/journal/jafcau. doi:10.1021/acs.jafc.1c02058.

Magistretti, P.J., Geisler, F.H., Schneider, J.S., Li, P.A., Fiumelli, H., Sipione, S., 2019. Gangliosides: treatment avenues in neurodegenerative disease. Front. Neurol. 1664-2295. 10. Available from: http://www.frontiersin.org/Neurology. doi:10.3389/fneur.2019.00859.

Makgobole, M.U., Onwubu, S., Khathi, A., Mpofana, N., Mkhwanazi, B., 2024. Cosmeceuticals from marine: the prospect of marine products in skin rejuvenation and care. J. Basic Appl. Sci. 2314-808X. 11 (1), 297–317. Available from: https://www.tandfonline.com/toc/teba20/current. doi:10.1080/2314808X.2024.2336307.

Menchinskaya, E.S., Chingizova, E.A., Pislyagin, E.A., Yurchenko, E.A., Klimovich, A.A., Zelepuga, E.A., Aminin, D. L., Avilov, S.A., Silchenko, A.S., 2024. Mechanisms of action of sea cucumber triterpene glycosides cucumarioside A0-1 and Djakonovioside A against human triple-negative breast cancer. Mar. Drugs. 1660-3397. 22 (10). Available from: http://www.mdpi.com/journal/marinedrugs. doi:10.3390/md22100474.

Menchinskaya, E.S., Pislyagin, E.A., Kovalchyk, S.N., Davydova, V.N., Silchenko, A.S., Avilov, S.A., Kalinin, V.I., Aminin, D.L., 2014. Antitumor activity of cucumarioside A2-2. Chemotherapy. 1421-9794. 59 (3), 181–191. doi:10.1159/000354156.

Minale, L., Riccio, R., Zollo, F., 1994. ChemInform abstract: steroidal oligoglycosides and polyhydroxysteroids from echinoderms. ChemInform. 0931-7597. 25 (45). doi:10.1002/chin.199445300.

Mititelu, M., Stanciu, G., Drăgănescu, D., Ioniță, A.C., Neacșu, S.M., Dinu, M., Stefan-van Staden, R.-I., Moroşan, E., 2022. Mussel shells, a valuable calcium resource for the pharmaceutical industry. Mar. Drugs. 1660-3397. 20 (1), 25. doi:10.3390/md20010025.

Mohamed, A.S., Mahmoud, S.A., Soliman, A.M., Fahmy, S.R., 2021. Antitumor activity of saponin isolated from the sea cucumber, holothuria arenicola against ehrlich ascites carcinoma cells in Swiss albino mice. Nat. Prod. Res. 1478-6427. 35(11), 1928–1932. Available from: https://www.tandf.co.uk/journals/titles/14786419.asp. doi:10.1080/14786419.2019.1644633.

Mohan, K., Ravichandran, S., Muralisankar, T., Uthayakumar, V., Chandirasekar, R., Rajeevgandhi, C., Rajan, D.K., Seedevi, P., 2019. Extraction and characterization of chitin from sea snail *Conus inscriptus* (Reeve, 1843). Int. J. Biol. Macromol. 1879-0003. 126, 555–560. Available from: https://www.elsevier.com/locate/ijbiomac. doi:10.1016/j.ijbiomac.2018.12.241.

Moreno-Tovar, R., Bucio, L., Thions, C., Tehuacanero-Cuapa, S., 2021. Thermal degradation and lifetime of β-chitin from *Dosidicus gigas* squid pen: effect of impact at 9.7 GPa and a comparative study with α-chitin. Carbohydr. Polym. 0144-8617. 251, 116987. doi:10.1016/j.carbpol.2020.116987.

Mutalipassi, M., Esposito, R., Ruocco, N., Viel, T., Costantini, M., Zupo, V., 2021. Bioactive compounds of nutraceutical value from fishery and aquaculture discards. Foods. 2304-8158. 10 (7), 1495. doi:10.3390/foods10071495.

Narasimman, V., Ramachandran, S., 2024. In silico analysis of low molecular weight sulfated chitosan from *Sepia brevimana* as potential inhibitors of white spot syndrome envelope proteins. Biomass Convers. Biorefin. 2190-6815. 14 (16), 18539–18550. doi:10.1007/s13399-023-03844-3.

Ngandjui, Y.A.T., Kereeditse, T.T., Kamika, I., Madikizela, L.M., Msagati, T.A.M., *et al.*, 2024. Nutraceutical and Medicinal Importance of Marine Molluscs. Marine drugs 22 (5), 201. https://www.mdpi.com/1660-3397/22/5/201. (Accessed 19 January 2025). doi:10.3390/md22050201.

Ngandjui, Y.A.T., Kereeditse, T.T., Kamika, I., Madikizela, L.M., Msagati, T.A.M., 2024. Nutraceutical and medicinal importance of marine Molluscs. Mar. Drugs. 1660-3397. 22 (5). Available from: http://www.mdpi.com/journal/marinedrugs. doi:10.3390/md22050201.

Nursid, M., Patantis, G., Dewi, A.S., Achmad, M.J., Sembodo, P.M., Estuningsih, S., 2021. Immunnostimulatory activity of *Holothuria atra* sea cucumber. Pharmacia. 0428-0296. 68 (1), 121–127. Available from: https://pharmacia.pensoft.net/. doi:10.3897/pharmacia.68.e58820.

Oh, R., Lee, M.J., Kim, Y.O., Nam, B.H., Kong, H.J., Kim, J.W., Park, J.Y., Seo, J.K., Kim, D.G., 2020. Myticusin-beta, antimicrobial peptide from the marine bivalve, *Mytilus coruscus*. Fish Shellfish Immunol. 1095-9947. 99, 342–352. Available from: http://www.elsevier.com/inca/publications/store/6/2/2/8/3/2/index.htt. doi:10.1016/j.fsi.2020.02.020.

Owczarzy, A., Kurasiński, R., Kulig, K., Rogóż, W., Szkudlarek, A., Maciążek-Jurczyk, M., 2020. Collagen - Structure, Properties and Application. doi:10.34821/ENG.BIOMAT.156.2020.17-23.

Parhi, R., 2020. Drug delivery applications of chitin and chitosan: a review. Environ. Chem. Lett. 1610-3661. 18 (3), 577–594. Available from: https://www.springer.com/journal/10311. doi:10.1007/s10311-020-00963-5.

Pezeshk, S., Ojagh, S.M., Rezaei, M., Shabanpour, B., 2019. Fractionation of protein hydrolysates of fish waste using membrane ultrafiltration: investigation of antibacterial and antioxidant activities. Probiotics Antimicrob. Proteins. 1867-1314. 11 (3), 1015–1022. Available from: http://www.springer.com/new+%26+forthcoming+titles+%28default%29/journal/12602. doi:10.1007/s12602-018-9483-y.

Pomin, V.H., Vignovich, W.P., Gonzales, A.V., Vasconcelos, A.A., Mulloy, B., 2019. Galactosaminoglycans: medical applications and drawbacks. Molecules (Basel, Switzerland). 1420-3049. 24 (15). doi:10.3390/molecules24152803.

Prié, V., 2019. Molluscs Encyclopedia of Caves, Third. Elsevier, France, pp. 725–731. Available from: https://www.sciencedirect.com/book/9780128141243. doi:10.1016/B978-0-12-814124-3.00087-X.

Prydz, K., 2015. Determinants of glycosaminoglycan (GAG) structure. Biomolecules. 2218-273X. 5 (3), 2003–2022. Available from: http://www.mdpi.com/2218-273X/5/3/2003/pdf. doi:10.3390/biom5032003.

Rahman, M.A., 2019. Collagen of extracellular matrix from marine invertebrates and its medical applications. Mar. Drugs. 1660-3397. 17 (2), 118. doi:10.3390/md17020118.

Rahman, M.A., Ismail, M., Parvez, M.S., Asadujjaman, M., Ashik, A.-A., Molla, M.H.R., 2022. Echinoderm fisheries: their culture, conservation, bioactive compounds and therapeutic applications. J. Biol. Stud. 2209-2560. 5 (3), 413–443. doi:10.62400/jbs.v5i3.7053.

Ramasamy, S., Manavalan, G., Santhanam, R., 2019. Biology and Ecology of Pharmaceutical Marine Mollusks, Vol. 2. doi:10.1201/9781351133395.

Ru, R., Chen, G., Liang, X., Cao, X., Yuan, L., Meng, M., 2023. Sea cucumber derived triterpenoid glycoside frondoside A: a potential anti-bladder cancer drug. Nutrients. 2072-6643. 15 (2), 378. doi:10.3390/nu15020378.

Saleuddin, S., Leys, S.P., Roer, R.D., Wilkie, I.C., 2024. Frontiers in Invertebrate Physiology: A Collection of Reviews: Volume 2: Crustacea, Vol. 2. Apple Academic Press, Canada. pp. 1–402, Available from: http://www.tandfebooks.com/doi/book/9781000900088. doi:10.1201/9781003405016.

Salvador, A., 2022. Fascinating Shells: An Introduction to 121 of the World's Most Wonderful Mollusks. University of Chicago Press. doi:10.7208/chicago/9780226819815.001.0001.

Sasaki, N., Itakura, Y., Toyoda, M., 2019. Gangliosides contribute to vascular insulin resistance. Int. J. Mol. Sci. 1422-0067. 20 (8), 1819. doi:10.3390/ijms20081819.

Sasaki, N., Toyoda, M., Ishiwata, T., 2021. Gangliosides as signaling regulators in cancer. Int. J. Mol. Sci. 1422-0067. 22 (10), 5076. doi:10.3390/ijms22105076.

Seo, H.D., Lee, J.Y., Park, S.H., Lee, E., Hahm, J.H., Ahn, J., Jang, A.R., An, S.H., Ha, J.H., No, K.T., Jung, C.H., 2024. Identification of novel anti-obesity saponins from the ovary of sea cucumber (*Stichopus japonicus*). Heliyon. 2405-8440. 10 (17). Available from https://www.sciencedirect.com/science/journal/24058440. doi:10.1016/j.heliyon.2024.e36943.

Sharath, B.O., Tiwari, R., Mal, S.S., Dutta, S., 2019. Straightforward synthesis of calcium levulinate from biomass-derived levulinic acid and calcium carbonate in egg-shells. Mater. Today Proc. 2214-7853. 17, 77–84. https://www.sciencedirect.com/journal/materials-today-proceedings. doi:10.1016/j.matpr.2019.06.403.

Shrestha, H., Bala, R., Arora, S., 2014. Lipid-based drug delivery systems. J. Pharm. 2090-9918. 2014, 1–10. doi:10.1155/2014/801820.

Smith, L.C., Arizza, V., Barela Hudgell, M.A., Barone, G., Bodnar, A.G., Buckley, K.M., Cunsolo, V., Dheilly, N.M., Franchi, N., Fugmann, S.D., Furukawa, R., Garcia-Arraras, J., Henson, J.H., Hibino, T., Irons, Z.H., Li, C., Lun, C. M., Majeske, A.J., Oren, M., Pagliara, P., Pinsino, A., Raftos, D.A., Rast, J.P., Samasa, B., Schillaci, D., Schrankel, C. S., Stabili, L., Stensväg, K., Sutton, E., 2018. Correction to: Echinodermata: the complex immune system in Echinoderms. In: Advances in Comparative Immunology. Springer Science and Business Media LLC, pp. E1. doi:10.1007/978-3-319-76768-0_32.

Sodhi, H., Panitch, A., 2021. Glycosaminoglycans in tissue engineering: a review. Biomolecules. 2218-273X. 11 (1), 29. doi:10.3390/biom11010029.

Song, W., Kong, X., Hua, Y., Li, X., Zhang, C., Chen, Y., 2020. Antioxidant and antibacterial activity and in vitro digestion stability of cottonseed protein hydrolysates. LWT. 0023-6438. 118, 108724. doi:10.1016/j.lwt.2019.108724.

Starikova, E., Mammedova, J., Ozhiganova, A., Lebedeva, A., Malashicheva, A., Semenova, D., Khokhlova, E., Mameli, E., Caporali, A., Wills, J., Sokolov, A., 2021. Protective role of *Mytilus edulis* hydrolysate in lipopolysaccharide-galactosamine acute liver injury. Front. Pharmacol. 1663-9812. 12. Available from: http://www.frontiersin.org/Pharmacology. doi:10.3389/fphar.2021.667572.

Thangaraj Jebarani, R., Muthukumarasamy, S., Theivasigamani, M., 2020. Physicochemical and functional characterization of chitosan from horn snail gastropod *Telescopium telescopium*. J. Appl. Pharm. Sci. 2231-3354. 11 (2). doi:10.7324/japs.2021.110207.

Thao, N.T.Thu, Wijerathna, H.M.S.M., Saravana Kumar, R., Choi, D., Dananjaya, S.H.S., Attanayake, A.P., 2021. Preparation and characterization of succinyl chitosan and succinyl chitosan nanoparticle film: in vitro and in vivo evaluation of wound healing activity. Int. J. Biol. Macromol. 0141-8130. 193 (Pt B), 1823–1834. doi:10.1016/j.ijbiomac.2021.11.015.

Thimmappa, R., Wang, S., Zheng, M., Misra, R.C., Huang, A.C., Saalbach, G., Chang, Y., Zhou, Z., Hinman, V., Bao, Z., Osbourn, A., 2022. Biosynthesis of saponin defensive compounds in sea cucumbers. Nat. Chem. Biol. 1552-4469. 18 (7), 774–781. Available from: http://www.nature.com/nchembio. doi:10.1038/s41589-022-01054-y.

Thron, C.D., Durant, R.C., Friess, S.L., 1964. Neuromuscular and cytotoxic effects of holothurin A and related saponins at low concentration levels. III. Toxicol. Appl. Pharmacol. 1096-0333. 6 (2), 182–196. doi:10.1016/0041-008X(64)90104-8.

Tkaczewska, J., 2020. Peptides and protein hydrolysates as food preservatives and bioactive components of edible films and coatings—A review. Trends Food Sci. Technol. 0924-2244. 106, 298–311. Available from: http://www.elsevier.com/wps/find/journaldescription.cws_home/601278/description#description. doi:10.1016/j.tifs.2020.10.022.

Torres, J., Costa, I., Peixoto, A.F., Silva, R., Lobo, J.M.S., Silva, A.C., 2023. Intranasal lipid nanoparticles containing bioactive compounds obtained from marine sources to manage neurodegenerative diseases. Pharmaceuticals. 1424-8247. 16 (2). Available from: https://www.mdpi.com/journal/pharmaceuticals. doi:10.3390/ph16020311.

Valcarcel, J., Novoa-Carballal, R., Pérez-Martín, R.I., Reis, R.L., Vázquez, J.A., 2017. Glycosaminoglycans from marine sources as therapeutic agents. Biotechnol. Adv. 0734-9750. 35 (6), 711–725. Available from: https://www.elsevier.com/inca/publications/store/5/2/5/4/5/5/index.htt. doi:10.1016/j.biotechadv.2017.07.008.

Varma, R., Vasudevan, S., 2020. Extraction, characterization, and antimicrobial activity of chitosan from horse mussel modiolus modiolus. ACS Omega. 2470-1343. 5 (32), 20224–20230. Available from: https://pubs.acs.org/journal/acsodf. doi:10.1021/acsomega.0c01903.

Vasconcelos, A.A., Pomin, V.H., 2017. The sea as a rich source of structurally unique glycosaminoglycans and mimetics. Microorganisms. 2076-2607. 5 (3). Available from: https://www.mdpi.com/2076-2607/5/3/51/pdf. doi:10.3390/microorganisms5030051.

Vien, L.T., Hanh, T.T.H., Quang, T.H., Cuong, N.X., Thao, D.T., Nam, N.H., Thung, D.C., Kiem, P.V., Minh, C.V., 2022. Polyhydroxylated steroid derivatives from the starfish *Pentaceraster regulus*. Nat. Prod. Res. 1478-6427. 36 (9), 2223–2229. Available from: https://www.tandf.co.uk/journals/titles/14786419.asp. doi:10.1080/14786419.2020.1826478.

Viruly, L., Suhartono, M.T., Nurilmala, M., Saraswati, S., Andarwulan, N., 2023. Identification and characterization of antimicrobial peptide (AMP) candidate from Gonggong sea snail (*Leavistrombus turturella*) extract. J. Food Sci. Technol. 09758402. 60 (1), 44–52. Available from https://www.springer.com/journal/13197. doi:10.1007/s13197-022-05585-z.

Wargasetia, T.L., Ratnawati, H., Widodo, N. 2020 Anticancer potential of Holothurin A, Holothurin B, and Holothurin B3 from the sea cucumber *Holothuria scabra*. In: AIP Conference Proceedings 2231 15517616 American Institute of Physics Inc. Indonesia doi:10.1063/5.0002552 http://scitation.aip.org/content/aip/proceeding/aipcp

Wargasetia, T.L., Widodo, 2017. Mechanisms of cancer cell killing by sea cucumber-derived compounds. Invest. New Drugs. 1573-0646. 35 (6), 820–826. Available from: https://www.wkap.nl/journalhome.htm/0167-6997. doi:10.1007/s10637-017-0505-5.

Wu, J., Guo, X., Liu, H., Chen, L., 2019. Isolation and comparative study on the characterization of guanidine hydrochloride soluble collagen and pepsin soluble collagen from the body of Surf clam shell (*Coelomactra antiquata*). Foods. 2304-8158. 8 (1), 11. doi:10.3390/foods8010011.

Xu, Z., Fan, F., Chen, H., Shi, P., Zhu, D., Yang, M., Wang, Z., Ei-Seedi, H.R., Du, M., 2021. Absorption and transport of a *Mytilus edulis* -derived peptide with the function of preventing osteoporosis. Food Funct. 2042-650X. 12 (5), 2102–2111. Available from: http://pubs.rsc.org/en/journals/journal/fo. doi:10.1039/d0fo02353a.

Yang, X.R., Qiu, Y.T., Zhao, Y.Q., Chi, C.F., Wang, B., 2019. Purification and characterization of antioxidant peptides derived from protein hydrolysate of the marine bivalve mollusk *Tergillarca granosa*. Mar. Drugs. 1660-3397. 17 (5). Available from: https://www.mdpi.com/1660-3397/17/5/251/pdf. doi:10.3390/md17050251.

Zhao, Y.C., Xue, C.H., Zhang, T.T., Wang, Y.M., 2018. Saponins from sea cucumber and their biological activities. J. Agric. Food Chem. 1520-5118. 66 (28), 7222–7237. Available from: http://pubs.acs.org/journal/jafcau. doi:10.1021/acs.jafc.8b01770.

Zhong, Z., Wang, Y., 2024. Overview. In: Ecophysiology and Ocean Acidification in Marine Mollusks. Elsevier BV, pp. 1–37. doi:10.1016/b978-0-443-15938-1.00001-5.

Zhukova, N.V., 2019. Fatty acids of marine mollusks: impact of diet, bacterial symbiosis and biosynthetic potential. Biomolecules. 2218-273X. 9 (12). Available from: https://www.mdpi.com/2218-273X/9/12/857/pdf. doi:10.3390/biom9120857.

PART III

Marine-derived nutritional and bioactive compounds

CHAPTER

5

Applications of marine polysaccharides

Sneha Sengupta, Soumajit Ghosh, Devargya Ganguly, and Gargi Goswami

Department of Life Sciences, GITAM (Deemed to be University), Visakhapatnam, Andhra Pradesh, India

5.1 Introduction

Oceans encompass more than one-third of the Earth and account for nearly 90% of the biosphere. The organisms inhabiting the marine environment are often exposed to diverse and extreme variations in temperature, pressure, salinity, pH, and light conditions (Ruocco et al., 2016). Marine organisms, which include a diverse array of bacteria, fungi, seaweeds, and microalgae, as well as larger nonmicroscopic organisms such as sponges, fishes, and crustaceans, can serve as a promising reservoir of bioactive compounds, as depicted in Fig. 5.1. These naturally derived molecules, including proteins, polysaccharides, and fatty acids, play a crucial role in their adaptation to the harsh and varying conditions of the marine environment and their survival (Aneiros and Garateix, 2004; Bergé and Barnathan, 2005; Laurienzo, 2010).

Among the diverse group of molecules, polysaccharides particularly play a significant role. Polysaccharides are carbohydrate polymers linked together by glycosidic bonds and can be classified as homopolysaccharides or heteropolysaccharides based on their composition (Ahmad et al., 2015; Mahmoud et al., 2021). Marine algae are a rich source of sulfated polysaccharides, namely, fucoidan, alginate, carrageenan, agarose, porphyran, and ulvan (Bhuyan et al., 2023; Patel et al., 2023). In addition, polysaccharides such as chitin, chitosan (CS), and oligosaccharides are derived from the exoskeletons of marine crustaceans (Younes and Rinaudo, 2015). Marine microorganisms also produce exopolysaccharides (EPSs) (Xiao and Zheng, 2016; Qi et al., 2022), which are biopolymers primarily composed of polysaccharides along with nucleic acids, lipids, and proteins. Polysaccharides derived from seaweeds are generally classified based on their source; for example, fucoidans are obtained from brown seaweeds, agaroids and carrageenans are

Harvesting the Sea
https://doi.org/10.1016/B978-0-443-29190-6.00009-X

FIG. 5.1 Diverse sources of marine polysaccharides: both macro- and microorganisms. *Source: Figure created with BioRender.*

sourced from red macroalgae, and ulvans are extracted from green algal species (Mišurcová et al., 2012).

Polysaccharides exhibit a wide range of structural features that aid in performing biological functions. Exopolysaccharides (EPSs) produced by microorganisms contribute to their survivability by assisting in withstanding extreme environmental conditions. Some EPSs act as bioprotective agents, offering antimicrobial and cryoprotective benefits, while others contribute to surface adhesion or improve nutrient availability, thereby supporting microbial growth and adaptation (Costa et al., 2018; Ibrahim et al., 2022). Polysaccharides in an algal system perform both structural functions and serve as energy reserves in metabolic processes (Mišurcová et al., 2012).

Marine polysaccharides possess immense potential for a wide range of applications across various fields, including biomedical, cosmeceutical, nutraceutical, pharmaceutical, and bioremediation (Fig. 5.2). This versatility can be attributed to their bioactive properties, along with key advantages such as biocompatibility, biodegradability, nontoxicity, cost-effectiveness, and abundance.

Recent advancements have highlighted the significance of marine polysaccharides across various disciplines. In the biomedical sector, polysaccharides are being extensively studied for their potential in novel drug discovery and for their anticancer, antioxidant, immunomodulatory, antidiabetic, and neuroprotective effects, owing to their structural versatility (De Jesus Raposo et al., 2015). In the cosmeceutical and nutraceutical industries, compounds such as fucoidan and carrageenan are highly regarded for their antioxidant, antiaging, moisturizing, wound healing, and immune-boosting properties (Ding et al., 2022). In addition, marine polysaccharides also demonstrate significant potential toward the promotion of environmental sustainability. Their ability to chelate heavy metals, degrade pollutants like dyes, and perform oil remediation accentuates their role in bioremediation strategies (Banerjee et al., 2021). The diverse applications of polysaccharides, coupled with their abundance and renewability, make them suitable alternatives for sustainable and eco-friendly applications. However, challenges such as efficient extraction methods, scalability,

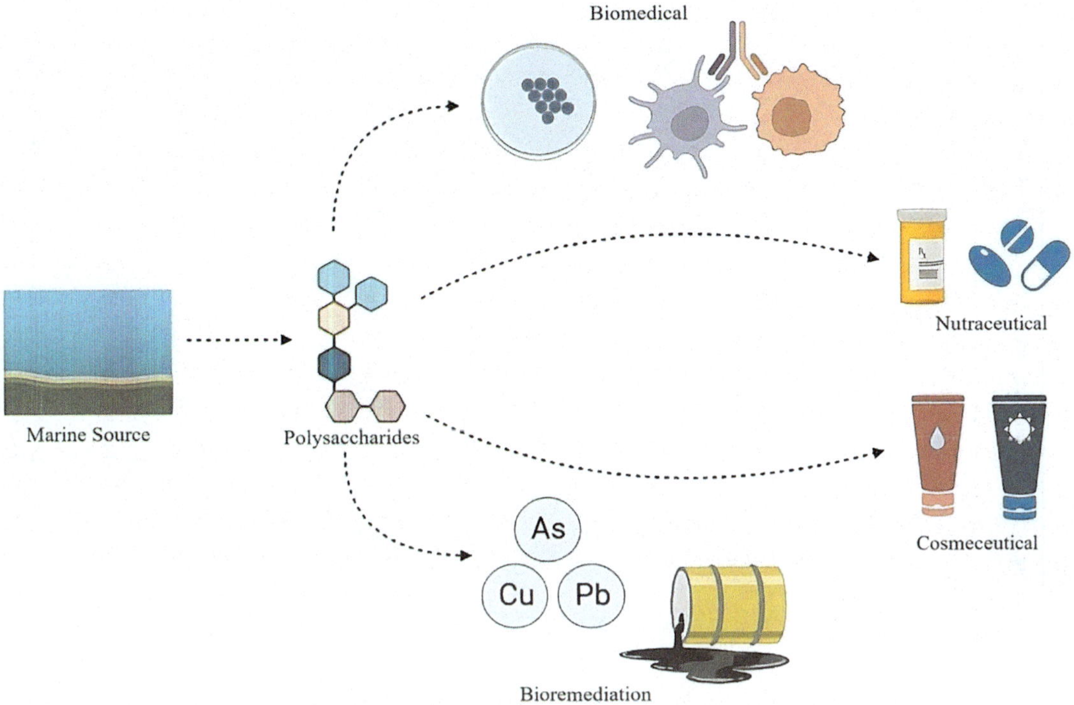

FIG. 5.2 Applications of marine polysaccharides across diverse fields. *Source: Figure created with BioRender.*

and regulatory compliance need to be addressed to unlock their full potential. This chapter provides a comprehensive summary of the pharmacological, cosmeceutical, nutraceutical, and bioremediation potential of marine polysaccharides derived from various sources, offering insights into their current applications and future prospects.

5.2 Different functions of polysaccharides

Polysaccharides in the marine habitat can be sourced from microorganisms, micro- and macroalgae, and crustaceans like crabs and prawns. The composition and glycosidic linkages of polysaccharides depend on the species (Table 5.1) and are influenced by the harvesting season and extraction process (Li et al., 2022). Macroalgae produce a variety of polysaccharides. Polysaccharides, namely, alginate, carrageenan, and agar, provide structural rigidity, while polysaccharides like laminarin and floridean starch function as energy reserves (Jeewon et al., 2023; Patel et al., 2023). On the contrary, microorganisms produce and secrete exopolysaccharides into their surroundings, which are generally present in slime layers or capsules and are not specific to cell types (Sutherland, 1982). Exopolysaccharides function as adhesives, promoting cellular associations and interactions. They help in creating a microenvironment to facilitate the transfer of genes and metabolites, ensuring the survival of the microorganism in nutrient-deficient conditions (Sutherland, 2001). Their role in

TABLE 5.1 List of polysaccharides obtained from marine bioresources and their respective biological activities.

Sl. no.	Organism	Composition	Molecular weight (kDa)	Biological activity	References
1.	*Vibrio alginolyticus* 364	Mannose, glucosamine, gluconic acid, galactosamine, and arabinose (5:9:3.4:0.5:0.8)	14.8	Antitumor activity	(Wang et al., 2021a)
2.	*Bacillus* sp. 11	Mannose, glucosamine, galacturonic acid, glucose, and xylose (1:2.58:0.68:0.13:3.09:1.41)	22.3	Anticancer activity	(Cao et al., 2018)
3.	*Microbacterium aurantiacum* FSW-25	Glucuronic acid, glucose, mannose, and fucose	7000	Antioxidant activity	(Sran et al., 2019b)
4.	*Rhodobacter johrii* CDR-SL 7Cii	Glucose, glucuronic acid, rhamnose, and galactose (3:1.5:0.25:0.25)	2000	Bioemulsification activity	(Sran et al., 2019a)
5.	*Bacillus licheniformis*	Fructose, fucose, glucose, galactosamine, and mannose (1.0:0.75:0.28:tr:tr) (where tr = trace)	1000	Cytotoxic property, antiviral activity, and immunomodulatory property	(Spanò et al., 2013)
6.	*Pantoea* sp. YU16-S3	Glucose, galactose, *N*-acetyl galactosamine, and glucosamine (1.9:1:0.4:0.02)	175	Wound healing property	(Sahana and Rekha, 2020)
7.	*Bacillus* sp. H5	Mannose, glucosamine, glucose, and galactose (1.00:0.02: 0.07:0.02)	89	Immunostimulatory property	(Wei et al., 2021)
8.	*Natronotalea sambharensis* sp. nov. AK103^T	Mannose, glucose, and glucuronic acid	4.6×10^6	Antioxidant activity	(Singh et al., 2019)
9.	*Bacillus cereus* KMS3-1	Mannose, xylose, glucose, and rhamnose (73.51:17.87:2.18:6.49)	–	Metal adsorption capacity	(Krishnamurthy et al., 2020)
10.	*Bacillus thuringiensis*	Fructose, galactose, xylose, glucose, rhamnose, and mannose (43.8:20:17.8:7.2:7.1:4.1)	–	Antioxidant activity and anticancer activity	(Ramamoorthy et al., 2018)
11.	*Pseudomonas stuzeri* 273	Glucosamine, rhamnose, and glucose (35.4:28.6:27.2)	190	Antibiofilms and antiinfection activity	(Wu et al., 2016)

12.	*Vibrio* sp. QY101	Galacturonic acid, glucuronic acid, rhamnose, and glucosamine	546	Antibacterial activity	(Jiang et al., 2011)
13.	*Halomonas* sp. 2E1	Mannose and glucose (3.76:1)	47	Immunomodulatory property	(Wang et al., 2021a)
14.	*Polaribacter* sp. SM1127	Rhamnose, fucose, glucuronic acid, mannose, galactose, glucose, and *N*-acetylglucosamine (0.8:7.4:21.4:23.4:17.3:1.6:28.0)	220	Antioxidant activity	(Sun et al., 2015)
15.	*Aureobasidium melanogenum* SCAU-266	Glucose, mannose, and galactose (97.30:1.9:0.08)	8	Immunomodulatory property	(Lin et al., 2022)
16.	*Alternaria* sp. SP-32	Mannose, glucose, and galactose (1.00:0.67:0.35)	27.4	Anticancer activity and antioxidant activity	(Chen et al., 2016)
17.	*Aspergillus versicolor* LCJ-5-4	Glucose and mannose (1.7:1.0)	7	Antioxidant activity	(Chen et al., 2012)
18.	*Hansfordia sinuosae*	Mannose, galactose, and glucose (96.1, 3.3, and 0.60)	22.5	Anticancer activity	(Li et al., 2018)
19.	*Porphyridium marinum*	Xylose, galactose, glucose, fucose, arabinose, and glucuronic acid (44–47:25–29:19–20:1:1–2:4–5)	1400–550	Antibacterial activity, antibiofilms, and anticancer activity	(Gargouch et al., 2021)
20.	*Tetraselmis suecica (Kylin)*	Arabinose, ribose, mannose, galacturonic acid, galactose, glucose, and glucuronic acid (5.23:0.83:6.64:0.1:25.27:35.46:21.47)	–	Antioxidant activity and anticancer activity	(Parra-Riofrío et al., 2020)

protection against salinity and temperature extremes, as well as potential predators, is another essential function (Nichols et al., 2005; Krüger et al., 2008). Marine prokaryotes protect themselves from antibiotics through the production of EPS. In addition, charged organic molecules and inorganic ions can be concentrated by exopolysaccharides (Maugeri et al., 2002).

Microorganisms in general colonize both abiotic surfaces like metal, suspended particles, and concrete, as well as biotic surfaces such as seaweeds, aided by the overproduction of extracellular polymeric substances, which play a critical role in the establishment and stability of these communities (Elsakhawy, 2017). The exopolysaccharides secreted by them act as the structural foundation of the biofilm matrix and serve a multifunctional role. They provide several advantages, including: (1) stabilization of extracellular enzyme activity, (2) facilitation of spatial organization among different species within the biofilm, (3) enhanced nutrient acquisition, (4) defense against predation, and (5) protection from toxins and antibiotics (Allison et al., 1998; Dang and Lovell, 2000).

The marine habitat comprises various nutritional sources, available in different concentrations (Singh et al., 2024). Microorganisms use numerous survival strategies under starvation. For instance, the production of proteins specific to starvation, such as proteolytic enzymes, that allow bacterial cells to scavenge nutrients (Mårdén et al., 1987; Albertson et al., 1990; Baijal et al., 2024; Zerbib et al., 2024). Other survival strategies for starvation-induced cells may include the adherence of the cells to hydrophobic surfaces, aiding the cells to spread over a broad area and scavenge nutrients (Hermansson et al., 1987). The S9 strain of *Pseudomonas* sp. has been reported to produce extracellular polysaccharides during the initial phase of starvation, which reduces cellular adhesion to hydrophobic surfaces and facilitates detachment (Wrangstadh et al., 1986).

The polysaccharide particle formation converts dissolved organic carbon into particulate organic carbon during phytoplankton blooms through the aggregation kinetics process. Transparent exopolymer particles (TEPs) represent a class of large transparent particles found in seawater and diatom cultures. These particles are formed from dissolved exopolymers, primarily exopolysaccharides, exuded by phytoplankton and bacteria. TEP can influence oceanic processes, including food web dynamics, carbon cycling, microbial activity, and particulate flux (Elsakhawy, 2017). While particle aggregation, which leads to the production of marine detritus, has been traditionally attributed to physicochemical flocculation, experiments using killed-bacterial cultures have demonstrated that active bacterial activity is essential for the formation of these aggregates (Kranck and Milligan, 1980; Biddanda, 1985). Transmission electron microscopy studies have demonstrated that bacterial extracellular polysaccharides are key drivers of marine aggregate formation (Liu et al., 2004; Huang et al., 2022). Similarly, extracellular materials also facilitate the attachment of bacteria to decomposing seaweeds (Biddanda, 1985). These aggregates sink to the ocean floor, thus facilitating the transport of nutrients and energy and bridging the gap of nutrient availability between surface and deep waters. The coagulation of individual particles into rapidly settling aggregates, often referred to as the "biological pump" plays a crucial role in this process (Engel et al., 2004). Another study by Engel identified a linear relationship between CO_2 concentrations and the production of TEPs, suggesting that TEP production is closely tied to theoretical CO_2 uptake rates (Engel, 2002). This natural aggregation process parallels the flocculation mechanism in wastewater treatment, where particle aggregation and sedimentation in settling tanks result in sludge formation (Elsakhawy, 2017). Studies have shown that nanoscale fibrils, rich in polysaccharides, are common colloidal particles in the aquatic systems (Gupta and Diwan, 2017; Bhatia et al., 2021). These fibrils provide a structural matrix for the development of larger aggregates and effectively scavenge metal ions from the surrounding water (Elsakhawy, 2017).

Extreme marine environments, such as deep-sea hydrothermal vents, cold seeps, coastal hot springs, polar regions, and hypersaline ponds, are potentially rich sources of unknown and uncultivated bacteria. These groups of bacteria adopt specialized metabolic pathways to survive through such harsh living conditions. These metabolic pathways enable them to produce bioactive compounds like exopolysaccharides more efficiently compared to other microorganisms (Laurienzo, 2010). Many of the extracellular polysaccharides produced by these extremophiles possess unique properties. Halophilic bacteria are particularly noteworthy since the exopolysaccharides produced by them have numerous industrial applications (Margesin and Schinner, 2001). Moreover, these groups of bacteria also exhibit significant osmotic tolerance, allowing polysaccharide production at higher sugar concentrations, a crucial factor for developing economically viable polysaccharide production processes (Mehta et al., 2014). Moreover, marine microbial polysaccharides are often characterized by the presence of glucuronic acid, galacturonic acid, amino sugars, and pyruvate with potential applications in pharmaceuticals, as food additives, and in industrial waste treatment processes (Wang et al., 2012).

The polysaccharide complex present in the macroalgal cell walls provides it with structural support and protection against environmental stress (Jönsson et al., 2020). Studies have also reported the role of the polysaccharides in cellular interaction and processes (Pomin and Mourão, 2008; Vilanova et al., 2009) and their role in pathogen prevention by a protective polysaccharide barrier (Aziz et al., 2003). The production and composition of these polysaccharides vary among species of the same genus due to the geographical location and algal physiological state (Pérez et al., 2016; Zammuto et al., 2022).

Brown algae are rich in polysaccharides such as fucoidan, alginate, laminarin, and cellulose. Each of these polysaccharides serves distinct physiological functions critical for algal survival in marine environments (Deniaud-Bouët et al., 2014). For instance, fucoidans play a crucial role in regulating water and ion retention in the extracellular matrices, thus helping prevention of dehydration and maintenance of osmotic balance in cells (Deniaud-Bouët et al., 2017). But, alginates contribute to the mechanical strength of these organisms (Zhao et al., 2012), enabling them to withstand the physical stresses of their marine environment. Laminarin, a relatively short polymer, serves as the primary energy reserve (Graiff et al., 2016). Rhodophyceae, commonly known as red algae, encompasses over 7000 species. Their cell wall is mainly composed of cellulose, carrageenan, and agar. Carrageenans are generally classified into three types: kappa (κ), iota (ι), and lambda (λ), depending on the number of ester-sulphate groups (Tuvikene, 2021). Commercially, κ- and ι-carrageenans are utilized as gel-forming agents, while λ-carrageenan is used as a thickening agent (Pacheco-Quito et al., 2020). Green algae, or Chlorophyceae, primarily consist of the polysaccharide ulvan. Ulvan is composed of sulfated rhamnose, uronic acids (glucuronic acid and iduronic acid), and xylose and constitutes nearly 29% of their dry cell weight (Brading et al., 1954; Lahaye and Robic, 2007).

5.3 Biomedical applications of marine polysaccharides

Marine polysaccharides have unique chemical structures and exhibit distinct anticancer, antimicrobial, antioxidant, and immunomodulatory properties, which make them suitable candidates for biomedical applications (Table 5.2). Furthermore, these polysaccharides are characterized by low molecular weight and high biocompatibility, thus enhancing their therapeutic applications.

5.3.1 Anticancer activities reported from marine polysaccharides

The anticancer properties exhibited by diverse marine macroalgal, microalgal, bacterial, and fungal species have been extensively highlighted in research (Fedorov et al., 2013). A novel polysaccharide, EPS364 obtained and purified from the marine bacterium *Vibrio alginolyticus* 364, was reported to inhibit liver cancer cell lines Bel-7402 and Huh7.5. EPS364 induced apoptosis in the cancer cells by increasing reactive oxygen species (ROS) generation in mitochondria, eventually leading to mitochondrial membrane depolarization (Wang et al., 2021b). Another exopolysaccharide, EPS11 sourced from *Bacillus* sp. 11, was found to exhibit antimetastatic activity against liver cancer cells tested in the cell lines Huh7.5, Bel-7402, and HepG2. EPS11 inhibited the collagen I/β1-integrin signaling pathway, thus showing antimetastatic activity (Liu et al., 2021). Similarly, Yahya et al. reported the *Brevundimonas subvibrioides* exopolysaccharide MSA1 which upregulated TLRs (Toll-like receptors) and proapoptotic proteins such as BAD, BAX, BID, and BIM while downregulating the prosurvival protein BCL-W pathway to inhibit the HepG2 cell line. An exopolysaccharide from the marine fungal species *Hansfordia sinuosae* was also reported by Li et al. (2018). In the HeLa cell line, this exopolysaccharide decreased mitochondrial membrane potential and increased ROS levels and caspase-3 expression. The activity of the polysaccharide was tested in the human cancer cell lines HeLa and MCF-7.

Low-molecular-weight λ-carrageenan derived from the macroalga *Chondrus ocellatus* demonstrated antitumor activity against Sarcoma-180 (S180) tumor cells, mainly through the upregulation of TNF-α levels (Zhou et al., 2005). A κ-carrageenan variant isolated from *Kappaphycus striatum* has also demonstrated comparable antitumor effects in S180 tumor cell-bearing murine models. These effects were linked to enhanced macrophage phagocytic activity, increased natural killer (NK) cell activity, and elevated levels of serum IL-2 and TNF-α (Yuan et al., 2006). Fucoidan extracted from the alga *Undaria pinnatifida* has been reported for its antitumor effect against the mouse hepatocarcinoma Hca-F cell line (Teng et al., 2015). Exposure of Hca-F cells to fucoidan resulted in suppressed HIF-1α expression and nuclear translocation, alongside increased TIMP-1 protein levels. This effectively reduced metastasis and lymphangiogenesis in mouse hepatocarcinoma.

Polysaccharides produced by marine animals have also been studied for their anticancer properties. A typical example is the SEP polysaccharide, derived from the eggs of the sea urchin *Strongylocentrotus nudus* (Zhang et al., 2017a). The isolated polysaccharide inhibited hepatocellular carcinoma growth in mice by enhancing the secretion of cytokines IL-2, TNF-α, and IFN-γ, and activating lymphocytes and macrophages (Wang et al., 2011).

TABLE 5.2 Structures of some polysaccharides obtained from marine bioresources.

Sl. no.	Organism	Source type	Possible structure of the compound	Biological activity	References
1.	*Strongylocentrotus nudus*	Echinoderm	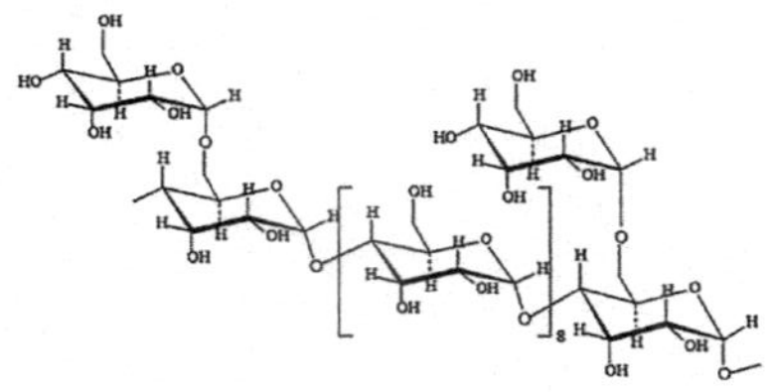	Anticancer activity and immunomodulator property	(Wang et al., 2011)
2.	*Kjellmaniella crassifolia*	Macroalga	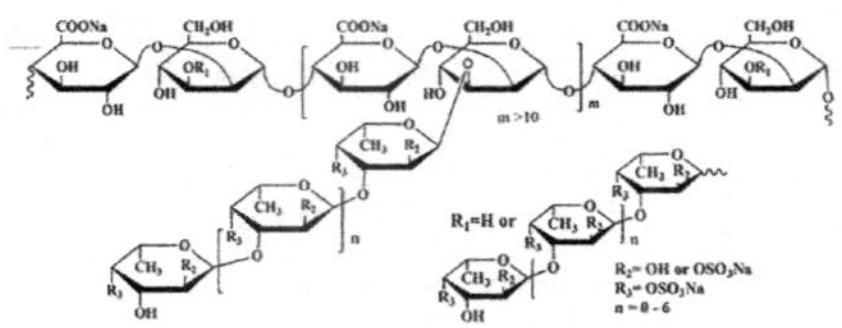	Antiviral activity against influenza A virus	(Wang et al., 2017)
3.	*Bacillus subtilis*	Bacterium	β-Glu-(1→4)-[[β-Glu-(1→4)]$_5$-β-Rha-(1→4)-β-Ara-(1→4)-β-Ara-(1→4)-β-Ara-(1→4)-β-Glu-(1→]$_n$	Antioxidant activity, antiinflammatory activity, cytotoxicity property, and anti-Alzheimer's activity	(Abdel-Wahab et al., 2022)
4.	*Tichocarpus crinitus*	Macroalga	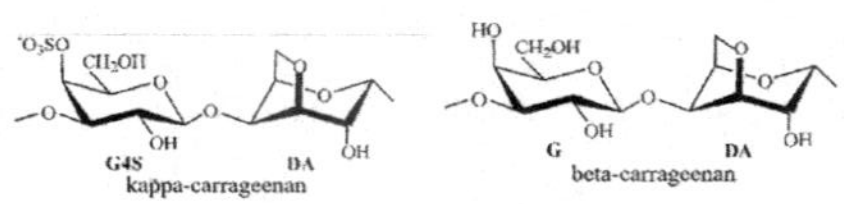	Immunomodulatory property	(Cicinskas et al., 2020)
5.	*Arca subcrenata*	Mollusk	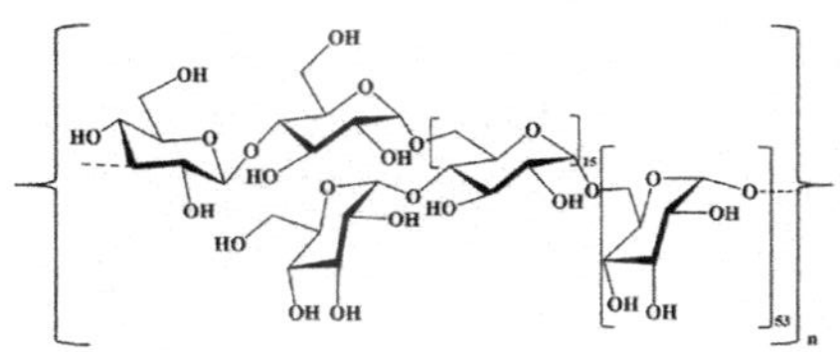	Immunomodulatory property	(Shi et al., 2021)

Continued

TABLE 5.2 Structures of some polysaccharides obtained from marine bioresources—cont'd

Sl. no.	Organism	Source type	Possible structure of the compound	Biological activity	References
6.	*Alteromonas* sp. *JL2810*	Bacterium	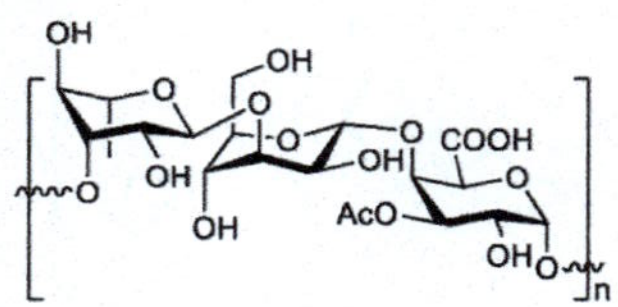	Metal adsorption capacity	(Zhang et al., 2017a)
7.	*Pseudoalteromonas* sp.	Bacterium	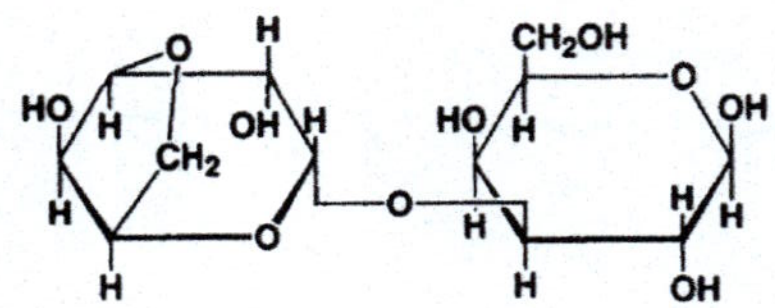	Moisturizing effect and whitening effect	(Kobayashi et al., 1997)
8.	*Ascophyllum nodosum*	Macroalga	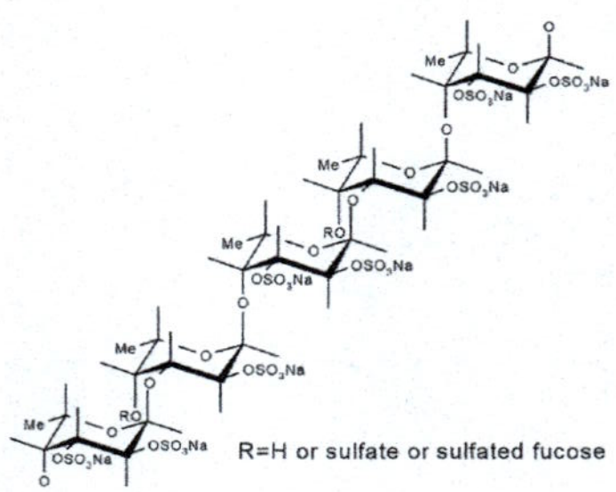	Antiaging activity	(Senni et al., 2006)
9.	*Bacillus* sp.	Bacterium	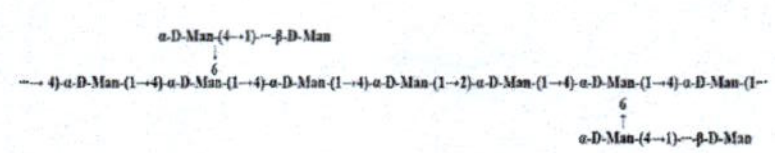	Immunomodulatory property	(Wei et al., 2021)

10.	*Penicillium janthinellum* N29	Fungus	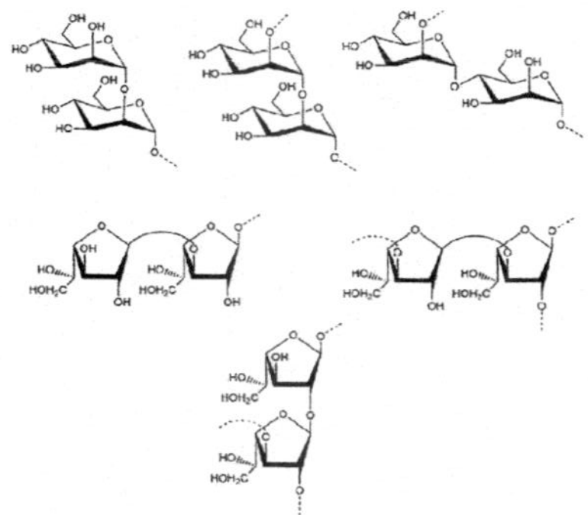	Antidiabetic activity	(Shao et al., 2023)
11.	*Pseudoaltermonas* strainSM20310	Bacterium	α-Man*p*-(1→6)-α-Man*p* 1 ↓ 2 →6)-α-Man*p*-(1→6)-α-Man*p*-(1→2)-α-Man*p*-(1→	Cryoprotective effect	(Liu et al., 2013)
12.	*Aspergillus ochraceus*	Fungus	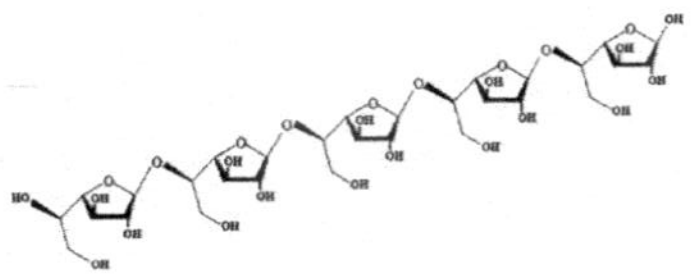	–	(Guo et al., 2014)
13.	*Aspergillus versicolor* SCAU141	Fungus	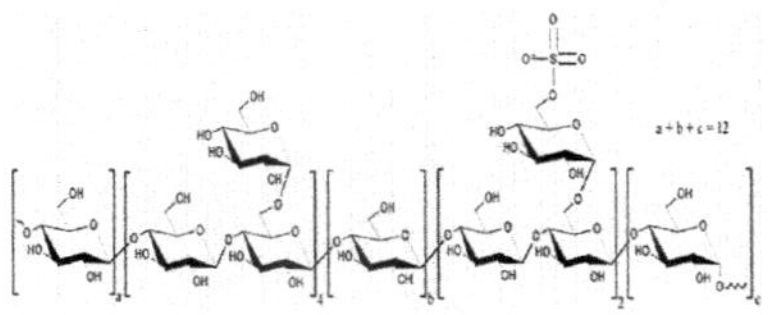	Immunomodulatory property	(Jeewon et al., 2023; Wu et al., 2023)

Marine polysaccharides represent a promising area for future research in the development of targeted and safer cancer treatments. Investigating their molecular mechanisms and clinical applications may lead to novel and effective cancer therapies.

5.3.2 Antimicrobial activities reported from marine polysaccharides

The growing population of multidrug-resistant pathogenic strains highlights the need to explore the vast array of understudied bioactive compounds that have potential as antimicrobials. Marine polysaccharides exhibit broad-spectrum antimicrobial activity against various pathogenic microorganisms. Advanced research into their structure–activity relationships and synergistic effects with existing antibodies could accelerate their therapeutic applications.

The marine bacterium *Enterobacter* sp. strain ACD2 was reported to produce EPS with the potential to inhibit the growth of *Escherichia coli* and *Staphylococcus aureus* (Almutairi and Helal, 2021). A study conducted by Aullybux and his group has reported two exopolysaccharides, namely, EPS-15 (levan) and EPS-37 (curdlan), isolated from *Halomonas* sp. and *Alcaligenes faecalis*, respectively, with the potential to inhibit methicillin-resistant *Staphylococcus aureus*, *Streptococcus pneumoniae*, and *Campylobacter jejuni* (Aullybux et al., 2019). The authors suggested that the antimicrobial properties can be attributed to the presence of functional groups that act as metal chelators and reduce the amount of essential metal ions required for the survival of the pathogens. Fucoidan, a sulfated polysaccharide sourced from the brown algae *Sargassum wightii*, exhibited significant antimicrobial activity against human bacterial pathogens, *Vibrio cholerae* and *Salmonella typhi* (Marudhupandi and Kumar, 2013).

Microalgal polysaccharides have also been reported to exhibit antiviral activity. For example, polysaccharides extracted from the microalgae *Porphyridium* sp. have strong inhibitory effects against alpha-herpes viruses, Herpes simplex virus types 1 and 2 (HSV-1 and HSV-2), and Varicella zoster virus (Huleihel et al., 2001). The polysaccharide prevents the entry of the viral particle into the host cell, thus avoiding replication of the viral particles and eventually interrupting the infection cycle. Similarly, exopolysaccharides sourced from *Porphyridium purpureum*—a red microalga—exhibit antiviral activity against the Vaccinia virus. These bioactive molecules restrict the viral membrane glycoproteins, thus preventing viral attachment and entry into host cells (Radonic et al., 2010). Lee and his group reported the extraction of fucoidan from *U. pinnatifida*, which exhibited broad-spectrum antiviral properties against various enveloped viruses, including Herpes simplex virus types 1 and 2 (HSV-1, HSV-2), human cytomegalovirus, and Influenza A virus, primarily through interference with virus-host cell attachment mechanisms (Lee et al., 2004). Furthermore, fucoidan derived from the brown alga *Kjellmaniella crassifolia* demonstrated anti-influenza A virus (IAV) activity through two mechanisms: (1) inhibition of viral neuraminidase activity to prevent virus release and (2) reduction of IAV endocytosis through suppression of the EGFR (Epidermal Growth Factor Receptor) signaling pathway (Wang et al., 2017).

5.3.3 Antioxidant activities reported from marine polysaccharides

Aerobic organisms generate ROS such as hydroxyl radicals (·OH), superoxide anion radicals ($O_2\cdot^-$), nitroxide radicals (NO·), peroxyl radicals ($ROO\cdot^-$), and hydrogen peroxide (H_2O_2). These ROS are produced as by-products of metabolic pathways, particularly cellular respiration. However, an imbalance or overproduction of these ROS can damage lipids, proteins, and DNA in human cells, thereby contributing to the development of various diseases. Extracellular polysaccharide substances derived from marine sources often exhibit potent antioxidant activity due to their distinct structural characteristics and functional groups that facilitate ROS scavenging.

The exopolysaccharide Mi-25, produced by *Microbacterium aurantiacum* strain FSW-25, exhibited a 2,2-diphenyl-1-picrylhydrazyl (DPPH) free radical scavenging rate of 80% ± 2.21% at a concentration of 1 mg/mL. This free radical scavenging rate surpassed the performance of commercial xanthan, which had a scavenging rate of 65%±1.51% (Sran et al., 2019b). The antioxidant activity of EPS Mi-25 thus demonstrates its potential for application in the pharmaceutical and healthcare sectors.

Similarly, EPSR4—produced by the marine *Bacillus subtilis* strain AG4—exhibited remarkable antioxidant properties, with DPPH free radical scavenging activity reaching 97.6% at a 1500 μg/mL concentration (IC_{50} − 300 μg/mL) and hydrogen peroxide scavenging activity of 64.8% at 1500 μg/mL (IC_{50} = 1500 μg/mL, 30 min) (Abdel-Wahab et al., 2022). The antioxidant activity of EPSR4, together with its additional anticancer, antiinflammatory, and anti-Alzheimer effects, reinforces its potential applicability in pharmaceutical and health-related fields. Furthermore, the EPS derived from the deep-sea bacterium *Zunongwangia profunda* SM-A87 exhibited a concentration-dependent enhancement in free radical scavenging, achieving 48.5 ± 1.5%, 58.7 ± 0.8%, and 27.2 ± 0.6% scavenging of DPPH, hydroxyl, and superoxide radicals, respectively, at a concentration of 10.0 mg/mL. These values were significantly higher compared to hyaluronic acid (HA), which showed minimal DPPH scavenging and achieved only 19.1 ± 0.1% hydroxyl and 15.2 ± 0.2% superoxide radical scavenging rates at the same concentration (Sun et al., 2014).

Macroalgal polysaccharides have also demonstrated antioxidant properties. Crude polysaccharide extract SLBS11P obtained from *Chnoospora minima* showed DPPH and alkyl radical scavenging activities, with IC_{50} values of 89.51 ± 17.00 and 106.80 ± 0.66 μg/mL, respectively. SLBS11P also demonstrated a strong ability to scavenge intracellular ROS in response to oxidative stress induced by H_2O_2 and AAPH (2,2′-azobis(2-amidinopropane) dihydrochloride), effectively reducing oxidative damage within cells (Fernando et al., 2017). Li and his group reported the antioxidant potential of a sulfated polysaccharide, fucoidan, obtained from *Nemacystus decipiens*. This polysaccharide was tested for DPPH radical scavenging and hydroxyl radical scavenging activities, achieving IC_{50} values of 3.96 and 4.12 mg/mL, respectively (Li et al., 2017). Polysaccharides extracted from *Gracilaria rubra* also demonstrated strong antioxidant properties, as evidenced by their radical scavenging activity experiments. At a concentration of 2.5 mg/mL, GRPS-1-1 showed DPPH radical scavenging activity of 41.59 ± 1.65%, ABTS (2,2′-azino-bis(3-ethylbenzothiazoline-6-sulfonic acid) radical scavenging activity of 59.01 ± 2.26%, superoxide anion radical scavenging activity of 64.78 ± 1.29%, and hydroxyl radical scavenging activity of 69.26 ± 1.04%. In addition, other polysaccharide fractions, GRPS-2-1 and GRPS-3-1, also demonstrated notable antioxidant potential (Di et al., 2017).

Polysaccharides sourced from microalgae are also known for their effective free radical scavenging activity. Sulfated polysaccharides WSPN (Sulfated polysaccharide from *Navicula* sp. in white wavelength), RSPN (Sulfated polysaccharide from *Navicula* sp. in red wavelength), and BSPN (Sulfated polysaccharide from *Navicula* sp. in blue wavelength), extracted from *Navicula* sp., showed DPPH radical scavenging activity, with WSPN having the lowest IC_{50}

value of 238 mg/mL, compared to BSPN—326 mg/mL and RSPN—3066 mg/mL (Fimbres-Olivarria et al., 2018). Extracellular polysaccharides of *Graesiella* sp. were reported to have high hydroxyl radical scavenging activity of 65.2% ± 4.7 at 2 mg/mL, and the IC_{50} value was reported to be 0.87 mg/mL. The extracted polysaccharide was also stated to have moderate DPPH and superoxide radical scavenging ability of 43.2 ± 1.4% and 15.2 ± 2.4% at 2 mg/mL, respectively (Trabelsi et al., 2016).

Marine animals, particularly marine invertebrates, are yet another rich source of antioxidants. The sulfated polysaccharide—HfP, obtained from the sea cucumber *Holothuria fuscogliva*, possesses strong antioxidant properties. The hydroxyl radical scavenging activity (EC_{50}) of HfP was determined to be 3.74 mg/mL, comparable to that of vitamin C (2.72 mg/mL), thereby highlighting its effective antioxidant potential. HfP displayed better superoxide radical scavenging activity with an EC_{50} value of 0.0378 mg/mL, significantly outperforming vitamin C (EC_{50} = 0.93 mg/mL) (Li et al., 2017). Hence, it can be concluded that HfP is a promising natural antioxidant that could be used to improve health products. Polysaccharides extracted from another sea cucumber, *Phyllophorus proteus* (PPP), were also studied and reported to have antioxidant properties. At the concentration of 4 mg/mL, PPP-2 exhibited DPPH, hydroxyl, superoxide, and ABTS radical scavenging rates of 30.90%, 27.99%, 99.66%, and 58.84%, respectively (Qin et al., 2018). These findings emphasize the significant role of polysaccharides derived from sea cucumbers in antioxidant applications.

5.3.4 Immunomodulatory activities reported from marine polysaccharides

Marine polysaccharides also demonstrate strong immunomodulatory properties, positioning them as promising candidates for use in immunotherapy, vaccine development, and the treatment of autoimmune diseases. A highly branched extracellular polysaccharide, EPSm, was extracted from *Halomonas* sp. 2E1, isolated from the South China Sea, and studied by Wang et al. (2021a). The researchers reported immunostimulatory effects through the activation of the MAPK and NF-κB pathways, resulting in increased production of NO, COX-2, TNF-α, IL-1β, and IL-6. Likewise, the exopolysaccharide, α-mannan, isolated from *Sphingobacterium* sp. IITKGP-BTPF3, has been reported to remarkably reduce nitric oxide (NO) production in LPS-treated mouse macrophages, RAW 264.7 cells (Chatterjee et al., 2018). This finding highlights its potential application as an immunomodulatory agent. The polysaccharide, PEP, obtained from the Antarctic bacterium *Pseudoalteromonas* sp. S-5, has been reported as a biological response modifier. The PEP polysaccharide has demonstrated immunomodulatory activities by activating macrophages, enhancing phagocytosis (133% increase relative to untreated cells), upregulating NO production (44.5% increase compared to control), and stimulating the release of pro-inflammatory cytokines such as TNF-α and IL-1β (Bai et al., 2012). Lin and his group documented an extracellular polysaccharide, AUM-1, with inflammation-promoting activity, isolated from the fungus *Aureobasidium melanogenum*, through the upregulation of cytokines COX2, iNOS, TNF-α, IL-1β, IL-6, and IL-10. In studies using RAW 264.7 cells, AUM-1 also exhibited ferroptosis-related immunomodulatory properties by modulating the expression of key proteins COX-2, SLC7A11, GPX4, ACSL4, and FTH1 (Lin et al., 2022).

Fucoidan has also been extensively studied with respect to its immunomodulatory properties. Fucoidans derived from *U. pinnatifida* have been reported to delay neutrophil apoptosis by upregulating the expression of the antiapoptotic protein Mcl-1 and downregulating activated caspase-3 levels. In addition, the *U. pinnatifida* fucoidan can induce the synthesis of IL-6, IL-8, and TNF-α from neutrophils in an AKT-dependent manner (Jin and Yu, 2015).

Fucoidan obtained from the kelp *Macrocystis pyrifera* has been documented and reported (Zhang et al., 2015b). Fucoidan was reported to delay neutrophil apoptosis and facilitate the maturation and activation of NK and splenic dendritic cells (DCs), consequently enhancing T-cell immune response and memory T-cell generation. Fucoidan from *Fucus vesiculosus* has also been reported for its role in the upregulation of CD40, CD80, and CD86 expression and the production of IL-6, IL-12, and TNF-α in spleen conventional DCs (Jin et al., 2014a). Carrageenans, particularly λ- and κ-carrageenans, extracted from red algae, can modulate immune responses via various pathways (Amin et al., 2020; Tiasto et al., 2022). The polysaccharide κ-carrageenan has been reported to enhance immune responses in S180-bearing mice through the upregulation of macrophage phagocytosis and the promotion of antibody secretion from spleen cells. In addition, it has been associated with increased production of cytokines, including IL-2 and TNF-α levels in murine serum (Yuan et al., 2006). Carrageenan extracted from *Tichocarpus crinitus* has also demonstrated immunostimulatory effects in murine models, significantly elevating the levels of proinflammatory cytokines such as IFN-γ, IL-12, and IL-1β (Cicinskas et al., 2020).

Marine mollusks have been a source of bioactive polysaccharides with immunomodulatory properties (Montuori et al., 2022). The water-soluble polysaccharide ASPG-2, extracted from *Arca subcrenata*, has been reported to promote macrophage polarization while enhancing NO secretion. It also activates critical signaling pathways, including MAPK and NF-κB (Shi et al., 2021). Similarly, the polysaccharide MP (*Mytilus coruscus* derived Polysaccharide), has demonstrated immunostimulatory effects by promoting the proliferation of RAW264.7 macrophage cells and modulating inflammatory cytokines such as TNF-α, IL-6, and IL-10. In addition, MP was found to inhibit the NF-κB signaling pathway (Xiang et al., 2021).

5.3.5 Other activities properties reported from marine polysaccharides

Marine polysaccharides have demonstrated promise in addressing the underlying mechanisms of chronic diseases such as diabetes, Alzheimer's, and other degenerative conditions. These diseases are commonly associated with oxidative stress, persistent inflammation, and metabolic imbalance. Due to their bioactive properties, biocompatibility, and structural flexibility, marine polysaccharides present a viable alternative for therapeutic applications in managing these conditions.

5.3.5.1 *Antidiabetic properties reported from marine polysaccharides*

Low-molecular-weight fucoidan (LMWF), isolated from *Laminaria japonica*, was tested on Goto-Kakizaki type 2 diabetic rats, a widely used model for studying diabetic vascular complications. The polysaccharide resulted in the upregulation of endothelial nitric oxide synthase phosphorylation and NO production, both of which are critical in the maintenance of vascular health. LMWF treatment has also been shown to restore endothelium-dependent relaxation in various vascular tissues, including the aorta, mesenteric arteries, and paw arteries. This effectively counteracts diabetes-induced vascular dysfunction, highlighting the potential of LMWF as a therapeutic candidate for managing diabetic vascular complications (Cui et al., 2014). Similarly, Liang and his group reported an LMWF, extracted from *L. japonica*, which demonstrated superior antidiabetic effects compared to the commercially available drug probucol in diabetes-induced Sprague-Dawley rats. Both LMWF and probucol effectively reduced triglycerides (TG) and low-density lipoprotein cholesterol (LDL-C), but LMWF demonstrated superior efficacy at a higher dose (100 mg/kg/day), significantly lowering total cholesterol (TC) in comparison with probucol. In addition, LMWF significantly lowered ROS levels, a major trigger of diabetes-related complications, and restored antioxidant enzyme activities, effects that were not observed in the probucol-treated groups (Liang et al., 2016).

5.3.5.2 Neuroprotective properties reported from marine polysaccharides

Fucoidans extracted from brown algae have been recognized for their neuroprotective potential. Studies have reported the potential of fucoidan to safeguard the brain from ischemia-reperfusion injury through the action of the MAPK signaling cascade. Fucoidan treatment led to a decrease in the levels of inflammatory cytokines IL-1, IL-6, myeloperoxidase (MPO), and TNF-α; oxidative stress-related proteins superoxide dismutase (SOD) and MDA; and apoptosis-related proteins such as p-p53, Bax, and Bcl-2 (Che et al., 2017). Fucoidans, specifically SC1, SC2, and SC3, extracted from *Sargassum crassifolium*, exhibited both neuroprotective and antioxidant properties on H_2O_2-induced rat pheochromocytoma PC-12 cells. All three fucoidan extracts—SC1, SC2, and SC3—exhibited notable DPPH and ABTS radical scavenging activity. The extracts improved cell viability, reduced apoptosis, and effectively preserved mitochondrial membrane potential. SC2 was found to be the most effective, enhancing cell viability by 2.28 times, reducing apoptotic markers, and preserving mitochondrial function more effectively than the other extracts (Yang et al., 2017). This suggests that fucoidans could be promising candidates for preventing neurodegenerative damage associated with oxidative stress (Yang et al., 2017). Heteropolysaccharides extracted from four other *Sargassum* sp. have also been reported for their neuroprotective properties (Jin et al., 2014b).

Sulfated agaran-SA-Gc, obtained from *Gracilaria cornea*, exhibited neuroprotective activity in Wistar rats treated with the neurotoxin 6-hydroxydopamine (6-OHDA). The SA-Gc-treated group rejuvenated locomotor performance and resulted in a decrease in lipid peroxidation levels in the cerebral areas of the rats. In addition, the treated group also showed downregulated expression of inflammatory genes such as p65, iNOS, and IL-1β. Therefore SA-Gc could potentially be used against neurodegenerative disorders (Souza et al., 2017).

Polysaccharide derived from *Spirulina platensis* was reported for its neuroprotective effects in a Parkinson's disease model using C_{57}BL/6J mice treated with intraperitoneal injections of 1-methyl-4-phenyl-1,2,3,6-tetrahydropyridine (MPTP). Pretreatment with the polysaccharide resulted in the preservation of dopaminergic neurons in the substantia nigra region. Furthermore, the polysaccharide led to the upregulation of antioxidant enzyme activity, including SOD and glutathione peroxidase (GSH-Px), in both the serum and midbrain region (Zhang et al., 2015a). Another sulfated polysaccharide—SJP, extracted from the sea cucumber *Stichopus japonicus*, was investigated for its neuroprotective property against 6-hydroxydopamine (6-OHDA)-treated SH-SY5Y cells. The *Stichopus japonicus* polysaccharide-SJP was found to promote SOD activity while minimizing the concentrations of malondialdehyde (MDA) and ROS. Moreover, the polysaccharide was shown to regulate the Bax/Bcl-2 ratio, inhibit the activation of phosphorylated p53, and suppress the NF-κB signaling pathway. This, in turn, suppressed the expression of inducible nitric oxide synthase and downregulated NO production (Cui et al., 2016).

5.4 Cosmeceutical applications of marine polysaccharides

Skin plays a protective role in the human body. As we age, the skin becomes thinner and less elastic, leading to dryness and the formation of wrinkles (Resende et al., 2021). Skincare is essential for maintaining skin health and appearance while enhancing its natural protective barrier (Bedoux et al., 2014). Polysaccharides extracted from marine microorganisms are widely known for their significant bioactive properties. Exopolysaccharides can effectively bind to positively charged biomolecules through ionic interactions, as they carry a negative charge due to the presence of functional groups like carboxyl, phosphate, sulfate, and pyruvate. Glycosaminoglycans, a linear polysaccharide present in mammalian tissue, interact with various proteins and play a key role in physiological processes such as adhesion, migration, proliferation, and differentiation (Perez et al., 2023), all of which are crucial for skin repair. Polysaccharides in cosmetic formulations are also utilized to preserve the structural integrity and health of the skin, owing to their moisturizing, soothing, and antiwrinkle properties, along with their whitening effects and ultraviolet (UV) protection (Yao and Baojun, 2022).

5.4.1 Moisturizing effects reported from marine polysaccharides

The dermis is primarily made up of fibroblasts, embedded within a matrix of collagen and elastic fibers, along with an extracellular matrix abundant in glycoproteins, HA, and glycosaminoglycans (GAGs). Water is essential for maintaining skin functionality, and GAGs play a major role in retaining the water content of the skin, thereby contributing to hydration and structural integrity (Bedoux et al., 2014). External factors such as UV rays, weather conditions, and air pollutants like smoke can weaken the skin, causing it to lose its natural moisturizing ability (Pimentel et al., 2018). The thickness of the algal cell wall, combined with the mucus layer formed by the secreted polysaccharides, plays an important role in maintaining cellular hydration. This protective barrier helps retain moisture, ensuring the hydration of the cells in challenging environmental conditions (Stoyneva-Gärtner et al., 2020).

Wang and his group extracted polysaccharides from five different marine algae: *Saccharina japonica, Enteromorpha linza, Bryopsis plumosa, Codium fragile*, and *Porphyra haitanensis*, and examined their moisture absorption and retention abilities (Wang et al., 2013). This study investigated the relationship between chemical structure (including sulfated groups, monosaccharides, and molecular weight) and the adsorption capacity of the extracted polysaccharides. The researchers found that the sulfated group acted as the main active site for water retention. They also discovered that low-molecular-weight polysaccharides enhance moisture absorption and retention. In this case, the extracted polysaccharide from brown algae with a low molecular weight exhibited superior moisturizing properties compared to HA. Moreover, Shao and colleagues reported that a polysaccharide extracted from *Sargassum horneri* exhibited enhanced moisture absorption and retention compared to propanediol and glycerin. They also found that the moisture retention capacity of the polysaccharide improved as its molecular weight decreased (Shao et al., 2015).

The exopolysaccharide (EPS) produced by *Polaribacter* sp. SM1127, an Arctic kelp-associated bacterium, demonstrated exceptional moisturizing properties that exceed those of commercially available HA, a common ingredient in the cosmetic industry. Furthermore, it provided protective effects on human dermal fibroblast cells at low temperatures, emphasizing its potential as an effective moisturizing agent in the cosmeceutical industry (Sun et al., 2015). Similarly, the EPS from *Vibrio* sp. CNCM I-4277 has been shown to stimulate the synthesis of HA in skin cells, further boosting its potential for improving skin hydration (González et al., 2016). In addition, Li and his group isolated an agarase-producing psychrophilic strain from Antarctica, which was subsequently identified as *Pseudoalteromonas* sp. NJ21 (Li et al., 2015). This strain is capable of enzymatically hydrolyzing agar to produce neoagarobiose. According to Kobayashi and his group, neoagarobiose demonstrates superior water absorption and retention properties compared to conventional hydrating agents like collagen, chitosan, and glycerol (Kobayashi et al., 1997). Interestingly, a novel strain of *Pseudoalteromonas* sp. was isolated from the polar regions by Martins and his group. The intracellular extracts of the strain are incorporated into cosmetic formulations and marketed as RefirMAR by BIOALVO. RefirMAR has been shown to reduce wrinkle depth by up to 23% and increase skin hydration by up to 64% after 28 days of use (Martins et al., 2014).

The sulfated polysaccharide extracted from *Porphyridium cruentum* increases the expression of glycosaminoglycans (GAGs) around keratinocytes in the epidermal layer. To evaluate the impact of the sulfated polysaccharide, the polysaccharide was topically applied to reconstructed epidermal skin. This treatment induced the strengthening of the junction between the dermis and epidermis. Studies have shown that the application of this polysaccharide improves the skin barrier, enhances moisturization, and helps regulate osmotic parameters, offering protection against external stresses such as UV radiation and pollution. In addition, its application improves the visual and sensory qualities of the skin by reducing roughness in dry skin (Ghibaudo et al., 2015).

5.4.2 Antiaging effects reported from marine polysaccharides

Skin aging is a natural process characterized by the development of folds and ridges, influenced by both endogenous and exogenous factors (Chaudhary et al., 2020). These factors collectively alter the structure and appearance of the skin (Ganceviciene et al., 2012). Aging skin cells experience a reduced ability to combat oxidative stress due to the loss of ROS scavenging properties, leading to wrinkles, skin laxity, abnormal pigmentation, and dryness (Brunt and Burgess, 2018).

A novel marine bacterial EPS derived from *Alteromonas infernus* (Raguénès et al., 1997) and *Vibrio diabolicus* (Raguenes et al., 1997) was isolated from a hydrothermal vent. Courtois and his group studied the modulatory effect of these two isolated EPS on the immune complement system using normal human serum (Courtois et al., 2014). These polysaccharides are naturally sulfated, allowing them to neutralize/depolymerize free radicals and undergo sulfation. Sulfated glycosaminoglycans are known for suppressing the immune complement system in vitro. The results suggested that these two low-molecular-weight EPS can inhibit the complement system via the classical pathway, contributing to their protective effects on cells. In addition, they have the ability to regenerate skin and bone cells, making them suitable for antiaging applications.

Borel and his group extracted EPS-229 from *Alteromonas macleodii* ferment, which contains natural sugars, uronic acid, and sulfates (Borel et al., 2017). EPS-229 demonstrates remarkable antiaging properties by protecting the skin from UV-induced oxidative stress and chelating heavy metals. It also prevents lipid peroxidation and structural damage caused by environmental pollutants, thereby safeguarding the skin. In addition, it creates a protective barrier against fine particles, promoting a healthier skin appearance. These beneficial properties make EPS-229 a valuable asset in the cosmetic industry, as it helps mitigate the effects of urban pollution on the skin.

A novel polysaccharide was isolated from the marine bacterium *Alteromonas* sp. JL2810 by Zhang et al. (2017b). This polysaccharide extract was combined with carnosine, sodium hyaluronate, and a tripeptide to create a new product. A study involving thirty-three healthy female subjects aged 45–65 years showed that the formulation significantly improved skin viscoelasticity and hydration in the deeper layers of the skin. In addition, the results revealed a reduction in skin sagging, smoother skin with improved texture, and enhanced hydration of the superficial epidermis. The formulation was found to be well-tolerated, safe, and effective in addressing skin aging (Garre et al., 2017).

The antioxidant effects of complex polysaccharides extracted from *Chlorella vulgaris*, which primarily consist of rhamnose, ribose, arabinose, xylose, 2-deoxy-D-glucose, mannose, glucose, galactose, and glucosamine, were tested on the model organism *Caenorhabditis elegans*. The extracted polysaccharides were found to enhance the activity of two key antioxidant enzymes—total SOD and catalase—both of which are associated with stress resistance. Furthermore, these polysaccharides significantly increased the organism's resistance to oxidative stress induced by hydrogen peroxide (Gui et al., 2019).

The level of MDA serves as a biomarker for lipid peroxidation, providing insights into oxidative stress within cells. In addition, intracellular antioxidant enzyme activities, such as SOD and GSH-Px, are assessed along with the measurement of total antioxidant capacity (TAOC) to evaluate the overall oxidative balance and the efficacy of cellular defense mechanisms against free radical-induced damage. A fraction of the sulfated galactans (F1) isolated from *P. haitanensis* was evaluated for its effects in mice. When treated at a slightly higher dose, F1 decreased MDA production, resulting in lower lipid peroxidation. In addition, treatment with F1 also resulted in increased activity of SOD and GSH-Px. The highest TAOC was observed in all tests when F1 was administered at a dose of 100 mg/kg (Zhang et al., 2004).

Fucoidan, extracted from the brown alga *Ascophyllum nodosum*, has the potential to protect skin tissues from degradation. It reduces the activity of matrix metalloproteinases (MMPs) (metalloproteinase-2 and metalloproteinase-3) and promotes their complex formation with tissue inhibitors of metalloproteinases, which ultimately prevents the degradation of the extracellular matrix. In the presence of fucoidan, the damaging effects of human leukocyte elastase are minimized, helping to maintain an intact dermoepidermal junction. In addition, when fucoidan is preincubated with skin cells, it completely inhibits the degradation of the skin's elastic network (Senni et al., 2006).

Galactofucan, extracted from the brown alga *Saccharina longicruris*, has been shown to enhance fibroblast migration and proliferation. When galactofucan undergoes depolymerization and is tested on dermal fibroblasts, it affects both the growth rate and apoptosis of the fibroblasts. In addition, it stimulates the synthesis of MMPs and collagen, which can counteract age-related degradation of the extracellular matrix and support wound healing. These properties make galactofucan an ideal candidate for use in the cosmetic industry (Rioux et al., 2013).

5.4.3 UV protection (sunscreen) effect reported from marine polysaccharides

UV radiation from the Sun plays a key role in the initiation and progression of various skin disorders, including skin cancer. While moderate UV exposure aids in the conversion of 7-dehydrocholesterol to cholecalciferol (vitamin D) in the skin, excessive UV exposure can have harmful effects. Prolonged and high levels of UV radiation can cause cellular damage, oxidative stress, and inflammation, which, in turn, heightens the risk of photoaging, immune suppression, and the development of skin malignancies (Battistin et al., 2020). Prolonged exposure of human skin to UVA (320–400 nm) and UVB (280–320 nm) radiation can result in both acute and chronic skin damage. UVB radiation, in particular, is known to cause DNA damage and increase the production of ROS, which accelerates skin photoaging. This occurs by triggering inflammatory signaling pathways and upregulating the expression of extracellular matrix remodeling proteins, including collagens and MMPs.

Previous research has demonstrated that UVB radiation raises matrix metalloproteinase-1 (MMP-1) levels by activating intracellular signaling transduction pathways. This elevation in MMP-1 disrupts collagen synthesis within the extracellular matrix, contributing to skin photoaging (Kim et al., 2010). Fucoidan, an algal polysaccharide, has been found to mitigate the photoaging process. A study conducted by Moon and his group investigated the effects of varying fucoidan concentrations on human fibroblast cells, which were subsequently exposed to UV radiation (Moon et al., 2009). The results demonstrated that fucoidan effectively inhibited the expression of MMP-1 by suppressing the extracellular signal-regulated kinase pathway. These findings suggest that fucoidan could act as a protective agent against UV-induced skin damage. Furthermore, fucoidan treatment led to an increase in type I procollagen mRNA and protein expression compared to the control cells.

A polysaccharide secreted by the strain HYD657, belonging to *Alteromonas macleodii* subsp. Fijiensis (CNCM I-1627T) (Cambon-Bonavita et al., 2002), is marketed under the name Abyssine. This strain was isolated in 1987 from a hydrothermal vent along the East Pacific Rise, known as biovar Deepsane. Abyssine effectively protects Langerhans cells, which are particularly sensitive to UV radiation and play a major role in the immune defense system of human skin. This bioactive ingredient is utilized in cosmeceuticals to soothe and reduce irritation in sensitive skin, particularly in response to chemical, mechanical, and UVB-induced damage (Alain and Atsushi, 2006).

Kim and his group conducted a study to investigate the protective effects of low-molecular-weight fucoidan against UVB-induced photoaging (Kim et al., 2018). The commercially available LMWF was procured from Glucan Corp. Ltd. (Busan, Korea) and used in this study. In this experiment, a mouse model was topically treated with varying concentrations of LMWF daily for 15 weeks while being exposed to UVB radiation. The results showed a significant reduction in wrinkle formation and cutaneous edema compared to the control group. In addition, the control group exhibited increased skin MPO activity, indicating enhanced neutrophil recruitment to UVB-induced skin lesions. The LMWF-treated group showed elevated levels of IL-10, an antiinflammatory cytokine. The LMWF treatment also inhibited MDA (a marker of lipid peroxidation) and superoxide anion (indicating oxidative stress), while boosting glutathione levels, a key antioxidant. This analysis suggests that LMWF may be an effective skin-protective agent for preventing photoaging.

5.4.4 Other properties reported from marine polysaccharides

A study was conducted by Veeraperumal and his group to evaluate the wound healing activities of a polysaccharide extracted from *Gracilaria lemaneiformis* (Veeraperumal et al., 2020). The study revealed that the extracted polysaccharide promoted the proliferation and migration of human keratinocytes through the activation of the PI3K/aPKC signaling pathway. The activation promoted the progression of the S phase in the cell cycle, increased the epithelial layer thickness, and stimulated collagen deposition. Consequently, the polysaccharide accelerated wound closure, indicating its potential as a wound healing product.

Resilience by Estée Lauder contains extracts from the Caribbean Sea whip, *Pseudopterogorgia elisabethae*, which is rich in pseudopterosins, the bioactive components. Pseudopterosins are tricyclic diterpene glycosides known for their potent antiinflammatory and analgesic properties. They inhibit eicosanoid biosynthesis by targeting phospholipase A2 (PLA2) and 5-lipoxygenase, effectively soothing skin irritation and inflammation. In addition, pseudopterosins inhibit human neutrophil PLA2 activity, helping to prevent skin damage caused by chemicals and sunlight. These properties make them ideal for use in cosmeceutical formulations (Martins et al., 2014; Silva et al., 2017).

SeaCode by Lipotech contains glycoproteins (GPs) and glucidic exopolymers (EPS) derived through biotechnological fermentation of *Pseudoalteromonas* sp. from Antarctic intertidal waters. GPs play a key role in various biological processes, including cell–cell communication, stress recovery, and the structural reinforcement of the dermis and epidermis, making them highly valuable for skincare products. GPs also enhance cellular protein maintenance, act as protective and lubricating agents, and serve as carriers for vitamins and hormones, further supporting skin health and wellness (Gottschalk, 1972; Ivatt, 2012).

Cationic polysaccharides derived from seaweed, such as chitosan, are highly effective film-forming agents. Their ability to bind strongly to negatively charged proteins found in human skin and hair makes them particularly useful in hair damage repair products and gel-based styling formulations. Nonionic polysaccharides, like hydroxy methylcellulose, are commonly used in nail products as film-forming agents and thickeners (Kanlayavattanakul and Lourith, 2015).

Polysaccharides derived from marine sources are renewable, biodegradable, and biocompatible, offering unique functional and bioactive properties that can be explored for innovative applications in the cosmetic skincare industry.

5.5 Nutraceutical applications of marine polysaccharides

Nutraceuticals are food products or ingredients that provide health benefits similar to pharmaceutical agents, beyond basic nutrition. Research has highlighted marine-derived compounds as a rich source of bioactive nutraceuticals, including polyunsaturated fatty acids, polysaccharides, polyphenols, bioactive peptides, and carotenoids. These compounds are known to contribute to the prevention and management of various health complications. Their bioactive properties make marine-derived compounds promising candidates for addressing chronic diseases and modulating the immune system (Pallela, 2014; Ruocco et al., 2016). Algal species serve as an abundant source of bioactive polysaccharides, which have garnered significant research attention due to their versatile applications. In the food and beverage industry, these polysaccharides are commonly used as stabilizers and texture-modifying agents. In addition, algal polysaccharides are recognized for their health-promoting properties, including antioxidant, antibacterial, antidiabetic, immunomodulatory, antiinflammatory, and anticancer activities (Fernando et al., 2019). These biological activities of algal polysaccharides are influenced by key characteristics, such as molecular weight, branching patterns, charge density, hydrophobicity, and the types of monomers and bonds they contain. In addition to their bioactive properties, algal polysaccharides also function as dietary fibers and prebiotics. They are resistant to digestion in the upper gastrointestinal tract but are fermentable in the lower gut. This fermentation process promotes the growth of beneficial gut microbiota, supporting overall gut health and contributing to improved digestive and metabolic functions (Shang et al., 2018).

5.5.1 Antioxidant properties

Oxidative stress, in general, results from an imbalance between the production of ROS and the ability of the body to neutralize them. This imbalance can trigger abnormal enzyme activity and damage cellular macromolecules. Oxidative stress can be linked to the development of various health conditions, including cardiovascular diseases, chronic inflammation, diabetes, neurodegenerative disorders, and cancer. Antioxidants, such as chitosan and its derivatives, help mitigate oxidative damage by disrupting the chain reactions of radical oxidation, thereby protecting cells and tissues from harm (Ngo et al., 2015; Šimat et al., 2020).

Chitosamin, a high-molecular-weight chitosan dietary supplement, was assessed in healthy individuals for its antioxidant effects. The supplement helped reduce lipid hydroperoxides and uremic toxins in the gastrointestinal tract, effectively preventing oxidative stress in systemic circulation (Anraku et al., 2011). Fucoidan extracted from *L. japonica*

(Wang et al., 2008) and *U. pinnatifida* (Mak et al., 2013) has also demonstrated antioxidant activities. The presence of electron-withdrawing groups enhances the antioxidant activity of low-molecular-weight fucoidan by altering its polarity or activating hydrogen atoms on anomeric carbons. In contrast, cations like amino groups exhibit weaker oxidation resistance due to their inability to activate hydrogen atoms. The position of substituted groups also plays a critical role in determining antioxidant effectiveness. A positive correlation between sulfate content and antioxidant capacity, together with the ratio of sulfate content to fucose, significantly influences hydroxyl radical scavenging activity (Wang et al., 2008, 2010). In addition to its ROS scavenging ability, fucoidan has also been reported to inhibit amyloid-β-induced toxicity by reducing amyloid-β accumulation (Wang et al., 2018). Alginate exhibits antioxidant properties owing to the presence of phenolic compounds and flavonoids. These molecules effectively neutralize free radicals, thus highlighting the potential of alginate as a natural antioxidant (Metha et al., 2024).

5.5.2 Antiobesity effect

Chitosan and its derivatives have been reported to possess antiobesity properties (Hu et al., 2016; Chiu et al., 2017; Inanli et al., 2020). As a dietary supplement, chitosan has been shown to reduce TC and LDL-C levels in plasma, as well as TG levels in both plasma and the liver. This reduction in lipid levels is attributed to chitosan's ability to bind to dietary lipids and bile acids, along with its capacity to inhibit pancreatic lipase activity. Together, these mechanisms help decrease the absorption of fats in the gastrointestinal tract, contributing to improved lipid profiles (Muanprasat and Chatsudthipong, 2017). Low-molecular-weight chitosan has been reported to enhance plasma levels of ATP and 5-HT by influencing intestinal microflora through oral administration. This modulation of gut microbial populations is believed to play a role in the antiobesity effects of surface-deacetylated chitin nanofibers. Furthermore, studies have demonstrated its effectiveness in reducing weight gain in overweight individuals, further supporting its potential as a weight management supplement (Kaats et al., 2006; Bondiolotti et al., 2007; Azuma et al., 2014).

Fucoidan also has the ability to inhibit lipogenesis (the formation of fat) and enhance lipolysis (the breakdown of fat), making it a favorable supplement for antiobesity applications (Wang et al., 2019). By regulating fat metabolism, fucoidan helps reduce fat accumulation and promotes fat breakdown, contributing to weight management and overall metabolic health. Fucoidan extracted from the sea cucumber *Acaudina molpadioides* has demonstrated antiobesity effects, both in vitro and in vivo, by downregulating the expression of SREBP-1c (sterol regulatory element-binding protein 1c) (Xu et al., 2014). This regulation of SREBP-1c, a key transcription factor involved in lipid biosynthesis, helps reduce fat accumulation and promotes healthier lipid metabolism, contributing to its potential as an antiobesity agent. Alginate, known for its ability to promote a healthy gut microbiome, is emerging as a potential functional food ingredient (Shang et al., 2018). Low-molecular-weight alginates have been shown to stimulate the growth of beneficial gut microbes, including *Bifidobacterium* spp. and *Lactobacillus* spp. (Ramnani et al., 2012). In addition, alginate has the ability to slow gastric emptying, which enhances feelings of fullness and can help reduce food intake (Hu et al., 2016). Chitooligosaccharides (COS) have gained attention in the nutraceutical field for their antidiabetic effects (Liu et al., 2007), ability to lower cholesterol levels (Kil-Nam et al., 2005), and inhibition of adipogenesis (Cho et al., 2008). Both chitosan and COS are commonly used in the food industry as additives and supplements, particularly for weight management and reducing serum lipid levels (Mhurchu et al., 2004).

5.5.3 Anticancer activity

The consumption of fucoidan, as food and dietary supplements, is regulated with a daily limit of 250 mg (Fitton et al., 2019). Studies have reported the anticancer activity of Fucoidan and its derivatives by inducing apoptosis in various cell lines, namely, 5637 human bladder cancer cells (Han et al., 2017), human breast (MCF-7) and colon (HCT15) cancer cells (Palanisamy et al., 2018), and colon adenocarcinoma (Caco-2) cells (Narayani et al., 2019). In addition, a fucoidan obtained from a brown alga also demonstrated inhibition of human melanoma and colon cancer cells (Ermakova et al., 2011).

Laminarin, a low-molecular-weight (~5 kDa) polysaccharide, generally obtained from the brown algal species *Laminaria*, *Saccharina*, *Ascophyllum*, *Fucus*, and *Cystoseira*, is a β-glucan storage consisting of (1,3)-β-D-glucan chains with occasional β-(1,6)-intrachain linkages (Kadam et al., 2015; Sanjeewa et al., 2017). Laminarin has been reported to possess anticancer properties, such as the promotion of apoptotic cell death, inhibition of angiogenesis, and suppression of colony formation in cancer cells (Zargarzadeh et al., 2020).

5.5.4 Antiallergic effect

Allergic and hypersensitive reactions can be classified into distinct categories based on their underlying mechanisms: Type I, also known as the anaphylactic response; Type II, involving cytotoxic-mediated reactions; and Type III, characterized by immune complex formation and associated responses (Choudhary et al., 2021). Hypersensitivity may also lead to consequential clinical conditions, including asthma, erythema, and anaphylactic shock. The polyphenolic extract from *Scytosiphon* species was investigated for its antiallergenic properties and compared to the commonly used antiallergic drug disodium cromoglycate. The study reported that the *Scytosiphon* extract, with an IC_{50} value of 1.13 mg/mL, exhibited superior antiallergic activity compared to disodium cromoglycate (Chen et al., 2015).

5.5.5 Antiinsomnia effect

Insomnia, a prevalent sleep disorder, affects approximately 10%–15% of individuals as a chronic condition, with an additional 25%–35% experiencing temporary or occasional episodes. One of the key molecular targets for developing sleep-enhancing therapies is the GABA receptor, which plays a crucial role in regulating sleep mechanisms (Johnston, 2015). Phlorotannins, such as dieckol, eckol, triphlorethol-A, and eckstolonol, derived from the seaweed *Ecklonia cava*, have been reported to promote sleep in mice. This effect is attributed to their ability to positively modulate the GABA-benzodiazepine receptor, a key target in sleep regulation, and has applications in sedatives and hypnotics (Cho et al., 2012). Triphlorethol-A has also demonstrated a reduction in sleep latency in a dose-dependent manner in C57BL/6N mice, with its sleep-inducing effect comparable to zolpidem, a widely used hypnotic drug (Yoon and Cho, 2018).

5.6 Bioremediation

Environmental pollution, exacerbated by rapid urbanization and industrialization, poses a serious threat to both ecosystems and human well-being. These pollutants are not only persistent in the environment but are also highly toxic at low concentrations and tend to bioaccumulate through food chains. Traditional remediation methods, such as electroplating, ion exchange, precipitation, and membrane-based technologies, have been widely employed to address pollution. However, their effectiveness is often hindered by high costs, which limit their scalability and sustainable application (Mathivanan et al., 2021). Bioremediation leverages biological systems to mitigate these pollutants. Microbial systems have long been recognized for their bioremediation capabilities (Velkova et al., 2018). More recently, research has highlighted the potential of marine polysaccharides as effective agents for bioremediation, offering a promising alternative for addressing environmental pollution (Jeewon et al., 2023).

5.6.1 Heavy metal bioremediation

Heavy metals, including chromium (Cr), cadmium (Cd), lead (Pb), arsenic (As), mercury (Hg), and silver (Ag), are among the most persistent and hazardous environmental pollutants. Marine polysaccharides have shown the ability to sequester these positively charged metal ions through strong interactions, facilitated by their negatively charged functional groups. Polysaccharides are rich in ionizable functional groups, such as sulfate units, uronic acids, carboxyl, hydroxyl, and phosphate groups, which facilitate multiple mechanisms—like ion exchange, complexation, and precipitation—that effectively bind and remove heavy metals from contaminated environments (Gupta and Diwan, 2017).

The marine ecosystem serves as an excellent source of polysaccharides with promising applications in bioremediation. Polysaccharides from *Enterobacter cloacae,* isolated from the West Coast of India, have demonstrated notable chelating and biosorptive properties against cadmium and copper metal ions. At a concentration of 100 mg/L, the polysaccharide exhibited a chelation efficiency of 65% for cadmium and 20% for copper. In addition, the polysaccharide was also able to bioabsorb cobalt, *albeit* with much lower efficacy (Iyer et al., 2005). Another marine bacterium, *Pseudoalteromonas* sp., known for producing an exopolysaccharide (EPS) Bsi20310, has been shown to enhance Fe(III) coagulation performance through mechanisms such as neutralization, bridging, and sweeping. These processes improve floc aggregation, leading to more effective pollutant removal (Zhou et al., 2010). In another study, four bacterial strains capable of producing polysaccharides were isolated from marine hydrothermal vents. The isolated strains were identified as *Exiguobacterium aquaticum, Mammaliicoccus sciuri, Micrococcus luteus,* and *Jeotgalicoccus huakuii.* These isolates exhibited metal ion removal efficiencies of arsenic (84%), cadmium (85%), copper (98%), and nickel (91%), respectively (Ahamed et al., 2024). Exopolysaccharides produced by *Bacillus cereus* strain KMS3-1 have been reported to exhibit maximum adsorption capacities of 54.05 mg/g for Cd(II), 71.42 mg/g for Cu(II), and

78.74 mg/g for Pb(II) (Mathivanan et al., 2021). A pH-responsive exopolysaccharide, isolated from the marine bacterium *Alteromonas* sp. JL2810, exhibited notable biosorption capabilities for heavy metals. The exopolysaccharide demonstrated maximum biosorption capacities of 140.8 ±8.2 mg/g for Cu^{2+} and 226.3±3.3 mg/g for Ni^{2+} at a pH of 5.0. For Cr^{6+}, the biosorption capacity peaked at 215.2 ± 5.1 mg/g at pH 5.5. The functional groups primarily responsible for the adsorption process included O–H, C=O, and C–O–C, which facilitated strong interactions with heavy metal ions. This highlights the potential of JL2810-derived exopolysaccharides as effective agents for the bioremediation of metal-contaminated environments (Zhang et al., 2017a). *Halomonas maura* secretes an exopolysaccharide that exhibits the ability to bind lead, achieving a maximum efficiency of 46.4 mg/g of the polysaccharide. In addition to lead, this exopolysaccharide is also capable of binding to other metal ions, including Cu^{2+}, Ag^{+}, Mn^{2+}, and Zn^{2+}, demonstrating its broad-spectrum metal-binding potential (Arias et al., 2003). Another exopolysaccharide derived from *Salipiger mucosus* strain $A3^T$ exhibited metal ion chelation capabilities, effectively binding copper, lead, and cobalt at capacities of 15.7, 43.5, and 8.7 mg/g of EPS, respectively (Llamas et al., 2010).

Similarly, EPS from *Enterobacter* sp. strain YU16-RN5 has been reported for its absorptive affinities for various metal ions such as cadmium, mercury, strontium, copper, arsenic, and uranium. Among the metals tested, cadmium showed the highest absorption capacity at 1555 mg/g, followed by mercury at 218 mg/g (Dhanya and Athmika, 2021). Two bacterial isolates, *Bacillus* sp. MC3B-22 and *Microbacterium* sp. MC3B-10, have been reported for their biosorption capacities for cadmium, with absorption rates of 141.1 and 97.1 mg/g, respectively (Camacho-Chab et al., 2018). In another instance, EPS from *Pseudoalteromonas* sp. SM9913, isolated from the Gulf of the Yellow Sea (China), has been reported for its ability to bind to several metal ions, including Fe^{2+}, Zn^{2+}, Cu^{2+}, and Co^{2+}, at rates of 85%, 58%, 52%, and 48%, respectively (Qin et al., 2007; Parrilli et al., 2021).

Marine polysaccharides not only offer eco-friendly and cost-effective solutions for heavy metal remediation but also demonstrate remarkable adaptability to diverse environmental conditions, such as variations in pH, salinity, and temperature. Advancements in understanding the molecular mechanisms behind metal–polysaccharide interactions could lead to the development of customized polysaccharides designed for specific pollutants. Moreover, integrating polysaccharide-based bioremediation with existing technologies, such as constructed wetlands and biofilters, has the potential to further enhance both the efficiency and sustainability of environmental cleanup efforts.

5.6.2 Oil pollution remediation

Marine ecosystems are frequently exposed to crude oil pollution caused by accidental spills, offshore drilling activities, and transportation activities. Such contamination poses severe threats to marine biodiversity, water quality, and human livelihoods. Traditional cleanup methods, such as chemical dispersants, skimmers, and sorbents, often have limited efficacy and may lead to secondary pollution. In recent years, biotechnological approaches utilizing marine-derived exopolysaccharides (EPSs) have gained attention as promising alternatives for oil spill mitigation. These EPS compounds offer surfactant-like properties, low toxicity, and inherent biodegradability, making them an eco-friendly and effective solution for addressing oil pollution (Poli et al., 2010).

These polysaccharides act by reducing the interfacial tension between oil and water through a process called emulsification, leading to the formation of microdroplets that are more accessible to hydrocarbon-degrading microorganisms (Banerjee et al., 2021). For instance, the exopolysaccharide (EPS) produced by *Pseudomonas furukawaii* strain PPS-19, isolated from an oil-polluted site at Paradip Port, Odisha, India, was capable of emulsifying and degrading crude oil, achieving 89.52% degradation within 5 days (Vandana, 2021). EPS produced by *Halomonas eurihalina* strain H-28 exhibited remarkable emulsifying properties with crude oil, outperforming conventional surfactants like Tween 20, Tween 80, and Triton X-100 (Martínez-Checa and Toledo, 2002). Another marine bacterium, *Hahella chejuensis*, produces an exopolysaccharide (EPS) that demonstrates remarkable emulsifying capabilities, achieving 89.2% efficiency in emulsifying kerosene. This performance surpasses that of commercial emulsifying agents like xanthan gum, gellan gum, and sodium alginate (Yim et al., 2004). EPS produced by the Antarctic *Pseudomonas* sp. ID1 demonstrated superior emulsification properties when tested against various food oils, including olive oil, sunflower oil, and corn oil. This EPS outperformed commercially available emulsifiers such as xanthan gum and arabic gum, showcasing its potential for use in food industry applications as a more efficient and natural emulsifying agent. The EPS has also been reported for its emulsifying capability against n-hexadecane (Carrión et al., 2015). In another study, two bacteria, *Alteromonas* and *Thalassospira*, isolated from the Gulf of Mexico, demonstrated the ability to degrade n-alkanes and polycyclic aromatic hydrocarbons (PAHs) (Bacosa et al., 2018). *Klebsiella pneumoniae* (KY494861), capable of producing EPS, was isolated from the eastern coastal regions of Tamil Nadu, India, and was tested for its potential to degrade PAHs such as naphthalene, anthracene, fluorene, and acenaphthene. The strain demonstrated maximum degradation rates of 74.31% for anthracene, 67.28% for acenaphthene, 62.48% for fluorene, 57.84% for naphthalene, and 55.85% for mixed PAHs, thus highlighting its role in the remediation of PAH-contaminated environments (Premnath et al., 2021). In addition, two bacterial species—namely, *Bacterium* Ex-DG74 and *Sporosarcina halophila*—have been reported to produce EPS with the ability to degrade toluene (Abari et al., 2012). These findings highlight the critical role of marine-derived EPS in the efficient bioremediation of oil pollution across a range of marine environments.

5.6.3 Dye bioremediation

Polysaccharides have also been effectively used to flocculate dyes present in wastewater from the textile, paper, and tannery industries (Banerjee et al., 2021). *Alteromonas* sp. CGMCC 10612 has been reported to effectively flocculate a variety of dyes, including Congo red, direct black, and methylene blue. The strain demonstrated impressive removal efficiencies of 98.5%, 97.9%, and 72.7%, respectively. The removal efficiency of the polysaccharide-rich bioflocculant extract derived from *Alteromonas* sp. CGMCC 10612 was found to increase with the rising concentration of the bioflocculant (Chen et al., 2017). Another exopolysaccharide, extracted from the marine cyanobacterium *Cyanothece* sp. ATCC 51142, has been studied for its removal efficiency against three dyes, namely, Porocion Golden Yellow (greater than 90%), Porocion Brilliant Blue (71%), and Porocion Red Brown (80%) (Shah et al., 1999). The EPS extracted from the halophilic bacterium *Exiguobacterium* sp. VK1 has shown maximal bioabsorptive capability of 79.1% against malachite green dye at pH 6 and 40°C. This efficiency was further increased to 97.61% following enhanced EPS production (Kalpana et al., 2020). Alginates derived from *Arthrospira platensis* have been documented for their remarkable ability to remove Ismate Violet 2R dye from wastewater, achieving a removal efficiency of 75.7% (Alprol et al., 2021).

Ulvan, a sulfated polysaccharide extracted from *Ulva lactuca*, has been evaluated for its dye removal capabilities. The study revealed that ulvan successfully removed 86.8% of methylene blue dye within just 30 min. This dye removal efficiency is largely attributed to the ionic interactions between the carboxyl groups present in ulvan and the positively charged methylene blue molecules (Ghazy et al., 2024). Commercially available iota-carrageenan has shown significant effectiveness in removing acidic dyes, including Alphanol Fast Blue. Among the various carrageenan variants tested, CAR4, which contains a higher nitrogen content, demonstrated the highest adsorption capacity for the acid dye. To evaluate the reusability of CAR4 as an adsorbent, five consecutive adsorption/desorption cycles were conducted. The results indicated that the removal efficiency of CAR4 remained consistent, with only a 1% variation, showcasing its stability and potential for repeated use in dye removal processes (Abdellatif and Abdellatif, 2020).

Marine polysaccharides, especially extracellular polysaccharides, are emerging as a promising and sustainable solution in the field of bioremediation, offering highly effective methods to address a range of environmental pollutants. As research continues to explore the vast potential of marine-derived EPS, these biopolymers are proving their effectiveness not only in pollutant removal but also for their eco-friendly, biodegradable, and sustainable properties. Moving forward, the integration of marine EPS into bioremediation strategies could transform pollution management, presenting a greener, more efficient alternative to traditional, often expensive methods.

5.7 Conclusion

Marine polysaccharides, derived from both macro- and microorganisms, hold significant potential for a wide range of applications across industries such as biomedical, cosmeceutical, nutraceutical, pharmaceutical, and environmental remediation. Their unique structural diversity and physicochemical properties, combined with beneficial traits like biocompatibility, biodegradability, nontoxicity, renewability, eco-friendliness, abundance, and cost-effectiveness, position them as a sustainable and versatile alternative to synthetic polymers. These advantages make marine polysaccharides highly promising for various industrial uses, aligning with global initiatives focused on sustainability and a circular economy. Moreover, advanced characterization tools, such as omics technologies and computational modeling, are shedding light on the structure–activity relationships of these polysaccharides. These insights can facilitate research on the chemical modification of marine polysaccharides, optimizing their biological properties and improving their therapeutic efficacy while minimizing risks for patients.

Despite the promising potential of marine polysaccharides, their industrial applications remain largely confined to laboratory-scale research. The transition from research to commercial-scale utilization is hindered by several challenges. Key obstacles include the lack of sustainable and efficient extraction techniques, difficulties in scaling up production, maintaining bioactivity during process development, and ensuring cost-effectiveness. In addition, there is a shortage of high-throughput screening studies and clinical trials, which further limits the integration of these polysaccharides into commercial products. Addressing these challenges is crucial for unlocking the full potential of marine polysaccharides in various industrial sectors. Developing efficient extraction methods, optimizing production processes, and conducting more extensive clinical trials will be essential to ensure their feasibility for large-scale commercialization. By overcoming these barriers, marine polysaccharides can become a sustainable and valuable resource across industries, ranging from biomedical applications to environmental remediation.

The vast biodiversity of marine ecosystems offers a unique opportunity to discover novel polysaccharides with distinct properties and diverse applications. By exploring previously untapped marine organisms, researchers can uncover polysaccharides with potential for innovation across various industrial sectors. However, fully realizing this potential requires significant advancements in scientific methodologies. To harness the functional and bioactive properties of these marine polysaccharides, it is crucial to develop sophisticated techniques for their separation, purification, and structural analysis. These advancements will not only allow researchers to identify and characterize novel polysaccharides but also enable the optimization of their extraction and processing methods for large-scale applications. With these improvements, marine polysaccharides can be effectively utilized in industries such as pharmaceuticals, nutraceuticals, cosmeceuticals, and environmental remediation, ultimately contributing to more sustainable and eco-friendly solutions.

Acknowledgement

Devargya Ganguly and Gargi Goswami acknowledge the financial support from the Science and Engineering Research Board, India (File No. SRG/2021/001256). Soumajit Ghosh, Sneha Sengupta, and Devargya Ganguly are thankful to Dr. Gargi Goswami for her constant guidance and support.

Author Contributions

Gargi Goswami was involved in conceptualization and designing of the contents of the book chapter and critical revision of the book chapter. Soumajit Ghosh, Sneha Sengupta, and Devargya Ganguly drafted the book chapter. All the authors have read and approved the final draft of this book chapter.

Competing Interests

The authors state that the work was carried out without the existence of any financial or commercial associations that could be interpreted as a potential conflict of interest.

References

Abari, A.H., Emtiazi, G., Ghasemi, S.M., 2012. The role of exopolysaccharide, biosurfactant and peroxidase enzymes on toluene degradation by bacteria isolated from marine and wastewater environments. J. Microbiol. 2008-4161. 5 (3), 479–485. Available from: http://jjmicrobiol.com/?page=download&file_id=8835. doi:10.5812/jjm.3554.

Abdel-Wahab, B.A., El-Kareem, H.F.A., Alzamami, A., Fahmy, C.A., Elesawy, B.H., Mahmoud, M.M., Ghareeb, A., Askary, A.E., Nahas, H.H.A., Attallah, N.G.M., Altwaijry, N., Saied, E.M., 2022. Novel exopolysaccharide from marine *Bacillus subtilis* with broad potential. Metabolites. 2218-1989. 12 (8), 715. Available from: https://www.ncbi.nlm.nih.gov/pmc/articles/PMC9413097. doi:10.3390/metabo12080715.

Abdellatif, F.H.H., Abdellatif, M.M., 2020. Bio-based i-carrageenan aerogels as efficient adsorbents for heavy metal. Cellulose (London, England). 0969-0239. 27 (1), 441–453. doi:10.1007/s10570-019-02818-x.

Ahamed, J.M., Dahms, H.U., Huang, Y.L., 2024. Heavy metal tolerance, and metal biosorption by exopolysaccharides produced by bacterial strains isolated from marine hydrothermal vents. Chemosphere. 18791298. 351. Available from: https://www.sciencedirect.com/science/journal/00456535. doi:10.1016/j.chemosphere.2024.141170.

Ahmad, N.H., Mustafa, S., Man, Y.B.C., 2015. Microbial polysaccharides and their modification approaches: a review. Int. J. Food Prop. 1094-2912. 18 (2), 332–347. doi:10.1080/10942912.2012.693561.

Alain, T., Atsushi, T., 2006. The applications and functions of new exopolysaccharide 'Deepsane' from. Fragrance J. 34, 61–64.

Albertson, N.H., Nyström, T., Kjelleberg, S., 1990. Exoprotease activity of two marine bacteria during starvation. Appl. Environ. Microbiol. 0099-2240. 56 (1), 218–223. Available from: https://www.ncbi.nlm.nih.gov/pmc/articles/PMC183279. doi:10.1128/aem.56.1.218-223.1990.

Allison, D.G., Ruiz, B., SanJose, C., Jaspe, A., Gilbert, P., 1998. Extracellular products as mediators of the formation and detachment. FEMS Microbiol. Lett. 0378-1097. 167 (2), 179–184. doi:10.1111/j.1574-6968.1998.tb13225.x.

Almutairi, M.H., Helal, M.M.I., 2021. Biological and microbiological activities of isolated *Enterobacter* sp. J. King Saud Univ. Sci. 1018-3647. 33 (2), 101328. doi:10.1016/j.jksus.2020.101328.

Alprol, A.E., Heneash, A.M.M., Ashour, M., Abualnaja, K.M., Alhashmialameer, D., Mansour, A.T., Sharawy, Z.Z., Abu-Saied, M.A., Abomohra, A.E.-F., 2021. Potential applications of Arthrospira platensis lipid-free biomass in bioremediation of organic dye from industrial textile effluents and its influence on Marine Rotifer (*Brachionus plicatilis*). Materials. 1996-1944. 14 (16), 4446. Available from: https://www.ncbi.nlm.nih.gov/pmc/articles/PMC8400522. doi:10.3390/ma14164446.

Amin, M.L., Mawad, D., Dokos, S., Koshy, P., Martens, P.J., Sorrell, C.C., 2020. Immunomodulatory properties of photopolymerizable fucoidan and carrageenans. Carbohydr. Polym. 0144-8617. 230 (115691), 115691. Available from: https://www.ncbi.nlm.nih.gov/pubmed/31887956. doi:10.1016/j.carbpol.2019.115691.

Aneiros, A., Garateix, A., 2004. Bioactive peptides from marine sources: pharmacological properties and. J. Chromatogr. B Analyt. Technol. Biomed. Life Sci. 1570-0232. 803 (1), 41–53. Available from: https://www.ncbi.nlm.nih.gov/pubmed/15025997. doi:10.1016/j.jchromb.2003.11.005.

Anraku, M., Fujii, T., Kondo, Y., Kojima, E., Hata, T., Tabuchi, N., Tsuchiya, D., Goromaru, T., Tsutsumi, H., Kadowaki, D., Maruyama, T., Otagiri, M., Tomida, H., 2011. Antioxidant properties of high molecular weight dietary chitosan in vitro. Carbohydr. Polym. 0144-8617. 83 (2), 501–505. doi:10.1016/j.carbpol.2010.08.009.

Arias, S., Del Moral, A., Ferrer, M.R., Tallon, R., Quesada, E., Béjar, V., 2003. Mauran, an exopolysaccharide produced by the halophilic bacterium. Extremophiles. 1431-0651. 7 (4), 319–326. Available from: https://www.ncbi.nlm.nih.gov/pubmed/12910391. doi:10.1007/s00792-003-0325-8.

Aullybux, A.A., Puchooa, D., Bahorun, T., Jeewon, R., 2019. Phylogenetics and antibacterial properties of exopolysaccharides from. Ann. Microbiol. 1590-4261. 69 (9), 957–972. doi:10.1007/s13213-019-01487-2.

Aziz, A., Poinssot, B., Daire, X., Adrian, M., Bézier, A., Lambert, B., Joubert, J.-M., Pugin, A., 2003. Laminarin elicits defense responses in grapevine and induces protection. Mol. Plant Microbe Interact. 0894-0282. 16 (12), 1118–1128. Available from: https://www.ncbi.nlm.nih.gov/pubmed/14651345. doi:10.1094/MPMI.2003.16.12.1118.

Azuma, K., Ifuku, S., Osaki, T., Okamoto, Y., Minami, S., 2014. Preparation and biomedical applications of chitin and chitosan nanofibers. J. Biomed. Nanotechnol. 1550-7033. 10 (10), 2891–2920. Available from: https://www.ncbi.nlm.nih.gov/pubmed/25992423. doi:10.1166/jbn.2014.1882.

Bacosa, H.P., Kamalanathan, M., Chiu, M.-H., Tsai, S.-M., Sun, L., Labonté, J.M., Schwehr, K.A., Hala, D., Santschi, P. H., Chin, W.-C., Quigg, A., 2018. Extracellular polymeric substances (EPS) producing and oil degrading. PloS One. 1932-6203. 13 (12), e0208406. doi:10.1371/journal.pone.0208406.

Bai, Y., Zhang, P., Chen, G., Cao, J., Huang, T., Chen, K., 2012. Macrophage immunomodulatory activity of extracellular polysaccharide (PEP). Int. Immunopharmacol. 1567-5769. 12 (4), 611–617. doi:10.1016/j.intimp.2012.02.009.

Baijal, K., Abramchuk, I., Herrera, C.M., Mah, T.-F., Trent, M.S., Lavallée-Adam, M., Downey, M., 2024. Polyphosphate kinase regulates LPS structure and polymyxin resistance. PLoS Biol. 1544-9173. 22 (3), e3002558. Available from: https://www.ncbi.nlm.nih.gov/pmc/articles/PMC10962826. doi:10.1371/journal.pbio.3002558.

Banerjee, A., Sarkar, S., Govil, T., González-Faune, P., Cabrera-Barjas, G., Bandopadhyay, R., Salem, D.R., Sani, R.K., 2021. Extremophilic exopolysaccharides: biotechnologies and wastewater. Front. Microbiol. 1664-302X. 12, 721365. Available from: https://www.ncbi.nlm.nih.gov/pmc/articles/PMC8417407. doi:10.3389/fmicb.2021.721365.

Battistin, M., Dissette, V., Bonetto, A., Durini, E., Manfredini, S., Marcomini, A., Casagrande, E., Brunetta, A., Ziosi, P., Molesini, S., Gavioli, R., Nicoli, F., Vertuani, S., Baldisserotto, A., 2020. A new approach to UV protection by direct surface functionalization of. Nanomaterials (Basel, Switzerland). 2079-4991. 10 (2), 231. Available from: https://www.ncbi.nlm.nih.gov/pmc/articles/PMC7075142. doi:10.3390/nano10020231.

Bedoux, G., Hardouin, K., Burlot, A.S., Bourgougnon, N., 2014. Bioactive components from seaweeds. Adv. Bot. Res. 345–378. doi:10.1016/b978-0-12-408062-1.00012-3.

Bergé, J.-P., Barnathan, G., 2005. Fatty acids from lipids of marine organisms: molecular biodiversity, roles. Adv. Biochem. Eng. Biotechnol. 49–125. doi:10.1007/b135782.

Bhatia, S.K., Gurav, R., Choi, Y.K., Choi, T.R., Kim, H.J., Song, H.S., Lee, S.M., Park, S.L., Lee, H.S., Kim, Y.G., Ahn, J., Yang, Y.H., 2021. Bioprospecting of exopolysaccharide from marine *Sphingobium yanoikuyae* BBL01: production, characterization, and metal chelation activity. Bioresour. Technol. 1873-2976. 324. Available from: www.elsevier.com/locate/biortech. doi:10.1016/j.biortech.2021.124674.

Bhuyan, P.P., Nayak, R., Patra, S., Abdulabbas, H.S., Jena, M., Pradhan, B., 2023. Seaweed-derived sulfated polysaccharides; The new age chemopreventives: a. Cancers. 2072-6694. 15 (3). Available from: https://www.ncbi.nlm.nih.gov/pmc/articles/PMC9913163. doi:10.3390/cancers15030715.

Biddanda, B.A., 1985. Microbial synthesis of macroparticulate matter. Mar. Ecol. Prog. Ser. 20, 241–251.

Bondiolotti, G., Bareggi, S.R., Frega, N.G., Strabioli, S., Cornelli, U., 2007. Activity of two different polyglucosamines, L112 and FF45, on body weight. Eur. J. Pharmacol. 0014-2999. 567 (1-2), 155–158. Available from: https://www.ncbi.nlm.nih.gov/pubmed/17482159. doi:10.1016/j.ejphar.2007.03.054.

Borel, M., Lamarque, E., Loing, E., 2017. Unique natural exopolysaccharides for biomimetic protective effect against. J. Cosmet. Sci. 68, 126–132.

Brading, J.W.E., Georg-Plant, M.M.T., Hardy, D.M., 1954. The polysaccharide from the alga *Ulva lactuca*. Purification, hydrolysis. J. Chem. Soc. 0368-1769. 319. doi:10.1039/jr9540000319.

Brunt, E.G., Burgess, J.G., 2018. The promise of marine molecules as cosmetic active ingredients. Int. J. Cosmet. Sci. 0142-5463. 40 (1), 1–15. doi:10.1111/ics.12435.

Camacho-Chab, J., Castañeda-Chávez, M., Chan-Bacab, M., Aguila-Ramírez, R., Galaviz-Villa, I., Bartolo-Pérez, P., Lango-Reynoso, F., Tabasco-Novelo, C., Gaylarde, C., Ortega-Morales, B., 2018. Biosorption of cadmium by non-toxic extracellular polymeric substances. Int. J. Environ. Res. Public Health. 1661-7827. 15 (2), 314. doi:10.3390/ijerph15020314.

Cambon-Bonavita, M.-A., Raguénès, G., Jean, J., Vincent, P., Guezennec, J., 2002. A novel polymer produced by a bacterium isolated from a deep-sea. J. Appl. Microbiol. 1364-5072. 93 (2), 310–315. Available from: https://www.ncbi.nlm.nih.gov/pubmed/12147080. doi:10.1046/j.1365-2672.2002.01689.x.

Cao, R., Jin, W., Shan, Y., Wang, J., Liu, G., Kuang, S., Sun, C., 2018. Marine bacterial polysaccharide EPS11 inhibits cancer cell growth via. Mar. Drugs. 1660-3397. 16 (3), 85. doi:10.3390/md16030085.

Carrión, O., Delgado, L., Mercade, E., 2015. New emulsifying and cryoprotective exopolysaccharide from Antarctic. Carbohydr. Polym. 0144-8617. 117, 1028–1034. Available from: https://www.ncbi.nlm.nih.gov/pubmed/25498731. doi:10.1016/j.carbpol.2014.08.060.

Chatterjee, S., Mukhopadhyay, S.K., Gauri, S.S., Dey, S., 2018. Sphingobactan, a new α-mannan exopolysaccharide from Arctic. Int. Immunopharmacol. 1567-5769. 60, 84–95. doi:10.1016/j.intimp.2018.04.039.

Chaudhary, M., Khan, A., Gupta, M., 2020. Skin ageing: pathophysiology and current market treatment approaches. Curr. Aging Sci. 1874-6098. 13 (1), 22–30. Available from: https://www.ncbi.nlm.nih.gov/pmc/articles/PMC7403684. doi:10.2174/1567205016666190809161115.

Che, N., Yijie, M., Xin, Y., 2017. Protective role of fucoidan in cerebral ischemia-reperfusion injury. Biomol. Ther. 1976-9148. 25 (3), 272–278. Available from: https://www.ncbi.nlm.nih.gov/pmc/articles/PMC5424637. doi:10.4062/biomolther.2016.098.

Chen, Y., Lin, H., Zhenxing, L., Mou, Q., 2015. The anti-allergic activity of polyphenol extracted from five marine algae. J. Ocean Univ. China. 1672-5182. 14 (4), 681–684. doi:10.1007/s11802-015-2601-5.

Chen, Y., Mao, W.-J., Yan, M.-X., Liu, X., Wang, S.-Y., Xia, Z., Xiao, B., Cao, S.-J., Yang, B.-Q., Jie, L., 2016. Purification, chemical characterization, and bioactivity of an. Mar. Biotechnol. (New York, N.Y.). 1436-2228. 18 (3), 301–313. Available from: https://www.ncbi.nlm.nih.gov/pubmed/27153822. doi:10.1007/s10126-016-9696-6.

Chen, Y., Mao, W., Yang, Y., Teng, X., Zhu, W., Xiaohui, Q., Chen, Y., Zhao, C., Hou, Y., Wang, C., Li, N., 2012. Structure and antioxidant activity of an extracellular polysaccharide from. Carbohydr. Polym. 0144-8617. 87 (1), 218–226. Available from: https://www.ncbi.nlm.nih.gov/pubmed/34662953. doi:10.1016/j.carbpol.2011.07.042.

Chen, Z., Zhipeng, L., Liu, P., Liu, Y., Wang, Y., Qingbiao, L., He, N., 2017. Characterization of a novel bioflocculant from a marine bacterium and its. BMC Biotechnol. 1472-6750. 17 (1), doi:10.1186/s12896-017-0404-z.

Chiu, C.-Y., Chang, T.-C., Liu, S.-H., Chiang, M.-T., 2017. The regulatory effects of fish oil and chitosan on hepatic lipogenic. J. Food Drug Anal. 1021-9498. 25 (4), 919–930. doi:10.1016/j.jfda.2016.11.015.

Cho, E.-J., Rahman, A., Kim, S.-W., Baek, Y.-M., Hwang, H.-J., Oh, J.-Y., Hwang, H.-S., Lee, S.-H., Yun, J.-W., 2008. Chitosan oligosaccharides inhibit adipogenesis in 3T3-L1 adipocytes. J. Microbiol. Biotechnol. 18, 80–87.

Cho, S., Yang, H., You-Jin Jeon, C.J.L., Jin, Y.-H., Baek, N.-I., Kim, D., Kang, S.-M., Yoon, M., Yong, H., Shimizu, M., Han, D., 2012. Phlorotannins of the edible brown seaweed *Ecklonia cava* Kjellman induce. Food Chem. 0308-8146. 132 (3), 1133–1142. doi:10.1016/j.foodchem.2011.08.040.

Choudhary, B., Chauhan, O.P., Mishra, A., 2021. Edible seaweeds: a potential novel source of bioactive metabolites and nutraceuticals with human health benefits. Front. Mar. Sci. 2296-7745. 8. doi:10.3389/fmars.2021.740054.

Cicinskas, E., Kalitnik, A.A., Karetin, Y.A., Ram, M.S.G.M., Achary, A., Kravchenko, A.O., 2020. Immunomodulating properties of carrageenan from *Tichocarpus crinitus*. Inflammation. 1573-2576. 43 (4), 1387–1396. Available from: www.wkap.nl/journalhome.htm/0360-3997. doi:10.1007/s10753-020-01216-x.

Costa, O.Y.A., Raaijmakers, J.M., Kuramae, E.E., 2018. Microbial extracellular polymeric substances: ecological function and impact on soil aggregation. Front. Microbiol. 1664-302X. 9. doi:10.3389/fmicb.2018.01636.

Courtois, A., Berthou, C., Guézennec, J., Boisset, C., Bordron, A., Stover, C.M., 2014. Exopolysaccharides isolated from hydrothermal vent bacteria can modulate the complement system. PLoS One. 1932-6203. 9 (4), e94965. doi:10.1371/journal.pone.0094965.

Cui, C., Cui, N., Wang, P., Song, S., Liang, H., Aiguo, J., 2016. Neuroprotective effect of sulfated polysaccharide isolated from sea cucumber Stichopus japonicus on 6-OHDA-induced death in SH-SY5Y through inhibition of MAPK and NF-κB and activation of PI3K/Akt signaling pathways. Biochem. Biophys. Res. Commun. 0006-291X. 470 (2), 375–383. Available from: https://www.sciencedirect.com/science/article/pii/S0006291X16300353. doi:10.1016/j.bbrc.2016.01.035.

Cui, W., Zheng, Y., Zhang, Q., Wang, J., Wang, L., Yang, W., Guo, C., Gao, W., Wang, X., Luo, D., 2014. Low-molecular-weight fucoidan protects endothelial function and ameliorates basal hypertension in diabetic Goto-Kakizaki rats. Lab. Investig. 0023-6837. 94 (4), 382–393. doi:10.1038/labinvest.2014.12.

Dang, H., Lovell, C.R., 2000. Bacterial primary colonization and early succession on surfaces in marine waters as determined by amplified rRNA gene restriction analysis and sequence analysis of 16S rRNA genes. Appl. Environ. Microbiol. 0099-2240. 66 (2), 467–475. doi:10.1128/AEM.66.2.467-475.2000.

De Jesus Raposo, M.F.D.J., De Morais, A.M.B., De Morais, R.M.S.C., 2015. Marine polysaccharides from algae with potential biomedical applications. Mar. Drugs. 1660-3397. 13 (5), 2967–3028. Available from: https://www.mdpi.com/1660-3397/13/5/2967/pdf. doi:10.3390/md13052967.

Deniaud-Bouët, E., Hardouin, K., Potin, P., Kloareg, B., Hervé, C., 2017. A review about brown algal cell walls and fucose-containing sulfated polysaccharides: cell wall context, biomedical properties and key research challenges. Carbohydr. Polym. 0144-8617. 175, 395–408. doi:10.1016/j.carbpol.2017.07.082.

Deniaud-Bouët, E., Kervarec, N., Michel, G., Tonon, T., Kloareg, B., Hervé, C., 2014. Chemical and enzymatic fractionation of cell walls from Fucales: insights into the structure of the extracellular matrix of brown algae. Ann. Bot. 0305-7364. 114 (6), 1203–1216. doi:10.1093/aob/mcu096.

Dhanya, B.E., Athmika, P.D.R., 2021. Characterization of an exopolysaccharide produced by *Enterobacter* sp. YU16-RN5 and its potential to alleviate cadmium induced cytotoxicity in vitro. 3 Biotech. 2190-5738. 11 (12). Available from: https://www.springer.com/journal/13205. doi:10.1007/s13205-021-03034-w.

Di, T., Chen, G., Sun, Y., Shiyi, O., Zeng, X., Hong, Y., 2017. Antioxidant and immunostimulating activities in vitro of sulfated polysaccharides isolated from *Gracilaria rubra*. J. Funct. Foods. 1756-4646. 28, 64–75. doi:10.1016/j.jff.2016.11.005.

Ding, J., Baochuan, W., Chen, L., 2022. Application of marine microbial natural products in cosmetics. Front. Microbiol. 1664-302X. 13. doi:10.3389/fmicb.2022.892505.

Elsakhawy, T., 2017. Marine microbial polysaccharides environmental role and applications: an overview. Env. Biodivers. Soil Secur.. 2536-9423. doi:10.21608/jenvbs.2017.1053.1004.

Engel, A., 2002. Direct relationship between CO_2 uptake and transparent exopolymer particles production in natural phytoplankton. J. Plankton Res. 1464-3774. 24 (1), 49–53. doi:10.1093/plankt/24.1.49.

Engel, A., Thoms, S., Riebesell, U., Rochelle-Newall, E., Zondervan, I., 2004. Polysaccharide aggregation as a potential sink of marine dissolved organic carbon. Nature. 0028-0836. 428 (6986), 929–932. doi:10.1038/nature02453.

Ermakova, S., Sokolova, R., Kim, S.M., Um, B.H., Isakov, V., Zvyagintseva, T., 2011. Fucoidans from brown seaweeds *Sargassum hornery, Eclonia cava, Costaria costata*: structural characteristics and anticancer activity. Appl. Biochem. Biotechnol. 1559-0291. 164 (6), 841–850. doi:10.1007/s12010-011-9178-2.

Fedorov, S., Ermakova, S., Zvyagintseva, T., Stonik, V., 2013. Anticancer and cancer preventive properties of marine polysaccharides: some results and prospects. Mar. Drugs. 1660-3397. 11 (12), 4876–4901. doi:10.3390/md11124876.

Fernando, I.P.S., Kim, K.N., Kim, D., Jeon, Y.J., 2019. Algal polysaccharides: potential bioactive substances for cosmeceutical applications. Crit. Rev. Biotechnol. 1549-7801. 39 (1), 99–113. Available from: https://www.tandfonline.com/loi/ibty20. doi:10.1080/07388551.2018.1503995.

Fernando, I.P.S., Sanjeewa, K.K.A., Samarakoon, K.W., Lee, W.W., Kim, H.S., Kim, E.A., Gunasekara, U.K., Abeytunga, D.T.U., Nanayakkara, C., De Silva, E.D., Lee, H.S., Jeon, Y.J., 2017. FTIR characterization and antioxidant activity of water soluble crude polysaccharides of Sri Lankan marine algae. Algae. 2093-0860. 32 (1), 75–86. Available from: https://www.e-algae.kr. doi:10.4490/algae.2017.32.12.1.

Fimbres-Olivarria, D., Carvajal-Millan, E., Lopez-Elias, J.A., Martinez-Robinson, K.G., Miranda-Baeza, A., Martinez-Cordova, L.R., Enriquez-Ocaña, F., Valdez-Holguin, J.E., 2018. Chemical characterization and antioxidant activity of sulfated polysaccharides from *Navicula* sp. Food Hydrocoll. 0268-005X. 75, 229–236. Available from: https://www.journals.elsevier.com/food-hydrocolloids. doi:10.1016/j.foodhyd.2017.08.002.

Fitton, J.H., Stringer, D.N., Park, A.Y., Karpiniec, S.S., 2019. Therapies from fucoidan: new developments. Mar. Drugs. 1660-3397. 17 (10). Available from: https://www.mdpi.com/1660-3397/17/10/571/pdf. doi:10.3390/md17100571.

Ganceviciene, R., Liakou, A.I., Theodoridis, A., Makrantonaki, E., Zouboulis, C.C., 2012. Skin anti-aging strategies. Dermato-Endocrinol. 1938-1980. 4 (3). Available from: https://www.landesbioscience.com/journals/dermatoendocrinology/2012DE0432.pdf?nocache=497028977. doi:10.4161/derm.22804Lithuania.

Gargouch, N., Elleuch, F., Karkouch, I., Tabbene, O., Pichon, C., Gardarin, C., Rihouey, C., Picton, L., Abdelkafi, S., Fendri, I., Laroche, C., 2021. Potential of exopolysaccharide from *Porphyridium marinum* to contend with bacterial proliferation, biofilm formation, and breast cancer. Mar. Drugs. 1660-3397. 19 (2), 66. doi:10.3390/md19020066.

Garre, A., Martinez-Masana, G., Piquero-Casals, J., Granger, C., 2017. Redefining face contour with a novel anti-aging cosmetic product: an open-label, prospective clinical study. Clin. Cosmet. Investig. Dermatol.. 1178-7015. 10, 473–482. doi:10.2147/CCID.S148597.

Ghazy, A.R., El-Sheekh, M.M., Ghazy, R., 2024. Structure, laser scattering, optical properties, adsorption, and dye removal studies by Ulvan extracted from the green seaweed *Ulva lactuca*. Egypt J. Mol. Liq. 0167-7322. 410. Available from: https://www.sciencedirect.com/science/journal/01677322. doi:10.1016/j.molliq.2024.125567.

Ghibaudo, M., Baltenneck, C., Minondo, A., Couturier, L.T., Potter, A., 2015. A natural sulfated polysaccharide from *Porphyridium cruentum* red microalgae—a potential mimetic of heparan sulfate and its modulation of the epidermal biological activity of human reconstructed skin.

González, R.D., Astals, A.S., Courtois, A., Thollas, B., 2016. Exopolysaccharide for the Treatment And/or Care of the Skin, Mucous Membranes And/or Nails. (Patent No. US 2016/0045423 A1).

Gottschalk, A., 1972. Glycoproteins: Their Composition, Structure and Function, Elsevier Publishing Company.

Graiff, A., Ruth, W., Kragl, U., Karsten, U., 2016. Chemical characterization and quantification of the brown algal storage compound laminarin—a new methodological approach. J. Appl. Phycol. 0921-8971. 28 (1), 533–543. doi:10.1007/s10811-015-0563-z.

Gui, J., Tong, W., Huang, S., Liang, X., Fang, Z., Wang, W., Zhang, Y., 2019. Effects of *Chlorella vulgaris* polysaccharides accumulation on growth characteristics of *Trachemys scripta* elegans. Int. J. Biol. Macromol. 0141-8130. 141, 1304–1313. doi:10.1016/j.ijbiomac.2019.08.248.

Guo, S., Mao, W., Yan, M., Zhao, C., Li, N., Shan, J., Lin, C., Liu, X., Guo, T., Guo, T., Wang, S., 2014. Galactomannan with novel structure produced by the coral endophytic fungus *Aspergillus ochraceus*. Carbohydr. Polym. 0144-8617. 105 (1), 325–333. doi:10.1016/j.carbpol.2014.01.079.

Gupta, P., Diwan, B., 2017. Bacterial exopolysaccharide mediated heavy metal removal: a review on biosynthesis, mechanism and remediation strategies. Biotechnol. Rep. 2215-017X. 13, 58–71. doi:10.1016/j.btre.2016.12.006.

Han, M.H., Lee, D.S., Jeong, J.W., Hong, S.H., Choi, I.W., Cha, H.J., Kim, S., Kim, H.S., Park, C., Kim, G.Y., Moon, S. K., Kim, W.J., Choi, Y.H., 2017. Fucoidan induces ROS-dependent apoptosis in 5637 human bladder cancer cells by downregulating telomerase activity via inactivation of the PI3K/Akt signaling pathway. Drug Dev. Res. 1098-2299. 78 (1), 37–48. Available from: https://onlinelibrary.wiley.com/journal/10.1002/(ISSN)1098-2299. doi:10.1002/ddr.21367.

Hermansson, M., W, J.G., Kjelleberg, S., 1987. Frequency of antibiotic and heavy metal resistance, pigmentation, and plasmids in bacteria of the marine air-water interface. Appl. Environ. Microbiol. 53 (10), 2338–2342. doi:10.1128/aem.53.10.2338-2342.1987.

Hu, X., Tao, N., Wang, X., Xiao, J., Wang, M., 2016. Marine-derived bioactive compounds with anti-obesity effect: a review. J. Funct. Foods. 1756-4646. 21, 372–387. doi:10.1016/j.jff.2015.12.006.

Huang, X., Wang, S., Fu, L., Fang, F., Guo, J., 2022. Characterization of extracellular polymeric substances by microscopic imaging techniques in wastewater biotreatment: a review. Environ. Eng. Sci. 1557-9018. 39 (6), 493–511. Available from: www.liebertonline.com/ees. doi:10.1089/ees.2021.0199.

Huleihel, M., Ishanu, V., Tal, J., Arad, S., 2001. Antiviral effect of red microalgal polysaccharides on *Herpes simplex* and *Varicella zoster* viruses. J. Appl. Phycol. 0921-8971. 13 (2), 127–134. doi:10.1023/A:1011178225912.

Ibrahim, H.A.H., Elhassayeb, H.E.A., El-Sayed, W.M.M., 2022. Potential functions and applications of diverse microbial exopolysaccharides in marine environments. J. Genet. Eng. Biotechnol. 2090-5920. 20 (1). Available from: https://www.springer.com/journal/43141. doi:10.1186/s43141-022-00432-2.

Inanli, A.G., Tümerkan, E.T.A., Abed, N.E., Regenstein, J.M., Özogul, F., 2020. The impact of chitosan on seafood quality and human health: a review. Trends Food Sci. Technol. 0924-2244. 97, 404–416. Available from: https://www.elsevier.com/wps/find/journaldescription.cws_home/601278/description#description. doi:10.1016/j.tifs.2020.01.029.

Ivatt, R.J., 2012. The Biology of Glycoproteins, Vol. 1. Springer, New York, NY, pp. 449. doi:10.1007/978-1-4684-7464-0.

Iyer, A., Mody, K., Jha, B., 2005. Biosorption of heavy metals by a marine bacterium. Mar. Pollut. Bull. 0025-326X. 50 (3), 340–343. doi:10.1016/j.marpolbul.2004.11.012.

Jeewon, R., Aullybux, A.A., Puchooa, D., Nazurally, N., Alrefaei, A.F., Zhang, Y., 2023. Marine microbial polysaccharides: an untapped resource for biotechnological applications. Mar. Drugs. 1660-3397. 21 (7). Available from: http://www.mdpi.com/journal/marinedrugs. doi:10.3390/md21070420.

Jiang, P., Jingbao, L., Han, F., Duan, G., Xinzhi, L., Yuchao, G., Wengong, Y., Hensel, M., 2011. Antibiofilm activity of an exopolysaccharide from marine bacterium *Vibrio* sp. QY101. PLoS One. 1932-6203. 6 (4), e18514. doi:10.1371/journal.pone.0018514.

Jin, J.-O., Zhang, W., Jiang-Yuan, D., Wong, K.-W., Oda, T., Qing, Y., 2014a. Fucoidan can function as an adjuvant in vivo to enhance dendritic cell maturation and function and promote antigen-specific T cell immune responses. PLoS One 9 (6), e99396. doi:https://doi.org/10.1371/journal.pone.0099396.

Jin, J.O., Yu, Q., 2015. Fucoidan delays apoptosis and induces pro-inflammatory cytokine production in human neutrophils. Int. J. Biol. Macromol. 1879-0003. 73 (1), 65–71. Available from: www.elsevier.com/locate/ijbiomac. doi:10.1016/j.ijbiomac.2014.10.059.

Jin, W., Zhang, W., Wang, J., Yao, J., Xie, E., Liu, D., Duan, D., Zhang, Q., 2014b. A study of neuroprotective and antioxidant activities of heteropolysaccharides from six *Sargassum* species. Int. J. Biol. Macromol. 0141-8130. 67, 336–342. Available from: https://www.sciencedirect.com/science/article/pii/S0141813014001846. doi:10.1016/j.ijbiomac.2014.03.031.

Johnston, G.A.R., 2015. Flavonoid nutraceuticals and ionotropic receptors for the inhibitory neurotransmitter GABA. Neurochem. Int. 1872-9754. 89, 120–125. Available from: www.elsevier.com/locate/neuint. doi:10.1016/j.neuint.2015.07.013.

Jönsson, M., Allahgholi, L., Sardari, R.R.R., Hreggviðsson, G.O., Karlsson, E.N., 2020. Extraction and modification of macroalgal polysaccharides for current and next-generation applications. Molecules. 1420-3049. 25 (4), 930. doi:10.3390/molecules25040930.

Kaats, G.R., Michalek, J.E., Preuss, H.G., 2006. Evaluating efficacy of a chitosan product using a double-blinded, placebo-controlled protocol. J. Am. Coll. Nutr. 1541-1087. 25 (5), 389–394. doi:10.1080/07315724.2006.10719550.

Kadam, S.U., Tiwari, B.K., O'Donnell, C.P., 2015. Extraction, structure and biofunctional activities of laminarin from brown algae. Int. J. Food Sci. Technol. 1365-2621. 50 (1), 24–31. doi:10.1111/ijfs.12692.

Kalpana, R., Maheshwaran, M., Vimali, E., Soosai, M.R., Shivamathi, C.S., Moorthy, I.G., Ashokkumar, B., Varalakshmi, P., 2020. Decolorization of textile dye by halophilic *Exiguobacterium* sp.VK1: biomass and exopolysaccharide (EPS) enhancement for bioremediation of malachite green. ChemistrySelect. 2365-6549. 5 (28), 8787–8797. Available from: https://onlinelibrary.wiley.com/journal/10.1002/(ISSN)2365-6549. doi:10.1002/slct.202001648.

Kanlayavattanakul, M., Lourith, N., 2015. Biopolysaccharides for skin hydrating cosmetics. Polysaccharides 1867–1892. doi:10.1007/978-3-319-16298-0_29.

Kil-Nam, K., Joo, E.-S., Kyu-Il, K., Se-Kwon, K., Hyun-Pyl, Y., You-Jin, J., 2005. Effect of chitosan oligosaccharides on cholesterol level and antioxidant enzyme activities in hypercholesterolemic rat. J. Korean Soc. Food Sci. Nutr. 1226-3311. 34 (1), 36–41. doi:10.3746/jkfn.2005.34.1.036.

Kim, C., Ryu, H.C., Kim, J.H., 2010. Low-dose UVB irradiation stimulates matrix metalloproteinase-1 expression via a BLT2-linked pathway in HaCaT cells. Exp. Mol. Med. 2092-6413. 42 (12), 833–841. Available from: https://www.e-emm.org/article/article_files/EMM-42-12-05.pdf. doi:10.3858/emm.2010.42.12.086.

Kim, Y.I., Oh, W.S., Song, P.H., Yun, S., Kwon, Y.S., Lee, Y.J., Ku, S.K., Song, C.H., Oh, T.H., 2018. Anti-photoaging effects of low molecular-weight fucoidan on ultraviolet B-irradiated mice. Mar. Drugs. 1660-3397. 16 (8). Available from: http://www.mdpi.com/journal/marinedrugs. doi:10.3390/md16080286.

Kobayashi, R., Takisada, M., Suzuki, T., Kirimura, K., Usami, S., 1997. Neoagarobiose as a novel moisturizer with whitening effect. Biosci. Biotechnolo. Biochem. 1347-6947. 61 (1), 162–163. doi:10.1271/bbb.61.162.

Kranck, K., Milligan, T., 1980. Macroflocs: production of marine snow in the laboratory. Mar. Ecol. Prog. Ser. 0171-8630. 3, 19–24. doi:10.3354/meps003019.

Krishnamurthy, M., Uthaya, C.J., Thangavel, M., Annadurai, V., Rajendran, R., Gurusamy, A., 2020. Optimization, compositional analysis, and characterization of exopolysaccharides produced by multi-metal resistant *Bacillus cereus* KMS3-1. Carbohydr. Polym. 0144-8617. 227, 115369. doi:10.1016/j.carbpol.2019.115369.

Krüger, M., Blumenberg, M., Kasten, S., Wieland, A., Känel, L., Klock, J.H., Michaelis, W., Seifert, R., 2008. A novel, multi-layered methanotrophic microbial mat system growing on the sediment of the Black Sea. Environ. Microbiol. 1462-2920. 10 (8), 1934–1947. doi:10.1111/j.1462-2920.2008.01607.x.

Lahaye, M., Robic, A., 2007. Structure and functional properties of ulvan, a polysaccharide from green seaweeds. Biomacromolecules. 1525-7797. 8 (6), 1765–1774. doi:10.1021/bm061185q.

Laurienzo, P., 2010. Marine polysaccharides in pharmaceutical applications: an overview. Mar. Drugs. 1660-3397. 8 (9), 2435–2465. Available from: https://www.mdpi.com/1660-3397/8/9/2435/pdf. doi:10.3390/md8092435.

Lee, J.B., Hayashi, K., Hashimoto, M., Nakano, T., Hayashi, T., 2004. Novel antiviral fucoidan from sporophyll of *Undaria pinnatifida* (Mekabu). Chem. Pharm. Bull. 1347-5223. 52 (9), 1091–1094. Available from: https://www.jstage.jst.go.jp/article/cpb/52/9/1091/_pdf. doi:10.1248/cpb.52.1091Japan.

Li, G.Y., Luo, Z.C., Yuan, F., Yu, X.B., 2017. Combined process of high-pressure homogenization and hydrothermal extraction for the extraction of fucoidan with good antioxidant properties from Nemacystus decipients. Food Bioprod. Process. 0960-3085. 106, 35–42. Available from: https://www.elsevier.com/wps/find/journaldescription.cws_home/713883/description#description. doi:10.1016/j.fbp.2017.08.002.

Li, H., Cao, K., Cong, P., Liu, Y., Cui, H., Xue, C., 2018. Structure characterization and antitumor activity of the extracellular polysaccharide from the marine fungus *Hansfordia sinuosae*. Carbohydr. Polym. 0144-8617. 190, 87–94. doi:10.1016/j.carbpol.2018.02.077.

Li, J., Qiushi, H., Yuquan, L., Yuan, X., 2015. Purification and characterization of cold-adapted beta-agarase from an Antarctic psychrophilic strain. Braz. J. Microbiol. 1678-4405. 46 (3), 683–690. doi:10.1590/S1517-838246320131289.

Li, J., Zhixiao, H., Liang, Y., Peng, T., Zhong, H., 2022. Insights into algal polysaccharides: a review of their structure, depolymerases, and metabolic pathways. J. Agric. Food Chem. 0021-8561. 70 (6), 1749–1765. doi:10.1021/acs.jafc.1c05365.

Liang, Z., Zheng, Y., Wang, J., Zhang, Q., Ren, S., Liu, T., Wang, Z., Luo, D., 2016. Low molecular weight fucoidan ameliorates streptozotocin-induced hyper-responsiveness of aortic smooth muscles in type 1 diabetes rats. J. Ethnopharmacol. 0378-8741. 191, 341–349. Available from: https://www.sciencedirect.com/science/article/pii/S0378874116304147. doi:10.1016/j.jep.2016.06.054.

Lin, Y., Yang, J., Luo, L., Zhang, X., Deng, S., Chen, X., Li, Y., Bekhit, A.E.D.A., Xu, B., Huang, R., 2022. Ferroptosis related immunomodulatory effect of a novel extracellular Polysaccharides from marine fungus *Aureobasidium melanogenum*. Mar. Drugs. 1660-3397. 20 (5). Available from: https://www.mdpi.com/1660-3397/20/5/332/pdf?version=1652924606. doi:10.3390/md20050332.

Liu, B., Liu, W.S., Han, B.Q., Sun, Y.Y., 2007. Antidiabetic effects of chitooligosaccharides on pancreatic islet cells in streptozotocin-induced diabetic rats. World J. Gastroenterol. 1007-9327. 13 (5), 725–731. Available from: https://www.wjgnet.com/1007-9327. doi:10.3748/wjg.v13.i5.725.

Liu, G., Liu, R., Shan, Y., Sun, C., 2021. Marine bacterial exopolysaccharide EPS11 inhibits migration and invasion of liver cancer cells by directly targeting collagen I. J. Biol. Chem. 0021-9258. 297 (4), 101133. doi:10.1016/j.jbc.2021.101133.

Liu, S.B., Chen, X.L., He, H.L., Zhang, X.Y., Xie, B.B., Yu, Y., Chen, B., Zhou, B.C., Zhang, Y.Z., 2013. Structure and ecological roles of a novel exopolysaccharide from the Arctic sea ice bacterium *Pseudoalteromonas* sp. strain SM20310. Appl. Environ. Microbiol. 1098-5336. 79 (1), 224–230. Available from: https://aem.asm.org/content/79/1/224.full.pdf. doi:10.1128/AEM.01801-12China.

Liu, Y.Q., Liu, Y., Tay, J.H., 2004. The effects of extracellular polymeric substances on the formation and stability of biogranules. Appl. Microbiol. Biotechnol. 0175-7598. 65 (2), 143–148. Available from: https://www.springer.de/link/service/journals/00253/index.htm. doi:10.1007/s00253-004-1657-8.

Llamas, I., Mata, J.A., Tallon, R., Bressollier, P., Urdaci, M.C., Quesada, E., Béjar, V., 2010. Characterization of the exopolysaccharide produced by *Salipiger mucosus* A3T, a halophilic species belonging to the Alphaproteobacteria, isolated on the Spanish Mediterranean seaboard. Mar. Drugs. 1660-3397. 8 (8), 2240–2251. Available from: https://www.mdpi.com/1660-3397/8/8/2240/pdf. doi:10.3390/md8082240.

Mahmoud, Y.A.G., El-Naggar, M.E., Abdel-Megeed, A., El-Newehy, M.H., 2021. Recent advancements in microbial polysaccharides: synthesis and applications. Polymers. 20734360. 13 (23). Available from: https://www.mdpi.com/2073-4360/13/23/4136/pdf. doi:10.3390/polym13234136.

Mak, W., Hamid, N., Liu, T., Lu, J., White, W.L., 2013. Fucoidan from New Zealand *Undaria pinnatifida*: monthly variations and determination of antioxidant activities. Carbohydr. Polym. 0144-8617. 95 (1), 606–614. doi:10.1016/j.carbpol.2013.02.047.

Mårdén, P., Nyström, T., Kjelleberg, S., 1987. Uptake of leucine by a marine Gram-negative heterotrophic bacterium during exposure to starvation conditions. FEMS Microbiol. Ecol. 0168-6496. 3 (4), 233–241. doi:10.1111/j.1574-6968.1987.tb02361.x.

Margesin, R., Schinner, F., 2001. Potential of halotolerant and halophilic microorganisms for biotechnology. Extremophiles. 1431-0651. 5 (2), 73–83. doi:10.1007/s007920100184.

Martínez-Checa, F., Toledo, F., 2002. Yield production, chemical composition, and functional properties of emulsifier H28 synthesized by *Halomonas eurihalina* strain H-28 in media containing various hydrocarbons. Appl. Microbiol. Biotechnol. 0175-7598. 58 (3), 358–363. doi:10.1007/s00253-001-0903-6.

Martins, A., Vieira, H., Gaspar, H., Santos, S., 2014. Marketed marine natural products in the pharmaceutical and cosmeceutical industries: tips for success. Mar. Drugs. 1660-3397. 12 (2), 1066–1101. doi:10.3390/md12021066.

Marudhupandi, T., Kumar, T.T.A., 2013. Antibacterial effect of fucoidan from *Sargassum wightii* against the chosen human bacterial pathogens. Int. Curr. Pharm. J. 2224-9486. 2 (10), 156–158. doi:10.3329/icpj.v2i10.16408.

Mathivanan, K., Chandirika, J.U., Mathimani, T., Rajaram, R., Annadurai, G., Yin, H., 2021. Production and functionality of exopolysaccharides in bacteria exposed to a toxic metal environment. Ecotoxicol. Environ. Saf. 0147-6513. 208, 111567. doi:10.1016/j.ecoenv.2020.111567.

Maugeri, T.L., Gugliandolo, C., Caccamo, D., Panico, A., Lama, L., Gambacorta, A., Nicolaus, B., 2002. A halophilic thermotolerant Bacillus isolated from a marine hot spring able to produce a new exopolysaccharide. Biotechnol. Lett. 0141-5492. 24 (7), 515–519. doi:10.1023/A:1014891431233.

Mehta, A., Sidhu, C., Pinnaka, A.K., Choudhury, A.R., 2014. Extracellular polysaccharide production by a novel osmotolerant marine strain of *Alteromonas macleodii* and its application towards biomineralization of silver. PLoS One. 19326203. 9 (6). Available from: https://www.plosone.org/article/fetchObject.action?uri=info%3Adoi%2F10.1371%2Fjournal.pone.0098798&representation=PDF. doi:10.1371/journal.pone.0098798.

Metha, C., Pawar, S., Suvarna, V., 2024. Recent advancements in alginate-based films for active food packaging applications. Sustain. Food Technol. 2753-8095. 2 (5), 1246–1265. Available from: https://www.rsc.org/journals-books-databases/about-journals/sustainable-food-technology/. doi:10.1039/d3fb00216k.

Mhurchu, C.N., Poppitt, S.D., McGill, A.-T., Leahy, F.E., Bennett, D.A., Lin, R.B., Ormrod, D., Ward, L., Strik, C., Rodgers, A., 2004. The effect of the dietary supplement, Chitosan, on body weight: a randomised controlled trial in 250 overweight and obese adults. Int. J. Obes. 0307-0565. 28 (9), 1149–1156. doi:10.1038/sj.ijo.0802693.

Mišurcová, L., Škrovánková, S., Samek, D., Ambrožová, J., Machů, L., 2012. Health benefits of algal polysaccharides in human nutrition. Adv. Food Nutr. Res. 75–145. doi:10.1016/b978-0-12-394597-6.00003-3.

Montuori, E., de Pascale, D., Lauritano, C., 2022. Recent discoveries on marine organism immunomodulatory activities. Mar. Drugs. 1660-3397. 20 (7). doi:10.3390/md20070422.

Moon, H.J., Lee, S.H., Ku, M.J., Yu, B.C., Jeon, M.J., Jeong, S.H., Stonil, V.A., Zvyagintseva, T.N., Ermakova, S.P., Lee, Y.H., 2009. Fucoidan inhibits UVB-induced MMP-I promoter expression and down regulation of type I procollagen synthesis in human skin fibroblasts. Eur. J. Dermatol. 19524-013. 19 (2), 129–134. Available from: https://www.john-libbey-eurotext.fr/e-docs/00/04/46/E1/vers_alt/VersionPDF.pdf. doi:10.1684/ejd.2008.0611.

Muanprasat, C., Chatsudthipong, V., 2017. Chitosan oligosaccharide: biological activities and potential therapeutic applications. Pharmacol. Ther. 01637258. 170, 80–97. doi:10.1016/j.pharmthera.2016.10.013.

Narayani, S.S., Saravanan, S., Ravindran, J., Ramasamy, M.S., Chitra, J., 2019. In vitro anticancer activity of fucoidan extracted from *Sargassum cinereum* against Caco-2 cells. Int. J. Biol. Macromol. 1879-0003. 138, 618–628. Available from: www.elsevier.com/locate/ijbiomac. doi:10.1016/j.ijbiomac.2019.07.127.

Ngo, D.H., Vo, T.S., Ngo, D.N., Kang, K.H., Je, J.Y., Pham, H.N.D., Byun, H.G., Kim, S.K., 2015. Biological effects of chitosan and its derivatives. Food Hydrocoll. 0268-005X. 51, 200–216. Available from: https://www.journals.elsevier.com/food-hydrocolloids. doi:10.1016/j.foodhyd.2015.05.023.

Nichols, C.A.M., Guezennec, J., Bowman, J.P., 2005. Bacterial exopolysaccharides from extreme marine environments with special consideration of the Southern Ocean, sea ice, and deep-sea hydrothermal vents: a review. Mar. Biotechnol. 1436-2228. 7 (4), 253–271. doi:10.1007/s10126-004-5118-2.

Pacheco-Quito, E.M., Ruiz-Caro, R., Veiga, M.D., 2020. Carrageenan: drug delivery systems and other biomedical applications. Mar. Drugs. 16603397. 18 (11). Available from: https://www.mdpi.com/journal/marinedrugs. doi:10.3390/md18110583.

Palanisamy, S., Vinosha, M., Manikandakrishnan, M., Anjali, R., Rajasekar, P., Marudhupandi, T., Manikandan, R., Vaseeharan, B., Prabhu, N.M., 2018. Investigation of antioxidant and anticancer potential of fucoidan from *Sargassum polycystum*. Int. J. Biol. Macromol. 1879-0003. 116, 151–161. Available from: www.elsevier.com/locate/ijbiomac. doi:10.1016/j.ijbiomac.2018.04.163.

Pallela, R., 2014. Nutraceutical and pharmacological implications of marine carbohydrates. In: Kim, S.-K., (Ed.), Marine Carbohydrates: Fundamentals and Applications, Part B. Elsevier, 183–195. doi:10.1016/b978-0-12-800268-1.00009-3.

Parra-Riofrío, G., García-Márquez, J., Casas-Arrojo, V., Uribe-Tapia, E., Abdala-Díaz, R.T., 2020. Antioxidant and cytotoxic effects on tumor cells of Exopolysaccharides from *Tetraselmis suecica* (Kylin) butcher grown under autotrophic and heterotrophic conditions. Mar. Drugs. 1660-3397. 18 (11), 534. doi:10.3390/md18110534.

Parrilli, E., Tedesco, P., Fondi, M., Tutino, M.L., Giudice, A.L., de Pascale, D., Fani, R., 2021. The art of adapting to extreme environments: the model system Pseudoalteromonas. Phys. Life Rev. 1571-0645. 36, 137–161. Available from: https://www.elsevier.com/inca/publications/store/6/8/0/8/3/5/index.htt. doi:10.1016/j.plrev.2019.04.003.

Patel, A.K., Vadrale, A.P., Singhania, R.R., Michaud, P., Pandey, A., Chen, S.J., Chen, C.W., Dong, C.D., 2023. Algal polysaccharides: current status and future prospects. Phytochem. Rev. 1572-980X. 22 (4), 1167–1196. Available from: https://www.springer.com/journal/11101. doi:10.1007/s11101-021-09799-5.

Pérez, M.J., Falqué, E., Domínguez, H., 2016. Antimicrobial action of compounds from marine seaweed. Mar. Drugs. 1660-3397. 14 (3). Available from: https://www.mdpi.com/1660-3397/14/3/52/pdf. doi:10.3390/md14030052.

Perez, S., Makshakova, O., Angulo, J., Bedini, E., Bisio, A., de Paz, J.L., Fadda, E., Guerrini, M., Hricovini, M., Hricovini, M., Lisacek, F., Nieto, P.M., Pagel, K., Paiardi, G., Richter, R., Samsonov, S.A., Vivès, R.R., Nikitovic, D., Blum, S.R., 2023. Glycosaminoglycans: what remains to be deciphered? JACS Au. 2691-3704. 3 (3), 628–656. Available from: https://pubs.acs.org/page/jaaucr/about.html. doi:10.1021/jacsau.2c00569.

Pimentel, F., Alves, R., Rodrigues, F., Oliveira, M.P.P., 2018. Macroalgae-derived ingredients for cosmetic industry—an update. Cosmetics. 2079-9284. 5 (1), 2. doi:10.3390/cosmetics5010002.

Poli, A., Anzelmo, G., Nicolaus, B., 2010. Bacterial Exopolysaccharides from extreme marine habitats: production, characterization and biological activities. Mar. Drugs. 1660-3397. 8 (6), 1779–1802. doi:10.3390/md8061779.

Pomin, V.H., Mourão, P.A.S., 2008. Structure, biology, evolution, and medical importance of sulfated fucans and galactans. Glycobiology. 1460-2423. 18 (12), 1016–1027. doi:10.1093/glycob/cwn085.

Premnath, N., Mohanrasu, K., Rao, R.G.R., Dinesh, G.H., Prakash, G.S., Arivalagan Pugazhendhi, J.J., Govarthanan, M., Kumar, P., Arun, A., 2021. Effect of C/N substrates for enhanced extracellular polymeric substances (EPS) production and poly cyclic aromatic hydrocarbons (PAHs) degradation. Environ. Pollut. 0269-7491. 275, 116035. doi:10.1016/j.envpol.2020.116035.

Qi, M., Zheng, C., Wenhui, W., Guangli, Y., Wang, P., 2022. Exopolysaccharides from marine microbes: source, structure and application. Mar. Drugs. 1660-3397. 20 (8), 512. doi:10.3390/md20080512.

Qin, G., Zhu, L., Chen, X., Wang, P.G., Zhang, Y., 2007. Structural characterization and ecological roles of a novel exopolysaccharide from the deep-sea psychrotolerant bacterium *Pseudoalteromonas* sp. SM9913. Microbiology. 1350-0872. 153 (5), 1566–1572. doi:10.1099/mic.0.2006/003327-0.

Qin, Y., Yuan, Q., Zhang, Y., Jialu, L., Zhu, X., Zhao, L., Wen, J., Liu, J., Zhao, L., Zhao, J., 2018. Enzyme-assisted extraction optimization, characterization and antioxidant activity of polysaccharides from sea cucumber *Phyllophorus proteus*. Molecules. 1420-3049. 23 (3), 590. doi:10.3390/molecules23030590.

Radonic, A., Thulke, S., Achenbach, J., Kurth, A., Vreemann, A., 2010. Anionic polysaccharides from phototrophic microorganisms exhibit antiviral. J. Antivir. Antiretrovir. 1948-5964. 02 (04), doi:10.4172/jaa.1000023.

Raguenes, G., Christen, R., Guezennec, J., Pignet, P., Barbier, G., 1997. *Vibrio diabolicus* sp. nov., a new polysaccharide-secreting organism isolated from a deep-sea hydrothermal vent polychaete annelid, *Alvinella pompejana*. Int. J. Syst. Bacteriol. 0020-7713. 47 (4), 989–995. doi:10.1099/00207713-47-4-989.

Raguénès, G.H.C., Peres, A., Ruimy, R., Pignet, P., Christen, R., Loaec, M., Rougeaux, H., Barbier, G., Guezennec, J. G., 1997. *Alteromonas infernus* sp. Nov., a new polysaccharideproducing bacterium isolated from a deep-sea hydrothermal vent. J. Appl. Microbiol. 1364-5072. 82 (4), 422–430. Available from: https://academic.oup.com/jambio. doi:10.1046/j.1365-2672.1997.00125.xFrance.

Ramamoorthy, S., Gnanakan, A., Lakshmana, S.S., Meivelu, M., Jeganathan, A., 2018. Structural characterization and anticancer activity of extracellular polysaccharides from ascidian symbiotic bacterium *Bacillus thuringiensis*. Carbohydr. Polym. 0144-8617. 190, 113–120. doi:10.1016/j.carbpol.2018.02.047.

Ramnani, P., Chitarrari, R., Tuohy, K., Grant, J., Hotchkiss, S., Philp, K., Campbell, R., Gill, C., Rowland, I., 2012. In vitro fermentation and prebiotic potential of novel low molecular weight polysaccharides derived from agar and alginate seaweeds. Anaerobe. 1075-9964. 18 (1), 1–6. doi:10.1016/j.anaerobe.2011.08.003.

Resende, D.I.S.P., Ferreira, M., Magalhães, C., Lobo, J.M.S., Sousa, E., Almeida, I.F., 2021. Trends in the use of marine ingredients in anti-aging cosmetics. Algal Res. 2211-9264. 55. Available from: https://www.sciencedirect.com/science/journal/aip/22119264. doi:10.1016/j.algal.2021.102273.

Rioux, L.E., Moulin, V., Beaulieu, M., Turgeon, S.L., 2013. Human skin fibroblast response is differentially regulated by galactofucan and low molecular weight galactofucan. Bioact. Carbohydr. Diet. Fibre. 2212-6198. 1 (2), 105–110. doi:10.1016/j.bcdf.2013.03.004.

Rongfeng, L., Huahua, Y., Yue, Y., Liu, S., Xing, R., Chen, X., Pengcheng, L., 2017. Sulfated polysaccharides with antioxidant and anticoagulant activity from the sea cucumber *Holothuria fuscogliva*. Chin. J. Oceanol. Limnol. 0254-4059. 35 (4), 763–769. doi:10.1007/s00343-017-5339-7.

Ruocco, N., Costantini, S., Guariniello, S., Costantini, M., 2016. Polysaccharides from the marine environment with pharmacological, cosmeceutical and nutraceutical potential. Molecules. 1420-3049. 21 (5), 551. doi:10.3390/molecules21050551.

Sahana, T.G., Rekha, P.D., 2020. A novel exopolysaccharide from marine bacterium Pantoea sp. YU16-S3 accelerates cutaneous wound healing through Wnt/β-catenin pathway. Carbohydr. Polym. 0144-8617. 238, 116191. doi:10.1016/j.carbpol.2020.116191.

Sanjeewa, K.K.A., Lee, J.S., Kim, W.S., Jeon, Y.J., 2017. The potential of brown-algae polysaccharides for the development of anticancer agents: an update on anticancer effects reported for fucoidan and laminaran. Carbohydr. Polym. 0144-8617. 177, 451–459. Available from: https://www.elsevier.com/wps/find/journaldescription.cws_home/405871/description#description. doi:10.1016/j.carbpol.2017.09.005.

Senni, K., Gueniche, F., Foucault-Bertaud, A., Igondjo-Tchen, S., Fioretti, F., Colliec-Jouault, S., Durand, P., Guezennec, J., Godeau, G., Letourneur, D., 2006. Fucoidan a sulfated polysaccharide from brown algae is a potent modulator of connective tissue proteolysis. Arch. Biochem. Biophys. 0003-9861. 445 (1), 56–64. doi:10.1016/j.abb.2005.11.001.

Shah, V., Garg, N., Madamwar, D., 1999. Exopolysaccharide production by a marine cyanobacterium *Cyanothece* sp.: application in dye removal by its gelation phenomenon. Appl. Biochem. Biotechnol. 0273-2289. 82 (2), 81–90. doi:10.1385/ABAB:82:2:81.

Shang, Q., Jiang, H., Cai, C., Hao, J., Guoyun, L., Guangli, Y., 2018. Gut microbiota fermentation of marine polysaccharides and its effects on intestinal ecology: an overview. Carbohydr. Polym. 0144-8617. 179, 173–185. doi:10.1016/j.carbpol.2017.09.059.

Shao, P., Chen, X., Sun, P., 2015. Improvement of antioxidant and moisture-preserving activities of *Sargassum horneri* polysaccharide enzymatic hydrolyzates. Int. J. Biol. Macromol. 0141-8130. 74, 420–427. doi:10.1016/j.ijbiomac.2014.12.021.

Shao, Z., Tian, Y., Liu, S., Chu, X., Mao, W., 2023. Anti-diabetic activity of a novel exopolysaccharide produced by the mangrove endophytic fungus *Penicillium janthinellum* N29. Mar. Drugs. 1660-3397. 21 (5), 270. doi:10.3390/md21050270.

Shi, H., Sixue, B., Hang, L., Jianhuan, L., Chunlei, L., Rongmin, Y., Song, L., Zhu, J., 2021. Purification and characterization of a novel mixed-linkage α,β-d-glucan from Arca subcrenata and its immunoregulatory activity. Int. J. Biol. Macromol. 0141-8130. 182, 207–216. Available from: https://www.sciencedirect.com/science/article/pii/S0141813021007467. doi:10.1016/j.ijbiomac.2021.03.196.

Silva, T.C., Pereira, D.M., Valentão, P., Andrade, P.B., 2017. Phospholipase A2 inhibitors of marine origin. In: Andrade, P.B., Valentão, P., Pereira, D.M., (Eds.), Natural Products Targeting Clinically Relevant Enzymes. Wiley-VCH Verlag GmbH & Co. KGaA, Weinheim, Germany, pp. 69–92. doi:10.1002/9783527805921.ch4.

Šimat, V., Elabed, N., Kulawik, P., Ceylan, Z., Jamroz, E., Yazgan, H., Čagalj, M., Regenstein, J.M., Özogul, F., 2020. Recent advances in marine-based nutraceuticals and their health benefits. Mar. Drugs. 1660-3397. 18 (12), 627. doi:10.3390/md18120627.

Singh, R.P., Kaur, H., Gupta, S., Singh, A., Singh, Y., Singh, H., Randhawa, M.K., Panesar, P.S., 2024. Exopolysaccharides from marine microbial resources. In: Bhatia, S.K., Panesar, P.S., Mehariya, S., (Eds.), Microbial Exopolysaccharides: Production and Applications. CRC Press, India, 164–193. Available from: https://www.tandfebooks.com/doi/book/9781003859987. doi:10.1201/9781003342687-8.

Singh, S., Sran, K.S., Pinnaka, A.K., Choudhury, A.R., 2019. Purification, characterization and functional properties of exopolysaccharide from a novel halophilic *Natronotalea sambharensis* sp. nov. Int. J. Biol. Macromol. 1879-0003. 136, 547–558. Available from: www.elsevier.com/locate/ijbiomac. doi:10.1016/j.ijbiomac.2019.06.080.

Souza, R.B., Frota, A.F., Sousa, R.S., Cezario, N.A., Santos, T.B., Souza, L.M.F., Coura, C.O., Monteiro, V.S., Filho, G. C., Vasconcelos, S.M.M., da Cunha, R.M.S., Aguiar, L.M.V., Benevides, N.M.B., 2017. Neuroprotective effects of sulphated agaran from marine alga *Gracilaria cornea* in rat 6-hydroxydopamine Parkinson's disease model: behavioural, neurochemical and transcriptional alterations. Basic Clin. Pharmacol. Toxicol. 1742-7843. 120 (2), 159–170.

Spanò, A., Gugliandolo, C., Lentini, V., Maugeri, T.L., Anzelmo, G., Poli, A., Nicolaus, B., 2013. A novel EPS-producing strain of *Bacillus licheniformis* isolated from a shallow vent Off Panarea Island (Italy). Curr. Microbiol. 1432-0991. 67 (1), 21–29. doi:10.1007/s00284-013-0327-4.

Sran, K.S., Bisht, B., Mayilraj, S., Choudhury, A.R., 2019b. Structural characterization and antioxidant potential of a novel anionic exopolysaccharide produced by marine *Microbacterium aurantiacum* FSW-25. Int. J. Biol. Macromol. 1879-0003. 131, 343–352. Available from: www.elsevier.com/locate/ijbiomac. doi:10.1016/j.ijbiomac.2019.03.016.

Sran, K.S., Sundharam, S.S., Krishnamurthi, S., Choudhury, A.R., 2019a. Production, characterization and bio-emulsifying activity of a novel thermostable exopolysaccharide produced by a marine strain of *Rhodobacter johrii* CDR-SL 7Cii. Int. J. Biol. Macromol. 1879-0003. 127, 240–249. Available from: www.elsevier.com/locate/ijbiomac. doi:10.1016/j.ijbiomac.2019.01.045.

Stoyneva-Gärtner, M., Uzunov, B., Gärtner, G., 2020. Enigmatic microalgae from aeroterrestrial and extreme habitats in cosmetics: the potential of the untapped natural sources. Cosmetics. 2079-9284. 7 (2), 27. doi:10.3390/cosmetics7020027.

Sun, M.L., Liu, S.B., Qiao, L.P., Chen, X.L., Pang, X., Shi, M., Zhang, X.Y., Qin, Q.L., Zhou, B.C., Zhang, Y.Z., Xie, B.B., 2014. A novel exopolysaccharide from deep-sea bacterium *Zunongwangia profunda* SM-A87: low-cost fermentation, moisture retention, and antioxidant activities. Appl. Microbiol. Biotechnol. 1432-0614. 98 (17), 7437–7445. Available from: https://www.springer.de/link/service/journals/00253/index.htm. doi:10.1007/s00253-014-5839-8.

Sun, M.L., Zhao, F., Shi, M., Zhang, X.Y., Zhou, B.C., Zhang, Y.Z., Chen, X.L., 2015. Characterization and biotechnological potential analysis of a new exopolysaccharide from the Arctic marine bacterium *Polaribacter* sp. SM1127. Sci. Rep. 2045-2322. 5. Available from: www.nature.com/srep/index.html. doi:10.1038/srep18435.

Sutherland, I.W., 1982. Biosynthesis of microbial exopolysaccharides. Adv. Microb. Physiol. 0065-2911. 23 (C), 79–150. doi:10.1016/S0065-2911(08)60336-7.

Sutherland, I.W., 2001. Biofilm exopolysaccharides: a strong and sticky framework. Microbiology. 1350-0872. 147 (1), 3–9. Available from: https://mic.sgmjournals.org. doi:10.1099/00221287-147-1-3.

Teng, H., Yang, Y., Wei, H., Liu, Z., Liu, Z., Yanhong, M., Gao, Z., Hou, L., Zou, X., 2015. Fucoidan suppresses hypoxia-induced lymphangiogenesis and lymphatic metastasis in mouse hepatocarcinoma. Mar. Drugs. 1660-3397. 13 (6), 3514–3530. doi:10.3390/md13063514.

Tiasto, V.A., Goncharov, N.V., Zhidkov, A.O., Zhidkov, M.E., Khotimchenko, Y.S., 2022. κ- and λ-Carrageenans from marine alga *Chondrus armatus* exhibit anticancer in vitro activity in human gastrointestinal cancers models. Mar. Drugs. 1660-3397. 20 (12). doi:10.3390/md20120741.

Trabelsi, L., Chaieb, O., Mnari, A., Abid-Essafi, S., Aleya, L., 2016. Partial characterization and antioxidant and antiproliferative activities of the aqueous extracellular polysaccharides from the thermophilic microalgae *Graesiella* sp. BMC Complement. Altern. Med. 1472-6882. 16 (1). Available from: https://www.biomedcentral.com/bmccomplementalternmed/. doi:10.1186/s12906-016-1198-6.

Tuvikene, R., 2021. Carrageenans Handbook of Hydrocolloids. Elsevier, pp. 767–804. doi:10.1016/b978-0-12-820104-6.00006-1.

Vandana, S.D., 2021. Structural and mechanical characterization of biofilm-associated bacterial polymer in the emulsification of petroleum hydrocarbon. 3 Biotech. 2190-572X. 11 (5). doi:10.1007/s13205-021-02795-8.

Veeraperumal, S., Qiu, H.M., Zeng, S.S., Yao, W.Z., Wang, B.P., Liu, Y., Cheong, K.L., 2020. Polysaccharides from *Gracilaria lemaneiformis* promote the HaCaT keratinocytes wound healing by polarised and directional cell migration. Carbohydr. Polym. 0144-8617. 241. Available from: https://www.elsevier.com/wps/find/journal-description.cws_home/405871/description#description. doi:10.1016/j.carbpol.2020.116310.

Velkova, Z., Kirova, G., Stoytcheva, M., Kostadinova, S., Todorova, K., Gochev, V., 2018. Immobilized microbial biosorbents for heavy metals removal. Eng. Life Sci. 1618-0240. 18 (12), 871–881. doi:10.1002/elsc.201800017.

Vilanova, E., Coutinho, C.C., Mourao, P.A.S., 2009. Sulfated polysaccharides from marine sponges (Porifera): an ancestor cell-cell adhesion event based on the carbohydrate-carbohydrate interaction. Glycobiology. 0959-6658. 19 (8), 860–867. doi:10.1093/glycob/cwp059.

Wang, J., Jin, W., Hou, Y., Niu, X., Zhang, H., Zhang, Q., 2013. Chemical composition and moisture-absorption /retention ability of polysaccharides extracted from five algae. Int. J. Biol. Macromol. 0141-8130. 57, 26–29. doi:10.1016/j.ijbiomac.2013.03.001.

Wang, J., Zhang, Q., Zhang, Z., Song, H., Pengcheng, L., 2010. Potential antioxidant and anticoagulant capacity of low molecular weight fucoidan fractions extracted from *Laminaria japonica*. Int. J. Biol. Macromol. 0141-8130. 46 (1), 6–12. doi:10.1016/j.ijbiomac.2009.10.015.

Wang, J., Zhang, Q., Zhang, Z., Zhien, L., 2008. Antioxidant activity of sulfated polysaccharide fractions extracted from *Laminaria japonica*. Int. J. Biol. Macromol. 0141-8130. 42 (2), 127–132. doi:10.1016/j.ijbiomac.2007.10.003.

Wang, M., Wang, H., Tang, Y., Kang, D., Gao, Y., Mengyun, K., Dou, J., Tao, X., Zhou, C., 2011. Effective inhibition of a *Strongylocentrotus nudus* eggs polysaccharide against hepatocellular carcinoma is mediated via immunoregulation in vivo. Immunol. Lett. 0165-2478. 141 (1), 74–82. doi:10.1016/j.imlet.2011.08.001.

Wang, Q., Wei, M., Zhang, J., Yue, Y., Ning, W., Geng, L., Sun, C., Zhang, Q., Wang, J., 2021a. Structural characteristics and immune-enhancing activity of an extracellular polysaccharide produced by marine *Halomonas* sp. 2E1. Int. J. Biol. Macromol. 0141-8130. 183, 1660–1668. doi:10.1016/j.ijbiomac.2021.05.143.

Wang, W., Jiandong, W., Zhang, X., Hao, C., Zhao, X., Jiao, G., Shan, X., Tai, W., Guangli, Y., 2017. Inhibition of influenza A virus infection by fucoidan targeting viral neuraminidase and cellular EGFR pathway. Sci. Rep. 2045-2322. 7 (1), doi:10.1038/srep40760.

Wang, W., Wang, S.X., Guan, H.S., 2012. The antiviral activities and mechanisms of marine polysaccharides: an overview. Mar. Drugs. 16603397. 10 (12), 2795–2816. Available from: https://www.mdpi.com/1660-3397/10/12/2795/pdf. doi:10.3390/md10122795.

Wang, X., Kaixuan, Y., Zhao, Y., 2018. Fucoidan inhibits amyloid-β-induced toxicity in transgenic *Caenorhabditis elegans* by reducing the accumulation of amyloid-β and decreasing the production of reactive oxygen species. Food Funct. 2042-6496. 9 (1), 552–560. doi:10.1039/C7FO00662D.

Wang, Y., Liu, G., Liu, R., Wei, M., Zhang, J., Sun, C., 2021b. EPS364, a novel deep-sea bacterial exopolysaccharide, inhibits liver cancer cell growth and adhesion. Mar. Drugs. 1660-3397. 19 (3), 171. doi:10.3390/md19030171.

Wang, Y., Xing, M., Cao, Q., Aiguo, J., Liang, H., Song, S., 2019. Biological activities of fucoidan and the factors mediating its therapeutic effects: a review of recent studies. Mar. Drugs. 1660-3397. 17 (3), 183. doi:10.3390/md17030183.

Wei, M., Geng, L., Wang, Q., Yue, Y., Wang, J., Ning, W., Wang, X., Sun, C., Zhang, Q., 2021. Purification, characterization and immunostimulatory activity of a novel exopolysaccharide from *Bacillus* sp. H5. Int. J. Biol. Macromol. 0141-8130. 189, 649–656. doi:10.1016/j.ijbiomac.2021.08.159.

Wrangstadh, M., Conway, P.L., Kjelleberg, S., 1986. The production and release of an extracellular polysaccharide during starvation of a marine *Pseudomonas* sp. and the effect thereof on adhesion. Arch. Microbiol. 1432-072X. 145 (3), 220–227. doi:10.1007/BF00443649.

Wu, K., Yiyang, L., Lin, Y., Baojun, X., Yang, J., Mo, L., Huang, R., Zhang, X., 2023. Structural characterization and immunomodulatory activity of an exopolysaccharide from marine-derived *Aspergillus* versicolor SCAU141. Int. J. Biol. Macromol. 0141-8130. 227, 329–339. doi:10.1016/j.ijbiomac.2022.12.127.

Wu, S., Liu, G., Jin, W., Xiu, P., Sun, C., 2016. Antibiofilm and anti-infection of a marine bacterial exopolysaccharide against *Pseudomonas aeruginosa*. Front. Microbiol. 1664-302X. 7. doi:10.3389/fmicb.2016.00102.

Xiang, X.-W., Wang, R., Yao, L.-W., Zhao, Y.-F., Sun, P.-L., Zheng, B., Chen, Y.-F., 2021. Anti-inflammatory effects of *Mytilus coruscus* polysaccharide on RAW264.7 cells and DSS-induced colitis in mice. Mar. Drugs. 1660-3397. 19 (8). doi:10.3390/md19080468.

Xiao, R., Zheng, Y., 2016. Overview of microalgal extracellular polymeric substances (EPS) and their applications. Biotechnol. Adv. 0734-9750. 34 (7), 1225–1244. doi:10.1016/j.biotechadv.2016.08.004.

Xu, H., Wang, J., Chang, Y., Jie, X., Wang, Y., Long, T., Xue, C., 2014. Fucoidan from the sea cucumber *Acaudina molpadioides* exhibits anti-adipogenic activity by modulating the Wnt/β-catenin pathway and down-regulating the SREBP-1c expression. Food Funct. 2042-6496. 5 (7), 1547–1555. doi:10.1039/C3FO60716J.

Yang, W.-N., Chen, P.-W., Huang, C.-Y., 2017. Compositional characteristics and in vitro evaluations of antioxidant and neuroprotective properties of crude extracts of fucoidan prepared from compressional puffing-pretreated *Sargassum crassifolium*. Mar. Drugs. 1660-3397. 15 (6). doi:10.3390/md15060183.

Yao, Y., Baojun, X., 2022. Skin health promoting effects of natural polysaccharides and their potential application in the cosmetic industry. Polysaccharides. 2673-4176. 3 (4), 818–830. doi:10.3390/polysaccharides3040048.

Yim, J.H., Kim, S.J., Aan, S.H., Lee, H.K., 2004. Physicochemical and rheological properties of a novel emulsifier, EPS-R, produced by the marine bacterium *Hahella chejuensis*. Biotechnol. Bioprocess Eng. 1226-8372. 9 (5), 405–413. Available from: https://www.springerlink.com/content/1226-8372. doi:10.1007/BF02933066.

Yoon, M., Cho, S., 2018. Triphlorethol A, a dietary polyphenol from seaweed, decreases sleep latency and increases non-rapid eye movement sleep in mice. Mar. Drugs. 1660-3397. 16 (5), 139. doi:10.3390/md16050139.

Younes, I., Rinaudo, M., 2015. Chitin and chitosan preparation from marine sources. Structure, properties and applications. Mar. Drugs. 1660-3397. 13 (3), 1133–1174. doi:10.3390/md13031133.

Yuan, H., Song, J., Xuegang, L., Ning, L., Dai, J., 2006. Immunomodulation and antitumor activity of κ-carrageenan oligosaccharides. Cancer Lett. 0304-3835. 243 (2), 228–234. Available from: https://www.sciencedirect.com/science/article/pii/S0304383505010402. doi:10.1016/j.canlet.2005.11.032.

Zammuto, V., Rizzo, M.G., Spanò, A., Spagnuolo, D., Di Martino, A., Morabito, M., Manghisi, A., Genovese, G., Guglielmino, S., Calabrese, G., Capparucci, F., Gervasi, C., Nicolò, M.S., Gugliandolo, C., 2022. Effects of crude polysaccharides from marine macroalgae on the adhesion and biofilm formation of *Pseudomonas aeruginosa* and *Staphylococcus aureus*. Algal Res. 2211-9264. 63. Available from: https://www.sciencedirect.com/science/journal/aip/22119264. doi:10.1016/j.algal.2022.102646.

Zargarzadeh, M., Amaral, A.J.R., Custódio, C.A., Mano, J.F., 2020. Biomedical applications of laminarin. Carbohydr. Polym. 0144-8617. 232. Available from: https://www.elsevier.com/wps/find/journaldescription.cws_home/405871/description#description. doi:10.1016/j.carbpol.2019.115774.

Zerbib, E., Levin, R., Gur, E., 2024. Tag recycling in the pup-proteasome system is essential for *Mycobacterium smegmatis* survival under starvation conditions. Mol. Microbiol. 0950-382X. 122 (4), 504–513. doi:10.1111/mmi.15312.

Zhang, F., Jian, L., Zhang, J.-G., Xie, J.-X., 2015a. Protective effects of a polysaccharide from *Spirulina platensis* on dopaminergic neurons in an MPTP-induced Parkinson's disease model in C57BL/6J mice. Neural Regen. Res. 1673-5374. 10 (2). Available from: https://journals.lww.com/nrronline/fulltext/2015/10020/protective_effects_of_a_polysaccharide_from.29.aspx.

Zhang, M., Liu, Y., Jingwen, L., Mengyun, K., Jie, Y., Dou, J., Wang, H., Zhou, C., 2017a. A polysaccharide component from *Strongylocentrotus nudus* eggs inhibited hepatocellular carcinoma in mice by activating T lymphocytes. Oncol. Lett. 1792-1074. 13 (3), 1847–1855. doi:10.3892/ol.2017.5624.

Zhang, Q., Ning, L., Liu, X., Zhao, Z., Zhien, L., Zuhong, X., 2004. The structure of a sulfated galactan from *Porphyra haitanensis* and its in vivo antioxidant activity. Carbohydr. Res. 0008-6215. 339 (1), 105–111. doi:10.1016/j.carres.2003.09.015.

Zhang, W., Oda, T., Yu, Q., Jin, J.O., 2015b. Fucoidan from *Macrocystis pyrifera* has powerful immune-modulatory effects compared to three other fucoidans. Mar. Drugs. 1660-3397. 13 (3), 1084–1104. Available from: https://www.mdpi.com/1660-3397/13/3/1084/pdf. doi:10.3390/md13031084.

Zhang, Z., Cai, R., Zhang, W., Yingnan, F., Jiao, N., 2017b. A novel exopolysaccharide with metal adsorption capacity produced by a marine bacterium *Alteromonas* sp. JL2810. Mar. Drugs. 1660-3397. 15 (6), 175. doi:10.3390/md15060175.

Zhao, X., Bafang, L., Xue, C., Sun, L., 2012. Effect of molecular weight on the antioxidant property of low molecular weight alginate from *Laminaria japonica*. J. Appl. Phycol. 0921-8971. 24 (2), 295–300. doi:10.1007/s10811-011-9679-y.

Zhou, G., Xin, H., Sheng, W., Sun, Y., Li, Z., Xu, Z., 2005. In vivo growth-inhibition of S180 tumor by mixture of 5-Fu and low molecular λ-carrageenan from *Chondrus ocellatus*. Pharmacol. Res. 1043-6618. 51 (2), 153–157. doi:10.1016/j.phrs.2004.07.003.

Zhou, W., Shen, B., Meng, F., Liu, S., Zhang, Y., 2010. Coagulation enhancement of exopolysaccharide secreted by an Antarctic sea-ice bacterium on dye wastewater. Sep. Purif. Technol. 1383-5866. 76 (2), 215–221. doi:10.1016/j.seppur.2010.10.011.

CHAPTER

6

Marine-derived lipids and fatty acids

Marcia Mazzuca[a,b], *Luis Fernando Flores*[c,g], *Brenda Juana Flores*[c,d,g], *and Gabriela Fabiana Malanga*[e,f]

[a]Institute of Biological Sciences of Patagonia, National Scientific and Technical Research Council of Argentina - National University of Patagonia San Juan Bosco, Comodoro Rivadavia, Chubut, Argentina

[b]Department of Chemistry, Faculty of Natural Sciences and Health Sciences, National University of Patagonia San Juan Bosco, Comodoro Rivadavia, Chubut, Argentina

[c]Agricultural and Environmental Biotechnology Club, Villa Regina, Argentina

[d]School of Production, Technology, and Environment, National University of Río Negro, Villa Regina, Argentina

[e]Faculty of Pharmacy and Biochemistry, University of Buenos Aires, Ciudad Autónoma de Buenos Aires, Argentina

[f]Institute of Biochemistry and Molecular Medicine, National Scientific and Technical Research Council of Argentina - University of Buenos Aires, Ciudad Autónoma de Buenos Aires, Argentina

[g]Wikimedians of Jujuy, San Salvador de Jujuy, Jujuy, Argentina

6.1 Introduction

The marine environment, which covers over 70% of the Earth's surface, represents an extraordinary reservoir of biodiversity and untapped resources. It is home to a vast array of biological organisms, ranging from microscopic algae to massive cetaceans, each playing essential roles in global ecological balance and biogeochemical cycles (Flores et al., 2025b). Among the numerous bioactive compounds derived from marine sources, lipids and fatty acids (FAs) stand out due to their diverse chemical structures and biological functions. These compounds are integral to cellular and systemic processes, contributing to energy storage, maintaining cell membrane integrity, and regulating signaling pathways. Their functionality extends beyond physiological roles, offering significant therapeutic, nutritional, and industrial benefits (Jayawardena et al., 2025; Conde et al., 2025; Parrish, 2025).

Marine-derived lipids exhibit a wide range of chemical structures, including phospholipids (PLs), glycolipids, triglycerides, and sterols (Cretton et al., 2023). FAs, in particular, are

Harvesting the Sea
https://doi.org/10.1016/B978-0-443-29190-6.00011-8

fundamental to the composition of these lipids, influencing their physicochemical properties and bioactivities. The nutritional and therapeutic potential of marine lipids is well documented, with numerous studies underscoring their beneficial impacts on human health. These findings have fueled the growing interest in incorporating marine-derived lipids into functional foods, nutraceuticals, and pharmaceutical formulations (Wei et al., 2025). For instance, polyunsaturated fatty acids (PUFAs), such as eicosapentaenoic acid (EPA) and docosahexaenoic acid (DHA), have garnered considerable attention due to their roles in maintaining cardiovascular health, cognitive function, and antiinflammatory responses. Other bioactive lipids, including PLs, sphingolipids, and sterols, have shown promise in therapeutic applications ranging from anticancer therapies to antimicrobial treatments (Xue et al., 2025; Parrish, 2025; Uthaiah et al., 2025).

Despite their immense potential, marine lipids and FAs remain underutilized. Several factors contribute to this, including the challenges associated with their extraction, purification, and sustainable production. In recent years, sustainability has become a central concern in the exploration of marine resources (Sanjeewa et al., 2024). Overexploitation of marine ecosystems poses significant threats to biodiversity and the long-term availability of these valuable resources. Consequently, researchers and industry stakeholders are prioritizing the development of sustainable methods for lipid production. This includes utilizing nontraditional sources such as marine microorganisms, which offer high lipid yields and can be cultivated under controlled conditions (Bastardie et al., 2024; Ogidi and Akpan, 2022). Technological innovations are also playing a key role in unlocking the potential of marine lipids. Novel extraction techniques, such as supercritical fluid extraction (SFE) and other green extraction methods, are enhancing lipid recovery efficiency while minimizing environmental impact. Lipidomics tools, including mass spectrometry (MS) and nuclear magnetic resonance (NMR) spectroscopy, are enabling detailed characterization of lipid profiles, which facilitates the identification of structure–activity relationships essential for their industrial application (Windarsih et al., 2024; Ying et al., 2024). Moreover, advances in bioprocess engineering and genetic manipulation are addressing these barriers, enabling the development of innovative methods for the efficient and eco-friendly harvesting of marine-derived lipids. Furthermore, the integration of marine biotechnology and biorefinery concepts is opening new avenues for utilizing marine biomass and by-products, transforming marine waste streams into valuable lipid-based products, and aligning with the principles of a circular bioeconomy (Jayawardena et al., 2025; Xue et al., 2025; Ahmad et al., 2025; Debbarma et al., 2025).

This chapter provides a comprehensive overview of marine-derived lipids and FAs, emphasizing their diversity, sources, and applications across various industries. It examines emerging technologies and methodologies for sustainable lipid recovery and commercialization. The exploration and utilization of these biomolecules require a multidisciplinary approach, integrating advancements in marine biology, biochemistry, bioprocess engineering, and environmental sciences. In this context, the chapter explores the vast potential of marine lipids, offering insights into the innovations driving their sustainable development and industrialization, while underscoring the importance of integrating scientific innovation with ecological responsibility to fully harness the potential of marine lipids in a sustainable manner.

6.2 Marine lipids and FAs

The marine environment remains one of the least exploited biological resources, yet it hosts a vast array of organisms capable of producing diverse bioactive metabolites adapted to extreme habitat conditions (Flores et al., 2025b). Covering 71% of the Earth's surface, marine ecosystems support nearly half of global biodiversity, encompassing organisms that thrive under extreme variations in temperature, salinity, pressure, and light intensity. These adaptations have led to the evolution of unique biological strategies essential for survival in highly competitive environments (Galasso et al., 2023; Rotter et al., 2021).

Among the bioactive compounds derived from marine organisms are polysaccharides, bioactive peptides, polyphenols, carotenoids, phytosterols, saponins, and polyunsaturated lipids, which can be sourced from algae, plants, animals, sponges, mollusks, and microorganisms such as microalgae and cyanobacteria. These molecules exhibit a wide range of pharmacological activities, including anticancer, antiinflammatory, antimicrobial, and anticoagulant properties, making them highly relevant to biomedical research and the nutraceutical industry (Flores et al., 2025b). In addition, marine lipids represent a particularly significant class of compounds, not only for their health benefits but also for their potential in biofuel production, positioning them as a key focus of multidisciplinary research on marine resource valorization (Hosseini et al., 2022; Lauritano, 2024; Rotter et al., 2021).

Lipids constitute a group of biomolecules with notable structural and functional diversity, whose common characteristic is their greater solubility in nonpolar organic solvents than in water, facilitating their extraction (Cretton et al., 2023). From a biochemical perspective, lipids share a common biosynthetic origin, as they are derived from the acetate pathway and are biosynthetically related to FAs. In biological systems, natural lipids are mainly composed of triglycerides, while other lipids, such as mono- and diglycerides, free FAs, PLs, sterols, terpenoids, and long-chain alcohols, are typically present in smaller proportions (Wang and Guo, 2025).

Lipids are also classified as simple or complex. Complex lipids are defined by the presence of one or more ester groups, which can be hydrolyzed under acidic or basic aqueous conditions to generate FAs and alcohols (Navalho et al., 2025). FAs, typically linear, are classified based on several criteria: the degree of unsaturation (saturated, monounsaturated, or polyunsaturated), the length of the carbon chain (short, medium, or long), and their dietary essentiality (essential or nonessential) (Xue et al., 2025).

PUFAs are further subdivided into two main classes, n-3 (n-3 PUFAs) and n-6 (n-6 PUFAs), depending on the position of the first double bond relative to the methyl (–CH3) end of the carbon chain. For instance, marine lipids, especially omega-3 polyunsaturated fatty acids (n-3 PUFAs), such as EPA and DHA, stand out for their health benefits. Since the human body does not produce DHA and EPA in significant amounts, it is essential to consume them through seafood or supplements based on fish oil, microalgae, and other marine invertebrates (Xue et al., 2025; Abril et al., 2023; Ahmad et al., 2025; Hosseini et al., 2022).

Lipids from marine species exhibit a high proportion of long-chain n-3 PUFAs (greater than 20 carbon atoms) (LC n-3 PUFAs), found predominantly in PLs rather than triglycerides (Mititelu et al., 2025). Among the LC n-3 PUFAs present in marine lipids, EPA (EPA; 20:5n-3), DHA (DHA; 22:6n-3), and arachidonic acid (ARA; 20:4n-6) are particularly noteworthy. In fish oils, LC n-3 PUFAs are primarily located in the sn-2 position, which is crucial for maintaining homeostasis in marine organisms, whereas in marine mammal oils, such as those from seals and whales, they are mainly found in the sn-1 and sn-3 positions. Moreover, the highly unsaturated structure of these LC-PUFAs contributes to the predominance of lipids in oil form, with lower melting points compared to saturated FAs of similar length. This is essential for maintaining the fluidity of cell membranes and their ability to adapt to extreme environments. The position of these FAs within the molecule determines their structure, physicochemical properties, and functionality of the lipid (Arts and Kohler, 2009; Beppu et al., 2017; Hosseini et al., 2022; Wang and Guo, 2025).

The high content of conjugated double bonds in molecules such as long-chain PUFAs or carotenoid pigments makes them chemically unstable. This renders them particularly susceptible to oxidative degradation caused by factors such as oxygen, light, heat, and metal ions (Ali et al., 2022; Wang et al., 2024). Oxidation of n-3 PUFAs results in the formation of fatty acid hydroperoxides, which then further oxidize to form lipid peroxides and generate free radicals (Wang et al., 2017), leading to a loss of biological activity and even toxicity. To mitigate these limitations, the encapsulation of PUFAs and other bioactive compounds has emerged as an effective strategy. This process creates protective barriers that enhance the chemical stability, water solubility, and functionality of these compounds, while also reducing unpleasant odors and protecting them against oxidation. Micro- and nanoencapsulation systems, designed with specific properties such as controlled release or the delivery of multiple compounds, represent innovative solutions for preserving the quality of marine oils and optimizing their applications in food, pharmaceuticals, and nutraceuticals (Arenas-Jal et al., 2020).

Fig. 6.1 shows representative examples of each type of unsaturated fatty acids. Omega-3 FAs, EPA and DHA, are essential in the diets of animals and marine fish, as they cannot

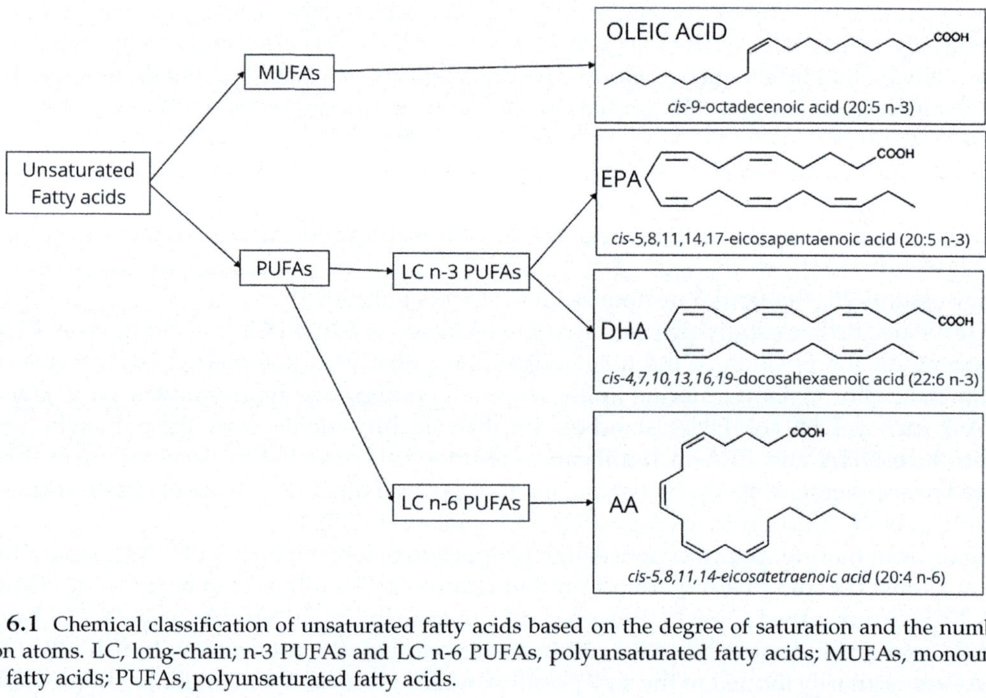

FIG. 6.1 Chemical classification of unsaturated fatty acids based on the degree of saturation and the number of carbon atoms. LC, long-chain; n-3 PUFAs and LC n-6 PUFAs, polyunsaturated fatty acids; MUFAs, monounsaturated fatty acids; PUFAs, polyunsaturated fatty acids.

synthesize them. Their supplementation ensures healthy growth and improves the quality of derived products. PUFA-rich microalgae, such as those from the *Schizochytrium* genus, offer an alternative to fish oil in aquaculture, enhancing the lipid profiles of fatty fish, pork, lamb, poultry, and eggs. In poultry, these microalgae increase DHA content in meat and eggs without altering their sensory qualities (Nagarajan et al., 2021).

6.3 Marine sources of lipids and FAs

Generally, the fatty acid composition of marine organisms varies according to the species' biology and environmental conditions, including water temperature, salinity, and the type and availability of food, which depend on geographic region and seasonality (Senadheera et al., 2023).

Fish is the most consumed marine species worldwide and is fundamental to the global economy. It is considered a healthy food alternative because it contains minerals, proteins, vitamins, and essential unsaturated FAs. The primary dietary source of LC n-3 PUFAs is fish (Tacon et al., 2025). Marine fish lipid content varies according to season, fish species, and diet. For instance, anchovy contains 21% EPA and 9% DHA, while sardines have a combined 16% of these FAs. Generally speaking, fatty fish such as tuna, mackerel, sardine, and salmon have a higher content of n-3 PUFAs (EPA and DHA) compared to lean fish such as pollock, cod, and hake. This difference arises because fatty fish store lipids in their flesh, whereas in lean fish, the liver serves as the primary lipid reservoir. Shark liver oil contains squalene, a triterpene crucial in cholesterol synthesis. This bioactive compound has been extensively studied for its antioxidant, antilipidemic, cardioprotective, anticancer, and antiaging properties. Furthermore, squalene has dermatological applications, where it serves as a protective agent against oxidative processes in the skin (Abril et al., 2023; Zhang et al., 2024).

Table 6.1 summarizes the total lipid content and fatty acid profiles of selected fish species that represent the main catches in various geographic locations around the world, including *Merluccius hubbsi* to represent the South Atlantic. It shows that, except for Atlantic herring, most commercial marine fish cuts are either lean (lipid content below 1%) or semilean (lipid content between 1% and 7%). However, the proportion of PUFAs is significant, especially in Alaska pollock.

Marine mammals, such as seals and whales, are significant sources of docosapentaenoic acid. Although fish and marine mammals are inefficient at synthesizing EPA and DHA, they accumulate these FAs through trophic webs, primarily from primary producers such as marine phytoplankton (Hamilton et al., 2020; Hosseini et al., 2022; Senadheera et al., 2023).

Marine invertebrates, including sponges, corals, and mollusks, host symbiotic microbial communities that play a key role in the biosynthesis of structurally complex lipids. Shellfish such as squid, cuttlefish, and octopus, although having low lipid content (1%–2%), are particularly rich in LC n-3 PUFAs, as well as fat-soluble vitamins A, D3, and E. Similarly, crustaceans, including shrimp, crabs, and prawns, are valuable sources of biologically active lipids,

TABLE 6.1 Total fat (g × 100 g^{-1} on a wet basis) and fatty acid profiles of different fish species around the world, selected based on global marine capture fisheries production (FAO, 2024), with the inclusion of ***Merluccius hubbsi*** to represent the South Atlantic.

Location	Finfish	Product	Total Fat	SFA	MUFA	n-3 PUFA	Reference
North Pacific	Alaska Pollock, *Gadus chalcogrammus*	Fillet	0.4–0.5	20.4 – 20.8	4.1–5.9	73.1–75.1	(Hintermeister Clara, 2017)
South Pacific	Anchoveta, *Engraulis* ringens	Dark muscle	1.5–7.7	33.2–46.0	19.7–26.8	25.3–38.8	(Albrecht-Ruiz and Salas-Maldonado, 2015)
North Atlantic	Atlantic herring, *Clupea harengus*	Fillett	6–16	32.6–35.9	24.5–17.1	38.7–43.2	(Wu et al., 2022)
	Blue whiting, *Micromesistius poutassou*	Full body	3.9	31.3	43.2	21.6	(Egerton et al., 2020)
Southwest Atlantic	Argentine hake, *Merluccius hubbsi*	Fillet	0.4	24.0	30.1	35.8	(Méndez and González, 1997)
Atlantic and Mediterranean	European pilchard, *Sardina pilchardus*	Edible portion	5.9	35.1	28.8	24.4	(Mathieu-Resuge et al., 2023)
Tropical and subtropical oceans	Skipjack tuna, *Katsuwonus pelamis*	Flesh	0.4	36.2	14.7	42.0	(Mahaliyana, 2015)
	Yellow fish Tuna, *Thunnus albacares*	Muscle	1.9	45.0	27.0	21.6	(Peng et al., 2013)

The intervals represent the seasonal variation, showing the highest and lowest lipid percentages reported in studies.

containing beneficial proportions of both saturated and unsaturated fats (Paulose and Chakraborty, 2024).

Sponges are abundant in biotechnologically significant secondary metabolites, such as sterols, terpenoids, and polyketides. Mollusks, including clams, oysters, scallops, and mussels, contain lipids with distinct biological properties. For example, by-products from *Mytilus edulis* are rich in valuable lipids like PUFAs, terpenoids, and steroids, positioning them as potential raw materials for functional foods and alternative medicines (Senadheera et al., 2023).

In recent years, marine zooplankton has emerged as an alternative source of LC-PUFAs, with notable examples, including Antarctic krill (*Euphausia superba*) and the copepod *Calanus finmarchicus*. In krill, LC-PUFAs are primarily found in PLs, which account for 30%–65% of the total lipid content. However, the exploitation of these resources raises ecological concerns due to potential impacts on marine food webs. While krill oil is marketed as a human nutritional supplement, its high cost limits its use in other applications, such as aquaculture feeds (Teng et al., 2023).

Crustacean heads, particularly from species such as shrimp, crabs, and prawns, contain hepatopancreas, a rich source of fat-soluble vitamins A, D, and E, which are essential for human metabolism. These vitamins offer significant health benefits: vitamin A supports vision and immunity, vitamin D regulates calcium metabolism, and vitamin E functions as an antioxidant, promoting muscular and reproductive development. In addition, crustaceans provide bioactive PLs, including phosphatidylcholine, phosphatidylethanolamine, and sphingomyelin, which are crucial for health and nutrition. Furthermore, crustaceans are notable sources of astaxanthin, astaxanthin esters, zeaxanthin, and PUFAs. These biologically active pigments are accumulated through the consumption of primary producers like phytoplankton (Senadheera et al., 2023; Xavier and Chattopadhyay, 2023).

Although some studies suggest lower yields of astaxanthin from crustacean by-products compared to freshwater microalgae such as *Haematococcus pluvialis* (which yields 27–50 $mg \cdot g^{-1}$ dry weight), crustaceans remain an important source of carotenoids with significant bioeconomic value. Moreover, the extraction of carotenoids from these by-products presents an exciting opportunity from a circular economy perspective, as it supports the valorization of waste materials, thereby contributing to the sustainability of marine industries. However, marine by-products, including crustaceans, also contain cholesterol, and excessive consumption has been associated with hypercholesterolemia. To mitigate this concern, an innovative process has been developed using beta-cyclodextrin to adsorb cholesterol during oil extraction from Pacific white shrimp heads. This process effectively reduces cholesterol by 95%, decreases oxidative substances, and increases astaxanthin levels by 86% compared to crude oils (Xavier and Chattopadhyay, 2023).

Macroalgae, or seaweeds, are particularly rich in long-chain PUFAs and bioactive sterols that demonstrate cardiovascular and antiinflammatory benefits. Notably, brown algae (Phaeophyceae) produce fucoxanthin-containing lipids, which have been shown to exhibit antiobesity and neuroprotective effects. While seaweeds typically have lower lipid contents than terrestrial plants like sunflower or soybean, they serve as important sources of carotenoids, which play a key role in photosynthesis and cellular protection. Marine organisms, including microalgae, are especially rich in carotenoids, and their production can be enhanced under stress conditions or through the addition of plant hormones in controlled cultures (Alsenani et al., 2019; Ma et al., 2018; Patel et al., 2021).

Despite the potential benefits, concerns about the ecological impact of large-scale harvesting of macroalgae and marine invertebrates have spurred the development of sustainable aquaculture systems and biotechnological innovations aimed at ensuring a reliable lipid supply without compromising marine biodiversity.

Microalgae, though situated at the base of the food chain, play a pivotal role in the biochemical carbon cycle and are among the leading lipid producers, surpassing their terrestrial counterparts by up to tenfold (Xin et al., 2024). Recent research has intensified interest in microalgal lipids across various industries, including food, pharmaceuticals, and cosmetics, with a focus on understanding the metabolic pathways involved in solar energy conversion, lipid biosynthesis, and catabolism, while also addressing technical challenges in lipid production. Their high photosynthetic efficiency and ability to accumulate valuable lipids make them an attractive and sustainable source for lipid extraction. Lipid synthesis in microalgae is influenced by various environmental factors, including CO_2 concentration, light intensity, temperature, and nutrients, alongside key genetic factors that regulate critical metabolic pathways. Recent omics studies have identified essential genes involved in these pathways, providing deeper insight into the molecular mechanisms regulating lipid accumulation. Certain microalgal strains, such as *Nannochloropsis* and *Schizochytrium*, have been recognized as prolific producers of omega-3 FAs (Xin et al., 2024).

In addition, marine bacteria and fungi synthesize structurally unique lipids, such as glycolipids and PLs, which possess potent antiinflammatory and immunomodulatory effects. Marine fungi, in particular, have gained attention as promising sources of secondary metabolites, including lipids and related compounds like polyketides, terpenoids, sterols, and their derivatives. These compounds exhibit significant biological and pharmacological activities, including antimicrobial, anticancer, and antioxidant properties, and have the potential to interact with various tissues and cells (Barzkar et al., 2024; Parrish, 2025).

Table 6.2 lists the main carotenoids found in marine organisms and their sources. Among these, many marine microalgae are notable sources of fucoxanthin. In contrast, most astaxanthin-producing microalgae are freshwater species. Astaxanthin is renowned for its exceptional antioxidant capacity, surpassing many conventional antioxidants, and for its higher safety profile compared to its synthetic counterpart. Its applications include the development of immunomodulatory, antiinflammatory, and cardiovascular disease treatments, as well as its use in antiaging cosmetic products. Recent research has also explored its potential as a supplement in oncology treatments (Xavier and Chattopadhyay, 2023).

TABLE 6.2 Chemical structures of the most relevant carotenoid pigments present in marine organisms, sources, and yields.

Carotenoid	Structure	Marine sources (mg·g^{-1} DW)
Astaxanthin	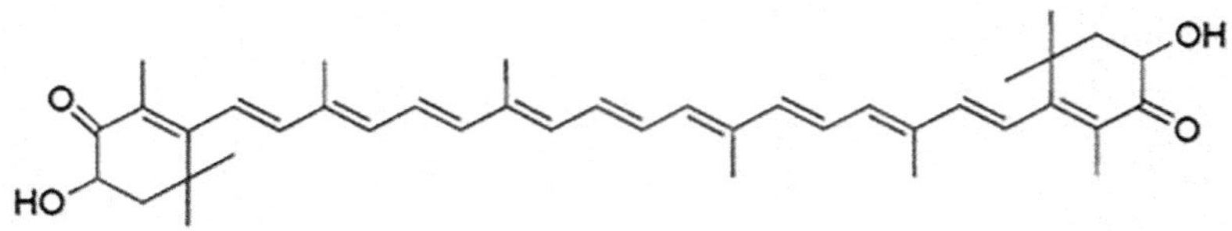	Crustacean waste: *Pleoticus muelleri* (0.3); *Parapenaeopsis stylifera* (0.15); *Pandalus borealis (0.15); Penaeus monodon (0.1); Farfantepenaeus paulensis (0.05); Vannamei (0.05)* (Cretton et al., 2021)
Fucoxanthin	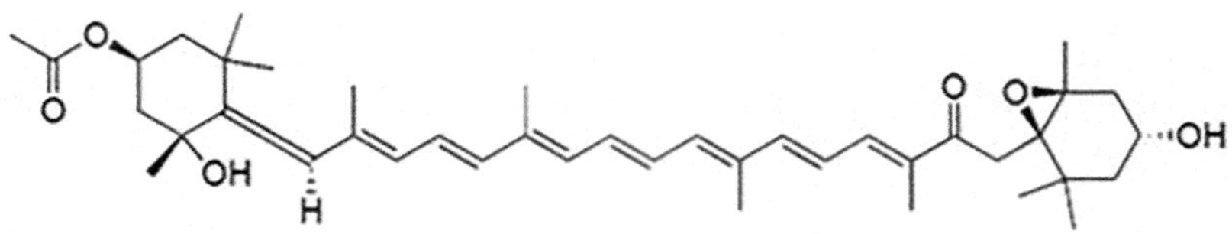	Seaweeds; *Phaeodactylum tricornutum (15–71); Odontella aurita (14–15); Chaetoceros calcitrans (5.13); Isochrysis galbana T-ISO (3.2); Nitzschia laevis (2.4); Isochrysis galbana (2.2); Chaetoceros calcitrans (2.3)* (Sun et al., 2023)
β-carotene	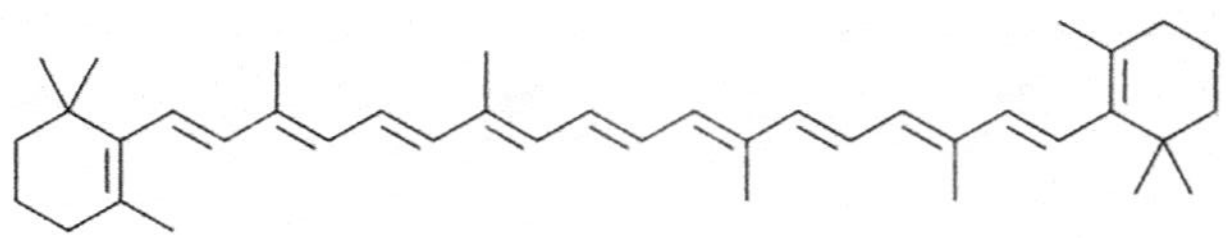	*Green alga Caulerpa racemosa (0.2)* (Kurniawan et al., 2023). Microalgae (e.g., *Spirulina, Dunaliella,* and *Chlorella*) (Alasalvar et al., 2002)
Canthaxanthin	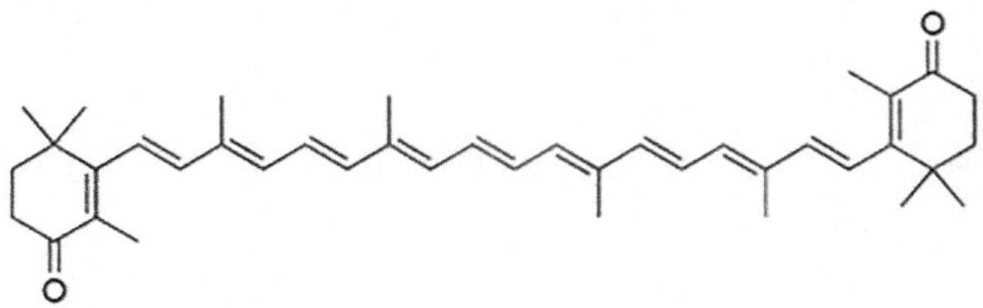	*Green alga Caulerpa racemosa (015)* (Kurniawan et al., 2023). Coelastrella sp. F50 *Sea urchin, starfish, holothurians, crustaceans, algae, and cyanobacteria* (Galasso et al., 2017)

Continued

TABLE 6.2 Chemical structures of the most relevant carotenoid pigments present in marine organisms, sources, and yields—cont'd

Carotenoid	Structure	Marine sources (mg·g^{-1} DW)
Lutein	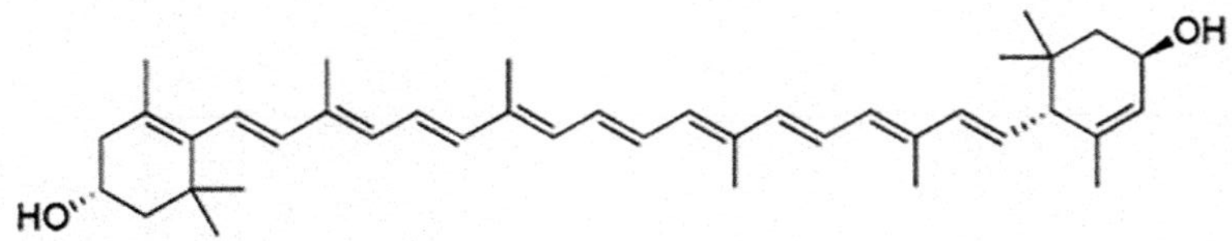	Microalgae: *Dunaliella salina, Chlorella species, seagrass, coral, and shellfish* (Galasso et al., 2017)
Zeaxanthin	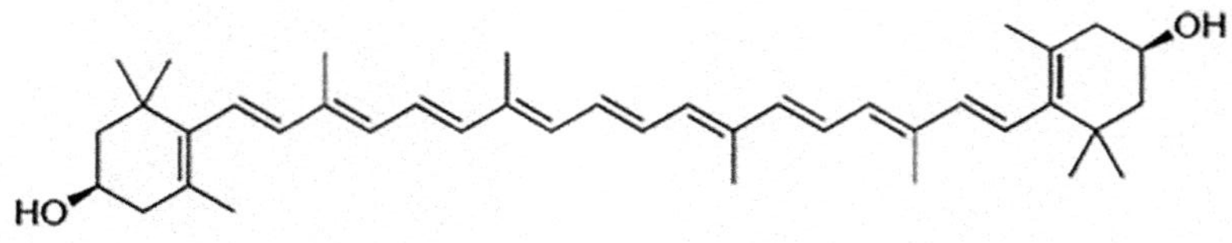	Microalgae*: Dunaliella salina; Seagrass: Posidonia oceanica, Cymodocea nodosa, Zostera noltii, and Halophila stipulacea;* Fungi-like protists*: Thraustochytrium strains ONC-T18 and CHN-1, Thraustochytriidae* sp. *AS4-A1 (Ulkenia* sp.*), and Aurantiochytrium* sp. *KH105;* Corals (due to their associations with cyanobacteria) and shellfish (Galasso et al., 2017).

6.4 Aquaculture-based marine lipids production

Aquaculture, encompassing the cultivation of aquatic organisms such as fish, mollusks, crustaceans, aquatic plants, and micro- and macroalgae, is key to providing protein-rich foods and easing the pressure on overexploited wild fisheries. It is one of the most dynamic sectors of the global food system. Government policies significantly influence the growth and geographical distribution of aquaculture, including species selection, management practices, addressing diseases, antibiotic use, and optimizing nutrition and infrastructure development (Naylor et al., 2023).

Global aquaculture has grown significantly over the past two decades, meeting key sustainability goals. This expansion, along with efforts to improve production efficiency, is essential for meeting the rising global demand for protein and strengthening food security, as long as it is done responsibly. Advances in processing technology are improving water quality, production efficiency, and animal welfare. The integration of artificial intelligence is transforming the sector by optimizing system performance, monitoring fish behavior, improving feed delivery, and enhancing disease detection. Aquaculture's ability to adapt to different environmental conditions also makes it more resilient to climate change. Practices such as aquaponics, biofloc systems, integrated multi-trophic aquaculture, and recirculating aquaculture systems align with circular economy principles, reinforcing sustainability (Boyd et al., 2020; Gupta et al., 2024; Tucciarone et al., 2024).

Fish and shellfish are rich in omega-3 FAs, which are essential for human health (Dunshea et al., 2024). Long-chain omega-3 polyunsaturated fatty acids (LC-PUFAs), particularly EPA and DHA, are critical for human health but must be obtained through dietary sources, as they are not efficiently synthesized in the body. Marine and vegetable oils are the primary commercial sources of these FAs. However, optimizing their production involves strategies such as selective breeding, improving the understanding of cultured species' biology and genetics, and optimizing culture conditions. Despite these efforts, the ability of food webs to biosynthesize sufficient LC-PUFAs like EPA and DHA for human consumption remains limited. Hernando et al. (2021) demonstrated that the bivalve *Limnoperna fortunei*, when fed *Microcystis aeruginosa* at high temperatures, significantly altered its FA profile, increasing EPA and ARA concentrations. These findings highlight the potential for enhancing PUFA production through dietary interventions (Hernando et al., 2021).

Marine fish traditionally meet their dietary requirements for EPA and DHA through fish oil and fish meal (Parrish, 2024), but concerns over limited supply and environmental impacts have led to research into alternative sources. Issues such as the vegan diet trend, conservation of FAs in farmed fish, and heavy metal contamination further drive this search. Microalgae, particularly thraustochytrids, present a sustainable alternative. These micro-heterotrophic organisms, which thrive without light, offer high oil yields and can treat wastewater and waste gases, reducing aquaculture's environmental footprint (Kumari et al., 2024). However, conventional microalgae cultivation methods for EPA and DHA production are often uneconomical. Innovative approaches, including supplementation with FA-rich extracts, antioxidants from natural sources like astaxanthin (Cretton et al., 2021), and cocultivation with other algal species, show promise in enhancing omega-3 production. Astaxanthin supplementation has been shown to improve the reproductive efficiency, meat quality, and survival rates of farmed fish and crustaceans, and its benefits extend to terrestrial animals, where it enhances oxidative stability and reproductive health (Babar et al., 2025; Elbahnaswy and Elshopakey, 2024; Gao et al., 2020).

6.5 Marine by-products as a source of lipids and FAs

FAO's 2024 reports highlight that marine capture fisheries continue to be the primary source of aquatic animals globally, reaching a record of 80 million tons in 2022. Of this total, 85% consisted of finfish, such as anchoveta, tuna, and Alaska pollock, while 15% comprised cephalopods like squids and octopuses, as well as crustaceans like lobsters (FAO, 2024). Despite technological advancements in fishing production chains, including improved catch methods and enhanced fish quality, biomass losses persist at critical stages. These losses are primarily due to bycatch, the discarding of low-commercial-value species, inefficient handling practices, and industrial processes that only utilize muscle tissue, leaving valuable by-products underexploited (FAO, 2024).

By-products from marine food processing, once considered low-value materials, are often relegated to uses such as fertilizers or are utilized in the tanning and leather industries. In other cases, these by-products are either directed to animal feed or biofuels, further limiting their potential value, or they are incinerated and discarded, leading to increased financial costs, energy consumption, and environmental impacts in their management. Unfortunately, most of these by-products are discarded as waste, contributing to environmental pollution and the spread of pathogens (Flores et al., 2025a). According to an FAO report, approximately 9.1 million tons of fish waste are discarded annually. This issue has become a global concern, also impacting the sustainability of aquaculture. Consequently, waste management is increasingly governed by strict environmental regulations, adding to the growing costs for the aquatic food industry (Kandyliari et al., 2020; Mkadem and Kaanane, 2024).

However, these by-products also represent a valuable opportunity for the sector. Marine animal co-products—such as bones, heads, shells, skin, and viscera from fish, mollusks, and crustaceans—are materials obtained during seafood processing that are often underutilized or discarded. For example, fish processing generates by-products like heads, bones, skin, and viscera, which can account for up to 15% of the fish's weight when scaled and gutted, and up to 70% when filleted. These by-products also constitute a significant source of bioactive lipids, including essential FAs and omega-3 (EPA and DHA), which are frequently wasted, representing about one-third of the global omega-3 PUFA production from both aquaculture and capture fisheries. Given their ability to reduce cardiovascular diseases, metabolic syndrome, and inflammation, as well as improve cognitive function, these co-products hold tremendous potential for developing resources to address health challenges associated with modern lifestyles (Kandyliari et al., 2020; Monteiro et al., 2024).

Fishing and aquaculture by-products, including heads, shells, fins, and visceral parts (e.g., liver and roe), can account for 30%–70% of the total catch biomass, depending on the organism type and processing level. For example, in tuna canning, 50%–70% of the tuna is discarded, while in shrimp and prawn processing, 40%–70% and 30%–60% of the total weight, respectively, are discarded. These residues can contain up to 19% lipids in fish and

12% in crustaceans, with a high concentration of LC n-3 PUFAs. In addition, they may include fat-soluble pigments such as carotenoids and vitamins A and D. This lipid composition represents a valuable opportunity for the valorization of marine residues, positioning the transformation of fishery by-products into functional ingredients as a key strategy in the circular economy. This approach not only reduces waste from the fishing industry but also maintains ecological balance, enhances sector competitiveness, and promotes economic development in coastal communities. Furthermore, it supports achieving the Sustainable Development Goals, including responsible production and consumption and marine ecosystem conservation (Cretton et al., 2021; FAO, 2024; Peñarubia et al., 2023).

In addition, the growing interest in valorizing fishery by-products is reflected in the increasing production of marine oils, positioning them as a sustainable alternative. Fish oil is a nutritious and digestible source for fish and other aquatic organisms, particularly due to its high content of LC n-3 PUFAs, such as EPA and DHA, which are essential for animal diets. In 2022 approximately 53% of global fish oil production came from by-products, emphasizing their strategic importance. While fish oil is primarily used in aquaculture, demand has driven prices up, leading to more strategic use in critical stages of aquaculture, such as breeding and final fattening, while its application in grow-out diets has declined. According to the Marine Ingredients Organization, 74% of fish oil production in 2021 was allocated to aquaculture, with 16% used for human consumption and 10% for other uses, such as pet food and biofuels. Fish oil, as the primary natural source of EPA and DHA, is linked to various health benefits, including improved cardiovascular health and the prevention of conditions such as asthma, attention-deficit/hyperactivity disorder in children, and Alzheimer's in the elderly. However, the sustainability of fish oil production largely relies on the use of by-products, which do not directly compete with human consumption, offering a low-cost alternative for high-quality oils. Nonetheless, in some developing countries, such as those in West Africa, whole fish are increasingly being allocated for oil and meal production for export, potentially compromising local food security by diverting resources that could be used to feed local populations. This has significant implications for nutrition and livelihoods (FAO, 2024; Mohanty et al., 2023; Thiao and Bunting, 2022).

Beyond aquaculture and human nutrition, waste fish oil (WFO) is emerging as a promising raw material for biodiesel production. The global energy demand, coupled with concerns over fossil fuel depletion and environmental pollution, has spurred research into alternative fuels. WFO presents an attractive option due to its high lipid content, superior engine performance, lower emissions, and enhanced fuel efficiency compared to fossil fuels. WFO biodiesel has several advantages, including a high cetane number, low viscosity, and reduced sulfur and aromatic compound content. As a low-cost, readily available lipid source, WFO provides a compelling alternative to petroleum diesel. However, further research is needed to evaluate the long-term performance of fish oil-based biodiesel and its viability for widespread use (Kumar et al., 2024).

Shrimp-derived oils offer additional benefits due to their rich composition of bioactive compounds, such as carotenoids (β-carotene and astaxanthin), fat-soluble vitamins (A, D, and E), PLs, and cholesterol. These oils are primarily extracted from shrimp head (cephalothorax) residues, which contain lipid levels ranging from 1% to 8% in most shrimp species (Mohanty et al., 2023). In cold-water species like the Argentinean Red shrimp *Pleoticus muelleri*, lipid levels can reach up to 11%. The high lipid content in Argentine shrimp is particularly notable, being comparable to that of the Pacific species *Litopenaeus vannamei* (11.9%), which has potential as a food ingredient. These residues also contain elevated levels of EPA (144 mg/100 g) and DHA (214 mg/100 g), as well as carotenoids (10.6 mg/100 g). Importantly, not all crustaceans are significant sources of EPA and DHA, as some species lack these essential compounds, making it crucial to assess local species for their nutritional potential. In addition, ethanolic extracts from the heads and shells of *P. muelleri* have been shown to enhance fish feed quality, enriching it with n-3 PUFAs and natural antioxidants. The development and application of technologies for commercializing these by-products can improve sustainability within the fisheries sector, promote a circular economy, and open new avenues for innovation and growth. In this context, Argentine shrimp serves as a prime example of how regional characteristics of lipid-rich species can contribute to local economic development and promote sustainable aquaculture. This approach not only reduces biomass waste but also generates high-value-added products (Cretton et al., 2021; FAO, 2024; Gómez-Estaca et al., 2017).

The circular bioeconomy, which promotes converting waste into products that integrate with or feed back into other production chains, optimizes resource use and mitigates the negative impacts of biomass losses (Xia et al., 2024). Given that the lipid composition of marine organisms is influenced by factors such as species, nutritional status, sex, age, diet, and geographic region, it is crucial to account for regional specificities to maximize their valorization. In addition, it is essential to ensure that valorization processes are environmentally sustainable and safe for human health. This includes assessing potential adverse impacts associated with handling by-products, such as microbial contamination, oxidation of compounds, and the presence of allergens. In this regard, strategies like the use of antioxidants, controlled drying, and monitoring of contaminants are vital to ensuring the quality of derived products and maximizing their efficiency within the circular bioeconomy framework. Standardizing data collection methods on a global scale is also necessary (FAO, 2024; Zhang et al., 2024).

Improved utilization is crucial not only from environmental, ecological, social, and economic perspectives but also for safeguarding consumer health and ensuring food security (FAO, 2024). Transforming fishing residues into value-added products enhances human nutrition and health, mitigates environmental pollution, generates economic revenue, and alleviates pressure on live aquatic resources, promoting sustainability. These actions align with the goals of the 2030 Agenda for Sustainable Development, which aims to reduce losses and waste in fishing value chains. To optimize efficiency, precise quantitative data are essential to assess losses at each stage of the process. However, the transition to a circular bioeconomy in this sector faces significant technological and regulatory challenges. Collaboration between industry, academia, and policymakers will be essential to drive research, develop innovative technologies, and ensure long-term sustainability. In this context, the marine by-product valorization model represents not only a path toward greater sustainability but also an integrated strategy for developing more resilient and sustainable regional economies.

Recent studies highlight various approaches to turning fish and seafood processing residues into value-added compounds, categorized by organism type, residue type, and extracted compound (e.g., omega-3 FAs, astaxanthin) (Table 6.3).

TABLE 6.3 Transformation of seafood by-products into added-value products.

Category of by-products	Species	By-products	Valorized products	References
Fish	Atlantic herring, *Clupea harengus*	Co-product mix heads, frames, skin, and viscera	Fish oil	Aitta et al., 2021
	Atlantic cod, *Gadus morhua*	Head, viscera, and skeletal frames	Fish oil	Shahidi et al., 1991
	Tuna *Euthinnus* sp.	Viscera, eyes, liver		Djamaludin et al. (2023)
	Merluza (*Merluccius hubbsi*)	Liver	Fish oil	Cretton et al. (2016)
Cephalopod	Argentinean shortfin squid (*Illex argentinus*)	Viscera	Oil	(Rodríguez et al., 2021)
Crustacean	Argentinean red shrimp *Pleoticus muelleri*	Cephalotorax, shell	Lipidic extract (astaxanthin, EPA, DHA)	Cretton et al., 2021
	White shrimp *Lithopenaeus vannamei*	Cephalotorax	Lipidic extract (astaxanthin, EPA, DHA)	Gómez-Guillén et al. (2018)
	Black tiger shrimp *Penaeus monodon*	Cephalothorax, shell, tails	Astaxanthin	Roy et al. (2020)

6.6 Advances in marine lipids characterization

Recent advancements in the structural and functional characterization of marine lipids have significantly expanded our understanding of their complexity and biological roles. Their unique structural features contribute to their functional diversity, with many exhibiting promising bioactivities for pharmaceutical, nutritional, and industrial applications. As research methodologies advance, the integration of high-resolution analytical techniques has provided new insights into the molecular composition and bioactive potential of these lipids.

Advancements in omics technologies have greatly facilitated the exploration of marine-derived lipids, enabling the identification of novel lipid metabolites. Recent progress in lipidomics, high-throughput screening, and computational bioinformatics has significantly advanced the characterization and industrial application of these compounds. Lipidomics, in particular, has emerged as an essential tool for investigating bioactive lipids in the marine environment. This approach allows for the analysis and identification of lipid species through advanced chromatography and MS techniques. In addition, lipidomics enhances fractionation and extraction processes by considering lipid polarity, making it an essential method for discovering compounds with antimicrobial properties and other therapeutic potential (Alves et al., 2020), as shown in Fig. 6.2.

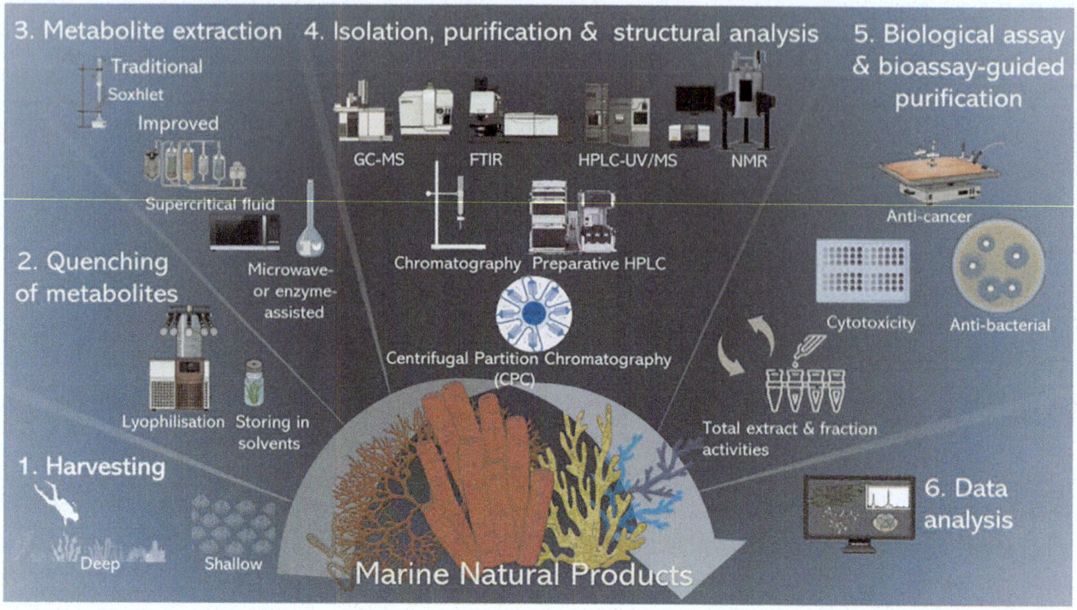

FIG. 6.2 Schematic overview of the process for uncovering marine metabolites. *Adopted from Jayawardena, T.U., Merindol, N., Liyanage, N.S., Awwad, F., Desgagné-Penix, I., 2025. Marine specialized metabolites: unveiling Nature's chemical treasures from the deep blue. TrAC Trends Anal. Chem. 183. 10.1016/j.trac.2024.118097.*

MS, evaporative light scattering detection, NMR spectroscopy, and fluorescence detection have revolutionized lipidomics by providing detailed structural insights into marine lipids and their functional properties. These analytical tools enable the rapid screening of marine lipid extracts for bioactive molecules, accelerating the transition from discovery to industrial application (Xue et al., 2025; Park et al., 2024).

MS remains a cornerstone in lipidomics, allowing for precise identification and quantification of marine lipid species. The development of soft ionization techniques, such as electrospray ionization (ESI) and matrix-assisted laser desorption/ionization (MALDI), has enabled the characterization of complex lipid mixtures without fragmentation, preserving structural integrity during analysis. ESI-MS has been particularly useful in profiling the PL and glycolipid compositions of marine microalgae, revealing novel lipid species with roles in stress adaptation and environmental resilience. MALDI-MS, on the other hand, has facilitated the spatial mapping of lipid distributions and the functional compartmentalization of lipids in different biological systems (Jayawardena et al., 2025; Xue et al., 2025; Barzkar et al., 2024).

NMR spectroscopy provides complementary insights into the structural properties and dynamics of marine lipids. Two-dimensional NMR techniques, such as correlation spectroscopy (COSY) and nuclear Overhauser effect spectroscopy (NOESY), have been applied to determining lipid stereochemistry and elucidating the three-dimensional conformations of complex lipid molecules (Alves et al., 2020; Barzkar et al., 2024).

oil. These advancements are crucial for ensuring an eco-friendly and reliable supply of marine lipids, benefiting industries such as pharmaceuticals, nutraceuticals, and biofuels (Patel et al., 2021).

6.8 Marine lipids commercial development

The commercial potential of microalgae-derived compounds has attracted significant investment and industry focus over recent years. In today's competitive global market, microalgae offer a sustainable and scalable source of high-value bioactive lipids that can be harnessed for diverse commercial applications. With their unique lipid profiles and robust production of nutraceutical ingredients, microalgae represent an emerging frontier for companies seeking to enhance their product portfolios in the food, pharmaceutical, cosmetic, and animal feed industries.

The market for microalgae-derived products is already showing impressive growth. Current estimates project the global microalgae industry to expand from approximately USD 11.8 billion to an estimated USD 25.4 billion by 2033. This growth, corresponding to a compound annual growth rate (CAGR) of 8%, is driven by increasing consumer demand for healthy, sustainable, and natural products. In particular, microalgae are positioned as a premium source for functional ingredients that contribute not only to improved nutritional profiles but also to the commercial appeal of finished products. With >50% of cultivated microalgal biomass designated for human consumption, the commercial opportunities in health foods and omega-3 production are robust, with health foods generating about USD 2.5 billion and omega-3 products contributing roughly USD 1.5 billion to the overall market, which approaches USD 6.5 billion in total value and 7.5 million tons in annual production (Uthaiah et al., 2025).

From a regional perspective, market penetration varies significantly, presenting targeted opportunities for strategic investments. North America leads the market with a 25.7% consumer base, followed by India with a 14.8% share and Germany with 5.3%. European markets are growing steadily at a CAGR of 5.8%, while Asia is expected to experience a higher growth rate of 7.4% between 2023 and 2030. North American markets are forecasted to reach a value of USD 670.58 million by 2030, and emerging regions such as the Middle East and Africa, with an expected CAGR of 4.6%, will offer additional revenue streams by achieving market values of approximately USD 36.80 million (Gallego et al., 2025). These regional trends not only highlight where market demand is strongest but also offer insights into where companies might focus their expansion efforts.

In the commercial sphere, two microalgal species have proven particularly valuable: *Spirulina* and *Chlorella*. *Spirulina* is cultivated at large scales in key markets such as India, China, and the United States due to its high protein content and its capacity for producing functional TAGs. These compounds are integrated into products aimed at health maintenance and disease prevention. On the contrary, *Chlorella vulgaris*, which has received formal approval from the European Food Safety Authority, is renowned for its rich biochemical profile. This microalga's portfolio includes ingredients for food, nutraceuticals, pharmaceuticals, cosmetics, natural dyes, aquaculture feeds, biofertilizers, and even biodiesel. Market forecasts suggest that the global *Chlorella* sector will grow from USD 292.55 million in 2023 to an estimated USD 506.99 million by 2031, marking a CAGR of 6.30% (Jadhav, 2024; Vladic et al., 2025). Commercial strategies increasingly favor these species because their established production processes and recognized market value reduce the risks associated with large-scale investments.

Nannochloropsis, a recognized microalga for its high protein content and elevated omega-3 levels, is another cornerstone in the commercial microalgae industry. In 2022, the market value for *Nannochloropsis* was approximately USD 8.5 million. It is expected to grow at a CAGR of 7.1% from 2023 to 2030 as demand for omega-3-rich products continues to rise. The growing global nutraceutical market, which is projected to surge from USD 209 billion in 2017 to USD 373 billion in 2025, has further propelled the adoption of *Nannochloropsis* across multiple regions, including Europe, North America, Asia-Pacific, Latin America, and the Middle East and Africa. Major industry players, such as Reed Mariculture, Cellana, and Algatech, have invested heavily in the commercialization of *Nannochloropsis* biomass, integrating it into products that range from animal feed supplements to wellness products (Uthaiah et al., 2025).

From a market segmentation perspective, the commercialization of microalgae-derived compounds can be broadly categorized into several key sectors. In the nutraceutical and dietary supplement segment, products enriched with omega-3 FAs, astaxanthin, and fucoxanthin are experiencing robust demand, driven by consumer trends toward health and wellness. The functional food sector is also a major beneficiary, as microalgae-based ingredients are incorporated into beverages, bakery products, and dairy alternatives to enhance nutritional content and appeal. In addition, the cosmetic industry is capitalizing on the antioxidant and antiinflammatory properties of microalgae-derived compounds by formulating products that cater to the growing consumer demand for natural and effective skincare solutions. High-value extraction processes have been central to unlocking the commercial potential of microalgae. Astaxanthin, a red–orange carotenoid from the xanthophyll family, stands out as one of the most commercially valuable compounds derived from microalgae. Owing to its potent antioxidant properties, which surpass those of vitamin C, β-carotene, and lutein, astaxanthin is widely used in dietary supplements, functional foods, and cosmetic formulations. In 2023 the global astaxanthin market was valued at USD 2.34 billion and is forecast to grow at a CAGR of 17.1% between 2024 and 2030. Companies such as Cyanotech, Algatech, and AstaReal have capitalized on this trend by pioneering the production of natural astaxanthin sourced primarily from *H. pluvialis*. Although *H. pluvialis* faces challenges such as slow growth rates and low mixotrophic capacity, the emergence of alternative microalgal sources like *Chromochloris zofingiensis*, renowned for its rapid growth and high biomass yield, offers promising commercial alternatives. Synthetic astaxanthin currently dominates >95% of the feed market; however, natural astaxanthin remains preferred for human consumption due to its superior bioactivity and safety profile. This market segment commands premium prices, ranging between USD 10,000 and 15,000 per kilogram (Tambat et al., 2025; Elbahnaswy and Elshopakey, 2024).

Fucoxanthin, another high-value carotenoid, has also carved out a significant niche in the commercial market. Predominantly sourced from brown microalgae, fucoxanthin is valued for its role as a natural pigment and its diverse biological activities, including antiobesity and antiinflammatory properties. The market for fucoxanthin was valued at USD 209.45 million and is projected to reach USD 294.72 million by 2030. While its principal application lies in the food industry as a natural pigment, fucoxanthin is increasingly finding its way into cosmetic and pharmaceutical products, particularly in formulations designed for skin and hair care, as well as in supplements aimed at addressing metabolic disorders such as obesity and depression (Yordi et al., 2024).

Marine-derived PLs have emerged as another commercially attractive product line. Their unique physicochemical properties, which facilitate dispersion, stabilization, emulsification, and wetting, make them indispensable in the formulation of a wide range of products, including food, pharmaceuticals, and cosmetics. Traditionally, PLs are sourced from soy, egg yolks, or brain tissues. However, soy lecithin, the most common commercial source, has limitations due to its high saturated fatty acid content and insufficient levels of essential unsaturated FAs like EPA and DHA. Consequently, research and commercial investment have shifted toward marine-derived PLs, which are naturally enriched with ω-3 PUFAs. An illustrative commercial application is the production of Lyprinol, a product derived via supercritical CO_2 extraction of a neutral oil from freeze-dried mussel powder. This product, formulated with olive oil and tocopherols, comprises a complex mixture of triglycerides, free FAs, sterols, and minor carotenoids. Clinical trials have confirmed its antiinflammatory benefits, making it a commercially viable remedy for conditions such as asthma and inflammatory bowel disease (Morel et al., 2024; Park et al., 2024).

Investment trends in microalgae-based industries have been robust and are expected to continue as companies expand their production capabilities. Key markets, such as the United States, Mexico, Canada, China, India, Germany, France, Australia, and Japan, are at the forefront of these investments. The substantial capital inflows into microalgae cultivation and processing underscore the sector's potential for high returns. As consumer preferences increasingly favor natural, sustainable, and health-enhancing products, the commercial viability of microalgae continues to strengthen. Companies are strategically positioning themselves to capture market share by developing innovative products that integrate microalgae-derived ingredients into functional foods, dietary supplements, and cosmeceuticals (Jadhav, 2024; Tambat et al., 2025; Uthaiah et al., 2025).

The commercial landscape for microalgae-based products is further enhanced by strong regulatory support in key markets. Regulatory bodies across the globe are increasingly recognizing the safety and efficacy of microalgae-derived ingredients, leading to favorable policies that encourage market entry and product innovation. The approval of *C. vulgaris* by the European Food Safety Authority, for example, has paved the way for its widespread commercial adoption. Similarly, robust regulatory frameworks in North America and Asia are fostering an environment of trust and transparency, which is critical for attracting both consumer confidence and investor interest. In conclusion, the commercial landscape for microalgae-derived compounds is characterized by rapid growth, significant investment, and a dynamic interplay between technology and market demand. With strong projections for market expansion, robust regional growth, and a continuously evolving technological base, microalgae offer a versatile and high-value solution for industries seeking to innovate and differentiate themselves in a competitive global market. The integrated value chain—from cultivation to final product formulation—presents multiple avenues for value addition and commercial success. As companies worldwide increasingly prioritize sustainable, health-focused, and natural products, the commercial potential of microalgae-derived compounds is poised to become a cornerstone of modern industrial strategies (Jadhav, 2024; Vladic et al., 2025).

6.9 Challenges

As lipidomics research progresses, there is an urgent need to ensure that lipid discovery and production practices adhere to sustainability principles and minimize ecological impact. The demand for bioactive lipids must be balanced with responsible bioprospecting efforts, avoiding overexploitation of marine resources. A concerted effort is required to establish ethical frameworks for marine bioprospecting, ensuring that lipid extraction processes do not compromise ecosystem health or biodiversity. Moreover, the application of high-throughput screening, lipidomics, and metabolic engineering could facilitate the identification and optimization of novel lipid-producing marine species, but these technologies require significant investment in research and development (FAO, 2024; Thiao and Bunting, 2022).

Lipidomics also faces several key challenges that hinder the identification and comprehensive understanding of lipids in complex biological matrices. A primary challenge is the accurate identification of specific lipids, many of which remain elusive in these intricate biological systems. The detection of a wide range of lipid species is often limited by their low abundance or chemical complexity. Furthermore, the identification of novel lipids from marine organisms is particularly difficult due to their intricate interactions with ecological environments, which often lead to rapid dilution and transformation of metabolites. The study of exuded lipids, especially those released into aquatic environments, is fraught with difficulties related to their dispersion and low concentrations. To overcome these issues, more refined isolation techniques, such as microscale approaches, are necessary. These methods should be optimized for both sensitivity and the ability to work with limited sample material, which remains a significant constraint in lipidomics research (Jayawardena et al., 2025; Fernandes and Cordeiro, 2021).

One of the major barriers to advancing lipidomics, particularly for marine bioactive lipids, is the lack of comprehensive and accessible databases for marine lipid species. Current lipidomics databases often underrepresent marine-specific lipids, making it difficult for researchers to rapidly identify and dereplicate novel compounds. The absence of reference spectra for many marine lipids complicates automated data analysis and necessitates the labor-intensive creation of in-house spectral libraries. This underscores the need for the development of an open-access, rigorously curated database that includes spectrometric data for marine lipids, which would significantly enhance the efficiency of lipid discovery and streamline the research process. In addition, advancements in lipidomics methodologies are crucial to improving detection sensitivity, resolution, and reproducibility, ensuring more accurate and comprehensive profiling of lipid species. Including both known and newly identified marine lipids in widely used lipidomics databases could provide an essential resource for accelerating lipid discovery (Jayawardena et al., 2025; Xue et al., 2025).

The scaling up of lipid production from marine organisms for both research and commercial applications presents its own set of challenges. Large-scale cultivation of marine microorganisms for lipid production requires the optimization of growth conditions and the development of cost-effective bioprocessing strategies. Moreover, the extraction and purification of lipids from marine organisms, especially for industrial use, demand optimized techniques that not only maximize yield but also preserve the bioactivity of the lipids. While laboratory-scale extraction and recovery techniques have demonstrated high efficiency, translating these methods into industrial-scale production is often challenging due to differences in reaction kinetics, mass transfer rates, and process economics. Factors such as bioreactor design, lipid solubility, and the optimization of harvesting techniques must be carefully addressed to ensure consistent lipid production at a commercial scale. In addition, the handling of marine biomass presents logistical challenges, as many lipid-rich marine organisms, such as microalgae and krill, have high water content and require rapid processing to prevent degradation (Getachew et al., 2024; Hemavathy et al., 2025; Senadheera et al., 2023).

Another major challenge is the efficiency of lipid extraction and recovery. Conventional solvent-based extraction methods, such as hexane or ethanol extraction, are widely used but suffer from several drawbacks, including incomplete lipid recovery, solvent toxicity, and environmental hazards associated with solvent disposal. Alternative extraction techniques, such as SFE using CO_2, have demonstrated potential for improving lipid yield while reducing environmental impact. However, these techniques require expensive equipment, complex operational control, and high energy inputs, making them less feasible for large-scale applications. Enzymatic-assisted extraction is another emerging approach that uses hydrolytic enzymes to break down cellular structures and release lipids more efficiently. Although this method enhances lipid yield and purity, it is highly dependent on enzyme specificity, reaction conditions, and cost-effectiveness, all of which require further optimization (Getachew et al., 2024; Hemavathy et al., 2025; Senadheera et al., 2023). As shown in Fig. 6.4, which presents a SWOT (Strengths, Weaknesses, Opportunities, and Threats) analysis, green extraction techniques are emerging as sustainable and safer alternatives, with strong potential despite current challenges in cost and scalability.

Additional key obstacles in scaling up marine lipid production, particularly from microalgae, are consumer acceptance. The strong marine-like flavor and dark green color of microalgae-derived products often reduce their sensory appeal, with higher concentrations further intensifying these characteristics. This trade-off between nutritional benefits and sensory limitations remains a major concern for the functional food industry. In addition, strict food safety regulations and limited availability of space for large-scale cultivation hinder the expansion of microalgae-based products. While closed photobioreactors provide controlled conditions for optimized growth, their high installation and operational costs make them less feasible for large-scale production. As a result, open-pond systems are the preferred method due to their lower costs, despite challenges related to contamination and environmental fluctuations (Jadhav, 2024).

The oxidative instability of marine lipids presents a significant barrier to effective production and recovery. PUFAs, particularly long-chain omega-3 FAs, are highly prone to oxidation due to their multiple double bonds, leading to the formation of lipid peroxides and free radicals. Oxidative degradation not only reduces the nutritional and therapeutic value of

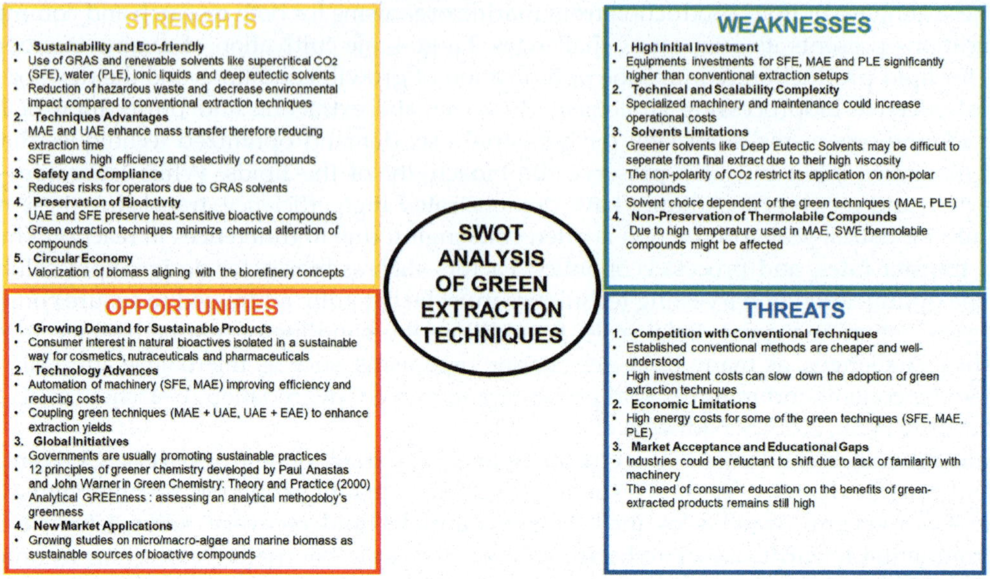

FIG. 6.4 SWOT (Strengths, Weaknesses, Opportunities, and Threats) of green extraction techniques. *From Kopp, G., Lauritano, C., 2025. Greener extraction solutions for microalgal compounds. Mar. Drugs 23 (7). doi:10.3390/md23070269.*

marine lipids but also affects their sensory properties, resulting in undesirable flavors and odors. Strategies to mitigate oxidation include the addition of natural antioxidants, such as tocopherols and polyphenols, during processing, as well as the use of low-temperature storage and inert gas flushing. However, these approaches add to production costs and require precise optimization to maintain lipid quality without affecting bioavailability (Rincó n-Cervera et al., 2019; Snipe et al., 2024).

Environmental concerns associated with marine lipid production further complicate large-scale implementation. The extraction and processing of fish oil and other marine lipids generate significant waste, including protein-rich residues, water-soluble by-products, and organic solvents. The improper disposal of these waste streams can lead to environmental pollution and ecological imbalance. Developing integrated biorefinery models that utilize waste streams for secondary applications, such as biofertilizers, animal feed, or biofuels, can help minimize waste and improve process sustainability. However, implementing such circular economy approaches requires technological advancements, regulatory support, and economic feasibility assessments (Monteiro et al., 2024; Hemavathy et al., 2025).

Moreover, regulatory constraints pose another major challenge in marine lipid production and recovery. Different regions impose strict guidelines on the use of marine-derived lipids in food, pharmaceuticals, and cosmetics, requiring compliance with safety standards related to contaminants, heavy metals, and microbial contamination. The presence of persistent organic pollutants, such as polychlorinated biphenyls and dioxins, in marine lipid sources necessitates extensive purification and quality control measures. While emerging purification techniques, such as activated carbon filtration and molecular distillation, have improved the removal of contaminants, they also contribute to higher production costs and energy consumption (Hosseini et al., 2022; Mathieu-Resuge et al., 2023; Parrish, 2025).

6.10 Future perspectives

To address the mentioned challenges, collaboration among academic institutions, biotechnology companies, and regulatory bodies will be essential. These partnerships can drive the development of more efficient production systems, sustainable extraction technologies, and robust regulatory guidelines. Harvesting strategies, such as ecosystem-based fisheries management, quota-based extraction, and real-time monitoring of marine biodiversity, will be necessary to balance lipid production with conservation efforts (Jayawardena et al., 2025).

As marine lipids continue to demonstrate their vast potential across diverse scientific and industrial domains, it is imperative to adopt sustainable sourcing practices and innovative processing technologies. The convergence of marine biotechnology and health sciences offers exciting prospects for the next generation of lipid-based solutions, reinforcing the significance of marine lipids in advancing human health and sustainable development (Monteiro net al., 2024; Parrish, 2025). Continued interdisciplinary collaboration and investment in cutting-edge research will be essential in unlocking the full potential of marine lipids, ensuring their responsible utilization for future generations. In this regard, the future of marine lipid research lies in the continued integration of advanced analytical techniques, biotechnological innovations, and computational modeling. The development of precision lipidomics will enable personalized nutrition and targeted therapeutic interventions, optimizing lipid formulations for specific health conditions (Xue et al., 2025).

One of the most pressing areas of future research is the development of sustainable and alternative sources of marine lipids. Traditionally, fish oil has been the dominant source of omega-3 FAs, but overfishing and environmental concerns have accelerated efforts to explore alternative lipid-producing organisms. Microalgae have emerged as the most promising alternative, given their ability to synthesize EPA and DHA without contributing to fish stock depletion. However, large-scale microalgal lipid production remains limited by high production costs, slow biomass accumulation, and technological challenges associated with cultivation, harvesting, and extraction (Motta et al., 2025). Future advancements in microalgal biotechnology will focus on strain improvement, optimization of photobioreactors, and the integration of closed-loop cultivation systems that minimize resource inputs and maximize lipid yields. The use of genetically modified microalgae, engineered for enhanced lipid biosynthesis through CRISPR-based metabolic modifications, is expected to become a key research area, though regulatory and ethical concerns surrounding genetically modified organisms will need to be addressed. In addition, emerging sources such as marine fungi and extremophilic microorganisms from deep-sea environments hold potential for the discovery of novel lipid biosynthetic pathways, warranting further exploration (Gallego et al., 2025; Hemavathy et al., 2025; Nagarajan et al., 2021).

Biotechnological advancements will play a critical role in shaping the future of marine lipid production. Genome editing and synthetic biology approaches will be applied to optimize lipid metabolism in microalgae, marine bacteria, and heterotrophic yeast.

CRISPR–Cas9, RNA interference, and metabolic flux analysis will allow for the targeted enhancement of lipid biosynthetic pathways, increasing the efficiency of EPA and DHA production. In addition, advancements in systems biology and computational modeling will enable researchers to fine-tune metabolic networks, improving lipid yield and composition. Fermentation-based production of marine lipids, particularly through engineered strains, is another promising avenue that could reduce reliance on wild fish stocks while offering a scalable and cost-effective approach (Guo et al., 2024; Becker et al., 2025; Filipe et al., 2025). Future research will focus on optimizing fermentation conditions, substrate utilization, and downstream processing to make microbial lipid production commercially viable. As new sources of marine lipids emerge, including genetically engineered microalgae and microbial fermentation-derived lipids, regulatory agencies will need to establish safety, traceability, and quality assurance standards. The harmonization of international regulations governing marine lipid purity, contaminant levels, and sustainability certifications will be essential for ensuring consumer confidence and market expansion. Future policy initiatives may include incentives for sustainable lipid production, stricter environmental regulations for marine resource utilization, and investment in research consortia focused on marine biotechnology innovations (Jayawardena et al., 2025; Hemavathy et al., 2025).

In addition, future research should emphasize the use of alternative extraction techniques, such as supercritical CO_2 extraction, ultrasound-assisted extraction, and enzymatic hydrolysis. In addition, emerging extraction techniques using deep eutectic solvents and ionic liquids are being explored for their potential to enhance lipid recovery while minimizing environmental impact. The integration of these techniques into biorefineries that utilize multiple extraction and purification methods will be essential for optimizing lipid recovery efficiency (Xue et al., 2025; Debbarma et al., 2025).

Novel antioxidant stabilization strategies are needed, including the incorporation of marine-derived polyphenols, flavonoids, and carotenoids to protect lipids from oxidative degradation. In addition, advances in nanoencapsulation will enable better control over lipid oxidation, extending shelf life and improving bioavailability. The use of lipid-based delivery systems, such as liposomes, nanostructured lipid carriers, and self-emulsifying lipid formulations, will further enhance the stability and absorption of marine lipids in functional foods and pharmaceuticals (Eladl et al., 2024; Wang et al., 2024).

In addition, artificial intelligence (AI) and machine learning are expected to play a growing role in lipid research, accelerating the discovery of novel bioactive compounds and improving predictive models for lipid functionality. AI-based lipidomics and metabolomics analyses will allow researchers to identify novel lipid compounds and predict their bioactive properties and could also be used to streamline operations in marine lipid biorefineries (Li et al., 2025; Xue et al., 2025).

6.11 Conclusion

The exploration, production, and valorization of marine lipids have advanced significantly, revealing their potential for applications across diverse industries while highlighting critical sustainability challenges. Despite their high nutritional and bioactive value, ensuring sustainable production with minimal environmental impact remains a key priority. Future advancements in marine lipid research and industry will depend on integrating biotechnological innovations, green extraction technologies, metabolic engineering, and artificial intelligence within circular bioeconomy frameworks. AI-driven modeling can enhance lipid extraction efficiency, optimize cultivation conditions for microalgae and microbial systems, and predict the bioactivity of novel marine lipids, accelerating their industrial application. In addition, the valorization of marine by-products and waste streams from fisheries, aquaculture, and marine biotechnology will be essential for reducing waste, improving resource efficiency, and minimizing environmental footprints. As novel lipid sources emerge and cutting-edge biotechnological processes evolve, international regulatory frameworks must adapt to ensure safety, traceability, and sustainability. Through continued investment in research, innovation, and policy development, marine lipids will remain integral to the bioeconomy, supporting human health, industrial progress, and the sustainable utilization of marine resources.

AI disclosure

During the preparation of this work, the author(s) used ChatGPT to apply it to some paragraphs to improve readability. After using this tool/service, the author(s) reviewed and edited the content as needed and take(s) full responsibility for the content of the publication.

Acknowledgements

MM acknowledges financial support from the National Agency of Science and Scientific Promotion under the former Ministry of Science, Technology, and Innovation of Argentina for funding the project Grant No. B17 Iniciativa Pampa Azul. LFF and BJF acknowledge financial support from the former Ministry of Science, Technology, and Innovation of Argentina for funding the project "Club de Biotecnología Agrícola y Ambiental" at Universidad Nacional de Río Negro, Argentina (Project 103B/UNRN-40-A-1025). Access to scholarly resources provided by The Wikipedia Library of the Wikimedia Foundation is also acknowledged.

References

Abril, A.G., Carrera, M., Pazos, M., 2023. Immunomodulatory effect of marine lipids on food allergy. Front. Nutr. 2296-861X. 10, 1254681. Available from: https://www.frontiersin.org/articles/10.3389/fnut.2023.1254681/full. doi:10.3389/fnut.2023.1254681.

Ahmad, S., Singh, A., Akram, W., Upadhyay, A., Abrol, G.S., 2025. Algal lipids: a review on current status and future prospects in food processing. J. Food Sci. 0022-1147, 1750-3841, 90 (1), e17618. Available from: https://ift.onlinelibrary.wiley.com/doi/10.1111/1750-3841.17618. doi:10.1111/1750-3841.17618.

Aitta, E., Marsol-Vall, A., Damerau, A., Yang, B., 2021. Enzyme-assisted extraction of fish oil from whole fish and by-products of Baltic herring (Clupea harengus membras). Foods 10 (8), 1811. doi:10.3390/foods10081811.

Alasalvar, C., Shahidi, F., Quantick, P., 2002. Food and health applications of marine nutraceuticals: a review. In: Alasalvar, C., Taylor, T., (Eds.), Seafoods—Quality, Technology and Nutraceutical Applications. Berlin, Heidelberg: Springer, pp. 175–204. doi:10.1007/978-3-662-09836-3_15.

Albrecht-Ruiz, M., Salas-Maldonado, A., 2015. Chemical composition of light and dark muscle of peruvian anchovy (*Engraulis ringens*) and its seasonal variation. J. Aquat. Food Prod. Technol. 1049-8850, 1547-0636, 24 (2), 191–196. Available from: https://www.tandfonline.com/doi/abs/10.1080/10498850.2012.762705. doi:10.1080/10498850.2012.762705.

Ali, A., Rehman, A., Jafari, S.M., Ranjha, M.M.A.N., Shehzad, Q., Shahbaz, H.M., Khan, S., Usman, M., Łukasz Kowalczewski, P., Jarzębski, M., Xia, W., 2022. Effect of co-encapsulated natural antioxidants with modified starch on the oxidative stability of β-carotene loaded within nanoemulsions. Appl. Sci. 2076-3417. 12 (3), 1070. Available from: https://www.mdpi.com/2076-3417/12/3/1070. doi:10.3390/app12031070.

Alsenani, F., Wass, T.J., Ruijuan, M., Eltanahy, E., Netzel, M.E., Schenk, P.M., 2019. Transcriptome-wide analysis of *Chlorella* reveals auxin-induced carotenogenesis pathway in green microalgae. Algal Res. 2211-9264. 37, 320–335. Available from: https://linkinghub.elsevier.com/retrieve/pii/S2211926418306817. doi:10.1016/j.algal.2018.12.002.

Alves, E., Dias, M., Lopes, D., Almeida, A., Do Rosário Domingues, M., Rey, F., 2020. Antimicrobial lipids from plants and marine organisms: an overview of the current state-of-the-art and future prospects. Antibiotics 2079-6382. 9 (8), 441. Available from: https://www.mdpi.com/2079-6382/9/8/441. doi:10.3390/antibiotics9080441.

Arenas-Jal M., Suñé-Negre J.M., García-Montoya E., 2020. Encarna García-Montoya, An overview of microencapsulation in the food industry: opportunities, challenges, and innovations. Eur. Food Res. Technol. 1438-2377, 1438-2385, 246 (7), 1371–1382. Available from: https://link.springer.com/10.1007/s00217-020-03496-x. doi:10.1007/s00217-020-03496-x.

Arts, M.T., Kohler, C.C., 2009. Health and condition in fish: the influence of lipids on membrane competency and immune response. In: Kainz, M., Brett, M.T., Arts, M.T. (Eds.), Lipids in Aquatic Ecosystems. Springer New York, New York, pp. 237–256. Available from: https://link.springer.com/10.1007/978-0-387-89366-2_10.

Babar, M., Buzdar, J.A., Zaheer, A., Nizam-ud-din, N.-U.-D., Mustafa, G., Khan, B.A., Hanif, M., Asghar, T., Qadeer, A., 2025. Carotenoids as a nutraceutical and health-promoting dietary supplement for human and animals: an updated review. Tradit. Med. Res. 2413-3973. 10 (3), 14. Available from: https://www.tmrjournals.com/article.html?J_num=1&a_id=3380. doi:10.53388/TMR20240831001.

Babu, S.S., Gondi, R., Vincent, G.S., JohnSamuel, G.C., Jeyakumar, R.B., 2022. Microalgae biomass and lipids as feedstock for biofuels: sustainable biotechnology strategies. Sustainability 2071-1050. 14 (22), 15070. Available from: https://www.mdpi.com/2071-1050/14/22/15070. doi:10.3390/su142215070.

Barzkar, N., Sukhikh, S., Babich, O., 2024. Study of marine microorganism metabolites: new resources for bioactive natural products. Front. Microbiol. 1664-302X. 14, 1285902. Available from: https://www.frontiersin.org/articles/10.3389/fmicb.2023.1285902/full. doi:10.3389/fmicb.2023.1285902.

Bastardie, F., Salvany, L., Cooper, A.M., Carvalho, N., 2024. A roadmap to reduce the risk of overexploiting EU marine living resources in a changing ocean. Front. Mar. Sci. 11. doi:10.3389/fmars.2024.1352500

Becker, K.W., Devresse, Q., Prieto-Mollar, X., Hinrichs, K.-U., Engel, A., 2025. Mixed-layer lipidomes suggest offshore transport of energy-rich and essential lipids by cyclonic eddies. Commun. Earth Environ. 6 (1). doi:10.1038/s43247-025-02152-0.

Behera, B., Unpaprom, Y., Ramaraj, R., Maniam, G.P., Govindan, N., Paramasivan, B., 2021. Integrated biomolecular and bioprocess engineering strategies for enhancing the lipid yield from microalgae. Renew. Sustain. Energy Rev. 1364-0321. 148, 111270. Available from: https://www.sciencedirect.com/science/article/pii/S1364032121005578. doi:10.1016/j.rser.2021.111270.

Beppu, F., Yasuda, K., Okada, A., Hirosaki, Y., Okazaki, M., Gotoh, N., 2017. Comparison of the distribution of unsaturated fatty acids at the sn-2 position of phospholipids and triacylglycerols in marine fishes and mammals. J. Oleo Sci. 1345-8957, 1347-3352, 66 (11), 1217–1227. Available from: https://www.jstage.jst.go.jp/article/jos/66/11/66_ess17132/_article. doi:10.5650/jos.ess17132.

Boyd, C.E., D'Abramo, L.R., Glencross, B.D., Huyben, D.C., Juarez, L.M., Lockwood, G.S., McNevin, A.A., Tacon, A.G.J., Teletchea, F., Tomasso, J.R., Tucker, C.S., Valenti, W.C., 2020. Achieving sustainable aquaculture: historical and current perspectives and future needs and challenges. J. World Aquac. Soc. 51 (3), 578–633. doi:10.1111/jwas.12714.

Chungjatupornchai, W., Areerat, K., Fa-Aroonsawat, S., 2019. Increased triacylglycerol production in oleaginous microalga *Neochloris oleoabundans* by overexpression of plastidial lysophosphatidic acid acyltransferase. Microb. Cell Fact. 1475-2859. 18 (1), 53. doi:10.1186/s12934-019-1104-2.

Conde, T., Lopes, D., Pinho, M., Melo, T., Coelho, N., Rodrigues, A.M.C., Pereira, H., Domingues, P., Domingues, R., 2025. Unraveling the lipidome of Pavlova gyrans, a natural reservoir of bioactive lipids for biotechnological applications. J. Appl. Phycol. doi:10.1007/s10811-024-03420-w.

Cretton, M., Malanga, G., Sobczuk, T.M., Mazzuca, M., 2021. Lipid fraction from industrial crustacean waste and its potential as a supplement for the feed industry: a case study in Argentine Patagonia. Waste Biomass Valorization

1877-2641, 1877-265X, 12 (5), 2311–2319. Available from: https://link.springer.com/10.1007/s12649-020-01162-7. doi:10.1007/s12649-020-01162-7.

Cretton, M., Malanga, G., Sobczuk, T.M., Mazzuca, M., 2023. Marine lipids as a source of high-quality fatty acids and antioxidants. Food Rev. Int. 39 (8), 4941–4964. doi:10.1080/87559129.2022.2042555.

Cretton, M., Rost, E., Mazzuca Sobczuk, T., Mazzuca, M., 2016. Variation in the proximate composition and fatty acid profile recovered from Argentine hake (Merluccius hubbsi) waste from Patagonia. Grasas Aceites 67 (1), e122. Available from: https://grasasyaceites.revistas.csic.es/index.php/grasasyaceites/article/view/1588.

Debbarma, S., Deb, S., Yadav, N.K., Kashyap, S., Debbarma, P., Chrisolite, B., Yirang, Y., Meena, D.K., Singh, S.K., Choudhury, T.G., Patel, A.B., 2025. Waste not, want not: unlocking the innovative potential of organic and eco-friendly insect and algal resources for future aquaculture. Aquac. Int. 0967-6120, 1573-143X, 33 (2), 158. Available from: https://link.springer.com/10.1007/s10499-024-01814-8. doi:10.1007/s10499-024-01814-8.

Djamaludin, H., Sulistiyati, T.D., Chamidah, A., Nurashikin, P., Roifah, M., Notonegoro, H., Ferdian, P.R., 2023. Quality and fatty acid profiles of fish oil from tuna by-products extracted using a dry-rendering method. Biodiversitas 24, (11). doi:10.13057/biodiv/d241131.

Dos Reis, G.A., Rodrigues, C., Wiatek, A.M., de Melo Pereira, G.V., de Carvalho, J.C., José Martinez-burgos, W., Karp, S.G., Soccol, V.T., Letti, L.A.J., Soccol, C.R., 2024. Exploring the potential of Thraustochytrids and other microorganisms for sustainable omega-3 production: a comprehensive review of patents, perspectives, and scale-up strategies. Syst. Microbiol. Biomanuf. 2662-7663. 4 (2), 448–462. doi:10.1007/s43393-023-00213-z.

Dunshea, F.R., Sutcliffe, M., Suleria, H.A.R., Giri, S.S., 2024. Global issues in aquaculture. Anim. Front. 14 (4), 3–5. doi:10.1093/af/vfae027.

Egerton, S., Mannion, D., Culloty, S., Whooley, J., Stanton, C., Ross, R.P., 2020. The proximate composition of three marine pelagic fish: blue whiting (*Micromesistius poutassou*), boarfish (*Capros* aper) and Atlantic herring (*Clupea harengus*). Ir. J. Agric. Food Res. 2009-9029. 59 (1). Available from: https://scienceopen.com/hosted-document?doi=10.15212/ijafr-2020-0112. doi:10.15212/ijafr-2020-0112.

Eladl, S.N., Elnabawy, A.M., Eltanahy, E.G., 2024. Recent biotechnological applications of value-added bioactive compounds from microalgae and seaweeds. Bot. Stud. 1999-3110. 65 (1), 28. Available from: https://as-botanicalstudies.springeropen.com/articles/10.1186/s40529-024-00434-y. doi:10.1186/s40529-024-00434-y.

Elbahnaswy, S., Elshopakey, G.E., 2024. Recent progress in practical applications of a potential carotenoid astaxanthin in aquaculture industry: a review. Fish Physiol. Biochem. 0920-1742, 1573-5168, 50 (1), 97–126. Available from: https://link.springer.com/10.1007/s10695-022-01167-0. doi:10.1007/s10695-022-01167-0.

FAO, 2024. The state of world fisheries and aquaculture 2024 – blue transformation in action. FAO, Rome, Italy, 264. Available from: https://doi.org/10.4060/cd0683en.

Fernandes, T., Cordeiro, N., 2021. Microalgae as sustainable biofactories to produce high-value lipids: biodiversity, exploitation, and biotechnological applications. Mar. Drugs 1660-3397. 19 (10), 573. Available from: https://www.mdpi.com/1660-3397/19/10/573. doi:10.3390/md19100573.

Filipe, H.A.L., Moreira, A.F., Miguel, S.P., Coutinho, P., 2025. Microalgae lipid membrane models: a computational biophysics characterization. Algal Res. 85. doi:10.1016/j.algal.2024.103884

Flores, L.F., Flores, B.J., Osorio-Gonzalez, C.S., Saini, R., Brar, S.K., Lahiri, D., 2025b. Marine-based bioactive compounds in healthcare and wellness industries. In: Lahiri, D., Nag, M., Bhattacharya, D., Pati, S., Sarkar, T., (Eds.), Bioactive Ingredients for Healthcare Industry Volume 1: Extraction Strategies, Stability and Medicinal Properties. Springer Nature Singapore, Singapore, pp. 75–104. doi:10.1007/978-981-96-3663-1_5.

Flores, L.F., Osorio-Gonzalez, C.S., Saini, R., Brar, S.K., 2024. Renewable Residues as Feedstock for Drop-in Biofuel Production. Springer Nature, Switzerland, Cham, pp. 41–74. doi:10.1007/978-3-031-61637-2_3.

Flores, L.F., Osorio-Gonzalez, C.S., Saini, R., Brar, S.K., 2025a. Streamlined ultrasonication-assisted lipid extraction and direct transesterification of Rhodosporidium toruloides 1588 single-cell oil. Process Biochem. 157, 1–9. doi:10.1016/j.procbio.2025.05.006.

Galasso, C., Corinaldesi, C., Sansone, C., 2017. Carotenoids from marine organisms: Biological functions and industrial applications. Antioxidants 6 (4), 96. doi:10.3390/antiox6040096.

Galasso, C., Ruocco, N., Mutalipassi, M., Barra, L., Costa, V., Giommi, C., Dinoi, A., Genovese, M., Pica, D., Romano, C., Greco, S., Pennesi, C., 2023. Marine polysaccharides, proteins, lipids, and silica for drug delivery systems: a review. Int. J. Biol. Macromol. 0141-8130. 253, 127145. Available from: https://linkinghub.elsevier.com/retrieve/pii/S0141813023040424. doi:10.1016/j.ijbiomac.2023.127145.

Gallego, I., Medic, N., Pedersen, J.S., Ramasamy, P.K., Robbens, J., Vereecke, E., Romeis, J., 2025. The microalgal sector in Europe: towards a sustainable bioeconomy. New Biotechnol. 1871-6784. 86, 1–13. Available from: https://www.sciencedirect.com/science/article/pii/S1871678425000020. doi:10.1016/j.nbt.2025.01.002.

Gao, S., Runhua, L., Heng, N., Chen, Y., Wang, L., Zheng, L., Guo, Y., Sheng, X., Wang, X., Xing, K., Hemin, N., Xiaolong, Q., 2020. Effects of dietary supplementation of natural astaxanthin from Haematococcus pluvialis on antioxidant capacity, lipid metabolism, and accumulation in the egg yolk of laying hens. Poult. Sci. 0032-5791. 99 (11), 5874–5882. Available from: https://linkinghub.elsevier.com/retrieve/pii/S003257912030568X. doi:10.1016/j.psj.2020.08.029.

Getachew, A.T., Jacobsen, C., Sørensen, A.-D.M., 2024. Supercritical CO_2 for efficient extraction of high-quality starfish (*Asterias rubens*) oil. J. Supercrit. Fluids 0896-8446. 206, 106161. Available from: https://linkinghub.elsevier.com/retrieve/pii/S089684462300325X. doi:10.1016/j.supflu.2023.106161.

Gómez-Estaca, J., Calvo, M.M., Álvarez-Acero, I., Montero, P., Gómez-Guillén, M.C., 2017. Characterization and storage stability of astaxanthin esters, fatty acid profile and α-tocopherol of lipid extract from shrimp (*L. vannamei*) waste with potential applications as food ingredient. Food Chem. 0308-8146. 216, 37–44. Available from: https://linkinghub.elsevier.com/retrieve/pii/S0308814616312468. doi:10.1016/j.foodchem.2016.08.016.

Gómez-Guillén, M.C., Montero, P., López-Caballero, M.E., Baccan, G.C., Gómez-Estaca, J., 2018. Bioactive and technological functionality of a lipid extract from shrimp (L. vannamei) cephalothorax. LWT 89, 704–711. doi:10.1016/j.lwt.2017.11.052.

Gomma, A.E., Lee, S.-K., Sun, S.M., Yang, S.H., Chung, G., 2015. Improvement in oil production by increasing malonyl-CoA and glycerol-3-phosphate pools in *Scenedesmus quadricauda*. Indian J. Microbiol. 0046-8991. 55 (4), 447–455. doi:10.1007/s12088-015-0546-4.

Guo, W., Weng, Y., Wenkai, M., Chang, C., Gao, Y., Huang, X., Zhang, F., 2024. Improving lipid content in the diatom *Phaeodactylum tricornutum* by the knockdown of the enoyl-CoA hydratase using CRISPR interference. Curr. Issues Mol. Biol. 46 (10), 10923–10933. doi:10.3390/cimb46100649.

Gupta, S., Makridis, P., Henry, I., Velle-George, M., Ribicic, D., Bhatnagar, A., Skalska-Tuomi, K., Daneshvar, E., Ciani, E., Persson, D., Netzer, R., 2024. Recent developments in recirculating aquaculture systems: a review. Aquac. Res. 2024. doi:10.1155/are/6096671

Hamilton, H.A., Newton, R., Auchterlonie, N.A., Müller, D.B., 2020. Systems approach to quantify the global omega-3 fatty acid cycle. Nat. Food 2662-1355. 1 (1), 59–62. Available from: https://www.nature.com/articles/s43016-019-0006-0. doi:10.1038/s43016-019-0006-0.

Hemavathy, R.V., Ragini, Y.P., Shruthi, S., Ranjani, S., Subhashini, S., Thamarai, P., 2025. Biofuel production from marine macroalgae: pathways, technologies, and sustainable energy solutions. Ind. Crops Prod. 0926-6690. 224, 120282. Available from: https://www.sciencedirect.com/science/article/pii/S0926669024022593. doi:10.1016/j.indcrop.2024.120282.

Hernando, M., De Troch, M., de la Rosa, F., Giannuzzi, L., 2021. Fatty acid response of the invasive bivalve *Limnoperna* fortunei fed with *Microcystis aeruginosa* exposed to high temperature. In: Comparative Biochemistry and Physiology Part - C: Toxicology and Pharmacology, Vol. 240, 108925. doi:10.1016/j.cbpc.2020.108925.

Hintermeister Clara, A., 2017. Nutritional Composition Changes in Alaska Pollock (*Gadus chalcogrammus*) Fillets during Season A and Season B. Oregon State University, Master of Science (M.S.). Food Science and Technology. https://ir.library.oregonstate.edu/concern/graduate_thesis_or_dissertations/6t053n239. doi:10.7267/6t053n239

Hosseini, S.F., Rezaei, M., McClements, D.J., 2022. Bioactive functional ingredients from aquatic origin: a review of recent progress in marine-derived nutraceuticals. Crit. Rev. Food Sci. Nutr. 1040-8398, 1549-7852, 62 (5), 1242–1269. Available from: https://www.tandfonline.com/doi/full/10.1080/10408398.2020.1839855. doi:10.1080/10408398.2020.1839855.

Imbs, A.B., Dembitsky, V.M., 2023. Coral lipids. Mar. Drugs 1660-3397. 21 (10), 539. Available from: https://www.mdpi.com/1660-3397/21/10/539. doi:10.3390/md21100539.

Jadhav, H.B., 2024. Functional triacylglycerols from microalgae and their use in the formulation of functional foods—Review. Food Chem. Adv. 4. doi:10.1016/j.focha.2024.100695

Jayawardena, T.U., Merindol, N., Liyanage, N.S., Awwad, F., Desgagné-Penix, I., 2025. Marine specialized metabolites: unveiling Nature's chemical treasures from the deep blue. TrAC Trends Anal. Chem. 183. doi:10.1016/j.trac.2024.118097

Kandyliari, A., Mallouchos, A., Papandroulakis, N., Golla, J.P., Lam, T.T., Sakellari, A., Karavoltsos, S., Vasiliou, V., Kapsokefalou, M., 2020. Nutrient composition and fatty acid and protein profiles of selected fish by-products. Foods 2304-8158. 9 (2), 190. Available from: https://www.mdpi.com/2304-8158/9/2/190. doi:10.3390/foods9020190.

Khoo, K.S., Ahmad, I., Chew, K.W., Iwamoto, K., Bhatnagar, A., Show, P.L., 2023. Enhanced microalgal lipid production for biofuel using different strategies including genetic modification of microalgae: a review. Prog. Energy Combust. Sci. 0360-1285. 96, 101071. Available from: https://www.sciencedirect.com/science/article/pii/S0360128523000011. doi:10.1016/j.pecs.2023.101071.

Kumar, S.S., Mageswari, S.D.U., Sakthivel, T.G., Umaeswari, P., 2024. A review on biofuel production from fish waste oil: review of biofuel production from fish waste oil. J. Environ. Nanotechnol. 2319-5541, 2279-0748, 13 (2), 294–297. Available from: https://nanoient.org/journals/index.php/jent/article/view/1039. doi:10.13074/jent.2024.06.241540.

Kumari, A., Pabbi, S., Tyagi, A., 2024. Recent advances in enhancing the production of long chain omega-3 fatty acids in microalgae. Crit. Rev. Food Sci. Nutr. 64 (29), 1040-8398. Available from: https://www.tandfonline.com/doi/full/10.1080/10408398.2023.2226720.

Kurniawan, R., Nurkolis, F., Taslim, N.A., Subali, D., Surya, R., Gunawan, W.B., Alisaputra, D., Mayulu, N., Salindeho, N., Kim, B., 2023. Carotenoids composition of green algae caulerpa racemosa and their antidiabetic, anti-obesity, antioxidant, and anti-inflammatory properties. Molecules 28 (7), 3267.

Lari, Z., Moradi-kheibari, N., Ahmadzadeh, H., Abrishamchi, P., Moheimani, N.R., Murry, M.A., 2016. Bioprocess engineering of microalgae to optimize lipid production through nutrient management. J. Appl. Phycol. 1573-5176. 28 (6), 3235–3250. doi:10.1007/s10811-016-0884-6.

Lauritano, C., 2024. Trends in study and application of marine resources. Biologia Marina Mediterranea 3034-8471. 28 (1), 9–14. Available from: https://www.biologiamarinamediterranea.it/index.php/metis/article/view/91.

Li, H., Chen, L., Zhang, F., Cai, Z., 2025. Graph-learning-based machine learning improves prediction and cultivation of commercial-grade marine microalgae Porphyridium. Bioresour. Technol. 416. doi:10.1016/j.biortech.2024.131728

Ma R., Thomas-Hall S.R., Chua E.T., Eltanahy E., Netzel M.E., Netzel G., Lu Y., Schenk P.M., 2018. LED power efficiency of biomass, fatty acid, and carotenoid production in Nannochloropsis microalgae. Bioresour. Technol. 0960-8524. 252, 118–126. Available from: https://linkinghub.elsevier.com/retrieve/pii/S0960852417322411. doi:10.1016/j.biortech.2017.12.096.

Mahaliyana, A.S., 2015. Nutritional composition of skipjack tuna (*Katsuwonus pelamis*) caught from the oceanic waters around Sri Lanka. Am. J. Food Nutr. 3 (4), 106–111. doi:10.12691/ajfn-3-4-3.

Mathieu-Resuge, M., Grand, F.L., Brosset, P., Lebigre, C., Soudant, P., Vagner, M., Pecquerie, L., 2023. Fany Sardenne, Red muscle of small pelagic fishes' fillets are high-quality sources of essential fatty acids. J. Food Compos. Anal. 0889-1575. 120, 105304. Available from: https://linkinghub.elsevier.com/retrieve/pii/S0889157523001783. doi:10.1016/j.jfca.2023.105304.

Méndez, E., González, R.M., 1997. Seasonal changes in the chemical and lipid composition of fillets of the Southwest Atlantic hake (*Merluccius hubbsi*). Food Chem. 0308-8146. 59 (2), 213–217. Available from: https://linkinghub.elsevier.com/retrieve/pii/S0308814696002257. doi:10.1016/S0308-8146(96)00225-7.

Mititelu, M., Lupuliasa, D., Neacşu, S.M., Olteanu, G., Sebastian Busnatu, Ş., Mihai, A., Popovici, V., Măru, N., Constanța Boroghină, S., Mihai, S., Ioniță-Mîndrican, C.-B., Scafa-Udriște, A., 2025. Polyunsaturated fatty acids and human health: a key to modern nutritional balance in association with polyphenolic compounds from food sources. Foods 14 (1). doi:10.3390/foods14010046.

Mkadem, H., Kaanane, A., 2024. A comprehensive review on marine by-products use for the recovery of value-added products. Curr. Opin. Green Sustain. Chem. 2452-2236. 50, 100972. Available from: https://linkinghub.elsevier.com/retrieve/pii/S2452223624000932. doi:10.1016/j.cogsc.2024.100972.

Mohanty, B.P., Ganguly, S., Mahanty, A., Mitra, T., Mohanty, S., Das, B.K., 2023. Production of Omega-3 Capsules from Fish Offal: Recycling of Resources for Sustainable Production and Consumption Advances in Fish Processing Technologies, Apple Academic Press.

Monteiro J., Domingues M., Calado R., 2024. Marine animal co-products—how improving their use as rich sources of health-promoting lipids can foster sustainability. Mar. Drugs 1660-3397. 22 (2), 73. Available from: https://www.mdpi.com/1660-3397/22/2/73. doi:10.3390/md22020073.

Morel, J., Catchpole, O., Moreno, T., Lagutin, K., MacKenzie, A., Fenton, T., Williams, A.M., 2024. Extraction of neutral lipids and phospholipids from marine biomasses using subcritical and supercritical fluids. J. Supercrit. Fluids 0896-8446. 206, 106160. Available from: https://linkinghub.elsevier.com/retrieve/pii/S0896844623003248. doi:10.1016/j.supflu.2023.106160.

Motta, D.N., Torres, V.G.S., de Queiroz Hermino, V.L., Neves, S.M.B., de Oliveira, L.V.V., Carvalho, R.P., da Silva, E. P., 2025. Challenges of Microalgae Production: From the Laboratory to Large Scale. Springer Nature, Switzerland, Cham, pp. 11–21. doi:10.1007/978-3-031-84282-5_2.

Nagarajan, D., Varjani, S., Lee, D.-J., Chang, J.-S., 2021. Sustainable aquaculture and animal feed from microalgae – nutritive value and techno-functional components. Renew. Sustain. Energy Rev. 1364-0321. 150, 111549. Available from: https://linkinghub.elsevier.com/retrieve/pii/S1364032121008273. doi:10.1016/j.rser.2021.111549.

Navalho, S., Ferrer-Ledo, N., Barbosa, M.J., Varela, J., 2025. Nannochloropsis lipids and polyunsaturated fatty acids: potential applications and strain improvement. Mar. Drugs 23 (3). doi:10.3390/md23030128.

Naylor, R., Fang, S., Fanzo, J., 2023. A global view of aquaculture policy. Food Policy 116. doi:10.1016/j.foodpol.2023.102422

Ogidi, O.I., Akpan, U.M., 2022. Aquatic Biodiversity Loss: Impacts of Pollution and Anthropogenic Activities and Strategies for Conservation. Springer Nature, Singapore, pp. 421–448. doi:10.1007/978-981-19-3326-4_16.

Park, J.-S., Roy, V.C., Han, J.-M., Yoo, S.H., Kim, G.-D., Chun, B.-S., 2024. Green extraction of marine phospholipids from conger eel by-product using SC-CO_2: lipidomic profiles and biological activities. Sustain. Chem. Pharm. 2352-5541. 42, 101742. Available from: https://linkinghub.elsevier.com/retrieve/pii/S2352554124003176. doi:10.1016/j.scp.2024.101742.

Parrish, C.C., 2024. Thraustochytrids and algae as sustainable sources of long-chain omega-3 fatty acids for aquafeeds. Sustainability (Switzerland 16 (21). doi:10.3390/su16219142.

Parrish, C.C., 2025. Production, transport, fate and effects of lipids in the marine environment. Mar. Drugs 1660-3397. 23 (2), 52. Available from: https://www.mdpi.com/1660-3397/23/2/52. doi:10.3390/md23020052.

Patel, A., Karageorgou, D., Katapodis, P., Sharma, A., Rova, U., Christakopoulos, P., Matsakas, L., 2021. Bioprospecting of thraustochytrids for omega-3 fatty acids: a sustainable approach to reduce dependency on animal sources. Trends Food Sci. Technol. 0924-2244. 115, 433–444. Available from: https://www.sciencedirect.com/science/article/pii/S0924224421004180. doi:10.1016/j.tifs.2021.06.044.

Paulose, S.K., Chakraborty, K., 2024. Nutritional properties of commercially important cephalopods from the south-west coast of India: nutritional qualities of cephalopod mollusks. Indian J. Fish. 0970-6011. 71 (3). Available from: https://epubs.icar.org.in/index.php/IJF/article/view/143304. doi:10.21077/ijf.2024.71.3.143304-14.

Peñarubia, O., Toppe, J., Ahern, M., Ward, A., Griffin, M., 2023. How value addition by utilization of tilapia processing by-products can improve human nutrition and livelihood. Rev. Aquac. 1753-5123, 1753-5131, 15 (S1), 32–40. Available from: https://onlinelibrary.wiley.com/doi/10.1111/raq.12737. doi:10.1111/raq.12737.

Peng, S., Chen, C., Shi, Z., Wang, L., 2013. Amino acid and fatty acid composition of the muscle tissue of yellowfin tuna (*Thunnus albacares*) and bigeye tuna (*Thunnus obesus*). J. Food Nutr. Res. 1 (4), 42–45. doi:10.12691/jfnr-1-4-2.

Rincón-Cervera, M.A., González-Barriga, V., Valenzuela, R., López-Arana, S., Romero, J., Valenzuela, A., 2019. Profile and distribution of fatty acids in edible parts of commonly consumed marine fishes in Chile. Food Chem. 0308-8146. 274, 123–129. Available from: https://linkinghub.elsevier.com/retrieve/pii/S0308814618315255. doi:10.1016/j.foodchem.2018.08.113.

Rodríguez, A., Trigo, M., Aubourg, S.P., Medina, I., 2021. Optimisation of healthy-lipid content and oxidative stability during oil extraction from squid (Illex argentinus) viscera by green processing. Mar. Drugs 19 (11), 616. doi:10.3390/md19110616.

Rotter, A., Barbier, M., Bertoni, F., Bones, A.M., Cancela, M.L., Carlsson, J., Carvalho, M.F., Cegłowska, M., Chirivella-Martorell, J., Dalay, M.C., Cueto, M., Dailianis, T., Deniz, I., Díaz-Marrero, A.R., Drakulovic, D., Dubnika, A., Edwards, C., Einarsson, H., Erdoğan, A., Eroldoğan, O.T., Ezra, D., Fazi, S., FitzGerald, R.J., Gargan, L.M., Gaudêncio, S.P., Udovič, M.G., Ivošević Denardis, N., Jónsdóttir, R., Kataržytė, M., Klun, K., Kotta, J., Ktari, L., Ljubešić, Z., Lukić Bilela, L., Mandalakis, M., Massa-Gallucci, A., Matijošytė, I., Mazur-Marzec, H., Mehiri, M., Laurentius Nielsen, S., Novoveská, L., Overlingė, D., Perale, G., Ramasamy, P., Rebours, C., Reinsch, T., Reyes, F., Rinkevich, B., Robbens, J., Röttinger, E., Rudovica, V., Sabotič, J., Safarik, I., Talve, S., Tasdemir, D., Schneider, X.T., Thomas, O.P., Toruńska-Sitarz, A., Varese, G.C., Vasquez, M.I., 2021. The essentials of marine biotechnology. Front. Mar. Sci. 2296-7745. 8, 629629. Available from: https://www.frontiersin.org/articles/10.3389/fmars.2021.629629/full. doi:10.3389/fmars.2021.629629.

Roy, V.C., Getachew, A.T., Cho, Y.-J., Park, J.-S., Chun, B.-S., 2020. Recovery and bio-potentialities of astaxanthin-rich oil from shrimp (Penaeus monodon) waste and mackerel (Scomberomous niphonius) skin using concurrent supercritical CO_2 extraction. J. Supercrit. Fluids 159, 104773. doi:10.1016/j.supflu.2020.104773.

Sanjeewa, K.K.A., Jayawardena, T.U., Herath, K.H.I.N.M., Jeon, Y.-J., 2024. The role of seaweeds in blue bioeconomy. The Role of Seaweeds in Blue Bioeconomy 1–353. doi:10.2174/97898152236441240101.

Senadheera, T.R.L., Hossain, A., Shahidi, F., 2023. Marine bioactives and their application in the food industry: a review. Appl. Sci. 2076-3417. 13 (21), 12088. Available from: https://www.mdpi.com/2076-3417/13/21/12088. doi:10.3390/app132112088.

Shahidi, F., Naczk, M., Pegg, R.B., Synowiecki, J., 1991. Chemical composition and nutritional value of processing discards of cod (Gadus morhua). Food Chem. 42 (2), 145–151. doi:10.1016/0308-8146(91)90030-R.

Snipe, R.M.J., Brelis, B., Kappas, C., Young, J.K., Eishold, L., Chui, J.M., Vatvani, M.D., Nigro, G.M.D., Hamilton, D.L., Convit, L., Carr, A., Condo, D., 2024. Omega-3 long chain polyunsaturated fatty acids as a potential treatment for reducing dysmenorrhoea pain: systematic literature review and meta-analysis. Nutr. Diet. 1446-6368, 1747-0080, 81 (1), 94–106. Available from: https://onlinelibrary.wiley.com/doi/10.1111/1747-0080.12835. doi:10.1111/1747-0080.12835.

Sun, H., Yang, S., Zhao, W., Kong, Q., Zhu, C., Fu, X., et al., 2023. Fucoxanthin from marine microalgae: A promising bioactive compound for industrial production and food application. Crit. Rev. Food Sci. Nutr. 63 (26), 7996–8012. doi:10.1080/10408398.2022.2054932.

Tacon, A.G.J., Levy, J., Coelho, R., Machado, T.M., Neiva, C.R.P., Lemos, D.E.L., 2025. Fish for health: role of fish in global food and animal protein supply. Rev. Fish. Sci. Aquac. 33 (2), 319–333. doi:10.1080/23308249.2024.2420917.

Tambat, V.S., Singhania, R.R., Patel, A.K., Chen, C.-W., Michaud, P., Dong, C.-D., 2025. Advancing sustainable astaxanthin-lipid biorefineries: robust two-stage phytohormone-driven bioprocess in Chromochloris zofingiensis. Bioresour. Technol. Rep. 29. doi:10.1016/j.biteb.2025.102022

Teng, X.-N., Wang, S.-C., Zeb, L., Dong, Y.-S., Xiu, Z.-L., 2023. Two-step enzymolysis of antarctic krill for simultaneous preparation of value-added oil and enzymolysate. Mar. Drugs 1660-3397. 21 (1), 47. Available from: https://www.mdpi.com/1660-3397/21/1/47. doi:10.3390/md21010047.

Thiao, D., Bunting, S.W., 2022. Socio-economic and Biological Impacts of the Fish-based Feed Industry for sub-Saharan Africa, Vol. 1236. FAO; WorldFish; University of Greenwich, Rome, Italy, pp. 120. Available from: https://www.fao.org/fishery/en/publication/280341.

Tucciarone, I., Secci, G., Contiero, B., Parisi, G., 2024. Sustainable aquaculture over the last 30 years: an analysis of the scientific literature by the Text Mining approach. Rev. Aquac. 16 (4), 2064–2076. doi:10.1111/raq.12950.

Udaypal, U., Goswami, R.K., Verma, P., 2025. Transforming dairy effluent into valuable resources: harnessing microalgae for sustainable production of nutraceuticals and pharmaceuticals. Process Biochem. 150, 342–356. doi:10.1016/j.procbio.2025.01.026.

Uthaiah, N.M., Venkataramareddy, S.R., Sheikh, A.Y., 2025. Oleaginous microalgae Nannochloropsis sp. as a potential source of EPA and bioactive metabolites: bottlenecks in the downstream processing and unravelling the challenges in its food, feed and therapeutic application. Algal Res. 2211-9264. 86, 103897. Available from: https://linkinghub.elsevier.com/retrieve/pii/S2211926425000062. doi:10.1016/j.algal.2025.103897.

Vladic, J., Radman, S., Zizak, Z., Besu, I., Jerkovic, I., Speranza, L.G., Hala, A.F., Kovacevic, S., Perreira, H., Gouveia, L., 2025. Clean production of microalgae high-value lipid fraction: influence of different pretreatments on chemical and cytotoxic profiles of Chlorella vulgaris supercritical extracts and life cycle assessment. J. Clean. Prod. 491. doi:10.1016/j.jclepro.2025.144823

Wang, C., Guo, M., 2025. Lipids and Their Impacts on Health Functional Foods. Elsevier, pp. 149–183. Available from: https://linkinghub.elsevier.com/retrieve/pii/B9780443191008000051.

Wang, W., Yang, H., Johnson, D., Gensler, C., Decker, E., Zhang, G., 2017. Chemistry and biology of ω-3 PUFA peroxidation-derived compounds. Prostaglandins Other Lipid Mediat. 1098-8823. 132, 84–91. Available from: https://linkinghub.elsevier.com/retrieve/pii/S1098882316301678. doi:10.1016/j.prostaglandins.2016.12.004.

Wang, Y., Rehman, A., Jafari, S.M., Shehzad, Q., Yu, L., Yijia, S., Gangcheng, W., Jin, Q., Zhang, H., Suleria, H.A. R., Wang, X., 2024. Micro/nano-encapsulation of marine dietary oils: a review on biomacromolecule-based delivery systems and their role in preventing cardiovascular diseases. Int. J. Biol. Macromol. 0141-8130. 261, 129820. Available from: https://linkinghub.elsevier.com/retrieve/pii/S0141813024006238. doi:10.1016/j.ijbiomac.2024.129820.

Wei, W., Cheng, Z., Wang, X., Chang, M., Xie, D., Sun, C., Guo, X., Bin, L., 2025. The potential of marine oils as functional foods: nutritional composition, health benefits, applications, and future perspectives. Food Rev. Int. doi:10.1080/87559129.2025.2528083.

Windarsih, A., Irnawati, S., Warmiko, H.D., Alam, L.P.M., Utami, I.D., Rohman, A., Indrianingsih, A.W., 2024. Lipidomics Analysis of Different Marine Fish Oils Using Untargeted Liquid Chromatography–Orbitrap High-Resolution Mass Spectrometry and Chemometrics. Chromatographia 87 (4), 203–214. doi:10.1007/s10337-024-04312-4.

Wu, H., Forghani, B., Abdollahi, M., Undeland, I., 2022. Five cuts from herring (*Clupea harengus*): comparison of nutritional and chemical composition between co-product fractions and fillets. Food Chem. X 2590-1575 16, 100488. Available from: https://linkinghub.elsevier.com/retrieve/pii/S2590157522002863. doi:10.1016/j.fochx.2022.100488.

Xavier, K.A.M., Chattopadhyay, K., 2023. Bioactive and Functional Nutraceuticals from Shrimp Processing Wastes Advances in Fish Processing Technologies, Apple Academic Press.

Xia, F.L.W., Supri, S., Djamaludin, H., Nurdiani, R., Seng, L.L., Yin, K.W., Rovina, K., 2024. Turning waste into value: extraction and effective valorization strategies of seafood by-products. Waste Manag. Bull. 2949-7507. 2 (3), 84–100. Available from: https://linkinghub.elsevier.com/retrieve/pii/S2949750724000579. doi:10.1016/j.wmb.2024.06.008.

Xin, Y., Shan, W., Miao, C., Tao, X., Yandu, L., 2024. Towards lipid from microalgae: products, biosynthesis, and genetic engineering. Life 2075-1729. 14 (4), 447. Available from: https://www.mdpi.com/2075-1729/14/4/447. doi:10.3390/life14040447.

Xue, J., Huixiang, W., Lijun, G., Weibo, L., Wang, H., Mao, P., Liao, J., Zeng, X., Wang, S., Jiang, L., Liang, J., Huang, J., Wang, Z., Shen, Q., 2025. Towards a safer food chain: recent advances in multi-technology based lipidomics application to food quality and safety. Trends Food Sci. Technol. 156. doi:10.1016/j.tifs.2024.104859

Ying, X., Xinyang, L., Lukai, M., Wang, F., Xiao, G., Benjakul, S., Liu, D., 2024. Lipidomics-based study of lipid quality of fish by-products: a case study of sea bass (Lateolabrax japonicus). Int. J. Food Sci. Technol. 59 (5), 3097–3105. doi:10.1111/ijfs.17051.

Yordi, E.G., Pérez Martínez, A., Radice, M., Scalvenzi, L., Abreu-Naranjo, R., Uriarte, E., Santana, L., Matos, M.J., 2024. Seaweeds as source of bioactive pigments with neuroprotective and/or anti-neurodegenerative activities: astaxanthin and fucoxanthin. Mar. Drugs 1660-3397. 22 (7), 327. Available from: https://www.mdpi.com/1660-3397/22/7/327. doi:10.3390/md22070327.

Zhang, Z.-X., Ying-Shuang, X., Zi-Jia, L., Lu-Wei, X., Wang, M., Ying-Feng, L., Guo, D.-S., Sun, X.-M., Huang, H., 2024. Turning waste into treasure: a new direction for low-cost production of lipid chemicals from Thraustochytrids. Biotechnol. Adv. 0734-9750. 73, 108354. Available from: https://linkinghub.elsevier.com/retrieve/pii/S073497502400048X. doi:10.1016/j.biotechadv.2024.108354.

CHAPTER

7

Proteins and amino acids isolated from marine organisms

D.M.C.B. Dissanayake[a], *Y.T. Senaweera*[b], *and P. Wijethunga*[a]

[a]Department of Biological Sciences, Faculty of Applied Sciences, Rajarata University of Sri Lanka, Mihintale, Sri Lanka

[b]Faculty of Graduate Studies, General Sir John Kotelawala Defense University, Rathmalana, Sri Lanka

7.1 Introduction

The marine ecosystem, which covers more than 70% of the Earth's surface, harbors an extraordinary diversity of life forms, including microorganisms, algae, invertebrates, and vertebrates. Many of these produce novel and unique bioactive compounds, which remain largely untapped but have immense potential for pharmaceutical, nutraceutical, and biotechnological applications. From Ecteinascidin-743, the first marine-derived anticancer drug extracted from the *Caribbean tunicate* (Molinski et al., 2009), various forms of drug candidates, including antibiotics (Subramani and Aalbersberg, 2012) and omega-3 fatty acids (Suleria et al., 2016), have been developed over the past few decades. Beyond medicine, marine bioactive compounds have applications in cosmetics, agriculture, and food preservation (Im et al., 2022; Gunasekara et al., 2025). Therefore marine organisms represent a vast and largely untapped resource for novel bioactive compounds, and ongoing research and innovation in marine biotechnology will be crucial to unlocking their full potential for human health and industrial applications.

7.2 Diversity of marine organisms

The marine ecosystem is a rich reservoir of biodiversity, comprising over 80% of lifeforms on Earth and serving as a crucial source of proteins and amino acids essential for numerous biological processes. It hosts a vast array of organisms, ranging from microscopic bacteria and fungi to complex vertebrates such as fish and sea mammals, which play a huge role in the

Harvesting the Sea
https://doi.org/10.1016/B978-0-443-29190-6.00012-X

production of biologically active compounds such as polysaccharides, peptides, and lipids. These bioactive compounds exhibit diverse health benefits, significantly contributing to functional food and pharmaceutical applications (Walquist et al., 2024; Zheng et al., 2025). This plays a crucial role in both ecological sustenance and human nutrition, establishing a vital link between marine systems and terrestrial life (Kim, 2013). Thus understanding marine organisms' capabilities further emphasizes their importance in global protein production.

7.2.1 Marine microorganism

The diversity of marine organisms, particularly bacteria, fungi, and microalgae, plays a significant role in the ecosystem. Some of these species live in various habitats, such as stressful habitats with cold, lightless, and highpressure conditions, and produce fascinating and structurally complex natural products. The water column of the oceans contains approximately 10^6 bacterial cells per milliliter (Hagström et al., 2002). In contrast to terrestrial macroorganisms, these microorganisms represent promising natural product sources. They have the capability to produce substantial amounts of secondary metabolites in a sustainable and costeffective manner through large-scale cultivation and fermentation (Debbab et al., 2010). In addition, marine bacteria and fungi have been identified as sources of novel proteins and amino acids, which could be vital for developing new pharmaceuticals and nutraceuticals. Furthermore, microalgae are essential for nutrient assimilation and participate in the nitrogen cycle, converting nitrogen into usable forms for other organisms, which is crucial for sustaining marine productivity (Fernando et al., 2023).

The role of marine bacteria and fungi in protein synthesis and nutrient cycling is vital for maintaining the health and functionality of ecosystems. Furthermore, they also play a key role in decomposing organic matter, which releases essential nutrients back into the marine environment. This includes the conversion of nitrogenous waste into usable forms through nitrification and denitrification by bacteria, as well as the decaying process by fungi. Both processes increase nutrient availability within marine food webs, contributing to a higher trophic level in the marine ecosystem. Recent studies highlight the intricate relationships within microbial communities, where phytoplankton and associated bacteria collaborate in metabolic processes, significantly impacting nutrient regeneration and protein production (Peng et al., 2024) (Table 7.1).

Microalgae are microscopic organisms rich in high protein content, constituting up to 70% of their dry weight in certain species, such as *Arthrospira platensis* and *Chlorella vulgaris*. This protein richness is complemented by a comprehensive amino acid profile, encompassing all essential amino acids for human nutrition (Lucakova et al., 2022). The levels of amino acids in microalgae, such as isoleucine, valine, lysine, tryptophan, methionine, threonine, and histidine, are similar to or even higher than those found in protein-rich sources, such as eggs and soy (Lima et al., 2025), and these values align closely with the amino acid profiles recommended by the Food and Agriculture Organization (FAO) and the World Health Organization (WHO) for human consumption (García-Encinas et al., 2025; *Protein and Amino Acid Requirements in Human Nutrition : Report of a Joint WHO/FAO/UNU Expert Consultation*, 2007). In addition, microalgae proteins are abundant in nonessential amino acids such as aspartic acid and glutamic acid, which collectively represent 8%–12% of the total amino acids (García-Encinas et al., 2025). However, the digestibility of proteins can be influenced by the robustness

TABLE 7.1 Protein/Amino Acid Production by Marine Microorganisms (This table summarizes the bioactive compounds derived from various marine bacteria and fungi, their applications in medicine and biotechnology, and the types of proteins and amino acids they produce, highlighting the significant biochemical contributions of marine microorganisms to health and industry

Microorganism	Protein/Amino Acids Produced	Uses
***Actinomycetes* (CNQ-418)**	Various enzymes including proteases	Antimicrobial against methicillin-resistant (Wang et al., 2020)
***Brevibacillus laterosporus* PNG276**	Novel lipopeptide named Tauraamide was identified. It contains two **D amino acids** (amino acids with a specific stereochemistry) and is acylated at the N-terminus of the peptide chain	less effective against multidrug-resistant (Desjardine et al., 2007)
***Cladosporium* sp. F14**	Polypeptides with antimicrobial properties	Protection against biofouling, antibacterial activity against certain bacterial strains, such as *Loktanella hongkongensis* and *Micrococcus luteus* (Qi et al., 2009)
***Cosmospora sp.* SF-5060**	Produces various secondary metabolites, including proteins like Cyclic tetrapeptides	Treatment for Type 2 diabetes and obesity (Seo et al., 2009)
Sponge-associated fungi *Petriella* sp	Numerous proteins with potential bioactivities	Antitumor activity (Proksch, 2008)
Micromonospora	Possible production of enzymes and peptides	Antitumor activity (Hirsch and Valdés, 2010)

of the microalgal cell walls; species with thicker cell walls may require processing techniques such as mechanical disruption to enhance protein bioavailability (Mosibo et al., 2024). Despite these considerations, the balanced amino acid composition and substantial protein content position microalgae as a promising, sustainable alternative to traditional animal and plant-based protein sources, with potential applications in the food, feed, and nutraceutical industries (García-Encinas et al., 2025).

7.2.2 Marine invertebrates

Marine invertebrates also represent a diverse and vital component of ocean ecosystems, showcasing a wide range of species that produce valuable proteins and amino acids. From sponges to mollusks, these organisms are essential for understanding both marine biology and the possible health advantages of their distinct molecular components. Marine invertebrates can be classified into several key taxonomic groups, each exhibiting unique characteristics and capabilities. Porifera (sponges) are known to produce various bioactive compounds, including collagen-like proteins that have applications in tissue engineering (Ganesan et al., 2020) and bioactive peptides with promising anticancer effects (Calcabrini et al., 2017). Cnidaria encompasses organisms such as corals and jellyfish that produce peptides with notable antioxidant and antiinflammatory properties (Sudhakar and Nazeer, 2015). Similarly, the protein hydrolysate from *Todarodes pacificus* demonstrates antiinflammatory properties, acting on tumor necrosis factor, showcasing the therapeutic potential of marinederived compounds (Zhang et al., 2016). *Annelida* (marine worms) are recognized for their unique proteins that can facilitate wound healing and possess immunomodulatory effects (Ganesan et al., 2020). Mollusca, which include clams, squids, and oysters, are another rich source of proteins, particularly with antioxidant and antimicrobial activities, thereby making them significant for both culinary and medicinal uses. Finally, *Echinodermata* also produce unique peptides that have shown potential applications in cancer treatment and tissue regeneration (Ganesan et al., 2020). The following figure highlights proteins and peptides produced and their diverse biological activities.

7.2.3 Marine vertebrates

The marine ecosystem contains a remarkable variety of vertebrates, which provide high-quality proteins and essential amino acids necessary for human health, making them a valuable part of a balanced diet (Tan et al., 2024). This group includes a diverse range of organisms, such as fish, marine mammals, sea turtles, sharks, rays, and lampreys, all of which are known for their high protein content and essential amino acids that are vital for human health. Fish, for example, typically contain between 16% and 21% proteins, along with eight essential amino acids such as lysine, methionine, and cysteine (Khalil et al., 2019), and serve as a primary source of nutrition for billions of people worldwide. They have myriad applications, ranging from fresh fillets and canned products to processed items such as fish oil and protein supplements (Bank et al., 2022). Fish proteins offer excellent functional attributes such as foaming capacity (Zamorano-Apodaca et al., 2020), emulsification, gelation, and stabilization (He et al., 2022). Besides fish and mammals, other marine vertebrates, such as sharks and rays, typically contain around 20% protein and contribute to human nutrition in various capacities (Lima et al., 2025; Bank et al., 2022). Tables 7.2 and 7.3 summarizes the protein content and uses of major marine vertebrates.

TABLE 7.2 NutritionalNutritional overview and culinary uses of marine vertebrates (This table summarizes the protein content, nutritional benefits, and culinary applications of various marine vertebrates, highlighting their importance in global food systems and the need for sustainable harvesting practices).

Marine Vertebrates	Protein Content (Approximate)	Uses	References
Fish	16% - 21% of body weight	Food source (fillets, canned, dried), protein supplements, fish oil, culinary applications	(The State of World Fisheries and Aquaculture 2020)
Marine Mammals	Varies (e.g., seals ~26.7% protein)	Traditional food source for Indigenous populations, oil production, medicinal uses	(Siddiqui et al., 2024)
Sea Turtles	Varies (14.78 %–42.08 % protein)	Traditional foods in some cultures, research uses	(Ren et al., 2026)
Sharks	~3.64% protein	Culinary uses (some cultures), production of cartilage supplements	(Agustin et al., 2025)
Rays	~20% protein	Food source in some regions, used for traditional dishes	(Bilgin, 2025)
Lampreys	~15% - 20% protein	Culinary uses in specific cultures, fishing bait	(McCauley et al., 2015)

TABLE 7.3 Summary of Marine-Derived Proteins and Amino Acids: Extraction, Purification, and Characterization Methods.

Category	Method/technique	Key features	Advantages	Limitations	References
Extraction methods	**Traditional**				
	Solid-liquid extraction (SLE)	Uses solvents like hexane, methanol, water	Simple Well-established methodology	Solvent dependency Environmental concerns	((Juel and Juul, 2024) (Karabulut et al., 2024)
	Advanced				
	Enzyme-assisted extraction (EAE)	Use specific enzymes to break cell walls	Higher yield Maintains protein integrity	Enzyme cost process optimization is needed	(Penggalih et al., 2023)
	Microwave-assisted extraction	Uses microwave energy for heating	Reduced extraction time Increased yield	Equipment cost Heat sensitivity	(Karabulut et al., 2024)
	Ultrasound-assisted extraction	Uses sound waves to disrupt cells	Improved extraction efficiency	Equipment cost Scale-up challenges	(Panchout et al., 2013)
	Pulsed electric field (PEF)	Uses electric pulses to disrupt cell membranes	Selective extraction Rapid processing	Technical complexity Energy consumption	(Musaimi and Jaradat, 2024)
	Supercritical fluid extraction	Uses CO_2 under specific pressure/temperature	Environmentally friendly High selectivity	High pressure requirement Operating costs	(Wang et al., 2017b)
Purification techniques	Membrane filtration	Uses pressure to separate molecules by size	High throughput Comparable to chromatography	Membrane fouling Pore size limitations	(Nunn and Keil, 2006)
Purification techniques	High-performance liquid chromatography	Separates compounds underhigh pressure	High resolution Analytical capabilities	High cost Technical expertise required	(Moore et al., 2014)
	Ion exchange chromatography	Separates based on charge	Effective for protein isolation Good resolution	Buffer requirement Sample preparation	(Moore et al., 2014)

Continued

TABLE 7.3 Summary of Marine-Derived Proteins and Amino Acids: Extraction, Purification, and Characterization Methods—cont'd

Category	Method/technique	Key features	Advantages	Limitations	References
	Electrophoresis (two-dimensional gel electrophoresis)	Separates proteins by isoelectric point and mass	High resolution Good for proteomics	Low throughput Labor intensive	Aldairi, 2020, Barzkar et al., 2023,
Structural characterization	X-ray crystallography	Determines 3D structure through X-ray diffraction	Atomic resolution Detailed structure	Requires crystallization Static structure	(Quitério et al., 2022)
	NMR spectroscopy	Analyzes protein structure in solution	Dynamic information No crystals needed	Size limitation Complex data analysis	(Quitério et al., 2022)
	Circular dichroism (CD)	Analyzes secondary structure	Rapid analysis Conformational changes	Lower resolution Limited information	(Bolje and Gobec, 2021)
	FTIR spectroscopy	Identify functional groups	Fast analysis Minimal sample preparation	Limited structural details	(Bolje and Gobec, 2021)

7.2.4 Seaweed and microalgae

Marine seaweed and macroalgae represent a diverse group of organisms that play a crucial role in marine ecosystems and human nutrition. Based on their pigmentation, they can be broadly classified into three main groups: red algae (Rhodophyta), green algae (Chlorophyta), and brown algae (Phaeophyceae) (Dissanayake et al., 2024). Each group exhibits unique characteristics and contributes to a wide range of bioactive compounds, including proteins and amino acids, which are essential for human health (Bikker et al., 2020).

7.2.4.1 Diversity of marine microalgae

Red algae (Rhodophyta): This group includes the *Porphyra*, *Gracilaria*, and *Chondrus* species. Red algae are known for their high protein content and nutritional value. Research has shown that red algae often provide complete profiles of essential amino acids crucial for human dietary needs. For example, species of Porphyra have been found to contain significant amounts of methionine and tryptophan, which are considered limiting amino acids in various seaweeds. This highlights the importance of dietary complementarity with other protein sources (Machado et al., 2020).

Green algae (Chlorophyta): Green algae species, such as *Ulva* sp., are recognized for their high protein content, which can range from 70% to 86% in algal protein concentrates. The proteins extracted from green algae are readily digestible, with studies showing that they can achieve proteolysis rates of at least 89% during simulated gastrointestinal digestion (Kazir et al., 2019). The protein quality in green algae can vary widely, but, similar to red algae, they also provide a substantial amount of essential amino acids. Studies suggest that green seaweed can contribute nearly 50% of essential amino acids from their total amino acid composition, making them a compelling protein source for both humans and livestock (Machado et al., 2020).

Brown algae (Phaeophyceae): This group includes economically significant species like *Laminaria* and *Fucus*. Brown algae offer a rich protein content and accumulate vitamins and minerals, making them nutritionally important. Studies indicate that certain species of brown algae outshine terrestrial protein sources in terms of protein levels, which can benefit dietary intake when integrated into food systems (Machado et al., 2020). Some of the known brown algae, like *Ascophyllum nodosum* and *Saccharina latissima,* are known for their high protein levels, ranging from 9.8% to 14.9%, enabling potential applications in food industries (Bikker et al., 2020). Marine organisms thus serve as an underutilized source of protein and nutrition due to their rich amino acid profiles and eco-friendly cultivation potential, marking them as a valuable addition to global food security strategies.

7.3 Unique characteristics of marine derived protein and amino acid

Marine organisms have developed complex metabolic pathways to survive in various conditions, including extreme and highly competitive marine environments, resulting in the production of unique bioactive compounds, including proteins, peptides, and amino acids (Harnedy and Fitzgerald, 2011). These biomolecules often display enhanced stability, resistance to degradation, and specialized functionality, making them highly sought after in the development of novel pharmaceuticals, nutraceuticals, and biotechnological industries (Kim, 2013). In addition, these marinederived proteins and amino acids exhibit various biological activities, including anticancer, antiproliferative, anticoagulant, antibacterial, antifungal, and antitumor properties (Pangestuti et al., 2018; Harnedy and Fitzgerald, 2011). However, further research and exploration in this field are needed to fully understand the unique properties of these proteins and amino acids (Pangestuti et al., 2018) (Figs. 7.1–7.3).

7.4 Potential applications: Pharmaceuticals and nutraceuticals

Marine proteins and peptides derived from various aquatic organisms are unique in their structural and functional characteristics, enabling potential applications in drug development, nutraceuticals, and biotechnology. As pharmaceutics, marine-derived proteins and peptides have demonstrated remarkable potential in addressing various health challenges. For instance, Plitidepsin, a compound isolated from the marine tunicate *Aplidium albicans*, has shown potent anticancer activity by inhibiting protein synthesis and inducing apoptosis in cancer cells, particularly in multiple myeloma (Leisch et al., 2019). Similarly, didemnin B, another tunicate-derived peptide, has exhibited strong antitumor properties (Suarez-Jimenez et al., 2012).

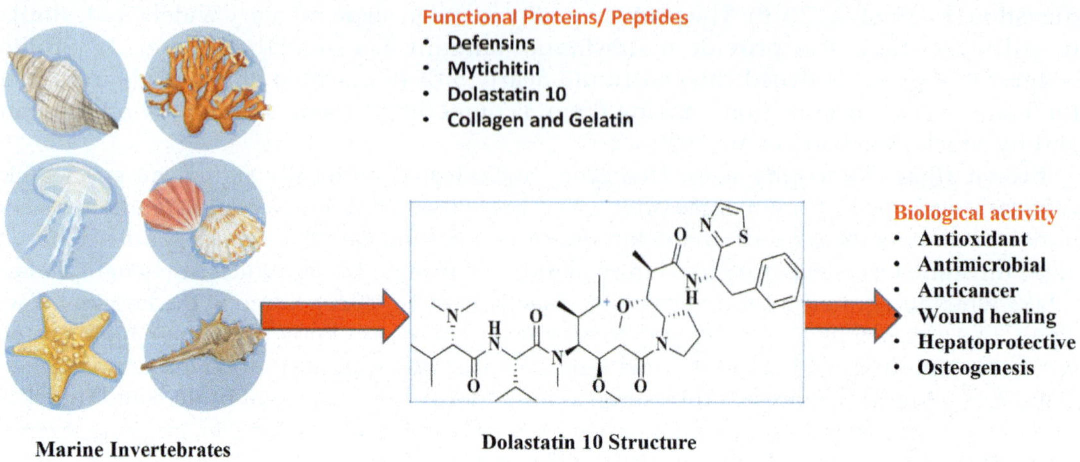

FIG. 7.1 Functional protein and peptides from marine invertebrate(Ganesan et al., 2020)

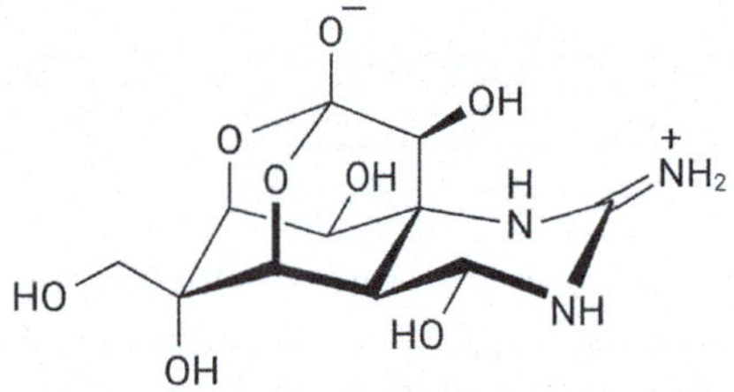

FIG. 7.2 A Structure of an aminoglycoside antibiotic with a six dash membered ring and multiple hydroxyl groups.

Marine-derived antimicrobial agents, such as antibiotics from marine actinomycetes, have proven effective against antibiotic-resistant bacteria (Subramani and Aalbersberg, 2012). In contrast, marine lectins and sulfated polysaccharides have demonstrated antiviral activity, offering potential treatments for viral infections (Hans et al., 2021). In addition, marine-derived compounds such as fish collagen peptides and fucoidan from brown seaweed have shown significant antiinflammatory and wound healing properties, making them valuable in regenerative medicine (Gómez-Guillén et al., 2011; Wijesinghe and Jeon, 2012).

In the nutraceutical sector, marine proteins and peptides have demonstrated considerable potential for improving overall health and preventing chronic diseases. Fish-derived peptides have exhibited antihypertensive effects, offering natural alternatives for managing blood pressure and cardiovascular health (Walquist et al., 2024). Marine collagen hydrolysates, extracted from fish and shellfish, have shown promising effects in alleviating osteoarthritis symptoms and enhancing joint function (Bourdon et al., 2021). In addition, marine protein hydrolysates have been found to regulate glucose metabolism, which could benefit individuals managing diabetes or seeking weight control solutions (Nasab et al., 2020). These findings emphasize the potential of marine-derived proteins and peptides in the pharmaceutical and nutraceutical industries, highlighting the necessity for further research to investigate their full spectrum of bioactivities and therapeutic uses.

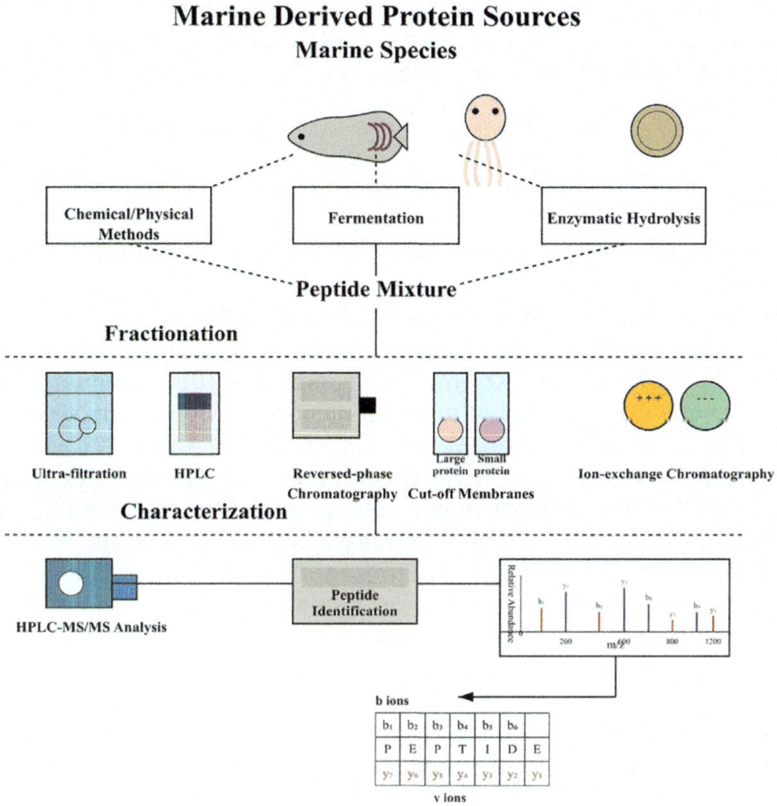

FIG. 7.3 A flowchart illustrating the process of obtaining and analyzing marine-derived protein sources.

7.5 Different types of proteins and amino acids

7.5.1 Structural proteins

Structural proteins form a skeleton or contribute to the mechanical properties of a living organism (LaPelusa and Kaushik, 2025). Collagen is regarded as one of the most useful biomaterials (Lee et al., 2001); however, extracting collagen from terrestrial animals is difficult, costly, and time-consuming (Rahman and Silva, 2022). Recently, marine organisms have emerged as a promising alternative source of collagen. For instance, fibrillar collagens isolated from jellyfish species (*Aurelia aurita, Cotylorhiza tuberculata, Pelagia noctiluca,* and *Rhizostoma pulmo*) exhibit similar biological effects to mammalian type I collagen (Addad et al., 2011). Collagen isolated from sole fish (*Aseraggodes umbratilis*) skin is also found to be similar to commercially available calf skin type I collagen (Arumugam et al., 2018). Furthermore, cartilaginous fish (*Mustelus mustelus*) collagen incorporated into chitosan-based biofilms exhibited potential UV barrier properties and antioxidant activity (Ben Slimane and Sadok, 2018). In addition, type I collagen extracted from snakehead scales has been used to create collagen patches that promote the growth of blood vessels and lymphatic vessels, suggesting that collagen isolated from marine organisms is a better alternative to existing collagens isolated from terrestrial animals.

7.5.2 Enzymes

To thrive in complex marine environments, marine organisms have evolved diverse enzyme systems that serve as valuable sources of novel biocatalysts (Zhang and Kim, 2010). Among these, proteases from marine bacteria demonstrate remarkable adaptability. For instance, enzymes isolated from *Bacillus subtilis* from marine sediments exhibit high proteolytic activity and tolerate extreme salt conditions (El-Hassayeb and Aziz, 2016). Cold-adapted enzymes are another hallmark of marine microbes. A lipase-producing bacterium from the Bay of Bengal retains 90% activity at 10°C (Arora, 2013), while *Aspergillus awamori* thrives at 40°C and pH 7 (Basheer et al., 2011). The lipolytic potential of Mediterranean *Bacillus cereus* is particularly notable, achieving 94.7% oil/grease removal in wastewater (Hassan et al., 2018). The enzymes of *Oceanobacillus* sp. show antimicrobial promise, as demonstrated by biofilm disruption against food pathogens (Seghal Kiran et al., 2014). Furthermore, *Pyrococcus furiosus*, isolated from volcanic marine sediments, grows at 100°C and produces a DNA polymerase that surpasses Taq polymerase (Cox and King, 2013).

7.5.3 Venom- and toxin-derived proteins

Marine ecosystems harbor more venomous species than terrestrial environments (Xie et al., 2016). These marine organisms produce a remarkable array of secondary metabolites, primarily as chemical defense mechanisms (Tubaro et al., 2012). Most of these bioactive compounds originate from dinoflagellates, though diatoms and cyanobacteria also contribute to this chemical diversity. Notably, while most marine toxins are derived from microalgae, tetrodotoxins, a neurotoxin, represent an exception with bacterial origins. This rich source of natural products has attracted growing scientific interest in pharmaceutical development (Díaz-Gómez et al., 2024). Tetrodotoxin ingestion leads to a variety of side effects and, in some cases, death; these outcomes often depend on the time since ingestion and the dose administered (Harish and Erwin, 2023).

7.6 Health benefits of marine-derived proteins and amino acids

Marine-derived peptides have emerged as a highly promising class of bioactive molecules, attracting significant scientific interest due to their diverse therapeutic properties. These are defined as short polymers of fewer than 50 amino acids, with a molecular weight below 10 kDa (Diao and Meibohm, 2013).

7.6.1 Antihypertensive activity

Hypertension is a major global health challenge because of its high prevalence and resulting cardiovascular disease and chronic kidney disease, and it is regarded as the leading preventable risk factor for premature death and disability worldwide. Specific marine peptides, such as SpirPep1 isolated from *A. platensis* (spirulina), inhibit angiotensin-converting enzyme (ACE), suggesting the possibility of developing it as an antihypertensive medication (Jo et al., 2024). Similarly, the peptide EVMAGNLYPG extracted from blue mussel (*Mytilus edulis*) (Je et al., 2004) and the peptide AEYLCEAC isolated from oyster (*Crassostrea gigas*) (Chen et al., 2022) also showed notably higher ACE inhibitory activities.

7.6.2 Antimicrobial activity

From 1940 to 1990, the discovery of new antibiotics effectively countered emerging resistance. However, in recent years, this approach has become less effective as resistance has grown faster than the development of new antibiotics (Livermore et al., 2011). A range of bioactive peptides derived from aquatic organisms demonstrates potent antibacterial properties. The peptide KLVDASHRLATGDVAVRA isolated from *Spirulina platensis* showed inhibitory concentrations of 8 mg/mL against *Escherichia coli* and 16 mg/mL against *Staphylococcus aureus* (Sun et al., 2016). Similarly, the peptide KLENCNYAVELGK from *Limnospira maxima* produced inhibition zones of 27 mm for *E. coli* and 14 mm for *S. aureus* (R. et al., 2022). The peptide NNS5-6 (DGGGKM PTTVVY) isolated from *Paenibacillus thiaminolyticus* demonstrated minimum inhibitory concentration and minimum bactericidal concentration values of 4–8 μg/mL against *Pseudomonas aeruginosa* and *Klebsiella pneumoniae* (Sermkaew et al., 2024). In addition, peptides isolated from marine invertebrates have also shown remarkable antibacterial efficacy. *Arenicola marina* produces arenicin, which exerts a membranolytic mechanism against Gram-negative strains (Orlov et al., 2019), and AmBRI-44a, which is specifically active against a narrow spectrum of Gram-positive bacteria (Safronova et al., 2024).

Several marine peptides also exhibit antifungal effects. *Nostoc calcicola* produces Nostophycin, which displays antifungal activity against *C. albicans* and *A. niger* (Gupta and Vyas, 2021). Aurelin from *Aurelia aurita* belongs to the ShKT domain family, sharing similarities with defensins and K+ channel-blocking toxins, indicating its functional potential (Ovchinnikova et al., 2006). In crustaceans, scyreprocin from *Scylla paramamosain* displays broad-spectrum antifungal activity, including disruption of fungal membrane integrity and rapid fungicidal kinetics (Yang et al., 2020). Some peptides, such as JFP-2826 isolated from *Rhopilema esculentum*, exhibit activity against both bacteria and fungi.

7.6.3 Antioxidant activity

Living organisms have complex defense mechanisms to protect against oxidative damage by reducing reactive species through enzymatic and nonenzymatic antioxidants. However, an increase in oxidative stress leads to an increased demand for other antioxidant substances (Lorenzo et al., 2018). Marine-derived peptides have gained significant attention due to their potent antioxidant properties. Specifically, *C. vulgaris* exhibits radical scavenging activity when treated with pepsin and promod enzymes, with a 2,2-diphenyl-1-picrylhydrazyl radical scavenging activity (Yaghoubzadeh and Safari, 2025). Similarly, peptides such as EP3, EP4, EP5, and EP6 isolated from *Eucheuma cottonii*, showed strong scavenging activities and were found to significantly protect human umbilical vein endothelial cells from oxidative damage, enhancing the levels of antioxidant enzymes like superoxide dismutase and glutathione peroxidase (Sun et al., 2022). In addition, *S. platensis* peptides, fermented with microbial strains such as *Thermus thermophilus* and *Saccharomyces cerevisiae*, exhibited inhibitory activity against myeloperoxidase, highlighting their potential in reducing inflammation and oxidative stress (Wei et al., 2024).

7.6.4 Neuroprotective effects

Neurodegenerative disorders are predicted to overtake cancer as the second leading cause of death among the elderly population in the future. Therefore neuroprotective drugs, which are safe and effective, are in high demand. Due to the possible side effects of synthetic drugs, there is a significant demand for naturally derived compounds with neuroprotective effects (Pangestuti and Kim, 2011). The peptides, including InhVJ from *Heteractis crispa*, were tested in a neurotoxicity model on neuroblastoma Neuro-2a cells and displayed a neuroprotective effect by reducing oxidative stress, showing their potential for neurodegenerative diseases (Sintsova et al., 2021). Furthermore, peptides, including HCIQ2c1 and HCIQ4c7 from *Heteractis magnifica*, modulated the purinergic receptor P2X7 in Parkinson's disease models (Kvetkina et al., 2022). In addition, peptides such as MoVIA and MoVIB from *Conus moncuri* have shown analgesic properties, reversing pain behavior in a rat neuropathic pain model (Sousa et al., 2018) indicating the possibility of using them in pain management (Keizer et al., 2003).

7.6.5 Immunomodulatory effects

Peptides isolated from marine organisms have been found to possess immunomodulatory characteristics that boost the innate immune response, effectively eliminate infections, and directly target and destroy pathogens (Kang et al., 2019). These peptides show notable immunomodulatory effects through various mechanisms. Peptides HIAEEADRK and AEQAESDKK from tuna trimmings (*Thunnus albacares*) significantly enhanced phagocytosis and cytokine secretion in RAW264.7 macrophages by the upregulation of the toll-like receptor (TLR)-2/4-nuclear factor (NF)-κB signaling pathway (Cai et al., 2022). A marine oligopeptide from chum salmon (*Oncorhynchus keta*) increased lymphocyte proliferation, natural killer cell activity, $CD4^+$ T-cell population, and secretion of proinflammatory cytokines in a mouse model, indicating both Th1 and Th2 immune activation (Yang et al., 2011). In addition, the peptide Myticin C obtained from *Mytilus galloprovincialis* showed antiviral activity and attracted hemocytes due to its immune defense and cell signaling properties (Balseiro et al., 2011). Furthermore, Piscidin 3 and 4, extracted from *Oreochromis niloticus*, provoked inflammatory responses against *Vibrio vulnificus* infection (Lin et al., 2016).

7.7 Isolation and characterization of marine-derived proteins and amino acids

The isolation and characterization of proteins derived from marine sources primarily involve the purification of secondary metabolites (Cheung et al., 2015; Fernando et al., 2023; Kendler et al., 2024; Okeke et al., 2024). Extraction systems for separating proteins and amino acids have evolved from traditional solid-liquid distillation to advanced methods like microwave-assisted extraction, enhancing efficiency and sustainability (Karabulut et al., 2024; Penggalih et al., 2023; Ren et al., 2021). In addition, protein purification is a crucial process that involves techniques such as chromatography and membrane filtration. These methods are essential for achieving the high cleanliness standards required for thorough research and development, as well as for commercial applications (Feng et al., 2024; Nunn and Keil, 2006). Finally, the fundamental description of marine-derivative proteins is crucial for understanding their capabilities and potential applications. Techniques such as X-ray crystallography, nuclear magnetic resonance (NMR) spectroscopy, and diverse spectroscopic methods help reveal their three-dimensional structures, highlighting their roles in food technology and pharmaceuticals (Juel et al., 2024; Quitério et al., 2022). However, challenges persist in optimizing the extraction and purification process to ensure the purity and specific characteristics of the proteins, which are critical for their efficiency in various applications.

7.7.1 Extraction methods

Marine protein extraction systems can be categorized into traditional and advanced methods, each with specific characteristics. Common traditional techniques, like mechanical processing, synthetic solvent extraction, and heat treatments, offer simplicity and cost efficiency but generally yield lower protein quantities along with reduced bioactivity and selectivity (Karabulut et al., 2024). Advanced techniques, such as enzymatic hydrolysis, ultrasound-assisted extraction, supercritical fluid distillation, pulsed energetic field technology, and microwave-assisted extraction, provide higher yields and better preservation of bioactive properties, though they demand more specific equipment and expertise. The choice of extraction method mainly depends on the marine source, manufacturing scale, and desired protein characteristics and application (Juel and Juul, 2024).

7.7.2 Traditional extraction methods

Solid-liquid extraction (SLE) is the primary method for extracting proteins from marine sources, using a solvent containing hexane, methanol, chloroform, ethanol, acetone, and water to dissolve target proteins (Karabulut et al., 2024). Although SLE is widely used due to its effectiveness, it requires the deliberate optimization of factors such as heat, pressure, extraction time, solvent-to-solid ratio, and pH to maximize protein yield and quality. The double screw press extraction is another primary extraction method, which applies both condensation and shearing forces together to physically rupture cells, contributing to higher yields on an industrial scale. Importantly, these conventional methods continuously evolve for improved efficiency, solvent recovery, and reduced environmental impact in commercial marine protein processing (Farooq et al., 2024).

7.7.3 Advanced extraction methods

Advanced marine protein extraction technologies have evolved beyond conventional systems with supercritical fluid extraction. These technologies offer environmentally friendly approaches with greater solvating efficiencies and vapor diffusion properties, effectively infiltrating marine molds without leaving toxic residues. However, their use is limited by high equipment costs and the complexity of polar protein compounds, despite exceptional results with nonpolar elements (Musaimi and Jaradat, 2024; Careri et al., 2001; Macías-Sánchez et al., 2008; Mendes et al., 2003). Pulsed electric field extraction is another advanced method in which short, extreme-voltage, energetic vibrations are applied to discharge intracellular proteins outside. This helps conserve thermally delicate bioactive peptides and enzymes and protects their fundamental purity, preserving their inherent functionality (Wang et al., 2017b; Zbinden et al., 2013). Ultrasound-assisted extraction is also a versatile and unique method that applies localized extreme temperatures, pressures, and liquid jets to disrupt cell walls. This method offers several advantages, including a significant reduction in processing time and solvent consumption, while simultaneously enhancing extraction yields (Grosso et al., 2015; Juel and Juul, 2024; Panchout et al., 2013).

7.7.4 Purification techniques

Marine protein purification employs advanced methods for optimal purity and yield. Membrane filtration, especially ultrafiltration, effectively collects protein hydrolysates, comparable to traditional chromatography methods (Aldairi, 2020; Wang et al., 2017b). Various chromatographic techniques, including continuous and periodic counter-current chromatography, are commonly used for purifying medically valuable proteins and monoclonal antibodies (Feng et al., 2024; Moore et al., 2014). High-performance liquid chromatography is a key method for isolating and purifying bioactive peptides (Barzkar et al., 2023; Wang et al., 2017b), often used together with mass spectrometry to enhance analysis. Ion exchange chromatography is another method that has also become progressively essential for protein separation, detection, and constitutional determination from marine sources. Modular bed-supported cassette chromatography also allows flexibility with interchangeable cassettes for media, improving mass transfer and flow dispersion for affinity, ion exchange, and hydrophobic chromatography. This supports faster optimization and scaling of purification methods. In addition, integrating these chromatographic procedures with two-dimensional gel electrophoresis serves as a sophisticated purification tool kit for marine-derived proteins and amino acids (Aldairi, 2020).

7.7.5 Structural characterization

The structural characterization of marine-derived proteins is crucial for understanding their hereditary characteristics and potential applications (Quitério et al., 2022). Advanced analytical technologies, such as X-ray crystallography, NMR spectroscopy, and electron microscopy, are used to elucidate their molecular structure and functionality (Quitério et al., 2022; Suyama et al., 2011). NMR spectroscopy is particularly important for discovering the native structure of proteins, contributing crucial insights into structural dynamics and conformational changes under various physical surroundings. Specialized NMR methods, such as multiplicity-altered heteronuclear distinct quantum equating patterns, are useful for understanding protein behavior under varying environments (Quitério et al., 2022; Vemulapalli et al., 2023). In addition, other spectroscopic methods, including circular dichroism (CD) and Fourier transform infrared (FTIR) spectroscopy, provide crucial insights into protein derivatives. CD spectroscopy reveals complete protein folding arrangements and conformational changes, while FTIR identifies functional groups and molecular interactions within the protein structure through characteristic absorption patterns in the infrared spectrum. Altogether, these techniques create a comprehensive toolkit for analyzing the complex structural features of marine-derived proteins, linking their biological activities and possible biotechnological uses (Derkach et al., 2022).

7.8 Challenges and future perspectives on proteins and amino acids isolated from marine organisms

The marine ecosystem is a vast reservoir of bioactive compounds, particularly proteins and amino acids with therapeutic and industrial potential. As the global population surges, the demand for proteins across diverse sectors, including food, textiles, biotechnology, cosmetics, and pharmaceuticals, is projected to double by 2050, thereby necessitating the exploration of novel and sustainable protein sources (Kopilovic et al., 2023). According to the literature, high levels of variability in protein content in seaweeds have been reported due to seasons, environmental factors, or harvesting locations (Magnusson et al., 2019; Pliego-Cortés et al., 2019), indicating that similar quantity and quality cannot be maintained throughout the year. Generally, higher amounts of proteins are reported in winter, such as in *Gracilaria tikvahiae* (Madden et al., 2012) collected in winter (~30% dw) compared to spring (~17% dw). This variability can lead to uncertainties for industries reliant on seaweed, as they may not always have a reliable supply. Nevertheless, cultivated seaweeds are a solution for reducing the variability seen in wild populations but are still mostly harvested from the wild because of the limitations in providing efficient conditions (Grote, 2019). This indicates a significant opportunity for the industry to shift toward more sustainable and green aquaculture practices, which could ensure a steady and reliable supply of marine organisms while contributing to higher protein yields. The effective isolation of proteins and amino acids from marine organisms, such as seaweed, is also challenging due to their complex biological structures (Weinbreck et al., 2004). Therefore advanced technical approaches should be adopted, targeting effective and easy protein extraction compatible with large-scale industrial settings. Finally, a careful selection and combination of purification techniques are always associated with the specific characteristics of the proteins and their intended applications (Pliego-Cortés et al., 2020). Therefore the purification of proteins and amino acids also represents a challenge, especially for novel proteins that have not been previously described.

7.9 Conclusion

Marine-derived organisms contain various bioactive proteins and amino acids essential for developing commercial products in the cosmetic, nutraceutical, and pharmaceutical industries. Notably, some organisms offer unique bioactive proteins and amino acids, exhibiting antiinflammatory, antioxidant, and anticancer properties, making them invaluable for drug discovery and commercial product development. The sustainable and green extraction, along with efficient isolation and purification of these compounds, will minimize ecological impact as marine ecosystems face increasing threats from climate change and anthropogenic activities. In addition, advances in analytical instrumentation and the integration of multiomics approaches will enhance the development of a comprehensive database for amino acid-derived compounds and their bioactivities, addressing industrial needs while ensuring their preservation for future generations.

AI disclosure

During the preparation of this work, the author(s) used Grammarly to edit grammar. After using this tool/service, the author(s) reviewed and edited the content as needed and take(s) full responsibility for the content of the publication.

References

Addad, S., Exposito, J.Y., Faye, C., Ricard-Blum, S., Lethias, C., 2011. Isolation, characterization and biological evaluation of jellyfish collagen for use in biomedical applications. Mar. Drugs. 1660-3397. 9 (6), 967–983. Available from: http://www.mdpi.com/1660-3397/9/6/967/pdf. doi:10.3390/md9060967.

Agustin, T., Nursyam, H., Firdaus, M., Rifa'i, M., Ekaputri, I.D., Mahmiah, 2025. The characterization of blue shark (Prionace glauca) cartilage potential as nature-derived drug material. IOP Conf. Ser.: Earth Environ. Sci. 1473, 012049. doi:10.1088/1755-1315/1473/1/012049.

Aldairi, A.F., 2020. Evaluation of various methodologies used in purification of biologically active carbohydrates derived from marine life. Biomed. J. Sci. Tech. Res. 27 (4). doi:10.26717/BJSTR.2020.27.004531.

Arora, P.K., 2013. *Staphylococcus lipolyticus* sp. nov., a new cold-adapted lipase producing marine species. Ann. Microbiol. 18692044. 63 (3), 913–922. doi:10.1007/s13213-012-0544-2.

Arumugam, G.K.S., Sharma, D., Balakrishnan, R.M., Ettiyappan, J.B.P., 2018. Extraction, optimization and characterization of collagen from sole fish skin. Sustain. Chem. Pharm. 23525541. 9, 19–26. Available from: http://www.journals.elsevier.com/sustainable-chemistry-and-pharmacy. doi:10.1016/j.scp.2018.04.003.

Balseiro, P., Falcó, A., Romero, A., Dios, S., Martínez-López, A., Figueras, A., Estepa, A., Novoa, B., Zamboni, D.S., 2011. Mytilus galloprovincialis myticin c: a chemotactic molecule with antiviral activity and immunoregulatory properties. PLoS ONE. 1932-6203. 6 (8), e23140. doi:10.1371/journal.pone.0023140.

Bank, M.S., Duarte, C.M., Sonne, C., 2022. Intergovernmental panel on blue foods in support of sustainable development and nutritional security. Environ. Sci. Technol. 1520-5851. 56 (9), 5302–5305. Available from: http://pubs.acs.org/journal/esthag. doi:10.1021/acs.est.2c00119.

Barzkar, N., Sukhikh, S., Babich, O., Maran, B.A.V., Jahromi, S.T., 2023. Marine collagen: purification, properties and application. Front. Mar. Sci. 2296-7745. 10. Available from: https://www.frontiersin.org/journals/marine-science#. doi:10.3389/fmars.2023.1245077.

Basheer, S.M., Chellappan, S., Beena, P.S., Sukumaran, R.K., Elyas, K.K., Chandrasekaran, M., 2011. Lipase from marine Aspergillus awamori BTMFW032: production, partial purification and application in oil effluent treatment. New Biotechnol. 1876-4347. 28 (6), 627–638. doi:10.1016/j.nbt.2011.04.007.

Ben Slimane, E., Sadok, S., 2018. Collagen from cartilaginous fish by-products for a potential application in bioactive film composite. Mar. Drugs. 1660-3397. 16 (6), 211. doi:10.3390/md16060211.

Bikker, P., Stokvis, L., van Krimpen, M.M., van Wikselaar, P.G., Cone, J.W., 2020. Evaluation of seaweeds from marine waters in NorthWestern Europe for application in animal nutrition. Anim. Feed Sci. Technol. 03778401. 263, 114460. doi:10.1016/j.anifeedsci.2020.114460.

Bilgin, Ö., 2025. Comparison of Nutritional Quality of Amino Acid in Elasmobranchii Species: Thornback Ray, Raja clavata (Rajidae) and Common Stingray, Dasyatis pastinaca (Dasyatidae) with Implications to Human Health. Thalassas: An International Journal of Marine Sciences, Vol. 41. doi:10.1007/s41208-025-00935-y.

Bolje, A. and Gobec, S., 2021. Analytical Techniques for Structural Characterization of Proteins in Solid Pharmaceutical Forms: An Overview. Pharmaceutics, 13 (4), 534. doi:10.3390/pharmaceutics13040534.

Bourdon, B., Contentin, R., Cassé, F., Maspimby, C., Oddoux, S., Noël, A., Legendre, F., Gruchy, N., Galéra, P., 2021. Marine collagen hydrolysates downregulate the synthesis of pro-catabolic and pro-inflammatory markers of osteoarthritis and favor collagen production and metabolic activity in equine articular chondrocyte organoids. Int. J. Mol. Sci. 1422-0067. 22 (2), 580. doi:10.3390/ijms22020580.

Cai, B., Chen, H., Wan, P., Luo, L., Ye, Z., Huang, J., Chen, D., Pan, J., 2022. Isolation and identification of immunomodulatory peptides from the protein hydrolysate of tuna trimmings (*Thunnas albacares*). LWT. 0023-6438. 164, 113614. doi:10.1016/j.lwt.2022.113614.

Calcabrini, C., Catanzaro, E., Bishayee, A., Turrini, E., Fimognari, C., 2017. Marine sponge natural products with anticancer potential: an updated review. Mar. Drugs. 1660-3397. 15 (10), 310. doi:10.3390/md15100310.

Careri, M., Furlattini, L., Mangia, A., Musci, M., Anklam, E., Theobald, A., von Holst, C., 2001. Supercritical fluid extraction for liquid chromatographic determination of carotenoids in Spirulina Pacifica algae: a chemometric approach. J. Chromatogr. A. 00219673. 912 (1), 61–71. doi:10.1016/s0021-9673(01)00545-3.

Chen, H., Chen, Y., Zheng, H., Xiang, X., Xu, L., 2022. A novel angiotensin-I-converting enzyme inhibitory peptide from oyster: simulated gastro-intestinal digestion, molecular docking, inhibition kinetics and antihypertensive effects in rats. Front. Nutr. 2296-861X. 9. doi:10.3389/fnut.2022.981163.

Cheung, R., Ng, T., Wong, J., 2015. Marine peptides: bioactivities and applications. Mar. Drugs. 1660-3397. 13 (7), 4006–4043. doi:10.3390/md13074006.

Cox, P.A., King, S., 2013. Bioprospecting. In: Encyclopedia of Biodiversity, Second. Elsevier Inc., United States, pp. 588–599. Available from http://www.sciencedirect.com/science/book/9780123847201. doi:10.1016/B978-0-12-384719-5.00350-6.

Debbab, A., Aly, A.H., Lin, W.H., Proksch, P., 2010. Bioactive compounds from marine bacteria and fungi: minireview. Microb. Biotechnol. 1751-7915. 3 (5), 544–563. doi:10.1111/j.1751-7915.2010.00179.x.

Derkach, S.R., Kuchina, Y.A., Kolotova, D.S., Petrova, L.A., Volchenko, V.I., Glukharev, A.Y., Grokhovsky, V.A., 2022. Properties of protein isolates from marine hydrobionts obtained by isoelectric solubilisation/precipitation: influence of temperature and Processing time. Int. J. Mol. Sci. 1422-0067. 23 (22). Available from: http://www.mdpi.com/journal/ijms. doi:10.3390/ijms232214221.

Desjardine, K., Pereira, A., Wright, H., Matainaho, T., Kelly, M. and Andersen, R.J., (2007). Tauramamide, a Lipopeptide Antibiotic Produced in Culture by Brevibacillus laterosporus Isolated from a Marine Habitat: Structure Elucidation and Synthesis. J. Nat. Prod. 70 (12), 1850–1853. doi:10.1021/np070209r.

Diao, L., Meibohm, B., 2013. Pharmacokinetics and pharmacokinetic–pharmacodynamic correlations of therapeutic peptides. Clin. Pharmacokinet. 0312-5963. 52 (10), 855–868. doi:10.1007/s40262-013-0079-0.

Díaz-Gómez, J.L., Martín-Estal, I., Rivera-Aboytes, E., Gaxiola-Muñíz, R.A., Puente-Garza, C.A., García-Lara, S., Castorena-Torres, F., 2024. Biomedical applications of synthetic peptides derived from venom of animal origin: a systematic review. Biomed. Pharmacother. 1950-6007. 170. Available from: https://www.sciencedirect.com/science/journal/07533322. doi:10.1016/j.biopha.2023.116015.

Dissanayake, D.S., Sanjeewa, K.K.A., Jayawardena, T.U., 2024. What is seaweed? General facts about seaweeds. In: The Role of Seaweeds in Blue Bioeconomy. Bentham Science Publishers, Sri Lanka, pp. 13–24. Available from: https://benthambooks.com/book/9789815223644/. doi:10.2174/9789815223644124010004.

El-Hassayeb, H.E.A., Aziz, S.M.Z.A., 2016. Screening, production and industrial application of protease enzyme from marine bacteria. Int. J. Curr. Microbiol. Appl. Sci. 23197692. 5 (7), 863–874. doi:10.20546/ijcmas.2016.507.099.

Farooq, S., Ahmad, M.I., Zheng, S., Ali, U., Li, Y., Shixiu, C., Zhang, H., 2024. A review on marine collagen: sources, extraction methods, colloids properties, and food applications. Collagen Leather. 2731-6998. 6 (1). Available from: https://jlse.springeropen.com/. doi:10.1186/s42825-024-00152-y.

Feng, L., Wu, Y., Han, Y., Yao, X., Li, Q., Liu, M., Cao, Y., 2024. Structural characteristics, functional properties and nutritional value of walnut protein by limited enzymatic hydrolysis. LWT. 0023-6438. 197, 115923. doi:10.1016/j.lwt.2024.115923.

Fernando, I.P.S., Jayawardena, T.U., Wu, J., 2023. Marine proteins and peptides: production, biological activities, and potential applications. Food Innov. Adv. 2836-774X. 2 (2), 69–84. Available from: https://www.maxapress.com/article/doi/10.48130/FIA-2023-0009. doi:10.48130/FIA-2023-0009.

Food and Agriculture Organization (FAO), 2020. The State of World Fisheries and Aquaculture. Rome: FAO. doi:10.4060/ca9229en.

Ganesan, A.R., Mohanram, M.S.G., Balasubramanian, B., Kim, I.H., Seedevi, P., Mohan, K., Kanagasabai, S., Arasu, M.V., Al-Dhabi, N.A., Ignacimuthu, S., 2020. Marine invertebrates' proteins: a recent update on functional property. J. King Saud Univ. Sci. 1018-3647. 32 (2), 1496–1502. Available from: http://www.sciencedirect.com/science/journal/10183647. doi:10.1016/j.jksus.2019.12.003.

García-Encinas, J.P., Ruiz-Cruz, S., Juárez, J., Ornelas-Paz, J.D.J., Toro-Sánchez, C.L.D., Márquez-Ríos, E., 2025. Proteins from microalgae: nutritional, functional and bioactive properties. Foods. 2304-8158. 14 (6). Available from: www.mdpi.com/journal/foods. doi:10.3390/foods14060921.

Gómez-Guillén, M.C., Giménez, B., López-Caballero, M.E., Montero, M.P., 2011. Functional and bioactive properties of collagen and gelatin from alternative sources: a review. Food Hydrocoll. 0268-005X. 25 (8), 1813–1827. doi:10.1016/j.foodhyd.2011.02.007.

Grosso, C., Valentão, P., Ferreres, F., Andrade, P., 2015. Alternative and efficient extraction methods for marine-derived compounds. Mar. Drugs. 1660-3397. 13 (5), 3182–3230. doi:10.3390/md13053182.

Grote, B., 2019. Recent developments in aquaculture of *Palmaria palmata* (Linnaeus) (Weber & Mohr 1805): cultivation and uses. Rev. Aquac. 1753-5123. 11 (1), 25–41. doi:10.1111/raq.12224.

Gunasekara, D.M.N.M., Wang, L., Herath, K.H.I.N.M., Sanjeewa, K.K.A., 2025. Cosmeceutical applications of phlorotannins from brown seaweeds. Phycology. 2673-9410. 5 (2), 15. doi:10.3390/phycology5020015.

Gupta, V., Vyas, D., 2021. Antimicrobial effect of a cyclic peptide Nostophycin isolated from wastewater cyanobacteria, *Nostoc calcicola*. Curr. Bot. 2220-4822. 94–101. doi:10.25081/cb.2021.v12.6612.

Hagström, A., Pommier, T., Rohwer, F., Simu, K., Stolte, W., Svensson, D., Zweifel, U.L., 2002. Use of 16S ribosomal DNA for delineation of marine bacterioplankton species. Appl. Environ. Microbiol. 68 (7), 3628–3633. doi:10.1128/AEM.68.7.3628-3633.2002.

Hans, N., Malik, A., Naik, S., 2021. Antiviral activity of sulfated polysaccharides from marine algae and its application in combating COVID-19: mini review. Bioresour. Technol. Rep. 2589014X. 13, 100623. doi:10.1016/j.biteb.2020.100623.

Harnedy, P.A., Fitzgerald, R.J., 2011. Bioactive proteins, peptides, and amino acids from macroalgae. J. Phycol. 1529-8817. 47 (2), 218–232. doi:10.1111/j.1529-8817.2011.00969.x.

Hassan, S.W.M., Latif, H.H.A.E., Ali, S.M., 2018. Production of cold-active lipase by free and immobilized marine Bacillus cereus HSS: application in wastewater treatment. Front. Microbiol. 1664-302X. 9. Available from: www.frontiersin.org/Microbiology. doi:10.3389/fmicb.2018.02377.

He, D., Yi, X., Xia, G., Liu, Z., Zhang, X., Li, C., Shen, X., 2022. Effects of fish oil on the gel properties and emulsifying stability of myofibrillar proteins: a comparative study of tilapia, hairtail and squid. LWT. 0023-6438. 161, 113373. doi:10.1016/j.lwt.2022.113373.

Hirsch, A.M., and Valdés, M., 2010. Micromonospora: An important microbe for biomedicine and potentially for biocontrol and biofuels. Soil Biol. Biochem. 42 (4), 536–542. doi:10.1016/j.soilbio.2009.11.023.

Im, S.T., Jang, Y., Park, S., Mun, H., Kim, D.S., Lee, D.S., Lee, J.M., Yim, M.J., Kim, J.Y., Kim, H.S., Ko, S.C., Jung, W.K., Lee, S.H., 2022. In vitro screening of extracts from 38 marine animal resources for novel cosmeceutical activities. Fish. Aquat. Sci. 22341757. 25 (6), 327–334. Available from https://www.e-fas.org/info/about. doi:10.47853/FAS.2022.e30.

Je, J.Y., Park, P.J., Kwon, J.Y., Kim, S.K., 2004. A novel angiotensin I converting enzyme inhibitory peptide from Alaska pollack (Theragra chalcogramma) frame protein hydrolysate. J. Agric. Food Chem. 0021-8561. 52 (26), 7842–7845.

Jo, D.M., Khan, F., Park, S.K., Ko, S.C., Kim, K.W., Yang, D., Kim, J.Y., Oh, G.W., Choi, G., Lee, D.S., Kim, Y.M., 2024. From sea to lab: angiotensin i-converting enzyme inhibition by marine peptides—mechanisms and applications. Mar. Drugs. 1660-3397. 22 (10). Available from: http://www.mdpi.com/journal/marinedrugs. doi:10.3390/md22100449.

Juel, N., Juul, L., 2024. Hartono tanambell, trine kastrup dalsgaard, extraction and purification of seaweed protein from Ulva sp.—challenges to overcome. LWT. 0023-6438. 198, 115944. doi:10.1016/j.lwt.2024.115944.

K., Harish, K., Erwin, 2023. Tetrodotoxin Toxicity, StatPearls Publishing.

Kang, H.K., Lee, H.H., Seo, C.H., Park, Y., 2019. Antimicrobial and immunomodulatory properties and applications of marine-derived proteins and peptides. Mar. Drugs. 1660-3397. 17 (6). Available from: https://www.mdpi.com/1660-3397/17/6/350/pdf. doi:10.3390/md17060350.

Karabulut, G., Purkiewicz, A., Goksen, G., 2024. Recent developments and challenges in algal protein and peptide extraction strategies, functional and technological properties, bioaccessibility, and commercial applications. Compr. Rev. Food Sci. Food Saf. 1541-4337. 23 (3). doi:10.1111/1541-4337.13372.

Kazir, M., Abuhassira, Y., Robin, A., Nahor, O., Luo, J., Israel, A., Golberg, A., Livney, Y.D., 2019. Extraction of proteins from two marine macroalgae, Ulva sp. and Gracilaria sp., for food application, and evaluating digestibility, amino acid composition and antioxidant properties of the protein concentrates. Food Hydrocoll. 0268005X. 87, 194–203. Available from: https://www.journals.elsevier.com/food-hydrocolloids. doi:10.1016/j.foodhyd.2018.07.047.

Keizer, D.W., West, P.J., Lee, E.F., Yoshikami, D., Olivera, B.M., Bulaj, G., Norton, R.S., 2003. Structural basis for tetrodotoxin-resistant sodium channel binding by μ-conotoxin SmIIIA. J. Biol. Chem. 278 (47), 46805–46813. doi:10.1074/jbc.M309222200.

Kendler, S., Sasidharan, A., Rustad, T., 2024. Extraction of proteinaceous components and biominerals from cold water fish fileting side streams: a review. Front. Sustain. Food Syst. 2571-581X. 7. doi:10.3389/fsufs.2023.1331113.

Khalil, H.S., Mansour, A.T., Goda, A.M.A., Omar, E.A., 2019. Effect of selenium yeast supplementation on growth performance, feed utilization, lipid profile, liver and intestine histological changes, and economic benefit in meagre, Argyrosomus regius, fingerlings. Aquaculture. 0044-8486. 501, 135–143. Available from: http://www.journals.elsevier.com/aquaculture/. doi:10.1016/j.aquaculture.2018.11.018.

Kim, S., 2013. Marine Proteins and Peptides: Biological Activities and Applications. Wiley. doi:10.1002/9781118375082.

Kopilovic, B., Valente, A.I., Ferreira, A.M., Almeida, M.R., Tavares, A.P.M., Freire, M.G., Coutinho, J.A.P., 2023. Towards the sustainable extraction and purification of non-animal proteins from biomass using alternative solvents. RSC Sustain. 2753-8125. 1 (6), 1314–1331. doi:10.1039/d3su00062a.

Kvetkina, A., Pislyagin, E., Menchinskaya, E., Yurchenko, E., Kalina, R., Kozlovskiy, S., Kaluzhskiy, L., Menshov, A., Kim, N., Peigneur, S., Tytgat, J., Ivanov, A., Ayvazyan, N., Leychenko, E., Aminin, D., 2022. Kunitz-type peptides from sea anemones protect neuronal cells against parkinson's disease inductors via inhibition of ROS production and ATP-induced P2X7 receptor activation. Int. J. Mol. Sci. 1422-0067. 23 (9), 5115. doi:10.3390/ijms23095115.

LaPelusa, A., Kaushik, R., (2025). Physiology, Proteins. StatPearls Publishing. https://www.ncbi.nlm.nih.gov/books/NBK555990/.

LaPelusa, A., Kaushik, R., 2025. Physiology.

Lee, C.H., Singla, A., Lee, Y., 2001. Biomedical applications of collagen. Int. J. Pharm. 0378-5173. 221 (1-2), 1–22. doi:10.1016/S0378-5173(01)00691-3.

Leisch, M., Egle, A., Greil, R., 2019. Plitidepsin: a potential new treatment for relapsed/refractory multiple myeloma. Future Oncol. 1479-6694. 15 (2), 109–120. doi:10.2217/fon-2018-0492.

Lima, N.P., Maciel, G.M., Pinheiro, D.F., Ribeiro, I.S., Lima, N.F., Liviz, C.D.A.M., Pedro, A.C., Haminiuk, C.W.I., 2025. Innovative protein sources from freshwater and marine environments - A comprehensive review. Meas. Food. 2772-2759. 17. Available from: https://www.sciencedirect.com/science/journal/27722759. doi:10.1016/j.meafoo.2025.100215.

Lin, W.-C., Chang, H.-Y., Chen, J.-Y., 2016. Electrotransfer of the tilapia piscidin 3 and tilapia piscidin 4 genes into skeletal muscle enhances the antibacterial and immunomodulatory functions of *Oreochromis niloticus*. Fish Shellfish Immunol. 1050-4648. 50, 200–209. doi:10.1016/j.fsi.2016.01.034.

Livermore, D.M., Blaser, M., Carrs, O., Cassell, G., Fishman, N., Guidos, R., Levy, S., Powers, J., Norrby, R., Tillotson, G., Davies, R., Projan, S., Dawson, M., Monnet, D., Keogh-Brown, M., Hand, K., Garner, S., Findlay, D., Morel, C., Wise, R., Bax, R., Burke, F., Chopra, I., Czaplewski, L., Finch, R., Livermore, D., Piddock, L.J.V., White, T., 2011. Discovery research: the scientific challenge of finding new antibiotics. J. Antimicrob. Chemother. 0305-7453. 66 (9), 1941–1944. doi:10.1093/jac/dkr262.

Lorenzo, J.M., Munekata, P.E.S., Gómez, B., Barba, F.J., Mora, L., Pérez-Santaescolástica, C., Toldrá, F., 2018. Bioactive peptides as natural antioxidants in food products – a review. Trends Food Sci. Technol. 0924-2244. 79, 136–147. Available from: http://www.elsevier.com/wps/find/journaldescription.cws_home/601278/description#description. doi:10.1016/j.tifs.2018.07.003.

Lucakova, S., Branyikova, I., Hayes, M., 2022. Microalgal proteins and bioactives for food, feed, and other applications. Appl. Sci. 2076-3417. 12 (9), 4402. doi:10.3390/app12094402.

Machado, M., Machado, S., Pimentel, F.B., Freitas, V., Alves, R.C., Oliveira, M.B.P.P., 2020. Amino acid profile and protein quality assessment of macroalgae produced in an integrated multi-trophic aquaculture system. Foods. 2304-8158. 9 (10). Available from: https://www.mdpi.com/2304-8158/9/10/1382. doi:10.3390/foods9101382.

Macías-Sánchez, M.D., Serrano, C.M., Rodríguez Rodríguez, M., de la Ossa, E.M., Lubián, L.M., Montero, O., 2008. Extraction of carotenoids and chlorophyll from microalgae with supercritical carbon dioxide and ethanol as cosolvent. J. Sep. Sci. 1615-9306. 31 (8), 1352–1362. doi:10.1002/jssc.200700503.

Madden, M., Mitra, M., Ruby, D., Schwarz, J., 2012. Seasonality of selected nutritional constituents of edible delmarva seaweeds. J. Phycol. 0022-3646. 48 (5), 1289–1298. doi:10.1111/j.1529-8817.2012.01207.x.

Magnusson, M., Glasson, C.R.K., Vucko, M.J., Angell, A., Neoh, T.L., de Nys, R., 2019. Enrichment processes for the production of high-protein feed from the green seaweed *Ulva ohnoi*. Algal Res. 2211-9264. 41. Available from: http://www.sciencedirect.com/science/journal/aip/22119264. doi:10.1016/j.algal.2019.101555.

McCauley, D.W., Docker, M.F., Whyard, S., Li, W., 2015. Lampreys as diverse model organisms in the Genomics era. BioScience 65, 1046–1056. doi:10.1093/biosci/biv139.

Mendes, R.L., Nobre, B.P., Cardoso, M.T., Pereira, A.P., Palavra, A.F., 2003. Supercritical carbon dioxide extraction of compounds with pharmaceutical importance from microalgae. Inorg. Chim. Acta. 0020-1693. 356, 328–334. Available from: http://www.journals.elsevier.com/inorganica-chimica-acta/. doi:10.1016/S0020-1693(03)00363-3.

Molinski, T.F., Dalisay, D.S., Lievens, S.L., Saludes, J.P., 2009. Drug development from marine natural products. Nat. Rev. Drug Discov. 1474-1784. 8 (1), 69–85. doi:10.1038/nrd2487.

Moore, E., Harvey, H., Faux, J., Goodlett, D., Nunn, B., 2014. Electrophoretic extraction and proteomic characterization of proteins buried in marine sediments. Chromatography. 2227-9075. 1 (4), 176–193. doi:10.3390/chromatography1040176.

Mosibo, O.K., Ferrentino, G., Udenigwe, C.C., 2024. Microalgae proteins as sustainable ingredients in novel foods: recent developments and challenges. Foods. 2304-8158. 13 (5). Available from: www.mdpi.com/journal/foods. doi:10.3390/foods13050733.

Musaimi, O.A., Jaradat, D.M.M., 2024. Advances in therapeutic peptides separation and purification. Separations. 22978739. 11 (8). Available from: www.mdpi.com/journal/separations. doi:10.3390/separations11080233.

Nasab, S.B., Homaei, A., Pletschke, B.I., Salinas-Salazar, C., Castillo-Zacarias, C., Parra-Saldívar, R., 2020. Marine resources effective in controlling and treating diabetes and its associated complications. Process Biochem. 1359-5113. 92, 313–342. Available from: www.elsevier.com/inca/publications/store/4/2/2/8/5/7. doi:10.1016/j.procbio.2020.01.024.

Nunn, B.L., Keil, R.G., 2006. A comparison of non-hydrolytic methods for extracting amino acids and proteins from coastal marine sediments. Mar. Chem. 0304-4203. 98 (1), 31–42. doi:10.1016/j.marchem.2005.06.005.

Okeke, E.S., Okagu, I.U., Chukwudozie, K., Ezike, T.C., Ezeorba, T.P.C., 2024. Marine-derived bioactive proteins and peptides: a review of current knowledge on anticancer potentials, clinical trials, and future prospects. Nat. Prod. Commun. 1555-9475. 19 (3). Available from: https://journals.sagepub.com/home/NPX. doi:10.1177/1934578X241239825.

Orlov, D.S., Shamova, O.V., Eliseev, I.E., Zharkova, M.S., Chakchir, O.B., Antcheva, N., Zachariev, S., Panteleev, P.V., Kokryakov, V.N., Ovchinnikova, T.V., Tossi, A., 2019. Redesigning arenicin-1, an antimicrobial peptide from the marine Polychaeta Arenicola marina, by strand rearrangement or branching, substitution of Specific residues, and backbone linearization or cyclization. Mar. Drugs. 1660-3397. 17 (6). Available from: https://www.mdpi.com/1660-3397/17/6/376/pdf. doi:10.3390/md17060376.

Ovchinnikova, T.V., Balandin, S.V., Aleshina, G.M., Tagaev, A.A., Leonova, Y.F., Krasnodembsky, E.D., Men'shenin, A.V., Kokryakov, V.N., 2006. Aurelin, a novel antimicrobial peptide from jellyfish *Aurelia aurita* with structural features of defensins and channel-blocking toxins. Biochem. Biophys. Res. Commun. 1090-2104. 348 (2), 514–523. doi:10.1016/j.bbrc.2006.07.078.

Panchout, F., Letendre, J., Bultelle, F., Denier, X., Rocher, B., Chan, P., Vaudry, D., Durand, F., 2013. Comparison of protein-extraction methods for gills of the Shore Crab, *Carcinus maenas* (L.), and application to 2DE. J. Biomol. Tech. 1524-0215. 24 (4), jbt.13-2404-002. doi:10.7171/jbt.13-2404-002.

Pangestuti, R., Kim, S.K., 2011. Neuroprotective effects of marine algae. Mar. Drugs. 1660-3397. 9 (5), 803–818. Available from: http://www.mdpi.com/1660-3397/9/5/803/pdf. doi:10.3390/md9050803.

Pangestuti, R., Siahaan, E.A., Kim, S.K., 2018. Photoprotective substances derived from marine algae. Mar. Drugs. 1660-3397. 16 (11). Available from: https://www.mdpi.com/1660-3397/16/11/399/pdf. doi:10.3390/md16110399.

Peng, X., Amend, A.S., Baltar, F., Blanco-Bercial, L., Breyer, E., Burgaud, G., Cunliffe, M., Edgcomb, V.P., Grossart, H. P., Mara, P., Masigol, H., Pang, K.L., Retter, A., Roberts, C., van Bleijswijk, J., Walker, A.K., Whitner, S., 2024.

Planktonic marine fungi: a review. J. Geophys. Res. Biogeosci. 2169-8961. 129 (3). Available from: https://agupubs.onlinelibrary.wiley.com/loi/21698961. doi:10.1029/2023JG007887.

Penggalih, M.H.S.T., Praditya, G.N., Rizqiansyah, C.Y., Setyawardani, A., Purnomo, A.F., Maulana, R.A., Gunawan, W.B., Subali, D., Kurniawan, R., Mayulu, N., Taslim, N.A., Hardinsyah, H., Sutanto, Y.S., Nurkolis, F., 2023. Marine-derived protein: peptide bioresources for the development of nutraceuticals for improved athletic performance. Front. Sports Act. Living. 26249367. 5. Available from www.frontiersin.org/journals/sports-and-active-living. doi:10.3389/fspor.2023.1281397.

Pliego-Cortés, H., Bedoux, G., Boulho, R., Taupin, L., Freile-Pelegrín, Y., Bourgougnon, N., Robledo, D., 2019. Stress tolerance and photoadaptation to solar radiation in Rhodymenia pseudopalmata (*Rhodophyta*) through mycosporine-like amino acids, phenolic compounds, and pigments in an integrated multi-trophic aquaculture system. Algal Res. 2211-9264. 41, 101542. doi:10.1016/j.algal.2019.101542.

Pliego-Cortés, H., Wijesekara, I., Lang, M., Bourgougnon, N., Bedoux, G., 2020. Current knowledge and challenges in extraction, characterization and bioactivity of seaweed protein and seaweed-derived proteins. In: Advances in Botanical Research, Vol. 95. Academic Press Inc., France, pp. 289–326. Available from: http://www.elsevier.com/wps/find/bookdescription.cws_home/675801/description#description. doi:10.1016/bs.abr.2019.11.008. 00652296.

Proksch, P., *et al* ., 2008. Sponge-associated fungi and their bioactive compounds: the Suberites case. botm, 51 (3) 209–218. doi:10.1515/BOT.2008.014.

Qi, S., Xu, Y., Xiong, H., Qian, P. and Zhang, S., 2009. Antifouling and antibacterial compounds from a marine fungus Cladosporium sp. F14. World J. Microbiol. Biotechnol. 25(3), 399–406. doi:10.1007/s11274-008-9904-2.

Quitério, E., Grosso, C., Ferraz, R., Delerue-Matos, C., Soares, C., 2022. A critical comparison of the advanced extraction techniques applied to obtain health-promoting compounds from seaweeds. Mar. Drugs. 1660-3397. 20 (11), 677. doi:10.3390/md20110677.

R. S., D. M., M.G. M.N., N. V., N. T. and Kim J., 2022. An Investigation of Pepsin Hydrolysate of Short Antibacterial Peptides Derived from Limnospira Sp. Appl. Biochem. Biotechnol. 194 (11), 5580–5593. doi:10.1007/s12010-022-04023-2.

Rahman, A., Silva, T.H., 2022. Collagens from marine organisms towards biomedical applications. Mar. Drugs. 1660-3397. 20 (3), 170. doi:10.3390/md20030170.

Ren, L.K., Guo, Q.Q., Yang, Y., Liu, X.F., Guan, H.N., Chen, F.L., Bian, X., Zhang, X.M., Zhang, N., 2021. Structural characterization and functional properties of CNPP, a byproduct formed during CPP preparation. J. Food Sci. 1750-3841. 86 (5), 1845–1860. Available from: http://onlinelibrary.wiley.com/journal/10.1111/(ISSN)1750-3841. doi:10.1111/1750-3841.15717.

Ren, M., Lu, J., Li, Z., Ye, T., Song, Y., Lu, H., Lai, N., Xiang, X., Yang, L., Lin, L., Jiang, S., Wang, Y., 2026. Turtle shell as a health food ingredient: a review of global utilization, nutrients, health benefits, and functional food potential. Trends Food Sci. Technol. 169, 105574. doi:10.1016/j.tifs.2026.105574.

Safronova, V.N., Panteleev, P.V., Kruglikov, R.N., Bolosov, I.A., Finkina, E.I., Ovchinnikova, T.V., 2024. Novel BRICHOS-related defensin-like antimicrobial peptide from the marine polychaeta *Arenicola* marina. Russ. J. Bioorg. Chem. 1068-1620. 50 (3), 629–643. doi:10.1134/s1068162024030087.

Seghal Kiran, G., Lipton, A.N., Kennedy, J., Dobson, A.D.W., Selvin, J., 2014. A halotolerant thermostable lipase from the marine bacterium Oceanobacillus sp. PUMB02 with an ability to disrupt bacterial biofilms. Bioengineered. 2165-5979. 5 (5), 305–318. doi:10.4161/bioe.29898.

Seo, C., Sohn, J.H., Oh, H., Kim, B.Y., and Ahn, J.S., 2009. Isolation of the protein tyrosine phosphatase 1B inhibitory metabolite from the marine-derived fungus Cosmospora sp. SF-5060. Bioorg. Med. Chem. Lett. 19 (21), 6095–6097. doi:10.1016/j.bmcl.2009.09.025.

Sermkaew, N., Atipairin, A., Krobthong, S., Aonbangkhen, C., Yingchutrakul, Y., Uchiyama, J., Songnaka, N., 2024. Unveiling a new antimicrobial peptide with Efficacy against *P. aeruginosa* and *K. pneumoniae* from mangrove-derived *Paenibacillus* thiaminolyticus NNS5-6 and genomic analysis. Antibiotics. 2079-6382. 13 (9), 846. doi:10.3390/antibiotics13090846.

Siddiqui, S.A., Baruah, S., Wu, Y.S., Yuansah, S.C., Castro-Muñoz, R., Szymkowiak, A., Kulawik, P., 2024. Investigating the sustainability, utilisation, consumption and conservation of sea mammals – A systematic review. Sustain. Prod. Consum. 46, 400–417. doi:10.1016/j.spc.2024.03.003.

Sintsova, O., Gladkikh, I., Monastyrnaya, M., Tabakmakher, V., Yurchenko, E., Menchinskaya, E., Pislyagin, E., Andreev, Y., Kozlov, S., Peigneur, S., Tytgat, J., Aminin, D., Kozlovskaya, E., Leychenko, E., 2021. Sea anemone kunitz-type peptides demonstrate neuroprotective activity in the 6-hydroxydopamine induced neurotoxicity model. Biomedicines. 2227-9059. 9 (3), 283. doi:10.3390/biomedicines9030283.

Sousa, S.R., McArthur, J.R., Brust, A., Bhola, R.F., Rosengren, K.J., Ragnarsson, L., Dutertre, S., Alewood, P.F., Christie, M.J., Adams, D.J., Vetter, I., Lewis, R.J., 2018. Novel analgesic ω-conotoxins from the vermivorous cone snail *Conus moncuri* provide new insights into the evolution of conopeptides. Sci. Rep. 2045-2322. 8 (1). Available from: www.nature.com/srep/index.html. doi:10.1038/s41598-018-31245-4.

Suarez-Jimenez, G.M., Burgos-Hernandez, A., Ezquerra-Brauer, J.M., 2012. Bioactive peptides and depsipeptides with anticancer potential: sources from marine animals. Mar. Drugs. 1660-3397. 10 (5), 963–986. Available from: http://www.mdpi.com/1660-3397/10/5/963/pdf. doi:10.3390/md10050963.

Subramani, R., Aalbersberg, W., 2012. Marine actinomycetes: an ongoing source of novel bioactive metabolites. Microbiol. Res. 0944-5013. 167 (10), 571–580. doi:10.1016/j.micres.2012.06.005.

Sudhakar, S., Nazeer, R.A., 2015. Structural characterization of an Indian squid antioxidant peptide and its protective effect against cellular reactive oxygen species. J. Funct. Foods. 1756-4646. 14, 502–512. Available from: http://www.elsevier.com/wps/find/journaldescription.cws_home/717426/description#description. doi:10.1016/j.jff.2015.02.028.

Suleria, H.A.R., Masci, P., Gobe, G., Osborne, S., 2016. Current and potential uses of bioactive molecules from marine processing waste. J. Sci. Food Agric. 10970010. 96 (4), 1064–1067. Available from: http://onlinelibrary.wiley.com/journal/10.1002/(ISSN)1097-0010. doi:10.1002/jsfa.7444.

Sun, K.L., Gao, M., Wang, Y.Z., Li, X.R., Wang, P., Wang, B., 2022. Antioxidant peptides from protein hydrolysate of Marine red algae *Eucheuma* cottonii: preparation, identification, and cytoprotective mechanisms on H_2O_2 oxidative damaged HUVECs. Front. Microbiol. 1664302X. 13. Available from: https://www.frontiersin.org/journals/microbiology#. doi:10.3389/fmicb.2022.791248.

Sun, Y., Chang, R., Li, Q., Li, B., 2016. Isolation and characterization of an antibacterial peptide from protein hydrolysates of Spirulina platensis. Eur. Food Res. Technol. 1438-2377. 242 (5), 685–692. doi:10.1007/s00217-015-2576-x.

Suyama, T.L., Gerwick, W.H., McPhail, K.L., 2011. Survey of marine natural product structure revisions: a synergy of spectroscopy and chemical synthesis. Bioorg. Med. Chem. 1464-3391. 19 (22), 6675–6701. doi:10.1016/j.bmc.2011.06.011.

Tan, K., Zhang, H., Zheng, H., 2024. Carotenoid content and composition: a special focus on commercially important fish and shellfish. Crit. Rev. Food Sci. Nutr. 1040-8398. 64 (2), 544–561. doi:10.1080/10408398.2022.2106937.

Tubaro, A., Sosa, S., Hungerford, J., 2012. Toxicology and diversity of marine toxins. In: Veterinary Toxicology. Elsevier Inc, Italy, pp. 896–934. Available from: http://www.sciencedirect.com/science/book/9780123859266. doi:10.1016/B978-0-12-385926-6.00080-6.

Vemulapalli, S.P.B., Griesinger, C., Dittmar, T., 2023. Expanding the limits of structural characterization of marine dissolved organic matter using nonuniform sampling frequency-reversed edited HSQC NMR. Anal. Chem. 15206882. 95 (39), 14770–14776. Available from: http://pubs.acs.org/journal/ancham. doi:10.1021/acs.analchem.3c02923.

Walquist, M.J., Eilertsen, K.E., Elvevoll, E.O., Jensen, I.J., 2024. Marine-derived peptides with anti-hypertensive properties: prospects for pharmaceuticals, supplements, and functional food. Mar. Drugs. 16603397. 22 (4). Available from: http://www.mdpi.com/journal/marinedrugs. doi:10.3390/md22040140.

Wang, C., Lu, Y., and Cao, S., 2020. Antimicrobial compounds from marine actinomycetes. Arch. Pharm. Res. 43 (7), 677–704. doi:10.1007/s12272-020-01251-0.

Wang, X., Yu, H., Xing, R., Li, P., 2017b. Characterization, preparation, and purification of marine bioactive peptides. Biomed. Res. Int. 2314-6133. 2017, 1–16. doi:10.1155/2017/9746720.

Wei, Y.F., Wang, J.Q., Luo, Y.J., Zhao, X.F., Zou, L.H., Qiu, X.F., Nie, S.P., Liu, H.Y., 2024. Isolation, purification and characterization of antioxidant peptides from Spirulina platensis extracts co-fermentation with Thermus thermophilus HB27 and *Saccharomyces cerevisiae* CH006. LWT. 0023-6438. 205. Available from: https://www.sciencedirect.com/science/journal/00236438. doi:10.1016/j.lwt.2024.116497.

Weinbreck, F., Nieuwenhuijse, H., Robijn, G.W., De Kruif, C.G., 2004. Complexation of whey proteins with carrageenan. J. Agric. Food Chem. 0021-8561. 52 (11), 3550–3555. doi:10.1021/jf034969t.

Wijesinghe, W.A.J.P., Jeon, Y.J., 2012. Enzyme-assistant extraction (EAE) of bioactive components: a useful approach for recovery of industrially important metabolites from seaweeds: a review. Fitoterapia. 1873-6971. 83 (1), 6–12. doi:10.1016/j.fitote.2011.10.016.

World Health Organization, Food and Agriculture Organization of the United Nations, United Nations University. 2007. Protein and amino acid requirements in human nutrition: Report of a joint WHO/FAO/UNU expert consultation. WHO Technical Report Series 935. Geneva: WHO.

Xie, B., Li, X., Lin, Z., Ruan, Z., Wang, M., Liu, J., Tong, T., Li, J., Huang, Y., Wen, B., Sun, Y., Shi, Q., 2016. Prediction of toxin genes from chinese yellow catfish based on transcriptomic and proteomic sequencing. Int. J. Mol. Sci. 1422-0067. 17 (4), 556. doi:10.3390/ijms17040556.

Yaghoubzadeh, Z., Safari, R., 2025. Extraction of bioactive peptides from *Chlorella vulgaris* using enzymatic hydrolysis: a green natural antioxidant. Int. J. Pept. Res. Ther. 1573-3904. 31 (2). doi:10.1007/s10989-025-10692-4.

Yang, R., Wang, J., Liu, Z., Pei, X., Han, X., Li, Y., 2011. Antioxidant effect of a marine oligopeptide preparation from Chum Salmon (*Oncorhynchus keta*) by enzymatic hydrolysis in radiation injured mice. Mar. Drugs. 1660-3397. 9 (11), 2304–2315. doi:10.3390/md9112304.

Yang, Y., Chen, F., Chen, H.Y., Peng, H., Hao, H., Wang, K.J., 2020. A novel antimicrobial peptide scyreprocin from mud crab *Scylla paramamosain* showing potent antifungal and anti-biofilm activity. Front. Microbiol. 1664-302X. 11. Available from: https://www.frontiersin.org/journals/microbiology#. doi:10.3389/fmicb.2020.01589.

Zamorano-Apodaca, J.C., García-Sifuentes, C.O., Carvajal-Millán, E., Vallejo-Galland, B., Scheuren-Acevedo, S.M., Lugo-Sánchez, M.E., 2020. Biological and functional properties of peptide fractions obtained from collagen hydrolysate derived from mixed by-products of different fish species. Food Chem. 1873-7072. 331. Available from: www.elsevier.com/locate/foodchem. doi:10.1016/j.foodchem.2020.127350.

Zbinden, M.D.A., Sturm, B.S.M., Nord, R.D., Carey, W.J., Moore, D., Shinogle, H., Stagg-Williams, S.M., 2013. Pulsed electric field (PEF) as an intensification pretreatment for greener solvent lipid extraction from microalgae. Biotechnol. Bioeng. 1097-0290. 110 (6), 1605–1615. doi:10.1002/bit.24829.

Zhang, C., Kim, S.K., 2010. Research and application of marine microbial enzymes: status and prospects. Mar. Drugs. 1660-3397. 8 (6), 1920–1934. Available from: http://www.mdpi.com/1660-3397/8/6/1920/pdf. doi:10.3390/md8061920.

Zhang, Y., Dai, B., Deng, Y., Zhao, Y., 2016. In vitro anti-inflammatory and antioxidant activities and protein quality of high hydrostatic pressure treated squids (*Todarodes pacificus*). Food Chem. 0308-8146. 203, 258–266. doi:10.1016/j.foodchem.2016.02.072.

Zheng, Y., Huang, C., Zhou, X., Sanjeewa, K.K.A., Mao, X., Wang, L., 2025. Health benefits of the blooming seaweed Ulva prolifera: a review. Aquaculture 598. Available from: https://www.sciencedirect.com/science/article/pii/S0044848624014972. doi:10.1016/j.aquaculture.2024.742035.

CHAPTER

8

Marine-based vitamins and future potentials

Nalin Medagoda[a] *and Bopitige Udith Ravindu Wijemanna*[b]

[a]Department of Marine Life Science, Jeju National University, Jeju, South Korea

[b]Department of Animal Science, Iowa State University, Ames, IA, United States

8.1 Introduction

Vitamins are a group of organic molecules essential in small quantities (micro or milligrams per day) for animals that are not synthesized in adequate amounts to meet bodily requirements for maintaining normal growth, health, and physiological functions (Zempleni et al., 2013; Webster-Gandy et al., 2020). These micronutrients do not provide animals with energy or structural functions but act as cofactors for enzymes, biological antioxidants, visual pigments (photoreceptive cofactors), hormones, blood clotting, and calcium balance. The animal body can synthesize a few vitamins, such as vitamin C, in small quantities but still needs to obtain them from the diet in adequate amounts to fulfill bodily requirements. Insufficient dietary vitamin intake, impaired absorption, and metabolic dysfunctions can lead to vitamin deficiency conditions, which can result in physiological, biochemical, and behavioral changes and pharmacological concerns. The Food and Nutrition Board of the Institute of Medicine has established dietary reference intake values (μg, mg/day) for nutrients, including each vitamin, with an estimated average requirement, recommended daily allowance (RDA), adequate intake, and upper intake levels for healthy males and females, including pregnant and lactating women, across different age categories. Those data can be accessed through the following link: https://nap.nationalacademies.org/catalog/11537/dietary-reference-intakes-the-essential-guide-to-nutrient-requirements. The intake of large doses of vitamins can lead to toxic conditions, especially for fat-soluble vitamins, since they are bound to fat and deposited in tissues, unlike water-soluble vitamins, which are readily excreted through urine. Moreover, some vitamins are single compounds, but vitamins A, D, E, and K, as well as individual vitamins of the B complex, are groups of compounds collectively known as vitamers. Each vitamer exhibits similar biological functions due to structural similarity or is converted into a similar final active biological molecule through metabolic

Harvesting the Sea
https://doi.org/10.1016/B978-0-443-29190-6.00014-3

activation in the body. According to the literature, there are four vitamers for vitamin A: retinol, retinal, retinoic acid, retinoids, and the provitamin A; three vitamers for vitamin B6: pyridoxol, pyridoxal, and pyridoxamine; and three vitamers for vitamin D: calcitriol, ergocalciferol, and cholecalciferol. In addition, there are eight vitamers for vitamin E: four tocopherols and four tocotrienols, and three vitamers for vitamin K: phylloquinone, menaquinones, and menadiones. Based on the polarity of the vitamin molecules or most of their vitamers, vitamins are divided into two categories: water-soluble or fat-soluble. Lipid-soluble vitamins can be obtained through fatty foods due to the nature of their solubility. For instance, fish oils are rich in vitamins A and D.

As alternative sources for traditional terrestrial vitamin sources, marine sources can be identified as future potential functional ingredients to extract a wide variety of vitamins. Marine microalgae, like *Tetraselmis* and *Pavlova*, are rich in vitamin A (alpha-carotene), B, C, E, and K, as well as other antioxidants, and macroalgae (seaweeds) are rich in water-soluble vitamins and vitamin D (Senadheera et al., 2023). Fish and fish by-products are primary sources of both water- and fat-soluble vitamins, including vitamins A, B, D, E, and K. However, vitamin C is not abundant in marine fish due to the lack of the L-gulonolactone oxidase enzyme in fish, which regulates a vital step in the vitamin C synthesis pathway. Shellfish are well-known sources of vitamins E, B6, and B12. Therefore understanding the history, physiological properties, molecular structure, absorption, transportation and storage, biological significance, dietary requirements, natural synthesis, extraction methods, and quantification techniques of each vitamin is essential.

8.2 Fat-soluble vitamins

8.2.1 Vitamin A

McCollum and Davis identified vitamin A in 1913 at the University of Wisconsin (McCollum and Davis, 1913). Mainly, two categories of vitamers can be identified owing to their similar characteristics: retinoids (retinol, retinal, and retinoic acid) and provitamin A (α, β, and γ-carotenoids), which are orange to yellow-colored crystals (Fig. 8.1). Among them, retinol is the most biologically active form of vitamin A, which is oxidized to retinal or retinoic acid or esterified with fatty acids such as acetic and palmitic acids due to the specific chemical reactivity of the 15th carbon. One international unit (IU) of retinol, retinyl acetate, and retinyl palmitate equals 0.30, 0.344, and 0.55 μg, respectively. Plant pigments, carotenes, are tail-conjoined retinoids widely known as provitamin A. The enzymatic conversion by β-carotene 15,15′-dioxygenase and retinaldehyde reductase is essential to convert provitamin A into active retinoids in the intestine. Retinyl esters in food are broken down to retinol by pancreatic lipase, carboxylic ester lipase, and retinyl ester hydrolases and absorbed into enterocytes through facilitated diffusion and passive uptake. The retinol bound to retinol-binding protein type II in the enterocytes is re-esterified by lecithin-retinol acyltransferase or acylated coenzyme A and transported to the bloodstream by the lymphatic system (Newcomer and Ong, 2000). Vitamin A absorption is comparatively high in the intestine, and 80%–90% of the total ingested vitamin A is absorbed (Blaner et al., 2016). Tissues uptake

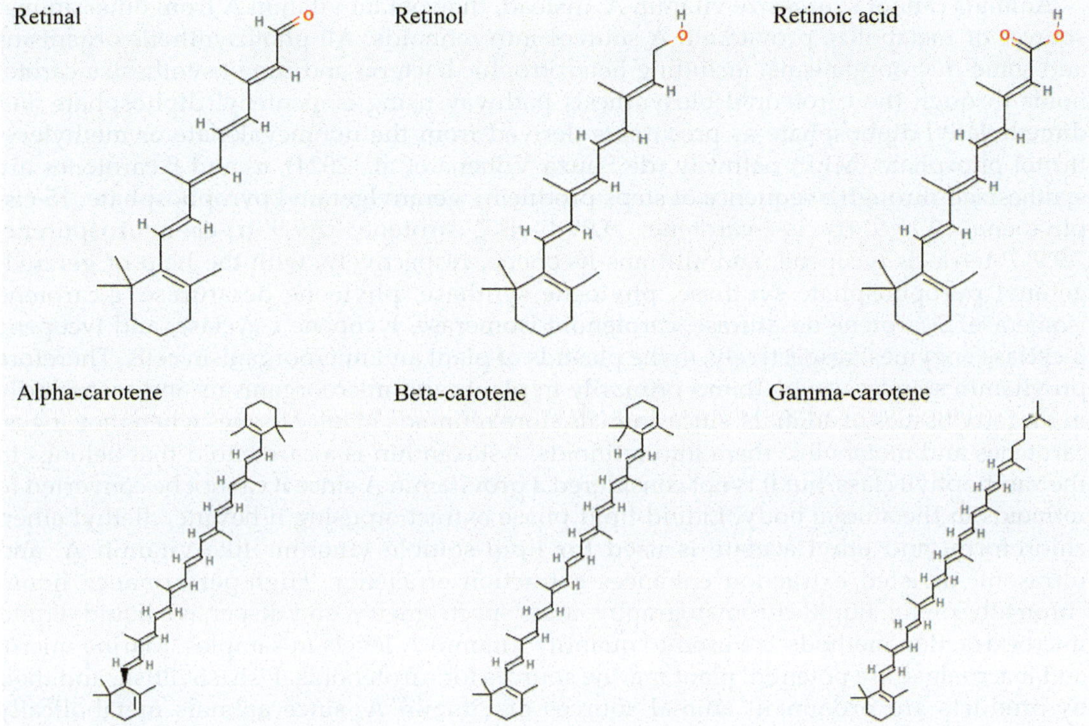

FIG. 8.1 Molecular structures of retinoids.

retinyl esters from chylomicrons, and most of the esters are stored in the hepatocytes and stellate cells in the liver, which hold 50%–80% of the total body vitamin A (Blomhoff et al., 1990).

Vitamin A is essential for vision, reproductive performance, and epithelial cell differentiation. 11-cis-retinal is bound to specific lysyl residues of the rhodopsin and iodopsin visual pigments in rod and cone retina cells, maintaining normal vision. In addition, retinoids influence epithelial cell differentiation by binding to nuclear chromatin to signal transcriptional processes, acting as analogs to steroid hormones. Retinoic acid enhances growth as a growth hormone in pituitary cells, acting synergistically with the thyroid hormone. Vitamin A is essential for bone mineralization, reproduction, embryogenesis, hematopoiesis, and proper immune function. The RDA for males and females in the 1–3-, 4–8-, and 9–13-year age categories is 300, 400, and 600 μg retinol activity equivalent (RAE) per day, respectively, while the RDA for males and females above 14 years is 900 and 700 μg RAE per day. During pregnancy and lactation, vitamin A requirements increase, and the RDAs for pregnant women below and above 18 years are 750 and 770 μg RAE per day, but the RDA for lactating women below and above 18 years is 1200 and 1300 μg RAE per day, respectively.

Animals cannot synthesize vitamin A; instead, they obtain vitamin A from other animal sources or metabolize provitamin A sources into retinoids. All photosynthetic organisms and some microorganisms, including heterotrophic bacteria and fungi, synthesize carotenoids through the carotenoid biosynthesis pathway using isopentenyl diphosphate and dimethylallyl diphosphate as precursors derived from the nonmevalonate or methylerythritol phosphate (MEP) pathway (de Souza Vilhena et al., 2024). α- and β-carotenes are synthesized through a sequence of steps producing geranylgeranyl pyrophosphate, 15-cis-phytoene, 9,15,9′-tri-cis-ζ-carotene, 9,9′-di-cis-ζ-carotene, 7,9,9′-tri-cis-neurosporene, 7,9,9′,7′-tetra-cis-lycopene, and all-trans-lycopene, respectively, with the help of geranylgeranyl pyrophosphate synthase, phytoene synthase, phytoene desaturase, ζ-carotene isomerase, ζ-carotene desaturase, carotenoid isomerase, lycopene ε-cyclase, and lycopene β-cyclase enzymes, respectively, in the plastids of plant and microorganism cells. Therefore provitamin sources can be found primarily in plants and microorganisms and secondarily in the fatty tissues of animals, since animals store retinoids in fatty tissues when they ingest carotenes and metabolize them into retinoids. Astaxanthin is a carotenoid that belongs to the xanthophyll class, but it is not considered a provitamin A since it cannot be converted to retinoids in the animal body. Liquid-lipid phase extraction using n-hexane, diethyl ether, chloroform, and ethyl acetate is used for lipid-soluble vitamins like vitamin A, and ultrasonic-assisted extraction enhances extraction efficiency. High-performance liquid chromatography, liquid chromatography–mass spectrometry, and dispersive liquid–liquid microextraction methods are used to quantify vitamin A levels in samples. Marine micro- and macroalgae are potential plant marine sources for carotenoids. Fish, shellfish, and their by-products are prominent animal sources of vitamin A, since animals metabolically convert ingested carotenoids into retinoids (Tables 8.1 and 8.2).

8.2.2 Vitamin D

Vitamin D belongs to a fat-soluble vitamin group of steroids known for preventing rickets in humans. Two major vitamers of vitamin D are ergocalciferol (vitamin D2) and cholecalciferol (vitamin D3), which are derived under UVB irradiation from their precursors: plant-derived ergosterol and 7-dehydrocholesterol, synthesized in animals (Fig. 8.2). The deficiency of this vitamin (rickets and osteomalacia) has been reported since ancient times, and the relationship between rickets and sunlight was identified in the late 1800s. McCollum's laboratory identified an antirachitic factor in fish oil, later named vitamin D (McCollum et al., 1921). The structure of synthetic vitamin D2 was identified by Windaus et al. (1932) and was named ergocalciferol. The same scientist identified the structure of the naturally derived vitamin D in cod liver oil, cholecalciferol, in 1936. However, the mode of action of the vitamin was identified 30 years after the structures were discovered. One IU equals 0.025 mg of vitamin D2 or vitamin D3. 7-Dehydrocholesterol is a precursor for vitamin D3, which is synthesized in skin sebaceous glands from cholesterol. UVB irradiation (UVB light between 290 and 310 nm, with peak synthesis occurring between 295 and 297 nm wavelengths) converts 7-dehydrocholesterol into provitamin D3, and the resulting provitamin D3 is reabsorbed into the epidermis layers; the highest concentrations are observed in the stratum basale and stratum spinosum. Following the cleavage of the B ring, rotation of the A ring

TABLE 8.1 Marine vitamin A sources and available contents.

Marine source	Form	Content	References
Fish, shellfish, and echinoderms			
Cod—raw fillet	Retinol	8–12 μg/100 g	(Ball, 2005)
Cod—oil	Retinol	9009 μg/100 g	(Ball, 2005)
Sardines	Retinol	6 μg/100 g	(Ball, 2005)
Tuna—whole	Retinol equivalent	80–830 μg/100 g	(Ball, 2005)
Tuna—liver	Vitamin A	7.83 ± 0.18 μg/g	(Ball, 2005)
Halibut—raw fillet	Retinol	15–47 μg/100 g	(Booth et al., 1992)
Halibut—oil	Retinol	15,000 μg/100 g	(Booth et al., 1992)
Requiem shark	Retinol	143,000 μg/100 g	(Booth et al., 1992)
Herring, raw	Retinol	44 μg/100 g	(Booth et al., 1992)
Dogfish (*Squalus acanthias*) liver oil	Retinol	66.8 mg/100 g oil	(Sunarya et al., 1996)
Thick shell mussel (*Mytilus coruscus*)	Retinol	3.61 ± 0.89 mg/kg	(Xing et al., 2023)
	Retinyl palmitate	2.41 ± 0.87–10.1 ± 1.6 mg/100 g	(Xing et al., 2023)
	β-Carotene	1.18 ± 0.80–4.52 ± 0.50 mg/100 g	(Xing et al., 2023)
Brown shrimp (*Crangon crangon*)	Vitamin A	537.0 ± 0.6 μg/100 g^{-1}	(Merdzhanova et al., 2014)
Antarctic Krill (*Euphasia superba*)	Vitamin A	16.40–28.55 mg/100 g	(Xie et al., 2017)
Pacific oyster (*Crassostrea gigas*)	Retinol	1.84 ± 0.90–4.59 ± 1.37 mg/100 g	(Xing et al., 2023)
	Retinyl palmitate	1.08 ± 0.052–2.88 ± 0.67 mg/100 g	(Xing et al., 2023)
	β-Carotene	1.25 ± 0.59–2.76 ± 0.58 mg/100 g	(Xing et al., 2023)
Farrer's scallop (*Azumapecten farreri*)	Retinol	1.25 ± 0.72–3.37 ± 1.03 mg/100 g	(Xing et al., 2023)
	Retinyl palmitate	2.61 ± 0.92–3.99 ± 0.69 mg/100 g	(Xing et al., 2023)
	β-Carotene	2.72 ± 0.51–10.1 ± 0.4 mg/100 g	(Xing et al., 2023)
Blood clam (*Scapharca subcrenata*)	Retinyl palmitate	4.23 ± 0.38–7.34 ± 0.74 mg/100 g	(Xing et al., 2023)
	β-Carotene	3.28 ± 0.64–40.6 ± 1.7 mg/100 g	(Xing et al., 2023)
Razor clam (*Sinonovacula constricta*)	Retinyl palmitate	7.32 ± 1.79–6.87 ± 0.37 mg/100 g	(Xing et al., 2023)
	β-Carotene	3.05 ± 0.51–29.4 ± 0.9 mg/100 g	(Xing et al., 2023)
Juvenile Octopus (*Octopus officinalis*)	Retinol	2.2 ± 0.8 μg/g (dry weight)	(Villanueva et al., 2009)

Continued

TABLE 8.1 Marine vitamin A sources and available contents—cont'd

Marine source	Form	Content	References
Macro algae			
Hypnea musciformis	Vitamin A	1.25±0.47 μM	(Balamurugan et al., 2013)
Gracilaria bursapastoris	Vitamin A	1.31±0.36 μM	(Yildiz et al., 2011)
Ulva rigida C.	Vitamin A	0.91±0.047 μM	(Yildiz et al., 2012)
Jania rubens	Retinol acetate	0.043±0.005 μg/g	(Caf et al., 2019)
Laurencia obtusa	Retinol acetate	0.15±0.05 μg/g	(Caf et al., 2019)
Corallina elongata	Retinol acetate	0.12±0.02 μg/g	(Caf et al., 2019)
Sargassum vulgare	Retinol acetate	0.005±0.001 μg/g	(Caf et al., 2019)
Cystoseira barbata	Retinol acetate	0.003±0.001 μg/g	(Caf et al., 2019)
Zooplankton			
Zooplankton	Retinol	0.2 μg/g (dry weight)	(van der Meeren et al., 2008)
Rotifer	Retinol	0.2 μg/g (dry weight)	(van der Meeren et al., 2008)
Microalgae			
Nannochloropsis gaditana	Chlorophyll *a*	3.71±0.10	(Camacho-Rodríguez et al., 2015
	Chlorophyll *b*	1.72±0.04	(Camacho-Rodríguez et al., 2015)
	β-Carotene	100 mg/100 g	(Di Lena et al., 2019)
Isochrysis galbana	β-Carotene	53 mg/100 g	(Di Lena et al., 2019)
Porphiridium cruentum	β-Carotene	53 mg/100 g	(Di Lena et al., 2019)
Phaeodactylum tricornutum	β-Carotene	34 mg/100 g	(Di Lena et al., 2019)
Tetraselmis suecica	β-Carotene	43 mg/100 g	(Di Lena et al., 2019)
	Vitamin A	4280 μg/g (dry basis)	(De Roeck-Holtzhauer et al., 1991)
Isochrysis galbana	Vitamin A	270 μg/g (dry basis)	(De Roeck-Holtzhauer et al., 1991)
Chaetoceros calcitrant	Vitamin A	970 μg/g (dry basis)	(De Roeck-Holtzhauer et al., 1991)
Pavlova lutheri	Vitamin A	262 μg/g (dry basis)	(De Roeck-Holtzhauer et al., 1991)
Dunaliella tertiolecta	β-Carotene	0.62 mg/100 g	(Diprat et al., 2017)
Heterochlorella luteoviridis	β-Carotene	0.50 mg/100 g	(Diprat et al., 2017)
Scenedesmus bijuga	β-Carotene	166 mg/100 g	(Fernandes et al., 2020)
Chlorella sorokiniana	β-Carotene	156 mg/100 g	(Fernandes et al., 2020)

TABLE 8.2 Chemical structures of the most relevant carotenoid pigments present in marine organisms, sources and yields.

Carotenoid	Structure, Marine Sources ($mg.g^{-1}$ DW)	
Astaxanthin	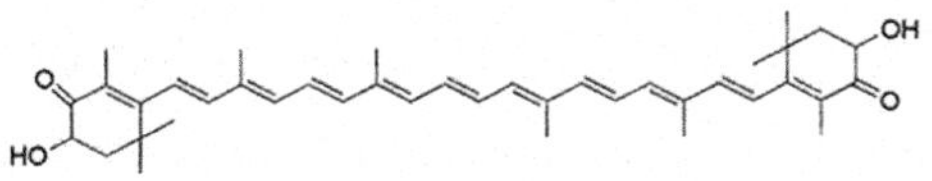	Crustacean waste: *P. Muelleri* (0.3); *Parapenaeopsis stylifera* (0.15); *Pandalus borealis (0.15); Penaeus monodon (0.1): Farfantepenaeus paulensis (0.05); vannamei (0.05)* (Cretton et al. 2021)
Fucoxanthin	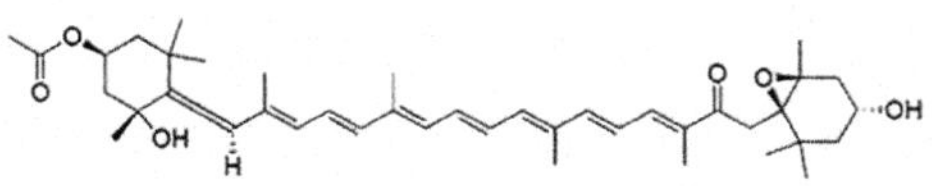	Seaweeds; *Phaeodactylum tricornutum (15-71); Odontella aurita (14-15); Chaetoceros calcitrans (5.13); Isochrysis galbana T-ISO (3.2); Nitzchia laevis (2.4); Isochrysis galbana (2.2); Chaetoceros calcitrans (2.3)*
b-carotene	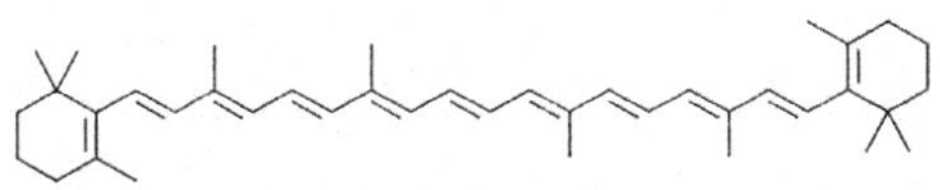	*Green alga Caulerpa racemosa (0,2)* (Kurniawan et al. (2023)). Microalgae (e.g. Spirulina, Dunaliela and Chlorella) (Alasalvar et al. (2002))
Canthaxanthin	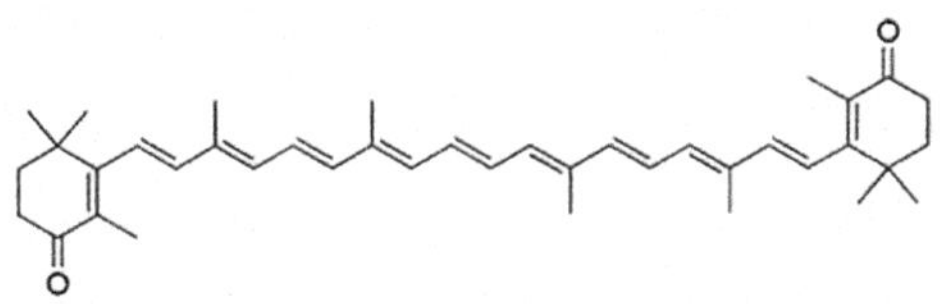	*Green alga Caulerpa racemosa (015)* (Kurniawan et al. (2023)). Coelastrella sp F50 *Sea urchin, starfish, holoturians, Crustaceans, algae and Cyanobacteria* (Galasso et al., 2017)
Lutein	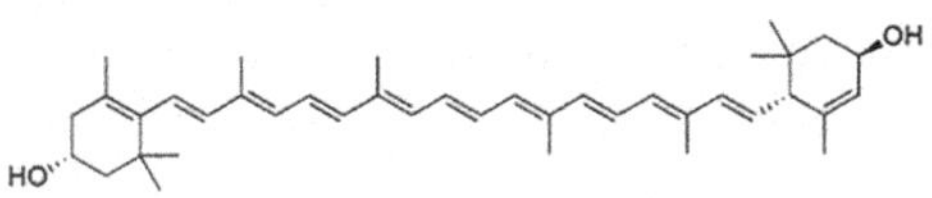	Microalgae: *Dunadelia salina, Chlorella species, Seagrass, coral and shelfishs* (Galasso et al., 2017)
Zeaxanthin	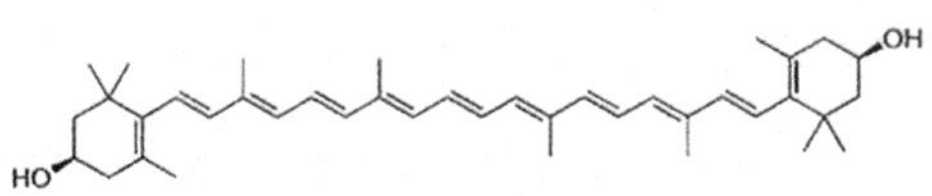	Microalgae: *Dunaliella salina; Seagrass: Posidonia oceanica, Cymodocea nodosa, Zostera noltii and Halophila stipulacea;* Fungi-like protists: *Thraustochytrium strains ONC-T18 and CHN-1, Thraustochytriidae sp. AS4-A1 (Ulkenia sp.) and Aurantiochytrium sp. KH105;* Corals (due to their associations with cyanobacteria) and shellfishes (Galasso et al., 2017)

under UV irradiation, and isomerization of the *cis* form to *the trans* form, vitamin D3 is synthesized. Apart from the natural synthesis of vitamin D under UV irradiation, vitamin D is chemically synthesized from cholesterol on a commercial scale. The molecular weights of cholecalciferol and ergocalciferol are 384.65 and 396.6 g/mol, respectively, and their melting points are around 84°C–85°C.

In the intestine, dietary vitamin D is absorbed into the enterocytes through dietary lipids attaching to the mixed bile salts and then incorporated into the chylomicrons at the enterocytes. Vitamin D is transported in the circulatory system in two ways: incorporated into chylomicrons and attached to vitamin D-binding Gc-protein, known as transcalciferin.

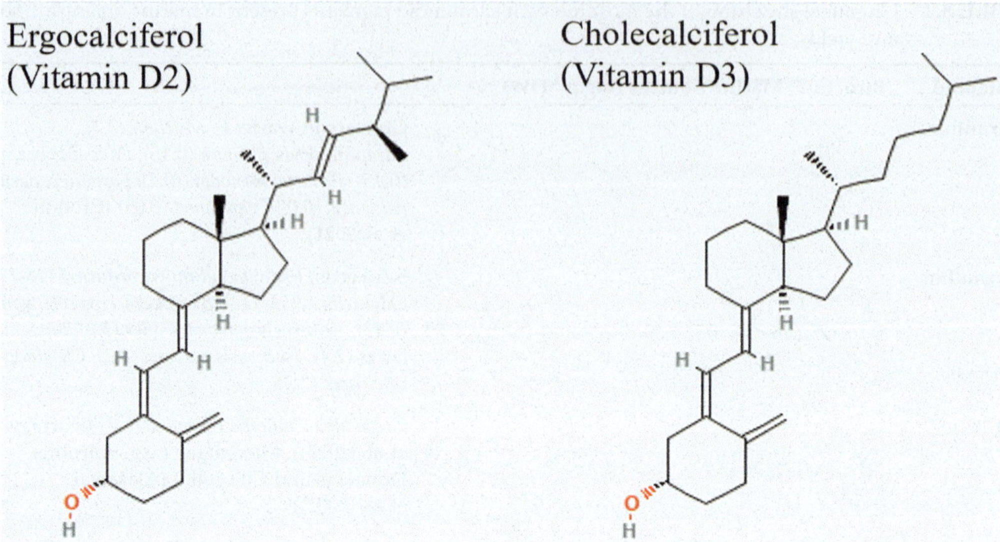

FIG. 8.2 Molecular Structures of Vitamin D.

Vitamin D synthesized in the skin is also transported by binding to transcalciferin due to its high affinity. Compared to other vitamins, vitamin D is not stored in the liver, but higher concentrations are observed in adipose tissues and plasma (Mawer et al., 1972). During the metabolism of vitamin D, it undergoes several hydroxylation processes in the body. Cholecalciferol 25-hydroxylase converts cholecalciferol into 25-hydroxycholecalciferol in the liver. It is further hydroxylated into 24,25-hydroxycalcidiol and 1,25-hydroxycalcidiol (calcitriol) by calcidiol 24-hydroxylase and calcidiol 1-α-hydroxylase in the kidneys. Both calcidiol and calcitriol are catabolized to calcitetrol and then to calcitroic acid, and the final product is excreted in bile or urine.

Vitamin D plays a significant role in maintaining plasma calcium levels by increasing intestinal calcium absorption, minimizing calcium excretion through reabsorption at the renal tubule, and metabolizing calcium from bones. Calcitriol regulates gene expression by binding to and activating nuclear receptors (Freedman, 1999). This receptor belongs to the zinc finger protein family, similar to other steroid hormone receptors that show a higher affinity for calcitriol and ercalcitriol.

Vitamin D synthesis depends on exposure to ultraviolet B irradiation, and environmental factors affect irradiation, such as season, time of day, latitude, and skin properties, including the intensity of melanin pigmentation (Holick, 1996). Ergocalciferol is limited in nature compared to ergosterol and can be found in sun-dried plant materials. Except for Ca^{++} homeostasis, vitamin D is involved in cell differentiation, including osteoclast and keratinocyte differentiation, maturation of leukemic cells into macrophages, hair follicle development, and suppression of adipocyte development. Vitamin D stimulates insulin production from β cells in the pancreas, modulates membrane fluidity by altering fatty acid composition, and affects mitochondrial energy production by regulating isocitrate lyase and malate synthase enzymes and influencing oxidative phosphorylation. Moreover, calcitriol regulates the immune system by affecting the proliferation, differentiation, and immune response of monocytes and lymphocytes. 1,25-Calcidiol suppresses inflammation by regulating 1,25-calcidiol-dependent cytokine production and inhibits parathyroid hormone gene expression and DNA synthesis in epidermal cells. Vitamin D intake has been shown to enhance cardiovascular health, reduce the risk of bone fractures, and lower the risk of developing autoimmune diseases such as rheumatoid arthritis, multiple sclerosis, type 1 and type 2 diabetes, and inflammatory skin conditions. Healthy vitamin D levels in the body reduce the risk of prostate, colon, and breast cancers. However, vitamin D deficiency conditions lead to rickets, osteomalacia, osteoporosis, and musculoskeletal pain. The recommended average intake of vitamin D for 0–50-year-old males and females, as well as pregnant and lactating women, is 5 μg/day. For people between 51 and 70 years, it is 10 μg/day, and for those over 70, it is 15 μg/day. Considering the lipophilic nature of vitamin D, organic solvent extraction methods such as saponification are used to extract vitamin D from samples. During the selection of analytical methods, several factors such as the stability of the vitamin D metabolite, levels of vitamin D in the sample, the solubility of vitamin D, isomerization and isobaric interference, response to environmental factors such as light and temperature, and specific biological activity and affinity of the metabolite to vitamin D binding proteins are considered. In clinical studies, 25-hydroxycalcidiol is frequently analyzed due to its longer half-life in biological tissues. The bioavailability of vitamin D in food or supplements is evaluated by rat bioassays as explained in the Association of Official Analytical Chemists International Method 936.14, though there are several constraints.

Primarily, vitamin D is abundant in marine animal flesh and oils such as cod, eel, salmon, trout, herring, and tuna. The origin of vitamin D in marine fish is unclear due to insufficient exposure to UVB irradiation in their habitats (Table 8.3). It is suggested that fish acquire vitamin D by consuming plankton and algae and accumulate it through food chains. Microalgae and plankton near the water surface synthesize vitamin D due to their higher exposure to UV irradiation compared to marine animals.

8.2.3 Vitamin E

Evans and Bishop (1922) identified vitamin E and named it Factor X. In 1938 Fernholz identified the chemical structure of α-tocopherol. Vitamers of vitamin E were identified in the early 1960s. Vitamin E (parent compound tocol, 2-methyl-2-(4′,8′,12′-trimethyltridecyl)-chroman-6-ol) is a plant-derived yellowish, viscous, and fat-soluble compound comprised of a chromanol ring and a C16 carbon side chain bound to the C2 position of the chromanol ring. Two different groups of vitamin E can be identified according to the saturation of side chains. Tocopherols contain a saturated phytyl side chain, and tocotrienols have an isoprenoid side chain containing three double bonds at the C3, C7, and C11 positions. Four different forms of tocopherols and tocotrienols are classified based on the degree and position of methylation on the chromanol ring. The α-configuration contains three methyl groups attached to the C5, C7, and C8 positions of the chromanol ring. The β configuration contains two methyl groups at the C5 and C8 positions, while the methyl groups of γ tocopherol are located at the C7 and C8 positions. The δ configuration contains only a single methyl group at the C8 position. Eight stereoisomers per tocopherol can be identified according to the arrangement of three asymmetric chiral carbons at

TABLE 8.3 Marine vitamin D sources and available contents.

Marine source	Marine source	Content	Reference
Fish, shellfish, and echinoderms			
Eel	Vitamin D3	25.6 μg/100g	Lamberg-Allardt, 2006
Herring	Vitamin D3	15.4 μg/100g	Lamberg-Allardt, 2006
Fish oil from anchovy leftovers	Vitamin D3	81.5 μg/kg oil	Scurria et al., 2020
Salmon	Vitamin D3	12.4 μg/100g	Scurria et al., 2020
Tuna	Vitamin D3	7.2 μg/100g	Scurria et al., 2020
Cod	Vitamin D3	7.0 μg/100g	Scurria et al., 2020
Cod	Vitamin D3	5 ng/g wet weight	Ostermeyer and Schmidt, 2006
Cod	Provitamin D3	97 ng/g wet weight	Ostermeyer and Schmidt, 2006
Cod -liver	Vitamin D3	44 μg/100g	Brustad et al., 2003
Cod - liver oil	Vitamin D3	122 μg/100g	Brustad et al., 2003
Rainbow trout	Vitamin D3	7.5 ± 1.7 μg/100 g raw sample	Ložnjak and Jakobsen, 2018
	Vitamin D2	<0.1 μg/100 g raw sample	
Sole	Vitamin D3	2.37 ± 0.04 μg/100g	Byrdwell et al., 2013
Halibut	Vitamin D3	5.82 ± 0.01 μg/100g	Byrdwell et al., 2013
Rockfish	Vitamin D3	6.16 ± 0.03 μg/100g	Byrdwell et al., 2013
Sardine - Canned	Vitamin D3	3.68 ± 0.02 μg/100g	Byrdwell et al., 2013
Swordfish	Vitamin D3	9.86 ± 0.54 μg/100g	Byrdwell et al., 2013
Pollack (*Pollachius pollachius*)	Vitamin D3	13 ng/g wet weight	Ostermeyer and Schmidt, 2006
	Provitamin D3	138 ng/g wet weight	
Turbot (*Psetta maxima*)	Vitamin D3	23 ng/g wet weight	Ostermeyer and Schmidt, 2006
	Provitamin D3	164 ng/g wet weight	
Horse Mackerel, *Trachurus trachurus*	Vitamin D3	62 ng/g wet weight	Ostermeyer and Schmidt, 2006
	Provitamin D3	247 ng/g wet weight	
Mackerel, *Scomber scombrus*	Vitamin D3	39 ng/g wet weight	Ostermeyer and Schmidt, 2006
	Provitamin D3	1725 ng/g wet weight	
Pilchard, *Sardina pilchardus*	Vitamin D3	212 ng/g wet weight	Ostermeyer and Schmidt, 2006
	Provitamin D3	646 ng/g wet weight	
Clam - Steamed	Vitamin D3	0.41 ± 0.03 μg/100g	Byrdwell et al., 2013

TABLE 8.3 Marine vitamin D sources and available contents—cont'd

Marine source	Marine source	Content	Reference
Mussel	Vitamin D3	0.20 ± 0.17 μg/100g	Byrdwell et al., 2013
Black Mussel	Vitamin D3	14.8 ± 1.0 μg/100g	Merdzhanova et al., 2014
Oyster	Vitamin D3	0.07 ± 0.02 μg/100g	Byrdwell et al., 2013
Scallops	Vitamin D3	0.12 ± 0.07 μg/100g	Byrdwell et al., 2013
Veined rapa whelk (*Rapana venosa*)	Vitamin D3	3.90 ± 0.40 μg/100g	Merdzhanova et al., 2014
Blue Crab, Steamed	Vitamin D3	0.05 ± 0.03 μg/100g	Byrdwell et al., 2013
Shrimp (unspecified variety)	Vitamin D3	0.53 ± 0.09 μg/100g	Byrdwell et al., 2013
Brown shrimp	Vitamin D3	12.99 ± 1.2 μg/100g	Merdzhanova et al., 2014
Phytoplankton			
Cyanobacteria *Arthrospira maxima*	Vitamin D3	0.004 μg/g DW	Ljubic et al., 2020
Chlorella minutissima	Vitamin D3	0.004 μg/g DW	Ljubic et al., 2020
Tetraselmis sp. CS-362	Vitamins D2 or D3	0.35 μg/g DW	Brown et al., 1999
Tetraselmis suecica	Vitamin D2	14 μg/g DW	De Roeck-Holtzhauer et al., 1991
Stichococcus sp. CS-92	Vitamins D2 or D3	0.35 μg/g DW	Brown et al., 1999
Rhodomonas salina	Vitamin D3	0.004 μg/g DW	Ljubic et al., 2020
Skeletonema costatum	Vitamin D2	11 μg/g DW	De Roeck-Holtzhauer et al., 1991
Isochrysis galbana	Vitamin D2	5 μg/g DW	De Roeck-Holtzhauer et al., 1991
Pavlova lutheri	Vitamin D2	39 μg/g DW	De Roeck-Holtzhauer et al., 1991
Pavlova pinguis	Vitamins D2 or D3	0.35 μg/g DW	Brown et al., 1999
Nannochloropsis sp. CS-246	Vitamins D2 or D3	0.35 μg/g DW	Brown et al., 1999
Nannochloropsis oceanica	Vitamin D3	0.48 μg/g DW	Ljubic et al., 2020
Phytoplankton (Chlorophyceae, Cyanophyceae, Bacillariophyceae, Euglenineae)	Provitamin D2	3889.0 ng/g (dry basis)	Rao and Raghuramulu, 1996
	Provitamin D3	23,581.0 ng/g (dry basis)	
	Vitamin D2	52.5 ng/g (dry basis)	
	Vitamin D3	803.5 ng/g (dry basis)	

Continued

TABLE 8.3 Marine vitamin D sources and available contents—cont'd

Marine source	Marine source	Content	Reference
Zooplankton			
Rotifers	Cholecalciferol	0.9 μg/g (DW)	van der Meeren et al., 2008
Artemia	Cholecalciferol	0.7 μg/g (DW)	van der Meeren et al., 2008
Zooplankton (Protozoans, Rotifers, planktonic forms of crustaceans)	Provitamin D2	7167.0 ng/g (dry basis)	Rao and Raghuramulu, 1996
	Provitamin D3	46,238.0 ng/g (dry basis)	
	Vitamin D2	724.0 ng/g (dry basis)	
	Vitamin D3	2717.0 ng/g (dry basis)	

positions 2 of the chromanol ring and 4 and 8 of the carbon side chain. For instance, RRR-, RSR-, RRS-, RSS-, SRR-, SSR-, SRS-, and SSS-α-tocopherol are eight α-tocopherol stereoisomers. Among them, only RRR-α-tocopherol is the naturally occurring form, and synthetic α-tocopherol (*all-rac*-α-T) is a racemic mixture of equal parts of each stereoisomer. Tocotrienols have only two stereoisomers due to double bonds in the carbon side chain, which lacks chiral carbons, and a single chiral stereocenter at the 2-position of the chromanol ring. Ester forms of vitamin E, such as acetate, nicotinate, succinate, and phosphate, are less prone to oxidation under heat, light, and alkaline conditions. Therefore these esters are commonly used in commercial applications. Based on biological assays in vitamin E-deficient rats, the biological activity of vitamin E widely varies depending on chemical structure, stereoisomeric configuration, and relative biological effectiveness of its vitamers (Bieri and Evarts, 1974). RRR-α-tocopherol possesses the highest biological activity, making it the most biologically significant form of vitamin E. Ninety percent of plasma vitamin E is α-tocopherol due to liver vitamin E metabolism and α-tocopherol binding protein. Different forms of vitamin E are not interconvertible in humans; therefore they act differentially in metabolism (Colombo, 2010).

Vitamin E is absorbed in the gut through passive diffusion, dietary lipids, and receptor-mediated transporters such as scavenger receptor-B1 and Niemann-Pick C1-like protein-1. The absorption of vitamin E in the intestine heavily depends on the presence of dietary lipids, functional bile, and pancreatic esterase for the formation of micelles and for maintaining the stability of micelles. Vitamin E is primarily concentrated in lysozyme and Golgi apparatus membranes in nonadipose tissues. In adipose tissues, which contain approximately 90% of the body's vitamin E, α-tocopherol is deposited in the bulk lipid phase, forming a fixed pool that metabolizes slowly over a long period of time (Traber and Kayden, 1987).

Vitamin E has long been known as a lipid-soluble, chain-breaking antioxidant in cell membranes and fatty tissues (Traber and Atkinson, 2007). Later, it was categorized under vitamins due to its role in certain functions, including testicular atrophy and necrotizing myopathy, which synthetic antioxidants cannot alleviate. Beyond its activity as an antioxidant, vitamin E is involved in cell signaling, platelet aggregation, interaction or inhibition of protein kinase C, cell proliferation, and modulation of gene expression (Giasuddin and Diplock, 1981; Yu et al., 1999; Wang and Quinn, 2000). Vitamin E stabilizes biological systems by scavenging free radicals as a lipid-soluble antioxidant and maintaining redox balance. Free radicals are unstable and highly reactive molecules with unpaired electrons generated due to natural cellular processes such as energy production, immune responses, autoxidation of heme proteins, and other enzymatic reactions, or external factors like UV irradiation, toxins, and pollutants. Numerous reactive oxygen species (ROS) can be identified in animal tissues, such as superoxide, hydrogen peroxide, hydroxyl, phenoxyl, singlet oxygen, alkoxyl, hydroperoxide, hydroperoxyl, semiquinone, carbonate, hypochlorous groups, hypobromous acid, hypothiocyanous acid, chloramines, bromamines, nitric oxide, peroxynitrite, and nitrogen dioxide. ROS are not considered entirely harmful since they are essential for combating pathogens, cell signaling, and metabolic reactions. However, excessive levels of ROS can induce cell damage and apoptosis. Apart from antioxidant enzymes, including superoxide dismutase, catalase, and glutathione peroxidase, antioxidants play a significant role in reducing the cell damage caused by ROS. Among them, vitamin E scavenges ROS, especially peroxyl radicals, in the lipid phase due to the lipophilic property of vitamin E. Lipid peroxidation occurs in three steps: chain initiation, chain propagation, and chain termination. When reactive species attack lipids, lipid radicals are formed. The second step begins with a lipid radical reacting with molecular oxygen, producing a lipid peroxyl radical. The lipid peroxyl radical abstracts a hydrogen atom from other lipids, forming lipid hydroperoxide and another lipid radical, which can propagate the reaction continuously. Chain termination can occur due to two reactions. One is when the resultant lipid radical reacts with another lipid radical, forming a stable nonradical product, a lipid dimer.

Saponification, solvent extraction, molecular distillation, supercritical fluid extraction, adsorption chromatography, enrichment from deodorizer distillates, crystallization, and membrane purification are techniques used for Vitamin E extraction (Parrish and Waltking, 2009; Ball, 1988; Bourgeois, 1992; AOAC International, 2005). Except for oils, the extraction of vitamin E from food or plant samples is quite complicated, and one or more of the above extraction techniques can be useful. The high-performance liquid chromatography method can detect eight vitamers of vitamin E using a fluorescence detector within a 25-minute runtime (Ye et al., 1998; Chun et al., 2006). Ultraviolet and evaporative light scattering detectors also provide precise and versatile data. The ultra-performance convergence chromatography technique is an advanced method that can identify tocopherol and α-tocotrienol in tocotrienol-rich fractions. Mass spectrometry coupled with gas or liquid chromatography greatly enhances the precision and accuracy of Vitamin E analysis to detect 10–1000 ng/mL concentrations (Kiyose et al., 1999; Indyk, 1988; Cooper et al., 1997; Delgado-Zamarreño et al., 2004). A Coulochem electrode array system has been developed to achieve well-resolved analyte separation (Gamache et al., 2024).

Adequate intake levels for infants aged 0–6 and 7–12 months are 4 and 5 mg/day, respectively, with no defined upper limit. The estimated average requirements for the age categories 1–3, 4–8, 9–13, and 14 years and older are 5, 6, 9, and 12 mg/day, respectively. The corresponding recommended dietary allowances are 6, 7, 11, and 15 mg/day, with upper limits of 200, 300, 600, 800, and 1000 mg/day, respectively. Estimated average requirements and recommended dietary allowances for pregnant women are 12 and 15 mg/day, and those levels for lactating women are 16 and 19 mg/day, respectively (Table 8.4; Figs. 8.3 and 8.4).

TABLE 8.4 Vitamin E sources and their quantities.

Marine source	Form	Content	References
Fish, shellfish, and echinoderms			
Cod—liver	α-Tocopherol	5.3–45.03 mg/g tissue	(Malone et al., 2004)
Cod—by-products	α-Tocopherol	0.25–1.69 mg/g tissue	(Malone et al., 2004)
Cod—viscera	α-Tocopherol	0.63–15.74 mg/g tissue	(Malone et al., 2004)
Ling—liver	α-Tocopherol	9.27–18.73 mg/g tissue	(Malone et al., 2004)
Saithe—liver	α-Tocopherol	4.0–25.64 mg/g tissue	(Malone et al., 2004)
Haddock—liver	α-Tocopherol	7.23–90.9 mg/g tissue	(Malone et al., 2004)
Chub mackerel	Vitamin E	1.3 mg/100 g of edible portion	(Dias et al., 2003)
Sardine	Vitamin E	110.1 mg/kg	(Nartea et al., 2023)
Salmon	Vitamin E	4.0 mg/100 g of edible portion	(Dias et al., 2003)
Tuna	Vitamin E	0.64 mg/100 g of edible portion	(Dias et al., 2003)
Red seabream	Vitamin E	0.70 mg/100 g of edible portion	(Dias et al., 2003)
Eel	Vitamin E	2.4 mg/100 g of edible portion	(Dias et al., 2003)
Grey mullet	α-Tocopherol	1112.2 ± 40.3 μg/100 g wet weight	(Stancheva and Dobreva, 2013)
Bonito	α-Tocopherol	121.4 ± 0.6 μg/100 g wet weight	(Stancheva and Dobreva, 2013)
Rainbow trout	α-Tocopherol	121.4 ± 0.6 μg/100 g wet weight	(Stancheva and Dobreva, 2013)
Juvenile Octopus (*Octopus officinalis*)	Total tocopherols	343.3 ± 59.6 μg/g dry weight	(Villanueva et al., 2009)
Squid—Pacific (*Loligo opalescens*)	Total tocopherols	3.0 ± 0.4 mg/100 g wet weight	(Torrinha et al., 2014)
Squid—Atlantic (*Loligo gahi*)	Total tocopherols	1.6 ± 0.3 mg/100 g wet weight	(Torrinha et al., 2014)
Squid—Indian (*Loligo duvaucelii*)	Total tocopherols	0.76 ± 0.41 mg/100 g wet weight	(Torrinha et al., 2014)
Antarctic Krill (*Euphasia superba*)	α-Tocopherol	157.0 ± 17.3 pmol/mg wet tissue	(Dunlap et al., 2002)
Sea cucumber	Vitamin E	21.77–34.8 mg/kg wet weight	(Li et al., 2020)
Sea urchin	α-Tocopherol	17.2–34.5 μg/specimen	(Tsushlma et al., 1997)
Macroalgae			
Ulva spp.	α-Tocopherol	9.10 ± 0.50 μg/g	(Durmaz et al., 2008)
Ulva rigida	Vitamin E	19.70 mg/kg	(Taboada et al., 2010)
Ulva rigida C.	α-Tocopherol	147.00 ± 0.28 mg/100 g	(Yildiz et al., 2012)
Ulva lactuca	Vitamin E	0.06 mg%	(Debbarma et al., 2016)

TABLE 8.4 Vitamin E sources and their quantities—cont'd

Marine source	Form	Content	References
Codium fragile	Total vitamin E	1617.6 6 10.3 mg/g lipid	(Ortiz et al., 2009)
Laminaria digitata	α-Tocopherol	0.275 mg/kg	(MacArtain et al., 2007)
Undaria pinnatifida	α-Tocopherol	1.39 mg/kg	(MacArtain et al., 2007)
Cystoseira spp.	α-Tocopherol	17.10 ± 0.10 μg/g	(Durmaz et al., 2008)
Sargassum spp.	Vitamin E	0.49 mg%	(Debbarma et al., 2016)
Macrocystis pyrifera	Total vitamin E	1457.2 6 11.4 mg/g lipid	(Ortiz et al., 2009)
Zostera spp.	α-Tocopherol	9.90 ± 0.10 μg/g	(Durmaz et al., 2008)
Gracilaria bursapastoris	α-Tocopherol	57.0 ± 38.0 mg/100 g	(Yildiz et al., 2011)
Gracileria edulis	Vitamin E	1.02 mg%	(Debbarma et al., 2016)
Hypnea musciformis	Vitamin—E	0.41 ± 0.13 mg/g	(Balamurugan et al., 2013)
Porphyra umbilicalis	α-Tocopherol	0.114 mg/kg	(MacArtain et al., 2007)
Gracilaria chilensis	Total vitamin E	391.9 6 9.7 mg/g lipid	(Ortiz et al., 2009)
Phytoplankton			
Chlorella stigmatophora	Vitamin E	669.0 mg/kg dry weight	(Fabregas and Herrero, 1990)
Dunaliella tertiolecta	Vitamin E	116.3 mg/kg dry weight	(Fabregas and Herrero, 1990)
Chateoceros calcitrans	Vitamin E	1628 μg/g dry matter	(De Roeck-Holtzhauer et al., 1991)
Skeletonema costatum	Vitamin E	108 μg/g dry matter	(De Roeck-Holtzhauer et al., 1991)
Isochrysis galbana	Vitamin E	117 μg/g dry matter	(De Roeck-Holtzhauer et al., 1991)
Tetraselmis suecica	Vitamin E	6323 μg/g dry matter	(De Roeck-Holtzhauer et al., 1991)
Tetraselmis suecica	Vitamin E	421.8 mg/kg dry weight	(Fabregas and Herrero, 1990)
Pavlova lutheri	Vitamin E	345 μg/g dry matter	(De Roeck-Holtzhauer et al., 1991)
Nannochloropsis oculata	α-Tocopherol	506–2325 μg/g dry weight	(Durmaz, 2007)
Isoehrysis galbana	Vitamin E	58.2 mg/kg dry weight	(Fabregas and Herrero, 1990)
Zooplankton			
Copepods	Total tocopherol	114 ± 61.3 μg/g dry weight	(van der Meeren et al., 2008)
Artemia	Total tocopherol	571.0 μg/g dry weight	(van der Meeren et al., 2008)
Common zooplankton	Total tocopherol	114.0 μg/g dry weight	(van der Meeren et al., 2008)
Rotifers	Total tocopherol	513.1 μg/g dry weight	(van der Meeren et al., 2008)
Marine fungi, archaea, and bacteria			
Spirulina platensis	α-Tocopherol	120 mg/kg	(Becker, 2013)

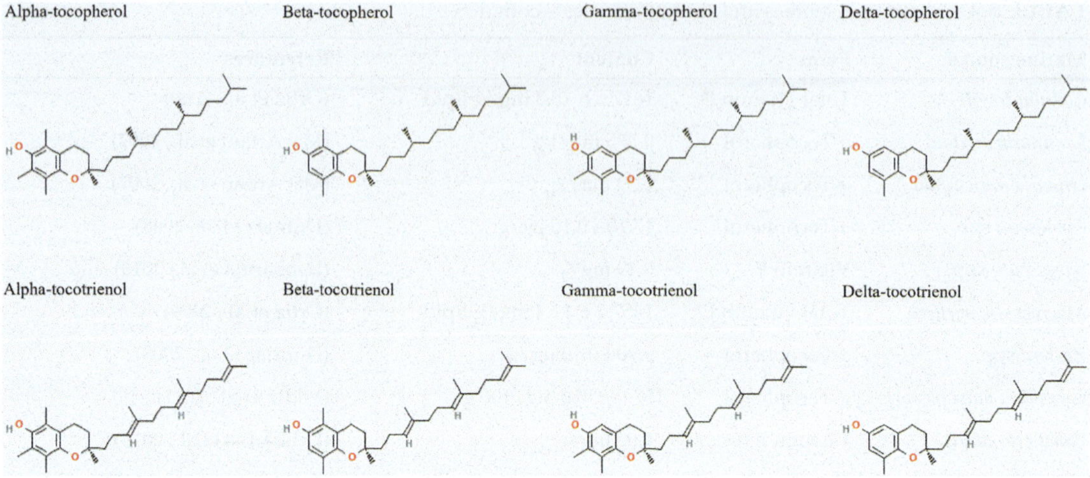

FIG. 8.3 Molecular structures of tocopherols and tocotrienols.

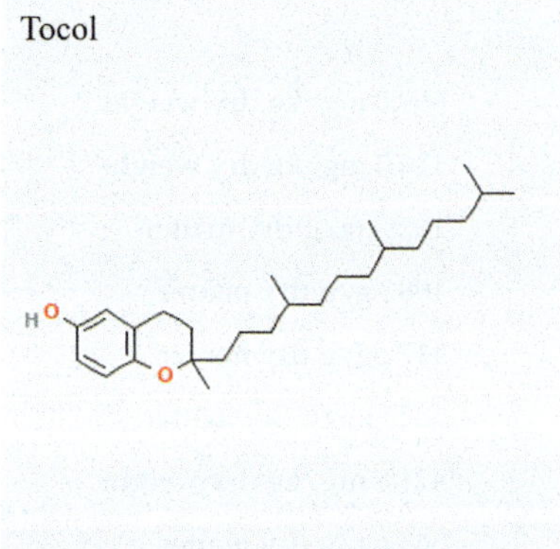

FIG. 8.4 The molecular structure of tocol.

8.2.4 Vitamin K

Three significant vitamers of vitamin K have been identified: one in green plants, another in microbes, and the third is synthetically produced. Vitamin K plays a crucial role in blood clotting. In 1935 Henrik Dam and his group discovered this essential fat-soluble factor and named it vitamin K, derived from the German word "Koagulation" (Dam, 1935). Later, they isolated vitamin K as a yellowish oil from alfalfa leaf extract. Green leafy materials contain phylloquinone (vitamin K1), a phytyl side chain attached to the third carbon atom of the 2-methyl-1,4-naphthoquinone ring structure (MacCorquodale et al., 1939) (Fig. 8.5). Phylloquinone is an electron receptor in the thylakoids, playing a key role in photosynthesis. Bacteria synthesize menaquinones (vitamin K2), which are structurally distinct from phylloquinone due to their differing side chains (Bentley and Meganathan, 1982). According to the

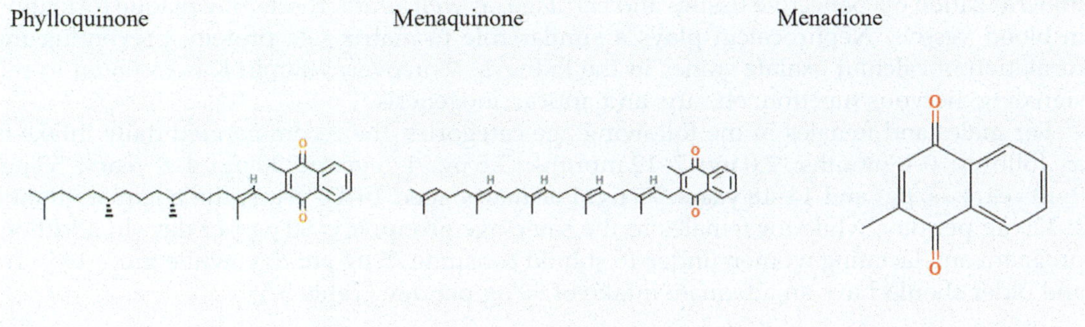

FIG. 8.5 Molecular structures of different forms of vitamin K.

number of isoprenoid units in their side chain attached to the third carbon atom of the core ring, vitamin K2 consists of various isoprenologs. The side chain of menaquinone consists of isoprenyl units, which can vary from 4 to 13. Predominant forms of menaquinones contain 6–10 isoprenoid units. Vitamin K3, also known as menadione, is the simpler form of vitamin K, consisting solely of the 2-methyl-1,4-naphthoquinone ring. Menadione sodium bisulfite, a derivative commonly used for commercial applications such as feed supplements, is stabilized by adding sodium bisulfite. In the presence of excess sodium bisulfite, vitamin K3 crystallizes in the matrix. Overall, vitamin K is termed as a group of molecules containing the 2-methyl-1,4-naphthoquinone ring, and derivatives of this compound exhibit antihemorrhagic properties. When animals ingest phylloquinone or menadione, they are metabolically converted to menaquinone in the body by dealkylating the phytyl side chain of phylloquinone into menadione and then realkylating menadione to menaquinone by adding an isoprenoid side chain. In the body, menaquinones are recycled: quinone reductase reduces menaquinone to dihydroxy vitamin K, and then vitamin K γ-glutamyl carboxylase oxidizes dihydroxy vitamin K into vitamin K 2,3-epoxide in the endoplasmic reticulum and Golgi apparatus. At the final step, vitamin K 2,3-epoxide is reduced back to menaquinone by epoxide reductase. Coumarin-type anticoagulants, such as warfarin, 2-phylloquinone derivatives, and 2,3,5,6-tetrachloro-pyridinol, inhibit the thiol-dependent process that regenerates the reduced form of vitamin K. This results in the accumulation of 2,3-epoxide metabolites. Vitamin K γ-glutamyl carboxylase uses CO_2 as a carboxyl precursor, and O_2 is needed to convert dihydroquinone to vitamin K 2,3-epoxide during the γ-carboxylation of glutamyl residues of vitamin K-dependent blood clotting factors. In blood clot formation, carboxylated glutamyl residues of clotting proteins, membrane phospholipids of blood platelets, and vascular and endothelial cells are bridged via calcium ions (Ca^{2+}). Prothrombin (Factor II), proconvertin (Factor VII), Christmas factor (Factor IX), Stuart factor (Factor X), protein C, protein M, protein S, and protein Z are vitamin K-dependent blood clotting regulatory proteins.

In addition to blood clotting, vitamin K is crucial in bone mineralization. Osteocalcin, periostin, and protein S are vitamin K-dependent proteins containing γ -carboxyglutamate residues. Osteocalcin is synthesized in osteoblasts, and this process is regulated by calcitriol. It binds to Ca^{+2} and hydroxyapatite and ensures that minerals are deposited properly within the bone matrix, preventing abnormal calcification. Matrix Gla protein prevents abnormal mineralization of connective tissues and cartilage, as well as atherosclerotic plaque formation in blood vessels. Nephrocalcin plays a similar role to matrix Gla protein, preventing the formation of calcium oxalate stones in the kidneys. Moreover, vitamin K is essential in cell signaling, nervous function, obesity, and anticarcinogenesis.

For males and females in the following age categories, the recommended daily intake is as follows: 0–6 months: 2.0 μg; 7–12 months: 2.5 μg; 1–3 years: 30 μg; 4–8 years: 55 μg; 9–13 years: 60 μg; and 14–18 years: 75 μg. For males aged 19–70 years, the adequate intake is 120 μg per day, while for females in the same age group, it is 90 μg per day. In addition, pregnant and lactating women under 18 should consume 75 μg per day, while those 19 years and older should have an adequate intake of 90 μg per day (Table 8.5).

TABLE 8.5 Marine vitamin K sources and available contents.

Marine source	Form	Content	References
Fish, shellfish, and echinoderms			
Anglerfish	Phylloquinone	0.3–0.4 ng/g wet weight	(Ostermeyer and Schmidt, 2001)
Cod	Phylloquinone	0.2–0.7 ng/g wet weight	(Ostermeyer and Schmidt, 2001)
	Menaquinone	1.0–2.4 ng/g wet weight	(Ostermeyer and Schmidt, 2001)
Common dab	Phylloquinone	0.1 ng/g wet weight	(Ostermeyer and Schmidt, 2001)
	Menaquinone	0.5 ng/g wet weight	(Ostermeyer and Schmidt, 2001)
Flounder	Phylloquinone	0.1–1.9 ng/g wet weight	(Ostermeyer and Schmidt, 2001)
	Menaquinone	0.5–1.6 ng/g wet weight	(Ostermeyer and Schmidt, 2001)
Haddock	Phylloquinone	0.2–0.3 ng/g wet weight	(Ostermeyer and Schmidt, 2001)
	Menaquinone	0.9 ng/g wet weight	(Ostermeyer and Schmidt, 2001)
Hake	Phylloquinone	0.1 ng/g wet weight	(Ostermeyer and Schmidt, 2001)
	Menaquinone	0.5 ng/g wet weight	(Ostermeyer and Schmidt, 2001)
Lemon sole	Phylloquinone	0.2 ng/g wet weight	(Ostermeyer and Schmidt, 2001)
	Menaquinone	0.9 ng/g wet weight	(Ostermeyer and Schmidt, 2001)
Megrim	Menaquinone	0.3 ng/g wet weight	(Ostermeyer and Schmidt, 2001)
Plaice	Phylloquinone	0.5–0.9 ng/g wet weight	(Ostermeyer and Schmidt, 2001)
	Menaquinone	0.25–1.5 ng/g wet weight	(Ostermeyer and Schmidt, 2001)
Saithe	Phylloquinone	0.2–0.35 ng/g wet weight	(Ostermeyer and Schmidt, 2001)
	Menaquinone	0.8–1.6 ng/g wet weight	(Ostermeyer and Schmidt, 2001)
Turbot	Menaquinone	0.6 ng/g wet weight	(Ostermeyer and Schmidt, 2001)
Whiting	Phylloquinone	0.1–0.5 ng/g wet weight	(Ostermeyer and Schmidt, 2001)
	Menaquinone	0.6–1.1 ng/g wet weight	(Ostermeyer and Schmidt, 2001)

TABLE 8.5 Marine vitamin K sources and available contents—cont'd

Marine source	Form	Content	References
Herring	Phylloquinone	0.4–0.8 ng/g wet weight	(Ostermeyer and Schmidt, 2001)
	Menaquinone	1.7–3.7 ng/g wet weight	(Ostermeyer and Schmidt, 2001)
Mackerel	Phylloquinone	6.1–7.5 ng/g wet weight	(Ostermeyer and Schmidt, 2001)
	Menaquinone	1.1–1.8 ng/g wet weight	(Ostermeyer and Schmidt, 2001)
Bass	Phylloquinone	0.3 ng/g wet weight	(Ostermeyer and Schmidt, 2001)
	Menaquinone	5.1 ng/g wet weight	(Ostermeyer and Schmidt, 2001)
Sprat	Phylloquinone	9.7 ng/g wet weight	(Ostermeyer and Schmidt, 2001)
	Menaquinone	11.5 ng/g wet weight	(Ostermeyer and Schmidt, 2001)
Fish powder	Vitamin K	150 µg/g dry matter	(De Roeck-Holtzhauer et al., 1991)
Shrimp	Vitamin K	1 µg/100 g tissue	(Venugopal and Gopakumar, 2017)
Lobster	Vitamin K	0.1 µg/100 g tissue	(Venugopal and Gopakumar, 2017)
Crab	Vitamin K	0.1 µg/100 g tissue	(Venugopal and Gopakumar, 2017)
Crayfish	Vitamin K	–µg/100 g tissue	(Venugopal and Gopakumar, 2017)
Oyster	Vitamin K	100 µg/100 g tissue	(Venugopal and Gopakumar, 2017)
Clam	Vitamin K	–µg/100 g tissue	(Venugopal and Gopakumar, 2017)
Scallop	Vitamin K	–µg/100 g tissue	(Venugopal and Gopakumar, 2017)
Mussel	Vitamin K	–µg/100 g tissue	(Venugopal and Gopakumar, 2017)
Cuttlefish	Vitamin K	0.1–0.2 µg/100 g tissue	(Venugopal and Gopakumar, 2017)
Macroalgae			
Cystoseira barbata	Vitamin K2	0.82 ± 0.1 µg/g	(Caf et al., 2019)
Sargassum vulgare	Vitamin K2	0.8 ± 0.1 µg/g	(Caf et al., 2019)
Jania rubens	Vitamin K2	0.25 ± 0.05 µg/g	(Caf et al., 2019)
Laurencia obtusa	Vitamin K2	0.8 ± 0.08 µg/g	(Caf et al., 2019)
Corallina elongata	Vitamin K2	0.25 ± 0.05 µg/g	(Caf et al., 2019)
Sargassum muticum	Vitamin K	750 µg/g dry matter	(De Roeck-Holtzhauer et al., 1991)
Microalgae			
Tetraselmis suecica	Vitamin K	28 µg/g dry matter	(De Roeck-Holtzhauer et al., 1991)
Isochrysis galbana	Vitamin K	8 µg/g dry matter	(De Roeck-Holtzhauer et al., 1991)
Pavlova lutheri	Vitamin K	6.5 µg/g dry matter	(De Roeck-Holtzhauer et al., 1991)
Skeletonema costatum	Vitamin K	5.5 µg/g dry matter	(De Roeck-Holtzhauer et al., 1991)

8.3 Water-soluble vitamins

8.3.1 Vitamin B complex

8.3.1.1 Thiamine

With the widespread use of rice polishing machines, beriberi, a peripheral nervous system disease, became a significant problem in the 19th century. It was found that the discarded rice polishings might contain the antiberiberi factor, later discovered as thiamine. Thiamine, a water-soluble vitamin, was the first to be identified and isolated; therefore it is known as vitamin B1. Earlier, thiamine was known as aneurine since it was utilized as a cure for polyneuritis in animals. However, with the advancement of chemistry, it was found that there are both sulfur and amino groups in the molecule; hence, it was named thiamine. The structure of thiamine includes a pyrimidine ring and a thiazole ring, which are linked through a methylene bridge, chemically known as 3-[(4-amino-2-methyl-5-pyrimidinyl) methyl]-5-(2-hydroxyethyl)-4-methylthiazolium ($C_{12}H_{18}ON_4SCl_2$). Two primary forms of thiamine are utilized in the food and pharmaceutical industries: thiamine hydrochloride and thiamine mononitrate (Bender, 2003) (Fig. 8.6).

Through evolution, many organisms have lost their ability to synthesize thiamine since it is readily available in natural diets (plants and animals) and exogenous sources, which may reduce energy costs and metabolic requirements for thiamine biosynthesis (Mrowicka et al., 2023). It is observed that 76% of the known 400 marine bacteria found in the ocean possess genetic material required for the de novo synthesis of thiamine (*Pseudomonas, Bacillus, Streptomyces*), while most of the phototrophic phytoplankton are thiamine auxotrophs, which depend on exogenous thiamine sources (Croft et al., 2006). The autotrophic phytoplankton, heterotrophic bacteria, and diatoms serve as the primary producers of thiamine in aquatic environments, and these organisms help to maintain the thiamine reservoir, supporting the entire marine food web. The dissolved thiamine concentration in the marine environment is in the picomolar range; therefore organisms are unable to obtain thiamine directly from seawater (Hylander et al., 2024; Suffridge et al., 2020). Autotrophic phytoplankton, diatoms, and bacteria are consumed by organisms at a higher trophic level: zooplankton, and

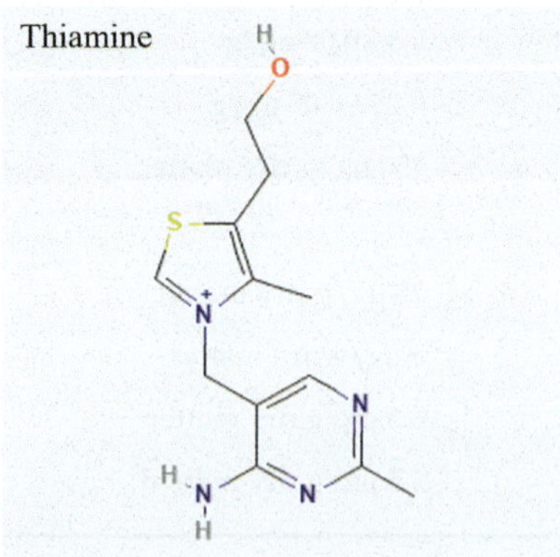

FIG. 8.6 The molecular structure of thiamine.

subsequently by planktivorous fish, which in turn are preyed upon by piscivorous fish, mammals, and birds. Fish and shellfish are a source of thiamine; however, their concentration is low compared to other B vitamins and is highly dependent on their diet.

It is no secret that seaweed is an excellent source of vitamins (Table 8.6). A study by Aktar et al., 2025 showed that the thiamine concentration of several seaweeds ranged from 0.324 to 2.326 mg/100 g, with *Enteromorpha clathrata* seaweed recording the highest concentration of thiamine, followed by *Gracilaria* sp. Recent studies have shown that marine sediments may serve as a potential source of thiamine since they contain the highest cellular density and diversity compared to any other environment on the planet. These microbial communities actively create a benthic flux of thiamine to the overlying water column. Whole genome sequencing has revealed that marine sediments contain many potential thiamine prototrophs (Monteverde et al., 2015). However, due to complex culturing processes, the dynamics of thiamine synthesis by most microorganisms are still unknown. Therefore measuring thiamine concentrations in sediment pore waters could be a novel directive for determining the sources of vitamins in the marine environment. This could be useful as a potent source of thiamine in human and animal nutrition, as well as for developing sustainable ways of cultivation and extraction.

Thiamine is required by all the cells and tissues of the human body and is found in the brain, heart, liver, kidneys, and skeletal muscles in higher concentrations (Mrowicka et al., 2023). Dietary thiamine phosphates are hydrolyzed by phosphatases in the intestine and absorbed mostly in the duodenum and proximal jejunum by active transport as free thiamine (Manzetti et al., 2014). This free thiamine is esterified with one or several phosphate groups. Some free thiamine is esterified into thiamine monophosphate (TMP) in the intestinal mucosa. Thiamine diphosphate (TDP) and thiamine triphosphate (TTP) are primarily synthesized in the liver by the phosphorylation of free thiamine and TMP, and all other tissues can synthesize TDP and TTP. Conversely, TMP emerges as a result of the sequential hydrolysis of TDP and TTP (Bender, 2003). The studies of Peters (1936) established the coenzymatic role of thiamine in the oxidative decarboxylation of pyruvate. The bioactive form, TDP, serves as a crucial cofactor of several essential enzymes in organisms: (1) oxidative decarboxylation of oxo-acids (pyruvate dehydrogenase and 2-oxoglutarate dehydrogenase complexes) in energy-yielding pathways, (2) branched-chain oxo-acid dehydrogenase in the catabolism of amino acids (leucine, isoleucine, and valine), and (3) transketolase in the pentose phosphate pathway for carbohydrate metabolism. TTP, synthesized by the phosphorylation of TDP in the brain and skeletal muscles, has a crucial role in neuronal function, beyond its coenzymatic role. It acts as a phosphate donor to activate a chloride channel in the nerve membrane and other membrane proteins in nerve and synaptosome preparations (Bettendorff et al., 1993).

Thiamine is essential in the oxidative defense mechanisms and the biosynthesis of cell constituents such as neurotransmitters and nucleic acid precursors (pentoses). Its deficiency in diets may hinder crucial biochemical and metabolic processes in the body: mitochondrial dysfunction, impaired glucose and energy metabolism, lactic acidosis, insufficient DNA synthesis, and impaired nerve signal transmission (Abdou and Hazell, 2015; Wijemanna and Lee, 2024). In addition, thiamine deficiencies may result in disease conditions in humans, namely, chronic peripheral neuropathy (beriberi), acute pernicious beriberi (Shoshin Beriberi), Wernicke–Korsakoff syndrome, and several neurodegenerative diseases such as Alzheimer's

TABLE 8.6 Marine vitamin B1 sources and available contents.

Marine source	Form	Content	References
Fish, shellfish, and echinoderms			
Haddock	Thiamine	0.02 mg/100 g	U.S. Department of Agriculture
European anchovy	Thiamine	0.055 mg/100 g	U.S. Department of Agriculture
Bluefish	Thiamine	0.058 mg/100 g	U.S. Department of Agriculture
Butterfish	Thiamine	0.12 mg/100 g	U.S. Department of Agriculture
Atlantic cod	Thiamine	0.076 mg/100 g	U.S. Department of Agriculture
Pacific cod	Thiamine	0.033 mg/100 g	U.S. Department of Agriculture
Atlantic croaker	Thiamine	0.075 mg/100 g	U.S. Department of Agriculture
Cusk	Thiamine	0.042 mg/100 g	U.S. Department of Agriculture
Flatfish (flounder and sole)	Thiamine	0.022 mg/100 g	U.S. Department of Agriculture
Grouper	Thiamine	0.07 mg/100 g	U.S. Department of Agriculture
Halibut (Atlantic and Pacific)	Thiamine	0.05 mg/100 g	U.S. Department of Agriculture
Halibut (Greenland)	Thiamine	0.06 mg/100 g	U.S. Department of Agriculture
Atlantic herring	Thiamine	0.092 mg/100 g	U.S. Department of Agriculture
Atlantic mackerel	Thiamine	0.176 mg/100 g	U.S. Department of Agriculture
Mahimahi	Thiamine	0.02 mg/100 g	U.S. Department of Agriculture
Milkfish	Thiamine	0.013 mg/100 g	U.S. Department of Agriculture
Monkfish	Thiamine	0.025 mg/100 g	U.S. Department of Agriculture
Mullet	Thiamine	0.09 mg/100 g	U.S. Department of Agriculture
Atlantic Ocean perch	Thiamine	0.04 mg/100 g	U.S. Department of Agriculture
Atlantic pollock	Thiamine	0.047 mg/100 g	U.S. Department of Agriculture
Florida pompano	Thiamine	0.56 mg/100 g	U.S. Department of Agriculture
Ocean pout	Thiamine	0.08 mg/100 g	U.S. Department of Agriculture
Sablefish	Thiamine	0.1 mg/100 g	U.S. Department of Agriculture
Atlantic salmon	Thiamine	0.226 mg/100 g	U.S. Department of Agriculture
Chinook salmon	Thiamine	0.054 mg/100 g	U.S. Department of Agriculture
Chum salmon	Thiamine	0.08 mg/100 g	U.S. Department of Agriculture
Pink salmon	Thiamine	0.08 mg/100 g	U.S. Department of Agriculture
Sockeye salmon	Thiamine	0.132 mg/100 g	U.S. Department of Agriculture
Scup	Thiamine	0.11 mg/100 g	U.S. Department of Agriculture
American shad	Thiamine	0.15 mg/100 g	U.S. Department of Agriculture
Swordfish	Thiamine	0.075 mg/100 g	U.S. Department of Agriculture

TABLE 8.6 Marine vitamin B1 sources and available contents—cont'd

Marine source	Form	Content	References
Bluefin tuna	Thiamine	0.241 mg/100 g	U.S. Department of Agriculture
Skipjack tuna	Thiamine	0.033 mg/100 g	U.S. Department of Agriculture
Yellowfin tuna	Thiamine	0.118 mg/100 g	U.S. Department of Agriculture
European turbot	Thiamine	0.066 mg/100 g	U.S. Department of Agriculture
Alaska king crab	Thiamine	0.043 mg/100 g	U.S. Department of Agriculture
Blue crab	Thiamine	0.08 mg/100 g	U.S. Department of Agriculture
Queen crab	Thiamine	0.08 mg/100 g	U.S. Department of Agriculture
Cray fish (mixed species)	Thiamine	0.045 mg/100 g	U.S. Department of Agriculture
Northern lobster	Thiamine	0.02 mg/100 g	U.S. Department of Agriculture
Spiny lobster (mixed species)	Thiamine	0.007 mg/100 g	U.S. Department of Agriculture
Abalone (mixed species)	Thiamine	0.19 mg/100 g	U.S. Department of Agriculture
Clam (mixed species)	Thiamine	0.015 mg/100 g	U.S. Department of Agriculture
Cuttlefish (mixed species)	Thiamine	0.009 mg/100 g	U.S. Department of Agriculture
Blue mussel	Thiamine	0.16 mg/100 g	U.S. Department of Agriculture
Common octopus	Thiamine	0.03 mg/100 g	U.S. Department of Agriculture
Eastern oyster	Thiamine	0.105 mg/100 g	U.S. Department of Agriculture
Pacific oyster	Thiamine	0.067 mg/100 g	U.S. Department of Agriculture
Whelk	Thiamine	0.026 mg/100 g	U.S. Department of Agriculture
Scallop (mixed species)	Thiamine	0.007 mg/100 g	U.S. Department of Agriculture
Squid (mixed species)	Thiamine	0.002 mg/100 g	U.S. Department of Agriculture
Macro algae			
Ulva lactuca	Thiamine	109 ± 18 mg/100 g, DW	(Cadar et al., 2023)
Enteromorpha intestinalis	Thiamine	3.95 ± 0.52 mg/100 g, DW	(Cadar et al., 2023)
Cladophora vagabunda	Thiamine	4.16 ± 0.25 mg/100 g, DW	(Cadar et al., 2023)
Enteromorpha compressa	Thiamine	0.56 mg/100 g	(Murata and Nakazoe, 2001)
Undaria pinnatifida	Thiamine	0.3 mg/100 g	(Murata and Nakazoe, 2001)
Hizikia fusiformis	Thiamine	0.01 mg/100 g	(Murata and Nakazoe, 2001)
Laminaria saccharina	Thiamine	0.48 mg/100 g	(Murata and Nakazoe, 2001)
Porphyra complex	Thiamine	1.15 mg/100 g	(Murata and Nakazoe, 2001)
Sargassum muticum	Thiamine	4 μg/g	(De Roeck-Holtzhauer et al., 1991)
Botryocladia leptopoda	Thiamine	2.34 ± 0.02 μg/g, DW	(Gajalakshmi et al., 2018)

Continued

TABLE 8.6 Marine vitamin B1 sources and available contents—cont'd

Marine source	Form	Content	References
Ascophyllum nodosum	Thiamine	0.216 mg/8 g, DW	(MacArtain et al., 2007)
Laminaria digitata	Thiamine	0.011 mg/8 g, DW	(MacArtain et al., 2007)
Undaria pinnatifida	Thiamine	0.403 mg/8 g, DW	(MacArtain et al., 2007)
Porphyra umbilicalis	Thiamine	0.077 mg/8 g, DW	(MacArtain et al., 2007)
Palmaria palmata	Thiamine	0.024 mg/8 g, DW	(MacArtain et al., 2007)
Ulva spp.	Thiamine	0.06 mg/8 g, DW	(MacArtain et al., 2007)
Microalgae			
Tertraselmis sp. CS-362	Thiamine	109 ± 18 μg/g	(Brown et al., 1999)
Pavlova pinguis	Thiamine	36 ± 5 μg/g	(Brown et al., 1999)
Stichococcus	Thiamine	29 ± 2 μg/g	(Brown et al., 1999)
Skeletonema costatum	Thiamine	710 μg/g	(De Roeck-Holtzhauer et al., 1991)
Chaetoceros calcitrans	Thiamine	655 μg/g	(De Roeck-Holtzhauer et al., 1991)
Tetraselmis suecica	Thiamine	627 μg/g	(De Roeck-Holtzhauer et al., 1991)
Isochrysis galbana	Thiamine	462 μg/g	(De Roeck-Holtzhauer et al., 1991)
Pavlova lutheri	Thiamine	290 μg/g	(De Roeck-Holtzhauer et al., 1991)
Nannochloropsis	Thiamine	70 μg/g	(Del Mondo et al., 2020)
Dunaliella	Thiamine	9–29 μg/g	(Del Mondo et al., 2020)
Chlorella stigmatophora	Thiamine	14.6 mg/kg	(Fabregas and Herrero, 1990)

disease, Parkinson's disease, and Huntington's disease. In acute deficiencies, lesions in the brain or spinal cord may occur (Bender, 2003). Deficiency diseases are primarily common among alcoholics with high energy intake, people with higher physical activity and low food intake, and in the presence of thiaminases and thiamine antagonists in some microorganisms, foods (plants and raw fish), and thermostable compounds in foods (polyphenols) (Smith et al., 2021). Moreover, it has been found that thiamine deficiency is associated with severe anorexia (an eating disorder), which may correlate with impaired transketolase activity in the intestinal mucosa (O'Brien et al., 2025).

The dietary thiamine requirement of organisms mainly depends on the total dietary energy intake. Generally, requirements are calculated based on dietary energy intake, assuming 40% of energy comes from fat. The dietary thiamine intake would be somewhat higher for diets low in fat and higher in protein and carbohydrates. Reference intakes show that adults consuming >2000 kcal/day must consume 0.5 mg of thiamine per 1000 kcal. In the event of fasting, 0.8 mg of thiamine per day is required to metabolize endogenous energy-yielding substrates (Bender, 2003). There has been no evidence of thiamine toxicity during higher dietary intake since thiamine absorption is limited (2.5 mg per dose), and the excess is filtered by the kidneys and excreted through urine. However, high parenteral doses have resulted in anaphylactic shock in humans. Hypersensitivity and contact dermatitis have been reported in medical workers handling large quantities of thiamine (Calogiuri, 2021).

8.3.1.2 Riboflavin

Riboflavin, or vitamin B2, was initially isolated from milk whey as a pigment called lactochrome and is a water-soluble, heat-stable vitamin (Northrop-Clewes and Thurnham, 2012). However, it is light-sensitive, and direct exposure to sunlight or fluorescent light destroys significant amounts of riboflavin in food (mainly milk) due to photolysis. The chemical structure of riboflavin (7,8-dimethyl-10-ribitylisoalloxazine) consists of a flavin isoalloxazine ring conjugated with ribitol sugar alcohol (Fig. 8.7). Upon phosphorylation by flavokinase, it serves as a precursor for two physiologically important coenzyme forms: flavin adenine dinucleotide (FAD) and flavin mononucleotide (Pinto and Rivlin, 2013). In the intestinal lumen, nucleotide diphosphatase and other phosphatases hydrolyze FAD and riboflavin phosphates to release free riboflavin, which is then absorbed in the proximal small intestine through an active sodium-dependent saturable mechanism (Malik et al., 2023).

The role of riboflavin has been identified as a redox coenzyme in energy-yielding reactions via the respiratory chain, acting as a catalyst in the transfer of electrons, and the flavin coenzymes function as electron carriers in many oxidation and reduction reactions (Malik et al., 2023). Moreover, riboflavin is involved in the metabolism of carbohydrates, lipids, proteins, and ketone bodies. Riboflavin promotes the normal growth of organisms and is involved in synthesizing red blood cells, glycogen, and steroids. Riboflavin coenzymes are involved in converting pyridoxine and folic acid to their coenzyme forms and transforming tryptophan to niacin. It has been found that riboflavin has a function in maintaining circadian rhythms in organisms, acting as the prosthetic group of the cryptochromes that are sensitive to blue light in the eye (Pinto and Zempleni, 2016).

Free riboflavin is found in milk and eggs, bound to specific binding proteins, and most of the riboflavin in food is found as flavin coenzymes (FAD mainly) bound to enzymes. While terrestrial sources such as meat, fish, cereals, fruits, and vegetables are considered excellent sources of riboflavin (Pallotta, 2019), the marine environment plays an indispensable role in the synthesis of riboflavin. Riboflavin is mainly biosynthesized in plants and microorganisms, employing guanosine triphosphate and ribulose 5-phosphate as precursors (Table 8.7).

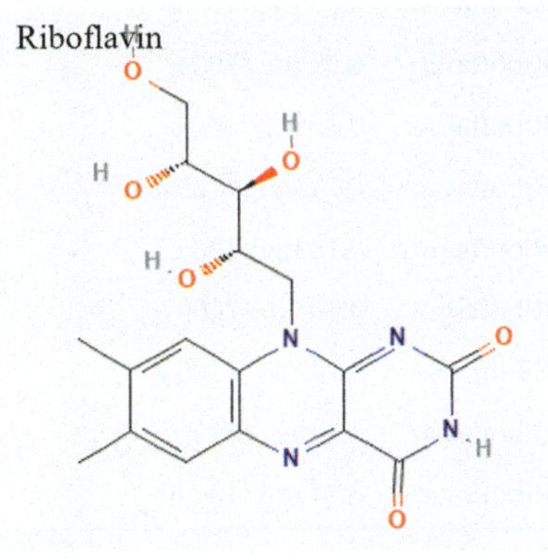

FIG. 8.7 The molecular structure of riboflavin.

TABLE 8.7 Marine vitamin B2 sources and available contents.

Marine source	Form	Content	References
Fish, shellfish, and echinoderms			
Haddock—raw	Riboflavin	0.057 mg/100 g	U.S. Department of Agriculture
European anchovy—raw	Riboflavin	0.256 mg/100 g	U.S. Department of Agriculture
Bluefish—raw	Riboflavin	0.08 mg/100 g	U.S. Department of Agriculture
Butterfish	Riboflavin	0.15 mg/100 g	U.S. Department of Agriculture
Atlantic cod	Riboflavin	0.065 mg/100 g	U.S. Department of Agriculture
Pacific cod	Riboflavin	0.045 mg/100 g	U.S. Department of Agriculture
Atlantic croaker—raw	Riboflavin	0.095 mg/100 g	U.S. Department of Agriculture
Cusk—raw	Riboflavin	0.133 mg/100 g	U.S. Department of Agriculture
Flatfish (flounder and sole)—raw	Riboflavin	0.02 mg/100 g	U.S. Department of Agriculture
Grouper—raw	Riboflavin	0.005 mg/100 g	U.S. Department of Agriculture
Halibut (Atlantic and Pacific)	Riboflavin	0.03 mg/100 g	U.S. Department of Agriculture
Halibut (Greenland)	Riboflavin	0.08 mg/100 g	U.S. Department of Agriculture
Atlantic herring	Riboflavin	0.233 mg/100 g	U.S. Department of Agriculture
Atlantic mackerel	Riboflavin	0.312 mg/100 g	U.S. Department of Agriculture
Mahimahi	Riboflavin	0.07 mg/100 g	U.S. Department of Agriculture
Milkfish	Riboflavin	0.054 mg/100 g	U.S. Department of Agriculture
Monkfish	Riboflavin	0.06 mg/100 g	U.S. Department of Agriculture
Mullet	Riboflavin	0.08 mg/100 g	U.S. Department of Agriculture
Atlantic Ocean perch	Riboflavin	0.048 mg/100 g	U.S. Department of Agriculture
Atlantic pollock	Riboflavin	0.185 mg/100 g	U.S. Department of Agriculture
Florida pompano	Riboflavin	0.12 mg/100 g	U.S. Department of Agriculture
Ocean pout	Riboflavin	0.06 mg/100 g	U.S. Department of Agriculture
Sablefish	Riboflavin	0.09 mg/100 g	U.S. Department of Agriculture
Atlantic salmon	Riboflavin	0.38 mg/100 g	U.S. Department of Agriculture
Chinook salmon	Riboflavin	0.113 mg/100 g	U.S. Department of Agriculture
Chum salmon	Riboflavin	0.18 mg/100 g	U.S. Department of Agriculture
Pink salmon	Riboflavin	0.105 mg/100 g	U.S. Department of Agriculture
Sockeye salmon	Riboflavin	0.207 mg/100 g	U.S. Department of Agriculture
Scup	Riboflavin	0.1 mg/100 g	U.S. Department of Agriculture
American shad	Riboflavin	0.24 mg/100 g	U.S. Department of Agriculture

TABLE 8.7 Marine vitamin B2 sources and available contents—cont'd

Marine source	Form	Content	References
Swordfish	Riboflavin	0.053 mg/100 g	U.S. Department of Agriculture
Bluefin tuna	Riboflavin	0.251 mg/100 g	U.S. Department of Agriculture
Skipjack tuna	Riboflavin	0.1 mg/100 g	U.S. Department of Agriculture
Yellowfin tuna	Riboflavin	0.115 mg/100 g	U.S. Department of Agriculture
European turbot	Riboflavin	0.08 mg/100 g	U.S. Department of Agriculture
Alaska king crab	Riboflavin	0.043 mg/100 g	U.S. Department of Agriculture
Blue crab	Riboflavin	0.04 mg/100 g	U.S. Department of Agriculture
Queen crab	Riboflavin	0.2 mg/100 g	U.S. Department of Agriculture
Cray fish (mixed species)	Riboflavin	0.032 mg/100 g	U.S. Department of Agriculture
Northern lobster	Riboflavin	0.014 mg/100 g	U.S. Department of Agriculture
Spiny lobster (mixed species)	Riboflavin	0.046 mg/100 g	U.S. Department of Agriculture
Abalone (mixed species)	Riboflavin	0.1 mg/100 g	U.S. Department of Agriculture
Clam (mixed species)	Riboflavin	0.04 mg/100 g	U.S. Department of Agriculture
Cuttlefish (mixed species)	Riboflavin	0.91 mg/100 g	U.S. Department of Agriculture
Blue mussel	Riboflavin	0.21 mg/100 g	U.S. Department of Agriculture
Common octopus	Riboflavin	0.04 mg/100 g	U.S. Department of Agriculture
Eastern oyster	Riboflavin	0.065 mg/100 g	U.S. Department of Agriculture
Pacific oyster	Riboflavin	0.233 mg/100 g	U.S. Department of Agriculture
Whelk	Riboflavin	0.107 mg/100 g	U.S. Department of Agriculture
Scallop (mixed species)	Riboflavin	0.015 mg/100 g	U.S. Department of Agriculture
Squid (mixed species)	Riboflavin	0.412 mg/100 g	U.S. Department of Agriculture
Macro algae			
Ulva lactuca	Riboflavin	0.99 ± 0.07 mg/100 g, DW	(Cadar et al., 2023)
Enteromorpha intestinalis	Riboflavin	0.97 ± 0.02 mg/100 g, DW	(Cadar et al., 2023)
Cladophora vagabunda	Riboflavin	0.89 ± 0.06 mg/100 g, DW	(Cadar et al., 2023)
Sargassum muticum	Riboflavin	65 μg/g	(De Roeck-Holtzhauer et al., 1991)
Botryocladia leptopoda	Riboflavin	1.18 ± 0.17 μg/g, DW	(Gajalakshmi et al., 2018)
Ascophyllum nodosum	Riboflavin	0.058 mg/8 g, DW	(MacArtain et al., 2007)
Laminaria digitata	Riboflavin	0.011 mg/8 g, DW	(MacArtain et al., 2007)
Undaria pinnatifida	Riboflavin	0.936 mg/8 g, DW	(MacArtain et al., 2007)

Continued

TABLE 8.7 Marine vitamin B2 sources and available contents—cont'd

Marine source	Form	Content	References
Porphyra umbilicalis	Riboflavin	0.274 mg/8 g, DW	(MacArtain et al., 2007)
Palmaria palmata	Riboflavin	0.080 mg/8 g, DW	(MacArtain et al., 2007)
Ulva spp.	Riboflavin	0.030 mg/8 g, DW	(MacArtain et al., 2007)
Enteromorpha compressa	Riboflavin	1.90 mg/100 g	(Murata and Nakazoe, 2001)
Undaria pinnatifida	Riboflavin	1.15 mg/100 g	(Murata and Nakazoe, 2001)
Hizikia fusiformis	Riboflavin	0.14 mg/100 g	(Murata and Nakazoe, 2001)
Laminaria saccharina	Riboflavin	0.37 mg/100 g	(Murata and Nakazoe, 2001)
Porphyra complex	Riboflavin	3.40 mg/100 g	(Murata and Nakazoe, 2001)
Microalgae			
Tertraselmis sp. CS-362	Riboflavin	26 ± 4 μg/g	(Brown et al., 1999)
Pavlova pinguis	Riboflavin	50 ± 10 μg/g	(Brown et al., 1999)
Stichococcus	Riboflavin	25 μg/g	(Brown et al., 1999)
Skeletonema costatum	Riboflavin	37 μg/g	(De Roeck-Holtzhauer et al., 1991)
Chaetoceros calcitrans	Riboflavin	12 μg/g	(De Roeck-Holtzhauer et al., 1991)
Tetraselmis suecica	Riboflavin	42 μg/g	(De Roeck-Holtzhauer et al., 1991)
Isochrysis galbana	Riboflavin	14 μg/g	(De Roeck-Holtzhauer et al., 1991)
Pavlova lutheri	Riboflavin	6 μg/g	(De Roeck-Holtzhauer et al., 1991)
Nannochloropsis	Riboflavin	22–25 μg/g	(Del Mondo et al., 2020)
Dunaliella	Riboflavin	9–31.2 μg/g	(Del Mondo et al., 2020)
Chlorella stigmatophora	Riboflavin	19.6 mg/kg	(Fabregas and Herrero, 1990)

In marine environments, Achromobacter species contribute significantly to riboflavin biosynthesis (Wang et al., 2024). Aktar et al., 2025) observed that the riboflavin content in several seaweed species varied between 0.065 and 0.891 mg/100 g, with *Glacilaria* sp. recording the highest concentration. Sánchez-Machado et al. (2004) observed that red marine algae have higher thiamine and riboflavin contents than brown algae. Furthermore, they explained that processing seaweed can affect the riboflavin content; canning reduced the riboflavin content more compared to drying. Pérez-García et al. (2022) established that *Corynebacterium glutamicum* is a potential host for riboflavin synthesis, utilizing brown seaweed (*Laminaria hyperborea*)-derived carbohydrates as the feedstock. Compared to chemical methods, the utilization of fermentation techniques has many advantages in riboflavin synthesis, such as easy recovery, lower energy requirements, and less waste produced. It has been found that the marine yeast strain isolated from seawater, *Candida membranifaciens* subsp. flavinogenie W14-3, has the potential to synthesize riboflavin through fermentation in the presence of xylose in the culture media (Wang et al., 2008).

8.3.1.3 *Niacin*

Niacin is a colorless, water-soluble vitamin and was initially recognized as the nicotinic acid produced through the oxidation of nicotine in 1867. Later, in the 1940s, it was discovered that niacin deficiency is the etiology of pellagra, and it was recommended that staple foods be fortified with niacin (Fig. 8.8). Nicotinic acid and nicotinamide are considered the two compounds having the biological action of the vitamin, and both forms function similarly (Chand and Savitri, 2016). Niacin is primarily found in tissues as nicotinamide nucleotides, and meat is an excellent source of free nicotinamide. In the intestinal lumen, NADs are hydrolyzed into nicotinamide, and several intestinal bacteria have the ability to deamidate dietary nicotinamide. Both nicotinic acid and nicotinamide are actively absorbed from the small intestine through a sodium-dependent, saturable process (Rasti et al., 2020).

Nicotinamide is the reactive moiety of NAD (nicotinamide adenine dinucleotide) and NADP (nicotinamide adenine dinucleotide phosphate), coenzymes involved in many oxidation and reduction reactions that act as proton and electron carriers and are crucial in energy production. The oxidized form, NAD+, serves as the electron acceptor in energy-yielding processes, and the resultant NADH is oxidized by the electron transport chain in mitochondria. $NADP^+$ is reduced to NADPH in the pentose phosphate pathway (Bender and Bender, 1986). Moreover, niacin is a substrate for nonredox signaling pathways, including gene expression and DNA repair, contributing to cell cycle regulation and cell death, and is also essential for the development and survival of neurons in the central nervous system (Gasperi et al., 2019; Aguilera-Méndez et al., 2012). NAD also plays a significant role in the metabolism of macronutrients: carbohydrates, protein, and fat. Nicotinic acid has antihyperlipidemic properties and is pharmacologically used as a hypolipidemic agent that helps lower cholesterol levels and other lipids in the blood. Moreover, niacin is used as a therapeutic agent to modify high LDL-C and low HDL-C, high lipoprotein levels, and hypertriglyceridemia, which are risk factors for cardiovascular diseases (Bender, 2003).

Animal and plant-based foodstuffs naturally contain niacin; specifically, in raw meat and fish, it is found as NAD(P) in higher concentrations than in processed food (Çatak, 2019). Seafood is an excellent source of niacin. Other than fish, invertebrates play a huge role in vitamin nutrition. Šimat (2021) reported that invertebrates such as sea cucumbers are a good source of thiamine, riboflavin, and niacin. According to Majewski and Lebiedzińska (2014), niacin content in several seafoods ranged from 0.62 to 1.15 mg per 100 g, with oysters having the highest concentration (Table 8.8). The same authors concluded that the inclusion of

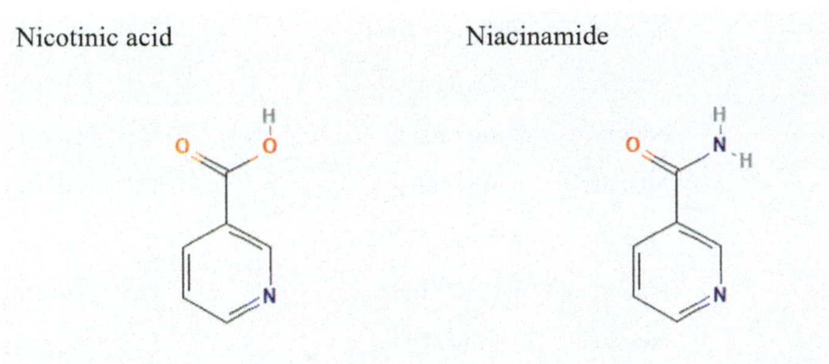

FIG. 8.8 Molecular structures of different forms of niacin.

TABLE 8.8 Marine vitamin B3 sources and available contents.

Marine source	Form	Content	References
Fish, shellfish, and echinoderms			
Haddock	Niacin	3.36 mg/100 g	U.S. Department of Agriculture
European anchovy	Niacin	14 mg/100 g	U.S. Department of Agriculture
Bluefish	Niacin	5.95 mg/100 g	U.S. Department of Agriculture
Butterfish	Niacin	4.5 mg/100 g	U.S. Department of Agriculture
Atlantic cod	Niacin	2.06 mg/100 g	U.S. Department of Agriculture
Pacific cod	Niacin	1.1 mg/100 g	U.S. Department of Agriculture
Atlantic croaker	Niacin	4.2 mg/100 g	U.S. Department of Agriculture
Cusk	Niacin	2.69 mg/100 g	U.S. Department of Agriculture
Flatfish (flounder and sole)	Niacin	1.04 mg/100 g	U.S. Department of Agriculture
Grouper	Niacin	0.313 mg/100 g	U.S. Department of Agriculture
Halibut (Atlantic and Pacific)	Niacin	6.51 mg/100 g	U.S. Department of Agriculture
Halibut (Greenland)	Niacin	1.5 mg/100 g	U.S. Department of Agriculture
Atlantic herring	Niacin	3.22 mg/100 g	U.S. Department of Agriculture
Atlantic mackerel	Niacin	9.08 mg/100 g	U.S. Department of Agriculture
Mahimahi	Niacin	6.1 mg/100 g	U.S. Department of Agriculture
Milkfish	Niacin	6.44 mg/100 g	U.S. Department of Agriculture
Monkfish	Niacin	2.1 mg/100 g	U.S. Department of Agriculture
Mullet	Niacin	5.2 mg/100 g	U.S. Department of Agriculture
Atlantic Ocean perch	Niacin	1 mg/100 g	U.S. Department of Agriculture
Atlantic pollock	Niacin	3.27 mg/100 g	U.S. Department of Agriculture
Florida pompano	Niacin	3 mg/100 g	U.S. Department of Agriculture
Ocean pout	Niacin	2.1 mg/100 g	U.S. Department of Agriculture
Sablefish	Niacin	4 mg/100 g	U.S. Department of Agriculture
Atlantic salmon	Niacin	7.86 mg/100 g	U.S. Department of Agriculture
Chinook salmon	Niacin	8.42 mg/100 g	U.S. Department of Agriculture
Chum salmon	Niacin	7 mg/100 g	U.S. Department of Agriculture
Pink salmon	Niacin	8 mg/100 g	U.S. Department of Agriculture
Sockeye salmon	Niacin	8.51 mg/100 g	U.S. Department of Agriculture
Scup	Niacin	4.1 mg/100 g	U.S. Department of Agriculture
American shad	Niacin	8.4 mg/100 g	U.S. Department of Agriculture
Swordfish	Niacin	7.76 mg/100 g	U.S. Department of Agriculture

TABLE 8.8 Marine vitamin B3 sources and available contents—cont'd

Marine source	Form	Content	References
Bluefin tuna	Niacin	8.65 mg/100 g	U.S. Department of Agriculture
Skipjack tuna	Niacin	15.4 mg/100 g	U.S. Department of Agriculture
Yellowfin tuna	Niacin	18.5 mg/100 g	U.S. Department of Agriculture
European turbot	Niacin	2.2 mg/100 g	U.S. Department of Agriculture
Alaska king crab	Niacin	1.1 mg/100 g	U.S. Department of Agriculture
Blue crab	Niacin	2.7 mg/100 g	U.S. Department of Agriculture
Queen crab	Niacin	2.5 mg/100 g	U.S. Department of Agriculture
Cray fish (mixed species)	Niacin	1.87 mg/100 g	U.S. Department of Agriculture
Northern lobster	Niacin	1.59 mg/100 g	U.S. Department of Agriculture
Spiny lobster (mixed species)	Niacin	4.24 mg/100 g	U.S. Department of Agriculture
Abalone (mixed species)	Niacin	1.5 mg/100 g	U.S. Department of Agriculture
Clam (mixed species)	Niacin	0.35 mg/100 g	U.S. Department of Agriculture
Cuttlefish (mixed species)	Niacin	1.22 mg/100 g	U.S. Department of Agriculture
Blue mussel	Niacin	1.6 mg/100 g	U.S. Department of Agriculture
Common octopus	Niacin	2.1 mg/100 g	U.S. Department of Agriculture
Eastern oyster	Niacin	1.27 mg/100 g	U.S. Department of Agriculture
Pacific oyster	Niacin	2.01 mg/100 g	U.S. Department of Agriculture
Whelk	Niacin	1.05 mg/100 g	U.S. Department of Agriculture
Scallop (mixed species)	Niacin	0.703 mg/100 g	U.S. Department of Agriculture
Squid (mixed species)	Niacin	2.18 mg/100 g	U.S. Department of Agriculture
Macro algae			
Ulva lactuca	Niacin	2.97 ± 0.28 mg/100 g, DW	(Cadar et al., 2023)
Enteromorpha intestinalis	Niacin	1.84 ± 0.45 mg/100 g, DW	(Cadar et al., 2023)
Cladophora vagabunda	Niacin	2.59 ± 0.32 mg/100 g, DW	(Cadar et al., 2023)
Sargassum muticum	Niacin	20 μg/g	(De Roeck-Holtzhauer et al., 1991)
Botryocladia leptopoda	Niacin	12.40 ± 0.17 μg/g, DW	(Gajalakshmi et al., 2018)
Laminaria digitata	Niacin	4.896 mg/8 g, DW	(MacArtain et al., 2007)
Undaria pinnatifida	Niacin	7.198 mg/8 g, DW	(MacArtain et al., 2007)
Porphyra umbilicalis	Niacin	0.761 mg/8 g, DW	(MacArtain et al., 2007)
Palmaria palmata	Niacin	0.800 mg/8 g, DW	(MacArtain et al., 2007)
Ulva spp.	Niacin	8 mg/8 g, DW	(MacArtain et al., 2007)

Continued

TABLE 8.8 Marine vitamin B3 sources and available contents—cont'd

Marine source	Form	Content	References
Microalgae			
Skeletonema costatum	Niacin	511 μg/g	(De Roeck-Holtzhauer et al., 1991)
Chaetoceros calcitrans	Niacin	25 μg/g	(De Roeck-Holtzhauer et al., 1991)
Tetraselmis suecica	Niacin	1410 μg/g	(De Roeck-Holtzhauer et al., 1991)
Isochrysis galbana	Niacin	2690 μg/g	(De Roeck-Holtzhauer et al., 1991)
Pavlova lutheri	Niacin	955 μg/g	(De Roeck-Holtzhauer et al., 1991)
Nannochloropsis	Niacin	0.12 μg/g	(Del Mondo et al., 2020)
Dunaliella	Niacin	10 μg/g	(Del Mondo et al., 2020)
Chlorella stigmatophora	Niacin	82.5 mg/kg	(Fabregas and Herrero, 1990)

shellfish in diets may enhance the overall health of humans due to the interplay among a wide variety of nutrients and its significant therapeutic potential.

Considering the role of nicotinamide nucleotides in energy-yielding metabolic processes, the niacin requirements are established based on energy expenditure. Based on depletion/repletion studies, the recommended average niacin requirement is 5.5 mg per 1000 kcal. However, it has been observed that the average dietary intake of proteins will provide sufficient tryptophan, which is then converted to niacin, satisfying its daily requirement, assuming that the diet comprises 15% energy from protein and provides 14 g of tryptophan per kg of dietary protein (assumption: 60 mg of tryptophan is equivalent to 1 mg of niacin) (Bender, 2003). deficiency has been recognized as the nutritional etiology for pellagra. It is characterized by dermatitis or severe burns on the skin exposed to sunlight, especially on the face and areas not exposed to sunlight but subject to pressure, such as elbows, knees, ankles, and wrists. Other symptoms, such as inflammation of the mouth and tongue, diarrhea, and dementia, have been observed in pellagra patients and could be fatal if untreated. Increased supplementation of niacin or tryptophan will treat the condition. Niacin deficiency also leads to slower metabolism and reduced cold tolerance. Moreover, niacin deficiency affects psychological health, showing signs of poor concentration, irritability, and anxiety (Redzic et al., 2020; Meyer-Ficca and Kirkland, 2016).

8.3.1.4 Pantothenic acid

Pantothenic acid, also known as vitamin B5, is water-soluble, ubiquitous in nature, and abundant in a wide range of foodstuffs. It was first discovered in 1931 by the chemist Roger J. Williams, who observed the ability of an acidic substance to stimulate the growth of *Saccharomyces cerevisiae* strains. The structure of pantothenic acid consists of a molecule of beta-alanine linked to pantoic acid via an amide link and is chemically known as 3-[(2,4-dihydroxy-3,3-dimethylbutanoyl)amino]propanoic acid, which is yellow in color (Gonzalez-Lopez et al., 2016) (Fig. 8.9). Pantothenic acid is crucial for organisms; it is the metabolically active component

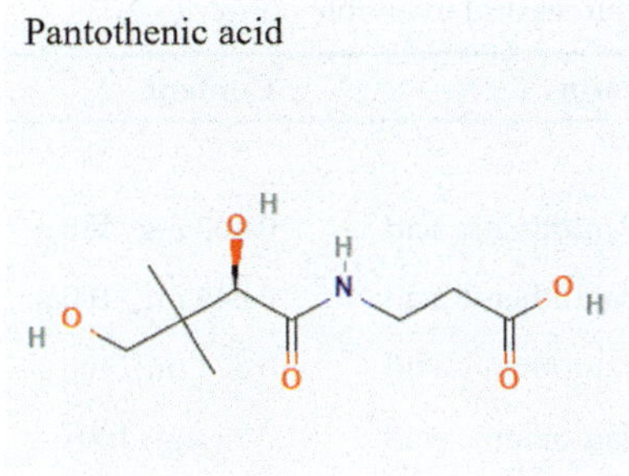

FIG. 8.9 The molecular structure of pantothenic acid.

in coenzyme A (CoA) and acyl carrier proteins (ACP) that participate in numerous biological pathways. CoA regulates the metabolism of carbohydrates, proteins, and lipids; fatty acid synthesis and oxidation; amino acid catabolism; synthesis of cholesterol, acetylcholine, and bile acids; and regulates gene expression. ACP is important in the synthesis of fatty acids, serving as a component of fatty acid synthase (Gonzalez-Lopez et al., 2016; Hrubša et al., 2022).

Most bacteria, fungi, and plants synthesize pantothenic acid, which is found everywhere, unlike other vitamins extracted from a single rich source; hence the name pantothenic acid, derived from the Greek word panthos, meaning "from all sides." Unlike other organisms (plants and bacteria), mammals and several microorganisms cannot perform de novo synthesis of pantothenic acid from oxo-isovalerate and aspartate enzymatically. Therefore it is essential to obtain it from external sources (Rucker and Bauerly, 2013). Islam et al. (2022) reported that seafoods are an excellent source of pantothenic acid (Table 8.9). According to their findings, the pantothenic content ranged from 0.14 ± 0.01 mg/100 g in fusiformis to 10.19 ± 0.46 mg/100 g in salted pollack roe.

Due to the wide availability of pantothenic acid and the contribution of intestinal bacteria toward the requirement, vitamin B5 deficiency diseases are rare among humans. However, due to the importance of the CoA molecule in various metabolic functions, it is vital to all living beings. Its deficiency can result in a wide array of consequences, including neurological, gastrointestinal, and immunological disorders, coupled with reduced feed intake and poor development. Some deficiency symptoms are comparable to those of other vitamin B deficiencies. A lack of CoA production may obstruct energy metabolism, leading to exhaustion, lethargy, and irritation. In addition, numbness, paresthesia, and muscle cramping may occur due to the inhibition of acetylcholine synthesis, which hinders nerve signal transmission (Bender, 2003). Animal trials have shown that vitamin B5 deficiency affects normal metabolism: loss of fur color in black and brown rats, impairment of peroxisomal oxidation of fatty acids affecting lipid metabolism, lack of steroid synthesis, and poor ability to respond to physical and metabolic stresses in rats (Riordan et al., 2012). It has been observed that pantothenic acid deficiency may cause hypoglycemia due to poor acylation of insulin receptors, which increases insulin sensitivity. Furthermore, it has been found that vitamin B5 deficiency increases susceptibility to infections compared to well-nourished animals, and most of the metabolic disorders occur due to defective metabolism of an acyl-CoA derivative. Thus an average intake of 2–7 mg of pantothenic acid is recommended,

TABLE 8.9 Marine vitamin B5 sources and available contents.

Marine source	Form	Content	References
Fish, shellfish, and echinoderms			
Haddock	Pantothenic acid	0.403 mg/100 g	U.S. Department of Agriculture
European anchovy	Pantothenic acid	0.645 mg/100 g	U.S. Department of Agriculture
Bluefish	Pantothenic acid	0.828 mg/100 g	U.S. Department of Agriculture
Butterfish	Pantothenic acid	0.75 mg/100 g	U.S. Department of Agriculture
Atlantic cod	Pantothenic acid	0.153 mg/100 g	U.S. Department of Agriculture
Pacific cod	Pantothenic acid	0.294 mg/100 g	U.S. Department of Agriculture
Atlantic croaker	Pantothenic acid	0.75 mg/100 g	U.S. Department of Agriculture
Cusk—raw	Pantothenic acid	0.279 mg/100 g	U.S. Department of Agriculture
Flatfish (flounder and sole)	Pantothenic acid	0.185 mg/100 g	U.S. Department of Agriculture
Grouper	Pantothenic acid	0.75 mg/100 g	U.S. Department of Agriculture
Halibut (Atlantic and Pacific)	Pantothenic acid	0.343 mg/100 g	U.S. Department of Agriculture
Halibut (Greenland)	Pantothenic acid	0.25 mg/100 g	U.S. Department of Agriculture
Atlantic herring	Pantothenic acid	0.645 mg/100 g	U.S. Department of Agriculture
Atlantic mackerel	Pantothenic acid	0.856 mg/100 g	U.S. Department of Agriculture
Mahimahi	Pantothenic acid	0.75 mg/100 g	U.S. Department of Agriculture
Milkfish	Pantothenic acid	0.75 mg/100 g	U.S. Department of Agriculture
Monkfish	Pantothenic acid	0.15 mg/100 g	U.S. Department of Agriculture
Mullet	Pantothenic acid	0.76 mg/100 g	U.S. Department of Agriculture
Atlantic Ocean perch	Pantothenic acid	0.272 mg/100 g	U.S. Department of Agriculture
Atlantic pollock	Pantothenic acid	0.358 mg/100 g	U.S. Department of Agriculture
Florida pompano	Pantothenic acid	0.75 mg/100 g	U.S. Department of Agriculture
Ocean pout	Pantothenic acid	0.15 mg/100 g	U.S. Department of Agriculture
Sablefish	Pantothenic acid	0.75 mg/100 g	U.S. Department of Agriculture
Atlantic salmon	Pantothenic acid	1.66 mg/100 g	U.S. Department of Agriculture
Chinook salmon	Pantothenic acid	0.75 mg/100 g	U.S. Department of Agriculture
Chum salmon	Pantothenic acid	0.75 mg/100 g	U.S. Department of Agriculture
Pink salmon	Pantothenic acid	1.03 mg/100 g	U.S. Department of Agriculture
Sockeye salmon	Pantothenic acid	1.07 mg/100 g	U.S. Department of Agriculture
Scup	Pantothenic acid	0.75 mg/100 g	U.S. Department of Agriculture
American shad	Pantothenic acid	0.75 mg/100 g	U.S. Department of Agriculture

TABLE 8.9 Marine vitamin B5 sources and available contents—cont'd

Marine source	Form	Content	References
Swordfish	Pantothenic acid	0.35 mg/100 g	U.S. Department of Agriculture
Bluefin tuna	Pantothenic acid	1.05 mg/100 g	U.S. Department of Agriculture
Skipjack tuna	Pantothenic acid	0.42 mg/100 g	U.S. Department of Agriculture
Yellowfin tuna	Pantothenic acid	0.28 mg/100 g	U.S. Department of Agriculture
European turbot	Pantothenic acid	0.57 mg/100 g	U.S. Department of Agriculture
Alaska king crab	Pantothenic acid	0.35 mg/100 g	U.S. Department of Agriculture
Blue crab	Pantothenic acid	0.35 mg/100 g	U.S. Department of Agriculture
Queen crab	Pantothenic acid	0.35 mg/100 g	U.S. Department of Agriculture
Cray fish (mixed species)	Pantothenic acid	0.57 mg/100 g	U.S. Department of Agriculture
Northern lobster	Pantothenic acid	1.45 mg/100 g	U.S. Department of Agriculture
Spiny lobster (mixed species)	Pantothenic acid	0.35 mg/100 g	U.S. Department of Agriculture
Abalone (mixed species)	Pantothenic acid	3 mg/100 g	U.S. Department of Agriculture
Clam (mixed species)	Pantothenic acid	0.148 mg/100 g	U.S. Department of Agriculture
Cuttlefish (mixed species)	Pantothenic acid	0.5 mg/100 g	U.S. Department of Agriculture
Blue mussel	Pantothenic acid	0.5 mg/100 g	U.S. Department of Agriculture
Common octopus	Pantothenic acid	0.5 mg/100 g	U.S. Department of Agriculture
Eastern oyster	Pantothenic acid	0.157 mg/100 g	U.S. Department of Agriculture
Pacific oyster	Pantothenic acid	0.5 mg/100 g	U.S. Department of Agriculture
Whelk	Pantothenic acid	0.208 mg/100 g	U.S. Department of Agriculture
Scallop (mixed species)	Pantothenic acid	0.215 mg/100 g	U.S. Department of Agriculture
Squid (mixed species)	Pantothenic acid	0.5 mg/100 g	U.S. Department of Agriculture
Microalgae			
Dunaliella	Pantothenic acid	5–13.2 μg/g	(Del Mondo et al., 2020)
Chlorella stigmatophora	Pantothenic acid	21.4 mg/kg	(Fabregas and Herrero, 1990)
Tetraselmis suecica	Pantothenic acid	37.7 mg/kg	(Fabregas and Herrero, 1990)
Isochrysis galbana	Pantothenic acid	9.1 mg/kg	(Fabregas and Herrero, 1990)
Dunaliella tertiolecta	Pantothenic acid	13.2 mg/kg	(Fabregas and Herrero, 1990)

considering individual variations and its wide availability in various food sources. Different forms of pantothenic acid, such as calcium pantothenate and pantetheine, are used in pharmacological formulations to treat different disease conditions affecting the skin, hair, liver, and cardiac muscle.

8.3.1.5 *Pyridoxin*

Pyridoxine, commonly known as vitamin B6, is a water-soluble vitamin naturally present in a wide range of foodstuffs. In 1938 vitamin B6 was isolated and crystallized by Samuel Lepkovsky, and later, it was identified as a pyrimidine derivative and given the name 3-hydroxy-4,5-dihydroxy-methyl-2-methyl-pyridine (Alturfan, 2023) (Fig. 8.10). Vitamin B6 has several natural forms (vitamers): pyridoxine, pyridoxal, pyridoxamine, and their respective 5′-phosphate esters, which are interconverted in biological systems (Malik et al., 2023). The biologically active form, pyridoxal phosphate (PLP), serves as a crucial coenzyme in numerous enzymatic activities, including the metabolism of macronutrients: carbohydrates, proteins, and lipids (Wilson et al., 2019). Among other things, PLP is involved in multiple metabolic reactions: decarboxylation, transamination, replacement, elimination, racemization, and beta-group interconversion. Vitamin B6 participates in the metabolism of amino acids, nucleic acids, the modulation of steroid hormone action, cellular signaling, and the regulation of gene expression (Dakshinamurti et al., 2014). Moreover, vitamin B6 is involved in synthesizing neurotransmitters, histamine synthesis, and hemoglobin synthesis and function. PLP is found in both eukaryotes (yeast) and prokaryotes (bacteria). However, humans or higher organisms do not synthesize this compound (Al-Shekaili et al., 2021). The phosphorylated forms of vitamin B6 are

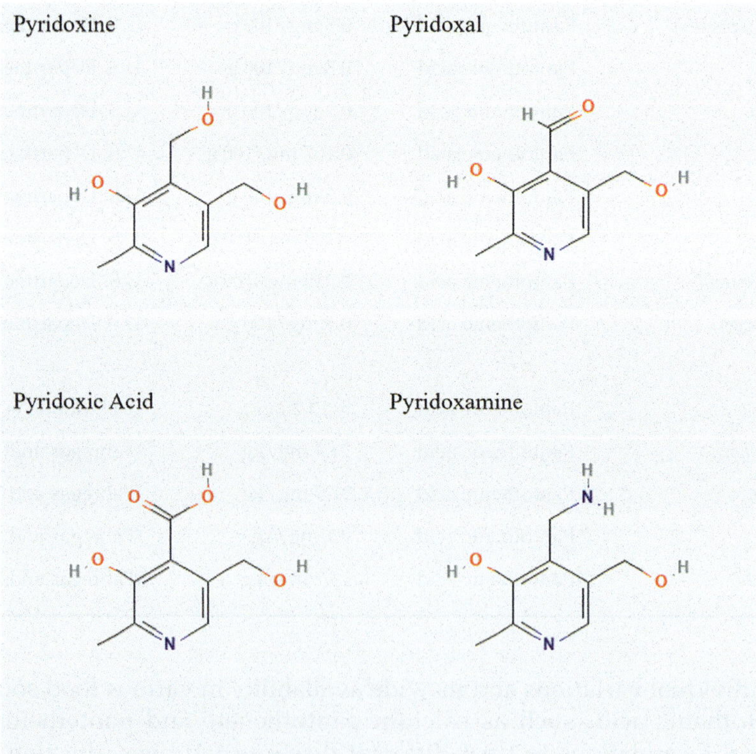

FIG. 8.10 Molecular structures of different forms of pyridoxine.

hydrolyzed by phosphatase in the intestinal mucosa. Then, these dephosphorylated forms are absorbed by facilitated diffusion, and most of the ingested pyridoxine is released into the portal circulation as pyridoxal.

In marine environments, vitamin B6 is mainly synthesized by Actinobacteria and Firmicutes (Wang et al., 2024). Islam et al. (2022) reported that vitamin B6 content in seafood varied from 0.01 ± 0.00 mg/100 g in seersucker to 1.39 ± 0.04 mg/100 g in salted pollack roe (Table 8.10). Vitamin B6 is widely spread among foodstuffs, and gut microorganisms contribute significantly to satisfying the vitamin B6 requirement. Therefore vitamin B6 deficiency is rare among humans. However, children have been diagnosed with vitamin B6 deficiency, which is associated with hyperexcitability of the central nervous system and convulsions. Moreover, pyridoxine deficiency tends to show symptoms similar to niacin deficiency, such as hyperexcitability, polyneuritis, and skin lesions, when patients are treated with drugs containing pyridoxal antagonists. Based on recent depletion/repletion studies, and considering the role of vitamin B6 in amino acid metabolism, it has been suggested that a dietary intake of 18–20 μg of pyridoxine per 1 g of protein is appropriate. Pyridoxine is clinically used in treating B6 hypovitaminosis, prophylaxis, the side effects

TABLE 8.10 Marine vitamin B6 sources and available contents.

Marine source	Form	Content	References
Fish, shellfish, and echinoderms			
Haddock	Vitamin B6	0.281 mg/100 g	U.S. Department of Agriculture
European anchovy	Vitamin B6	0.143 mg/100 g	U.S. Department of Agriculture
Bluefish	Vitamin B6	0.402 mg/100 g	U.S. Department of Agriculture
Butterfish	Vitamin B6	0.3 mg/100 g	U.S. Department of Agriculture
Atlantic cod	Vitamin B6	0.245 mg/100 g	U.S. Department of Agriculture
Pacific cod	Vitamin B6	0.117 mg/100 g	U.S. Department of Agriculture
Atlantic croaker	Vitamin B6	0.3 mg/100 g	U.S. Department of Agriculture
Cusk	Vitamin B6	0.387 mg/100 g	U.S. Department of Agriculture
Flatfish (flounder and sole)	Vitamin B6	0.098 mg/100 g	U.S. Department of Agriculture
Grouper	Vitamin B6	0.3 mg/100 g	U.S. Department of Agriculture
Halibut (Atlantic and Pacific)	Vitamin B6	0.548 mg/100 g	U.S. Department of Agriculture
Halibut (Greenland)	Vitamin B6	0.42 mg/100 g	U.S. Department of Agriculture
Atlantic herring	Vitamin B6	0.302 mg/100 g	U.S. Department of Agriculture
Atlantic mackerel	Vitamin B6	0.399 mg/100 g	U.S. Department of Agriculture
Mahimahi	Vitamin B6	0.4 mg/100 g	U.S. Department of Agriculture
Milkfish	Vitamin B6	0.423 mg/100 g	U.S. Department of Agriculture

Continued

TABLE 8.10 Marine vitamin B6 sources and available contents—cont'd

Marine source	Form	Content	References
Monkfish	Vitamin B6	0.24 mg/100 g	U.S. Department of Agriculture
Mullet	Vitamin B6	0.425 mg/100 g	U.S. Department of Agriculture
Atlantic Ocean perch	Vitamin B6	0.073 mg/100 g	U.S. Department of Agriculture
Atlantic pollock	Vitamin B6	0.287 mg/100 g	U.S. Department of Agriculture
Florida pompano	Vitamin B6	0.2 mg/100 g	U.S. Department of Agriculture
Ocean pout	Vitamin B6	0.24 mg/100 g	U.S. Department of Agriculture
Sablefish	Vitamin B6	0.3 mg/100 g	U.S. Department of Agriculture
Atlantic salmon	Vitamin B6	0.818 mg/100 g	U.S. Department of Agriculture
Chinook salmon	Vitamin B6	0.4 mg/100 g	U.S. Department of Agriculture
Chum salmon	Vitamin B6	0.4 mg/100 g	U.S. Department of Agriculture
Pink salmon	Vitamin B6	0.611 mg/100 g	U.S. Department of Agriculture
Sockeye salmon	Vitamin B6	0.731 mg/100 g	U.S. Department of Agriculture
Scup	Vitamin B6	0.3 mg/100 g	U.S. Department of Agriculture
American shad	Vitamin B6	0.4 mg/100 g	U.S. Department of Agriculture
Swordfish	Vitamin B6	0.543 mg/100 g	U.S. Department of Agriculture
Bluefin tuna	Vitamin B6	0.455 mg/100 g	U.S. Department of Agriculture
Skipjack tuna	Vitamin B6	0.85 mg/100 g	U.S. Department of Agriculture
Yellowfin tuna	Vitamin B6	0.933 mg/100 g	U.S. Department of Agriculture
European turbot	Vitamin B6	0.21 mg/100 g	U.S. Department of Agriculture
Alaska king crab	Vitamin B6	0.15 mg/100 g	U.S. Department of Agriculture
Blue crab	Vitamin B6	0.15 mg/100 g	U.S. Department of Agriculture
Queen crab	Vitamin B6	0.15 mg/100 g	U.S. Department of Agriculture
Cray fish (mixed species)	Vitamin B6	0.076 mg/100 g	U.S. Department of Agriculture
Northern lobster	Vitamin B6	0.104 mg/100 g	U.S. Department of Agriculture
Spiny lobster (mixed species)	Vitamin B6	0.15 mg/100 g	U.S. Department of Agriculture
Abalone (mixed species)	Vitamin B6	0.15 mg/100 g	U.S. Department of Agriculture
Clam (mixed species)	Vitamin B6	0.01 mg/100 g	U.S. Department of Agriculture
Cuttlefish (mixed species)	Vitamin B6	0.15 mg/100 g	U.S. Department of Agriculture
Blue mussel	Vitamin B6	0.05 mg/100 g	U.S. Department of Agriculture
Common octopus	Vitamin B6	0.36 mg/100 g	U.S. Department of Agriculture
Eastern oyster	Vitamin B6	0.06 mg/100 g	U.S. Department of Agriculture

TABLE 8.10 Marine vitamin B6 sources and available contents—cont'd

Marine source	Form	Content	References
Pacific oyster	Vitamin B6	0.05 mg/100 g	U.S. Department of Agriculture
Whelk	Vitamin B6	0.342 mg/100 g	U.S. Department of Agriculture
Scallop (mixed species)	Vitamin B6	0.073 mg/100 g	U.S. Department of Agriculture
Squid (mixed species)	Vitamin B6	0.056 mg/100 g	U.S. Department of Agriculture
Macro algae			
Ulva lactuca	Vitamin B6	0.58 ± 0.12 mg/100 g, DW	(Cadar et al., 2023)
Enteromorpha intestinalis	Vitamin B6	0.53 ± 0.66 mg/100 g, DW	(Cadar et al., 2023)
Cladophora vagabunda	Vitamin B6	0.69 ± 0.06 mg/100 g, DW	(Cadar et al., 2023)
Botryocladia leptopoda	Vitamin B6	0.33 ± 0.12 μg/g, DW	(Gajalakshmi et al., 2018)
Ascophyllum nodosum	Vitamin B6	0.001 mg/8 g, DW	(MacArtain et al., 2007)
Laminaria digitata	Vitamin B6	0.513 mg/8 g, DW	(MacArtain et al., 2007)
Undaria pinnatifida	Vitamin B6	0.259 mg/8 g, DW	(MacArtain et al., 2007)
Porphyra umbilicalis	Vitamin B6	0.119 mg/8 g, DW	(MacArtain et al., 2007)
Palmaria palmata	Vitamin B6	0.002 mg/8 g, DW	(MacArtain et al., 2007)
Microalgae			
Tertraselmis sp. CS-362	Vitamin B6	5.8 ± 0.4 μg/g	(Brown et al., 1999)
Pavlova pinguis	Vitamin B6	8.4 μg/g	(Brown et al., 1999)
Stichococcus	Vitamin B6	17 μg/g	(Brown et al., 1999)
Skeletonema costatum	Vitamin B6	134 μg/g	(De Roeck-Holtzhauer et al., 1991)
Chaetoceros calcitrans	Vitamin B6	4 μg/g	(De Roeck-Holtzhauer et al., 1991)
Tetraselmis suecica	Vitamin B6	155 μg/g	(De Roeck-Holtzhauer et al., 1991)
Isochrysis galbana	Vitamin B6	183 μg/g	(De Roeck-Holtzhauer et al., 1991)
Pavlova lutheri	Vitamin B6	4 μg/g	(De Roeck-Holtzhauer et al., 1991)
Nannochloropsis	Vitamin B6	3.6 μg/g	(Del Mondo et al., 2020)
Dunaliella	Vitamin B6	2.2–4 μg/g	(Del Mondo et al., 2020)
Chlorella stigmatophora	Vitamin B6	1.9 mg/kg	(Fabregas and Herrero, 1990)

caused by isoniazid therapy to treat tuberculosis, polyneuritis, dermatitis, gestational toxicosis, and impaired hepatic function. Several animal studies have shown that overconsumption of vitamin B6 may cause toxicity in animals, acting as a neurotoxin, causing peripheral neuropathy, muscle weakness, and loss of balance.

8.3.1.6 Biotin

Biotin is a nonprotein (prosthetic) group that covalently binds to bicarbonate-dependent carboxylase enzymes, including acetyl-CoA carboxylase, pyruvate carboxylase, propionyl CoA carboxylase, and methylcrotonyl CoA carboxylase. Biotin is synthesized by bacteria, yeast, fungi, algae, and some plant species and is abundant in animal and plant-based food sources. A synthetic form of biotin was commercially produced using the Hoffmann-La Roche process. Primarily, it exists in free and protein-bound forms. A higher amount of biotin is microbiologically derived and exists in free forms. Biotin is bound via an amide bond to the lysine amino acid of proteins, and biotinidases are involved in releasing biotin by hydrolyzing the amide bond. Biotin deficiency is known to be extremely rare, and under induced conditions, symptoms such as dermatitis, lassitude, anorexia, muscle pain, alopecia, conjunctivitis, and abnormalities of the central nervous system are exhibited in humans.

Biotin is a sulfur-containing, water-soluble vitamin characterized by a ureido ring fused with a thiophene ring and a pentanoic acid side chain. The chemical structure of biotin was confirmed in the 1940s, and its chemical formula and IUPAC name are $C_{10}H_{16}O_3N_2S$ (MW = 244.3) and *cis*-hexahydro-2-oxo-1*H*-thieno[3,4-*d*]imidazole-4-pentanoic acid, respectively (Fig. 8.11) (Vigneaud, 1942). Biotin crystallizes as colorless needles, and the melting point is 232°C. This molecular structure includes three asymmetric chiral carbon molecules, which result in eight stereoisomers. Among them, only *d*-biotin exhibits biological activity.

Little is known about how protein-bound biotin is digested, and its bioavailability varies widely depending on the presence of biotin-binding inhibiting factors, such as avidin. Avidin, a protein found in raw egg white, has been shown to bind to biotin, preventing its absorption. However, studies have shown that intravenously administered avidin-bound biotin is biologically active, suggesting that avidin inhibits intestinal biotin absorption. Individuals who consumed raw egg whites over long periods have exhibited symptoms of biotin deficiency. Protein-bound biotin is not directly released as free biotin during intestinal enzymatic digestion because the amide bond between biotin and lysyl residues resists breakdown by gastrointestinal proteases. However, proteases release biocytin and biotin-containing peptides. Biotinidase in pancreatic juice, intestinal mucosa, and plasma further

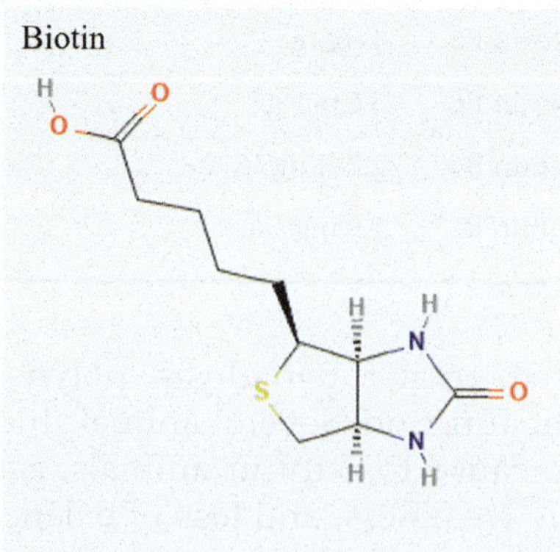

FIG. 8.11 The molecular structure of biotin.

breaks down biocytin into free biotin and biotinyl peptides, and free biotin is absorbed across the brush-border membrane of the enterocytes of the proximal small intestine and at the colon actively via Na^+-dependent multivitamin transporters and passive diffusion. In addition to biotin, Na^+-dependent multivitamin transporters transport pantothenic acid, lipoic acid, anticonvulsants, ethanol, and acetaldehyde across cell membranes. Therefore these components are considered biotin absorption inhibitors.

The same Na^+ ion-dependent transportation process involves loading biotin from enterocytes into plasma. Biotin circulates in plasma as free biotin, biotinyl 5-adenylate, biotinyl-CoA, protein-bound biotin, bis-nor-biotins, biotin sulfoxide, biotin sulfone, and unidentified derivatives. Biotin enters hepatocytes via Na^+ ion-dependent multivitamin transporters through an energy-requiring process. Na^+ ion-dependent multivitamin transporters facilitate biotin transport to the fetus, where fetal plasma levels are significantly higher than maternal levels. Biotin concentration in milk also exceeds maternal plasma, indicating active transport. Newly absorbed biotin is primarily stored in the liver at concentrations of 800–3000 ng/g, most of which is in mitochondrial acetyl-CoA carboxylase (Combs Jr and McClung, 2016). However, these stores are poorly mobilized during deprivation, leading to reductions in plasma biotin levels. Biotin is rapidly excreted as free biotin, bis-nor-biotin, bis-nor-biotin methyl ketone, biotin sulfone, tetra-nor-biotin-L-sulfoxide, and other derivatives in the urine, with 60.6%–95% of an oral dose excreted in rats within 24 h (Zempleni et al., 2013; Combs Jr and McClung, 2016).

The key role of biotin is to act as a prosthetic group for carboxylase enzymes involved in the carboxylation reactions of fatty acid metabolism, gluconeogenesis, and amino acid metabolism. During carboxylation, a carboxyl group binds to biotin, forming carboxy biotin, and then the carboxyl group is transferred to the acceptor subsite of the specific carboxylation molecule. Pyruvate carboxylase facilitates the addition of bicarbonate to pyruvate, forming oxaloacetic acid for gluconeogenesis and replenishing the tricarboxylic acid cycle. Furthermore, it supports the formation of citrate for fatty acid synthesis. Acetyl-CoA carboxylase-1 enzyme incorporates bicarbonate into acetyl-CoA to form malonyl-CoA in fatty acid synthesis and elongation. Acetyl-CoA carboxylase-2 enzyme inhibits fatty acid oxidation in mitochondria. Propionyl-CoA carboxylase converts propionyl-CoA into methylmalonyl-CoA by adding a bicarbonate group. Methylmalonyl-CoA is subsequently transformed into succinyl-CoA, which enters the TCA cycle for energy production or glucose synthesis. 3-Methylcrotonoyl-CoA carboxylase is involved in the degradation of leucine, a ketogenic amino acid, by catalyzing the conversion of 3-methylcrotonoyl-CoA into intermediates. In the absence of sufficient biotin, leucine degradation is diverted through an alternative pathway, leading to the production of 3-hydroxyisovaleric acid, which is excreted in the urine. This indicates a disruption in normal amino acid metabolism.

Adequate daily biotin intake levels for males and females are 5, 6, 8, 12, 20, 25, and 30 μg/day for ages 0–6 months, 7–12 months, 1–3 years, 4–8 years, 9–13 years, 14–18 years, and 19 years and older, respectively. Pregnant and lactating women require 30 and 35 μg/day, respectively.

Biotin is synthesized in microorganisms through a multi-step process (Streit and Entcheva, 2003; Lin and Cronan, 2011). The synthesis begins with the condensation of pimeloyl-CoA with alanine. This reaction is catalyzed by keto-aminopelargonic acid synthase. Nitrogen is incorporated into keto–amino pelargonic acid through transamination, a unique role for

S-adenosylmethionine, which is typically a methyl donor. A carboxylation reaction using CO_2 and ATP forms the ureido ring, yielding dethiobiotin. The sulfur insertion step, catalyzed by biotin synthase, involves the insertion of sulfur into dethiobiotin. This reaction requires iron-sulfur clusters, NADPH, and S-adenosylmethionine. Intestinal bacteria contribute significantly to biotin synthesis and overall biotin nutrition. It has been shown that bacterial fermentation of substrates such as cellulose or sorbitol can alleviate biotin deficiency in experimental animals.

8.3.1.7 Folic acid

Folic acid (vitamin B9) is a water-soluble vitamin composed of several vitamers. Folic acid refers to the oxidized form of the vitamin, but various folic acid derivatives in different oxidative states are collectively referred to as folic acid. Microbes and plants can synthesize folic acid, but other animals, including humans, cannot synthesize it. Therefore humans depend on a frequent and adequate supply of dietary folic acid. Folic acid deficiency develops faster than vitamin B12 deficiency due to its fast metabolism and storage method. Folic acid was identified in liver extract and yeast extract, which were used as treatments for megaloblastic anemia. It was then isolated from spinach leaves in 1941, and its name was derived from the Latin word "folium," as folic acid was abundant in leafy vegetables (Mitchell et al., 1941). It was chemically synthesized in 1945, and its role in DNA synthesis, megaloblastic anemia, and other metabolic activities has been identified since the 1950s (Donnelly, 2001). The relationship between neural tube disorders and folic acid was identified in the 1990s, and fortification of cereals and grain products was mandated in the United States and several other countries in 1998 (Minns, 1996; Bishai and Nalubola, 2002). Basic folic acid consists of 6-methylpterin, p-aminobenzoic acid, and glutamic acid. Polyglutamated folic acid can be formed by binding glutamate groups (up to more than eight groups), and their affinity to enzymes and cellular retention increases with the length of the glutamate chain. Therefore folyl monoglutamate is scarcely found in food. In the cytosol, dihydrofolate reductase reduces folic acid to dihydrofolate and tetrahydrofolate in two steps, with the expense of two NADPH molecules. Single-carbon fragments such as methyl, formyl, methylene, and formamino groups can bind to N5, N10, or between N5 and N10 nitrogen atoms of tetrahydrofolate, forming 5-methyl-, 5-formyl, 10-formyl-, 5,10-methylene-, 5,10-methenyl-, and 5-formamino-tetrahydrofolic acid, respectively (Fig. 8.12). Serine, glycine, choline, histidine, and formic acid donate one-carbon groups to folic acid. Predominant folate forms in food are 5-methyl- and 10-formyl-tetrahydrofolates, which function as one-carbon transporters in single-carbon metabolism. Considering its stability against oxidation, 5-formyltetrahydrofolate (also known as folinic acid) is used in pharmaceutical applications.

Food-derived folates are absorbed in the small intestine. Most folates in food are found in their reduced polyglutamate form. Before they can be absorbed, polyglutamate must be broken down into mono- or diglutamate forms. This process is facilitated by enzymes known as folyl γ-glutamyl carboxypeptidases, often referred to as folyl conjugases or pteroylpolyglutamate hydrolase, and its activity is optimal at pH 5–6 (Chandler et al., 1986). This enzyme is widely distributed in the small intestine, inside enterocytes and brush borders, and likely represents two distinct enzymes in these two locations. This enzyme is a zinc-dependent metalloenzyme, and zinc deficiency can impair enzyme activity and cause an

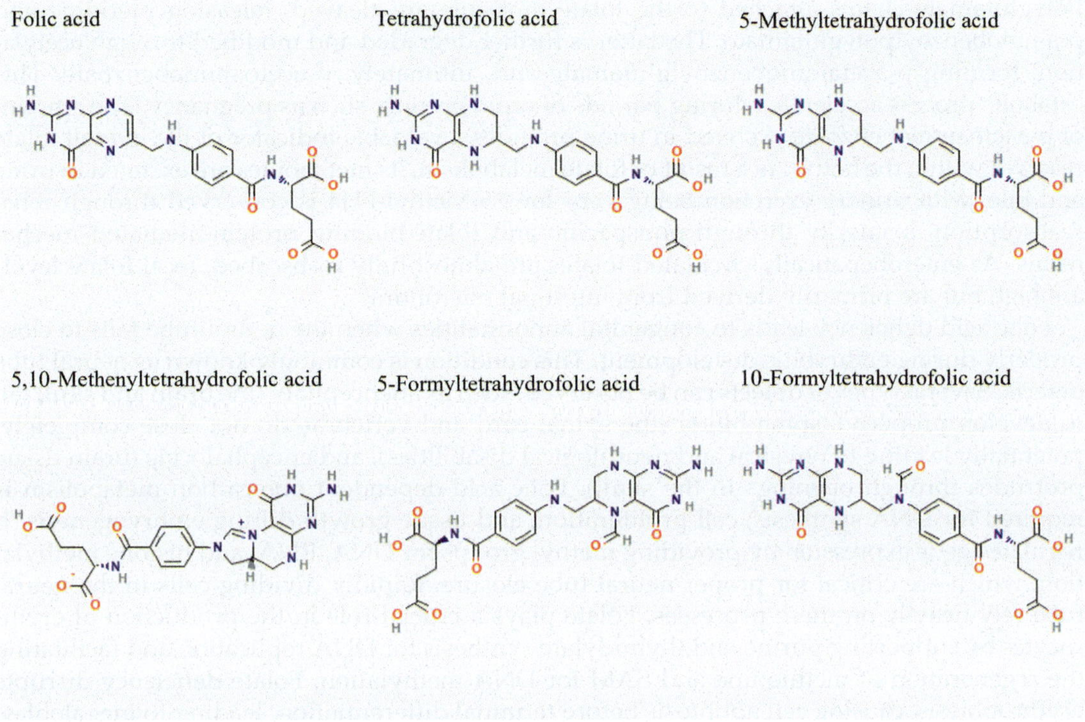

FIG. 8.12 Molecular structures of different forms of folic acid.

impairment of absorption. Free folates derived from food and biliary folates are absorbed into enterocytes by a receptor- or carrier-mediated active absorption mechanism. Participating receptors in active absorption are proton-coupled folate transporters, reduced folate carriers, and folate receptors. Passive absorption of folic acid can account for 20%–30% of high folic acid intakes (Combs Jr and McClung, 2016). Absorbed folic acid can reconjugate with glutamic acid to avoid being pumped out of the cell by folylpolyglutamate synthetase. Absorbed folate is exported directly or after methylation through organic anion transporters to the portal circulation. These folate pumps do not transport conjugated folates, which are deconjugated by γ-glutamyl hydrolase in cells. A mixture of folates, including different oxidative forms of folic acid, is transported, and the liver converts folates to 5-methyl tetrahydrofolates or 10-formyl tetrahydrofolates after being taken up into the liver. Plasma folate concentrations in humans range between 10 and 30 nmol/L, but erythrocyte concentrations range between 50 and 100 nmol/L (Combs Jr and McClung, 2016). Somatic cells take up monoglutamyl folic acids similar to enterocytes through proton-coupled folate transporters, reduced folate carriers, and folate receptors. Total body folate levels vary from 5 to 10 mg, and their distribution in tissues widely varies. Brain tissue contains lower folate levels, and the liver serves as the reservoir for half of the body's folate amount. Rapidly dividing cells contain higher levels of 10-formyl tetrahydrofolates, while slowly dividing cells contain higher levels of 5-methyl tetrahydrofolates. Folate within tissues is broken down in a multi-step process.

Polyglutamate chains attached to the folate molecule are cleaved, releasing pteridine and p-aminobenzoylpolyglutamate. The latter is further degraded and modified through acetylation, forming p-acetaminobenzoylglutamate and, ultimately, p-acetoaminobenzoate. This catabolic process accelerates during periods of rapid growth, such as pregnancy. The amount of p-acetoaminobenzoate excreted in urine provides a valuable indicator of the overall folate reserves within the body. As a result of folate metabolism, its metabolites are excreted in urine and bile, with urinary excretion being very low. 5-Methyl-FH4 is conserved through renal reabsorption, primarily through nonspecific and folate-binding protein-mediated mechanisms. As enterohepatically circulated folates are almost fully reabsorbed, fecal folate levels are high but are primarily derived from intestinal microflora.

Folic acid deficiency leads to congenital abnormalities when the neural tube fails to close properly during embryonic development. This condition is commonly known as neural tube defects. Several types of defects can be observed, such as anencephaly (the brain and skull fail to develop properly), spina bifida (the spinal cord and vertebrae do not close completely, potentially leading to physical and neurological disabilities), and encephalocele (brain tissue protrudes through openings in the skull). Folic acid-dependent one-carbon metabolism is required for DNA synthesis, cell proliferation, and tissue growth during embryogenesis. It regulates gene expression by providing methyl groups for DNA, RNA, and histone methylation, which are critical for proper neural tube closure. Rapidly dividing cells in the neural tube rely heavily on these processes. Folate plays a crucial role in the production of erythrocytes by supporting purine and thymidylate synthesis for DNA replication and facilitating the regeneration of methionine and SAM for DNA methylation. Folate deficiency disrupts erythropoiesis, causing cell apoptosis before terminal differentiation, leading to megaloblastic anemia. Food fortification and dietary intake at adequate levels reduce the risk of these conditions, including congenital heart defects, cancer risk, and vascular diseases. Numerous efforts are being made globally to enhance the folate content in foods through bioprocessing and bioengineering research (Table 8.11).

8.3.1.8 Cobalamins

Cobalamin (vitamin B12) deficiency, or pernicious anemia, was first identified in 1824, with clinical symptoms detailed in 1855 (Addison, n.d.). Large-sized bold cells (megaloblastic cells) were observed in the blood count of a patient with pernicious anemia in 1880. Key milestones include the discovery in 1925 that raw liver cured anemia in dogs (Whipple and Robscheit-Robbins, 1925) and humans (Minot and Murphy, 1926). Later developments included the isolation, determination of the structure, and naming of vitamin B12 in 1948, the identification of cyanocobalamin and hydroxocobalamin in 1949, the establishment of its structure and microbial synthesis in 1955, and its chemical synthesis in 1973 (Rickes et al., 1948; Woodward, 1973; Hodgkin et al., 1956). Dorothy Hodgkin was awarded the Nobel Prize in 1964 for determining its structure through X-ray crystallography.

Cobalamin, vitamin B12, or 5′,6′-dimethylbenzimidazolylcobamide, is an organometallic compound with a unique corrin ring structure surrounding a central cobalt atom. This corrin ring is linked to a lower ligand comprising a phosphate group, ribose, and 5,6-dimethylbenzimidazole, and an upper ligand that varies, forming distinct cobalamin forms such as cyanocobalamin (cyanide group), methylcobalamin (methyl group), 5′-deoxyadenosylcobalamin

TABLE 8.11 Vitamin B9 sources and their quantities.

Marine source	Form	Content	References
Fish, shellfish, and echinoderms			
Axillary bream	Vitamin B9	17 μg/100 g edible portion	(Dias et al., 2003)
Black scabbardfish	Vitamin B9	8.3 μg/100 g edible portion	(Dias et al., 2003)
Blue-spotted seabream	Vitamin B9	7.0 μg/100 g edible portion	(Dias et al., 2003)
Chub mackerel	Vitamin B9	14 μg/100 g edible portion	(Dias et al., 2003)
Cod	Vitamin B9	8.1 μg/100 g edible portion	(Dias et al., 2003)
Common seabream	Vitamin B9	10 μg/100 g edible portion	(Dias et al., 2003)
Common sole	Vitamin B9	10 μg/100 g edible portion	(Dias et al., 2003)
Eel	Vitamin B9	9.3 μg/100 g edible portion	(Dias et al., 2003)
Gilthead bream	Vitamin B9	9.4 μg/100 g edible portion	(Dias et al., 2003)
Hake	Vitamin B9	13 μg/100 g edible portion	(Dias et al., 2003)
Horse mackerel	Vitamin B9	13 μg/100 g edible portion	(Dias et al., 2003)
Mackerel	Vitamin B9	15 μg/100 g edible portion	(Dias et al., 2003)
Red seabream	Vitamin B9	15 μg/100 g edible portion	(Dias et al., 2003)
Sardine	Vitamin B9	15 μg/100 g edible portion	(Dias et al., 2003)
Seabass	Vitamin B9	9.5 μg/100 g edible portion	(Dias et al., 2003)
Skate	Vitamin B9	12 μg/100 g edible portion	(Dias et al., 2003)
Swordfish	Vitamin B9	14 μg/100 g edible portion	(Dias et al., 2003)
Rainbow trout	Vitamin B9	14 μg/100 g edible portion	(Dias et al., 2003)
Tuna	Vitamin B9	8.3 μg/100 g edible portion	(Dias et al., 2003)
Shrimp	Vitamin B9	4–12 μg/100 g tissue	(Venugopal and Gopakumar, 2017)
Lobster	Vitamin B9	11–12 μg/100 g tissue	(Venugopal and Gopakumar, 2017)
Crab	Vitamin B9	44–51 μg/100 g tissue	(Venugopal and Gopakumar, 2017)
Crayfish	Vitamin B9	44 μg/100 g tissue	(Venugopal and Gopakumar, 2017)
Clam	Vitamin B9	29 μg/100 g tissue	(Venugopal and Gopakumar, 2017)
Scallop	Vitamin B9	0 μg/100 g tissue	(Venugopal and Gopakumar, 2017)
Oyster	Vitamin B9	7 μg/100 g tissue	(Venugopal and Gopakumar, 2017)
Mussel	Vitamin B9	16–76 μg/100 g tissue	(Venugopal and Gopakumar, 2017)
Cuttle fish	Vitamin B9	16–46 μg/100 g tissue	(Venugopal and Gopakumar, 2017)
Cuttlefish	Vitamin B9	8.8 μg/100 g edible portion	(Dias et al., 2003)

Continued

TABLE 8.11 Vitamin B9 sources and their quantities—cont'd

Marine source	Form	Content	References
Octopus	Vitamin B9	8.1 μg/100 g edible portion	(Dias et al., 2003)
Squid	Vitamin B9	5 μg/100 g tissue	(Venugopal and Gopakumar, 2017)
Macroalgae			
Ulva rigida	Vitamin B9	1.08 g/kg	(Taboada et al., 2010)
Botryocladia leptopoda	Vitamin B9	1.55 ± 0.01 μg/g dry weight	(Gajalakshmi et al., 2018)
Ascophyllum nodosum	Vitamin B9	3.648 mg/8 g dry portion	(MacArtain et al., 2007)
Undaria pinnatifida	Vitamin B9	0.528 mg/8 g dry portion	(MacArtain et al., 2007)
Porphyra umbilicalis	Vitamin B9	1.003 mg/8 g dry portion	(MacArtain et al., 2007)
Palmaria palmata	Vitamin B9	0.021 mg/8 g dry portion	(MacArtain et al., 2007)
Ulva spp.	Vitamin B9	0.012 mg/8 g dry portion	(MacArtain et al., 2007)
Phytoplankton			
Porphyridium	Vitamin B9	5.39 ng/g dry weight	(Del Mondo et al., 2020)
Isochrysis	Vitamin B9	3 ng/g dry weight	(Del Mondo et al., 2020)
Pavlova	Vitamin B9	23 ng/g dry weight	(Del Mondo et al., 2020)
Nannochloropsis	Vitamin B9	17–22 ng/g dry weight	(Del Mondo et al., 2020)
Microchloropsis	Vitamin B9	43.6 ng/g dry weight	(Del Mondo et al., 2020)
Tetraselmis	Vitamin B9	3–20 ng/g dry weight	(Del Mondo et al., 2020)
Chlorella	Vitamin B9	3.1–34 ng/g dry weight	(Del Mondo et al., 2020)
Dunaliella	Vitamin B9	0.4–53.7 ng/g dry weight	(Del Mondo et al., 2020)
Picochlorum	Vitamin B9	64.7 ng/g dry weight	(Del Mondo et al., 2020)

(adenosyl group), and hydroxocobalamin (hydroxyl group or water molecule) (Fig. 8.13). Cyanocobalamin is the most chemically stable form of cobalamin; therefore it is used in pharmaceutical applications and food fortification but lacks a specific role in the body as it is. Light exposure and processing can degrade the molecular structure, often converting active compounds to hydroxocobalamin. In the body, these inactive forms are converted to methylcobalamin or 5′-deoxyadenosylcobalamin, which are biologically active. In addition, cobalamin analogs (glutathionylcobalamin, sulfitocobalamin, and nitritocobalamin) are found in the body, with their origins and impacts remaining under study.

Vitamin B12 is exclusively synthesized by bacteria, and it is found in foods derived from bacterial fermentation or in animal tissues that accumulate the vitamin through their diet or gut microflora. There are a number of vitamin B12-rich dietary sources, including dairy products, meats (especially liver and muscle), fermented food products, eggs, fish, and shellfish. Some algae and seaweed may contain vitamin B12 due to bacterial contamination, but this is not synthesized naturally in plants or animals. However, plant-based sources often

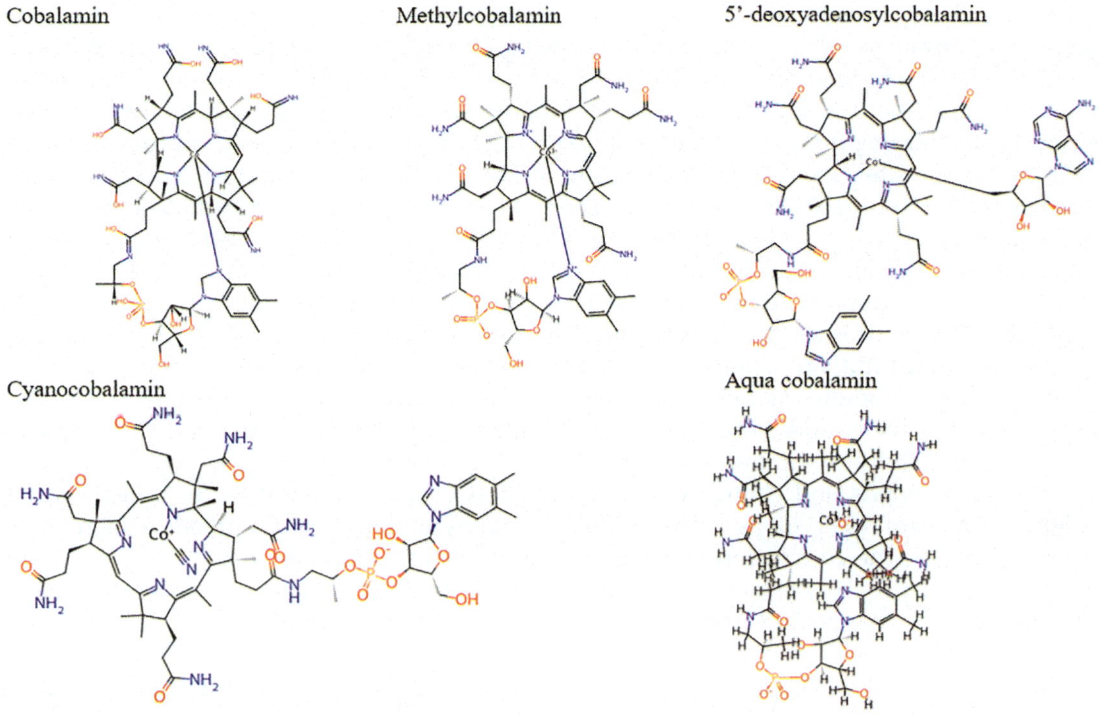

FIG. 8.13 Molecular structures of cobalamins.

contain inactive vitamin B12 analogs with no nutritional value for humans. Fortified foods like cereals, flour, and soy products can provide vitamin B12 for strict vegetarians, but naturally fermented plant-based foods generally lack significant amounts.

Vitamin B12 in food exists bound to proteins. It is primarily absorbed from food through a multi-step process involving various proteins and enzymes (Hippe, 1970). Initially, dietary B12 is released from food proteins by the action of gastric acid and pepsin in the stomach. Once released from the proteins, it binds to salivary haptocorrin, a glycoprotein present in the saliva that protects vitamin B12 from digestive enzymes. In the duodenum, the vitamin B12–haptocorrin complex is hydrolyzed by pancreatic trypsin, allowing it to bind to the intrinsic factor, a glycoprotein secreted by the parietal cells of the stomach wall. The intrinsic factor-vitamin B12 complex protects vitamin B12 from intestinal bacteria and enhances its stability by preventing side-chain modification of the corrin ring. This complex travels to the ileum, where cubilin and the receptor-associated protein are located, mediating receptor-mediated endocytosis (Moestrup and Verroust, 2001). This internalization process is calcium-dependent and requires near-neutral pH. Following endocytosis, vitamin B12 is released from the intrinsic factor in the enterocytes by the degradation of lysosomal proteases. Vitamin B12 is then bound to transcobalamin II, a transport protein that facilitates the secretion of B12 into the portal circulation. Intrinsic factor itself is degraded and does not enter the bloodstream. In addition to absorption from the diet, vitamin B12 undergoes enterohepatic circulation, where bile excreted in the ileum is reabsorbed. As a result of this recycling system, the body is able to conserve B12 and maintain normal levels of the vitamin, slowing down the development of vitamin B12 deficiency. A very small amount, <1%, is passively diffused into enterocytes. Once transported to the bloodstream, the majority of vitamin B12 binds to transcobalamin, a protein crucial for its transport. This complex binds to CD320 receptors located on the surface of cells in storage sites, including the liver and kidney, initiating endocytosis and facilitating the release of vitamin B12 inside the cell. Then, it is utilized for various biological functions. In the human liver, 60% of the total body vitamin B12 is stored (Combs and McClung, 2017). The half-life of vitamin B12 in the liver is 350–400 days in humans (Combs and McClung, 2017). The predominant vitamin B12 form in plasma is methylcobalamin, but in the liver and other tissues, it is adenosylcobalamin.

Cobalamin naturally functions as a coenzyme in essential biochemical reactions, including the conversion of homocysteine to methionine, methylmalonyl-CoA to succinyl-CoA, reductive dehalogenation, and radical-mediated S-adenosylmethionine processes. The unique biochemical activity of vitamin B12 comes from the cobalt ion at its core, which can form cobalt-carbon bonds and change oxidation states (Co(I), Co(II), Co(III)). In mitochondria, adenosylcobalamin serves as a coenzyme for methylmalonyl-CoA mutase, which plays a crucial role in the conversion of methylmalonyl-CoA to succinyl-CoA, a key step in the breakdown of propionate derived from odd-chain fatty acids and certain amino acids. A deficiency in vitamin B12 impairs the function of enzymes, leading to the accumulation of methylmalonic acid. This disruption affects the tricarboxylic acid cycle and can result in metabolic issues such as methylmalonic aciduria and the synthesis of abnormal odd-chain and branched-chain fatty acids.

In the cytoplasm, methylcobalamin facilitates the folate-dependent methylation of homocysteine to methionine, acting as a coenzyme for methionine synthase. In this process, vitamin B12 acts as a methyl carrier, with methylcobalamin being regenerated through methylation by methyltetrahydrofolate, which simultaneously produces tetrahydrofolate. Methionine is not only vital for protein synthesis but also acts as a precursor for S-adenosylmethionine, the universal methyl donor involved in diverse methylation reactions, including DNA, RNA, and protein methylation.

Reductive dehalogenation, primarily occurring in specific bacteria, involves the removal of halogen atoms from organic compounds coupled with electron transfer. Vitamin B12-dependent reductive dehalogenases catalyze this reaction through the formation of a cobalt-halogen bond. These processes aid in environmental bioremediation, particularly in detoxifying halogenated pollutants in soil and water.

Vitamin B12 is involved in one-carbon metabolism. A deficiency of vitamin B12 can cause neural tube defects similar to folic acid deficiency, as explained above. However, vitamin B12 may play a role in the residual incidence of neural tube defects not prevented by folate supplementation. Research indicates that lower vitamin B12 levels in the amniotic fluid of neural tube defect pregnancies, despite comparable maternal serum levels, suggest a limitation in the maternal capacity to transfer sufficient vitamin B12 to the fetus.

Vitamin B12 deficiency leads to megaloblastic anemia, characterized by the release of immature erythrocytes, low white cell and platelet counts, and hyper-segmented neutrophils. This condition arises from impaired DNA synthesis due to a methylation defect. Pernicious anemia, a specific form of megaloblastic anemia caused by vitamin B12 deficiency, often results from the autoimmune destruction of intrinsic factor. Neurological degeneration linked to vitamin B12 deficiency involves demyelination, likely due to impaired methylation of myelin proteins and cytokine imbalances. Insufficient vitamin B12 levels can cause neurodegeneration in glial cells, myelin, and the neural interstitium due to abnormal methylmalonic acid incorporation into neuronal lipids, reduced choline synthesis, and increased inflammatory cytokines.

Estimated average requirements of vitamin B12 for males and females in the age groups of 1–3 years, 4–8 years, 9–13 years, and 14 years and older are 0.7, 1.0, 1.5, and 2.0 µg/day, respectively. RDAs for the same age groups are 0.9, 1.2, 1.8, and 2.4 µg/day, respectively. Adequate intake levels have been defined for the 0–6-month-old and 7–12-month-old age groups, and these levels are 0.4 and 0.5 µg/day, respectively. For pregnant and lactating women, estimated average requirements are 2.2 and 2.4 2.4 µg/day, respectively, and RDAs are 2.4 and 2.8 µg/day, respectively.

Vitamin B12 is a corrinoid, a structurally essential molecule crucial for regulating various biochemical pathways in prokaryotes, animals, and humans; only a subset of archaea and bacteria can synthesize these molecules de novo. This pathway begins with the synthesis of δ-aminolevulinic acid, and this molecule can lead to two pathways, including the Shemin pathway (which starts from glycine and succinyl CoA, (Shemin and Kikuchi, 1958) and the C5 pathway (which starts from glutamic acid, Jahn et al., 1992). δ-aminolevulinic acid is converted to uroporphyrinogen III through several enzyme-dependent intermediate steps, including condensation to form porphobilinogen and the polymerization and cyclization of porphobilinogen into uroporphyrinogen III. The corrinoid pathway begins with uroporphyrinogen III, forming precorrin 2. Depending on oxygen availability, precorrin 2 undergoes different processes, such as aerobic and anaerobic pathways. In the aerobic pathway, adenosylcobyric(III) acid is synthesized through a series of intermediate steps, progressing sequentially through precorrin 2, precorrin 3A, precorrin 3B, precorrin 4, precorrin 5, precorrin 6A, precorrin 6B, precorrin 8, hydrogenobyrinic acid, hydrogenobyrinic acid a,c-diamide, cob(II)yrinic acid a,c-diamide, and adenosylcob(III)yrinic acid a,c-diamide (Heldt et al., 2005).

The anaerobic pathway of vitamin B12 biosynthesis differs from the aerobic pathway primarily in the timing of cobalt insertion and ring contraction (Hazra et al., 2015; Moore et al., 2013). In this pathway, cobalt is inserted into precorrin-2 by distinct cobalt chelatases, forming cobalt-factor II, followed by factor II oxidation and methylation, ring contraction, lactone opening, and double bond formation between C1 and C19. Subsequent steps include C1 carbon methylation, reduction of the C18–C19 double bond, methylation and substitution at the C5 and C12 carbon atoms, methyl migration, amidation of side chains, and finally, adenosylation to complete the process. As the final step, a nucleotide loop is attached to the adenosylcob(III)yrinic acid. The nucleotide attachment loop involves synthesizing an activated α-nucleotide by transferring the phosphoribosyl moiety of nicotinate mononucleotide onto a base via phosphoribosyltransferase, followed by its linkage to adenosylcobinamide phosphate through cobalamin synthase, completing the cobalamin molecule.

Vitamin B12 is contained at a lower level in most biological samples, usually <10 µg/100 g (Watanabe and Bito, 2018) (Table 8.12). Numerous methods have been optimized to

TABLE 8.12 Vitamin B12 sources and their quantities.

Marine source	Form	Content	References
Fish, shellfish, and echinoderms			
Blue-spotted seabream	Vitamin B12	1.3 μg/100 g	(Dias et al., 2003)
Chub mackerel	Vitamin B12	14 μg/100 g	(Dias et al., 2003)
Cod	Vitamin B12	0.95 μg/100 g	(Dias et al., 2003)
Common seabream	Vitamin B12	2.1 μg/100 g	(Dias et al., 2003)
Common sole	Vitamin B12	0.94 μg/100 g	(Dias et al., 2003)
Hake	Vitamin B12	0.47 μg/100 g	(Dias et al., 2003)
Horse mackerel	Vitamin B12	7.2 μg/100 g	(Dias et al., 2003)
Octopus	Vitamin B12	0.83 μg/100 g	(Dias et al., 2003)
Sardine	Vitamin B12	10 μg/100 g	(Dias et al., 2003)
Tuna	Vitamin B12	2.4 μg/100 g	(Dias et al., 2003)
Wreckfish	Vitamin B12	0.7 μg/100 g	(Dias et al., 2003)
Shrimp	Vitamin B12	1.2–1.9 μg/100 g tissue	(Venugopal and Gopakumar, 2017)
Lobster	Vitamin B12	3.1 μg/100 g tissue	(Venugopal and Gopakumar, 2017)
Crayfish	Vitamin B12	2.1 μg/100 g tissue	(Venugopal and Gopakumar, 2017)
Crab	Vitamin B12	7.3–11.5 μg/100 g tissue	(Venugopal and Gopakumar, 2017)
Mussel	Vitamin B12	8–24 μg/100 g tissue	(Venugopal and Gopakumar, 2017)
Oyster	Vitamin B12	15–29 μg/100 g tissue	(Venugopal and Gopakumar, 2017)
Clam	Vitamin B12	1.3–100 μg/100 g tissue	(Venugopal and Gopakumar, 2017
Scallop	Vitamin B12	1.3–1.5 μg/100 g tissue	(Venugopal and Gopakumar, 2017)
Cuttle fish	Vitamin B12	3–5.4 μg/100 g tissue	(Venugopal and Gopakumar, 2017)
Cuttlefish	Vitamin B12	1.2 μg/100 g	(Dias et al., 2003)
Squid	Vitamin B12	1.3 μg/100 g tissue	(Venugopal and Gopakumar, 2017)
Squid	Vitamin B12	1.1 μg/100 g	(Dias et al., 2003)
Macroalgae			
Sargassum muticum	Vitamin B12	10 μg/g dry matter	(De Roeck-Holtzhauer et al., 1991)
Laminaria digitata	Vitamin B12	2.84 mg/g	(MacArtain et al., 2007)
Undaria pinnatifida	Vitamin B12	14.8 mg/g	(MacArtain et al., 2007)
Porphyra umbilicalis	Vitamin B12	12.9 mg/g	(MacArtain et al., 2007)
Palmaria palmata	Vitamin B12	1.840 mg/8 g dry portion	(MacArtain et al., 2007)
Ulva rigida	Vitamin B12	60 μg/kg	(Taboada et al., 2010)

TABLE 8.12 Vitamin B12 sources and their quantities—cont'd

Marine source	Form	Content	References
Phytoplankton and bacteria			
Pavlova lutheri	Vitamin B12	1162 μg/g dry matter	(De Roeck-Holtzhauer et al., 1991)
Skeletonema costatum	Vitamin B12	117 μg/g dry matter	(De Roeck-Holtzhauer et al., 1991)
Isochrysis galbana	Vitamin B12	89 μg/g dry matter	(De Roeck-Holtzhauer et al., 1991)
Tetraselmis suecica	Vitamin B12	9 μg/g dry matter	(De Roeck-Holtzhauer et al., 1991)
Chaetoceros calcitrans	Vitamin B12	8 μg/g dry matter	(De Roeck-Holtzhauer et al., 1991)
Spirulina platensis	Vitamin B12	80 mg/g	(Becker, 2013)
Chlorella pyrenoidosa	Vitamin B12	0	(Becker, 2013)
Senedesmus quadricauda	Vitamin B12	396 mg/g	(Becker, 2013)

determine vitamin B12 levels in biological, pharmaceutical, and food samples, such as microbiological assays, radio-ligand binding assays, and polarographic, spectrophotometric, electrophoretic, and chromatography techniques (Rothenberg, 1963; Quadros et al., 1988; AOAC International, 1993; Muhammad et al., 1993; Li and Chen, 2000). AOAC International methods describe microbial assays to determine vitamin B12 levels in samples employing *Lactobacillus delbrueckii* (*leichmannii*) ATCC 7830, which depends on vitamin B12 for growth. It is part of the Official Methods of Analysis of AOAC International (section 45.2.02). Radio-ligand binding assays are used to analyze vitamin B12 in blood, tissues, and food (Quadros et al., 1988). Binding proteins such as transcobalamin I, transcobalamin II, and haptocorrin have been used in this assay, and the porcine intrinsic factor is commonly used as the binding protein in clinical applications. Chemiluminescence, fluorometric, spectrophotometric, solid-phase extraction, and capillary electrophoresis methods have been published to quantify vitamin B12 levels in samples; however, these methods are moderately recommended for vitamin B12 analysis due to their lower sensitivity and selectivity. Purification and concentration steps are applied in liquid chromatography to address limitations in liquid chromatography, especially for natural products and high-concentration pharmaceutical formulations. Proteases and amylases are utilized in buffered solutions to liberate bound vitamin B12 from food or biological samples. In addition, heat treatments, sonication, and acidic precipitation are used to liberate vitamin B12 from bound forms. A cyanide source (potassium or sodium cyanide) is then added to convert the liberated cobalamin into cyanocobalamin, enabling the measurement of total cobalamin content as cyanocobalamin. Pharmaceuticals often require minimal preparation as they can be solubilized in buffer solutions (e.g., trichloro acetic acid, Na_2SO_4 containing NaEDTA, and phosphate buffer). Extracted solutions should be protected from ultraviolet and visible light and alkaline conditions to avoid the decomposition of cyanocobalamin. Chromatography methods have been widely utilized for the separation, identification, and quantification of vitamin B12 (and its forms). Earlier methods like thin-layer chromatography and gas-liquid chromatography offered means to analyze vitamin B12; however, high-performance liquid chromatography provides a more versatile and efficient approach (Luo et al., 2006). Reversed-phase chromatography on C18 columns with various buffers forms the foundation of most liquid chromatography methods for vitamin B12 assays. High-performance liquid chromatography coupled with ultraviolet or fluorescent detection techniques is promising in vitamin B12 analysis due to its precision and ability to work with complex samples (Li and Chen, 2000). For detection, 361 and 550 nm are frequently used in ultraviolet and visible light ranges, and excitation and emission wavelengths used in fluorescent detection are 275 and 305 nm, respectively (Alsberg et al., 2001; Viñas et al., 2003; Li and Chen, 2000).

8.3.2 Vitamin C

Vitamin C (ascorbic acid, $C_6H_8O_6$, 2-oxo-l-theo-hexono-4-lactone-2,3-enediol) is a water-soluble vitamin found in both animals and plants that functions as an antioxidant and a cofactor for a wide array of enzymatic and hormonal processes in animals (Fig. 8.14). It contributes to the production of biomolecules (carnitine, neurotransmitters, and connective tissues) and influences iron metabolism (absorption, transport, and storage). Vitamin C is a reducing sugar, which has a near-planar five-membered ring. Ascorbic acid has two chiral centers, resulting in four stereoisomers. It undergoes one-electron oxidation to form monodehydroascorbate (semidehydro-l-ascorbic acid) free radical, which can be further oxidized to dehydroascorbic acid by the nonenzymatic disproportionation of the monodehydroascorbate free radical. A key chemical property of ascorbic acid is the ability of dehydroascorbic acid (the oxidized form) to be enzymatically or nonenzymatically reduced back to ascorbic acid via monodehydroascorbate (Buettner and Jurkiewicz, 1996; Bors and Buettner, 1997). If dehydroascorbic acid is not reduced, it undergoes irreversible hydrolysis in an enzymatic reaction pathway. A small percentage of ascorbic acid exists in the form of dehydroascorbic acid, and this proportion can increase during oxidative stress. Dehydroascorbic acid has a short in vivo half-life, potentially as brief as 2 min, and is rapidly converted to 2,3-diketogulonate, a process

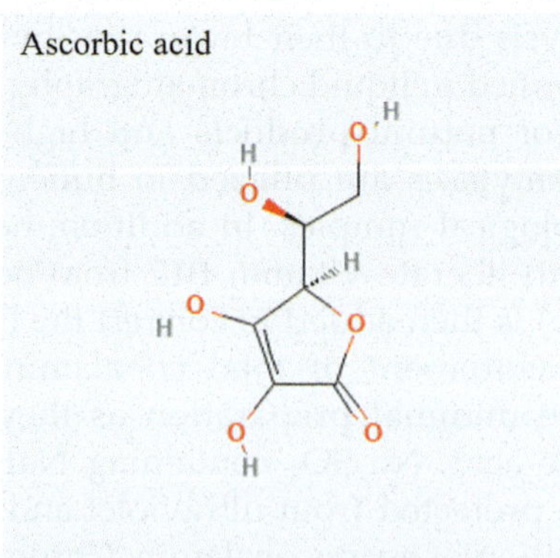

FIG. 8.14 The molecular structure of ascorbic acid.

facilitated by bicarbonate and aldonolactonase enzymes present in the livers of most mammals (Koshiishi and Imanari, 1997). Diketogulonate degrades into various products, including L-xylose, L-xylonate, L-lyxonate, oxalate, L-threonate, glycogen, and CO_2, with products varying based on the presence of enzymes and the conditions. In addition, the proton on oxygen-3 is acidic (pK1 = 4.17), which contributes to the acidic nature of ascorbic acid.

Humans do not synthesize vitamin C due to the lack of the L-gulonolactone oxidase enzyme, which catalyzes the final step of vitamin C biosynthesis from glucose (Nishikimi and Yagi, 1991). Therefore adequate vitamin C intake is essential to avoid deficiency conditions. Vitamin C is absorbed in the intestine with an efficiency of 70%–90% when dietary intake ranges from 30 to 180 mg/day. However, absorption decreases to approximately 50% when the intake increases to 1000 mg/day. Excessive intakes of >3000 mg/day can cause diarrhea and other gastrointestinal disturbances. However, there is no evidence suggesting that vitamin C is toxic or carcinogenic, but it can result in renal stone formation due to its metabolite, oxalic acid. Unabsorbed vitamin C in the intestine serves as a substrate for gut bacteria. Vitamin C is absorbed in the intestine through different mechanisms. Passive absorption can be observed in animals that do not synthesize vitamin C. It is actively absorbed depending on its oxidized forms, in which ascorbic acid is absorbed by sodium-dependent ascorbate transporters, and dehydroascorbic acid is efficiently absorbed by glucose transporters. Dehydroascorbic acid is converted to ascorbic acid soon after being absorbed into enterocytes. Human cells saturate at a dietary intake of vitamin C of about 100 mg/day, and plasma accumulates excess vitamin C. Ascorbic acid is absorbed into cells following the same mechanism, with sodium-dependent ascorbate transporter-1 uptaking ascorbic acid in the liver and kidney epithelial cells, while sodium-dependent ascorbate transporter-2 is responsible for absorbing ascorbic acid in the brain, heart, lungs, placenta, and exocrine and endocrine tissues. Vitamin C is distributed and accumulated in nearly all body tissues. When plasma is saturated with vitamin C, excess vitamin C is filtered by the glomeruli of the kidney and excreted in urine.

Ascorbic acid serves as a chain-breaking antioxidant that protects lipids, DNA, and proteins from oxidative damage, which is caused by reducing ROS, including superoxide, peroxyl radicals, aqueous peroxyl radicals, singlet oxygen, ozone, peroxynitrite, nitrogen dioxide, nitroxide radicals, and hypochlorous acid. Ascorbic acid undergoes single-electron oxidation, forming the ascorbyl radical (a one-electron oxidation product), a less reactive intermediate that breaks down into ascorbate and dehydroascorbate (a two-electron oxidation product). Oxidized products can be restored within the body through chemical regeneration by reduced glutathione or through enzymatic processes involving glutathione- or NADPH-dependent reductases. It reacts with peroxyl radicals, preventing lipid peroxidation and preserving vitamin E by reducing the α-tocopheroxyl radical. It protects DNA from oxidative damage, preventing DNA mutations that may lead to cancers or congenital disabilities. Vitamin C supplementation combined with vitamin E reduces 8-oxo-deoxyguanosine levels, a marker of DNA oxidation. Vitamin C reduces protein oxidation, maintaining proper protein folding and disulfide bond formation in the endoplasmic reticulum. It interacts with the ascorbate/dehydroascorbate system, supporting redox reactions essential for protein stability and function. This property of vitamin C is important for the degradation of 4-hydroxyphenylpyruvate and the desaturation of fatty acids forming double bonds. The concentrations of glutathione and ascorbic acid are closely related, and they mutually spare each other and work synergistically to protect cells from oxidative damage. Ascorbic acid (dehydroascorbic acid) is recycled back to its reduced form by glutathione, maintaining the antioxidant capacity of vitamin C. Their interdependence is based on recycling mechanisms and enzymatic pathways, emphasizing their role as regulators of redox balance. As a biochemical redox agent, ascorbic acid can donate electrons during the hydroxylation of proline and lysine residues in the unfolded procollagen chain for collagen synthesis. Lysyl hydroxylase, prolyl 4-hydroxylase, and prolyl 3-hydroxylase enzymes catalyze this process, which depends on O_2, Fe^{+2}, and ascorbic acid. Hydroxylation is essential to form a stable triple-helical structure of collagen. Severe vitamin C deficiency can lead to scurvy, a condition caused by impaired collagen synthesis. This deficiency results in delayed wound healing, fragile blood vessels, bone and cartilage abnormalities, reduced iron absorption, and weakened immune function. Ascorbic acid acts as a cofactor for two iron-containing enzymes (trimethyl lysine dioxygenase and 4-trimethylaminobutyraldehyde dioxygenase) that help produce carnitine, a molecule essential for transporting fatty acids into mitochondria to generate cellular energy. Furthermore, it is involved in drug, steroid, and tyrosine metabolism.

The brain and adrenal glands store a significant amount of ascorbic acid, which acts as a cofactor for dopamine β-hydroxylase, an enzyme essential for converting dopamine to norepinephrine. The dopamine β-hydroxylase enzyme requires copper ions in a reduced state for its function, and ascorbic acid helps maintain the reduced state of these copper ions. Furthermore, ascorbic acid plays a similar role in supporting α-amidases, which are essential for the α-amidation of hormones and hormone-releasing factors. These include hormones involved in growth and metabolism (releasing factors for growth hormone, corticotrophin, thyrotropin, and calcitonin), stress and water regulation (pro-ACTH, vasopressin, and oxytocin), pigmentation (melanotropins), and digestion (cholecystokinin and gastrin).

Daily reference intakes for vitamin C are comparatively higher than those of other vitamins. According to the Institute of Medicine of the National Academies, the recommended daily intakes of vitamin C for males and females in the 1–3, 4–8, and 9–13 age groups are 15, 25, and 45 mg/day, respectively. For males in the age groups of 14–18 and 19 years and older, the recommended intakes are 75 and 90 mg/day, while for females in these age groups, they are 65 and 75 mg/day, respectively. For pregnant women in the age groups of under 18 and 19 years or older, the recommended daily vitamin C intakes are 80 and 85 mg/day, while for lactating women in those age groups, they are 115 and 120 mg/day, respectively. Tolerable upper limits are significantly higher than the recommended daily intake values. The upper limits for the 1–3, 4–8, 9–13, 14–18, and 19 or older age groups are 400, 650, 1200, 1800, and 2000 mg/day, respectively.

Certain properties of ascorbic acid must be considered when selecting an extraction medium or analytical method, such as pH, light, temperature, oxygen, and metals or enzymes like ascorbic acid oxidase in the medium (Spínola et al., 2014). The light absorption properties of L-ascorbic acid depend on the pH of the medium and the ionic form of the molecule (Malacaria and Furia, 2023). Although L-ascorbic acid itself does not fluoresce, it can be derivatized with O-phenylenediamine to produce a fluorescent product, which is useful in chemical and liquid chromatography methods. Effective vitamin C extraction and stabilization require tailored methods depending on the sample type and analytical needs. Common extraction media used for vitamin C include metaphosphoric acid, ethylenediaminetetraacetic acid, acetic acid, sulfuric acid, ethanol, and acetone. Trichloroacetic and perchloric acids have been used to extract tissue-bound vitamin C from animal tissue samples for high-pressure liquid chromatography. Dry materials should be blended with the relevant extractant, centrifuged (15,000 × g for 4 min), and filtered (0.45 μm) to prepare the extraction. During the past decades, analytical methods for vitamin C have rapidly advanced from guinea pig bioassay to more sophisticated methods, including spectrophotometric, fluorometric, electrochemical, and chemiluminescence methods (Dawbarn, 1945; Mori et al., 1998; Chaichi and Alijanpour, 2013; Škugor Rončević et al., 2022; Arya et al., 2000). Key factors include maintaining an acidic environment, preventing oxidation, addressing matrix-specific challenges, proper sample handling, rapid processing, and using appropriate stabilizers, which are critical for accurate analysis (Zaporozhets and Krushinskaya, 2002; Ottaway, 2012; Moser and Bendich, 1990). Rapid enzymatic assay, flow-injection spectrophotometric determination, microplate assay, and liquid chromatography with fluorescent detectors are suitable analytical methods to determine vitamin C levels in food or tissue samples (Casella et al., 1989; Sánchez-Mata et al., 2000; Kleszczewski and Kleszczewska, 2002; Stevens et al., 2006). The reducing ability of ascorbic acid is commonly used to develop colorimetric methods to determine vitamin C levels, and 2,4-dinitrophenylhydrazine became the preferred method for analyzing vitamin C in tissues. Dehydroascorbic acid is converted to ascorbic acid prior to redox reaction-based analysis (Davies et al., 2007). High-performance liquid chromatography coupled with ultraviolet and electrochemical detection is used to detect forms of vitamin C accurately (Gazdik et al., 2008).

Ascorbic acid is not synthesized in some species, including primates, bats, guinea pigs, and teleosts, due to an evolutionary lack of L-gulonolactone oxidase, which is essential to catalyze L-gulonolactone in the ascorbic acid synthesis pathway (Birney et al., 1976; Drouin et al., 2011). Biosynthesis of ascorbic acid in yeast, algae, higher plants, and mammals follows different enzymatic pathways (Smirnoff et al., 2004). L-Ascorbic acid is synthesized in animals from D-glucose via D-glucuronic acid, L-gulonic acid, L-gulonolactone, and 2-keto-L-gulonolactone. In plants, it starts from D-glucose-6-phosphate and progresses via D-fructose-6-phosphate, D-mannose-6-phosphate, D-mannose-L-phosphate, guanosine diphosphate-D-mannose, guanosine diphosphate-L-galactose, L-galactose-L-phosphate, L-galactose, and L-galactono-1,4-lactone to synthesize L-ascorbic acid. Yeast synthesizes D-erythroascorbate, a C5 analog, using D-arabinose and D-arabinono-1,4-lactone (Linster and Van Schaftingen, 2007; Duque et al., 2022) (Table 8.13).

TABLE 8.13 Marine vitamin C sources and available content.

Marine source	Form	Content	References
Fish, shellfish, and echinoderms			
Parrot fish—liver	Vitamin C	31.9–109.6 mg/kg tissue	(Wang et al., 2003)
Korean rockfish	Ascorbic acid	14.9–60.6 mg/kg tissue	(Lee et al., 1998)
Sprat larvae	Vitamin C	428–487 μg/g wet weight	(Hapette and Poulet, 1990)
European sea bass	Vitamin C	580–3580 μg AA/g dry weight	(Merchie et al., 1997)
Turbot	Vitamin C	500–2600 μg AA/g dry weight	(Merchie et al., 1997)
Large yellow croaker	Vitamin C	0.0–27.1 μg/g wet weight	
Red sea bream—liver	Vitamin C	47–187 μg/g wet weight	(Medagoda et al., 2024)
Kuruma shrimp (*Marsupenaeus japonicus*)	Vitamin C	1.3–22.3 μg/g	(Moe et al., 2004)
Abalone (*Haliotis tuberculata*)	Vitamin C	55.6–141 μg/g wet weight	(Mai, 1998)
Abalone (*Haliotis discus hannai*)	Vitamin C	38–68.6 μg/g wet weight	(Mai, 1998)
Macro algae			
Sargassum muticum	Vitamin C	560 μg/g dry matter	(De Roeck-Holtzhauer et al., 1991)
Laminaria digitata	Vitamin C	2.84 mg/g	(MacArtain et al., 2007)
Undaria pinnatifida	Vitamin C	0.528 mg/g	(MacArtain et al., 2007)
Laminaria saccharina	Vitamin C	25 mg/100 g	(Murata and Nakazoe, 2001)
Porphyra complex	Vitamin C	100 mg/100 g	(Murata and Nakazoe, 2001)
Porphyra umbilicalis	Vitamin C	1.00 mg/g	(MacArtain et al., 2007)
Palmaria palmata	Vitamin C	5.520 mg/8 g dry portion	(MacArtain et al., 2007)
Gracilaria changgi	Vitamin C	28.5 mg/100 g	(Norziah and Ching, 2000)
Gracilaria bursapastoris	L-Ascorbic acid	21.6 ± 8.0 mg/100 g	(Yildiz et al., 2011)
Hypnea musciformis	Vitamin C	0.72 ± 0.12 mg/g	(Balamurugan et al., 2013)
Botryocladia leptopoda	Vitamin C	5.45 ± 0.18 μg/g dry weight	(Gajalakshmi et al., 2018)
Ulva rigida C.	L-Ascorbic acid	46.00 ± 0.17 mg/100 g	(Yildiz et al., 2012)
Ulva rigida	Ascorbic acid	94.20 mg/kg	(Taboada et al., 2010)
Enteromorpha intestinalis	Ascorbic acid	136.16 ± 0.85 mg/100 g dry weight	(Cadar et al., 2023)
Enteromorpha compressa	Vitamin C	40 mg/100 g	(Murata and Nakazoe, 2001)

TABLE 8.13 Marine vitamin C sources and available content—cont'd

Marine source	Form	Content	References
Caulerpa cottonii	Vitamin C	35.3 ± 0.01 mg/100 g wet weight	(Matanjun et al., 2009)
Caulerpa lentillifera	Vitamin C	34.7 ± 0.02 mg/100 g wet weight	(Matanjun et al., 2009)
Caulerpa polycystum	Vitamin C	34.5 ± 0.01 mg/100 g wet weight	(Matanjun et al., 2009)
Cladophora vagabunda	Ascorbic acid	149.66 ± 0.58 mg/100 g dry weight	(Cadar et al., 2023)
Zooplankton			
Rotifers	Ascorbic acid	220.1 μg/g dry matter	(van der Meeren et al., 2008)
Cysts of artemia	Vitamin C	160–517 μg/g dry weight	(Dabrowski, 1991;
Artemia	Ascorbic acid	361.3–530.6 μg/g dry weight	(van der Meeren et al., 2008)
Mixed zooplankton	Vitamin C	15.16–39.07 μg/g dry matter	(Mitra et al., 2007)
Copepods	Ascorbic acid	476.6–552.9 μg/g dry matter	(van der Meeren et al., 2008)
Calanus helgolandicus	Vitamin C	80–538 μg/g wet weight	(Poulet et al., 1989)
Temora longicornis	Vitamin C	125–368 μg/g wet weight	(Poulet et al., 1989)
Acartia clausi	Vitamin C	30–260 μg/g wet weight	(Poulet et al., 1989)
Nauplii (*A. clausi, T, longicornis*)	Vitamin C	201.5–234.5 μg/g wet weight	(Hapette and Poulet, 1990)
Ldolea metallica	Vitamin C	272.58 μg/g wet weight	(Hapette and Poulet, 1990)
Cavolinia inflexa	Vitamin C	147.04 μg/g wet weight	(Hapette and Poulet, 1990)
Microalgae			
Pavlova lutheri	Vitamin C	837 μg/g dry matter	(De Roeck-Holtzhauer et al., 1991)
Isochrysis galbana	Vitamin C	772 μg/g dry matter	(De Roeck-Holtzhauer et al., 1991)
Tetraselmis suecica	Vitamin C	498 μg/g dry matter	(De Roeck-Holtzhauer et al., 1991)
Chaetoceros calcitrans	Vitamin C	121 μg/g dry matter	(De Roeck-Holtzhauer et al., 1991)
Skeletonema costatum	Vitamin C	59 μg/g dry matter	(De Roeck-Holtzhauer et al., 1991)
Spirulina platensis	Vitamin C	80 mg/g	(Becker, 2013)
Chlorella pyrenoidosa	Vitamin C	0	(Becker, 2013)
Senedesmus quadricauda	Vitamin C	396 mg/g	(Becker, 2013)

References

Abdou, E., Hazell, A.S., 2015. Thiamine deficiency: an update of pathophysiologic mechanisms and future therapeutic considerations. Neurochem. Res. 40, 353–361. doi:10.1007/s11064-014-1430-z.

Addison, T., n.d. On the Constitutional and Local Effects of Disease of the Suprarenal Capsules. Samuel Highley.

Aguilera-Méndez, A., Fernández-Lainez, C., Ibarra-González, I., Fernandez-Mejia, C., 2012. V. R. Preedy (ed.), B Vitamins and Folate: Chemistry, Analysis, Function and Effects, The Royal Society of Chemistry, pp. 108–126.

Aktar, F., Rahaman, M.N., Simol, H.A., Soma, M.A., Sultan, M.Z., 2025. Quantification and evaluation of thiamine and riboflavin in marine resources of the Bay of Bengal. Biomed. Pharmacology J. 18 (2), 1381–1388.

Al-Shekaili, H.H., van Karnebeek, C., Leavitt, B.R., 2021. Vitamin B6 and related inborn errors of metabolism. In: B-Complex Vitamins-Sources, Intakes and Novel Applications, IntechOpen. doi:10.5772/intechopen.99751.

Alasalvar, C., Taylor, K., Zubcov, E., Shahidi, F., Alexis, M., 2002. Differentiation ofcultured and wild sea bass (Dicentrarchus labrax): total lipid content, fatty acid andtrace mineral composition. Food Chem. 79, 145–150.

Alsberg, T., Minten, J., Haglund, J., Törnqvist, M., 2001. Determination of hydroxyalkyl derivatives of cobalamin (vitamin B12) using reversed phase high performance liquid chromatography with electrospray tandem mass spectrometry and ultraviolet diode array detection. Rapid Commun. Mass Spectrom. 09514198. 15 (24), 2438–2445. Available from: http://www.onlinelibrary.wiley.com/journal/10.1002/(ISSN)1097-0231. doi:10.1002/rcm.527.

Alturfan, A.A., 2023. Water-soluble Vitamins. A Guide to Vitamins and Their Effects on Diseases. Cambridge Scholars Publishing, Newcastle upon Tyne, UK, pp. 71–74.

AOAC International. 2005. Official methods of analysis. 18th edn. Arlington, VA: AOAC International.

AOAC International, 1993. Report of the AOAC International task force on Methods for Nutrient Labeling Analyses. J. AOAC Int. 76, 180A.

Arya, S.P., Mahajan, M., Jain, P., 2000. Non-spectrophotometric methods for the determination of vitamin C. Anal. Chim. Acta 417 (1), 1–14.

Balamurugan, M., Selvam, G.G., Thinakaran, T., Sivakumar, K., 2013. Biochemical study and GC-MS analysis of *Hypnea musciformis* (Wulf.) Lamouroux. Am.-Eurasian J. Sci. Res. 8 (3), 117–123.

Ball, G.F., 1988. Fat-soluble Vitamin Assays in Food Analysis: A Comprehensive Review, pp. xii+–326.

Ball, G.F., 2005. Vitamins in foods: analysis, bioavailability, and stability. CRC press.

Becker, E.W., 2013. Microalgae for human and animal nutrition. Handbook of microalgal culture: applied phycology and biotechnology, pp. 461–503.

Bender, D.A., 2003. Nutritional Biochemistry of the Vitamins, Cambridge University Press.

Bender, D.A., Bender, A.E., 1986. Niacin and Tryptophan Metabolism: The Biochemical Basis of Niacin Requirements and Recommendations.

Bentley, R., Meganathan, R., 1982. Biosynthesis of vitamin K (menaquinone) in bacteria. Microbiol. Rev. 0146-0749. 46 (3), 241–280. doi:10.1128/mmbr.46.3.241-280.1982.

Bettendorff, L., Kolb, H.A., Schoffeniels, E., 1993. Thiamine triphosphate activates an anion channel of large unit conductance in neuroblastoma cells. J. Membr. Biol. 136, 281–288. doi:10.1007/BF00233667.

Bieri, J.G., Evarts, R.P., 1974. Vitamin E activity of γ-tocopherol in the rat, chick and hamster. J. Nutr. 104 (7), 850–857.

Birney, E.C., Jenness, R., Ayaz, K.M., 1976. Inability of bats to synthesise L-ascorbic acid. Nature 260 (5552), 626–628.

Bishai, D., Nalubola, R., 2002. The history of food fortification in the United States: its relevance for current fortification efforts in developing countries. Econ. Dev. Cult. Change. 0013-0079. 51 (1), 37–53. doi:10.1086/345361.

Blaner, W.S., Li, Y., Brun, P.J., Yuen, J.J., Lee, S.A., Clugston, R.D., 2016. Vitamin A absorption, storage and mobilization. Subcell. Biochem. 03060225. 81, 95–125. Available from: http://www.springer.com/series/6515. doi:10.1007/978-94-024-0945-1_4.

Blomhoff, R., Green, M.H., Berg, T., Norum, K.R., 1990. Transport and storage of vitamin A. Science 00368075. 250 (4977), 399–404. Available from: http://www.sciencemag.org.

Booth, S.L., Johns, T., Kuhnlein, H.V., 1992. Natural food sources of vitamin A and provitamin A. Food Nutr. Bull. 14 (1), 1–15.

Bors, W., Buettner, G.R., 1997. The Vitamin C Radical and Its Reactions.

Bourgeois, C., 1992. Determination of Vitamin E: Tocopherols and Tocotrienols.

Brown, M.R., Mular, M., Miller, I., Farmer, C., Trenerry, C., 1999. The vitamin content of microalgae used in aquaculture. J. Appl. Phycol. 11 (3), 247–255. doi:10.1023/A:10080.

Brustad, M., Sandanger, T., Wilsgaard, T., Aksnes, L., Lund, E., 2003. Change in plasma levels of vitamin D after consumption of cod-liver and fresh cod-liver oil as part of the traditional north Norwegian fish dish "Molje". Int. J. Circumpolar Health 62 (1), 40–53.

Buettner, G.R., Jurkiewicz, B.A., 1996. Catalytic metals, ascorbate and free radicals: combinations to avoid. Radiat. Res. 00337587. 145 (5), 532–541. doi:10.2307/3579271.

Byrdwell, W.C., Horst, R.L., Phillips, K.M., Holden, J.M., Patterson, K.Y., Harnly, J.M., Exler, J., 2013. Vitamin D levels in fish and shellfish determined by liquid chromatography with ultraviolet detection and mass spectrometry. J. Food Compos. Anal. 30 (2), 109–119.

Cadar, E., Negreanu-Pirjol, T., Sirbu, R., Dragan, A.M.L., Negreanu-Pirjol, B.S., Axente, E.R., Ionescu, A.M., 2023. Biocompounds from green algae of romanian Black Sea coast as potential nutraceuticals. Processes 11 (6), 1750.

Caf, F., Özdemir, N.Ş., Yılmaz, Ö., Durucan, F., Ak, İ.L.K.N.U.R., 2019. Fatty acid and lipophilic vitamin composition of seaweeds from Antalya and Çanakkale (Turkey). Grasas Y Aceites 70 (3), e312.

Calogiuri, G., 2021. Hypersensitivity to Vitamins, Bentham Science Publishers.

Camacho-Rodríguez, J., Cerón-García, M.C., Fernández-Sevilla, J.M., Molina-Grima, E., 2015. The influence of culture conditions on biomass and high value product generation by *Nannochloropsis gaditana* in aquaculture. Algal Res. 11, 63–73.

Casella, L., Gullotti, M., Marchesini, A., Petrarulo, M., 1989. Rapid enzymatic method for vitamin C assay in fruits and vegetables using peroxidase. J. Food Sci. 0022-1147. 54 (2), 374–375. doi:10.1111/j.1365-2621.1989.tb03084.x.

Çatak, J., 2019. Determination of niacin profiles in some animal and plant based foods by high performance liquid chromatography: association with healthy nutrition. J. Anim. Sci. Technol. 61 (3), 138.

Chaichi, M.J., Alijanpour, S.O., 2013. Determination of vitamin C in drugs using of an optimized novel TCPO–Amplex red–gold/silver alloy nanoparticles–H_2O_2 chemiluminescence method by the Box–Behnken design. J. Lumin. 00222313. 134, 195–200. doi:10.1016/j.jlumin.2012.08.048.

Chand, T., Savitri, B., 2016. Vitamin B3, niacin. In: Industrial Biotechnology of Vitamins, Biopigments, and Antioxidants, pp. 41–65.

Chandler, C.J., Wang, T.T.Y., Halsted, C.H., 1986. Pteroylpolyglutamate hydrolase from human jejunal brush-borders—purification and characterization. J. Biol. Chem. 261, 928–933.

Chun, J., Lee, J., Ye, L., Exler, J., Eitenmiller, R.R., 2006. Tocopherol and tocotrienol contents of raw and processed fruits and vegetables in the United States diet. J. Food Compos. Anal. 08891575. 19 (2-3), 196–204. doi:10.1016/j.jfca.2005.08.001.

Colombo, M.L., 2010. An update on vitamin E, tocopherol and tocotrienol—perspectives. Molecules 15 (4), 2103–2113.

Combs Jr G.F., McClung, J.P., 2016. The vitamins: fundamental aspects in nutrition and health. Academic press.

Combs, G.F., McClung, J.P., 2017. The Vitamins: Fundamental Aspects in Nutrition and Health the Vitamins: Fundamental Aspects in Nutrition and Health. Elsevier, United States, pp. 1–612. Available from: https://www.elsevier.com/books/the-vitamins/combs-jr/978-0-323-90473-5. doi:10.1016/B978-0-12-802965-7.01001-7.

Cooper, J., Thadwal, R., Cooper, M., 1997. Determination of vitamin E in human plasma by highperformance liquid chromatography. J. Chromatogr. B 690, 355.

Cretton, M., Malanga, G., Mazzuca Sobczuk, T., Mazzuca, M., 2021. Lipid fraction from industrial crustacean waste and its potential as a supplement for the feed industry: a case study in Argentine Patagonia. Waste Biomass Valor. 12 (5), 2311–2319.

Croft, M.T., Warren, M.J., Smith, A.G., 2006. Algae need their vitamins. Eukaryot. Cell 5, 1175–1183.

Dabrowski, K., 1991. Some aspects of ascorbate metabolism in developing embryos of the brine shrimp. Can. J. Fish. Aquat. Sci. 48, 1–3.

Dakshinamurti, S., Dakshinamurti, K., Zempleni, J., Suttie, J.W., Gregory, J.F., Stover, P.J., 2014. Vitamin B6. Handb. Vitam., 4.

Dam, H., 1935. The antihaemorrhagic vitamin of the chick: occurrence and chemical nature. Nature. 0028-0836. 135 (3417), 652–653. doi:10.1038/135652b0.

Davies, M.B., Partridge, D.A., Austin, J.A., 2007. Vitamin C: Its Chemistry and Biochemistry, Royal Society of Chemistry.

Dawbarn, M.C., 1945. The determination of vitamin C by the growth of Guinea-pigs. Aust. J. Exp. Biol. Med. Sci. 0818-9641. 23 (4), 281–287. doi:10.1038/icb.1945.45.

De Roeck-Holtzhauer, Y., Quere, I., Claire, C., 1991. Vitamin analysis of five planktonic microalgae and one macroalga. J. Appl. Phycol. 3, 259–264.

de Souza Vilhena, K.D.S., da Silva, M.P., de Santana Botelho, A., Cardoso, F.J.B., Ferreira, O.O., de Oliveira, M.S., de Aguiar Andrade, E.H., 2024. Biosynthesis of carotenoids in plants and microorganisms. Carotenoids Trends Adv. 37.

Debbarma, J., Rao, B.M., Murthy, L.N., Mathew, S., Venkateshwarlu, G., Ravishankar, C.N., 2016. Nutritional profiling of the edible seaweeds *Gracilaria edulis*, *Ulva lactuca* and *Sargassum* sp. Indian J. Fish. 63 (3), 81–87.

Del Mondo, A., Smerilli, A., Sané, E., Sansone, C., Brunet, C., 2020. Challenging microalgal vitamins for human health. Microb. Cell Fact. 19, 1–23.

Delgado-Zamarreño, M.M., Bustamente-Rangel, M., Sánchez-Pérez, A., Carabias-Martínez, R., 2004. Pressurized liquid extraction prior to liquid chromatography with electrochemicaldetection for the analysis of vitamin E isomers in seeds and nuts. J. Chromatogr. A 1056, 249.

Di Lena, G., Casini, I., Lucarini, M., Lombardi-Boccia, G., 2019. Carotenoid profiling of five microalgae species from large-scale production. Food Res. Int. 120, 810–818.

Dias, M.G., Sanchez, M.V., Bartolo, H., Oliveira, L., 2003. Vitamin content of fish and fish products consumed in Portugal. Electronic Journal of Environmental, Agricultural and Food Chemistry 2 (4), 510–515.

Diprat, A., Menegol, T., Boelter, J., Zmozinski, A., Vale, M., Rodrigues, E., Rech, R., 2017. Chemical composition of microalgae *Heterochlorella luteoviridis* and *Dunaliella tertiolecta* with emphasis on carotenoids. J. Sci. Food Agric. 97 (10).

Donnelly, J.G., 2001. Folic acid. Crit. Rev. Clin. Lab. Sci. 10408363. 38 (3), 183–223. doi:10.1080/20014091084209.

Drouin, G., Godin, J.R., Pagé, B., 2011. The genetics of vitamin C loss in vertebrates. Curr. Genomics 12 (5), 371–378.

Dunlap, W.C., Fujisawa, A., Yamamoto, Y., Moylan, T.J., Sidell, B.D., 2002. Notothenioid fish, krill and phytoplankton from Antarctica contain a vitamin E constituent (α-tocomonoenol) functionally associated with cold-water adaptation. Comp. Biochem. Physiol. B Biochem. Mol. Biol. 133 (3), 299–305.

Duque, P., Vieira, C.P., Bastos, B., Vieira, J., 2022. The evolution of vitamin C biosynthesis and transport in animals. BMC Ecol. Evol. 14726785. 22 (1). Available from: https://www.bmcevolbiol.biomedcentral.com/. doi:10.1186/s12862-022-02040-7.

Durmaz, Y., 2007. Vitamin E (α-tocopherol) production by the marine microalgae *Nannochloropsis oculata* (Eustigmatophyceae) in nitrogen limitation. Aquaculture 272 (1-4), 717–722.

Durmaz, Y., Duyar, H.A., Gokpinar, S., Taskaya, L., Ogretmen, Y.O., Bandarra, N.M., Nunes, M.L., 2008. Fatty acids, α-tocopherol and total pigment contents of *Cystoseira* spp., *Ulva* spp. and *Zostera* spp. from Sinop Bay (Turkey). Int. J. Nat. Eng. Sci. 2 (3), 111–114.

Evans, H.M., Bishop, K.S., 1922. On the existence of a hitherto unrecognized dietary factor essential for reproduction. Science 56, 650–651.

Fabregas, J., Herrero, C., 1990. Vitamin content of four marine microalgae. Potential use as source of vitamins in nutrition. J. Ind. Microbiol. Biotechnol. 5 (4), 259–263.

Fernandes, A.S., Petry, F.C., Mercadante, A.Z., Jacob-Lopes, E., Zepka, L.Q., 2020. HPLC-PDA-MS/MS as a strategy to characterize and quantify natural pigments from microalgae. Curr. Res. Food Sci. 3, 100–112.

Freedman, L.P., 1999. Increasing the complexity of coactivation in nuclear receptor signaling. Cell. 00928674. 97 (1), 5–8. Available from: https://www.sciencedirect.com/journal/cell. doi:10.1016/S0092-8674(00)80708-4.

Gajalakshmi, D., Shettu, N., Murugesan, S., 2018. Vitamin content of marine red alga *Botryocladia leptopoda* (J. Agardh) Kylin Potential use as a source of vitamins in nutrition. Int. J. Life Sci. Res. 6 (2), 263–268.

Galasso, C., Corinaldesi, C., Sansone, C., 2017. Carotenoids from marine organisms: Biological functions and industrial applications. Antioxidants 6 (4), 96.

Gamache, P.H., Cabe, D.R.M., Parvez, H., Parvez, S., Acworth, I.N., 2024. The measurement of markers of oxidative damage, anti oxidants and related compounds using HPLC and coulometric array analysis. In: Coulometric Electrode Array Detectors for HPLC. CRC Press, United States, pp. 99–126. Available from: https://www.taylorfrancis.com/books/edit/10.1201/9780429070303/coulometric-electrode-array-detectors-hplc-naoi-acworth-parvez-hasan-parvez.

Gasperi, V., Sibilano, M., Savini, I., Catani, M.V., 2019. Niacin in the central nervous system: an update of biological aspects and clinical applications. Int. J. Mol. Sci. 20 (4), 974.

Gazdik, Z., Zitka, O., Petrlova, J., Adam, V., Zehnalek, J., Horna, A., Reznicek, V., Beklova, M., Kizek, R., 2008. Determination of vitamin C (ascorbic acid) using high performance liquid chromatography coupled with electrochemical detection. Sensors. 1424-8220. 8 (11), 7097–7112. doi:10.3390/s8117097.

Giasuddin, A.S.M., Diplock, A.T., 1981. The influence of vitamin E on membrane lipids of mouse fibroblasts in culture. Arch. Biochem. Biophys. 00039861. 210 (1), 348–362. doi:10.1016/0003-9861(81)90198-3.

Gonzalez-Lopez, J., Aliaga, L., Gonzalez-Martinez, A., Martinez-Toledo, M.V., 2016. Pantothenic Acid, in Industrial Biotechnology of Vitamins, Biopigments, and Antioxidants. Chichester: Wiley, pp. 67–102. doi:10.1002/9783527681754.ch4.

Gonzalez-Lopez, J., Aliaga, L., Gonzalez-Martinez, A., Martinez-Toledo, M.V., 2016. Pantothenic acid. In: Industrial Biotechnology of Vitamins, Biopigments, and Antioxidants, pp. 67–101

Hapette, A.M., Poulet, S.A., 1990. Variation of vitamin C in some common species of marine plankton. Marine ecology progress series. Oldendorf 64 (1), 69–79.

Hazra, A.B., Han, A.W., Mehta, A.P., Mok, K.C., Osadchiy, V., Begley, T.P., Taga, M.E., 2015. Anaerobic biosynthesis of the lower ligand of vitamin B12. Proc. Natl. Acad. Sci. U. S. A. 10916490. 112 (34), 10792–10797. Available from: http://www.pnas.org/content/112/34/10792.full.pdf. doi:10.1073/pnas.1509132112.

Heldt, D., Lawrence, A.D., Lindenmeyer, M., Deery, E., Heathcote, P., Rigby, S.E., Warren, M.J., 2005. Aerobic synthesis of vitamin B12: ring contraction and cobalt chelation. Biochem. Soc. Trans. 0300-5127. 33 (4), 815–819. doi:10.1042/bst0330815.

Hippe, E., 1970. Changes in stokes radius on binding of vitamin B12 to human intrinsic factor and transcobalamins. Biochim. Biophys. Acta Gen. Subj. 03044165. 208 (2), 337–339. doi:10.1016/0304-4165(70)90255-2.

Hodgkin, D.C., et al, 1956. Structure of vitamin B12. Nature 178, 64.

Holick, M.F., 1996. Vitamin D and bone health. J. Nutr. 126, 1159S–64S.

Oemonom, Hrubša, M., Siatka, T., Nejmanová, I., Vopršalová, M., Kujovská Krčmová, L., Matoušová, K., 2022. Biological properties of vitamins of the B-complex, part 1: vitamins B1, B2, B3, and B5. Nutrients 14 (3), 484.

Hylander, S., Farnelid, H., Fridolfsson, E., Hauber, M.M., Todisco, V., Ejsmond, M.J., Lindehoff, E., 2024. Thiamin (vitamin B1, thiamine) transfer in the aquatic food web from lower to higher trophic levels. PLoS One 19 (12), e0308844. doi:10.1371/journal.pone.0308844.

Indyk, H.E., 1988. Simplified saponification procedure for the routine determination of total vitamin E in dairy products, foods, and tissues by high-performance liquid chromatography. Analyst 113, 1217.

Islam, M.A., Park, E., Jeong, B., Gwak, Y.J., Kim, J., Hong, W.H., Chun, J., 2022. Validation of vitamin B5 (pantothenic acid) and B6 (pyridoxine, pyridoxal, and pyridoxamine) analyses in seafood. J. Food Compos. Anal. 109, 104518.

Jahn, D., Verkamp, E., Soll, D., 1992. Glutamyl-transfer RNA: a precursor of heme and chlorophyll biosynthesis. Trends Biochem. Sci. 17 (215), 218.

Kiyose, C., Kaneko, K., Muramatsu, R., Ueda, T., Igarashi, O., 1999. Simultaneous determination of RRR- and SRR-α-tocopherols and their quinones in rat plasma and tissues by using chiral high-performance liquid chromatography. Lipids. 00244201. 34 (4), 415–422. doi:10.1007/s11745-999-0380-x.

Kleszczewski, T., Kleszczewska, E., 2002. Flow injection spectrophotometric determination of l-ascorbic acid in biological matters. J. Pharm. Biomed. Anal. 07317085. 29 (4), 755–759. doi:10.1016/s0731-7085(02)00127-9.

Koshiishi, I., Imanari, T., 1997. Measurement of ascorbate and dehydroascorbate contents in biological fluids. Anal. Chem. 0003-2700. 69 (2), 216–220. doi:10.1021/ac960704k.

Kurniawan, R., Nurkolis, F., Taslim, N.A., Subali, D., Surya, R., Gunawan, W.B., *et al.*, 2023. Carotenoids composition of green algae Caulerpa racemosa and their antidiabetic, anti-obesity, antioxidant, and anti-inflammatory properties. Molecules 28 (7), 3267.

Lamberg-Allardt, C., 2006. Vitamin D in foods and as supplements. Prog. Biophys. Mol. Biol. 92 (1), 33–38.

Lee, K.J., Kim, K.-W., Bal, S.C., 1998. Effects of different dietary levels of L-ascorbic acid on growth and tissue vitamin C concentration in juvenile Korean rockfish, *Sebastes schlegeli* (Hilgendorf). Aquac. Res. 29, 237–244.

Li, B., Wang, L., Wang, J., Sun, Y., Wang, S., Jiang, L., Huang, B., 2020. Requirement of vitamin E of growing sea cucumber *Apostichopus japonicus* Selenka. Aquac. Res. 51 (3), 1284–1292.

Li, H.B., Chen, F., 2000. Determination of B12 in pharmaceutical preparations by a highly sensitive fluorimetric method, Fres. J. Anal. Chem. 368, 836.

Lin, S., Cronan, J.E., 2011. Closing in on complete pathways of biotin biosynthesis. Mol. Biosyst. 17422051. 7 (6), 1811–1821. doi:10.1039/c1mb05022b.

Linster, C.L., Van Schaftingen, E., 2007. Vitamin C: biosynthesis, recycling and degradation in mammals. FEBS J. 17424658. 274 (1), 1–22. doi:10.1111/j.1742-4658.2006.05607.x.

Ljubic, A., Jacobsen, C., Holdt, S.L., Jakobsen, J., 2020. Microalgae nannochloropsis oceanica as a future new natural source of vitamin D3. Food Chem. 320, 126627. doi:10.1016/j.foodchem.2020.126627.

Ložnjak, P., Jakobsen, J., 2018. Stability of vitamin D3 and vitamin D2 in oil, fish and mushrooms after household cooking. Food Chem. 254, 144–149.

Luo, X., Chen, B., Ding, L., Tang, F., Yao, S., 2006. HPLC-ESI-MS analysis of vitamin B12 in food products and in multivitamins-multimineral tablets. Anal. Chim. Acta. 00032670. 562 (2), 185–189. doi:10.1016/j.aca.2006.01.073.

MacArtain, P., Gill, C., Brooks, M., Campbell, R., Rowland, I.R., 2007. Nutritional value of edible seaweeds. Nutr. Res. 65, 535–543.

MacCorquodale, D.W., Cheney, L.C., Binkley, S.B., Holcomb, W.F., McKee, R.W., Thayer, S.A., Doisy, E.A., 1939. The constitution and synthesis of vitamin K1. J. Biol. Chem. 00219258. 131 (1), 357–370. doi:10.1016/s0021-9258(18)73508-4.

Mai, K., 1998. Comparative studies on the nutrition of two species of abalone, *Haliotis tuberculata* L. and *Haliotis discus hannai* Ino.: VII. Effects of dietary vitamin C on survival, growth and tissue concentration of ascorbic acid. Aquaculture 161 (1-4), 383–392.

Majewski, M., Lebiedzińska, A., 2014. The evaluation of selected shellfish as a source of niacin in nutrition and therapy of modern human. Pol. Ann. Med. 21 (1), 14–19.

Malacaria, L., Furia, E., 2023. Wide pH range potentiometric and spectrophotometric investigation into the acidic constants of quercetin, luteolin and l-ascorbic acid in aqueous media. Appl. Sci. 2076-3417. 13 (2), 776. doi:10.3390/app13020776.

Malik, D., Narayanasamy, N., Pratyusha, V.A., Thakur, J., Sinha, N., 2023. Water-soluble vitamins. In: Textbook of Nutritional Biochemistry. Springer Nature Singapore, Singapore, pp. 291–389.

Malone, C., Shaw, N.B., Kerry, J.P., 2004. Effect of season on vitamin E, fatty acid profile, and nutritional value of fish by-products from cod, saithe, ling and haddock species caught in southern Irish coastal waters. J. Aquat. Food Prod. Technol. 13 (3), 127–149.

Manzetti, S., Zhang, J., van der Spoel, D., 2014. Thiamin function, metabolism, uptake, and transport. Biochemistry 53 (5), 821–835.

Matanjun, P., Mohamed, S., Mustapha, N.M., Muhammad, K., 2009. Nutrient content of tropical edible seaweeds, *Eucheuma cottonii, Caulerpa lentillifera* and *Sargassum polycystum*. J. Appl. Phycol. 21, 75–80.

Mawer, E.B., Backhouse, J., Holman, C.A., Lumb, G.A., Stanbury, S.W., 1972. The distribution and storage of vitamin D and its metabolites in human tissues Clin. Sci. 43 (3), 413–431.

McCollum, E.V., Davis, M., 1913. The necessity of certain lipins in the diet during growth. J. Biol. Chem. 15, 167–175.

McCollum, E.V., Simmonds, N., Kinney, M., Shipley, P.G., Park, E.A., 1921. Studies on experimental rickets. XVII. The effects of diets deficient in calcium and in fat-soluble A in modifying the histological structure of the bones. Am. J. Epidemiol. discussion 279. doi: 10.1093/aje/141.4.280 141 (4), 280–296.

Medagoda, N., Kim, J.S., Song, J.W., Lee, K.J., 2024. Citrus by-product is a potential alternative to replace synthetic ascorbic acid in the diet of juvenile red seabream, *Pagrus major*. Fish Aquat. Sci. 27 (11), 744–752. doi:10.47853/FAS.2024.e69.

Merchie, G., Lavens, P., Sorgeloos, P., 1997. Optimization of dietary vitamin C in fish and crustacean larvae: a review. Aquaculture 155 (1-4), 165–181.

Merdzhanova, A., Dobreva, D.A., Stancheva, M., Makedonski, L., 2014. Fat soluble vitamins and fatty acid composition of wild Black Sea mussel, rapana and shrimp. Ovidius Univ. Ann. Chem. 25 (1), 15–23.

Meyer-Ficca, M., Kirkland, J.B., 2016. Niacin. Adv. Nutr. 7 (3), 556–558.

Minns, R.A., 1996. Folic acid and neural tube defects. Spinal Cord. 1362-4393. 34 (8), 460–465. doi:10.1038/sc.1996.79.

Minot, G.R., Murphy, W.P., 1926. Treatment of pernicious anemia by a special diet. JAMA. 23768118. 87 (7), 470–476. doi:10.1001/jama.1926.02680070016005.

Mitchell, H.K., Snell, E.E., Williams, R.J., 1941. The concentration of "folic acid". J. Am. Chem. Soc. 63 (8), 2284.

Mitra, G., Mukhopadhyay, P.K., Ayyappan, S., 2007. Biochemical composition of zooplankton community grown in freshwater earthen ponds: nutritional implication in nursery rearing of fish larvae and early juveniles. Aquaculture 272 (1-4), 346–360.

Moe, Y.Y., Koshio, S., Teshima, S.I., Ishikawa, M., Matsunaga, Y., Panganiban Jr, A., 2004. Effect of vitamin C derivatives on the performance of larval kuruma shrimp, *Marsupenaeus japonicus*. Aquaculture 242 (1-4), 501–512.

Moestrup, S.K., Verroust, P.J., 2001. Megalin- and cubulin-mediated endocytosis of protein-bound vitamins, lipids, and hormones in polarized epithelia. Annu. Rev. Nutr. 21, 407.

Monteverde, D.R., Gómez-Consarnau, L., Cutter, L., Chong, L., Berelson, W., Sañudo-Wilhelmy, S.A., 2015. Vitamin B1 in marine sediments: pore water concentration gradient drives benthic flux with potential biological implications. Front. Microbiol. 6, 434.

Moore, S.J., Lawrence, A.D., Biedendieck, R., Deery, E., Frank, S., Howard, M.J., Warren, M.J., 2013. Elucidation of the anaerobic pathway for the corrin component of cobalamin (vitamin B12). Proc. Natl. Acad. Sci. U. S. A. 110 (37), 14906–14911.

Mori, K., Kidawara, M., Iseki, M., Umegaki, C., Kishi, T., 1998. A simple fluorometric determination of vitamin C. Chem. Pharm. Bull. 0009-2363. 46 (9), 1474–1476. doi:10.1248/cpb.46.1474.

Moser, U., Bendich, A., 1990. Chapter 5: vitamin C. In: Handbook of Vitamins, Machlin, L.J. Ed., Marcel Dekker, New York.

Mrowicka, M., Mrowicki, J., Dragan, G., Majsterek, I., 2023. The importance of thiamine (vitamin B1) in humans. Biosci. Rep. 43 (10), BSR20230374. doi:10.1042/BSR20230374.

Muhammad, K., Briggs, D., Jones, G., 1993. The appropriateness of using cyanocobalamin as calibration standards in *Lactobacillus leichmanni* ATCC 7830 assay of vitamin B12. Food Chem. 48, 427.

Murata, M., Nakazoe, J.I., 2001. Production and use of marine algae in Japan. Jpn. Agric. Res. Q. 35 (4), 281–290.

Nartea, A., Ismaiel, L., Frapiccini, E., Falcone, P.M., Pacetti, D., Frega, N.G., Colella, S., 2023. Impact of modern oven treatments on lipid oxidation and vitamin E content of fillets from sardine (*Sardina pilchardus*) at different reproductive cycle phases. Antioxidants 12 (6), 1312.

Newcomer, M.E., Ong, D.E., 2000. Plasma retinol binding protein: structure and function of the prototypic lipocalin. Biochim. Biophys. Acta Protein Struct. Mol. Enzymol. 01674838. 1482 (1-2), 57–64. doi:10.1016/S0167-4838(00)00150-3.

Nishikimi, M., Yagi, K., 1991. Molecular basis for the deficiency in humans of gulonolactone oxidase, a key enzyme for ascorbic acid biosynthesis. Am. J. Clin. Nutr. 00029165. 54 (6), 1203S. doi:10.1093/ajcn/54.6.1203s.

Northrop-Clewes, C.A., Thurnham, D.I., 2012. The discovery and characterization of riboflavin. Ann. Nutr. Metab. 61 (3), 224–230.

Norziah, M.H., Ching, C.Y., 2000. Nutritional composition of edible seaweed *Gracilaria changgi*. Food Chem. 68 (1), 69–76.

O'Brien, M., Quirke, R., Gowan, R., McNicholas, F., 2025. Prevalence of thiamine deficiency in anorexia nervosa: a systematic review and narrative synthesis. Eur. Child Adolesc. Psychiatry 34 (4), 1281–1293.

Ortiz, J., Uquiche, E., Robert, P., Romero, N., Quitral, V., Llantén, C., 2009. Functional and nutritional value of the Chilean seaweeds *Codium fragile*, *Gracilaria chilensis* and *Macrocystis pyrifera*. Eur. J. Lipid Sci. Technol. 111 (4), 320–327.

Ostermeyer, U., Schmidt, T., 2001. Determination of vitamin K in the edible part of fish by high-performance liquid chromatography. Eur. Food Res. Technol. 212 (4), 518–528.

Ostermeyer, U., Schmidt, T., 2006. Vitamin D and provitamin D in fish: determination by HPLC with electrochemical detection. Eur. Food Res. Technol. 222 (3), 403–413.

Ottaway, P.B., 2012. The Technology of Vitamins in Food, Springer Science & Business Media.

Pallotta, M.L., 2019. Flavin cofactors FMN and FAD, the biologically active forms of riboflavin and healthy life. EC Nutr. 14 (8), 614–615.

Parrish, D.B., Waltking, A.E., 2009. Determination of vitamin e in foods—a review. Crit. Rev. Food Sci. Nutr. 0099-0248. 13 (2), 161–187. doi:10.1080/10408398009527288.

Pérez-García, F., Klein, V.J., Brito, L.F., Brautaset, T., 2022. From brown seaweed to a sustainable microbial feedstock for the production of riboflavin. Front. Bioeng. Biotechnol. 10, 863690.

Peters, R.A., 1936. Pyruvic acid oxidation in brain: Vitamin B_1 and the pyruvate oxidase in pigeon's brain. Biochem. J. 30 (12), 2206–2218. doi:10.1042/bj0302206.

Pinto, J., Rivlin, R., 2013. Riboflavin (vitamin B2). Handb. Vitam. 5, 191–266.

Pinto, J.T., Zempleni, J., 2016. Riboflavin. Adv. Nutr. 7 (5), 973–975.

Poulet, S.A., Hapette, A.M., Cole, R.B., Tabet, J.C., 1989. Vitamin C in marine copepods. Limnol. Oceanogr. 34 (7), 1325–1331.

Quadros, E.V., Rothenberg, S.P., Polu, S., 1988. A specific radioimmunoassay for 5′-deoxyadenosyl cobalamin in serum. Br. J. Haematol. 69, 551.

Rao, D.S., Raghuramulu, N., 1996. Food chain as origin of vitamin D in fish. Comp. Biochem. Physiol. A Physiol. 114 (1), 15–19.

Rasti, G., Simonet, N.G., Vaquero, A., 2020. Niacin. In: Principles of Nutrigenetics and Nutrigenomics. Academic Press. pp. 287–293.

Redzic, S., Hashmi, M.F., Gupta, V., 2020. Niacin Deficiency.

Rickes, E.L., Brink, N.G., Koniuszy, F.R., Wood, T.R., Folkers, K., 1948. Crystalline vitamin B12. Science. 00368075. 107 (2781), 396–397. doi:10.1126/science.107.2781.396.

Riordan, H.D., Mikirova, N., Taylor, P.R., Feldkamp, C.A., Casciari, J.J., 2012. The effects of a primary nutritional deficiency (Vitamin B study). Food Nutr. Sci. 3 (9), 1238–1244.

Rothenberg, S.P., 1963. Radioassay of serum vitamin B12 by quantitating the competition between CO57B12 and unlabeled B12 for the binding sites of intrinsic factor. J. Clin. Investig. 00219738. 42, 1391–1398. doi:10.1172/JCI104823.

Rucker, R.B., Bauerly, K., 2013. Pantothenic acid. In: Handbook of Vitamins Chicago, 5.

Sánchez-Machado, D.I., López-Cervantes, J., López-Hernández, J., Paseiro-Losada, P., 2004. Simultaneous determination of thiamine and riboflavin in edible marine seaweeds by high-performance liquid chromatography. J. Chromatogr. Sci. 42 (3), 117–120.

Sánchez-Mata, M.C., Cámara-Hurtado, M., Díez-Marqués, C., Torija-Isasa, M.E., 2000. Comparison of high-performance liquid chromatography and spectrofluorimetry for vitamin C analysis of green beans (*Phaseolus vulgaris* L.). Eur. Food Res. Technol. 14382385. 210 (3), 220–225. Available from: http://www.link.springer.de/link/service/journals/00217/index.htm. doi:10.1007/PL00005516.

Scurria, A., Lino, C., Pitonzo, R., Pagliaro, M., Avellone, G., Ciriminna, R., 2020. Vitamin D3 in fish oil extracted with limonene from anchovy leftovers. Chem. Data Collect. 25, 100311.

Senadheera, T.R.L., Hossain, A., Shahidi, F., 2023. Marine bioactives and their application in the food industry: a review. Appl. Sci. 13 (21), 12088. doi:10.3390/app132112088.

Shemin, D., Kikuchi, G., 1958. Enzymatic synthesis of δ-aminolevulinic acid. Ann. N. Y. Acad. Sci. 17496632. 75 (1), 122–128. doi:10.1111/j.1749-6632.1958.tb36857.x.

Šimat, V., 2021. Nutraceuticals and pharmaceuticals from marine fish and invertebrates. Mar. Drugs 19 (7), 401.

Škugor Rončević, I., Skroza, D., Vrca, I., Kondža, A.M., Vladislavić, N., 2022. Development and optimization of electrochemical method for determination of vitamin C. Chemosensors. 22279040. 10 (7). Available from: https://www.mdpi.com/journal/chemosensors. doi:10.3390/chemosensors10070283.

Smirnoff, N., Running, J.A., Gatzek, S., 2004. Ascorbate biosynthesis a diversity of pathways. In: Vitamin C. Taylor & Francis. pp. 8–33.

Smith, T.J., Johnson, C.R., Koshy, R., Hess, S.Y., Qureshi, U.A., Mynak, M.L., Fischer, P.R., 2021. Thiamine deficiency disorders: a clinical perspective. Ann. N. Y. Acad. Sci. 1498 (1), 9–28.

Spínola, V., Llorent-Martínez, E.J., Castilho, P.C., 2014. Determination of vitamin C in foods: current state of method validation. J. Chromatogr. A 1369, 2–17.

Stancheva, M., Dobreva, D.A., 2013. Bulgarian marine and freshwater fishes as a source of fat-soluble vitamins for a healthy human diet. Foods 2 (3), 332–337.

Stevens, R., Buret, M., Garchery, C., Carretero, Y., Causse, M., 2006. Technique for rapid, small-scale analysis of vitamin C levels in fruit and application to a tomato mutant collection. J. Agric. Food Chem. 0021-8561. 54 (17), 6159–6165. doi:10.1021/jf061241e.

Streit, W.R., Entcheva, P., 2003. Biotin in microbes, the genes involved in its biosynthesis, its biochemical role and perspectives for biotechnological production. Appl. Microbiol. Biotechnol. 0175-7598. 61 (1), 21–31. doi:10.1007/s00253-002-1186-2.

Suffridge, C.P., Bolaños, L.M., Bergauer, K., Worden, A.Z., Morré, J., Behrenfeld, M.J., Giovannoni, S.J., 2020. Exploring vitamin B1 cycling and its connections to the microbial community in the North Atlantic Ocean. Front. Mar. Sci. 7, 606342.

Sunarya, H., Hole, M., Taylor, K.D.A., 1996. Methods of extraction composition and stability of vitamin A and other components in dogfish (*Squalus acanthias*) liver oil. Food Chem. 55 (3), 215–220.

Taboada, C., Millán, R., Míguez, I., 2010. Composition, nutritional aspects and effect on serum parameters of marine algae Ulva rigida. J. Sci. Food Agric. 90 (3), 445–449.

Torrinha, A., Gomes, F., Oliveira, M., Cruz, R., Mendes, E., Delerue-Matos, C., Morais, S., 2014. Commercial squids: characterization, assessment of potential health benefits/risks and discrimination based on mineral, lipid and vitamin E concentrations. Food Chem. Toxicol. 67, 44–56.

Traber, M.G., Atkinson, J., 2007. Vitamin E, antioxidant and nothing more. Free Radic. Biol. Med. 08915849. 43 (1), 4–15. doi:10.1016/j.freeradbiomed.2007.03.024.

Traber, M.G., Kayden, H.J., 1987. Tocopherol distribution and intracellular localization in human adipose tissue. Am. J. Clin. Nutr. 00029165. 46 (3), 488–495. doi:10.1093/ajcn/46.3.488.

Tsushlma, M., Kawakami, T., Mine, M., Matsuno, T., 1997. The role of carotenoids in the development of the sea urchin Pseudocentrotus depressus. Invertebr. Reprod. Dev. 32 (2), 149–153.

van der Meeren, T., Olsen, R.E., Hamre, K., Fyhn, H.J., 2008. Biochemical composition of copepods for evaluation of feed quality in production of juvenile marine fish. Aquaculture 274 (2-4), 375–397.

Venugopal, V., Gopakumar, K., 2017. Shellfish: nutritive value, health benefits, and consumer safety. Compr. Rev. Food Sci. Food Saf. 16 (6), 1219–1242.

Vigneaud, V.D., 1942. The structure of biotin. Science. 00368075. 96 (2499), 455–461. doi:10.1126/science.96.2499.455.

Villanueva, R., Escudero, J.M., Deulofeu, R., Bozzano, A., Casoliva, C., 2009. Vitamin A and E content in early stages of cephalopods and their dietary effects in *Octopus vulgaris* paralarvae. Aquaculture 286 (3-4), 277–282.

Viñas, P., López-Erroz, C., Balsalobre, N., Hernández-Córdoba, M., 2003. Speciation of cobalamins in biological samples using liquid chromatography with diode-array detection. Chromatographia. 0009-5893. 58 (1-2), 5–10. doi:10.1365/s10337-003-0003-0.

Wang, L., Chi, Z.M., Wang, X.H., L., J., Chi, Z., Guo, N., 2008. Isolation and characterization of *Candida membranifaciens* subsp. flavinogenie W14-3, a novel riboflavin-producing marine yeast. Microbiol. Res. 163, 255–266.

Wang, L., Zhao, H., Bi, R., et al., 2024. Roles and sources of B vitamins in the marine ecosystem. Rev. Fish Biol. Fish. (34), 111–130. doi:10.1007/s11160-023-09818-y.

Wang, X., Kim, K.W., Bai, S.C., Huh, M.D., Cho, B.Y., 2003. Effects of the different levels of dietary vitamin C on growth and tissue ascorbic acid changes in parrot fish (*Oplegnathus fasciatus*). Aquaculture 215 (1-4), 203–211.

Wang, X., Quinn, P.J., 2000. The location and function of vitamin E in membranes (Review). Mol. Membr. Biol. 09687688. 17 (3), 143–156. Available from: http://www.tandfonline.com/loi/imbc20. doi:10.1080/09687680010000311.

Watanabe, F., Bito, T., 2018. Vitamin B 12 sources and microbial interaction. Exp. Biol. Med. 1535-3702. 243 (2), 148–158. doi:10.1177/1535370217746612.

Webster-Gandy, J., Madden, A., Holdsworth, M., Eds.. 2020. Oxford Handbook of Nutrition and Dietetics. Oxford University Press.

Whipple, G.H., Robscheit-Robbins, F.S., 1925. Blood regeneration in severe anemia. Am. J. Physiol. 0002-9513. 72 (3), 395–407. doi:10.1152/ajplegacy.1925.72.3.395.

Wijemanna, U., Lee, K.J., 2024. Dietary thiamine requirement and its effects on growth and innate immunity of Pacific white shrimp (Penaeus vannamei). Aquac. Int. 32 (3), 2999–3016. doi:10.1007/s10499-024-01372-y.

Wilson, M.P., Plecko, B., Mills, P.B., Clayton, P.T., 2019. Disorders affecting vitamin B6 metabolism. J. Inherit. Metab. Dis. 42 (4), 629–646.

Windaus, A., Linsert, O., Lüttringhaus, A., Weidlich, G., 1932. Über das krystallisierte vitamin D2. Justus Liebigs Ann. Chem. 492 (1), 226–241.

Woodward, R.B., 1973. The total synthesis of vitamin B12. Pure Appl. Chem. 1365-3075. 33 (1), 145–178. doi:10.1351/pac197333010145.

Xie, D., Jin, J., Sun, J., Liang, L., Wang, X., Zhang, W., Jin, Q., 2017. Comparison of solvents for extraction of krill oil from krill meal: lipid yield, phospholipids content, fatty acids composition and minor components. Food Chem. 233, 434–441.

Xing, L., Gao, Q., Huang, X., Xue, Y., Ding, L., Li, J., Dong, P., 2023. Simultaneous determination of six vitamin A substances by multi-wavelength HPLC and application to different marine shellfishes. J. Food Compos. Anal. 115, 105039.

Ye, L., Landen, W.O., Lee, J., Eitenmiller, R.R., 1998. Vitamin E content of margarine and reduced fat products using a simplified extraction procedure and HPLC determination. J. Liq. Chromatogr. Relat. Technol. 10826076. 21 (8), 1227–1238. Available from: http://www.tandf.co.uk/journals/titles/10826076.asp. doi:10.1080/10826079808006596.

Yildiz, G., Celikler, S., Vatan, O., Dere, S., 2012. Determination of the anti-oxidative capacity and bioactive compounds in green seaweed *Ulva rigida* C. Agardh. Int. J. Food Prop. 15 (6), 1182–1189. doi:10.1080/10942912.2010.517341.

Yildiz, G., Vatan, Ö., Çelikler, S., Dere, Ş., 2011. Determination of the phenolic compounds and antioxidative capacity in red algae *Gracilaria bursa-pastoris*. Int. J. Food Prop. 14 (3), 496–502. doi:10.1080/10942910903256949.

Yu, W., Simmons-Menchaca, M., Gapor, A., Sanders, B.G., Kline, K., 1999. Induction of apoptosis in human breast cancer cells by tocopherols and tocotrienols. Nutr. Cancer. 01635581. 33 (1), 26–32. Available from: http://www.tandfonline.com/toc/hnuc20/current. doi:10.1080/01635589909514744.

Zaporozhets, O.A., Krushinskaya, E.A., 2002. Determination of ascorbic acid by molecular spectroscopic techniques. Zh. Anal. Khim. 57 (4), 343–355.

Zempleni, J., Suttie, J.W., Gregory III, J.F., Stover, P.J., 2013. Handbook of vitamins. CRC Press.

CHAPTER

9

Marine pigmented bacteria associated with seaweeds and their potential industrial applications

Malith Chalitha Lalaksha Jayasinghe

Department of Botany, University of Sri Jayewardenepura, Nugegoda, Sri Lanka

9.1 Introduction to marine pigmented bacteria associated with seaweeds

Earth's surfaces are dominated by marine ecosystems, with a striking 71% Earth surface coverage, which hosts 80% of the total life present on Earth (Santhanam, 2020). Marine microbes represent over 95% of the total marine biomass, including bacteria, where the density of bacteria per liter counts up to one billion bacteria (Abida et al., 2013); however, the diversity and known species of marine bacteria are still underexplored (Ameen et al., 2021). Seaweeds (so-called marine macroalgae) play crucial roles in marine ecosystem dynamics by serving as primary producers, oxygen producers, agents for carbon sequestration and storage, and substrates for different organisms (by means of shelter, protection, and/or breeding grounds). Moreover, they also serve as "marine ecosystem engineers," as they can be utilized as biological agents for biomonitoring and bioremediation (Chan et al., 2006; Harley et al., 2012; Hurd et al., 2022; Ould and Caldwell, 2022). The associations between marine bacteria and seaweeds have gained significant interest among the scientific community recently (Egan et al., 2013; Hollants et al., 2013; Singh and Reddy, 2014), especially with the concept of the "seaweed–bacteria holobiont" (Egan et al., 2013). The seaweed–bacteria holobiont concept explains the relationship between seaweed and associated phycoplane bacteria, which act together as a single unit to assist and regulate macroalgal health, performance, and resilience (see Fig. 12.1). The phycoplane is the living surface of marine seaweeds (Diab and Metwalli, 1982; Raghukumar, Nagarkar and Raghukumar, 1992), and it harbors highly diverse microbial communities (Egan et al., 2013; Singh and Reddy, 2014), including pigmented bacteria (Ganesh Kumar et al., 2013; Suresh et al., 2015; Serratia marcescens P25, A New Strain Isolated From The Phycoplane of the Red Marine Alga Punctaria sp Produced Potent

https://doi.org/10.1016/B978-0-443-29190-6.00013-1

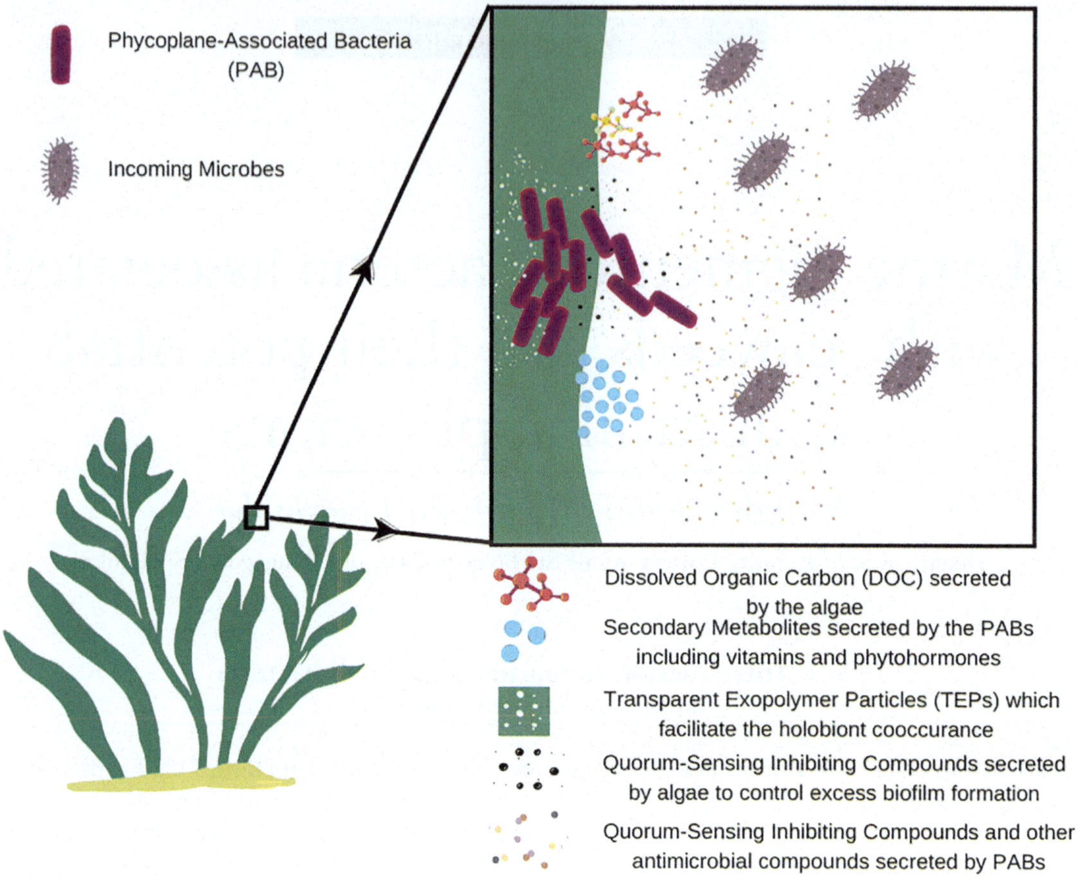

FIG. 9.1 Seaweed–bacteria holobiont concept.

Biosurfactant Used for Enhancing the Bioremediation of Spent Motor Oil-Polluted Soil, 2017; Lunggani et al., 2020), which are ecologically significant and notable for their bioactive pigment profiles (Soliev et al., 2011) (Fig. 9.1).

Among the other bacteria, pigmented bacteria have great potential in many aspects, especially regarding the availability of natural pigments. These pigments can be used in various industrial applications, including the food, beverage, and dairy industries, the textile industry, the paper industry, the cosmetic industry, agriculture, water science, and technology. These natural bacterial pigments also offer bioactive properties such as antioxidant properties, antibacterial properties, ultraviolet (UV)-blocking properties, anticancer properties, provitamins, probiotic properties, antiproliferative properties, and immunosuppressive properties, making them valuable essentials in the above-mentioned applications and also making them good candidates for the pharmaceutical industry as well (Joux et al., 1999; Stafsnes et al., 2010; Malik et al., 2012; Pawar et al., 2015; Nigam and Luke, 2016; Dzeha et al., 2019; Ramesh et al., 2019; Paillïè-Jiménez et al., 2020). Due to the adverse effects caused by artificial pigments, such as environmental pollution, toxicological side effects, oncogenicity, and teratogenic properties, natural pigments are becoming more in demand. Changes in consumer attitudes toward eco-friendly natural pigment consumption have made natural pigments more favorable over synthetic pigments (Malik et al., 2012; Nigam and Luke, 2016). When it comes to natural pigments, microbial and plant pigments are more prominent. However, plant pigments have limitations such as instability against heat, adverse pH, and light, low water solubility, and unavailability throughout the year in sufficient amounts (Nigam and Luke, 2016). This makes microbial pigments the leading candidate for replacing synthetic pigments due to their properties, such as stability to light, heat, and pH, as well as fermentation advantages, including cheaper production, easier extraction, no lack of raw materials, higher yields through strain improvements, no seasonal variations (Malik et al., 2012), and, moreover, easy scalability for mass production. Some challenges regarding microbial pigments include difficulties in downstream processes for extracting sufficiently concentrated pure forms of pigments, instability against UV light and high temperatures, and long processes involving organic chemicals, which may lead to degradation or undesirable isomer formation (Nigam and Luke, 2016).

9.2 Overview of taxonomic groups of seaweed-associated bacteria and their importance

Seaweeds host a wide array of associated bacteria from different gram-negative phyla such as *Proteobacteria, Bacteroidetes, Planctomycetes,* and *Chloroflexota* (mostly gram-negative) and gram-positive phyla such as *Firmicutes* and *Actinobacteria* (Martin et al., 2014; Singh and Reddy, 2014) and facilitate the complex interaction between phycoplane bacteria and themselves (Egan et al., 2013). These seaweed-associated bacteria play vital roles in the growth and development of seaweeds, especially by providing key nutrients such as CO_2, nitrogen from nitrogen fixation, vitamins, phytohormones, thiamine derivatives, and siderophores; morphogenesis of seaweeds, particularly in *Ulva* species and in *Monostroma oxyspermum* through the morphogenesis factor known as thallusin; nutrient cycling; induction of spore liberation in *Acrochaetium*; zoospore settlement of *Ulva* species; acting as antifouling agents (especially *Vibrio* spp. and *Pseudomonas* spp.) and antimicrobial agents (from the phyla *Proteobacteria, Actinobacteria, Bacteroidetes,* and *Firmicutes*); and secretion of many bioactive compounds such as haliangicin, violacein, pelagiomicin A, korormicin, macrolactins, and chlorophyll *d* (Armstrong et al., 2001; Joint et al., 2007; Weinberger et al., 2007; Egan et al., 2013; Singh and Reddy, 2014; Singh et al., 2015; Ren et al., 2022; Li et al., 2023; Sterling et al., 2023). In addition, these pigmented bacteria contribute to nutrient and organic matter cycling and also serve as indicators of environmental changes in marine ecosystems (Paerl et al., 2003; Giovannoni and Stingl, 2005; Thompson and Polz, 2014). Therefore it is evident that the phycoplane and the ambient environment act as microcosms to shape the community dynamics and functional processes between algae, macroalgae, and microbial communities. Understanding these phycoplane-associated bacteria (PAB) provides a good approach to understanding the role of pigmented bacteria in these associated communities and the intricate network of marine ecosystems.

9.3 Understanding the phycoplane and ambient microenvironment and its role in algal–bacterial interactions

The phycoplane can be defined as all the surfaces of seaweeds that are exposed to the outer environment, thus providing substrates for surrounding organisms and facilitating interactions between them and each other. Algae are the primary producers of marine ecosystems, and they secrete exudates to the phycoplane and the ambient microenvironment (Kouzuma and Watanabe, 2015), attracting beneficial microbiota and facilitating their growth and development. In return, these microbial communities, including pigmented bacteria, express their gratitude to algae in various ways (which are described in Section 1) that stimulate and facilitate algae growth, development, and survival. Most importantly, these interactions seem to be specific to several genera of bacteria and several algal groups. As an example, morphogenesis induction in the algae *Monostroma oxyspermum* is restricted to the interaction between bacterial communities in the phycoplane from the groups *Cytophaga*, *Flavobacterium*, and *Bacteroides* (Matsuo et al., 2005; Egan et al., 2013). In addition, it has also been recorded that there is a close association between diatoms and bacteria from the phyla *Proteobacteria* (e.g., *Roseobacter* sp. and *Sulfitobacter* sp.) and *Bacteroidetes* (e.g., *Cytophaga* sp. and *Flavobacterium* sp.) (Amin et al., 2012; Sterling et al., 2023). Mainly, these interactions are of three types: nutrient exchange, signal transduction, and gene transfer (Kouzuma and Watanabe, 2015; Ren et al., 2022; Li et al., 2023).

When it comes to nutrient exchange interactions, marine algae excrete dissolved organic carbon, and PAB assimilate and decompose a significant amount of it. Algae are able to control the microbial population through nutrient exchange interactions by limiting nutrients and light. In addition, PAB degrades and decomposes unprotected or damaged cells of algae and dead algae. PAB is also able to produce nitrogen required for algal growth, vitamin B12, which eukaryotic algae are auxotrophic for, phytohormones such as indole acetic acid (IAA) abscisic acid, cytokinins, ethylene, gibberellins, siderophores, and thiamin derivatives.

When it comes to signal transduction interactions, it was discussed earlier how chemical signals generated by the PAB affect the morphogenesis of the algae. Diatoms produce an intercellular osmolyte known as dimethylsulfoniopropionate (DMSP) for PAB, *Acropora cytherea*, and *Tridacna maxima*. DMSP concentrations can be modulated with respect to species assemblage complexity. Algae are able to produce specific signal chemicals that interrupt the quorum sensing (QS) between bacteria, thus controlling the excessive formation of biofilms on the phycoplane. The PAB of *Colpomenia sinuosa* is able to produce quorum-sensing inhibiting compounds to limit the colonization of other bacteria. In addition, algae are also able to produce volatile halogenated compounds and fatty acids, which can be considered antimicrobial compounds to control bacterial populations on the phycoplane. As an example, algae producing chlorellin can affect both gram-positive and gram-negative species such as *Staphylococcus aureus*, *B. subtilis*, *Escherichia coli*, and *Pseudomonas aeruginosa*. However, some algae are able to secrete transparent exopolymer particles, which facilitate the cooccurrence of the PAB and algae. Acyl homoserine lactones, which are QS compounds secreted by bacteria, interfere with *Ulva zoospore* germination and growth. Some bacteria from *Shewanella*, *Streptomyces*, and *Bacillus* taxa are able to produce algicidal metabolites.

Finally, the third interaction type is horizontal gene transfer between bacteria and algae. This is supported by findings such as evidence of horizontal gene transfer in the *Seminavis robusta* chloroplast genome, close relations of *Dinoflagellate Minicircles* genome genes to *Algoriphagus* sp. and *Cytophaga* sp. bacterial genes, and also diatom ornithine–urea cycle enzyme genes of bacterial origin.

When analyzing these interactions, bacterial and algal interactions can fall into all sorts of categories, such as mutualism, commensalism, and parasitism. However, all these symbiotic relationships reflect the fact that most of the time, PAB and associated algae act as a single unit to get things done, especially in survival against harsh conditions, pathogen attacks, etc., and therefore act as holobionts (Egan et al., 2013; Ghaderiardakani et al., 2022; Ren et al., 2022).

It is now evident that bioactive compounds, signal transduction molecules, and other metabolites in the phycoplane and ambient microenvironment act as dynamic intermediaries of algal–bacterial interactions.

9.4 Mechanisms of communication and signaling between algae and bacteria

Algae and associated bacteria employ different mechanisms to communicate and exchange signals. QS is one of the mechanisms bacteria use to communicate with each other and also with other species of bacteria, but not only with bacteria, also with associated algae (Dow, 2021). To take advantage of the QS mechanisms, organisms need to have two sets of synthase genes: one set of genes to synthesize QS signal molecules, also known as autoinducers, and another set of synthase genes to synthesize the QS signal receptor molecules. Autoinducers are diverse in their chemical structure, but mainly there are two categories they can fall into, such as N-acyl-homoserine lactone and AQ. When there is a certain amount of autoinducers that fulfill the minimum concentration to start a signal cascade, QS signal receptors recognize the presence of autoinducer molecules and start to express relevant genes. In this way, QS mechanisms are able to dictate gene expression and physiological processes according to the signals created by the interaction between algae and associated bacteria. Understanding these mechanisms will help provide better insights into the relationship between PAB and associated algae.

9.5 Factors influencing the composition and diversity of the phycoplane community

Various internal factors, such as the biological, physical, and chemical properties of the phycoplane and associated organisms, and external factors, such as environmental conditions and seasonal modulations, may have a negative or positive effect on the phycoplane and ambient microenvironment communities (Egan et al., 2013). The phycoplane is normally rich in oxygen, a by-product of algal photosynthesis, and thus facilitates the growth of aerobic microbes such as ammonium-oxidizing bacteria. However, having higher levels of oxygen in the phycoplane can cause detrimental effects, especially by forming reactive oxygen species (ROS). Therefore there is a section on the microbes' mechanisms to overcome this oxidative stress in the phycoplane. Algal exudates and the composition of algal cell walls also shape the phycoplane microbial communities when these act as carbon sources and other nutrient sources for the PAB, but algal cell wall degradation by the PAB is tightly controlled by the host algae. Metabolites secreted by both members of the holobiont, such as antifouling agents, antibacterial compounds, and different signal molecules, alter the composition and diversity of the phycoplane community as well. External surface physical characteristics for attachment, as well as adaptations for attachment of PAB, such as having lectins (more studies are needed with macroalgal communities), mannose-sensitive hemagglutinin-pili (e.g., *P. tunicata* attachment to *U. lactuca*), and curli-proteinaceous fiber for attachment, are also important in influencing microbial communities in the phycoplane. Most importantly, the competition for the resources available in the phycoplane and ambient microenvironment exerts selective pressure and is therefore critical in influencing phycoplane communities. In addition, external factors such as available light intensity, temperature, and salinity, and internal factors such as interactions of these seaweed–bacteria holobionts with other species, such as viruses and fungi, also determine the composition and diversity of the phycoplane communities.

9.6 Types of marine bacterial pigments

Marine pigmented bacteria have diverse pigment profiles with vibrant to dull colors, ranging from red, violet, yellow, brown, black, etc. (see Table 9.1) (Velmurugan et al., 2020). Most importantly, these pigments exhibit various bioactive properties, as shown in Table 9.1 (Soliev et al., 2011; Nawaz et al., 2021; Velmurugan et al., 2020; Panintingjati Brotosudarmo et al., 2021).

TABLE 9.1 Different pigments isolated from marine bacteria, their bioactive properties, and bacterial strains responsible for pigment production.

Pigment	Activity	Bacterial strains
Undecylprodigiosin	Anticancer	*Streptomyces rubber*
Cycloprodigiosin	Immunosuppressant; anticancer; antimalarial	*Pseudoalteromonas denitrificans*
Heptyl prodigiosin	Antiplasmodial	α-Proteobacteria
2-(p-Hydroxybenzyl) prodigiosin	Cytotoxic	*Pseudoalteromonas rubra* sp.
Prodigiosin	Antibacterial; anticancer; algicidal; cytotoxic; antiinflammatory; antiviral; UV protection	*Pseudoalteromonas rubra* strains PS1 and SB14
		Hahella chejuensis
		Zooshikella sp.

TABLE 9.1 Different pigments isolated from marine bacteria, their bioactive properties, and bacterial strains responsible for pigment production—cont'd

Pigment	Activity	Bacterial strains
		Streptomyces sp.
		Vibryoruber DSM 14379
		Actinomyces sp.
		Serratiarubidaea sp. RAM_Alex
Astaxanthin (carotene)	Antioxidant; anticancer	*Agrobacterium aurantiacum*
		Pseudoalteromonas luteoviolacea
		Pontibacterkorlensis sp. AG6
Decaprenoxanthin (carotenoid)	Antioxidant	*Arthrobacter* sp. P40
Decaprenoxanthin monoglucoside (carotenoid)	Antioxidant	*Arthrobacter* sp. P40
Decaprenoxanthin diglucoside (carotenoid)	Antioxidant	*Arthrobacter* sp. P40
2′-Hydroxy-astaxanthin (carotenoid)	Antioxidant	*Brevundimonas* sp. strain N-5
2,2′-Dihydroxy-astaxanthin (carotenoid)	Antioxidant	*Brevundimonas* scallop
		Brevundimonas sp. strain N-5
Dehydroflexixanthin (carotenoid)	Antioxidant	*Rhodopirellula rubra* LF2^T
2′-Isopentenyldehy-drosaproxanthin (carotenoid)	Antioxidant	*Rhodopirellula rubra* LF2^T
Saproxanthin (carotenoid)	Antioxidant	*Rhodopirellula rubra* LF2^T
		Rubinisphaera brasiliensis Gr7
		Flavobacteriaceae sp.
Myxol (carotenoid)	Antioxidant	*Flavobacteriaceae* sp.
Lycopene (carotenoid)	Antitumor	*Arthrobacter* sp. G2O
Zeaxanthin (carotenoid)	Nitric oxide scavenging, inhibition of lipid peroxidation, DPPH radical scavenging activities	*Muricauda* aquimarina sp.
		Muricauda olearia sp.
Zeaxanthin sulfate (carotenoid)	Antioxidant	*Erythrobacterflavusi* KJ5

Continued

TABLE 9.1 Different pigments isolated from marine bacteria, their bioactive properties, and bacterial strains responsible for pigment production—cont'd

Pigment	Activity	Bacterial strains
Violacein	Antibiotic; antiprotozoan; anticancer; antimicrobial; UV protection	*Pseudoalteromonas tunicata* *Pseudoalteromonas* sp. 520P1 *Collimonas* sp. CT *Pseudoalteromonas amylolytica* *Chromobacterium violaceum* *Janthinobacterium lividum*
Phenazine	Antibiotic; antiviral; insecticidal agent; anticancerous; antiprotozoal agents	*Pseudomonas* sp. *Actinomycetes* sp. *Pelagibacter* sp. *Vibrio* sp.
Methyl saphenate (phenazine derivative)	Antibiotic	*Pseudonocardia* sp. B6273
Phenazine derivatives	Cytotoxic	*Bacillus* sp.
Pyocyanin and pyorubrin	Antibacterial	*Pseudomonasaeruginosa*
Phenazine-1-carboxylic acid (PCA)	Antibiotic; anticancer	*Pseudomonasaeruginosa* *Pseudomonasaeruginosa* sp. GS-33
5,10-Dihydrophencomycin methyl ester	Antibiotic	*Streptomycete* sp.
Fridamycin D, himalomycin A, himalomycin B	Antibacterial	*Streptomycete* sp. B6921
Chinikomycin A and chinikomycin B, manumycin A	Anticancer	*Streptomycete* sp. M045
Tambjamines (BE-18591, pyrroles and their synthetic analogs)	Antibiotic; anticancer	*Pseudoalteromonas tunicata*
Melanins	Protection from UV irradiation; antioxidative; anticancer; antibacterial; antiviral; antiquorum sensing	*Vibrio cholera* *Shewanella colwelliana* *Alteromonas nigrifaciens* *Cellulophaga tyrosinoxydans* *Bacillus* sp.

TABLE 9.1 Different pigments isolated from marine bacteria, their bioactive properties, and bacterial strains responsible for pigment production—cont'd

Pigment	Activity	Bacterial strains
		Pseudomonas stutzeri
		Providencia sp.
		Salinicoccus roseus
		Planococcus maritimus
		Staphylococcus sciuri
		Pseudolateromonas lipolytica
		Pseudomonas stutzeri
		Nocardiopsis sp.
		Streptomyces sp.
Poly melanin	Antibiotic resistance	*Leclercia* sp. BTCZ22
Scytonemin	Protection from UV irradiation, antiinflammatory, antiproliferative	*Cyanobacteria*
Tryptanthrin	Antibiotic	*Cytophaga/Flexibacteria* AM13,1 strain
Quinone (common)	Antiviral; anticancer; antimicrobial; insecticidal	*Streptomyces* sp., XMA39
		Streptomyces sp., 182SMLY
Bisantraquinone 1 (quinone)	Antibacterial; anticancer	*Ecteinascidia turbinada*
Fridamycins H (quinone)	Antitrypanosomal	*Actinokineospora spheciospongiae* sp. nov.
Fridamycins I (Quinone)	Antitrypanosomal	*Actinokineospora spheciospongiae* sp. nov.
N-acetyl-N demethylmayamycin (quinone)	Antibacterial	*Streptomyces* sp. 182SMLY
Mersaquinone (quinone)	Antibacterial	*Streptomyces* sp.
Streptoanthraquinone A (quinone)	Antibacterial	*Streptomyces* sp. 182SMLY

9.7 UV-blocking pigments in pigmented bacteria

9.7.1 Types of UV-blocking pigments produced by bacteria

Bacterial pigments are secondary metabolites with color (Narsing Rao et al., 2017). There are several types of bacterial pigments responsible for blocking or absorbing UV. As mentioned in Section 1.2 of this review, melanins, mycosporines, scytonemin, and carotenoids are the major groups of these pigments. In addition, prodigiosin and violacein have UV-blocking properties as well (Borić et al., 2011; Abboud and Arment, 2013).

Bacterial melanins are heteropolymeric pigments that are polymerized reactive quinones formed by the oxidation of hydroxylated aromatic compounds (Pavan et al., 2020). Bacterial melanins play important roles not only in stress conditions caused by UV radiation (UVR) but also in toxic heavy metal stresses and oxidative stresses. In addition, melanins are important to bacteria in several ways, such as contributing to their virulence by reducing host–pathogen defense susceptibility, increasing iron bioavailability during biofilm formation (as seen in the marine bacterium *Pseudoalteromonas lipolytica*), providing a selective advantage in stress conditions, etc. Mycosporines are molecules that have a cyclohexanone or cyclohexenimine chromophore linked to an amino acid or amino alcohol (Colabella et al., 2015). They are water-soluble molecules as well. In addition to their UV protection ability, mycosporines are important for their antioxidant ability, high photostability, nitrogen storage, role as photosynthetic accessory pigments, osmotic stress protection, thermal protection, and other functions. Scytonemin is a lipid-soluble yellow–brown pigment mainly found in the cyanobacterial extracellular polysaccharide sheath, which is composed of indolic and phenolic subunits (Shailendra et al., 2010). An olefinic carbon atom acts as the linkage between these two subunits, which is quite unique to naturally occurring compounds. In addition to UV protection, scytonemin also shows promising potential for antiinflammatory and antiproliferative activities.

Carotenoids are natural pigments of isoprenoid origin, having at least six conjugated double bonds in their polyene chain, resulting in a red to yellow coloration range (López et al., 2021). These carotenoids are diverse among the categories, with some containing only hydrocarbons in their structure and others having additional oxygen functional groups in their structure. Even when it comes to the carbon number, carotenoids exhibit variations, such as C30, C40, C45, C50, and even <30 carbons. In addition to photoprotective activities, carotenoids offer a vast range of bioactive properties due to their different varieties, including antioxidant, antiinflammatory, and anticancer properties.

Prodigiosin is a tripyrrole molecule that can form linear or ring derivatives (Han et al., 2021). It consists of pyrrole (ring A), 3-methoxypyrrole (ring B), and 2-methyl-3-pentylpyrrole (ring C). In addition to its UV-blocking abilities, prodigiosin exhibits bioactivities such as antibacterial, antiinsect, antimalaria, antialgae, antitumor, and immunosuppressive activities.

Violacein is a purple-colored dimeric indole derivative composed of 2-pyrrolidone, 5-hydroxyindole, and oxindole (August et al., 2000; Durán et al., 2007). In addition to UV protection, it exhibits activities such as antiparasitic, antimicrobial, antiviral, antitumoral, antibiotic, and antiprotozoan properties.

9.7.2 Biosynthesis and regulation of UV-protective pigments

Understanding the biosynthesis and regulation of the above-mentioned pigments is important in bioproduction, synthetic biology, and their applications (Gao and Garcia-Pichel, 2011). Most of the melanin biosynthesis pathways are related to tyrosine metabolism (Pavan et al., 2020). It is mediated by tyrosinases, which catalyze the associated hydroxylation (monophenols) and oxidation reactions (O-diphenols to O-quinones). *Streptomyces* sp. has melC, which is a two-cistron operon that encodes tyrosinase and chaperone-like proteins to incorporate metal into apotyrosinase. Transcription factors involved in aerial mycelial development regulate gene expression and melanin production in actinobacteria. However, in some other bacteria, like *Marinomonas mediterranea,* the polyphenol oxidase B (ppoB) operon is responsible for melanogenesis. In *Modestobacter versicolor*, melanin production is induced by exposure to UVB and oligotrophy, but low nitrogen availability prevents its production. Melanogenesis in fungi involves polyketide intermediates and several other genes responsible for the synthesis of HN reductase, polyketide synthase, and scytalone dehydratase.

The synthesis of mycosporin-like amino acids is associated with the shikimate pathway (Colabella et al., 2015). Demethyldeoxygadusol synthase and demethyldeoxygadusol methyltransferase genes are involved in mycosporin-like amino acid synthesis, and the synthesis pathway branches from central metabolism rather than the shikimate pathway. However, core genes involved in mycosporin synthesis, such as those related to amino acid synthesis, are conserved. Exposure to UVB radiation can enhance mycosporin-like amino acid production. UVB photoreceptors (with pterin as the chromophore) are implicated in the UVB-dependent synthesis in *Chlorogleopsis* spp. Furthermore, osmotic stress, nitrogen and sulfur availability, and visible light can affect the biosynthesis of mycosporin-like amino acids.

Scytonemin biosynthesis in *Nostoc punctiforme* ATCC 29133 involves a gene cluster with 18 genes, including the key gene ScyD (Shailendra et al., 2010). The synthesis of scytonemin is involved in aromatic amino acid biosynthesis, and building blocks are provided by the downstream genes. ScyA, ScyB, and ScyC have been the key characterized enzymes. Gene ScyB is responsible for the conversion of tryptophan to indole-3-pyruvic acid (I3P). Another precursor, p-hydroxyphenylpyruvic acid (HPP), is also synthesized by the oxidation of prephenate by a putative prephenate dehydrogenase. Gene ScyA is responsible for the catalysis of the condensation of I3P and HPP to form a key intermediate, while gene ScyC mediates its transformation to scytonemin. UVA radiation exposure induces the expression of the scytonemin gene cluster.

The synthesis of carotenoids starts with the mevalonate pathway by isomerizing isopentenyl pyrophosphate into dimethylallyl pyrophosphate (López et al., 2021). The pathway of 2-methyl-D-erythritol-4-phosphate is also active to add additional C5 units. C30 and C40 carotenoids have two different pathways of synthesis after forming common geranyl diphosphate with the help of prenyl transferase (CrtE). The addition of two C15-farnesyl pyrophosphate units and two C20-geranylgeranyl diphosphate units gives rise to C30 and C40 carotenoids, respectively. Then, these molecules go through different addition, elimination, substitution, or rearrangement reactions to produce a variety of carotenoids.

Biosynthesis of prodigiosin takes place through a bifurcated pathway (Han et al., 2021). Two key intermediates, which are 2-methyl-3-n-amylpyrrole (MAP) and 4-methoxy-2,2′-bipyrrole-5-carbaldehyde (MBC), are formed from 2-octenoyl CoA and L-proline, respectively, and eventually condense to form prodigiosin. *Serratia marcescens* employs the Pig cluster (Pig B, Pig D, and Pig E for MAP biosynthesis; Pig A and Pig F-Pig N for MBC biosynthesis; and Pig C for condensing MAP and MBC) genes, while *Streptomyces coelicolor* employs Red cluster genes for prodigiosin biosynthesis. Under anaerobic conditions, higher temperatures (above 30°C), glucose, ATP, and increased phosphate levels decrease or inhibit prodigiosin production, while higher light intensities, thiamine, and iron increase its synthesis.

The biosynthesis of violacein takes place with the help of the flavoenzyme l-tryptophan oxidase (VioA), which converts l-tryptophan into the corresponding α-imine oxidatively (Sánchez et al., 2006; Füller et al., 2016). The formed prodeoxyviolaceinic acid is converted by the VioB enzyme into deoxyviolacein, and the VioE enzyme adds two oxygens to form 5′-hydroxydeoxyviolacein. Finally, it is converted into violacein with the help of the VioD and VioC enzymes. The VioABCD gene cluster is responsible for violacein biosynthesis.

9.7.3 Mechanisms of UV protection and photoprotection by bacterial pigments

Bacterial pigments employ different mechanisms to fight against UV and photodamage. Mycosporines act as UVR-absorbing pigments and dissipate absorbed energy, thus preventing UV damage (Oren and Gunde-Cimerman, 2007). Mycosporines have the special ability to transfer absorbed energy into heat and dissipate it without creating ROS. Moreover, some mycosporines have antioxidant activity to scavenge free radicals formed from UV damage. Melanin also absorbs UVR (Cordero and Casadevall, 2020; López et al., 2021; Pavan et al., 2020), and absorbed energy is dissipated as heat without giving ROS much chance to form. Melanin also has the ability to scavenge or neutralize ROS. Scytonemin, on the contrary, has the ability to absorb short-wavelength UVR and dissipate the absorbed energy through thermal deexcitation (Rastogi et al., 2015). In addition, it has shown the ability to minimize the formation of ROS and thymine dimers caused by UVR. Carotenoids also function in the same manner to provide photoprotection and scavenge ROS caused by photodamage (Sies and Stahl, 2004; Reis-Mansur et al., 2019). It is now evident that these pigments provide a UV-protective shield for bacteria and also associate with algae in mutualism to maintain their viability and cellular integrity. Both prodigiosin and violacein show the same types of mechanisms that the previously described pigments exhibit.

9.8 Antioxidant properties of pigments in pigmented bacteria

9.8.1 ROS and oxidative stress in marine environments

ROS such as hydrogen peroxide (HOOH), superoxide (O_2 -), and hydroxyl radical (HO·) are highly reactive molecules that can be either cellular metabolic by-products (from photosynthesis, cellular respiration, and different enzymatic activities: xanthine oxidase, NADPH oxidases, and cytochrome P450) or produced by the influence of external factors such as UVR, temperature, chemical contaminants, and other pollutants (Zinser, 2018). These produced ROS are important to marine organisms in ways such as cellular signaling in growth, cell differentiation, apoptosis, and biofilm formation in bacteria, defense against pathogens and other invaders, and sometimes adaptations for environmental stresses. However, when there is an imbalance between ROS production and antioxidant activity against them, or in other words, if there is excessive production of ROS, it becomes the cause of oxidative stress for marine organisms. Oxidative stress negatively affects them by damaging their DNA and other nucleic matter, proteins, lipids, and other important biomolecules, causing impaired physiological and biochemical processes, cellular dysfunction, mutations, and even death. In general, it impacts the growth and development, reproduction, immunity, and overall fitness of marine organisms, destroys biodiversity, and can even lead to the extinction of marine communities.

Marine organisms employ different strategies to overcome oxidative stress, which can be classified as enzymatic and nonenzymatic methods (Shalaby, 2013; Zinser, 2018). In the enzymatic method, enzymes involved in mitigating oxidative stress are catalases, superoxide dismutase, thioredoxins, glutathione peroxidase, and glutathione S-transferase. Nonenzymatic methods employ vitamins (water-soluble vitamin C and fat-soluble vitamins A, E, and K), minerals (zinc and selenium), low molecular weight compounds (glutathione and uric acid), cofactors (coenzyme Q_{10}), organosulfur compounds (allium, allyl sulfide, indoles), different phenolic metabolites (tocopherol), and bioactive pigments (carotenoids, melanin, etc.), which is the interesting aspect of this review.

9.8.2 Role of pigments in antioxidant defense mechanisms

As given in Table 9.1, it is now understood that many bacterial pigments have the ability to act as antioxidants. Antioxidants have the ability to scavenge or neutralize free radicals, such as ROS, thus providing protection against the damage caused by free radicals (Lü et al., 2010). Bacterial pigments are able to do this by donating electrons to the free radicals to neutralize their unpaired electrons and thus stabilize the free radicals (Azman et al., 2018). Most importantly, by doing this, the stability of antioxidant agents does not change, so they do not become reactive. Another way is to break the chain reactions by reacting with starters or intermediates and stopping the progression of the reaction by neutralizing them (Burton, 1989; Gulcin, 2020). Another important mechanism they follow is to chelate metal ions (by melanin), such as lead, iron, and copper, which can cause the formation of free radicals (Sajjan et al., 2013; Cai et al., 2015). In addition, carotenoids have the ability to quench singlet oxygen to provide antioxidative protection (Shimidzu et al., 1996; Cantrell et al., 2003).

9.9 Biotechnological applications and potential of bacterial pigments

9.9.1 Potential applications of UV-blocking pigments in skincare and sunscreens

Pigments with UV-blocking activity that originate from bacteria have a promising future in skincare and sunscreen formulations. Bacterial pigments, such as violacein and prodigiosin, have already shown positive responses toward UV blocking when applied with commercially available sunscreens as additives (Suryawanshi et al., 2015). On the contrary, melanin shows great potential for use in sunscreens as a direct active sunscreen ingredient (Geng et al., 2008). As it is now understood that many bacterial pigments show UV-blocking activities, applying them to sunscreen formulations and evaluating their efficacy has become a novel trend in relevant fields of study.

9.9.2 Exploitation of bacterial pigments for natural UV filters

UV filters are the active ingredients present in a sunscreen formulation. These UV filters can be UV absorbers (natural or artificial) and inorganic particulates (zinc oxide and titanium dioxide) (Manaia et al., 2013). However, most of them are not derived from nature, so synthetic UV filters have detrimental effects on humans, animals, and the environment (Jesus et al., 2022). Apart from their photoinstability and related consequences, the direct negative effects of consumption on human health can include acting as endocrine disruptors by penetrating the skin and entering the blood, as well as causing allergic reactions and cytotoxicity. In addition, they can negatively affect men's fertility as well. Apart from humans, they have negative effects on animals, especially marine organisms such as fish, crustaceans, corals, algae, mussels, dolphins, and sea urchins. Ecosystems such as estuaries, lakes, seawater, rivers, and sediments contain higher amounts of accumulated nondegradable UV filters, causing sediment contamination and promoting bioaccumulation in organisms, which leads to various issues. However, by employing naturally derived, nontoxic, degradable UV filters such as bacterial pigments, there is a promising future for their use in sunscreen formulations as UV filters.

9.9.3 Antioxidant properties and potential therapeutic applications of pigments

Throughout this literature review, the bioactive properties of bacterial pigments are discussed, and it is evident that their importance in the pharmaceutical industry is significant. Being nontoxic, noncarcinogenic, biodegradable, stable, bioavailable, biocompatible, and also easy to upscale and use, with simple techniques for pigment extraction, bioactive pigments of bacterial origin show great potential in therapeutic applications, including antimicrobial, anticancer, cytotoxic, antiobesity, antiinflammatory, antiangiogenic, and antioxidant pharmaceutical formulations (Numan et al., 2018). Moreover, they can also be used in drug delivery, photoacoustic imaging, photothermal therapy, photodynamic therapy, and wound healing (Manivasagan et al., 2018).

9.9.4 Bacterial pigments in the textile industry and other related industries

Bacterial pigments can be used in the textile industry as a colorant with additional properties such as antimicrobial, antiodor, and UV blocking, and in other industries such as colored candle making, the paper industry, the soap industry, the pencil case pouch industry, ballpoint pen ink, and highlighter pen ink industries, etc. (Venil et al., 2013; Metwally et al., 2021; Kramar and Kostic, 2022). Synthetic dyes mainly originate from nonrenewable sources like fossil oil, are environmentally unfriendly, and pose a risk to human health as well. Using microbial pigments as dyes solves these problems and In addition addresses stability issues, seasonal variation effects, extraction difficulties, etc., that other natural pigments pose. In addition, microbial pigments have the ability to undergo chemical modifications according to industrial needs and are therefore a good prospect as an alternative to artificial dyes.

9.10 Environmental implications and future perspectives

9.10.1 Ecological significance of UV-blocking and antioxidant pigments in marine ecosystems

Bioactive pigments, in the context of UV-blocking and antioxidant capabilities in bacteria, have given an upper hand to the bacteria and associated symbiotic seaweeds in extreme conditions. Thus their existence, rather than extinction, has shaped the marine ecosystem's food chains, community structure, community dynamics, and fitness, thereby facilitating the continuation of life on Earth (Soliev et al., 2011). Without these bioactive pigments, the whole marine ecosystem could fall apart, and only a few species might survive, ultimately destroying the diversity of the marine ecosystem. Therefore it is important to understand their significance in maintaining balance in the ecosystem and to extend our understanding through research and studies to gain more insights and conserve them.

9.10.2 Implications of climate change and increased UVR on seaweed–bacterial interactions

After the establishment of the Montreal Protocol in 1987, the recovery of the depleted stratospheric ozone layer is effectively in action, but it is a slow process that can take decades (Häder et al., 2014; Williamson et al., 2014). However, the uncontrolled negative human intervention on global climate has caused new serious considerations, partnering with UVR that has sneaked through the recovering ozone layer (Häder et al., 2014; Williamson et al., 2014). Global warming, due to various human activities, has caused the melting of glaciers, thus increasing sea levels and exposing masked dissolved organic matter (DOM) in dissolved ice to UVR. These additional DOMs, added to the ocean due to global warming, may form increased amounts of ROS through various photochemical reactions, thereby increasing oxidative stress (Häder et al., 2014; Williamson et al., 2014). Marine environments already face oxidative stress, but this additional excessive oxidative stress can exceed the limits of control for marine pigmented bacteria, thereby limiting their protection for themselves or symbiotic seaweed. This can lead to the extinction of current species and interactions among species in marine ecosystems, with only a few surviving due to their capability to overcome this excessive oxidative stress. The same applies to seaweed–bacterial interactions, thus limiting the diversity of interactions and only facilitating those that can adapt to this situation. This will result in the loss of current marine ecosystem diversity and facilitate the dominance of a few species.

9.10.3 Future research directions and areas of investigation

The current understanding of algae and bacteria is mainly based on the interactions between phytoplankton and bacteria, rather than on seaweed and bacteria interactions. There are far fewer studies on macroalgae and bacterial interactions, especially the mutualistic bacteria on the phycoplane of macroalgae. Macroalgae-related industries are widespread all over the world, but the utilization of knowledge on algae–bacterial interactions for the betterment of these industries has yet to be addressed. Most of these associated bacteria are yet to be discovered, and their metabolite profiles, with unlimited biotechnological applications, remain unexplored. Pigmented bacteria, their pigment profiles, their roles, their pigment production pathways, and pigment bioactive mechanisms to mitigate oxidative stress still require further research. The real effect of UVR and how UVR has structured the marine community composition, including algal and bacterial communities, over a period of time, is an interesting topic to shed light on. There should be more developments in the methods that effectively evaluate the UV-blocking properties of compounds.

References

Serratia marcescens P25, A New Strain Isolated From The Phycoplane of the Red Marine Alga Punctaria sp. Produced Potent Biosurfactant Used for Enhancing the Bioremediation of Spent Motor Oil-Polluted Soil, 2017. Int. J. Sci. Res. 23197064. 6 (1), 1034–1047. doi:10.21275/ART20164145.

Abboud, A.N., Arment, A., 2013. The protective effects of the violacein pigment against UV-C irradiation in *Chromobacterium violaceum*. Ohio J. Sci. 00300950. 111 (2-5), 28–32.

Abida, H., Ruchaud, S., Rios, L., Humeau, A., Probert, I., De Vargas, C., Bach, S., Bowler, C., 2013. Bioprospecting marine plankton. Mar. Drugs. 1660-3397. 11 (11), 4594–4611. doi:10.3390/md11114594.

Ameen, F., AlNadhari, S., Al-Homaidan, A.A., 2021. Marine microorganisms as an untapped source of bioactive compounds. Saudi J. Biol. Sci. 1319562X. 28 (1), 224–231. Available from: http://www.ksu.edu.sa/sites/Colleges/CollegeofScinces/BotanyDepartment/SJBS/Default.aspx. doi:10.1016/j.sjbs.2020.09.052.

Amin, S.A., Parker, M.S., Armbrust, E.V., 2012. Interactions between diatoms and bacteria. Microbiol. Mol. Biol. Rev. 10985557. 76(3), 667–684. Available from: http://mmbr.asm.org/content/76/3/667.full.pdf+html. doi:10.1128/MMBR.00007-12United.

Ann Cantrell, D.J.M., Truscott, T.G., Rancan, F., Böhm, F., 2003. Singlet oxygen quenching by dietary carotenoids in a model membrane environment. Arch. Biochem. Biophys. 00039861. 412 (1), 47–54. doi:10.1016/s0003-9861(03)00014-6.

Armstrong, E., Yan, L., Boyd, K.G., Wright, P.C., Burgess, J.G., 2001. The symbiotic role of marine microbes on living surfaces. Hydrobiologia 461 (1–3), 37–40.00188158. doi:10.1023/A:1012756913566.

August, P.R., Grossman, T.H., Minor, C., Draper, M.P., MacNeil, I.A., Pemberton, J.M., Call, K.M., Holt, D., Osburne, M.S., 2000. Sequence analysis and functional characterization of the violacein biosynthetic pathway from *Chromobacterium violaceum*. J. Mol. Microbiol. Biotechnol. 14641801. 2 (4), 513–519.

Azman, A.-S., Mawang, C.-I., Abubakar, S., 2018. Bacterial pigments: the bioactivities and as an alternative for therapeutic applications. Nat. Prod. Commun. 1934-578X. 13 (12). doi:10.1177/1934578x1801301240.

Borić, M., Danevčič, T., Stopar, D., 2011. Prodigiosin from Vibrio sp. DSM 14379; a new UV-protective pigment. Microb. Ecol. 00953628. 62 (3), 528–536. doi:10.1007/s00248-011-9857-0.

Burton, G.W., 1989. Antioxidant action of carotenoids. J. Nutr. 00223166. 119 (1), 109–111. doi:10.1093/jn/119.1.109.

Cai, X., Yang, Q., Wang, S., 2015. Antioxidant and hepatoprotective effects of a pigment-protein complex from *Chlorella vulgaris* on carbon tetrachloride-induced liver damage in vivo. RSC Adv. 20462069. 5 (116), 96097–96104. Available from: http://pubs.rsc.org/en/journals/journalissues. doi:10.1039/c5ra17544e.

Chan, C.X., Ho, C.L., Phang, S.M., 2006. Trends in seaweed research. Trends Plant Sci. 13601385. 11 (4), 165–166. doi:10.1016/j.tplants.2006.02.003.

Chandralata Raghukumar, S.N., Raghukumar, S., 1992. Association of thraustochytrids and fungi with living marine algae. Mycol. Res. 09537562. 96 (7), 542–546. doi:10.1016/s0953-7562(09)80978-7.

Colabella, F., Moline, M., Libkind, D., 2015. UV sunscreens of microbial origin: mycosporines and mycosporine- like aminoacids. Recent Pat. Biotechnol. 18722083. 8 (3), 179–193. doi:10.2174/1872208309666150102104520.

Cordero, R.J.B., Casadevall, A., 2020. Melanin. Curr. Biol. 18790445. 30 (4), R142 Available from: http://www.elsevier.com/journals/current-biology/0960-9822. doi:10.1016/j.cub.2019.12.042.

Diab, A., Metwalli, A., 1982. Thermophilic bacillus species associated with phaeophyceae in the Arabian gulf shore water at Kuwait. Zentralblatt Für Mikrobiologie. 02324393. 137 (3), 197–206. doi:10.1016/s0232-4393(82)80027-9.

Dow, L., 2021. How do quorum-sensing signals mediate algae–bacteria interactions? Microorganisms. 2076-2607. 9 (7), 1391. doi:10.3390/microorganisms9071391.

Durán, N., Justo, G.Z., Ferreira, C.V., Melo, P.S., Cordi, L., Martins, D., 2007. Violacein: properties and biological activities. Biotechnol. Appl. Biochem. 0885-4513. 48 (3), 127–133. doi:10.1042/ba20070115.

Dzeha, T., Nyiro, C., Kardasopoulos, D., Mburu, D., Mwafaida, J., Hall, M.J., Burgess, J.G., 2019. UV resistance of bacteria from the Kenyan Marine cyanobacterium Moorea producens. MicrobiologyOpen. 2045-8827. 8 (4). doi:10.1002/mbo3.697.

Egan, S., Harder, T., Burke, C., Steinberg, P., Kjelleberg, S., Thomas, T., 2013. The seaweed holobiont: understanding seaweed–bacteria interactions. FEMS Microbiol. Rev. 1574-6976. 37 (3), 462–476. doi:10.1111/1574-6976.12011.

Füller, J.J., Röpke, R., Krausze, J., Rennhack, K.E., Daniel, N.P., Blankenfeldt, W., Schulz, S., Jahn, D., Moser, J., 2016. Biosynthesis of violacein, structure and function of L-tryptophan oxidase VioA from chromobacterium violaceum. J. Biol. Chem. 1083351X. 291 (38), 20068–20084. Available from: http://www.jbc.org/content/291/38/20068.full.pdf. doi:10.1074/jbc.M116.741561.

Ganesh Kumar, C., Sahu, N., Narender Reddy, G., Prasad, R.B.N., Nagesh, N., Kamal, A., 2013. Production of melanin pigment from Pseudomonas stutzeri isolated from red seaweed Hypnea musciformis. Lett. Appl. Microbiol. 02668254. 57 (4), 295–302. doi:10.1111/lam.12111.

Gao, Q., Garcia-Pichel, F., 2011. Microbial ultraviolet sunscreens. Nat. Rev. Microbiol. 17401534. 9 (11), 791–802. doi:10.1038/nrmicro2649.

Geng, J., Tang, W., Wan, X., Zhou, Q., Wang, X.-J., Shen, P., Lei, T.-.C., Chen, X.-.D., 2008. Photoprotection of bacterial-derived melanin against ultraviolet A–induced cell death and its potential application as an active sunscreen. J. Eur. Acad. Dermatol. Venereol. 0926-9959. 22 (7), 852–858. doi:10.1111/j.1468-3083.2007.02574.x.

Ghaderiardakani, F., Langhans, L., Kurbel, V.B., Fenizia, S., Wichard, T., 2022. Metabolite profiling reveals insights into the species-dependent cold stress response of the green seaweed holobiont Ulva (*Chlorophyta*). Environ. Exp. Bot. 00988472. 200, 104913. doi:10.1016/j.envexpbot.2022.104913.

Giovannoni, S.J., Stingl, U., 2005. Molecular diversity and ecology of microbial plankton. Nature. 14764687. 437 (7057), 343–348. Available from: http://www.nature.com/nature/index.html. doi:10.1038/nature04158.

Gulcin, İ., 2020. Antioxidants and antioxidant methods: an updated overview. Arch. Toxicol. 14320738. 94 (3), 651–715. Available from: https://www.link.springer.de/link/service/journals/00204/index.htm. doi:10.1007/s00204-020-02689-3.

Häder, D.-P., Williamson, C.E., Wängberg, S.-Å., Rautio, M., Rose, K.C., Gao, K., Helbling, E.W., Sinha, R.P., Worrest, R., 2014. Effects of UV radiation on aquatic ecosystems and interactions with other environmental factors. Photochem. Photobiol. Sci. 1474-905X. 14 (1), 108–126. doi:10.1039/c4pp90035a.

Han, R., Xiang, R., Jinglin, L., Wang, F., Wang, C., 2021. High-level production of microbial prodigiosin: a review. J. Basic Microbiol. 0233-111X. 61 (6), 506–523. doi:10.1002/jobm.202100101.

Harley, C.D.G., Anderson, K.M., Demes, K.W., Jorve, J.P., Kordas, R.L., Coyle, T.A., Graham, M.H., 2012. EFfects of climate change on global seaweed communities. J. Phycol. 15298817. 48 (5), 1064–1078. doi:10.1111/j.1529-8817.2012.01224.x.

Hollants, J., Leliaert, F., De Clerck, O., Willems, A., 2013. What we can learn from sushi: a review on seaweed-bacterial associations. FEMS Microbiol. Ecol. 01686496. 83 (1), 1–16. doi:10.1111/j.1574-6941.2012.01446.x.

Hurd, C.L., Law, C.S., Bach, L.T., Britton, D., Hovenden, M., Paine, E.R., Raven, J.A., Tamsitt, V., Boyd, P.W., 2022. Forensic carbon accounting: assessing the role of seaweeds for carbon sequestration. J. Phycol. 15298817. 58 (3), 347–363. Available from: http://onlinelibrary.wiley.com/journal/10.1111/(ISSN)1529-8817. doi:10.1111/jpy.13249.

Jesus, A., Sousa, E., Cruz, M., Cidade, H., Lobo, J., Almeida, I., 2022. UV filters: challenges and prospects. Pharmaceuticals. 1424-8247. 15 (3), 263. doi:10.3390/ph15030263.

Joint, I., Tait, K., Wheeler, G., 2007. Cross-kingdom signalling: exploitation of bacterial quorum sensing molecules by the green seaweed Ulva. Philos. Trans. R. Soc. Lond. B Biol. Sci. 09628436. 362 (1483), 1223–1233. Available from: http://rstb.royalsocietypublishing.org/. doi:10.1098/rstb.2007.2047.

Joux, F., Jeffrey, W.H., Lebaron, P., Mitchell, D.L., 1999. Marine bacterial isolates display diverse responses to UV-B radiation. Appl. Environ. Microbiol. 0099-2240. 65 (9), 3820–3827. doi:10.1128/aem.65.9.3820-3827.1999.

Kouzuma, A., Watanabe, K., 2015. Exploring the potential of algae/bacteria interactions. Curr. Opin. Biotechnol. 18790429. 33, 125–129. Available from: http://www.elsevier.com/locate/copbio. doi:10.1016/j.copbio.2015.02.007.

Kramar, A., Kostic, M.M., 2022. Bacterial secondary metabolites as biopigments for textile dyeing. Textiles. 26737248. 2 (2), 252–264. Available from: https://www.mdpi.com/journal/textiles. doi:10.3390/textiles2020013.

Li, S.N., Zhang, C., Li, F., Ren, N.Q., Ho, S.H., 2023. Recent advances of algae-bacteria consortia in aquatic remediation. Crit. Rev. Environ. Sci. Technol. 15476537. 53 (3), 315–339. Available from: https://www.tandf.co.uk/journals/titles/10643389.asp. doi:10.1080/10643389.2022.2052704.

López, G.-D., Álvarez-Rivera, G., Carazzone, C., Ibáñez, E., Leidy, C., Cifuentes, A., 2021. Bacterial carotenoids: extraction, characterization, and applications. Crit. Rev. Anal. Chem. 1040-8347. 53 (6), 1239–1262. doi:10.1080/10408347.2021.2016366.

Lü, J.M., Lin, P.H., Yao, Q., Chen, C., 2010. Chemical and molecular mechanisms of antioxidants: experimental approaches and model systems. J. Cell. Mol. Med. 15821838. 14 (4), 840–860. doi:10.1111/j.1582-4934.2009.00897.x.

Lunggani, A.T., Purwantisari, S., Jannah, S.N., 2020. Characterization of yellow pigmented bacteria associated with *Gracilaria* sp. Advance Sustainable Science, Engineering and Technology. 27154211. 2 (2). Available from: https://journal.upgris.ac.id/index.php/asset/article/view/7041/pdf. doi:10.26877/asset.v2i2.7041.

Malik, K., Tokas, J., Goyal, S., 2012. Microbial pigments: a review. Int. J. Microbial Res. Technol. 1, 361–365.

Manaia, E.B., Kaminski, R.C.K., Corrêa, M.A., Chiavacci, L.A., 2013. Inorganic UV filters. Braz. J. Pharm. Sci. 21759790. 49 (2), 201–209. Available from: http://www.scielo.br/pdf/bjps/v49n2/02.pdf. doi:10.1590/S1984-82502013000200002.

Manivasagan, P., Bharathiraja, S., Moorthy, M.S., Mondal, S., Seo, H., Lee, K.D., Oh, J., 2018. Marine natural pigments as potential sources for therapeutic applications. Crit. Rev. Biotechnol. 0738-8551. 38 (5), 745–761. doi:10.1080/07388551.2017.1398713.

Martin, M., Portetelle, D., Michel, G., Vandenbol, M., 2014. Microorganisms living on macroalgae: diversity, interactions, and biotechnological applications. Appl. Microbiol. Biotechnol. 0175-7598. 98 (7), 2917–2935. doi:10.1007/s00253-014-5557-2.

Matsuo, Y., Imagawa, H., Nishizawa, M., Shizuri, Y., 2005. Isolation of an algal morphogenesis inducer from a marine bacterium. Science. 0036-8075. 307 (5715), 1598. doi:10.1126/science.1105486.

Metwally, R.A., Sikaily, A.E., El-Sersy, N.A., Ghozlan, H.A., Sabry, S.A., 2021. Antimicrobial activity of textile fabrics dyed with prodigiosin pigment extracted from marine *Serratia rubidaea* RAM_Alex bacteria. Egyptian Journal of Aquatic Research. 20903278. 47 (3), 301–305. Available from: http://www.elsevier.com/journals/egyptian-journal-of-aquatic-research/1687-4285. doi:10.1016/j.ejar.2021.05.004.

Narsing Rao, M.P., Xiao, M., Li, W.J., 2017. Fungal and bacterial pigments: secondary metabolites with wide applications. Front. Microbiol. 1664302X. (8). Available from: http://journal.frontiersin.org/article/10.3389/fmicb.2017.01113/full. doi:10.3389/fmicb.2017.01113.

Nawaz, A., Chaudhary, R., Shah, Z., Dufossé, L., Fouillaud, M., Mukhtar, H., Ul Haq, I., 2021. An overview on industrial and medical applications of bio-pigments synthesized by marine bacteria. Microorganisms. 2076-2607. 9 (1), 11. doi:10.3390/microorganisms9010011.

Nigam, P.S., Luke, J.S., 2016. Food additives: production of microbial pigments and their antioxidant properties. Curr. Opin. Food Sci. 22147993. 7, 93–100. Available from: http://www.journals.elsevier.com/current-opinion-in-food-science/. doi:10.1016/j.cofs.2016.02.004.

Numan, M., Bashir, S., Mumtaz, R., Tayyab, S., Rehman, N.U., Khan, A.L., Shinwari, Z.K., Al-Harrasi, A., 2018. Therapeutic applications of bacterial pigments: a review of current status and future opportunities. 3 Biotech. 2190-572X. 8 (4). doi:10.1007/s13205-018-1227-x.

Oren, A., Gunde-Cimerman, N., 2007. Mycosporines and mycosporine-like amino acids: UV protectants or multipurpose secondary metabolites? FEMS Microbiol. Lett. 15746968. 269 (1), 1–10. doi:10.1111/j.1574-6968.2007.00650.x.

Ould, E., Caldwell, G.S., 2022. The potential of seaweed for carbon capture. CAB Rev. 17498848. 2022. Available from: https://www.cabidigitallibrary.org/doi/epdf/10.1079/cabireviews202217009. doi:10.1079/cabireviews202217009.

Paerl, H.W., Dyble, J., Moisander, P.H., Noble, R.T., Piehler, M.F., Pinckney, J.L., Steppe, T.F., Twomey, L., Valdes, L. M., 2003. Microbial indicators of aquatic ecosystem change: current applications to eutrophication studies. FEMS Microbiol. Ecol. 01686496. 3 (46), 233–246. http://femsec.oxfordjournals.org/. doi:10.1016/S0168-6496(03)00200-9.

Paillié-Jiménez, M.E., Stincone, P., Brandelli, A., 2020. Natural pigments of microbial origin. Front. Sustain. Food Syst. 2571581X. 4. Available from: https://www.frontiersin.org/journals/sustainable-food-systems#. doi:10.3389/fsufs.2020.590439.

Panintingjati Brotosudarmo, T.H., Limantara, L., Pringgenies, D., 2021. Recent exploration of bioactive pigments from marine bacteria. ScienceAsia. 1513-1874. 47 (3), 265. doi:10.2306/scienceasia1513-1874.2021.050.

Pavan, M.E., López, N.I., Pettinari, M.J., 2020. Melanin biosynthesis in bacteria, regulation and production perspectives. Appl. Microbiol. Biotechnol. 14320614. 104 (4), 1357–1370. Available from: https://www.link.springer.de/link/service/journals/00253/index.htm. doi:10.1007/s00253-019-10245-y.

Pawar, R., Mohandass, C., Sivaperumal, E., Sabu, E., Rajasabapathy, R., Jagtap, T., 2015. Epiphytic marine pigmented bacteria: a prospective source of natural antioxidants. Braz. J. Microbiol. 16784405. 46 (1), 29–39. Available from: http://www.scielo.br/pdf/bjm/v46n1/1517-8382-bjm-46-01-0029.pdf. doi:10.1590/S1517-838246120130353.

Ramesh, C., Vinithkumar, N., Kirubagaran, R., 2019. Marine pigmented bacteria: a prospective source of antibacterial compounds. Journal of Natural Science, Biology and Medicine. 0976-9668. 10 (2), 104. doi:10.4103/jnsbm.jnsbm_201_18.

Rastogi, R.P., Sonani, R.R., Madamwar, D., 2015. Cyanobacterial sunscreen scytonemin: role in photoprotection and biomedical research. Appl. Biochem. Biotechnol. 15590291. 176 (6), 1551–1563. Available from: http://www.springer.com/humana+press/journal/12010. doi:10.1007/s12010-015-1676-1.

Reis-Mansur, M.C.P.P., Cardoso-Rurr, J.S., Silva, J.V.M.A., de Souza, G.R., Cardoso, V.D.S., Mansoldo, F.R.P., Pinheiro, Y., Schultz, J., Balottin, L.B.L., da Silva, A.J.R., Lage, C., Dos Santos, E.P., Rosado, A.S., Vermelho, A. B., 2019. Carotenoids from UV-resistant antarctic Microbacterium sp. LEMMJ01. Sci. Rep. 20452322. 9 (1). Available from: https://www.nature.com/srep/index.html. doi:10.1038/s41598-019-45840-6.

Ren, C., Liu, Z., Wang, X., Qin, S., 2022. The seaweed holobiont: from microecology to biotechnological applications. Microb. Biotechnol. 1751-7915. 15 (3), 738–754. doi:10.1111/1751-7915.14014.

Sajjan, S.S., Anjaneya, O., Kulkarni, G.B., Nayak, A.S., Mashetty, S.B., Karegoudar, T.B., 2013. Properties and functions of melanin pigment from *Klebsiella* sp. J. Microbiol. Biotechnol. 1598642X. 41 (1), 60–69. doi:10.4014/kjmb.1210.10002.

Sánchez, C., Braña, A.F., Méndez, C., Salas, J.A., 2006. Reevaluation of the violacein biosynthetic pathway and its relationship to indolocarbazole biosynthesis. ChemBioChem. 1439-4227. 7 (8), 1231–1240. doi:10.1002/cbic.200600029.

Santhanam, R., 2020. Biology and Ecology of Venomous Marine Cnidarians Biology and Ecology of Venomous Marine Cnidarians. Springer Singapore, India, pp. 1–343. Available from https://link.springer.com/book/10.1007/978-981-15-1603-0. doi:10.1007/978-981-15-1603-0.

Shailendra, P.S., Sunita, K., Rajesh, P.R., Kanchan, L.S., Richa, Rajeshwar, P.S., 2010. Photoprotective and biotechnological potentials of cyanobacterial sheath pigment, scytonemin. Afr. J. Biotechnol. 1684-5315. 9 (5), 581–588. doi:10.5897/AJB09.019.

Shalaby, A., 2013. Antioxidant compounds, assays of determination and mode of action. Afr. J. Pharm. Pharmacol. 1996-0816. 7 (10), 528–539. doi:10.5897/AJPP2013.3474.

Shimidzu, N., Goto, M., Miki, W., 1996. Carotenoids as singlet oxygen quenchers in marine organisms. Fish. Sci. 09199268. 62 (1), 134–137. Available from: http://www.springer.com/life+sci/zoology/journal/12562?detailsPage=contactPublishing. doi:10.2331/fishsci.62.134.

Sies, H., Stahl, W., 2004. Carotenoids and UV protection. Photochem. Photobiol. Sci. 14749092. 3 (8), 749–752. doi:10.1039/b316082c.

Singh, R.P., Kumari, P., Reddy, C.R.K., 2015. Antimicrobial compounds from seaweeds-associated bacteria and fungi. Appl. Microbiol. Biotechnol. 14320614. 99 (4), 1571–1586. Available from: https://www.link.springer.de/link/service/journals/00253/index.htm. doi:10.1007/s00253-014-6334-y.

Singh, R.P., Reddy, C.R.K., 2014. Seaweed-microbial interactions: key functions of seaweed-associated bacteria. FEMS Microbiol. Ecol. 15746941. 88 (2), 213–230. Available from: http://www.wiley.com/bw/journal.asp?ref=0168-6496. doi:10.1111/1574-6941.12297.

Soliev, A.B., Hosokawa, K., Enomoto, K., 2011. Bioactive pigments from marine bacteria: applications and physiological roles. Evid. Based Complement. Alternat. Med. 17414288. 2011. doi:10.1155/2011/670349.

Stafsnes, M.H., Josefsen, K.D., Kildahl-Andersen, G., Valla, S., Ellingsen, T.E., Bruheim, P., 2010. Isolation and characterization of marine pigmented bacteria from Norwegian coastal waters and screening for carotenoids with UVA-blue light absorbing properties. J. Microbiol. 1225-8873. 48 (1), 16–23. doi:10.1007/s12275-009-0118-6.

Sterling, A.R., Holland, L.Z., Bundy, R.M., Burns, S.M., Buck, K.N., Chappell, P.D., Jenkins, B.D., 2023. Potential interactions between diatoms and bacteria are shaped by trace element gradients in the Southern Ocean. Front. Mar. Sci. 2296-7745. 9. doi:10.3389/fmars.2022.876830.

Suresh, M., Renugadevi, B., Brammavidhya, S., Iyapparaj, P., Anantharaman, P., 2015. Antibacterial activity of red pigment produced by *Halolactibacillus alkaliphilus* MSRD1—an isolate from seaweed. Appl. Biochem. Biotechnol. 0273-2289. 176 (1), 185–195. doi:10.1007/s12010-015-1566-6.

Suryawanshi, R.K., Patil, C.D., Borase, H.P., Narkhede, C.P., Stevenson, A., Hallsworth, J.E., Patil, S.V., 2015. Towards an understanding of bacterial metabolites prodigiosin and violacein and their potential for use in commercial sunscreens. Int. J. Cosmet. Sci. 0142-5463. 37 (1), 98–107. doi:10.1111/ics.12175.

Thompson, J.R., Polz, M.F., 2014. Dynamics of vibrio populations and their role in environmental nutrient cycling. Am. Soc. Microbiol. 190–203. doi:10.1128/9781555815714.ch13.

Velmurugan, P., Venil, C.K., Ravi, A.V., Dufossé, L., 2020. Marine bacteria is the cell factory to produce bioactive pigments: a prospective pigment source in the ocean. Front. Sustain. Food Syst. 2571-581X. 4. doi:10.3389/fsufs.2020.589655.

Venil, C.K., Zakaria, Z.A., Ahmad, W.A., 2013. Bacterial pigments and their applications. Process Biochem. 13595113. 48 (7), 1065–1079. doi:10.1016/j.procbio.2013.06.006.

Weinberger, F., Beltran, J., Correa, J.A., Lion, U., Pohnert, G., Kumar, N., Steinberg, P., Kloareg, B., Potin, P., 2007. Spore release in *Acrochaetium* sp. (Rhodophyta) is bacterially controlled. J. Phycol. 00223646. 43 (2), 235–241. doi:10.1111/j.1529-8817.2007.00329.x.

Williamson, C.E., Zepp, R.G., Lucas, R.M., Madronich, S., Austin, A.T., Ballaré, C.L., Norval, M., Sulzberger, B., Bais, A.F., McKenzie, R.L., Robinson, S.A., Häder, D.P., Paul, N.D., Bornman, J.F., 2014. Solar ultraviolet radiation in a changing climate. Nat. Clim. Change. 17586798. 4 (6), 434–441. Available from: http://www.nature.com/nclimate/index.html. doi:10.1038/nclimate2225.

Zinser, E.R., 2018. The microbial contribution to reactive oxygen species dynamics in marine ecosystems. Environ. Microbiol. Rep. 17582229. 10 (4), 412–427. Available from: http://onlinelibrary.wiley.com/journal/10.1111/(ISSN)1758-2229. doi:10.1111/1758-2229.12626.

CHAPTER

10

Bioactive phenolic compounds identified from marine organisms

Dineth Nagahwatta and Nisansala Madushani Liyanage

Department of Medicine, University of Alberta, Edmonton, AB, Canada

10.1 Introduction

Marine polyphenolic compounds (MPCs) are a diverse group of secondary metabolites characterized by the presence of one or more hydroxyl groups attached to an aromatic ring. These compounds are widely distributed in nature and play essential roles in various biological processes, including defense mechanisms against environmental stressors such as UV radiation, pathogen invasion, and oxidative stress. While terrestrial plants have been extensively studied as sources of phenolic compounds, marine organisms have gained increasing attention due to their unique structural diversity and potent biological activities (Nagahawatta et al., 2024a).

The marine environment is one of the richest sources of novel bioactive compounds, offering a largely untapped reservoir of chemical diversity. Marine ecosystems, including deep-sea habitats, coral reefs, and coastal regions, provide unique conditions that drive the biosynthesis of structurally distinct and biologically active phenolic compounds. These compounds have been identified in a wide range of marine organisms, including macroalgae (seaweeds), microalgae, sponges, corals, mollusks, and marine microorganisms (Ghallab et al., 2024). The structural complexity of marine-derived phenolics is often influenced by the extreme environmental conditions in which these organisms thrive, such as high salinity, varying pressure, low temperatures, and limited nutrients. As a result, marine phenolic compounds often exhibit enhanced bioactivity compared to their terrestrial counterparts (Greer et al., 2020). Among marine-derived phenolics, phlorotannins from brown algae, bromophenols from sponges, and flavonoids from marine microorganisms have demonstrated significant pharmacological properties (Martins et al., 2019). These compounds have been widely studied for their antioxidant, antiinflammatory, antimicrobial, and anticancer activities. Moreover, the increasing demand for natural bioactive compounds in functional foods and dietary supplements has further driven research into marine-derived phenolics (Prates, 2025).

Harvesting the Sea
https://doi.org/10.1016/B978-0-443-29190-6.00025-8

Despite their promising potential, several challenges hinder the large-scale exploitation of marine phenolic compounds. Issues related to the extraction and purification of these compounds, low natural yields, and the sustainable harvesting of marine resources need to be addressed (Chandrasekhar et al., 2024). Advances in biotechnology, including microbial fermentation, synthetic biology, and metabolic engineering, offer promising solutions to overcome these limitations. In addition, further research is required to elucidate the mechanisms of action, pharmacokinetics, and safety profiles of marine phenolic compounds to facilitate their clinical translation.

This chapter aims to provide a comprehensive overview of the bioactive phenolic compounds identified from marine organisms. It explores their sources, structural diversity, and pharmacological properties, highlighting their potential applications in various industries. Furthermore, challenges associated with the extraction, production, and commercialization of marine phenolics are discussed, along with future perspectives for their development as therapeutic agents.

10.2 Sources of marine phenolic compounds

10.2.1 Chemical classification of marine phenolic compounds

MPCs possess unique structural characteristics, such as halogenation, sulfation, and complex dimerization, which are rarely observed in terrestrial phenolics. These structural modifications enhance their stability, bioavailability, and bioactivity in extreme marine environments. According to Nagahawatta et al. (2024b), the major classification of MPCs is summarized in Fig. 10.1 (Nagahawatta et al., 2024a).

Marine flavonoids are a group of polyphenolic compounds produced by various marine organisms such as algae, seagrasses, and mangroves. Although they share structural similarities with terrestrial flavonoids, they are adapted to the unique conditions of the marine environment (Martins et al., 2019). The chemical properties and characteristics of these flavonoids are influenced by several factors. The core structure of marine flavonoids consists of two

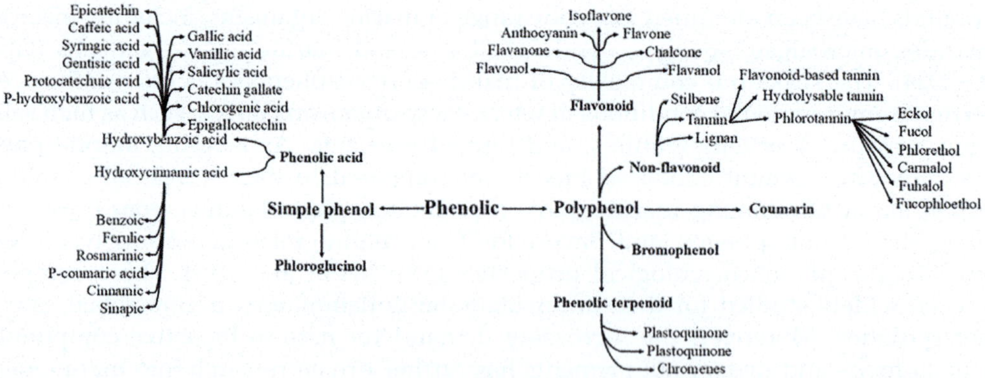

FIG. 10.1 Chemical classification of marine phenolic compounds.

phenyl rings (A and B) connected by a three-carbon bridge, forming a heterocyclic ring known as the C6–C3–C6 structure. The A ring typically features a benzoyl moiety, while the B ring is usually a cinnamoyl moiety. Variations in the structure and degree of hydroxylation on these rings, particularly the B ring, play a crucial role in the antioxidant potential of flavonoids. A higher number of hydroxyl groups enhances the capacity to scavenge free radicals (Martins et al., 2019). Many marine flavonoids undergo glycosylation, where sugar molecules attach to the flavonoid structure. This modification affects their solubility, stability, and bioavailability. In addition, methylation of hydroxyl groups can occur, increasing the lipophilicity of the flavonoids and influencing their interaction with biological membranes. Unique to marine flavonoids are uncommon substitutions, such as sulfate, chlorine, and amino groups, which can significantly impact their biological activities (Chen et al., 2025). The solubility of flavonoids is influenced by the extent of glycosylation and methylation. Glycosylated flavonoids are more water-soluble, while methylated flavonoids are more lipophilic. Marine flavonoids are also sensitive to environmental factors like light, temperature, and pH. Glycosylation can improve their stability against oxidative degradation (Del Mar Ribas-taberner et al., 2025). The antioxidant activity of flavonoids primarily stems from their ability to donate hydrogen atoms or electrons to neutralize free radicals. The presence of an ortho-dihydroxy configuration on the B ring is particularly important for this activity. Furthermore, flavonoids can interact with various biological targets, such as enzymes, receptors, and nucleic acids, contributing to their antiinflammatory, anticancer, and cardioprotective effects (Martins et al., 2019).

The bioavailability of marine flavonoids is influenced by their chemical structure, metabolism, and interaction with gut microbiota. Glycosylation can enhance absorption in the human digestive system. Marine flavonoids are categorized into six major groups, each with distinct chemical characteristics. According to Nagahawatta et al. (2024b), the major groups of flavonoids are summarized in Table 10.1 (Nagahawatta et al., 2024a).

Delving into the extensive chemical diversity of marine polyphenolic compounds (MPCs) reveals that nonflavonoid constituents form a remarkable and underexplored segment of

TABLE 10.1 Major groups of marine flavonoids.

Group	Structure	Structural characterization	Common source	Primary bioactivity
Flavones		Double bond between C2 and C3, no hydroxyl group at C3	Marine plants	Antioxidant
Flavonols		Hydroxyl group at C3	Seaweeds	AntiinflammatoryCardioprotective
Flavanones		No double bond between C2 and C3 (saturated C chain)	Various marine algae	AntioxidantAntiinflammatory
Isoflavones		B ring attached at C3 position	Marine fungi and bacteria	EstrogenicAnticancer
Anthocyanidins		Glycosylated forms of anthocyanins	Some marine plants	Pigmentation, antioxidant
Flavanols		Catechins, often found as part of larger tannin molecules	Green algae, seaweeds	Antioxidant, anticarcinogenic

marine natural products. These compounds, including phlorotannins, marine tannins, and sulfated polyphenols, are structurally distinct from flavonoids and exhibit unique bioactivities. These bioactivities are vital for the ecological roles they play in marine environments and their potential therapeutic applications (Sadeghi et al., 2024).

Phenolic acids are simple phenolic compounds that often serve as precursors for more complex polyphenols. They typically feature one carboxylic acid group attached to an aromatic ring with one or more hydroxyl groups. Examples include hydroxybenzoic and hydroxycinnamic acids, which are known for their antioxidant properties. With a C6–C1 structure, they effectively neutralize free radicals through radical scavenging mechanisms (Jimenez-Lopez et al., 2021).

Tannins are high-molecular-weight polyphenolic compounds that can be classified into hydrolyzable and condensed tannins. Known for their astringent properties, tannins have the ability to bind and precipitate proteins (Saini et al., 2024). Stilbenes are characterized by a 1,2-diphenylethylene nucleus. Compounds like resveratrol, produced by plants in response to stress, have demonstrated antiinflammatory and cardioprotective effects. With a C6–C2–C6 structure, stilbenes are less common than other polyphenols but are known for their potential health benefits, including antiinflammatory and anticancer activities (De Luca and Lauritano, 2020).

Lignans are formed through the dimerization of two phenylpropanoid units, resulting in a C6–C3–C6 structure. Examples such as secoisolariciresinol and matairesinol offer health benefits, including anticancer, antiinflammatory, and antioxidant properties (Lu et al., 2020). Overall, marine polyphenols, including nonflavonoids, possess unique structures that contribute to their biological activities. Their structural complexity and diversity arise from the diverse environmental conditions in which marine organisms thrive, leading to the production of distinctive polyphenolic compounds (Matulja et al., 2022). For instance, marine algae produce phlorotannins, a type of nonflavonoid polyphenol synthesized via the acetate–malonate pathway, also known as the polyketide pathway. Phlorotannins, formed through the polymerization of phloroglucinol units, have been observed to possess antiinflammatory and antioxidative properties. According to Nagahawatta et al. (2024b), major nonflavonoid groups are summarized in Table 10.2.

TABLE 10.2 Major groups of marine nonflavonoids.

Group	Structural characterization	Common source	Primary bioactivity
Phenolic acids	C6–C1 structure, presence of a carboxylic acid group and hydroxyl groups on an aromatic ring	Seaweeds, marine fungi	Antioxidant, radical scavenging
Tannins	High molecular weight, ability to bind and precipitate proteins	Marine plants, algae	Antioxidant
Stilbenes	C6–C2–C6 structure, 1,2-diphenylethylene nucleus	Mangroves, marine sponges	Antiinflammatory, cardioprotective
Lignans	C6–C3–C6 structure, dimerization of two phenylpropanoid units	Marine algae, seagrasses	Antioxidant, anticancer
Phlorotannins	Polymers of phloroglucinol units, varying degrees of polymerization	Brown algae	Antioxidant, antiinflammatory

10.3 Biosynthesis of marine phenolic compounds

The biosynthesis of MPCs involves complex pathways that are similar to those found in terrestrial plants but with some unique adaptations to the marine environment. Three main pathways are involved in the production of these compounds: the shikimate pathway, the pentose phosphate pathway (PPP), and the phenylpropanoid pathway (Dewick, 1984).

10.4 Shikimate pathway

The shikimate pathway is a crucial metabolic route found in microorganisms, plants, and some protozoans, but notably absent in animals. This seven-step pathway links carbohydrate metabolism to the biosynthesis of aromatic compounds, including the aromatic amino acids phenylalanine, tyrosine, and tryptophan, as well as numerous secondary metabolites. The pathway converts two primary metabolites, phosphoenolpyruvate (PEP) from glycolysis and erythrose 4-phosphate (E4-P) from the PPP, into chorismate, the branch-point compound for aromatic amino acid synthesis and many other aromatic secondary metabolites (Shende et al., 2024).

The shikimate pathway involves seven key enzymatic reactions. The process begins with DAHP synthase (DHS) catalyzing the condensation of PEP and E4-P. This is followed by reactions catalyzed by 3-dehydroquinate synthase, 3-dehydroquinate dehydratase, shikimate dehydrogenase, and shikimate kinase, which phosphorylates shikimate to shikimate-3-phosphate. 5-enolpyruvylshikimate-3-phosphate (EPSP) synthase then couples shikimate-3-phosphate with PEP, and finally, chorismate synthase produces the pathway's end product, chorismate (Maeda and Dudareva, 2012).

Regulation of the shikimate pathway differs between microorganisms and plants. In microorganisms, the pathway is tightly controlled through feedback inhibition by aromatic amino acids and repression of the first enzyme, DHS (Weaver and Herrmann, 1997; Jiang and Zhang, 2016). However, plant regulation is more complex and less understood. No clear physiological feedback inhibitor has been identified in plants, and regulation likely occurs primarily at the genetic level. Isoenzymic forms of pathway enzymes in plants show variable expression based on environmental conditions and plant organs (Tzin and Galili, 2010). In plants, the shikimate pathway is localized in plastids. cDNAs of higher plants encode proteins with amino-terminal signal sequences for plastid import, and plastids are believed to be the exclusive site for chorismate biosynthesis. This subcellular localization is crucial for the pathway's function and regulation in plant metabolism (Moon et al., 2008).

The importance of the shikimate pathway in plant metabolism cannot be overstated. It directs bulk carbon flow toward the biosynthesis of aromatic amino acids and serves as a precursor for numerous aromatic phytochemicals. These compounds play critical roles in plant physiology and adaptation, and many are widely used as nutraceuticals, pharmaceuticals, and biomaterials. The pathway's intermediates also act as branch points for other metabolic pathways, further emphasizing its central role in plant biochemistry (Moon et al., 2008).

The absence of the shikimate pathway in animals highlights its evolutionary importance and makes it a potential target for antimicrobial and herbicidal compounds. For instance, glyphosate, a widely used herbicide, specifically inhibits EPSP synthase in this pathway. This unique characteristic of the shikimate pathway has significant implications for biotechnology, including the development of glyphosate-tolerant transgenic plants and the potential for metabolic engineering to enhance the production of aromatic compounds (Goyal, 2018).

10.5 The pentose phosphate pathway

The PPP is a fundamental metabolic route that operates parallel to glycolysis, primarily in the cytosol of cells. This pathway serves crucial anabolic functions by generating NADPH, pentoses (five-carbon sugars), and ribose-5-phosphate, a vital precursor for nucleotide synthesis. The PPP consists of two distinct phases: the oxidative phase, which converts glucose-6-phosphate to ribulose-5-phosphate while generating NADPH, and the nonoxidative phase, which interconverts various sugar phosphates to produce ribose-5-phosphate and other metabolic intermediates (Shetty and Wahlqvist, 2004).

Key enzymes in the oxidative phase include glucose-6-phosphate dehydrogenase (G6PDH), which catalyzes the first and rate-limiting step, followed by 6-phosphogluconolactonase and 6-phosphogluconate dehydrogenase. The nonoxidative phase involves enzymes such as ribose-5-phosphate isomerase, transketolase, and transaldolase. These enzymes work in concert to facilitate the complex sugar-phosphate interconversions characteristic of this phase (Marchiosi et al., 2020).

Regulation of the PPP is primarily achieved at the level of G6PDH, the pathway's rate-limiting enzyme. The NADPH/NADP+ratio plays a crucial role, with high NADPH levels inhibiting G6PDH and low levels promoting pathway activity. Hormonal control also influences the pathway, with insulin stimulating G6PDH activity, while glucagon inhibits it. Transcriptional regulation, particularly through factors like NRF2, controls enzyme expression, especially under conditions of oxidative stress. In addition, substrate availability, specifically higher glucose-6-phosphate levels, can stimulate the pathway's activity (Marchiosi et al., 2020). In terms of subcellular localization, the PPP primarily occurs in the cytoplasm of most cells. However, in plants, a significant portion of the pathway's steps takes place within plastids, highlighting the pathway's adaptability to different cellular environments and metabolic needs.

The importance of the PPP in cellular metabolism cannot be overstated. It plays a critical role in maintaining redox balance by generating NADPH for biosynthetic reactions and antioxidant defense mechanisms. The pathway is essential for nucleotide synthesis, providing ribose-5-phosphate for nucleic acid formation. It also contributes to amino acid synthesis by producing erythrose-4-phosphate, a precursor for aromatic amino acids. The PPP's ability to allow interconversion between different sugar phosphates is crucial for carbon homeostasis within the cell. In rapidly proliferating cells, such as cancer cells exhibiting the Warburg effect, the PPP supports accelerated growth by providing necessary building blocks. Moreover, the pathway is integral to the cellular response to oxidative stress, playing a vital role in maintaining reduced glutathione levels (Stincone et al., 2015).

The versatility and significance of the PPP in cellular metabolism make it a critical area of study for understanding various physiological and pathological processes. Its involvement in cancer metabolism, oxidative stress responses, and metabolic disorders underscores its importance in both basic science and medical research. As our understanding of this pathway continues to evolve, it promises to provide valuable insights into cellular function and potential therapeutic targets for a range of conditions.

10.5.1 The phenylpropanoid pathway

The phenylpropanoid pathway is a vital metabolic route in plants, serving as a rich source of diverse secondary metabolites. This pathway begins with the amino acid phenylalanine and leads to the biosynthesis of numerous important compounds, including lignin, flavonoids, coumarins, and lignans. The pathway consists of three main steps that redirect carbon flow from primary metabolism: the deamination of phenylalanine by phenylalanine ammonia-lyase (PAL) to form cinnamic acid, the hydroxylation of cinnamic acid by cinnamic acid 4-hydroxylase (C4H) to produce p-coumaric acid, and the activation of p-coumaric acid to p-coumaroyl-CoA by 4-coumarate-CoA ligase (4CL). These initial steps constitute the general phenylpropanoid pathway, which is ubiquitous in plants. From p-coumaroyl-CoA, the pathway branches into various specialized routes, leading to the synthesis of different classes of phenylpropanoids (Vogt, 2010).

Regulation of the phenylpropanoid pathway occurs at multiple levels, including transcriptional regulation (with MYB transcription factors playing a crucial role), posttranscriptional and posttranslational modifications, metabolon formation (where enzymes in the pathway may organize into complexes to optimize biosynthesis and regulate metabolic flux), and transport mechanisms (involving vesicle trafficking, glutathione S-transferase-supported transport, and membrane transporters such as MATE and ABC proteins). This multi-layered regulation allows plants to finely tune their phenylpropanoid metabolism in response to various environmental stimuli and developmental cues (Ando et al., 2013). The phenylpropanoid pathway plays vital roles in plant development, stress tolerance, and defense against pathogens. Its products contribute to various physiological functions, including UV protection, antioxidant activity, antimicrobial defense, structural support through lignin biosynthesis, and pollinator attraction via flavonoids and anthocyanins. The pathway is highly responsive to environmental stimuli, allowing plants to adapt to changing conditions and threats. Recent research has focused on re-engineering the phenylpropanoid pathway to enhance the production of valuable compounds for industrial applications, such as pharmaceuticals, colorants, and food nutrients (Singh et al., 2023).

Understanding the complex regulation and diverse functions of the phenylpropanoid pathway continues to be an active area of research, with significant implications for improving crop resilience, nutritional value, and the production of high-value plant-derived compounds. As our knowledge of this pathway expands, it offers exciting possibilities for biotechnological applications in agriculture, medicine, and industry, highlighting the importance of continued investigation into this fundamental aspect of plant metabolism.

10.6 Extraction, isolation, and purification methods

10.6.1 Isolation strategies

Extracting and purifying these bioactive compounds effectively is crucial for their application in pharmaceuticals, nutraceuticals, and cosmeceuticals. These industries rely on high-purity polyphenolic compounds due to their remarkable therapeutic potential, including antiaging properties, cardiovascular protection, and immune system enhancement. In addition, marine-derived polyphenols have shown promising applications in functional foods, dietary supplements, and dermatological formulations due to their superior bioavailability and unique mechanisms of action. However, their commercial utilization faces challenges such as low natural abundance, complex extraction procedures, and potential structural degradation during processing (Stengel and Connan, 2015). Addressing these challenges requires the development of optimized extraction and purification methodologies that preserve bioactivity while ensuring scalability and sustainability. The exploration of eco-friendly and high-efficiency techniques is essential to unlock the full potential of marine polyphenols and facilitate their widespread industrial applications.

10.7 Conventional extraction methods

Conventional extraction methods for marine polyphenols primarily involve solvent-based techniques that rely on the principles of solubility and diffusion. These methods, though widely used, often face limitations such as low efficiency, extended extraction times, and potential degradation of sensitive bioactive compounds. The most common conventional extraction techniques include:

10.7.1 Maceration

Maceration is one of the simplest and oldest extraction techniques. It involves soaking marine biomass in a suitable solvent, allowing the diffusion of polyphenols into the solvent over time. This method is inexpensive and does not require sophisticated equipment; however, it is time-consuming and may result in the coextraction of unwanted compounds (Carreira-Casais et al., 2021).

10.7.2 Soxhlet extraction

Soxhlet extraction is a widely used technique for marine polyphenols, providing continuous solvent percolation through the biomass. This method utilizes a Soxhlet apparatus, which repeatedly circulates the solvent, ensuring a high degree of compound recovery. However, the prolonged exposure to heat and solvent can degrade thermolabile polyphenols and increase energy consumption (Catarino et al., 2019).

10.7.3 Decoction

Decoction involves boiling the marine sample in water or an aqueous solvent mixture to extract polyphenols. This method is commonly used for water-soluble polyphenols but may lead to hydrolysis or degradation of heat-sensitive compounds (Saudagar and Saokar, 2019).

10.7.4 Percolation

Percolation is an advanced version of maceration, where the solvent is continuously passed through the marine biomass in a controlled manner. This technique enhances extraction efficiency by maintaining a dynamic solvent-to-sample interaction but requires precise optimization of flow rates to prevent excessive dilution (Maheswari and Babu, 2022).

10.8 Advanced extraction methods

To overcome the limitations of conventional methods, several advanced extraction techniques have been developed to enhance efficiency, reduce solvent use, and minimize environmental impact. These methods optimize extraction parameters such as temperature, pressure, and solvent selection to improve the yield and purity of marine polyphenolic compounds.

10.8.1 Ultrasound-assisted extraction

Ultrasound-assisted extraction (UAE) utilizes high-frequency sound waves to create cavitation in the solvent, leading to cell wall disruption and enhanced polyphenol release. This method offers several advantages, including shorter extraction time, lower solvent consumption, and improved efficiency. UAE is particularly effective for extracting polyphenols from marine algae and sponges, as it enhances mass transfer while preserving the structural integrity of the compounds (Chemat et al., 2011).

10.8.2 Microwave-assisted extraction

Microwave-assisted extraction (MAE) applies electromagnetic radiation to rapidly heat the solvent and biomass, accelerating the extraction process. MAE is advantageous due to its energy efficiency, rapid heating, and ability to extract thermolabile polyphenols without significant degradation. The selection of an appropriate solvent and optimization of microwave power are crucial to achieving high yields and purity (Yuan et al., 2018).

10.8.3 Supercritical fluid extraction

Supercritical fluid extraction (SFE) employs supercritical CO_2 as a solvent to extract polyphenolic compounds. This technique is environmentally friendly, as CO_2 is nontoxic, nonflammable, and easily removed postextraction. SFE provides high selectivity and efficiency while avoiding the use of organic solvents. The addition of co-solvents, such as ethanol, enhances the solubility of polar polyphenols, improving extraction yields (Liu et al., 2011).

10.8.4 Enzyme-assisted extraction

Enzyme-assisted extraction (EAE) utilizes specific enzymes, such as cellulases and proteases, to break down the cell walls of marine organisms, facilitating polyphenol release. This method enhances extraction efficiency while maintaining the bioactivity of polyphenols. EAE is particularly useful for extracting polyphenols from complex marine matrices, such as brown algae, where polysaccharides form barriers to solvent penetration (Gligor et al., 2019).

10.8.5 Pressurized liquid extraction

Pressurized liquid extraction (PLE), also known as accelerated solvent extraction, involves applying high pressure and temperature to improve the solubility and mass transfer of polyphenols. This method reduces solvent consumption and extraction time while increasing yield. PLE is particularly suitable for marine-derived polyphenols with moderate thermal stability (Herrero et al., 2015).

10.8.6 Ionic liquid-based extraction

Ionic liquids (ILs) are nonvolatile, thermally stable solvents that can selectively dissolve marine polyphenols. IL-based extraction (ILE) minimizes solvent toxicity and improves extraction efficiency. By tuning the properties of ILs, researchers can enhance selectivity and yield for targeted polyphenolic compounds (Ventura et al., 2017).

10.8.7 Deep eutectic solvent extraction

Deep eutectic solvents (DESs) are an emerging class of green solvents composed of natural components, such as organic acids and amino acids. These solvents provide a tunable and biodegradable alternative to conventional organic solvents. DES extraction has shown promising results for marine polyphenols due to its enhanced solubility and eco-friendly nature (Amador et al., 2025).

10.8.8 Comparative analysis of extraction methods

Selecting an appropriate extraction method for marine polyphenolic compounds is crucial for maximizing yield, preserving bioactivity, and ensuring sustainability. Each extraction technique has distinct advantages and limitations based on factors such as extraction efficiency, environmental impact, selectivity, energy consumption, and scalability.

Conventional methods like maceration and decoction consume large amounts of solvents but yield lower efficiency, whereas advanced methods such as UAE, MAE, and SFE offer higher yields with lower solvent consumption (Getachew et al., 2020). Traditional

TABLE 10.3 Comparative analysis of the effectiveness of extraction methods.

Extraction method	Efficiency	Solvent consumption	Environmental impact	Selectivity	Scalability	Cost
Maceration	Low	High	Moderate	Low	High	Low
Soxhlet extraction	High	High	Moderate	Moderate	Moderate	Medium
Decoction	Moderate	High	High	Low	High	Low
Percolation	Moderate	Moderate	Moderate	Moderate	High	Medium
Ultrasound-assisted (UAE)	High	Low	Low	High	High	Medium
Microwave-assisted (MAE)	High	Low	Low	High	High	Medium
Supercritical fluid (SFE)	Very High	Very Low	Very low	Very high	High	High
Enzyme-assisted (EAE)	High	Low	Very low	High	High	Medium
Pressurized liquid (PLE)	High	Low	Low	High	High	Medium
Ionic liquid (ILE)	Very High	Very Low	Low	Very High	High	High
Deep eutectic solvent (DES)	High	Very Low	Very low	High	High	Medium

solvent-based techniques pose environmental concerns due to organic solvent disposal, whereas SFE, EAE, and DES extraction are eco-friendly alternatives. Advanced methods such as ILE and SFE offer superior selectivity, isolating specific polyphenolic compounds while reducing impurities (Chemat et al., 2012). Methods like percolation, UAE, and MAE are highly scalable for industrial applications, while SFE and ILE, despite their high efficiency, require specialized equipment and incur higher costs. Conventional methods are low-cost but less efficient, while techniques like SFE and ILE provide high purity but involve higher operational costs (Ang et al., 2022; Azmir et al., 2013). The choice of extraction method depends on the specific characteristics of marine polyphenols, intended application, and industrial feasibility. Hybrid approaches, combining different green techniques, may offer optimal solutions for sustainable and efficient extraction. A comparative analysis of conventional and advanced extraction methods is presented in Table 10.3.

10.9 Marine macroalgae as a source of marine phenolic compounds

Marine macroalgae, commonly known as seaweeds, are photosynthetic organisms classified into three major groups: brown algae (Phaeophyceae), red algae (Rhodophyceae), and green algae (Chlorophyceae). These organisms produce a wide range of bioactive

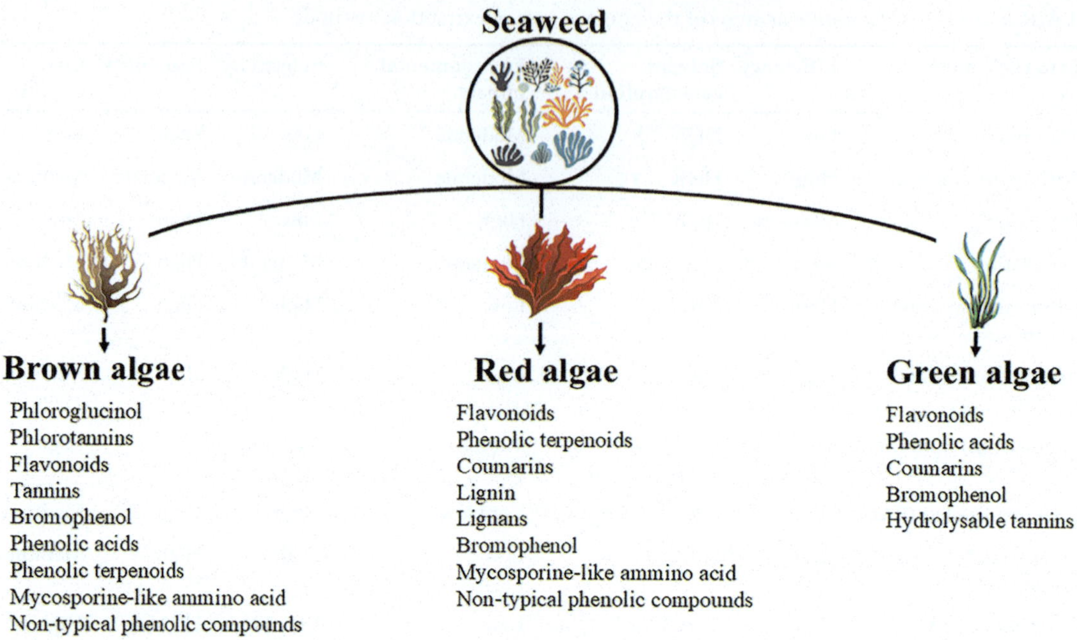

FIG. 10.2 Major phenolic compounds found in macroalgae species.

compounds, including phenolic compounds, to adapt to the harsh marine environment (Nagahawatta et al., 2024b). Marine algae-derived phenolics exhibit potent antioxidant, antiinflammatory, and antimicrobial properties, making them valuable candidates for pharmaceutical and nutraceutical applications. Fig. 10.2 shows the major MPCs reported from macroalgae species.

Brown algae are particularly notable for their high content of phlorotannins, a unique group of polyphenolic compounds formed by the polymerization of phloroglucinol units (1,3,5-trihydroxybenzene) and serving as primary defense compounds against herbivory, UV stress, and microbial infections. These compounds have been widely studied for their antioxidant, antimicrobial, antiinflammatory, and neuroprotective properties (Imbs and Zvyagintseva, 2018). Species such as *Ecklonia cava*, *Fucus vesiculosus*, and *Sargassum* spp. are known to contain high levels of phlorotannins, which contribute to their strong free radical-scavenging activity (Wang et al., 2012; Yong et al., 2009). Phlorotannins have also shown promising effects in reducing inflammation, inhibiting bacterial growth, and protecting neuronal cells from oxidative stress. These phlorotannins exhibit remarkable structural diversity, which is reflected in their various subclasses, including fucophlorethols, eckol-type phlorotannins, carmalols, fuhalols, and branched phlorotannins. As mentioned earlier, these compounds are polymerized derivatives of phloroglucinol and are primarily linked through aryl-aryl, ether, or phenyl ether bonds, contributing to their structural complexity and bioactivity (Fernando et al., 2022).

Fuhalols are a subclass of hydroxylated phlorotannins with additional –OH groups and ether bonds. These compounds exhibit enhanced free radical scavenging capacity, making them particularly effective in protecting algae from extreme environmental conditions, such as high UV exposure and oxidative stress. Their strong antioxidant activity suggests potential applications in various fields, though specific cosmetic applications require further validation (Catarino et al., 2022). Diphlorethohydroxycarmalol, found in *Ishige okamurae*, is a specific phlorotannin compound that has shown promise in enzymatic inhibition studies, particularly against acetylcholinesterase. This highlights the potential of certain phlorotannins in addressing neurodegenerative disorders (Heo and Jeon, 2009). Branched phlorotannins, such as fucofurotriphlorethol found in species like *Fucus vesiculosus*, demonstrate complex structural arrangements that influence their solubility and bioactivity. The degree of branching and polymerization can affect properties such as antioxidant capacity, with higher molecular weight compounds often showing increased activity up to a certain point (Catarino, 2021). The structural diversity of phlorotannins significantly influences their bioactivities, though the relationship is complex and not always directly correlated. Factors such as the degree of polymerization, specific linkage types, and hydroxylation patterns all contribute to the compounds' properties. This structural variety highlights the potential of brown algae as a rich source of bioactive compounds for pharmaceutical, nutraceutical, and cosmeceutical industries, warranting continued research to fully understand and utilize these valuable natural products.

Red algae produce bromophenols, flavonoids, and phenolic acids, which contribute to their antimicrobial and cytotoxic activities. Species such as *Gracilaria* and *Gelidium* have been reported to contain bromophenol derivatives with potent antibacterial and antifungal properties. These compounds have been tested against various human pathogens and have demonstrated significant inhibitory effects. Some bromophenols have also been identified as potential anticancer agents due to their ability to induce apoptosis in tumor cells (Aziz et al., 2021).

Green algae, including *Ulva* and *Codium* species, contain flavonoids, tannins, and phenolic acids. These compounds are known for their strong antioxidant and antiinflammatory activities, making them valuable for potential therapeutic applications. In addition to their free radical-scavenging properties, green algal phenolics have been shown to promote wound healing by stimulating fibroblast activity, which is crucial in skin regeneration and repair (El-Bilawy et al., 2022).

10.10 Seagrass as marine phenolic compounds

Seagrasses are marine flowering plants that grow in shallow coastal waters, forming extensive underwater meadows. They play a crucial role in marine ecosystems by stabilizing sediments, providing habitat for marine organisms, and serving as a primary food source for many herbivorous species (Orth et al., 2006). In addition to their ecological significance, seagrasses have been recognized as a valuable source of marine phenolic compounds (MPCs). These bioactive compounds contribute to the survival and adaptation of seagrasses in harsh marine environments, providing defense against herbivory, UV radiation, microbial infections, and oxidative stress. Due to their potent biological activities, seagrass-derived phenolic compounds have gained attention for their potential applications in pharmaceuticals, nutraceuticals, and cosmetics.

Seagrasses synthesize a diverse range of phenolic compounds, including flavonoids, tannins, phenolic acids, and lignins. These compounds act as natural antioxidants and antimicrobial agents, offering protection against various environmental stressors.

10.10.1 Flavonoids

Flavonoids are among the most abundant phenolic compounds found in seagrasses. These compounds have been identified in several seagrass species, including *Zostera marina*, *Posidonia oceanica*, *Enhalus acoroides*, *Halophila spinulosa*, and *Thalassia testudinum*. Common flavonoids in seagrasses include luteolin, apigenin, and chrysoeriol, as well as their glycosylated and sulfated derivatives. Sulfated flavonoids are particularly prevalent in seagrasses, with sulfated flavones characterized in genera such as *Zostera*, *Thalassia*, *Halophila*, *Enhalus*, and *Phyllospadix* (Orth et al., 2006; Botes et al., 2025).

These molecules exhibit strong antioxidant properties, helping seagrasses combat oxidative stress caused by various environmental factors. However, their role extends beyond UV protection and environmental stress response. Flavonoids in seagrasses are also involved in regulating plant growth, forming stable complexes with other compounds, and facilitating ecological adaptation to marine habitats (Botes et al., 2025). Interestingly, the flavonoid content and distribution vary among different seagrass species and plant parts. For instance, significantly higher total flavonoid content has been observed in seagrass leaves compared to rhizomes, with *Halophila spinulosa* leaves containing the highest recorded levels (Santoso et al., 2023). The antioxidant activity of seagrass extracts is often strongly correlated with their total phenolic content, which includes flavonoids and other phenolic compounds.

In the context of climate change, flavonoids are likely to play a crucial role in helping seagrasses adapt to new stresses, such as increased sulfate levels, changes in salt concentration, shifts in herbivore distributions, and ocean acidification. In addition, the interplay between flavonoid and simple phenolic levels may serve as a useful indicator for assessing the health of seagrass meadows in response to changing environmental conditions.

10.10.2 Tannins, phenolic acids, and lignins

Seagrasses are known to produce condensed tannins, which serve as chemical deterrents against herbivores and microbial pathogens. Tannins interfere with protein digestion in herbivorous marine organisms, reducing their feeding efficiency. In addition to their antiherbivory function, tannins also possess strong antioxidant and antimicrobial activities (Choi et al., 2009).

Hydroxycinnamic acids, such as caffeic acid, ferulic acid, and p-coumaric acid, are commonly found in seagrasses. These compounds contribute to the structural integrity of the plant cell walls and provide defense against microbial infections. Studies have demonstrated that seagrass-derived phenolic acids exhibit significant antibacterial and antifungal properties (Mateos et al., 2020). Seagrasses contain lignin-derived phenolics, which play a crucial role in structural support and defense against environmental stressors. Lignins enhance the mechanical strength of seagrass tissues, making them more resistant to physical damage and microbial degradation.

10.11 Marine sponges as a source of marine phenolic compounds

Although marine sponges are a rich source of highly bioactive compounds, there is limited literature on the extraction and characterization of their polyphenolic constituents. Traditionally, methanol and dichloromethane have been employed as extraction solvents. Recent advancements have enabled the identification of novel phenolic compounds using techniques such as high-resolution atmospheric pressure chemical ionization mass spectrometry, high-resolution electrospray ionization mass spectrometry, and nuclear magnetic resonance spectroscopy. Among these, bisabolenes represent a particularly intriguing class of sponge-derived polyphenols. These compounds exhibit a distinctive C-7 absolute stereochemistry, with all sponge-derived bisabolenes characterized by a 7S configuration, in contrast to the 7R configuration commonly found in their marine and terrestrial counterparts (Cichewicz et al., 2005). Among the most studied are bromophenols, which are distinguished by their halogenated aromatic rings. Bromophenols isolated from sponges such as *Aplysina cavernicola* and *Dysidea fragilis* have demonstrated potent antioxidant capacities by scavenging free radicals and inhibiting lipid peroxidation. The presence of bromine atoms is believed to enhance both the stability and reactivity of these molecules, contributing to heightened antibacterial and antifungal activities. Several bromophenols have shown efficacy against pathogenic bacteria and marine biofouling organisms, supporting their role as chemical defenses in sponge ecology (Rodríguez et al., 2023).

Avarol and avarone, two hydroquinone derivatives isolated from the Mediterranean sponge *Dysidea avara*, exemplify the antiviral and cytotoxic properties of sponge-derived polyphenols. Avarol inhibits HIV-1 replication and demonstrates cytotoxicity toward a range of human cancer cell lines, including leukemia and neuroblastoma. In addition, avarone has been investigated for its ability to block cell proliferation and induce apoptotic pathways in malignant cells. The unique redox behavior and aromatic structure of these compounds are thought to underlie their effectiveness in interfering with cellular metabolism and viral replication (Schneider et al., 2025). Plakortin-like polyphenols, found in sponges of the genus *Plakortis*, are polyether derivatives with promising antimalarial and cytotoxic effects. Studies have shown that plakortin can disrupt cellular respiration in *Plasmodium falciparum*, the causative agent of malaria, by generating reactive oxygen species and interfering with oxidative stress pathways within parasite-infected cells. Nakaoic acid, a polyphenolic molecule from *Dysidea nakaoensis*, acts as a strong radical scavenger, reducing oxidative damage in biological systems. Its antioxidant capabilities highlight its potential role in mitigating age-related and degenerative disorders that are linked to oxidative stress (Rodríguez et al., 2023).

Halocyamines, bisphenol derivatives from sponges belonging to the *Haliclona* species, have attracted attention for their robust antibacterial activity. These compounds effectively inhibit the growth of Gram-positive and Gram-negative bacteria, including clinical strains resistant to conventional antibiotics. This activity is crucial for sponges, as it assists in managing microbial populations on their surfaces and provides a competitive advantage in densely populated marine environments (Schneider et al., 2025). Sesquiterpene phenols such as (+)-curcuphenol and (+)-curcudiol—isolated from *Didiscus aceratus*—have exhibited selective cytotoxicity against certain cancer cell lines and also show antioxidant properties.

The structural complexity and bioactivity of these sponge-derived phenolics offer considerable promise in drug development, especially as sources for new chemotherapeutic and antiinfective agents (Wright et al., 1987). Collectively, the bioactivities of polyphenolic compounds from marine sponges—spanning antioxidant, antibacterial, antiviral, and cytotoxic effects—reflect both their ecological functions in the marine environment and their potential utility in biotechnology and medicine. Their structural diversity and potency underscore marine sponges as a rich resource for natural product chemistry and drug discovery research (Kim et al., 2025).

10.12 Conclusions

Marine ecosystems represent a rich and largely untapped reservoir of chemically diverse phenolic compounds with extraordinary biological potential. Shaped by millions of years of evolution under extreme conditions, such as high salinity, pressure, UV radiation, and intense microbial competition, marine organisms have developed complex metabolic pathways that give rise to structurally unique phenolics. These often exhibit rare modifications, including halogenation (e.g., bromination, chlorination), sulfation, glycosylation, and complex dimerization. Such structural features, uncommon in terrestrial analogs, contribute to the remarkable stability, enhanced bioavailability, and potent bioactivities of marine phenolics.

This chapter has highlighted the structural diversity and ecological origins of marine phenolic compounds, including flavonoids, phlorotannins, bromophenols, stilbenes, lignans, tannins, and phenolic acids sourced from algae, sponges, corals, seagrasses, and, notably, marine microorganisms such as fungi and bacteria. Advances in natural product chemistry have been accompanied by the development of increasingly efficient extraction and purification techniques, such as UAE, MAE, SFE, and solid-phase extraction. These approaches now enable the isolation of complex phenolic structures with high specificity and yield, which is essential for downstream applications.

Biologically, marine phenolic compounds demonstrate a broad spectrum of pharmacological activities, including antioxidant, antiinflammatory, antimicrobial, antiviral, and anticancer effects. These effects are mediated through diverse and often synergistic mechanisms, such as scavenging reactive oxygen and nitrogen species, inhibiting key enzymes, modulating signaling pathways, and influencing gene expression related to inflammation, apoptosis, and oxidative stress. In addition, some marine phenolics have shown potential in modulating gut microbiota, further expanding their therapeutic relevance. These multifunctional properties underscore their potential as lead compounds in the development of pharmaceuticals, functional foods, and cosmeceuticals aimed at managing chronic lifestyle diseases such as cancer, diabetes, obesity, cardiovascular disease, and neurodegenerative disorders.

Despite their promise, the large-scale application of marine phenolics remains constrained by several challenges: low natural abundance, complex extraction processes, and concerns over the sustainability of marine biomass harvesting. However, emerging biotechnological tools—including microbial fermentation, synthetic biology, and metabolic engineering—offer sustainable alternatives for production. Moreover, hybrid extraction and purification strategies continue to improve compound yield and purity, enhancing feasibility for industrial use.

For marine phenolics to transition from promising bioactives to clinically viable agents, further research is essential. Critical areas include elucidating their mechanisms of action, pharmacokinetics, metabolism, and long-term safety. Interdisciplinary collaboration across marine biotechnology, analytical chemistry, pharmacology, and systems biology will be key to unlocking their full potential.

References

Amador, S., Martins, A., Matias, M., Pedrosa, R., Pinteus, S., 2025. Deep eutectic systems: a game changer for marine bioactives recovery. Mar. Drugs. 1660-3397. 23 (5), 211. doi:10.3390/md23050211.

Ando, E., Ohnishi, M., Wang, Y., Matsushita, T., Watanabe, A., Hayashi, Y., Fujii, M., Ma, J.F., Inoue, S.I., Kinoshita, T., 2013. TWIN SISTER OF FT, GIGANTEA, and CONSTANS have a positive but indirect effect on blue light-induced Stomatal opening in *Arabidopsis*. Plant Physiol. 15322548. 162 (3), 1529–1538. Available from: http://www.plantphysiol.org/content/162/3/1529.full.pdf+html. doi:10.1104/pp.113.217984.

Ang, M.B.M.Y., Deang, A.B.G., Chiao, Y.-H., Aquino, R.R., Millare, J.C., Huang, S.-H., Tsai, H.-A., Lee, K.-R., 2022. Integrating nanoclay intercalated with interlayers of cationic surfactant into thin-film nanocomposite nanofiltration membranes to improve performance and antifouling property. Sep. Purif. Technol. 13835866. 285, 120360. doi:10.1016/j.seppur.2021.120360.

Aziz, E., Batool, R., Khan, M.U., Rauf, A., Akhtar, W., Heydari, M., Rehman, S., Shahzad, T., Malik, A., Mosavat, S.H., Plygun, S., Shariati, M.A., 2021. An overview on red algae bioactive compounds and their pharmaceutical applications. J. Complement. Integr. Med. 1553-3840. 17 (4). doi:10.1515/jcim-2019-0203.

Azmir, J., Zaidul, I.S.M., Rahman, M.M., Sharif, K.M., Mohamed, A., Sahena, F., Jahurul, M.H.A., Ghafoor, K., Norulaini, N.A.N., Omar, A.K.M., 2013. Techniques for extraction of bioactive compounds from plant materials: a review. J. Food Eng. 02608774. 117 (4), 426–436. doi:10.1016/j.jfoodeng.2013.01.014.

Botes, J., Ma, X., Chang, J., Van de Peer, Y., Berger, D.K., 2025. Flavonoids and anthocyanins in seagrasses: implications for climate change adaptation and resilience. Front. Plant Sci. 1664-462X. 15. doi:10.3389/fpls.2024.1520474.

Carreira-Casais, A., Otero, P., Garcia-Perez, P., Garcia-Oliveira, P., Pereira, A.G., Carpena, M., Soria-Lopez, A., Simal-Gandara, J., Prieto, M.A., 2021. Benefits and drawbacks of ultrasound-assisted extraction for the recovery of bioactive compounds from marine algae. Int. J. Environ. Res. Public Health. 1660-4601. 18 (17), 9153. doi:10.3390/ijerph18179153.

Catarino, M.D., 2021. Florotaninos da Alga Fucus Vesiculosus: extração, Caracterização Estrutural e Efeitos ao Longo do Trato Gastrointestinal, pp. 175.

Catarino, M.D., Pires, S.M.G., Silva, S., Costa, F., Braga, S.S., Pinto, D.C.G.A., Silva, A.M.S., Cardoso, S.M., 2022. Overview of phlorotannins' constituents in fucales. Mar. Drugs. 1660-3397. 20 (12), 754. doi:10.3390/md20120754.

Catarino, M.D., Silva, A.M.S., Mateus, N., Cardoso, S.M., 2019. Optimization of phlorotannins extraction from fucus vesiculosus and evaluation of their potential to prevent metabolic disorders. Mar. Drugs. 16603397. 17 (3). Available from: https://www.mdpi.com/1660-3397/17/3/162/pdf. doi:10.3390/md17030162.

Chandrasekhar, T., Riazunnisa, K., Lakshmi, D.V., Prasanna, V.A., Bramhachari, P.V., 2024. Exploration of bioactive functional molecules from marine algae: challenges and applications in nutraceuticals. In: Marine Bioactive Molecules for Biomedical and Pharmacotherapeutic Applications. Springer Nature, India, pp. 187–196. Available from: https://link.springer.com/book/9789819967704. doi:10.1007/978-981-99-6770-4_10.

Chemat, F., Vian, M.A., Cravotto, G., 2012. Green extraction of natural products: concept and principles. Int. J. Mol. Sci. 14220067. 13 (7), 8615–8627. Available from: http://www.mdpi.com/1422-0067/13/7/8615/pdf. doi:10.3390/ijms13078615.

Chemat, F., Zill-E-Huma, Khan, M.K., 2011. Applications of ultrasound in food technology: processing, preservation and extraction. Ultrason. Sonochem. 13504177. 18 (4), 813–835. https://www.elsevier.com/inca/publications/store/5/2/5/4/5/1. doi:10.1016/j.ultsonch.2010.11.023.

Chen, W., Li, Z., Liu, T., Lyu, Q., Liu, W., 2025. Discovery and characterization of CcrSult1: a novel flavonoid sulfotransferase from the red alga chondrus crispus. J. Agric. Food Chem. 0021-8561. 73 (22), 14008–14019. doi:10.1021/acs.jafc.5c02656.

Choi, H.-G., Lee, J.-H., Park, H.-H., Sayegh, F.A.Q., 2009. Antioxidant and antimicrobial activity of *Zostera marina* L. extract. ALGAE. 1226-2617. 24 (3), 179–184. doi:10.4490/algae.2009.24.3.179.

Cichewicz, R.H., Clifford, L.J., Lassen, P.R., Cao, X., Freedman, T.B., Nafie, L.A., Deschamps, J.D., Kenyon, V.A., Flanary, J.R., Holman, T.R., Crews, P., 2005. Stereochemical determination and bioactivity assessment of (S)-(+)-curcuphenol dimers isolated from the marine sponge *Didiscus aceratus* and synthesized through laccase biocatalysis. Bioorg. Med. Chem. 09680896. 13 (19), 5600–5612. doi:10.1016/j.bmc.2005.06.020.

De Luca, D., Lauritano, C., 2020. In silico identification of type III PKS chalcone and stilbene synthase homologs in marine photosynthetic organisms. Biology. 20797737. 9 (5). Available from: https://www.mdpi.com/2079-7737/9/5/110/pdf. doi:10.3390/biology9050110.

Del Mar Ribas-taberner, M., Mir-Rossello, P.M., Gil, L., Sureda, A., Capó, X., 2025. Potential use of marine plants as a source of bioactive compounds. Molecules. 1420-3049. 30 (3), 485. doi:10.3390/molecules30030485.

Dewick, P.M., 1984. The biosynthesis of shikimate metabolites. Nat. Prod. Rep. 02650568. 1 (5), 451–469. doi:10.1039/NP9840100451.

El-Bilawy, E.H., Al-Mansori, A.-N.A., Alotibi, F.O., Al-Askar, A.A., Arishi, A.A., Teiba, I.I., Sabry, A.E.-N., Elsharkawy, M.M., Heflish, A.A., Behiry, S.I., Abdelkhalek, A., 2022. Antiviral and antifungal of Ulva fasciata extract: HPLC analysis of polyphenolic compounds. Sustainability. 2071-1050. 14 (19), 12799. doi:10.3390/su141912799.

Fernando, I.P.S., Lee, W.W., Ahn, G., 2022. Marine algal flavonoids and phlorotannins; an intriguing frontier of biofunctional secondary metabolites. Crit. Rev. Biotechnol. 15497801. 42 (1), 23–45. Available from: http://www.tandfonline.com/loi/ibty20. doi:10.1080/07388551.2021.1922351.

Getachew, A.T., Jacobsen, C., Holdt, S.L., 2020. Emerging technologies for the extraction of marine phenolics: opportunities and challenges. Mar. Drugs. 16603397. 18 (8). Available from: https://www.mdpi.com/1660-3397/18/8/389. doi:10.3390/MD18080389.

Ghallab, D.S., Ibrahim, R.S., Mohyeldin, M.M., Shawky, E., 2024. Marine algae: a treasure trove of bioactive anti-inflammatory compounds. Mar. Pollut. Bull. 0025326X. 199, 116023. doi:10.1016/j.marpolbul.2023.116023.

Gligor, O., Mocan, A., Moldovan, C., Locatelli, M., Crişan, G., Ferreira, I.C.F.R., 2019. Enzyme-assisted extractions of polyphenols – a comprehensive review. Trends Food Sci. Technol. 09242244. 88, 302–315. doi:10.1016/j.tifs.2019.03.029.

Goyal, M., 2018. Structural modeling of shikimate pathway enzymes for herbicide and drug development: a review. J. Entomol. Zool. Stud. 6 (2), 785–790.

Greer, A.T., Lehrter, J.C., Binder, B.M., Nayak, A.R., Barua, R., Rice, A.E., Cohen, J.H., McFarland, M.N., Hagemeyer, A., Stockley, N.D., Boswell, K.M., Shulman, I., deRada, S., Penta, B., 2020. High-resolution sampling of a broad marine life size spectrum reveals differing size- and composition-based associations with physical oceanographic structure. Front. Mar. Sci. 22967745. 7. Available from: https://www.frontiersin.org/journals/marine-science#. doi:10.3389/fmars.2020.542701.

Heo, S.J., Jeon, Y.J., 2009. Evaluation of diphlorethohydroxycarmalol isolated from Ishige okamurae for radical scavenging activity and its protective effect against H_2O_2-induced cell damage. Process Biochem. 13595113. 44 (4), 412–418. doi:10.1016/j.procbio.2008.12.005.

Herrero, M., Del Pilar Sánchez-camargo, A., Cifuentes, A., Ibáñez, E., 2015. Plants, seaweeds, microalgae and food by-products as natural sources of functional ingredients obtained using pressurized liquid extraction and supercritical fluid extraction. TrAC Trends Anal. Chem. 01659936. 71, 26–38. doi:10.1016/j.trac.2015.01.018.

Imbs, T.I., Zvyagintseva, T.N., 2018. Phlorotannins are polyphenolic metabolites of brown algae. Russian J. Mar. Biol. 1063-0740. 44 (4), 263–273. doi:10.1134/s106307401804003x.

Jiang, M., Zhang, H., 2016. Engineering the shikimate pathway for biosynthesis of molecules with pharmaceutical activities in E. coli. Curr. Opin. Biotechnol. 18790429. 42, 1–6. Available from: http://www.elsevier.com/locate/copbio. doi:10.1016/j.copbio.2016.01.016.

Jimenez-Lopez, C., Pereira, A.G., Lourenço-Lopes, C., Garcia-Oliveira, P., Cassani, L., Fraga-Corral, M., Prieto, M.A., Simal-Gandara, J., 2021. Main bioactive phenolic compounds in marine algae and their mechanisms of action supporting potential health benefits. Food Chem. 03088146. 341 (Pt 2), 128262. doi:10.1016/j.foodchem.2020.128262.

Kim, J.A., Choi, S.S., Lim, J.K., Kim, E.S., 2025. Unlocking marine treasures: isolation and mining strategies of natural products from sponge-associated bacteria. Nat. Prod. Rep. 14604752. 42 (7), 1195–1225. Available from: http://pubs.rsc.org/en/journals/journal/np. doi:10.1039/d5np00013k.

Liu, J., Lin, S., Wang, Z., Wang, C., Wang, E., Zhang, Y., Liu, J., 2011. Supercritical fluid extraction of flavonoids from *Maydis stigma* and its nitrite-scavenging ability. Food Bioprod. Process. 09603085. 89 (4), 333–339. doi:10.1016/j.fbp.2010.08.004.

Lu, P., Wang, W., Zhang, G., Li, W., Jiang, A., Cao, M., Zhang, X., Xing, K., Peng, X., Yuan, B., Feng, Z., Arora, P.K., 2020. Isolation and characterization marine bacteria capable of degrading lignin-derived compounds. PLOS ONE. 1932-6203. 15 (10), e0240187. doi:10.1371/journal.pone.0240187.

Maeda, H., Dudareva, N., 2012. The shikimate pathway and aromatic amino acid biosynthesis in plants. Annu. Rev. Plant Biol. 15452123. 63, 73–105. doi:10.1146/annurev-arplant-042811-105439.

Maheswari, V., Babu, P.A.S., 2022. Phlorotannin and its derivatives, a potential antiviral molecule from brown seaweeds, an overview. Russian J. Mar. Biol. 16083377. 48 (5), 309–324. Available from: https://www.springer.com/journal/11179. doi:10.1134/S1063074022050169India.

Marchiosi, R., Dos Santos, W.D., Constantin, R.P., de Lima, R.B., Soares, A.R., Finger-Teixeira, A., Mota, T.R., de Oliveira, D.M., de Paiva Foletto-felipe, M., Abrahão, J., Ferrarese-Filho, O., 2020. Biosynthesis and metabolic actions of simple phenolic acids in plants. Phytochem. Rev. 1568-7767. 19 (4), 865–906. doi:10.1007/s11101-020-09689-2.

Martins, B.T., da Silva, M.C., Pinto, M., Cidade, H., Kijjoa, A., 2019. Marine natural flavonoids: chemistry and biological activities. Nat. Prod. Res. 14786427. 33 (22), 3260–3272. Available from: https://www.tandf.co.uk/journals/titles/14786419.asp. doi:10.1080/14786419.2018.1470514.

Mateos, R., Pérez-Correa, J.R., Domínguez, H., 2020. Bioactive properties of marine phenolics. Mar. Drugs. 16603397. 18 (10). Available from: https://www.mdpi.com/1660-3397/18/10/501. doi:10.3390/md18100501.

Matulja, D., Vranješević, F., Markovic, M.K., Kraljević Pavelić, S., Marković, D., 2022. Anticancer activities of marine-derived phenolic compounds and their derivatives. Molecules. 1420-3049. 27 (4), 1449. doi:10.3390/molecules27041449.

Moon, C., Kim, S., Kim, J., Hyun, J.W., Lee, N.H., Park, J.W., Shin, T., 2008. Protective effect of phlorotannin components phloroglucinol and eckol on radiation-induced intestinal injury in mice. Phytother. Res. 0951-418X. 22 (2), 238–242. doi:10.1002/ptr.2298.

Nagahawatta, D.P., Liyanage, N.M., Jayawardena, T.U., Jayawardhana, H.H.A.C.K., Jeong, S.-H., Kwon, H.-J., Jeon, Y.-J., 2024b. Role of marine natural products in the development of antiviral agents against SARS-CoV-2: potential and prospects. Mar. Life Sci. Technol. 2662-1746. 6 (2), 280–297. doi:10.1007/s42995-023-00215-9.

Nagahawatta, D.P., Liyanage, N.M., Jayawardena, T.U., Jeon, Y.-J., 2024a. Marine polyphenols in cardiovascular health: unraveling structure–activity relationships, mechanisms, and therapeutic implications. Int. J. Mol. Sci. 1422-0067. 25 (15), 8419. doi:10.3390/ijms25158419.

Orth, R.J., Carruthers, T.J.B., Dennison, W.C., Duarte, C.M., Fourqurean, J.W., Heck, K.L., Hughes, A.R., Kendrick, G. A., Kenworthy, W.J., Olyarnik, S., Short, F.T., Waycott, M., Williams, S.L., 2006. A global crisis for seagrass ecosystems. BioScience. 00063568. 56 (12), 987–996. doi:10.1641/0006-3568(2006)56[987:AGCFSE]2.0.CO;2.

Prates, J.A.M., 2025. Applications of bioactive compounds from marine microalgae in health, cosmetics, and functional foods. Appl. Sci. 2076-3417. 15 (11), 6144. doi:10.3390/app15116144.

Rodríguez, P.F., Murillo-González, L., Rodríguez, E., Pérez, A.M., 2023. Marine phenolic compounds: sources, commercial value, and biological activities. In: Marine Phenolic Compounds: Science and Engineering. Elsevier, Costa Rica, 47–86. Available from: https://www.sciencedirect.com/book/9780128235898. doi:10.1016/B978-0-12-823589-8.00005-4.

Sadeghi, A., Rajabiyan, A., Nabizade, N., Nezhad, N.M., Zarei-Ahmady, A., 2024. Seaweed-derived phenolic compounds as diverse bioactive molecules: a review on identification, application, extraction and purification strategies. Int. J. Biol. Macromol. 01418130. 266 (Pt 1), 131147. doi:10.1016/j.ijbiomac.2024.131147.

Saini, R., Ali, M.I., Pant, M., Warghane, A., 2024. Current status of potential antiviral drugs derived from plant, marine, and microbial sources. Anti-Infect. Agents. 22113533. 22 (2), 61–73. Available from: https://www.eurekaselect.com/download_file. doi:10.2174/0122113525272349231210055403.

Santoso, J., Purwaningsih, S., Ramadhan, W., Noveliyana, Y., 2023. Bioactive compounds, phenol content and antioxidant activity of tropical seagrass Halodule pinifolia. Coast. Ocean J. 2549-8223. 7 (2), 62–75. doi:10.29244/coj.v7i1.47784.

Saudagar, R.B., Saokar, S., 2019. Anti-inflammatory natural compounds from herbal and marine origin. J. Drug Deliv. Ther. 2250-1177. 9 (3), 669–672. doi:10.22270/jddt.v9i3.2906.

Schneider, A., Dutra, F., Silva, S.M., Maraschin, M., 2025. Phenolic compounds from marine invertebrates: bioprospecting and chemical synthesis. Stud. Nat. Prod. Chem. 15725995. 84, 495–524. Available from: https://www.sciencedirect.com/science/bookseries/15725995. doi:10.1016/B978-0-443-23752-2.00017-7.

Shende, V.V., Bauman, K.D., Moore, B.S., 2024. The shikimate pathway: gateway to metabolic diversity. Nat. Prod. Rep. 14604752. 41 (4), 604–648. Available from: http://pubs.rsc.org/en/journals/journal/np. doi:10.1039/d3np00037k.

Shetty, K., Wahlqvist, M., 2004. A model for the role of the proline-linked pentosephosphate pathway in phenolic phytochemical biosynthesis and mechanism of action for human health and environmental applications. Asia Pac. J. Clin. Nutr. 13 (1), 1–24.

Singh, P., Singh, A., Choudhary, K.K., 2023. Revisiting the role of phenylpropanoids in plant defense against UV-B stress. Plant Stress. 2667064X. 7. Available from:https://www.journals.elsevier.com/plant-stress. doi:10.1016/j.stress.2023.100143.

Stengel, D.B., Connan, S., 2015. Natural products from marine algae. Preface. Methods in Molecular Biology (Clifton, N.J.). 19406029. 1308, v.

Stincone, A., Prigione, A., Cramer, T., Wamelink, M.M.C., Campbell, K., Cheung, E., Olin-Sandoval, V., Grüning, N., Krüger, A., Alam, M.T., Keller, M.A., Breitenbach, M., Brindle, K.M., Rabinowitz, J.D., Ralser, M., 2015. The return of metabolism: biochemistry and physiology of the pentose phosphate pathway. Biol. Rev. 1464-7931. 90 (3), 927–963. doi:10.1111/brv.12140.

Tzin, V., Galili, G., 2010. New Insights into the shikimate and aromatic amino acids biosynthesis pathways in plants. Mol. Plant. 17529867. 3 (6), 956–972. doi:10.1093/mp/ssq048.

Ventura, S.P.M., Silva, F.A.E., Quental, M.V., Mondal, D., Freire, M.G., Coutinho, J.A.P., 2017. Ionic-liquid-mediated extraction and separation processes for bioactive compounds: past, present, and future trends. Chem. Rev. 15206890. 117 (10), 6984–7052. Available from: http://pubs.acs.org/journal/chreay. doi:10.1021/acs.chemrev.6b00550.

Vogt, T., 2010. Phenylpropanoid Biosynthesis. Mol. Plant. 16742052. 3 (1), 2–20. doi:10.1093/mp/ssp106.

Wang, T., Jónsdóttir, R., Liu, H., Gu, L., Kristinsson, H.G., Raghavan, S., Ólafsdóttir, G., 2012. Antioxidant capacities of phlorotannins extracted from the brown algae fucus vesiculosus. J. Agric. Food Chem. 0021-8561. 60 (23), 5874–5883. doi:10.1021/jf3003653.

Weaver, L.M., Herrmann, K.M., 1997. Dynamics of the shikimate pathway in plants. Trends Plant Sci. 13601385. 2 (9), 346–351. Available from: https://www.elsevier.com/inca/publications/store/3/0/9/6/0/index.htt. doi:10.1016/S1360-1385(97)84622-5.

Wright, A.E., Pomponi, S.A., McConnell, O.J., Kohmoto, S., McCarthy, P.J., 1987. (+)-curcuphenol and (+)-curcudiol, sesquiterpene phenols from shallow and deep water collections of the marine sponge didiscus flavus. J. Nat. Prod. 15206025. 50 (5), 976–978. doi:10.1021/np50053a042.

Yong, L., Qian, Z.-J., Ryu, B., Lee, S.-H., Kim, M.-M., Kim, S.-K., 2009. Chemical components and its antioxidant properties in vitro: an edible marine brown alga, *Ecklonia cava*. Bioorg. Med. Chem. 09680896. 17 (5), 1963–1973. doi:10.1016/j.bmc.2009.01.031.

Yuan, Y., Zhang, J., Fan, J., Clark, J., Shen, P., Li, Y., Zhang, C., 2018. Microwave assisted extraction of phenolic compounds from four economic brown macroalgae species and evaluation of their antioxidant activities and inhibitory effects on α-amylase, α-glucosidase, pancreatic lipase and tyrosinase. Food Res. Int. 18737145. 113, 288–297. Available from: https://www.elsevier.com/inca/publications/store/4/2/2/9/7/0. doi:10.1016/j.foodres.2018.07.021.

PART IV

Marine-derived functional food types

CHAPTER

11

Seaweeds based on food products

H.D. Tharushi Wijerathne[a], *Kim Hyun-Soo*[a], *Dineth Nagahwatta*[b], *Thilina U. Jayawardena*[b], *and K.K. Asanka Sanjeewa*[c]

[a]Department of Seafood Science and Technology, Institute of Marine Industry, Gyongsang National University, South Korea

[b]Department of Medicine, University of Alberta, Edmonton, AB, Canada

[c]Department of Biosystems Technology, Faculty of Technology, University of Sri Jayewardenepura, Homagama, Sri Lanka

11.1 Introduction to seaweed

Seaweed refers to macroscopic marine algae and is an essential living resource from the ocean. Unlike terrestrial plants, seaweed possesses photosynthetic ability, utilizing pigments, sunlight, and seawater nutrients to produce food through photosynthesis. Ecologically, seaweeds play a crucial role in coastal protection by reducing erosion. They also contribute to primary production for human consumption and help regulate biodiversity in ocean ecosystems by providing nursery habitats for marine organisms, supplying food for a variety of herbivores, and offering shelter from carnivorous predators (Cotas et al., 2023). In addition, seaweeds absorb and store carbon dioxide, aiding in the reduction of greenhouse gases in the atmosphere, which has been found to be useful in carbon sequestration (Hurd et al., 2022) (see Box 15.1 for the definition of carbon sequestration). Seaweeds are nonflowering plants that lack true roots, stems, or leaves. They typically grow in shallow waters, attaching themselves to rocks, stones, corals, or other solid substrates for support (Fig. 11.1). Generally, seaweeds can be classified into two categories: edible and non-edible. Edible seaweeds are widely used in various food-related industries worldwide. In contrast, non-edible seaweeds can be cultivated in coastal regions to help mitigate contamination, including heavy metal pollution (Hossain et al., 2022).

Harvesting the Sea
https://doi.org/10.1016/B978-0-443-29190-6.00016-7

Box 15.1 Definition of carbon sequestration

Carbon sequestration refers to the long-term storage of carbon-containing molecules for over 100 years. In the context of carbon dioxide removal for climate mitigation, this CO_2 originates from the atmosphere. On land, trees play a significant role in sequestering carbon within their woody biomass. Similarly, seaweeds have been proposed to serve an analogous function in ocean carbon sequestration.

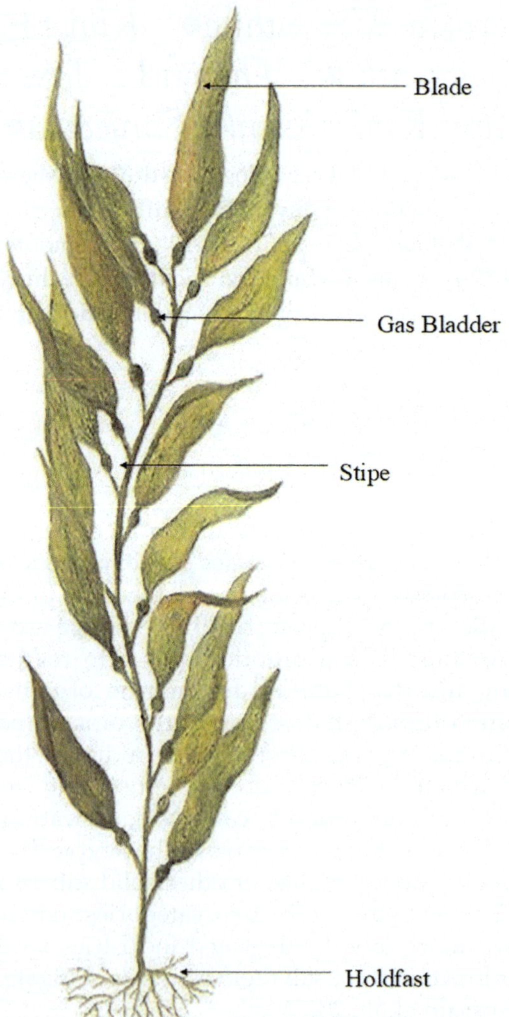

FIG. 11.1 Morphology of seaweed. This figure shows the anatomy of a seaweed plant. Seaweeds are nonflowering plants without true roots, stems, or leaves.

11.2 Classification of seaweeds

Based on biochemical, physiological, and electron microscopic image studies, seaweeds have photosynthetic pigments. Based on these pigments or appearance, seaweeds can be categorized into three main groups: red algae (Rhodophyta), green algae (Chlorophyta), and brown algae (Phaeophyceae under the phylum Ochrophyta). The three types of seaweeds can be classified based on their classes as Rhodophyceae, Chlorophyceae, and Phaeophyceae, respectively, for red, green, and brown seaweeds.

Red seaweeds can thrive in deep seawater primarily due to their high concentration of phycoerythrins, which efficiently absorb light wavelengths ranging from 450 to 570 nm. Phycoerythrins are classified into three main types: B-phycoerythrin ($\lambda max = 565$ nm, 546 nm, with a shoulder at 499 nm), C-phycoerythrin ($\lambda max = 565$ nm), and R-phycoerythrin ($\lambda max = 565$ nm, 498 nm, with a shoulder or peak at 540 nm). Among these, R-phycoerythrin is the most abundant phycobiliprotein found in red algae (Cotas et al., 2020). Red seaweeds also contain low amounts of chlorophyll *a*, β-carotene, and xanthophylls (Freitas et al., 2022). Phycoerythrin, the primary pigment responsible for the red color of red algae, reflects red light, absorbs blue light, and transfers energy to chlorophyll *a* (French and Young, 1952). However, red seaweeds are not always strictly red; in some cases, they may appear purple or even brownish-red (Thangavel and Sridevi, 2015). Green seaweeds are characterized by eukaryotic algae that are multicellular, photosynthetic, and oxygenic, based on their high amount of chlorophyll (Aina et al., 2022). Green seaweeds have a high content of chlorophyll "a" and chlorophyll "b" (Orosa et al., 2000; Chen et al., 2023) and low amounts of β-carotene and xanthophylls (Freitas et al., 2022). Brown seaweeds are actively grown in cold-temperature water. The presence of fucoxanthin contributes to the brownish color of algae. Brown seaweeds are reported to carry a number of bioactive secondary metabolites compared to other seaweeds (Aina et al., 2022).

11.3 Seaweeds reproduction

The life cycle and reproductive strategies of seaweeds are complex, often involving alternation of generations with distinct morphological phases. Depending on the species, some seaweeds exhibit an annual life cycle, completing their growth and reproduction within a single year, while others are perennial, persisting for multiple years or even decades. Reproduction in seaweeds may occur through both sexual and asexual mechanisms, with sexual reproduction involving the fusion of gametes and subsequent formation of a zygote, followed by sporophyte and gametophyte phases in many taxa (Fig. 11.2). Asexual reproduction occurs through processes such as fragmentation or division (Liu et al., 2017). In the fragmentation process, small fragments are broken off from the parental plants and grow separately. These plants are genetically identical to the parental plants (Takahashi and Mikami, 2017).

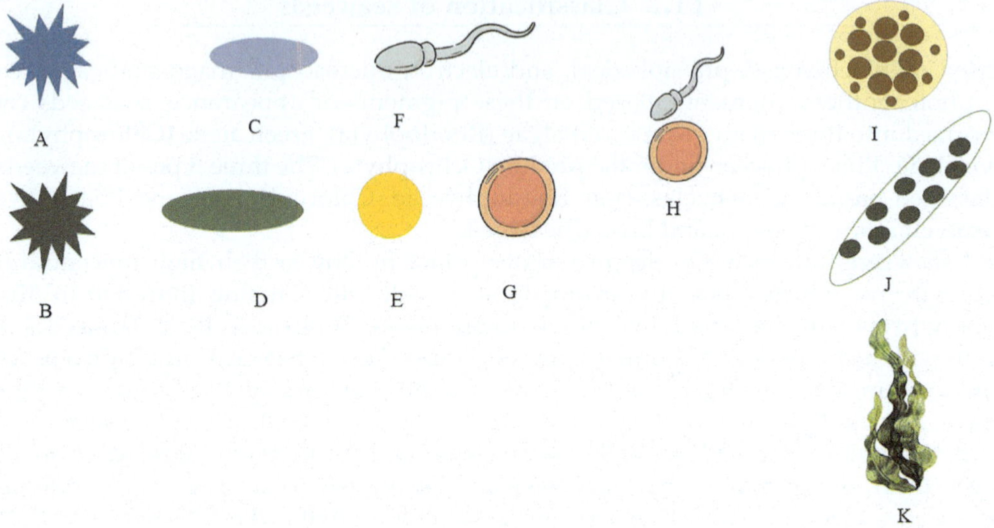

FIG. 11.2 Sexual reproduction of algae. (A) male zoospore, (B) female zoospore, (C) male gametophyte, (D) antheridium, (E) female gametophyte, (F) sperm, (G) egg, (H) fertilization, (I) zygote, (J) sporophyte, (K) seaweed. This figure illustrates the process of sexual reproduction, where male and female reproductive cells (gametes) bind together. The cycle begins with the sporophyte stage, which consists of separate male and female plants known as gametophytes. These gametophytes produce sperm and eggs separately; male gametophytes produce sperm, while female gametophytes produce eggs. The sperm and eggs may either be retained within the plant body or released into the surrounding marine environment. When fertilization occurs, the sperm and egg unite to form a zygote. This zygote then develops into a new sporophyte, continuing the reproductive cycle.

11.4 Global utilization of different seaweed species for food production

The global utilization of different seaweed species in food products offers food security and sustainability. Canada, China, Korea, and Japan have already used these seaweeds for their food production (Table 11.1).

TABLE 11.1 Global utilization of different seaweed species for food production.

Country	Rhodophyta	Chlorophyta	Ochrophyta (Class-Phaeophyceae)	Reference
Canada	*Mazzaella splenden* *Palmaria hecatensis* *Palmaria mollis* *Palmaria palmata*	*Ulva* spp. *Enteromorpha intestinalis*	*Alaria marginata* *Egregia menziesii* *Laminaria groenlandica*	(Thangavel and Sridevi, 2015; Xu et al., 2023)
China	*Eucheuma gelatine* *Porphyrah aitanensis* *Porphyra yezoensis*	*Ulva lactica*	*Laminaria japonicaSargassum* *Sargassum* spp.	(Battershill, 2024; Thangavel and Sridevi, 2015)

TABLE 11.1 Global utilization of different seaweed species for food production—cont'd

Country	Rhodophyta	Chlorophyta	Ochrophyta (Class-Phaeophyceae)	Reference
Hawaii	*Asparagopsis taxiformis* *Gracilaria parvispora*	*Codium muelleri* *Enteromorpha* spp.	*Sargassum polycystum*	(Thangavel and Sridevi, 2015)
Indonesia	*Asparagopsis taxiformis,* *Caloglossa adnata,* *Gracilaria eucheumoides*	*Enteromorpha intestinalis* *Codium tenue* *Codium tomentosum*	*Sargassum aquifolium* *Sargassum polycystum*	(Thangavel and Sridevi, 2015; Patel et al., 2017)
Japan	*Eucheuma cartilagineum* *Gracilaria bursa-pastoris* *Meristotheca papulosa*	*Monostroma nitidum* *Ulva lactica* *Enteromorpha intestinalis*	*Alaria crassifolia* *Cladosiphone okamuranus* *Ecklonia cava*	(Thangavel and Sridevi, 2015; Morais et al., 2020)
Korea	*Chondria crassicaulis* *Gelidium anansii* *Nemalion vericulare* *Porphyra* spp. (gim) *Gelidium amansii* (wu-mu-ga-sa-ri) *Gracilaria verrucosa* (go-si-rae-gi) *Gloiopeltis tenax* (pul-ga-sa-ri)	*Capsosiphon fulvescens* *Codium fragile* *Enteromorpha clathrata* *Capsosiphon fulvescens* *Caulerpa lentillifera* *Enteromorpha linza*	*Ecklonia stolonifera* *Ecklonia cava* *Hizikia fusiforme* (Tot) *Laminaria japonica* (Da-si-ma) *Sargassum fulvellum* (Mo–Ja-ban)	(Thangavel and Sridevi, 2015; Sanjeewa, Lee and Jeon, 2018; Sanjeewa and Jeon, 2018; Michalak et al., 2022)
Philippines	*Acanthophora spicifera* *Kappaphycus alvarezii* *Hypnea pannosa*	*Caulerpa racemosa* *Codium edule* *Codium bartletti*	*Hydroclathrus clathratus*	(Thangavel and Sridevi, 2015)
Vietnam	*Acanthophora spicifera* *Gracilaria coronopifolia* *Gracilaria coronopifolia*	*Ulva reticulate* *Ulva lactuca* *Caulerpa racemosa*	*Sargassum crassifolium* *Sargassum graminifolium* *Sargassum mcclurei*	(Probst, 2015; Thangavel and Sridevi, 2015; Xu et al., 2023)

11.5 Nutritional composition of edible seaweeds

Seaweed is among the most abundant marine species and has a superior nutritional composition compared to terrestrial plants (Abdel-Kareem and ElSaied, 2021). The nutritional value of seaweeds mostly depends on growth conditions (Probst, 2015). In addition, seaweeds serve as an important marine biological resource, rich in proteins, amino acids, fatty acids, and dietary fibers. Green seaweeds, in particular, are valuable sources of lipids and proteins. Furthermore, seaweeds also contain significant amounts of carbohydrates, pigments, and other bioactive compounds (Xu et al., 2023). The protein, dietary fiber, carbohydrate, and lipid compositions reported from different edible seaweeds are summarized in Table 11.2.

TABLE 11.2 The nutrient profiles of selected edible seaweed species (% of dry weight).

Species	Protein (%)	Dietary fiber (%)	Carbohydrate (%)	Lipid (%)	Reference
Ulva pertusa	20–26	–	47.0	–	(Probst, 2015)
Ulva lactuca	8.65–25	29–55	36–43	6.0–1.6	(Xu et al., 2023)
Ulva reticulata	17–20	65.7	50–58	1.7–2.3	(Xu et al., 2023)
Laminaria digitata	8–15	36-37	48.0	1.0	(Probst, 2015)
Undaria pinnatifida	9.4	48.2	54.5	4.8	(Battershill, 2024)
Saccharina japonica	7–8	10-41	51.9	1.0–1.9	(Probst, 2015)
Palmaria palmata	8–35	29–46	46–56	0.7–3.0	(Morais et al., 2020)
Alaria esculenta	9–20	42.9	46–51	1.0–2.0	(Morais et al., 2020)

11.6 Major types of seaweed-based food products

11.6.1 Nori

Nori is the Japanese term for edible seaweed derived from species of red algae belonging to the genus *Porphyra* (class Rhodophyceae). Commonly cultivated and consumed varieties include *Porphyra yezoensis* and *Porphyra tenera*, which are widely used in Japanese cuisine, particularly in sushi preparation and as dried seaweed sheets (Sinurat et al., 2022). These red seaweed species occur naturally, growing attached to rocks in the intertidal zone along shallow coastal shorelines, predominantly in the Pacific Ocean, the North Atlantic, and the Irish Sea (Sari et al., 2019). Nori is one of the most extensively domesticated Asian seaweed varieties, cultivated for centuries in Japan using traditional aquaculture methods, and today produced worldwide (Fig. 11.3). It is considered a nutrient-rich food, rich in essential vitamins (A and C), minerals (calcium and iodine), dietary fiber, protein, and omega-3 fatty acids. These nutrients contribute to various health benefits, including supporting immune function, promoting digestive health, maintaining thyroid function, reducing oxidative stress through its antioxidant compounds, and contributing to cardiovascular health by lowering LDL cholesterol and triglyceride levels (Sinurat et al., 2022; Rajapakse and Kim, 2011). In culinary applications, nori is most commonly used to wrap sushi and rice balls, as an ingredient in Japanese rice crackers, and as flavored snack sheets (Fig. 11.4). Owing to its low-calorie content, high nutritional value, and bioactive compounds, nori is increasingly recognized as both a traditional food and a functional food. Some studies also suggest that regular nori consumption may contribute to the reduction of diastolic blood pressure (Michalak et al., 2022; Wahyudi et al., 2022). Nori is also a good source of omega-3 fatty acids, which promote heart health by reducing and stabilizing LDL cholesterol and triglyceride levels. With its low-calorie content, nori supports a healthy lifestyle. In addition, its high iodine content plays a crucial role in maintaining thyroid health (Smyth, 2021). Studies suggest that nori consumption may also help lower diastolic blood pressure levels in children (Wada et al., 2021).

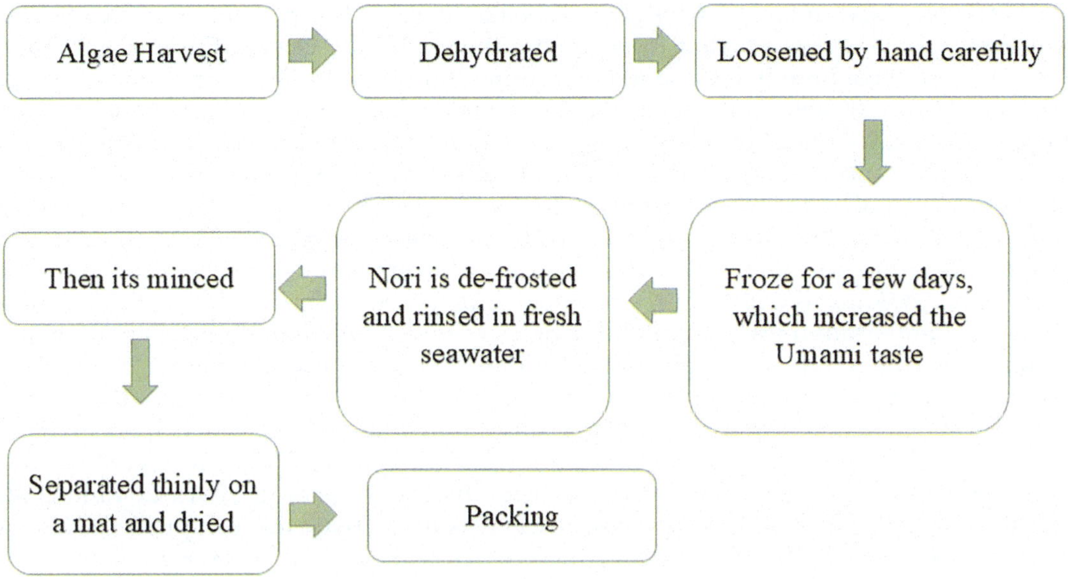

FIG. 11.3 Processing of porphyra (Nori): from harvest to drying. This figure illustrates the process of harvesting and processing Porphyra algae. First, the algae are harvested and dehydrated, followed by freezing for a few days to enhance the umami taste. After defrosting, the nori is rinsed in fresh seawater, minced, and evenly spread thinly on a mat. Finally, it is dried to complete the process.

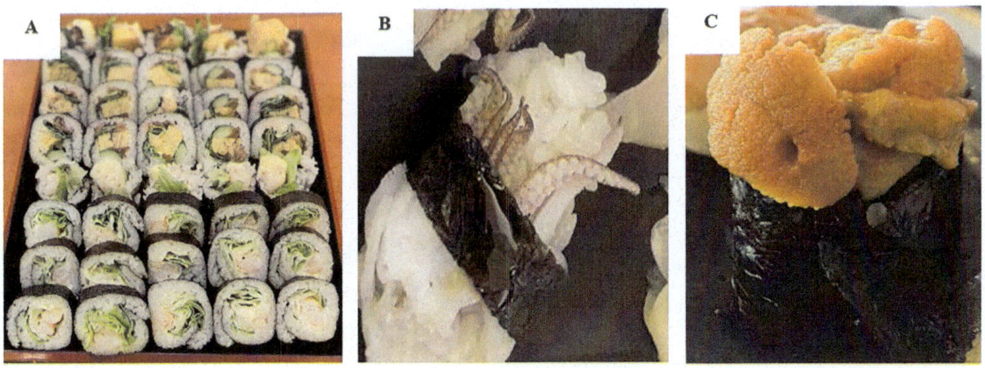

FIG. 11.4 (A) Salmon sushi roll wrapped with a nori sheet, which is a common type of sushi. (B) Cuttlefish wrapped with a nori sheet. (C) Tobiko sushi, which is fish eggs wrapped with a nori sheet and has lots of health benefits.

11.6.2 Wakame

Wakame (*Undaria pinnatifida*) is a species of kelp and belongs to the class Phaeophyceae and the family Alariaceae (Taboada et al., 2013; Epstein and Smale, 2017). Wakame is an edible seaweed that is nutritious and a good source of various minerals and vitamins. It has been cultivated in Japan and Korea (Taboada et al., 2013). Wakame is native to cold, temperate coasts and originated from the Northwest Pacific Ocean (Epstein and Smale, 2017). It is most commonly used in making soups and salads because it provides a unique taste and texture (Prabhasankar et al., 2009). *U. pinnatifida is rich in* minerals and vitamins such as A, C, E, iodine, manganese, magnesium, and calcium. Moreover, *U. pinnatifida* is low in calories and rich in nutrients (Taboada et al., 2013). Wakame possesses bioactive properties, including anticancer, antidiabetes, antihypertensive, antiviral, and antiobesity effects (Chan et al., 2023). *U. pinnatifida* may discourage the absorption and assimilation of cholesterol in the intestine. As a consequence, it lowers serum cholesterol levels (Hata et al., 2001). Wakame is a rich source of essential minerals such as iodine and manganese, which play important roles in maintaining proper thyroid function. Iodine, in particular, is required for the synthesis of thyroid hormones, which regulate growth, metabolism, protein synthesis, and cellular repair (Aakre et al., 2023; Taboada et al., 2013; Aakre et al., 2023). Wakame is also rich in n-3 polyunsaturated fatty acids, which are responsible for antihyperlipidemic, antiallergic, and antithrombotic actions. These fatty acids also help lower blood pressure and reduce the risk of heart disease (Hata et al., 2001). Commercially, four main forms of wakame are available: fresh wakame, dried wakame, pickled wakame, and wakame powder (Fig. 11.5). Among these, dried wakame and wakame powder are the most widely distributed and popular products in the market due to their extended shelf life, ease of storage, and convenience in culinary applications (Prabhasankar et al., 2009; Coniglio et al., 2021).

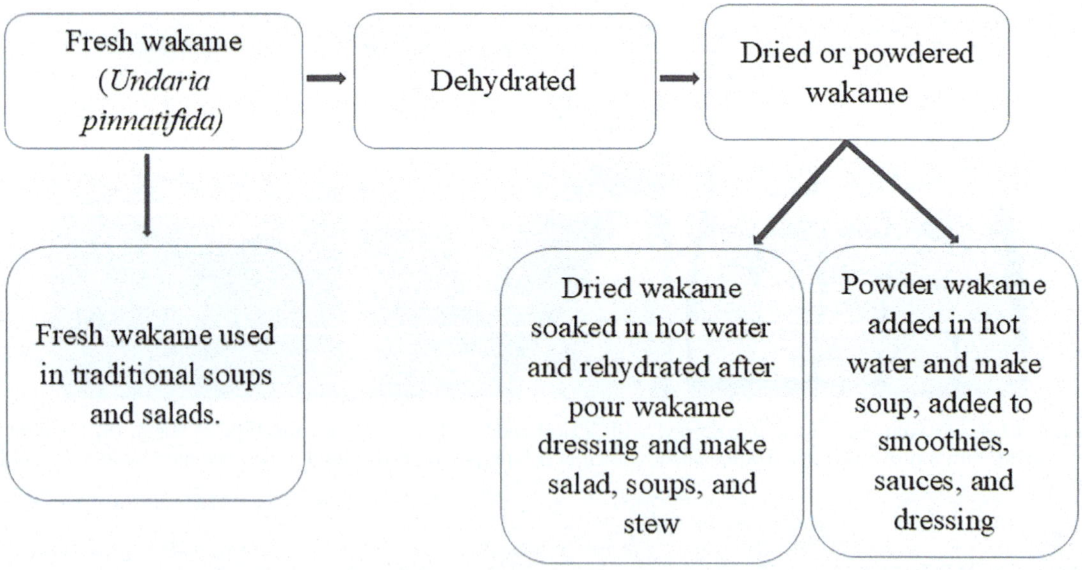

FIG. 11.5 Fresh and powdered wakame processing procedure. This figure illustrates two wakame processing techniques. Fresh wakame is commonly used in traditional soups and salads, while dehydrated wakame is dried, ground into powder, and utilized as a seasoning for salads and soups.

11.6.3 Dulse

The *Palmaria palmata,* commonly called dulse, is a common red seaweed in the genus Palmaria and the class Florideophyceae. It is found in cold environments along the rocky northern coasts of the Atlantic and Pacific Oceans (Stévant et al., 2023). Dulse is an edible seaweed, widely consumed as a snack or a food supplement in France, the United States, and Canada (Mouritsen et al., 2013). It also contains a high level of antioxidants. High amounts of antioxidants protect our body from free radicals that can damage cells and cause oxidative stress (Sato et al., 2019). Its anti-inflammatory properties help with skin issues, allergies, and pain (Millan-Linares et al., 2019). Dulse is a nutrient-rich red seaweed containing vitamins B1 (thiamine), B2 (riboflavin), and B12 (cobalamin). Vitamin B12 is essential for nerve function and red blood cell production; however, vegetarians and vegans are often at risk of B12 deficiency, as animal-derived foods are the primary dietary sources of this vitamin. Incorporating dulse into plant-based diets can help address this nutritional gap. In addition to B vitamins, dulse is a valuable source of iodine, magnesium, calcium, potassium, and vitamin C, contributing to overall mineral and antioxidant intake (Lee et al., 2017). Proteins and amino acids are present, but the protein content of dulse is one of the major reasons for its commercial usage in both food and feed applications (Stévant et al., 2023). Normally, people enjoy the taste of dulse, which can be eaten as a raw salad garnish or dried snack, as well as in dietary or medical supplements in its dried or powdered form (Fig. 11.6) (Stévant et al., 2023).

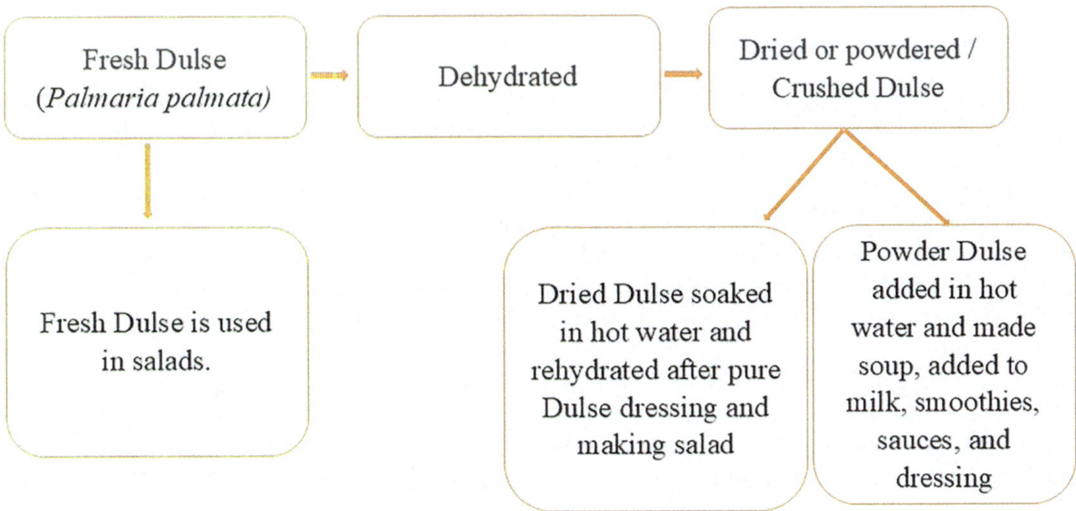

FIG. 11.6 Dried and powdered dulse-making procedure. This figure illustrates the process of making dried and powdered dulse. Fresh dulse is commonly used in salads, while dehydrated dulse is dried and ground into a powder for use in soups.

11.6.4 Kombu

Kombu (*Saccharina japonica*) is an edible kelp in the family Laminariaceae and is mostly eaten by East Asians. It is called "dasima" in Korea, "Hokkaido" in Japan, and "Haidai" in China (Chan et al., 2023). Kombu displays pharmacological properties, including anticoagulant, anticancer, hypolipidemic, antiobesity, vascular protective, antioxidant, antimicrobial, anti-inflammatory, and neuroprotective activities (Amorim et al., 2012; Nishimura et al., 2019). Kombu and wakame contain bioactive metabolites like fucoxanthin, fucoidan, and fucosterol (Zarekarizi et al., 2019; Zarekarizi et al., 2019). Kombu exerts an anti-obesity effect by reducing body fat and improving adiponectin levels (Nishimura et al., 2019). Kelps are rich in amino acids and vitamins like iodine, zinc, copper, and vitamin E (Barandiaran et al., 2024). They are also a rich source of minerals, dietary fibers, and fucoxanthin-like carotenoids (red-to-red-orange). Carotenoids represent antioxidants, angiogenesis, and anti-inflammatory properties. Frequent intake of kombu has been shown to reduce the serum triglyceride levels of subjects with abnormally high serum triglyceride levels (Nishiumi et al., 2020). Kombu is normally found in East Asian markets and health food stores. These kelp strips are covered with a white powder from natural salts. Before cooking, it is recommended to wipe them with a cloth to remove the powder (Zarekarizi et al., 2019). Kombu is used to make vegetarian broth, stew, salad, and dashi. Kombu is also used to soften beans. The amino acids in kombu help soften beans and make them more digestible. These seaweeds are rich in glutamate and become tender during cooking. Kombu is used as a seasoning condiment and in tea (Chan et al., 2023). Dried, crushed kombu is used as a salt substitute and is sprinkled over grains and tofu (Zarekarizi et al., 2019). It is also used as a traditional and functional food preservative (Peñalver et al., 2020).

11.6.5 Kelp—alginates

Alginates are produced from algae and bacteria, normally by brown algae (*Laminaria, Sargassum*, and kelp). The commercial source of alginate species is kelp (Macrocystis) (Belattmania et al., 2020). Alginate is present in the cell walls of brown seaweeds, and after the extraction process, it is possible to collect the alginate. Alginates are naturally occurring anionic polysaccharides found in brown seaweeds. In novel food industries, alginate has been used to film and coat some fruits and vegetables to improve the quality of the product, enhance the product's appearance, and extend the product's shelf life (Puscaselu et al., 2020). Alginate-based edible films and coatings are used for packaging applications because they reduce product dehydration and control respiration (Parreidt et al., 2018). Alginate products include ammonium alginate, calcium alginate, potassium alginate, propylene glycol alginate, and sodium alginate (Qin, 2018). Alginates play major roles in making different types of foods such as dairy products, mayonnaise, beer, pasta, noodles, meat, bakery products, and artificial fruits (Fig. 11.7). Major properties of alginates are summarized in Fig. 11.7.

11.6.5.1 Food applications of alginate

Alginate is a valuable source of hydrocolloids and bioactive compounds (Abka Khajouei et al., 2021). It has also been used to investigate the effects of high hydrostatic pressure pretreatment on alginate extraction (Bojorges et al., 2023). The biocompatibility of alginate

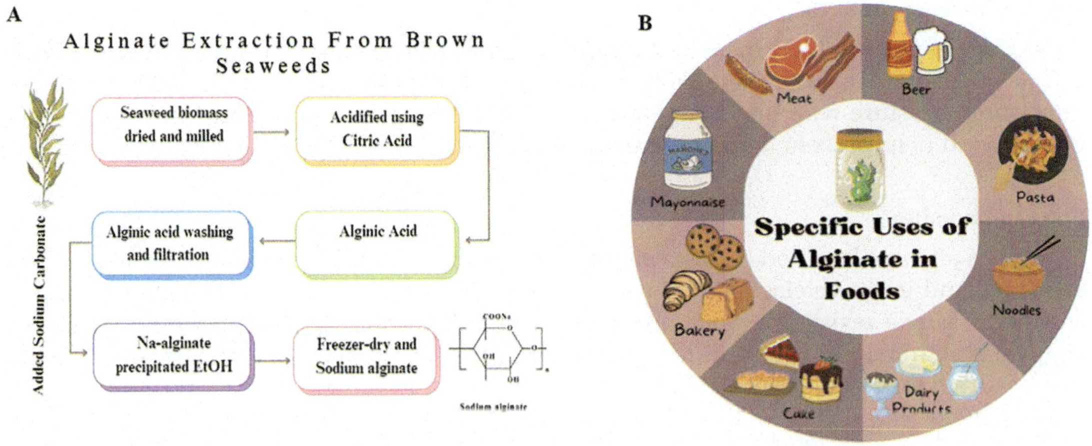

FIG. 11.7 (A) Alginate extraction process. (B) Specific uses of alginate in food products. This figure illustrates (A) the alginate extraction process from brown algae, beginning with the acidification of the algae biomass using citric acid, followed by alginic acid washing and filtration. Sodium carbonate is then added to precipitate sodium alginate, which is collected through freeze-drying. (B) The diverse applications of alginate in food products, including pasta, noodles, cakes, dairy products, beer, meat, and mayonnaise.

makes it useful for human body weight management and obesity control (Jensen et al., 2012). It increases satiety, reduces calorie intake, alters gut microbiota, delays gastric emptying, and slows nutrient absorption in the small intestine (Zhou et al., 2023). Alginate helps maintain blood sugar levels (Puscaselu et al., 2020) and is also used in oral insulin administration (Bácskay et al., 2024). In addition, alginate plays a role in controlling cholesterol levels (Idota et al., 2016). It can aid colonic health and intestinal absorption, lowering blood lipid levels, triglycerides, and low-density lipoprotein cholesterol, thereby reducing the risk of cardiovascular disease (Guo et al., 2020). Alginate also exhibits anti-inflammatory, antibacterial, antioxidant, and prebiotic properties (Lukova et al., 2024).

Dairy products

Alginate is used in making dairy products such as yogurt, ice cream, and cheese because it has stable emulsions and crosslinking properties. Alginate is convenient at diverse viscosities and gelling properties and offers stability to foods at both high and low temperatures (Pournaki et al., 2024). The accumulation of alginate gel depends on H^{+} or Ca^{2+}. Alginate forms gels at low temperatures. In addition, alginate stabilizes foods and beverages, especially ice cream, to reduce ice crystal formation during freezing. This makes the ice cream smoother and more palatable (Sharifi, Golestan and Sharifzadeh Baei, 2013).

Engineered foods

Engineered foods, or genetically modified foods such as shrimp substitutes, are utilized for Alginate (ALG). Alginate is used to bind proteins and re-engineer food. In addition, sodium alginate, as an immunostimulant, has been incorporated into shrimp feed (Santos, 2018).

Meat and bakery products

Alginate is used to bind meat and can form a soft, elastic texture (Puscaselu et al., 2020). Sodium alginate enhances the quality of bakery products by reducing the baking rate and improving moisture resistance. Alginates also increase cake swelling, improve texture and volume, and help prevent aging (Tabara et al., 2016).

Pasta and noodles

Sodium alginate is added to pasta and noodle products to increase their toughness, reduce breakage, and improve elasticity (Lee et al., 2008). It also reduces stickiness after steaming, enhances storage resistance, and improves taste. In addition, when combined with flour, alginate is particularly useful in low-gluten products, as alginate helps maintain structure and texture (Hong et al., 2021).

Beer

Sodium alginate acts as a stabilizer. Then, alginate stabilizes beer foam, increases transparency, and extends the shelf life. Also, it helps remove tannins and nitrogenous substances from wine (Nedović et al., 2001).

11.6.6 Agar-agar

Agar, also known as agar-agar, is produced from red algae, primarily from species such as *Gelidium* and *Gracilaria* (Pandya, 2022). *Gelidium* belongs to the Gelidiaceae family, while *Gracilaria* is part of the Gracilariaceae family. The origins of agar trace back to Japan, where it was first used in 1658. It was extracted from Gracilaria, an agarophyte seaweed, which was traditionally used in food products (Padmesh and Singh, 2021). Agar is a jelly-like substance composed of polysaccharides obtained from red seaweeds. Agar is primarily used as a gelling agent, thickener, and stabilizer (Qin, 2018). Agar consists of a mixture of agarose and agaropectin. Food-grade agar is refined to remove agaropectin, leaving pure agarose (Pandya, 2022). Agar plays a role in gel formation, synthesis control, emulsification, texture enhancement, reduction of sugar bloom, fiber content improvement, yield increase, and production cost reduction (Menon, 2011). Agar also possesses several functional properties, such as gelling ability, thermal stability, pH range adaptability, and transparency (Phan et al., 2005). In addition, agar-based food has been found to demonstrate antioxidant properties (Reshma et al., 2022). Furthermore, agar is high in soluble fibers, which contribute to digestion and overall gut health. It also supports probiotic consumption in humans (Praiboon et al., 2023). Furthermore, agar-based foods can aid in weight loss, as agar is considered a calorie-free food source from marine environments (Louala and Lamri-Senhadji, 2019). The fibers in agar help regulate carbohydrate absorption in the human gut, which may contribute to reduced blood glucose levels (Maeda et al., 2005). Agar fiber also reduces fat absorption and helps manage blood cholesterol levels (Louala and Lamri-Senhadji, 2019).

11.6.6.1 Food applications of agar

Mitsumame

Mitsumame is a traditional Japanese dessert with various ingredients served in a bowl, typically refreshing and sweet (Menon, 2011). The dessert often includes cubes of agar jelly, fruit (such as cherries, mandarin oranges, or peaches), sweet beans (like red bean paste or cooked adzuki beans), and sometimes mochi or other chewy elements. Mitsumame is typically topped with syrup, often a sweet, clear sugar syrup or a thicker, slightly sweetened syrup. It is commonly enjoyed in Japan during warm weather as a cool and light treat (Faudon and Savage, 2014).

Canned meat and sausage

Seaweed agar is widely used in the production of canned meat and sausages for its gelling and binding properties. In canned meat products, agar forms a gel matrix that helps hold the meat together, providing a firmer texture and retaining moisture during the canning process, which prevents the meat from becoming dry or tough (Menon, 2011). In addition, agar aids in emulsifying fat and water, ensuring a uniform product and improving shelf life by stabilizing the texture and appearance over time (Öztürk-Kerimoğlu, 2021). In sausages, agar enhances texture and firmness, particularly in low-fat or reduced-fat formulations, by maintaining shape and preventing the separation of ingredients. It also helps retain moisture, ensuring sausages remain juicy and tender after cooking or reheating. Moreover, agar binds fat and other components in sausages, contributing to a smooth and consistent texture without the need for excessive fats or fillers. These applications highlight agar's role in improving the texture, stability, and quality of both canned meat and sausages (Nouban et al., 2017).

Bakery products

Agar incorporation in bakery products can modify starch properties by lowering the pasting temperature and increasing starch susceptibility to enzymatic degradation by amylase. This leads to improvements in crust characteristics, color, and texture of the baked goods. The functional effects of agar vary depending on the type of flour used and the specific baking methods employed. In addition, agar can serve as a fat substitute, making it valuable in gluten-free and celiac-friendly bakery formulations. It is also utilized as a glazing or icing agent within the bakery industry (Menon, 2011). Nowadays, cookies are made using canola oil and agar. Hydrogels are formed from rapeseed oil solidified by agar, and they have good texture, color, and crispness (Banaś et al., 2024).

Green grass jelly drink

Agar can be utilized in green grass jelly drink production to reduce syneresis, a process where water leaks from the gel due to the formation of new polymer bonds. Popular green grass jelly is commonly used in preparing various drinks and desserts. The use of agar as a gelling agent could improve its physicochemical properties (pH, color, and Brix) and leaching properties (Hassan et al., 2024)

Pudding and yogurt

Agar is primarily used as a stabilizing ingredient. Its main function is to act as a thickening agent, which also allows for measuring the quality of yogurt. Adding agar to yogurt helps release moisture and reduces syneresis (Malaka, Ningrum and Hajrawati, 2020). In addition, agar is commonly used in making coconut pudding and chocolate jelly pudding (Martins, 2024). Novel Days has also developed a sugar-free jelly using leafy vegetables and agar as a functional food (Amer and El-Rahman, 2023).

11.6.7 Seaweed salad

Edible seaweed is rich in essential minerals and vitamins, including vitamins A, C, and E, as well as iodine, manganese, magnesium, and calcium. Due to these nutrients, seaweed salads are low in calories, nutrient-dense, and help regulate cholesterol levels in the human body (Taboada et al., 2013). In addition, seaweed salads possess various beneficial properties, such as enhancing beauty and offering medicinal and pharmacological benefits, including antioxidant, antidiabetic, anticancer, antifungal, and anti-inflammatory effects (Pati et al., 2016). Moreover, the consumption of seaweed salads is now recognized as beneficial for managing type 2 diabetes, prediabetes, osteoporosis, and obesity (Brownlee, 2011).

11.6.7.1 Lato

Lato or Ensaladang Lato is a type of Filipino salad. The main ingredient is edible seaweed. This type of seaweed is sometimes called sea grapes (Atias and Sciences, 2016). In a bowl, combine vinegar, pepper, sugar, tomato, onion, and salt with the lato and mix. Let it stay for 10 min and serve (Harsha Mohan et al., 2023).

11.6.7.2 Guso

The Philippines has commercial seaweed cultivation farms of guso called *Kappaphycus*. But commonly, guso is referred to as *Eucheuma* spp. *Eucheuma spinosum* is used to make fish balls and salads (Damongilala et al., 2021). *Kappaphycus alvarezii* is used to prepare salads, sausages, flat rice noodles, and yellow alkaline noodles, and they are also used as functional foods (Matos et al., 2024).

11.6.7.3 Agar-agar salad

Agar-agar strips are used to make salads (Zuberi and Bengisu, 2019). Southern China and Taiwan commonly use these agar strips in salads. Cut the agar-agar into 2–3-in. strips and soak it in water for at least 1 h. Then, mix a spice mixture with the salad, adding salad dressing (Kaláb and Larocque, 1996).

11.6.7.4 Hijiki

Sargassum fusiforme is a brownish-green seaweed found on the coasts of Japan and Korea (Mise et al., 2019). Korean people also know it as tot (Afonso et al., 2019). Hijiki is a traditional food and part of the Japanese people's daily diet. This hijiki is used steamed, boiled, marinated in soy or fish sauce, cooked in oil, or added to soup (Zheng et al., 2013).

11.7 Future direction

The future diversification of seaweed-based food products will develop snacks like chips and crackers, using powder-formed seaweed to produce these snacks. Agar jelly is used to make sweet candy, and we can also introduce smoothies or energy drink supplements. Vegetarian people can introduce seaweed-based alternatives for the development of seaweed-based meat products and egg replacements for the vegan market. Seaweed is rich in high amounts of bioactive compounds like antioxidants and omega-3 fatty acids. Nowadays, seaweed food products are used as functional foods.

Seaweed is rich in minerals, vitamins, and iodine. Seaweed contains prebiotics that promote human gut health. Seaweed is used to make processed foods and low-sodium content alternatives to enhancers to reduce salt content. Finally, seaweed is called a carbon-neutral food and promotes sustainable and low-carbon food sources.

11.8 Conclusion

Seaweed-based food products represent a world of nutrition and sustainability, rich in nutrient-dense food production. Based on the diverse array of health benefits, seaweed-based food products are being introduced as part of a sustainable diet. As the global population grows, the demand for protein and other nutrients is rapidly increasing. Due to resource limitations, traditional sources like vegetables and meat may not be sustainable for long-term consumption. Nowadays, seaweed has become an integral part of the future food landscape.

References

Aakre, I., Tveit, I.B., Myrmel, L.S., Fjære, E., Ballance, S., Rosendal-Riise, H., 2023. Bioavailability of iodine from a meal consisting of sushi and a wakame seaweed salad—A randomized crossover trial. Food Sci. Nutr. 11 (12), 7707–7717. doi:10.1002/fsn3.3689.

Abdel-Kareem, M.S.M., ElSaied, A.A.F., 2021. Global seaweeds diversity. In: Handbook of Algal Biofuels: Aspects of Cultivation, Conversion, and Biorefinery, pp. 39–55. doi:10.1016/B978-0-12-823764-9.00001-7.

Abka Khajouei, R., Keramat, J., Hamdami, N., Ursu, A.-V., Delattre, C., Gardarin, C., Lecerf, D., Desbrières, J., Djelveh, G., Michaud, P., 2021. Effect of high voltage electrode discharge on the physicochemical characteristics of alginate extracted from an Iranian brown seaweed (*Nizimuddinia zanardini*). Algal Res. 22119264. 56, 102326. doi:10.1016/j.algal.2021.102326.

Afonso, N.C., Catarino, M.D., Silva, A.M.S., Cardoso, S.M., 2019. Brown macroalgae as valuable food ingredients. Antioxidants 8 (9). doi:10.3390/antiox8090365.

Aina, O., Bakare, O.O., Daniel, A.I., Gokul, A., Beukes, D.R., Fadaka, A.O., Keyster, M., Klein, A., 2022. Seaweed-derived phenolic compounds in growth promotion and stress alleviation in plants. Life 12 (10). doi:10.3390/life12101548.

Amer, S.A., Abd El-Rahman, H.S.M., 2023. Development and evaluation of free sugar jelly made with leafy vegetables as a functional food. Carpathian J. Food Sci. Technol. 15 (3), 26–46. doi:10.34302/crpjfst/2023.15.3.3.

Amorim, K., Lage-Yusty, M.A., López-Hernández, J., 2012. Short communication: changes in bioactive compounds content and antioxidant activity of seaweed after cooking processing. CYTA J. Food. 19476345. 10 (4), 321–324. doi:10.1080/19476337.2012.658871.

Asanka Sanjeewa, K.K., Jeon, Y.-J., 2018. Edible brown seaweeds: a review. J. Food Bioact. 2637-8752. 37–50. doi:10.31665/jfb.2018.2139.

Asanka Sanjeewa, K.K., Lee, W., Jeon, Y.-J., 2018. Nutrients and bioactive potentials of edible green and red seaweed in Korea. Fish. Aquat. Sci. 21 (1). doi:10.1186/s41240-018-0095-y.

Atias, J.R., Sciences, P., 2016. Rediscovery of naturally occurring seagrape Caulerpa lentillifera from the Gulf of Mannar and its mariculture. Curr. Sci. 97, 1418–1420.

Bácskay, I., Papp, B., Pártos, P., Budai, I., Pető, Á., Fehér, P., Ujhelyi, Z., Kósa, D., 2024. Formulation and evaluation of insulin-loaded sodium-alginate microparticles for oral administration. Pharmaceutics. 1999-4923. 16 (1), 46. doi:10.3390/pharmaceutics16010046.

Banaś, K., Piwowar, A., Harasym, J., 2024. Agar-rapeseed oil hydroleogels as solid fat substitute in short-bread cookies. Food Hydrocoll. 151. 10.1016/j.foodhyd.2024.109889

Barandiaran, L.N., Taylor, V.F., Karagas, M.R., 2024. Exposure to iodine, essential and non-essential trace element through seaweed consumption in humans. Sci. Rep. 20452322. 14 (1). Available from: https://www.nature.com/srep/. doi:10.1038/s41598-024-64556-w.

Battershill, Z.V., 2024. Nutritional profiling of five New Zealand seaweeds – a preliminary assessment. Front. Mar. Sci. 22967745. 11. Available from: https://www.frontiersin.org/journals/marine-science#. doi:10.3389/fmars.2024.1410005.

Belattmania, Z., Kaidi, S., Atouani, S.E., Katif, C., Bentiss, F., Jama, C., Reani, A., Sabour, B., Vasconcelos, V., 2020. Isolation and FTIR-ATR and 1H NMR characterization of alginates from the main alginophyte species of the Atlantic coast of Morocco. Molecules. 1420-3049. 25 (18), 4335. doi:10.3390/molecules25184335.

Bojorges, H., Martínez-Abad, A., Martínez-Sanz, M., Dolores Rodrigo, M., Vilaplana, F., López-Rubio, A., José Fabra, M., 2023. Structural and functional properties of alginate obtained by means of high hydrostatic pressure-assisted extraction. Carbohydr. Polym. 01448617. 299, 120175. doi:10.1016/j.carbpol.2022.120175.

Brownlee, I.A., 2011. The physiological roles of dietary fibre. Food Hydrocoll. 25 (2), 238–250. doi:10.1016/j.foodhyd.2009.11.013.

Chan, E.W.C., Kezuka, M., Chan, H.T., Wong, S.K., 2023. The health-promoting properties of seaweeds: clinical evidence based on Wakame and Kombu. J. Nat. Remedies 23 (3), 687–698. doi:10.18311/jnr/2023/30820.

Chen, Z., Wu, W., Wen, Y., Zhang, L., Wu, Y., Farid, M.S., El-Seedi, H.R., Capanoglu, E., Zhao, C., 2023. Recent advances of natural pigments from algae. Food Prod. Process. Nutr. 5 (1). doi:10.1186/s43014-023-00155-y.

Coniglio, D., Bianco, M., Ventura, G., Calvano, C.D., Losito, I., Cataldi, T.R.I., 2021. Lipidomics of the edible brown alga wakame (*Undaria pinnatifida*) by liquid chromatography coupled to electrospray ionization and tandem mass spectrometry. Molecules 26 (15). doi:10.3390/molecules26154480.

Cotas, J., Gomes, L., Pacheco, D., Pereira, L., 2023. Ecosystem services provided by seaweeds. Hydrobiology 2 (1), 75–96. doi:10.3390/hydrobiology2010006.

Cotas, J., Leandro, A., Pacheco, D., Gonçalves, A.M.M., Pereira, L., 2020. A comprehensive review of the nutraceutical and therapeutic applications of red seaweeds (*Rhodophyta*). Life 10 (3). doi:10.3390/life10030019.

Damongilala, L.J., Wewengkang, D.S., Losung, F., Tallei, T.E., 2021. Phytochemical and antioxidant activities of eucheuma spinosum as natural functional food from north sulawesi waters, Indonesia. Pak. J. Biol. Sci. 24 (1), 132–138. doi:10.3923/pjbs.2021.132.138.

Epstein, G., Smale, D.A., 2017. Undaria pinnatifida: a case study to highlight challenges in marine invasion ecology and management. Ecol. Evol. 7 (20), 8624–8642. doi:10.1002/ece3.3430.

Faudon, S., Savage, G., 2014. Manufacture of a low oxalate mitsumame-type dessert using rhubarb juice and calcium salts. Food Nutr. Sci. 2157-944X. 05 (17), 1621–1627. doi:10.4236/fns.2014.517174.

Freitas, M.V., Pacheco, D., Cotas, J., Mouga, T., Afonso, C., Pereira, L., 2022. Red seaweed pigments from a biotechnological perspective. Phycology 2 (1), 1–29. doi:10.3390/phycology2010001.

French, C.S., Young, V.K., 1952. The fluorescence spectra of red algae and the transfer of energy from phycoerythrin to phycocyanin and chlorophyll. J. Gen. Physiol. 35 (6), 873–890.

Guo, L., Goff, H.D., Xu, F., Liu, F., Ma, J., Chen, M., Zhong, F., 2020. The effect of sodium alginate on nutrient digestion and metabolic responses during both in vitro and in vivo digestion process. Food Hydrocoll. 0268005X. 107, 105304. doi:10.1016/j.foodhyd.2019.105304.

Harsha Mohan, E, Madhusudan, S., Baskaran, R., 2023. The sea lettuce Ulva sensu lato: future food with health-promoting bioactives. Algal Res. 71. 10.1016/j.algal.2023.103069

Hassan, Fadhlina, Nor, Sheikh, Yusof, Azman, 2024. Enhancing physicochemical and leaching properties of grass jelly formulation by incorporation of gelatine. Food Res. 8 (2), 371–381. doi:10.26656/fr.2017.8(2).195.

Hata, Y., Nakajima, K., Uchida, J.-I., Hidaka, H., Nakano, T., 2001. Clinical effects of brown seaweed, *Undaria pinnatifida* (wakame), on blood pressure in hypertensive subjects. J. Clin. Biochem. Nutr. 30, 43–53. doi:10.3164/jcbn.30.43.

Hong, T., Zhang, Y., Xu, D., Wu, F., Xu, X., 2021. Effect of sodium alginate on the quality of highland barley fortified wheat noodles. LWT. 00236438. 140, 110719. doi:10.1016/j.lwt.2020.110719.

Hossain, M.S., Liyana, E., Sifat, S.A.-D., Ameen, F., Ullah, M.A., Jolly, Y.N., Quraishi, S.B., Hossain, M., Salleh, S., Akter, S., Hossain, M.A., Mukhlish, M.Z.B., Elliott, M., 2022. Trace element bioaccumulation in edible red seaweeds (*Rhodophyta*): a risk assessment for consumers. Environ. Pollut. 307. 10.1016/j.envpol.2022.119560

Hurd, C.L., Law, C.S., Bach, L.T., Britton, D., Hovenden, M., Paine, E.R., Raven, J.A., Tamsitt, V., Boyd, P.W., 2022. Forensic carbon accounting: assessing the role of seaweeds for carbon sequestration. J. Phycol. 58 (3), 347–363. doi:10.1111/jpy.13249.

Idota, Y., Kogure, Y., Kato, T., Ogawa, M., Kobayashi, S., Kakinuma, C., Yano, K., Arakawa, H., Miyajima, C., Kasahara, F., Ogihara, T., 2016. Cholesterol-lowering effect of calcium alginate in rats. Biol. Pharm. Bull. 0918-6158. 39 (1), 62–67. doi:10.1248/bpb.b15-00503.

Jensen, M.G., Kristensen, M., Astrup, A., 2012. Effect of alginate supplementation on weight loss in obese subjects completing a 12-wk energy-restricted diet: a randomized controlled trial. Am. J. Clin. Nutr. 19383207. 96 (1), 5–13. Available from: http://www.ajcn.org/content/96/1/5.full.pdf+html. doi:10.3945/ajcn.111.025312Denmark.

Kaláb, M., Larocque, G., 1996. Suitability of agar gel encapsulation of milk and cream for electron microscopy. LWT 29 (4), 368–371. doi:10.1006/fstl.1996.0056.

Lee, D., Nishizawa, M., Shimizu, Y., Saeki, H., 2017. Anti-inflammatory effects of dulse (*Palmaria palmata*) resulting from the simultaneous water-extraction of phycobiliproteins and chlorophyll a. Food Res. Int. 100, 514–521. doi:10.1016/j.foodres.2017.06.040.

Lee, S., Bae, I.Y., Jung, J., Jang, K., Kim, Y.W., Lee, H.G., 2008. Physicochemical, textural and noodle-making properties of wheat dough containing alginate. J. Texture Stud. 0022-4901. 39 (4), 393–404. doi:10.1111/j.1745-4603.2008.00149.x.

Liu, X., Bogaert, K., Engelen, A.H., Leliaert, F., Roleda, M.Y., De Clerck, O., 2017. Seaweed reproductive biology: environmental and genetic controls. Bot. Mar. 60 (2), 89–108. doi:10.1515/bot-2016-0091.

Louala, S., Lamri-Senhadji, M., 2019. Beneficial effects of low-calorie-carbohydrate/high-agar diet on cardiometabolic disorders associated with non-alcoholic fatty liver disease in obese rats. Prev. Nutr. Food Sci. 24 (4), 400–409. doi:10.3746/pnf.2019.24.4.400.

Lukova, P., Kokova, V., Baldzhieva, A., Murdjeva, M., Katsarov, P., Delattre, C., Apostolova, E., 2024. Alginate from *Ericaria crinita* possesses antioxidant activity and attenuates systemic inflammation via downregulation of pro-inflammatory cytokines. Mar. Drugs. 1660-3397. 22 (11), 482. doi:10.3390/md22110482.

Maeda, H., Yamamoto, R., Hirao, K., Tochikubo, O., 2005. Effects of agar (kanten) diet on obese patients with impaired glucose tolerance and type 2 diabetes. Diabetes, Obes. Metab. 1462-8902. 7 (1), 40–46. doi:10.1111/j.1463-1326.2004.00370.x.

Malaka, R., Murphi Ningrum, E.H., 2020. Yoghurt syneresis with addition of agar as stabilizer. Hasanuddin J. Anim. Sci 2 (1), 43–50.

Martins, A.C., 2024. The effect of sugar and jelly concentration on the quality of young coconut pudding (Cocos Nucifera L.). Teknol. Pangan dan Ilmu Pertan. 2, 1–6. doi:10.59581/jtpip-widyakarya.v2i1.2859.

Matos, Cardoso, Serralheiro, Bandarra, Afonso, 2024. Seaweed bioactives potential as nutraceuticals and functional ingredients: a review. J. Food Compos. Anal. 133. 10.1016/j.jfca.2024.106453

Menon, V.V., 2011. Seaweed Polysaccharides - Food Applications. In: Handbook of Marine Macroalgae: Biotechnology and Applied Phycology. John Wiley and Sons, India, pp. 541–555. Available from: http://onlinelibrary.wiley.com/book/10.1002/9781119977087. doi:10.1002/9781119977087.ch36.

Michalak, I., Tiwari, R., Dhawan, M., Alagawany, M., Farag, M.R., Sharun, K., Emran, T.B., Dhama, K., 2022. Antioxidant effects of seaweeds and their active compounds on animal health and production–a review. Vet. Q. 42 (1), 48–67. doi:10.1080/01652176.2022.2061744.

Millan-Linares, M.C., Martin, M.E., Rodriguez, N.M., Toscano, R., Claro, C., Bermudez, B., Pedroche, J., Millan, F., La Paz, S.M.-D., 2019. Nutraceutical extract from dulse (*Palmaria palmata* L.) inhibits primary human neutrophil activation. Mar. Drugs 17 (11). doi:10.3390/md17110610.

Mise, N., Ohtsu, M., Ikegami, A., Mizuno, A., Cui, X., Kobayashi, Y., Nakagi, Y., Nohara, K., Yoshida, T., Kayama, F., 2019. Hijiki seaweed consumption elevates levels of inorganic arsenic intake in Japanese children and pregnant women. Food Addit. Contam. Part A Chem. Anal. Control. Expo. Risk Assess. 36 (1), 84–95. doi:10.1080/19440049.2018.1562228.

Morais, T., Inácio, A., Coutinho, T., Ministro, M., Cotas, J., Pereira, L., Bahcevandziev, K., 2020. Seaweed potential in the animal feed: a review. J. Mar. Sci. Eng. 2077-1312. 8 (8), 559. doi:10.3390/jmse8080559.

Mouritsen, O.G., Dawczynski, C., Duelund, L., Jahreis, G., Vetter, W., Schröder, M., 2013. On the human consumption of the red seaweed dulse (*Palmaria palmata* (L.) Weber & Mohr). J. Appl. Phycol. 09218971. 25 (6), 1777–1791. doi:10.1007/s10811-013-0014-7.

Nedović, V.A., Obradović, B., Leskošek-Čukalović, I., Trifunović, O., Pešić, R., Bugarski, B., 2001. Electrostatic generation of alginate microbeads loaded with brewing yeast. Process Biochem. 37 (1), 17–22. doi:10.1016/S0032-9592(01)00172-8.

Nishimura, M., Sugawara, M., Kudo, M., Kinoshita, Y., Yoshino, H., Nishihira, J., 2019. Effects of daily intake of Harudori-kombu: a randomized, double-blind, placebo-controlled, parallel-group study. Funct. Foods Health Dis. 21603855. 9 (4), 205–223. Available from: https://ffhdj.com/index.php/ffhd/article/view/594/1154. doi:10.31989/ffhd.v9i4.594.

Nishiumi, S., Izumi, Y., Kobayashi, T., Yoshida, M., 2020. Possible involvement of lipids in the effectiveness of kombu in individuals with abnormally high serum triglyceride levels. J. Nutr. Sci. Vitaminol. 0301-4800. 66 (2), 185–190. doi:10.3177/jnsv.66.185.

Nouban, F., Abazid, M., Abazid, F., 2017. Plastic degrading fungi Trichoderma viride and Aspergillus nomius isolated from local landfill soil in Medan. Iopscience.Iop.Org 8, 68–74. doi:10.1088/1755-1315.

Orosa, Torres, Fidalgo, Abalde, 2000. Production and analysis of secondary carotenoids in green algae. J. Appl. Phycol. 12 (3-5), 553–556. doi:10.1023/A:1008173807143.

Öztürk-Kerimoğlu, B., 2021. A promising strategy for designing reduced-fat model meat emulsions by utilization of pea protein-agar agar gel complex. Food Struct. 29. 10.1016/j.foostr.2021.100205

Padmesh, S., Singh, A., 2021. Agars: properties and applications. In: Polysaccharides: Properties and Applications, pp. 75–93. doi:10.1002/9781119711414.ch5.

Pandya, Y., 2022. Agar-agar extraction, structural properties and applications: a review. Pharma Innov. J. 6, 1151–1157.

Parreidt, T.S., Müller, K., Schmid, M., 2018. Alginate-based edible films and coatings for food packaging applications. Foods. 23048158. 7 (10). Available from: https://www.mdpi.com/2304-8158/7/10. doi:10.3390/foods7100170.

Patel, R., Pandya, K., Jasrai, R.T., Brahmbhatt, N., 2017. A review: scope of utilizing seaweed as a biofertilizer in agriculture. Int. J. Adv. Res. 23205407. 5 (7), 2046–2054. doi:10.21474/IJAR01/4941.

Pati, M.P., Sharma, S.D., Nayak, L., Panda, C.R., 2016. Uses of seaweed and its application to human welfare: a review. Int. J. Pharm. Pharm. Sci. 09751491. 8 (10), 12–20. Available from: http://innovareacademics.in/journals/index.php/ijpps/article/download/12740/7307. doi:10.22159/ijpps.2016v8i10.12740.

Peñalver, R., Lorenzo, J.M., Ros, G., Amarowicz, R., Pateiro, M., Nieto, G., 2020. Seaweeds as a functional ingredient for a healthy diet. Mar. Drugs 18 (6). doi:10.3390/md18060301.

Phan, T.D., Debeaufort, F., Luu, D., Voilley, A., 2005. Functional properties of edible agar-based and starch-based films for food quality preservation. J. Agric. Food Chem. 00218561. 53 (4), 973–981. doi:10.1021/jf040309s.

Pournaki, S.K., Aleman, R.S., Hasani-Azhdari, M., Marcia, J., Yadav, A., Moncada, M., 2024. Current review: alginate in the food applications. J. 2571-8800. 7 (3), 281–301. doi:10.3390/j7030016.

Prabhasankar, Ganesan, Bhaskar, Hirose, Stephen, N., Gowda, L.R., Hosokawa, Miyashita, 2009. Edible Japanese seaweed, wakame (*Undaria pinnatifida*) as an ingredient in pasta: chemical, functional and structural evaluation. Food Chem. 115 (2), 501–508. doi:10.1016/j.foodchem.2008.12.047.

Praiboon, J., Chantorn, S., Krangkratok, W., Choosuwan, P., La-ongkham, O., 2023. Evaluating the prebiotic properties of agar oligosaccharides obtained from the red alga gracilaria fisheri via enzymatic hydrolysis. Plants. 2223-7747. 12 (23), 3958. doi:10.3390/plants12233958.

Probst, Y., 2015. A review of the nutrient composition of selected rubus berries. Nutr. Food Sci. 17586917. 45 (2), 242–254. Available from: http://www.emeraldinsight.com/info/journals/nfs/nfs.jsp. doi:10.1108/NFS-07-2014-0063.

Puscaselu, R., Gutt, G., Amariei, S., 2020. The use of edible films based on sodium alginate in meat product packaging: an eco-friendly alternative to conventional plastic materials. Coatings. 20796412. 10 (2). Available from: https://res.mdpi.com/d_attachment/coatings/coatings-10-00166/article_deploy/coatings-10-00166-v2.pdf. doi:10.3390/coatings10020166.

Qin, Y., 2018. Seaweed bioresources. In: Bioactive Seaweeds for Food Applications: Natural Ingredients for Healthy Diets, pp. 3–24. doi:10.1016/B978-0-12-813312-5.00001-7.

Rajapakse, N., Kim, S.-K., 2011. Nutritional and digestive health benefits of seaweed. Adv. Food Nutr. Res. 64, 17–28. doi:10.1016/B978-0-12-387669-0.00002-8.

Reshma, B.S., Aavula, T., Narasimman, V., Ramachandran, S., Essa, M.M., Qoronfleh, M.W., Anand, V., 2022. Antioxidant and antiaging properties of agar obtained from brown seaweed *Laminaria digitata* (Hudson) in D-galactose-induced Swiss albino mice. Evid. Based Complement. Alternat. Med. 1741-4288. 2022, 1–16. doi:10.1155/2022/7736378.

Santos, H.M., 2018. Effects of sodium alginate - fed Pacific white shrimps, *Litopenaeus vannamei*, on toll - like receptors and *Vibrio alginolyticus* infection, 1–9. doi: 10.1111/are.13989.

Sari, Rahardjanto, Husamah, Purwanti, Permana, Fauzi, 2019. The formulation of artificial nori with the base mixture ingredients of *Gracilaria* sp. and *Arenga pinnata* (Wurmb) Merr. using the natural colorant from *Pleomele angustifolia* (Medik.) N.E. Br. IOP Conf. Ser. Earth Environ. Sci. 276 (1). doi:10.1088/1755-1315/276/1/012013.

Sato, N., Furuta, T., Takeda, T., Miyabe, Y., Ura, K., Takagi, Y., Yasui, H., Kumagai, Y., Kishimura, H., 2019. Antioxidant activity of proteins extracted from red alga dulse harvested in Japan. J. Food Biochem. 17454514. 43 (2). Available from: http://onlinelibrary.wiley.com/journal/10.1111/(ISSN)1745-4514. doi:10.1111/jfbc.12709.

Sharifi, A., Golestan, L., Baei, M.S., 2013. Studying the enrichment of ice cream with alginate nanoparticles including Fe and Zn salts. J. Nanoparticles. 2314-484X. 2013, 1–5. doi:10.1155/2013/754385.

Sinurat, E., Fransiska, D., Utomo, B.S.B., Subaryono, Nurhayati, Sihono, 2022. Characteristics of nori-like product prepared from seaweeds growing in Indonesia. J. Aquat. Food Prod. Technol. 31 (6), 525–535. doi:10.1080/10498850.2022.2077677.

Smyth, P.P.A., 2021. Iodine, Seaweed, and the Thyroid. Eur. Thyroid J. 10 (2), 101–108. doi:10.1159/000512971.

Stévant, P., Søndergaard Schmedes, P., Gall, L.L., Wegeberg, S., Dumay, J., Rebours, C., 2023. Concise review of the red macroalga dulse, *Palmaria palmata* (L.) Weber & Mohr. J. Appl. Phycol. 0921-8971. 35 (2), 523–550. doi:10.1007/s10811-022-02899-5.

Tabara, A., Miyajima, C., Moki, N., Kasahara, F., Seguchi, M., 2016. Improvement of bread making properties by the addition of alginates. Food Sci. Technol. Res. 1344-6606. 22 (1), 145–151. doi:10.3136/fstr.22.145.

Taboada, C., Millan, R., Miguez, I., 2013. Evaluation of marine algae *Undaria pinnatifida* and *Porphyra purpurea* as a food supplement: composition, nutritional value and effect of intake on intestinal, hepatic and renal enzyme activities in rats. J. Sci. Food Agric. 93 (8), 1863–1868. doi:10.1002/jsfa.5981.

Takahashi, M., Mikami, K., 2017. Oxidative stress promotes asexual reproduction and apogamy in the red seaweed *Pyropia yezoensis*. Front. Plant Sci. 8 (January). doi:10.3389/fpls.2017.00062.

Thangavel, P., Sridevi, G., 2015. Environmental Sustainability: Role of Green Technologies. Springer, India, pp. 1–324. Available from. doi:10.1007/978-81-322-2056-5.

Wada, K., Tsuji, M., Nakamura, K., Oba, S., Nishizawa, S., Yamamoto, K., Watanabe, K., Ando, K., Nagata, C., 2021. Effect of dietary nori (dried laver) on blood pressure in young Japanese children: an intervention study. J. Epidemiol. 31 (1), 37–42. doi:10.2188/jea.JE20190176.

Wahyudi, V.A., Harini, N., Manshur, H.A., Wachid, M., Aini, A.N., 2022. Study of protein concentrate from flying fish roe filament and its application for nutrified rice-corn milk. Curr. Res. Nutr. Food Sci. 23220007. 10 (2), 766–776. Available from: https://www.foodandnutritionjournal.org/. doi:10.12944/CRNFSJ.10.2.29.

Xu, J., Liao, W., Liu, Y., Guo, Y., Jiang, S., Zhao, C., 2023. An overview on the nutritional and bioactive components of green seaweeds. Food Prod. Process. Nu. 2661-8974. 5 (1). doi:10.1186/s43014-023-00132-5.

Zarekarizi, A., Hoffmann, L., Burritt, D., 2019. Approaches for the sustainable production of fucoxanthin, a xanthophyll with potential health benefits. J. Appl. Phycol. 15735176. 31 (1), 281–299. Available from: https://www.wkap.nl/journalhome.htm/0921-8971. doi:10.1007/s10811-018-1558-3.

Zheng, T., Liu, C., Yang, J., Liu, Q., Li, J., 2013. Hijiki seaweed (*Hizikia fusiformis*): nutritional value, safety concern and arsenic removal method. Adv. Mater. Res. 634-638 (1), 1247–1252. doi:10.4028/www.scientific.net/AMR.634-638.1247.

Zhou, X., Peng, Z., Liao, Y., Li, D., Xu, S., Wen, Y., Gao, J., Qi, X., Zhang, X., Feng, L., Zhang, H., Hao, X., Wang, Q., Liu, L., Yang, W., 2023. Weight reduction effect of alginate associated with gut microbiota and bile acids: a double-blind and randomized trial. J. Funct. Foods. 17564646. 108, 105774. doi:10.1016/j.jff.2023.105774.

Zuberi, H.S., Bengisu, M., 2019. Agar-based adaptable DIY materials. FME Trans. 47 (3), 442–451. doi:10.5937/fmet1903442H.

CHAPTER

12

Marine polysaccharides based on food products

Navindu Dinara Gajanayaka[a,b], *Minthari Sakethanika Bandara*[a,b], *and Chulhong Oh*[a,b]

[a]Jeju Bio Research Center, Korea Institute of Ocean Science and Technology (KIOST), Jeju-si, Republic of Korea

[b]Department of Marine Technology & Convergence Engineering (Marine Biotechnology), KIOST School, University of Science and Technology, Daejeon, Republic of Korea

12.1 Introduction

With the advent of global demand for sustainable and health-promoting food alternatives, marine polysaccharides have emerged as crucial and versatile ingredients, shaping the future of functional food products. Marine polysaccharides are a category of complex carbohydrates with macromolecular polymers (Samalens et al., 2022). Marine polysaccharides are among the most abundant substances in the ocean and include seaweed-oriented (ulvan, alginate, carrageenan, fucoidan, and laminarin), animal-oriented (chitosan and hyaluronic acid), and marine microorganism-oriented polymers (Samalens et al., 2022; Sun et al., 2021; Jeewon et al., 2023). In recent years, these natural polymers have gained attention due to their structural and functional diversity, becoming vital across various industries, particularly the food industry (Althagbi et al., 2020; Yosri et al., 2021).

Marine polysaccharides are distinguished from terrestrial polysaccharides by their distinctive molecular structures and properties, which are influenced by the unique marine environment. These distinctive functional and biological advantages create potential applications (Althagbi et al., 2020). Compared to terrestrial ecosystems, marine ecosystems exist in a salinized buffer system with water pressure, high salinity, low temperature, low dissolved oxygen, and insufficient light conditions (Wang et al., 2023c). Thus marine polysaccharides are composed of rare sugars such as rhamnose, fucose, and xylose, which are less abundant in terrestrial-derived polysaccharides (Podolean et al., 2022). These rare sugars contribute to many aspects in terms of biotechnological applications. For instance, rhamnose can be used as an alternative for the synthesis of meat-like flavor (Wong et al., 1983) and as a medication

Harvesting the Sea
https://doi.org/10.1016/B978-0-443-29190-6.00015-5

for gastric ulcers (Lahaye and R, 2007). Furthermore, marine polysaccharides are abundant in sulfate groups, which lead to enhanced properties such as hydrophilicity, ionic interactions, and biological activities (Berezhnaya et al., 2024; Augusto et al., 2024a, 2024a).

With the rapid development of world food technology, marine-derived alternatives are emerging as crucial components in creating sustainable, nutritional, and innovative food products (Augusto et al., 2024a). In particular, marine polysaccharides are used to improve the consistency and texture of food products (Han et al., 2022). In recent years, chitosan, alginate, agar, and other cheap and readily available marine polysaccharides have gained more attention in the application of the food industry (Han et al., 2022). Sodium alginate, a natural linear polymer derived from brown algae, has been used in the food industry due to its nontoxicity, biodegradability, and biocompatibility (Wang et al., 2022; Montes et al., 2022). Moreover, alginate shows a unique ability to form a strong, heat-stable gel in the presence of various divalent cations (Ca^{2+}, Mg^{2+}) (Burdick and Stevens, 2005). Beyond the food industry, alginate is a preferred material for drug delivery systems in the pharmaceutical industry (Abourehab et al., 2022). Chitosan, with its abundance of reactive groups like amine and hydroxyl groups, along with its antibacterial properties, is regarded as an ideal raw material for food preservation (Han et al., 2022; Wu et al., 2022). The versatile properties of these marine-derived polysaccharides, such as the ability to enhance the texture and shelf life of foods, highlight their potential application within the food sector. With the advancement of research perspectives, these eco-friendly materials are expected to play an important role in creating sustainable, functional, and innovative food solutions for the future.

Marine polysaccharides play a crucial role in the marine ecosystem while contributing eco-friendly solutions for various human-oriented applications. In marine ecosystems, these polymers provide structural support to maintain ecological stability across the ecosystem and serve as a source of nutrients and energy for marine organisms (Ahmed et al., 2024). For instance, cellulose and ulvan in green algae, agarose in red algae, and laminarin in brown algae offer structural support by maintaining cellular integrity, while also sustaining marine food webs (Ahmed et al., 2024). Beyond these ecological significances, marine polysaccharides provide an undeniable contribution to sustainable resource utilization. Unlike terrestrial agriculture, marine resource cultivation does not depend on freshwater or arable land. Furthermore, fertilization or additional nutrients are generally unnecessary, and farming strategies can deliver both economic and ecological benefits (Sultana et al., 2023). Therefore seaweed cultivation is ideal for addressing the rising global demand for food. In addition, seaweed cultivation contributes significantly to carbon sequestration by absorbing atmospheric CO_2 and mitigating greenhouse gas emissions (Hossain et al., 2020b). This dual-function role of resource provision and climate mitigation highlights the complex environmental advantages of marine polysaccharides.

The potential application in the circular economy further strengthens the importance of marine polysaccharides. For instance, in the seafood industry, most of the crustacean shells are discarded as waste. However, these can be converted into high-value-added polysaccharides like chitosan, which can be utilized in food preservation and biodegradable packaging (Duan et al., 2019; Riseh et al., 2024). Moreover, marine polysaccharide-based biodegradable packaging materials can serve as an eco-friendly alternative in the biomedical and food industries (Behrooznia and Nourmohammadi, 2024). The transformation of marine waste into value-added products not only benefits the environment but also contributes to a sustainable economic approach in resource efficiency (Vidal et al., 2022).

In recent years, global attention to sustainable and innovative approaches has significantly shifted the trends in the food industry (Yıkmış et al., 2024). This global trend has also highlighted the role of marine-derived polysaccharides in food applications (Yıkmış et al., 2024). Advancements in extraction processing (Arias et al., 2023), an increase in demand for sustainable ingredients (Sabu Mathew et al., 2024), and expanding applications in novel marine-based food products (Błaszak et al., 2018) are key shifting patterns in the marine-based food industry. The advancement of green technologies, such as enzymatic hydrolysis (Huang et al., 2024a) and supercritical CO_2 extraction (Zhao et al., 2021), has enhanced the production of high-purity polysaccharides and oligosaccharides with different biological properties (Sacramento et al., 2022). These advancements contribute to minimizing environmental impact and reducing chemical waste (Mishra et al., 2023). Moreover, the bioavailability of the final products enhances their desirability for both consumers and industries. The expansion of marine-derived ingredients in novel food products is also a rapidly growing trend. For instance, seaweed extracts and polysaccharides like fucoidan can be combined with different beverages, which can influence gut health and immunity (Mensah et al., 2023). Marine seaweeds, such as *Sargassum* and *Ulva*, can serve as viable alternatives to traditional tea, offering potential health benefits. These seaweeds are rich in bioactive compounds, including antioxidants, polysaccharides, and vitamins, which can contribute to improved human health when incorporated into dietary routines (Sangma et al., 2023).

Combination of unique characteristics, environmental relevance, and potential application in different industries, marine polysaccharides offer an unparalleled perspective for innovative, sustainable, and health-promoting food solutions. Moreover, their growing adaptations address not only the positive impacts of science and technology but also their vital contribution to global consumption trends and environmental imperatives. This chapter further explores how these biopolymers are shaping the future of food innovation, promoting health benefits, and addressing the growing need for eco-friendly solutions in the food industry. Furthermore, we discuss the application of novel technological approaches, along with their associated challenges and future trends, in relation to marine polysaccharides and their use in functional food products.

12.2 Diversity of marine polysaccharides

Marine ecosystems are a rich source of bioactive compounds, including polysaccharides, which have contributed to a wide range of industrial applications due to their structural and functional properties (Sanjeewa, 2024). These polysaccharides are considered a crucial source of functional components with tremendous roles in human health advancement and the reduction of disease risk vulnerability (Senadheera et al., 2023). Therefore recently, a huge and growing interest has been observed in studies related to the development of marine-derived functional dietary supplements (Shahidi and Ambigaipalan, 2015). The discovery and incorporation of marine-derived bioactive compounds into food products are relatively recent, especially when compared to bioactive compound sources from terrestrial origins. Many underexplored marine bioactive sources possess high nutritional values (Senadheera et al., 2023). A wide range of sources, including marine algae, marine animals, and marine

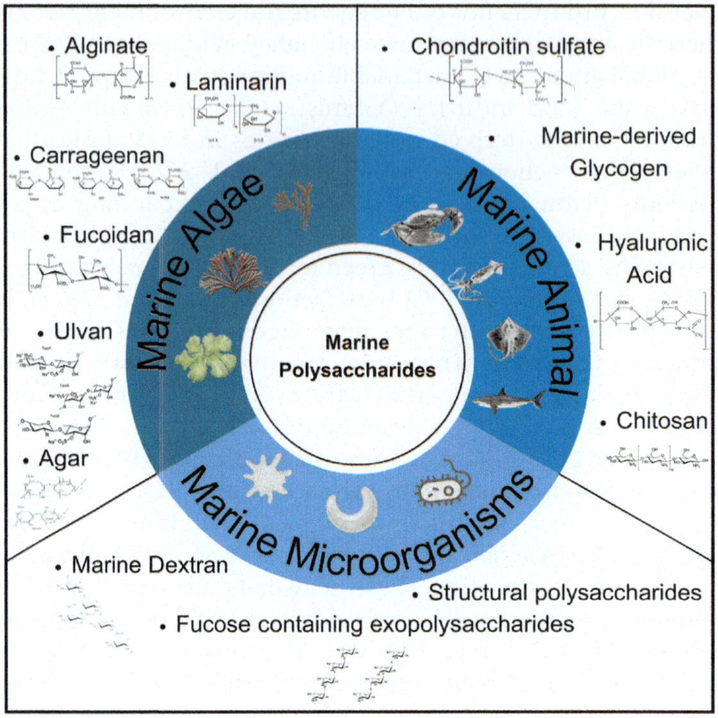

FIG. 12.1 Diversity and distribution of common marine polysaccharides and their structures. This figure illustrates the various types of marine polysaccharides, their sources, and structural compositions.

microorganism-derived polysaccharides, are contributing to the functional food industry (Ghosh et al., 2022; Hossain et al., 2020b). The diversity of common polysaccharides and their structural composition is illustrated in Fig. 12.1.

12.2.1 Marine algae-derived polysaccharides

Seaweeds, or marine macroalgae, representing around 90% of all marine plants (Senadheera et al., 2023), are the most significant contributors to marine polysaccharide-derived food products. Macro algae can be classified into three main categories based on their primary pigment, including Rhodophyta (red algae), Phaeophyta (brown algae), and Chlorophyta (green algae) (Deepika et al., 2022). Major algal polysaccharides, such as agar and carrageenan, derived from red algae, alginates derived from brown algae, and ulvan derived from green algae, are mainly used in the food-related industry.

Generally, agar is found in the cell wall matrix of red seaweeds belonging to the *Gelidiales* and *Gracilariales* orders. In the production of agar, due to its abundance and easier manipulation, *Gelidium tenuistipitata* is considered the most economical macro algae (Torres et al., 2019). Agar is a linear polysaccharide consisting of alternating (1,3)-linked D-galactose and (1,4)-linked 3,6-anhydro-L-galactose units. It can be slightly substituted with sulfate, methyl, or pyruvate groups (Lee et al., 2016). The two main components are agarose, a neutral linear polysaccharide, and agaropectin, an acidic polysaccharide containing other substitutes (sulfates, pyruvic acids, and D-glucuronic acid). Various agarose-derived molecules can be produced through chemical or enzymatic degradation. Hydrolysis products such as agarooligosaccharides, neoagarooligosaccharides, neoagarobiose, and 3,6-anhydro-L-galactose (L-AHG) have demonstrated biological activities (Yu et al., 2019). The properties of agar can vary with the type of species, geographical environment, and season (Torres et al., 2019; Sousa et al., 2012; Vergara-Rodarte et al., 2010). Furthermore, the process of extraction and the methods of pre- and postharvest storage also affect the properties of agar (Torres et al., 2019; Rhein-Knudsen et al., 2017). The market value is determined by the quality of the agar, which is affected by its molecular arrangement, as well as its chemical and physical properties (Riquelme et al., 2023; Arvizu-Higuera et al., 2007; Liao et al., 2021).

Carrageenan is a linear hydrophilic sulfated polysaccharide derived from red seaweed. It is composed of alternating units of D-galactose and 3,6-anhydrogalactose, linked by β-1,3 and α-1,4 glycosidic bonds. According to the average molecular size, the sulfated and ester group positioning, and the presence of a 3,6-anhydro ring in the α-linked galactose, carrageenan can be classified into three main categories: κ-carrageenan, ι-carrageenan, and λ-carrageenan. K-carrageenan, ι-carrageenan, and λ-carrageenan have one, two, and three sulfate groups per disaccharide, respectively (Fontes-Candia et al., 2020b). Furthermore, these polysaccharides are Generally Recognized As Safe (GRAS) compounds for use in the food industry (Necas and Bartosikova, 2013). The genera *Iridaea, Furcellaria, Kappaphycus, Chondrus, Hypnea, Eucheuma,* and *Gigartina* are widely used for the obtainment of carrageenan for industrial purposes. *Kappaphycus alvarezii* and *Eucheuma denticulatum* are commonly used for κ-carrageenan and ι-carrageenan.

Alginates, fucoidans, and laminarin are polysaccharides extracted from brown algae. Laminarin and fucoidan are widely used for their biological properties (Karuppusamy et al., 2022; Levy-Ontman et al., 2023), while alginates are utilized in the food and medical industries due to their physical properties, such as stabilizing, emulsifying, and thickening (Khajouei et al., 2018; Dobrincic et al., 2020; Ponce and Stortz, 2020). Alginates are composed of β-d-mannuronic (M) acid and α-l-guluronic acid (G) units, linked by (1,4) bonds and adopting a pyranose ring structure. These units are organized into uniform (MM or GG) or mixed (MG or GM) blocks, resulting in a wide variety of structures, molecular weights, and physicochemical properties (Khajouei et al., 2018; Hentati et al., 2018). Commercial alginates are derived from several seaweed species, including *Laminaria hyperborea, Laminaria digitata, Macrocystis pyrifera, Ascophyllum nodosum, Ecklonia maxima, Saccharina japonica, Lessonia nigrescens, Durvillea antarctica,* and various *Sargassum* spp. (Fernando et al., 2020). The molecular weight and the arrangement of monomers within the chain determine the physical and chemical properties of alginates (Fernando et al., 2020). The properties of alginates can also be influenced by environmental factors. Alginates extracted from relatively warm water environments tend to exhibit stronger gel formation, while those extracted from cold water environments generally have lower viscosity (Sari-Chmayssem et al., 2015; Phillips et al., 1988).

Fucoidan, a sulfated, water-soluble, complex polysaccharide, is widely found in the cell walls of brown algae. The primary backbone of fucoidans is composed of fucose, with sulfate groups linked to it, as well as acetyl groups and other monosaccharides such as glucose, mannose, galactose, and xylose (Kopplin et al., 2018; Yu et al., 2021; Wang et al., 2020). The composition and complexity of brown seaweeds exhibit significant diversity across different species (Anisha et al., 2022). Fucoidan exhibits diverse biological properties that make it applicable across various fields, including the food industry (Mensah et al., 2023). The molecular mass and sulfate content play a critical role in determining the type and extent of biological activity (Chen et al., 2021).

Compared to other algal polysaccharides, laminarin is one of the most underexploited polysaccharides (Karuppusamy et al., 2022). Laminarin is a branched, water-soluble polysaccharide made up of (1–3)-β-D-glucan with β (1–6)-linkages, and it can be extracted from brown algae. Laminarin has been extracted from various brown macroalgal species, such as *Laminaria digitata*, *Saccharina latissima*, *Laminaria japonica*, *Ecklonia kurome*, and *Eisenia bicyclis*, with smaller amounts found in *Ascophyllum*, *Fucus*, and *Undaria* species from both Asian and European regions (Garcia-Vaquero et al., 2018; Karuppusamy et al., 2022; Zargarzadeh et al., 2020).

Ulvan is a mostly linear, acidic, water-soluble sulfated polysaccharide extracted from the cell wall of green algae (Robic et al., 2009). The source of extraction (type of *Ulva* sp.), harvesting season, extraction, and purification method can alter the composition of ulvan (Li et al., 2023a). Ulvan is mainly composed of four types of monosaccharides: a high amount of sulfated rhamnose, followed by glucuronic acid, xylose, and iduronic acid. Due to the relatively high diversity of its composition, the structure of ulvan is considered a more complex polysaccharide compared to other algal polysaccharides such as agarose, alginate, and carrageenan (Li et al., 2023a). Moreover, the rare sugar composition of ulvan polysaccharides exhibits a broader range of unique characteristics compared to other algal polysaccharides (Wang et al., 2023).

12.2.2 Marine animal-derived polysaccharides

Chitin, a linear, cationic, alkaline, and natural fiber-forming polysaccharide, is found in the exoskeletons of crustaceans like crabs and shrimp, with crab shells being the primary source of both chitin and its derivative, chitosan (Bose et al., 2023; Wang et al., 2023c). Chitin is considered the most abundant biopolymer after cellulose (Samalens et al., 2022). Chitin is found in arthropods like shrimp, lobster, crab, and insects, as well as in mollusks such as squid pens. In addition, various unicellular eukaryotic organisms in marine environments produce chitin, including ciliates, cnidospora, rhizopods, diatoms, yeasts, and fungi (Zhang et al., 2021; Shao et al., 2023; Samalens et al., 2022). Annually, nearly 60,000–80,000 tons of shellfish waste are accumulated from the food industry as a by-product (Divya and Jisha, 2018). The utilization and application of this waste can play a significant role in resource recycling and reducing environmental pollution. Chitosan (β-1,4-glucosamine and *N*-acetyl-D-glucosamine) is produced by deacetylation of chitin and by fermentation with specific microorganisms (Tan et al., 2020), and it is considered the only naturally occurring alkaline amino polysaccharide (Zhang et al., 2022a). The amount of deacetylation depends on the source of chitin (Haripriya and Vijayakrishna, 2024). The biodegradation of chitin and chitosan is greatly influenced by their degree of deacetylation (Samalens et al., 2022). The key properties of chitosan are determined by the presence of three distinct functional groups (primary –OH, secondary –OH, and –NH2) and its ability to dissolve in water under acidic pH conditions (Duan et al., 2019).

12.2.3 Marine microorganism-derived polysaccharides

Exopolysaccharides (EPS) are considered natural high molecular weight polymers produced and secreted by microbes. Generally, microbes secrete these compounds into their surrounding environment to establish and maintain the functional and structural integrity of microbial biofilms (Ibrahim et al., 2022a). EPS have diverse functional properties with significant potential in the food sector (Korcz and Varga, 2021). Compared to marine algal and animal-derived polysaccharides, marine microorganism-derived polysaccharides and their potential applications are underexplored.

Concordio-Reis et al. (2021) have recorded six *Alteromonas* strains found in French Polynesia atolls that are capable of producing EPS. According to their records, the EPS are heteropolysaccharides containing neutral monosaccharides such as glucose, galactose, mannose, and rhamnose, as well as uronic acids. Moreover, these EPS exhibit high thermal tolerance, with a degradation temperature above 260°C, and possess notable rheological properties. They can form viscous aqueous solutions with shear-thinning behavior and are capable of forming strong gels in the presence of Fe^{2+} or under alkaline conditions with Mg^{2+}, Cu^{2+}, and Ca^{2+} (Concordio-Reis et al., 2021). Another study has documented that *Alteromonas australica* is capable of producing EPS with notable antioxidant properties, demonstrating its potential application as an effective emulsifier (Li et al., 2022).

Fucose-containing exopolysaccharides (FcEPS) are high molecular weight polysaccharides secreted by bacteria, exhibiting diverse biological properties such as antioxidant, antiinflammatory, antiviral, and prebiotic activities. These bioactivities make FcEPS highly suitable for applications in functional foods and nutraceuticals. Moreover, their production through bacterial fermentation offers a convenient and sustainable approach (Xiao et al., 2022; Gan et al., 2020). *Pseudoalteromonas* sp. YU16-DR3A produces an exopolysaccharide (EPS-DR3A) with a monosaccharide composition of fucose, erythrotetrose, ribose, and glucose. EPS-DR3A demonstrates several functional properties that make it highly relevant for the food industry. These include emulsifying activity, with an excellent ability to form stable emulsions with hydrocarbons and food oils; rheological behavior, exhibiting pseudoplastic properties suitable for use as a thickening or stabilizing agent; and thermal stability, with a high thermal degradation temperature (249°C) and melting point (322°C). In addition, EPS-DR3A displays notable antioxidant properties, making it an excellent candidate for applications in food packaging and as a functional ingredient in food products (Dhanya et al., 2022).

12.3 Technological applications in food products

The utilization of marine polysaccharides in the food industry has transformed trends in functional and health-oriented food products. These biopolymers, derived from marine algae, crustaceans, and marine microorganisms, are widely applied in the food sector due to their diverse physiological benefits, such as thickening, gelling, stabilizing, and encapsulating (Ghosh et al., 2022). This section highlights the technological applications of marine polysaccharides originating from different sources as thickening and gelling agents, emulsifiers, stabilizers, encapsulation materials, and texture modifiers. Emerging trends in their applications are summarized in Tables 12.1 and 12.2. Marine polysaccharides such as agar, alginate, and carrageenan are well formulated as thickening and gelling agents.

TABLE 12.1 The marine seaweed-oriented polysaccharides, physical and biochemical properties, and potential applications in the food sector.

Species	Polysaccharide type	Activity and property	Application	Reference
Brown algae	Alginate	Enhances physicochemical, textural, sensory, and rheological properties. Antioxidant and antiinflammatory properties Stabilizer Improves creaming ability	Yogurt Food packaging	(Bankole et al., 2023; Elbayomi et al., 2021) (Akshaya and Nathanael, 2024)
Jania adhaerens	Sulfated xylogalactan	Antioxidant	Natural additive	(Hentati et al., 2022)
	Alginate Carrageenan	Gelling property Edible gels	Food packing, Food coatings, 3D-printed personalized diets for health treatments	(Pascuta et al., 2022)
	Carrageenan, Agar, Alginate	Thickening, gelling, emulsifying	Food additives	(Liao et al., 2021)
	Alginate and chitosan	Moisture and water retention, barrier for gas concentration, delays ripening, antimicrobial, antioxidant	Edible film/coating Fruits (guava, pear, blueberries), vegetables (cucumber, capsicum, mushroom)	(Nair et al., 2020)
	Alginate Oligosaccharides	Immunomodulatory, antimicrobial, antioxidant, prebiotic, antihypertensive, antidiabetic, antitumor, anticoagulant, and other bioactivities	Prebiotics, feed supplements for aquaculture, poultry, and swine, cryoprotectants for frozen foods, postharvest treatments	(Liu et al., 2019)
	Alginate, pectin	Gelling agent	Jelly candy	(De Avelar and Efraim, 2020)
	Alginate	Encapsulation agent maintains cell viability, enhances survival under heat treatments, and simulated gastrointestinal conditions.	Dairy products, fermented foods	(Mahmoud et al., 2020)

TABLE 12.1 The marine seaweed-oriented polysaccharides, physical and biochemical properties, and potential applications in the food sector—cont'd

Species	Polysaccharide type	Activity and property	Application	Reference
	Alginate	Encapsulation, delivery system for nutraceuticals	Functional foods (lycopene-loaded emulsion gels)	(Shu et al., 2023)
	Alginate	Formation of hydrogel particles with shear-thinning behavior and increased viscosity.	3D-printed hydrogel particles for delivery systems in food applications	(Rysenaer et al., 2023)
	Sodium alginate	Edible film production, shelf-life enhancement	Wrapping materials for chocolate, Indian vegetable puff	(Singh et al., 2020b)
	Agar, alginate, carrageenan	Stabilizing, gelling, thickening	Bakery products (for better loaf volume), dairy products (to control viscosity, phase separation, etc.), toppings, icings	(Kumar et al., 2021)
	Carrageenan	Emulsifiers, stabilizers, thickening agents, gelling agents, antiviral, antibiotic, antifungal, and antitumor		(Li et al., 2023a)
	Carrageenan	Gelling agents, stabilizers, and emulsifiers improve Shelf-life and taste, enhancing the appearance of food products	Dairy products, ice cream, jams, jellies, and confections	(Liu et al., 2021a)
Ulva rigida	Ulvan	Texture enhancement	Beef sausage	(Alboofetileh et al., 2024)
Ulva lactuca	Ulvan	Antimicrobial activity, antifouling properties, hydrogel formation	Antimicrobial packaging and antifouling properties can be used in food processing equipment to prevent biofilm formation, ensuring hygiene	(Ibrahim et al., 2022b)
Ulva lactuca	Ulvan	Antifungal activity Fungicidal inhibition	Raw or processed tomato sauce (to increase shelf life)	(El Fayoumy et al., 2022)
Macroalgae		Antimicrobial properties	Meat Dairy Beverages Fish Bakery products packing	(Cabral et al., 2021)
	Ulvan oligosaccharides	Antidiabetic properties		(Gajanayaka et al., 2024)
	Fucoidan	Antioxidant activity, antiinflammatory activity	Functional foods edible films	(Zhang et al., 2022b)

Continued

TABLE 12.1 The marine seaweed-oriented polysaccharides, physical and biochemical properties, and potential applications in the food sector—cont'd

Species	Polysaccharide type	Activity and property	Application	Reference
S. japonica	Fucoidan	Antitumor activity, immunomodulatory effect		(Cotas et al., 2021)
Chondrus crispus	Carrageenan	Cholesterol-lowering activity	Thickening, gelling, and stabilizing agent	(Chudasama et al., 2021)
	Xylooligosaccharides	Prebiotic effect	Sugar and fat replacers in bakery and dairy products, the meat industry, and food packaging film	(Lin et al., 2024)

TABLE 12.2 The marine animal-oriented polysaccharides, physical and biochemical properties, and potential applications in the food sector.

Species	Source	Polysaccharide type	Activity and property	Reference
Raja pulchra	Head and axial cartilage	Chondroitin sulfate	Antiobesity	(Li et al., 2019)
	Cartilage		Antioxidant	(Li et al., 2021b)
Haliotis discus hannai ino	Gonad	Sulfated polysaccharides	Antiobesity	(Ai et al., 2018)
			Anticoagulant	(Zhao et al., 2016)
Abalone	Viscera	Polysaccharides	Antioxidant	(Guo et al., 2020)
Stichopus japonicus		Sulfated polysaccharides	Antiobesity	(Zhu et al., 2018)
	Body wall	Chondroitin sulfate	Antidiabetics	(Gong et al., 2021)
	Degutted flesh	Neutral polysaccharide		(Gong et al., 2022)
Holothuria nobilis		Fucosylated glycosaminoglycan	Anticoagulant	(Li et al., 2021a)
Holothuria fuscopunctata	Body wall	Fucosylated chondroitin sulfate		(Gao et al., 2020)
Holothuria floridana	Body wall	Fucosylated chondroitin sulfate	Anticoagulant	(Farrag et al., 2023)
Pentacta pygmaea	Body wall	Fucosylated chondroitin sulfate	Anticoagulant	(Dwivedi et al., 2023)
Stichopus japonicus		Fucosylated chondroitin sulfate	Antihyperlipidemic	(Zhu et al., 2019)

TABLE 12.2 The marine animal-oriented polysaccharides, physical and biochemical properties, and potential applications in the food sector—cont'd

Species	Source	Polysaccharide type	Activity and property	Reference
Rhopilema esculentum	Skin	Polysaccharides	Antiinflammatory	(Cao et al., 2021)
			bowel disease	
Ostrea rivularis		Glycosaminoglycan	Antihyperlipidemic	(Kong et al., 2021)
Crassostrea gigas	Deshelled flesh	Polysaccharide	Antiobesity	(Ma et al., 2022b)
			Antioxidant	(Ma et al., 2022a)
Asterias rollestoni		Glucan		(Zhang et al., 2013)
Mussel		Glucan		(Xiang et al., 2022)
Amussium pleurinectus		Hyaluronan		(Kanchana et al., 2013)
Scylla tranquebarica	Shell	Chitin	Antioxidant	(Sreeja et al., 2023)
Mackerel fish		Glycosaminoglycan	Antiobesity	(Rysenaer et al., 2023)
Belone belone	Skin, head, and bone	Glycosaminoglycan	Anticoagulant	(Chikha et al., 2023)
Sciaena umbra	Skin	Chondroitin sulfate		(Bougatef et al., 2018)
		Dermatan sulfate		
Shark	Cartilage	Chondroitin sulfate	Antioxidant	(Chen et al., 2022a)
Hippocampus mohnikei		Polysaccharide	Antiinflammatory	(Cui et al., 2022)
Cuttlefish	Skin and muscle	Sulfated polysaccharide	Antioxidant and antibacterial	(Jridi et al., 2019) (Chattopadhyay et al., 2019)
Shell fish		Chitosan	Antibacterial	(Sotong, 2019; Chattopadhyay et al., 2019)
			Coating barrier Cross-linking ability emulsifier	
Squid	Squid mantle, tentacle, ovary, ink, and visceral organs	Chitooligosaccharide	Gel formation enhances texture	(Singh et al., 2020a)
Asian green mussel (Perna viridis)		Sulfated glycosaminoglycan	Antiinflammatory	(Pai et al., 2024)

12.3.1 Agar

In the United States, agar is classified as a "GRAS" food additive, and it is also approved in Europe as a food additive with the code E406 (Torres et al., 2019). Agar is a nontoxic polysaccharide composed of agarose and agaropectin, which contributes to the diverse physicochemical and rheological properties of agar (Thiviya et al., 2023; Liao et al., 2021). Agarose, the gelling component, is a neutral polysaccharide and is able to form a transparent gel by the aggregation of molecules through hydrogen bonding (Thiviya et al., 2023). Its rheological properties can be enhanced by incorporating sugars such as sucrose, glucose, or trehalose, or by combining it with other polysaccharides, such as alginates, locust bean gum, or xanthan gum (Alba and Kontogiorgos, 2018). Incorporation of sugar promotes a higher gelation temperature and stronger gel formation. Moreover, agar has the ability to form a thermoreversible strong gel that melts when heated above 85°C (depending on the algal species) and reforms its gel structure upon cooling, without requiring any additional cations (Olatunji, 2020). Another key characteristic of agar is its high gelling hysteresis (40–60°C), which refers to the significant temperature gap between its gelling point (around 38°C) and melting point (approximately 85°C) (Kocira et al., 2021). This remarkable feature contributes to versatile applications in the food industry. For instance, it is used in traditional Asian cuisines, canned meats, confectionery jellies, and sparkling products such as marshmallows, nougats, and toffees (Liao et al., 2021; Anderson et al., 1993). In addition, the higher melting point of agar results in more thermostable products compared to gelatin, making it suitable for preparing bakery fillings (Anderson et al., 1993). Its natural taste and gel-forming ability without the need for additional supplements make agar a versatile choice in confectionery, particularly for jelly candies (Burey et al., 2009). Agar has also been mixed with fruit pulps for the production of structured fruits, which are considered fruit bars. These are cost-effective, environmentally sustainable to produce, and help reduce waste by extending the shelf life of fruits while maintaining their nutritional value and visual appeal (Reis., 2019; Da Costa et al., 2020). Kaneda and Shibata have reported a low-oil mayonnaise achieved by substituting oil with agar microgels (Khalil et al., 2018). Recently, researchers discovered an enzymatic conversion of D-galactose, a sugar found in agar, into D-tagatose, a rare and low-calorie sweetener. This breakthrough positions agar as a valuable resource for producing sweeteners in the food industry (Jeong et al., 2020).

12.3.2 Carrageenan

Carrageenan cannot directly disperse in oil. However, it can form a stable mixture called emulsion gels by combining a carrageenan solution with a proportion of oil through a simple emulsification process (Iqbal et al., 2019). A study showed that the gelation process and the types of gel networks formed by various commercial carrageenan grades differ significantly (Fontes-Candia et al., 2020). In addition, these characteristics can be modified by the addition of a cation. The inclusion of K^+ was found to result in the formation of stronger κ-carrageenan and λ-carrageenan hydrogels. Conversely, while Ca^{2+} was more effective than K^+ at inducing the formation of ι-carrageenan hydrogels (needing lower salt concentrations), the strength of these hydrogels was not significantly impacted by the type of salt. This indicates that the resultant supramolecular structures were likely similar (Fontes-Candia et al., 2020b).

κ-carrageenan stands out as the most extensively used among the three types due to its superior gel-forming capabilities, followed by ι- and λ-carrageenan (Humayun et al., 2023; Levy-Ontman et al., 2023).

In the 1950s, carrageenan was first used as a stabilizer in ice creams and chocolate milk. Eventually, it found its way into other products like pudding, condensed milk, and toothpaste (Hotchkiss et al., 2016). To meet the growing demand for high-value food products, carrageenan applications have been investigated in various innovative technologies. Carrageenan has shown potential in areas such as encapsulation, plant-based meat products, and 3D/4D printing, where it can be used as a wall material, an edible sheet composite, a texturing agent, and food ink (Udo et al., 2023). Carrageenan is also used to make thin films through a process called solvent casting. These films are neat and brittle but have weak physical properties, such as low strength, poor barrier protection, and limited water resistance. Because of this, they are mostly used in food packaging (Mavelil-Sam et al., 2023; Ferdiansyah et al., 2023). Adding plasticizers like glycerol and crosslinking calcium ions, when combined with other biopolymers, can enhance the physical properties of the films. Furthermore, incorporating functional bioactive compounds into the film-forming solution significantly improves their UV protection (Kokkuvayil Ramadas et al., 2024). Carrageenan-based food packaging films are reported to have significant potential in extending the shelf life of packaged foods like meat, fish, and bananas (Roy et al., 2021; Khojah, 2020). In addition, smart color-changing freshness indicators are effective in monitoring the real-time freshness of meat, fish, and milk products (Janseerat et al., 2024).

12.3.3 Alginate

Alginate is a soluble major dietary fiber polymer found in brown algae (Ghosh et al., 2022). Alginate is widely used in the food industry as a gelling and thickening agent (Fertah, 2017). The strength of a gel depends on the molecular weight of the polymer (Thiviya et al., 2023). The strength of the gel increases up to a molecular size of 2.4×10^5, after which it becomes independent of further increases in molecular weight (Lavrentieva et al., 2021). Alginate can form gels with various divalent cations through ion-induced ionization and cross-linking. The resulting gels exhibit diverse structures and properties depending on the specific cross-linked cation (Nezamdoost-Sani et al., 2023). For instance, the binding site for Ca^{2+} is a cavity created by the connection of adjacent guluronic acid residues, which interact with the carboxyl groups on the sodium alginate backbone, forming a spatial structure resembling an "eggshell" (Yan et al., 2024). Mg, Mn, Al, and Fe are also used as divalent cations to form various types of alginate-derived gels (Borchard et al., 2005; Aruna Dhanapal et al., 2012). The ratio of guluronic acid (G) to mannuronic acid (M) residues determines the molecular weight of the polymer, which ultimately affects the thickening properties of the structure (Yan et al., 2024). The ratio $M/G<1$, capable of forming rigid thicken gel (Yan et al., 2024). The ratio $M/G>1$ is able to form soft and flexible gels (Yan et al., 2024). Furthermore, external factors such as pH and ionic strength also able to determine the viscosity of the alginate solution (Yan et al., 2024).

This water-insoluble gel is used to coat edible vegetables and fruits (Rhim, 2004). Alginate is able to form transparent and glossy-looking films that are impermeable to fat and oils (Mahajan et al., 2017; Rhim, 2004; Kocira et al., 2021). Furthermore, alginate possesses distinctive colloidal properties, including stabilizing, thickening, suspending, film-forming, and coating capabilities, as well as gel-forming and emulsion-stabilizing functions (Aruna Dhanapal et al., 2012). Alginate can be used in meat and fish processing as a film cover to prevent mass loss and degradation of color and texture, and as a binder in restructured meat and fish products, where it holds meat or fish particles together into cohesive portions (Gheorghita et al., 2020). Spherification, a technique in molecular gastronomy, uses alginate to create caviar-like spheres with liquid interiors, enhancing both the visual appeal and sensory experience of food (Sivakumaran and Prabodhani, 2018). Moreover, alginate can serve as a gluten substitute in gluten-free bakery products, offering both structure and moisture retention, such as in gluten-free bread or pasta (Lux Nee Bantleon et al., 2023).

12.3.4 Fucoidan

Fucoidan, extracted from *Fucus vesiculosus* and *Undaria pinnatifida*, is an approved macroalgal product by the European Union for use in food and dietary supplements as a novel food (Anisha et al., 2022). Fucoidans are also present in edible species like *Laminaria japonica* and *Cladosiphon okamuranus* (Wells et al., 2017). Fucoidan-based active biopolymer food packaging films exhibit improved physical properties, including reduced water vapor and oxygen permeability, enhanced light penetration capabilities, better thermal stability, and increased film thickness (Anisha et al., 2022). The brown macroalga *C. okamuranus*, which is commercially cultivated, is utilized for industrial-scale fucoidan production. Its distinctive rheological properties have facilitated its commercialization in Japan, where it is included in drinks and tablets as a nutritional supplement (D et al., 2020). The ability of fucoidan to form a viscous gel without solidifying makes it an ideal emulsifier (Hifney et al., 2016). Moreover, fucoidan is suitable for commercial applications as a stabilizing, thickening, and water-holding agent in the food market, cosmetics, and various other industries, such as in stabilizing protein oil-in-water emulsions (Hifney et al., 2016; Tako, 2003).

12.3.5 Laminarin

Recent studies have explored the improvement of texture and the delay in digestion achieved by physically mixing nonstarch polysaccharides with native starch in noodle dough production (Feng et al., 2020a, 2020b). Similarly, laminarin-mixed wheat noodle has demonstrated improved cooking performance and enhanced texture (Wei et al., 2023). Laminarin can be utilized as a dietary fiber, offering potential health benefits for humans (Sterner and Gröndahl, 2021; Devillé et al., 2004).

12.3.6 Ulvan

According to recent records, a few attempts have been made to produce edible film from ulvan. Ulvan, extracted from *Ulva lactuca*, was studied under various extraction conditions to evaluate its suitability for producing ulvan-based films (Guidara et al., 2019, 2020). Ulvan has the potential to formulate thermoreversible gel with calcium and boron ions under alkaline conditions, which can be applicable as a stable, stabilized colloidal formulator in the food industry (Morelli et al., 2019). Due to its amphiphilic nature, ulvan can effectively serve as a food preservative, a rheology modifier, and a quality enhancer for various food products (Cindana Mo'o et al., 2020; Alipour et al., 2018). Moreover, ulvan is used in food technology, acting as a prebiotic in yogurt and altering properties in fish products (Alipour et al., 2018; Shalaby and Amin, 2019).

12.3.7 Chitosan

Chitosan is a versatile, nontoxic, biodegradable, biocompatible, and GRAS-recommended biopolymer. Due to these properties, interest in chitosan applications in the food industry has increased in recent years (Hu and Ganzle, 2019). Industrially, chitosan is used as a coating agent for preservation due to its various biologically advantageous properties. Moreover, novel technological approaches, such as the incorporation of bioceramic materials into chitosan, can enhance its biological properties and its industrial applications (Aktug et al., 2021). The film formation property of chitosan is determined by its molecular and chemical characteristics. Chitosan molecules possess numerous hydroxyl and amino groups, which can form hydrogen bonds with water molecules, enabling chitosan to create a colloidal solution in water. The gradual evaporation of water enhances the hydrogen bonds among chitosan molecules and ultimately forms a film with specific elasticity and strength (Wang et al., 2024). The formation process of chitosan film favors high solution concentration and lower pH (Riseh et al., 2023). In application, the chitosan coating on fruits forms a transparent film that helps prevent water evaporation, weight loss, and surface deformations. In addition, the chitosan film facilitates oxygen transport, reduces metabolism, and ultimately extends the shelf life of the fruit (Wang et al., 2024). Moreover, the moisturizing property of chitosan films (Chaiwong et al., 2020) helps retain the moisture in foods, thereby preserving their quality. The proven characteristics, such as inhibition of microorganism growth, reduction of lipid oxidation rate, and protein degradation rate, enhance the application of chitosan in the food industry (Liu et al., 2022; Zhang et al., 2022a). However, the limited water and alkaline solubility of chitosan restricts its applications as a food preservative (Zhang et al., 2022a). Chitooligosaccharides exhibit enhanced biological properties compared to chitosan due to the presence of more functional groups and low molecular weights. Nevertheless, the diminished antimicrobial activity of chitooligosaccharides is a key factor limiting their application in the food industry. Recently, chemically mediated modification methods have successfully improved the solubility of chitosan and enhanced its biological properties, including antibacterial activity (Zhang et al., 2022a). Therefore chemical modification can enhance the physical and biological properties of chitosan, making it beneficial for use in the food industry. Recently, several studies have explored the properties of chitosan as a stabilizer in classic emulsions, multilayer emulsions, and Pickering emulsions (Sharkawy et al., 2020; Tabatabaei et al., 2022). The emulsifying properties can be improved by modifying the chemical structure through bonding with compounds like stearic acid (Atarian et al., 2019), myristic acid (Hosseini et al., 2019), and capric acid (Yang et al., 2023), thereby expanding its potential application in the food industry. These modifications not only improve its performance in food preservation and texture enhancement but also contribute to the development of more sustainable and bioactive food packaging solutions.

12.4 Potential biological properties and related health benefits as functional foods

In recent years, consumers have become more concerned about the relationship between diet and health. Over the past decade, marine-derived resources have emerged as one of the most capable natural resources for bioactive carbohydrates. This wide diversity has captured the attention of scientists due to their significant biological activities and vast potential in biomedical, technological, and functional food applications (Paz Otero et al., 2023). Compared to terrestrial polysaccharides, marine polysaccharides demonstrate diverse and unique properties that are highly beneficial to human health. Their distinctive characteristics are shaped by the survival challenges created by extreme marine environments, such as low light, high pressure, and high salinity. The activity of polysaccharides is affected by various factors such as their composition, glycosidic linkages, and molecular weight. Polysaccharides with different monosaccharide compositions display unique biological activities (Huang et al., 2024b). For instance, glucose-based polysaccharides like β-glucans can boost the immune system's overall function by activating macrophages. On the contrary, galactose-based polysaccharides, such as galactooligosaccharides, enhance intestinal barrier integrity by influencing tight junction proteins. The bioactivity and physicochemical properties of polysaccharides are influenced by their monosaccharide composition and glycosidic bond types, whether linear or branched. Linear polysaccharides often exhibit greater viscosity and stability, while branched polysaccharides tend to provide improved solubility and bioavailability (Huang et al., 2024b; Yang et al., 2022). With reference to their diverse bioactivities and nutritional benefits, marine-based polysaccharides hold significant potential to positively impact human health, food security, and, most importantly, affordable nutritional diets (Xie et al., 2023). Most animal-derived polysaccharides function as prebiotics, supporting the gut microbiome and contributing to the structural integrity of the gut (Guan et al., 2024; Krangkratok et al., 2023). In recent years, marine polysaccharides have gained recognition for their wide range of physiological functions, such as antiobesity and anticoagulant effects. These polysaccharides play a crucial role in maintaining health by regulating metabolites and supporting the immune system (Guan et al., 2024). Moreover, these polysaccharides can function as dietary fibers, which are nondigestible polymers that physiologically contribute to human health (Shang et al., 2018). Dietary fibers contribute to increasing stool bulk, promoting colonic fermentation, lowering blood glucose levels after meals, and reducing cholesterol levels before meals (Delzenne et al., 2020).

Fucoidan extracts from *Undaria pinnatifida* and *Fucus vesiculosus* have gained regulatory approvals for use in foods and dietary supplements across various global jurisdictions. The US FDA granted GRAS status to these extracts, permitting daily consumption of up to 250 mg. Similarly, the European Commission approved these extracts as novel foods under EU regulation, deeming them equivalent to their seaweed sources and allowing the same daily intake. In Canada and Australia, several listed medicines containing fucoidan extracts have been approved, with Australia specifically recognizing these extracts as listable components tied to their respective seaweed sources (Fitton et al., 2019). Moreover, different studies have confirmed the anticancer effects of fucoidans against many cell lines, including HepG2, HL-60, HCT-116, and Raji (Arumugam et al., 2019; Bittkau et al., 2019). A diet enriched with 6% *Sargassum horneri* mitigated high-fat diet-induced obesity in mice by improving related metabolic disorders and reducing diabetes and hepatic steatosis through inhibition of pancreatic lipase activity and decreased intestinal lipid absorption (Murakami et al., 2021). In addition, seaweed consumption has been identified as a protective factor against breast cancer. A daily intake of 5 g of *Undaria* sp. led to a 50% reduction in urinary urokinase-type plasminogen activator receptor concentrations, potentially explaining Japan's comparatively lower postmenopausal breast cancer rates (Teas et al., 2013). Seaweed polysaccharides, due to their unique structures, target key therapeutic pathways in inflammatory bowel disease, including proinflammatory cytokines, NF-kB, and intestinal epithelial cells, making them capable candidates for developing drugs, dietary supplements, and functional foods (Besednova et al., 2020). It has recently been proven that cardiovascular diseases, heart malfunctions, hypertension, and strokes are generally connected with the gut microbiome. Marine algal polysaccharides help maintain gut microbiome homeostasis, promote the production of functional metabolites and short-chain fatty acids, and impact bile acid signaling, all of which contribute to cardiovascular disease prevention and management (Cheong et al., 2022). This highlights their potential as sustainable sources for functional foods and nutraceuticals targeting cardiovascular diseases. Fucose and fucose oligosaccharides are essential ingredients in infant formula powder, as they mimic the prebiotic effects of human milk oligosaccharides (Xiao et al., 2022). Recent studies have demonstrated that combining traditional and innovative polysaccharide extraction methods can significantly enhance the functional properties of various marine polysaccharides, including their antioxidant, antitumor, and immunomodulatory activities (Wu et al., 2022).

12.5 Novel technological approaches and modification techniques

Chitin nanomaterials offer a wide range of applications in the food industry. Nanochitin is frequently utilized as an organic solid particle emulsifier, known as a Pickering emulsion, and as a nanofiller to improve the mechanical strength, barrier effectiveness, and thermal stability of biodegradable food films. In addition, nanochitin acts as a saltiness enhancer, inhibits starch retrogradation, and functions as a dietary fiber in various food products (Ngasotter et al., 2023). An emulsion is a combination of immiscible liquids that require stabilization to prevent separation. This is typically achieved using surfactants, which reduce surface tension and maintain stability. Alternatively, solid colloid particles can serve as stabilizers by accumulating at the oil-water interface, preventing droplet coalescence and forming what is known as "Pickering emulsions" (Ngasotter et al., 2023). A mixture of chitin nanocrystals and fucoidan is able to reduce the size of the emulsion droplet and is more stable in oil-water Pickering emulsions (Liu et al., 2021b). Similarly, the combination of cellulose and chitin nanoparticles produced a Pickering emulsion with enhanced creaming stability and exceptional resistance to droplet coalescence (Lv et al., 2021). Moreover, chitin nanocrystals can be used in food packaging applications. Chitin nanocrystals, extracted from shrimp shells, were combined with carboxymethyl cellulose and grapefruit seed extract to create advanced composite films, which can be used as a food packaging material. The addition of chitin nanocrystals enhanced UV blocking, mechanical strength, and reduced water vapor permeability, while grapefruit seed extract imparted strong antibacterial effects against foodborne pathogens such as *E. coli* and L. *monocytogenes* (Oun and Rhim, 2020). Excessive salt consumption is linked to various health problems, including obesity, high blood pressure, cardiovascular diseases, and stomach cancer (Lucarini et al., 2021). Therefore recent attention toward alternative solutions has rapidly increased. A mixture of NaCl with chitin nanomaterials is able to significantly increase salty taste compared to a solution without chitin nanomaterials (Tsai et al., 2019). Inflammatory bowel disease is a chronic condition that requires lifelong management and treatment. Marine polysaccharides such as chitosan, alginate, chondroitin sulfate, and hyaluronic acid have been used to create multifunctional nanocarriers that facilitate the targeted delivery of polyphenols to the colon, which can be utilized as bioactive components in functional food applications (Tie and Tan, 2022).

Chemical, physical, and biological modifications of polysaccharides are widely used to enhance specific properties, making them more suitable for applications in the food industry. However, in comparison with physical and biological modifications, chemical modifications are more productive and impactful (Han et al., 2022). For instance, chemical modification techniques such as oxidation, copolymerization, and esterification can target specific sites within the alginate polysaccharide chain. These modifications improve various properties, such as antioxidant activity, thickening, and stabilizing capabilities, thereby broadening their potential applications (Park et al., 2021; Dai et al., 2018; Han et al., 2022; Khoshdouni Farahani et al., 2022).

Pressurized liquid extraction (PLE) is a recently developed, eco-friendly technique utilized for extracting components from seaweeds. This method employs high temperatures and pressure combined with a solvent to effectively isolate compounds from solid matrices (Perez-Vazquez et al., 2023). Most importantly, this technique enables the extraction of target components in a short time frame, using a low amount of solvent while achieving a high yield. In addition, combining PLE with other innovative methods can enhance both the quality and specificity of the extraction process (Perez-Vazquez et al., 2023). A recent study proved that an edible film produced by chitosan and algal extract derived from *H. elongata* and *P. palmata* effectively inhibited the development of mesophilic and psychrophilic microorganisms. This film was able to maintain the pH level and hydrophilic activity without altering the sensorial characteristics such as flavor, taste, color, or smell (Albertos et al., 2019). Alginate-based nanoemulsions containing *Zataria multiflora* Boiss essential oil have shown effective results in extending the shelf life of rainbow trout fillets. The results indicate that nanoemulsion solutions can extend the shelf life by more than 4 days compared to coarse emulsions (Khanzadi et al., 2020).

Advancements in novel bioinformatics technologies and computational tools have significantly impacted marine polysaccharide-derived oligo products and their diverse bioactive applications. For instance, in the past decade, xylooligosaccharides have gained attention for their unique chemical structure and prebiotic effects. Xylooligosaccharides show potential applications in the food industry, most notably as sugar and fat replacers and as cryoprotectants. However, their production remains challenging due to variations in bioactivity associated with their degree of polymerization. Consequently, innovative bioinformatics solutions, such as genome sequencing and gene mining applications, are crucial for addressing the challenges of identifying and designing novel xylanases (Lin et al., 2024).

Recently, the combination of marine polysaccharides, chitosan, and EPS produced by two *Alteromonas* sp. has been shown to form homogeneous, transparent, and colorless films, demonstrating excellent compatibility between these polysaccharides. The formed film exhibited low gloss values (5–10 GU), resulting in a nonglossy and rough surface. Moreover, this film exhibited hydrophobicity and superior mechanical properties, making it a potential active packaging material for the food industry. In addition, its ability to form water-resistant and robust films ensures its application in protecting food products and maintaining their freshness (Zarandona et al., 2020).

3D food printing is a novel technology used to craft food products with specific characteristics and properties according to the consumer's desire (Donn et al., 2022). Building upon the advancements in 3D food printing, the emerging technology of 4D food printing has been extensively developed, introducing dynamic and transformative food applications (Chen et al., 2022b). These technologies enable the design of functional foods with predetermined nutrient compositions by utilizing a layer-by-layer deposition of desired compounds (Derossi et al., 2024). Marine polysaccharides such as fucoidan and alginate are able to enhance the function of quinoa protein-derived emulsion gels, which are used for 3D printing applications. Both polysaccharides improved the structural integrity of these gels by enhancing electrostatic interactions, hardness, freeze-thaw stability, reducing swelling ratio, and water holding capacity. In addition, these polysaccharides are able to enhance the viscoelasticity of the gels, enabling better stability and quality during 3D printing (Zhao et al., 2024).

A food-grade delivery system represents a cutting-edge approach in encapsulation technology within the food industry. This innovation ensures the protection of bioactive compounds while enabling their targeted release within the gastrointestinal tract. Furthermore, the encapsulation process enhances the solubility and absorption of compounds at the target site, thereby improving the bioavailability of specific compounds (Han et al., 2022b; Lim et al., 2020). Marine polysaccharides are extensively utilized as food-grade encapsulation materials because of their nontoxic nature (Han et al., 2022b; Wei et al., 2020).

12.6 Future perspectives and challenges

Over the last few decades, the application of marine-derived polysaccharides has become widely renowned, driven by multiple contributing aspects, particularly their health-promoting abilities and diverse potential applications. However, due to several factors, such as the extraction and utilization of polysaccharides, the complexity of the polymer, diverse solubility, environmental contamination, and the formulation of products with enduring stability, the application of these polymers is significantly challenging (Augusto et al., 2024a; Getachew et al., 2020; Wan et al., 2020). In addition, the limited exploration of various biochemical properties of marine polysaccharides remains a significant challenge (Wan et al., 2020).

To date, several types of extraction methods, such as conventional water extraction, ultrasound-assisted extraction (UAE), microwave-assisted extraction (MAE), enzyme-assisted extraction (EAE), PLE, and supercritical fluid extraction, have been used for marine polysaccharide extraction. Each type of extraction presents both advantages and disadvantages, which are crucial for industrial and specific applications (Jonsson et al., 2020). The pretreatment and purification of marine polysaccharides, under distinct and standardized extraction protocols, are essential for maintaining both their structural integrity and biological properties. Advanced and innovative assisted extraction techniques, such as UAE, MAE, PAE, and EAE, provide higher extraction efficiency while reducing operation and extraction time, energy, and cost requirements (Gomez et al., 2020). However, the use of novel approaches in marine polysaccharide extraction has not yet been widely adopted in industry and remains limited to laboratory settings. Therefore a proper, well-defined, and robust approach needs to be established for both the production and consumption of marine-derived polysaccharides. Only then will producers be able to innovate and scale up novel approaches in the future.

Environmental contamination has recently emerged as a critical factor of concern. Recent industrial and anthropogenic activities have contributed to the accumulation and spread of various toxic compounds in marine environments. Although the consumption of marine algal polysaccharides offers health benefits, there are concerns in various sectors regarding potential risks associated with toxic elements, such as iodine, cadmium, mercury, and lead (Schepers et al., 2020; Filippini et al., 2021). Recent studies have been conducted to assess the amount of several potentially health-vulnerable heavy metals, such as arsenic, cadmium, lead, and mercury, in marine algae (Duinker et al., 2020; Panebianco et al., 2021; Roleda et al., 2019). However, the levels of these metals vary based on the dry weight, species type, location, and season of sample collection, as well as other environmental and growth conditions (Roleda et al., 2019). The European Union has established a cadmium limit of 3 mg/kg (wet weight) for marine algae, while Germany has set a national iodine limit of 20 mg/kg (dry weight) due to the vulnerability of overconsumption. In contrast, Centre d'Etude et de Valorization des Algues (CEVA) in France has proposed a higher iodine limit of 2000 mg/kg (dry weight), although no standardized iodine limit exists across the EU despite recognizing the health risks associated with excessive iodine intake from algal products (Blikra et al., 2021). Therefore greater attention must be given to the extraction and purification processes to optimize yield in the future.

To date, most of the bioactive properties identified from marine sources remain at the laboratory or clinical stage. Therefore there should be an advancement in the application and production sector to bring algal raw polysaccharides as an innovative and more beneficial product to the market (Mohammed et al., 2021). The production of the desired product often relies heavily on biomass availability. Consequently, for industrial applications, large-scale production and achieving the desired carbohydrate composition are highly dependent on optimizing the biomass growth environment (Patel et al., 2022). Therefore achieving cost-effective production and application remains a significant challenge. As industrial applications advance, companies are increasingly inclined to explore innovative technologies, such as high-performance drying equipment (high capacity, low temperature), which offer energy efficiency and ensure high-quality outcomes (Blikra et al., 2021). Furthermore, establishing offshore industrial facilities presents significant challenges due to physical risks, the potential for biofouling on infrastructure and equipment, and the high initial costs involved (Van Den Burg et al., 2020).

Novel scientific approaches, such as the chemical modification of polysaccharide structures, are highly efficient for food applications, including processing and preservation, in terms of their effectiveness and impact. However, challenges such as the toxicity of catalysts, additional purification steps, and study limitations regarding regioselective modifications are hindering practical applications. Therefore the development of green solvents, catalysts, and crosslinking agents for sustainable and more practical reactions will be even more crucial in the future (Han et al., 2022a).

The use of marine polysaccharides as prebiotics is often considered challenging due to their complex structures and the potential to form antibacterial compounds. These properties can interfere with the growth and colonization of natural lactic acid bacteria under natural conditions (Fabris et al., 2020). Despite these challenges, advancements in structural modification, such as enzymatic and fermentation processes, have offered a considerable contribution to overcoming the above limitations. Modification of complex polysaccharides enhances their compatibility with beneficial microbiota by promoting the selective growth of gut microbes while minimizing antimicrobial effects (Ren et al., 2024). For instance, porphyran, a complex and sulfated nondigestible polysaccharide, limits bioaccessibility. However, fermentation using microorganisms can convert these polysaccharides into bioactive oligosaccharides, enhancing their nutritional value and health benefits, including antioxidant, antidiabetic, and antiinflammatory properties. Consequently, these fermented products support the development of innovative functional foods, such as sauces, beverages, and health-enhancing ingredients (Ghagane et al., 2024; Wang et al., 2023b; Yang et al., 2024).

Marine polysaccharides, particularly those derived from algae, have long been interconnected with traditional and cultural diets in various regions of the world, especially in the Asian region. Numerous in vitro and in vivo studies have confirmed the significant health benefits of these polysaccharides, strengthening their potential as functional food ingredients. This growing body of evidence has encouraged increased research interest in exploring marine polysaccharides for applications in the functional food and pharmaceutical industries. However, despite this interest and notable advancements, there remain critical gaps in our understanding. In particular, the structural characteristics of marine polysaccharides during digestion and the mechanisms by which fermentation metabolism induces their prebiotic and bioactive effects in humans are not fully examined (Zheng et al., 2020). A deeper understanding of the physicochemical properties of marine polysaccharides, especially their transformations during digestion and fermentation, is essential to fully realize their potential as prebiotics and health-promoting compounds.

According to the 2024 World Bank report, the global population facing severe food insecurity is projected to reach 956 million by 2028. This prediction could approach a billion in a pessimistic economic scenario, particularly if central banks are unable to manage inflation and resort to additional monetary tightening, resulting in slowed economic growth (Bank, 2024). Innovative approaches have the potential to tackle this global challenge. Therefore advancing current technologies and adopting practical strategies should be priorities for both governments and the private sector. Undoubtedly, addressing the challenges associated with marine polysaccharide-based processes to achieve cost-effective industrial applications could pave the way for a sustainable future and a circular global bioeconomy.

12.7 Conclusion

The rapidly growing global population is driving an exponential increase in food demand. To address this global issue, sustainable and robust solutions are urgently required. The ocean, a vast and diverse ecosystem, hosts a wide range of organisms spanning from prokaryotes to eukaryotes, offering immense potential for innovative approaches to meet global food needs. Marine-derived polysaccharides from algae, animals, and microorganisms offer both nutritional benefits and a sustainable, economical solution to meet global food demand. These polysaccharides contribute to a healthy diet by providing functional properties such as prebiotic effects, antioxidant activity, antidiabetic potential, and antiinflammatory benefits. In addition, their remarkable physical properties, including gel-forming ability, viscosity and thickening, water retention, film-forming capability, emulsification, thermal stability, and pseudoplastic behavior, open up a wide range of applications in the food industry. These applications include their use as texturizers, stabilizers, emulsifiers, edible films and coatings, encapsulating agents, bakery product enhancers, food packaging materials, food additives, and low-calorie sweeteners, making them indispensable for innovative and functional food product development. However, advancing processing and modification technologies, addressing challenges related to environmental contamination, and expanding industrial and technological applications require greater focus in the future to maximize the potential benefits and profitability of marine polysaccharides in the food sector. Further advancements in technology are essential to develop more sustainable and eco-friendly approaches to meet the growing global demand. By bridging these gaps, marine-derived polysaccharides hold immense potential to revolutionize the food sector while contributing to a healthier and more sustainable future.

AI disclosure

During the preparation of this work, the author(s) used ChatGPT and Perplexity AI to improve the readability of sentences. After using these tools/services, the author(s) reviewed and edited the content as needed and take(s) full responsibility for the content of the publication.

References

Abourehab, M.A.S., Rajendran, R.R., Singh, A., Pramanik, S., Shrivastav, P., Ansari, M.J., Manne, R., Amaral, L.S., Deepak, A., 2022. Alginate as a Promising Biopolymer in Drug Delivery and Wound Healing: a Review of the State-of-the-Art. Int. J. Mol. Sci. 14220067. 23 (16). Available from: http://www.mdpi.com/journal/ijms. doi:10.3390/ijms23169035.

Ahmed, N., Sheikh, M.A., Ubaid, M., Chauhan, P., Kumar, K., Choudhary, S., 2024. Comprehensive exploration of marine algae diversity, bioactive compounds, health benefits, regulatory issues, and food and drug applications. Meas. Food. 27722759. 14. Available from: https://www.sciencedirect.com/science/journal/27722759. doi:10.1016/j.meafoo.2024.100163.

Ai, C., Duan, M., Ma, N., Sun, X., Yang, J., Wen, C., Sun, Y., Zhao, N., Song, S., 2018. Sulfated polysaccharides from pacific abalone reduce diet-induced obesity by modulating the gut microbiota. J. Funct. Foods 47, 211–219.

Akshaya, S., Nathanael, A.J., 2024. A Review on Hydrophobically Associated Alginates: approaches and Applications. ACS Omega. 24701343. 9 (4), 4246–4262. Available from: http://pubs.acs.org/journal/acsodf. doi:10.1021/acsomega.3c08619.

Aktug, S.L., Durdu, S., Kalkan, S., Cavusoglu, K., Usta, M., 2021. In vitro biological and antimicrobial properties of chitosan-based bioceramic coatings on zirconium. Nature Research, Turkey Scientific Reports. 20452322. 11 (1). Available from: www.nature.com/srep/index.html. doi:10.1038/s41598-021-94502-z.

Alba, K., Kontogiorgos, V., 2018. Seaweed Polysaccharides (Agar, Alginate Carrageenan) Encyclopedia of Food Chemistry. Elsevier, United Kingdom, pp. 240–250. Available from http://dx.doi.org/10.1016/B978-0-08-100596-5.21587-4. doi:10.1016/B978-0-08-100596-5.21587-4.

Albertos, I., Martin-Diana, A.B., Burón, M., Rico, D., 2019. Development of functional bio-based seaweed (Himanthalia elongata and Palmaria palmata) edible films for extending the shelflife of fresh fish burgers. Food Packag. Shelf Life. 22142894. 22, 100382. doi:10.1016/j.fpsl.2019.100382.

Alboofetileh, M., Jeddi, S., Mohammadzadeh, B., Noghani, F., Kamali, S., 2024. Exploring the impact of enzyme-extracted Ulvan from Ulva rigida on quality characteristics and oxidative stability of beef sausage during refrigerated storage. J. Food Sci. Technol. 0022–1155. doi:10.1007/s13197-024-06153-3.

Alipour, H.J., Rezaei, M., Shabanpour, B., Tabarsa, M., 2018. Effects of sulfated polysaccharides from green alga Ulva intestinalis on physicochemical properties and microstructure of silver carp surimi. Food Hydrocoll. 0268005X. 74, 87–96. Available from: https://www.journals.elsevier.com/food-hydrocolloids. doi:10.1016/j.foodhyd.2017.07.038.

Althagbi, H.I., Alarif, W.M., Al-Footy, K.O., Abdel-Lateff, A., 2020. Marine-derived macrocyclic alkaloids (MDMAs): chemical and biological diversity. Mar. Drugs. 16603397. 18 (7). Available from: https://www.mdpi.com/1660-3397/18/7/368. doi:10.3390/MD18070368.

Anderson, M.A.R.K., Baird, J.O.H.N.K., Barkalow, D.A.V.I.D.G., Bemiller, J.A.M.E.S.N., Brigand, G., Chen, -C.H.Y.I.-C.H.E.N.G., Clare, K.E.N.N.E.T.H., Debelder, A.T.H.O.N.Y.N., Dea, I.A.I.N.C.M., Desmarais, A.J., Feddersen, R. O.B.E.R.T.L., Grover, J.O.S.E.P.H.A., Harada, A.K.I.R.A., Harada, T.O.K.U.Y.A., Kang, K.S., Karl, C.U.R.T., Magnuson, K.E.N.N.E.T.H., Maier, H.E.L.M.U.T., Mitsuhashi, M.A.S.A.K.A.Z.U., Pettitt, D.J., Rolin, C.L.A.U.S., Selby, H.H., Sutherland, I.A.N.W., Terasaki, M.A.M.O.R.U., Therkelsen, G.E.O.R.G.H., Thorp, S.T.E.P.H.E.N.N., Towle, G.O.R.D.O.N.A., Tsujisaka, Y.O.S.H.I.O., Whistler, R.O.Y.L., Wint, R.F., 1993. CONTRIBUTORS, Vol. v. Elsevier BV. doi:10.1016/b978-0-08-092654-4.50003-6.

Anisha, G.S., Padmakumari, S., Patel, A.K., Pandey, A., Singhania, R.R., 2022. Fucoidan from Marine Macroalgae: biological Actions and Applications in Regenerative Medicine, Drug Delivery Systems and Food Industry. Bioengineering. 23065354. 9 (9). Available from: www.mdpi.com/journal/bioengineering. doi:10.3390/bioengineering9090472.

Arias, A., Feijoo, G., Moreira, M.T., 2023. Macroalgae biorefineries as a sustainable resource in the extraction of value-added compounds. Algal Res., 69.

Arumugam, P., Arunkumar, K., Sivakumar, L., Murugan, M., Murugan, K., 2019. Anticancer effect of fucoidan on cell proliferation, cell cycle progression, genetic damage and apoptotic cell death in HepG2 cancer cells. Toxicol. Rep. 22147500. 6, 556–563. doi:10.1016/j.toxrep.2019.06.005.

Arvizu-Higuera, D.L., Rodríguez-Montesinos, Y.E., Murillo-Álvarez, J.I., Muñoz-Ochoa, M., Hernández-Carmona, G., 2007. Effect of alkali treatment time and extraction time on agar from Gracilaria vermiculophylla. J. Appl. Phycol. 20, 515–519.

Atarian, M., Rajaei, A., Tabatabaei, M., Mohsenifar, A., Bodaghi, H., 2019. Formulation of Pickering sunflower oil-in-water emulsion stabilized by chitosan-stearic acid nanogel and studying its oxidative stability. Carbohydr. Polym. 210, 47–55.

Augusto, A., Lemos, M., Silva, S., 2024a. Exploring Marine-Based Food Production: the Challenges for a Sustainable and Fast Biotechnology-Based Development. Appl. Sci. 14.

Bangladeshb, Anderkillah, Chattogram, Bangladesh., Sams Creation,.

Bank, W., Food security trends for 2024 and beyond. (2024).

Bankole, A.O., Irondi, E.A., Awoyale, W., Ajani, E.O., 2023. Application of natural and modified additives in yogurt formulation: types, production, and rheological and nutraceutical benefits. Front. Nutr. 2296861X. 10. Available from: http://journal.frontiersin.org/journal/nutrition. doi:10.3389/fnut.2023.1257439.

Behrooznia, Z., Nourmohammadi, J., 2024. Polysaccharide-based materials as an eco-friendly alternative in biomedical, environmental, and food packaging. Giant. 26665425. 19, 100301. doi:10.1016/j.giant.2024.100301.

Berezhnaya, Y.D., Kazachenko, A.S., Kazachenko, A.S., Malyar, Y.N., Borovkova, V.S., 2024. Sulfation of Various Polysaccharide Structures: different Methods and Perspectives. Federation Chemistry (Switzerland. 26248549. 6 (4), 640–665. Available from: www.mdpi.com/journal/chemistry. doi:10.3390/chemistry6040038.

Besednova, N.N., Zaporozhets, T.S., Kuznetsova, T.A., Makarenkova, I.D., Kryzhanovsky, S.P., Fedyanina, L.N., Ermakova, S.P., 2020. Extracts and marine algae polysaccharides in therapy and prevention of inflammatory diseases of the intestine. Federation Marine Drugs. 16603397. 18 (6). Available from: https://www.mdpi.com/1660-3397/18/6/289/pdf. doi:10.3390/md18060289.

Bittkau, K.S., Dorschmann, P., Blumel, M., Tasdemir, D., Roider, J., Klettner, A., Alban, S., 2019. Comparison of the Effects of Fucoidans on the Cell Viability of Tumor and Non-Tumor Cell Lines. Mar. Drugs 17.

Błaszak, B., Gozdecka, G., Shyichuk, A., 2018. Carrageenan as a functional additive in the production of cheese and cheese-like products [pdf]. Acta Sci. Pol. Technol. Aliment. 17, 107–116.

Blikra, M.J., Altintzoglou, T., Løvdal, T., Rognså, G., Skipnes, D., Skåra, T., Sivertsvik, M., Fernández, E.N., 2021. Seaweed products for the future: using current tools to develop a sustainable food industry. Trends Food Sci. Technol. 09242244. 118, 765–776. Available from: http://www.elsevier.com/wps/find/journaldescription.cws_home/601278/description#description. doi:10.1016/j.tifs.2021.11.002.

Borchard, W., Kenning, A., Kapp, A., Mayer, C., 2005. Phase diagram of the system sodium alginate/water: a model for biofilms. Int. J. Biol. Macromol. 01418130. 35 (5), 247–256. doi:10.1016/j.ijbiomac.2005.02.006.

Bose, I., Nousheen, S.R., Yaduvanshi, P., Sharma, S., Chandel, V., Biswas, D., 2023. Unveiling the Potential of Marine Biopolymers: sources, Classification, and Diverse Food Applications. Materials. 1996-1944. 16 (13), 4840. doi:10.3390/ma16134840.

Bougatef, H., Krichen, F., Capitani, F., Amor, I.B., Maccari, F., Mantovani, V., Galeotti, F., Volpi, N., Bougatef, A., Sila, A., 2018. Chondroitin sulfate/dermatan sulfate from corb (Sciaena umbra) skin: purification, structural analysis and anticoagulant effect. Carbohydr. Polym. 01448617. 196, 272–278. Available from: http://www.elsevier.com/wps/find/journaldescription.cws_home/405871/description#description. doi:10.1016/j.carbpol.2018.05.019.

Burdick, J.A., Stevens, M.M., 2005. 11 - Biomedical Hydrogels Biomaterials, Artificial Organs and Tissue Engineering, Woodhead Publishing.

Burey, P., Bhandari, B.R., Rutgers, R.P.G., Halley, P.J., Torley, P.J., 2009. Confectionery Gels: a Review on Formulation, Rheological and Structural Aspects. Int. J. Food Prop. 1094-2912. 12 (1), 176–210. doi:10.1080/10942910802223404.

Cabral, E.M., Oliveira, M., Mondala, J.R.M., Curtin, J., Tiwari, B.K., Garcia-Vaquero, M., 2021. Antimicrobials from seaweeds for food applications. Mar. Drugs. 16603397. 19 (4). Available from: http://www.mdpi.com/journal/marinedrugs. doi:10.3390/md19040211.

Cao, Y., Gao, J., Zhang, L., Qin, N., Zhu, B., Xia, X., 2021. Jellyfish skin polysaccharides enhance intestinal barrier function and modulate the gut microbiota in mice with DSS-induced colitis. Food Funct. 2042-6496. 12 (20), 10121–10135. doi:10.1039/d1fo02001c.

Chaiwong, N., Leelapornpisid, P., Jantanasakulwong, K., Rachtanapun, P., Seesuriyachan, P., Sakdatorn, V., 2020. Noppol Leksawasdi, Yuthana Phimolsiripol, Antioxidant and Moisturizing Properties of Carboxymethyl Chitosan with Different Molecular Weights. Polymers. 2073-4360. 12 (7), 1445. doi:10.3390/polym12071445.

Chattopadhyay, K., Xavier, K.A.M., Layana, P., Balange, A.K., Nayak, B.B., 2019. Chitosan hydrogel inclusion in fish mince based emulsion sausages: effect of gel interaction on functional and physicochemical qualities. Int. J. Biol. Macromol. 134, 1063–1069.

Chen, C.Y., Wang, S.H., Huang, C.Y., Dong, C.D., Huang, C.Y., Chang, C.C., Chang, J.S., 2021. Effect of molecular mass and sulfate content of fucoidan from Sargassum siliquosum on antioxidant, anti-lipogenesis, and anti-inflammatory activity. J. Biosci. Bioeng. 13474421. 132 (4), 359–364. Available from: http://www.elsevier.com/wps/find/journaldescription.cws_home/505516/description#description. doi:10.1016/j.jbiosc.2021.06.005.

Chen, J., Chen, X., Li, J., Luo, B., Fan, T., Li, R., Liu, X., Song, B., Jia, X., Zhong, S., 2022a. Preparation and Characterization of Nano-Selenium Decorated by Chondroitin Sulfate Derived from Shark Cartilage and Investigation on Its Antioxidant Activity. Mar. Drugs, 20.

Chen, Y., Zhang, M., Sun, Y., Phuhongsung, P., 2022. Improving 3D/4D printing characteristics of natural food gels by novel additives: a review. Food Hydrocoll. 0268005X. 123, 107160. doi:10.1016/j.foodhyd.2021.107160.

Cheong, K.L., Yu, B., Chen, J., Zhong, S., 2022. A Comprehensive Review of the Cardioprotective Effect of Marine Algae Polysaccharide on the Gut Microbiota. Foods. 23048158. 11 (22). Available from: www.mdpi.com/journal/foods. doi:10.3390/foods11223550.

Chikha, S.B., Bougatef, H., Capitani, F., Amor, I.B., Maccari, F., Gargouri, J., Sila, A., Volpi, N., Bougatef, A., 2023. Composition and Anticoagulant Potential of Chondroitin Sulfate and Dermatan Sulfate from Inedible Parts of

Garfish (Belone belone). Foods. 23048158. 12 (21). Available from: www.mdpi.com/journal/foods. doi:10.3390/foods12213887.

Chudasama, N.A., Sequeira, R.A., Moradiya, K., Prasad, K., 2021. Seaweed polysaccharide based products and materials: an assessment on their production from a sustainability point of view. Molecules. 14203049. 26 (9). Available from: https://www.mdpi.com/1420-3049/26/9/2608/pdf. doi:10.3390/molecules26092608.

Cindana Mo'o, F.R., Wilar, G., Devkota, H.P., 2020. Nasrul Wathoni, Ulvan, a Polysaccharide from Macroalga Ulva sp.: a Review of Chemistry, Biological Activities and Potential for Food and Biomedical Applications. Appl. Sci. 2076-3417. 10 (16), 5488. doi:10.3390/app10165488.

Concordio-Reis, P., Alves, V.D., Moppert, X., Guezennec, J., Freitas, F., Reis, M.A.M., 2021. Characterization and Biotechnological Potential of Extracellular Polysaccharides Synthesized by Alteromonas Strains Isolated from French Polynesia Marine Environments. Mar. Drugs 19.

Cotas, J., Pacheco, D., Gonçalves, A.M.M., Silva, P., Carvalho, L.G., Pereira, L., 2021. Seaweeds' nutraceutical and biomedical potential in cancer therapy: a concise review. J. Cancer Metastasis Treat. 2454-2857. 2021. doi:10.20517/2394-4722.2020.134.

Cui, M., Wang, Y., Liu, K., 2022. Structure Analysis and Anti-Inflammatory Effect of a Polysaccharide from Hippocampus mohnikei. Chem. Biodivers. 19, e202100739.

D, P., S, G.-P., J, C., Amm, G., L, P., 2020. Fucoidan - a valuable source from the ocean to pharmaceutical. Frontiers in Drug, Chemistry and Clinical Research, 3.

Da Costa, J.N., Leal, A.R., Nascimento, L.G.L., Rodrigues, D.C., Muniz, C.R., Figueiredo, R.W., Mata, P., Noronha, J. P., De Sousa, P.H.M., 2020. Texture, microstructure and volatile profile of structured guava using agar and gellan gum. Int. J. Gastronomy Food Sci. 18784518. 20. Available from: www.journals.elsevier.com/international-journal-of-gastronomy-and-food-science. doi:10.1016/j.ijgfs.2020.100207.

Dai, L., Zhan, X., Wei, Y., Sun, C., Mao, L., Mcclements, D.J., Gao, Y., 2018. Composite zein - propylene glycol alginate particles prepared using solvent evaporation: characterization and application as Pickering emulsion stabilizers. Food Hydrocoll. 85, 281–290.

De Avelar, M.H.M., Efraim, P., 2020. Alginate/pectin cold-set gelation as a potential sustainable method for jelly candy production. LWT. 00236438. 123. Available from: https://www.journals.elsevier.com/lwt. doi:10.1016/j.lwt.2020.109119.

Deepika, C., Ravishankar, G.A., Rao, A.R., 2022. Potential Products from Macroalgae: an Overview. In: Ranga Rao, A., Ravishankar, G.A. (Eds.), In. *Sustainable Global Resources Of Seaweeds Volume 1: Bioresources, cultivation, trade and multifarious applications*. Springer International Publishing, Cham.

Delzenne, N.M., Olivares, M., Neyrinck, A.M., Beaumont, M., Kjølbæk, L., Larsen, T.M., Benítez-Páez, A., Romaní-Pérez, M., Garcia-Campayo, V., Bosscher, D., Sanz, Y., Van Der Kamp, J.W., 2020. Nutritional interest of dietary fiber and prebiotics in obesity: lessons from the MyNewGut consortium. Churchill Livingstone. Belgium Clin. Nutr. 15321983. 39 (2), 414–424. Available from: http://www.elsevier-international.com/journals/clnu/. doi:10.1016/j.clnu.2019.03.002.

Derossi, A., Spence, C., Corradini, M.G., Jekle, M., Fahmy, A.R., Caporizzi, R., Devahastin, S., Moses, J.A., Le-Bail, A., Zhou, W., Zhang, M., Bhandari, B., Severini, C., 2024. Personalized, digitally designed 3D printed food towards the reshaping of food manufacturing and consumption. Npj Sci. Food. 23968370. 8 (1). Available from: https://www.nature.com/npjscifood/about. doi:10.1038/s41538-024-00296-5.

Devillé, C., Damas, J., Forget, P., Dandrifosse, G., Peulen, O., 2004. Laminarin in the dietary fibre concept. J. Sci. Food Agric. 0022-5142. 84 (9), 1030–1038. doi:10.1002/jsfa.1754.

Dhanapal, K.A.V.I.T.H.A.V.R.A.J.A.M.A.N.I., Banu, Y.A.Z.H.I.N.I., S, M., Edible films from Polysaccharides Food Science and Quality Management. (2012).

Dhanya, B.E., Prabhu, A., Rekha, P.D., 2022. Extraction and characterization of an exopolysaccharide from a marine bacterium. Int. Microbiol. 25, 285–295.

Divya, K., Jisha, M.S., 2018. Chitosan nanoparticles preparation and applications. Environ. Chem. Lett. 1610-3653. 16 (1), 101–112. doi:10.1007/s10311-017-0670-y.

Dobrincic, A., Balbino, S., Zoric, Z., Pedisic, S., Bursac Kovacevic, D., Elez Garofulic, I., Dragovic-Uzelac, V., 2020. Advanced Technologies for the Extraction of Marine Brown Algal Polysaccharides. Mar. Drugs 18.

Donn, P., Prieto, M.A., Mejuto, J.C., Cao, H., Simal-Gandara, J., 2022. Functional foods based on the recovery of bioactive ingredients from food and algae by-products by emerging extraction technologies and 3D printing. Food Biosci. 22124306. 49. Available from: http://www.journals.elsevier.com/food-bioscience/. doi:10.1016/j.fbio.2022.101853.

Duan, C., Meng, X., Meng, J., Khan, M.I.H., Dai, L., Khan, A., An, X., Zhang, J., Huq, T., Ni, Y., 2019. Chitosan as A Preservative for Fruits and Vegetables: a Review on Chemistry and Antimicrobial Properties. J. Bioresour. Bioprod. 4, 11–21.

Duinker, A., Kleppe, M., Fjære, E., Biancarosa, I., Heldal, H.E., Dahl, L., Lunestad, B.T., Knowledge update on macroalgae food and feed safety-based on data generated in the period 2014-2019 by the Institute of Marine Research. (2020).

Dwivedi, R., Sharma, P., Eilts, F., Zhang, F., Linhardt, R.J., Tandon, R., Pomin, V.H., 2023. Anti-SARS-CoV-2 and anticoagulant properties of Pentacta pygmaea fucosylated chondroitin sulfate depend on high molecular weight structures. Glycobiology. 14602423. 33 (1), 75–85. Available from: http://glycob.oxfordjournals.org/. doi:10.1093/glycob/cwac063.

El Fayoumy, R.A., El-Sheekh, M.M., Abu Ahmed, S.E., 2022. Potential of Ulvan Polysaccharide from Ulva lactuca as Antifungal Against Some Foodborne Fungi Isolated from Spoiled Tomato Sauce Cans. J. Aquat. Food Prod. Technol. 31, 658–671.

Elbayomi, S.M., Wang, H., Tamer, T.M., You, Y., 2021. Enhancement of antioxidant and hydrophobic properties of alginate via aromatic derivatization: preparation, characterization, and evaluation. Polymers. 20734360. 13 (15). Available from: https://www.mdpi.com/2073-4360/13/15/2575/pdf. doi:10.3390/polym13152575.

Fabris, M., Abbriano, R.M., Pernice, M., Sutherland, D.L., Commault, A.S., Hall, C.C., Labeeuw, L., Mccauley, J.I., Kuzhiuparambil, U., Ray, P., Kahlke, T., Ralph, P.J., 2020. Emerging Technologies in Algal Biotechnology: toward the Establishment of a Sustainable, Algae-Based Bioeconomy. Front. Plant Sci. 1664462X. 11. Available from: https://www.frontiersin.org/journals/plant-science. doi:10.3389/fpls.2020.00279.

Farrag, M., Dwivedi, R., Sharma, P., Kumar, D., Tandon, R., Pomin, V.H., Lahmann, M., 2023. Structural requirements of Holothuria floridana fucosylated chondroitin sulfate oligosaccharides in anti-SARS-CoV-2 and anticoagulant activities. PLOS ONE. 1932-6203. 18 (5), e0285539. doi:10.1371/journal.pone.0285539.

Feng, Y.Y., Mu, T.H., Zhang, M., Ma, M.M., 2020a. Effects of different polysaccharides and proteins on dough rheological properties, texture, structure and in vitro starch digestibility of wet sweet potato vermicelli. Int. J. Biol. Macromol. 148, 1–10.

Feng, Y.Y., Mu, T.H., Zhang, M., Ma, M.M., 2020b. Effects of ionic polysaccharides and egg white protein complex formulations on dough rheological properties, structure formation and in vitro starch digestibility of wet sweet potato vermicelli. Int. J. Biol. Macromol. 149, 1170–1179.

Ferdiansyah, R., Abdassah, M., Zainuddin, A., Rachmaniar, R., Chaerunisaa, A.Y., 2023. Effects of Alkaline Solvent Type and pH on Solid Physical Properties of Carrageenan from Eucheuma cottonii. Gels. 23102861. 9 (5). Available from: www.mdpi.com/journal/gels. doi:10.3390/gels9050397.

Fernando, I.P.S., Lee, W.W., Han, E.J., Ahn, G., 2020. Alginate-based nanomaterials: fabrication techniques, properties, and applications. Chem. Eng. J. 13858947. 391. Available from: www.elsevier.com/inca/publications/store/6/0/1/2/7/3/index.htt. doi:10.1016/j.cej.2019.123823.

Fertah, M., 2017. Chapter 2 - Isolation and Characterization of Alginate from Seaweed Seaweed Polysaccharides, Elsevier.

Filippini, M., Baldisserotto, A., Menotta, S., Fedrizzi, G., Rubini, S., Gigliotti, D., Valpiani, G., Buzzi, R., Manfredini, S., Vertuani, S., 2021. Heavy metals and potential risks in edible seaweed on the market in Italy. Chemosphere. 00456535. 263, 127983. doi:10.1016/j.chemosphere.2020.127983.

Fitton, H.J., Stringer, D.S., Park, A.Y., Karpiniec, S.N., 2019. Therapies from Fucoidan: new Developments. Mar. Drugs 17.

Fontes-Candia, C., Ström, A., Gómez-Mascaraque, L.G., López-Rubio, A., Martínez-Sanz, M., 2020. Understanding nanostructural differences in hydrogels from commercial carrageenans: combined small angle X-ray scattering and rheological studies. Algal Res. 22119264. 47. Available from: http://www.sciencedirect.com/science/journal/aip/22119264. doi:10.1016/j.algal.2020.101882.

Fontes-Candia, C., Ström, A., Lopez-Sanchez, P., López-Rubio, A., Martínez-Sanz, M., 2020b. Rheological and structural characterization of carrageenan emulsion gels. Algal Res. 47.

Gajanayaka, N.D., Jo, E., Bandara, M.S., Marasinghe, S.D., Bawkar, C., Lee, Y.J., Park, G.H., Oh, C., Lee, Y., 2024. Characterisation of High Alkaline-Tolerant Novel Ulvan Lyase from Pseudoalteromonas agarivorans: potential Applications of Enzyme Derived Oligo-Ulvan as Anti-Diabetic Agent. Mar. Drugs. 16603397. 22 (12). Available from: http://www.mdpi.com/journal/marinedrugs. doi:10.3390/md22120577.

Gan, L., Xiaoguang, L., Wang, H., Peng, B., Tian, Y., 2020. Structural characterization and functional evaluation of a novel exopolysaccharide from the moderate halophile Gracilibacillus sp. SCU50. Int. J. Biol. Macromol. 01418130. 154, 1140–1148. doi:10.1016/j.ijbiomac.2019.11.143.

Gao, N., Chen, R., Mou, R., Xiang, J., Zhou, K., Li, Z., Zhao, J., 2020. Purification, structural characterization and anticoagulant activities of four sulfated polysaccharides from sea cucumber Holothuria fuscopunctata. Int. J. Biol. Macromol. 01418130. 164, 3421–3428. doi:10.1016/j.ijbiomac.2020.08.150.

Garcia-Vaquero, M., Rajauria, G., Tiwari, B., Sweeney, T., O'doherty, J., 2018. Extraction and Yield Optimisation of Fucose, Glucans and Associated Antioxidant Activities from Laminaria digitata by Applying Response Surface Methodology to High Intensity Ultrasound-Assisted Extraction. Mar. Drugs. 1660-3397. 16 (8), 257. doi:10.3390/md16080257.

Getachew, A.T., Jacobsen, C., Holdt, S.L., 2020. Emerging Technologies for the Extraction of Marine Phenolics: opportunities and Challenges. Mar. Drugs 18.

Ghagane, S.C., Shetti, P.P., Vidyashree, J.S., Kumbar, V.M., Nadaf, R., Patil, D., 2024. Evaluation of in vitro antidiabetic and anti-microbial activity of edible red seaweed Pyropia vietnamensis. Discov. Appl. Sci. 30049261. 6 (9). Available from: https://link.springer.com/journal/42452. doi:10.1007/s42452-024-06126-3.

Gheorghita, R., Gutt, G., Amariei, S., 2020. The Use of Edible Films Based on Sodium Alginate in Meat Product Packaging: an Eco-Friendly Alternative to Conventional Plastic Materials. Coatings 10.

Ghosh, S., Sarkar, T., Pati, S., Kari, Z.A., Edinur, H.A., Chakraborty, R., 2022. Novel Bioactive Compounds From Marine Sources as a Tool for Functional Food Development. Front. Mar. Sci. 22967745. 9. Available from: https://www.frontiersin.org/journals/marine-science#. doi:10.3389/fmars.2022.832957.

Gomez, L.P., Alvarez, C., Zhao, M., Tiwari, U., Curtin, J., Garcia-Vaquero, M., Tiwari, B.K., 2020. Innovative processing strategies and technologies to obtain hydrocolloids from macroalgae for food applications. Carbohydr. Polym. 01448617. 248, 116784. doi:10.1016/j.carbpol.2020.116784.

Gong, P.X., Li, Q.Y., Wu, Y.C., Lu, W.Y., Zeng, J., Li, H.J., 2021. Structural elucidation and antidiabetic activity of fucosylated chondroitin sulfate from sea cucumber Stichopus japonicas. Carbohydr. Polym. 01448617. 262. Available from: http://www.elsevier.com/wps/find/journaldescription.cws_home/405871/description#description. doi:10.1016/j.carbpol.2021.117969.

Gong, P.X., Wu, Y.C., Liu, Y., Lv, S.Z., You, Y., Zhou, Z.L., Chen, X., Li, H.J., 2022. Structure and hypoglycemic effect of a neutral polysaccharide isolated from sea cucumber Stichopus japonicus. Int. J. Biol. Macromol. 18790003. 216, 14–23. Available from: www.elsevier.com/locate/ijbiomac. doi:10.1016/j.ijbiomac.2022.06.160.

Guan, X., Wang, F., Zhou, B., Sang, X., Zhao, Q., 2024. The nutritional function of active polysaccharides from marine animals: a review. Food Biosci. 22124292. 58, 103693. doi:10.1016/j.fbio.2024.103693.

Guidara, M., Yaich, H., Benelhadj, S., Adjouman, Y.D., Richel, A., Blecker, C., Sindic, M., Boufi, S., Attia, H., Garna, H., 2020. Smart ulvan films responsive to stimuli of plasticizer and extraction condition in physico-chemical, optical, barrier and mechanical properties. Int. J. Biol. Macromol. 18790003. 150, 714–726. Available from: www.elsevier.com/locate/ijbiomac. doi:10.1016/j.ijbiomac.2020.02.111.

Guidara, M., Yaich, H., Richel, A., Blecker, C., Boufi, S., Attia, H., Garna, H., 2019. Effects of extraction procedures and plasticizer concentration on the optical, thermal, structural and antioxidant properties of novel ulvan films. Int. J. Biol. Macromol. 135, 647–658.

Guo, S., Wang, J., Chuanbo, H., Wei, H., Ying, M., Xiong, H., 2020. Preparation and antioxidant activities of polysaccharides obtained from abalone viscera by combination of enzymolysis and multiple separation methods. J. Food Sci. 0022-1147. 85 (12), 4260–4270. doi:10.1111/1750-3841.15520.

Han, X., Zheng, Z., Chen, Y., Deng, Y., Qing, Y., Niu, F., Chen, Q., Pan, W., Wang, Y., 2022. Preparation, characterization and antibacterial activity of new ionized chitosan. Carbohydr. Polym. 01448617. 290, 119490. doi:10.1016/j.carbpol.2022.119490.

Haripriya, P., Vijayakrishna, K., 2024. Synthesis of poly(ionic liquid-OH) mediated deacetylated chitin and its hydrogels: a study on their applications in controlled release of paracetamol and urea. Int. J. Biol. Macromol. 01418130. 266 (Pt 2), 131230. doi:10.1016/j.ijbiomac.2024.131230.

Hentati, F., Tounsi, L., Pierre, G., Barkallah, M., Ursu, A.V., Hlima, H.B., Desbrières, J., Cerf, D.L., Fendri, I., Michaud, P., Abdelkafi, S., 2022. Structural Characterization and Rheological and Antioxidant Properties of Novel Polysaccharide from Calcareous Red Seaweed. Mar. Drugs. 16603397. 20 (9). Available from: http://www.mdpi.com/journal/marinedrugs. doi:10.3390/md20090546.

Hentati, F., Ursu, A., Pierre, G., Delattre, C., Trica, B., Abdelkafi, S., Djelveh, G., Dobre, T., Michaud, P., 2018. Handbook of Algal Technologies and Phytochemicals, CRC Press.

Hifney, A.F., Fawzy, M.A., Abdel-Gawad, K.M., Gomaa, M., 2016. Industrial optimization of fucoidan extraction from Sargassum sp. and its potential antioxidant and emulsifying activities. Food Hydrocoll. 0268005X. 54, 77–88. Available from: https://www.journals.elsevier.com/food-hydrocolloids. doi:10.1016/j.foodhyd.2015.09.022.

Hossain, M.S., Alamgir, M., Uddin, S.A., Msn, C. 2020b. *Seaweeds for Blue Economy in Bangladesh.*

Hosseini, E., Rajaei, A., Tabatabaei, M., Mohsenifar, A., Jahanbin, K., 2019. Preparation of Pickering Flaxseed Oil-in-Water Emulsion Stabilized by Chitosan-Myristic Acid Nanogels and Investigation of Its Oxidative Stability in Presence of Clove Essential Oil as Antioxidant. Food Biophys. 15, 216–228.

Hotchkiss, S., Brooks, M., Campbell, R., Philp, K., Trius, A., 2016. The Use of Carrageenan in Food Carrageenans: Sources and Extraction Methods, Molecular Structure, Bioactive Properties and Health Effects. Nova Science Publishers, Inc., Ireland, 229–243. Available from: https://www.novapublishers.com/catalog/product_info.php?products_id=58914.

Hu, Z., Ganzle, M.G., 2019. Challenges and opportunities related to the use of chitosan as a food preservative. J. Appl. Microbiol. 126, 1318–1331.

Huang, A., Wu, X., Lu, F., Liu, F., 2024a. Sustainable Production of Ulva Oligosaccharides via Enzymatic Hydrolysis: a Review on Ulvan Lyase. Foods 13.

Huang, R., Zhang, J., Xu, X., Sun, M., Xu, L., Kuang, H., Xu, C., Guo, L., 2024b. The multiple benefits of bioactive polysaccharides: from the gut to overall health. Trends Food Sci. Technol. 152.

Humayun, S., Premarathna, A.D., Rjabovs, V., Howlader, M.M., Darko, C.N.S., Mok, I.K., Tuvikene, R., 2023. Biochemical Characteristics and Potential Biomedical Applications of Hydrolyzed Carrageenans. Mar. Drugs. 16603397. 21 (5). Available from: http://www.mdpi.com/journal/marinedrugs. doi:10.3390/md21050269.

Ibrahim, H.A.H., Abou Elhassayeb, H.E., El-Sayed, W.M.M., 2022a. Potential functions and applications of diverse microbial exopolysaccharides in marine environments. J. Genet. Eng. Biotechnol. 20, 151.

Ibrahim, M.I.A., Amer, M.S., Ibrahim, H.A.H., Zaghloul, E.H., 2022. Considerable Production of Ulvan from Ulva lactuca with Special Emphasis on Its Antimicrobial and Anti-fouling Properties. Appl. Biochem. Biotechnol. 15590291. 194 (7), 3097–3118. Available from: https://www.springer.com/journal/12010. doi:10.1007/s12010-022-03867-y.

Iqbal, S., Xu, Z., Huang, H., Chen, X.D., 2019. Structuring of water-in-oil emulsions using controlled aggregation of polysaccharide in aqueous phases. J. Food Eng. 258, 34–44.

Janseerat, M.K., Reddy, C.S., Sharma, S., Roy, S., 2024. Anthocyanin-Based Natural Color Induced Intelligent Food Packaging Sensor: a Review. Curr. Food Sci. Technol. Rep. 2662-8473. 2 (2), 157–167. doi:10.1007/s43555-024-00021-z.

Jeewon, R., Aullybux, A.A., Puchooa, D., Nazurally, N., Alrefaei, A.F., Zhang, Y., 2023. Marine Microbial Polysaccharides: an Untapped Resource for Biotechnological Applications. Mar. Drugs. 16603397. 21 (7). Available from: http://www.mdpi.com/journal/marinedrugs. doi:10.3390/md21070420.

Jeong, D.W., Hyeon, J.E., Shin, S.K., Han, S.O., 2020. Trienzymatic Complex System for Isomerization of Agar-Derived d -Galactose into d -Tagatose as a Low-Calorie Sweetener. J. Agric. Food Chem. 15205118. 68 (10), 3195–3202. Available from: http://pubs.acs.org/journal/jafcau. doi:10.1021/acs.jafc.9b07536.

Jonsson, M., Allahgholi, L., Sardari, R.R.R., Hreggviethsson, G.O., Nordberg Karlsson, E., 2020. Extraction and Modification of Macroalgal Polysaccharides for Current and Next-Generation Applications. Molecules 25.

Jridi, M., Nasri, R., Marzougui, Z., Abdelhedi, O., Hamdi, M., Nasri, M., 2019. Characterization and assessment of antioxidant and antibacterial activities of sulfated polysaccharides extracted from cuttlefish skin and muscle. Int. J. Biol. Macromol. 123, 1221–1228.

Kanchana, S., Arumugam, M., Giji, S., 2013. Thangavel Balasubramanian, Isolation, characterization and antioxidant activity of hyaluronic acid from marine bivalve mollusc Amussium pleuronectus (Linnaeus, 1758). Bioact. Carbohydr. Diet. Fibre. 22126198. 2 (1), 1–7. doi:10.1016/j.bcdf.2013.06.001.

Karuppusamy, S., Rajauria, G., Fitzpatrick, S., Lyons, H., Mcmahon, H., Curtin, J., Tiwari, B.K., O'donnell, C., 2022. Biological Properties and Health-Promoting Functions of Laminarin: a Comprehensive Review of Preclinical and Clinical Studies. Mar. Drugs. 16603397. 20 (12). Available from: http://www.mdpi.com/journal/marinedrugs. doi:10.3390/md20120772.

Khajouei, R.A., Keramat, J., Hamdami, N., Ursu, A.V., Delattre, C., Laroche, C., Gardarin, C., Lecerf, D., Desbrières, J., Djelveh, G., Michaud, P., 2018. Extraction and characterization of an alginate from the Iranian brown seaweed Nizimuddinia zanardini. Int. J. Biol. Macromol. 18790003. 118, 1073–1081. Available from: www.elsevier.com/locate/ijbiomac. doi:10.1016/j.ijbiomac.2018.06.154.

Khalil, H.P.S.A., Lai, T.K., Tye, Y.Y., Rizal, S., Chong, E.W.N., Yap, S.W., Hamzah, A.A., Fazita, M.R.N., Paridah, M. T., 2018. A review of extractions of seaweed hydrocolloids: properties and applications. Express Polym. Lett. 1788618X. 12 (4), 296–317. doi:10.3144/expresspolymlett.2018.27.

Khanzadi, S., Keykhosravy, K., Hashemi, M., Azizzadeh, M., (2020).

Khojah, S.M., 2020. Bio-based Coating from Fish Gelatin, K-Carrageenan and Extract of Pomegranate Peels for Maintaining the Overall Qualities of Fish Fillet. J. Aquat. Food Prod. Technol. 15470636. 29 (8), 810–822. Available from: http://www.tandfonline.com/toc/wafp20/current. doi:10.1080/10498850.2020.1718261.

Khoshdouni Farahani, Z., Mousavi, M., Seyedain Ardebili, S.M., Bakhoda, H., 2022. Modification of sodium alginate by octenyl succinic anhydride to fabricate beads for encapsulating jujube extract. Curr. Res. Food Sci. 5, 157–166.

Kocira, A., Kozłowicz, K., Panasiewicz, K., Staniak, M., Szpunar-Krok, E., Hortyńska, P., 2021. Polysaccharides as Edible Films and Coatings: characteristics and Influence on Fruit and Vegetable Quality—A Review. Agronomy. 2073-4395. 11 (5), 813. doi:10.3390/agronomy11050813.

Kokkuvayil Ramadas, B., Rhim, J.W., S, R.O.Y., 2024. Recent Progress of Carrageenan-Based Composite Films in Active and Intelligent Food Packaging Applications. Polymers (Basel 16).

Kong, Y., Li, Y., Dai, Z.R., Qin, M., Fan, H.L., Hao, J.G., Zhang, C.X., Zhong, Q.P., Qi, C., Wang, P., 2021. Glycosaminoglycan from Ostrea rivularis attenuates hyperlipidemia and regulates gut microbiota in high-cholesterol diet-fed zebrafish. Food Sci. Nutr. 20487177. 9 (9), 5198–5210. Available from: http://onlinelibrary.wiley.com/journal/10.1002/%28ISSN%292048-7177. doi:10.1002/fsn3.2492.

Kopplin, G., Rokstad, A.M., Mélida, H., Bulone, V., Skjåk-Bræk, G., Aachmann, F.L., 2018. Structural Characterization of Fucoidan from Laminaria hyperborea: assessment of Coagulation and Inflammatory Properties and Their Structure-Function Relationship. ACS Appl. Bio Mater. 25766422. 1 (6), 1880–1892. Available from: http://pubs.acs.org/journal/aabmcb. doi:10.1021/acsabm.8b00436.

Korcz, E., Varga, L., 2021. Exopolysaccharides from lactic acid bacteria: techno-functional application in the food industry. Trends Food Sci. Technol. 09242244. 110, 375–384. doi:10.1016/j.tifs.2021.02.014.

Krangkratok, W., Chantorn, S., Choosuwan, P., Phomkaivon, N., La-Ongkham, O., Kosawatpat, P., Tamtin, M., Praiboon, J., 2023. Production of prebiotic ulvan-oligosaccharide from the green seaweed Ulva rigida by enzymatic hydrolysis. Biocatal. Agric. Biotechnol. 18788181. 54, 102922. doi:10.1016/j.bcab.2023.102922.

Kumar, A., Kumari, P., Gupta, K., Singh, M., Tomer, V., 2021. Recent Advances in Extraction, Techno-functional Properties, Food and Therapeutic Applications as Well as Safety Aspects of Natural and Modified Stabilizers. Food Rev. Int. 39, 2233–2276.

Lahaye, M., R, A., 2007. Structure and Functional Properties of Ulvan, a Polysaccharidefrom Green Seaweeds. Biomacromolecules 8, 1765–1774.

Lavrentieva, A., Pepelanova, I., Seliktar, D., 2021. Tunable Hydrogels: Smart Materials for Biomedical Applications, Springer Nature Switzerland AG.

Lee, W.-K., Lim, -Y.-Y., Leow, A.T.-C., Namasivayam, P., Abdullah, J.O., Ho, C.-L., 2016. Factors affecting yield and gelling properties of agar. J. Appl. Phycol. 29, 1527–1540.

Levy-Ontman, O., Abu-Galiyun, E., Huleihel, M., 2023. Studying the Relationship between the Antiviral Activity and the Structure of i-Carrageenan Using Ultrasonication. Int. J. Mol. Sci. 24.

Li, C., Tang, T., Du, Y., Jiang, L., Yao, Z., Ning, L., B, Z.H.U., 2023a. Ulvan and Ulva oligosaccharides: a systematic review of structure, preparation, biological activities and applications. Bioresour. Bioprocess. 10, 66.

Li, F., Hu, X., Sun, X., Li, H., Lu, J., Li, Y., Bao, M., 2022. Effect of fermentation pH on the structure, rheological properties, and antioxidant activities of exopolysaccharides produced by Alteromonas australica QD. Glycoconj. J. 39 (6), 773–787. doi:10.1007/s10719-022-10087-3. PMID: 36367683.

Li, S., Zhong, W., Pan, Y., Lin, L., Cai, Y., Mao, H., Zhang, T., Li, S., Chen, R., Zhou, L., Wang, W., Cui, Q., Yin, R., Huang, S., Zhao, J., 2021a. Structural characterization and anticoagulant analysis of the novel branched fucosylated glycosaminoglycan from sea cucumber Holothuria nobilis. Carbohydr. Polym. 269, 118290.

Li, W., Kobayashi, T., Moroi, S., Kotake, H., Ikoma, T., Saeki, H., Ura, K., Takagi 2019. Anti-obesity effects of chondroitin sulfate oligosaccharides from the skate Raja pulchra. Carbohydr. Polym. 214, 303–310.

Li, W., Terauchi, N., Meng, D., Miyamoto, N., Tsutsumi, N., Ura, K., Takagi, Y., 2021b. Antioxidant and fibroblast-activating activities of the by-product of skate chondroitin extractive production. Sustain. Chem. Pharm., 23.

Liao, Y.C., Chang, C.C., Nagarajan, D., Chen, C.Y., Chang, J.S., 2021. Algae-derived hydrocolloids in foods: applications and health-related issues. Bioengineered. 21655987. 12 (1), 3787–3801. Available from: http://www.tandfonline.com/loi/kbie20. doi:10.1080/21655979.2021.1946359.

Lim, H.P., Ho, K.W., Singh, C.K.S., Ooi, C.W., Tey, B.T., Chan, E.S., 2020. Pickering emulsion hydrogel as a promising food delivery system: synergistic effects of chitosan Pickering emulsifier and alginate matrix on hydrogel stability and emulsion delivery. Food Hydrocoll. 0268005X. 103. Available from: https://www.sciencedirect.com/science/journal/0268005X. doi:10.1016/j.foodhyd.2020.105659.

Lin, Y., Dong, Y., Xiangling, L., Cai, J., Cai, L., Zhang, G., 2024. Enzymatic production of xylooligosaccharide from lignocellulosic and marine biomass: a review of current progress, challenges, and its applications in food sectors. Int. J. Biol. Macromol. 01418130. 277 (Pt 3), 134014. doi:10.1016/j.ijbiomac.2024.134014.

Liu, F., Duan, G., Yang, H., 2023. Recent advances in exploiting carrageenans as a versatile functional material for promising biomedical applications. Int. J. Biol. Macromol. 235, 123787.

Liu, F., Hou, P., Zhang, H., Tang, Q., Xue, C., Li, R.W., 2021a. Food-grade carrageenans and their implications in health and disease. Compr. Rev. Food Sci. Food Saf. 20, 3918–3936.

Liu, J., Yang, S., Li, X., Yan, Q., Reaney, M.J.T., Jiang, Z., 2019. Alginate Oligosaccharides: production, Biological Activities, and Potential Applications. Compr. Rev. Food Sci. Food Saf. 18, 1859–1881.

Liu, T., Li, J., Tang, Q., Qiu, P., Gou, D., Zhao, J., 2022. Chitosan-Based Materials: an Overview of Potential Applications in Food Packaging. Foods, 11.

Liu, Z., Hu, M., Zhang, S., Jiang, L., Xie, F., Li, Y., 2021b. Oil-in-water Pickering emulsion stabilization with oppositely charged polysaccharide particles: chitin nanocrystals/fucoidan complexes. J. Sci. Food Agric. 101, 3003–3012.

Lucarini, M., Durazzo, A., Sette, S., Lombardi-Boccia, G., Santini, A., Strazzullo, P., 2021. Sodium Intake and Related Diseases. Int. J. Mol. Sci. 1422-0067. 22 (14), 7608. doi:10.3390/ijms22147608.

Lux Nee Bantleon, T., Spillmann, F., Reimold, F., Erdos, A., Lochny, A., Floter, E., 2023. Physical quality of gluten-free doughs and fresh pasta made of amaranth. Food Sci. Nutr. 11, 3213–3223.

Lv, S., Zhou, H., Bai, L., Rojas, O.J., Mcclements, D.J., 2021. Development of food-grade Pickering emulsions stabilized by a mixture of cellulose nanofibrils and nanochitin. Food Hydrocoll. 0268005X. 113. Available from: https://www.sciencedirect.com/science/journal/0268005X. doi:10.1016/j.foodhyd.2020.106451.

Ma, Y., Liu, X., Liu, D., Yin, Z., Yang, X., Zeng, M., 2022a. Oyster (Crassostrea gigas) Polysaccharide Ameliorates High-Fat-Diet-Induced Oxidative Stress and Inflammation in the Liver via the Bile Acid-FXR-AMPKalpha Pathway. J. Agric. Food Chem. 70, 8662–8671.

Ma, Y., Zhu, L., Ke, H., Jiang, S., Zeng, M., 2022. Oyster (Crassostrea gigas) polysaccharide ameliorates obesity in association with modulation of lipid metabolism and gut microbiota in high-fat diet fed mice. Int. J. Biol. Macromol. 18790003. 216, 916–926. Available from: www.elsevier.com/locate/ijbiomac. doi:10.1016/j.ijbiomac.2022.07.100.

Mahajan, B.V.C., Tandon, R., Kapoor, S., Sidhu, M.K., Natural coatings for shelf-life enhancement and quality maintenance of fresh fruits and vegetables- A Review. (2017).

Mahmoud, M., Abdallah, N.A., El-Shafei, K., Tawfik, N.F., El-Sayed, H.S., 2020. Survivability of alginate-microencapsulated Lactobacillus plantarum during storage, simulated food processing and gastrointestinal conditions. Heliyon. 24058440. 6 (3). Available from: http://www.journals.elsevier.com/heliyon/. doi:10.1016/j.heliyon.2020.e03541.

Mavelil-Sam, R., Ouseph, E.M., Morreale, M., Scaffaro, R., Thomas, S., 2023. Recent Developments and Formulations for Hydrophobic Modification of Carrageenan Bionanocomposites. Polymers. 20734360. 15 (7). Available from: http://www.mdpi.com/journal/polymers. doi:10.3390/polym15071650.

Mensah, E.O., Kanwugu, O.N., Panda, P.K., Adadi, P., 2023. Marine fucoidans: structural, extraction, biological activities and their applications in the food industry. Federation Food Hydrocolloids. 0268005X. 142. Available from: https://www.journals.elsevier.com/food-hydrocolloids. doi:10.1016/j.foodhyd.2023.108784.

Mishra, K., Siwal, S.S., Nayaka, S.C., Guan, Z., Thakur, V.K., 2023. Waste-to-chemicals: green solutions for bioeconomy markets. Sci. Total Environ. 18791026. 887. Available from: www.elsevier.com/locate/scitotenv. doi:10.1016/j.scitotenv.2023.164006.

Mohammed, A.S.A., Naveed, M., Jost, N., 2021. Polysaccharides; Classification, Chemical Properties, and Future Perspective Applications in Fields of Pharmacology and Biological Medicine (A Review of Current Applications and Upcoming Potentialities). J. Polym. Environ. 29, 2359–2371.

Montes, L., Santamaria, M., Garzon, R., Rosell, C.M., Moreira, R., 2022. Effect of the addition of different sodium alginates on viscoelastic, structural features and hydrolysis kinetics of corn starch gels. Food Biosci. 22124306. 47. Available from: http://www.journals.elsevier.com/food-bioscience/. doi:10.1016/j.fbio.2022.101628.

Morelli, A., Massironi, A., Puppi, D., Creti, D., Domingo Martinez, E., Bonistalli, C., Fabroni, C., Morgenni, F., Chiellini, F., 2019. Development of ulvan-based emulsions containing flavour and fragrances for food and cosmetic applications. Flavour Fragr. J. 34, 411–425.

Murakami, S., Hirazawa, C., Ohya, T., Yoshikawa, R., Mizutani, T., Ning, M., Moriyama, M., Ito, T., Matsuzaki, C., 2021. The Edible Brown Seaweed Sargassum horneri (Turner) C. Agardh Ameliorates High-Fat Diet-Induced Obesity, Diabetes, and Hepatic Steatosis in Mice. Nutrients. 2072-6643. 13 (2), 551. doi:10.3390/nu13020551.

Nair, M.S., Tomar, M., Punia, S., Kukula-Koch, W., Kumar, M., 2020. Enhancing the functionality of chitosan- and alginate-based active edible coatings/films for the preservation of fruits and vegetables: a review. Int. J. Biol. Macromol. 18790003. 164, 304–320. Available from: www.elsevier.com/locate/ijbiomac. doi:10.1016/j.ijbiomac.2020.07.083.

Necas, J., Bartosikova, L., 2013. Carrageenan: a review. Veterinární Medicína. 03758427. 58 (4), 187–205. doi:10.17221/6758-vetmed.

Nezamdoost-Sani, N., Khaledabad, M.A., Amiri, S., Khaneghah, A.M., 2023. Alginate and derivatives hydrogels in encapsulation of probiotic bacteria: an updated review. Food Biosci. 22124306. 52. Available from: http://www.journals.elsevier.com/food-bioscience/. doi:10.1016/j.fbio.2023.102433.

Ngasotter, S., Xavier, K.A.M., Meitei, M.M., Waikhom, D., Madhulika, J.P., Singh, S.K., 2023. Crustacean shell waste derived chitin and chitin nanomaterials for application in agriculture, food, and health – a review. Carbohydr. Polym. Technol. Appl. 26668939. 6. Available from: https://www.journals.elsevier.com/carbohydrate-polymer-technologies-and-applications. doi:10.1016/j.carpta.2023.100349.

Olatunji, O., 2020. Aquatic Biopolymers: Understanding Their Industrial Significance and Environmental Implications, Springer.

Oun, A.A., Rhim, J.W., 2020. Preparation of multifunctional carboxymethyl cellulose-based films incorporated with chitin nanocrystal and grapefruit seed extract. Int. J. Biol. Macromol. 18790003. 152, 1038–1046. Available from: www.elsevier.com/locate/ijbiomac. doi:10.1016/j.ijbiomac.2019.10.191.

Pai, A.A., Chakraborty, K., Dhara, S., Raj, A., Kariyil, B.J., R, A., 2024. Therapeutic potential of sulfated glycosaminoglycan from seafood Asian green mussel (Perna viridis): insights from an in vivo study. Food Biosci. 22124306. 61. Available from: https://www.sciencedirect.com/science/journal/22124292. doi:10.1016/j.fbio.2024.104837.

Panebianco, F., Nava, V., Giarratana, F., Gervasi, T., Cicero, N., 2021. Assessment of heavy- and semi-metals contamination in edible seaweed and dried fish sold in ethnic food stores on the Italian market. J. Food Compos. Anal. 08891575. 104, 104150. doi:10.1016/j.jfca.2021.104150.

Park, J., Nam, J., Yun, H., Jin, H.J., Kwak, H.W., 2021. Aquatic polymer-based edible films of fish gelatin crosslinked with alginate dialdehyde having enhanced physicochemical properties. Carbohydr. Polym. 01448617. 254. Available from: http://www.elsevier.com/wps/find/journaldescription.cws_home/405871/description#description. doi:10.1016/j.carbpol.2020.117317.

Pascuta, M.S., Varvara, R.A., Teleky, B.E., Szabo, K., Plamada, D., Nemeş, S.A., Mitrea, L., Martău, G.A., Ciont, C., Călinoiu, L.F., Barta, G., Vodnar, D.C., 2022. Polysaccharide-Based Edible Gels as Functional Ingredients: characterization, Applicability, and Human Health Benefits. Gels. 23102861. 8 (8). Available from: www.mdpi.com/journal/gels. doi:10.3390/gels8080524.

Patel, A.K., Vadrale, A.P., Singhania, R.R., Michaud, P., Pandey, A., Chen, S.-J., Chen, C.-W., Dong, C.-D., 2022. Algal polysaccharides: current status and future prospects. Phytochem. Rev. 22, 1167–1196.

Paz Otero, M.C., Garcia-Oliveira, P., Echave, J., Soria-Lopez, A., Garcia-Perez, P., Fraga-Corral, M., Cao, H., Nie, S., Jianbo Xiao, J.S.-G., Prieto, M.A., 2023. Seaweed polysaccharides: emerging extraction technologies, chemical modifications and bioactive properties. Crit. Rev. Food Sci. Nutr. 1040-8398. 63 (13), 1901–1929. doi:10.1080/10408398.2021.1969534.

Perez-Vazquez, A., Carpena, M., Barciela, P., Cassani, L., Simal-Gandara, J., Prieto, M.A., 2023. Pressurized Liquid Extraction for the Recovery of Bioactive Compounds from Seaweeds for Food Industry Application: a Review. Antioxidants. 20763921. 12 (3). Available from: www.mdpi.com/journal/antioxidants. doi:10.3390/antiox12030612.

Phillips, G.O., Williams, P.A., Wedlock, D.J. 1988. *Gums and Stabilisers for the Food Industry 4*.

Podolean, I., Coman, S.M., Bucur, C., Teodorescu, C., Kikionis, S., Ioannou, E., Roussis, V., Primo, A., Garcia, H., Parvulescu, V.I., 2022. Catalytic transformation of the marine polysaccharide ulvan into rare sugars, tartaric and succinic acids. Catal. Today. 09205861. 383, 345–357. Available from: http://www.sciencedirect.com/science/journal/09205861. doi:10.1016/j.cattod.2020.06.086.

Ponce, N.M.A., Stortz, C.A., 2020. A Comprehensive and Comparative Analysis of the Fucoidan Compositional Data Across the Phaeophyceae. Front. Plant Sci. 1664462X. 11. Available from: https://www.frontiersin.org/journals/plant-science. doi:10.3389/fpls.2020.556312.

Reis., F.R. (2019). Reports on the Processing of Exotic Fruits, Cham, Switzerland: Springer.

Ren, X., Cai, S., Zhong, Y., Tang, L., Xiao, M., Shuang, L., Zhu, C., Dongyu, L., Mou, H., Xiaodan, F., 2024. Marine-Derived Fucose-Containing Carbohydrates: review of Sources, Structure, and Beneficial Effects on Gastrointestinal Health. Foods. 2304-8158. 13 (21), 3460. doi:10.3390/foods13213460.

Rhein-Knudsen, N., Ale, M.T., Ajalloueian, F., Yu, L., Meyer, A.S., 2017. Rheological properties of agar and carrageenan from Ghanaian red seaweeds. Food Hydrocoll. 0268005X. 63, 50–58. Available from: https://www.journals.elsevier.com/food-hydrocolloids. doi:10.1016/j.foodhyd.2016.08.023.

Rhim, J.-W., 2004. Physical and mechanical properties of water resistant sodium alginate films. LWT. 00236438. 37 (3), 323–330. doi:10.1016/j.lwt.2003.09.008.

Riquelme, N., Savignones, C., López, A., Zúñiga, R.N., Arancibia, C., 2023. Effect of Gelling Agent Type on the Physical Properties of Nanoemulsion-Based Gels. Colloids Interfaces. 25045377. 7 (3). Available from: www.mdpi.com/journal/colloids/. doi:10.3390/colloids7030049.

Riseh, R.S., Vatankhah, M., Hassanisaadi, M., Kennedy, J.F., 2023. Chitosan-based nanocomposites as coatings and packaging materials for the postharvest improvement of agricultural product: a review. Carbohydr. Polym. 01448617. 309, 120666. doi:10.1016/j.carbpol.2023.120666.

Riseh, R.S., Vatankhah, M., Hassanisaadi, M., Shafiei-Hematabad, Z., Kennedy, J.F., 2024. Advancements in coating technologies: unveiling the potential of chitosan for the preservation of fruits and vegetables. Int. J. Biol. Macromol. 01418130. 254 (Pt 1), 127677. doi:10.1016/j.ijbiomac.2023.127677.

Robic, A., Gaillard, C., Sassi, J.-F., Lerat, Y., Lahaye, M., 2009. Ultrastructure of ulvan: a polysaccharide from green seaweeds. Biopolymers. 00063525. 91 (8), 652–664. doi:10.1002/bip.21195.

Roleda, M.Y., Marfaing, H., Desnica, N., Jónsdóttir, R., Skjermo, J., Rebours, C., Nitschke, U., 2019. Variations in polyphenol and heavy metal contents of wild-harvested and cultivated seaweed bulk biomass: health risk assessment and implication for food applications. Food Control 95, 121–134.

Roy, S., Ezati, P., Rhim, J.-W., 2021. Gelatin/Carrageenan-Based Functional Films with Carbon Dots from Enoki Mushroom for Active Food Packaging Applications. ACS Appl. Polym. Mater. 3, 6437–6445.

Rysenaer, V.B.J., Ahmadzadeh, S., Van Bockstaele, F., Ubeyitogullari, A., 2023. An extrusion-based 3D food printing approach for generating alginate-pectin particles. Curr. Res. Food Sci. 26659271. 6. Available from: https://www.journals.elsevier.com/current-research-in-food-science. doi:10.1016/j.crfs.2022.11.023.

Sabu Mathew, S., Jaiswal, A.K., Jaiswal, S., 2024. Carrageenan-based sustainable biomaterials for intelligent food packaging: a review. Carbohydr. Polym. 342, 122267.

Sacramento, M.M.A., Borges, J., Correia, F.J.S., Calado, R., Rodrigues, J.M.M., Patrício, S.G., Mano, J.F., 2022. Green approaches for extraction, chemical modification and processing of marine polysaccharides for biomedical applications. Front. Bioeng. Biotechnol. 22964185. 10. Available from: http://journal.frontiersin.org/journal/bioengineering-and-biotechnology#archive. doi:10.3389/fbioe.2022.1041102.

Samalens, F., Thomas, M., Claverie, M., Castejon, N., Zhang, Y., Pigot, T., Blanc, S., Fernandes, S.C.M., 2022. Progresses and future prospects in biodegradation of marine biopolymers and emerging biopolymer-based materials for sustainable marine ecosystems. Green Chem. 1463-9262. 24 (5), 1762–1779. doi:10.1039/d1gc04327g.

Sangma, S.A.L.N.A.M.C.H.I.J., Adithiyalakshmi, S., Sivakami., P.S., 2023. Formulation of Instant Tea from Seaweeds: sargassum wightii and Ulva lactuca. The Pharma Innovation Journal 12, 119–123.

Sanjeewa, 2024. Exploring the Blue Bioeconomy: Marine Bioresources and Sustainable Applications Exploring the Blue Bioeconomy: Marine Bioresources and Sustainable Applications. CRC Press, pp. 1–144. doi:10.1201/9781003477365.

Sari-Chmayssem, N., Taha, S., Mawlawi, H., Guégan, J.-P., Jeftić, J., Benvegnu, T., 2015. Extracted and depolymerized alginates from brown algae Sargassum vulgare of Lebanese origin: chemical, rheological, and antioxidant properties. J. Appl. Phycol. 28, 1915–1929.

Schepers, M., Martens, N., Tiane, A., Vanbrabant, K., Liu, H.B., Lütjohann, D., Mulder, M., Vanmierlo, T., 2020. Edible seaweed-derived constituents: an undisclosed source of neuroprotective compounds. Neural Regen. Res. 18767958. 15 (5), 790–795. Available from: http://www.nrronline.org/. doi:10.4103/1673-5374.268894.

Senadheera, T.R.L., Hossain, A., Shahidi, F., 2023. Marine Bioactives and Their Application in the Food Industry: a Review. Appl. Sci. 13.

Shahidi, F., Ambigaipalan, P., 2015. Novel functional food ingredients from marine sources. Curr. Opin. Food Sci. 22147993. 2, 123–129. doi:10.1016/j.cofs.2014.12.009.

Shalaby, M.S., Amin, H.H., 2019. Potential Using of Ulvan Polysaccharide from Ulva lactuca as a Prebiotic on Synbiotic Yogurt Production. J. Prob. Health 7, 208. doi:10.35248/2329-8901.19.7.208.

Shang, Q., Jiang, H., Cai, C., Hao, J., Guoyun, L., Guangli, Y., 2018. Gut microbiota fermentation of marine polysaccharides and its effects on intestinal ecology: an overview. Carbohydr. Polym. 01448617. 179, 173–185. doi:10.1016/j.carbpol.2017.09.059.

Shao, Z., Ampomah, O., Vieira, F.R.J., Dorrell, R.G., Shaoxuan, L., Tirichine, L., Bulone, V., Duan, D., Bowler, C., Wilmes, P., Busi, S., 2023. Characterization of a Marine Diatom Chitin Synthase Using a Combination of Meta-Omics. Genomics, and Heterologous Expression Approaches. mSystems. 2379-5077. 8 (2). doi:10.1128/msystems.01131-22.

Sharkawy, A., Barreiro, M.F., Rodrigues, A.E., 2020. Chitosan-based Pickering emulsions and their applications: a review. Carbohydr. Polym. 250, 116885.

Shu, J., Mcclements, D.J., Luo, S., Ye, J., Liu, C., 2023. Effect of internal and external gelation on the physical properties, water distribution, and lycopene encapsulation properties of alginate-based emulsion gels. Food Hydrocoll. 0268005X. 139. Available from: https://www.journals.elsevier.com/food-hydrocolloids. doi:10.1016/j.foodhyd.2023.108499.

Singh, A., Mittal, A., Benjakul, S., 2020a. Full Utilization of Squid Meat and Its Processing By-products: revisit. Food Rev. Int. 38, 455–479.

Singh, P., Baisthakur, P., Yemul, O.S., 2020b. Synthesis, characterization and application of crosslinked alginate as green packaging material. Heliyon 6, e03026.

Sivakumaran, K.E.E.R.T.H.A.N.A., Prabodhani, W., 2018. An overview of the applications molecular gastronomy in food industry. Int. J. Food Sci. Nutr. 3, 35–40.

Sotong, L.K.B.P., 2019. Antibacterial properties of chitosan edible films incorporated with musk lime extracts for the preservation of squids. Malaysian J. Anal. Sci. 23, 914–925.

Sousa, A.M.M., Morais, S., Abreu, M.H., Pereira, R., Sousa-Pinto, I., Cabrita, E.J., Delerue-Matos, C., Gonçalves, M.P., 2012. Structural, physical, and chemical modifications induced by microwave heating on native agar-like galactans. J. Agric. Food Chem. 15205118. 60 (19), 4977–4985. Available from: http://pubs.acs.org/journal/jafcau. doi:10.1021/jf2053542.

Sreeja, S.J., Tamilarutselvi, K., Tamilselvi, A., Kuttalam Pillai, S., Jasmin, K.J., Malini, M.M., 2023. Production of chitin and conversion into chitosan from crab (Scylla tranquebarica) shells and evaluation of its antioxidant activities. Biomass Convers. Biorefin. 14, 17193–17199.

Sterner, M., Gröndahl, F., 2021. Extraction of laminarin from Saccharina latissima seaweed using cross-flow filtration. J. Appl. Phycol. 0921-8971. 33 (3), 1825–1844. doi:10.1007/s10811-021-02398-z.

Sultana, F., Wahab, M.A., Nahiduzzaman, M., Mohiuddin, M., Iqbal, M.Z., Shakil, A., Mamun, A.A., Khan, M.S.R., Wong, L.L., Asaduzzaman, M., 2023. Seaweed farming for food and nutritional security, climate change mitigation and adaptation, and women empowerment: a review. Aquac. Fish. 2468550X. 8 (5), 463–480. Available from: www.keaipublishing.com/en/journals/aquaculture-and-fisheries/. doi:10.1016/j.aaf.2022.09.001.

Sun, Y., Ma, X., Hu, H., 2021. Marine Polysaccharides as a Versatile Biomass for the Construction of Nano Drug Delivery Systems. Mar. Drugs 19.

Tabatabaei, M., Ebrahimi, B., Rajaei, A., Movahednejad, M.H., Rastegari, H., Taghavi, E., Aghbashlo, M., Gupta, V. K., Lam, S.S., 2022. Producing submicron chitosan-stabilized oil Pickering emulsion powder by an electrostatic collector-equipped spray dryer. Carbohydr. Polym. 01448617. 294. Available from: http://www.elsevier.com/wps/find/journaldescription.cws_home/405871/description#description. doi:10.1016/j.carbpol.2022.119791.

Tako, M., 2003. Rheological Characteristics of Fucoidan Isolated from Commercially Cultured Cladosiphon okamuranus. Bot. Mar. 46 (5). doi:10.1515/bot.2003.047.

Tan, Y.N., Lee, P.P., Chen, W.N., 2020. Dual Extraction of Crustacean and Fungal Chitosan from a Single Mucor circinelloides Fermentation. Fermentation 6.

Teas, J., Vena, S., Cone, D.L., Irhimeh, M., 2013. The consumption of seaweed as a protective factor in the etiology of breast cancer: proof of principle. J. Appl. Phycol. 09218971. 25 (3), 771–779. doi:10.1007/s10811-012-9931-0.

Thiviya, P., Gamage, A., Liyanapathiranage, A., Makehelwala, M., Dassanayake, R.S., Manamperi, A., Merah, O., Mani, S., Koduru, J.R., Madhujith, T., 2023. Algal polysaccharides: structure, preparation and applications in food packaging. Elsevier Ltd, Sri Lanka Food Chemistry. 18737072. 405. Available from: www.elsevier.com/locate/foodchem. doi:10.1016/j.foodchem.2022.134903.

Tie, S., Tan, M., 2022. Current Advances in Multifunctional Nanocarriers Based on Marine Polysaccharides for Colon Delivery of Food Polyphenols. J. Agric. Food Chem. 0021-8561. 70 (4), 903–915. doi:10.1021/acs.jafc.1c05012.

Torres, M.D., Florez-Fernandez, N., Dominguez, H., 2019. Integral Utilization of Red Seaweed for Bioactive Production. Mar. Drugs 17.

Tsai, W.C., Wang, S.T., Chang, K.B., Tsai, M.L., 2019. Enhancing Saltiness Perception Using Chitin Nanomaterials. Polymers (Basel, 11).

Udo, T., Mummaleti, G., Mohan, A., Singh, R.K., Kong, F., 2023. Current and emerging applications of carrageenan in the food industry. Food Res. Int. 09639969. 173 (Pt 2), 113369. doi:10.1016/j.foodres.2023.113369.

Van Den Burg, S.W.K., Röckmann, C., Banach, J.L., Van Hoof, L., 2020. Governing Risks of Multi-Use: seaweed Aquaculture at Offshore Wind Farms. Front. Mar. Sci. 22967745. 7. Available from: https://www.frontiersin.org/journals/marine-science#. doi:10.3389/fmars.2020.00060.

Vergara-Rodarte, M.A., Hernández-Carmona, G., Rodríguez-Montesinos, Y.E., Arvizu-Higuera, D.L., Riosmena-Rodríguez, R., Murillo-Álvarez, J.I., 2010. Seasonal variation of agar from Gracilaria vermiculophylla, effect of alkali treatment time, and stability of its Colagar. J. Appl. Phycol. 09218971. 22 (6), 753–759. doi:10.1007/s10811-010-9516-8.

Vidal, J.L., Jin, T., Lam, E., Kerton, F., Moores, A., 2022. Blue is the new green: valorization of crustacean waste. Curr. Res. Green Sustain. Chem. 26660865. 5, 100330. doi:10.1016/j.crgsc.2022.100330.

Wan, X.Z., Ai, C., Chen, Y.H., Gao, X.X., Zhong, R.T., Liu, B., Chen, X.H., Zhao, C., 2020. Physicochemical Characterization of a Polysaccharide from Green Microalga Chlorella pyrenoidosa and Its Hypolipidemic Activity via Gut Microbiota Regulation in Rats. J. Agric. Food Chem. 15205118. 68 (5), 1186–1197. Available from: http://pubs.acs.org/journal/jafcau. doi:10.1021/acs.jafc.9b06282.

Wang, H., Cao, Z., Yao, L., Feng, T., Song, S., Sun, M., 2023. Insights into the Edible and Biodegradable Ulvan-Based Films and Coatings for Food Packaging. Foods. 2304-8158. 12 (8), 1622. doi:10.3390/foods12081622.

Wang, H., Zhu, B., Yao, Z., 2023b. Porphyran, porphyran oligosaccharides and porphyranase: source, structure, preparation methods and applications. Algal Res. 73.

Wang, J., Yuan, Y., Liu, Y., Xiang, L., Shengjun, W., 2024. Application of chitosan in fruit preservation: a review. Food Chem. X. 25901575. 23, 101589. doi:10.1016/j.fochx.2024.101589.

Wang, K., Sui, J., Gao, W., Bin, Y., Yuan, C., Guo, L., Bo Cui, -A.M.A.E.-A., 2022. Effects of xanthan gum and sodium alginate on gelatinization and gels structure of debranched pea starch by pullulanase. Food Hydrocoll. 0268005X. 130, 107733. doi:10.1016/j.foodhyd.2022.107733.

Wang, L., Jayawardena, T.U., Yang, H.W., Lee, H.G., Kang, M.C., Sanjeewa, K.K.A., Oh, J.Y., Jeon, Y.J., 2020. Isolation, Characterization, and Antioxidant Activity Evaluation of a Fucoidan from an Enzymatic Digest of the Edible Seaweed, Hizikia fusiforme. Antioxidants (Basel 9).

Wang, Z., Xu, Z., Yang, X., Li, M., Yip, R.C.S., Li, Y., Chen, H., 2023c. Current application and modification strategy of marine polysaccharides in tissue regeneration: a review. Biomater. Adv. 154, 213580.

Wei, Y., Yang, S., Zhang, L., Dai, L., Tai, K., Liu, J., Mao, L., Yuan, F., Gao, Y., Mackie, A., 2020. Fabrication, characterization and in vitro digestion of food grade complex nanoparticles for co-delivery of resveratrol and coenzyme Q10. Food Hydrocoll. 0268005X. 105, 105791. doi:10.1016/j.foodhyd.2020.105791.

Wei, Z., Yujia, O., Lan, X., Tang, J., Zheng, B., 2023. Effects of laminarin on the structural properties and in vitro digestion of wheat starch and its application in noodles. LWT. 00236438. 178, 114543. doi:10.1016/j.lwt.2023.114543.

Wells, M.L., Potin, P., Craigie, J.S., Raven, J.A., Merchant, S.S., Helliwell, K.E., Smith, A.G., Camire, M.E., Brawley, S.H., 2017. Algae as nutritional and functional food sources: revisiting our understanding. J. Appl. Phycol. 15735176. 29 (2), 949–982. Available from: www.wkap.nl/journalhome.htm/0921-8971. doi:10.1007/s10811-016-0974-5.

Wong, C.H., Mazenod, F.P., Whitesides, G.M., 1983. Chemical and enzymatic syntheses of 6-deoxyhexoses. Conversion to 2,5-dimethyl-4-hydroxy-2,3-dihydrofuran-3-one (Furaneol) and analogs. J. Org. Chem. 48, 3493–3497.

Wu, D.T., He, Y., Fu, M.X., Gan, R.Y., Hu, Y.C., Peng, L.X., Zhao, G., Zou, L., 2022. Structural characteristics and biological activities of a pectic-polysaccharide from okra affected by ultrasound assisted metal-free Fenton reaction. Food Hydrocoll. 0268005X. 122. Available from: https://www.journals.elsevier.com/food-hydrocolloids. doi:10.1016/j.foodhyd.2021.107085.

Xiang, X.W., Wang, R., Chen, H., Chen, Y.F., Shen, G.X., Liu, S.L., Sun, P.L., Chen, L., 2022. Structural characterization of a novel marine polysaccharide from mussel and its antioxidant activity in RAW264.7 cells induced by H2O2. Food Biosci. 22124306. 47. Available from: http://www.journals.elsevier.com/food-bioscience/. doi:10.1016/j.fbio.2022.101659.

Xiao, M., Ren, X., Ying, Y., Gao, W., Zhu, C., Sun, H., Kong, Q., Xiaodan, F., Mou, H., 2022. Fucose-containing bacterial exopolysaccharides: sources, biological activities, and food applications. Food Chem. X. 25901575. 13, 100233. doi:10.1016/j.fochx.2022.100233.

Xie, C., Lee, Z.J., Ye, S., Barrow, C.J., Dunshea, F.R., Suleria, H.A.R., 2023. A Review on Seaweeds and Seaweed-Derived Polysaccharides: nutrition, Chemistry, Bioactivities, and Applications. Food Rev. Int. 40, 1312 1347.

Yan, P., Lan, W., Xie, J., 2024. Modification on sodium alginate for food preservation: a review. Trends Food Sci. Technol. 143.

Yang, M., Zhou, D., Xiao, H., Xiaodan, F., Kong, Q., Zhu, C., Han, Z., Mou, H., 2022. Marine-derived uronic acid-containing polysaccharides: structures, sources, production, and nutritional functions. Trends Food Sci. Technol. 09242244. 122, 1–12. doi:10.1016/j.tifs.2022.02.013.

Yang, Y., Gupta, V.K., Amiri, H., Pan, J., Aghbashlo, M., Tabatabaei, M., Rajaei, A., 2023. Recent developments in improving the emulsifying properties of chitosan. Int. J. Biol. Macromol. 01418130. 239, 124210. doi:10.1016/j.ijbiomac.2023.124210.

Yang, Y., Lin, H., Fu, X., 2024. Fermentation of Pyropia spp. seaweed: a comprehensive review on processing conditions, biological activities and potential applications in the food industry. Crit. Rev. Food Sci. Nutr., 1–16.

Yıkmış, S., Türkol, M., Abdi, G., İmre, M., Alkan, G., Türk Aslan, S., Rabail, R., Aadil, R.M., 2024. Culinary trends in future gastronomy: a review. J. Agric. Food Res. 26661543. 18. Available from: https://www.sciencedirect.com/science/journal/26661543. doi:10.1016/j.jafr.2024.101363.

Yosri, N., Khalifa, S.A.M., Guo, Z., Xu, B., Zou, X., El-Seedi, H.R., 2021. Marine organisms: pioneer natural sources of polysaccharides/proteins for green synthesis of nanoparticles and their potential applications. Int. J. Biol. Macromol. 18790003. 193, 1767–1798. Available from: www.elsevier.com/locate/ijbiomac. doi:10.1016/j.ijbiomac.2021.10.229.

Yu, J., Li, Q., Wu, J., Yang, X., Yang, S., Zhu, W., Liu, Y., Tang, W., Nie, S., Hassouna, A., White, W.L., Zhao, Y., Lu, J., 2021. Fucoidan Extracted From Sporophyll of Undaria pinnatifida Grown in Weihai, China – chemical Composition and Comparison of Antioxidant Activity of Different Molecular Weight Fractions. Front. Nutr. 2296861X. 8. Available from: http://journal.frontiersin.org/journal/nutrition. doi:10.3389/fnut.2021.636930.

Yu, S., Yun, E.J., Kim, D.H., Park, S.Y., Kim, K.H., 2019. Anticariogenic Activity of Agarobiose and Agarooligosaccharides Derived from Red Macroalgae. J. Agric. Food Chem. 67, 7297–7303.

Zarandona, I., Estupiñán, M., Pérez, C., Alonso-Sáez, L., Guerrero, P., De La Caba, K., 2020. Chitosan Films Incorporated with Exopolysaccharides from Deep Seawater Alteromonas sp. Mar. Drugs. 16603397. 18 (9). Available from: https://www.mdpi.com/1660-3397/18/9/447. doi:10.3390/md18090447.

Zargarzadeh, M., Amaral, A.J.R., Custódio, C.A., Mano, J.F., 2020. Biomedical applications of laminarin. Carbohydr. Polym. 01448617. 232. Available from: http://www.elsevier.com/wps/find/journaldescription.cws_home/405871/description#description. doi:10.1016/j.carbpol.2019.115774.

Zataria multiflora Boiss essential oil as edible coatings and the impact on microbial quality of trout fillet. *Aquaculture Research*, 51, 873–881.

Zhang, B., Lan, W., Xie, J., 2022a. Chemical modifications in the structure of marine polysaccharide as serviceable food processing and preservation assistant: a review. Int. J. Biol. Macromol. 223, 1539–1555.

Zhang, W., Wang, J., Jin, W., Zhang, Q., 2013. The antioxidant activities and neuroprotective effect of polysaccharides from the starfish Asterias rollestoni. Carbohydr. Polym. 01448617. 95 (1), 9–15. doi:10.1016/j.carbpol.2013.02.035.

Zhang, X., Wei, Z., Xue, C., 2022b. Physicochemical properties of fucoidan and its applications as building blocks of nutraceutical delivery systems. Crit. Rev. Food Sci. Nutr. 62, 8935–8953.

Zhang, X., Yuan, J., Fuhua, L., Xiang, J., 2021. Chitin Synthesis and Degradation in Crustaceans: a Genomic View and Application. Mar. Drugs. 1660-3397. 19 (3), 153. doi:10.3390/md19030153.

Zhao, H.-K., Wei, X.-Y., Xie, Y.-M., 2021. Supercritical CO2 extraction, structural analysis and bioactivity of polysaccharide from Grifola frondosa. J. Food Compos. Anal. 102.

Zhao, J., Yang, J., Song, S., Zhou, D., Qiao, W., Zhu, C., Liu, S., Zhu, B., 2016. Anticoagulant Activity and Structural Characterization of Polysaccharide from Abalone (Haliotis discus hannai Ino) Gonad. Molecules. 1420-3049. 21 (6), 697. doi:10.3390/molecules21060697.

Zhao, K., Hao, Y., Guo, X., Chang, Y., Shen, X., 2024. Development, characterization and underling mechanism of 3D printable quinoa protein emulsion gels by incorporating of different polysaccharides for curcumin delivery. Int. J. Biol. Macromol. 01418130. 280 (Pt 1), 135648. doi:10.1016/j.ijbiomac.2024.135648.

Zheng, L.X., Chen, X.Q., Cheong, K.L., 2020. Current trends in marine algae polysaccharides: the digestive tract, microbial catabolism, and prebiotic potential. Int. J. Biol. Macromol. 151, 344–354.

Zhengguo, W., Tang, S., Deng, W., Luo, J., Wang, X., 2021. Antibacterial chitosan composite films with food-inspired carbon spheres immobilized AgNPs. Food Chem. 03088146. 363, 130342. doi:10.1016/j.foodchem.2021.130342.

Zhu, Z., Dong, X., Yan, C., Ai, C., Zhou, D., Yang, J., Zhang, H., Liu, X., Song, S., Xiao, H., Zhu, B., 2019. Structural Features and Digestive Behavior of Fucosylated Chondroitin Sulfate from Sea Cucumbers Stichopus japonicus. J. Agric. Food Chem. 67, 10534–10542.

Zhu, Z., Zhu, B., Sun, Y., Chunqing, A., Wang, L., Wen, C., Yang, J., Song, S., Liu, X., 2018. Sulfated Polysaccharide from Sea Cucumber and its Depolymerized Derivative Prevent Obesity in Association with Modification of Gut Microbiota in High-Fat Diet-Fed Mice. Mol. Nutr. Food Res. 1613-4125. 62 (23). doi:10.1002/mnfr.201800446.

CHAPTER

13

Marine proteins and fatty acids based on food products

Sanduni Umayangani[a], *R.C.N. Thilakarathna*[b], *Takagi Anna*[b], *Chi Yie Choong*[b], *Ishara De Silva*[c], *Yee-Ying Lee*[b,d], *Mayumi Silva*[c], *and Jayani Chandrapala*[c]

[a]Department of Biosystems Technology, Faculty of Technology, University of Sri Jayewardenepura, Homagama, Sri Lanka
[b]School of Science, Monash University Malaysia, Subang Jaya, Selangor, Malaysia
[c]School of Science, RMIT University, Bundoora, VIC, Australia
[d]Monash Industry Research Laboratory, Monash University Malaysia, Subang Jaya, Selangor, Malaysia

Marine-derived proteins, including hydrolysates and bioactive peptides, exhibit significant functional properties, including antioxidant, antihypertensive, and antimicrobial activities. They offer improved solubility, emulsification, and water holding capacity (WHC), making them suitable for applications in dairy, meat, and bakery products. The integration of marine lipids and proteins into food products enhances nutritional quality while addressing health concerns related to modern diets. However, challenges such as lipid oxidation, sensory acceptability, and sustainable sourcing require further research. This chapter explores the potential of marine fatty acids and proteins, their health benefits, functional properties, and applications in food systems, while highlighting future directions for sustainable utilization.

13.1 Marine-derived fatty acids

Fatty acids are essential macromolecules that play a crucial role in various physiological functions in humans. They are fundamental components of plasma cell membranes, serve as intracellular sources of ATP, act as bioactive compounds, and contribute to overall health, particularly by supporting cardiovascular well-being (Sakamoto et al., 2019). Fatty acids can be broadly classified as either saturated or unsaturated, with the latter further divided into monounsaturated fatty acids and polyunsaturated fatty acids (PUFAs).

Harvesting the Sea
https://doi.org/10.1016/B978-0-443-29190-6.00017-9

Although the human body can synthesize most fatty acids, it cannot produce α-linolenic acid and linoleic acid (LA), which are essential PUFAs required for the biosynthesis of eicosapentaenoic acid (EPA) and docosahexaenoic acid (DHA) (Mititelu et al., 2025; Kapoor et al., 2021).

Marine sources, including fish, mollusks, crustaceans, and algae, are the primary natural reservoirs of EPA and DHA. The lipid and fatty acid composition of these sources varies depending on species, gender, age, maturity stage, diet, and environmental factors such as temperature (Barta et al., 2021). Among marine sources, fatty fish such as salmon, herring, trout, and mackerel contain 10%–15% fat and are reported to have a higher omega-3 fatty acid content than lean fish like cod, bass, and flounder, which contain only 2%–5% fat (Araujo et al., 2021). Crustaceans, particularly krill, are also significant sources of DHA and EPA, and their omega-3 fatty acids are often considered more bioavailable than those derived from fish oils. Moreover, marine algae contribute to omega-3 fatty acid production, with species such as the marine oleaginous diatom *Fistulifera solaris* and *Crypthecodinium cohnii* producing up to 65% PUFA. Other microalgae, including Thraustochytrium and Schizochytrium, are also known for their ability to synthesize EPA and DHA (Barta et al., 2021). Given their significant benefits, marine-derived fatty acids have garnered increasing attention in both nutritional research and the food industry. The incorporation of omega-3-rich marine sources into the human diet is widely recommended to mitigate the risks of cardiovascular diseases (CVDs), neurodegenerative disorders, and chronic inflammation. However, the high demand for omega-3-rich seafood can contribute to overfishing, disrupting marine ecosystems. Therefore marine algae can offer a promising source of omega-3 as a more sustainable alternative. In addition, certain fish species may accumulate heavy metals, posing a potential health risk if consumed in large quantities, and may also trigger allergic reactions in some individuals. Another challenge is the susceptibility of omega-3 fatty acids to oxidation, which can lead to rancidity and loss of nutritional value if not properly processed or stored. To address these concerns, further research is needed to enhance the stability, bioavailability, and sustainability of fatty acid sources. Advanced techniques such as microencapsulation, nanoemulsion, and lipid oxidation prevention strategies should be explored to improve the shelf life and efficacy of marine-derived fatty acids. In addition, advancements in algae cultivation and biotechnology could provide more scalable and eco-friendly alternatives.

13.1.1 Marine-derived fatty acid sources

13.1.1.1 Fish

Fish is one of the most well-known marine sources of omega-3 polyunsaturated fatty acids (*n*-3 PUFA), particularly EPA (C20:5*n*-3) and DHA (C22:6*n*-3) (Alfio et al., 2021). The fatty acid composition in fish varies widely depending on factors such as species, season, body parts, and extraction methods (Balikci, 2024; Rahman et al., 2023; Zhang et al., 2020). However, certain species, including anchovy, herring, mackerel, salmon, sardine, and tuna, consistently demonstrate high EPA and DHA contents, with levels ranging from 3.72% to 10.91% and 11.82% to 38.40%, respectively (Table 13.1). These fatty acids have been extensively studied for their cardiovascular, antiinflammatory, and cognitive benefits (Elagizi

TABLE 13.1 Fatty acid composition (% of total fatty acids) in selected fish species.

FA (%)	Common name	Anchovy	Herring	Mackerel	Salmon	Sardine	Tuna
C16:0	Palmitic acid	28.89	23.58	19.00	19.07	23.75	23.10
C18:0	Stearic acid	5.68	3.95	6.10	4.61	4.45	10.50
C18:1*n*-9	Oleic acid	11.92	20.97	13.10	9.49	14.11	12.50
C18:2*n*-6	Linoleic acid	1.92	1.39	1.20	1.05	2.26	0.90
C20:4*n*-6	AA	2.12	1.48	1.40	0.70	0.89	4.75
C18:3*n*-3	ALA	1.07	0.72	0.80	0.67	1.95	0.19
C20:5*n*-3	EPA	9.02	5.85	9.70	8.28	10.91	3.72
C22:6*n*-3	DHA	25.24	11.82	15.40	20.50	22.88	38.40

AA, arachidonic acid; *ALA*, alpha-linolenic acid; *DHA*, docosahexaenoic acid; *EPA*, eicosapentaenoic acid; FA, fatty acids.
Modified from Biton-Porsmoguer, S., Bou, R., Lloret, E., Alcaide, M., Lloret, J., Cooke, S., 2020. Fatty acid composition and parasitism of European sardine (Sardina pilchardus) and anchovy (Engraulis encrasicolus) populations in the northern Catalan Sea in the context of changing environmental conditions. Conserv. Physiol. 2051–1434. 8 (1). https://doi.org/10.1093/conphys/coaa121; Colombo, S.M., Mazal, X., 2020. Investigation of the nutritional composition of different types of salmon available to Canadian consumers. J. Agric. Food Res. 26661543. 2, 100056. https://doi.org/10.1016/j.jafr.2020.100056; Stefanie M. Colombo, Xenna Mazal, Investigation of the nutritional composition of different types of salmon available to Canadian consumers . Journal of Agriculture and Food Research. 26661543. 2 (2020), 100056. https://doi.org/10.1016/j.jafr.2020.100056; Domingues, V.F., Quaresma, M., Sousa, S., Rosas, M., Ventoso, B., Nunes, M.L., Delerue-Matos, C., 2021. Evaluating the lipid quality of yellowfin tuna (Thunnus albacares) harvested from different oceans by their fatty acid signatures. Foods. 23048158. 10 (11). Available from: https://www.mdpi.com/2304-8158/10/11/2816/pdf. https://doi.org/10.3390/foods10112816; Egerton, S., Mannion, D., Culloty, S., Whooley, J., Stanton, C., Ross, R.P., 2020. The proximate composition of three marine pelagic fish: blue whiting (Micromesistius poutassou), boarfish (Capros aper) and Atlantic herring (Clupea harengus). Ir. J. Agric. Food Res. 07916833. 59 (1), 185–200. Available from: https://www.t-stor.teagasc.ie/handle/11019/2576. https://doi.org/10.15212/ijafr-2020-0112; Ferreira, I., Gomes-Bispo, A., Lourenço, H., Matos, J., Afonso, C., Cardoso, C., Castanheira, I., Motta, C., Prates, J.A.M., Bandarra, N.M., 2020. The chemical composition and lipid profile of the chub mackerel (Scomber colias) show a strong seasonal dependence: contribution to a nutritional evaluation. Biochimie. 61831638. 178, 181–189. Available from: http://www.elsevier.com/locate/biochimie. https://doi.org/10.1016/j.biochi.2020.09.022.

et al., 2021; Mora et al., 2022; Shibabaw, 2021). As a result, fish-derived lipids are increasingly being utilized in functional foods, nutraceuticals, and animal feeds (Moghadam et al., 2019; Raeisi et al., 2019; Wang et al., 2024). While the nutritional benefits of fish oil are well-established, its utilization presents several challenges. One significant challenge is the strong, unpleasant odor of fish oil, which limits its use in food products unless advanced processing techniques are employed to improve sensory acceptability (Güner et al., 2019). Furthermore, environmental concerns such as global warming and overfishing threaten the sustainability of fish oil production, jeopardizing marine biodiversity and the long-term availability of fish as a resource (Lotze et al., 2019; Shepon et al., 2022). Fish oils are also unsuitable for individuals following plant-based diets, further restricting their consumer base. These limitations have driven the search for alternative sources of *n*-3 PUFA that can provide comparable nutritional benefits.

13.1.1.2 *Algae*

Algae are photosynthetic eukaryotes that serve as primary producers in aquatic environments, forming the foundation of marine food chains (Vijayaram et al., 2024). They are broadly classified into two categories based on size: macroalgae and microalgae (Vijayaram et al., 2024). Macroalgae, commonly referred to as seaweed, are visible to the naked eye and include red (*Rhodophyta*), brown (*Phaeophyta*), and green (*Chlorophyta*), which are characterized by their pigmentation (Vijayaram et al., 2024). In contrast, microalgae are microscopic, unicellular organisms, often referred to as phytoplankton (Ahmed et al., 2023). They include blue-green algae (*Cyanophyta*), dinoflagellates (*Pyrrophyta*), diatoms, golden-brown algae (*Chrysophyta*), and microscopic green algae (*Chlorophyta*) (Visuddho et al., 2024). The lipid composition of macroalgae typically ranges from 0.2% to 8% of their dry weight, constraining its use as a substantial lipid source (Rey et al., 2020; Salehi et al., 2019). Conversely, microalgae exhibit a much higher lipid content, ranging from 20% to 50% of their dry weight, depending on species and cultivation conditions (Jadhav, 2024). Factors such as temperature, nutrient concentration, and pH significantly influence the quality and composition of lipids in microalgae (Jadhav, 2024). Microalgae lipids primarily consist of a mixture of C16 and C18 saturated (e.g., palmitic and stearic acids) and unsaturated (e.g., oleic acid and LA), with notable amounts of *n*-3 PUFA such as EPA and DHA when present (Table 13.2). These *n*-3 PUFAs are of particular interest due to their health benefits and the potential of microalgae as a sustainable alternative to fish oil. Bartek et al. (2021) emphasized that microalgae oil has a lower environmental footprint compared to fish oil, with reduced global warming potential, land use, and freshwater eutrophication. However, further research is needed to comprehensively characterize the lipid profiles of diverse species and optimize their large-scale production and utilization.

13.1.1.3 Shellfish

Shellfish are aquatic organisms characterized by shells or exoskeletons and are classified into two main groups of invertebrates: crustaceans and mollusks. Mollusks include bivalves such as mussels, oysters, and scallops, while crustaceans comprise species like crabs, shrimp, and lobsters. The lipid content in shellfish varies significantly, ranging from 0.43% to 25.00%, depending on species, body parts, and processing methods (Reksten et al., 2024). As shown in Table 13.3, palmitic acid is the predominant fatty acid in shellfish, while their fatty acid composition also highlights a high proportion of omega-3 fatty acids, particularly EPA and DHA, which together account for 72%–96% of the total omega-3 content across various species (Reksten et al., 2024). Moreover, the levels of omega-3 fatty acids typically exceed those of omega-6 fatty acids, signifying shellfish as a valuable source of long-chain PUFAs (Reksten et al., 2024). Among shellfish, krill, a small marine crustacean, has gained significant attention for its superior bioavailability of omega-3 fatty acids compared to fish oil (Pham et al., 2024). This advantage comes from its unique chemical composition, as the omega-3 fatty acids in krill oil are bound to phospholipids rather than triglycerides (Andraka et al., 2020). Phospholipid-bound omega-3 fatty acids are more water-soluble, thereby enhancing their absorption efficiency relative to the triglyceride-bound forms found in fish oil (Colletti et al., 2021). Consequently, krill oil is recognized as a highly desirable source of omega-3 fatty acids, particularly in dietary and nutraceutical applications (Hwang et al., 2022; Sánchez et al., 2021; Sun et al., 2023).

TABLE 13.2 Fatty acid composition (% of total fatty acids) in selected microalgae species.

FA (%)	Common name	*Arthrospira platensis*	*Chaetoceros ceratosporum*	*Chaetoceros gracilis*	*Isochrysus galbana*	*Nannochloropsis oculata*	*Pavlova lutheri*
C16:0	Palmitic acid	46.66	5.10	5.39	6.79	21.86	16.95
C18:0	Stearic acid	0.77	0.38	0.43	0.28	0.19	0.18
C18:1*n*-9	Oleic acid	1.89	0.43	0.59	9.34	3.10	0.20
C18:2*n*-6	Linoleic acid	21.59	0.83	1.19	5.86	4.05	0.70
C20:4*n*-6	AA	n.d.	1.58	2.11	0.12	3.74	0.32
C18:3*n*-3	ALA	0.04	0.10	0.10	n.d.	n.d.	2.64
C20:5*n*-3	EPA	n.d.	11.49	10.67	0.12	26.21	17.76
C22:6*n*-3	DHA	n.d.	1.41	0.78	8.76	0.52	7.61

AA, arachidonic acid; *ALA*, alpha-linolenic acid; *DHA*, docosahexaenoic acid; *EPA*, eicosapentaenoic acid; *FA*, fatty acids; *n.d.*, not detected.
Modified from Yang, Y., Lei, D., Hosokawa, M., Miyashita, K., 2020. Total lipids content, lipid class and fatty acid composition of ten species of microalgae. J. Oleo Sci. 1345-8957. 69 (10), 1181–1189. https://doi.org/10.5650/jos.ess20140.

TABLE 13.3 Fatty acid composition (% of total fatty acids) in selected shellfish species.

FA (%)	Common name	Clam	Cuttlefish	Krill	Mussel	Shrimp	Squid
C16:0	Palmitic acid	33.30	23.30	21.65	31.10	19.30	30.60
C18:0	Stearic acid	8.30	9.04	3.32	4.70	6.90	3.30
C18:1*n*-9	Oleic acid	4.40	6.40	13.34	1.60	14.00	2.50
C18:2*n*-6	Linoleic acid	1.70	2.40	1.39	2.40	1.50	0.35
C20:4*n*-6	AA	4.10	4.20	0.02	3.20	5.10	0.73
C18:3*n*-3	ALA	3.40	n.d.	0.66	0.75	n.d.	n.d.
C20:5*n*-3	EPA	11.40	16.50	15.97	18.80	15.00	n.d.
C22:6*n*-3	DHA	16.90	24.60	8.17	14.90	19.50	41.50

AA, arachidonic acid; *ALA*, alpha-linolenic acid; *DHA*, docosahexaenoic acid; *EPA*, eicosapentaenoic acid; *FA*, fatty acids; *n.d.*, not detected.
Modified from Mesa, M.D., Gil, F., Olmedo, P., Gil, A., 2021. Nutritional importance of selected fresh fishes, shrimps and mollusks to meet compliance with nutritional guidelines of N-3 LC-Pufa intake in Spain. Nutrients. 20726643. 13 (2), 1–14. Available from: https://www.mdpi.com/2072-6643/13/2/465/pdf. https://doi.org/10.3390/nu13020465; Xie, D., Gong, M., Wei, W., Jin, J., Wang, X., Wang, X., Jin, Q., 2019. Antarctic krill (Euphausia superba) oil: a comprehensive review of chemical composition, extraction technologies, health benefits, and current applications. Compr. Rev. Food Sci. Food Saf. 15414337. 18 (2), 514–534. Available from: http://onlinelibrary.wiley.com/journal/10.1111/(ISSN)1541-4337. https://doi.org/10.1111/1541-4337.12427.

13.1.1.4 *Sponges*

Sponges, members of the phylum *Porifera*, are immobile invertebrates capable of thriving in diverse environments, including marine and freshwater habitats across temperate, polar, and tropical areas (Gökalp et al., 2021; Varijakzhan et al., 2021). The major classes of sponges include *Calcarea*, *Demospongiae*, *Hexactinellida*, and *Homoscleromorpha*, which contribute significantly to marine ecosystems by creating habitats for benthic communities and supporting diverse microbial symbionts (Zhang et al., 2019). Sponges are known to produce a wide range of lipids, such as fatty acids, sterols, and lipid-soluble compounds, which constitute approximately 10.1%–12.3% of their dry weight (Mehbub et al., 2024; Radnaeva et al., 2020). These lipids are highly distinctive due to the contributions of their microbial symbionts, which lead to complex and unique fatty acid compositions. For example, studies on North Atlantic deep-sea demosponges reveal that bacterial fatty acids dominate the lipid profile, contributing 56%–79%, while sponge-specific fatty acids account for 11%–29% (de Kluijver et al., 2021). The fatty acids derived from sponges exhibit structural diversity. Bacterial-specific fatty acids are often branched and monoenic, with chain lengths shorter than 20 carbons (de Kluijver et al., 2021). In contrast, sponge-specific fatty acids tend to include long-chain varieties, with carbon chain lengths ranging from 24 to 29 (de Kluijver et al., 2021). Notable examples include demospongic acid (C26:3), nervonic acid (C24:1), and isomers of hexadecenoic acid (C16:1) and octadecenoic acid (C18:1) in *Baikalospongia intermedia*. In addition, microbial-specific fatty acids, such as *iso*-palmitic acid (*i*-C16:0), *iso*-stearic acid (*i*-C18:0), and oxy-stearic acid, further underscore the complexity of sponge lipids. However, the full spectrum of fatty acids in sponges is far from being comprehensively understood, with existing studies focusing on a limited number of species. Emerging evidence suggests that sponge-derived fatty acids may have significant bioactive properties. For instance, fatty acids isolated from sponges have been shown to inhibit biofilm formation in methicillin-susceptible and methicillin-resistant *Staphylococcus aureus* (Khan et al., 2023). Similarly, fatty acids from *Xestospongia testudinaria*, such as stearic acid (C18:0) and arachidic acid (C20:0), exhibited antibacterial activity against shrimp-pathogenic *Vibrio* species (Latifah et al., 2021). These findings prove the therapeutic potential of sponge-derived fatty acids, yet their development as a sustainable source for nutraceutical or pharmaceutical applications remains constrained. Further studies into their fatty acid compositions, biochemical pathways, symbiotic interactions, and safety profiles could establish sponges as innovative and sustainable sources of bioactive lipids.

13.1.2 Health benefits of marine-derived fatty acids

Marine-derived fatty acids, particularly omega-3 PUFAs like EPA and DHA, are essential nutrients recognized for their significant health benefits. Sourced primarily from fish and algae, these fatty acids play a crucial role in human health due to their limited biosynthesis in the human body, making dietary intake vital (Bafkar et al., 2024; Ghosh et al., 2022). Since the human body has a limited ability to convert ALA and LA into DHA and EPA, obtaining these essential fatty acids through dietary sources is highly beneficial. Marine-derived fatty acids remain the only natural and direct sources of EPA and DHA, making seafood a vital component of a heart-healthy diet (Sakamoto et al., 2019). Omega-3 fatty acids help manage chronic diseases linked to inflammation, oxidative stress, and lipid dysregulation. They alleviate conditions like rheumatoid arthritis and CVDs by modulating inflammatory mediators (Djuricic and Calder, 2021; Poggioli et al., 2023). In addition to their antiinflammatory effects, marine-derived fatty acids exhibit significant antioxidant activity, providing cellular protection against oxidative stress. This property is vital for mitigating the risks of diseases such as cancer and cardiovascular ailments (Zhong et al., 2019; Ghosh et al., 2022). The ability of marine compounds to enhance antioxidant defenses and their potential as substitutes for synthetic antioxidants underscores their therapeutic value in promoting health and preventing disease (Ebrahimi et al., 2023). As dietary guidelines recommend increased omega-3 intake, the exploration of marine-derived fatty acids continues to highlight their importance in public health and nutrition strategies aimed at preventing and managing chronic diseases (Dempsey et al., 2023; Laupsa-Borge et al., 2023).

13.1.2.1 *Antiinflammatory activity*

Inflammation is closely linked to oxidative stress and is a key factor in the development of health complications such as CVD and diabetes. Both EPA and DHA exhibit potent antiinflammatory effects. According to the literature, arachidonic acid (AA) is converted into eicosanoids, such as thromboxanes and prostaglandins, which stimulate the production of proinflammatory cytokines. These cytokines promote the release of lysosomal enzymes and the formation of free radicals, contributing to oxidative stress and inflammation. However, EPA supplementation can counteract this process by integrating into the phospholipid layer of cell membranes, thereby inhibiting the conversion of AA into eicosanoids. This, in turn, reduces the production of proinflammatory cytokines and mitigates oxidative stress. In addition, DHA plays a crucial role in regulating inflammatory pathways. It downregulates proinflammatory signaling cascades, such as the nuclear factor kappa-light-chain-enhancer of activated B cells (NF-κB) pathway, while upregulating anti-inflammatory intracellular signaling pathways, including peroxisome proliferator-activated receptor α/γ transcriptional activation. The NF-κB pathway is particularly significant, as it facilitates the translocation of NF-κB from the cytosol to the nucleus, where it induces the transcription of various genes involved in inflammation, immune response, and apoptosis (Banaszak et al., 2024).

Research has demonstrated that combining various marine extracts can yield greater antiinflammatory effects than individual components. For instance, studies involving the phlorotannins from Ecklonia cava and sulfated polysaccharides from *Sargassum horneri* have shown that their combination possesses superior antiinflammatory activity compared to either extract alone (Sanjeewa et al., 2019). This synergistic effect is thought to stem from the diverse bioactive properties of the compounds present in the extracts, which work in concert to inhibit key inflammatory mediators such as iNOS, COX-2, IL-1β, IL-6, and TNF-α in activated macrophage cells (Mahmoud et al., 2021).

13.1.2.2 *Antioxidant activity*

Oxidative stress contributes to serious health conditions like cancer and CVD. Omega-3 fatty acids, particularly DHA, combat oxidative stress by enhancing antioxidant enzyme activity, such as manganese-dependent superoxide dismutase, and reducing oxidative damage. They also boost nitric oxide (NO) production by stimulating endothelial NO synthase, promoting cardiovascular health through antiatherosclerotic effects, vasodilation, and reduced inflammation. However, oxidative stress can impair NO production, leading to endothelial dysfunction, vascular inflammation, and prothrombotic conditions. Omega-3s thus play a protective role against oxidative stress-related diseases (Oppedisano et al., 2020).

The antioxidant activity of marine-derived substances has garnered considerable attention due to the increasing recognition of oxidative stress as a contributing factor to various diseases, including cancer and cardiovascular ailments. Marine organisms, with their unique metabolic profiles and biological adaptations, are particularly promising sources of natural antioxidants. A systematic review evaluating the antioxidant potential of marine natural products found that compounds derived from algae, fungi, sponges, mollusks, and sea cucumbers exhibited significant antioxidant activity, consistently demonstrating their potential as substitutes for synthetic antioxidants and therapeutic agents for oxidative stress-related conditions (Ebrahimi et al., 2023).

Marine-derived bioactive compounds exhibit diverse and synergistic interactions in modulating antiinflammatory and antioxidant activities. Sesquiterpenoids, like lemnalol extracted from *Lemnalia cervicorni* and clovane compound 1 extracted from *Rumphella antipathies,* suppress key inflammatory mediators, with lemnalol inhibiting iNOS and COX-2 expression in macrophages, and clovane reducing superoxide anion generation in neutrophils, thereby mitigating oxidative stress. Similarly, diterpenoids such as 8-hydroxybriaranes derived from *Junceella fragilis* and excavatolide B derived from *Briareum excavatum* also target COX-2 and iNOS pathways. Steroidal compounds like solomonsterol A from *Theonella swinhoei* and ergosta-7,22-dien-3-ol from *Marthasterias glacialis* demonstrate antiarthritic effects and iNOS suppression, respectively, while steroid compound 5 from *Astropecten polyacanthus* enhances proinflammatory cytokines like IL-12p40, IL-6, and TNF-α. Fatty acid derivatives from *Octopus vulgaris* ink modulate cytokine secretion through JAK-STAT and PI3K-Akt pathways, suggesting dual roles in inflammation resolution and oxidative stress regulation (Chen et al., 2019; Ebrahimi et al., 2023; Hernández-Zazueta et al., 2021).

13.1.3 Functional properties of marine fatty acids in food formulation

13.1.3.1 Nutritional enhancement

Marine fatty acids, including EPA, DHA, and docosapentaenoic acid (DPA), are essential nutrients valued for their critical role in human nutrition. EPA and DHA are n-3 long-chain PUFAs classified as essential since the human body cannot efficiently synthesize them, necessitating dietary intake or supplementation (Jónasdóttir, 2019). EPA and DHA are commonly sourced from fatty fish such as salmon, mackerel, and sardines, along with seafood and algae. In marine foods, DHA and EPA are primarily found in the form of triacylglycerides (TAG) and bound with phospholipids (PL). (Zhang et al., 2019) demonstrated that the content and proportions of DHA/EPA-TAG and DHA/EPA-PL vary depending on the specific marine source. The Food and Agriculture Organization (FAO) of the United Nations recommends a daily intake of 250 mg/day of EPA plus DHA for adult males and nonpregnant/nonlactating adult females (WHO Expert Consultation on Rabies, 2005). Maintaining a balanced n-6 to n-3 ratio of 1:1 to 4:1 is essential for optimizing their health benefits (Ahmmed et al., 2020). However, modern diets often skew toward higher n-6 PUFA consumption, which disrupts this balance and impairs the utilization of omega-3 PUFAs in organs like the brain.

Marine omega-3 fatty acids exhibit significant healthful advantages. Direct dietary intake of EPA and DHA has been shown to increase their levels in the heart, liver, brain, and serum, underscoring the importance of adequate consumption (Zhang et al., 2019). DHA plays a crucial role in visual and brain functions by enhancing cell membrane fluidity, regulating enzyme activities, and improving signal transmission. These processes collectively support neurotransmitter production, serotonin absorption, and receptor function, all of which are essential for neural communication. Adequate DHA supplementation further acts to strengthen neural network functions and improve brain sensing, leading to better learning and memory abilities (Chen and Liu, 2020). Moreover, DHA is one of the most important PUFAs for infant development, and Chamorro et al. (2022) highlight the necessity of improving diet quality, specifically increasing DHA content in pregnant and lactating women. In addition, their cardioprotective effects are attributed to their ability to reduce inflammation and improve arterial compliance, thereby lowering the risks of CVDs (Donat-Vargas et al., 2020; Manson et al., 2019). For instance, supplementation with krill oil, a rich source of EPA and DHA, has been shown to decrease plasma concentrations of Low-Density Lipoprotein (LDL) cholesterol and triglycerides while increasing High-Density Lipoprotein (HDL) (Kim et al., 2022). Similarly, fish oil rich in DHA and EPA has demonstrated efficacy in lowering triglyceride levels, thereby reducing cardiovascular risk factors (Fard et al., 2019).

In food systems, omega PUFAs are often converted into colloidal forms such as liposomes, lipid droplets, or biopolymer particles to enhance their delivery and are often incorporated into foods and beverages (McClements and Öztürk, 2021). Advanced technologies, including encapsulation, microfluidization, and ultrasonication, are used to prepare the colloidal forms, mainly to serve a few purposes such as enhancing the stability of PUFAs, reducing lipid oxidation, and improving bioavailability (Shi et al., 2020). Therefore once consumed, omega PUFAs are readily absorbed by the body and play a critical role in improving overall health. Besides, omega-3 PUFAs, particularly EPA and DHA, when integrated into the phospholipid bilayer of cell membranes, improve their fluidity and permeability. This structural enhancement facilitates the absorption of fat-soluble vitamins (A, D, E, and K) and supports the activity of membrane-bound proteins, enzymes, and receptors essential for metabolic and immune processes.

13.1.3.2 *Enhancing antioxidant properties*

Antioxidants are compounds that scavenge and neutralize free radicals. Oxidation is a natural process in living organisms that generates free radicals and reactive oxygen species (ROS), which, in excess, can lead to oxidative stress, contributing to aging and chronic conditions such as CVDs and diabetes (Gulcin, 2020). Marine fatty acids, particularly omega-3 PUFAs like EPA and DHA, have shown significant antioxidant properties that help mitigate oxidative damage (Meital et al., 2019; Xiao et al., 2022). In addition to their well-known health benefits, EPA and DHA can enhance oxidative stability in food systems, reducing lipid oxidation, delaying rancidity, and enhancing shelf life. The antioxidative properties arise from the multiple double bonds that result in the susceptibility of EPA and DHA to oxidation, enabling EPA and DHA to act as antioxidants by neutralizing free radicals and breaking oxidative chain reactions (Cretton et al., 2023).

As an antioxidant, EPA and DPA are easily susceptible to oxidation. Hence, techniques such as encapsulation and emulsification, along with co-formulation with natural antioxidants like tocopherols and polyphenols, have been employed not only to protect against the degradation of EPA and DHA but also to enhance the antioxidant capacity in food systems. For example, encapsulated marine oils, such as krill oil, have shown improved oxidative stability and enhanced antioxidant capacity in food applications (Sánchez et al., 2021). Similarly, Rahim et al. (2022) reported that spray-dried microcapsules of fish oil effectively minimized the loss of EPA and DHA while maintaining low peroxide values, showcasing their potential for maintaining oxidative stability. Another promising approach involves combining tuna eye oil, rich in omega-3, with antioxidant-rich *Spirulina platensis* in supplement tablets. This combination not only improved oxidative stability, as indicated by an IC50 value of 119.63 ppm, but also maintained a high fatty acid content of 99.95%, demonstrating its potential to enhance antioxidant activity and product shelf life (Saeidi et al., 2016). These strategies not only preserve the functional properties of omega-3 PUFAs but also improve their integration into diverse food formulations.

EPA and DHA not only provide additional protection against oxidation but also further enhance the nutritional and functional value of omega-3-enriched food products. This dual functionality, offering both health benefits and improved food stability, makes marine fatty acids valuable components in the development of innovative food formulations. In addition, emerging trends in food technology, such as advanced packaging solutions like oxygen-absorbing or nitrogen-flushed packaging, complement the stability of omega-3 fatty acids by minimizing oxidative degradation during storage. These advancements, coupled with the incorporation of bioactive omega-3 metabolites, pave the way for the creation of functional food products with extended shelf life and enhanced antioxidant properties.

13.1.3.3 *Flavor enhancement*

Flavor, one of the most prominent sensory attributes, is significantly influenced by marine fatty acids, which contribute to the development of diverse aroma compounds. Often, the flavor is developed when lipids undergo degradative processes such as autoxidation, photo-oxidation, and enzymatic oxidation, producing odor-active volatiles, including aldehydes, ketones, and alcohols (Shahidi and Hossain, 2022). Aldehydes derived from PUFAs are particularly important. Key compounds such as hexanal, octenal, and nonenal are recognized for their significant role in flavor development. Among these, hexanal is notable for its high odor threshold, which ranges from 4 to 23 μg/L. These aldehydes create a spectrum of aromas, from fresh green to fatty green, with aldehydes accounting for approximately 77% of the total aroma profile in marine lipid-based foods (Francezon et al., 2021). A sensory wheel developed by Larssen et al. (2018) categorizes marine oil flavors with 184 descriptors covering aroma, flavor, and mouthfeel. Their findings suggest that minimal oxidation produces sensory characteristics such as *grass, butter, and acidic*, whereas advanced oxidation leads to undesirable *rancid, metallic*, and *fishy* aromas. The role of glycerophospholipids in enhancing flavor has also been extensively studied. Ahmmed et al. (2023) identified 72 volatile compounds in unrinsed mixed sturgeon surimi gels, showcasing the complexity of marine lipid-based flavors. Similarly, Nie et al. (2023) investigated flavor development in *Saccharina japonica*, a widely consumed brown algae in Asian cuisine. Their findings revealed that fermentation, hydrolysis, and oxidation are key mechanisms contributing to the distinctive flavor of this seaweed.

Fishy flavors often pose challenges in certain applications. Excessive fish oil supplementation in animal feed reduces the sensory acceptability of resulting products. To address this issue, Feng et al. (2020) suggested microalgae oil as a viable alternative to fish oil for enriching DHA in hens' eggs without negatively affecting flavor or aroma. Furthermore, advancements in encapsulation technologies, such as the coating matrix designed by Sun et al. (2022) using pea proteins and flaxseed gum, have successfully mitigated off-flavor formation by stabilizing algae oil droplets and preventing lipid oxidation.

The influence of cooking methods on the fatty acid composition and sensory quality of marine-based foods is highly significant. Temgoua et al. (2022) evaluated different cooking techniques for Atlantic salmon and found that steaming was the most effective method for delivering the highest levels of PUFAs. Both steaming and boiling also received high ratings for enhancing taste and odor, highlighting their potential to maximize sensory quality. Fermentation offers additional opportunities for flavor enhancement. For example, Lu et al. (2024) demonstrated that fermenting oysters generates characteristic fruity flavors by producing undecanone and 5-nonanone during lipid oxidation.

The flavor development and the resulting food applications of marine fatty acids are not confined to direct consumption. Their ability to produce unique aromas and enhance flavor complexity is helpful in functional food formulations. Value additions through encapsulation and emulsification enable marine oil to be incorporated into various food products without compromising the flavor or generating off-flavors. The interplay of lipid degradation, cooking methods, and technological interventions underscores the versatility of marine fatty acids in contributing to flavor development.

13.1.4 Application of marine fatty acids in different food industries

13.1.4.1 Dairy products

The incorporation of marine fatty acids into dairy products has garnered significant interest due to their widespread consumption and potential as vehicles for delivering omega-3 PUFAs. A few strategies have been employed to enhance the incorporation of marine fatty acids into dairy products, either via feeding regimens or direct incorporation into food systems.

Feeding marine fatty acids, particularly omega-3 PUFAs, to dairy animals such as cows and goats is an effective strategy for enhancing the nutritional profile of milk (Pajor et al., 2019, 2021, 2023; Nguyen et al., 2019). Marine-derived omega-3 PUFAs can be incorporated into the animals' diets through supplements like fish oil, algal oil, or encapsulated marine lipid formulations (Nguyen et al., 2019). These feed supplements alter the fatty acid composition of the animals' milk by increasing the omega-3 PUFA content, such as DHA, aligning with dietary recommendations to reduce saturated fatty acid intake and increase PUFAs (Urrutia et al., 2020). Mitguard et al. (2023) have reported that higher concentrations of omega-3 fatty acids in milk lead to better health outcomes in neurodevelopment, body composition, allergy, skin, and eczema in infants. These fatty acids are associated with a reduced risk of coronary heart disease and have antiinflammatory properties (Pajor et al., 2021, 2023). This dietary intervention also reduces the omega-6 to omega-3 ratio, aligning the milk's fatty acid profile with recommended nutritional standards (Hadrová et al., 2021). Besides, the inclusion of marine algae in dairy diets has been shown to decrease the somatic cell count and prevalence of mastitis pathogens in milk, indicating improved udder health in dairy goats (Pajor et al., 2021). This enhances not only the quality of milk but also its safety for consumption. Pajor et al. (2023) commented that although marine algae supplementation can alter the fatty acid composition, it does not negatively impact the sensory profile of milk, maintaining its flavor while enhancing its nutritional value. Furthermore, incorporating PUFA-enriched feeds in lactating cow diets reduced methane production and improved milk quality without affecting yield (Thanh et al., 2023). Adding rumen-protected marine-derived supplements is the most effective way to increase omega-3 fatty acid content in dairy products for human consumption. Diets supplemented with n-3 PUFAs have enhanced reproductive parameters in dairy cows, including improved conception rates and shortened calving-to-conception intervals (Pereira et al., 2022). These benefits are largely attributed to the positive influence of n-3 PUFAs on ovarian function and the uterine environment. Specifically, PUFA supplementation has been found to modulate reproductive hormones by enhancing estradiol and prostaglandin levels, which play critical roles in successful reproduction (Amini et al., 2024). This hormonal regulation not only supports improved reproductive efficiency but also positively impacts milk production by promoting better metabolic health and lactation performance. Enhanced reproductive outcomes reduce the interval between calving and subsequent lactation cycles, leading to more consistent milk yield (Amini et al., 2024).

An extensive study was conducted by Savatinova and Ivanova (2024) on the incorporation of marine fatty acids, especially fish oil, in various types of dairy products such as infant formula, natural drinkable yogurt, cream, and cheese. However, the main challenge in fortifying dairy products with omega-3 fatty acids is their high susceptibility to oxidation, which can lead to undesirable flavors and reduced nutritional value. Techniques like ultrasonication and high-pressure homogenization are commonly utilized to produce stable emulsions, while freeze drying and spray drying are employed to create stable microcapsules for long-term storage and improved oxidative stability. Gumus and Gharibzahedi (2021) highlighted that omega-3 fatty acids can be encapsulated in micro- or nanoemulsions to enhance their bioavailability and stability in yogurt, thereby improving its nutritional properties. Nosratpour et al. (2022) reported that fish oil and milk fat encapsulation into solid lipid particles promotes stability and reduces undesirable taste and odor in dairy products. On the other hand, microencapsulation has been shown to protect these fatty acids from oxidation, maintaining the physicochemical and sensory properties of bakery products (Dhiman et al., 2024). These innovative approaches demonstrate the potential for integrating omega-3 fatty acids into dairy and bakery products while maintaining or enhancing product quality and functionality.

13.1.4.2 *Bakery products*

Traditional bakery products are often high in trans fats, which have been associated with various health risks, including CVDs, obesity, and diabetes. Recent studies suggest that replacing these unhealthy fats with omega-3-rich oils can significantly enhance the nutritional quality of baked goods. Zaitseva (2022), Karantonis et al. (2022), and Dhiman et al. (2024) highlighted that omega-3-enriched bakery products not only improve health outcomes but also maintain the stability and concentration of omega-3 fatty acids even after exposure to high baking temperatures, preserving their health benefits during consumption. Moreover, Karantonis et al. (2022) reported that omega-3-enriched baked goods exhibit higher in vitro antithrombotic and antiinflammatory activity compared to traditional bakery products, further underscoring their potential health advantages. Consumer studies have shown positive reception of omega-3-enriched baked goods, with no significant differences in taste compared to conventional products (Karantonis et al., 2022). In addition, the incorporation of omega-3-rich oils has been found to improve the textural properties of bakery products. For instance, Punia et al. (2020) and Ponomareva et al. (2020) observed that these enriched products demonstrated increased specific volume and elasticity, enhancing the overall quality of buns and similar baked goods. These findings suggest that omega-3 fortification in bakery products offers a promising approach to improving both nutritional and sensory attributes.

13.1.4.3 *Meat products*

Marine fatty acids significantly enhance the quality of meat products. Marine fatty acids can be applied in many diverse modes, including incorporation through animal feed, direct enrichment of processed meat, and innovative techniques such as emulsions, encapsulations, and edible coatings. Incorporation of marine fatty acids in any of these approaches helps to improve not only the nutritional profiles but also intensify flavor, while simultaneously preventing oxidative degradation (marine fatty acids possess inherent antioxidant properties) (Valença et al., 2021; Sun et al., 2023). This has driven the meat industry to innovate, incorporating marine fatty acids to cater to the growing demand for functional and nutrient-rich products (Wang et al., 2023).

The nutritional quality of meat can be improved through the incorporation of marine fatty acids, either indirectly in animal feed or directly into meat products. Fatty acid-rich dietary supplements derived from marine sources, such as fish oil and microalgae, have been extensively studied for their potential to enrich meat products with beneficial omega-3 fatty acids, EPA, and DHA. Xu et al. (2021) revealed that when cattle were fed a diet supplemented with DHA-rich microalgae, the increased DHA and EPA levels enhanced the activity of antioxidant enzymes in the resultant beef. Furthermore, research by Al-Soufi et al. (2024) demonstrated the positive effects of microalgae diets on the fatty acid composition of rabbit meat. By increasing the proportion of omega-3 fatty acids, microalgae-enriched diets improved the nutritional profile of the meat, offering a healthier alternative to conventionally produced meat. In another instance, high levels of saturated fat in conventional processed meats pose a significant health concern. Marine fat, rich in PUFAs, provides an effective solution. Incorporating marine-derived oils directly into meat products reduces saturated fat content while increasing levels of health-promoting PUFAs. Gullón et al. (2020) reported a remarkable 70% reduction in saturated fat content and significant increases in DHA and EPA levels when beef patties were formulated with 1% algae oil. This dual benefit of reducing unhealthy fats while enriching meat with essential omega-3 fatty acids highlights the potential of marine oils as functional ingredients in meat processing. Beyond their nutritional benefits, marine fatty acids also play a critical role in altering the sensory characteristics of meat products. For instance, Xu et al. (2021) highlighted the influence of marine fatty acid-rich diets on the volatile composition of beef, identifying aldehydes derived from fatty acids as the main volatile component. These volatiles significantly enhance the beef's sensory properties, creating a desirable aroma and flavor profile that appeals to consumers. Moreover, Tu et al. (2020) discussed partially replacing animal fats with vegetable and marine oils to modify the flavor profile of processed meats. This approach addresses the challenge of maintaining the desired sensory attributes while improving the nutritional quality of the product. When blended with other oils, marine fats contribute to a balanced flavor profile, making them a valuable component in formulating healthier meat products.

However, the direct incorporation of fish oil in bulk into meat products often leads to an unfavorable fishy flavor. Therefore microencapsulation of fish oil using carrier matrices is preferred to mask these off-flavors, besides enhancing controlled release (Jamshidi et al., 2020). Advanced modification techniques, such as emulsifications and encapsulations, have been developed to optimize the delivery and stability of marine fatty acids in meat products (Pérez-Palacios et al., 2019). These techniques protect the sensitive fatty acids from oxidative degradation, ensuring their stability throughout processing and storage. Encapsulation, in particular, has gained significant attention for its ability to shield PUFAs from environmental factors while allowing their controlled release during cooking or digestion. Accordingly, Pourashouri et al. (2021) revealed the high stability of prefried nuggets using microencapsulated fish oil. Moreover, Solomando et al. (2021) investigated the efficacy of multilayer microcapsules of omega-3 fatty acids in dry-cured and cooked meat products with high oxidative stability. Emulsions and edible coatings also offer unique advantages in meat processing. Emulsion systems enable the uniform distribution of marine oils within the meat matrix (Pérez-Palacios et al., 2019). Emulsion hydrogels of oil mixtures, mainly algal oil, have successfully replaced animal fat in dry fermented foal sausage (Cittadini et al., 2022).

13.2 Marine-derived protein

Marine sources, including fish, mollusks, crustaceans, and algae, are rich in high-quality proteins. Most of these marine-derived proteins contain all 20 amino acids, including essential amino acids, and exhibit a higher biological value (BV) compared to plant-based protein sources (Raja et al., 2024). The muscle proteins of fish, mollusks, and crustaceans can be classified into three major categories: sarcoplasmic proteins, which are soluble in water or low-salt buffers; myofibrillar proteins, which are highly soluble in high-salt solutions and are crucial for muscle structure; and stromal proteins, which are structural connective tissue proteins. These proteins contribute to the total muscle protein content in the following proportions: 15%–35% (w/w) for sarcoplasmic proteins, 65%–75% (w/w) for myofibrillar proteins, and approximately 3% (w/w) for stromal proteins (Lionetto and Corcione, 2021).

Sarcoplasmic proteins primarily consist of enzymes involved in energy metabolism, such as aldolase and creatine kinase. They also contain pigmented proteins like myoglobin and hemoglobin, which contribute to oxygen transport. Myofibrillar proteins include actin and myosin, the key contractile proteins responsible for muscle movement. Other important proteins in this group include tropomyosin, troponin, actinin, desmin, nebulin, and C and M proteins, all of which play essential roles in muscle contraction and structural integrity. Stromal proteins mainly consist of collagen, followed by elastin and gelatin, which provide structural support to connective tissues (Alam et al., 2022).

Both microalgae and macroalgae serve as excellent sources of naturally occurring marine proteins. Macroalgae, commonly known as seaweeds, are particularly rich in phycobiliproteins and lectins, which possess antioxidant and immunomodulatory properties. Microalgae, on the other hand, are an abundant source of sulfur-containing amino acids such as methionine and cysteine, as well as high-quality proteins (Zhang et al., 2024b). The composition and bioavailability of these marine proteins vary depending on factors such as species type, growth stage, and environmental conditions in their natural habitat.

13.2.1 Marine-derived protein hydrolysates and peptides

Marine-derived protein hydrolysates and peptides refer to small protein products obtained from the breakdown of proteins naturally occurring in marine organisms. Peptides typically consist of 2–20 amino acids. Although most marine proteins remain inactive within the body of the species, the derived peptides exhibit a range of bioactive properties when extracted from marine sources. These bioactive properties include anticancer, antimicrobial, antiinflammatory, antihypertensive, and antioxidant effects (Cunha and Pintado, 2022). Today, these peptides are widely utilized in pharmaceuticals (Banday et al., 2024), the food industry (Penggalih et al., 2023), and cosmetics production (Rotter et al., 2024). The growing demand for these bioactive compounds has driven the fish protein hydrolysate market, which is expected to reach USD 558 million by 2025 (Nikoo et al., 2023).

There are several methods for producing protein hydrolysates and peptides, including enzymatic hydrolysis, subcritical water hydrolysis, microbial fermentation, autolysis, and chemical hydrolysis (acid and alkali hydrolysis). Among these, enzymatic hydrolysis and autolysis are the most eco-friendly and widely used, as they do not generate toxic by-products. Enzymatic hydrolysis employs commercially manufactured proteolytic enzymes to break down proteins and generate bioactive peptides in a controlled manner (Siddik et al., 2021). In contrast, autolysis utilizes endogenous proteolytic enzymes, such as pepsin, trypsin, chymotrypsin, and gastricsin, found in marine organisms. This process is more cost-effective than enzymatic hydrolysis, and peptide properties can be controlled by adjusting processing factors such as temperature, pH, and hydrolysis duration (Nikoo et al., 2021).

Chemical hydrolysis involves the cleavage of peptide bonds using either acidic or alkaline conditions: acid hydrolysis employs sulfuric acid (H_2SO_4) or hydrochloric acid (HCl), commonly for fish scales. Other acids, such as maleic acid, nitric acid, and oxalic acid, have also been used. However, this method requires high temperatures and pressure, leading to salt accumulation in the final product. The harsh conditions may degrade essential amino acids, such as tryptophan, reducing the protein hydrolysate's quality. Alkali hydrolysis uses calcium hydroxide, sodium hydroxide, or potassium hydroxide. Although milder than acid hydrolysis, it still generates toxic by-products and diminishes the functionality of the resultant peptides (Pavlicevic et al., 2020).

Microbial fermentation employs microorganisms, such as lactic acid bacteria, to break down proteins and generate bioactive peptides with antioxidant, antimicrobial, and antihypertensive properties. This method enhances peptide digestibility and is considered an environmentally friendly alternative (Marti-Quijal et al., 2020). Subcritical water hydrolysis involves heating water to temperatures between 100°C and 300°C under high pressure, ensuring it remains in a liquid state throughout the process. Under these conditions, the viscosity of water decreases, facilitating better penetration into marine biomass. The hydrolysis reaction is catalyzed by the formation of hydronium (H_3O^+) and hydroxide (OH^-) ions, which act as natural acid and base catalysts. This process does not require chemical additives or enzymes, making it a green technology that is environmentally safe (Melgosa et al., 2020). The efficiency of this method depends on factors such as the type of marine organism, the body part used, and specific processing conditions (Pratama et al., 2022).

Numerous marine-derived peptides with a variety of bioactive properties, including antimicrobial, antioxidative, antiviral, anticancer, and antiinflammatory activities, have been discovered so far. Fish bioactive peptides include YENGG, EGYPWN, YIVYPG, and WGDAGGYY peptides from tuna (antioxidative); MEVFVP and VSQLTR peptides from flounder fish (*Paralichthys olivaceus*) (antihypertensive); PLKKFRSIRREL and NITRQAYWQIHM peptides from striped murrel (*Channa striatus*) (antimicrobial); and YALPAH from half-fin anchovy (*Setipinna taty*) (antiproliferative) (Vásquez et al., 2023).

Some of the reported algae-derived peptides include FGMPLDR and MELVLR peptides from *Ulva intestinalis* (antioxidative), YGFVMPRSGLWFR from *S. platensis* (anticancer) (Okeke et al., 2024), Theonellamide G from the Red Sea sponge *Theonella swinhoei* (antifungal), and cyclic depsipeptides mirabamides A–D from the marine sponge *Siliquariaspongia mirabilis* (antiviral) (Penggalih et al., 2023). Some of the reported bioactive peptides obtained from crustaceans and mollusks include MEAP from blue mussel (*Mytilus edulis*) (anticoagulant), conotoxins from worm-hunting cone snails *Conus nux*, *Conus brunneus*, and *Conus princeps* (analgesic), PFAOP peptide from *Pinctada fucata* (antioxidant), and LLEYSL and LLEYSI peptides from oyster (*Crassostrea gigas*) (anti-HIV) (Cunha and Pintado, 2022).

Marine-derived protein hydrolysates and peptides have significant applications in pharmaceuticals, functional foods, and nutraceuticals due to their diverse bioactivities. Advances in extraction and hydrolysis techniques continue to improve the efficiency and sustainability of peptide production, positioning marine bioactive compounds as promising candidates for future health and industrial applications.

13.2.2 Marine-derived protein sources

13.2.2.1 *Fish*

Among a wide range of protein sources, fish play a crucial role in providing nutritional, economic, and cultural benefits to human society. Fish is widely recognized for its high nutritional quality, offering a readily digestible and bioavailable source of protein (Mogobe et al., 2024). The fish farming industry significantly contributes to global protein production, with applications extending beyond food into the cosmetic and pharmaceutical sectors. Fish protein is derived from three primary components: skin, flesh (muscle), and processing by-products such as fins, frames, heads, trimmings, and viscera. These components are commonly used to produce fish protein hydrolysates, which have diverse functional and nutritional applications (Khan et al., 2022). Fish proteins are particularly rich in essential amino acids, including tryptophan, cystine, lysine, methionine, and threonine, which contribute to their superior digestibility and nutritional value (Ramos-Miras et al., 2019). The digestibility of fish protein ranges from 77% to 98.7%, as determined by comparisons with the standard protein recommendations established by the World Health Organization (WHO). Moreover, the BV and protein efficiency ratio, which measure the amino acid profile and growth-supporting ability of proteins, are higher for fish than for other protein sources such as beef, pork, chicken, and milk (Mogobe et al., 2024).

The concentration, quality, and type of protein vary among fish species and across different organs. The protein content in fish ranges from 8.2% to 23.9%, depending on the species. Notably, *Thunnus albacares* contains the highest protein content at 23.9%, whereas *Harpadon nehereus* has the lowest at 8.2%. Several fish species have a protein content exceeding 20%, including *Mugil cephalus, Euthynnus affinis, Etroplus suratensis, Stolephorus waitei, Tenualosa ilisha*, and *Lates calcarifer* (Mohanty et al., 2019).

Fish muscles contain different types of proteins, classified into three major categories based on their solubility: structural proteins, sarcoplasmic proteins, and myofibrillar proteins. The predominant type in fish is striated muscle, which consists of muscle fibers connected by connective tissue. Fish muscle proteins differ from those found in land animals and other lower vertebrates, with significant variations in protein composition among species (Khan et al., 2022). Fish processing by-products are an excellent source of high-quality protein. While only 40% of fish-derived protein is directly consumed by humans, the remaining 60%, including skin, head, frames, fins, trimmings, roes, and viscera, is often considered waste despite its substantial protein content (Kandyliari et al., 2020). The efficient utilization of these by-products can contribute to sustainable food production and reduce environmental waste while providing valuable protein sources for human and animal nutrition.

13.2.2.2 *Algae*

Algae are rich in essential nutrients and bioactive compounds, including proteins, making them nutritionally complete food products or ingredients. Microalgae, in particular, have a high protein content, with some species boasting up to 70% protein by dry weight (Geada et al., 2021). Among the most widely recognized microalgae species for vegan protein sources are *Arthrospira platensis* (spirulina) and *Chlorella vulgaris*, which contain 50%–60% protein by dry weight. These species also provide substantial levels of essential amino acids and vitamin B12, making them valuable alternatives to traditional protein sources (Durdakova et al., 2024).

Although seaweeds generally have a lower protein content, ranging from 9% to 22%, certain red algae species have been reported to contain protein levels as high as 47% (Pereira et al., 2024). The widespread availability and rapid growth of seaweeds compensate for their lower protein content, making them a valuable and sustainable protein source. The quality of algae-derived proteins is considered superior to many plant-based proteins, including those from wheat, rice, and beans, but slightly inferior to animal proteins such as milk or meat (Ścieszka and Klewicka, 2019). However, algae proteins surpass those found in eggs, as they contain all essential amino acids in proportions that meet the FAO guidelines (Wu et al., 2023). Furthermore, algae are rich in aspartic acid, glutamic acid, alanine, and glycine, which contribute to various physiological functions. Their amino acid composition, particularly the presence of branched-chain amino acids, is comparable to that of soy and eggs, making algae a promising protein source for muscle health and overall nutrition (Wu et al., 2023).

Beyond their nutritional benefits, algae offer significant advantages over traditional plant-based protein sources. They exhibit rapid growth, have higher photosynthetic efficiency, and require minimal freshwater, reducing competition for arable land. With approximately 70% of the Earth's surface covered by oceans, the large-scale cultivation and harvesting of algae as a sustainable protein feedstock are highly feasible, reinforcing their potential as a future food source (Geada et al., 2021).

13.2.2.3 *Crustacean and mollusk*

Marine crustaceans, including crabs, shrimp, lobsters, prawns, and crayfish, are rich sources of biologically active compounds. Proteins derived from shrimp waste provide a well-balanced profile of essential (56.8%) and nonessential (43.2%) amino acids, making them valuable for nutritional applications (Senadheera et al., 2023). Hydrolysates obtained from shrimp shells have been explored for their potential use in functional foods and protein supplements, offering an eco-friendly solution for utilizing seafood processing by-products (Zhang et al., 2024a). Shrimp heads, which account for approximately 30%–50% of the total shrimp weight, are particularly rich in high-quality protein, containing 50%–65% protein on a dry-weight basis. Notably, the heads of Kuruma shrimp (*Marsupenaeus japonicus*), a major by-product of the seafood processing industry, represent an underutilized source of high-quality protein with potential applications in the food and health industries (Zhou et al., 2023).

Crustaceans are also known for their bioactive proteins, which exhibit antioxidant, antimicrobial, and therapeutic properties, making them valuable for food industry applications. Peptides derived from Izumi shrimp (*Plesionika izumiae*) and *Acetes chinensis* shrimp have demonstrated superior angiotensin-I-converting enzyme (ACE) inhibitory activity compared to terrestrial protein sources, highlighting their potential for developing functional foods for hypertension management (Senadheera et al., 2023). Moreover, the red king crab (*Paralithodes camtschaticus*) produces a novel antimicrobial peptide (AMP), positioning it as a promising biopreservative for food products (Molchanov et al., 2023). Invasive green crabs (*Carcinus maenas*) have also been explored for the production of carbohydrase-inhibitory peptides, which offer health benefits such as antihyperglycemic effects, making them suitable as functional food ingredients (Kang et al., 2019).

Mollusks, belonging to the phylum Mollusca, include oysters, clams, mussels, and scallops, and also contain bioactive proteins with significant health benefits. Peptides derived from mollusks, such as mussels (*Perna canaliculus, Mytilus edulis*), clams (*Ruditapes philippinarum*), and Asian clams (*Corbicula fluminea*), exhibit strong antioxidant properties. These bioactive compounds have potential applications in functional foods and nutraceuticals, helping to prevent oxidative stress and promote overall health (Senadheera et al., 2023; Wang et al., 2022; Xu et al., 2020).

13.2.2.4 Marine sponges

Marine sponges are primarily recognized for their medicinal properties and are also a potential source of valuable proteins, such as collagen and antioxidant peptides. Collagen has been successfully isolated from the marine sponge *Stylissa flabelliformis*, demonstrating significant antioxidant activity, with an IC_{50} value of 61.5 ± 2.132 ppm. This activity correlated with the protein content of the isolate, highlighting its potential as a bioactive compound (Sunarwidhi et al., 2021). In addition to collagen, marine sponges have been found to contain novel proteins with diverse biological functions. For instance, ilborin, a protein identified in marine sponges, plays a crucial role in energy metabolism and calcium binding. These findings underscore the complex biochemical functions of sponge-derived proteins and their potential applications in biotechnology (Borisenko et al., 2022). Although marine sponges are not yet widely utilized as a direct food source, their unique proteins and bioactive peptides present promising opportunities for use in health supplements and potentially in future functional food products. As research continues to uncover the full range of bioactive compounds within marine sponges, their applications in nutraceuticals and pharmaceuticals are expected to expand.

13.2.3 Health benefits of marine-derived proteins

Fish consumption accounts for approximately 20% of global animal protein intake, making it a vital dietary protein source. Fish proteins are highly nutritious, containing all 20 amino acids and exhibiting a high BV of around 75%, comparable to that of beef. The crude protein content in fish flesh ranges between 17% and 22% (w/w). Similarly, cephalopods contain 15%–25% protein relative to their total body weight, a composition comparable to that of beef, highlighting their potential as an alternative protein source (Raja et al., 2024).

Marine algae are also recognized as a rich source of protein, with their protein content varying by species. Red seaweed contains between 12.5% and 35.2% protein, green seaweed ranges from 9.6% to 23.3%, and brown seaweed contains 4.5%–16.8% protein. A significant portion of this protein fraction consists of essential amino acids, which typically make up approximately 30% of the total protein content. In addition, marine algae contain bioactive proteins such as glycoproteins, phycobiliproteins, and lectins, which exhibit various essential biological activities, including immune modulation and antioxidant properties (Echave et al., 2022).

Beyond their nutritional value, marine-derived proteins and peptides offer numerous health benefits due to their bioactive properties. These include ACE inhibition for blood pressure regulation, antioxidative effects that help combat oxidative stress, antiinflammatory properties that modulate immune responses, and antidiabetic effects that contribute to glucose homeostasis. As research into marine-derived proteins continues to expand, their applications in functional foods and nutraceuticals present promising opportunities for improving human health.

13.2.3.1 *Antihypertensive activity*

According to the WHO, hypertension is one of the leading causes of premature death, as it significantly increases the risk of CVDs, heart attacks, strokes, and arteriosclerosis. One of the primary factors contributing to hypertension is the production of angiotensin II, a potent vasoconstrictor. In the renin–angiotensin–aldosterone system, angiotensinogen is first cleaved by renin to produce angiotensin I. Angiotensin I is then converted into angiotensin II by ACE. The inhibition of this ACE-mediated conversion is a well-established strategy for managing hypertension. While ACE inhibitors are commonly prescribed for hypertension management, they can cause various side effects, including chronic cough, loss of taste, renal impairment, and angioneurotic edema. As a natural alternative, marine-derived ACE inhibitor peptides have gained attention for their potential to regulate blood pressure without the adverse effects associated with synthetic drugs (Cunha and Pintado, 2022).

These bioactive peptides have been extracted from a wide range of marine organisms, including fish, mollusks, crustaceans, and algae. Notable examples include the Alanine – Tryptophan – Tryptophan (AWW), Isoleucine – Tryptophan – Tryptophan (IWW), Tryptophan – Leucine (WL), Valine – Arginine – Proline (VRP), Isoleucine – Lysine – Proline (IKP), Leucine – Arginine – Proline (LRP), and Isoleucine – Arginine – Proline (IRP) peptides derived from *Rachycentron canadum* and bonito fish; the VECYGPNRPQF peptide from *C. vulgaris*; and the FQIN[M(O)]CILR and TGAPCR peptides from the microalga *Gracilariopsis lemaneiformis* (Cunha and Pintado, 2022). Moreover, the valyl-tyrosine (VY) peptide from *Litopenaeus vannamei* shrimp, isoleucyl–prolyl–proline from *C. gigas* oyster, isoleucyl–leucine–proline from *Portunus trituberculatus* crab, and leucine–aspartic acid from *Meretrix meretrix* clams have also demonstrated ACE inhibitory activity, making them promising candidates for natural antihypertensive therapies (Okeke et al., 2024).

13.2.3.2 *Antioxidant activity*

Oxidative stress is a physiological imbalance that occurs when the production of reactive ROS exceeds the body's antioxidant defense capacity. This imbalance can be triggered by various external and internal factors, including lifestyle behaviors such as smoking, excessive alcohol consumption, poor dietary habits, and chronic stress. These factors contribute to the generation of highly reactive free radicals. Excessive accumulation of these radicals leads to oxidative damage at the cellular and molecular levels, contributing to the pathogenesis of various chronic diseases, including cancer, CVDs, neurodegenerative disorders, and nephropathy. Antioxidants play a crucial role in mitigating oxidative stress by neutralizing excess free radicals or preventing their formation through scavenging mechanisms.

Peptides containing amino acid residues such as histidine, tyrosine, methionine, and cysteine (thiol-containing groups) exhibit strong antioxidant properties, as these residues enhance radical scavenging activity and metal ion chelation, thereby preventing oxidative damage (Ma et al., 2021). Numerous antioxidative peptides have been identified from marine organisms, demonstrating their potential as functional ingredients in food and pharmaceutical applications. Notable examples include the GIISHR peptide from *Mustelus griseus* (marine fish), the PFAOP peptide from *Pinctada fucata* (marine bivalve), and the HVGGCG peptide from *Oratosquilla woodmasoni* (marine squilla) (Penggalih et al., 2023). Moreover, the VECYGPNRPQF peptide from *C. vulgaris* and peptides derived from *Ishige okamurae* (marine microalga) have demonstrated strong antioxidative activity, further highlighting the potential of marine-derived peptides in combating oxidative stress and preventing related diseases (Cunha and Pintado, 2022). Given their bioavailability, stability, and potency, marine-derived antioxidative peptides offer a promising natural alternative to synthetic antioxidants.

13.2.3.3 *Antiinflammatory activity*

Inflammation is a localized tissue response to irritation, injury, or infection, triggered by the production of specific cytokines and chemokines. This inflammatory process can lead to increased oxidative stress, resulting in the overproduction of ROS and a reduction in antioxidant enzyme activity, including catalase, superoxide dismutase, and glutathione peroxidase. Controlling the rise of inflammatory factors during inflammation is crucial for preventing oxidative damage and associated health complications (Park et al., 2020).

Research has identified numerous marine-derived peptides with significant antiinflammatory properties, offering promising alternatives to conventional pharmaceuticals. These peptides can inhibit the production of proinflammatory cytokines and modulate key inflammatory pathways, such as the nuclear factor kappa-B (NF-κB) signaling pathway, thereby reducing chronic inflammation and supporting overall immune health (Barzkar et al., 2024; Rivera-Jiménez et al., 2022).

The literature further highlights several marine-derived peptide sources with potent antiinflammatory effects, including the novel NCWPFQGVPLGFQAPP peptide (NCW peptide) from *Marphysa sanguinea* (a marine polychaete), the Val–Glu–Cys–Tyr–Gly–Pro–Asn–Arg–Pro–Gln–Phe (chlorella-11) peptide from *C. vulgaris*, the Leu–Asn–Gly–Asp–Val–Trp peptide from *Chlorella ellipsoidea* (a microalga), and the Skipjack enzymatic peptide from *Katsuwonus pelamis* (a marine fish) (Penggalih et al., 2023). Recent studies have also explored the antinociceptive and antiinflammatory effects of various marine algae species. For instance, a methanol extract of *Bryothamnion triquetrum* exhibited significant antinociceptive and antiinflammatory activities in Swiss mice. These effects were evaluated using multiple models, including the acetic acid-induced writhing test, the hot-plate test, and tests for glutamate- and formalin-induced nociception. In addition, antiinflammatory effects were assessed through zymosan A-induced peritonitis analysis, further validating the potential of marine-derived compounds in inflammation management (Lee et al., 2013). Other notable marine sources with antiinflammatory activity include sulfated polysaccharide fractions from *Gracilaria caudata*, galactans from *Gelidium crinale*, and mucin-binding agglutinins from *Hypnea cervicornis*. These bioactive compounds have demonstrated the ability to regulate inflammatory responses, making them promising candidates for the development of novel antiinflammatory agents (Alencar et al., 2019).

13.2.3.4 *Antidiabetic activity*

The antidiabetic activity of marine-derived proteins has emerged as a significant area of research within the field of nutritional science, focusing on the therapeutic potential of bioactive compounds sourced from marine organisms. Marine proteins and peptides, primarily extracted from fish, shellfish, and algae, have demonstrated the ability to interact with biological mechanisms involved in glucose metabolism, offering promising alternatives for the prevention and management of diabetes mellitus, particularly type 2 diabetes and its related complications (Penggalih et al., 2023). Research suggests that marine-derived proteins play a significant role in enhancing insulin sensitivity, regulating glucose homeostasis, and inhibiting digestive enzymes that contribute to elevated blood sugar levels. These effects are linked to the modulation of incretin hormones, such as gastric inhibitory polypeptide (GIP) and glucagon-like peptide-1 (GLP-1), as well as the antiinflammatory and antioxidant properties of these proteins, which help combat oxidative stress associated with diabetes pathogenesis (Ovchinnikova and Shi, 2023).

A common therapeutic approach for managing diabetes mellitus involves suppressing carbohydrate digestion by inhibiting key enzymes such as α-glucosidase and α-amylase. Moreover, maintaining blood glucose levels can be achieved by regulating GIP and GLP-1 through the inhibition of dipeptidyl peptidase IV (DPP-IV). While synthetic drugs are widely used for these purposes, marine-derived proteins have also been reported to be effective in diabetes treatment, offering a natural alternative with potential health benefits (Cunha and Pintado, 2022). For example, trypsin hydrolysate from phycobiliproteins of *Spirulina* (*A. platensis*) has DPP-IV and α-amylase inhibitory activity (Ramos-Romero et al., 2021). Other reported marine-derived proteins with antidiabetic properties include DPP-IV inhibitor peptides from *Phaeodactylum tricornutum* and *Porphyridium purpureum* (microalgae), the Val–Glu–Cys–Tyr–Gly–Pro–Asn–Arg–Pro–Gln–Phe (chlorella-11) peptide from *C. vulgaris*, and the Leu–Asn–Gly–Asp–Val–Trp peptide from *C. ellipsoidea* (microalgae) (Penggalih et al., 2023).

13.2.3.5 *Anticancer activity*

Marine-derived proteins, particularly peptides, have emerged as a promising focus in cancer research due to their potential anticancer properties. With cancer being a leading cause of mortality worldwide, traditional treatment methods often present significant side effects, driving interest in alternative therapies sourced from natural products. These proteins, which exhibit unique structural characteristics and diverse mechanisms of action, can offer novel avenues for therapeutic development and cancer management, making them a significant area of investigation in the field of oncology (Seyed and Ayesha, 2021; Azemin et al., 2025).

The anticancer activity of marine-derived proteins encompasses various mechanisms, including immune modulation, apoptosis induction, and disruption of critical signaling pathways. For instance, AMPs derived from marine organisms have shown both direct cytotoxic effects against cancer cells and supportive roles in enhancing immune responses in patients with compromised systems (Seyed and Ayesha, 2021; Khalifa et al., 2019). Natural peptides derived from marine sources have demonstrated promising anticancer effects against various cancer types. For example, the pentapeptide ILYMP from *Cyclina sinensis* has been shown to induce apoptosis in prostate cancer cells (DU-145), while the LKEENRRRRD peptide from *Sepia esculenta* exhibits moderate antiproliferative effects on the prostate cancer cell line (PC-3). Similarly, the VECYGPNRPQF peptide from *C. vulgaris* inhibits the proliferation of gastric cancer (Adenocarcinoma Gastric Stomach (AGS)) cells (Okeke et al., 2024). In addition, the SCAP1 peptide from *Crassostrea cucullata* has demonstrated inhibitory effects on human colon carcinoma (HT-29), while a hydrophilic peptide from *Mytilus coruscus* has been reported to induce apoptosis in prostate cancer (PC-3), lung cancer (A549), and breast cancer (MDA-MB-231) (Cunha and Pintado, 2022).

13.2.4 Functional properties of marine protein in food formulations

13.2.4.1 *Solubility*

Protein solubility is still one of the most important properties to consider when proteins are being selected for specific food formulations and contributes to other functional properties, such as foaming and emulsification. Improved solubility, achieved through the breakdown of proteins into smaller peptides, enhances their dispersion in solutions and facilitates the formation of stable colloidal systems (Siddik et al., 2021).

Marine-derived proteins, particularly fish protein hydrolysates, are highly soluble across a wide pH range, which makes them suitable for various food industry applications (Ucak et al., 2021). Marine protein hydrolysates often outperform plant proteins in solubility, especially under varying pH conditions. Studies on protein isolates extracted from sources like yellowfin tuna roe have provided valuable insights into their behavior across different pH levels, highlighting their potential performance in food systems (Yoon et al., 2019). Algal proteins, such as *Chlorella protothecoides,* have shown exceptional solubility across a wide pH range of 2–12, exceeding 84% solubility. This solubility surpasses that of many food proteins, such as pea, casein, soluble meat proteins, and soy protein, which exhibit reduced solubility (≤9%) near their isoelectric point of pH 4–5 (Grossmann et al., 2019). This high solubility makes algal proteins suitable for use in food formulations at both acidic and neutral pH levels. In acidic pH ranges, marine collagen also shows high solubility. Collagen sourced from fish species such as Bigeye Tuna, Puffer Fish, and Lizardfish (Ahmed et al., 2019; Iswariya et al., 2018; Chen et al., 2021b) is known for its high solubility, which is linked to its lower thermal stability compared to mammalian collagen. With denaturation temperatures generally below 35°C (Ahmed et al., 2019), marine collagen is easy to process and offers a valuable alternative in food industrial applications.

13.2.4.2 *Water holding capacity*

The interactions between protein and water or oil are crucial in food systems as they influence the flavor, texture, and overall quality of the food product. The WHC pertains to protein functioning associated with hydration through protein–water interactions (Yoon et al., 2019). The solubility of fish proteins has a direct impact on their WHC, with an increase in solubility often leading to a decrease in water retention. However, despite this relationship, fish peptides demonstrate excellent WHC, ranging from 2.47 to 6.60 mL/g (Ucak et al., 2021). Marine-derived proteins provide unique advantages over plant-based and terrestrial proteins, particularly in terms of their WHC.

WHC of marine protein can be modified through various modification techniques, such as nanostarch modification, phosphate incorporation, or pH adjustments, to form stronger and more elastic gels with improved water retention, which is a crucial factor in food formulation. For instance, the addition of nanostarch improves the gel characteristics of grass carp myofibrillar protein, resulting in improved WHC and mechanical properties such as breaking force and elasticity (Li et al., 2022). Similarly, the incorporation of sodium tripolyphosphate into myofibrillar protein derived from mantis shrimp has been found to enhance WHC, leading to higher gel strength and elasticity (Chen et al., 2020).

Marine collagen sourced from species like chum salmon (skin) and Nile tilapia (skin) also demonstrates varying WHCs, enabling them to be utilized in a variety of food applications. According to Zheng et al. (2024), collagen derived from chum salmon skin is more suitable for sparkling beverages and cold-water food preservation, while Nile tilapia skin collagen offers better WHC and stability, making it suitable for solid food processing and the production of biofunctional materials. In summary, the significant WHC of marine proteins, along with their versatility and adaptability through various modification techniques, underscores their potential as valuable functional ingredients in food formulations. Their ability to enhance texture, shelf life, and overall product quality makes them a potential alternative to conventional protein sources.

13.2.4.3 *Emulsifying and foaming properties*

As many foods consist of lipidic and aqueous phases, the emulsifying properties of protein are crucial for food formulations. As proteins stabilize emulsions, they can protect lipid compounds from oxidation and enhance the overall nutritional value of the product. However, the effectiveness of protein-based emulsification depends on several factors, including the protein source, concentration, pH, structure, molecular weight, and adsorption behavior at the oil-water interface (Kim et al., 2020).

Proteins also play a significant role in food foams by creating an elastic and dense interfacial film that bridges two phases. Their ability to retain air within the food matrix improves desirable textural attributes such as crispiness, tenderness, foam stability, and airy texture (Geada et al., 2021). Algal proteins possess high emulsification capacity and foaming capacity, making them a promising functional food ingredient in various food formulations such as baked goods, milk and meat alternatives, beverages, soups and sauces, and ice cream and frozen desserts (Geada et al., 2021). Studies have shown that protein isolates derived from microalgae species such as *Arthrospira maxima, Nannochloropsis gaditana,* and *Tisochrysis impellucida* exhibit greater emulsification efficiency than whey protein isolates, achieving a similar minimum droplet size of approximately 300 nm (Bertsch et al., 2021). Furthermore, crude extracts from microalgae produce stable emulsions and foams, outperforming soy flour in terms of emulsifying capacity and foam overrun, further highlighting their functionality in food formulations. The adsorption behavior of microalgae protein isolates at oil-water interfaces is faster compared to egg white albumin and whey protein isolates (Bertsch et al., 2021).

Moreover, lyophilized protein extracts from *P. tricornutum* and *Chlorella sorokiniana* have shown significant interfacial activity at the oil-water interface, leading to the formation of stable emulsions with high salt stabilities (Ebert et al., 2019). In addition to microalgae, fish proteins also serve as natural emulsifiers, forming a stable emulsifying system that improves the texture and stability of meat products such as surimi sausages. The double protein system, composed of whey protein fibers and fish gelatin, significantly improved the emulsifying properties and enhanced the stability of the emulsion (Dang et al., 2021).

13.2.4.4 Antioxidative activity

Oxidation is a major deterioration reaction in food, especially in products with high protein and low-fat contents. It compromises the protein quality, functional properties, and nutritional value. Antioxidants, particularly radical scavenging compounds, have been widely studied and recognized as effective means to control oxidative reactions (Chuang et al., 2023). Antioxidative peptides play a dual role in food formulations by acting as additives to curb oxidation and to control, delay, or prevent oxidative stress-related disease conditions (Nwachukwu and Aluko, 2019).

Antioxidative peptides derived from fish sources have demonstrated a greater ability to prevent oxidation. Studies have reported that protein hydrolysate obtained from smooth hound viscera, cod backbones, unicorn seabass skin, bluefin tuna head, along with bioactive peptides derived from Alaska pollock frame, Yellowfin sole frame, jumbo squid skin, sardine heads and viscera, hoki frame, chum salmon skin, sea rotifer, black pomfret viscera, and squid by-products, possess strong antioxidant activities by inhibiting lipid oxidation and peroxide formation (Elavarasan and Shamasundar, 2022; Ucak et al., 2021). These peptides can contribute to extending the shelf life of food products by preventing oxidative deterioration.

In addition to fish-derived peptides, marine algae are also a promising source of antioxidative proteins and peptides. Proteins in red algae, the second most abundant component after polysaccharides, have gained attention for their potent antioxidant activities. For example, *Eucheuma cottonii*, an edible tropical red macroalga, produces natural antioxidants with significant bioactivities that can be applied in functional foods, as well as to develop biopharmaceutical antioxidant drugs with low toxicity (Sun et al., 2022). Similarly, *Porphyra haitanensis*, another marine red alga species, contains novel antioxidant proteins that could serve as natural food stabilizers, enhancing product stability and shelf life (Pan et al., 2021).

Furthermore, phycobiliproteins, found in cyanobacteria and red algae, have great potential in the food industry as antioxidant ingredients (Imchen and Singh, 2023). Protein concentrates derived from *Ulva* sp. and *Gracilaria* sp. exhibit antioxidant activity comparable to known protein isolates via the hydrogen atom transfer mechanism but are found to be 10–20 times more effective in the single electron transfer mechanism. Polyphenolic compounds present in these algal protein concentrates further contribute to their antioxidant activity. This research suggests that *Ulva* sp. and *Gracilaria* sp. are promising sustainable marine sources for human nutrition, offering benefits such as antioxidant properties, a rich essential amino acid content, and digestibility (Ucak et al., 2021). In summary, marine proteins and peptides, particularly those derived from fish and algae, offer a sustainable and effective means to prevent oxidation in food systems. Their ability to enhance the oxidative stability of food products while providing additional health benefits makes them valuable ingredients in the development of functional foods and nutraceuticals.

13.2.4.5 *Antimicrobial activity*

Proteins and peptides with antimicrobial properties can be used as biopreservatives, aiming to extend shelf life and improve the microbial stability of food systems. Natural marine proteins and peptides are also gaining attention as potential alternatives to synthetic preservatives due to their broad-spectrum activity and minimal adverse effects on human health. Their ability to inhibit spoilage microorganisms and foodborne pathogens makes them valuable ingredients in food formulations aimed at improving safety and quality (Olatunde et al., 2020; Ucak et al., 2021; Dai et al., 2023).

Fish are a rich source of AMPs, including histone-derived peptides, defensins, hepcidins, cathelicidins, and piscidins (Ucak et al., 2021). These compounds, generated through the hydrolysis of fish proteins, produce low molecular weight peptides, many of which possess potent antimicrobial properties. The effectiveness of these peptides depends on their amino acid composition and structural characteristics (Heffernan et al., 2021). The positively charged and amphipathic nature of AMPs enables them to bind to negatively charged bacterial membranes. This interaction can lead to membrane disruption through pore formation or intracellular targeting via macropinosomes, ultimately resulting in cell death. In addition, the hydrophobicity, flexibility, and stability of AMPs enhance their specificity and effectiveness against target microorganisms (Valero et al., 2020).

Several fish species, including *Pleuronectes americanus, Hippoglossoides platessoides, Hippoglossus hippoglossus,* and red sea bream, contain body parts rich in peptides with antimicrobial properties, such as pleurocidins (Ucak et al., 2021). A notable AMP sequence, Lys–Val–Glu–Ile–Val–Ala–Ile–Asn–Asp–Pro–Phe–Ile–Asp–Leu, identified in Protamex hydrolysates of Atlantic mackerel, exhibits antibacterial activity against *Listeria ivanovii* and *Listeria monocytogenes,* two common food spoilage microbes (Offret et al., 2019). Furthermore, peptide fragments produced by simulated gastrointestinal digestion of yellowfin tuna muscle have shown antimicrobial activity against both Gram-negative and Gram-positive bacteria, demonstrating their potential as functional ingredients in food formulation (Cerrato et al., 2020).

In addition to fish-derived AMPs, algae represent another promising source of bioactive compounds. The high protein content in algal protein and their unique amino acid profiles make them a better source for bioactive peptides with antimicrobial activity. Algal peptides are nontoxic and have no side effects upon consumption, offering a natural alternative to synthetic preservatives. Due to their smaller molecular size and structural simplicity, algal peptides exhibit excellent antimicrobial properties compared to larger protein molecules (Surendhiran et al., 2021). For instance, hydrolyzed protein fractions (>10 kDa) extracted from *Saccharina longicruris* have shown antibacterial properties against *S. aureus,* a major foodborne pathogen (Cabral et al., 2021). Similarly, lectin isolated from *Solieria filiformis* has shown antimicrobial activity against *P. aeruginosa, Enterobacter aerogenes, Serratia marcescens, Salmonella typhi, Klebsiella pneumoniae,* and *Proteus* sp. (Surendhiran et al., 2021).

Therefore the antimicrobial potential of marine proteins and peptides from both fish and algae offers an exciting opportunity for natural food preservation. Their ability to inhibit a wide spectrum of spoilage and pathogenic microorganisms makes them valuable additions to food formulations aimed at extending shelf life and improving microbial safety while maintaining a natural and health-conscious approach.

13.2.4.6 *Antifreeze property*

The formation of large ice crystals and recrystallization during food freezing can cause significant damage to food tissues, leading to deterioration in both sensory and nutritional quality. Therefore food products stored at low temperatures or transported within the cold chain are highly dependent on the protective measures of the freezing process (Dang et al., 2021).

Antifreeze proteins and peptides are a unique class of biomolecules that interact specifically with ice crystal planes in a structured manner. Their effectiveness in preventing ice formation and recrystallization depends on factors such as the type of protein involved and the temperature range in which they function. By binding to ice crystal gaps, antifreeze proteins inhibit nucleation and further growth of ice crystals, preventing the aggregation of ice crystals that could otherwise negatively affect the texture and quality of food products. These proteins have been identified in various cold-adapted marine organisms, including fish, microalgae, bacteria, and fungi (Ghalamara et al., 2022).

Incorporating antifreeze proteins into food formulations, even at low concentrations, has been shown to effectively suppress ice recrystallization, thereby enhancing the overall quality and nutritional profile of frozen foods. Studies have proven that antifreeze proteins can significantly reduce the size of ice crystals in frozen meats. Moreover, at higher concentrations, antifreeze proteins can be utilized in products that are not intended for freezing, where they further inhibit ice crystal growth and allow sensitive foods, like strawberries, to be stored at low temperatures with minimal quality degradation (Ghalamara et al., 2022). Antifreeze proteins are increasingly being utilized in low-temperature, ready-to-eat food products and desserts, such as ice cream. Their addition helps maintain the desired texture and flavor profiles by preventing ice crystal formation. Interestingly, no evidence has been reported to suggest that antifreeze proteins derived from fish or other marine sources act as food allergens. Furthermore, current studies indicate that antifreeze proteins are safe for human consumption, with no adverse health effects over both short- and long-term use (Zhou et al., 2024).

13.2.5 Application of marine protein in different food industries

Marine proteins are gaining recognition as valuable ingredients in food formulations due to their diverse functional properties, which play a crucial role in enhancing food quality. Extensive research has focused on extracting proteins from various marine sources and incorporating them into different food products. These studies have demonstrated significant improvements in the functional, sensory, and nutritional attributes of the final products. A summary of the applications of marine proteins across different food industries is presented in Fig. 13.1 and Table 13.4.

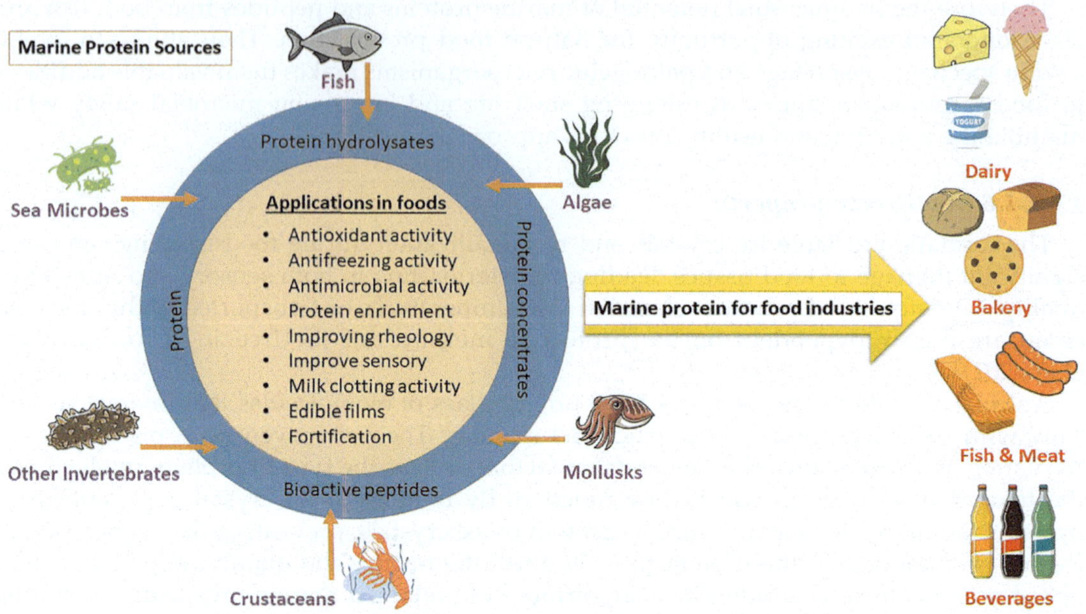

FIG. 13.1 Applications of marine proteins in the food industry.

TABLE 13.4 Studies on the applications of marine proteins in different food industries.

Food product	Marine protein source	Protein type	Application in food matrix	Key characteristics and beneficial outcomes	References
Cheese	*Gracilaria edulis* (microalge)	Protease	Milk clotting	Optimum activity at neutral pH and 50°C–70°C.	(Arbita, 2022)
				Higher activity toward α-, β-, and κ-caseins compared to calf rennet.	
				Most suitable for fresh cheeses, such as feta cheese	
Cheese	*Balistes capriscus* (grey triggerfish)	Gelatin	Edible film	Reduced yeast and mold growth.	Mezhoudi et al. (2022b).
				Decrease in moisture loss and lipid oxidation.	
				Improved cheese texture during storage	
Ice cream	*Spirulina platensis* (microalgae)	Biomass powder	Fortification	Increased protein with higher algal concentration.	(Boyanova et al., 2022)
				High overrun and enhanced melting point	

TABLE 13.4 Studies on the applications of marine proteins in different food industries—cont'd

Food product	Marine protein source	Protein type	Application in food matrix	Key characteristics and beneficial outcomes	References
Yogurt	Fish	Collagen peptides GPLGAAGP, GRDGEP, and MTGTQGEAGR	Fortification	Improved protein content, water holding capacity, and viscosity with a high peptide concentration. Higher antioxidant activity, ACE inhibitory, and DPP-IV	(Ayati et al., 2022).
Yogurt	*Cynoscion guatucupa* (stripped weakfish)	Protein hydrolysates	Fortification	Reduced syneresis and viscoelasticity. Improved antioxidant and antihypertensive activities. Microencapsulation helps in masking the fish flavor	(Lima et al., 2021)
Steamed bread	*Gadus morhua* (Atlantic cod)	Fish skin collagen peptide	Dough stability during freezing and antioxidant activity of bread	Improved yeast activity and reduced water migration in the dough. Improved antioxidant activity Refined bread structure Most suitable concentration is 1.5%	(Li et al., 2024)
Rye-wheat loaves	*Gadus morhua* (Baltic cod)	Fish protein	Fortification	Increased protein by 70.2% Appealing texture and taste	(Klyuchko et al., 2022)
Bread	*Hypophthalmichthys moltrix* (silver carp)	Fish protein powder	Substitute wheat flour	Improved bread stability Delayed staling Substitution at 3% shows the best characteristics	(Shekarabi and Shahbazi, 2022)
Gluten free bread	*Chlorella sorokiniana* (microalgae)	Biomass powder	Fortification	Enhanced protein content by <55% No undesirable effects on texture or volume	(Diprat et al., 2020)
Bread	*Microchloropsis gaditana, Tetraselmis chui, and Chlorella vulgaris* (microalgae)	Algae biomass	Protein enrichment and substitute wheat flour	Dough rheology and baking performance *Microchloropsis gaditana* shows better bread characteristics.	(Qazi et al., 2021)

Continued

TABLE 13.4 Studies on the applications of marine proteins in different food industries—cont'd

Food product	Marine protein source	Protein type	Application in food matrix	Key characteristics and beneficial outcomes	References
				Replacement of wheat flour with 12% microalgae increased the protein content and improved the protein quality of the bread	
Oat-based cookies	Bullseye and pacu fish	Protein isolates	Fortification	High essential amino acid content in 6% pacu protein isolate-enriched cookies. Reduced hardness in protein-enriched cookies	(Baraiya et al., 2024)
Biscuits	Yellowfin tuna *Thunnus albacares*	Protein isolates	Fortification	The biscuits with 2.7% protein isolate had similar physicochemical and sensory properties to the control biscuits without the protein isolate	(Huynh Nguyen Duy Bao and Dung, 2024)
Cookies	Snakehead fish *Channa striata*	Protein concentrate powder	Fortification	Increase protein and fat content. Enhanced appearance and texture	(Ikasari et al., 2020)
Biscuit	Scales/skin of fish	Collagen	Functional additive	Improved antioxidative properties Increased protein content Slightly reduced caloric values	(Kumar et al., 2019)
Cookies	*Piaractus brachypomus* (red-bellied pacu)	Protein hydrolysate	Fortification	Higher protein content Improved protein digestibility Improved antioxidative properties Reduced lipid oxidation	(Hashem et al., 2023)
Ground beef	*Colossoma macropomum* (blackfin pacu)	Protein hydrolysate	Antioxidant	Inhibited lipid oxidation Preserve color and freshness	(De Quadros et al., 2019)
Beef	*Trichiurus haumela* (hairtail surimi)	Antioxidative peptide fractions	Antioxidant	Delayed lipid oxidation Preserve color Improve Umami taste	(Wang et al., 2021)

TABLE 13.4 Studies on the applications of marine proteins in different food industries—cont'd

Food product	Marine protein source	Protein type	Application in food matrix	Key characteristics and beneficial outcomes	References
Tilapia fish	Fish	Collagen peptide	Antifreeze	Stable freshness Improved texture Reduced ice crystal damage	(Liu et al., 2024)
Fish fillets	*Pseudosciaena crocea* (yellow croaker)	Trypsin hydrolysate	Antifreeze	Delayed lipid oxidation Improve texture Preserve freshness	(Xu et al., 2023)
Red porgy fillets	*Prionace glauca* (blue shark)	Collagen	Edible coating	Reduced lipid oxidation Extended shelf life	(Liu et al., 2020)
Smooth-hound shark fillets	*Balistes capriscus* (gray triggerfish)	Gelatin	Edible coating	Higher firmness Extended shelf life Preserve color	(Mezhoudi et al., 2022a)
Fish Fillets	*Ctenopharyngodon idella* (grass carp)	Myofibrillar protein	Biofilm	Inhibited lipid oxidation	(Du et al., 2021)
Apple juice	*Sardina pilchardus* (sardine)	Protein	Antioxidant	Improved antioxidant activity by 80% Improved antihypertensive activity by 65%	(Rivero-Pino et al., 2020)
Kombucha	*Alaria esculenta* (algae)	Biomass	Functional ingredient	Higher protein content Improved antioxidant activity	(Healy et al., 2024)
Functional drink	*Cucumaria frondosa* (sea cucumber)	Protein hydrolysate	Functional ingredient	Improved antioxidant activity Better sensory properties	(Chen et al., 2021a)
Malted grain-based health beverage	*Thunnus albacares* (yellowfin tuna)	Protein hydrolysate	Functional ingredient	Improved sensory properties Improved nutritional properties 2.5% concentration gives best results	(Unnikrishnan et al., 2021)

A, alanine; *ACE*, angiotensin-converting enzyme; *D*, aspartic acid; *DPP-IV*, dipeptidyl peptidase-IV; *E*, glutamic acid; *G*, glycine; *M*, methionine; *P*, proline; *Q*, glutamine; *R*, arginine; *T*, threonine.

13.2.5.1 *Dairy industry*

Cheese products

Milk coagulation is a fundamental step in cheese production, achieved using milk-clotting enzymes. Enzymes derived from marine ecosystems offer unique advantages due to their adaptation to the harsh ocean environment, enabling them to function under a wide range of food processing conditions. Therefore marine sources such as fish, crustaceans, shrimp, sea mammals, algae, and marine microorganisms have been researched as potential alternative milk coagulants. Protease is the most essential enzyme in cheese production, as it plays a crucial role in milk coagulation, curd formation, and the enhancement of texture and flavor development. Marine proteases show functional capabilities across a wide range of pH and temperatures, broadening their applicability in cheese making. Many marine-derived proteases are found to have suitable coagulant activity values. Cheese produced using proteases from marine sources such as tuna, harp seal, and cod has shown chemical and sensory properties comparable to those of calf rennet-based cheese. Cheese produced using protease derived from harp seal exhibited slightly lower moisture content and reduced or no bitterness in the final product. In addition, different protein degradation patterns were observed, leading to the formation of unique polypeptides. However, there was no significant difference in sensory scores. Meanwhile, protease derived from the shrimp *Pleoticus muelleri* enhanced the texture of semi-soft cheeses, improving overall acceptability compared to commercial rennet (Arbita, 2022; Arbita and Zhao, 2024; Zhao et al., 2023).

The protease derived from *Gracilaria edulis*, a macroalgae, showed optimum milk-clotting activity at neutral pH with temperatures ranging between 50°C and 70°C. *G. edulis* proteases can effectively support commercial cheese-making procedures in the presence of calcium chloride and under the optimal enzyme-to-substrate ratios. It has been found that this protease displayed higher activity toward α-, β-, and κ-caseins compared to calf rennet, indicating its potential as a superior milk coagulant in cheese production (Arbita, 2022). High pasteurization temperatures lead to protein denaturation, resulting in the loss of natural water-binding properties in cheese. Marine-derived gelatin extracted from fish by-products, such as skin, bones, tendons, and scales, offers a solution by restoring this lost functionality. The inclusion of 0.5%–0.8% gelatin has been reported to reduce syneresis, enhance the spreading ability of blended cheese products, and provide a smooth texture, thereby improving the overall quality of cheese (Zhao et al., 2023).

Moreover, gelatin has also been shown to improve the viscosity of cottage cheese dressings, ensuring a uniform dispersion of ingredients and reducing fat content, all while maintaining desirable sensory attributes such as texture and flavor (Bose et al., 2023). Furthermore, marine-derived proteins have also been found to be applicable in preserving and improving the quality of cheese. Edible films made from triggerfish gelatin combined with *Moringa oleifera* extract have proven to be effective in reducing microbial spoilage and minimizing water loss in ricotta cheese (Mezhoudi et al., 2022b). Therefore the use of marine-derived proteins and enzymes in cheese production presents promising opportunities for improving the functional, sensory, and nutritional properties of cheese products. Their ability to withstand diverse processing conditions, coupled with their potential to replace traditional coagulants, makes them a highly valuable component in modern cheese manufacturing.

Ice cream products

Maintaining the quality and nutritional balance of ice cream is challenging due to its low protein content and high levels of fat and sugar. To address these challenges, various studies have been carried out to explore the incorporation of marine-derived proteins, focusing on improving both the nutritional composition and physical properties of ice cream. Microalgae proteins can notably contribute to the nutritional and functional properties of ice cream owing to their high content of essential amino acids (Yalcin et al., 2024).

The protein content of some microalgae species, such as *S. platensis*, is higher than that of eggs and soybeans, reaching 510–710 g/kg dry powder, compared to 132 g/kg in eggs and 370 g/kg in soybeans. In addition, these proteins have a high digestibility rate (87.45%–97.81%), making them a valuable ingredient in ice cream formulations (Dineshbabu et al., 2019). Microalgae, either in powder or microencapsulated forms, have been shown to enhance the protein content of ice cream while simultaneously improving its melting and overrun properties. Furthermore, microalgae act as natural stabilizers and emulsifiers, contributing to the overall stability and quality parameters (melting rate, whipping rate, volume, penetration value, overrun) of the product (Yalcin et al., 2024).

However, one of the primary challenges in ice cream production is the formation of ice crystals during the freezing step, which can degrade the original texture, resulting in a rough and undesirable mouthfeel. The addition of antifreezing peptides derived from marine proteins offers a promising solution to this issue. These peptides effectively inhibit ice crystal formation, preserving the smooth and creamy texture of ice cream even after thawing (Zhou et al., 2024). Therefore the incorporation of marine-derived proteins, particularly from microalgae, presents a viable approach to improve the nutritional profile and physical properties, enabling their use as valuable ingredients in the development of high-quality, functional ice cream formulations.

Yogurt products

Marine proteins offer significant potential in yogurt fortification, enhancing its functional and health-promoting properties without compromising its sensory and physicochemical characteristics. Recent studies have explored the incorporation of bioactive peptides derived from fish collagen, such as GPLGAAGP, GRDGEP, and MTGTQGEAGR, into yogurt formulations and have demonstrated significant functional benefits in concentrations ranging from 0.2 to 1.0 mg/mL. Fortification with these peptides has been shown to enhance antioxidant capacity, angiotensin-converting enzyme inhibitory activity, and dipeptidyl peptidase-IV inhibitory activity, contributing to potential health benefits. Notably, the addition of these peptides did not result in any detectable changes to the physicochemical or sensory properties of yogurt, ensuring consumer acceptability (Ayati et al., 2022).

In another study, fish protein hydrolysates derived from stripped weakfish (*Cynoscion guatucupa*) were incorporated into yogurt to further enhance its functional profile. These hydrolysates were produced using the enzyme Protamex and microencapsulated with maltodextrin to improve stability and bioavailability. Both free and microencapsulated hydrolysates improved the antioxidant and angiotensin-converting enzyme inhibitory activities of yogurt. Moreover, both forms of hydrolysates were effective in reducing syneresis, thereby helping to maintain the desirable gel-like structure of yogurt.

Notably, microencapsulation, in particular, enhanced rheological stability and reduced undesirable viscoelastic changes. The storage modulus (G') and loss modulus (G'') indicated a gel-like behavior in yogurt, with increased gel strength over the storage period (Lima et al., 2021).

Fortified yogurts containing marine proteins have thus demonstrated potential antihypertensive and antioxidant effects, highlighting their value as functional foods. However, to fully validate these health benefits and assess the bioavailability of the incorporated marine compounds, further in vivo research is required. Expanding this research could provide more insights into how marine protein fortification can be optimized for broader consumer markets.

13.2.5.2 *Bakery and confectionery*

Bread products

Marine-derived proteins have gained attention in the bakery industry due to their potential to enhance the nutritional and functional properties of bread. Among these, fish skin collagen peptides, as bioactive components, have been successfully utilized to improve the quality of frozen dough and steamed bread. It was found that the addition of fish skin collagen peptides reduced freezable water content and inhibited water migration, which enhances dough stability during freezing. In addition, these peptides mitigate freeze damage to gluten, preserving disulfide bond content and dough viscoelasticity. Furthermore, steamed bread enriched with fish skin collagen peptides, particularly at a concentration of 1.5%, demonstrated improved texture and higher antioxidant activity, highlighting the potential of fish skin collagen peptides in frozen dough applications and the development of functional bakery products (Li et al., 2024).

Moreover, the enrichment of rye-wheat loaves with a protein-mineral additive derived from Baltic cod (*Gadus morhua*) has shown significant nutritional and sensory improvements. The enriched bread exhibited a 47% increase in protein content and a 96.3% rise in mineral content, with the BV of the protein reaching 124%. The enriched bread maintained an appealing texture and taste, with a slight fishy aroma (Klyuchko et al., 2022). In another study, fish protein powder obtained from silver carp (*Hypophthalmichthys molitrix*) has been evaluated as a partial substitute for wheat flour in bread formulations. A substitution level of 3% yielded optimum results, achieving the highest development and stability times, improved farinograph quality, and delayed staling after 48 h without adversely affecting sensory characteristics. This suggests that fish protein powder is a viable alternative for partial wheat flour replacement in bread formulations (Shekarabi and Shahbazi, 2022).

In addition to fish-based proteins, microalgae proteins have been extensively studied for their fortification potential in bread production. The incorporation of microalgae proteins has been shown to significantly increase the protein content of wheat bread while strengthening the dough and enhancing bread volume, texture, and color (Diprat et al., 2020; Qazi et al., 2021, 2022). These improvements demonstrate the potential of microalgae proteins as sustainable and functional ingredients for the bakery industry. Moreover, all these advancements provide opportunities for developing healthier, protein-enriched bread while maintaining desirable sensory and structural attributes.

Biscuits and cookie products

Biscuits and cookies are widely consumed bakery products; however, they are typically low in protein and essential amino acids. Incorporating marine protein isolates, concentrates, and peptides into these products has demonstrated potential to enhance their nutritional value while preserving or improving their physicochemical and sensory properties.

Studies have demonstrated that marine protein isolates derived from bullseye and pacu fish processing waste can be incorporated into oat-based cookies at varying levels of up to 10%. Among these, cookies supplemented with 6% pacu protein isolates exhibited the highest total essential amino acid content, whereas those enriched with 4% bullseye protein isolates displayed higher levels of nonessential amino acids. Furthermore, these fortified cookies also displayed reduced hardness compared to control samples, highlighting their improved texture and nutritional quality (Baraiya et al., 2024). Similarly, yellowfin tuna fish protein isolates were incorporated into biscuits. No significant differences were observed in any sensory attributes at a 2.7% concentration, while only color and aroma were affected at higher concentrations (Huynh Nguyen Duy Bao and Dung, 2024).

Fish protein concentrate powder from snakehead fish (*Channa striata*) has also been explored for use in cookie formulations at 5% and 10% concentrations. The incorporation resulted in increased protein and fat content while enhancing the appearance and texture, resulting in soft yet crispy cookies that improved consumer acceptability. Notably, cookies with a 10% fish protein concentrate exhibited a protein content of 11.41% (Ikasari et al., 2020). Furthermore, marine collagen peptides were tested as functional additives in biscuit formulations at a concentration of up to 10%. The inclusion improved the antioxidative properties, increased the protein content, and slightly reduced the caloric values of the produced biscuits without significantly altering sensory attributes (Kumar et al., 2019). In addition, red-bellied pacu fish meat protein hydrolysate prepared through probiotic *Bacillus* fermentation, was incorporated into cookies. These protein hydrolysate-enriched cookies displayed higher protein content ranging from 11.76% to 14.40%, improved protein digestibility between 65.21% and 96.0%, and enhanced antioxidant properties compared to control cookies. Furthermore, the incorporation of fish meat protein hydrolysates contributed to reduced lipid oxidation during storage, thereby improving the shelf life of the product (Hashem et al., 2023). In conclusion, the use of marine-derived proteins in biscuits and cookies offers significant benefits in terms of nutritional enrichment, improved texture, and extended shelf life. These proteins provide a sustainable and functional alternative for developing healthier bakery products while maintaining desirable taste and sensory attributes.

13.2.5.3 *Meat and fish products*

Meat products

Marine-derived proteins and peptides have demonstrated significant potential in improving the quality, shelf life, and safety of fish and meat products. These applications focus on their antioxidant, cryoprotective, and antimicrobial properties, making them suitable for natural food preservation. Marine protein hydrolysates and peptides demonstrate significant antioxidant activity by scavenging free radicals, chelating metal ions, and acting as reducing agents, thus preventing lipid oxidation in food products. For instance, protein hydrolysate (Wang et al., 2021) obtained from blackfin pacu (*Colossoma macropomum*) has been applied to bovine ground beef, where it effectively inhibited lipid oxidation by 60.9% over a 7-day period. This treatment helped maintain the natural color and freshness of the product, proving the potential of marine-derived proteins as natural antioxidants in refrigerated meats (De Quadros et al., 2019). Similarly, antioxidative peptide fractions obtained from hairtail surimi hydrolysates delayed lipid oxidation and preserved beef color, making them effective for meat preservation. This is also a cultural advantage in East Asia, where consumers believe combining fish and meat flavors enhances the "umami" taste (Wang et al., 2021).

In addition to their antioxidative benefits, marine-derived antifreeze peptides play a crucial role in frozen meat storage. These peptides effectively prevent the formation of ice crystals, which helps preserve the texture, flavor, and nutritional value of meat products. Fish-derived antifreeze peptides have been widely used in frozen meat storage, where they inhibit ice crystal formation, maintain water-retention capacity, and preserve the nutritional stability of meat products during freezing and thawing. This functionality extends the shelf life of frozen meats and ensures retention of the product's original taste and nutritional value. Moreover, tilapia skin-derived antifreeze peptides have been found to offer protective effects on probiotic bacteria during freezing. They stabilize cell membranes and reduce enzyme activity loss, thereby maintaining the viability of probiotics in frozen products and improving the overall quality of functional meat-based formulations (Zeng et al., 2022). Therefore the integration of marine-derived proteins and peptides into meat products presents a promising approach for enhancing product stability, extending shelf life, and improving sensory characteristics. Their multifunctional properties, ranging from antioxidative to antifreezing effects, make them valuable natural alternatives for meat preservation in the food industry.

Fish products

Marine-derived proteins and peptides have shown great potential in improving the quality, shelf life, and safety of fish products through their antifreezing, antioxidant, and antimicrobial properties (Xu et al., 2023). Fish collagen peptides have been glycosylated with glucosamine hydrochloride, catalyzed by transglutaminase, to produce a transglutaminase-catalyzed glycosylation product, which has been investigated for its antifreezing effects on tilapia. The treated tilapia exhibited stable freshness with improved textural and rheological properties. Notably, hardness and elasticity, critical attributes of fish muscle texture, were significantly improved, reflecting reduced ice crystal damage and improved structural integrity. Therefore these findings suggest that glycosylated fish collagen peptide products can be used as new and effective green antifreeze agents in frozen foods (Liu et al., 2024). Furthermore, the cryoprotective effects of antifreeze peptides obtained from tilapia scales were evaluated on *Streptococcus thermophilus* during cold stress. The results indicated that these antifreeze peptides from tilapia scales effectively protected probiotic bacteria during freezing, suggesting a broader application in functional foods by reducing the damage caused by frozen conditions and temperature fluctuations (Chen et al., 2019).

In addition to their cryoprotective functions, marine peptides offer antioxidant properties that play a crucial role in preserving fish products. The trypsin hydrolysate of yellow croaker (*Pseudosciaena crocea*) protein exhibited significant antioxidant potential for protecting fish fillets from oxidative damage during freezing and storage. The application of these fish peptides as a treatment for fish fillets resulted in delayed oxidation and aggregation of myofibrillar proteins, which are vital for maintaining the structural integrity and texture of fish meat. Moreover, they maintained critical sensory and functional properties, including color, WHC, and texture, ensuring the fish fillets retained their freshness and appeal throughout storage (Xu et al., 2023).

Marine-based gelatin and collagen have also been successfully incorporated into edible coatings to enhance fish preservation. Chitosan-collagen composite coatings, produced from pepsin-soluble collagen extracted from blue shark (*Prionace glauca*) skin, reduced lipid oxidation in red porgy fillets stored at 4°C, extending shelf life while maintaining thiobarbituric acid values below critical thresholds. The presence of high collagen content in the coating suppressed lipid oxidation by forming a protective film that prevented oxygen penetration into the fillet (Liu et al., 2020). Similarly, an edible coating produced using gelatin extracted from the skin of gray triggerfish (*Balistes capriscus*) and enriched with *Moringa oleifera* leaf extract delayed the deterioration of smooth-hound shark (*Mustelus mustelus*) fillets during chilled storage. The coating helped preserve the color of the fillet surface and reduced undesirable texture changes, ensuring better firmness and consistency throughout storage (Mezhoudi et al., 2022a). In addition, biofilms combining fish myofibrillar proteins, chitosan, and rosemary extract have been developed to inhibit lipid oxidation and monitor freshness in fish fillets. These biofilms proved their potential as edible food packaging solutions for transportation by effectively reducing lipid oxidation and maintaining the quality of the fish fillet (Du et al., 2021). Another edible coating enriched with garden cress extract also enhanced the preservation of smooth-hound shark fillets during refrigeration, highlighting the role of marine-derived bioactive compounds in food packaging innovations (Salem et al., 2024).

Marine-derived peptides also possess remarkable antimicrobial properties, which are important for fish preservation due to the high susceptibility of fish products to microbial spoilage. Hemolymph from *Liagore rubromaculata* crabs exhibited potent antibacterial activity against foodborne pathogens like *S. typhi* and *Vibrio cholerae* (Olatunde et al., 2020). Similarly, low molecular weight peptides derived from salt-fermented shrimp (*Penaeus vannamei*) effectively inhibited the growth of *Vibrio parahaemolyticus* and *Escherichia coli,* highlighting their potential as natural antimicrobial agents for food preservation (Dai et al., 2023).

13.2.5.4 Beverage products

Marine-derived proteins and peptides have demonstrated significant potential in the beverage industry by offering functional, nutritional, and health-promoting benefits. These bioactive compounds have been widely studied for their ability to enhance antioxidant capacity, improve protein content, and provide therapeutic properties to various beverage formulations. One promising source is sardine (*Sardina pilchardus*) protein, which has been identified as a rich source of antioxidant and antihypertensive peptides. Incorporating these biopeptides into fish soup and apple juice, as sugar-free and high-sugar matrices, can enhance their bioactivity with improved antioxidant and antihypertensive activities by up to 80% and 65%, respectively, after sterilization in apple juice. The bioactivity of these peptides was preserved or even improved after 100 days of storage. Therefore the findings of this study underscore the potential of fish protein hydrolysates as nutraceuticals, emphasizing the importance of considering the composition of food matrices when designing functional beverage formulations (Rivero-Pino et al., 2020). In another study, *Alaria esculenta* was fermented using *Lactiplantibacillus plantarum* and kombucha symbiotic culture of bacteria and yeast (SCOBY). Results revealed that the resulting seaweed kombucha exhibited a higher protein content and enhanced antioxidant activity compared to traditional tea kombucha, highlighting the potential of marine algae in creating functional beverages with improved nutritional profiles (Healy et al., 2024).

Sea cucumber (*Cucumaria frondosa*) protein hydrolysates have also been explored for beverage applications, demonstrating enhanced antioxidant properties. These sea cucumber peptides have been integrated into functional beverages without compromising sensory or functional qualities. The study findings suggest that sea cucumber peptides could be utilized in the development of novel antiaging drinks, offering potential health benefits with a focus on longevity and wellness (Chen et al., 2021a).

Furthermore, bioactive peptides derived from yellowfin tuna red meat were used to formulate a malted grain-based health beverage. However, the bitter taste of peptides can compromise the quality when incorporated into beverages. Therefore bitterness and overall acceptability were evaluated. The acceptability of the beverage depended on the fish flavor and associated bitterness, which were influenced by the concentration of protein hydrolysate. At a concentration of 2.5%, tuna protein hydrolysate exhibited the most desirable sensory properties, with no detectable bitterness. A comprehensive analysis of the beverage's nutritional, functional, and antioxidant properties, along with its in vitro digestibility and storage stability under ambient conditions, further demonstrated its potential to meet consumer demands for health-promoting beverages (Unnikrishnan et al., 2021).

In summary, the incorporation of marine-derived proteins and peptides into beverage formulations presents an exciting opportunity to develop functional drinks with enhanced nutritional value and health benefits. Whether sourced from fish, seaweed, or sea cucumber, these bioactive compounds have shown great promise in improving beverage stability, antioxidant capacity, and overall consumer appeal, making them viable options for the expanding functional beverage market.

13.3 Conclusions and future directions

Marine-derived fatty acids and proteins play a crucial role in enhancing the nutritional and functional quality of food products. Omega-3 fatty acids, particularly EPA and DHA, contribute to cardiovascular health, antiinflammatory responses, and neurological well-being. Marine proteins, including hydrolysates and bioactive peptides, exhibit bioactive properties such as antioxidation, antimicrobial activity, and antihypertensive effects, making them valuable for both food and pharmaceutical applications. The utilization of marine lipids and proteins in dairy, bakery, and meat products presents opportunities to improve the health benefits of these foods while enhancing their sensory and stability attributes.

Despite their advantages, several challenges hinder the widespread application of marine-derived ingredients. These include the lipid and/or protein yields, the susceptibility of marine lipids to oxidation, off-flavors in food formulations, and environmental concerns related to overfishing and marine biodiversity. Technological advancements, such as encapsulation, nanoemulsions, and alternative marine sources, offer potential solutions to these limitations.

However, the future of marine-derived bioactives depends on sustainable sourcing. Advances in genetic engineering and biotechnology could enhance lipid and protein yields, making marine ingredients more viable for the food industry. As mentioned earlier, encapsulation techniques such as spray drying and liposomal delivery, along with natural antioxidants, are crucial for improving the stability and shelf life of omega-3 fatty acids. Sensory challenges in marine-based foods can be addressed through masking technologies and hybrid formulations that blend marine proteins with plant-based lipids. Expanding functional food applications, particularly in nutraceuticals, could aid in preventing CVDs and metabolic disorders. Sustainable processing technologies should focus on eco-friendly extraction and the utilization of seafood by-products. Further clinical and nutritional research is needed to establish dietary guidelines for omega-3 intake and proteins, ensuring sustainable food innovation while maximizing health benefits.

References

WHO Expert Consultation on Rabies, 2005.

Ahmed, R., Haq, M., Chun, B.S., 2019. Characterization of marine derived collagen extracted from the by-products of bigeye tuna (*Thunnus obesus*). Int. J. Biol. Macromol. 18790003. 135, 668–676. Available from: http://www.elsevier.com/locate/ijbiomac. doi:10.1016/j.ijbiomac.2019.05.213.

Ahmmed, M.K., Ahmmed, F., Tian, H., Carne, A., Bekhit, A.E.D., 2020. Marine omega-3 (n-3) phospholipids: a comprehensive review of their properties, sources, bioavailability, and relation to brain health. Compr. Rev. Food Sci. Food Saf. 15414337. 19 (1), 64–123. Available from: http://www.onlinelibrary.wiley.com/journal/10.1111/(ISSN)1541-4337. doi:10.1111/1541-4337.12510.

Ahmmed, M.K., Carne, A., Haizhou, W., Bekhit, A.E.-D.A., 2023. Navigating the depths of marine lipids: from extraction efficiency to flavour enhancement. Food Chem. X. 25901575. 20, 100958. doi:10.1016/j.fochx.2023.100958.

Al-Soufi, S., García, J., Nicodemus, N., Lorenzo, J.M., Cegarra, E., Muíños, A., Losada, A.P., Miranda, M., López-Alonso, M., 2024. Marine macroalgae in rabbit feed—effects on meat quality. Meat Sci. 03091740. 216. Available from: https://www.sciencedirect.com/science/journal/03091740. doi:10.1016/j.meatsci.2024.109584.

Alam, M.R., Shahid, M.A., Alimuzzaman, S., Khan, A.N., 2022. Sources, extractions and applications of bio-maker collagen—a review. Biomed. Eng. Adv. 26670992. 4. Available from: https://www.journals.elsevier.com/biomedical-engineering-advances. doi:10.1016/j.bea.2022.100064.

Alencar, P.O.C., Lima, G.C., Barros, F.C.N., Costa, L.E.C., Ribeiro, C.V.P.E., Sousa, W.M., Sombra, V.G., Abreu, C.M.W.S., Abreu, E.S., Pontes, E.O.B., Oliveira, A.C., de Paula, R.C.M., Freitas, A.L.P., 2019. A novel antioxidant sulfated polysaccharide from the algae *Gracilaria caudata*: in vitro and in vivo activities. Food Hydrocoll. 0268005X. 90, 28–34. doi:10.1016/j.foodhyd.2018.12.007.

Alfio, V.G., Manzo, C., Micillo, R., 2021. From fish waste to value: an overview of the sustainable recovery of omega-3 for food supplements. Molecules. 1420-3049. 26 (4), 1002. doi:10.3390/molecules26041002.

Amini, M.R., Naserian, A.A., Valizadeh, R., Dirandeh, E., Baghshahi, H., Razavi, S.A., 2024. The effects of saturated and polyunsaturated fatty acids on reproductive performance and reproductive hormonal changes in dairy cows at the transition period. World's Vet. J. 23224568. 14 (2), 184–193. Available from: https://www.science-line.com/. doi:10.54203/scil.2024.wvj23.

Andraka, J.M., Sharma, N., Marchalant, Y., 2020. Can krill oil be of use for counteracting neuroinflammatory processes induced by high fat diet and aging? Neurosci. Res. 18728111. 157, 1–14. Available from: http://www.elsevier.com/locate/neures. doi:10.1016/j.neures.2019.08.001.

Araujo, P., Truzzi, C., Belghit, I., Antonucci, M., 2021. The impact of seawater warming on fatty acid composition and nutritional quality indices of *Trematomus bernacchii* from the Antarctic region. Food Chem. 03088146. 365, 130500. doi:10.1016/j.foodchem.2021.130500.

Arbita, A., 2022. Biochemical, Milk-Clotting and Cheese Making Properties of Proteases from the Marine Macroalga *Gracilaria Edulis* (Ph.d. Thesis). University of New South Wales. doi:10.26190/unsworks/2003.

Arbita, A.A., Zhao, J., 2024. Milk clotting enzymes from marine resources and their role in cheese-making: a mini review. Crit. Rev. Food Sci. Nutr. 15497852. 64 (27), 10036–10047. Available from: https://www.tandfonline.com/loi/bfsn20. doi:10.1080/10408398.2023.2220030.

Ayati, S., Eun, J.B., Atoub, N., Mirzapour-Kouhdasht, A., 2022. Functional yogurt fortified with fish collagen-derived bioactive peptides: antioxidant capacity, ACE and DPP-IV inhibitory. J. Food Process. Preserv. 17454549. 46 (1). Available from: http://www.onlinelibrary.wiley.com/journal/10.1111/(ISSN)1745-4549. doi:10.1111/jfpp.16208.

Azemin, W.A., Rashid, S.A., Philip, N., Ali, A.M., Shamsir, M.S., 2025. Fish-derived hepcidins in cancer treatment: a scoping review. Beni-Suef Univ. J. Basic Appl. Sci. 23148543. 14 (1). Available from: https://www.springer.com/journal/43088. doi:10.1186/s43088-025-00599-w.

Bafkar, N., Zeraattalab-Motlagh, S., Jayedi, A., Shab-Bidar, S., 2024. Efficacy and safety of omega-3 fatty acids supplementation for anxiety symptoms: a systematic review and dose-response meta-analysis of randomized controlled trials. BMC Psychiatry. 1471-244X. 24 (1). doi:10.1186/s12888-024-05881-2.

Balikci, E., 2024. Influence of seasons and fish body parts on fatty acid profile and effect of seasons on proximate composition of *Anatolian khramulya* (*Capoeta tinca*) and *Colchic khramulya* (*Capoeta sieboldii*) captured from the Çekerek Dam in Yozgat, Turkey. J. Food Compos. Anal. 08891575. 132, 106267. doi:10.1016/j.jfca.2024.106267.

Banaszak, M., Dobrzyńska, M., Kawka, A., Górna, I., Woźniak, D., Przysławski, J., Drzymała-Czyż, S., 2024. Role of omega-3 fatty acids eicosapentaenoic (EPA) and docosahexaenoic (DHA) as modulatory and anti-inflammatory agents in noncommunicable diet-related diseases—reports from the last 10 years. Clin. Nutr. ESPEN. 24054577. 63, 240–258. doi:10.1016/j.clnesp.2024.06.053.

Banday, A.H., Azha, N.U., Farooq, R., Sheikh, S.A., Ganie, M.A., Parray, M.N., Mushtaq, H., Hameed, I., Lone, M.A., 2024. Exploring the potential of marine natural products in drug development: a comprehensive review. Phytochem. Lett. 18767486. 59, 124–135. Available from: https://www.sciencedirect.com/science/journal/18743900. doi:10.1016/j.phytol.2024.01.001.

Baraiya, K.G., Bojayanaik, M., Surasani, V.K.R., Chavan, S.D., Chavan, M.R., Lende, S.R., Alandur, V.S., 2024. Utilization of fish protein isolates to supplement oat-based cookies and assessment of end product quality. Environ. Sci. Pollut. Res. 16147499. 31 (53), 62201–62212. Available from: https://www.springer.com/journal/11356. doi:10.1007/s11356-023-27804-6.

Barta, D.G., Coman, V., Vodnar, D.C., 2021. Microalgae as sources of omega-3 polyunsaturated fatty acids: biotechnological aspects. Algal Res. 22119264. 58. Available from: http://www.sciencedirect.com/science/journal/aip/22119264. doi:10.1016/j.algal.2021.102410.

Bartek, L., Strid, I., Henryson, K., Junne, S., Rasi, S., Eriksson, M., 2021. Life cycle assessment of fish oil substitute produced by microalgae using food waste. Sustain. Prod. Consum. 23525509. 27, 2002–2021. doi:10.1016/j.spc.2021.04.033.

Barzkar, N., Sukhikh, S., Babich, O., 2024. Study of marine microorganism metabolites: new resources for bioactive natural products. Front. Microbiol. 1664-302X. 14. doi:10.3389/fmicb.2023.1285902.

Bertsch, P., Böcker, L., Mathys, A., Fischer, P., 2021. Proteins from microalgae for the stabilization of fluid interfaces, emulsions, and foams. Trends Food Sci. Technol. 09242244. 108, 326–342. doi:10.1016/j.tifs.2020.12.014.

Biton-Porsmoguer, S., Bou, R., Lloret, E., Alcaide, M., Lloret, J., Cooke, S., 2020. Fatty acid composition and parasitism of European sardine (*Sardina pilchardus*) and anchovy (*Engraulis encrasicolus*) populations in the northern Catalan Sea in the context of changing environmental conditions. Conserv. Physiol. 2051-1434. 8 (1). doi:10.1093/conphys/coaa121.

Borisenko, I., Daugavet, M., Ereskovsky, A., Lavrov, A., Podgornaya, O., 2022. Novel protein from larval sponge cells, ilborin, is related to energy turnover and calcium binding and is conserved among marine invertebrates. Open Biol. 2046-2441. 12 (2). doi:10.1098/rsob.210336.

Bose, I., Nousheen, S.R., Yaduvanshi, P., Sharma, S., Chandel, V., Biswas, D., 2023. Unveiling the potential of marine biopolymers: sources, classification, and diverse food applications. Materials. 1996-1944. 16 (13), 4840. doi:10.3390/ma16134840.

Boyanova, P., Gradinarska, D., Dobreva, V., Panayotov, P., Momchilova, M., Gabor Zsivanovits, P.M., Ivanov, G., Kostov, G., Dragoev, S., 2022. Effect of *Spirulina platensis* on the quality and antioxidants characteristics of ice cream. BIO Web Conf. 2117-4458. 45, 01009. doi:10.1051/bioconf/20224501009.

Cabral, E.M., Oliveira, M., Mondala, J.R.M., Curtin, J., Tiwari, B.K., Garcia-Vaquero, M., 2021. Antimicrobials from seaweeds for food applications. Mar. Drugs. 16603397. 19 (4). Available from: http://www.mdpi.com/journal/marinedrugs. doi:10.3390/md19040211.

Cerrato, A., Capriotti, A.L., Capuano, F., Cavaliere, C., Montone, A.M.I., Montone, C.M., Piovesana, S., Chiozzi, R.Z., Laganà, A., 2020. Identification and antimicrobial activity of medium-sized and short peptides from yellowfin tuna (*Thunnus albacares*) simulated gastrointestinal digestion. Foods. 23048158. 9 (9). Available from: https://www.mdpi.com/2304-8158/9/9/1185. doi:10.3390/foods9091185.

Chamorro, R., Bascuñán, K.A., Barrera, C., Sandoval, J., Puigrredon, C., Valenzuela, R., 2022. Reduced n-3 and n-6 PUFA (DHA and AA) concentrations in breast milk and erythrocytes phospholipids during pregnancy and lactation in women with obesity. Int. J. Environ. Res. Public Health. 1660-4601. 19 (4), 1930. doi:10.3390/ijerph 19041930.

Chen, F., Lin, L., Zhao, M., Zhu, Q., 2021a. Modification of *Cucumaria frondosa* hydrolysate through maillard reaction for sea cucumber peptide based-beverage. LWT. 00236438. 136, 110329. doi:10.1016/j.lwt.2020.110329.

Chen, J., Liu, H., 2020. Nutritional indices for assessing fatty acids: a mini-review. Int. J. Mol. Sci. 1422-0067. 21 (16), 5695. doi:10.3390/ijms21165695.

Chen, J., Ren, Y., Zhang, K., Xiong, Y.L., Wang, Q., Shang, K., Zhang, D., 2020. Site-specific incorporation of sodium tripolyphosphate into myofibrillar protein from mantis shrimp (*Oratosquilla oratoria*) promotes protein crosslinking and gel network formation. Food Chem. 03088146. 312, 126113. doi:10.1016/j.foodchem.2019.126113.

Chen, J., Wang, G., Yushuang, L., 2021b. Preparation and characterization of thermally stable collagens from the scales of lizardfish (*Synodus macrops*). Mar. Drugs. 1660-3397. 19 (11), 597. doi:10.3390/md19110597.

Chen, Y.Y., Fang, L.S., Chen, Y.H., Peng, B.R., Su, T.P., Huynh, T.H., Lin, F.Y., Hu, C.C., Lin, N.C., Wen, Z.H., Chen, J. J., Lee, C.Y., Wang, J.W., Sung, P.J., 2019. New 8-hydroxybriaranes from the Gorgonian coral *Junceella fragilis* (Ellisellidae). Mar. Drugs. 16603397. 17 (9). Available from: https://www.mdpi.com/1660-3397/17/9/534/pdf. doi:10.3390/md17090534.

Chuang, P., Tao, F., Qi, B., Xiang, H., Chen, S., Yang, X., 2023. A novel antioxidant protein of non-phycobiliprotein family derived from marine red alga *Porphyra haitanensis*. J. Sci. Food Agric. 0022-5142. 103 (11), 5277–5287. doi:10.1002/jsfa.12601.

Cittadini, A., Domínguez, R., Munekata, P.E.S., Pateiro, M., Sarriés, M.V., Lorenzo, J.M., 2022. Use of oil mixture emulsion hydrogels as partial animal fat replacers in dry-fermented foal sausages. Food Res. Int. 18737145. 161. Available from: http://www.elsevier.com/inca/publications/store/4/2/2/9/7/0. doi:10.1016/j.foodres.2022. 111881.

Colletti, A., Cravotto, G., Citi, V., Martelli, A., Testai, L., Cicero, A.F.G., 2021. Advances in technologies for highly active omega-3 fatty acids from krill oil: Clinical applications. Mar. Drugs 19 (6), 306. Available from: http://www.mdpi.com/journal/marinedrugs. doi:10.3390/md19060306.

Colombo, S.M., Mazal, X., 2020. Investigation of the nutritional composition of different types of salmon available to Canadian consumers. J. Agric. Food Res. 26661543. 2, 100056. doi:10.1016/j.jafr.2020.100056.

Cretton, M., Malanga, G., Sobczuk, T.M., Mazzuca, M., 2023. Marine lipids as a source of high-quality fatty acids and antioxidants. Food Rev. Int. 8755-9129. 39 (8), 4941–4964. doi:10.1080/87559129.2022.2042555.

Cunha, S.A., Pintado, M.E., 2022. Bioactive peptides derived from marine sources: biological and functional properties. Trends Food Sci. Technol. 09242244. 119, 348–370. Available from: http://www.elsevier.com/wps/find/journaldescription.cws_home/601278/description#description. doi:10.1016/j.tifs.2021.08.017.

Dai, J., Jin, R., Gao, J., Aweya, J.J., Lin, R., Guiling, L., Yang, S., 2023. Antibacterial activity and mechanism of peptide PV-Q5 against *Vibrio parahaemolyticus* and *Escherichia coli*, derived from salt-fermented *Penaeus vannamei*. Foods. 2304-8158. 12 (9), 1804. doi:10.3390/foods12091804.

Dang, D.S., Bastarrachea, L.J., Martini, S., Matarneh, S.K., 2021. Crystallization behavior and quality of frozen meat. Foods. 23048158. 10 (11). Available from: https://www.mdpi.com/2304-8158/10/11/2707/pdf. doi:10.3390/foods10112707.

de Kluijver, A., Nierop, K.G.J., Morganti, T.M., Bart, M.C., Slaby, B.M., Hanz, U., de Goeij, J.M., Mienis, F., Middelburg, J.J., 2021. Bacterial precursors and unsaturated long-chain fatty acids are biomarkers of North-

Atlantic deep-sea demosponges. PLoS One. 19326203. 16 (1). Available from: https://www.journals.plos.org/plosone/article/file?id=10.1371/journal.pone.0241095&type=printable. doi:10.1371/journal.pone.0241095.

De Quadros, C.D.C., Lima, K.O., Bueno, C.H.L., Fogaça, F.H.D.S., Rocha, M.D., Prentice, C., 2019. Evaluation of the antioxidant and antimicrobial activity of protein hydrolysates and peptide fractions derived from *Colossoma macropomum* and their effect on ground beef lipid oxidation. J. Aquat. Food Prod. Technol. 15470636. 28 (6), 677–688. Available from: http://www.tandfonline.com/toc/wafp20/current. doi:10.1080/10498850.2019.1628152.

Dempsey, M., Rockwell, M.S., Wentz, L.M., 2023. The influence of dietary and supplemental omega-3 fatty acids on the omega-3 index: a scoping review. Front. Nutr. 2296-861X. 10. doi:10.3389/fnut.2023.1072653.

Dhiman, A., Chopra, R., Singh, P.K., Homroy, S., Chand, M., Talwar, B., 2024. Amelioration of nutritional properties of bakery fat using omega-3 fatty acid-rich edible oils: a review. J. Sci. Food Agric. 10970010. 104 (6), 3175–3184. Available from: http://www.onlinelibrary.wiley.com/journal/10.1002/(ISSN)1097-0010. doi:10.1002/jsfa.13225.

Dineshbabu, G., Goswami, G., Kumar, R., Sinha, A., Das, D., 2019. Microalgae–nutritious, sustainable aqua- and animal feed source. J. Funct. Foods 62. doi:10.1016/J.JFF.2019.103545.

Diprat, A.B., Thys, R.C.S., Rodrigues, E., Rech, R., 2020. *Chlorella sorokiniana*: a new alternative source of carotenoids and proteins for gluten-free bread. LWT. 00236438. 134. Available from: https://www.journals.elsevier.com/lwt. doi:10.1016/j.lwt.2020.109974.

Djuricic, I., Calder, P.C., 2021. Beneficial outcomes of omega-6 and omega-3 polyunsaturated fatty acids on human health: an update for 2021. Nutrients. 2072-6643. 13 (7), 2421. doi:10.3390/nu13072421.

Domingues, V.F., Quaresma, M., Sousa, S., Rosas, M., Ventoso, B., Nunes, M.L., Delerue-Matos, C., 2021. Evaluating the lipid quality of yellowfin tuna (*Thunnus albacares*) harvested from different oceans by their fatty acid signatures. Foods. 23048158. 10 (11). Available from: https://www.mdpi.com/2304-8158/10/11/2816/pdf. doi:10.3390/foods10112816.

Donat-Vargas, C., Bellavia, A., Berglund, M., Glynn, A., Wolk, A., Åkesson, A., 2020. Cardiovascular and cancer mortality in relation to dietary polychlorinated biphenyls and marine polyunsaturated fatty acids: a nutritional-toxicological aspect of fish consumption. J. Intern. Med. 0954-6820. 287 (2), 197–209. doi:10.1111/joim.12995.

Du, H., Liu, C., Unsalan, O., Altunayar-Unsalan, C., Xiong, S., Manyande, A., Chen, H., 2021. Development and characterization of fish myofibrillar protein/chitosan/rosemary extract composite edible films and the improvement of lipid oxidation stability during the grass carp fillets storage. Int. J. Biol. Macromol. 01418130. 184, 463–475. doi:10.1016/j.ijbiomac.2021.06.121.

Durdakova, M., Kolackova, M., Ridoskova, A., Cernei, N., Pavelicova, K., Urbis, P., Richtera, L., Pelcova, P., Adam, V., Huska, D., 2024. Exploring the potential nutritional benefits of *Arthrospira maxima* and *Chlorella vulgaris*: a focus on vitamin B12, amino acids, and micronutrients. Food Chem. 03088146. 452, 139434. doi:10.1016/j.foodchem.2024.139434.

Ebert, S., Grossmann, L., Hinrichs, J., Weiss, J., 2019. Emulsifying properties of water-soluble proteins extracted from the microalgae: *chlorella sorokiniana* and *Phaeodactylum tricornutum*. Food Funct. 2042650X. 10 (2), 754–764. Available from: http://www.pubs.rsc.org/en/journals/journal/fo. doi:10.1039/c8fo02197j.

Ebrahimi, B., Baroutian, S., Jinyao, L., Zhang, B., Ying, T., Jun, L., 2023. Combination of marine bioactive compounds and extracts for the prevention and treatment of chronic diseases. Front. Nutr. 2296-861X. 9. doi:10.3389/fnut.2022.1047026.

Echave, J., Otero, P., Garcia-Oliveira, P., Munekata, P.E.S., Pateiro, M., Lorenzo, J.M., Simal-Gandara, J., Prieto, M.A., 2022. Seaweed-derived proteins and peptides: promising marine bioactives. Antioxidants. 20763921. 11 (1). Available from: https://www.mdpi.com/2076-3921/11/1/176/pdf. doi:10.3390/antiox11010176.

Egerton, S., Mannion, D., Culloty, S., Whooley, J., Stanton, C., Ross, R.P., 2020. The proximate composition of three marine pelagic fish: blue whiting (*Micromesistius poutassou*), boarfish (*Capros aper*) and Atlantic herring (*Clupea harengus*). Ir. J. Agric. Food Res. 07916833. 59 (1), 185–200. Available from: https://www.t-stor.teagasc.ie/handle/11019/2576. doi:10.15212/ijafr-2020-0112.

Elagizi, A., Lavie, C.J., O'keefe, E., Marshall, K., O'keefe, J.H., Milani, R.V., 2021. An update on omega-3 polyunsaturated fatty acids and cardiovascular health. Nutrients. 20726643. 13 (1), 1–12. Available from: https://www.mdpi.com/2072-6643/13/1/204/pdf. doi:10.3390/nu13010204.

Elavarasan, K., Shamasundar, B.A., 2022. Antioxidant properties of papain mediated protein hydrolysates from fresh water carps (*Catla catla*, *Labeo rohita* and *Cirrhinus mrigala*) and its application on inhibition of lipid oxidation in oil sardine mince during ice storage. J. Food Sci. Technol. 09758402. 59 (2), 636–645. Available from: http://www.springerlink.com/content/121580/. doi:10.1007/s13197-021-05053-0.

Fard, S.G., Wang, F., Sinclair, A.J., Elliott, G., Turchini, G.M., 2019. How does high DHA fish oil affect health? A systematic review of evidence. Crit. Rev. Food Sci. Nutr. 15497852. 59 (11), 1684–1727. Available from: https://www.tandfonline.com/loi/bfsn20. doi:10.1080/10408398.2018.1425978.

Feng, J., Long, S., Zhang, H.J., Wu, S.G., Qi, G.H., Wang, J., 2020. Comparative effects of dietary microalgae oil and fish oil on fatty acid composition and sensory quality of table eggs. Poult. Sci. 15253171. 99 (3), 1734–1743. Available from: https://www.journals.elsevier.com/poultry-science/. doi:10.1016/j.psj.2019.11.005.

Ferreira, I., Gomes-Bispo, A., Lourenço, H., Matos, J., Afonso, C., Cardoso, C., Castanheira, I., Motta, C., Prates, J. A.M., Bandarra, N.M., 2020. The chemical composition and lipid profile of the chub mackerel (*Scomber colias*) show a strong seasonal dependence: contribution to a nutritional evaluation. Biochimie. 61831638. 178, 181–189. Available from: http://www.elsevier.com/locate/biochimie. doi:10.1016/j.biochi.2020.09.022.

Francezon, N., Tremblay, A., Mouget, J.L., Pasetto, P., Beaulieu, L., 2021. Algae as a source of natural flavors in innovative foods. J. Agric. Food Chem. 15205118. 69 (40), 11753–11772. Available from: http://www.pubs.acs.org/journal/jafcau. doi:10.1021/acs.jafc.1c04409.

Geada, P., Moreira, C., Silva, M., Nunes, R., Madureira, L., Rocha, C.M.R., Pereira, R.N., Vicente, A.A., Teixeira, J.A., 2021. Algal proteins: production strategies and nutritional and functional properties. Bioresour. Technol. 18732976. 332. Available from: http://www.elsevier.com/locate/biortech. doi:10.1016/j.biortech.2021.125125.

Ghalamara, S., Silva, S., Brazinha, C., Pintado, M., 2022. Structural diversity of marine anti-freezing proteins, properties and potential applications: a review. Bioresour. Bioprocess. 2197-4365. 9 (1). doi:10.1186/s40643-022-00494-7.

Ghosh, S., Sarkar, T., Pati, S., Kari, Z.A., Edinur, H.A., Chakraborty, R., 2022. Novel bioactive compounds from marine sources as a tool for functional food development. Front. Mar. Sci. 22967745. 9. Available from: https://www.frontiersin.org/journals/marine-science#. doi:10.3389/fmars.2022.832957.

Gökalp, M., Mes, D., Nederlof, M., Zhao, H., de Goeij, J.M., Osinga, R., 2021. The potential roles of sponges in integrated mariculture. Rev. Aquac. 17535131. 13 (3), 1159–1171. Available from: http://www.onlinelibrary.wiley.com/journal/10.1111/(ISSN)1753-5131. doi:10.1111/raq.12516.

Grossmann, L., Hinrichs, J., Weiss, J., 2019. Solubility and aggregation behavior of protein fractions from the heterotrophically cultivated microalga *Chlorella protothecoides*. Food Res. Int. 09639969. 116, 283–290. doi:10.1016/j.foodres.2018.08.037.

Gulcin, İ., 2020. Antioxidants and antioxidant methods: an updated overview. Arch. Toxicol. 14320738. 94 (3), 651–715. Available from: http://www.link.springer.de/link/service/journals/00204/index.htm. doi:10.1007/s00204-020-02689-3.

Gullón, B., Gagaoua, M., Barba, F.J., Gullón, P., Zhang, W., Lorenzo, J.M., 2020. Seaweeds as promising resource of bioactive compounds: overview of novel extraction strategies and design of tailored meat products. Trends Food Sci. Technol. 09242244. 100, 1–18. Available from: http://www.elsevier.com/wps/find/journaldescription.cws_home/601278/description#description. doi:10.1016/j.tifs.2020.03.039.

Gumus, C.E., Gharibzahedi, S.M.T., 2021. Yogurts supplemented with lipid emulsions rich in omega-3 fatty acids: new insights into the fortification, microencapsulation, quality properties, and health-promoting effects. Trends Food Sci. Technol. 09242244. 110, 267–279. Available from: http://www.elsevier.com/wps/find/journaldescription.cws_home/601278/description#description. doi:10.1016/j.tifs.2021.02.016.

Güner, M., Yılmaz, E., Yüceer, Y., 2019. Off-odor removal from fish oil by adsorbent treatment with selected metal-organic frameworks. Flavour Fragr. J. 0882-5734. 34 (3), 163–174. doi:10.1002/ffj.3489.

Hadrová, S., Sedláková, K., Křížová, L., Malyugina, S., 2021. Alternative and unconventional feeds in dairy diets and their effect on fatty acid profile and health properties of milk fat. Animals. 2076-2615. 11 (6), 1817. doi:10.3390/ani11061817.

Hashem, A.M.A., Sakhare, S.D., Kudre, T.G., 2023. Effect of *Piaractus brachypomus* fish protein hydrolysate on physicochemical, sensory, and storage properties of cookies. Biocatal. Agric. Biotechnol. 18788181. 51. Available from: http://www.journals.elsevier.com/biocatalysis-and-agricultural-biotechnology/. doi:10.1016/j.bcab.2023.102761.

Healy, L., Zhu, X., Dong, G., Selli, S., Kelebek, H., Sullivan, C., Tiwari, U., Tiwari, B.K., 2024. Investigation into the use of novel pretreatments in the fermentation of *Alaria esculenta* by *Lactiplantibacillus plantarum* and kombucha SCOBY. Food Chem. 03088146. 442, 138335. doi:10.1016/j.foodchem.2023.138335.

Heffernan, S., Giblin, L., O'Brien, N., 2021. Assessment of the biological activity of fish muscle protein hydrolysates using in vitro model systems. Food Chem. 03088146. 359, 129852. doi:10.1016/j.foodchem.2021.129852.

Hernández-Zazueta, M.S., Luzardo-Ocampo, I., García-Romo, J.S., Noguera-Artiaga, L., Carbonell-Barrachina, Á.A., Taboada-Antelo, P., Campos-Vega, R., Rosas-Burgos, E.C., Burboa-Zazueta, M.G., Ezquerra-Brauer, J.M., Burgos-Hernández, A., 2021. Bioactive compounds from *Octopus vulgaris* ink extracts exerted anti-proliferative and anti-inflammatory effects in vitro. Food Chem. Toxicol. 18736351. 151. Available from: http://www.elsevier.com/locate/foodchemtox. doi:10.1016/j.fct.2021.112119.

Huynh Nguyen Duy Bao, D.B., Dung, N.N.T., 2024. Effect of the addition of fish protein isolate on biscuits' physicochemical and sensory properties. J. Fish. Sci. Technol. 1859-2252. 1, 031–037. doi:10.53818/jfst.01.2024.240.

Hwang, S.M., Kim, Y.U., Kim, J.K., Chun, Y.S., Kwon, Y.S., Ku, S.K., Song, C.H., 2022. Preventive and therapeutic effects of krill oil on obesity and obesity-induced metabolic syndromes in high-fat diet-fed mice. Mar. Drugs. 16603397. 20 (8). Available from: http://www.mdpi.com/journal/marinedrugs. doi:10.3390/md20080483.

Ikasari, D., Hastarini, E., Suryaningrum, T.D., 2020. Characteristics of cookies formulated with fish protein concentrate powder produced from snakehead fish (*Channa striata*) extraction by-product. E3S Web Conf. 22671242, 147. http://www.e3s-conferences.org/. doi:10.1051/e3sconf/202014703028.

Imchen, T., Singh, K.S., 2023. Marine algae colorants: antioxidant, anti-diabetic properties and applications in food industry. Algal Res. 22119264. 69 (102898). doi:10.1016/j.algal.2022.102898.

Iswariya, S., Velswamy, P., Uma, T.S., 2018. Isolation and characterization of biocompatible collagen from the skin of puffer fish (*Lagocephalus inermis*). J. Polym. Environ. 1566-2543. 26 (5), 2086–2095. doi:10.1007/s10924-017-1107-1.

Jadhav, H.B., 2024. Functional triacylglycerols from microalgae and their use in the formulation of functional foods—review. Food Chem. Adv. 2772753X. 4 (100695). doi:10.1016/j.focha.2024.100695.

Jamshidi, A., Cao, H., Xiao, J., Simal-Gandara, J., 2020. Advantages of techniques to fortify food products with the benefits of fish oil. Food Res. Int. 09639969. 137, 109353. doi:10.1016/j.foodres.2020.109353.

Jónasdóttir, H.S., 2019. Fatty acid profiles and production in marine phytoplankton. Mar. Drugs. 1660-3397. 17 (3), 151. doi:10.3390/md17030151.

Kandyliari, A., Mallouchos, A., Papandroulakis, N., Golla, J.P., Lam, T.K.T., Sakellari, A., Karavoltsos, S., Vasiliou, V., Kapsokefalou, M., 2020. Nutrient composition and fatty acid and protein profiles of selected fish by-products. Foods. 23048158. 9 (2). Available from: https://www.mdpi.com/2304-8158/9/2/190. doi:10.3390/foods9020190.

Kang, B., Myracle, A., Skonberg, D., Evaluation of in vitro anti-hyperglycemic effect of green crab hydrolysates derived by commercially available enzymes (P06-105-19). 2019. Curr. Dev. Nutr. 24752991. 3. nzz031.P06-105-19. doi:10.1093/cdn/nzz031.p06-105-19.

Kapoor, B., Kapoor, D., Gautam, S., Singh, R., Bhardwaj, S., 2021. Dietary polyunsaturated fatty acids (PUFAs): uses and potential health benefits. Curr. Nutr. Rep. 2161-3311. 10 (3), 232–242. doi:10.1007/s13668-021-00363-3.

Karantonis, H.C., Nasopoulou, C., Skalkos, D., 2022. Functional bakery snacks for the post-COVID-19 market, fortified with omega-3 fatty acids. Sustainability. 2071-1050. 14 (8), 4816. doi:10.3390/su14084816.

Khalifa, S.A.M., Elias, N., Farag, M.A., Chen, L., Saeed, A., Hegazy, M.E.F., Moustafa, M.S., El-Wahed, A.A., Al-Mousawi, S.M., Musharraf, S.G., Chang, F.R., Iwasaki, A., Suenaga, K., Alajlani, M., Göransson, U., El-Seedi, H.R., 2019. Marine natural products: a source of novel anticancer drugs. Mar. Drugs. 16603397. 17 (9). Available from: https://www.mdpi.com/1660-3397/17/9/491/pdf. doi:10.3390/md17090491.

Khan, N.A., Barthes, N., McCormack, G., O'Gara, J.P., Thomas, O.P., Boyd, A., 2023. Sponge-derived fatty acids inhibit biofilm formation of MRSA and MSSA by down-regulating biofilm-related genes specific to each pathogen. J. Appl. Microbiol. 13652672. 134 (8). Available from: https://academic.oup.com/jambio. doi:10.1093/jambio/lxad152.

Khan, S., Rehman, A., Shah, H., Aadil, R.M., Ali, A., Shehzad, Q., Ashraf, W., Yang, F., Karim, A., Khaliq, A., Xia, W., 2022. Fish protein and its derivatives: the novel applications, bioactivities, and their functional significance in food products. Food Rev. Int. 15256103. 38 (8), 1607–1634. Available from: http://www.tandf.co.uk/journals/titles/87559129.asp. doi:10.1080/87559129.2020.1828452.

Kim, O.K., Yun, J.M., Kim, D., Park, S.J., Lee, C., Go, E.B., Kim, J.S., Park, S.Y., Lee, J., 2022. Krill oil inhibits cholesterol synthesis and stimulated cholesterol excretion in hypercholesterolemic rats. Mar. Drugs. 16603397. 20 (10). Available from: http://www.mdpi.com/journal/marinedrugs. doi:10.3390/md20100609.

Kim, W., Wang, Y., Selomulya, C., 2020. Dairy and plant proteins as natural food emulsifiers. Trends Food Sci. Technol. 09242244. 105, 261–272. doi:10.1016/j.tifs.2020.09.012.

Klyuchko, N.Y., Pozdnyakova, D.A., Romazyaeva, I.R., Kovaleva, E.D., 2022. Using fish protein in innovative technologies of bakery and flour confectionery products. Ser. Fish. Ind. 2073-5529. 2022 (3), 98–105. doi:10.24143/2073-5529-2022-3-98-105.

Kumar, A., Elavarasan, K., Hanjabam, M.D., Binsi, P.K., Mohan, C.O., Zynudheen, A.A., K, A.K., 2019. Marine collagen peptide as a fortificant for biscuit: effects on biscuit attributes. LWT. 00236438. 109, 450–456. Available from: https://www.journals.elsevier.com/lwt. doi:10.1016/j.lwt.2019.04.052.

Larssen, W.E., Monteleone, E., Hersleth, M., 2018. Sensory description of marine oils through development of a sensory wheel and vocabulary. Food Res. Int. 09639969. 106, 45–53. doi:10.1016/j.foodres.2017.12.045.

Latifah, L.A., Soekamto, N.H., Tahir, A., 2021. New antibacterial activities of brominated C18 and C20 fatty acids isolated from marine sponge *Xestospongia testudinaria* against shrimp pathogenic bacteria. Rasayan J. Chem. 09760083. 14 (1), 460–465. Available from: https://rasayanjournal.co.in/admin/php/upload/3089_pdf.pdf. doi:10.31788/RJC.2021.1415999.

Laupsa-Borge, J., Grytten, E., Bohov, P., Bjørndal, B., Strand, E., Skorve, J., Nordrehaug, J.E., Berge, R.K., Rostrup, E., Mellgren, G., Dankel, S.N., Nygård, O.K., 2023. Sex-specific responses in glucose-insulin homeostasis and lipoprotein-lipid components after high-dose supplementation with marine n-3 PUFAs in abdominal obesity: a randomized double-blind crossover study. Front. Nutr. 2296-861X. (10). doi:10.3389/fnut.2023.1020678.

Lee, J.C., Hou, M.F., Huang, H.W., Chang, F.R., Yeh, C.C., Tang, J.Y., Chang, H.W., 2013. Marine algal natural products with anti-oxidative, anti-inflammatory, and anti-cancer properties. Cancer Cell Int. 14752867. 13 (1). Available from: http://www.cancerci.com/content/13/1/55. doi:10.1186/1475-2867-13-55Taiwan.

Li, P., Guo, J., Can, L., Yue, C., Wang, P., 2024. Effects of fish skin collagen peptide on physicochemical properties of frozen dough and quality and antioxidant activity of steamed bread. Food Hydrocoll. 0268005X. 157, 110420. doi:10.1016/j.foodhyd.2024.110420.

Li, X., Fan, M., Huang, Q., Zhao, S., Xiong, S., Yin, T., Zhang, B., 2022. Effect of micro- and nano-starch on the gel properties, microstructure and water mobility of myofibrillar protein from grass carp. Food Chem. 03088146. 366, 130579. doi:10.1016/j.foodchem.2021.130579.

Lima, K.O., da Rocha, M., Alemán, A., López-Caballero, M.E., Tovar, C.A., Gómez-Guillén, M.C., Montero, P., Prentice, C., 2021. Yogurt fortification by the addition of microencapsulated stripped weakfish (*Cynoscion guatucupa*) protein hydrolysate. Antioxidants. 20763921. 10 (10). Available from: https://www.mdpi.com/2076-3921/10/10/1567. doi:10.3390/antiox10101567.

Lionetto, F., Corcione, C.E., 2021. Recent applications of biopolymers derived from fish industry waste in food packaging. Polymers. 2073-4360. 13 (14), 2337. doi:10.3390/polym13142337.

Liu, J., Shibata, M., Qingbao, M., Liu, F., Lu, Q., Shan, Q., Hagiwara, T., Bao, J., 2020. Characterization of fish collagen from blue shark skin and its application for chitosan-collagen composite coating to preserve red porgy (*Pagrus major*) meat. J. Food Biochem. 0145-8884. 44 (8). doi:10.1111/jfbc.13265.

Liu, S., Zhang, L., Zhuyi, L., Chen, J., Zhang, Y., Yang, X., Chen, Q., Cai, H., Hong, P., Zhu, C., Zhong, S., 2024. The cryoprotective effect of an antifreeze collagen peptide complex obtained by enzymatic glycosylation on tilapia. Foods. 2304-8158. 13 (9), 1319. doi:10.3390/foods13091319.

Lotze, H.K., Tittensor, D.P., Bryndum-Buchholz, A., Eddy, T.D., Cheung, W.W.L., Galbraith, E.D., Barange, M., Barrier, N., Bianchi, D., Blanchard, J.L., Bopp, L., Büchner, M., Bulman, C.M., Carozza, D.A., Christensen, V., Coll, M., Dunne, J.P., Fulton, E.A., Jennings, S., Jones, M.C., Mackinson, S., Maury, O., Niiranen, S., Oliveros-Ramos, R., Roy, T., Fernandes, J.A., Schewe, J., Shin, Y.J., Silva, T.A.M., Steenbeek, J., Stock, C.A., Verley, P., Volkholz, J., Walker, N.D., Worm, B., 2019. Global ensemble projections reveal trophic amplification of ocean biomass declines with climate change. Proc. Natl. Acad. Sci. U. S. A. 10916490. 116 (26), 12907–12912. Available from: https://www.pnas.org/content/pnas/116/26/12907.full.pdf. doi:10.1073/pnas.1900194116.

Lu, X., Shi, M., Liu, L., Chen, Z., Xinxing, X., Feng, G., Zeng, M., 2024. Enhancement of flavor quality in oyster hydrolysate through fermentation with oyster-derived lactic acid bacteria. Food Biosci. 22124292. 62, 105231. doi:10.1016/j.fbio.2024.105231.

Ma, Y., Huang, K., Wu, Y., 2021. In vivo/in vitro properties of novel antioxidant peptide from Pinctada fucata. J. Microbiol. Biotechnol. 17388872. 31 (1), 33–42. Available from: http://www.jmb.or.kr/journal/view.html?doi=10.4014/jmb.2006.06002. doi:10.4014/JMB.2006.06002.

Mahmoud, A.M., Bin-Jumah, M., Abukhalil, M.H., 2021. Antiinflammatory natural products from marine algae. In: Gopi, S., Amalraj, A., Kunnumakkara, A., Thomas, S. (Eds.), Inflammation and Natural Products. Elsevier, Egypt, 173–201. Available from https://www.sciencedirect.com/book/9780128192184. doi:10.1016/B978-0-12-819218-4.00012-2.

Manson, J.A.E., Cook, N.R., Lee, I.M., Christen, W., Bassuk, S.S., Mora, S., Gibson, H., Albert, C.M., Gordon, D., Copeland, T., D'Agostino, D., Friedenberg, G., Ridge, C., Bubes, V., Giovannucci, E.L., Willett, W.C., Buring, J.E.,

2019. Marine n-3 fatty acids and prevention of cardiovascular disease and cancer. J. Med. 15334406. 380 (1), 23–32. Available from: http://www.nejm.org/medical-index. doi:10.1056/NEJMoa1811403.

Marti-Quijal, F.J., Remize, F., Meca, G., Ferrer, E., Ruiz, M.J., Barba, F.J., 2020. Fermentation in fish and by-products processing: an overview of current research and future prospects. Curr. Opin. Food Sci. 22147993. 31, 9–16. Available from: http://www.journals.elsevier.com/current-opinion-in-food-science/. doi:10.1016/j.cofs.2019.08.001.

McClements, D.J., Öztürk, B., 2021. Utilization of nanotechnology to improve the handling, storage and biocompatibility of bioactive lipids in food applications. Foods. 23048158. 10 (2), 1–17. Available from: https://www.mdpi.com/2304-8158/10/2/365/pdf. doi:10.3390/foods10020365.

Mehbub, M.F., Yang, Q., Cheng, Y., Franco, C.M.M., Zhang, W., 2024. Marine sponge-derived natural products: trends and opportunities for the decade of 2011-2020. Front. Mar. Sci. 22967745. 11. Available from: https://www.frontiersin.org/journals/marine-science#. doi:10.3389/fmars.2024.1462825.

Meital, L.T., Windsor, M.T., Perissiou, M., Schulze, K., Magee, R., Kuballa, A., Golledge, J., Bailey, T.G., Askew, C.D., Russell, F.D., 2019. Omega-3 fatty acids decrease oxidative stress and inflammation in macrophages from patients with small abdominal aortic aneurysm. Sci. Rep. 20452322. 9 (1). Available from: http://www.nature.com/srep/index.html. doi:10.1038/s41598-019-49362-z.

Melgosa, R., Trigueros, E., Sanz, M.T., Cardeira, M., Rodrigues, L., Fernández, N., Matias, A.A., Bronze, M.R., Marques, M., Paiva, A., Simões, P., 2020. Supercritical CO_2 and subcritical water technologies for the production of bioactive extracts from sardine (*Sardina pilchardus*) waste. J. Supercrit. Fluids. 08968446. 164. Available from: https://www.journals.elsevier.com/the-journal-of-supercritical-fluids. doi:10.1016/j.supflu.2020.104943.

Mesa, M.D., Gil, F., Olmedo, P., Gil, A., 2021. Nutritional importance of selected fresh fishes, shrimps and mollusks to meet compliance with nutritional guidelines of N-3 LC-Pufa intake in Spain. Nutrients. 20726643. 13 (2), 1–14. Available from: https://www.mdpi.com/2072-6643/13/2/465/pdf. doi:10.3390/nu13020465.

Mezhoudi, M., Salem, A., Abdelhedi, O., Fakhfakh, N., Debeaufort, F., Jridi, M., Zouari, N., 2022b. Edible films from triggerfish gelatin and *Moringa oleifera* extract: physical properties and application in wrapping ricotta cheese. J. Food Meas. Charact. 2193-4126. 16 (5), 3987–3997. doi:10.1007/s11694-022-01472-5.

Mezhoudi, M., Salem, A., Abdelhedi, O., Fakhfakh, N., Mabrouk, M., Khorchani, T., Debeaufort, F., Jridi, M., Zouari, N., 2022a. Development of active edible coatings based on fish gelatin enriched with *Moringa oleifera* extract: application in fish (*Mustelus mustelus*) fillet preservation. Food Sci. Nutr. 20487177. 10 (11), 3979–3992. Available from: http://www.onlinelibrary.wiley.com/journal/10.1002/%28ISSN%292048-7177. doi:10.1002/fsn3.2993.

Mitguard, S., Doucette, O., Miklavcic, J., 2023. Human milk polyunsaturated fatty acids are related to neurodevelopmental, anthropometric, and allergic outcomes in early life: a systematic review. J. Dev. Orig. Health Dis. 2040-1744. 14 (6), 763–772. doi:10.1017/s2040174423000454.

Mititelu, M., Lupuliasa, D., Neacşu, S.M., Olteanu, G., Busnatu, Ş.S., Mihai, A., Popovici, V., Măru, N., Boroghină, S. C., Mihai, S., Ioniță-Mîndrican, C.B., Scafa-Udriște, A., 2025. Polyunsaturated fatty acids and human health: a key to modern nutritional balance in association with polyphenolic compounds from food sources. Foods. 23048158. 14 (1). Available from: http://www.mdpi.com/journal/foods. doi:10.3390/foods14010046.

Moghadam, F.V., Pourahmad, R., Mortazavi, A., Davoodi, D., Azizinezhad, R., 2019. Use of fish oil nanoencapsulated with gum Arabic carrier in low fat probiotic fermented milk. Food Sci. Anim. Resour. 26360780. 39 (2), 309–323. Available from: https://www.kosfaj.org/download/download_pdf?pid=kosfa-39-2-309. doi:10.5851/kosfa.2019.e25.

Mogobe, O., Mazrui, N.M., Gondwe, M.J., Mosepele, K., Masamba, W.R.L., 2024. Nutrient composition of common fish species in the Okavango Delta: potential contribution to nutrition security. Environ. Dev. Sustain. 15732975. 26 (8), 19731–19753. Available from: https://www.springer.com/journal/10668. doi:10.1007/s10668-023-03434-3.

Mohanty, B.P., Mahanty, A., Ganguly, S., Mitra, T., Karunakaran, D., Anandan, R., 2019. Nutritional composition of food fishes and their importance in providing food and nutritional security. Food Chem. 18737072. 293, 561–570. Available from: http://www.elsevier.com/locate/foodchem. doi:10.1016/j.foodchem.2017.11.039.

Molchanov, V., Yegorov, A., Molchanov, M., Timchenko, A., Novikov, V., Novojilov, N., Timchenko, M., 2023. Novel antimicrobial peptide from the hepatopancreas of the red king crab. Int. J. Mol. Sci. 1422-0067. 24 (21), 15607. doi:10.3390/ijms242115607.

Mora, I., Arola, L., Caimari, A., Escoté, X., Puiggròs, F., 2022. Structured long-chain omega-3 fatty acids for improvement of cognitive function during aging. Int. J. Mol. Sci. 1422-0067. 23 (7), 3472. doi:10.3390/ijms23073472.

Nguyen, Q.V., Malau-Aduli, B.S., Cavalieri, J., Malau-Aduli, A.E.O., Nichols, P.D., 2019. Enhancing omega-3 long-chain polyunsaturated fatty acid content of dairy-derived foods for human consumption. Nutrients. 20726643. 11 (4). Available from: https://www.mdpi.com/2072-6643/11/4/743/pdf. doi:10.3390/nu11040743.

Nie, J., Xiaoting, F., Wang, L., Jiachao, X., Gao, X., 2023. Impact of *Monascus purpureus* fermentation on antioxidant activity, free amino acid profiles and flavor properties of kelp (*Saccharina japonica*). Food Chem. 03088146. 400, 133990. doi:10.1016/j.foodchem.2022.133990.

Nikoo, M., Regenstein, J.M., Noori, F., Gheshlaghi, S.P., 2021. Autolysis of rainbow trout (*Oncorhynchus mykiss*) by-products: enzymatic activities, lipid and protein oxidation, and antioxidant activity of protein hydrolysates. LWT. 00236438. 140, 110702. doi:10.1016/j.lwt.2020.110702.

Nikoo, M., Regenstein, J.M., Yasemi, M., 2023. Protein hydrolysates from fishery processing by-products: production, characteristics, food applications, and challenges. Foods. 2304-8158. 12 (24), 4470. doi:10.3390/foods12244470.

Nosratpour, M., Ma, J., Haritos, V., Wang, Y., 2022. The physicochemical and sensory characteristics of yoghurt fortified with encapsulated fish oil/milk fat. In: Proceedings of 2022 AOCS Annual Meeting & Expo. doi:10.21748/yhap5375.

Nwachukwu, I.D., Aluko, R.E., 2019. Structural and functional properties of food protein-derived antioxidant peptides. J. Food Biochem. 17454514. 43 (1), 1–13. Available from: http://www.onlinelibrary.wiley.com/journal/10.1111/(ISSN)1745-4514. doi:10.1111/jfbc.12761.

Offret, C., Fliss, I., Bazinet, L., Marette, A., Beaulieu, L., 2019. Identification of a novel antibacterial peptide from atlantic mackerel belonging to the GAPDH-related antimicrobial family and its in vitro digestibility. Mar. Drugs. 1660-3397. 17 (7), 413. doi:10.3390/md17070413.

Okeke, E.S., Okagu, I.U., Chukwudozie, K., Ezike, T.C., Ezeorba, T.P.C., 2024. Marine-derived bioactive proteins and peptides: a review of current knowledge on anticancer potentials, clinical trials, and future prospects. Nat. Prod. Commun. 15559475. 19 (3). Available from: https://www.journals.sagepub.com/home/NPX. doi:10.1177/1934578X241239825.

Olatunde, O.O., Benjakul, S., Yesilsu, A.F., 2020. Antimicrobial compounds from crustaceans and their applications for extending shelf-life of marine-based foods. Turk. J. Fish. Aquat. Sci. 13032712. 20 (8), 629–646. Available from: http://www.trjfas.org/pdf.php?&id=1474. doi:10.4194/1303-2712-v20_8_06.

Oppedisano, F., Macrì, R., Gliozzi, M., Musolino, V., Carresi, C., Maiuolo, J., Bosco, F., Nucera, S., Zito, M.C., Guarnieri, L., Scarano, F., Nicita, C., Coppoletta, A.R., Ruga, S., Scicchitano, M., Mollace, R., Palma, E., Mollace, V., 2020. The anti-inflammatory and antioxidant properties of n-3 PUFAs: their role in cardiovascular protection. Biomedicines. 22279059. 8 (9). Available from: https://www.res.mdpi.com/d_attachment/biomedicines/biomedicines-08-00306/article_deploy/biomedicines-08-00306.pdf. doi:10.3390/biomedicines8090306.

Ovchinnikova, T.V., Shi, Q., 2023. Editorial: omics-based identification, characterization, and validation of marine bioactive peptides. Front. Mar. Sci. 22967745. 10. Available from: https://www.frontiersin.org/journals/marine-science#. doi:10.3389/fmars.2023.1323847.

Pajor, F., Egerszegi, I., Steiber, O., Bodnár, Á., Póti, P., 2019. Effect of marine algae supplementation on the fatty acid profile of milk of dairy goats kept indoor and on pasture. J. Anim. Feed Sci. 1230-1388. 28 (2), 169–176. doi:10.22358/jafs/109955/2019.

Pajor, F., Egerszegi, I., Szűcs, Á., Póti, P., Bodnár, Á., 2021. Effect of marine algae supplementation on somatic cell count, prevalence of udder pathogens, and fatty acid profile of dairy goats' milk. Animals. 2076-2615. 11 (4), 1097. doi:10.3390/ani11041097.

Pajor, F., Várkonyi, D., Dalmadi, I., Pásztorné-Huszár, K., Egerszegi, I., Penksza, K., Póti, P., Bodnár, Á., 2023. Changes in chemical composition and fatty acid profile of milk and cheese and sensory profile of milk via supplementation of goats' diet with marine algae. Animals. 2076-2615. 13 (13), 2152. doi:10.3390/ani13132152.

Pan, C., Jingrong, M., Tao, F., Chenyang, J., Zhao, Y., Chen, S., Yang, X., 2021. Novel insight into the antioxidant proteins derived from laver (*Porphyra haitanensis*) by proteomics analysis and protein based bioinformatics. Food Biosci. 22124292. 42, 101134. doi:10.1016/j.fbio.2021.101134.

Park, Y.R., Park, C.I., Soh, Y., 2020. Antioxidant and anti-inflammatory effects of NCW peptide from Clam Worm (*Marphysa sanguinea*). J. Microbiol. Biotechnol. 17388872. 30 (9), 1387–1394. Available from: http://www.jmb.or.kr/journal/view.html?uid=5672&vmd=Full. doi:10.4014/jmb.2003.03050.

Pavlicevic, M., Maestri, E., Marmiroli, M., 2020. Marine bioactive peptides—an overview of generation, structure and application with a focus on food sources. Mar. Drugs. 1660-3397. 18 (8), 424. doi:10.3390/md18080424.

Penggalih, M.H.S.T., Praditya, G.N., Rizqiansyah, C.Y., Setyawardani, A., Purnomo, A.F., Maulana, R.A., Gunawan, W.B., Subali, D., Kurniawan, R., Mayulu, N., Taslim, N.A., Hardinsyah, H., Sutanto, Y.S., Nurkolis, F., 2023. Marine-derived protein: peptide bioresources for the development of nutraceuticals for improved athletic performance. Front. Sports Act. Living. 26249367. 5. Available from: http://www.frontiersin.org/journals/sports-and-active-living. doi:10.3389/fspor.2023.1281397.

Pereira, G., Simões, P., Bexiga, R., Silva, E., Mateus, L., Fernandes, T., Alves, S.P., Bessa, R.J.B., Lopes-da-Costa, L., 2022. Effects of feeding rumen-protected linseed fat to postpartum dairy cows on plasma n-3 polyunsaturated fatty acid concentrations and metabolic and reproductive parameters. J. Dairy Sci. 15253198. 105 (1), 361–374. Available from: http://www.elsevier.com/wps/find/journaldescription.cws_home/721317/description#description. doi:10.3168/jds.2021-20674.

Pereira, L., Cotas, J., Gonçalves, A.M., 2024. Seaweed proteins: a step towards sustainability? Nutrients. 20726643. 16 (8). doi:10.3390/nu16081123.

Pérez-Palacios, T., Ruiz-Carrascal, J., Solomando, J.C., Antequera, T., 2019. Strategies for enrichment in ω-3 fatty acids aiming for healthier meat products. Food Rev. Int. 15256103. 35 (5), 485–503. Available from: http://www.tandf.co.uk/journals/titles/87559129.asp. doi:10.1080/87559129.2019.1584817.

Pham, T.P.T., Hoang, T.V., Cao, P.T.N., Le, T.T.D., Ho, V.T.N., Vu, T.M.H., Le, T.H.T., Pham, H.T.X., Tran, T.T., Mafruhah, O.R., Pham, T.T.L., Hsieh, M.T., Ha, H.A., 2024. Comparison of omega-3 polyunsaturated fatty acids bioavailability in fish oil and krill oil: network meta-analyses. Food Chem. X. 25901575. (24). Available from: https://www.sciencedirect.com/science/journal/25901575. doi:10.1016/j.fochx.2024.101880.

Poggioli, R., Hirani, K., Jogani, V.G., Ricordi, C., 2023. Modulation of inflammation and immunity by omega-3 fatty acids: a possible role for prevention and to halt disease progression in autoimmune, viral, and age-related disorders. Eur. Rev. Med. Pharmacol. Sci. 22840729. 27 (15), 7380–7400. Available from: https://www.europeanreview.org/wp/wp-content/uploads/7380-7400.pdf. doi:10.26355/eurrev_202308_33310.

Ponomareva, E.I., Alehina, N.N., Gubareva, Y.P., Tereshchenko, D.A., 2020. The influence of the dosage of corn oil on the quality indicators of the dough and products with germinated flax seeds. DOAJ. doi:10.20914/2310-1202-2020-1-53-58.

Pourashouri, P., Shabanpour, B., Heydari, S., Raeisi, S., 2021. Encapsulation of fish oil by carrageenan and gum tragacanth as wall materials and its application to the enrichment of chicken nuggets. LWT. 00236438. 137, 110334. doi:10.1016/j.lwt.2020.110334.

Pratama, I.S., Putra, Y., Pangestuti, R., Kim, S.K., Siahaan, E.A., 2022. Bioactive peptides-derived from marine by-products: development, health benefits and potential application in biomedicine. Fish. Aquat. Sci. 22341757. 25 (7), 357–379. Available from: https://www.e-fas.org/info/about. doi:10.47853/FAS.2022.e33.

Punia, S., Sandhu, K.S., Dhull, S.B., Kaur, M., Siroha, A.K., 2020. Kinetic, rheological and thermal studies of flaxseed (*Linum usitatissiumum* L.) oil and its utilization. J. Food Sci. Technol. 09758402. 57 (11), 4014–4021. Available from: http://www.springerlink.com/content/121580/. doi:10.1007/s13197-020-04434-1.

Qazi, M.W., de Sousa, I.G., Nunes, M.C., Raymundo, A., 2022. Improving the nutritional, structural, and sensory properties of gluten-free bread with different species of microalgae. Foods. 23048158. 11 (3). Available from: https://www.mdpi.com/2304-8158/11/3/397/pdf. doi:10.3390/foods11030397.

Qazi, W.M., Ballance, S., Uhlen, A.K., Kousoulaki, K., Haugen, J.E., Rieder, A., 2021. Protein enrichment of wheat bread with the marine green microalgae *Tetraselmis chuii*—impact on dough rheology and bread quality. LWT. 00236438. 143. Available from: https://www.journals.elsevier.com/lwt. doi:10.1016/j.lwt.2021.111115.

Radnaeva, L.D., Bazarsadueva, S.V., Taraskin, V.V., Tulokhonov, A.K., 2020. First data on lipids and microorganisms of deepwater endemic sponge Baikalospongia intermedia and sediments from hydrothermal discharge area of the Frolikha Bay (North Baikal, Siberia). J. Great Lakes Res. 03801330. 46 (1), 67–74. Available from: http://www.elsevier.com. doi:10.1016/j.jglr.2019.09.021.

Raeisi, S., Ojagh, S.M., Quek, S.Y., Pourashouri, P., Salaün, F., 2019. Nano-encapsulation of fish oil and garlic essential oil by a novel composition of wall material: persian gum-chitosan. LWT. 00236438. 116. Available from: https://www.journals.elsevier.com/lwt. doi:10.1016/j.lwt.2019.108494.

Rahim, M.A., Imran, M., Khan, M.K., Ahmad, M.H., Nadeem, M., Khalid, W., Lorenzo, J.M., Afifi, M.M., AL-Farga, A., Aljobair, M.O., 2022. Omega-3 fatty acid retention and oxidative stability of spray-dried chia–fish-oil-prepared microcapsules. Processes. 22279717. 10 (11). Available from: http://www.mdpi.com/journal/processes. doi:10.3390/pr10112184.

Rahman, N., Hashem, S., Akther, S., Jothi, J.S., 2023. Impact of various extraction methods on fatty acid profile, physicochemical properties, and nutritional quality index of Pangus fish oil. Food Sci. Nutr. 20487177. 11 (8), 4688–4699. Available from: http://www.onlinelibrary.wiley.com/journal/10.1002/%28ISSN%292048-7177. doi:10.1002/fsn3.3431.

Raja, K., Suresh, K., Anbalagan, S., Ragini, Y.P., Kadirvel, V., 2024. Investigating the nutritional viability of marine-derived protein for sustainable future development. Food Chem. 18737072. 448. Available from: https://www.sciencedirect.com/science/journal/03088146. doi:10.1016/j.foodchem.2024.139087.

Ramos-Miras, J.J., Sanchez-Muros, M.J., Morote, E., Torrijos, M., Gil, C., Zamani-Ahmadmahmoodi, R., Rodríguez Martin, J.A., 2019. Potentially toxic elements in commonly consumed fish species from the western Mediterranean Sea (Almería Bay): bioaccumulation in liver and muscle tissues in relation to biometric parameters. Sci. Total Environ. 00489697. 671, 280–287. doi:10.1016/j.scitotenv.2019.03.359.

Ramos-Romero, S., Torrella, J.R., Pagès, T., Viscor, G., Torres, J.L., 2021. Edible microalgae and their bioactive compounds in the prevention and treatment of metabolic alterations. Nutrients. 20726643. 13 (2), 1–16. Available from: https://www.mdpi.com/2072-6643/13/2/563/pdf. doi:10.3390/nu13020563.

Reksten, A.M., Wiech, M., Aakre, I., Markhus, M.W., Nøstbakken, O.J., Hannisdal, R., Madsen, L., Dahl, L., 2024. Exploring the nutrient composition of various shellfish available in Norway and their role in providing key nutrients. J. Food Compos. Anal. 08891575. 128. Available from: https://www.sciencedirect.com/science/journal/08891575. doi:10.1016/j.jfca.2024.106003.

Rey, F., Cartaxana, P., Melo, T., Calado, R., Pereira, R., Abreu, H., Domingues, P., Sónia Cruz, M.R.D., 2020. Domesticated populations of *Codium tomentosum* display lipid extracts with lower seasonal shifts than conspecifics from the wild—relevance for biotechnological applications of this green seaweed. Mar. Drugs. 1660-3397. 18 (4), 188. doi:10.3390/md18040188.

Rivera-Jiménez, J., Berraquero-García, C., Pérez-Gálvez, R., García-Moreno, P.J., Espejo-Carpio, F.J., Guadix, A., Guadix, E.M., 2022. Peptides and protein hydrolysates exhibiting anti-inflammatory activity: sources, structural features and modulation mechanisms. Food Funct. 2042650X. 13 (24), 12510–12540. Available from: http://pubs.rsc.org/en/journals/journal/fo. doi:10.1039/d2fo02223k.

Rivero-Pino, F., Espejo-Carpio, F.J., Guadix, E.M., 2020. Evaluation of the bioactive potential of foods fortified with fish protein hydrolysates. Food Res. Int. 18737145. 137. Available from: http://www.elsevier.com/inca/publications/store/4/2/2/9/7/0. doi:10.1016/j.foodres.2020.109572.

Rotter, A., Varamogianni-Mamatsi, D., Pobirk, A.Z., Matjaž, M.G., Cueto, M., Díaz-Marrero, A.R., Jónsdóttir, R., Sveinsdóttir, K., Catalá, T.S., Romano, G., Guler, B.A., Atak, E., Zrimec, M.B., Bosch, D., Deniz, I., Gaudêncio, S. P., Grigalionyte-Bembič, E., Klun, K., Zidar, L., Rius, A.C., Baebler, Š., Lukić Bilela, L., Rinkevich, B., Mandalakis, M., 2024. Marine cosmetics and the blue bioeconomy: from sourcing to success stories. iScience. 25890042. 27 (12). Available from: https://www.sciencedirect.com/science/journal/25890042. doi:10.1016/j.isci.2024.111339.

Saeidi, K., Alirezalu, A., Akbari, Z., 2016. Evaluation of chemical constitute, fatty acids and antioxidant activity of the fruit and seed of sea buckthorn (*Hippophae rhamnoides* L.) grown wild in Iran. Nat. Prod. Res. 14786427. 30 (3), 366–368. Available from: http://www.tandf.co.uk/journals/titles/14786419.asp. doi:10.1080/14786419.2015.1057728.

Sakamoto, A., Saotome, M., Iguchi, K., Maekawa, Y., 2019. Marine-derived omega-3 polyunsaturated fatty acids and heart failure: current understanding for basic to clinical relevance. Int. J. Mol. Sci. 1422-0067. 20 (16), 4025. doi:10.3390/ijms20164025.

Salehi, B., Sharifi-Rad, J., Seca, A.M.L., Pinto, D.C.G.A., Michalak, I., Trincone, A., Mishra, A.P., Nigam, M., Zam, W., Martins, N., 2019. Current trends on seaweeds: looking at chemical composition, phytopharmacology, and cosmetic applications. Molecules. 14203049. 24 (22). Available from: https://www.mdpi.com/1420-3049/24/22/4182/pdf. doi:10.3390/molecules24224182.

Salem, A., Abdelhedi, O., Sebii, H., Taheur, F.B., Fakhfakh, N., Debeaufort, F., Jridi, M., Zouari, N., 2024. Enhancing the quality of refrigerated smooth-hound shark fillet with a fish gelatin coating enriched with garden cress extract. J. Aquat. Food Prod. Technol. 1049-8850. 33 (3), 247–262. doi:10.1080/10498850.2024.2320264.

Sánchez, C.A.O., Zavaleta, E.B., García, G.R.U., Solano, G.L., Díaz, M.P.R., 2021. Krill oil microencapsulation: antioxidant activity, astaxanthin retention, encapsulation efficiency, fatty acids profile, in vitro bioaccessibility and storage stability. LWT. 00236438. 147. Available from: http://www.elsevier.com/inca/publications/store/6/2/2/9/1/0/index.htt. doi:10.1016/j.lwt.2021.111476.

Sanjeewa, K.K.A., Fernando, I.P.S., Kim, S.-Y., Kim, W.-S., Ahn, G., Jee, Y., Jeon, Y.-J., 2019. *Ecklonia cava* (Laminariales) and *Sargassum horneri* (Fucales) synergistically inhibit the lipopolysaccharide-induced inflammation via blocking NF-κB and MAPK pathways. Algae. 1226-2617. 34 (1), 45–56. doi:10.4490/algae.2019.34.2.10.

Savatinova, M., Ivanova, M., 2024. Functional dairy products enriched with omega-3 fatty acids. Food Sci. Appl. Biotechnol. 26033380. 7 (1), 1–13. Available from: https://www.ijfsab.com/index.php/fsab/article/view/301. doi:10.30721/fsab2024.v7.i1.301.

Ścieszka, S., Klewicka, E., 2019. Algae in food: a general review. Crit. Rev. Food Sci. Nutr. 1040-8398. 59 (21), 3538–3547. doi:10.1080/10408398.2018.1496319.

Senadheera, T.R.L., Hossain, A., Shahidi, F., 2023. Marine bioactives and their application in the food industry: a review. Appl. Sci. 20763417. 13 (21). Available from: http://www.mdpi.com/journal/applsci/. doi:10.3390/app132112088.

Seyed, M.A., Ayesha, S., 2021. Marine-derived pipeline anticancer natural products: a review of their pharmacotherapeutic potential and molecular mechanisms. Future J. Pharm. Sci. 2314-7253. 7 (1). doi:10.1186/s43094-021-00350-z.

Shahidi, F., Hossain, A., 2022. Role of lipids in food flavor generation. Molecules. 1420-3049. 27 (15), 5014. doi:10.3390/molecules27155014.

Shekarabi, S.P.H., Shahbazi, M., 2022. Partial substitution of wheat flour with fish protein powder influences physicochemical, rheological, and sensory properties of the flour, dough, and bread. J. Aquat. Food Prod. Technol. 15470636. 31 (1), 2–18. Available from: http://www.tandfonline.com/toc/wafp20/current. doi:10.1080/10498850.2021.2008078.

Shepon, A., Makov, T., Hamilton, H.A., Müller, D.B., Gephart, J.A., Henriksson, P.J.G., Troell, M., Golden, C.D., 2022. Sustainable optimization of global aquatic omega-3 supply chain could substantially narrow the nutrient gap. Resour. Conserv. Recycl. 18790658. 181. Available from: http://www.elsevier.com/locate/resconrec. doi:10.1016/j.resconrec.2022.106260.

Shi, M., Ying, D.Y., Hlaing, M.M., Ye, J.H., Sanguansri, L., Augustin, M.A., 2020. Oxidative stability of spray dried matcha-tuna oil powders. Food Res. Int. 18737145. 132. Available from: http://www.elsevier.com/inca/publications/store/4/2/2/9/7/0. doi:10.1016/j.foodres.2020.109050.

Shibabaw, T., 2021. Omega-3 polyunsaturated fatty acids: anti-inflammatory and anti-hypertriglyceridemia mechanisms in cardiovascular disease. Mol. Cell. Biochem. 0300-8177. 476 (2), 993–1003. doi:10.1007/s11010-020-03965-7.

Siddik, M.A.B., Howieson, J., Fotedar, R., Partridge, G.J., 2021. Enzymatic fish protein hydrolysates in finfish aquaculture: a review. Rev. Aquac. 17535131. 13 (1), 406–430. Available from: http://www.onlinelibrary.wiley.com/journal/10.1111/(ISSN)1753-5131. doi:10.1111/raq.12481.

Solomando, J.C., Antequera, T., Martín, A., Perez-Palacios, T., 2021. Fish oil microcapsules as omega-3 enrichment strategy: changes in volatile compounds of meat products during storage and cooking. Foods. 2304-8158. 10 (4), 745. doi:10.3390/foods10040745.

Sun, K.L., Gao, M., Wang, Y.Z., Li, X.R., Wang, P., Wang, B., 2022. Antioxidant peptides from protein hydrolysate of marine red algae *Eucheuma cottonii*: preparation, identification, and cytoprotective mechanisms on H_2O_2 oxidative damaged HUVECs. Front. Microbiol. 1664302X. 13. Available from: https://www.frontiersin.org/journals/microbiology#. doi:10.3389/fmicb.2022.791248.

Sun, Q., Qiaoqiao, M., Yan, Y., Jiang, W., Lin, Y., Qian, L., Jiang, S., Zhao, Y., Zhou, Q., Miao, L., Liu, B., Xianping, G., 2023. Krill oil showed a good application potential as the dietary lipid source on the growth, physiology, nutritional quality, and flavor of the edible meat of the oriental river prawns (*Macrobrachium nipponense*). Aquac. Rep. 23525134. 32, 101721. doi:10.1016/j.aqrep.2023.101721.

Sunarwidhi, A.L., Rosyantari, A., Prasedya, E.S., Ardiana, N., Ilhami, B.T.K., Abidin, A.S., Ambana, Y., Kirana, I.A.P., Wirasisya, D.G., Anugrah, W., Fersiyana, R.D., Dewi, N.M.A.R., 2021. The correlation between total protein content and antioxidant activity of collagen isolated from a marine sponge Stylissa flabelliformis collected from

North Lombok Indonesia coast. IOP Conf. Ser. Earth Environ. Sci. 1755-1307. 913 (1), 012103. doi:10.1088/1755-1315/913/1/012103.

Surendhiran, D., Li, C., Cui, H., Lin, L., 2021. Marine algae as efficacious bioresources housing antimicrobial compounds for preserving foods—a review. Int. J. Food Microbiol. 18793460. 358. Available from: http://www.elsevier.com/locate/ijfoodmicro. doi:10.1016/j.ijfoodmicro.2021.109416.

Temgoua, N.S.F., Sun, Z., Okoye, C.O., Pan, H., 2022. Fatty acid profile, physicochemical composition, and sensory properties of Atlantic salmon fish (*Salmo salar*) during different culinary treatments. J. Food Qual. 17454557. 2022. Available from: https://www.hindawi.com/journals/jfq/. doi:10.1155/2022/7425142.

Thanh, L.P., Suksombat, W., Loor, J.J., Hang, T.T.T., 2023. Polyunsaturated fatty acids and rumen undegradable protein alter ruminal fermentation and milk fatty acid profiles in dairy cows. Arch. Anim. Nutr. 14772817. 77 (1), 58–76. Available from: https://www.tandfonline.com/loi/gaan20. doi:10.1080/1745039X.2023.2176150.

Tu, L., Xugan, W., Wang, X., Shi, W., 2020. Effects of fish oil replacement by blending vegetable oils in fattening diets on nonvolatile taste substances of swimming crab (*Portunus trituberculatus*). J. Food Biochem. 0145-8884. 44 (9). doi:10.1111/jfbc.13345.

Ucak, I., Afreen, M., Montesano, D., Carrillo, C., Tomasevic, I., Simal-Gandara, J., Barba, F.J., 2021. Functional and bioactive properties of peptides derived from marine side streams. Mar. Drugs. 16603397. 19 (2). Available from: https://www.mdpi.com/1660-3397/19/2/71/pdf. doi:10.3390/MD19020071.

Unnikrishnan, P., Kizhakkethil, B.P., George, J.C., Sivam, V., Panda, S.K., Ninan, G., Zynudheen, A.A., 2021. Characterization of health beverage fortified with peptides from yellowfin tuna. J. Aquat. Food Prod. Technol. 15470636. 30 (9), 1142–1158. Available from: http://www.tandfonline.com/toc/wafp20/current. doi:10.1080/10498850.2021.1974631.

Urrutia, O., Antonio Mendizabal, J., Alfonso, L., Soret, B., Insausti, K., Arana, A., 2020. Adipose tissue modification through feeding strategies and their implication on adipogenesis and adipose tissue metabolism in ruminants. Int. J. Mol. Sci. 1422-0067. 21 (9), 3183. doi:10.3390/ijms21093183.

Valença, R.D.L., Sobrinho, A.G.D.S., Borghi, T.H., Meza, D.A.R., Andrade, N.D., Silva, L.G., Bezerra, L.R., 2021. Performance, carcass traits, physicochemical properties and fatty acids composition of lamb's meat fed diets with marine microalgae meal (*Schizochytrium* sp.). Livest. Sci. 18711413. 243. Available from: http://www.elsevier.com/wps/find/journaldescription.cws_home/706547/description#description. doi:10.1016/j.livsci.2020.104387.

Valero, Y., Saraiva-Fraga, M., Costas, B., Guardiola, F.A., 2020. Antimicrobial peptides from fish: beyond the fight against pathogens. Rev. Aquac. 17535131. 12 (1), 224–253. Available from: http://www.onlinelibrary.wiley.com/journal/10.1111/(ISSN)1753-5131. doi:10.1111/raq.12314.

Varijakzhan, D., Loh, J.Y., Yap, W.S., Yusoff, K., Seboussi, R., Lim, S.H.E., Lai, K.S., Chong, C.M., 2021. Bioactive compounds from marine sponges: fundamentals and applications. Mar. Drugs. 16603397. 19 (5). Available from: https://www.mdpi.com/1660-3397/19/5/246/pdf. doi:10.3390/md19050246.

Vásquez, P., Cian, R.E., Drago, S.R., 2023. Marine Bioactive Peptides (Fishes, Algae, Cephalopods, Molluscs, and Crustaceans). Springer Science and Business Media LLC, pp. 1–30. doi:10.1007/978-3-030-81404-5_16-1.

Vijayaram, S., Ringø, E., Ghafarifarsani, H., Hoseinifar, S.H., Ahani, S., Chou, C.C., 2024. Use of algae in aquaculture: a review. Fishes. 24103888. 9 (2). Available from: http://www.mdpi.com/journal/fishes/. doi:10.3390/fishes9020063.

Visuddho, V., Halim, P., Helen, H., Muhar, A.M., Iqhrammullah, M., Mayulu, N., Surya, R., Tjandrawinata, R.R., Ribeiro, R.I.M.A., Tallei, T.E., Taslim, N.A., Kim, B., Syahputra, R.A., Nurkolis, F., 2024. Modulation of apoptotic, cell cycle, DNA repair, and senescence pathways by marine algae peptides in cancer therapy. Mar. Drugs. 16603397. 22 (8). Available from: http://www.mdpi.com/journal/marinedrugs. doi:10.3390/md22080338.

Wang, M., Zhou, J., Tavares, J., Pinto, C.A., Saraiva, J.A., Prieto, M.A., Cao, H., Xiao, J., Simal-Gandara, J., Barba, F.J., 2023. Applications of algae to obtain healthier meat products: a critical review on nutrients, acceptability and quality. Crit. Rev. Food Sci. Nutr. 15497852. 63 (26), 8357–8374. Available from: https://www.tandfonline.com/loi/bfsn20. doi:10.1080/10408398.2022.2054939.

Wang, P., Zhang, J., Tang, Y., Zhang, Z., Zhang, Y., Jiamiao, H., 2021. Purification and characterization of antioxidant peptides from hairtail surimi hydrolysates and their effects on beef color stability. J. Food Sci. 0022-1147. 86 (7), 2898–2909. doi:10.1111/1750-3841.15804.

Wang, Q., Liu, F.J., Wang, X.M., Zhao, G.H., Cai, D., Yu, J.H., Yin, F.W., Zhou, D.Y., 2022. Preparation and hepatoprotective activities of peptides derived from mussels (*Mytilus edulis*) and Clams (*Ruditapes philippinarum*). Mar. Drugs. 16603397. 20 (11). doi:10.3390/md20110719.

Wang, X., Liu, J., Cui, S., Wang, Z., Zihao, Y., Yifang, X., Tang, M., Chunxiu, L., Chen, C., Huang, Y., Huang, W., Zhang, Y., Zhang, Z., 2024. Effects of terrestrial animal fats replacing dietary fish oil on the growth performance, antioxidant capacity and lipid metabolism in juvenile *Scylla paramamosain*. Aquac. Rep. 23525134. 35, 102016. doi:10.1016/j.aqrep.2024.102016.

Wu, J.Y., Tso, R., Teo, H.S., Haldar, S., 2023. The utility of algae as sources of high value nutritional ingredients, particularly for alternative/complementary proteins to improve human health. Front. Nutr. 2296861X 10. Available from: http://www.journal.frontiersin.org/journal/nutrition. doi:10.3389/fnut.2023.1277343.

Xiao, B., Yuanyuan, L., Lin, Y., Lin, J., Zhang, L., Daren, W., Zeng, J., Jian, L., Wen Liu, J., Guiling, L., 2022. Eicosapentaenoic acid (EPA) exhibits antioxidant activity via mitochondrial modulation. Food Chem. 03088146. 373 (Pt A), 131389. doi:10.1016/j.foodchem.2021.131389.

Xie, D., Gong, M., Wei, W., Jin, J., Wang, X., Wang, X., Jin, Q., 2019. Antarctic krill (*Euphausia superba*) oil: a comprehensive review of chemical composition, extraction technologies, health benefits, and current applications. Compr. Rev. Food Sci. Food Saf. 15414337. 18 (2), 514–534. Available from: http://www.onlinelibrary.wiley.com/journal/10.1111/(ISSN)1541-4337. doi:10.1111/1541-4337.12427.

Xu, C., Zhang, S., Sun, B., Xie, P., Liu, X., Chang, L., Fushan, L., Zhang, S., 2021. Dietary supplementation with microalgae (*Schizochytrium* sp.) improves the antioxidant status, fatty acids profiles and volatile compounds of beef. Animals. 2076-2615. 11 (12), 3517. doi:10.3390/ani11123517.

Xu, H., Lv, S., Jiang, S., Lu, J., Lin, L., 2020. Radical scavenging activities of peptide from Asian clam (*Corbicula fluminea*) and its protective effects on oxidative damage induced by hydrogen peroxide in HepG2 cells. J. Food Biochem. 17454514. 44 (3). Available from: http://www.onlinelibrary.wiley.com/journal/10.1111/(ISSN)1745-4514. doi:10.1111/jfbc.13146.

Xu, Z., Zhu, Z., Maolin, T., Chang, J., Han, S., Han, L., Chen, H., Tan, Z., Ming, D., Li, T., 2023. Characterizations and the mechanism underlying cryoprotective activity of peptides from enzymatic hydrolysates of *Pseudosciaena crocea*. Foods. 2304-8158. 12 (4), 875. doi:10.3390/foods12040875.

Yalcin, B., Senem, K., Meral, L.-A., 2024. Investigation of the use of microalgae in ice cream formulation. ITU J. Food Sci. Technol. 2, 47–56.

Yang, Y., Lei, D., Hosokawa, M., Miyashita, K., 2020. Total lipids content, lipid class and fatty acid composition of ten species of microalgae. J. Oleo Sci. 1345-8957. 69 (10), 1181–1189. doi:10.5650/jos.ess20140.

Yoon, I.S., Lee, H.J., Kang, S.I., Park, S.Y., Kang, Y.M., Kim, J.S., Heu, M.S., 2019. Food functionality of protein isolates extracted from yellowfin tuna (*Thunnus albacares*) roe using alkaline solubilization and acid precipitation process. Food Sci. Nutr. 20487177. 7 (2), 412–424. Available from: http://www.onlinelibrary.wiley.com/journal/10.1002/%28ISSN%292048-7177. doi:10.1002/fsn3.793.

Zaitseva, L.V., 2022. Confectionery products with omega-3 fatty acids in the focus of a healthy lifestyle. Tovaroved. Prodovolstvenn. Tovarov. 5, 328–330. doi:10.33920/igt-01-2205-08.

Zeng, Y., Weinan, L., Liu, Y., Jiang, W., 2022. Antifreeze peptides preparation from tilapia skin and evaluation of its cryoprotective effect on *Lacticaseibacillus rhamnosus*. Foods. 2304-8158. 11 (6), 857. doi:10.3390/foods11060857.

Zhang, F., Jonas, L., Lin, H., Hill, R.T., 2019. Microbially mediated nutrient cycles in marine sponges. FEMS Microbiol. Ecol. 15746941. 95 (11), Available from http://www.femsec.oxfordjournals.org/. doi:10.1093/femsec/fiz155.

Zhang, J., Ahmmed, M.K., Regenstein, J.M., Wu, H., 2024a. Recent advances of recycling proteins from seafood by-products: industrial applications, challenges, and breakthroughs. Trends Food Sci. Technol. 09242244. 149. Available from: https://www.sciencedirect.com/science/journal/09242244. doi:10.1016/j.tifs.2024.104533.

Zhang, W., Boateng, I.D., Xu, J., 2024b. Novel marine proteins as a global protein supply and human nutrition: extraction, bioactivities, potential applications, safety assessment, and deodorization technologies. Trends Food Sci. Technol. 09242244. 143. Available from: https://www.sciencedirect.com/science/journal/09242244. doi:10.1016/j.tifs.2023.104283.

Zhang, X., Ning, X., Xiaoxiao, H., Sun, X., Xinjian, Y., Cheng, Y., Ri-Qing, Y., Yuping, W., Shu-Chien, A.C., 2020. Fatty acid composition analyses of commercially important fish species from the Pearl River Estuary, China. PLoS One. 1932-6203. 15 (1), e0228276. doi:10.1371/journal.pone.0228276.

Zhao, Y., Khalesi, H., Jun, H., Fang, Y., 2023. Application of different hydrocolloids as fat replacer in low-fat dairy products: ice cream, yogurt and cheese. Food Hydrocoll. 0268005X. 138, 108493. doi:10.1016/j.foodhyd.2023.108493.

Zheng, Y., Yushuang, L., Cong, K., Gao, X., Liu, Z., Chen, J., 2024. Comparison of structural and physicochemical characteristics of skin collagen from chum salmon (cold-water fish) and Nile tilapia (warm-water fish). Foods. 2304-8158. 13 (8), 1213. doi:10.3390/foods13081213.

Zhong, Q., Wei, B., Wang, S., Songze, K., Chen, J., Zhang, H., Wang, H., 2019. The antioxidant activity of polysaccharides derived from marine organisms: an overview. Mar. Drugs. 1660-3397. 17 (12), 674. doi:10.3390/md17120674.

Zhou, J., Feng, Q., Fu, H., Ren, K., Shang, W., Li, C., Zhang, X., MAI, T.T.N., He, Y., 2024. Current trends and perspectives on aquatic-derived protein: a focus on structure-technofunctional properties relationship and application for food preservation. Trends Food Sci. Technol. 09242244. 151. Available from: https://www.sciencedirect.com/science/journal/09242244. doi:10.1016/j.tifs.2024.104651.

Zhou, J., Han, Q., Koyama, T., Ishizaki, S., 2023. Preparation, purification and characterization of antibacterial and ACE inhibitory peptides from head protein hydrolysate of kuruma shrimp, *Marsupenaeus japonicus*. Molecules. 1420-3049. 28 (2), 894. doi:10.3390/molecules28020894.

CHAPTER

14

Marine-based food supplements and food additives

Uvin Eksith Senadheera[a,b] *and Jasintha Jayasanka*[a]

[a]Department of Biosystems Technology, Faculty of Technology, University of Sri Jayewardenepura, Homagama, Sri Lanka

[b]Faculty of Graduate Studies, University of Sri Jayewardenepura, Gangodawila, Nugegoda, Sri Lanka

14.1 Introduction

Marine ecosystems continue to be the birthplace of novel marine-based food additives and supplements for the development of new functional foods. Many studies have utilized biotechnologies for the discovery and synthesis of novel marine-based food additives and supplements to improve the chemical diversity and availability of beneficial marine bioactive compounds (Freitas et al., 2012; Barbosa et al., 2014). Marine ecosystems exhibit most of the underutilized biological resources and have a wide range of creatures with distinctive biological characteristics. In scientific literature, microalgae and algae are identified as excellent resources for marine-based food supplements and additives for applications in functional foods. According to Freitas et al. (2012) and Gallego et al. (2019), compounds extracted from marine organisms are of high interest due to the features of the marine ecosystem, which comprises highly varying degrees of salt, temperature, and illumination. In addition, natural marine molecules derived from algae and invertebrates have made noteworthy contributions to functional food development, and there is space for further diversification (Sigwart et al., 2020). Natural organic bioactive compounds such as polyethers, peptides, proteins, fatty acids, enzymes, and polysaccharides are among many potent and highly specialized bioactive chemicals synthesized by marine organisms. Proteins obtained from marine sources show excellent potential for functional components of foods due to their multifarious, distinctive, and significant characteristics, such as gel-forming ability, antibacterial action, and foaming capacity. Photosynthetic microalgae are one of the most well-studied and promising sources for the sustainable development of natural bioactive marine-based food supplements and additives (Centella et al., 2017; Mondo et al., 2021).

https://doi.org/10.1016/B978-0-443-29190-6.00018-0

The high diversity of the marine ecosystem is considered to be a good source of bioactive peptides, fish oils, fish proteins, microalgae, and macroalgae. Marine plants thrive and survive in hostile conditions by living close to each other. Marine plants have evolved with different physiological adaptations, such as the production of bioactive molecules for protection against grazers. Some marine plants comprise secondary metabolites with potential for food supplements and additives due to their effectiveness in preventing pathogens from growing (Pooja et al., 2016). Due to the elevated carotenoid and polyphenol levels, seaweed is now exploited as a source of food supplements and additives. Products obtained from the postharvesting of marine fish and seafood varieties are also a noteworthy source of marine-based food supplements and food additives. Fish oil [primarily omega-3 polyunsaturated fatty acids (PUFAs)], shark liver oil, chitin, chitosan, enzymes, peptides, vitamins (A, D, and E), and protein hydrolysates are some examples of marine-fish-based food supplements and additives (Shahidi and Ambigaipalan, 2015). According to Trang et al. (2021), marine organisms can synthesize novel compounds, such as glucosidase inhibitors, with potential applications for functional food development for diabetic medication. The high biodiversity and the abundance of estuarine mollusks, as well as their wide range of uses as food supplements and additives, have piqued the attention and interest of food technologists and dieticians for a long time. For example, *Gastropoda, Cephalopoda, Bivalvia,* and phyla *Mollusca* can be readily considered as sources for marine-based food supplements and additives (Chakraborty and Joy, 2020). The increased awareness of health benefits and positive perception of marine-based food supplements and additives has caused a global expansion of the development of such marine-based food supplements and additives. Figs. 14.1 and 14.2 showcase the

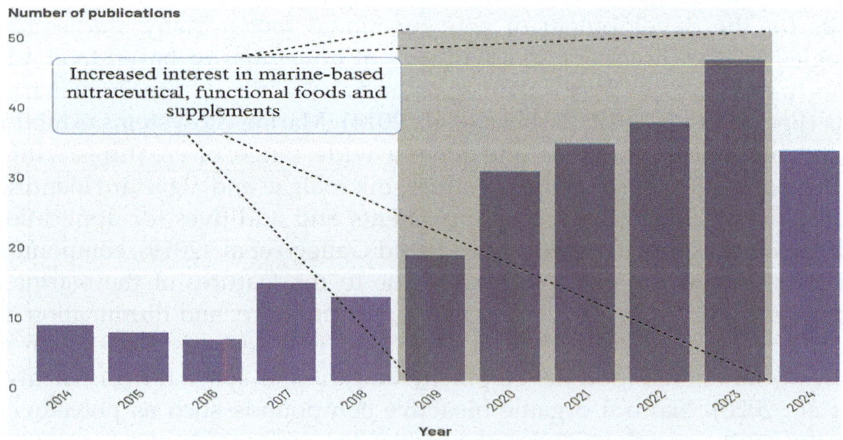

FIG. 14.1 The increased trend in the number of publications on marine-based food supplements and additives over the last decade. The search strategy comprised TITLE-ABS-SCHEMA with primary keywords: "marine-based," "food supplements," "food additives," "nutraceuticals," "food development," and "functional foods." Articles were obtained from a bibliometric survey in the SCOPUS and Web of Science databases. This highlights the recent trend of publications related to marine-based food supplements and food additives in major research databases.

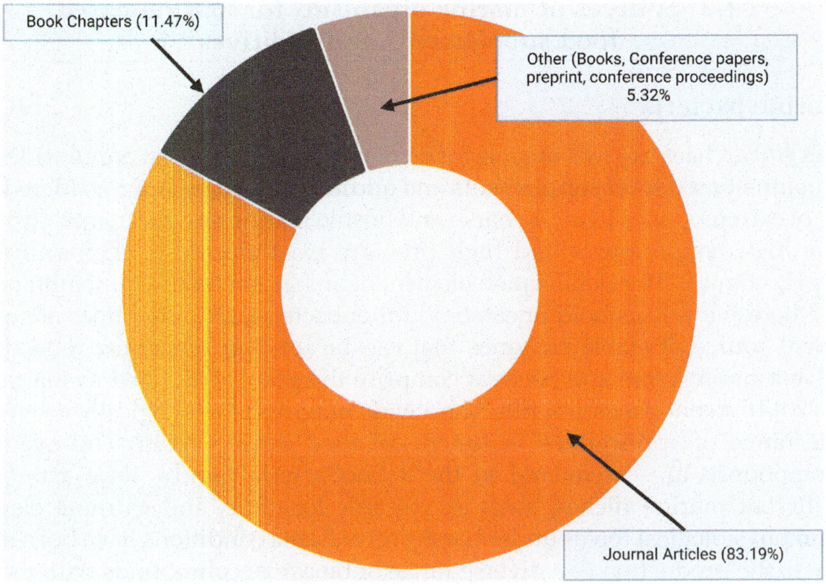

FIG. 14.2 The distribution of research publications on marine-based food supplements and additives over the last decade. The vast majority of publications represented in journal platforms highlight that the focus on marine-based food supplements and additives is mainstream among the research community in both academia and industry. Distribution of marine-based food supplements and additives research articles in publishing platforms.

bibliometric landscape of the marine-based food supplements and food additive research on a global scale.

When compared with terrestrial sources, marine organisms have evolved to synthesize a wide array of bioactive chemical compounds due to the diversity in their ecosystem. The value of functional food supplements and additives, in terms of reducing disease risk and promoting health, has long been acknowledged in the scientific literature (Shahidi and Ambigaipalan, 2015). This study classifies the biomolecules isolated from marine organisms into two categories: marine-based food supplements and marine-based food additives. Marine-based food supplements are used in functional foods to deliver health benefits to users (e.g., antiinflammatory defense and cardiovascular disease prevention) (Suleria et al., 2015). Marine-based food additives are used to enhance the texture, stability, and color of foods, thereby improving the overall quality and shelf life of food products (Karunarathne et al., 2020).

The primary objective of this book chapter is to provide a concise explanation of different aspects of marine-based food supplements and additives, including a precise overview of the marine sources for the extraction of marine-based food supplements and additives, the types of marine-based food supplements and additives, the health benefits of such marine-based food supplements and additives, and the primary uses of marine-based food supplements and additives in conventional food production.

14.2 Sources of marine organisms for marine-based food supplements and additives

14.2.1 Marine bacteria

Eubacteria and archaea are two notable forms of marine bacteria that are considered potential sources for marine-based food supplements and additives. Archaea are considered to comprise the majority of extremophiles living in harsh and hostile marine environments, such as deep-sea vents, where high temperatures and high pressure exist alongside chemically demanding conditions (pH, salinity). The stabilization element of archaea in their cell membranes is glycerol tetraethers, whereas it is hopanoid or catenoid for eubacteria. Extremophiles are considered to be an excellent source of stable enzymes that can be used in functional food development. Gram-positive actinomycetes and bacteria comprise the group identified as marine eubacteria (Kiuru et al., 2014). Actinobacteria synthesize a wide range of physiologically active compounds with a wide range of applications in functional food development. The majority of such bioactive compounds are discovered in the actinobacteria family. Rare actinobacteria are found in different marine niches, such as volcanic locations and extreme elevations. The metabolic and physiological forces under harsh and extreme conditions have been hypothesized to contribute to the production of a diverse range of bioactive compounds with the potential to be used as food supplements and additives in functional foods (Dhakal et al., 2017).

14.2.1.1 Proteobacteria

Numerous bioactive chemical compounds, such as ectoine, vibriobactin, thiomarinol, bromoalterochromide, eryostatins, pentabromopseudilin, solonamide, and violacein, have been identified from Gram-positive actinobacteria, filamentous fungal species, and Gram-negative bacteria, which have garnered less attention (Buijs et al., 2019; Speitling et al., 2007; Zhang et al., 2017; Carbonell, 2000). Nevertheless, genome mining studies reveal that the Gram-negative bacterium of the phylum Proteobacteria comprises gene clusters likely to encode beneficial compounds with the potential to be used as functional food ingredients. Even though Proteobacteria is the most prevalent phylum in pelagic marine ecosystems, few bioactive compounds have been discovered from this phylum with the potential to be used as functional food ingredients, such as supplements and additives (Buijs et al., 2019; Desriac et al., 2013).

14.2.1.2 Microalgae

Sea algae are unicellular or multicellular benthic marine plants that are macroscopic or microscopic in nature (Freile-Pelegrin and Robledo, 2014; Kadam et al., 2013). According to Balasubramaniam et al. (2021), microalgae have garnered attention as an excellent tool for the long-term production of a wide range of bioactive compounds with the potential to be used as functional foods, such as carotenoids, chlorophylls, phycobiliproteins, fatty acids, and vitamins. Food supplements and additives originating from microalgae sources have potential medicinal capabilities, such as anticancer, antioxidant, antifungal, antiviral, antimalarial, and antiinflammatory activities. Furthermore, microalgae act as natural sources of bioactive compounds for food supplements and additives, such as vital amino acids, enzymes, vitamins, carotenoids, PUFAs, fibers, and minerals. Microalgae can be used as a whole for the functional components of the organism to boost health and reduce disease occurrences (Matos et al., 2017). Minerals extracted from microalgal biomass are a key additive in functional foods, and the usage of enzymatic, supercritical, microwave, high-pressure, and ultrasonic solid–liquid and liquid–liquid extraction protocols can be successfully employed for profitable, simple, and long-term recovery of bioactive compounds for food supplements and additives (Menaa et al., 2021). According to Camacho et al. (2019), microalgae-based food supplements and additives have the potential to be beneficial in diabetes, gastric ulcers, constipation, anemia, and hypertension. Some of the unique properties of microalgae include: (1) the ability to be isolated from marine ecosystems and grown in biomass bioreactors, (2) a wide range of different strains and species with the potential to produce bioactive compounds to be used as food supplements and additives, (3) the usage of minimal arable land for microalgae mass production, (4) simple cultivation methods, and (5) the ability to thrive in adverse conditions, such as cold or hot springs and high salinity, which makes them an excellent candidate as a source for functional food development (Khoo et al., 2020; Gallego et al., 2019).

14.2.1.3 *Cyanobacteria*

According to Stincone and Brandelli (2020), unicellular and filamentous cyanobacteria cultivated in axenic cultures are still under exploration for bioactive compounds with the potential to be used as food supplements and additives. Screening for potential bioactive compounds has resulted in the identification of several compounds synthesized by cyanobacteria. Many of these compounds have significant biological functions and chemical conformations, as well as the ability to influence the biochemical processes of cells. The majority of cyanobacteria generate secondary metabolites that are not mandatory for their metabolism and growth. *Oscillatoria, Anabaena, Nostoc,* and *Microcystis* are some of the potential sources of cyanobacteria for the identification of marine-based food supplements and additives such as cryptophycins, dolastatin, and curacin A (Kiuru et al., 2014).

14.2.1.4 *Macroalgae and seaweed*

Seaweed or macroalgae are considered to be an untapped source of novel marine-based food supplements and additives, such as carbohydrates and peptides. Numerous bioactive compounds isolated from macroalgae demonstrate several biological actions, including antioxidant and antihypertensive properties, which can be effectively utilized in the development of functional food supplements and additives. Significantly, the peptides and polysaccharides extracted from macroalgae exhibit antitumor, antiinflammatory, antioxidant, anticoagulant, and immunostimulatory properties (Jesumani et al., 2019; Zhao et al., 2017; Lafarga et al., 2020). According to Harnedy and Fitzgerald (2011), several proteinaceous compounds, such as amino acids, linear peptides, proteins, and amino acid-like molecules, have been discovered in macroalgae with the potential to be used as marine-based food supplements and additives. Furthermore, marine macroalgae are recognized for their useful bioactive compounds, such as phlorotannins. This polyphenolic molecule can be used in a marine-based food supplement or directly as a food additive due to its antihyperlipidemic and antihyperglycemic characteristics. Because of their ability to mediate hyperlipidemia and antihyperglycemia, phlorotannins from marine macroalgae are being utilized in the development of functional foods for individuals with type 2 diabetes and cardiovascular diseases (Rani and Yadav, 2018). According to Tanna and Mishra (2018), seaweeds are being utilized as a source for the sustainable production of functional food additives and supplements owing to their medicinal properties. Seaweed proteins are primarily used as marine-based food supplements due to their nutrient-dense nature.

14.2.1.5 *Marine actinomycetes*

According to Mahapatra et al. (2019), marine actinomycetes are recognized for the biosynthesis of a variety of antibiotics and have the potential to be explored as a source of bioactive compounds for functional food development. It is essential to focus attention on the symbiotic marine actinomycetes with symbiotic relationships with other marine organisms such as cone snails, marine sponges (*Haliclona*, *Axinella polypoides*, etc.), invertebrates, and flora-like insects to screen novel bioactive compounds for food supplement and additive development.

14.2.1.6 *Marine invertebrates*

Marine invertebrates such as Echinodermata, mollusks, and crabs are being utilized as organic biopolymers. These biopolymers are rich in proteins and polysaccharides, with the potential to be used as nutraceuticals in marine-based food supplements and additives. According to Ganesan et al. (2020), biopolymeric compounds such as chitosan, glycosaminoglycans, chitin, and fucosylated chondroitin sulfate hold the potential to be used in marine-based food supplements and additives due to their excellent health benefits, ranging from wound healing and antimicrobial capacity to anticancer effects. Furthermore, gelatin and collagen are also commercially important food additives derived from edible marine invertebrates (Ganesan et al., 2019). Sponges are long-acknowledged marine invertebrates rich in bioactive compounds. Their unusual sterols might be an excellent precursor for developing marine-based food supplements and additives. Shrimps are one of the popular marine invertebrates consumed around the world and are rich in triglycerides, fatty acids, and carotenoids, with functional food applications alongside their therapeutic effects against cancer. Furthermore, marine invertebrates such as starfish, sea urchins, and sea cucumbers have also garnered attention as marine sources due to the presence of bioactive compounds such as asterosaponins, which are sterol derivatives with the potential for food supplement and additive development (John et al., 2018; Datta et al., 2015; Papon et al., 2021).

14.2.1.7 *Marine fishes*

According to Šimat (2021) and Šimat et al. (2021), current advancements in the field of nutraceuticals, functional food supplements, and additive development have opened novel avenues for the creation of marine-based food supplements and additives derived from marine fishes as a source. By-products such as peptides, collagen, and enzymes from marine fish are also being utilized to develop food supplements and additives with values extending beyond mere nutritional benefits, providing therapeutic effects as well. The presence of minerals such as iron, zinc, calcium, and selenium; vitamins such as A, D, E, B_6, B_3, and B_{12}; and unsaturated essential fatty acids makes marine fish an excellent source for marine-based food supplements and additives. Furthermore, peptides extracted from fermented marine fish hold the potential to be used as a food additive, alongside their therapeutic effects on viral infections, cancer, Alzheimer's disease, and varying degrees of hypertension (Suleria et al., 2015). Fish oil is also a widely used food supplement that heavily relies on marine fish sources for its production (Kari et al., 2022, 2021). Table 14.1 provides an overview of a diverse range of marine-based food supplements isolated from marine organisms and additives that exhibit beneficial properties in foods.

TABLE 14.1 An overview of a diverse range of marine-based food supplements isolated from marine organisms and additives with beneficial properties in foods.

Classification	Bioactive compounds	Marine source of isolation	Mechanism of bioaction	Application in food development	References
Protein and peptides	Gelatin, collagen, albumin	Bonito, tuna, sardine, prawn, oyster, cod, algae, mackerel, jumbo squid, krill, dried salmon, mussel, herring	Hypertensive, anticoagulant, and antioxidant	Production of edible coatings for meat products, clarifiers in low-fat meals, texturizers, thickeners, and stabilizers in low-fat meals	(Shahidi and Ambigaipalan, 2015; Suleria et al., 2015)
Minerals	Zn, Fe, I, and Mn	All marine organisms (the majority of seaweeds)	Food fortification	Pharmaceutical, food and nutraceutical manufacturing	(Jimenez-Lopez et al., 2020)
Marine-based enzymes	Lipases, pepsins, chymosins, gastricsins, serine, and cysteine proteases	Short-finned squid, mollusks, and crustaceans	Antiinflammatory and antioxidant activity	Removal of unfavorable color changes in food, squid and herring fermentation, and meat tenderizing	(Mateos et al., 2020)
Carbohydrates	Fucans, chitin, ascophyllan, polyuronides, chitosan, carrageenans, fucanoids, laminarin, glutathione, alginates, agar-agar	Macroalgae	Antiinflammatory, antithrombotic, antiviral, anticoagulant, bactericidal, fungicidal, and antitumor	Production of gelling agents, gel coatings for meat, acidification and clarification of fruit products, and edible protective film production	(Mayakrishnan et al., 2013; Wang et al., 2009)
Fatty acids	Omega-3 fatty acids	Algae, fungi, mackerel, sardine, krill, menhaden, cod, halibut, salmon, Pacific anchovy, whale blubber, seal blubber	Cancer prevention, retina and brain development enhancement, artery disease, coronary arthritis, inflammation, and diabetes prevention	Fish oil and capsule production, infant formula production, and livestock fortification	(Martins et al., 2013; Moi et al., 2018)
Pigments and other phenolic compounds	Xanthophylls, lycopene, and α-, β-, ε-carotene	Sea urchin, red snapper, shrimp, algae, mussel, lobster, squid, crayfish, starfish, trout, coral, tuna, sea cucumber, sea anemones, pink salmon, crab, octopus, sea bream, salmon, redfish, sponges, fungi.	Neurodegenerative and cardiovascular disease progression prevention, anticancer, antioxidant, antiinflammatory, and immunomodulatory characteristics	Food colorings and nutraceuticals	(Haoujar et al., 2019; Shahidi and Ambigaipalan, 2015; Šimat et al., 2020)

14.3 Marine-based food supplements utilized in conventional food, functional foods, and nutraceuticals development

14.3.1 Proteins and protein hydrolysate

Marine-based proteins are extensively used due to their nutritional and therapeutic properties in functional food supplements. Marine-based proteins are often utilized as a source of agar, edible alginates for foods, and dietary supplements. One of the most commonly used forms of marine proteins is seaweed (Pooja et al., 2016). Fish protein hydrolysates (FPHs), produced from the digestion of unused fish meat under the influence of proteolytic enzymes, are considered excellent marine-based food supplements due to their potential pharmaceutical and nutritional applications. Protein hydrolysates comprise a superior balance in amino acid composition with good digestion and rapid gastrointestinal uptake (Elias et al., 2008). According to Shahidi and Zhong (2008), FPH is regarded as an excellent food supplement, contributing to texture, water holding, gelling, and emulsification characteristics when incorporated into food. Furthermore, these FPHs can provide therapeutic effects such as antithrombotic, immunomodulatory, anticancer, and antihypertension benefits when added to foods.

Furthermore, the presence of antihypertensive peptides in FPHs is effective as a food supplement that can be incorporated into foods targeting hypertensive populations. Antihypertensive peptides are effective in preventing hypertensive conditions due to the inhibition of angiotensin-converting enzymes (ACEs) (Gu et al., 2011). ACE-inhibitory peptides can be found in FPHs extracted from the Atlantic salmon, cuttlefish, and sea cucumber (Amado et al., 2013).

According to Zhu et al. (2010), marine-based collagen peptides can be effectively used as food supplements and offer therapeutic protection against hypertension and diabetes by regulating the compounds involved in the pathogenesis of these conditions. Furthermore, marine protein hydrolysates from smooth round muscle and skipjack roe can be effectively used as functional food supplements for their potential antioxidant properties (Intarasirisawat et al., 2013; Wang et al., 2013). A study by Nalinanon et al. (2011) shows that the protein hydrolysate extracted from skipjack tuna pepsin can be effectively utilized as a functional food supplement with antioxidant effects. In addition, marine collagen peptide from chum salmon skin can be effectively used as a food supplement in targeted functional foods for populations with memory deficiencies (Pei et al., 2009). Zhu et al. (2010) also show that the protein hydrolysate of salmon skin has the potential to be used as a food supplement for populations with type 2 diabetes.

Color is believed to be the primary element perceived by the senses, and it has played an important role in the acceptance of foods for millennia, improving their actual appearance and quality (Ghosh et al., 2021a, 2021b). The microalgal pigment is a natural food coloring component that is used commercially. Carotene is found in large concentrations in some microalgae (in addition to beta-carotene). Other types of colors can also be observed in microalgae. Beta-carotene is used as a food pigment (most notably to give margarine its yellow color), as a food additive to enhance the color of fish flesh and egg yolks, and to improve the health and fertility of grain-fed cattle. Natural beta-carotene has physical features that distinguish it from synthetic beta-carotene (Aman Mohammadi et al., 2022). Natural beta-carotene is a far superior anticarcinogen and antiheart disease agent because it is fat-soluble. The role of microalgae as a food coloring source is limited, however, because algal-derived food coloring is not photostable and bleaches with cooking. Despite this limitation, the market for microalgae-derived food coloring has enormous potential. *Dunaliella salina* is cultivated to produce beta-carotene, a photosynthetic pigment. Beta-carotene is an orange color and a vitamin C supplement (Priyadarshani and Rath, 2012). Lutein is a high-value bioactive molecule that has significant industrial applications in aquaculture, poultry farming, food processing, and healthcare (Low et al., 2019).

14.3.1.1 *Mineral compounds*

High mineral content (Na, P, Mg, Ca, and K) can be obtained from marine sources, as well as trace minerals such as I, Mn, and Zn from sources such as seaweed (Kasmiati et al., 2022). The characteristic high mineral content of seaweed is due to its unique ability to retain inorganic compounds owing to the surface polysaccharides in the cells. A large variety of seaweeds are recommended to be taken as food supplements to meet the mineral demand for Ca and K (Imchen, 2021; Muñoz and Díaz, 2020). Furthermore, other marine organisms are also used as sources of minerals and incorporated as food supplements, such as some species of shellfish, which are rich in Se, I, Zn, Mg, and Ca (Davis and Gatlin, 1996). Utilization of marine-based Se in the diet reduces the risk of cancer (Wallace et al., 1988). Obtaining iodine supplements through seaweed instead of fortified salt iodine is an alternative for developing countries with poor iodine consumption (Aakre et al., 2023).

14.3.1.2 *Amino acids and marine peptides*

Water-soluble amino acids such as Gln, Arg, Gly, Pro, and Ala are abundantly found in the muscle tissues of marine organisms. Due to the quick deterioration of water-soluble amino acids, to retain these amino acids, seafood should be lightly prepared. Consumption of marine-based amino acids originating from the muscle tissue of marine fish in large quantities can boost human nutrition (Dewapriya and Kim, 2013; Romano et al., 2016). Marine-based peptides are isolated from algae, marine fish, and crustaceans. These bioactive marine peptides comprise 3–20 amino acid residues. The marine peptides used in functional food development exhibit antithrombotic, immunomodulatory, antioxidant, antihypertensive, and antibacterial activity (Hamed et al., 2015). According to Lee and Hur (2017), marine-based peptides are highly beneficial as antihypertensive compounds, which can be integrated into functional foods. Table 14.2 provides an overview of marine-based peptides with beneficial characteristics and potential for incorporation into functional foods as therapeutic action biomolecules.

14.3.1.3 *Omega-3 oils*

Omega-3 PUFAs are primarily composed of docosahexaenoic acid (DHA) and eicosapentaenoic acid (EPA). Docosapentaenoic acid is detected in very low concentrations in fish oil compared to EPA and DHA. DHA and EPA are biosynthesized by multicellular and unicellular marine plants (e.g., algae and phytoplankton) (Coupland, 2008; Avijit et al., 1999). There is critical evidence to support that omega-3 fatty acids possess beneficial implications for human health. In vitro and animal studies demonstrate that the utilization of omega-3 fatty acids as food supplements affects cardiovascular health, lipid profiles, eicosanoid biosynthesis, lipid composition, gene expression, and signaling cascades (Islam et al., 2018; Shahidi and

TABLE 14.2 An overview of marine-based peptides with beneficial characteristics for functional food development, intended for use as therapeutic molecules.

Type of marine-based peptide	Source of isolation	Mechanism of bioaction	References
Pipecolidepsin A	*Homophymia lamellosa*	Cytotoxic action on A549, MDA-MB-231	(Pelay-Gimeno et al., 2013)
Hemiasterlin	*Cymbastela, Hemiasterella*	NFF cytotoxicity and PC3 cytotoxicity	(Zheng et al., 2011)
Mycothiazole	*Spongia mycofijiensis*	Antihelminthic action	(Zheng et al., 2011)
Phakellistatin 16	*Phakellia* sp.	Cytotoxic action on P388 and BEL-7402	(Zhang et al., 2010)
Pembamide	*Cribrochalina*	Cytotoxic action on MDA-MB-231 and HT-29	(Urda et al., 2016)
Phakellistatin 15	*Phakellia* sp.	Antitumor activity	(Shaheen et al., 2016)
Carteritin A	*Stylissa carteri*	Cytotoxic action on HCT116	(Afifi et al., 2016)
Reniochalistatin E	*Reniochalina stalagmitis*	Cytotoxic action on MGC-803 and RPMI-8226	(Zhan et al., 2014)

Miraliakbari, 2004). The utilization of omega-3 fatty acids as food supplements is able to delay the onset of autoimmune, cancer, cardiovascular, and mental diseases (Wang et al., 2009). Omega-3 fatty acids primarily originate from marine mammals and fish and also come from marine algae. The white lean fish liver (halibut and cod), the body of fatty fish (herring and mackerel), marine mammal blubber (whales and seals), and krill are widely used for manufacturing omega-3 fatty oils available on the market (Shahidi, 2009). Omega-3 fatty oils can also be extracted from marine fungal and algal oils. DHA oils isolated from marine microalgae are commercially available as food supplements. Algal oils comprise predominantly one type of PUFA rather than a combination of different PUFA. This makes algal oils popular as a food supplement compared to fish oils, which often comprise a combination of PUFA (Santigosa et al., 2020; Sukenik et al., 1994). The Food and Drug Administration has concluded that food supplements with omega-3 fatty acids (EPA and/or DHA) up to 3 g/day are generally recommended as safe to consume (Tahergorabi et al., 2013; Pietrowski et al., 2011).

14.4 Marine-based food additives utilized in conventional food products, functional foods, and nutraceuticals development

14.4.1 Chitosan, chitosan oligosaccharides, and glucosamine

Chitin and chitosan polymers have received significant attention in the food development field. Chitin is subjected to partial deacetylation to form a wide range of derivatives, such as chitosan oligosaccharides, chitosan, and glucosamine (Satitsri and Muanprasat, 2020;

Tsvetkov et al., 2020). Chitin derivatives are extensively used in functional food development due to their increased water solubility. Chitosan derivatives are used as cholesterol-reducing and fat-reducing agents in foods. Nevertheless, the ability of chitin and chitin derivatives to reduce fat absorption is highly questioned in the field of nutrition science (Setiati et al., 2021). The biodegradability, noncytotoxicity, and biocompatibility make chitin and chitin derivatives widely used marine-based compounds in food products (Raafat and Sahl, 2009). According to Zhang et al. (2012), low-molecular-weight chitosan is used in high-fat food products to reduce cholesterol buildup in humans and to lower liver triacylglycerol and low-density lipoprotein levels. Furthermore, incorporating chitosan derivatives, such as chitooligosaccharides and low-molecular-weight chitosan, into functional foods enhances antitumor and antibacterial activity, as well as immunity in humans (Kuroiwa et al., 2008). According to Ngo et al. (2010), low-molecular-weight chitooligosaccharides can be used as food additives due to their antioxidant effect on live human cells.

Furthermore, due to the inhibitory action of chitooligosaccharides on oxidative damage to DNA caused by lymphoma U937 and their direct radical scavenging ability, chitooligosaccharides can be used as additives in nutraceuticals. The incorporation of chito-oligosaccharides enables nutritionists to formulate food products with dietary intake and body weight control, as well as blood lipid and glucose-lowering effects (Kumar et al., 2009; Ngo et al., 2010). The incorporation of glucosamine as a food additive enhances antiinflammatory responses and acts as a chondroprotective agent. Marine-based glucosamine is often provided as an oral food additive in the form of glucosamine sulfate to populations with osteoarthritis (Bottegoni et al., 2014; Nagaoka et al., 2010). Chitosan dietary fibers and tablets are already available as commercial products.

14.4.1.1 *Marine-based pigments*

Color enhancement is being used to improve the overall visual acceptance of food and quality (Ghosh et al., 2021a, 2021b). Marine-based microalgal pigments are extensively utilized for food coloring at a commercial level. Out of the marine-based pigments used in the food industry, carotenoids contribute to the yellow and orange coloring of food items. One of the primary uses of carotenoids (primarily β-carotene) is their use in making vitamin A precursors. Furthermore, β-carotene is utilized as a pigment in coloring margarine, improving the color of egg yolks and enhancing the color of fish flesh (Mohammadi et al., 2021). According to Priyadarshani and Rath (2012), the current role of pigments isolated from microalgae is limited.

Nevertheless, the demand for microalgae-based food coloring is gradually building. For example, β-carotene from *D. salina* can be used as a food coloring agent and a vitamin C supplement in foods (Priyadarshani and Rath, 2012). Marine-based carotenoids can inhibit lipase activity, reduce the gastrointestinal tract's fat absorption, and obstruct triacylglycerol absorption in food (Matsumoto et al., 2009). When used in food, fucoxanthin, a primary marine-based carotenoid in brown seaweed, enhances insulin regulation and resistance and lowers blood glucose levels (Miyashita et al., 2011). Furthermore, the use of fucoxanthin in functional food development inhibits human cancer cell growth. For example, fucoxanthin from the marine sea squirt *H. roretzi* suppresses human leukemia cell growth (HL-60), Caco-2 human colon cancer, and human breast cancer. In addition, fucoxanthin is extensively used in functional foods to control metabolic disorders and manage obesity (Miyashita, 2013; Miyashita et al., 2011).

14.4.1.2 Marine enzymes

Marine-based enzymes, vitamins, and minerals are effectively used for the transformation of molecules into useful ingredients. Enzymes such as chitinolytic enzymes, transglutaminase, polyphenol oxidase, and lipase are used in starch degradation (e.g., -1,4-glucanase) and have a marine biological origin. According to Zhang and Kim, 2012, marine enzymes possess superior catalytic properties compared to their terrestrial counterparts. Furthermore, marine enzymes are used as culinary additives. For example, gastric enzymes of marine fish are being used as rennet substitutes in cheese production, fish sauce production, fish protein concentrates, and hydrolysates. A considerable number of marine enzymes from shellfish and fish discards are identified as potential food supplements, such as acetylglucosaminidase, alkaline phosphatase, hyaluronidase, chitinase, and protease. For example, Hayet et al. (2010) have identified a low-molecular-weight serine–protease from *Sardinella aurita* viscera with collagenolytic ability, which can be used as a food supplement for populations with collagen-related diseases.

Furthermore, alkaline phosphatase, identified and characterized from shrimp, has the potential to be used as a food supplement for vitamin D deficiency, as well as a supplement for populations with tumors and hypophosphatasia. In addition to this, marine enzymes can be used to eradicate Gram-positive and Gram-negative bacteria on food items. For example, cold-active peptide enzymes from Arctic scallops can be used for the growth inhibition of Gram-positive and Gram-negative bacteria (Shahidi and Zhong, 2010). Table 14.3 gives an overview of marine enzymes with the potential to be used as functional food additives. Fig. 14.3 provides a comprehensive summary of marine-based food supplements and food additives discussed in the book chapter.

TABLE 14.3 An overview of marine-based enzymes with their specific bioaction in functional foods as additives.

Type of marine enzyme	Isolated source	Biological function in functional foods	References
α-Galactosidases	Yeasts	Oligosaccharides hydrolysis in soybean food products	(De Borba Gurpilhares et al., 2018)
Chitinase	*Alcaligenes* sp., *Streptomyces* sp.	Biotransformation of chitinous marine compounds to be usable in functional foods	(Jahromi and Barzkar, 2018)
Carrageenases	Marine proteobacteria and Bacteroidetes	Carrageenans' degradation by internal β-(1→4) glycosidic hydrolysis	(Ghanbarzadeh et al., 2018)
L-Arginine deiminase	*Aspergillus fumigatus*	Destroying auxotrophic cancer cells	(Toldrá and Kim, 2016)
Lipoxygenase	Trout gills and skin, gray mullet, sea urchin, corals	Polyunsaturated fatty acid oxygenation catalysis	(Ozogul et al., 2021)

TABLE 14.3 An overview of marine-based enzymes with their specific bioaction in functional foods as additives—cont'd

Type of marine enzyme	Isolated source	Biological function in functional foods	References
Xylanase	Filamentous fungi	The brewing of Xylan products in beverages and foods	(Wang et al., 2015)
Monooxygenase	*Dicentrarchus labrax*	Indicator molecule of aromatic hydrocarbon exposure in food production	(Debashish et al., 2005)
Protease	*Bacillus* sp.	Anticancer activity	(Salamone et al., 2011)
L-Arginase	*Penicillium chrysogenum*	L-arginine depletion and anticancer activity	(Toldra and Kim, 2017)
Collagenase	*Clostridium histolyticum*	Collagen hydrolysis in food	(Kim, 2013)
Phenol oxidase	*Perna viridis* (a marine mussel)	Phenolic compound oxidation in foods	(Debashish et al., 2005)

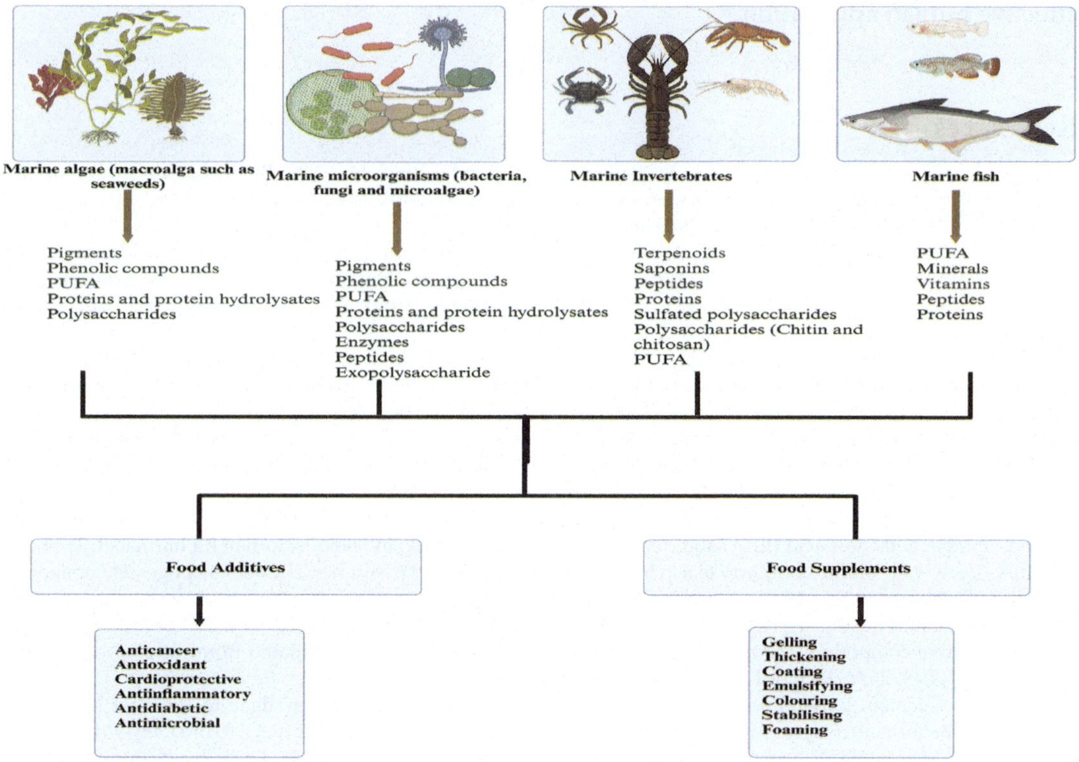

FIG. 14.3 Comprehensive summary of marine-based food supplements and food additives used as food supplements and food additives. The distribution of marine-based food supplements and food additives across marine resources.

14.5 Conclusion

Marine resources provide a wide range of biomolecules with beneficial uses, such as food supplements and additives, with expanding research focusing on their widespread integration into more food products. Among these marine-based food additives and supplements, omega-3 fatty acids have been proven to be highly effective in addressing various health concerns, which drives their integration into the human diet as a food additive or supplement for the conventional human diet. Food supplements and additives obtained from marine algae with a high concentration of antioxidants and omega-3 fatty acids have received increased attention for their unique ability in chronic cardiovascular disease prevention. Marine sponges, fungi, and bacteria remain primarily unexploited, and more attention is needed to isolate and characterize food supplements and additives, as they harbor unique biomolecules as a result of adaptation to extreme marine ecosystems. Extraction method optimization, bioassay, and biosafety assessment are essential in adequately harnessing these biomolecules as food supplements and additives. Future developments will require adequate clinical evaluations to translate already discovered and newly discovered biomolecules into functional foods and additives to ensure safe and effective human application.

References

Aakre, I., Tveit, I.B., Myrmel, L.S., Fjære, E., Ballance, S., Rosendal-Riise, H., 2023. Bioavailability of iodine from a meal consisting of sushi and a wakame seaweed salad—a randomized crossover trial. Food Sci. Nutr. 2048-7177. 11 (12), 7707–7717. doi:10.1002/fsn3.3689.

Afifi, A.H., El-Desoky, A.H., Kato, H., Mangindaan, R.E.P., De Voogd, N.J., Ammar, N.M., Hifnawy, M.S., Tsukamoto, S., 2016. Carteritins A and B, cyclic heptapeptides from the marine sponge *Stylissa carteri*. Tetrahedron Lett. 18733581. 57 (11), 1285–1288. Available from: http://www.journals.elsevier.com/tetrahedron-letters/. doi:10.1016/j.tetlet.2016.02.031.

Amado, I.R., Vázquez, J.A., González, M.P., Murado, M.A., 2013. Production of antihypertensive and antioxidant activities by enzymatic hydrolysis of protein concentrates recovered by ultrafiltration from cuttlefish processing wastewaters. Biochem. Eng. J. 1873295X. (76), 43–54. doi: 10.1016/j.bej.2013.04.009.

Aman Mohammadi, M., Ahangari, H., Mousazadeh, S., Hosseini, S.M., Dufossé, L., 2022. Microbial pigments as an alternative to synthetic dyes and food additives: a brief review of recent studies. Bioprocess Biosyst. Eng. 45 (1), 1–12. doi:10.1007/s00449-021-02621-8.

Avijit, H., Tripathi, S., Ghosh, A., 1999. Pharmacology and therapeutic potential of the n-3 polyunsaturated fatty acids, eicosapentaenoic acid (EPA) and docosahexaenoic acid (DHA) in fish oils. Indian J. Pharmacol. 31 (4), 247. https://www.ijp-online.com/article.asp?issn=0253-7613;year=1999;volume=31;issue=4;spage=247;epage=264;aulast=Avijit;type=0.

Balasubramaniam, V., Gunasegavan, R.D.N., Mustar, S., Lee, J.C., Noh, M.F.M., 2021. Isolation of industrial important bioactive compounds from microalgae. Molecules. 14203049. 26 (4). Available from: https://www.mdpi.com/1420-3049/26/4/943/pdf. doi:10.3390/molecules26040943.

Barbosa, M., Valentão, P., Andrade, P.B., 2014. Bioactive compounds from macroalgae in the new millennium: implications for neurodegenerative diseases. Mar. Drugs 12 (9), 4934–4972. doi:10.3390/md12094934.

Bottegoni, C., Muzzarelli, R.A.A., Giovannini, F., Busilacchi, A., Gigante, A., 2014. Oral chondroprotection with nutraceuticals made of chondroitin sulphate plus glucosamine sulphate in osteoarthritis. Carbohydr. Polym. 01448617. 109, 126–138. doi:10.1016/j.carbpol.2014.03.033.

Buijs, Y., Bech, P.K., Vazquez-Albacete, D., Bentzon-Tilia, M., Sonnenschein, E.C., Gram, L., Zhang, S.D., 2019. Marine Proteobacteria as a source of natural products: advances in molecular tools and strategies. Nat. Prod. Rep.

14604752. 36 (9), 1333–1350. Available from: http://www.pubs.rsc.org/en/journals/journal/np. doi:10.1039/c9np00020h.

Camacho, F., Macedo, A., Malcata, F., 2019. Potential industrial applications and commercialization of microalgae in the functional food and feed industries: a short review. Mar. Drugs 17 (6), 312. doi:10.3390/md17060312.

Carbonell, G., 2000. Clinical relevance and virulence factors of pigmented *Serratia marcescens*. FEMS Immunol. Med. Microbiol. 28 (2), 143–149. doi:10.1016/s0928-8244(00)00146-2.

Centella, M.H., Arévalo-Gallegos, A., Parra-Saldivar, R., Iqbal, H.M.N., 2017. Marine-derived bioactive compounds for value-added applications in bio- and non-bio sectors. J. Clean. Prod. 09596526. 168, 1559–1565. doi:10.1016/j.jclepro.2017.05.086.

Chakraborty, K., Joy, M., 2020. High-value compounds from the molluscs of marine and estuarine ecosystems as prospective functional food ingredients: an overview. Food Res. Int. 09639969. 137, 109637. doi:10.1016/j.foodres.2020.109637.

Coupland, K., 2008. Stearidonic acid: a plant produced omega-3 PUFA and a potential alternative for marine oil fatty acids. Lipid Technol. 0956-666X. 20 (7), 152–154. doi:10.1002/lite.200800045.

Datta, D., Talapatra, S.N., Swarnakar, S., 2015. Bioactive compounds from marine invertebrates for potential medicines—an overview. Int. Lett. Nat. Sci. 34, 42–61. doi:10.56431/p-i22ej9.

Davis, D.A., Gatlin, D.M., 1996. Dietary mineral requirements of fish and marine crustaceans. Rev. Fish. Sci. 10641262. 4 (1), 75–99. Available from: http://www.tandfonline.com/loi/brfs21#.V0fw6k1f3cs. doi:10.1080/10641269609388579.

De Borba Gurpilhares, D., Cinelli, L.P., Simas, N.K., Pessoa Jr, A., Sette, L.D., 2018. Marine prebiotics: polysaccharides and oligosaccharides obtained by using microbial enzymes. Food Chem. 280, 175–186. doi:10.1016/j.foodchem.2018.12.023.

Debashish, G., Malay, S., Barindra, S., Joydeep, M., 2005. Marine enzymes. Adv. Biochem. Eng. Biotechnol. 07246145. 96, 189–218. doi:10.1007/b135785.

Desriac, F., Jégou, C., Balnois, E., Brillet, B., Chevalier, P., Fleury, Y., 2013. Antimicrobial peptides from marine proteobacteria. Mar. Drugs. 1660-3397. 11 (10), 3632–3660. doi:10.3390/md11103632.

Dewapriya, P., Kim, S., 2013. Marine microorganisms: an emerging avenue in modern nutraceuticals and functional foods. Food Res. Int. 56, 115–125. doi:10.1016/j.foodres.2013.12.022.

Dhakal, D., Pokhrel, A.R., Shrestha, B., Sohng, J.K., 2017. Marine rare actinobacteria: isolation, characterization, and strategies for harnessing bioactive compounds. Front. Microbiol. 1664302X. 8. Available from: http://www.journal.frontiersin.org/article/10.3389/fmicb.2017.01106/full. doi:10.3389/fmicb.2017.01106.

Elias, R.J., Kellerby, S.S., Decker, E.A., 2008. Antioxidant activity of proteins and peptides. Crit. Rev. Food Sci. Nutr. 48 (5), 430–441. doi:10.1080/10408390701425615.

Freile-Pelegrin, Y., Robledo, D.R., 2014. Bioactive phenolic compounds from algae. In: Hernández-Ledesma, B., Herrero, M. (Eds.), Bioactive Compounds from Marine Foods: Plant and Animal Sources. Wiley, pp. 797–802.

Freitas, A.C., Rodrigues, D., Rocha-Santos, T.A.P., Gomes, A.M.P., Duarte, A.C., 2012. Marine biotechnology advances towards applications in new functional foods. Biotechnol. Adv. 07349750. 30 (6), 1506–1515. doi:10.1016/j.biotechadv.2012.03.006.

Gallego, R., Bueno, M., Herrero, M., 2019. Sub- and supercritical fluid extraction of bioactive compounds from plants, food-by-products, seaweeds and microalgae—an update. TrAC Trends Anal. Chem. 116, 198–213. doi:10.1016/j.trac.2019.04.030.

Ganesan, A.R., Guru, M.S., Balasubramanian, B., Mohan, K., Liu, W.C., Arasu, M.V., Al-Dhabi, N.A., Duraipandiyan, V., Ignacimuthu, S., Sudhakar, M., Seedevi, P., 2020. Biopolymer from edible marine invertebrates: a potential functional food. J. King Saud Univ. Sci. 32 (2), 1772–1777. doi:10.1016/j.jksus.2020.01.015.

Ganesan, A.R., Mohanram, M.S.G., Balasubramanian, B., Kim, I.H., Seedevi, P., Mohan, K., Kanagasabai, S., Arasu, M.V., Al-Dhabi, N.A., Ignacimuthu, S., 2019. Marine invertebrates' proteins: a recent update on functional property. J. King Saud Univ. Sci. 32 (2), 1496–1502. doi:10.1016/j.jksus.2019.12.003.

Ghanbarzadeh, M., Golmoradizadeh, A., Homaei, A., 2018. Carrageenans and carrageenases: versatile polysaccharides and promising marine enzymes. Phytochem. Rev. 17 (3), 535–571. doi:10.1007/s11101-018-9548-2.

Ghosh, S., Sarkar, T., Das, A., Chakraborty, R., 2021a. Micro and nanoencapsulation of natural colors: a holistic view. Appl. Biochem. Biotechnol. 0273-2289. 193 (11), 3787–3811. doi:10.1007/s12010-021-03631-8.

Ghosh, S., Sarkar, T., Das, A., Chakraborty, R., 2021b. Natural colorants from plant pigments and their encapsulation: an emerging window for the food industry. LWT 153, 112527. doi:10.1016/j.lwt.2021.112527.

Gu, R.-Z., Chen-Yue, L., Liu, W.-Y., Wei-Xue, Y., Cai, M.-Y., 2011. Angiotensin I-converting enzyme inhibitory activity of low-molecular-weight peptides from Atlantic salmon (*Salmo salar* L.) skin. Food Res. Int. 09639969. 44 (5), 1536–1540. doi:10.1016/j.foodres.2011.04.006.

Hamed, I., Özogul, F., Özogul, Y., Regenstein, J.M., 2015. Marine bioactive compounds and their health benefits: a review. Compr. Rev. Food Sci. Food Saf. 15414337. 14 (4), 446–465. Available from: http://www.onlinelibrary.wiley.com/journal/10.1111/(ISSN)1541-4337. doi:10.1111/1541-4337.12136Turkey.

Haoujar, I., Cacciola, F., Abrini, J., Mangraviti, D., Giuffrida, D., Majdoub, Y.O.E., Kounnoun, A., Miceli, N., Taviano, M.F., Mondello, L., Rigano, F., Senhaji, N.S., 2019. The contribution of carotenoids, phenolic compounds, and flavonoids to the antioxidative properties of marine microalgae isolated from Mediterranean Morocco. Molecules. 14203049. 24 (22). Available from: https://www.mdpi.com/1420-3049/24/22/4037/pdf. doi:10.3390/molecules24224037.

Harnedy, P.A., Fitzgerald, R.J., 2011. Bioactive proteins, peptides, and amino acids from macroalgae. J. Phycol. 15298817. 47 (2), 218–232. doi:10.1111/j.1529-8817.2011.00969.x.

Hayet, B.K., Rym, N., Ali, B., Sofiane, G., Moncef, N., 2010. Low molecular weight serine protease from the viscera of sardinelle (*Sardinella aurita*) with collagenolytic activity: purification and characterisation. Food Chem. 124 (3), 788–794. doi:10.1016/j.foodchem.2010.06.096.

Imchen, T., 2021. Nutritional value of marine seaweeds and their potential to serve as nutraceutical supplements. Pharmacogn. Mag. 75 (17), 399–405.

Intarasirisawat, R., Benjakul, S., Jianping, W., Visessanguan, W., 2013. Isolation of antioxidative and ACE inhibitory peptides from protein hydrolysate of skipjack (*Katsuwana pelamis*) roe. J. Funct. Foods. 17564646. 5 (4), 1854–1862. doi:10.1016/j.jff.2013.09.006.

Islam, M.S., Castellucci, C., Fiorini, R., Greco, S., Gagliardi, R., Zannotti, A., Giannubilo, S.R., Ciavattini, A., Frega, N. G., Pacetti, D., Ciarmela, P., 2018. Omega-3 fatty acids modulate the lipid profile, membrane architecture, and gene expression of leiomyoma cells. J. Cell. Physiol. 10974652. 233 (9), 7143–7156. Available from: http://www.onlinelibrary.wiley.com/journal/10.1002/(ISSN)1097-4652. doi:10.1002/jcp.26537.

Jahromi, S.T., Barzkar, N., 2018. Marine bacterial chitinase as sources of energy, eco-friendly agent, and industrial biocatalyst. Int. J. Biol. Macromol. 01418130. 120 (Pt B), 2147–2154. doi:10.1016/j.ijbiomac.2018.09.083.

Jesumani, V., Hong, D., Aslam, M., Pei, P., Huang, N., 2019. Potential use of seaweed bioactive compounds in skincare—a review. Mar. Drugs. 1660-3397. 17 (12), 688. doi:10.3390/md17120688.

Jimenez-Lopez, C., Pereira, A., Lourenço-Lopes, C., Garcia-Oliveira, P., Cassani, L., Fraga-Corral, M., Prieto, M., Simal-Gandara, J., 2020. Main bioactive phenolic compounds in marine algae and their mechanisms of action supporting potential health benefits. Food Chem. 341, 128262. doi:10.1016/j.foodchem.2020.128262.

John, B.A., Nelson, B.R., Sheikh, H.I., Cheung, S.G., Wardiatno, Y., Dash, B.P., Tsuchiya, K., Iwasaki, Y., Pati, S., 2018. A review on fisheries and conservation status of Asian horseshoe crabs. Biodivers. Conserv. 15729710. 27 (14), 3573–3598. Available from: http://www.springerlink.com/content/0960-3115. doi:10.1007/s10531-018-1633-8.

Kadam, S.U., Tiwari, B.K., O'Donnell, C.P., 2013. Application of novel extraction technologies for bioactives from marine algae. J. Agric. Food Chem. 61 (20), 4667–4675. doi:10.1021/jf400819p.

Kari, Z.A., Kabir, M.A., Dawood, M.A.O., Razab, M.K.A.A., Ariff, N.S.N.A., Sarkar, T., Pati, S., Edinur, H.A., Mat, K., Ismail, T.A., Wei, L.S., 2022. Effect of fish meal substitution with fermented soy pulp on growth performance, digestive enzyme, amino acid profile, and immune-related gene expression of African catfish (*Clarias gariepinus*). Aquaculture. 00448486. 546. Available from: http://www.journals.elsevier.com/aquaculture/. doi:10.1016/j.aquaculture.2021.737418.

Kari, Z.A., Kabir, M.A., Mat, K., Rusli, N.D., Razab, M.K.A.A., Ariff, N.S.N.A., Edinur, H.A., Rahim, M.Z.A., Pati, S., Dawood, M.A., Wei, L.S., 2021. The possibility of replacing fish meal with fermented soy pulp on the growth performance, blood biochemistry, liver, and intestinal morphology of African catfish (*Clarias gariepinus*). Aquac. Rep. 21, 100815. doi:10.1016/j.aqrep.2021.100815.

Karunarathne, S., Mendis, E., Kim, S., 2020. The potential of developing additives from marine bioresources for the food industry. Encycl. Mar. Biotechnol. 3085–3107. doi:10.1002/9781119143802.ch137.

Kasmiati, K., Syahrul, S., Badraeni, B., Rahmi, M.H., 2022. Proximate and mineral compositions of the green seaweeds *Caulerpa lentilifera* and *Caulerpa racemosa* from South Sulawesi Coast, Indonesia. IOP Conf. Ser. Earth Environ. Sci. 1755-1307. 1119 (1), 012049. doi:10.1088/1755-1315/1119/1/012049.

Khoo, K.S., Chew, K.W., Yew, G.Y., Leong, W.H., Chai, Y.H., Show, P.L., Chen, W.H., 2020. Recent advances in downstream processing of microalgae lipid recovery for biofuel production. Bioresour. Technol. 18732976. 304. Available from: https://www.elsevier.com/locate/biortech. doi:10.1016/j.biortech.2020.122996.

Kim, S.K., 2013. Seafood Processing By-products: Trends and Applications. Springer New York, South Korea, pp. 1–597. doi:10.1007/978-1-4614-9590-1.

Kiuru, P., D'Auria, M., Muller, C., Tammela, P., Vuorela, H., Yli-Kauhaluoma, J., 2014. Exploring marine resources for bioactive compounds. Planta Med. 0032-0943. 80 (14), 1234–1246. doi:10.1055/s-0034-1383001.

Kumar, S.G., Rahman, A.M., Lee, S.H., Hwang, H.S., Kim, H.A., Yun, J.W., 2009. Plasma proteome analysis for anti-obesity and anti-diabetic potentials of chitosan oligosaccharides in ob/ob mice. Proteomics. 16159861. 9 (8), 2149–2162. Available from: http://www3.interscience.wiley.com/cgi-bin/fulltext/122267007/PDFSTART. doi:10.1002/pmic.200800571.

Kuroiwa, T., Izuta, H., Nabetani, H., Nakajima, M., Sato, S., Mukataka, S., Ichikawa, S., 2008. Selective and stable production of physiologically active chitosan oligosaccharides using an enzymatic membrane bioreactor. Process Biochem. 44 (3), 283–287. doi:10.1016/j.procbio.2008.10.020.

Lafarga, T., Acién-Fernández, F.G., Garcia-Vaquero, M., 2020. Bioactive peptides and carbohydrates from seaweed for food applications: natural occurrence, isolation, purification, and identification. Algal Res. 48, 101909. doi:10.1016/j.algal.2020.101909.

Lee, S.Y., Hur, S.J., 2017. Antihypertensive peptides from animal products, marine organisms, and plants. Food Chem. 18737072. 228, 506–517. Available from: https://www.elsevier.com/locate/foodchem. doi:10.1016/j.foodchem.2017.02.039.

Low, K., Idris, A., Yusof, N.M., 2019. Novel protocol optimized for microalgae lutein used as food additives. Food Chem. 307, 125631. doi:10.1016/j.foodchem.2019.125631.

Mahapatra, G.P., Raman, S., Nayak, S., Gouda, S., Das, G., Patra, J.K., 2019. Metagenomics approaches in discovery and development of new bioactive compounds from marine actinomycetes. Curr. Microbiol. 77 (4), 645–656. doi:10.1007/s00284-019-01698-5.

Martins, D., Custódio, L., Barreira, L., Pereira, H., Ben-Hamadou, R., Varela, J., Abu-Salah, K., 2013. Alternative sources of n-3 long-chain polyunsaturated fatty acids in marine microalgae. Mar. Drugs. 1660-3397. 11 (7), 2259–2281. doi:10.3390/md11072259.

Mateos, R., Pérez-Correa, J.R., Domínguez, H., 2020. Bioactive properties of marine phenolics. Mar. Drugs 18 (10), 501. doi:10.3390/md18100501.

Matos, J., Cardoso, C., Bandarra, N.M., Afonso, C., 2017. Microalgae as healthy ingredients for functional food: a review. Food Funct. 2042-6496. 8 (8), 2672–2685. doi:10.1039/c7fo00409e.

Matsumoto, M., Hosokawa, M., Matsukawa, N., Hagio, M., Shinoki, A., Nishimukai, M., Miyashita, K., Yajima, T., Hara, H., 2009. Suppressive effects of the marine carotenoids, fucoxanthin and fucoxanthinol on triglyceride absorption in lymph duct-cannulated rats. Eur. J. Nutr. 49 (4), 243–249. doi:10.1007/s00394-009-0078-y.

Mayakrishnan, V., Kannappan, P., Abdullah, N., Ahmed, A.B.A., 2013. Cardioprotective activity of polysaccharides derived from marine algae: an overview. Trends Food Sci. Technol. 09242244. 30 (2), 98–104. doi:10.1016/j.tifs.2013.01.007.

Menaa, F., Wijesinghe, U., Thiripuranathar, G., Althobaiti, N.A., Albalawi, A.E., Khan, B.A., Menaa, B., 2021. Marine algae-derived bioactive compounds: a new wave of nanodrugs? Mar. Drugs. 16603397. 19 (9). Available from: https://www.mdpi.com/1660-3397/19/9/484/pdf. doi:10.3390/md19090484.

Miyashita, K., 2013. Anti-obesity therapy by food component: unique activity of marine carotenoid, fucoxanthin. Obes. Control Ther. Open Access. 23748354. 1 (1). doi:10.15226/2374-8354/1/1/00103.

Miyashita, K., Nishikawa, S., Beppu, F., Tsukui, T., Abe, M., Hosokawa, M., 2011. The allenic carotenoid fucoxanthin, a novel marine nutraceutical from brown seaweeds. J. Sci. Food Agric. 0022-5142. 91 (7), 1166–1174. doi:10.1002/jsfa.4353.

Mohammadi, M.A., Ahangari, H., Mousazadeh, S., Hosseini, S.M., Dufossé, L., 2021. Microbial pigments as an alternative to synthetic dyes and food additives: a brief review of recent studies. Bioprocess Biosyst. Eng. 45 (1), 1–12. doi:10.1007/s00449-021-02621-8.

Moi, I.M., Leow, A.T.C., Ali, M.S.M., Rahman, R.N.Z.R.A., Salleh, A.B., Sabri, S., 2018. Polyunsaturated fatty acids in marine bacteria and strategies to enhance their production. Appl. Microbiol. Biotechnol. 14320614. 102 (14), 5811–5826. Available from: http://www.link.springer.de/link/service/journals/00253/index.htm. doi:10.1007/s00253-018-9063-9.

Mondo, A.D., Smerilli, A., Ambrosino, L., Albini, A., Noonan, D.M., Sansone, C., Brunet, C., 2021. Insights into phenolic compounds from microalgae: structural variety and complex beneficial activities from health to nutraceutics. Crit. Rev. Biotechnol. 15497801. 41 (2), 155–171. Available from: http://www.tandfonline.com/loi/ibty20. doi:10.1080/07388551.2021.1874284.

Muñoz, I.L., Díaz, N.F., 2020. Minerals in edible seaweed: health benefits and food safety issues. Chile Crit. Rev. Food Sci. Nutr. 15497852. 62 (6), 1592–1607. Available from: https://www.tandfonline.com/loi/bfsn20. doi:10.1080/10408398.2020.1844637.

Nagaoka, I., Igarashi, M., Hua, J., Ju, Y., Yomogida, S., Sakamoto, K., 2010. Recent aspects of the anti-inflammatory actions of glucosamine. Carbohydr. Polym. 84 (2), 825–830. doi:10.1016/j.carbpol.2010.04.007.

Nalinanon, S., Benjakul, S., Kishimura, H., Shahidi, F., 2011. Functionalities and antioxidant properties of protein hydrolysates from the muscle of ornate threadfin bream treated with pepsin from skipjack tuna. Food Chem. 03088146. 124 (4), 1354–1362. doi:10.1016/j.foodchem.2010.07.089.

Ngo, D.-N., Kim, -M.O.O.N.-M.O.O., Qian, Z.H.O.N.G.-J.I., Jung, W.O.N.-K.Y.O., Lee, S.A.N.G.-H.O.O.N., Kim, S.E.-K.W.O.N., 2010. Free radical-scavenging activities of low molecular weight chitin oligosaccharides lead to antioxidant effect in live cells. J. Food Biochem. 01458884. 34 (1), 161–177. doi:10.1111/j.1745-4514.2009.00321.x.

Ozogul, F., Cagalj, M., Šimat, V., Ozogul, Y., Tkaczewska, J., Hassoun, A., Kaddour, A.A., Kuley, E., Rathod, N.B., Phadke, G.G., 2021. Recent developments in valorisation of bioactive ingredients in discard/seafood processing by-products. Trends Food Sci. Technol. 09242244. 116, 559–582. Available from: http://www.elsevier.com/wps/find/journaldescription.cws_home/601278/description#description. doi:10.1016/j.tifs.2021.08.007.

Papon, N., Copp, B.R., Courdavault, V., 2021. Marine drugs: biology, pipelines, current and future prospects for production. Biotechnol. Adv. 54, 107871. doi:10.1016/j.biotechadv.2021.107871.

Pei, X., Yang, R., Zhang, Z., Gao, L., Wang, J., Xu, Y., Zhao, M., Han, X., Liu, Z., Li, Y., 2009. Marine collagen peptide isolated from Chum Salmon (*Oncorhynchus keta*) skin facilitates learning and memory in aged C57BL/6J mice. Food Chem. 118 (2), 333–340. doi:10.1016/j.foodchem.2009.04.120.

Pelay-Gimeno, M., García-Ramos, Y., Jesús Martin, M., Spengler, J., Molina-Guijarro, J.M., Munt, S., Francesch, A.M., Cuevas, C., Tulla-Puche, J., Albericio, F., 2013. The first total synthesis of the cyclodepsipeptide pipecolideps in A. Nat. Commun. 20411723. 4. doi:10.1038/ncomms3352.

Pietrowski, B.N., Tahergorabi, R., Jaczynski, J., 2011. Dynamic rheology and thermal transitions of surimi seafood enhanced with ω-3-rich oils. Food Hydrocoll. 27 (2), 384–389. doi:10.1016/j.foodhyd.2011.10.016.

Pooja, K., Rani, S., Rana, V., Pal, G.K., 2016. Aquatic plants as a natural source of antimicrobial and functional ingredients. In: Bhanu, P. (Ed.), Functional and Preservative Properties of Phytochemicals. Elsevier, pp. 1–26.

Priyadarshani, I., Rath, B., 2012. Commercial and industrial applications of micro algae–a review. J. Algal Biomass Utln. 3 (4), 89–100.

Raafat, D., Sahl, H., 2009. Chitosan and its antimicrobial potential—a critical literature survey. Microb. Biotechnol. 1751-7915. 2 (2), 186–201. doi:10.1111/j.1751-7915.2008.00080.x.

Rani, V., Yadav, U.C.S., 2018. Functional food and human health. In: Rani, V., Yadav, U.C.S. (Eds.), Functional Food and Human Health. Springer Singapore, India, pp. 1–694. Available from: http://www.springer.com/in/book/9789811311222. doi:10.1007/978-981-13-1123-9.

Romano, G., Costantini, M., Sansone, C., Lauritano, C., Ruocco, N., Ianora, A., 2016. Marine microorganisms as a promising and sustainable source of bioactive molecules. Mar. Environ. Res. 128, 58–69. doi:10.1016/j.marenvres.2016.05.002.

Salamone, M., Cuttitta, A., Seidita, G., Tutino, L., Masullo, T., Rigogliuso, S., Bertuzzi, F., Ghersi, G., 2011. Proteolytic marine enzymes and their applications. Mar. Res, 1–12.

Santigosa, E., Constant, D., Prudence, D., Wahli, T., Verlhac-Trichet, V., 2020. A novel marine algal oil containing both EPA and DHA is an effective source of omega-3 fatty acids for rainbow trout (*Oncorhynchus mykiss*). J. World Aquac. Soc. 0893-8849. 51 (3), 649–665. doi:10.1111/jwas.12699.

Satitsri, S., Muanprasat, C., 2020. Chitin and chitosan derivatives as biomaterial resources for biological and biomedical applications. Molecules. 1420-3049. 25 (24), 5961. doi:10.3390/molecules25245961.

Setiati, R., Siregar, S., Wahyuningrum, D., Rinanti, A., 2021. Synthesis method of chitin become chitosan polymer from shrimp shells for enhanced oil recovery. IOP Conf. Ser. Earth Environ. Sci. 1755-1307. 737 (1), 012048. doi:10.1088/1755-1315/737/1/012048.

Shaheen, F., Ziaee, M.A., Ali, S.A., Simjee, S.U., Ahmed, A., Choudhary, M.I., 2016. The first solid-phase synthesis and structural studies on phakellistatin 15. Rec. Nat. Prod. 13076167. 10 (4), 397–406. Available from http://www.acgpubs.org/RNP/2016/Volume10/Issue%201/50-RNP-1503-049.pdf.

Shahidi, F., 2009. Nutraceuticals and functional foods: whole versus processed foods. Trends Food Sci. Technol. 09242244. 20 (9), 376–387. doi:10.1016/j.tifs.2008.08.004.

Shahidi, F., Ambigaipalan, P., 2015. Novel functional food ingredients from marine sources. Curr. Opin. Food Sci. 22147993. 2, 123–129. doi:10.1016/j.cofs.2014.12.009.

Shahidi, F., Miraliakbari, H., 2004. Omega-3 (n-3) fatty acids in health and disease: part 1—cardiovascular disease and cancer. J. Med. Food. 1096620X. 7 (4), 387–401. Available from: https://www.liebertonline.com/jmf. doi:10.1089/jmf.2004.7.387.

Shahidi, F., Zhong, Y., 2008. Bioactive peptides. J. AOAC Int. 1060-3271. 91 (4), 914–931. doi:10.1093/jaoac/91.4.914.

Shahidi, F., Zhong, Y., 2010. Novel antioxidants in food quality preservation and health promotion. Eur. J. Lipid Sci. Technol. 112 (9), 930–940. doi:10.1002/ejlt.201000044.

Sigwart, J.D., Blasiak, R., Jaspars, M., Jouffray, J., Tasdemir, D., 2020. Unlocking the potential of marine biodiscovery. Nat. Prod. Rep. 38 (7), 1235–1242. doi:10.1039/d0np00067a.

Šimat, V., 2021. Nutraceuticals and pharmaceuticals from marine fish and invertebrates. Mar. Drugs. 1660-3397. 19 (7), 401. doi:10.3390/md19070401.

Šimat, V., Čagalj, M., Skroza, D., Gardini, F., Tabanelli, G., Montanari, C., Hassoun, A., Ozogul, F., 2021. Sustainable sources for antioxidant and antimicrobial compounds used in meat and seafood products. Adv. Food Nutr. Res. 55–118. doi:10.1016/bs.afnr.2021.03.001.

Šimat, V., Elabed, N., Kulawik, P., Ceylan, Z., Jamroz, E., Yazgan, H., Čagalj, M., Regenstein, J.M., Özogul, F., 2020. Recent advances in marine-based nutraceuticals and their health benefits. Mar. Drugs. 16603397. 18 (12). Available from: https://www.mdpi.com/1660-3397/18/12/627. doi:10.3390/md18120627.

Speitling, M., Smetanina, O.F., Kuznetsova, T.A., Laatsch, H., 2007. Bromoalterochromides A and A′, unprecedented chromopeptides from a marine *Pseudoalteromonas maricaloris* strain KMM 636T. J. Antibiot. 00218820. 60 (1), 36–42. doi:10.1038/ja.2007.5.

Stincone, P., Brandelli, A., 2020. Marine bacteria as source of antimicrobial compounds. Crit. Rev. Biotechnol. 0738-8551. 40 (3), 306–319. doi:10.1080/07388551.2019.1710457.

Sukenik, A., Takahashi, H., Mokady, S., 1994. Dietary lipids from marine unicellular alagae enhance the amount of liver and blood omega-3 fatty acids in rats. Ann. Nutr. Metab. 38, 85–96.

Suleria, H.A.R., Osborne, S., Masci, P., Gobe, G., 2015. Marine-based nutraceuticals: an innovative trend in the food and supplement industries. Mar. Drugs. 16603397. 13 (10), 6336–6351. Available from: http://www.mdpi.com/1660-3397/13/10/6336/pdf. doi:10.3390/md13106336.

Tahergorabi, R., Beamer, S.K., Matak, K.E., Jaczynski, J., 2013. Chemical properties of ω-3 fortified gels made of protein isolate recovered with isoelectric solubilisation/precipitation from whole fish. Food Chem. 18737072. 139 (1-4), 777–785. doi:10.1016/j.foodchem.2013.01.077.

Tanna, B., Mishra, A., 2018. Metabolites unravel nutraceutical potential of edible seaweeds: an emerging source of functional food. Compr. Rev. Food Sci. Food Saf. 1541-4337. 17 (6), 1613–1624. doi:10.1111/1541-4337.12396.

Toldrá, F., Kim, S.K., 2016. Advances in Food and Nutrition Research: Marine Enzymes Biotechnology: Production and Industrial Applications, Part I-Production of Enzymes, Academic Press.

Toldra, F., Kim, S.K., 2017. Marine Enzymes Biotechnology: Production and Industrial Applications, Part III-Application of Marine Enzymes, vol. 80. Academic Press.

Trang, N.T.H., Tang, D.Y.Y., Chew, K.W., Linh, N.T., Hoang, L.T., Cuong, N.T., Yen, H.T., Thao, N.T., Trung, N.T., Show, P.L., Tuyen, D.T., 2021. Discovery of α-glucosidase inhibitors from marine microorganisms: optimization of culture conditions and medium composition. Mol. Biotechnol. 15590305. 63 (11), 1004–1015. Available from: http://www.springer.com/humana+press/journal/12033. doi:10.1007/s12033-021-00362-3.

Tsvetkov, Y.E., Paulovičová, E., Paulovičová, L., Farkaš, P., Nifantiev, N.E., 2020. Synthesis of biotin-tagged chitosan oligosaccharides and assessment of their immunomodulatory activity. Front. Chem. 22962646. 8. Available from: http://www.journal.frontiersin.org/journal/chemistry. doi:10.3389/fchem.2020.554732.

Urda, C., Pérez, M., Rodríguez, J., Jiménez, C., Cuevas, C., Fernández, R., 2016. Pembamide, a *N*-methylated linear peptide from a sponge *Cribrochalina* sp. Tetrahedron Lett. 00404039. 57 (30), 3239–3242. doi:10.1016/j.tetlet.2016.05.054.

Wallace, C.L., Watson, R.R., Watson, A.A., 1988. Reducing cancer risk with vitamins C, E, and selenium. Am. J. Health Promot. 3 (1), 5–32. doi:10.4278/0890-1171-3.1.5.

Wang, B., Gong, Y., Li, Z., Yu, D., Chi, C., Ma, J., 2013. Isolation and characterisation of five novel antioxidant peptides from ethanol-soluble proteins hydrolysate of spotless smoothhound (*Mustelus griseus*) muscle. J. Funct. Foods 6, 176–185. doi:10.1016/j.jff.2013.10.004.

Wang, T., Jónsdóttir, R., Ólafsdóttir, G., 2009. Total phenolic compounds, radical scavenging and metal chelation of extracts from Icelandic seaweeds. Food Chem. 116 (1), 240–248. doi:10.1016/j.foodchem.2009.02.041.

Wang, X., Luo, H., Yu, W., Ma, R., You, S., Liu, W., Hou, L., Zheng, F., Xie, X., Yao, B., 2015. A thermostable *Gloeophyllum trabeum* xylanase with potential for the brewing industry. Food Chem. 199, 516–523. doi:10.1016/j.foodchem.2015.12.028.

Zhan, K.-X., Jiao, W.-H., Yang, F., Jing, L., Wang, S.-P., Yu-Shan, L., Han, B.-N., Lin, H.-W., A–E, R., 2014. Cyclic peptides from the marine sponge *Reniochalina stalagmitis*. J. Nat. Prod. 0163-3864. 77 (12), 2678–2684. doi:10.1021/np5006778.

Zhang, C., Kim, S.K., 2012. Application of marine microbial enzymes in the food and pharmaceutical industries. In: Se-Kwon, K. (Ed.), Advances in Food and Nutrition Research, vol. 65. Academic Press, pp. 423–435. Available from: http://www.elsevier.com/wps/find/bookdescription.cws_home/704624/description#description. doi:10.1016/B978-0-12-416003-3.00028-7.

Zhang, H., Yi, Y., Yang, G., Hu, M., Cao, G., Yang, F., Lin, H., 2010. Proline-containing cyclopeptides from the Marine Sponge *Phakellia fusca*. J. Nat. Prod. 73 (4), 650–655. doi:10.1021/np9008267.

Zhang, J.J., Tang, X., Zhang, M., Nguyen, D., Moore, B.S., 2017. Broad-host-range expression reveals native and host regulatory elements that influence heterologous antibiotic production in Gram-negative bacteria. mBio. 21507511. 8 (5). Available from: http://www.mbio.asm.org/content/8/5/e01291-17.full.pdf. doi:10.1128/mBio.01291-17.

Zhang, J., Zhang, W., Mamadouba, B., Xia, W., 2012. A comparative study on hypolipidemic activities of high and low molecular weight chitosan in rats. Int. J. Biol. Macromol. 01418130. 51 (4), 504–508. doi:10.1016/j.ijbiomac.2012.06.018.

Zhao, C., Yang, C., Liu, B., Lin, L., Sarker, S.D., Nahar, L., Yu, H., Cao, H., Xiao, J., 2017. Bioactive compounds from marine macroalgae and their hypoglycemic benefits. Trends Food Sci. Technol. 72, 1–12. doi:10.1016/j.tifs.2017.12.001.

Zheng, L.-H., Wang, Y.-J., Sheng, J., Wang, F., Zheng, Y., Lin, X.-K., Sun, M., 2011. Antitumor peptides from marine organisms. Mar. Drugs. 1660-3397. 9 (10), 1840–1859. doi:10.3390/md9101840.

Zhu, C.-F., Li, G.-Z., Peng, H.-B., Zhang, F., Chen, Y., Li, Y., 2010. Effect of marine collagen peptides on markers of metabolic nuclear receptors in type 2 diabetic patients with/without hypertension. Biomed. Environ. Sci. 08953988. 23 (2), 113–120. doi:10.1016/s0895-3988(10)60040-2.

PART V

Therapeutic properties of functional foods incorporated with marine ingredients

CHAPTER

15

Impacts on noninfectious diseases

Susara Madduma Hewage

St. Boniface Hospital Research Centre, Winnipeg, MB, Canada

Department of Animal Science, University of Manitoba, Winnipeg, MB, Canada

15.1 Introduction

Noninfectious diseases refer to a group of health conditions that are not attributable to infectious agents like bacteria, viruses, or parasites. These diseases typically emerge from a multifaceted interaction of genetic susceptibility, environmental influences, and lifestyle choices. Common examples include chronic conditions such as cardiovascular disease (CVD), diabetes, fatty liver disease, chronic kidney disease (CKD), and cancer, which tend to develop progressively and can be affected by factors, including nutrition, physical activity, and exposure to harmful chemicals (Budreviciute et al., 2020). This category also encompasses autoimmune disorders, neurological diseases, and various mental health conditions, illustrating the broad spectrum of noninfectious diseases. Unlike infectious diseases, which can be spread from person to person, noninfectious diseases are generally not transmissible (Budreviciute et al., 2020). This distinction highlights the necessity for preventive approaches that concentrate on modifying risk factors and managing health rather than aiming for elimination.

Nonalcoholic fatty liver disease (NAFLD) and CKD represent major noncommunicable diseases that have garnered increasing attention due to their rising incidence and severe health implications. NAFLD is characterized by the abnormal accumulation of lipids within hepatocytes, independent of alcohol consumption. This condition is rapidly becoming a primary contributor to liver-related morbidity and mortality on a global scale (Wong et al., 2023). Its potential to progress to advanced liver diseases, such as nonalcoholic steatohepatitis (NASH), fibrosis, and cirrhosis, poses a significant risk of liver failure, which may require transplantation. In addition, NAFLD is commonly associated with several metabolic disorders, including obesity, type 2 diabetes mellitus (T2DM), and CVDs, complicating the management of affected individuals (Bonora and Targher, 2012). Concurrently, CKD is a progressive renal disorder that diminishes the kidneys' ability to effectively filter metabolic waste from the bloodstream. This deterioration can lead to numerous complications, including cardiovascular events, anemia, and disruptions in bone mineral metabolism.

Harvesting the Sea
https://doi.org/10.1016/B978-0-443-29190-6.10015-7

CKD impacts millions of individuals worldwide and remains a critical precursor to end-stage renal disease (ESRD), necessitating interventions such as dialysis or kidney transplantation (Francis et al., 2024).

Dietary intervention has emerged as an effective strategy for the management of various metabolic diseases, including NAFLD and CKD (Miller, 2020; Ingelfinger et al., 2017). Beyond the regulation of macronutrient and micronutrient intake, the incorporation of natural bioactive compounds—such as polyphenols, flavonoids, polyunsaturated fatty acids (PUFAs), vitamins, and other chemicals—has been shown to mitigate oxidative stress, chronic inflammation, hypertension, and dyslipidemia commonly associated with NAFLD and CKD (Banwo et al., 2021). These compounds possess antioxidant and antiinflammatory properties that can reduce disease-related complications and enhance overall metabolic health.

Marine resources represent a vital reservoir of bioactive compounds with significant potential for managing metabolic diseases. Marine organisms, including fish, shellfish, algae, and microorganisms, produce various chemical entities, including omega-3 fatty acids, proteins, peptides, polysaccharides, polyphenols, and terpenoids (Ghosh et al., 2022; Sharanagat et al., 2019). Omega-3 fatty acids, particularly eicosapentaenoic acid (EPA) and docosahexaenoic acid (DHA), derived primarily from fish, play a crucial role in reducing inflammation and improving cardiovascular health. Their consumption has been linked to lower risks of metabolic syndrome, which often coexists with conditions such as NAFLD and CKD (Lorente-Cebrián et al., 2013). In addition, many marine-derived bioactive compounds exhibit potent antioxidant properties, helping to combat oxidative stress, a significant contributor to the pathogenesis of metabolic disorders. For instance, certain seaweeds and marine-derived phenolic compounds scavenge free radicals and mitigate cellular damage. Moreover, marine sources can provide antiinflammatory agents that address the chronic inflammation seen in metabolic diseases (D'Orazio et al., 2012; Karthikeyan et al., 2022). Compounds such as fucoidans from brown seaweeds have demonstrated antiinflammatory effects, potentially improving insulin sensitivity and lipid metabolism (Wang et al., 2016). The unique structural diversity of marine bioactive compounds also presents opportunities for discovering novel therapeutic agents with mechanisms of action not found in terrestrial sources.

This book chapter aims to explore the health benefits of marine bioactive compounds in managing noninfectious diseases, with a particular focus on CKD and NAFLD. It seeks to review the existing literature on various marine-derived substances, such as omega-3 fatty acids, polyphenols, and polysaccharides, and their roles in mitigating the biochemical pathways associated with oxidative stress, inflammation, dyslipidemia, and hypertension, which are prevalent in CKD and NAFLD.

15.2 Nonalcoholic fatty liver disease

15.2.1 Definition of NAFLD

NAFLD is a comprehensive term that encompasses a range of liver disorders, including simple steatosis, NASH, cirrhosis, and hepatocellular carcinoma (HCC), in the absence of excessive alcohol intake (defined as <30 g per day for men and <20 g per day for women) (Burt et al., 2015). The hallmark and earliest histopathological feature of NAFLD is steatosis, which is characterized by the accumulation of fat, primarily in the form of triglycerides, in >5% of hepatocytes. Histologically, steatosis is identified by the presence of macrovesicular lipid vacuoles, which consist of both large and small lipid droplets within the liver tissue. During steatosis, lipid accumulation in hepatocytes causes the displacement of the nucleus toward the periphery of the cell, leading to the formation of a single large lipid droplet (Burt et al., 2015).

NASH represents a more advanced form of NAFLD, characterized by hepatic inflammation and fibrosis as a result of ongoing lipid accumulation in the liver. The histopathological features typical of NASH include lobular inflammation and hepatocyte ballooning, accompanied by macrovesicular steatosis (Byrne and Targher, 2015). The inflammation in the liver is identified histologically through the presence of inflammatory foci containing lymphocytes, eosinophils, leukocytes, and Kupffer cells (the resident macrophages of the liver) (Byrne and Targher, 2015). A key distinguishing feature of NASH, compared to simple steatosis, is the presence of hepatic ballooning, where hepatocytes become swollen due to cellular injury. Chronic inflammation and cellular damage activate hepatic stellate cells, which play a crucial role in the production and secretion of collagen (Michelotti et al., 2013). While NASH can potentially revert to simple steatosis, the accumulation of collagen leads to irreversible progression to advanced cirrhosis. Patients diagnosed with cirrhosis often exhibit comorbidities such as metabolic syndrome, diabetes, and obesity. In addition, the incidence of cirrhosis significantly increases the risk of developing HCC and portal hypertension, both of which are notable contributors to NAFLD-related mortality (Michelotti et al., 2013).

15.2.2 Prevalence and etiology of NAFLD

NAFLD is a primary contributor to chronic liver disease worldwide. It includes a range of conditions, from uncomplicated hepatic steatosis to NASH, which may advance to liver fibrosis and cirrhosis. Recent studies indicate a rising global prevalence of NAFLD, with a meta-analysis suggesting that approximately 32% of adults are affected by this condition (Cheemerla and Balakrishnan, 2021; Riazi et al., 2022). This trend correlates with the worldwide epidemics of obesity and diabetes (GBD 2015 Obesity Collaborators, 2017; Stefan and Cusi, 2022). NASH has emerged as the most rapidly increasing cause of HCC globally and is also the leading reason for liver transplantation in the United States (Younossi et al., 2021). The global prevalence of NAFLD is currently 30% and is on the rise, necessitating urgent and comprehensive strategies to enhance awareness and tackle all facets of NAFLD at local, regional, and global scales.

The etiology of NAFLD is intricately associated with obesity, T2DM, and dietary factors, all of which considerably influence the emergence and advancement of this condition (Wei et al., 2024). Obesity, characterized by an excessive accumulation of adipose tissue, is recognized as a significant risk factor for NAFLD, primarily due to its strong correlation with insulin resistance (Ko et al., 2023). Increased adiposity, particularly visceral fat accumulation, results in dysregulation of lipid metabolism and leads to heightened levels of free fatty acids in the circulation. This lipid overload fosters the deposition of triglycerides within hepatocytes, a process that not only promotes the development of simple hepatic steatosis but can also activate inflammatory cascades, progressing to NASH and potentially advancing to fibrosis and cirrhosis (Dietrich and Hellerbrand, 2014). Concurrently, T2DM contributes to an elevated risk of developing NAFLD; patients with this condition frequently exhibit severe insulin resistance and persistent hyperglycemia (Targher et al., 2021). Elevated serum glucose concentrations stimulate hepatic gluconeogenesis and de novo lipogenesis, further exacerbating hepatic fat accumulation (Targher et al., 2021). Furthermore, dietary patterns play a crucial role in this multifactorial etiology; contemporary diets characterized by high caloric intake, prevalent consumption of processed foods, and excessive sugars, particularly fructose, significantly contribute to obesity and the development of insulin resistance (Semmler et al., 2021). These dietary habits facilitate the deposition of excess fat in the liver and disrupt overall metabolic homeostasis, creating a self-reinforcing cycle that accelerates the progression of NAFLD (Martins, 2015). Elucidating the interplay among these factors is essential for effective prevention and intervention strategies against NAFLD's pathogenesis.

15.2.3 Pathophysiology of NAFLD

The pathophysiology of NAFLD can be characterized by a multifaceted interplay of lipid overload, oxidative stress, inflammation, and impaired lipid metabolism, which together culminates in liver injury and disease progression (Dietrich and Hellerbrand, 2014). Lipid overload is a primary feature of NAFLD, often resulting from excessive dietary intake of carbohydrates, particularly simple sugars such as fructose, and saturated fats (Arab et al., 2018). This excessive caloric consumption exceeds the liver's capacity to metabolize fats, leading to enhanced de novo lipogenesis and the subsequent accumulation of triglycerides within hepatocytes. The excessive storage of lipids can overwhelm normal metabolic pathways, resulting in steatosis and, if unchecked, advancing to more severe manifestations such as NASH.

As lipid accumulation in the liver escalates, the condition progresses into a state of oxidative stress, which is characterized by an imbalance between the generation of reactive oxygen species (ROS) and the hepatic capacity to neutralize these detrimental metabolic byproducts (Chen et al., 2020). The surplus of fatty acids within hepatocytes significantly enhances ROS production, leading to cellular damage manifested through processes such as lipid peroxidation, alterations in protein structure, and genetic damage to cellular DNA. This oxidative stress serves as a pivotal catalyst for inflammatory responses, as damaged hepatocytes release proinflammatory cytokines that activate innate immune pathways (Chen et al., 2020). Consequently, this activation leads to the infiltration of immune cells and the perpetuation of inflammation within the liver. The sustained inflammatory environment not only exacerbates hepatocellular injury but also facilitates the progression from simple hepatic steatosis to NASH, which signifies the beginning of fibrosis and potentially escalates to cirrhosis if the underlying factors remain unaddressed (Li et al., 2016b).

Impaired lipid metabolism is a crucial aspect of the pathophysiology of NAFLD, significantly undermining the liver's ability to maintain lipid homeostasis. The liver is essential for regulating fatty acid uptake, synthesis, oxidation, and export. However, in the context of NAFLD, fatty acid oxidation is notably compromised, primarily due to mitochondrial dysfunction (Pei et al., 2020). This dysfunction exacerbates hepatic lipid accumulation and is intricately linked to the dysregulation of several key metabolic pathways, including those involving peroxisome proliferator-activated receptors (PPARs) and sterol regulatory element-binding proteins (SREBPs) (Liss and Finck, 2017; Hewage et al., 2021). These pathways are vital for fatty acid synthesis and oxidation; thus their alteration leads to increased lipogenesis while simultaneously impairing the liver's ability to export lipids as very low-density lipoproteins (Pei et al., 2020). Consequently, this interferes with lipid export mechanisms and perpetuates the cycle of lipid overload and hepatic steatosis. Furthermore, the impairment of the AMP-activated protein kinase (AMPK) and Notch signaling pathways during NAFLD further complicates this dysregulation, contributing to metabolic disturbances and promoting the pathogenesis of steatosis and its progression to more severe liver injuries (Pajvani et al., 2013; Smith et al., 2016).

During NAFLD, multiple inflammatory signaling pathways are activated within the liver, each playing a crucial role in mediating the inflammatory response and contributing to hepatocellular damage. A central pathway involved is the nuclear factor kappa B (NF-κB) pathway, which is triggered by oxidative stress and the presence of proinflammatory cytokines. Upon activation, NF-κB regulates the expression of various inflammatory genes, producing cytokines such as tumor necrosis factor-alpha (TNF-α) and interleukin-6 (IL-6) (Muriel, 2009). Simultaneously, the c-Jun N-terminal kinase (JNK) pathway is also activated, primarily through signals associated with oxidative stress and the engagement of cytokines (Yan et al., 2020). Activation of JNK is linked to increased insulin resistance and promotes apoptosis in hepatocytes, further exacerbating liver inflammation. Furthermore, the toll-like receptor (TLR) pathway, particularly TLR4, is significantly activated, largely due to the translocation of bacterial products such as lipopolysaccharides (LPS) from the gut into the portal circulation (Song, 2022). The activation of TLR4 triggers downstream inflammatory signaling cascades, including the NF-κB pathway, which amplifies the inflammatory response. In addition, the engagement of the NLRP3 (NOD-like receptor family pyrin domain containing 3) inflammasome is notable, as it is activated by various cellular stressors and leads to the secretion of proinflammatory cytokines, including interleukin-1 beta (IL-1β) and interleukin-18 (Wan et al., 2016). Collectively, these activated pathways create a persistent inflammatory environment within the liver, facilitating the progression from simple steatosis to NASH and promoting fibrogenesis. Thus they highlight the intricate inflammatory mechanisms at play in NAFLD.

Understanding the pathophysiology of NAFLD involves recognizing the complex interplay between lipid overload, oxidative stress, and inflammation, alongside impaired lipid metabolism. These interconnected mechanisms converge to create a detrimental cycle that can lead to progressive liver disease. As researchers strive to elucidate these pathways further, targeting specific components within this multifactorial framework offers promise for developing therapeutic interventions.

15.3 Chronic kidney disease

15.3.1 Definition of CKD

CKD refers to a range of disorders that compromise the structure and/or function of the kidneys. Kidney Disease Improving Global Outcomes (KDIGO), an independent nonprofit foundation initiated by the United States National Kidney Foundation, defines CKD as "abnormalities of kidney structure or function, present for more than three months, with implications for health." It is clinically diagnosed through criteria, including a decrease in glomerular filtration rate (GFR) to <60 mL/min/1.73 m^2 and/or the presence of kidney damage indicators, such as albuminuria (the presence of albumin in the urine) greater than 30 mg/g. CKD is classified into five stages based on GFR levels, with associated implications for health, progression to complications, and overall prognosis (Cheung et al., 2021; Анемії, 2012). The disease can result from a variety of underlying causes, including obesity, diabetes, hypertension, and glomerulonephritis, and requires ongoing management to slow disease progression and prevent associated complications (Fig. 15.1).

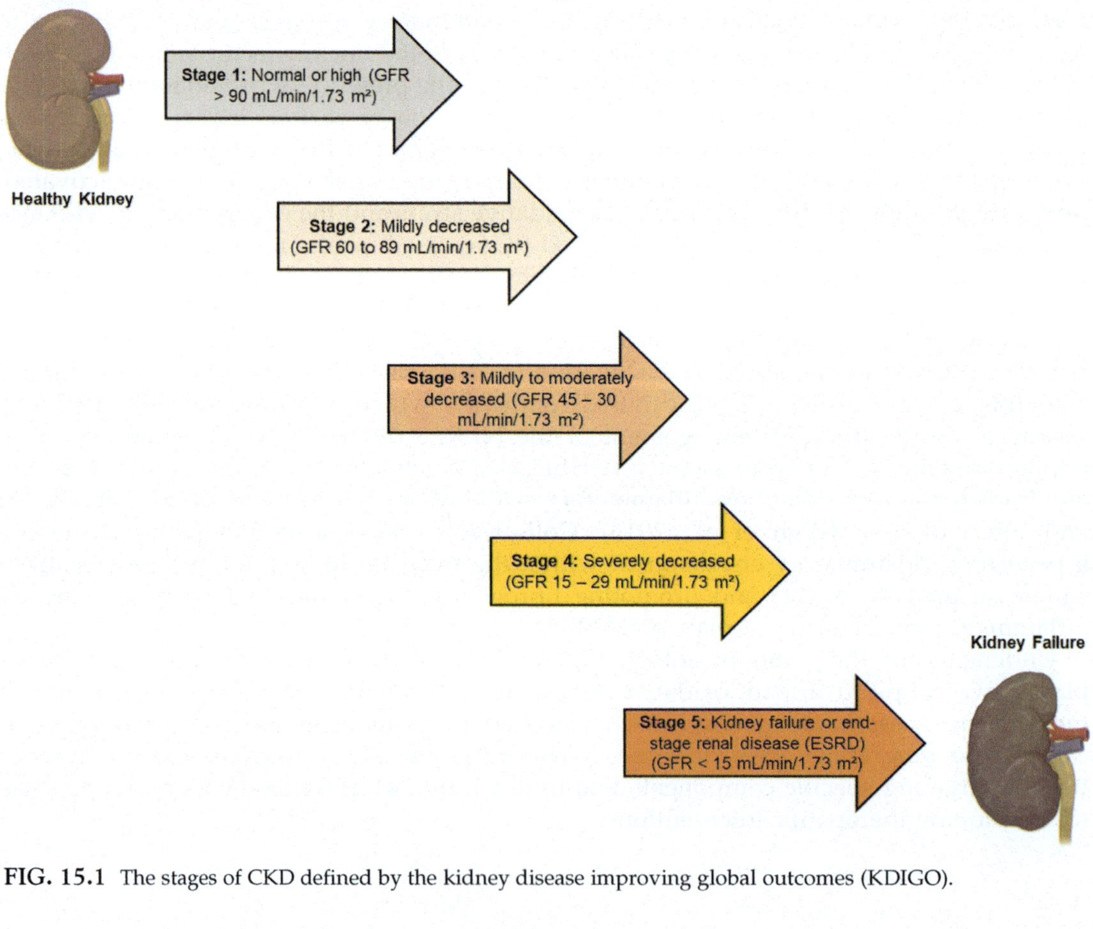

FIG. 15.1 The stages of CKD defined by the kidney disease improving global outcomes (KDIGO).

15.3.2 Prevalence and etiology of CKD

CKD poses a serious public health issue, with an estimated global prevalence of about 13.4%, ranging from 11.7% to 15.1%. It is projected that between 4.9 million and 7.1 million people are currently living with ESRD requiring renal replacement therapy. CKD significantly contributes to the overall burden of illness and death worldwide, especially due to its association with increased cardiovascular risks and its potential to progress to ESRD (Lv and Zhang, 2019).

Among the key mediators of CKD are obesity, diabetes mellitus, and hypertension, all of which significantly contribute to the disease's pathogenesis (Romagnani et al., 2017). Obesity leads to an increased risk of CKD through mechanisms such as the development of insulin resistance, altered lipid metabolism, and chronic inflammation, which can induce or exacerbate kidney damage (Nehus, 2018). Diabetes is a leading cause of CKD, primarily due to diabetic nephropathy, where prolonged hyperglycemia results in glomerular hyperfiltration, inflammation, and eventual fibrosis (Expert et al., 2018). Hypertension also plays a critical role, as sustained high blood pressure can cause direct injury to the renal vasculature and glomeruli, further impairing kidney function (Kaysen, 2007). The interplay among these conditions illustrates a complex relationship where obesity predisposes individuals to diabetes and hypertension, thereby increasing the overall burden of CKD (Zhang and Lerman, 2017). Addressing these key mediators through lifestyle modifications and management strategies is essential for preventing the onset and progression of CKD in at-risk populations.

15.3.3 Pathophysiology of CKD

CKD is a complex disorder that gradually diminishes kidney function, often progressing to ESRD, which may require treatments like dialysis or kidney transplantation. The underlying mechanisms of CKD involve a series of interconnected events that lead to the loss of nephrons, the functional units of the kidney, and subsequent kidney dysfunction. Various factors can initiate this process, including conditions such as lipid overload in obesity, diabetes, hypertension, glomerulonephritis, and systemic diseases like lupus. These factors trigger a cascade of events that result in kidney damage and disease progression (Mac Way et al., 2012).

A key player in the progression of CKD is hyperfiltration, which occurs in the remaining healthy nephrons after some have been lost or damaged. When nephrons become impaired, the remaining ones compensate by working harder, often with increased filtration rates to maintain overall kidney function. While this initial response can help, it ultimately leads to increased pressure within the glomeruli, the tiny filters within the kidney (Mac Way et al., 2012). Over time, this heightened pressure causes mechanical damage, resulting in glomerulosclerosis, where the tissue becomes scarred and loses its ability to filter blood effectively. This cycle of injury becomes self-perpetuating, with further nephron loss accelerating the decline in kidney function. Another critical factor in CKD is oxidative stress, which refers to the imbalance between the production of ROS and the body's ability to detoxify them (Daenen et al., 2019). The kidneys are particularly vulnerable to oxidative stress because of their high metabolic activity. CKD creates an environment rich in oxidative stress due to factors like inflammation and the accumulation of waste products. Increased ROS contributes to injury and death of kidney cells, leading to fibrosis within the nephron. Moreover, oxidative stress can damage the endothelium (the inner lining of blood vessels), disrupting kidney blood flow and further aggravating existing damage (Daenen et al., 2019).

Inflammation is also central to the development and worsening of CKD. Uremic toxins and metabolic changes associated with kidney dysfunction trigger the production of inflammatory substances in the body. These include cytokines like TNF-α and IL-6, which can create a persistent low-grade inflammatory state (Akchurin and Kaskel, 2015). As inflammation progresses, it leads to fibrosis (scarring) in the kidney tissue, which compounds the loss of functional nephrons. The body's immune response, involving various immune cells such as macrophages, only intensifies the inflammatory atmosphere, contributing to further kidney damage. The interplay between inflammation and fibrosis is crucial, as this interaction fosters a cycle of injury that continues to harm renal function over time (Akchurin and Kaskel, 2015). Metabolic disorders, often seen in CKD patients—especially those with obesity, diabetes, and hypertension—further complicate the situation. Obesity leads to chronic inflammation and metabolic disruption, which can increase stress on the kidneys. In diabetes, high blood sugar levels result in glomerular hyperfiltration and increased protein levels in urine, exacerbating kidney harm. Hypertension adds another layer of complexity by exerting additional pressure on the glomeruli, thereby worsening kidney injuries and accelerating nephron loss (Zhang and Lerman, 2017).

As CKD progresses toward ESRD, patients may experience a wide range of symptoms due to the build-up of waste products in the body, imbalances in electrolytes, and metabolic disturbances. This can lead to manifestations such as anemia (due to erythropoietin deficiency), fluid retention, and issues with blood clotting. The intricate interactions between hyperfiltration, oxidative stress, inflammation, metabolic dysregulation, and mineral imbalances all contribute to the worsening of kidney function. Therefore the pathophysiology of CKD is a complicated interplay of factors, each one playing a role in the disease's progression (Romagnani et al., 2017). Recognizing these underlying mechanisms highlights the importance of early detection and intervention in managing CKD effectively to preserve kidney function and improve patient outcomes.

15.4 Management of NAFLD and CKD: Importance of dietary intervention and marine bioactives

Currently, there are no Food and Drug Administration (FDA)-approved (US Food and Drug Administration) pharmacological treatments specifically for NAFLD; however, insulin sensitizers and glucagon-like peptide-1 receptor agonists are being employed in clinical practice to manage the condition (Cardoso et al., 2021). Similarly, CKD lacks a definitive cure, with management primarily focused on controlling its risk factors and addressing associated complications. In this context, angiotensin-converting enzyme (ACE) inhibitors and angiotensin receptor blockers (ARBs) are frequently prescribed to manage hypertension and reduce renal damage (Davison et al., 2019).

With the limitations of pharmacological treatments for various metabolic diseases, including NAFLD and CKD, there has been growing interest in the application of bioactive compounds as potential therapeutic agents. Bioactive compounds, which are naturally occurring substances found in foods, particularly in fruits, vegetables, whole grains, and marine sources, possess a variety of health-promoting properties, such as antiinflammatory, antioxidant, and insulin-sensitizing effects (Allan et al., 2023). These compounds, including polyphenols, flavonoids, omega-3 fatty acids, and other phytochemicals, can modulate metabolic pathways, enhance lipid metabolism, and reduce oxidative stress, thereby mitigating the progression of metabolic diseases. For instance, omega-3 fatty acids sourced from fish oil have demonstrated efficacy in improving liver fat levels and reducing inflammatory markers in NAFLD patients, while polyphenols, such as resveratrol and curcumin, exhibit protective effects on renal function through their antioxidant properties and modulation of inflammatory pathways (Barroso et al., 2023; Abenavoli et al., 2021). Given the rising prevalence of metabolic disorders, ongoing research focusing on the efficacy and mechanisms of these bioactive compounds is crucial for developing integrative strategies that complement traditional pharmacological approaches, ultimately enhancing patient outcomes and quality of life.

Marine bioactive compounds have attracted significant scientific interest as potential therapeutic agents for the management of NAFLD and CKD, attributable to their various biochemical properties and associated health benefits (Pangestuti and Kim, 2017). Among these compounds, omega-3 fatty acids, particularly EPA and DHA, have been extensively studied for their efficacy in decreasing liver lipid accumulation and attenuating inflammatory processes linked to NAFLD (Scorletti and Byrne, 2018). Clinical research indicates that omega-3 fatty acid supplementation can yield significant improvements in liver function parameters, reduction of hepatic steatosis, and lower levels of inflammatory markers, suggesting their potential as a noninvasive therapeutic strategy for individuals affected by NAFLD (Scorletti and Byrne, 2018; Lu et al., 2016). Other than lipid-lowering effects, marine bioactive compounds are known for modulating inflammation and oxidative stress, two key contributors to the progression of NASH and kidney damage (D'Orazio et al., 2012; Fernando et al., 2016; Ko et al., 2012). Marine peptides, derived from fish and shellfish, exhibit potent antioxidant properties that can protect renal tissues from oxidative damage, thereby reducing fibrosis and inflammation (Chen et al., 2019; Sheih et al., 2009). Furthermore, polysaccharides isolated from marine sources, such as fucoidan and alginate, have shown potential in promoting renal health by improving glucose metabolism and regulating lipid profiles, which are critical components in the management of CKD (Wang et al., 2016; Idota et al., 2018). The subsequent sections concentrate on elucidating a range of marine bioactive compounds that may serve as potential therapeutic candidates in the management of NAFLD and CKD.

15.5 Understanding marine compounds and their protective effects against NAFLD and CKD

15.5.1 Definition of marine bioactive compounds

Marine bioactive compounds are naturally occurring substances sourced from marine organisms such as fish, shellfish, algae, and microorganisms, which have been shown to offer significant health benefits. These compounds play a crucial role in influencing various

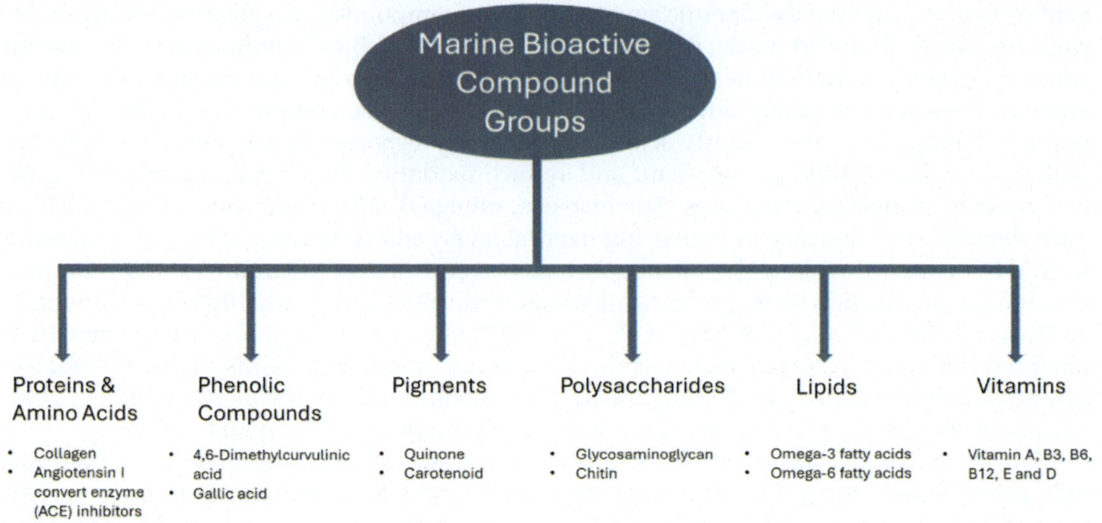

FIG. 15.2 Key marine bioactive compound groups based on their chemical structure.

biological processes, ultimately contributing to improved health and a reduced risk of numerous diseases (D'Orazio et al., 2012). The spectrum of marine bioactives includes several chemical classes, such as fatty acids, proteins, peptides, polysaccharides, antioxidants, and other phytochemicals. They are particularly noted for their capacity to enhance metabolic health, modulate inflammatory responses, counteract oxidative stress, and promote overall well-being (Sharanagat et al., 2019; Karthikeyan et al., 2022). Therefore researchers are increasingly exploring their potential applications in nutrition and therapeutic interventions. Fig. 15.2 summarizes major marine bioactive compound groups based on their chemical structure.

15.5.2 Oxidative stress

Oxidative stress is a pathological condition that arises when there is an imbalance between the production of ROS and the cellular antioxidant defense mechanism, ultimately leading to cellular damage (Sies et al., 2017). During normal cellular metabolism, particularly in the mitochondria, ROS are generated as by-products of the electron transport chain (ETC). These highly reactive molecules, which include free radicals such as superoxide anions (O_2^-), hydrogen peroxide (H_2O_2), and hydroxyl radicals (•OH), play dual roles in cellular physiology, serving as signaling molecules in various biological processes. However, excessive ROS production can overwhelm the cellular antioxidant defenses, which are comprised of enzymes like superoxide dismutase (SOD), catalase, and glutathione peroxidase, as well as nonenzymatic antioxidants such as vitamins C and E (Sies et al., 2017).

Under conditions such as lipotoxicity (due to excessive lipid accumulation), hyperglycemia, inflammation, or exposure to environmental toxins, the production of ROS can markedly increase, resulting in oxidative stress (Hauck and Bernlohr, 2016; Pouvreau et al., 2018; Münzel and Daiber, 2018). For example, during lipotoxicity, heightened lipid oxidation within the mitochondria leads to elevated ROS generation, which damages mitochondrial membranes and induces electron leakage. The resultant excessive ROS can interact with DNA, lipids, and proteins, resulting in oxidative modifications that disrupt cellular function. Oxidation of DNA may provoke mutations that facilitate the onset of carcinogenesis, while lipid peroxidation can deteriorate cell membrane integrity, ultimately disrupting cellular homeostasis and triggering apoptosis. In addition, oxidatively modified proteins may lose their functional capabilities or undergo conformational changes, which can contribute to the pathogenesis of various diseases (Hauck and Bernlohr, 2016). This cascade of oxidative damage underscores the critical need to understand the mechanisms behind ROS generation and its implications for cellular health and disease progression.

The consequences of oxidative stress are profound and widespread, playing a crucial role in the development of numerous chronic conditions, including NAFLD, CKD, CVD, neurodegenerative disorders, and metabolic syndromes (Rani et al., 2016; Singh et al., 2019). For instance, in diabetes, an increased concentration of glucose generates additional ROS, while in CVDs, oxidative stress can provoke endothelial dysfunction, a precursor to atherosclerosis (Rani et al., 2016; Victor et al., 2009). In addition, chronic inflammation, a common feature in many diseases, is characterized by the sustained production of ROS, thereby exacerbating oxidative damage (Pouvreau et al., 2018). The cumulative effect of oxidative stress underscores its critical role in cellular dysfunction, providing a compelling rationale for the development of antioxidant therapies aimed at mitigating ROS-induced damage and restoring redox balance in affected tissues (Sies et al., 2017). Therefore understanding the impact of oxidative stress on cellular metabolism is essential for elucidating the mechanisms underlying various diseases and for developing effective therapeutic interventions.

NAFLD pathogenesis is characterized by a state of oxidative stress resulting from an imbalance between ROS production and antioxidant capacity. Hepatic lipid accumulation, caused by increased dietary intake, de novo lipogenesis, and impaired mitochondrial fatty acid β-oxidation, triggers a cascade of events that create an oxidative stress environment (Gonzalez et al., 2020). Mitochondrial dysfunction in NAFLD disrupts ETC activity, increasing ROS production, predominantly H_2O_2 and $O_2\bullet^-$. This ROS generation is amplified by extra-mitochondrial sources, including endoplasmic reticulum oxidoreductase 1α within the endoplasmic reticulum and NADPH oxidases expressed in both hepatocytes and hepatic stellate cells (Sumida et al., 2013). Increased ROS levels cause oxidative damage to cellular macromolecules, thereby initiating inflammatory signaling (e.g., NF-κB, MAPK pathways), inducing hepatocyte apoptosis, and driving disease progression from NAFLD to NASH, potentially culminating in fibrosis and cirrhosis (Sumida et al., 2013). This pathogenesis involves a complex interplay between oxidative stress, endoplasmic reticulum (ER) stress, mitochondrial dysfunction, and innate immune responses, forming a positive feedback loop that sustains and exacerbates NAFLD progression. The impairment of endogenous antioxidant defenses further amplifies the overall oxidative burden within the liver (Liu et al., 2015).

Oxidative stress plays a pivotal role in the pathogenesis and progression of CKD. Its amplification begins in the early stages of CKD, contributing significantly to renal failure. Excessive production of ROS by various enzymatic systems (NADPH oxidase, lipoxygenase, xanthine oxidase, and mitochondrial respiratory chain), coupled with impaired antioxidant defense mechanisms, creates a prooxidant milieu. This imbalance leads to the oxidation of macromolecules, resulting in cellular damage and dysfunction (Duni et al., 2019). Directly, ROS are linked to podocyte damage, proteinuria, and the development of focal segmental glomerulosclerosis, as well as tubulointerstitial fibrosis. Vascular oxidative stress further impairs renal autoregulation by affecting the myogenic responses of afferent renal arterioles. ROS promotes inflammation via the formation of proinflammatory oxidized lipids and advanced oxidation protein products, while the activation of NF-κB in a prooxidant environment upregulates proinflammatory cytokines and recruits inflammatory cells. This vicious cycle sustains CKD progression, leading to various clinical manifestations, including diabetic nephropathy, IgA nephropathy, and polycystic kidney disease. The disruption of the delicate balance between prooxidant and antioxidant systems, therefore, constitutes a central mechanism underlying CKD pathogenesis (Duni et al., 2019).

The marine environment is a rich source of structurally diverse bioactive compounds, many of which possess significant antioxidant properties (Ghosh et al., 2022; Sharanagat et al., 2019). This abundance of natural antioxidants offers a promising avenue for developing novel therapeutic approaches to address the pathogenesis of NAFLD and CKD, two prevalent and increasingly interlinked chronic diseases characterized by substantial oxidative stress (Targher and Byrne, 2017). The multifaceted antioxidant mechanisms of these marine-derived molecules offer a potential advantage over conventional synthetic antioxidants. Several key classes of marine bioactive compounds demonstrate potent antioxidant activity. Phlorotannins, a group of polyphenolic compounds abundant in brown seaweeds, exemplify this capacity through multiple mechanisms. Their large molecular size and high number of hydroxyl groups enable effective scavenging of ROS, such as superoxide, hydroxyl, and peroxyl radicals, directly neutralizing their damaging potential. Moreover, phlorotannins exhibit potent metal-chelating capabilities, sequestering redox-active transition metals such as iron and copper (Shin et al., 2014). These metals catalyze the formation of highly reactive hydroxyl radicals via the Fenton reaction, a key driver of oxidative damage in both NAFLD and CKD (Zhao et al., 2023; Shah, 2004). By chelating these metals, phlorotannins effectively prevent the initiation of this destructive chain reaction.

Fucoidan, a sulfated polysaccharide predominantly found in brown algae, provides a further example of a marine compound with multifaceted antioxidant effects. Fucoidan's sulfated esters can directly interact with ROS, neutralizing their damaging effects (Qu et al., 2014). Beyond this direct scavenging action, fucoidan also demonstrates significant antiinflammatory properties. It inhibits the activity of various enzymes involved in inflammatory signaling cascades, effectively dampening the chronic inflammatory response often associated with NAFLD and CKD progression (Apostolova et al., 2020). This dual action is crucial, as oxidative stress and inflammation are closely intertwined and mutually reinforcing in the pathogenesis of these diseases. Fucoxanthin, a carotenoid pigment particularly abundant in brown seaweeds, exhibits potent antioxidant effects primarily due to its ability to effectively quench singlet oxygen, a highly reactive form of oxygen implicated in lipid peroxidation, a key driver of cellular damage in both the liver and kidney (Fung et al., 2013). Fucoxanthin's unique chemical structure allows it to interact with singlet oxygen in a way that effectively prevents its damaging reactions. Marine-derived peptides, another class of bioactive compounds, offer yet another avenue for antioxidant intervention. These peptides, often extracted from fish and shellfish, demonstrate a range of antioxidant mechanisms. Some directly scavenge ROS, while others may act by inhibiting the activity of enzymes involved in ROS generation, such as NADPH oxidase. Furthermore, some marine peptides exhibit antiinflammatory properties, thereby contributing to the overall reduction of oxidative stress.

Omega-3 PUFAs, abundantly present in fatty fish, represent a widely recognized class of marine-derived compounds with demonstrable antioxidant and antiinflammatory properties (Fernando et al., 2023). Omega-3 PUFAs, like EPA and DHA, exert their antiinflammatory effects primarily through their influence on lipid metabolism and eicosanoid synthesis. They reduce the production of proinflammatory eicosanoids while increasing the production of antiinflammatory mediators, thus effectively mitigating the inflammatory component of NAFLD and CKD (Banaszak et al., 2024). The incorporation of omega-3 PUFAs into the diet has shown promise in improving liver function tests and reducing the severity of hepatic steatosis in NAFLD (Jump et al., 2018). In CKD, the use of omega-3 PUFAs has shown potential in reducing inflammation and slowing the progression of kidney disease, although further research is needed (Saglimbene et al., 2020).

The potential of marine bioactive compounds to manage NAFLD and CKD pathogenesis extends beyond their direct antioxidant properties. Many also exhibit the capability to modulate key signaling pathways involved in the development and progression of these diseases. For instance, some marine compounds have shown promise in activating the Nrf2-Keap1 pathway, a crucial cellular defense mechanism against oxidative stress. Activation of this pathway leads to increased expression of antioxidant and cytoprotective enzymes, ultimately strengthening the body's inherent defense against oxidative damage (Nisar et al., 2024). Simultaneously, some compounds effectively suppress the NF-κB signaling pathway, a master regulator of inflammation, reducing the inflammatory burden significantly associated with NAFLD and CKD (Terracciano et al., 2006). In summary, the diverse array of antioxidant mechanisms exhibited by marine bioactive compounds offers a rich source of potential therapeutic agents for mitigating the pathogenesis of NAFLD and CKD. However, further research is essential to fully elucidate their mechanisms, optimize their delivery systems, and thoroughly assess their safety and efficacy in human clinical trials. The eventual integration of these promising compounds into effective management strategies holds significant potential for improving the treatment and prognosis of these widely prevalent diseases. Table 15.1 presents examples of marine bioactive compounds that exhibit antioxidant properties.

15.5.3 Inflammation

Inflammation is a crucial immune system response that protects the body from infection, injury, and harmful stimuli. It is characterized by a complex interplay of cellular and molecular components aimed at restoring homeostasis (Ward, 2010). Acute inflammation is typically a short-term process that culminates in the resolution of the inflammatory response once the initial trigger has been eliminated. However, chronic low-grade inflammation is increasingly recognized as a significant health concern, often persisting for months or even years, and is associated with a myriad of chronic diseases. This form of inflammation is characterized by a sustained, subclinical immune response marked by the continuous release of proinflammatory cytokines, immune cell activation, and tissue remodeling. Unlike acute inflammation, which presents with classic signs like redness and swelling, chronic low-grade

TABLE 15.1 Examples of marine bioactive compounds that exhibit antioxidant properties.

Marine bioactive group	Compound/ sequence	Antioxidant activity	References
Polyphenols	Phlorotannins	Direct ROS scavenging, enzyme inhibition: specific phlorotannins, such as phloroglucinol, eckol, and dieckol, inhibit enzymes involved in inflammatory signaling pathways (e.g., MKK4/SEK1-JNK-AP-1 cascade). Upregulation of protective factors: Eckol, for instance, upregulates the expression of protective factors like heme oxygenase-1 (HO-1) and increases levels of reduced glutathione, thereby enhancing the cell's ability to combat oxidative stress.	Shin et al. (2014)
Polyphenols	Fucoidan	Direct ROS scavenging, lower-molecular-weight fucoidans generally show stronger antioxidant capacity than higher molecular weight forms, likely due to better cell membrane penetration. The degree and pattern of sulfation significantly impact antioxidant activity.	Wang et al. (2019)
Carotenoid	Fucoxanthin	Direct ROS scavenging (ABTS, β-carotene bleaching) and reduction of ferric ions (FRAP).	Foo et al. (2017)
Peptides	VECYGPNRPQF	Scavenged hydroxyl, superoxide, peroxyl, DPPH, and ABTS radicals; prevent DNA damage.	Sheih et al. (2009), Sheih et al. (2010)
Peptides	NDAEYGICGF	Attenuating ethanol-induced cytotoxicity; inhibition of intracellular reactive oxygen species production by DCFH-DA assay.	Chen et al. (2019)
Peptides	PIIVYWK	Scavenged DPPH radical; ORAC activity.	Park et al. (2016)
Peptides	KTFCGRH	Scavenged DPPH radical; inhibition of linoleic acid peroxidation; protective effect against hydroxyl radicals.	Nazeer et al. (2012)
PUFAs	Eicosapentaenoic acid (EPA)	Direct ROS scavenging; increased total cellular antioxidant capacity (T-AOC), as measured by ABTS and FRAP assays, indicating an overall boost in the cell's antioxidant defense system. Increased the activity and mRNA expression of key antioxidant enzymes (SOD, CAT, GPX) at both the cellular and mitochondrial levels. Improved mitochondrial membrane potential (MMP) and reduced the ADP/ATP ratio, indicating enhanced mitochondrial function and energy production.	Xiao et al. (2022)
PUFAs	Docosahexaenoic acid (DHA)	Direct ROS scavenging; increased the activity and/or expression of various antioxidant enzymes (SOD, CAT, GPx) and related proteins (GSH, TRx). Oxidized DHA metabolites (e.g., HHE, neuroprostanes) may directly activate the Nrf2 transcription factor, leading to increased expression of antioxidant enzymes. DHA's incorporation into cell membranes may alter membrane fluidity and the activity of membrane-associated proteins involved in redox signaling, ultimately influencing antioxidant gene expression.	Borgonovi et al. (2023)

inflammation is often asymptomatic and subtle, making it challenging to detect without specific medical assessments (Ward, 2010).

Chronic inflammation is a fundamental mechanism underlying metabolic diseases, including obesity, type 2 diabetes, NAFLD, and CKD (Hotamisligil, 2006). Central to this inflammatory response is the activation of the NF-κB pathway, which triggers the expression of various proinflammatory cytokines. TNF-α plays a crucial role by promoting insulin resistance and facilitating lipolysis in adipocytes, linking inflammation directly to metabolic dysregulation. In addition, monocyte chemoattractant protein-1 (MCP-1) fosters the recruitment of monocytes to adipose tissues, enhancing local inflammation and contributing to tissue dysfunction. IL-6, often elevated in obesity, influences glucose metabolism and further exacerbates insulin resistance (Kunnumakkara et al., 2020). C-reactive protein (CRP), an acute-phase protein, serves as a clinical biomarker of systemic inflammation and is associated with increased cardiovascular risk in metabolic syndrome (Ridker and Silvertown, 2008). Together, these mediators create a feedback loop of inflammation, with NF-κB as a key regulatory hub, perpetuating metabolic disturbances (Kunnumakkara et al., 2020).

Chronic inflammation plays a major role in the pathogenesis of NAFLD. The progression of NAFLD is often interconnected with metabolic syndrome, obesity, and insulin resistance, where chronic low-grade inflammation serves as a common pathogenic link (Yan et al., 2020). The initial phase of NAFLD involves the accumulation of triglycerides in hepatocytes, primarily due to increased lipogenesis and decreased fatty acid oxidation. As the disease progresses, the presence of excess lipids triggers inflammatory pathways that contribute to hepatocyte injury and further disease advancement, leading to NASH, fibrosis, and potentially cirrhosis (Michelotti et al., 2013). At the cellular level, the activation of the NF-κB signaling pathway is central to the inflammatory response in NAFLD. During conditions of lipid overload, hepatocytes experience an accumulation of fatty acids that exceeds their metabolic capacity, leading to the incomplete metabolism of these lipids. This results in the release of toxic lipid intermediates, such as diacylglycerol (DAG) and ceramides, which, in turn, activate several inflammatory pathways, including the NF-κB signaling pathway. Activation of NF-κB leads to the production of proinflammatory cytokines, such as TNF-α and IL-6, exacerbating local inflammation within the liver and promoting a cycle of hepatocyte injury and inflammation (Jani et al., 2022; Hajduch et al., 2021). TNF-α, particularly, has been linked to the induction of insulin resistance, creating a feedback loop that exacerbates both systemic and hepatic metabolic dysfunction.

Moreover, the infiltration of immune cells, particularly macrophages, into the liver is a hallmark of chronic inflammation in NAFLD. These macrophages are often activated by damaged hepatocytes and lipid accumulation signals, further contributing to the inflammatory milieu (Peiseler et al., 2022). MCP-1 is a key chemokine involved in recruiting these immune cells, amplifying the inflammatory response, and perpetuating liver injury. Activated macrophages increase the secretion of proinflammatory cytokines and ROS generation, further damaging hepatocytes and leading to necrosis and fibrosis (Barreby et al., 2022). In addition to cytokine release, chronic inflammation is associated with the activation of hepatic stellate cells, which play a significant role in fibrosis development. In response to inflammatory mediators, these cells transform from a quiescent state to an activated form, producing extracellular matrix components that contribute to liver fibrosis (Wang et al., 2022).

The progression from simple steatosis to NASH and subsequently to liver fibrosis is intricately linked to the sustained inflammatory response within the liver. The systemic effects of chronic inflammation also extend to the liver. Elevated levels of inflammatory markers, such as CRP, are often found in individuals with NAFLD and serve as indicators of increased cardiovascular risk associated with the condition (Lizardi-Cervera et al., 2007). The liver's role in producing these inflammatory mediators places it at the nexus of metabolic and inflammatory diseases, highlighting the importance of targeting inflammation in the management of NAFLD.

Chronic inflammation plays a pivotal role in the pathogenesis and progression of CKD, influencing renal damage, fibrosis, and overall disease outcomes. In CKD, a persistent inflammatory state is predominantly manifested by an increase in proinflammatory cytokines, activation of immune cells, and renal tissue damage, all of which contribute to systemic and glomerular inflammation (Akchurin and Kaskel, 2015). One of the primary drivers of chronic inflammation in CKD is the dysregulation of the immune system, which is often exacerbated by factors such as metabolic syndrome and oxidative stress. As kidney function declines, the kidneys become less capable of filtering waste products, leading to their accumulation in the bloodstream. This retention of uremic toxins can stimulate inflammatory responses, activating resident immune cells like macrophages and interstitial fibroblasts within the kidney (Amdur et al., 2016). The activation of these cells results in the production of a variety of proinflammatory cytokines, such as TNF-α, IL-6, and CRP, which further propagate the inflammatory response (Amdur et al., 2016). The NF-κB signaling pathway is critically involved in the regulation of inflammatory processes in CKD. Upon activation, NF-κB facilitates the transcription of multiple proinflammatory genes, leading to an upsurge in the production of cytokines and chemokines that are known to mediate inflammation. This culminates in the recruitment of additional immune cells, creating a vicious cycle of inflammation and injury within the renal parenchyma. In addition, chronic activation of NF-κB has been implicated in the progression of renal fibrosis by promoting the activation of renal fibroblasts and enhancing the deposition of extracellular matrix components (Pedruzzi et al., 2012).

Another key aspect of chronic inflammation in CKD involves the role of endothelial dysfunction. The inflammatory milieu in CKD is associated with structural and functional alterations in the renal vasculature. Endothelial cells can become activated in response to proinflammatory cytokines, promoting the expression of adhesion molecules that facilitate the infiltration of leukocytes into the renal interstitium. This process exacerbates local inflammation and contributes to vascular calcification and impaired renal hemodynamics, further accelerating the decline in kidney function (Roumeliotis et al., 2020). The interplay between inflammation and oxidative stress is another crucial factor in the pathogenesis of CKD. Uremic toxins can induce oxidative stress by generating ROS, which, in turn, activate various signaling pathways, including NF-κB, reinforcing the inflammatory response (Harlacher et al., 2022). Moreover, oxidative stress can directly damage renal cells, promoting cellular apoptosis and necrosis. The combination of inflammation and oxidative stress creates a detrimental microenvironment that contributes to renal cell injury, fibrosis, and the overall decline in kidney function (Ebert et al., 2021). Clinical studies have demonstrated that persistent inflammation is associated with adverse outcomes in patients with CKD, including faster progression to ESRD and increased cardiovascular morbidity and mortality. Biomarkers of inflammation, such as elevated levels of IL-6, TNF-α, and CRP, have been correlated with the severity of CKD and can serve as prognostic indicators for disease progression (Amdur et al., 2016; Laclair et al., 2008).

Marine bioactive compounds have gained increasing attention due to their potent antiinflammatory properties, which can significantly impact health and disease management. These compounds include omega-3 fatty acids, polyphenols, peptides, and carotenoids, each playing distinct roles in modulating inflammatory responses and offering potential therapeutic benefits for various chronic inflammatory conditions, including CVDs, metabolic syndrome, and autoimmune disorders (D'Orazio et al., 2012). One of the most extensively studied classes of marine bioactive compounds is omega-3 fatty acids, particularly EPA and DHA. The antiinflammatory effects of omega-3 fatty acids are largely attributed to their ability to modify the production of proinflammatory mediators. For instance, omega-3 fatty acids can be enzymatically converted into resolvins and protectins, which are specialized proresolving mediators that actively promote the resolution of inflammation (Kwon, 2020). Furthermore, supplementation with omega-3 fatty acids has been shown to reduce the levels of proinflammatory cytokines such as TNF-α and IL-6, which are often elevated in chronic inflammatory states (Kang and Weylandt, 2008). Clinical studies have demonstrated that individuals who consume diets rich in omega-3 fatty acids exhibit lower serum levels of inflammatory markers, including CRP, thereby highlighting their role in managing systemic inflammation (Taha et al., 2022).

Seaweeds, representing a particularly prolific source of bioactive metabolites, produce a diverse range of compounds with antiinflammatory properties. Phlorotannins, a class of unique polyphenolic compounds found primarily in brown seaweeds, are particularly well-studied. Phlorofucofuroeckol A and B, isolated from *Eisenia bicyclis* and *Ecklonia stolonifera*, respectively, exemplify this activity by suppressing NO production and downregulating the expression of proinflammatory cytokines such as TNF-α, IL-1β, and IL-6. Simultaneously, they inhibit the activity of key inflammatory enzymes, including inducible nitric oxide synthase (iNOS) and cyclooxygenase-2 (COX-2), thereby targeting multiple facets of the inflammatory cascade (Kim et al., 2013; Yu et al., 2015). Dieckol, another prominent phlorotannin found in various *Ecklonia* species, further exemplifies this multitarget approach by effectively suppressing the production of NO, COX-2, and iNOS, thus simultaneously impeding multiple steps in the inflammatory response (Jung et al., 2009). Diphlorethohydroxycarmalol, isolated from *Ishige okamurae,* presents a different mechanism of action, demonstrating antiinflammatory properties through the downregulation of NF-κB and p38 MAPK signaling pathways, resulting in a decrease in the production of proinflammatory cytokines and NO (Kang et al., 2015). This highlights the complex mechanisms through which seaweed-derived compounds exert their effects. Other seaweed-derived compounds, such as octaphlorethol A (*Ishige foliacea*), various fucoidans (*Saccharina japonica, Colpomenia sinuosa, Fucus vesiculosus*), sulfated polysaccharides (*Sargassum fulvellum*), alginic acid (*Sargassum horneri*), laminarin (*Grifola frondosa*), fucosterol (*Padina boryana*), and epitaondiol (*Stypopodium flabelliforme*), contribute to this rich diversity of antiinflammatory agents. Their effects often involve the inhibition of key inflammatory enzymes (iNOS, COX-2), the modulation of critical immune regulatory processes (including the inhibition of NF-κB and MAPK pathways), and a consequent reduction in the levels of proinflammatory cytokines (TNF-α, IL-1β, IL-6, and others) and mediators (e.g., PGE2, NO) (Jaworowska and Murtaza, 2023).

Marine sponges, known for their extraordinary capacity to produce bioactive secondary metabolites, offer another significant source of antiinflammatory compounds. Stylissatin A and its derivatives, isolated from *Stylissa massa*, exhibit potent inhibitory effects on NO production in macrophages while maintaining minimal cytotoxicity, representing a desirable characteristic for therapeutic development (Kita et al., 2013). Dactylospongins A and B, derived from *Dactylospongia* species, effectively suppress the production of several proinflammatory cytokines (IL-6, IL-1β, IL-8, and PGE2) in THP-1 cells (Magri et al., 2023). Septosones A, isolated from *Dysidea septosa*, demonstrate antiinflammatory activity by inhibiting NF-κB activation and reducing macrophage migration, showcasing their impact on cellular processes crucial to the inflammatory response (Gui et al., 2019). In addition, 9,11-dihydrogracilin A, a terpene derived from *Dendrilla membranosa*, further exemplifies the multifaceted nature of sponge-derived compounds, inducing apoptosis in human mononuclear cells, downregulating key inflammatory signaling pathways (NF-κB, STAT, ERK), and reducing the levels of IL-6 and IL-10, while also affecting HaCaT cell migration and viability (Ciaglia et al., 2017). Furthermore, deacetylphylloketal from *Phyllospongia* species comprehensively inhibits the production and/or gene expression of NO, PGE2, IL-6, IL-1β, TNFα, iNOS, and COX-2 (Lee et al., 2019).

Beyond seaweeds and sponges, other marine organisms, including jellyfish, shellfish, and peanut worms, also contribute to this rich reservoir of bioactive compounds with antiinflammatory potential. Polysaccharides from jellyfish, such as *Rhopilema esculentum*, suppress myeloperoxidase activity, proinflammatory cytokines, and NO, suggesting broad-spectrum effects on inflammatory processes (Khursheed et al., 2023). Shellfish, including mussels, clams, and abalone, contain a variety of bioactive peptides and lipid fractions that exert antiinflammatory effects via various mechanisms, including the modulation of NF-κB and MAPK pathways, the inhibition of iNOS and COX-2, and the reduction of inflammatory cytokines (Ahmad et al., 2018; Kemp and Kwon, 2021). Extracts and peptides from peanut worms, such as *Sipunculus nudus* and *Phascolosoma esculenta*, demonstrate antiinflammatory effects in various models, reducing paw edema, modulating proinflammatory cytokine levels, and promoting wound healing (Zhang et al., 2011; Yang et al., 2019).

Marine sources represent a rich and largely unexplored source of novel antiinflammatory compounds with diverse mechanisms of action. Their ability to target multiple aspects of the inflammatory response, from the inhibition of key enzymes to the modulation of complex signaling pathways, showcases their potential as promising therapeutic agents for a wide range of inflammatory diseases, particularly metabolic diseases, including NAFLD and CKD. Table 15.2 presents examples of marine bioactive compounds that exhibit antiinflammatory properties.

15.5.4 Obesity and hyperlipidemia

Obesity and hyperlipidemia are closely interlinked conditions that significantly contribute to the burden of metabolic diseases globally. The prevalence of obesity has escalated dramatically over the past few decades. By 2022, an estimated 2.5 billion adults aged 18 years and

TABLE 15.2 Examples of marine bioactive compounds that exhibit antiproperties.

Compound/extract	Species	Activity	References
Phlorofucofuroeckol A	*Eisenia bicyclis*	Suppresses LPS-induced NO production.	Kim et al. (2013)
Phlorofucofuroeckol B	*Ecklonia stolonifera*	Inhibits TNF-α, IL-1β, and IL-6 secretion; downregulates COX-2 and iNOS; inhibits NF-κB and Akt/ERK/JNK pathways.	Yu et al. (2015)
Dieckol	*Ecklonia stolonifera*	Inhibits LPS-induced COX-2, iNOS, and NO production.	Jung et al. (2009)
Diphlorethohydroxycarmalol	*Ishige okamurae*	Downregulates NF-κB and p38 MAPK; reduces proinflammatory cytokines and NO.	Kang et al. (2015), Wang et al. (2020)
Octaphlorethol A	*Ishige foliacea*	Exhibits antiinflammatory activity via transcriptional regulation of NF-κB through MAPK.	Manzoor et al. (2013)
Fucoidan	*Saccharina japonica*	Reduces NO production; downregulates MAPK and NF-κB pathways.	Ni et al. (2020)
Sulfated polysaccharides	*Sargassum fulvellum*	Downregulates proinflammatory cytokines and reduces NO and NF-κB/MAPK pathways.	Saraswati et al. (2019)
Alginic acid	*Sargassum horneri*	Suppresses PGE2 production via COX-2 inhibition; decreases proinflammatory cytokines and NF-κB/MAPK pathways.	Sarithakumari et al. (2013)
Laminarin	*Grifola frondosa*	Inhibits NO and PGE2 production; suppresses proinflammatory cytokines and NF-κB.	Pramanik et al. (2024)
Fucosterol	*Padina boryana*	Inhibits NO and PGE2 production, suppresses proinflammatory cytokines, and NF-kB/MAPK pathways.	Jayawardena et al. (2020)
Lectin	*Bryothamnion triquetrum*	Inhibits neutrophil migration to inflammation sites.	Pereira-da-Silva et al. (2012)
Peptide fractions	*Ulva* spp.	Modulates TLR4 and the NF-κB/p38/JNK pathway.	Cian et al. (2018)
Azocinyl morpholinone	*Gracilaria opuntia*	Inhibits COX-2 and 5-LOX activity.	Makkar and Chakraborty (2018)
3-Hydroxy-4,7-megastigmadien-9-one	*Ulva pertusa*	Inhibits IL-12 p40, IL-6, and TNF-α production; blocks MAPKs and NF-κB pathways.	Ali et al. (2017)
Sargachromanol G	*Sargassum siliquastrum*	Reduces proinflammatory cytokines and suppresses NO and PGE2 production.	Yoon et al. (2012)
Bryostatin-1	*Bugula neritina*	Downregulates COX-2 mRNA expression; reduces hemorrhagic transformation; promotes the M2 macrophage phenotype; suppresses proinflammatory cytokines; increases IL-10.	Safaeinejad et al. (2018)
Peptide (EGLLGDVF)	*Perna viridis*	Inhibits proinflammatory cytokines; reduces NO and COX-2; downregulates iNOS and COX-2.	Joshi and Nazeer (2020)
Peptide (HKGQCC)	*Meretrix meretrix*	Inhibits NO, COX-2, TNF-α, and IL-1β in LPS-stimulated RAW 264.7 cells.	Joshi et al. (2021)

older were classified as overweight, with >890 million of these individuals living with obesity (Okunogbe et al., 2022). This rising trend in obesity is paralleled by an increase in hyperlipidemia, characterized by elevated levels of lipids (fats) in the blood, most notably triglycerides and low-density lipoprotein (LDL) cholesterol (Stewart et al., 2020). Hyperlipidemia is commonly found in individuals with obesity due to the altered metabolic processes associated with excess adipose tissue. The combination of these two conditions significantly enhances the risk of developing serious metabolic diseases, including type 2 diabetes, CVD, NAFLD, CKD, and metabolic syndrome (Stewart et al., 2020).

Obesity is increasingly recognized as a significant risk factor for the development and progression of NAFLD. Obese individuals have a 3.5-fold increased risk of developing NAFLD, and there is a positive relationship between body mass index (BMI) and NAFLD risk (Li et al., 2016a). The underlying mechanisms linking obesity to NAFLD are complex and multifactorial, with lipid overload and subsequent lipotoxicity playing central roles. In individuals with obesity, excess adipose tissue, particularly visceral fat, leads to an overproduction of free fatty acids that enter the liver in excessive amounts, resulting in lipid overload. This overflow of lipids promotes the accumulation of triglycerides in hepatocytes, contributing to the initial stages of fatty liver disease (Rojano et al., 2023). As the liver's capacity to store fat becomes overwhelmed, toxic lipid intermediates, such as DAGs and ceramides, accumulate, further exacerbating hepatocellular injury. These intermediates are known to activate proinflammatory signaling pathways, creating a state of cellular stress and dysfunction (Jani et al., 2022; Hajduch et al., 2021). Mitochondrial dysfunction plays a critical role in this context, as the overwhelmed liver cells struggle to metabolize the excessive lipid load. Impaired mitochondrial function leads to decreased fatty acid oxidation and increased production of ROS, contributing to oxidative stress (Bass and Merriman, 2005). In addition, oxidative stress serves as a catalyst for the progression of NAFLD, where it not only contributes to hepatocyte damage but also activates inflammatory pathways that progress simple fatty liver to cirrhosis (Peiseler et al., 2022).

Hyperlipidemia, often coexisting with obesity, exacerbates these pathological processes by enhancing the lipid burden on the liver and sustaining chronic inflammation. Elevated levels of circulating LDL cholesterol and triglycerides contribute to hepatic lipid accumulation, leading to a vicious cycle of lipotoxicity and cellular dysfunction. As the liver attempts to manage the excessive influx of lipids, the dysregulation of lipid metabolism may result in further hepatic inflammation and fibrosis. With ongoing inflammation and cellular damage, there is a progressive risk of developing advanced liver diseases, including cirrhosis and HCC (Phan et al., 2019). The interplay between obesity, hyperlipidemia, and NAFLD necessitates a comprehensive understanding and interventions targeted at limiting fat accumulation and addressing the metabolic dysfunction associated with these conditions.

Obesity has emerged as a significant risk factor for the development and progression of CKD, primarily through mechanisms involving lipid overload, lipotoxicity, mitochondrial dysfunction, and associated oxidative stress and inflammation (Jiang et al., 2023). The pathophysiological link between obesity and CKD is highlighted by the excessive accumulation of adipose tissue, which plays a crucial role in disrupting normal metabolic processes and contributing to renal injury. In obese individuals, there is an increase in free fatty acids released into circulation due to the enlargement of adipocytes. This lipid overload can overwhelm the kidneys, leading to their dysfunction (Jiang et al., 2023).

The kidneys are actively involved in lipid metabolism, and when exposed to excessive free fatty acids and toxic lipid intermediates, such as DAGs and ceramides, they become susceptible to lipotoxicity. These harmful lipid intermediates not only accumulate due to obesity but also disrupt metabolic homeostasis, thereby impairing kidney function (Jiang et al., 2023). The presence of such toxic lipids can induce cellular stress and trigger a cascade of pathological responses within renal cells, culminating in mitochondrial dysfunction. Mitochondria are essential for energy production and metabolic regulation, and their dysfunction in the context of obesity leads to decreased mitochondrial oxidative phosphorylation. This impairment results in decreased ATP production and, consequently, promotes the accumulation of ROS (Ren et al., 2023). The accumulation of ROS is a critical factor in the development of oxidative stress, which damages cellular components, including lipids, proteins, and DNA. In the kidneys, oxidative stress triggers inflammatory pathways, enhancing the production of proinflammatory cytokines such as TNF-α, IL-6, and TGF-β (Escasany et al., 2019). This chronic inflammatory state has profound implications for renal health, as it contributes to the fibrosis and scarring of kidney tissues, progressively leading to nephron loss and a decline in renal function. Furthermore, the coexistence of hyperlipidemia, often present in individuals with obesity, exacerbates lipid-induced renal injury by sustaining higher levels of circulating lipids, which further perpetuates lipotoxicity and oxidative stress (Escasany et al., 2019).

The interrelation between obesity, hyperlipidemia, and CKD exemplifies a vicious cycle. As kidney function deteriorates, the body's ability to clear excess lipids diminishes, leading to an exacerbation of lipid overload and subsequent renal injury (Ren et al., 2023). In addition, obesity is associated with the development of insulin resistance, which further complicates renal pathology by promoting a metabolic environment that favors inflammation and fibrosis. Insulin resistance can lead to increased levels of circulating glucose and free fatty acids, both of which are detrimental to renal function (Arabi et al., 2023). Recent studies have highlighted the importance of weight loss and metabolic interventions to improve renal outcomes in obese patients with CKD (Alkhatib et al., 2023). Moreover, pharmacological treatments targeting hyperlipidemia have also demonstrated potential benefits in preserving renal function (Zuzda et al., 2022; Harper and Jacobson, 2008).

Marine-derived bioactive compounds represent a promising therapeutic strategy for the management of obesity and hyperlipidemia. Extensive research demonstrates the lipid-lowering potential of various compounds derived from diverse marine sources, including algae, crustaceans, and fish. The observed effects are mediated through multiple interconnected mechanisms affecting key pathways involved in energy homeostasis and lipid metabolism (Hu et al., 2016). One significant mechanism involves the modulation of appetite. Alginate, a polysaccharide abundant in brown algae, induces satiety by slowing gastric emptying, delaying the sensation of hunger, and promoting feelings of fullness. This effect is partially attributable to the formation of alginate gels in the stomach, which increase distension and influence the release of gut hormones that regulate appetite (Odunsi et al., 2010).

Another key mechanism is the suppression of dietary fat absorption. Orlistat, a clinically used drug, inhibits pancreatic lipase, an enzyme responsible for fat digestion (Heck et al., 2000). Marine-derived compounds also target this pathway; for example, brominated lipids from sponges and certain saponins from sea cucumbers have demonstrated significant pancreatic lipase inhibitory activity in vitro. This inhibition reduces the amount of dietary fat absorbed, contributing to a reduction in calorie intake and potential weight loss (Liang et al., 2014; Guo et al., 2016). The crude Calanus oil and its wax esters significantly decreased fat pad deposition in animal models, though the precise mechanisms remain under investigation (Salma, 2015). Furthermore, phospholipids of EPA and DHA from sea cucumbers significantly inhibited body fat deposition, lowered serum triacylglycerol levels, and alleviated hepatic steatosis (Han et al., 2019).

The regulation of adipogenesis, the process of fat cell formation, is another target for marine antiobesity compounds. Numerous marine compounds have demonstrated the capacity to suppress adipogenesis. Fucoidan, a sulfated polysaccharide extracted from brown algae and sea cucumbers, inhibits the differentiation of preadipocytes into mature adipocytes by modulating the expression of key genes involved in adipogenesis, such as PPARγ and C/EBPα (Kim et al., 2009). Similarly, phlorotannins from brown algae, like Eckol and Dieckol, exhibit significant antiadipogenic effects via multiple pathways, including inhibition of PPARγ and C/EBPα, and the activation of AMPK (Ko et al., 2013). Furthermore, fucosylated chondroitin sulfate extracted from sea cucumbers exhibits potent antiadipogenic effects, also through modulation of the Wnt/β-catenin pathway and regulation of PPARγ and SREBP1 expression (Xu et al., 2015). Stimulation of energy expenditure plays a significant role in the management of obesity. Fucoxanthin, a carotenoid found in brown algae, is a potent stimulator of thermogenesis, the process of heat production, primarily through the activation of uncoupling proteins (UCPs), particularly UCP1 in brown adipose tissue (BAT). Although humans possess less BAT than rodents, fucoxanthin's ability to stimulate UCPs in other tissues also contributes to increased energy expenditure (Maeda et al., 2005). Interestingly, astaxanthin, another carotenoid found in algae and crustaceans, also significantly reduced body weight and improved lipid profiles in obese animal models through mechanisms involving modulation of insulin signaling (Bhuvaneswari and Anuradha, 2012).

Lipids, specifically PUFAs such as EPA and DHA, found abundantly in fish oil, also contribute to the antiobesity effects of marine compounds. Studies indicate that increased consumption of fish oil is associated with a lower BMI and reduced body fat (Lopez-Huertas, 2012). The mechanisms behind these effects are complex and involve several factors, including reduced appetite, improved lipid metabolism via modulation of lipoprotein synthesis, and increased fatty acid oxidation, the breaking down of fats for energy (Gray et al., 2013). Moreover, other marine compounds beyond the extensively studied ones, such as fucosterol (a sterol), various peptides from fish, and specific metabolites of carotenoids, also present antiobesity properties (Abdul et al., 2016; Mizushige et al., 2017; Bonet et al., 2020). These demonstrate inhibitory effects on adipogenesis or pancreatic lipase activity, though many remain understudied with regard to detailed mechanisms. Table 15.3 summarizes the antiobesity effects of the above-described marine compounds.

TABLE 15.3 Antiobesity effects of commonly used marine bioactive compounds.

Marine compound	Antiobesity effect	References
Alginate	Appetite control; increased satiety: Increases distension and influences the release of gut hormones that regulate appetite	Odunsi et al. (2010)
Brominated lipids (sponges)	Inhibits pancreatic lipase; reduces the amount of dietary fat absorbed, contributing to a reduction in calorie intake and potential weight loss	Liang et al. (2014)
Saponins (sea cucumbers)	Inhibits pancreatic lipase; reduces the amount of dietary fat absorbed, contributing to a reduction in calorie intake and potential weight loss	Guo et al. (2016)
EPA and DHA phospholipids (sea cucumber)	Reduces body fat, lowers serum triacylglycerols, and alleviates hepatic steatosis	Han et al. (2019)
EPA & DHA (Fish oil)	Lowers BMI; reduces body fat; improves lipid metabolism via modulation of lipoprotein synthesis and increased fatty acid oxidation	Lopez-Huertas (2012)
Fucoidan	Suppresses adipogenesis; inhibits the differentiation of preadipocytes by modulating the gene expression of PPARγ and C/EBPα	Kim et al. (2009)
Phlorotannins (Eckol, Dieckol)	Suppresses adipogenesis; inhibits PPARγ and C/EBPα, and the activation of AMPK	Ko et al. (2013)
Fucosylated chondroitin sulphate (FCS)	Suppresses adipogenesis; modulates the Wnt/β-catenin pathway and regulates PPARγ and SREBP1 expression	Xu et al. (2015)
Fucoxanthin	Stimulates thermogenesis; increases energy expenditure: activation of uncoupling protein 1 in brown adipose tissue	Maeda et al. (2005)
Astaxanthin	Reduces body weight; improves lipid profiles; reduces insulin resistance	Bhuvaneswari and Anuradha (2012)

15.5.5 Hypertension

Hypertension, often referred to as high blood pressure, is a significant public health concern affecting millions globally. According to estimates, approximately 1.28 billion people (aged 30–80) worldwide suffer from hypertension, which is a leading risk factor for CVDs, including heart attacks and strokes. The World Health Organization indicates that the prevalence of hypertension is increasing, particularly in low- and middle-income countries, where lifestyle changes and urbanization contribute to rising blood pressure levels (Kario et al., 2024). Hypertension is often termed a "silent killer" as it frequently presents without noticeable symptoms, leading many individuals to remain unaware of their condition (Oparil et al., 2018).

Hypertension is considered a key component of metabolic syndrome and is thought to have a significant association with NAFLD, which is supported by extensive evidence.

For instance, epidemiological studies indicate that the prevalence of NAFLD among patients with hypertension is about 49.5%, a figure that exceeds the prevalence found in the general population (Ma et al., 2021). At the molecular level, the pathophysiological interactions between hypertension and NAFLD involve the activation of proinflammatory pathways and oxidative stress. Increased systemic vascular resistance, often associated with hypertension, leads to endothelial dysfunction, which is characterized by impaired vasodilation and increased production of ROS (Nakagami, 2023). These ROS play a critical role in the progression of NAFLD by promoting hepatocyte injury, inflammation, and subsequent fibrosis. Moreover, the renin–angiotensin–aldosterone system (RAAS) is a key mediator in this association; activation of RAAS not only contributes to hypertension by inducing vascular constriction and sodium retention but also independently promotes hepatic steatosis and inflammation (Yuan et al., 2024).

Clinical studies have demonstrated a clear correlation between elevated blood pressure and the severity of hepatic steatosis, with hypertensive patients showing a higher prevalence of advanced degrees of liver disease, including NASH and fibrosis (Ryoo et al., 2014). This relationship positions hypertension as both a consequence and a perpetrator of liver dysfunction, thereby complicating the clinical management of patients with NAFLD. Furthermore, the coexistence of hypertension and NAFLD increases the risk of CVDs, highlighting the importance of comprehensive monitoring and management strategies (Ng et al., 2022). Addressing hypertension through lifestyle modifications and pharmacotherapy can thus have therapeutic benefits for liver health, ultimately aiming to interrupt the pathological cascade linking these two prevalent conditions.

Hypertension is recognized as a major risk factor for the development and progression of CKD, contributing to renal injury through various interrelated mechanisms. The relationship between hypertension and CKD is bidirectional; while high blood pressure can lead to kidney damage, CKD itself can also contribute to elevated blood pressure through a series of compensatory mechanisms (Hamrahian and Falkner, 2017). Prolonged hypertension causes structural and functional alterations in the kidneys, including nephrosclerosis, glomerulosclerosis, and tubulointerstitial fibrosis, which collectively impair renal function. The glomeruli become less efficient in filtering blood, resulting in proteinuria and a progressive decline in the GFR (Huan et al., 2015). From a pathophysiological perspective, the RAAS plays a pivotal role in mediating the effects of hypertension on kidney health. Hypertension activates the RAAS, leading to increased vascular resistance, sodium retention, and fluid overload. This not only exacerbates blood pressure but also contributes to kidney damage through heightened intraglomerular pressure and subsequent injury to glomerular capillaries. Moreover, hypertension generates oxidative stress and promotes inflammatory pathways that further contribute to the fibrotic processes in the kidneys (Huan et al., 2015).

Epidemiological studies have consistently shown a high prevalence of hypertension among individuals with CKD, with estimates, indicating that nearly 80%–85% of patients with advanced CKD also have elevated blood pressure (Horowitz et al., 2015). In fact, poorly controlled hypertension is associated with a faster decline in renal function and an increased risk of progression to ESRD. Managing blood pressure effectively has been shown to slow the progression of CKD, underscoring the importance of early diagnosis and intervention. Antihypertensive therapies, particularly those targeting the RAAS, such as ACE inhibitors and ARBs, have demonstrated protective effects on renal function and are often recommended as first-line treatments in patients with hypertension and CKD (Townsend and Taler, 2015; Levin et al., 2013). The association between hypertension and CKD is a crucial aspect of renal health, highlighting the need for vigilant monitoring and management of blood pressure in patients at risk. Addressing hypertension not only helps to preserve kidney function but also links to broader cardiovascular health, creating a comprehensive approach to patient care. This multifaceted relationship emphasizes the critical role of hypertension in the continuum of CKD and underscores the necessity for targeted interventions to improve outcomes for affected individuals.

Marine-derived compounds are showing increasing promise as potential antihypertensive agents, offering a diverse range of mechanisms to combat hypertension. Bioactive peptides, often derived from protein hydrolysates of various seaweed species, have shown promising results in multiple studies (Walquist et al., 2024). These peptides demonstrate potent ACE inhibitory activity, directly impacting the RAAS. The RAAS plays a crucial role in blood pressure regulation; by inhibiting ACE, these peptides prevent the conversion of angiotensin I to the potent vasoconstrictor angiotensin II, thus reducing peripheral vascular resistance and lowering blood pressure. Numerous studies have identified specific peptides from seaweed sources such as *Undaria pinnatifida*, *Pyropia yezoensis*, and *Ulva rigida*, exhibiting varying degrees of ACE inhibitory potency. The efficacy of these peptides in vivo is dependent on factors such as their resistance to gastrointestinal proteases and their absorption into the bloodstream (Walquist et al., 2024).

Beyond ACE inhibition, other marine compounds contribute to antihypertensive effects through diverse mechanisms. Phlorotannins, abundant in brown seaweeds, exhibit multiple beneficial properties. These polyphenolic compounds demonstrate ACE inhibitory activity but also contribute to vasodilation by acting as antioxidants and potentially modulating NO production (Brai et al., 2024). NO plays a crucial role in vascular tone and dilation. Studies highlight dieckol and phlorofucofuroeckol A, isolated from *Ecklonia cava* and *E. stolonifera* respectively, showcasing their ACE inhibitory properties and demonstrating the ability of some phlorotannins to promote NO release within endothelial cells (Wijesinghe et al., 2011; Jung et al., 2006). The exact mechanisms underlying their effects are still being elucidated, but these multifaceted actions contribute to their overall antihypertensive potential. Polysaccharides, particularly those from brown seaweeds like *Pelvetia canaliculata*, have also demonstrated antihypertensive effects. For instance, D-polymannuronic sulfate from *Pelvetia canaliculata* has shown acute and prophylactic hypotensive activity in vivo, potentially by increasing NO levels and reducing angiotensin II and endothelin-1 concentrations (Haibo et al., 2001). The precise mechanisms of action for these polysaccharides in blood pressure regulation require further research, but their effects on vascular tone and inflammation appear to play a significant role.

Marine-derived carotenoids, such as fucoxanthin from various brown algae, also exhibit potential benefits. Fucoxanthin demonstrates potent antioxidant properties, which can combat oxidative stress, a major contributor to hypertension-related vascular damage. While direct ACE inhibition has not been explicitly demonstrated for fucoxanthin, its antioxidant properties and influence on pathways regulating inflammation and vascular function may contribute indirectly to its antihypertensive potential (Menichini et al., 2020). The exploration of marine compounds for hypertension treatment is in its early stages, with much research still needed to fully understand the mechanisms of action and clinical efficacy of these diverse

TABLE 15.4 Examples of marine bioactive compounds with antihypertensive effects.

Compound	Source	Antihypertensive Effect	References
Peptides (IPP, VPP, FY, VY, IY)	*Undaria pinnatifida*	Direct ACE inhibition	Sato et al. (2002)
Peptide (AKYSY)	*Pyropia yezoensis*	Direct ACE inhibition	Allur Subramaniyan et al. (2021)
Dieckol	*Ecklonia cava*	Direct ACE inhibition. Antioxidant, potential NO modulation	Wijesinghe et al. (2011)
Phlorofucofuroeckol A	*Ecklonia stolonifera*	Direct ACE inhibition	Jung et al. (2006)
6,6′-Bieckol	*Ecklonia cava*	Direct ACE inhibition. Antioxidant properties, potential NO modulation, and in vivo effects observed	Ko et al. (2017)
D-Polymannuronic sulphate	*Pelvetia canaliculata*	Increases NO, lowers Ang II, and endothelin-1 levels	Haibo et al. (2001)
Peptide (ALLAGDPSVLEDR)	*Bangia fuscopurpurea*	Direct ACE inhibition	Wu et al. (2017)
Peptide (QVEY)	*Gracilariopsis lemaneiformis*	Direct ACE inhibition	Cao et al. (2017)
Peptide (GIIGPSGSP)	*Hippocampus abdominalis*	Direct ACE inhibition	Kim et al. (2019)
Peptide (IPVDM)	*Capros aper*	Direct ACE inhibition	Harnedy-Rothwell et al. (2020)

compounds. Further studies are necessary to establish structure–activity relationships, optimize extraction methods for enhanced bioactivity, and conduct comprehensive in vivo and clinical trials to validate the therapeutic potential of these promising marine-derived antihypertensive agents. Table 15.4 summarizes the marine bioactive compounds with antihypertensive effects.

15.6 Conclusion

Exploring alternative treatment options for metabolic diseases such as NAFLD and CKD is essential due to the limitations of current pharmacological interventions. As dietary interventions rise in significance, they present a promising strategy for managing these conditions. The incorporation of marine bioactive compounds into dietary practices has shown considerable potential in addressing major risk factors associated with NAFLD and CKD. These compounds, particularly omega-3 fatty acids, phenolic compounds, peptides, and polysaccharides, possess notable antioxidant, antiinflammatory, lipid-lowering, and antihypertensive properties that can significantly mitigate the underlying mechanisms driving these diseases.

While this chapter has focused primarily on the health benefits of marine bioactive compounds in the context of NAFLD and CKD, it does not exhaustively cover all relevant studies or potential compounds. Future research should aim to deepen our understanding of the mechanisms by which these compounds exert their beneficial effects, optimize their delivery methods, and validate their long-term safety and efficacy through clinical trials. In addition, addressing gaps in the understanding of the potential interactions between marine bioactives and existing treatments could further enhance therapeutic approaches for individuals afflicted by these growing health concerns. By advancing marine bioactive research, we can potentially transform dietary interventions into effective, integrative strategies for managing metabolic diseases like NAFLD and CKD, ultimately improving patient outcomes and quality of life.

AI disclosure

During the preparation of this work, the author(s) used Open AI to enhance the text. The Open AI platform was used to smoothen the texts and correct grammar mistakes. After using this tool/service, the author(s) reviewed and edited the content as needed and take(s) full responsibility for the content of the publication.

References

Abdul, Q.A., Choi, R.J., Jung, H.A., Choi, J.S., 2016. Health benefit of fucosterol from marine algae: a review. J. Sci. Food Agric. 1097-0010. 96 (6), 1856–1866. Available from: http://onlinelibrary.wiley.com/journal/10.1002/(ISSN)1097-0010. doi:10.1002/jsfa.7489.

Abenavoli, L., Larussa, T., Corea, A., Procopio, A.C., Boccuto, L., Dallio, M., Federico, A., Luzza, F., 2021. Dietary polyphenols and non-alcoholic fatty liver disease. Nutrients. 2072-6643. 13 (2), 494. doi:10.3390/nu13020494.

Ahmad, T.B., Liu, L., Kotiw, M., Benkendorff, K., 2018. Review of anti-inflammatory, immune-modulatory and wound healing properties of molluscs. J. Ethnopharmacol. 1872-7573. 210, 156–178. Available from: www.elsevier.com/locate/jethpharm. doi:10.1016/j.jep.2017.08.008.

Akchurin, O.M., Kaskel, F., 2015. Update on inflammation in chronic kidney disease. Blood Purif. 1421-9735. 39 (1-3), 84–92. www.karger.ch/journals/bpu/bpu_jh.htm. doi:10.1159/000368940.

Ali, I., Manzoor, Z., Koo, J.-E., Kim, J.-E., Byeon, S.-H., Yoo, E.-S., et al., 2017. 3-Hydroxy-4, 7-megastigmadien-9-one, isolated from Ulva pertusa, attenuates TLR9-mediated inflammatory response by down-regulating mitogen-activated protein kinase and NF-κB pathways. Pharm. Biol. 55 (1), 435–440.

Alkhatib, L., Diaz, L.A.V., Varma, S., Chowdhary, A., Bapat, P., Pan, H., Kukreja, G., Palabindela, P., Selvam, S.A., Kalra, K., 2023. Lifestyle modifications and nutritional and therapeutic interventions in delaying the progression of chronic kidney disease: a review. Cureus. 2168-8184. 15 (2). doi:10.7759/cureus.34572.

Allan, R.B.P.V., Murthy, K.S., Dable-Tupas, G., 2023. Bioactive compounds for metabolic diseases. In: Role of Nutrigenomics in Modern-day Healthcare and Drug Discovery. Elsevier BV, pp. 517–546. doi:10.1016/b978-0-12-824412-8.00012-6.

Allur Subramaniyan, S., Begum, N., Kim, S., Choi, Y., Nam, T.J., 2021. Biopeptides of Pyropia yezoensis and their potential health benefits: a review . Wolters Kluwer Medknow Publications. Asian Pac. J. Trop. Biomed. 2221-1691. 11 (9), 375–384. Available from: http://www.apjtb.org/aboutus.asp. doi:10.4103/2221-1691.321127.

Amdur, R.L., Feldman, H.I., Gupta, J., Yang, W., Kanetsky, P., Shlipak, M., Rahman, M., Lash, J.P., Townsend, R.R., Ojo, A., Roy-Chaudhury, A., Go, A.S., Joffe, M., He, J., Balakrishnan, V.S., Kimmel, P.L., Kusek, J.W., Raj, D.S., 2016. Inflammation and progression of CKD: the CRIC study. Clin. J. Am. Soc. Nephrol. 1555-905X. 11 (9), 1546–1556. Available from: http://cjasn.asnjournals.org/content/11/9/1546.full.pdf. doi:10.2215/CJN.13121215.

Apostolova, E., Lukova, P., Baldzhieva, A., Katsarov, P., Nikolova, M., Iliev, I., Peychev, L., Trica, B., Oancea, F., Delattre, C., Kokova, V., 2020. Immunomodulatory and anti-inflammatory effects of fucoidan: a review. Polymers. 2073-4360. 12 (10), 2338. doi:10.3390/polym12102338.

Arab, J.P., Arrese, M., Trauner, M., 2018. Recent insights into the pathogenesis of nonalcoholic fatty liver disease. Annu. Rev. Pathol. 1553-4014. 13, 321–350. Available from: http://www.annualreviews.org/journal/pathmechdis. doi:10.1146/annurev-pathol-020117-043617.

Arabi, T., Shafqat, A., Sabbah, B.N., Fawzy, N.A., Shah, H., Abdulkader, H., Razak, A., Sabbah, A.N., Arabi, Z., 2023. Obesity-related kidney disease: beyond hypertension and insulin-resistance. Front. Endocrinol. 1664-2392. 13. Available from: https://www.frontiersin.org/journals/endocrinology#. doi:10.3389/fendo.2022.1095211.

Banaszak, M., Dobrzyńska, M., Kawka, A., Górna, I., Woźniak, D., Przysławski, J., Drzymała-Czyż, S., 2024. Role of Omega-3 fatty acids eicosapentaenoic (EPA) and docosahexaenoic (DHA) as modulatory and anti-inflammatory agents in noncommunicable diet-related diseases – reports from the last 10 years. Clin. Nutr. ESPEN. 2405-4577. 63, 240–258. doi:10.1016/j.clnesp.2024.06.053.

Banwo, K., Olojede, A.O., Adesulu-Dahunsi, A.T., Verma, D.K., Thakur, M., Tripathy, S., Singh, S., Patel, A.R., Gupta, A.K., Aguilar, C.N., Utama, G.L., 2021. Functional importance of bioactive compounds of foods with potential health benefits: a review on recent trends. Food Biosci. 2212-4306. 43. Available from: http://www.journals.elsevier.com/food-bioscience/. doi:10.1016/j.fbio.2021.101320.

Barreby, E., Chen, P., Aouadi, M., 2022. Macrophage functional diversity in NAFLD—more than inflammation. Nat. Rev. Endocrinol. 1759-5029. 18 (8), 461–472. doi:10.1038/s41574-022-00675-6.

Barroso, L.N., Salarini, J., Leite, N.C., Villela-Nogueira, C.A., Dávalos, A., Carmo, M.D.G.T., Peres, W.A.F., 2023. Effect of fish oil supplementation on the concentration of miRNA-122, FGF-21 and liver fibrosis in patients with NAFLD: study protocol for a randomized, double-blind and placebo-controlled clinical trial. Clin. Nutr. ESPEN. 2405-4577. 57, 117–125. Available from: http://www.journals.elsevier.com/clinical-nutrition-espen. doi:10.1016/j.clnesp.2023.06.027.

Bass, N.M., Merriman, R.B., 2005. Fatty acid metabolism and lipotoxicity in the pathogenesis of NAFLD/NASH. In: Farrell, G.C., George, J., Hall, P., McCullough, A.J. (Eds.), NASH and Related Disorders. Blackwell Publishing, Malden, USA, pp. 109–122.

Bhuvaneswari, S., Anuradha, C.V., 2012. Astaxanthin prevents loss of insulin signaling and improves glucose metabolism in liver of insulin resistant mice. Can. J. Physiol. Pharmacol. 1205-7541. 90 (11), 1544–1552. doi:10.1139/y2012-119.

Bonet, M.L., Ribot, J., Galmés, S., Serra, F., Palou, A., 2020. Carotenoids and carotenoid conversion products in adipose tissue biology and obesity: pre-clinical and human studies. Biochim. Biophys. Acta Mol. Cell Biol. Lipids. 1388-1981. 1865 (11), 158676. doi:10.1016/j.bbalip.2020.158676.

Bonora, E., Targher, G., 2012. Increased risk of cardiovascular disease and chronic kidney disease in NAFLD. Nat. Rev. Gastroenterol. Hepatol. 1759-5045. 9 (7), 372–381. doi:10.1038/nrgastro.2012.79.

Borgonovi, S.M., Iametti, S., Di Nunzio, M., 2023. Docosahexaenoic acid as master regulator of cellular antioxidant defenses: a systematic review. Antioxidants. 2076-3921. 12 (6), 1283. doi:10.3390/antiox12061283.

Brai, A., Hasanaj, A., Vagaggini, C., Poggialini, F., Dreassi, E., 2024. Infesting seaweeds as a novel functional food: analysis of nutrients, antioxidants and ACE inhibitory effects. Int. J. Mol. Sci. 1422-0067. 25 (14), 7588. doi:10.3390/ijms25147588.

Budreviciute, A., Damiati, S., Sabir, D.K., Onder, K., Schuller-Goetzburg, P., Plakys, G., Katileviciute, A., Khoja, S., Kodzius, R., 2020. Management and prevention strategies for non-communicable diseases (NCDs) and their risk factors. Front. Public Health. 2296-2565. 8. Available from: http://journal.frontiersin.org/journal/public-health/section/public-health-education-and-promotion#archive. doi:10.3389/fpubh.2020.574111.

Burt, A.D., Lackner, C., Tiniakos, D.G., 2015. Diagnosis and assessment of NAFLD: definitions and histopathological classification. Semin. Liver Dis. 1098-8971. 35 (3), 207–220. Available from: http://www.thieme.de/sld/index.html. doi:10.1055/s-0035-1562942.

Byrne, C.D., Targher, G., 2015. NAFLD: a multisystem disease. J. Hepatol. 1600-0641. 62 (1), S47. Available from: http://www.sciencedirect.com/science/journal/01688278. doi:10.1016/j.jhep.2014.12.012.

Cao, D., Lv, X., Xu, X., Yu, H., Sun, X., Xu, N., 2017. Purification and identification of a novel ACE inhibitory peptide from marine alga Gracilariopsis lemaneiformis protein hydrolysate. Eur. Food Res. Technol. 1438-2377. 243 (10), 1829–1837. doi:10.1007/s00217-017-2886-2.

Cardoso, A.C., De figueiredo-mendes, C., Villela-Nogueira, C.A., 2021. Current management of NAFLD/NASH. Liver Int. 1478-3223. 41 (S1), 89–94. doi:10.1111/liv.14869.

Cheemerla, S., Balakrishnan, M., 2021. Global epidemiology of chronic liver disease. Clin. Liver Dis. 2046-2484. 17 (5), 365–370. doi:10.1002/cld.1061.

Chen, M.F., Zhang, Y.Y., Di He, M., Li, C.Y., Zhou, C.X., Hong, P.Z., Qian, Z.J., 2019. Antioxidant peptide purified from enzymatic hydrolysates of Isochrysis Zhanjiangensis and its protective effect against ethanol induced oxidative stress of HepG2 cells. Biotechnol. Bioprocess Eng. 1976-3816. 24 (2), 308–317. Available from: http://www.springerlink.com/content/1226-8372. doi:10.1007/s12257-018-0391-5.

Chen, Z., Tian, R., She, Z., Cai, J., Li, H., 2020. Role of oxidative stress in the pathogenesis of nonalcoholic fatty liver disease. Free Radic. Biol. Med. 0891-5849. 152, 116–141. doi:10.1016/j.freeradbiomed.2020.02.025.

Cheung, A.K., Chang, T.I., Cushman, W.C., Furth, S.L., Hou, F.F., Ix, J.H., Knoll, G.A., Muntner, P., Pecoits-Filho, R., Sarnak, M.J., Tobe, S.W., Tomson, C.R.V., Mann, J.F.E., 2021. KDIGO 2021 clinical practice guideline for the management of blood pressure in chronic kidney disease. Kidney Int. 1523-1755. 99 (3), S1. Available from: https://www.journals.elsevier.com/kidney-international. doi:10.1016/j.kint.2020.11.003.

Ciaglia, E., Malfitano, A., Laezza, C., Fontana, A., Nuzzo, G., Cutignano, A., Abate, M., Pelin, M., Sosa, S., Bifulco, M., Gazzerro, P., 2017. Immuno-modulatory and anti-inflammatory effects of dihydrogracilin A, a terpene derived from the marine sponge *Dendrilla* membranosa. Int. J. Mol. Sci. 1422-0067. 18 (8), 1643. doi:10.3390/ijms18081643.

Cian, R.E., Hernández-Chirlaque, C., Gámez-Belmonte, R., Drago, S.R., Sánchez de Medina, F., Martínez-Augustin, O., 2018. Green alga *Ulva* spp. hydrolysates and their peptide fractions regulate cytokine production in splenic macrophages and lymphocytes involving the TLR4-NFκB/MAPK pathways. Mar. Drugs. 1660-3397. 16 (7). Available from: http://www.mdpi.com/1660-3397/16/7/235/pdf. doi:10.3390/md16070235.

D'Orazio, N., Gammone, M.A., Gemello, E., De Girolamo, M., Cusenza, S., Riccioni, G., 2012. Marine bioactives: pharmacological properties and potential applications against inflammatory diseases. Mar. Drugs. 1660-3397. 10 (4), 812–833. Available from: http://www.mdpi.com/1660-3397/10/4/812/pdf. doi:10.3390/md10040812.

Daenen, K., Andries, A., Mekahli, D., Van Schepdael, A., Jouret, F., Bammens, B., 2019. Oxidative stress in chronic kidney disease. Pediatr. Nephrol. 0931-041X. 34 (6), 975–991. doi:10.1007/s00467-018-4005-4.

Davison, S.N., Tupala, B., Wasylynuk, B.A., Siu, V., Sinnarajah, A., Triscott, J., 2019. Recommendations for the care of patients receiving conservative kidney management focus on management of CKD and symptoms. Clin. J. Am. Soc. Nephrol. 1555-905X. 14 (4), 626–634. Available from: https://cjasn.asnjournals.org/content/clinjasn/14/4/626.full.pdf. doi:10.2215/CJN.10510917.

Dietrich, P., Hellerbrand, C., 2014. Non-alcoholic fatty liver disease, obesity and the metabolic syndrome. Best Pract. Res. Clin. Gastroenterol. 1521-6918. 28 (4), 637–653. doi:10.1016/j.bpg.2014.07.008.

Duni, A., Liakopoulos, V., Roumeliotis, S., Peschos, D., Dounousi, E., 2019. Oxidative stress in the pathogenesis and evolution of chronic kidney disease: untangling ariadne's thread. Int. J. Mol. Sci. 1422-0067. 20 (15), 3711. doi:10.3390/ijms20153711.

Ebert, T., Neytchev, O., Witasp, A., Kublickiene, K., Stenvinkel, P., Shiels, P.G., 2021. Inflammation and oxidative stress in chronic kidney disease and dialysis patients. Antioxid. Redox Signal. 1557-7716. 35 (17), 1426–1448. Available from: www.liebertonline.com/ars. doi:10.1089/ars.2020.8184.

Escasany, E., Izquierdo-Lahuerta, A., Medina-Gomez, G., 2019. Underlying Mechanisms of Renal Lipotoxicity in Obesity. Nephron. 1660-8151. 143 (1), 28–32. doi:10.1159/000494694.

Expert, C., Mcfarlane, P., Cherney, D., Gilbert, R.E., Senior, P., 2018. Diabetes Canada Clinical Practice Guidelines. Can. J. Diabetes 42.

Fernando, I.P.S., Jayawardena, T.U., Wu, J., 2023. Marine proteins and peptides: production, biological activities, and potential applications. Food Innov. Adv. 2836-774X. 2 (2), 69–84. Available from: https://www.maxapress.com/article/doi/10.48130/FIA-2023-0009. doi:10.48130/FIA-2023-0009.

Fernando, I.P.S., Kim, M., Son, K.-T., Jeong, Y., Jeon, Y.-J., 2016. Antioxidant activity of marine algal polyphenolic compounds: a mechanistic approach. J. Med. Food. 1096-620X. 19 (7), 615–628. doi:10.1089/jmf.2016.3706.

Foo, S.C., Yusoff, F.M., Ismail, M., Basri, M., Yau, S.K., Khong, N.M.H., Chan, K.W., Ebrahimi, M., 2017. Antioxidant capacities of fucoxanthin-producing algae as influenced by their carotenoid and phenolic contents. J. Biotechnol. 1873-4863. 241, 175–183. Available from: www.elsevier.com/locate/jbiotec. doi:10.1016/j.jbiotec.2016.11.026.

Francis, A., Harhay, M.N., Ong, A., Tummalapalli, S.L., Ortiz, A., Fogo, A.B., et al., 2024. Chronic kidney disease and the global public health agenda: an international consensus. Nat. Rev. Nephrol. 20 (7), 473–485.

Fung, A., Hamid, N., Lu, J., 2013. Fucoxanthin content and antioxidant properties of *Undaria pinnatifida*. Food Chem. 0308-8146. 136 (2), 1055–1062. doi:10.1016/j.foodchem.2012.09.024.

GBD 2015 Obesity Collaborators, 2017. Health effects of overweight and obesity in 195 countries over 25 years. N. Engl. J. Med. 0028-4793. 377 (1), 13–27. doi:10.1056/NEJMoa1614362.

Ghosh, S., Sarkar, T., Pati, S., Kari, Z.A., Edinur, H.A., Chakraborty, R., 2022. Novel bioactive compounds from marine sources as a tool for functional food development. Front. Mar. Sci. 2296-7745. 9. Available from: https://www.frontiersin.org/journals/marine-science#. doi:10.3389/fmars.2022.832957.

Gonzalez, A., Huerta-Salgado, C., Orozco-Aguilar, J., Aguirre, F., Tacchi, F., Simon, F., Cabello-Verrugio, C., Gasparovic, A.C., 2020. Role of oxidative stress in hepatic and extrahepatic dysfunctions during nonalcoholic fatty liver disease (NAFLD). Oxid. Med. Cell. Longev. 1942-0994. 2020, 1–16. doi:10.1155/2020/1617805.

Gray, B., Steyn, F., Davies, P.S.W., Vitetta, L., 2013. Omega-3 fatty acids: a review of the effects on adiponectin and leptin and potential implications for obesity management. Eur. J. Clin. Nutr. 0954-3007. 67 (12), 1234–1242. doi:10.1038/ejcn.2013.197.

Gui, Y.H., Jiao, W.H., Zhou, M., Zhang, Y., Zeng, D.Q., Zhu, H.R., Liu, K.C., Sun, F., Chen, H.F., Lin, H.W., 2019. Septosones A-C, in vivo anti-inflammatory meroterpenoids with rearranged carbon skeletons from the marine sponge *Dysidea* septosa. Org. Lett. 1523-7052. 21 (3), 767–770. Available from: http://pubs.acs.org/journal/orlef7. doi:10.1021/acs.orglett.8b04019.

Guo, L., Gao, Z., Zhang, L., Guo, F., Chen, Y., Li, Y., Huang, C., 2016. Saponin-enriched sea cucumber extracts exhibit an antiobesity effect through inhibition of pancreatic lipase activity and upregulation of LXR-β signaling. Pharm. Biol. 1388-0209. 54 (8), 1312–1325. doi:10.3109/13880209.2015.1075047.

Haibo, Z., Meiyu, G., Huashi, 2001. Effects of D-polymannuronic sulfate on serum nitric oxide levels and plasma endothelin-1 and angiotensin II Contents in renovascular hypertensive rats (series II). J. Ocean Univ. Qingdao 31 (2), 179–184.

Hajduch, E., Lachkar, F., Ferré, P., Foufelle, F., 2021. Roles of ceramides in non-alcoholic fatty liver disease. J. Clin. Med. 2077-0383. 10 (4), 792. doi:10.3390/jcm10040792.

Hamrahian, S.M., Falkner, B., 2017. Hypertension in chronic kidney disease. Advances in Experimental Medicine and Biology. 2214-8019. 956, 307–325. Available from: http://www.springer.com/series/5584. doi:10.1007/5584_2016_84.

Han, X.Q., Zhang, L.Y., Ding, L., Shi, H.H., Xue, C.H., Zhang, T.T., Wang, Y.M., 2019. Synergistic effect of sea cucumber saponins and EPA-enriched phospholipids on insulin resistance in high-fat diet-induced obese mice. Food Funct. 2042-650X. 10 (7), 3955–3964. Available from: http://pubs.rsc.org/en/journals/journal/fo. doi:10.1039/c9fo01147a.

Harlacher, E., Wollenhaupt, J., Baaten, C.C.F.M.J., Noels, H., 2022. Impact of uremic toxins on endothelial dysfunction in chronic kidney disease: a systematic review. Int. J. Mol. Sci. 1422-0067. 23 (1), 531. doi:10.3390/ijms23010531.

Harnedy-Rothwell, P.A., McLaughlin, C.M., O'Keeffe, M.B., Gouic, A.V.L., Allsopp, P.J., McSorley, E.M., Sharkey, S., Whooley, J., McGovern, B., O'Harte, F.P.M., FitzGerald, R.J., 2020. Identification and characterisation of peptides from a boarfish (*Capros aper*) protein hydrolysate displaying in vitro dipeptidyl peptidase-IV (DPP-IV) inhibitory and insulinotropic activity. Food Res. Int. 1873-7145. 131. Available from: www.elsevier.com/inca/publications/store/4/2/2/9/7/0. doi:10.1016/j.foodres.2020.108989.

Harper, C.R., Jacobson, T.A., 2008. Managing dyslipidemia in chronic kidney disease. J. Am. Coll. Cardiol. 0735-1097. 51 (25), 2375–2384. doi:10.1016/j.jacc.2008.03.025.

Hauck, A.K., Bernlohr, D.A., 2016. Oxidative stress and lipotoxicity. J. Lipid Res. 0022-2275. 57 (11), 1976–1986. doi:10.1194/jlr.r066597.

Heck, A.M., Yanovski, J.A., Calis, K.A., 2000. Orlistat, a new lipase inhibitor for the management of obesity. Pharmacotherapy. 0277-0008. 20 (3), 270–279. doi:10.1592/phco.20.4.270.34882.

Hewage, S.M., Prashar, S., Karmin, O., Siow, Y.L., 2021. Lingonberry improves non-alcoholic fatty liver disease by reducing hepatic lipid accumulation, oxidative stress and inflammatory response. Antioxidants. 2076-3921. 10 (4), 565. doi:10.3390/antiox10040565.

Horowitz, B., Miskulin, D., Zager, P., 2015. Epidemiology of hypertension in CKD. Adv. Chronic Kidney Dis. 1548-5595. 22 (2), 88–95. doi:10.1053/j.ackd.2014.09.004.

Hotamisligil, G.S., 2006. Inflammation and metabolic disorders. Nature. 1476-4679. 444 (7121), 860–867. doi:10.1038/nature05485.

Hu, X., Tao, N., Wang, X., Xiao, J., Wang, M., 2016. Marine-derived bioactive compounds with anti-obesity effect: a review. J. Funct. Foods. 1756-4646. 21, 372–387. doi:10.1016/j.jff.2015.12.006.

Huan, Y., Cohen, D.L., Townsend, R.R., 2015. Pathophysiology of hypertension in chronic kidney disease. In: Chronic Renal Disease. Elsevier Inc., United States, pp. 163–169. Available from: http://www.sciencedirect.com/science/book/9780124116023. doi:10.1016/B978-0-12-411602-3.00014-7.

Idota, Y., Kato, T., Shiragami, K., Koike, M., Yokoyama, A., Takahashi, H., Yano, K., Ogihara, T., 2018. Mechanism of suppression of blood glucose level by calcium alginate in rats. Biol. Pharm. Bull. 0918-6158. 41 (9), 1362–1366. doi:10.1248/bpb.b18-00155.

Ingelfinger, J.R., Kalantar-Zadeh, K., Fouque, D., 2017. Nutritional management of chronic kidney disease. N. Engl. J. Med. 0028-4793. 377 (18), 1765–1776. doi:10.1056/nejmra1700312.

Jani, S., Eira, D.D., Stefanovic, M., Ceddia, R.B., 2022. The ketogenic diet prevents steatosis and insulin resistance by reducing lipogenesis, diacylglycerol accumulation and protein kinase C activity in male rat liver. J. Physiol. 0022-3751. 600 (18), 4137–4151. doi:10.1113/jp283552.

Jaworowska, A., Murtaza, A., 2023. Seaweed derived lipids are a potential anti-inflammatory agent: a review. Int. J. Environ. Res. Public Health. 1660-4601. 20 (1), 730. doi:10.3390/ijerph20010730.

Jayawardena, T.U., Sanjeewa, K.K.A., Lee, H.G., Nagahawatta, D.P., Yang, H.W., Kang, M.C., Jeon, Y.J., 2020. Particulate matter-induced inflammation/oxidative stress in macrophages: fucosterol from *Padina boryana* as a potent protector, activated via NF-κB/MAPK pathways and Nrf2/HO-1 involvement. Mar. Drugs. 1660-3397. 18 (12). Available from: https://www.mdpi.com/1660-3397/18/12/628. doi:10.3390/md18120628.

Jiang, Z., Wang, Y., Zhao, X., Cui, H., Han, M., Ren, X., Gang, X., Wang, G., 2023. Obesity and chronic kidney disease. Am. J. Physiol. Endocrinol. Metab. 0193-1849. 324 (1), E24. doi:10.1152/ajpendo.00179.2022.

Joshi, I., Mohideen, H.S., Nazeer, R.A., 2021. A Meretrix meretrix visceral mass derived peptide inhibits lipopolysaccharide-stimulated responses in RAW264.7 cells and adult zebrafish model. Int. Immunopharmacol. 1567-5769. 90, 107140. doi:10.1016/j.intimp.2020.107140.

Joshi, I., Nazeer, R.A., 2020. EGLLGDVF: a novel peptide from green mussel perna viridis foot exerts stability and anti-inflammatory effects on LPS-stimulated RAW264.7 cells. Protein Pept. Lett. 1875-5305. 27 (9), 851–859. Available from: file:///C:/Users/304401/Downloads/002PPL%20(2).pdf. doi:10.2174/0929866527666200224111832.

Jump, D.B., Lytle, K.A., Depner, C.M., Sasmita Tripathy, 2018. Omega-3 polyunsaturated fatty acids as a treatment strategy for nonalcoholic fatty liver disease. Pharmacol. Ther. 0163-7258. 181, 108–125. doi:10.1016/j.pharmthera.2017.07.007.

Jung, H.A., Hyun, S.K., Kim, H.R., Choi, J.S., 2006. Angiotensin-converting enzyme I inhibitory activity of phlorotannins from *Ecklonia stolonifera*. Fish. Sci. 1444-2906. 72 (6), 1292–1299. doi:10.1111/j.1444-2906.2006.01288.x.

Jung, W.K., Heo, S.J., Jeon, Y.J., Lee, C.M., Park, Y.M., Byun, H.G., Choi, Y.H., Park, S.G., Choi, I.L.W., 2009. Inhibitory effects and molecular mechanism of dieckol isolated from marine brown alga on COX-2 and iNOS in microglial cells. J. Agric. Food Chem. 0021-8561. 57(10), 4439–4446. Available from: http://pubs.acs.org/doi/pdfplus/10.1021/jf9003913. doi: 10.1021/jf9003913 . doi:.

Kang, J.X., Weylandt, K.H., 2008. Modulation of inflammatory cytokines by omega-3 fatty acids. Subcell. Biochem. 0306-0225. 49, 133–143. Available from: http://www.springer.com/series/6515. doi:10.1007/978-1-4020-8831-5_5.

Kang, N.J., Han, S.C., Kang, G.J., Koo, D.H., Koh, Y.S., Hyun, J.W., Lee, N.H., Ko, M.H., Kang, H.K., Yoo, E.S., 2015. Diphlorethohydroxycarmalol inhibits interleukin-6 production by regulating NF-κB, STAT5 and SOCS1 in lipopolysaccharide-stimulated RAW264.7 cells. Mar. Drugs. 1660-3397. 13 (4), 2141–2157. Available from: http://www.mdpi.com/1660-3397/13/4/2141/pdf. doi:10.3390/md13042141.

Kario, K., Okura, A., Hoshide, S., Mogi, M., 2024. The WHO global report 2023 on hypertension warning the emerging hypertension burden in globe and its treatment strategy. Hypertens. Res. 0916-9636. 47 (5), 1099–1102. doi:10.1038/s41440-024-01622-w.

Karthikeyan, A., Joseph, A., Nair, B.G., 2022. Promising bioactive compounds from the marine environment and their potential effects on various diseases. J. Genet. Eng. Biotechnol. 1687-157X. 20 (1), 14. doi:10.1186/s43141-021-00290-4.

Kaysen, G.A., 2007. Hyperlipidemia in chronic kidney disease. Int. J. Artif. Organs. 0391-3988. 30 (11), 987–992. doi:10.1177/039139880703001107.

Kemp, D.C., Kwon, J.Y., 2021. Fish and shellfish-derived anti-inflammatory protein products: properties and mechanisms. Molecules. 1420-3049. 26 (11), 3225. doi:10.3390/molecules26113225.

Khursheed, M., Ghelani, H., Jan, R.K., Adrian, T.E., 2023. Anti-inflammatory effects of bioactive compounds from seaweeds, bryozoans, jellyfish, shellfish and peanut worms. Mar. Drugs. 1660-3397. 21 (10). Available from: http://www.mdpi.com/journal/marinedrugs. doi:10.3390/md21100524.

Kim, A.R., Lee, M.S., Choi, J.W., Utsuki, T., Kim, J.I., Jang, B.C., Kim, H.R., 2013. Phlorofucofuroeckol a suppresses expression of inducible nitric oxide synthase, cyclooxygenase-2, and pro-inflammatory cytokines via inhibition of nuclear factor-κB, c-Jun NH2-terminal kinases, and akt in microglial cells. Inflammation. 1573-2576. 36 (2), 259–271. doi:10.1007/s10753-012-9542-6.

Kim, H.S., Je, J.G., Ryu, B., Kang, N., Fernando, I.P.S., Jayawardena, T.U., Sanjeewa, K.K.A., Oh, J.Y., Lee, T.G., Jeon, Y.J., 2019. Antioxidant and angiotensin-I converting enzyme inhibitory peptides from Hippocampus abdominalis. Eur. Food Res. Technol. 1438-2385. 245 (2), 479–487. Available from: http://link.springer.de/link/service/journals/00217/index.htm. doi:10.1007/s00217-018-3179-0.

Kim, M.J., Chang, U.J., Lee, J.S., 2009. Inhibitory effects of fucoidan in 3T3-L1 adipocyte differentiation. Mar. Biotechnol. 1436-2228. 11 (5), 557–562. doi:10.1007/s10126-008-9170-1.

Kita, M., Gise, B., Kawamura, A., Kigoshi, H., 2013. Stylissatin A, a cyclic peptide that inhibits nitric oxide production from the marine sponge *Stylissa massa*. Tetrahedron Lett. 0040-4039. 54 (50), 6826–6828. doi:10.1016/j.tetlet.2013.10.003.

Ko, E., Yoon, E.L., Jun, D.W., 2023. Risk factors in nonalcoholic fatty liver disease. Clin. Mol. Hepatol. 2287-2728. 29 (Suppl), S79. doi:10.3350/cmh.2022.0398.

Ko, S.-C., Lee, M., Lee, J.-H., Lee, S.-H., Lim, Y., Jeon, Y.-J., 2013. Dieckol, Ecklonia cava, inhibits adipogenesis through AMP-activated protein kinase (AMPK) activation in 3T3-L1 preadipocytes. Environ. Toxicol. Pharmacol. 36(3), 1253–1260.

Ko, S.C., Kang, M.C., Kang, N., Kim, H.S., Lee, S.H., Ahn, G., Jung, W.K., Jeon, Y.J., 2017. Effect of angiotensin I-converting enzyme (ACE) inhibition and nitric oxide (NO) production of 6,6′-bieckol, a marine algal polyphenol and its anti-hypertensive effect in spontaneously hypertensive rats. Process Biochem. 1359-5113. 58, 326–332. Available from: www.elsevier.com/inca/publications/store/4/2/2/8/5/7. doi:10.1016/j.procbio.2017.04.014.

Ko, S.C., Kim, D., Jeon, Y.J., 2012. Protective effect of a novel antioxidative peptide purified from a marine Chlorella ellipsoidea protein against free radical-induced oxidative stress. Food Chem. Toxicol. 1873-6351. 50 (7), 2294–2302. doi:10.1016/j.fct.2012.04.022.

Kunnumakkara, A.B., Shabnam, B., Girisa, S., Harsha, C., Banik, K., Devi, T.B., Choudhury, R., Sahu, H., Parama, D., Sailo, B.L., Thakur, K.K., Gupta, S.C., Aggarwal, B.B., 2020. Inflammation, nf-κb, and chronic diseases: how are they linked? Crit. Rev. Immunol. 1040-8401. 40 (1), 1–39. Available from: http://www.dl.begellhouse.com/download/article/779eef5b714f5b81/CRI-33210.pdf. doi:10.1615/CritRevImmunol.2020033210.

Kwon, Y., 2020. Immuno-resolving ability of resolvins, protectins, and maresins derived from omega-3 fatty acids in metabolic syndrome. Mol. Nutr. Food Res. 1613-4133. 64 (4). Available from: http://onlinelibrary.wiley.com/journal/10.1002/(ISSN)1613-4133. doi:10.1002/mnfr.201900824.

Laclair, R., O'Neal, K., Ofner, S., Sosa, M.J., Labarrere, C.A., Moe, S.M., 2008. Precision of biomarkers to define chronic inflammation in CKD. Am. J. Nephrol. 0250-8095. 28 (5), 808–812. doi:10.1159/000135692.

Lee, S.M., Kim, N.H., Lee, S., Kim, Y.N., Heo, J.D., Jeong, E.J., Rho, J.R., 2019. Deacetylphylloketal, a new phylloketal derivative from a marine sponge, genus Phyllospongia, with potent anti-inflammatory activity in in vitro co-culture model of intestine. Mar. Drugs. 1660-3397. 17 (11). Available from: https://www.mdpi.com/1660-3397/17/11/634/pdf. doi:10.3390/md17110634.

Levin, A., Stevens, P.E., Bilous, R.W., Coresh, J., De Francisco, A.L.M., De Jong, P.E., Griffith, K.E., Hemmelgarn, B.R., Iseki, K., Lamb, E.J., Levey, A.S., Riella, M.C., Shlipak, M.G., Wang, H., White, C.T., Winearls, C.G., 2013. Kidney disease: improving global outcomes (KDIGO) CKD work group. KDIGO 2012 clinical practice guideline for the evaluation and management of chronic kidney disease. Kidney Int. Suppl. 2157-1716. 3 (1), 1–150. Available from: https://www.journals.elsevier.com/kidney-international-supplements. doi:10.1038/kisup.2012.73.

Li, L., Liu, D.W., Yan, H.Y., Wang, Z.Y., Zhao, S.H., Wang, B., 2016a. Obesity is an independent risk factor for non-alcoholic fatty liver disease: evidence from a meta-analysis of 21 cohort studies. Obes. Rev. 1467-789X. 17 (6), 510–519. Available from: http://www.wiley.com/bw/journal.asp?ref=1467-7881. doi:10.1111/obr.12407.

Li, S., Hong, M., Tan, H.-Y., Wang, N., Feng, Y., Reyes-Gordillo, K., 2016b. Insights into the role and interdependence of oxidative stress and inflammation in liver diseases. Oxid. Med. Cell. Longev. 1942-0900. 2016 (1). doi:10.1155/2016/4234061.

Liang, L.F., Wang, T., Cai, Y.S., He, W.F., Sun, P., Li, Y.F., Huang, Q., Taglialatela-Scafati, O., Wang, H.Y., Guo, Y.W., 2014. Brominated polyunsaturated lipids from the Chinese sponge *Xestospongia testudinaria* as a new class of pancreatic lipase inhibitors. Eur. J. Med. Chem. 1768-3254. 79, 290–297. Available from: http://www.journals.elsevier.com/european-journal-of-medicinal-chemistry/. doi:10.1016/j.ejmech.2014.04.003.

Liss, K.H.H., Finck, B.N., 2017. PPARs and nonalcoholic fatty liver disease. Biochimie. 6183-1638. 136, 65–74. Available from: www.elsevier.com/locate/biochimie. doi:10.1016/j.biochi.2016.11.009.

Liu, W., Baker, S.S., Baker, R.D., Zhu, L., 2015. Antioxidant mechanisms in nonalcoholic fatty liver disease. Curr. Drug Targets. 1873-5592. 16 (12), 1301–1314. Available from: http://www.benthamdirect.org/pages/all_b_by-publication.php. doi:10.2174/1389450116666150427155342.

Lizardi-Cervera, J., Chavez-Tapia, N.C., Pérez-Bautista, O., Ramos, M.H., Uribe, M., 2007. Association among C-reactive protein, fatty liver disease, and cardiovascular risk. Dig. Dis. Sci. 0163-2116. 52 (9), 2375–2379. doi:10.1007/s10620-006-9262-6.

Lopez-Huertas, E., 2012. The effect of EPA and DHA on metabolic syndrome patients: a systematic review of randomised controlled trials. Br. J. Nutr. 0007-1145. 107 (S2), S185. doi:10.1017/s0007114512001572.

Lorente-Cebrián, S., Costa, A.G.V., Navas-Carretero, S., Zabala, M., Martínez, J.A., Moreno-Aliaga, M.J., 2013. Role of omega-3 fatty acids in obesity, metabolic syndrome, and cardiovascular diseases: a review of the evidence. J. Physiol. Biochem. 1877-8755. 69 (3), 633–651. Available from: http://www.springerlink.com/content/121648/. doi:10.1007/s13105-013-0265-4.

Lu, W., Li, S., Li, J., Wang, J., Zhang, R., Zhou, Y., Yin, Q., Zheng, Y., Wang, F., Xia, Y., Chen, K., Liu, T., Lu, J., Zhou, Y., Guo, C., 2016. Effects of omega-3 fatty acid in nonalcoholic fatty liver disease: a meta-analysis. Gastroenterol. Res. Pract. 1687-6121. 2016, 1–11. doi:10.1155/2016/1459790.

Lv, J.C., Zhang, L.X., 2019. Prevalence and disease burden of chronic kidney disease. In: Advances in Experimental Medicine and Biology, Vol. 1165, Springer New York LLC, China, pp. 3–15. Available from: http://www.springer.com/series/5584. doi:10.1007/978-981-13-8871-2_1.

Ma, C., Yan, K., Wang, Z., Zhang, Q., Gao, L., Xu, T., Sai, J., Cheng, F., Du, Y., 2021. The association between hypertension and nonalcoholic fatty liver disease (NAFLD): literature evidence and systems biology analysis. Bioengineered. 2165-5979. 12 (1), 2187–2202. doi:10.1080/21655979.2021.1933302.

Mac Way, F., Lessard, M., Lafage-Proust, M.H., 2012. Pathophysiology of chronic kidney disease-mineral and bone disorder. Joint Bone Spine. 1778-7254. 79 (6), 544–549. doi:10.1016/j.jbspin.2012.09.014.

Maeda, H., Hosokawa, M., Sashima, T., Funayama, K., Miyashita, K., 2005. Fucoxanthin from edible seaweed, *Undaria pinnatifida*, shows antiobesity effect through UCP1 expression in white adipose tissues. Biochem. Biophys. Res. Commun. 0006-291X. 332 (2), 392–397. doi:10.1016/j.bbrc.2005.05.002.

Magri, A.M.P., Avanzi, I.R., Vila, G.T., Granito, R.N., Estadella, D., Jimenez, P.C., Ribeiro, A.M., Renno, A.C.M., 2023. Anti-inflammatory effects of compounds Extracted from marine sponges: a systematic review. Anti-Inflamm. Anti-Allergy Agents Med. Chem. 1875-614X. 22 (3), 164–197. Available from: https://www.eurekaselect.com/article/136445. doi:10.2174/0118715230272152231106094727.

Makkar, F., Chakraborty, K., 2018. Previously undescribed antioxidative azocinyl morpholinone alkaloid from red seaweed *Gracilaria opuntia* with anti-cyclooxygenase and lipoxygenase properties. Nat. Prod. Res. 1478-6419. 32 (10), 1150–1160. doi:10.1080/14786419.2017.1326041.

Manzoor, Z., Mathema, V.B., Chae, D., Kang, H.K., Yoo, E.S., Jeon, Y.J., Koh, Y.S., 2013. Octaphlorethol a inhibits the CPG-induced inflammatory response by attenuating the mitogen-activated protein kinase and nf-κb pathways. Biosci., Biotechnol. Biochem. 1347-6947. 77 (9), 1970–1972. Available from: https://www.jstage.jst.go.jp/article/bbb/77/9/77_130299/_pdf. doi:10.1271/bbb.130299.

Martins, I., 2015. Unhealthy nutrigenomic diets accelerate NAFLD and adiposity in global communities. J. Mol. Genet. Med. 9 (1), 1–11.

Menichini, D., Alrais, M., Liu, C., Xia, Y., Blackwell, S.C., Facchinetti, F., Sibai, B.M., Longo, M., 2020. Maternal supplementation of inositols, fucoxanthin, and hydroxytyrosol in pregnant murine models of hypertension. Am. J. Hypertens. 1941-7225. 33 (7), 652–659. Available from: https://academic.oup.com/ajh/issue. doi:10.1093/ajh/hpaa041.

Michelotti, G.A., Machado, M.V., Diehl, A.M., 2013. NAFLD, NASH and liver cancer. Nat. Rev. Gastroenterol. Hepatol. 1759-5053. 10 (11), 656–665. doi:10.1038/nrgastro.2013.183.

Miller, E.F., 2020. Nutrition management strategies for nonalcoholic fatty liver disease: treatment and prevention. Clin. Liver Dis. 2046-2484. 15 (4), 144–148. Available from: http://aasldpubs.onlinelibrary.wiley.com/hub/journal/10.1002/(ISSN)2046-2484/. doi:10.1002/cld.918.

Mizushige, T., Komiya, M., Onda, M., Uchida, K., Hayamizu, K., Kabuyama, Y., 2017. Fish protein hydrolysate exhibits anti-obesity activity and reduces hypothalamic neuropeptide Y and agouti-related protein mRNA expressions in rats. Biomed. Res. 1880-313X. 38 (6), 351–357. Available from: https://www.jstage.jst.go.jp/article/biomedres/38/6/38_351/_pdf/-char/en. doi:10.2220/biomedres.38.351.

Münzel, T., Daiber, A., 2018. Environmental stressors and their impact on health and disease with focus on oxidative stress. Antioxid. Redox Signal. 1523-0864. 28 (9), 735–740. doi:10.1089/ars.2017.7488.

Muriel, P., 2009. NF- κ B in liver diseases: a target for drug therapy. J. Appl. Toxicol. 0260-437X. 29 (2), 91–100. doi:10.1002/jat.1393.

Nakagami, H., 2023. Mechanisms underlying the bidirectional association between nonalcoholic fatty liver disease and hypertension. Hypertens. Res. 0916-9636. 46 (2), 539–541. doi:10.1038/s41440-022-01117-6.

Nazeer, R.A., Kumar, N.S.S., Ganesh, R.J., 2012. In vitro and in vivo studies on the antioxidant activity of fish peptide isolated from the croaker (*Otolithes ruber*) muscle protein hydrolysate. Peptides. 0196-9781. 35 (2), 261–268. doi:10.1016/j.peptides.2012.03.028.

Nehus, E., 2018. Obesity and chronic kidney disease. Curr. Opin. Pediatr. 1040-8703. 30 (2), 241–246. doi:10.1097/mop.0000000000000586.

Ng, C.H., Wong, Z.Y., Chew, N.W.S., Chan, K.E., Xiao, J., Sayed, N., Lim, W.H., Tan, D.J.H., Loke, R.W.K., Tay, P.W.L., Yong, J.N., Kong, G., Huang, D.Q., Wang, J.W., Chan, M., Dalakoti, M., Tamaki, N., Noureddin, M., Siddiqui, M.S., Sanyal, A.J., Muthiah, M., 2022. Hypertension is prevalent in non-alcoholic fatty liver disease and increases all-cause and cardiovascular mortality. Front. Cardiovasc. Med. 2297-055X. 9. Available from: www.frontiersin.org/journals/cardiovascular-medicine. doi:10.3389/fcvm.2022.942753.

Ni, L., Wang, L., Fu, X., Duan, D., Jeon, Y.J., Xu, J., Gao, X., 2020. In vitro and in vivo anti-inflammatory activities of a fucose-rich fucoidan isolated from *Saccharina japonica*. Int. J. Biol. Macromol. 1879-0003. 156, 717–729. Available from: www.elsevier.com/locate/ijbiomac. doi:10.1016/j.ijbiomac.2020.04.012.

Nisar, M.F., Li, M., Xu, J., Wan, C., 2024. Anti-diabetic effects of marine natural products through redox modulation via Nrf2/HO-1 cytoprotective pathways. Front. Mar. Sci. 2296-7745. 11. Available from: https://www.frontiersin.org/journals/marine-science#. doi:10.3389/fmars.2024.1438955.

Odunsi, S.T., Vázquez-Roque, M.I., Camilleri, M., Papathanasopoulos, A., Clark, M.M., Wodrich, L., Lempke, M., McKinzie, S., Ryks, M., Burton, D., Zinsmeister, A.R., 2010. Effect of alginate on satiation, appetite, gastric function, and selected gut satiety hormones in overweight and obesity. Obesity. 1930-739X. 18 (8), 1579–1584. doi:10.1038/oby.2009.421.

Okunogbe, A., Nugent, R., Spencer, G., Powis, J., Ralston, J., Wilding, J., 2022. Economic impacts of overweight and obesity: current and future estimates for 161 countries. BMJ Glob. Health. 2059-7908. 7 (9), e009773. doi:10.1136/bmjgh-2022-009773.

Oparil, S., Acelajado, M.C., Bakris, G.L., Berlowitz, D.R., Cífková, R., Dominiczak, A.F., et al., 2018. Hypertension. Nat. Rev. Dis. Primers 4.

Pajvani, U.B., Qiang, L., Kangsamaksin, T., Kitajewski, J., Ginsberg, H.N., Accili, D., 2013. Inhibition of Notch uncouples Akt activation from hepatic lipid accumulation by decreasing mTorc1 stability. Nat. Med. 1546-170X. 19 (8), 1054–1060. doi:10.1038/nm.3259.

Pangestuti, R., Kim, S.-K., 2017. Bioactive peptide of marine origin for the prevention and treatment of non-communicable diseases. Mar. Drugs. 1660-3397. 15 (3), 67. doi:10.3390/md15030067.

Park, S.Y., Kim, Y.S., Ahn, C.B., Je, J.Y., 2016. Partial purification and identification of three antioxidant peptides with hepatoprotective effects from blue mussel (*Mytilus edulis*) hydrolysate by peptic hydrolysis. J. Funct. Foods. 1756-4646. 20, 88–95. Available from: http://www.elsevier.com/wps/find/journaldescription.cws_home/717426/description#description. doi:10.1016/j.jff.2015.10.023.

Pedruzzi, L.M., Stockler-Pinto, M.B., Leite, M., Mafra, D., 2012. Nrf2-keap1 system versus NF-κB: the good and the evil in chronic kidney disease? Biochimie. 6183-1638. 94 (12), 2461–2466. doi:10.1016/j.biochi.2012.07.015.

Pei, K., Gui, T., Kan, D., Feng, H., Jin, Y., Yang, Y., Zhang, Q., Du, Z., Gai, Z., Wu, J., Li, Y., Kharbanda, K., 2020. An overview of lipid metabolism and nonalcoholic fatty liver disease. Biomed. Res. Int. 2314-6133. 2020 (1). doi:10.1155/2020/4020249.

Peiseler, M., Schwabe, R., Hampe, J., Kubes, P., Heikenwälder, M., Tacke, F., 2022. Immune mechanisms linking metabolic injury to inflammation and fibrosis in fatty liver disease—novel insights into cellular communication circuits. J. Hepatol. 0168-8278. 77 (4), 1136–1160. doi:10.1016/j.jhep.2022.06.012.

Pereira-da-Silva, G., Carvalho, F.C., Roque-Barreira, M.C., 2012. Neutrophil activation induced by plant lectins: modulation of inflammatory processes. Inflamm. Allergy Drug Targets. 1871-5281. 11 (6), 433–441. doi:10.2174/187152812803589985.

Phan, J., Ng, V., Sheinbaum, A., French, S., Choi, G., Kabany, M.E., Durazo, F., Saab, S., Tong, M., Busuttil, R., Han, S.-H., 2019. Hyperlipidemia and nonalcoholic Steatohepatitis predispose to hepatocellular carcinoma development without Cirrhosis. J. Clin. Gastroenterol. 0192-0790. 53 (4), 309–313. doi:10.1097/mcg.0000000000001062.

Pouvreau, C., Dayre, A., Butkowski, E.G., De Jong, B., Jelinek, H.F., 2018. Inflammation and oxidative stress markers in diabetes and hypertension. J. Inflamm. Res. 1178-7031. 11, 61–68. Available from: https://www.dovepress.com/getfile.php?fileID=40582. doi:10.2147/JIR.S148911.

Pramanik, S., Singh, A., Abualsoud, B.M., Deepak, A., Nainwal, P., Sargsyan, A.S., Bellucci, S., 2024. From algae to advancements: laminarin in biomedicine. RSC Adv. 2046-2069. 14 (5), 3209–3231. Available from: http://pubs.rsc.org/en/journals/journal/ra. doi:10.1039/d3ra08161c.

Qu, G., Liu, X., Wang, D., Yuan, Y., Han, L., 2014. Isolation and characterization of fucoidans from five brown algae and evaluation of their antioxidant activity. J. Ocean Univ. China. 1672-5182. 13 (5), 851–856. doi:10.1007/s11802-014-2260-y.

Rani, V., Deep, G., Singh, R.K., Palle, K., Yadav, U.C.S., 2016. Oxidative stress and metabolic disorders: pathogenesis and therapeutic strategies. Life Sci. 1879-0631. 148, 183–193. Available from: www.elsevier.com/locate/lifescie. doi:10.1016/j.lfs.2016.02.002.

Ren, L., Cui, H., Wang, Y., Ju, F., Cai, Y., Gang, X., Wang, G., 2023. The role of lipotoxicity in kidney disease: from molecular mechanisms to therapeutic prospects. Biomed. Pharmacother. 0753-3322. 161, 114465. doi:10.1016/j.biopha.2023.114465.

Riazi, K., Azhari, H., Charette, J.H., Underwood, F.E., King, J.A., Afshar, E.E., Swain, M.G., Congly, S.E., Kaplan, G. G., Shaheen, A.-A., 2022. The prevalence and incidence of NAFLD worldwide: a systematic review and meta-analysis. Lancet Gastroenterol. Hepatol. 2468-1253. 7 (9), 851–861. doi:10.1016/s2468-1253(22)00165-0.

Ridker, P.M., Silvertown, J.D., 2008. Inflammation, C-reactive protein, and atherothrombosis. J. Periodontol. 0022-3492. 79 (8), 1544–1551. http://www.joponline.org/doi/pdf/10.1902/jop.2008.080249. doi:10.1902/jop.2008.080249.

Rojano, A., Sena, E., Manzano-Nuñez, R., Pericàs, J.M., Ciudin, A., 2023. NAFLD as the metabolic hallmark of obesity. Intern. Emerg. Med. 1970-9366. 18 (1), 31–41. Available from: https://www.springer.com/journal/11739. doi:10.1007/s11739-022-03139-x.

Romagnani, P., Remuzzi, G., Glassock, R., Levin, A., Jager, K.J., Tonelli, M., Massy, Z., Wanner, C., Anders, H.J., 2017. Chronic kidney disease. Nat. Rev. Dis. Primers. 2056-676X. 3. Available from: http://www.nature.com/nrdp/. doi:10.1038/nrdp.2017.88.

Roumeliotis, S., Mallamaci, F., Zoccali, C., 2020. Endothelial dysfunction in chronic kidney disease, from biology to clinical outcomes: a 2020 update. J. Clin. Med. 2077-0383. 9 (8), 2359. doi:10.3390/jcm9082359.

Ryoo, J., Suh, Y.J., Shin, H.C., Cho, Y.K., Choi, J., Park, S.K., 2014. Clinical association between non-alcoholic fatty liver disease and the development of hypertension. J. Gastroenterol. Hepatol. 0815-9319. 29 (11), 1926–1931. doi:10.1111/jgh.12643.

Safaeinejad, F., Bahrami, S., Redl, H., Niknejad, H., 2018. Inhibition of inflammation, suppression of matrix metalloproteinases, induction of neurogenesis, and antioxidant property make bryostatin-1 a therapeutic choice for multiple sclerosis. Front. Pharmacol. 1663-9812. 9. doi:10.3389/fphar.2018.00625.

Saglimbene, V.M., Wong, G., van Zwieten, A., Palmer, S.C., Ruospo, M., Natale, P., Campbell, K., Teixeira-Pinto, A., Craig, J.C., Strippoli, G.F.M., 2020. Effects of omega-3 polyunsaturated fatty acid intake in patients with chronic kidney disease: systematic review and meta-analysis of randomized controlled trials. Clin. Nutr. 1532-1983. 39 (2), 358–368. Available from: http://www.elsevier-international.com/journals/clnu/. doi:10.1016/j.clnu.2019.02.041.

Salma, W., 2015. Anti-Obesity and Anti-Hypertensive Action of Calanus Oil. The Arctic University of Norway, Norway.

Saraswati,, Giriwono, P.E., Iskandriati, D., Tan, C.P., Andarwulan, N., 2019. Sargassum seaweed as a source of anti-inflammatory substances and the potential insight of the tropical species: a review. Mar. Drugs. 1660-3397. 17 (10). Available from: https://www.mdpi.com/1660-3397/17/10/590/pdf. doi:10.3390/md17100590.

Sarithakumari, C.H., Renju, G.L., Kurup, G.M., 2013. Anti-inflammatory and antioxidant potential of alginic acid isolated from the marine algae, Sargassum wightii on adjuvant-induced arthritic rats. Inflammopharmacology. 1568-5608. 21 (3), 261–268. doi:10.1007/s10787-012-0159-z.

Sato, M., Oba, T., Yamaguchi, T., Nakano, T., Kahara, T., Funayama, K., Kobayashi, A., Nakano, T., 2002. Antihypertensive effects of hydrolysates of wakame *(Undaria pinnatifida)* and their angiotensin-I-converting enzyme inhibitory activity. Ann. Nutr. Metab. 0250-6807. 46 (6), 259–267. doi:10.1159/000066495.

Scorletti, E., Byrne, C.D., 2018. Omega-3 fatty acids and non-alcoholic fatty liver disease: evidence of efficacy and mechanism of action. Mol. Aspects Med. 1872-9452. 64, 135–146. Available from: www.elsevier.com/locate/mam. doi:10.1016/j.mam.2018.03.001.

Semmler, G., Datz, C., Reiberger, T., Trauner, M., 2021. Diet and exercise in NAFLD/NASH: beyond the obvious. Liver Int. 1478-3223. 41 (10), 2249–2268. doi:10.1111/liv.15024.

Shah, S.V., 2004. Oxidants and iron in chronic kidney disease. Kidney Int. 0085-2538. 66 (91), S50. doi:10.1111/j.1523-1755.2004.09108.x.

Sharanagat, V.S., Singla, V., Singh, L., 2019. Bioactive compounds from marine sources. In: Technological Processes for Marine Foods, from Water to Fork. Informa UK Limited, pp. 23–60. doi:10.1201/9780429425271-3.

Sheih, I.C., Fang, T.J., Wu, T.K., Lin, P.H., 2010. Anticancer and antioxidant activities of the peptide fraction from algae protein waste. J. Agric. Food Chem. 0021-8561. 58 (2), 1202–1207. Available from: http://pubs.acs.org/doi/pdfplus/10.1021/jf903089m. doi:10.1021/jf903089mTaiwan.

Sheih, I.C., Wu, T.K., Fang, T.J., 2009. Antioxidant properties of a new antioxidative peptide from algae protein waste hydrolysate in different oxidation systems. Bioresour. Technol. 0960-8524. 100 (13), 3419–3425. doi:10.1016/j.biortech.2009.02.014.

Shin, T., Ahn, M., Hyun, J.W., Kim, S.H., Moon, C., 2014. Antioxidant marine algae phlorotannins and radioprotection: a review of experimental evidence. Acta Histochem. 1618-0372. 116 (5), 669–674. Available from: www.urbanfischer.de/journals/actahist/acthist.htm. doi:10.1016/j.acthis.2014.03.008.

Sies, H., Berndt, C., Jones, D.P., 2017. Oxidative stress. Annu. Rev. Biochem. 86, 715–748.

Singh, A., Kukreti, R., Saso, L., Kukreti, S., 2019. Oxidative stress: a key modulator in neurodegenerative diseases. Molecules. 1420-3049. 24 (8), 1583. doi:10.3390/molecules24081583.

Smith, B.K., Marcinko, K., Desjardins, E.M., Lally, J.S., Ford, R.J., Steinberg, G.R., 2016. Treatment of nonalcoholic fatty liver disease: role of AMPK. Am. J. Physiol. Endocrinol. Metab. 1522-1555. 311 (4), E730. Available from: http://ajpendo.physiology.org/content/311/4/E730full.pdf. doi:10.1152/ajpendo.00225.2016.

Song, Z., 2022. The role of gut–liver axis in gut microbiome dysbiosis associated NAFLD and NAFLD-HCC. Biomedicines 10(3), 524.

Stefan, N., Cusi, K., 2022. A global view of the interplay between non-alcoholic fatty liver disease and diabetes. Lancet Diabetes Endocrinol. 2213-8587. 10 (4), 284–296. doi:10.1016/s2213-8587(22)00003-1.

Stewart, J., McCallin, T., Martinez, J., Chacko, S., Yusuf, S., 2020. Hyperlipidemia. Pediatr. Rev. 0191-9601. 41 (8), 393–402. doi:10.1542/pir.2019-0053.

Sumida, Y., Niki, E., Naito, Y., Yoshikawa, T., 2013. Involvement of free radicals and oxidative stress in NAFLD/NASH. Free Radic. Res. 1071-5762. 47 (11), 869–880. doi:10.3109/10715762.2013.837577.

Taha, A.M., Shaarawy, A.S., Omar, M.M., Abouelmagd, K., Shalma, N.M., Alhashemi, M., Ahmed, H.M., Allam, A. H., Abd-ElGawad, M., 2022. Effect of omega-3 fatty acids supplementation on serum level of C-reactive protein in patients with COVID-19: a systematic review and meta-analysis of randomized controlled trials. J. Transl. Med. 1479-5876. 20 (1). Available from: https://translational-medicine.biomedcentral.com. doi:10.1186/s12967-022-03604-3.

Targher, G., Byrne, C.D., 2017. Non-alcoholic fatty liver disease: an emerging driving force in chronic kidney disease. Nat. Rev. Nephrol. 1759-507X. 13 (5), 297–310. Available from: http://www.nature.com/nrneph/archive/index.html. doi:10.1038/nrneph.2017.16.

Targher, G., Corey, K.E., Byrne, C.D., Roden, M., 2021. The complex link between NAFLD and type 2 diabetes mellitus—mechanisms and treatments. Nat. Rev. Gastroenterol. Hepatol. 1759-5053. 18 (9), 599–612. Available from: http://www.nature.com/nrgastro/index.html. doi:10.1038/s41575-021-00448-y.

Terracciano, S., Aquino, M., Rodriquez, M., Monti, M.C., Casapullo, A., Riccio, R., Gomez-Paloma, L., 2006. Chemistry and biology of anti-inflammatory marine natural products: molecules interfering with cyclooxygenase, NF-kB and other unidentified targets. Curr. Med. Chem. 0929-8673. 13 (16), 1947–1969. doi:10.2174/092986706777585095.

Townsend, R.R., Taler, S.J., 2015. Management of hypertension in chronic kidney disease. Nat. Rev. Nephrol. 1759-507X. 11 (9), 555–563. Available from: http://www.nature.com/nrneph/archive/index.html. doi:10.1038/nrneph.2015.114.

Victor, V., Rocha, M., Sola, E., Banuls, C., Garcia-Malpartida, K., Mijares, A.H., 2009. Oxidative stress, endothelial dysfunction and atherosclerosis. Curr. Pharm. Des. 1381-6128. 15 (26), 2988–3002. doi:10.2174/138161209789058093.

Walquist, M.J., Eilertsen, K.E., Elvevoll, E.O., Jensen, I.J., 2024. Marine-derived peptides with anti-hypertensive properties: prospects for pharmaceuticals, supplements, and functional food. Mar. Drugs. 1660-3397. 22 (4). Available from: http://www.mdpi.com/journal/marinedrugs. doi:10.3390/md22040140.

Wan, X., Xu, C., Yu, C., Li, Y., 2016. Role of NLRP3 inflammasome in the progression of NAFLD to NASH. Can. J. Gastroenterol. Hepatol. 2291-2789. 2016, 1–7. doi:10.1155/2016/6489012.

Wang, J., Hu, S., Jiang, W., Song, W., Cai, L., Wang, J., 2016. Fucoidan from sea cucumber may improve hepatic inflammatory response and insulin resistance in mice. Int. Immunopharmacol. 1567-5769. 31, 15–23. doi:10.1016/j.intimp.2015.12.009.

Wang, L., Kim, H.S., Je, J.-G., Oh, J.Y., Kim, Y.-S., Cha, S.-H., 2020. Protective effect of diphlorethohydroxycarmalol isolated from Ishige okamurae against particulate matter-induced skin damage by regulation of NF-κB, AP-1, and MAPKs signaling pathways in vitro in human dermal fibroblasts. Molecules 25(5), 1055.

Wang, M., Li, L., Xu, Y., Du, J., Ling, C., 2022. Roles of hepatic stellate cells in NAFLD: from the perspective of inflammation and fibrosis. Front. Pharmacol. 1663-9812. 13. doi:10.3389/fphar.2022.958428.

Wang, Y., Xing, M., Cao, Q., Ji, A., Liang, H., Song, S., 2019. Biological activities of fucoidan and the factors mediating its therapeutic effects: a review of recent studies. Mar. Drugs. 1660-3397. 17 (3), 183. doi:10.3390/md17030183.

Ward, P.A., 2010. Acute and Chronic Inflammation. Cambridge University Press (CUP), pp. 1–16. doi:10.1017/cbo9781139195737.002.

Wei, S., Wang, L., Evans, P.C., Xu, S., 2024. NAFLD and NASH: etiology, targets and emerging therapies. Drug Discov. Today. 1359-6446. 29 (3), 103910. doi:10.1016/j.drudis.2024.103910.

Wijesinghe, W.A.J.P., Ko, S.C., Jeon, Y.J., 2011. Effect of phlorotannins isolated from *Ecklonia cava* on angiotensin I-converting enzyme (ACE) inhibitory activity. Nutr. Res. Pract. 2005-6168. 5 (2), 93–100. Available from: http://www.ncbi.nlm.nih.gov/pmc/articles/PMC3085808/pdf/nrp-5-93.pdf. doi:10.4162/nrp.2011.5.2.93.

Wong, V.W.S., Ekstedt, M., Wong, G.L.H., Hagström, H., 2023. Changing epidemiology, global trends and implications for outcomes of NAFLD. J. Hepatol. 1600-0641. 79 (3), 842–852. Available from: http://www.sciencedirect.com/science/journal/01688278. doi:10.1016/j.jhep.2023.04.036.

Wu, Q., Cai, Q.F., Yoshida, A., Sun, L.C., Liu, Y.X., Liu, G.M., Su, W.J., Cao, M.J., 2017. Purification and characterization of two novel angiotensin I-converting enzyme inhibitory peptides derived from R-phycoerythrin of red algae (Bangia fusco-purpurea). Eur. Food Res. Technol. 1438-2385. 243 (5), 779–789. Available from: http://link.springer.de/link/service/journals/00217/index.htm. doi:10.1007/s00217-016-2792-z.

Xiao, B., Li, Y., Lin, Y., Lin, J., Zhang, L., Wu, D., Zeng, J., Li, J., Wen Liu, J., Li, G., 2022. Eicosapentaenoic acid (EPA) exhibits antioxidant activity via mitochondrial modulation. Food Chem. 03088146. 373 (Pt A), 131389. doi:10.1016/j.foodchem.2021.131389.

Xu, H., Wang, J., Zhang, X., Li, Z., Wang, Y., Xue, C., 2015. Inhibitory effect of fucosylated chondroitin sulfate from the sea cucumber Acaudina molpadioides on adipogenesis is dependent on Wnt/β-catenin pathway. J. Biosci. Bioeng. 1389-1723. 119 (1), 85–91. doi:10.1016/j.jbiosc.2014.05.026.

Yan, F.-J., Wang, X., Wang, S.-E., Hong, H.-T., Lu, J., Ye, Q., Zheng, Y.-L., Wang, Y.-J., 2020. C-Jun/C7ORF41/NF-κB axis mediates hepatic inflammation and lipid accumulation in NAFLD. Biochem. J. 0264-6021. 477 (3), 691–708. doi:10.1042/bcj20190799.

Yang, Z., Pan, Y., Chen, J., Zhang, H., Wei, H., Zufang, W., Liu, L., 2019. Anti-inflammatory, anti-oxidative stress effect of *Phascolosoma esculenta* oligosaccharides on *Escherichia coli*-induced sepsis mice. Food Sci. Biotechnol. 1226-7708. 28 (6), 1871–1879. doi:10.1007/s10068-019-00620-w.

Yoon, W.J., Heo, S.J., Han, S.C., Lee, H.J., Kang, G.J., Kang, H.K., Hyun, J.W., Koh, Y.S., Yoo, E.S., 2012. Anti-inflammatory effect of sargachromanol G isolated from Sargassum siliquastrum in RAW 264.7 cells. Arch. Pharm. Res. 1976-3786. 35 (8), 1421–1430. doi:10.1007/s12272-012-0812-5.

Younossi, Z.M., Stepanova, M., Ong, J., Trimble, G., AlQahtani, S., Younossi, I., Ahmed, A., Racila, A., Henry, L., 2021. Nonalcoholic Steatohepatitis is the most rapidly increasing indication for liver transplantation in the United States. Clin. Gastroenterol. Hepatol. 1542-7714. 19 (3), 580. Available from: http://www.elsevier.com/inca/publications/store/6/7/2/7/4/3/index.htt. doi:10.1016/j.cgh.2020.05.064.

Yu, D.K., Lee, B., Kwon, M., Yoon, N., Shin, T., Kim, N.G., Choi, J.S., Kim, H.R., 2015. Phlorofucofuroeckol B suppresses inflammatory responses by down-regulating nuclear factor κB activation via Akt, ERK, and JNK in LPS-stimulated microglial cells. Int. Immunopharmacol. 1878-1705. 28 (2), 1068–1075. Available from: www.elsevier.com/locate/intimp. doi:10.1016/j.intimp.2015.08.028.

Yuan, M., He, J., Hu, X., Yao, L., Chen, P., Wang, Z., Liu, P., Xiong, Z., Jiang, Y., Li, L., 2024. Hypertension and NAFLD risk: insights from the NHANES 2017–2018 and Mendelian randomization analyses. Chin. Med. J. 0366-6999. 137 (4), 457–464. doi:10.1097/cm9.0000000000002753.

Zhang, C.X., Dai, Z.R., Cai, Q.X., 2011. Anti-inflammatory and anti-nociceptive activities of *Sipunculus nudus* L. extract. J. Ethnopharmacol. 1872-7573. 137 (3), 1177–1182. doi:10.1016/j.jep.2011.07.039.

Zhang, X., Lerman, L.O., 2017. The metabolic syndrome and chronic kidney disease. Transl. Res. 1878-1810. 183, 14–25. Available from: http://www.sciencedirect.com/science/journal/19315244. doi:10.1016/j.trsl.2016.12.004.

Zhao, M., Jin, Z., Xia, C., Chen, S., Zeng, L., Qin, S., He, Q., 2023. Inhibition of free heme-catalyzed Fenton-like reaction prevents non-alcoholic fatty liver disease by hepatocyte-targeted hydrogen delivery. Biomaterials. 0142-9612. 301, 122230. doi:10.1016/j.biomaterials.2023.122230.

Zuzda, K., Grycuk, W., Małyszko, J., Małyszko, J., 2022. Kidney and lipids: novel potential therapeutic targets for dyslipidemia in kidney disease? Expert Opin. Ther. Targets. 1472-8222. 26 (11), 995–1009. doi:10.1080/14728222.2022.2161887.

Анемії, К., 2012. KDIGO clinical practice guideline for anemia in chronic kidney disease. Kidney Int. 2, 279.

CHAPTER

16

The impact of marine organisms on infectious diseases

D.M.N.M. Gunasekara[a], *Lei Wang*[b,c], *K.K. Asanka Sanjeewa*[a], *and K.H.I.N.M. Herath*[d]

[a]Department of Biosystems Technology, Faculty of Technology, University of Sri Jayewardenepura, Homagama, Sri Lanka
[b]State Key Laboratory of Marine Food Processing & Safety Control, College of Food Science and Engineering, Ocean University of China, Qingdao, China
[c]Sanya Oceanographic Institution, Ocean University of China, Sanya, China
[d]Department of Biosystems Engineering, Faculty of Agriculture and Plantation Management, Wayamba University of Sri Lanka, Makandura, Sri Lanka

16.1 Introduction

Infectious diseases are illnesses caused by pathogenic organisms such as viruses, bacteria, fungi, parasites, and, rarely, prions (Agrebi and Larbi, 2020). Many of them resolve without complications, but some can cause lasting damage, serious and life-threatening complications, and even lead to pandemic situations. In the 20th century, one of the deadliest pandemics in human history was reported as the Spanish flu, caused by the H1N1 influenza A virus, which infected over a quarter of the world's population (Brüssow, 2022). COVID-19 [severe acute respiratory syndrome coronavirus 2 (SARS-CoV-2)] has recently caused a pandemic that spread to >200 countries, with infections occurring in nearly 6,310,635 individuals within the global population (Tabish, 2020). Infectious diseases can be classified into three types: established, reemerging, and newly emerging (Bhowmik, 2023). Tuberculosis is a prominent bacterial infectious disease and an example of an established infectious disease that has been prevalent for a prolonged period, during which its patterns of incidence, morbidity, and mortality have become well-understood and predictable (Wellington Macêdo Viana and Torres, 2022). Reemerging illnesses are diseases that were formerly controlled but have resurfaced due to factors like drug resistance, environmental changes, weakened public health efforts, or globalization. These diseases can appear in new forms (e.g., methicillin-resistant *Staphylococcus aureus*)

https://doi.org/10.1016/B978-0-443-29190-6.00021-0

or reappear in areas where they were previously eradicated (e.g., polio in Africa), creating significant challenges for public health (Biswas et al., 2023). Furthermore, infections that are known to infect humans for the first time, such as the SARS-CoV virus and Nipah virus, are categorized as newly emerging infectious diseases (Al-Salihi and Khalaf, 2021; Haider et al., 2020).

Infectious diseases can be transmitted between hosts through various mechanisms, primarily direct human-to-human contact, via intermediate animal hosts, and through contaminated food, water, or soil (Prabhu, 2023). These diseases can exhibit clear and noticeable signs and symptoms (symptomatic) or may not show any visible or noticeable symptoms despite the individual being infected (asymptomatic) (Dutta et al., 2024). Symptoms of infectious diseases depend on the type of illness. Viral and bacterial infections commonly present symptoms such as fever, chills, congestion, cough, fatigue, muscle aches, headache, and gastrointestinal issues like diarrhea, nausea, and vomiting (Bhowmik, 2023; Derakhshan-Nezhad, 2025). Numerous types of medicines, such as antibiotics, antivirals, antifungals, and antiparasitic drugs, have existed in human society probably as long as sickness itself to combat infections. While these medications play a critical role in managing infections, their overuse or misuse can result in the development of drug resistance, the onset of adverse side effects, and a reduction in therapeutic effectiveness over time (Muteeb et al., 2023). Notably, multidrug-resistant pathogens (simultaneously resistant to three or more types of drugs used in clinical settings) have become increasingly difficult to control effectively. This has emerged as a significant challenge in clinical treatment. Currently, drug-resistant diseases are responsible for approximately 700,000 deaths globally each year, and this number could rise to 10 million annually by 2050 if action is not taken (Lin et al., 2022). However, with the advent of the modern pharmaceutical industry, biochemical approaches to preventing and treating diseases have acquired a new level of prominence and comprehensive substance, which have the potential to significantly improve patient-related outcomes and enhance public health. Over 50% of the several drugs currently used as therapeutic agents have originated from natural sources, and this percentage is much higher for anticancer and antimicrobial treatment agents (Naeem et al., 2022). Even though drug discovery from global biodiversity is under continuous exploration, the majority of it remains unexplored, which could lead to an important source of novel therapeutics (Chaachouay and Zidane, 2024).

The ocean remains the largest unexplored biological resource, covering approximately 70% of the Earth's surface (Sunagawa et al., 2020). Its broad spectrum of untapped diversity offers sustainable solutions to a number of globally pressing issues. The extreme variations in environmental gradients, including physical, chemical, and hydrological parameters such as temperature, pressure, light intensity, salinity, and nutrients, lead to incredible diversity in chemical space and the biological activities of marine ecosystems (Rotter et al., 2021). Marine bioresources encompass all living organisms in the marine environment, such as marine plants, animals, microorganisms, and compounds derived from these organisms (Collins et al., 2020). These biological organisms and their biosynthesized metabolites with significant biological activities represent valuable resources (Fig. 16.1) (Nguyen et al., 2020; Daniotti and Re, 2021). Furthermore, diverse biological functions are associated with polysaccharides, pigments, polyphenols, polyketides, peptides, proteins, steroids, and essential fatty acids, which are widely studied for their extensive activities that can be applied across a wide range of industries (Ghosh et al., 2022). Oceans provide unlimited biological resources for developing therapeutic drugs, including enzymes, antibiotics, and biopolymer substances derived from various marine sources for

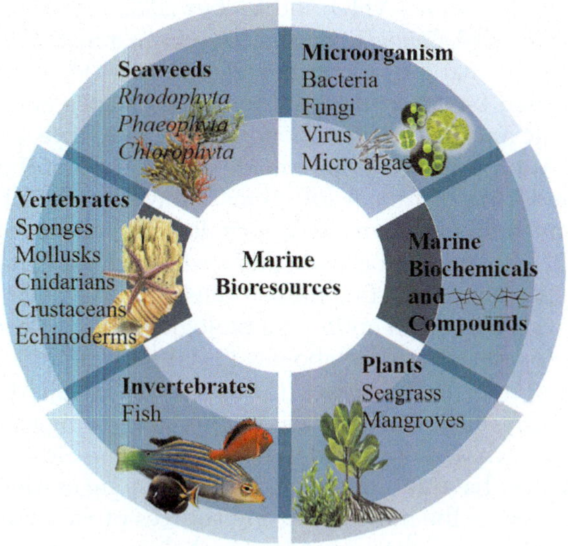

FIG. 16.1 Different types of marine bioresources.

treating numerous human diseases (Lyu et al., 2021). These therapeutics may exhibit a range of beneficial pharmaceutical activities, such as antibacterial, antiviral, neuroprotective, anticancer, and antiinflammatory effects (Karthikeyan et al., 2022).

16.2 Important impact of marine bioresource properties related to infectious diseases

16.2.1 Antifungal activity of marine bioresources for fungal infectious diseases

Fungi have been recognized as a highly diverse eukaryotic group in the world, where they spread across almost all habitats (Deng et al., 2023; Mankar et al., 2022). Being important decomposers, they are essential for many ecosystems in the degradation of organic matter, playing a fundamental role in ecological and biogeochemical processes. They are also highly beneficial to human society, particularly in the medical industry, where they provide essential drugs like antibiotics, immunosuppressants, vitamins, steroids, and alkaloids (Roy et al., 2021). However, various pathogenic fungal infections are common in humans, plants, and animals, primarily due to their adaptability, spread to new regions, vulnerability of populations, immunocompromised individuals, misuse of antifungal drugs, drug-resistant strains, and climate changes (Jafarlou, 2024). Particularly in individuals with weakened immune systems, such as those with cancer, diabetes, and HIV/AIDS, several forms of fungi, including molds, yeasts, and poisonous mushrooms, cause diseases ranging from mild to severe (Firacative, 2020). Some of the important fungal infections in humans, including Candidiasis, Aspergillosis, Cryptococcosis, Histoplasmosis, and Mucormycosis, can cause symptoms ranging from minor to severe. According to the highest-ranked fungal pathogens in the Fungal Priority Pathogens List, released by the World Health Organization (WHO) in October 2022, *Cryptococcus neoformans, Candida auris, Aspergillus fumigatus,* and *Candida albicans* have been ranked as the four main clinical isolates of fungal pathogens that are in critical stages (W.H. Organization, 2022.). They have been primarily documented based on their antifungal resistance and mortality rates, lack of evidence-based diagnostic and treatment options, annual incidence, and associated complications (Zhao et al., 2023). Antifungal agents have made significant contributions to medical care by enabling the control and cure of fungal infections. However, the physiological similarities between fungal cells and human cells make it a challenging endeavor to develop antifungal agents (Fisher et al., 2022). Therefore, it is necessary to discover new fungal-specific targets that can eliminate fungal cells without affecting human cells.

There is currently much interest in antifungal biologically active compounds derived from marine organisms and/or secondary metabolites (SMs) that can efficiently act on molecular targets involved in fungal infectious diseases (Cardoso et al., 2020). Marine bacteria and fungi isolated from various marine bioresources, including sediments, algae, seawater, fish, and most marine vertebrates, have been demonstrated to produce unique SMs with specific and diverse chemical structures that may have antifungal properties. Among marine organisms, vertebrates and sponges are recognized as the most promising sources of SMs with antifungal properties (Cardoso et al., 2020). Furthermore, they provide a rich habitat for endophytic fungi and bacteria that have a great ability to produce unique and diverse bioactive metabolites, which serve as a defense mechanism and possess antifungal properties (Wong Chin et al., 2021). The fungus *Penicillium* cf. *montanense,* which forms a symbiotic relationship with the sponge *Xestospongia exigua,* produces a compound known as xestodecalactone B. The isolated compound xestodecalactone B has shown significant antifungal activity against *C. albicans,* which is the causative agent of different forms of candidiasis, most commonly at concentrations of 20 μM or higher. Another extract obtained from the marine sponge *Hyrtios erectus* exhibits significant antifungal effects against several clinical fungal species, primarily *Aspergillus flavus, C. albicans,* and *Candida tropicalis.* These organisms produce diverse compounds, such as steroids, flavonoids, terpenoids, tannins, and coumarins, which have potential as lead pharmaceutical agents for developing new treatments (El-Bondkly et al., 2021).

Marine bacteria are highly effective microorganisms with antifungal activity. Ieodoglycolipid and Ieodoglucomide C are glycolipids that have both been isolated from the aquatic bacterium *Bacillus licheniformis.* It was found that they exhibit potent antifungal activity [minimum inhibitory concentration (MIC) values of 0.02–0.03 μM] against human fungal pathogens, including *C. albicans, Aspergillus niger, Colletotrichum acutatum, Rhizoctonia solani,* and *Botrytis cinerea* (Thawabteh et al., 2023). Meanwhile, Pedein A, isolated from the myxobacterium *Chondromyces pediculatus,* has demonstrated significant inhibitory effects on filamentous fungi such as *Trichoderma koningii, B. cinerea, Rhizopus arrhizus,* and *Pythium debaryanum.* Moreover, marine fungi have been identified as a greater antifungal resource compared to terrestrial fungi (Jakubczyk and Dussart, 2020). Didymellamide A, derived from the marine fungus *Stagonosporopsis cucurbitacearum,* and sclerotide B, obtained from *Aspergillus sclerotiorum* PT06-1, have shown growth inhibitory activity against *C. albicans* with MICs of 1.6–3.1 μg/mL and 3.5 μM, respectively. Seaweeds are another abundant marine bioresource that represents antimicrobial properties for the development of novel pharmaceutical products. *Ecklonia cava, Pelvetia siliquosa, Eisenia arborea, Eisenia bicyclis, Ishige okamurae, Ecklonia kurome,* and *Ecklonia stolonifera* have been reported as brown seaweeds containing high levels of phlorotannins (phenolic compounds) (Negara et al., 2021). The antifungal activity of phlorotannins has been summarized by (Negara et al., 2021). According to the studies, fucofuroeckol-A and dieckol, isolated from *E. bicyclis* and *E. cava*, respectively, exhibit antifungal activity against dermal fungal pathogens, namely, *C. albicans*, *Epidermophyton floccosum*, and *Trichophyton mentagrophytes*. These phlorotannin compounds have the potential to inhibit dimorphic complexes, reduce virulence, and decrease the capacity of fungal cells to invade host cells (Negara et al., 2021).

16.2.2 Antibacterial activity of marine bioresources for bacterial infectious diseases

Bacteria are widespread microorganisms that can colonize even in hot springs and glaciers (Das et al., 2021). Generally, they exist in millions inside and outside the human body, with a huge diversity of different species. They contribute to various aspects of physiological functions, including digestion, immune system support, maintaining gut health, and significantly to the overall health of the body (Dekaboruah et al., 2020). There are a few (<1% of all bacteria types) that cause negative impacts on human health as a result of their reproduction and survival processes. Antibiotics are drugs designed to kill pathogenic bacteria or inhibit their growth without harming the host. They were one of the greatest medical advances of the 20th century, revolutionizing the treatment of bacterial infections (Wang et al., 2020). However, despite their initial success, antibiotics are rapidly becoming less effective due to antibiotic resistance, with even common infections becoming untreatable and once again deadly (Wang et al., 2020). This problem is driven by the misuse and overuse of antibiotics. A growing number of antibiotic-resistant bacteria (ARB) leads to a huge economic burden for almost all countries in the modern world. This is because numerous bacterial infections are becoming harder to treat due to their increased resistance to conventional antibiotics (Li et al., 2023). To address this threat to human health, the WHO compiled a list of priority pathogens, including *Enterococcus faecium, S. aureus, Klebsiella pneumoniae, Acinetobacter baumannii, Pseudomonas aeruginosa*, and *Enterobacter* spp., collectively known as ESKAPE. These pathogens alone are responsible for causing > 50% infection-related bacterial deaths globally (Table 16.1) (Soni et al., 2024).

ARB can survive even when exposed to one or more antibacterial agents. Multidrug resistance has been regarded as the acquired resistance to at least one drug from three or more different classes of antibiotics. This phenomenon poses a serious threat to public health and requires immediate action, including the development of novel antibiotics and better stewardship of existing ones. The molecular mechanisms that underlie antimicrobial resistance are prerequisites. Several major mechanisms can be identified in Fig. 16.2, including bacterial efflux systems, reduced cell membrane permeability, antibiotic inactivation by enzymes, target modification, and target protection (Sun and Chen, 2024; Bharathi and Lee, 2024).

The development of new antibiotics is essential to preserving the effectiveness of therapies for common and life-threatening bacterial infections. Current efforts have increased in the research and development of novel antibiotics, along with alternative therapies or resistance inhibitors, by targeting drug resistance mechanisms. The marine environment provides valuable, diverse natural compounds such as alkaloids, sterols, lactones, peptides, polyketides, terpenoids, fatty acids, phenols, halogenated compounds, and nucleosides, which are derived from various bioresources (Patnaik and Das, 2024). These compounds have the

TABLE 16.1 The updated bacterial priority pathogens list (BPPL) released by the World Health Organization (WHO) on May 17, 2024.

Priority level	Bacteria	Resistance
Critical priority	*Acinetobacter baumannii*	Carbapenem-resistant
	Enterobacterales	Third-generation cephalosporin-resistant
	Enterobacterales	Carbapenem-resistant
	Mycobacterium tuberculosis	Rifampicin-resistant
High priority	*Salmonella Typhi*	Fluoroquinolone-resistant
	Shigella spp.	Fluoroquinolone-resistant
	Enterococcus faecium	Vancomycin-resistant
	Pseudomonas aeruginosa	Carbapenem-resistant
	Nontyphoidal Salmonella	Fluoroquinolone-resistant
	Neisseria gonorrhoeae	Third-generation cephalosporin- and/or fluoroquinolone-resistant
	Staphylococcus aureus	Methicillin-resistant
Medium priority	Group A streptococci	Macrolide-resistant
	Streptococcus pneumoniae	Macrolide-resistant
	Haemophilus influenzae	Ampicillin-resistant
	Group B streptococci	Penicillin-resistant

Source: https://www.who.int/publications/i/item/9789240093461.

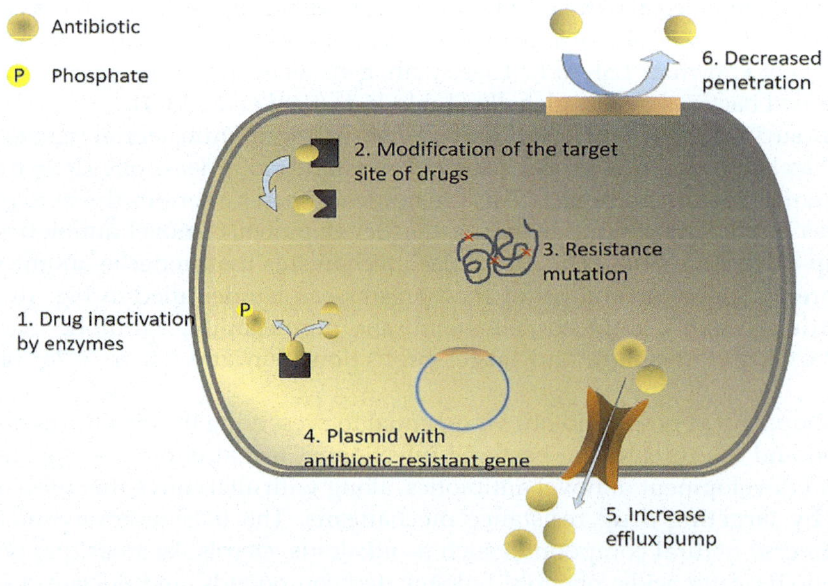

FIG. 16.2 Mechanism of antimicrobial drug resistance: analysis on six potential occasions.

ability to combat numerous drug-resistant bacteria, making them a very promising resource in the search for novel antimicrobial agents, especially against multidrug-resistant pathogenic microorganisms (Barbosa et al., 2020; Nweze et al., 2020).

Marine seaweeds (macroalgae) are valuable bioresources rich in bioactive compounds such as carrageenans, alginates, fucoidans, and phenols, which are widely used in the pharmaceutical industry (Salleh and Rashid, 2024; Gunasekara et al., 2025). One group of phenolic compounds, known as phlorotannins, is particularly abundant in brown seaweeds and has gained significant attention due to its strong antibiotic and cytotoxic properties (Lemesheva et al., 2023; Herath et al., 2025). Phlorotannin extracts from various brown seaweed species, including *Desmarestia aculeata*, *Fucus vesiculosus*, and *Ectocarpus siliculosus*, exhibited the lowest MICs against a range of pathogenic bacteria, with values ranging from 4 to 25 μg/mL. Notably, > 98% of *Escherichia coli* cells died after only a few hours (~5 h) of exposure to phlorotannin preparations in a 0.9% NaCl solution (Lemesheva et al., 2023). Another study performed by Mofeed et al. (2022) investigated the antibacterial activity of three different seaweeds: *Amphiroa anceps* (red seaweed), *Corallina officinalis* (red seaweed), and *Sargassum filipendula* (brown seaweed) against three pathogenic bacteria (*Salmonella typhimurium, S. aureus,* and *Escherichia coli*). Organic extracts of *A. anceps* (tetradecanoic acid, phytone, pentadecanoic acid), *C. officinalis* (thiosemicarbazide, tetradecanoic acid, phytol, pentadecanoic acid, *hexa*-decane), and *S. filipendula* showed bacterial inhibition activity at the highest percentages. Surprisingly, the *C. officinalis* extract (100 μg/mL) inhibited 100% of *S. aureus* cells (Mofeed et al., 2022). The review by Liu et al. (2024) summarized 141 different terpenoid compounds, studied recently (between 2019 and 2024), that are primarily derived from sponges, red algae, fungi, bacteria, soft corals, and marine sediments. These include 20 triterpenes, 34 meroterpenoids, 39 diterpenes, and 48 sesquiterpenes, all of which exhibit antibacterial and/or antifungal activities. According to a comparison of each compound with different pathogenic bacteria (test strains: *B. cereus, S. aureus, E. coli, Pseudomonas* sp.), the activities are relatively evenly distributed and do not show a particular preference for any specific skeletal type (Liu et al., 2024). Marine peptides derived from different bioresources, including Porifera, Cnidaria, Chordata, and Echinodermata, as well as algae, fungi, and bacteria, serve as effective pharmacophores because of their selectivity and high binding affinity to carry out signaling processes in cells. Callyaerins A–F, derived from the marine sponge *Callyspongia aerizusa,* and Salinamide A and B—produced by *Streptomyces* species—a marine bacterium isolated from the surface of the jellyfish *Cassiopeia xamachana*—have exhibited antibacterial activity against Gram-positive and Gram-negative bacteria (Sable et al., 2017).

16.2.3 Antiprotozoal activity of marine bioresources for protozoal infectious diseases

Protozoal diseases are a type of parasitic infection in humans caused by single-celled organisms called protozoa (Vaisusuk and Saijuntha, 2021). They are most common in tropical and subtropical countries, owing to serious infectious diseases, and are responsible for significant mortality and morbidity risks in cases of infection. There are various protozoal pathogens that lead to fatal or life-threatening infections, capable of affecting humans and a wide range of animals, namely, Plasmodium, Trypanosoma, Cryptosporidium, Toxoplasma, and Giardia (Hu et al., 2022). Malaria is the primary infectious disease caused by the apicomplexan parasite *Plasmodium falciparum*, which is transmitted via the bite of infected female Anopheles mosquitoes that carry Plasmodium parasites (Cite). According to the annual report of the WHO on global trends in malaria control and elimination, published on December 11, 2024, global incidences of malaria increased by 11 million in 2023 compared to 2022, resulting in the incidence rate rising from 58.6 to 60.4 cases per 1000 people at risk (Venkatesan, 2025). Furthermore, leishmaniasis and trypanosomiasis are other protozoal infections caused by >20 leishmania species and *Toxoplasma gondii*. They are leading causes of morbidity and mortality, affecting a significant number of people and costing emerging economies billions of dollars per year. By 2023, around 25 commonly used drugs for treating protozoal infections, most of which were discovered between the 1930s and 1970s (with the exception of artemisinin and its derivatives), were available. A few of these drugs can effectively treat these infections under specific conditions (Sykes and Papich, 2022). Nevertheless, in the modern era, these drugs exhibit significant limitations, including poor absorption, toxicity, low therapeutic index, extensive resistance, and long treatment durations. Furthermore, there are no commercial vaccines for protozoans yet, except for the RTS,S/AS01 (RTS,S) vaccine, which was recently approved by WHO for the prevention of *P. falciparum* malaria in children at the highest risk of severe disease (Mumtaz et al., 2023). These limitations underscore the urgent need for the development of novel compounds that are more effective and reduce both the cost and time required during the early stages of drug development.

Marine organisms are a rich source of antiparasitic compounds, such as peptides, polyketides, alkaloids, terpenoids, terpenes, steroids, coumarins, fatty acid derivatives, and lactones (Nweze et al., 2021). This study screened 36 studies on the derivatives of invertebrates and microorganisms. Alkaloids, terpenoids, and sterols are the most common types extracted from sponges that belong to the Xestospongia, Pseudoceratina, Dysidea, Hyrtios, and Monanchora genera. These compounds have shown strong antiparasite activities against *P. falciparum*, *Trypanosoma*, and *Leishmania* species. Notably, pentacyclic and bromotyrosine alkaloids have shown strong antimalarial properties, and some of these compounds show synergy with existing drugs. Sesquiterpenoids with strong antimalarial effects include avarol and avarone. Furthermore, compounds like saringosterol and kaimanol that are derived from marine sponges exhibit a moderate level of antimalarial action (Zhang et al., 2023). An additional study was conducted by Varijakzhan et al. (2021) on the antimalarial activity of the New Caledonian deep-sea sponge. The ethanol extract, namely, alisiaquinone A, alisiaquinone B, alisiaquinone C, meroterpenes, and alisiaquinol, has been shown to exhibit antimalarial activity against protein farnesyltransferase and the enzyme plasmodium kinase Pfnek-1, which can be used as inhibitors to treat malaria (Varijakzhan et al., 2021).

When discussing marine microorganisms (fungi and bacteria) with antiparasitic properties, bacterial compounds such as staurosporine and marinopyrrole A exhibit activity against parasites like *Toxoplasma gondii* and *Acanthamoeba* (Zhang et al., 2023). In addition, hydroxyquinoline molecules with antimalarial and anti-*Trypanosoma cruzi* activities are produced by *P. aeruginosa*. Compounds like harzialactone A, which is effective against Leishmania and other parasites, are produced by marine fungi such as *Paecilomyces* sp. A substance with potent antimalarial properties, astechrome, is also produced by *Aspergillus terreus* (Zhang et al., 2023). Marine cyanobacteria (blue–green algae) produce different antimicrobial peptides, which are linear, depsi, and cyclic in their structures. Ikoamide, a kind of linear peptide derived from *Okeania,* has demonstrated strong activity against the *P. falciparum* 3D7 strain (Iwasaki et al., 2020). Similarly, Iheyamides A exhibit moderate activity against *Trypanosoma brucei rhodesiense* (IC_{50} = 1.5 μM) and *T. b. brucei* (Kurisawa et al., 2020). The growth of *T. b. rhodesiense* was also inhibited by a cyclic decapeptide isolated from *Leptolyngbya* sp. A polyhydroxy macrolide compound, palstimolide A, isolated from cyanobacteria, has also shown very high antiplasmodial activity against the Dd2 strain of *P. falciparum* (Keller et al., 2020).

16.2.4 Antiviral activity of marine bioresources for virus infections

Over the centuries, people have suffered from viral infectious diseases such as Human Immunodeficiency Virus (HIV), Herpes simplex virus (HSV), respiratory syncytial virus, influenza, dengue virus, and Ebola virus, which have killed millions of people worldwide, and most recently, coronavirus disease 2019 (COVID-19) (Dharsan et al., 2020). Although many preventive strategies have been used up to now, the availability of efficient therapies for viral infectious diseases is still limited (Vazquez-Munoz, 2023). Most commonly, anti-HIV, antiflu, and anti-Ebola drugs are facing resistance from these viruses, stimulating the search for new efficient therapeutic molecules (Ma et al., 2021). The new infectious diseases caused by viruses, such as COVID-19, remain a challenge (Wang and Alexander, 2021). Furthermore, drug failure in human trials is a general process that requires work and attention. Promising results are expected through the emergence of many new technologies. Growing knowledge about viruses and rapidly developing techniques and tools provides greater help in the development of new drugs with antiviral activities (Kausar et al., 2021). More than 1200 new natural products are reported annually from a range of marine sources, including microbes, algae, corals, and sponges. Some bioactive SMs derived from these sources are currently being intensively used for their potent, strong antiviral properties. Over the last few decades, the demand for novel antivirals has increased as different forms of infectious diseases caused by emerging or reemerging viruses continue to pose a significant threat to humanity.

Natural resources, particularly from fungi, bacteria, marine algae, and mangroves, have shown promising antiviral potential against influenza virus infections. Fungal species like *Cladosporium sphaerospermum* produce the compound Cladosin C, which targets Influenza A (H1N1), while *Penicillium chrysogenum* PJX-17, sourced from mangrove sediment, yields Sorbicatechols A and B, inhibiting H1N1 in Madin-Darby canine kidney cells (Eichberg et al., 2022). In addition, *Aspergillus sydowii* ZSDS1-F6 produces Cordyol C and Diorcinol, which target Influenza A (H3N2) (Riccio et al., 2020). Marine bacterial species such as *Leptolyngbya,* which produces chlorinated compounds, exhibit antiviral activity against Influenza A and B by inhibiting neuraminidase activity and viral replication. Specifically, an ethyl acetate extract rich in sulfated polysaccharides has been shown to act against two influenza strains, A (H1N1)pdm09-WT and A(H3N2)-WT. *Gyrodinium impudicum,* producing sulfated polysaccharide p-KG03, targets Influenza A (H1N1 and H3N2) by preventing viral attachment to cell receptors and internalization through virus–cell fusion. Further, marine algae like *Ulva clathrata* and *E. cava* produce Phlorofucofuroeckol, which inhibits neuraminidase activity across various Influenza A strains (H1N1, H3N2, H9N2). In addition, mangrove plants such as *Xylocarpus moluccensis* (producing Khayanolides) and *Sonneratia paracaseolaris* (producing Triterpenoids) inhibit the cytopathic effects of Influenza A (H1N1).

Marine bioresources-based antiviral peptides (AVPs) are short chains of amino acids derived from marine microbes, vertebrates, and invertebrates that possess antiviral properties. AVPs are primarily dedicated to treating enveloped viruses at their fundamental stages of the life cycle, such as entry, synthesis, or assembly, against human viruses, including influenza viruses, HIV, hepatitis C virus (HCV), HSVs, and SARS-CoV-2 (Sukmarini, 2022). The mechanisms of antiviral activity of AVPs include direct binding, where AVPs like papuamide A, isolated from the marine sponge *Cacospongia mycofijiensis*, exhibit an immediate virucidal effect by inhibiting HIV-1 through a viral-membrane-targeting mechanism. Similarly, Stellettapeptines A and B from *Stelletta* sp. target HIV-1 viral entry, ultimately leading to viral inactivation. In addition, myticin C from *Mytilus galloprovincialis* affects both HSV-1 and HSV-2 by targeting the intracellular phase of viral replication. For HCV, Cyclo(l-Tyr-l-Pro) diketopiperazine from *Aspergillus versicolor* inhibits the NS3-4A protease, which plays a critical role in the replication of HCV. According to a study conducted by Teng et al. (2020), antiviral compounds derived from marine microorganisms have been classified into eight biosynthetic categories, interpreting their contribution to antiviral activities as follows: alkaloids (11%), quinones (7%), peptides (35%), polyketones (18%), pyrones (10%), sterols (3%), terpenoids (6%), and others (10%). Among these, peptides are the most prominent group of SMs with potentially useful activities, especially against H1N1 (Teng et al., 2020).

Macroalgae, or seaweed, are valuable marine bioresources that have demonstrated antiviral activity in many studies. Alginate compounds derived from brown algae, most commonly in *Callophyllis variegata, Cryptonemia crenulata, Schizymenia binderi,* and *Agardhiella tenera,* have exhibited activity against HIV, HBV, and IAV by targeting reverse transcriptase during viral replication and transcription. In addition, 8,8′-bieckol (a phloroglucinol derivative) is a phlorotannin compound extracted from the brown seaweed *E. cava.* This compound has been shown to block viral penetration and inhibit the reverse transcriptase activity of HIV-1. Red seaweed species such as *A. tenera, C. variegata, C. crenulata,* and *S. binderi* have demonstrated antiviral effects against HIV-1, HIV-2, HAV-1, HSV-2, and DENV. Another red seaweed extract, Galactan, can prevent not only virus entry, but also inhibit their replication and transcription (Sansone et al., 2020).

COVID-19 is the recently reported respiratory disease caused by the strain of coronavirus, SARS-CoV-2 (Geahchan et al., 2021). The spread of COVID-19 was rapid across the globe, causing serious concerns and showing a 0.5%–1% mortality rate (Muralidar et al., 2020). This single-stranded ribonucleic acid virus infects the respiratory system by attaching to angiotensin-converting enzyme 2 protein receptors on the surface of host cells (Wan et al., 2020). This virus has four structural proteins: spike (S), nucleocapsid (N), membrane (M), and envelope (E), which can function as antigens. These antigens may produce neutralizing antibodies in the human body and boost CD4+/8+ T-cell responses. The therapy regimens for this condition are based on these mechanisms of action (Ratha et al., 2021). Though a number of researchers have attempted to produce effective therapeutics, drugs, and vaccines, no preventative vaccinations or viable therapeutics against SARS-CoV-2 infection have been developed to date (Rahman et al., 2022). The world is still uncertain about COVID-19's spread and protection against it. As a result, additional research is needed to find a long-term solution to this devastating condition. Marine compounds have served as milestones in the search for new or repurposed lead drugs that are effective against viruses. Compounds produced from marine organisms, such as phenols, alkaloids, lignans, flavonoids, and peptides, have been shown to inhibit SARS-CoV-2 in various ways (Fig. 16.3). In silico, in vitro, and clinical analytical

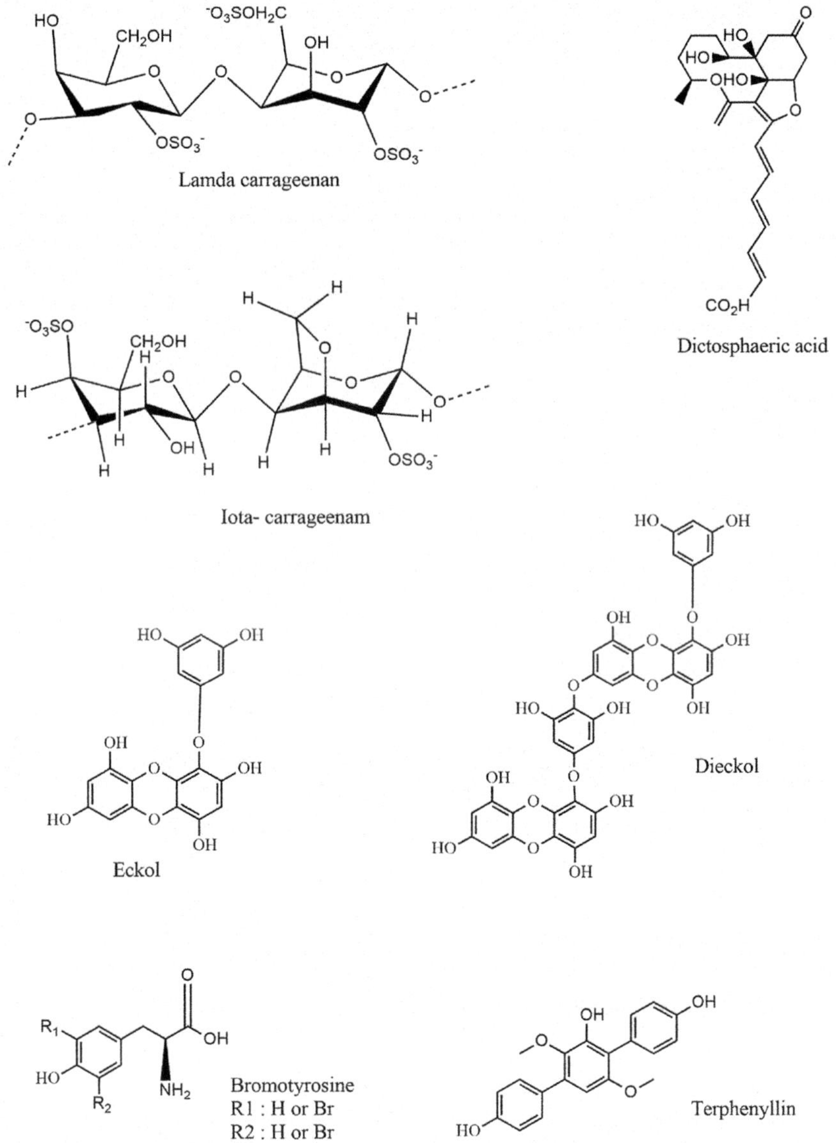

FIG. 16.3 Chemical structure of some selected compounds from marine bioresources with SARS-CoV-2 inhibitory properties.

approaches have been investigated to study effects on virus pathogenesis stages, mainly viral entry and penetration, replication of the viral nucleic acid, inflammatory and immunostimulatory effects, and both structural and nonstructural proteins (Table 16.2) (Morokutti-Kurz et al., 2021).

Marine sponges have demonstrated highly specialized defensive mechanisms against SARS-CoV-2 viral particles and are capable of rapidly filtering out these virus particles. Bromotyrosine compounds derived from demosponges have been investigated as potential

TABLE 16.2 Anti-SARS-COV-2 marine compounds.

Source	Marine compound	Target	References
Marine red algae	Lambda Carrageenan, Iota-carrageenan	S protein	(Rahman et al., 2022; Morokutti-Kurz et al., 2021; Cai et al., 2020; Jang et al., 2021; Bansal et al., 2021)
Brown algae *Sargassum spinuligerum*	Phlorotannins	M^{pro}	(Gurung et al., 2020)
Cyanobacteria, brown algae (*Saccharina japonica*)	Sulfated polysaccharides	S protein	(Nagle et al., 2020)
Marine sponges	Bromotyrosines	S protein	(Binnewerg et al., 2020; Drechsel et al., 2020; Binnewerg et al., 2020)
Scleractinia-associated organisms	Terphenyllin, Tirandamycin A	M^{pro}	(Zahran et al., 2020)
Marine Sponge *Desmapsamma anchorata*	Chimyl alcohol	M^{pro}	(Hamoda et al., 2021) (Khalili et al., 2020)
Green algae *Dictyosphaeria versluyii*	Dictyosphaeric acid A	TMPRSS2-S PPIs, hACE2-S	(Senapati et al., 2021)
Brown algae *Ecklonia cava*	Dieckol and eckol	$3CL^{pro}$	(Arunkumar et al., 2022) (Alqathama et al., 2022)
Marine sponge *Axinella* cf. *corrugat*	Esculetin-4-carboxylicacid methyl ester	M^{pro}	(Coelho et al., 2020) (Hamoda et al., 2021)
Soft coral *Formosan gorgonian Briareum*	Excavatolide M	TMPRSS2-S PPIs, hACE2-S	(Rahman et al., 2020)
Fish and other seafood, algae	Omega-3 fatty acids	IL-1β, IL-TNF-α, COX, LOX pathway	(Morokutti-Kurz et al., 2021; Hathaway et al., 2020)

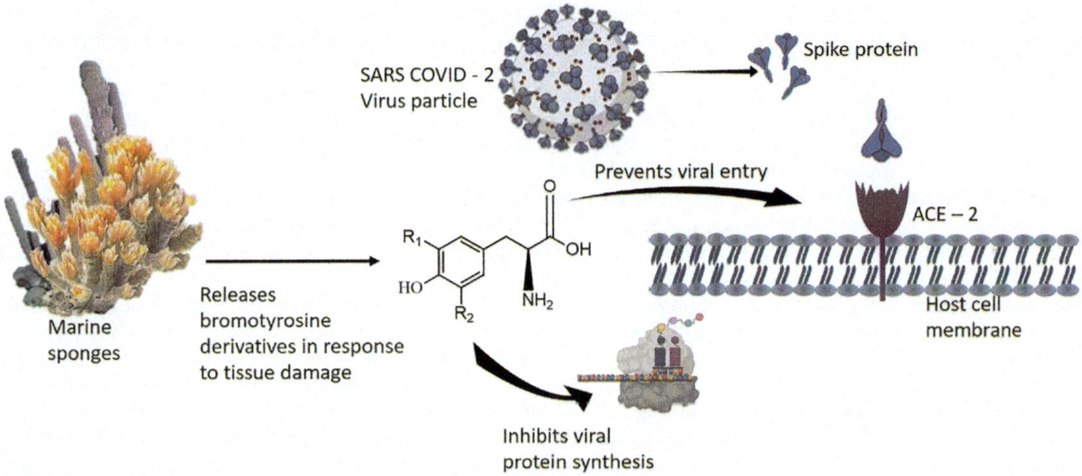

FIG. 16.4 Overview of SARS-CoV-2 infection and bromotyrosine, a marine compound that has antiviral properties.

blockers of SARS-CoV-2 infection by targeting and inhibiting the spike proteins of the virus, ultimately terminating viral proliferation inside the host cell (Fig. 16.4) (Geahchan et al., 2022).

16.3 Conclusion

Infectious diseases caused by bacteria, fungi, viruses, and parasites have affected millions of people globally and have become a major socioeconomic threat. Pharmaceutical industries are seeking new possibilities to overcome the rapid increase in pathogenic infections and the growing issue of drug resistance. Their priority is to focus on marine bioresources in developing alternative drugs because their vast biodiversity provides a broad spectrum of biological activities and chemical structures that allow for the consistent and effective control of pathogen-caused diseases. Many research studies have proven that the contribution of marine bioresources is sustainable and eco-friendly in terms of disease treatment and prevention. In essence, marine bioresources present a promising frontier in the fight against infectious diseases, but further in vivo, in vitro, and in silico experiments are needed to unlock their full potential for drug-like properties while preserving both the health of the oceans and public health, which leads to pharmaceutical advancement.

References

Agrebi, S., Larbi, A., 2020. Use of artificial intelligence in infectious diseases. In: Artificial Intelligence in Precision Health: From Concept to Applications. Elsevier, Tunisia, 415–438. Available from: https://www.sciencedirect.com/book/9780128171332. doi:10.1016/B978-0-12-817133-2.00018-5.

Al-Salihi, K.A., Khalaf, J.M., 2021. The emerging SARS-CoV, MERS-CoV, and SARS-CoV-2: an insight into the viruses zoonotic aspects. Vet. World. 22310916. 14 (1), 190–199. Available from: http://www.veterinaryworld.org/Vol.14/January-2021/25.pdf. doi:10.14202/VETWORLD.2021.190-199.

Alqathama, A.A., Ahmad, R., Alsaedi, R.B., Alghamdi, R.A., Abkar, E.H., Alrehaly, R.H., Abdalla, A.N., 2022. The vital role of animal, marine, and microbial natural products against COVID-19. Pharm. Biol. 17445116. 60 (1), 509–524. Available from: https://www.tandfonline.com/loi/iphb20. doi:10.1080/13880209.2022.2039215.

Arunkumar, M., Gunaseelan, S., Aravind, M.K., Mohankumar, V., Anupam, P., Harikrishnan, M., Siva, A., Ashokkumar, B., Varalakshmi, P., 2022. Marine algal antagonists targeting 3CL protease and spike glycoprotein of SARS-CoV-2: a computational approach for anti-COVID-19 drug discovery. J. Biomol. Struct. Dyn. 15380254. 40 (19), 8961–8988. Available from: http://www.tandfonline.com/loi/tbsd20. doi:10.1080/07391102.2021.1921032.

Bansal, S., Jonsson, C.B., Taylor, S.L., Figueroa, J.M., Dugour, A.V., Palacios, C., Vega, J.C., 2021. Lota-carrageenan and xylitol inhibit SARS-CoV-2 in cell culture. PLoS One 16 (11), e0259943. doi:10.1371/journal.pone.0259943.

Barbosa, F., Pinto, E., Kijjoa, A., Pinto, M., Sousa, E., 2020. Targeting antimicrobial drug resistance with marine natural products. Int. J. Antimicrob. Agents. 18727913. 56 (1). Available from: https://www.elsevier.com/locate/ijantimicag. doi:10.1016/j.ijantimicag.2020.106005.

Bharathi, D., Lee, J., 2024. Recent advances in marine-derived compounds as potent antibacterial and antifungal agents: a comprehensive review. Mar. Drugs. 16603397. 22 (8). Available from: http://www.mdpi.com/journal/marinedrugs. doi:10.3390/md22080348.

Bhowmik, A.J.A.O.H.M.R., 2023. Reviews, role of diagnostic procedures in managing human Bacterial infections: a comprehensive overview. Arch. Hematol. Case Rep. Rev. 8, 8–019.

Binneweg, B., Schubert, M., Voronkina, A., Muzychka, L., Wysokowski, M., Petrenko, I., Djurović, M., Kovalchuk, V., Tsurkan, M., Martinovic, R., Bechmann, N., Fursov, A., Ivanenko, V.N., Tabachnick, K.R., Smolii, O.B., Joseph, Y., Giovine, M., Bornstein, S.R., Stelling, A.L., Tunger, A., Schmitz, M., Taniya, O.S., Kovalev, I.S., Zyryanov, G.V., Guan, K., Ehrlich, H., 2020. Marine biomaterials: biomimetic and pharmacological potential of cultivated *Aplysina aerophoba* marine demosponge. Mater. Sci. Eng. 09284931. C (109), 110566. doi:10.1016/j.msec.2019.110566.

Biswas, S., Khuntia, H., Bal, M., Pati, S., Ranjit, M., 2023. Emerging and re-emerging diseases: a narrative review. One Health Bull. 2773-0344. 3 (1), 20. doi:10.4103/2773-0344.389607.

Brüssow, H., 2022. The beginning and ending of a respiratory viral pandemic-lessons from the Spanish flu. Microb. Biotechnol. 17517915. 15 (5), 1301–1317. Available from: http://onlinelibrary.wiley.com/journal/10.1111/(ISSN)1751-7915. doi:10.1111/1751-7915.14053.

Cai, Y., Xu, W., Gu, C., Cai, X., Qu, D., Lu, L., Xie, Y., Jiang, S., 2020. Griffithsin with a broad-spectrum antiviral activity by binding glycans in viral glycoprotein exhibits strong synergistic effect in combination with a pan-coronavirus fusion inhibitor targeting SARS-CoV-2 Spike S2 subunit. Virol. Sin. 1995820X. 35 (6), 857–860. Available from: http://www.springerlink.com/content/120859/. doi:10.1007/s12250-020-00305-3.

Cardoso, J., Gonçalves Nakayama, D., Sousa, E., Pinto, E., 2020. Marine-derived compounds and prospects for their antifungal application. Molecules. 1420-3049. 25 (24), 5856. doi:10.3390/molecules25245856.

Chaachouay, N., Zidane, L., 2024. Plant-derived natural products: a source for drug discovery and development. Drugs Drug Candidates. 2813-2998. 3 (1), 184–207. doi:10.3390/ddc3010011.

Chin, J.M.W., Puchooa, D., Bahorun, T., Jeewon, R., 2021. Antimicrobial properties of marine fungi from sponges and brown algae of Mauritius. Mycology. 21501211. 12 (4), 231–244. Available from: http://www.tandfonline.com/toc/tmyc20/current. doi:10.1080/21501203.2021.1895347.

Coelho, C., Gallo, G., Campos, C.B., Hardy, L., Würtele, M., Bogyo, M., 2020. Biochemical screening for SARS-CoV-2 main protease inhibitors. PLOS ONE. 1932-6203. 15 (10), e0240079. doi:10.1371/journal.pone.0240079.

Collins, J.E., Vanagt, T., Huys, I., Vieira, H.J.F.I.M.S., 2020. Marine bioresource development–stakeholder's challenges, implementable actions, and business models. Front. Mar. Sci. 7. doi:10.3389/fmars.2020.00062.

Daniotti, S., Re, I., 2021. Marine biotechnology: challenges and development market trends for the enhancement of biotic resources in industrial pharmaceutical and food applications. a statistical analysis of scientific literature and business models. Mar. Drugs. 16603397. 19 (2). Available from: http://www.mdpi.com/journal/marinedrugs. doi:10.3390/md19020061.

Das, S., Sherpa, M.T., Najar, I.N., Thakur, N., 2021. Hot and cold bacteria of Sikkim: biodiversity and enzymology. In: Bioprospecting of Enzymes in Industry, Healthcare and Sustainable Environment. Singapore, India, 269–289. Available from: https://link.springer.com/book/10.1007/978-981-33-4195-1. doi:10.1007/978-981-33-4195-1_13.

Dekaboruah, E., Suryavanshi, M.V., Chettri, D., Verma, A.K., 2020. Human microbiome: an academic update on human body site specific surveillance and its possible role. Arch. Microbiol. 1432072X. 202 (8), 2147–2167. Available from: https://link.springer.de/link/service/journals/00203/index.htm. doi:10.1007/s00203-020-01931-x.

Deng, W., Zhang, F., Fornacca, D., Yang, X.Y., Xiao, W., 2023. Those nematode-trapping fungi that are not everywhere: hints towards soil microbial biogeography. J. Microbiol. 19763794. 61 (5), 511–523. Available from: https://www.springer.com/journal/12275. doi:10.1007/s12275-023-00043-7.

Derakhshan-Nezhad, E., 2025. An overview of viral respiratory infections. J. Med. Bacteriol. 13. doi:10.18502/jmb.v13i1.18050.

Dharsan, R., Geetha, R.V., Lakshmi, T., 2020. Emerging and re-emerging virus. Indian J. Forensic Med. Toxicol. 09739130. 14 (4), 4515–4521. Available from: http://www.ijfmt.com/issues.html. doi:10.37506/ijfmt.v14i4.12353.

Drechsel, A., Helm, J., Ehrlich, H., Pantovic, S., Bornstein, S.R., Bechmann, N., 2020. Anti-tumor activity vs. normal cell toxicity: therapeutic potential of the bromotyrosines aerothionin and homoaerothionin in vitro. Mar. Drugs. 16603397. 18 (5). Available from: https://www.mdpi.com/1660-3397/18/5/236. doi:10.3390/md18050236.

Dutta, S., Dutta, P., Samanta, G., 2024. Modelling disease transmission through asymptomatic carriers: a societal and environmental perspective. Int. J. Dyn. Control. 2195-268X. 12 (8), 3100–3122. doi:10.1007/s40435-024-01387-7.

Eichberg, J., Maiworm, E., Oberpaul, M., Czudai-Matwich, V., Lüddecke, T., Vilcinskas, A., Hardes, K., 2022. Antiviral potential of natural resources against influenza virus infections. Viruses. 19994915. 14 (11). Available from: http://www.mdpi.com/journal/viruses/. doi:10.3390/v14112452.

El-Bondkly, E.A.M., El-Bondkly, A.A.M., El-Bondkly, A.A.M., 2021. Marine endophytic fungal metabolites: a whole new world of pharmaceutical therapy exploration. Heliyon. 24058440. 7 (3). Available from: http://www.journals.elsevier.com/heliyon/. doi:10.1016/j.heliyon.2021.e06362.

Firacative, C., 2020. Invasive fungal disease in humans: are we aware of the real impact? Mem. Inst. Oswaldo Cruz. 16788060. 115 (9), 1–9. Available from: https://memorias.ioc.fiocruz.br/article/10760/0430-invasive-fungal-disease-in-humans-are-we-aware-of-the-real-impact. doi:10.1590/0074-02760200430.

Fisher, M.C., Alastruey-Izquierdo, A., Berman, J., Bicanic, T., Bignell, E.M., Bowyer, P., Bromley, M., Brüggemann, R., Garber, G., Cornely, O.A., Gurr, S.J., Harrison, T.S., Kuijper, E., Rhodes, J., Sheppard, D.C., Warris, A., White, P.L., Xu, J., Zwaan, B., Verweij, P.E., 2022. Tackling the emerging threat of antifungal resistance to human health. Nat. Rev. Microbiol. 17401534. 20 (9), 557–571. Available from: http://www.nature.com/nrmicro/index.html. doi:10.1038/s41579-022-00720-1.

Geahchan, S., Ehrlich, H., Rahman, A., 2022. A short overview: marine resources as potential interventions for the Omicron SARS-CoV-2 variant. COVID. 26738112. 2 (4), 501–512. Available from: https://www.mdpi.com/journal/covid/about. doi:10.3390/covid2040037.

Geahchan, S., Ehrlich, H., Rahman, M.A., 2021. The anti-viral applications of marine resources for COVID-19 treatment: an overview. Mar. Drugs. 16603397. 19 (8). Available from: https://www.mdpi.com/1660-3397/19/8/409/pdf. doi:10.3390/md19080409.

Ghosh, S., Sarkar, T., Pati, S., Kari, Z.A., Edinur, H.A., Chakraborty, R., 2022. Novel bioactive compounds from marine sources as a tool for functional food development. Front. Mar. Sci. 22967745. 9. Available from: https://www.frontiersin.org/journals/marine-science#. doi:10.3389/fmars.2022.832957.

Gunasekara, D.M.N.M., Wang, L., Herath, K.H.I.N.M., Sanjeewa, K.K.A., 2025. Cosmeceutical applications of phlorotannins from brown seaweeds. Phycology. 2673-9410. 5 (2), 15. doi:10.3390/phycology5020015.

Gurung, A.B., Ali, M.A., Lee, J., Farah, M.A., Al-Anazi, K.M., 2020. Unravelling lead antiviral phytochemicals for the inhibition of SARS-CoV-2 Mpro enzyme through in silico approach. Life Sci. 18790631. 255. Available from: https://www.elsevier.com/locate/lifescie. doi:10.1016/j.lfs.2020.117831.

Haider, N., Rothman-Ostrow, P., Osman, A.Y., Arruda, L.B., Macfarlane-Berry, L., Elton, L., Thomason, M.J., Yeboah-Manu, D., Ansumana, R., Kapata, N., Mboera, L., Rushton, J., McHugh, T.D., Heymann, D.L., Zumla, A., Kock, R.A., 2020. COVID-19—zoonosis or emerging infectious disease? Front. Public Health 8. doi:10.3389/fpubh.2020.596944.

Hamoda, A.M., Fayed, B., Ashmawy, N.S., El-Shorbagi, A.N.A., Hamdy, R., Soliman, S.S.M., 2021. Marine sponge is a promising natural source of anti-SARS-CoV-2 scaffold. Front. Pharmacol. 16639812. 12. Available from: http://www.frontiersin.org/Pharmacology. doi:10.3389/fphar.2021.666664.

Hathaway, D., Pandav, K., Patel, M., Riva-Moscoso, A., Singh, B.M., Patel, A., Min, Z.C., Singh-Makkar, S., Sana, M. K., Sanchez-Dopazo, R., Desir, R., Fahem, M.M.M., Manella, S., Rodriguez, I., Alvarez, A., Abreu, R., 2020. Omega 3 fatty acids and COVID-19: a comprehensive review. Infect. Chemother. 2093-2340. 52 (4), 478. doi:10.3947/ic.2020.52.4.478.

Herath, K.H.I.N.M., Nagahawatta, D.P., Wang, L., Asanka Sanjeewa, K.K., 2025. The role of phlorotannins to treat inflammatory diseases. Chemistry. 2624-8549. 7 (3), 77. doi:10.3390/chemistry7030077.

Hu, R.S., Hesham, A.E.L., Zou, Q., 2022. Machine learning and its applications for protozoal pathogens and protozoal infectious diseases. Front. Cell. Infect. Microbiol. 22352988. 12. Available from: http://www.frontiersin.org/Cellular_and_Infection_Microbiology/archive. doi:10.3389/fcimb.2022.882995.

Iwasaki, K., Iwasaki, A., Sumimoto, S., Matsubara, T., Sato, T., Nozaki, T., Saito-Nakano, Y., Suenaga, K., 2020. Ikoamide, an antimalarial lipopeptide from an Okeania sp. Marine Cyanobacterium. J. Nat. Prod. 15206025. 83 (2), 481–488. Available from: http://pubs.acs.org/journal/jnprdf. doi:10.1021/acs.jnatprod.9b01147.

Jafarlou, M., 2024. Unveiling the menace: a thorough review of potential pandemic fungal disease. Front. Fungal Biol. 26736128. 5. Available from: https://www.frontiersin.org/journals/fungal-biology. doi:10.3389/ffunb.2024.1338726.

Jakubczyk, D., Dussart, F., 2020. Selected fungal natural products with antimicrobial properties. Molecules. 14203049. 25 (4). Available from: https://www.mdpi.com/1420-3049/25/4/911/pdf. doi:10.3390/molecules25040911.

Jang, Y., Shin, H., Lee, M.K., Kwon, O.S., Shin, J.S., Kim, Y.I., Kim, C.W., Lee, H.R., Kim, M., 2021. Antiviral activity of lambda-carrageenan against influenza viruses and severe acute respiratory syndrome coronavirus 2. Sci. Rep. 20452322. 11 (1). Available from: https://www.nature.com/srep/index.html. doi:10.1038/s41598-020-80896-9.

Karthikeyan, A., Joseph, A., Nair, B.G., 2022. Promising bioactive compounds from the marine environment and their potential effects on various diseases. J. Genet. Eng. Biotechnol. 1687-157X. 20 (1), 14. Available from: https://www.sciencedirect.com/science/article/pii/S1687157X23009411. doi:10.1186/s43141-021-00290-4.

Kausar, S., Khan, F.S., Ur Rehman, M.I.M., Akram, M., Riaz, M., Rasool, G., Hamid Khan, A., Saleem, I., Shamim, S., Malik, A., 2021. A review: mechanism of action of antiviral drugs. Int. J. Immunopathol. Pharmacol. 20587384. 35. Available from: http://iji.sagepub.com/. doi:10.1177/20587384211002621.

Keller, L., Siqueira-Neto, J.L., Souza, J.M., Eribez, K., LaMonte, G.M., Smith, J.E., Gerwick, W.H., 2020. Palstimolide a: a complex polyhydroxy macrolide with antiparasitic activity. Molecules. 1420-3049. 25 (7), 1604. doi:10.3390/molecules25071604.

Khalili, J.S., Zhu, H., Mak, N.S.A., Yan, Y., Zhu, Y., 2020. Novel coronavirus treatment with ribavirin: groundwork for an evaluation concerning COVID-19. J. Med. Virol. 10969071. 92 (7), 740–746. Available from: http://onlinelibrary.wiley.com/journal/10.1002/(ISSN)1096-9071. doi:10.1002/jmv.25798.

Kurisawa, N., Iwasaki, A., Jeelani, G., Nozaki, T., Suenaga, K., Iheyamides, A.-C., 2020. Antitrypanosomal linear peptides isolated from a marine Dapis sp. Cyanobacterium. J. Nat. Prod. 15206025. 83 (5), 1684–1690. Available from: http://pubs.acs.org/journal/jnprdf. doi:10.1021/acs.jnatprod.0c00250.

Lemesheva, V., Islamova, R., Stepchenkova, E., Shenfeld, A., Birkemeyer, C., Tarakhovskaya, E., 2023. Antibacterial, antifungal and algicidal activity of phlorotannins, as principal biologically active components of ten species of brown algae. Federation Plants. 22237747. 12 (4). Available from: https://www.mdpi.com/journal/plants. doi:10.3390/plants12040821.

Li, T., Wang, Z., Guo, J., de la Fuente-nunez, C., Wang, J., Han, B., Tao, H., Liu, J., Wang, X., 2023. Bacterial resistance to antibacterial agents: mechanisms, control strategies, and implications for global health. Sci. Total Environ. 00489697. 860, 160461. doi:10.1016/j.scitotenv.2022.160461.

Lin, J., Du, F., Long, M., Li, P., 2022. Limitations of phage therapy and corresponding optimization strategies: a review. Molecules. 14203049. 27 (6). Available from: https://www.mdpi.com/1420-3049/27/6/1857/pdf. doi:10.3390/molecules27061857.

Liu, X., Xin, J., Sun, Y., Zhao, F., Niu, C., Liu, S., 2024. Terpenoids from marine sources: a promising avenue for new antimicrobial drugs. Mar. Drugs. 16603397. 22 (8). Available from: http://www.mdpi.com/journal/marinedrugs. doi:10.3390/md22080347.

Lyu, C., Chen, T., Qiang, B., Liu, N., Wang, H., Zhang, L., Liu, Z., 2021. CMNPD: a comprehensive marine natural products database towards facilitating drug discovery from the ocean. Nucleic Acids Res. 13624962. 49 (1), D509. Available from: https://academic.oup.com/nar/issue. doi:10.1093/nar/gkaa763.

Ma, Y., Frutos-Beltrán, E., Kang, D., Pannecouque, C., De Clercq, E., Menéndez-Arias, L., Liu, X., Zhan, P., 2021. Medicinal chemistry strategies for discovering antivirals effective against drug-resistant viruses. Chem. Soc. Rev. 14604744. 50 (7), 4514–4540. Available from: http://pubs.rsc.org/en/journals/journal/cs. doi:10.1039/d0cs01084g.

Mankar, S., Vikhe, S., Pawar, T., Bhawar, S., 2022. Current perspective of fungi-a review. Asian J. Res. Pharm. Sci. 12 (3), 251–257. doi:10.52711/2231-5659.2022.00044.

Mofeed, J., Deyab, M., Mohamed, A., Moustafa, M., Negm, S., El-Bilawy, E., 2022. Antimicrobial activities of three seaweeds extract against some human viral and bacterial pathogens. Biocell. 16675746. 46 (1), 247–261. Available from: https://www.techscience.com/biocell/v46n1. doi:10.32604/biocell.2022.015966.

Morokutti-Kurz, M., Fröba, M., Graf, P., Große, M., Grassauer, A., Auth, J., Schubert, U., Prieschl-Grassauer, E., Polyak, S.J., 2021. Iota-carrageenan neutralizes SARS-CoV-2 and inhibits viral replication in vitro. PLOS ONE. 1932-6203. 16 (2), e0237480. doi:10.1371/journal.pone.0237480.

Mumtaz, H., Nadeem, A., Bilal, W., Ansar, F., Saleem, S., Khan, Q.A., Tango, T., Farkouh, C., Belay, N.F., Verma, R., Farkouh, M., Saqib, M., 2023. Acceptance, availability, and feasibility of RTS, S/AS01 malaria vaccine: a review. Immun. Inflam. Dis. 20504527. 11 (6). Available from: http://onlinelibrary.wiley.com/journal/10.1002/(ISSN)2050-4527. doi:10.1002/iid3.899.

Muralidar, S., Ambi, S.V., Sekaran, S., Krishnan, U.M., 2020. The emergence of COVID-19 as a global pandemic: understanding the epidemiology, immune response and potential therapeutic targets of SARS-CoV-2. Biochimie. 61831638. 179, 85–100. Available from: https://www.elsevier.com/locate/biochimie. doi:10.1016/j.biochi.2020.09.018.

Muteeb, G., Rehman, M.T., Shahwan, M., Aatif, M., 2023. Origin of antibiotics and antibiotic resistance, and their impacts on drug development: a narrative review. Pharmaceuticals. 14248247. 16 (11). Available from: https://www.mdpi.com/journal/pharmaceuticals. doi:10.3390/ph16111615.

Naeem, A., Hu, P., Yang, M., Zhang, J., Liu, Y., Zhu, W., Zheng, Q., 2022. Natural products as anticancer agents: current status and future perspectives. Molecules. 14203049. 27 (23). Available from: http://www.mdpi.com/journal/molecules. doi:10.3390/molecules27238367.

Nagle, V., Gaikwad, M., Pawar, Y., Dasgupta, S., 2020. Marine red alga Porphyridium sp. as a source of sulfated polysaccharides (sps) for combating against COVID-19. Preprints. doi:10.13140/RG.2.2.35640.90883.

Negara, B.F.S.P., Sohn, J.-H., Kim, J.-S., Choi, J.-S., 2021. Antifungal and larvicidal activities of phlorotannins from brown seaweeds. Mar. Drugs. 1660-3397. 19 (4), 223. doi:10.3390/md19040223.

Nguyen, T.T., Heimann, K., Zhang, W., 2020. Protein recovery from underutilised marine bioresources for product development with nutraceutical and pharmaceutical bioactivities. Mar. Drugs. 1660-3397. 18 (8), 391. doi:10.3390/md18080391.

Nweze, J.A., Mbaoji, F.N., Huang, G., Li, Y., Yang, L., Zhang, Y., Huang, S., Pan, L., Yang, D., 2020. Antibiotics development and the potentials of marine-derived compounds to stem the tide of multidrug-resistant pathogenic bacteria, fungi, and protozoa. Mar. Drugs. 16603397. 18 (3). Available from: https://www.mdpi.com/1660-3397/18/3/145. doi:10.3390/md18030145.

Nweze, J.A., Mbaoji, F.N., Li, Y.M., Yang, L.Y., Huang, S.S., Chigor, V.N., Eze, E.A., Pan, L.X., Zhang, T., Yang, D.F., 2021. Potentials of marine natural products against malaria, leishmaniasis, and trypanosomiasis parasites: a review of recent articles. Infect. Dis. Poverty. 20499957. 10 (1). Available from: http://www.idpjournal.com/. doi:10.1186/s40249-021-00796-6.

Patnaik, S., Das, P., Marine metabolites in disease management. In: Advances in Medical Technologies and Clinical Practice. IGI Global. 2024. 10.4018/979-8-3693-5878-8

Prabhu, S.R., 2023. Infectious and communicable diseases: an overview. In: Textbook of General Pathology for Dental Students. Science and Business Media LLC, 63–72. doi:10.1007/978-3-031-31244-1_9.

Rahman, M.M., Islam, M.R., Shohag, S., Hossain, M.E., Shah, M., Khan Shuvo, S., Khan, H., Chowdhury, M.A.R., Bulbul, I.J., Hossain, M.S., Sultana, S., Ahmed, M., Akhtar, M.F., Saleem, A., Rahman, M.H., 2022. Retracted article: multifaceted role of natural sources for COVID-19 pandemic as marine drugs. Environ. Sci. Pollut. Res. 0944-1344. 29 (31), 46527–46550. doi:10.1007/s11356-022-20328-5.

Rahman, N., Basharat, Z., Yousuf, M., Castaldo, G., Rastrelli, L., Khan, H., 2020. Virtual screening of natural products against type II transmembrane serine protease (TMPRSS2), the priming agent of coronavirus 2 (SARS-COV-2). Molecules. 14203049. 25 (10). Available from: https://www.mdpi.com/1420-3049/25/10/2271/pdf. doi:10.3390/molecules25102271.

Ratha, S.K., Renuka, N., Rawat, I., Bux, F., 2021. Prospective options of algae-derived nutraceuticals as supplements to combat COVID-19 and human coronavirus diseases. Nutrition. 0899-9007. 83, 111089. Available from https://www.sciencedirect.com/science/article/pii/S0899900720303725. doi:10.1016/j.nut.2020.111089.

Riccio, G., Ruocco, N., Mutalipassi, M., Costantini, M., Zupo, V., Coppola, D., de Pascale, D., Lauritano, C., 2020. Ten-year research update review: antiviral activities from marine organisms. Biomolecules. 2218273X. 10 (7), 1–36. Available from: https://www.mdpi.com/2218-273X/10/7/1007/pdf. doi:10.3390/biom10071007.

Rotter, A., Barbier, M., Bertoni, F., Bones, A.M., Cancela, M.L., Carlsson, J., Carvalho, M.F., Cegłowska, M., Chirivella-Martorell, J., Dalay, M.C., Cueto, M., Dailianis, T., Deniz, I., Díaz-Marrero, A.R., Drakulovic, D., Dubnika, A., Edwards, C., Einarsson, H., Erdoğan, A., Eroldoğan, O.T., Ezra, D., Fazi, S., FitzGerald, R.J., Gargan, L.M., Gaudêncio, S.P., Udovič, M.G., Ivošević Denardis, N., Jónsdóttir, R., Kataržytė, M., Klun, K., Kotta, J., Ktari, L., Ljubešić, Z., Lukić Bilela, L., Mandalakis, M., Massa-Gallucci, A., Matijošytė, I., Mazur-Marzec, H., Mehiri, M., Laurentius Nielsen, S., Novoveská, L., Overlingė, D., Perale, G., Ramasamy, P., Rebours, C., Reinsch, T., Reyes, F., Rinkevich, B., Robbens, J., Röttinger, E., Rudovica, V., Sabotič, J., Safarik, I., Talve, S., Tasdemir, D., Schneider, X.T., Thomas, O.P., Toruńska-Sitarz, A., Varese, G.C., Vasquez, M.I., 2021. The Essentials of Marine Biotechnology. Front. Mar. Sci. 2296-7745. 8. doi:10.3389/fmars.2021.629629.

Roy, A., Ahuja, S., Garg, S., 2021. Fungal secondary metabolites: biological activity and potential applications. In: Fungal Biology. Springer Science and Business Media LLC, 159–188. doi:10.1007/978-3-030-60659-6_7.

Sable, R., Parajuli, P., Jois, S., 2017. Peptides, peptidomimetics, and polypeptides from marine sources: a wealth of natural sources for pharmaceutical applications. Mar. Drugs. 16603397. 15 (4). Available from: http://www.mdpi.com/1660-3397/15/4/124/pdf. doi:10.3390/md15040124.

Salleh, K.M., Rashid, N.F.A., 2024. Potential seaweed-derived bioactive compounds for pharmaceutical applications. In: Marine Biomass: Biorefinery, Bioproducts and Environmental Bioremediation. De Gruyter, Malaysia, 297–318. Available from https://www.degruyter.com/document/doi/10.1515/9783111353951/html. doi:10.1515/9783111353951-013.

Sansone, C., Brunet, C., Noonan, D.M., Albini, A., 2020. Marine algal antioxidants as potential vectors for controlling viral diseases. Antioxidants. 20763921. 9 (5). Available from: https://www.mdpi.com/2076-3921/9/5/392/pdf. doi:10.3390/antiox9050392.

Senapati, S., Banerjee, P., Bhagavatula, S., Kushwaha, P.P., Kumar, S., 2021. Contributions of human ACE2 and TMPRSS2 in determining host–pathogen interaction of COVID-19. J. Genet. 09737731. 100 (1). Available from: http://www.springerlink.com/content/120421/. doi:10.1007/s12041-021-01262-w.

Soni, J., Sinha, S., Pandey, R., 2024. Understanding bacterial pathogenicity: a closer look at the journey of harmful microbes. Front. Microbiol. 1664302X. (15). Available from https://www.frontiersin.org/journals/microbiology#. doi:10.3389/fmicb.2024.1370818.

Sukmarini, L., 2022. Antiviral peptides (AVPs) of marine origin as propitious therapeutic drug candidates for the treatment of human viruses. Molecules. 1420-3049. 27 (9), 2619. doi:10.3390/molecules27092619.

Sun, S., Chen, X., 2024. Mechanism-guided strategies for combating antibiotic resistance. World J. Microbiol. Biotechnol. 15730972. 40 (10). Available from: https://www.springer.com/journal/11274. doi:10.1007/s11274-024-04106-8.

Sunagawa, S., Acinas, S.G., Bork, P., Bowler, C., Acinas, S.G., Babin, M., Bork, P., Boss, E., Bowler, C., Cochrane, G., de Vargas, C., Follows, M., Gorsky, G., Grimsley, N., Guidi, L., Hingamp, P., Iudicone, D., Jaillon, O., Kandels, S., Karp-Boss, L., Karsenti, E., Lescot, M., Not, F., Ogata, H., Pesant, S., Poulton, N., Raes, J., Sardet, C., Sieracki, M., Speich, S., Stemmann, L., Sullivan, M.B., Sunagawa, S., Wincker, P., Eveillard, D., Gorsky, G., Guidi, L., Iudicone, D., Karsenti, E., Lombard, F., Ogata, H., Pesant, S., Sullivan, M.B., Wincker, P., de Vargas, C., 2020. Tara Oceans: towards global ocean ecosystems biology. Nat. Rev. Microbiol. 1740-1526. 18 (8), 428–445. doi:10.1038/s41579-020-0364-5.

Sykes, J.E., Papich, M.G., 2022. Antiprotozoal Drugs. In: Greene's Infectious Diseases of the Dog and Cat, Fifth Edition. Elsevier, United States, 140–148. Available from https://www.sciencedirect.com/book/9780323509343. doi:10.1016/B978-0-323-50934-3.00012-4.

Tabish, S.A., 2020. Covid-19 pandemic: emerging perspectives and future trends. J. Public Health Res. 2279-9036. 9 (1). Available from: https://www.jphr.2020.1786. doi:10.4081/jphr.2020.1786.

Teng, Y.F., Xu, L., Wei, M.Y., Wang, C.Y., Gu, Y.C., Shao, C.L., 2020. Recent progresses in marine microbial-derived antiviral natural products. Arch. Pharm. Res. 19763786. 43 (12), 1215–1229. Available from: http://www.springerlink.com/content/0253-6269. doi:10.1007/s12272-020-01286-3.

Thawabteh, A.M., Swaileh, Z., Ammar, M., Jaghama, W., Yousef, M., Karaman, R., Bufo, S.A., Scrano, L., 2023. Antifungal and antibacterial activities of isolated marine compounds. Toxins. 20726651. 15 (2). Available from: http://www.mdpi.com/journal/toxins/. doi:10.3390/toxins15020093.

Vaisusuk, K., Saijuntha, W., 2021. Intestinal protozoa: their role as human pathogens and zoonoses. In: Parasitology Research Monographs. Springer Science and Business Media LLC, 35–61. doi:10.1007/978-3-030-71161-0_3.

Varijakzhan, D., Loh, J.Y., Yap, W.S., Yusoff, K., Seboussi, R., Lim, S.H.E., Lai, K.S., Chong, C.M., 2021. Bioactive compounds from marine sponges: fundamentals and applications. Mar. Drugs. 16603397. 19 (5). Available from: https://www.mdpi.com/1660-3397/19/5/246/pdf. doi:10.3390/md19050246.

Vazquez-Munoz, R., 2023. Infectious diseases: overview, challenges, and perspectives. In: Smart Nanomaterials to Combat the Spread of Viral Infections: Advanced Strategies for the Prevention of Viral Infections. Elsevier, United States, 1–21. Available from: https://www.sciencedirect.com/book/9780323991483. doi:10.1016/B978-0-323-99148-3.00010-8.

Venkatesan, P., 2025. WHO world malaria report 2024. Lancet Microbe. 2666-5247. 6 (4). Available from: https://doi.org/10.1016/j.lanmic.2025.101073. doi:10.1016/j.lanmic.2025.101073.

W.H. Organization, 2022. WHO Fungal Priority Pathogens List to Guide Research. World Health Organization.

Wan, Y., Shang, J., Graham, R., Baric, R., Li, F.J.J.V., 2020. Receptor recognition by novel coronavirus from Wuhan: ananalysis based on decade-long structural studies of SARS. J. Virol. 94. doi:10.1128/jvi.00127-20.

Wang, C.H., Hsieh, Y.H., Powers, Z.M., Kao, C.Y., 2020. Defeating antibiotic-resistant bacteria: exploring alternative therapies for a post-antibiotic era. Int. J. Mol. Sci. 14220067. 21 (3). Available from: https://www.mdpi.com/1422-0067/21/3/1061/pdf. doi:10.3390/ijms21031061.

Wang, L., Alexander, C.A., 2021. Covid-19 compared with other viral diseases: novelties, progress, and challenges. Electron. J. Gen. Med. 25163507. 18 (1), 1–12. Available from: https://www.ejgm.co.uk/download/sars-cov-2-rapid-detection-system-using-electric-current-spectrum-for-human-exhaled-air-samples-8944.pdf. doi:10.29333/ejgm/8575.

Wellington Macêdo Viana, J., Torres, R.P., 2022. Resistance mechanisms of mycobacterium tuberculosis in the tuberculosis pathogenicity in humans: an overview. Revista Interfaces: Saúde, Hum. E Tecnol. 2317434X. 10 (2), 1431–1440. doi:10.16891/2317-434x.v10.e2a2022.pp1431-1440.

Zahran, E.M., Albohy, A., Khalil, A., Ibrahim, A.H., Ahmed, H.A., El-Hossary, E.M., Bringmann, G., Abdelmohsen, U.R., 2020. Bioactivity potential of marine natural products from scleractinia-associated microbes and in silico anti-SARS-COV-2 evaluation. Mar. Drugs. 16603397. 18 (12). Available from: https://www.mdpi.com/1660-3397/18/12/645. doi:10.3390/md18120645.

Zhang, M., Zhang, Q., Zhang, Q., Cui, X., Zhu, L., 2023. Promising antiparasitic natural and synthetic products from marine invertebrates and microorganisms. Mar. Drugs. 16603397. 21 (2). Available from: http://www.mdpi.com/journal/marinedrugs. doi:10.3390/md21020084.

Zhao, Q., Chen, Y., Huang, W., Zhou, H., Zhang, W., 2023. Drug-microbiota interactions: an emerging priority for precision medicine. Signal Transduct. Target. Ther. **8**, 1–27. doi:10.1038/s41392-023-01619-w.

CHAPTER

17

Impacts of marine metabolites on gut health

Hiruni S. Kumarasinghe[a], *Thiruni Perera*[a], *Kalpa W. Samarakoon*[b], and *Thilina L. Gunathilaka*[c]

[a]Department of Interdisciplinary Graduate Program in Advanced Convergence Technology and Science, Jeju National University, Jeju, Republic of Korea
[b]Institute for Combinatorial Advanced Research and Education (KDU-CARE), General Sir John Kotelawala Defence University, Ratmalana, Sri Lanka
[c]Department of Basic Science and Social Science for Nursing, Faculty of Nursing, University of Colombo, Colombo, Sri Lanka

17.1 Introduction

The marine ecosystem serves as a valuable source of novel bioactive compounds with various therapeutic benefits, including anticancer, antiinflammatory, antimicrobial, and gut health. The concept of gut health has gained significant attention recently in both scientific research and the food industry as a marker of overall well-being. A scientific committee researching gut health has proposed five primary criteria to define gut health, which are: effective digestion and absorption of food, absence of gastrointestinal (GI) illness, normal and stable intestinal microbiota, effective immune status of the GI tract, and overall well-being (Bischoff, 2011).

The lower GI tract comprises the vast majority of the small intestine and the entire large intestine. From a structural perspective, it consists of a stacked monolayer of developed epithelial cells, with columnar enterocytes forming the core structure. These intestinal epithelial cells' layer serves as the major physical and biochemical defense barrier against foreign substances such as toxins, allergens, and bacteria. A functioning intestinal epithelial barrier not only creates a barrier to prevent harmful bacteria from entering but also encourages beneficial microbial species, thus maintaining a stable gut microbiota. In contrast, the microbiota can impact the integrity of the barrier by affecting the innate immune response, demonstrating a reciprocal interaction. The intestinal barrier and gut microbiota work together to maintain gut health, as any disruptions in barrier integrity or microbial

Harvesting the Sea
https://doi.org/10.1016/B978-0-443-29190-6.00020-9

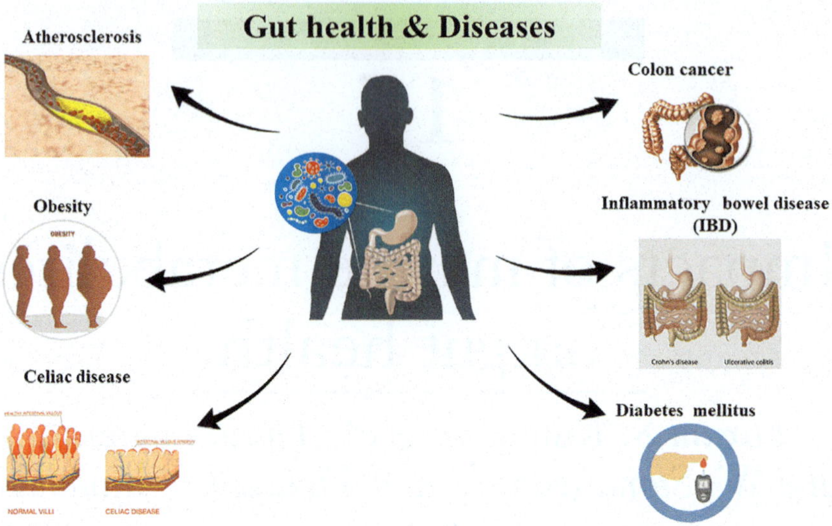

FIG. 17.1 Graphical illustration of gut health-related diseases. The impact of gut health on various diseases, including atherosclerosis, obesity, celiac disease, colon cancer, inflammatory bowel disease (IBD), and diabetes mellitus.

composition can lead to gastrointestinal diseases (GIDs) (Wan et al., 2019). Some common GIDs include irritable bowel syndrome (IBS), inflammatory bowel disease (IBD) (including Crohn's disease and ulcerative colitis), leaky gut syndrome, colorectal cancer (CRC), and gastroesophageal reflux disease (Fig. 17.1) (Jabłońska and Mrowiec, 2023).

The marine ecosystem, which is abundant in bioactive compounds, presents an exciting opportunity to influence gut health. Marine-derived metabolites, including polysaccharides, polyphenols, omega-3 fatty acids, and marine peptides, provide significant potential in enhancing gut microbiota, alleviating inflammation, and promoting intestinal integrity. This chapter covers the effects of marine metabolites on gut health, detailing their mechanisms of action and therapeutic potential.

17.2 Major sources of marine metabolites

Marine-derived compounds exhibit unique structures and functionalities, offering tremendous potential in therapeutic applications. Below are key classes of marine metabolites, with specific emphasis on their impacts on gut health (Fig. 17.2).

17.2.1 Polysaccharides

Polysaccharide bioactive macromolecules derived from marine plants, animals, and microorganisms have great potential in therapeutic applications, particularly for regulating metabolic syndrome via intracellular communication and molecular signaling. Furthermore, because these marine-derived polysaccharides are resistant to digestion by human enzymes,

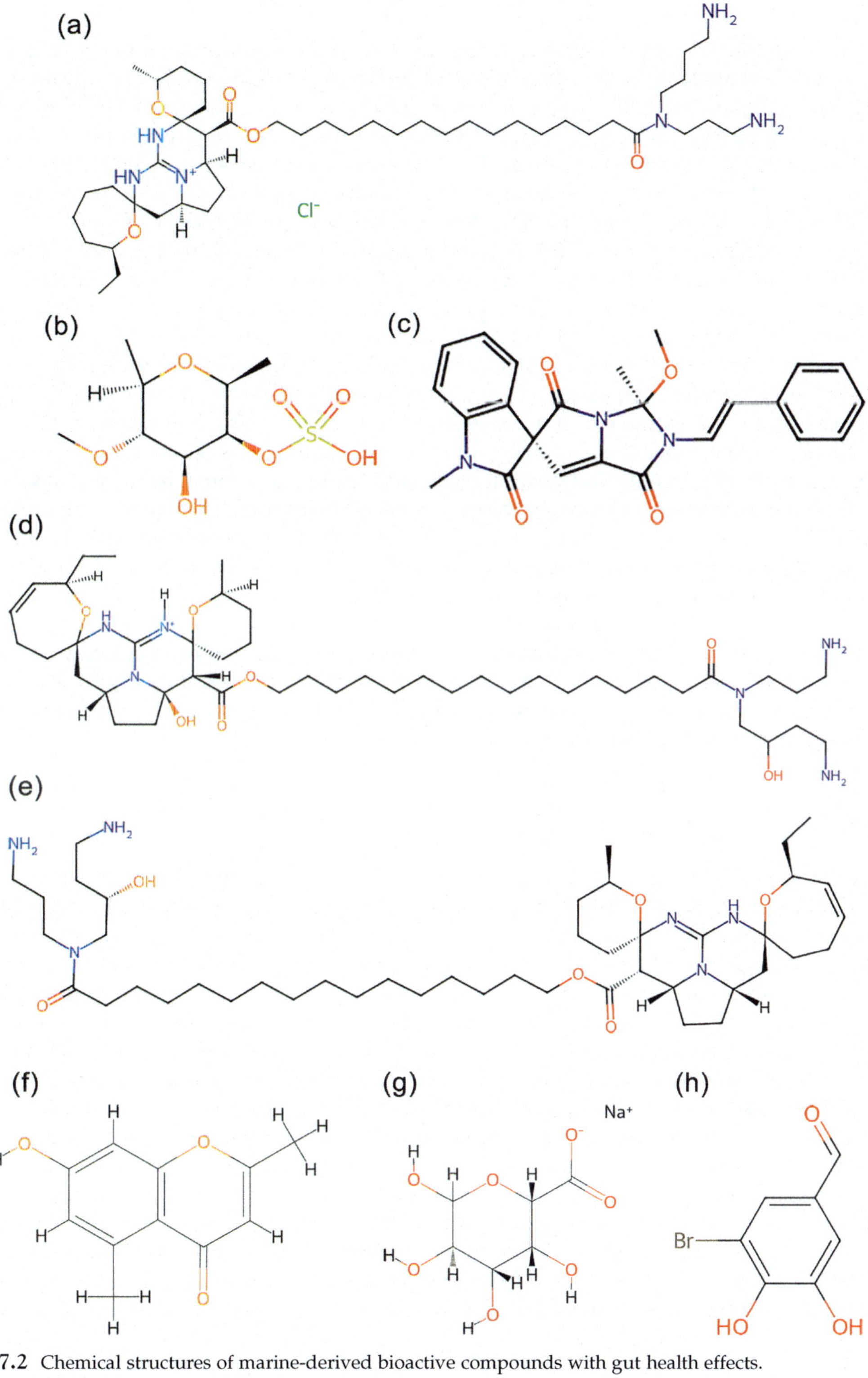

FIG. 17.2 Chemical structures of marine-derived bioactive compounds with gut health effects.

they promote the growth of beneficial gut bacteria that generate metabolites (Ullah et al., 2021). Fucoidans are large molecular weight, fucose-based, sulfated polysaccharides found in brown macroalgae. Fucoidans differ in structure and chemical composition between algal species and typically consist of a (1→3)-linked α-L-fucopyranosyl backbone, with some, including both (1→3)- and (1→4)-linked α-L-fucopyranosyl structures (Fig. 17.2). The four primary categories of brown algae that produce these polysaccharides are *Chordariales, Laminariales, Ectocarpales,* and *Fucales.* Despite variances in chemical makeup, fucoidans found in these algae have comparable backbone structures, allowing them to interact with different proteins and perform a variety of biological functions (Yang and Lim, 2021). Fucoidan promotes intestinal health, particularly through its prebiotic properties. As a prebiotic, fucoidan—a sulfated polysaccharide with an L-fucose composition—escapes digestion by stomach and pancreatic enzymes and enters the large intestine intact, where it has therapeutic effects. Fucoidan promotes the growth of beneficial bacteria in the large intestine, including *Lactobacillus, Bifidobacterium longum,* and *Bifidobacterium bifidum,* resulting in a balanced microbiota that enhances host health (Kim and Wijesekara, 2013). Alginate oligosaccharides (AOSs) are indigestible polysaccharides generated from hydrolyzed alginate that have been shown to have prebiotic effects and support gut health. AOS, made up of β-D-mannuronic and α-L-guluronic acids, regulates the gut microbiota and promotes the generation of beneficial short-chain fatty acids (SCFAs), which are necessary for a balanced intestinal environment. According to studies, AOS promotes intestinal health by inhibiting proinflammatory cytokines, scavenging reactive oxygen species (ROS), reducing intestinal epithelial cell death, decreasing intestinal permeability, and restoring microbial balance. These qualities highlight AOS's importance in gut health and intestinal homeostasis, although more research is needed to identify its effects on intestinal dysfunction in low birth weight mammals (Ghallab et al., 2024).

17.2.2 Polyphenols and flavonoids

Marine polyphenols are compounds consisting of several phenolic rings interlinked by diverse chemical bonds, which may be further modified by groups such as sugars or sulfates. A distinctive characteristic of marine polyphenols is their capacity to create complex aggregates, referred to as "tannins," via intermolecular interactions, including hydrogen bonding and hydrophobic interactions. These tannins possess distinct physical and chemical properties, including increased solubility and stability. The polyphenol 5-bromo-3,4-dihydroxybenzaldehyde, isolated from *Polysiphonia morrowii,* has been reported to exhibit antiadipogenic activity by inhibiting intracellular lipid accumulation and triglyceride levels in 3T3-L1 preadipocytes (Ko et al., 2019).

Flavonoids are low-molecular-weight secondary metabolites found in terrestrial plants and marine organisms. Since the discovery of the first marine flavonoids in 1966, around 100 distinct flavonoids have been identified. Recent studies have emphasized the extensive spectrum of beneficial biological actions demonstrated by marine flavonoids, making them more widely recognized and explored for possible applications (Martins et al., 2019). Altechromone A, also known as 2,5-dimethyl-7-hydroxychromone, is a hydroxyketone derived from the marine-derived fungus *Penicillium chrysogenum* (XY-14-0-4). In a study evaluating its potential for treating IBD, Altechromone A demonstrated significant antiinflammatory activity in various zebrafish models, including those induced by $CuSO_4$, tail cutting, and LPS (Li et al., 2024).

17.2.3 Omega-3 fatty acids

Unsaturated fatty acids (UFAs), comprising monounsaturated and polyunsaturated fatty acids (PUFAs), are vital bioactive substances acknowledged for their considerable health advantages. UFAs have a minimum of one double bond in their hydrocarbon chain, which enhances their functional and structural diversity (Lu et al., 2022). Marine ecosystems, especially fish oils, marine sponges, sea cucumbers, algae, coral, and microorganism species, constitute a significant natural source of UFAs, providing a sustainable and important source of these bioactive compounds (Lu et al., 2019). Several studies have shown that omega-3 PUFAs improve gut health by altering the composition of the microbiome. For example, omega-3 supplementation has been linked to an increase in the abundance of antiobesogenic bacteria in the gut, potentially aiding in the prevention and treatment of metabolic disorders. Specific alterations identified include a decline in *Faecalibacterium*, a decrease in proinflammatory bacteria, and a rise in Bacteroidetes and butyrate-producing bacteria from the *Lachnospiraceae* family. Changes in microbiota composition are expected to lead to improved gut health and enhanced synthesis of antiinflammatory chemicals, such as SCFAs (Lu et al., 2019).

17.2.4 Marine proteins and peptides

Proteins play a vital role in human health because dietary proteins are known to contain compounds with physiological activity that improve body functioning after digestion, which are essential to human nutrition. Microbial fermentation, enzymatic hydrolysis, or solvent extraction are utilized to produce bioactive peptides, which are beneficial compounds derived from proteins (Fernando et al., 2023). Peptides derived from marine organisms serve a variety of biological functions, including appetite regulation. A low-molecular-weight peptide isolated from shrimp head protein hydrolysates has shown a significant increase in cholecystokinin (CCK) from gut endocrine cells. CCK levels are frequently used as biomarkers to assess satiety in vivo (Cudennec et al., 2008). C-phycocyanin (C-PC) is a water-soluble, light-harvesting protein found in cyanobacteria, red algae, and some cryptomonads. A study on the effects of C-PC-derived peptides found that they significantly improved TNBS (Triple-Negative Breast Cancer)-induced colitis in a zebrafish model. Colitis is a complex, incurable type of chronic IBD that affects millions of individuals worldwide (Xu et al., 2024a).

17.2.5 Marine alkaloids and terpenes

Marine alkaloids are nitrogen-dense secondary metabolites synthesized by marine organisms, including sponges, bacteria, fungi, cyanobacteria, algae, and tunicates. Marine-derived alkaloids are important natural compounds characterized by varied structures and biological activity. Almost 40% of the 800 chemicals that were isolated from cyanobacteria were discovered to be alkaloids, making them an especially rich source. Their remarkable ability to fix nitrogen, which promotes alkaloid production, is responsible for this massive yield (Yu et al., 2024). The marine-derived alkaloids such as crambescidin 800, crambescidin 816, and ptilomycalin are reported to possess anticancer activity against CRC. These compounds are derived from the marine sponge Crambe, the French Polynesian *Monanchora n.* sp. sponge, and sponges *Ptilocaulis spiculifer* and *Hemimycale* sp., respectively (Singh et al., 2025; Sfecci et al., 2016). Moreover, Cyanogramide ($AC1_4$), a novel alkaloid isolated from the marine-derived *Actinoalloteichus cyanogriseus*, has shown promising effects in IBD treatment (Xu et al., 2024b)

Terpenes are primary and secondary metabolites derived from the five-carbon isoprene unit. The combining and alteration of these isoprene units yield an extensive variety of structures, each with unique chemical and biological properties (Ruzicka, 1953). Recent studies on the methanolic extract of the Red Sea-derived soft coral *Sarcophyton glaucum* have shown significant enhancements in metabolic markers associated with obesity and metabolic syndrome. The extract lowers glucose, insulin resistance (HOMA-IR), cholesterol, triglycerides, LDL-C, fetuin A and B, and PTP1B levels (Tammam et al., 2024). Furthermore, the expression of genes such as UCP1 and PPARGC1A was dramatically increased, indicating better metabolic performance. These findings imply that *S. glaucum*'s bioactive components, notably terpenes, may play an important role in obesity management (Tammam et al., 2024).

17.3 Impact of marine metabolites on human gut health and associated mechanisms of actions

Marine metabolites, which originate in marine animals such as algae, fish, and crustaceans, have emerged as intriguing bioactive substances with significant gut health benefits. These metabolites, which include PUFAs, peptides, polyphenols, carotenoids, and other bioactive compounds, have a variety of mechanisms for modulating the gut microbiota, reducing inflammation, enhancing antioxidant defenses, and supporting intestinal barrier integrity. These processes work together to keep the gut healthy and prevent the onset or progression of a variety of GI illnesses, such as IBD, IBS, and dysbiosis (Hamed et al., 2015) (Fig. 17.3).

17.3.1 Antiinflammatory activity

One of the most well-known methods by which marine metabolites improve intestinal health is through their antiinflammatory properties. Chronic inflammation in the gut plays an important role in the development and progression of many GIDs (Anwer et al., 2025). Marine-derived bioactive compounds, especially omega-3 PUFAs like eicosapentaenoic acid (EPA) and docosahexaenoic acid (DHA), have been shown to modulate inflammatory responses by reducing the production of proinflammatory cytokines, such as tumor necrosis factor-alpha (TNF-α) and interleukins (Wall et al., 2010; Calder, 2015). Omega-3 fatty acids

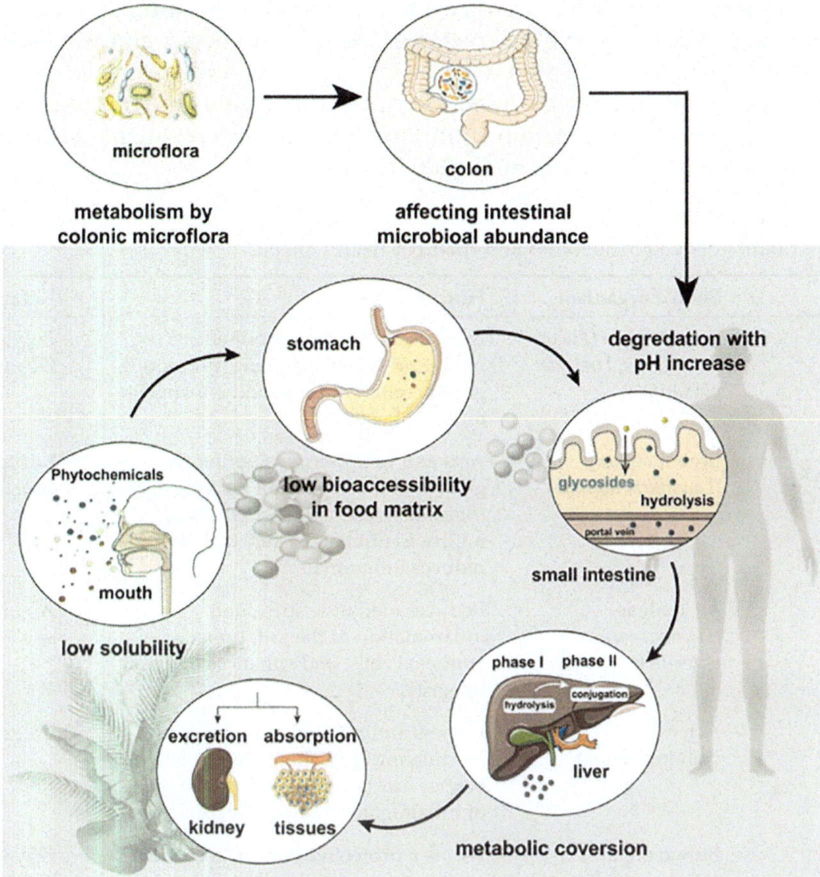

FIG. 17.3 Marine metabolites from algae, fish, and crustaceans affect gut health via a variety of metabolic pathways. Following consumption, these bioactive chemicals are digested, hydrolyzed, and metabolically converted, altering their bioavailability. Some metabolites are absorbed and used in tissues, while others enter the colon and alter gut microbiota, reduce inflammation, boost antioxidant defenses, and maintain intestinal barrier integrity. These processes aid in preventing gastrointestinal illnesses like IBD, IBS, and dysbiosis.

prevent the activation of critical inflammatory pathways, including the nuclear factor-kappa B (NF-κB) pathway, which is essential for the inflammatory response (Calder, 2015).

Marine metabolites help to prevent chronic gut inflammation by decreasing inflammatory mediators, encouraging tissue repair, and improving intestinal health. This antiinflammatory activity is especially useful in disorders like IBD, where uncontrolled inflammation causes tissue damage and reduces gut barrier function (Calder, 2015). The gut has one of the most extensive immune systems in humans, providing both physical (barrier function) and biochemical defenses against invading foodborne pathogens and other threats. The proinflammatory response works by first being activated as the main reaction to intruders, followed by the recruitment or activation of the amplification process (Evidente et al., 2003). As a natural response to invading antigens, cells develop into several subtypes of immune cells and migrate to the affected areas. This entire process steadily accumulates and strengthens throughout the first few days, at the expense of significant energy and nutrient intake. The GI epithelial cells, which are coated with a variety of ligand receptors, are continually exposed to harmful substances or infections that can easily and regularly activate the immune response. As a result, certain immunological disorders, such as IBD and autoimmune illnesses, may arise (Evidente et al., 2003) (Table 17.1).

TABLE 17.1 Marine-derived metabolites and their gut health effects.

Metabolite	Source organism	Findings	References
Fucoidan	Brown algae (*Fucus vesiculosus, Undaria pinnatifida*)	Exhibits antiinflammatory effects by inhibiting NF-κB signaling, promoting gut microbiota health, and enhancing mucosal healing.	Bulya et al. (2024), Oliyaei et al. (2025)
Laminarin	Brown algae (*Laminaria digitata*)	Acts as a prebiotic, promoting the growth of beneficial gut bacteria (*Bifidobacterium* and *Lactobacillus*); improves gut barrier function, and reduces inflammation.	Karuppusamy et al. (2022)
Astaxanthin	Microalgae (*Haematococcus pluvialis*)	Reduces oxidative stress and inflammation in the gut, protects epithelial cells, and enhances mucosal integrity.	Ambati et al. (2014)
Omega-3 Fatty Acids	Fish oil (Salmon, Tuna)	Possess antiinflammatory properties by modulating eicosanoids, improve gut barrier function, and reduce the severity of inflammatory bowel diseases.	Calder (2015)
Alginate	Brown algae (*Macrocystis pyrifera*)	Forms a protective barrier over the mucosa; promotes healing and reduces irritation in gastrointestinal conditions like mucositis.	Oliyaei et al. (2025)
Chitosan	Shellfish (Shrimp, Crab)	Exhibits antimicrobial activity against gut pathogens, promotes wound healing, and supports gut barrier function.	Yan et al. (2021)
Marine peptides	Fish (Cod, Tuna)	Demonstrate immunomodulatory effects by reducing proinflammatory cytokines; enhance intestinal epithelial repair.	Ahmad et al. (2008)
Carrageenan	Red algae (*Chondrus crispus*)	Potential prebiotic effects: supports the growth of beneficial gut microbiota and improves stool consistency.	Liu et al. (2015)
Squalene	Deep-sea shark liver oil	Possesses antioxidant and antiinflammatory properties; may support gut health by modulating immune responses and protecting against oxidative stress.	Gao et al. (2022)

TABLE 17.1 Marine-derived metabolites and their gut health effects—cont'd

Metabolite	Source organism	Findings	References
Eicosapentaenoic acid (EPA)	Marine algae (*Nannochloropsis*)	Exhibits antiinflammatory effects, modulates gut microbiota composition, and enhances the production of beneficial metabolites like short-chain fatty acids.	Tojjari et al. (2023)
Docosahexaenoic acid (DHA)	Marine algae (*Schizochytrium*)	Supports gut health by modulating inflammatory pathways and maintaining mucosal integrity; influences gut microbiota composition favorably.	Tojjari et al. (2023)
Phlorotannins	Brown algae (*Ecklonia cava*)	Exhibit antioxidant and antiinflammatory properties; may protect against gut inflammation and support overall gut health.	Kim et al. (2023)
Spirulina polysaccharides	Blue–green algae (*Spirulina platensis*)	Possess immunomodulatory and antioxidant properties; may enhance gut health by promoting beneficial microbiota and protecting against oxidative stress.	Pulito et al. (2020)

Antiinflammation, which prevents excessive inflammation, is required to maintain immunological homeostasis. Butyrate has been extensively investigated as an antiinflammatory agent that regulates cytokine synthesis, kinase activity, and immune-associated signaling pathways. Previously documented methods include the upregulation of immunosuppressive IL-10, G-protein-coupled receptors (GPRs), notably GPR41, and the suppression of several immune mediators, including toll-like receptor-engaged IL-12/23p40, nuclear factor (NF)-κB, and histone deacetylase (Siddiqui and Cresci, 2021). IBD, including ulcerative colitis and Crohn's disease, is a common chronic relapsing inflammatory disease characterized by a genetic antiinflammation deficit. It is thought that in IBD patients, the sensing capability of antiinflammatory mediators such as butyrate is reduced. However, supplementing with more butyrate-loaded dietary fibers may help to reduce the severity of these symptoms (Chen and Vitetta, 2020).

17.3.2 Antioxidant activity

In addition to their antiinflammatory characteristics, marine metabolites are strong antioxidants (Kim et al., 2018). The gut is constantly exposed to ROS, which are produced during normal metabolic processes and by environmental variables such as nutrition and pollution. These ROS have the potential to significantly harm intestinal cells, alter gut barrier integrity, and worsen inflammatory diseases (Chueh et al., 2014). Marine-derived antioxidants, such as astaxanthin, a carotenoid found in algae and shellfish, and polyphenols from marine plants, are effective free radical scavengers. These bioactive secondary metabolites prevent oxidative stress by neutralizing ROS, which helps the intestinal epithelium maintain its structural integrity (Vladkova et al., 2022). Astaxanthin, in particular, is known for its ability to prevent lipid peroxidation while also protecting cellular membranes from oxidative damage. Marine metabolites use antioxidant mechanisms to protect gut epithelial cells, maintain intestinal lining health, and prevent conditions such as "leaky gut," which occurs when the intestinal barrier is compromised, allowing harmful substances to enter the bloodstream (Vladkova et al., 2022).

17.3.3 Gut microbiome modulation

Marine metabolites have an impact on gut health by altering the gut flora. A diverse and balanced microbiota is essential for gut health because it regulates immune responses, produces beneficial metabolites such as SCFAs, and maintains gut barrier function (Gunathilaka and Kumarasinghe, 2025). Marine-derived bioactive chemicals, such as fucoidans from brown algae, have been proven to boost the growth of beneficial gut bacteria like *Bifidobacterium* and *Lactobacillus* while inhibiting harmful species (Shang et al., 2016). This specific microbiota modification contributes to a more favorable microbial balance, which can lower gut inflammation and increase the synthesis of SCFAs, particularly butyrate, which acts as an energy source for colonocytes and helps to maintain the intestinal barrier (Nogal et al., 2021a, 2021b). Marine metabolites promote a healthy microbiota, which helps to maintain gut homeostasis and prevent dysbiosis, a microbial imbalance linked to a variety of GI illnesses.

17.3.4 Gut barrier function

Marine metabolites improve gut barrier function, which is an important part of maintaining intestinal health. The gut barrier protects the bloodstream from dangerous bacteria, poisons, and other substances while allowing necessary nutrients to be absorbed (Gunathilaka and Kumarasinghe, 2025). A damaged gut barrier, often known as a "leaky gut," can cause systemic inflammation and immunological dysfunction. Marine bioactives, particularly omega-3 fatty acids, play an important role in improving intestinal barrier integrity by influencing the production of tight junction proteins, which are responsible for the gut lining's selective permeability (Calder, 2015). These tight junctions are crucial for preventing hazardous chemicals from entering the bloodstream. Furthermore, marine metabolites enhance mucus production, which contributes to the intestine's protective barrier. This improved gut barrier function not only defends against hazardous substances but also helps to maintain immunological tolerance, thereby reducing the development of chronic inflammatory disorders (Li et al., 2008).

Dietary fibers and SCFAs are associated with epithelial barrier function. However, butyrate has a bidirectional influence on epithelial barrier function (Nogal et al., 2021a, 2021b). According to previous studies, diarrhea and antibiotic-induced dysbiosis drastically reduce the number of commensal bacteria responsible for butyrate synthesis. Decreased commensal bacteria result in a lack of accessible butyrate as an energy source for colonocytes, interrupting enterocyte refreshment (O'Riordan et al., 2022). In humans, supplementing with particular fibers can enrich butyrate-producing bacteria in the stomach; when butyrate production increases, the diarrhea state changes and improves. Butyrate has also been shown to increase barrier function by nourishing TJs (Tight-Junctions) such as beta-defensin, cingulin, ZO-1, and ZO-2 proteins in chicken HD11 macrophage cells, primary monocytes, bone marrow cells, and jejunal and cecal explants (Beane et al., 2021). Another study employing the in vitro Caco-2 human epithelial colorectal adenocarcinoma cell line demonstrated that butyrate might compensate for *Campylobacter jejuni* barrier damage (Ney et al., 2023).

Furthermore, butyrate-induced intracellular signaling pathways associated with barrier function have been investigated. For example, it has been claimed that butyrate improves TJs by activating the Akt/mTOR pathways and replenishing ATP (Yan and Ajuwon, 2017). Another study found that dietary butyrate improves barrier function by inhibiting IL-10 receptor-dependent claudin-2 (Beane et al., 2021). The IL-10 receptor consists of two ligand-binding subunits: the alpha subunit (IL-10RA) and the beta subunit (IL-10RB). When IL-10 binds to IL10R, it activates the JAK-STAT signaling pathway, with STAT3 activation being important for antiinflammatory action. Butyrate was discovered to increase epithelial barrier function by IL-10RA suppression of claudin-2, which regulates paracellular channels for small cations and water, inducing diarrhea via the leak-flux mechanism (Yan and Ajuwon, 2017)

17.3.5 Immune system modulation

The gut-associated lymphoid tissue (GALT), which is responsible for maintaining immunological homeostasis in the GI tract, also plays an important role in the action of marine compounds. The GALT is inextricably linked to the gut microbiota, and its proper function is critical for preventing both hyperactive immune responses, which can lead to inflammatory disorders, and underactive responses, which can increase vulnerability to infection (Bemark et al., 2024). Marine metabolites, particularly omega-3 fatty acids, have been shown to influence the immune system by increasing the development of regulatory T cells (Tregs), which are essential for dampening overactive immunological responses and avoiding autoimmunity (Shukla et al., 2019). Furthermore, marine bioactives can help reduce proinflammatory cytokine release and improve immune cell function, including macrophages and dendritic cells, which play a role in immune response initiation. Marine metabolites modulate immunological responses in this way to help maintain immune tolerance and prevent the development of inflammatory diseases in the gut (Ho et al., 2023).

The gut–brain axis, which connects the GI and central neurological systems, is another important avenue through which marine metabolites influence overall health. Immune cells in the stomach produce cytokines and neurotransmitters, which can have a significant impact on brain function, mood, and cognitive ability. Marine-derived bioactive chemicals, such as omega-3 fatty acids, regulate neurotransmitter synthesis, including serotonin, which is important for mood and behavior regulation. Marine metabolites can improve mental health and well-being by altering the gut-brain axis, offering effects that extend beyond the GI tract (Chaudhry et al., 2023).

Marine metabolites' combined effects on gut inflammation, oxidative stress, microbial balance, immunological modulation, and gut barrier integrity make them effective agents for enhancing gut health and preventing disease. Marine-derived bioactives are prospective treatment possibilities for a variety of GI illnesses, including IBD, IBS, and dysbiosis, due to their ability to lower chronic inflammation, counteract oxidative damage, increase microbial diversity, and improve the intestinal barrier (Siddiqui and Cresci, 2021). Furthermore, their extensive health advantages apply to other systemic illnesses, such as metabolic diseases, cardiovascular health, and neurodegenerative disorders, providing a comprehensive approach to enhancing health through natural, marine-based therapies (Siddiqui and Cresci, 2021).

17.4 Impact of marine metabolites on specific GI conditions and their therapeutic applications

17.4.1 Celiac disease

Celiac disease (CD) is an autoimmune condition caused by the intake of gluten, a protein present in wheat, barley, and rye, in genetically predisposed individuals. The disorder causes inflammation, villous atrophy, and malabsorption in the small intestine, leading to symptoms such as diarrhea, bloating, fatigue, and nutritional deficits (Caio et al., 2019a, 2019b). Currently, the primary treatment is a strict gluten-free diet for the rest of one's life. However, adhering to such a diet is difficult, and accidental gluten consumption remains a possibility (Caio et al., 2019a, 2019b). Thus other treatment techniques are being investigated, including the use of marine-derived metabolites (Caio et al., 2019a, 2019b).

Marine creatures like algae, sponges, and fungi produce bioactive metabolites with distinct chemical structures and methods of action, which have enormous potential for treating a variety of ailments, including CD (Jiménez-González et al., 2023). Marine polysaccharides, peptides, and secondary metabolites have antiinflammatory, antioxidant, and immunomodulatory effects that may help with the pathophysiology of CD. Fucoidans, sulfated polysaccharides from brown seaweeds, have been shown to reduce inflammation by blocking critical pathways like NF-κB (Sanjeewa et al., 2019). These pathways are essential in the inflammatory cascade seen in CD. Fucoidans could help to reduce intestinal inflammation and restore gut integrity (Elbandy, 2023).

Another potential chemical is laminarin, a polymer found in brown algae and renowned for its prebiotic effects. Laminarin stimulates the growth of good gut bacteria like *Bifidobacterium* and *Lactobacillus*, which may improve gut barrier function and reduce gluten-induced inflammation (Gunathilaka and Kumarasinghe, 2025). Bioactive peptides from marine sources can reduce proinflammatory cytokines like TNF-α and IL-6, which are often high in celiac patients (Karuppusamy et al., 2022). Antioxidants produced from marine sources, such as astaxanthin found in microalgae, may also have therapeutic properties. Astaxanthin decreases oxidative stress in the stomach, which contributes to epithelium damage in CD (Barkia et al., 2019). These substances, which protect intestinal cells and modulate immunological responses, provide a supplement to standard dietary control (Ambati et al., 2014).

While marine metabolites show promise in controlling CD, further research is needed to understand their precise processes, efficacy, and safety in clinical settings. Advances in biotechnology and marine pharmacology may allow for the creation of novel nutraceuticals and therapies that supplement gluten-free diets and improve the quality of life for celiac patients.

17.4.2 Inflammatory bowel disease

IBD, also known as Crohn's disease and ulcerative colitis, is characterized by chronic inflammation of the GI tract, which is frequently associated with gut homeostasis abnormalities, immunological dysregulation, and intestinal microbiota changes (Lee et al., 2018). Marine-derived bioactive chemicals, such as lipopolysaccharides and sulfated polysaccharides, have demonstrated promising therapeutic effects in IBD models by reducing inflammation and restoring gut health. Studies using in vitro, in vivo, and clinical techniques emphasize the potential of these marine metabolites in treating IBD due to their multiple modes of action (Repici et al., 2024).

Repici et al. (2024) found that a lipopolysaccharide extract from the green algae *Ulva pertusa* had antiinflammatory effects in a mouse model of IBD induced by dextran sulfate sodium (DSS). The extract lowered proinflammatory factors in blood and colon tissues, including NF-κB, MAPK, IL-4, IL-6, and IL-10, after oral dosing. Furthermore, *Ulva pertusa* extract had a protective impact on colon tissue architecture, preserving tight junction proteins such as occludin, claudin-1, and mucin, which are necessary for intestinal barrier integrity. The treatment also restored the disturbed gut microbiota induced by DSS, demonstrating the extract's potential to modulate both immunological and microbial homeostasis, which are critical elements in IBD management (Repici et al., 2024).

Polysaccharides produced from diverse marine algae have also shown antiinflammatory properties (Asanka Sanjeewa et al., 2021). Polysaccharides from the red algae *Digenea simplex* were found to possess inhibitory properties against inflammation caused by histamine, 5-hydroxytryptamine, and bradykinin, as well as carrageenan-induced edema (Pereira et al., 2014). These polysaccharides reduced neutrophil infiltration into inflamed tissues, resulting in lower amounts of TNF-α and IL-1β in peritoneal fluid. Such findings demonstrate their efficacy in targeting proinflammatory pathways and, hence, reducing inflammation in IBD (Batista et al., 2020). Similarly, sulfated polysaccharides isolated from the green alga *Caulerpa mexicana* (Cm-SPs) showed antiinflammatory and analgesic activity. Cm-SPs prevented histamine-induced paw edema in mice, indicating that histamine is a target for their antiswelling activities. Furthermore, Cm-SPs inhibited carrageenan-induced edema and myeloperoxidase activity, demonstrating their wide-ranging antiinflammatory pathways (Carneiro et al., 2014).

The gastroprotective benefits of marine sulfated polysaccharides go beyond inflammation to include oxidative stress and prostaglandin pathways. In a study on ethanol-induced stomach injury, Cm-SPs were found to dramatically lower oxidative stress indicators and protect gastric tissue, implying potential applications for IBD-related gastric problems. These findings highlight the versatility of marine polysaccharides in targeting a variety of inflammatory pathways, making them interesting options for IBD treatment (Repici et al., 2024).

Fucoidan, a long-chain sulfated polysaccharide produced from brown algae, has also garnered interest for its potential therapeutic uses in IBD. In a DSS-induced mouse model of colitis, fucoidan was administered orally and intraperitoneally, which reduced weight loss, diarrhea, and blood in the stool while preserving colon tissue architecture and lowering inflammation markers (Lean et al., 2015). Notably, oral treatment was more effective, most likely due to the limited absorption and tissue distribution of the large fucoidan molecule, which ranges in size from 5 to 1000 kDa. This research underscores the value of customized delivery systems in improving treatment outcomes. Fucoidan's efficacy was further demonstrated in a study by Lean et al., who found that a fucoidan-polyphenol complex (Maritech Synergy) and depyrogenated fucoidan reduced DSS-induced acute colitis. These formulations improved disease-related pathology, highlighting fucoidan's potential to act on numerous therapeutic targets concurrently (Bulya et al., 2024).

Given the complexity of IBD pathophysiology, which includes several inflammatory mediators, cytokine networks, and disease sites, marine-derived bioactives offer a distinct advantage due to their broad-spectrum activity. Compounds such as fucoidan and algal polysaccharides inhibit several proinflammatory pathways, alter gut microbiota composition, and restore intestinal barrier integrity. Their natural origin and minimal toxicity add to their promise as sustainable alternatives or supplemental therapies to traditional IBD treatments.

17.4.3 Colon inflammation and cancer

Colon inflammation and cancer are serious health concerns that are frequently linked to chronic inflammatory states and dysregulated cellular communication (Porter et al., 2021). Marine-derived bioactive compounds, such as peptides, polysaccharides, lipids, and secondary metabolites, have shown remarkable promise in reducing colon inflammation and combating CRC via a variety of mechanisms, including signaling pathway modulation, apoptosis induction, and immune response regulation (Chapkin et al., 2007; Sanjeewa et al., 2017).

Curacin-A, a peptide produced from the cyanobacterium *Lyngbya majuscula,* has demonstrated powerful antiproliferative effects on tumor cell lines, including kidney, colon, and breast cancer. Elisidepsin (PM02734), a peptide isolated from the mollusk *Elysia rufescens*, is being evaluated for the treatment of advanced solid tumors and metastatic gastroesophageal cancer (Gehret et al., 2011). Elisidepsin inhibits the AKT/mTOR signaling pathway and activates death-associated protein kinases, affecting cancer cells' survival strategies. These findings demonstrate the therapeutic potential of marine peptides in targeting specific signaling pathways involved in tumor growth and metastasis (Ahmed et al., 2021).

Fucoidan has been shown to promote apoptosis in patients with metastatic CRC by inhibiting the p38 MAPK and PI3K/AKT signaling pathways and activating the ERK 1/2 MAPK pathway. These dual actions allow for the selective killing of cancer cells while protecting healthy tissues (Jin et al., 2021). Furthermore, the efficacy of fucoidan in CRC treatment is increased when delivered orally, as this route ensures targeted action in the GI system, overcoming the limitations posed by its massive molecular structure (Jin et al., 2021).

Lipids, such as omega-3 fatty acids found in krill oil and fish oil, have also shown promise in the treatment of CRC. Bioactive fatty acids, such as EPA and DHA, have been demonstrated to suppress cell proliferation and increase apoptosis in CRC cell lines HCT-15, SW-480, and Caco-2 (Tojjari et al., 2023). These effects are mediated by mitochondria-dependent apoptotic pathways, which change mitochondrial membrane potential and cause programmed cell death. The use of these fatty acids in dietary therapies for colon cancer highlights their potential as natural chemopreventive agents (Tojjari et al., 2023).

Secondary metabolites from marine sources, such as quinones and hydroquinones, provide an additional layer of anticancer activity. For example, 2-amino-6-hydroxy-[1,4]-benzoquinone accumulates ROS, which induce apoptosis in a variety of cancer types, including gastric and colorectal tumors (Veríssimo et al., 2021). Compounds such as GS03 and the bishydroquinone renieramycin M affect mitochondrial apoptotic pathways, increasing the therapeutic efficacy against CRC and gastric cancer by inducing cell death. Similarly, bioactive extracts from marine species such as *Jaspis stellifera*, *Axinyssa* sp., and *Cystoseira usneoides* have been demonstrated to slow the spread of colon cancer by modifying critical pathways such as the AKT signaling pathway (Pinkhien et al., 2016).

Saponins derived from both terrestrial and marine sources have antiinflammatory and anticancer properties (Manoharan and Perumal, 2022). α-Hederin from *Hedera helix* L. targets CRC through ROS-mediated pathways, while *Panax ginseng* saponins promote AMPK pathways to limit tumor growth (Sun et al., 2019). These findings highlight the potential synergistic relationship between marine and plant-derived compounds in the development of integrated cancer treatments (Elekofehinti et al., 2021). The link between chronic inflammation and cancer development is well established, with inflammatory mediators playing an important role in tumor initiation and progression (Fernandes et al., 2024). Marine-derived bioactives target inflammation-associated signaling pathways such as PI3K/AKT, MAPK, and Wnt/β-catenin (Pulito et al., 2020), offering a multifaceted approach to treating colon inflammation and CRC. These chemicals not only lower inflammation, but also boost immune responses, stimulate apoptosis, and restore intestinal homeostasis (Zhao et al., 2021).

17.4.4 Mucositis

Mucositis is a devastating side effect of cancer treatments, such as chemotherapy and radiation, that is characterized by inflammation and ulceration of the GI tract's mucosa (Pulito et al., 2020). Symptoms include pain, trouble eating, and a higher risk of infection, all of which have a substantial impact on patients' quality of life (Calder, 2015). Current therapies mostly focus on symptom management, such as oral rinses, analgesics, and nutritional assistance, but there is no universal cure. Marine-derived metabolites have recently emerged as interesting options for mucositis treatment due to their significant antiinflammatory, antioxidant, and regenerative characteristics (Pulito et al., 2020).

Moreover, fucoidans suppress proinflammatory cytokines such as TNF-α and IL-6, which are increased in mucositis (Flórez-Fernández et al., 2023). Furthermore, these chemicals promote epithelial healing and reduce oxidative stress, both of which are essential for preserving mucosal integrity. Fucoidans have been shown in studies to expedite wound healing in the mucosa, making them a prospective treatment for mucositis (Flórez-Fernández et al., 2023).

Another type of bioactive chemical with antiinflammatory characteristics is marine-derived omega-3 fatty acids, which include EPA and DHA (Calder, 2013). These fatty acids regulate the generation of lipid mediators, such as prostaglandins and leukotrienes, lowering the intensity of inflammation in the GI tract. Clinical trials have shown that omega-3 supplementation can reduce the frequency and severity of chemotherapy-induced mucositis (Calder, 2015).

Astaxanthin, a carotenoid derived from marine microalgae, is an effective antioxidant that protects mucosal cells from oxidative damage caused by ROS produced during cancer treatment. Astaxanthin can reduce oxidative stress and promote faster healing of the mucosal barrier (Ambati et al., 2014). Similarly, bioactive peptides originating from marine species such as fish and shellfish have immunomodulatory properties, increasing the generation of antiinflammatory cytokines and decreasing epithelial damage (Oliyaei et al., 2025).

Another intriguing marine chemical is alginate, a polymer derived from brown algae that acts as a protective barrier on the mucosal surface, reducing irritation and encouraging healing. Its mucoadhesive characteristics can also assist in preserving moisture and preventing subsequent infections, resulting in symptom relief for mucositis patients (Oliyaei et al., 2025).

Despite promising preclinical and early clinical evidence, the use of marine metabolites in mucositis treatment requires additional validation in large-scale trials. Advances in marine biotechnology and nutraceutical research may result in the development of novel, effective medications to supplement traditional treatments and enhance outcomes for mucositis patients.

17.5 Concluding remarks and future directions

The marine environment, which makes up >70% of the planet's surface, is still a relatively unexplored but extremely promising source of bioactive substances that could greatly improve gut health. Polysaccharides, polyphenols, omega-3 fatty acids, peptides, and other secondary metabolites originating from marine sources have shown promise in modifying the composition of the gut microbiota, enhancing the function of the intestinal barrier, controlling immunological responses, and reducing inflammation. Marine bioactives are excellent candidates for creating innovative functional foods, nutraceuticals, and medicinal agents targeted at enhancing GI health because of these advantageous qualities.

There are still a number of obstacles to overcome, despite the mounting evidence of marine bioactives' positive effects on gut health. The incomplete knowledge of the precise molecular mechanisms driving their interactions with host physiology and gut microbiota is one of the main research gaps. Although many substances have been demonstrated to increase the number of good bacteria or inhibit harmful strains, further research is required to clarify how they directly affect microbial metabolism, signaling cascades, and host immunological responses. In addition, as factors, including digestion, absorption, and structural changes during metabolism, may affect the efficacy of marine-derived chemicals in the gut, more research is needed to determine their bioavailability, stability, and metabolic fate.

The conversion of preclinical results into clinical applications is another crucial area that needs focus. Despite the fact that a large number of in vitro and in vivo investigations have demonstrated the therapeutic potential of marine bioactives, carefully planned human clinical trials are required to confirm their effectiveness and safety in practical settings. Furthermore, genetic variables and individual differences in gut microbiota composition may affect how people react to marine-derived bioactives, requiring tailored strategies for future studies and product development.

The sustainable extraction, purification, and large-scale manufacturing of marine bioactives present additional challenges from an industrial standpoint. Since overharvesting marine resources can have negative environmental effects, sustainable sourcing practices, including aquaculture, synthetic biology, and marine biotechnology, are essential. The yield, purity, and utility of these molecules could be enhanced by advancements in bioprocessing technologies, such as fermentation and enzymatic alterations, while maintaining environmentally friendly production practices.

To thoroughly evaluate the mechanisms, effectiveness, and safety of marine-derived bioactives, future research should concentrate on an integrated approach that blends in vitro, in vivo, and in silico studies. To close the gap between fundamental research and real-world applications, multidisciplinary partnerships between marine biologists, microbiologists, dietitians, and pharmaceutical specialists will be essential. Marine bioactives have the potential to transform gut health therapies and ultimately improve human well-being if marine biotechnology and gut microbiome studies continue to progress.

References

Ahmad, I., Zahin, M., Aqil, F., Hasan, S., Khan, M.S.A., Owais, M., 2008. Bioactive compounds from *Punica granatum, Curcuma longa* and Zingiber officinale and their therapeutic potential. Drugs Future. 0377-8282. 33 (4), 329–346. doi:10.1358/dof.2008.033.04.1186159.

Ahmed, S., Mirzaei, H., Aschner, M., Khan, A., Al-Harrasi, A., Khan, H., 2021. Marine peptides in breast cancer: therapeutic and mechanistic understanding. Biomed. Pharmacother. 1950-6007. 142. Available from: https://www.journals.elsevier.com/biomedicine-and-pharmacotherapy. doi:10.1016/j.biopha.2021.112038.

Ambati, R.R., Moi, P.S., Ravi, S., Aswathanarayana, R.G., 2014. Astaxanthin: sources, extraction, stability, biological activities and its commercial applications—A review. Mar. Drugs. 1660-3397. 12 (1), 128–152. Available from: http://www.mdpi.com/1660-3397/12/1/128/pdf. doi:10.3390/md12010128.

Anwer, E.K.E., Ajagbe, M., Sherif, M., Musaibah, A.S., Mahmoud, S., ElBanbi, A., Abdelnaser, A., 2025. Gut Microbiota secondary metabolites: key roles in GI tract cancers and infectious diseases. Biomedicines 13 (1). doi:10.3390/biomedicines13010100.

Asanka Sanjeewa, K.K., Herath, K.H.I.N.M., Yang, H.-W., Choi, C.S., Jeon, Y.-J., 2021. Anti-inflammatory mechanisms of fucoidans to treat inflammatory diseases: a review. Mar. Drugs 19 (12). doi:10.3390/md19120678.

Barkia, I., Saari, N., Manning, S.R., 2019. Microalgae for high-value products towards human health and nutrition. Mar. Drugs 17. doi:10.3390/md17050304

Batista, J.A., Magalhães, D.D.A., Sousa, S.G., Ferreira, J.D.S., Pereira, C.M.C., Lima, J.V.D.N., de Albuquerque, I.F., Bezerra, N.L.S.D., de Brito, T.V., Monteiro, C.E.D.S., Franco, A.X., Di Lenardo, D., Oliveira, L.A., Feitosa, J.P.D.A., de Paula, R.C.M., Barros, F.C.N., Freitas, A.L.P., de Oliveira, J.S., Vasconcelos, D.F.P., Soares, P.M.G., Barbosa, A. L.D.R., 2020. Polysaccharides derived from *Morinda citrifolia* Linn reduce inflammatory markers during experimental colitis. J. Ethnopharmacol. 1872-7573. 248. Available from: www.elsevier.com/locate/jethpharm.. doi:10.1016/j.jep.2019.112303.

Beane, K.E., Redding, M.C., Wang, X., Pan, J.H., Le, B., Cicalo, C., Jeon, S., Kim, Y.J., Lee, J.H., Shin, E.C., Li, Y., Zhao, J., Kim, J.K., 2021. Effects of dietary fibers, micronutrients, and phytonutrients on gut microbiome: a review. Appl. Biol. Chem. doi:10.1186/s13765-021-00605-6.

Bemark, M., Pitcher, M.J., Dionisi, C., Spencer, J., 2024. Gut-associated lymphoid tissue: a microbiota-driven hub of B cell immunity. Trends Immunol. 45 (3), 211–223. doi:10.1016/j.it.2024.01.006.

Bischoff, S.C., 2011. 'Gut health': a new objective in medicine? BMC Med. 1741-7015. 9. Available from: http://www.biomedcentral.com/1741-7015/9/24. doi:10.1186/1741-7015-9-24.

Bulya, E.T., Alqubelat, R.S., Rao, A.R., 2024. Nutritional Benefits of Fucoidan and Its Applications (A Review). Unchanged. 1608-330X. 50 (6), 2107–2133. Available from: https://www.springer.com/journal/11171. doi:10.1134/S1068162024060335.

Caio, G., Volta, U., Sapone, A., Leffler, D.A., De Giorgio, R., Catassi, C., Fasano, A., 2019a. Celiac disease: a comprehensive current review. BMC Med. 1741-7015. 17 (1). Available from: http://www.biomedcentral.com/bmcmed/. doi:10.1186/s12916-019-1380-z.

Caio, G., Volta, U., Sapone, A., Leffler, D.A., De Giorgio, R., Catassi, C., 2019b. Celiac disease: a comprehensive current review. BMC Med. 17 (1). doi:10.1186/s12916-019-1380-z.

Calder, P.C., 2013. Omega-3 polyunsaturated fatty acids and inflammatory processes: nutrition or pharmacology? Br. J. Clin. Pharmacol. 75 (3), 645–662. doi:10.1111/j.1365-2125.2012.04374.x.

Calder, P.C., 2015. Marine omega-3 fatty acids and inflammatory processes: effects, mechanisms and clinical relevance. Biochim. Biophys. Acta Mol. Cell Biol. Lipids. 1879-2618. 1851 (4), 469–484. Available from: www.elsevier.com/locate/bbalip.. doi:10.1016/j.bbalip.2014.08.010.

Carneiro, J.G., Rodrigues, J.A.G., Vanderlei, E.D.S.O., Souza, R.B., Quinderé, A.L.G., Coura, C.O., de Araújo, I.W.F., Chaves, H.V., Bezerra, M.M., Benevides, N.M.B., 2014. Peripheral antinociception and anti-inflammatory effects of sulphated polysaccharides from the alga Caulerpa mexicana. Basic Clin. Pharmacol. Toxicol. 1742-7843. 115 (4), 335–342. Available from: http://www.wiley.com/bw/journal.asp?ref=1742-7835. doi:10.1111/bcpt.12234.

Chapkin, R.S., McMurray, D.N., Lupton, J.R., 2007. Colon cancer, fatty acids and anti-inflammatory compounds. Curr. Opin. Gastroenterol. 0267-1379. 23 (1), 48–54. doi:10.1097/MOG.0b013e32801145d7.

Chaudhry, T.S., Senapati, S.G., Gadam, S., Mannam, H.P.S.S., Voruganti, H.V., Abbasi, Z., Abhinav, T., Challa, A.B., Pallipamu, N., Bheemisetty, N., Arunachalam, S.P., 2023. The impact of microbiota on the gut–brain axis: examining the complex interplay and implications. J. Clin. Med. 2077-0383. 12 (16). Available from: https://www.mdpi.com/journal/jcm. doi:10.3390/jcm12165231.

Chen, J., Vitetta, L., 2020. Butyrate in inflammatory bowel disease therapy. Gastroenterology. 0016-5085. 158 (5), 1511. doi:10.1053/j.gastro.2019.08.064.

Chueh, P.J., Liang, R.Y., Lee, Y.H., Zeng, Z.M., Chuang, S.M., 2014. Differential cytotoxic effects of gold nanoparticles in different mammalian cell lines. J. Hazard. Mater. 1873-3336. 264, 303–312. Available from: www.elsevier.com/locate/jhazmat.. doi:10.1016/j.jhazmat.2013.11.031.

Cudennec, B., Ravallec-Plé, R., Courois, E., Fouchereau-Peron, M., 2008. Peptides from fish and crustacean by-products hydrolysates stimulate cholecystokinin release in STC-1 cells. Food Chem. 0308-8146. 111 (4), 970–975. doi:10.1016/j.foodchem.2008.05.016.

Elbandy, M., 2023. Anti-Inflammatory effects of marine bioactive compounds and their potential as functional Food ingredients in the prevention and treatment of neuroinflammatory disorders. Molecules. 1420-3049. 28 (1), 2. doi:10.3390/molecules28010002.

Elekofehinti, O.O., Iwaloye, O., Olawale, F., Ariyo, E.O., 2021. Saponins in cancer treatment: current progress and future prospects. Pathophysiology. 1873-149X. 28 (2), 250–272. Available from: https://www.mdpi.com/1873-149X/28/2/17/pdf. doi:10.3390/pathophysiology28020017.

Evidente, V.G.H., Caviness, J.N., Adler, C.H., 2003. Case studies in movement disorders. Semin. Neurol. 23 (3), 277–284. doi:10.1055/s-2003-814739.

Fernandes, Q., Inchakalody, V.P., Bedhiafi, T., Mestiri, S., Taib, N., Uddin, S., Merhi, M., Dermime, S., 2024. Chronic inflammation and cancer; the two sides of a coin. Life Sci. 338. doi:10.1016/j.lfs.2023.122390

Fernando, I.P.S., Jayawardena, T.U., Wu, J., 2023. Marine proteins and peptides: production, biological activities, and potential applications. Food Innov. Adv. 2836-774X. 2 (2), 69–84. Available from: https://www.maxapress.com/article/doi/10.48130/FIA-2023-0009. doi:10.48130/FIA-2023-0009.

Flórez-Fernández, N., Vaamonde-García, C., Torres, M.D., Buján, M., Muíños, A., Muiños, A., Lamas-Vázquez, M. J., Meijide-Faílde, R., Blanco, F.J., Domínguez, H., 2023. Relevance of the extraction stage on the anti-inflammatory action of fucoidans. Pharmaceutics. 1999-4923. 15 (3). Available from: http://www.mdpi.com/journal/pharmaceutics. doi:10.3390/pharmaceutics15030808.

Gao, Y., Ma, X., Zhou, Y., Li, Y., Xiang, D., 2022. Dietary supplementation of squalene increases the growth performance of early-weaned piglets by improving gut microbiota, intestinal barrier, and blood antioxidant capacity. Front. Vet. Sci. 2297-1769. 9. doi:10.3389/fvets.2022.995548.

Gehret, J.J., Gu, L., Gerwick, W.H., Wipf, P., Sherman, D.H., Smith, J.L., 2011. Terminal alkene formation by the thioesterase of curacin A biosynthesis: structure of a decarboxylating thioesterase. J. Biol. Chem. 1083-351X. 286 (16), 14445–14454. Available from: http://www.jbc.org/content/286/16/14445.full.pdf+html. doi:10.1074/jbc.M110.214635.

Ghallab, D.S., Ibrahim, R.S., Mohyeldin, M.M., Shawky, E., 2024. Marine algae: a treasure trove of bioactive anti-inflammatory compounds. Mar. Pollut. Bull. 1879-3363. 199. Available from: https://www.sciencedirect.com/science/journal/0025326X. doi:10.1016/j.marpolbul.2023.116023.

Gunathilaka, T.L., Kumarasinghe, H.S., 2025. Seaweeds as silent healers: insight into the pharmacological potential of marine algal metabolites., Gunathilaka, T.L., Kumarasinghe, H.S.. Cambridge Scholars Publishing, Newcastle upon Tyne, NE6 2PA, UK. doi:10.13140/RG.2.2.17805.86245.

Hamed, I., Özogul, F., Özogul, Y., Regenstein, J.M., 2015. marine bioactive compounds and their health benefits: a review. Compr. Rev. Food Sci. Food Saf. 1541-4337. 14 (4), 446–465. Available from: http://onlinelibrary.wiley.com/journal/10.1111/(ISSN)1541-4337. doi:10.1111/1541-4337.12136.

Ho, T.L., Lee, J., Ahn, S.Y., Lee, D.H., Song, W.J., Kang, I., Ko, E.J., 2023. Immunostimulatory effects of marine algae extracts on in vitro antigen-presenting cell activation and in vivo immune cell recruitment. Food Sci. Nutr. 2048-7177. 11 (10), 6560–6570. Available from: onlinelibrary.wiley.com/journal/10.1002/%28ISSN%292048-7177. doi:10.1002/fsn3.3605.

Jabłońska, B., Mrowiec, S., 2023. Gastrointestinal disease: new diagnostic and therapeutic approaches. Biomedicines. 2227-9059. 11 (5). Available from: www.mdpi.com/journal/biomedicines. doi:10.3390/biomedicines11051420.

Jiménez-González, C., Agrasar, A.M.T., Mallo, F., Rúa, M.L., Fuciños, C., 2023. Red seaweed proteins: valuable marine-origin compounds with encouraging applications. Algal Res. 2211-9264. 75. Available from: http://www.sciencedirect.com/science/journal/aip/22119264. doi:10.1016/j.algal.2023.103262.

Jin, J.O., Chauhan, P.S., Arukha, A.P., Chavda, V., Dubey, A., Yadav, D., 2021. The therapeutic potential of the anticancer activity of fucoidan: current advances and hurdles. Mar. Drugs. 1660-3397. 19 (5). Available from: https://www.mdpi.com/1660-3397/19/5/265/pdf. doi:10.3390/md19050265.

Karuppusamy, S., Rajauria, G., Fitzpatrick, S., Lyons, H., McMahon, H., Curtin, J., Tiwari, B.K., O'Donnell, C., 2022. Biological properties and health-promoting functions of Laminarin: a comprehensive review of preclinical and clinical studies. Mar. Drugs. 1660-3397. 20 (12). Available from: http://www.mdpi.com/journal/marinedrugs. doi:10.3390/md20120772.

Kim, H.-S., Lee, W., Lee, J.-H., Asanka Sanjeewa, K.K., Fernando, I.P.S., Ko, S.-C., Lee, S.-H., Kim, Y.-T., Jeon, Y.-J., 2018. Purification and Identification of an antioxidative peptide from digestive enzyme hydrolysis of Cutlassfish muscle. J. Aquat. Food Prod. Technol. 27 (8), 934–944. doi:10.1080/10498850.2018.1508532.

Kim, S.K., Wijesekara, I., 2013. Marine-derived peptides: development and health prospects. In: Marine Proteins and Peptides: Biological Activities and Applications. John Wiley and Sons, South Korea, pp. 1–3. Available from: http://onlinelibrary.wiley.com/book/10.1002/9781118375082. doi:10.1002/9781118375082.ch1.

Kim, Y.M., Kim, H.Y., Jang, J.T., Hong, S., 2023. Preventive effect of Ecklonia cava extract on DSS-induced Colitis by elevating intestinal barrier function and improving pathogenic inflammation. Molecules. 1420-3049. 28 (24). Available from: http://www.mdpi.com/journal/molecules. doi:10.3390/molecules28248099.

Ko, S.C., Ding, Y., Kim, J., Ye, B.R., Kim, E.A., Jung, W.K., Heo, S.J., Lee, S.H., 2019. Bromophenol (5-bromo-3,4-dihydroxybenzaldehyde) isolated from red alga *Polysiphonia morrowii* inhibits adipogenesis by regulating expression of adipogenic transcription factors and AMP-activated protein kinase activation in 3T3-L1 adipocytes. Phytother. Res. 1099-1573. 33 (3), 737–744. Available from: http://onlinelibrary.wiley.com/journal/10.1002/(ISSN)1099-1573. doi:10.1002/ptr.6266.

Lean, Q.Y., Eri, R.D., Fitton, J.H., Patel, R.P., Gueven, N., 2015. Fucoidan extracts ameliorate acute colitis. PLoS One. 1932-6203. 10 (6). Available from: http://www.plosone.org/article/fetchObject.action?uri=info:doi/10.1371/journal.pone.0128453&representation=PDF. doi:10.1371/journal.pone.0128453.

Lee, S.H., Kwon, J.E., Cho, M.L., 2018. Immunological pathogenesis of inflammatory bowel disease. Intest. Res. 2288-1956. 16 (1), 26–42. Available from: http://www.irjournal.org/journal/download_pdf.php?spage=26&volume=16&number=1. doi:10.5217/ir.2018.16.1.26.

Li, L., Huang, J., Feng, L., Xu, L., Lin, H., Liu, K., Li, X., Wang, R., 2024. Altechromone A ameliorates inflammatory bowel disease by inhibiting NF-κB and NLRP3 pathways. Mar. Drugs. 1660-3397. 22 (9), 410. doi:10.3390/md22090410.

Li, Q., Zhang, Q., Zhang, M., Wang, C., Zhu, Z., Li, N., Li, J., 2008. Effect of n-3 polyunsaturated fatty acids on membrane microdomain localization of tight junction proteins in experimental colitis. FEBS J. 1742-464X. 275 (3), 411–420. doi:10.1111/j.1742-4658.2007.06210.x.

Liu, J., Kandasamy, S., Zhang, J., Kirby, C.W., Karakach, T., Hafting, J., Critchley, A.T., Evans, F., Prithiviraj, B., 2015. Prebiotic effects of diet supplemented with the cultivated red seaweed *Chondrus crispus* or with fructo-oligosaccharide on host immunity, colonic microbiota and gut microbial metabolites. BMC Complement. Altern. Med. 1472-6882. 15 (1). Available from: http://www.biomedcentral.com/bmccomplementalternmed/. doi:10.1186/s12906-015-0802-5.

Lu, X.T., Wang, Y.D., Zhu, T.T., Zhu, H.L., Liu, Z.Y., 2022. Dietary fatty acids and risk of non-alcoholic steatohepatitis: a national study in the United States. Front. Nutr. 9. doi:10.3389/fnut.2022.952451

Lu, Y., Chen, Y., Wu, Y., Hao, H., Liang, W., Liu, J., Huang, R., 2019. Marine unsaturated fatty acids: structures, bioactivities, biosynthesis and benefits. RSC Adv. 2046-2069. 9 (61), 35312–35327. Available from: http://pubs.rsc.org/en/journals/journal/ra. doi:10.1039/c9ra08119d.

Manoharan, S., Perumal, E., 2022. Potential role of marine bioactive compounds in cancer signaling pathways: a review. Eur. J. Pharmacol. 936. doi:10.1016/j.ejphar.2022.175330

Martins, B.T., da Silva, M.C., Pinto, M., Cidade, H., Kijjoa, A., 2019. Marine natural flavonoids: chemistry and biological activities. Nat. Prod. Res. 1478-6427. 33 (22), 3260–3272. Available from: www.tandf.co.uk/journals/titles/14786419.asp.. doi:10.1080/14786419.2018.1470514.

Ney, L.M., Wipplinger, M., Grossmann, M., Engert, N., Wegner, V.D., Mosig, A.S., 2023. Short chain fatty acids: key regulators of the local and systemic immune response in inflammatory diseases and infections. Open Biol. 2046-2441. 13 (3). Available from: http://rsob.royalsocietypublishing.org/. doi:10.1098/rsob.230014.

Nogal, A., Valdes, A.M., Menni, C., 2021a. The role of short-chain fatty acids in the interplay between gut microbiota and diet in cardio-metabolic health. Gut Microbes. 1949-0984. 13 (1), 1–24. Available from: http://www.tandfonline.com/toc/kgmi20/current. doi:10.1080/19490976.2021.1897212.

Nogal, A., Valdes, A.M., Menni, C., 2021b. The role of short-chain fatty acids in the interplay between gut microbiota and diet in cardio-metabolic health. Gut Microbes. 13 (1). doi:10.1080/19490976.2021.1897212.

O'Riordan, K.J., Collins, M.K., Moloney, G.M., Knox, E.G., Aburto, M.R., Fülling, C., Morley, S.J., Clarke, G., Schellekens, H., Cryan, J.F., 2022. Short chain fatty acids: microbial metabolites for gut-brain axis signalling. Mol. Cell. Endocrinol. 1872-8057. 546. Available from: www.elsevier.com/locate/mce.. doi:10.1016/j.mce.2022.111572.

Oliyaei, N., Zekri, S., Iraji, A., Oliyaei, A., Tanideh, R., Mussin, N.M., Tamadon, A., Tanideh, N., 2025. Health benefits of algae and marine-derived bioactive metabolites for modulating ulcerative colitis symptoms. J. Funct. Foods. 1756-4646. 125, 106690. doi:10.1016/j.jff.2025.106690.

Pereira, J.G., Mesquita, J.X., Aragão, K.S., Franco, A.X., Souza, M.H.L.P., Brito, T.V., Dias, J.M., Silva, R.O., Medeiros, J.V.R., Oliveira, J.S., Abreu, C.M.W.S., De Paula, R.C.M., Barbosa, A.L.R., Freitas, A.L.P., 2014. Polysaccharides isolated from Digenea simplex inhibit inflammatory and nociceptive responses. Carbohydr. Polym. 0144-8617. 108 (1), 17–25. Available from: http://www.elsevier.com/wps/find/journaldescription.cws_home/405871/description#description. doi:10.1016/j.carbpol.2014.01.105.

Pinkhien, T., Maiuthed, A., Chamni, S., Suwanborirux, K., Saito, N., Chanvorachote, P., 2016. Bishydroquinone renieramycin M induces apoptosis of human lung cancer cells through a mitochondria-dependent pathway. Anticancer Res. 1791-7530. 36 (12), 6327–6333. Available from: http://ar.iiarjournals.org/content/36/12/6327.full.pdf. doi:10.21873/anticanres.11229.

Porter, R.J., Arends, M.J., Churchhouse, A.M.D., Din, S., 2021. Inflammatory bowel disease-associated colorectal cancer: translational risks from mechanisms to medicines. J. Crohns Colitis. 15 (12), 2131–2141. doi:10.1093/ecco-jcc/jjab102.

Pulito, C., Cristaudo, A., Porta, C.L., Zapperi, S., Blandino, G., Morrone, A., Strano, S., 2020. Oral mucositis: the hidden side of cancer therapy. J. Exp. Clin. Cancer Res. 1756-9966. 39 (1). Available from: http://www.jeccr.com/. doi:10.1186/s13046-020-01715-7.

Repici, A., Hasan, A., Capra, A.P., Scuderi, S.A., Paterniti, I., Campolo, M., Ardizzone, A., Esposito, E., 2024. Marine algae and deriving biomolecules for the management of inflammatory bowel diseases: potential clinical therapeutics to decrease gut inflammatory and oxidative stress markers? Mar. Drugs. 1660-3397. 22 (8). Available from: http://www.mdpi.com/journal/marinedrugs. doi:10.3390/md22080336.

Ruzicka, 1953. The isoprene rule and the biogenesis of terpenic compounds. Experientia 9, 357–367. doi:10.1007/bf02167631.

Sanjeewa, K.K.A., Lee, J.-S., Kim, W.-S., Jeon, Y.-J., 2017. The potential of brown-algae polysaccharides for the development of anticancer agents: an update on anticancer effects reported for fucoidan and laminaran. Carbohydr. Polym. 177, 451–459. doi:10.1016/j.carbpol.2017.09.005.

Sanjeewa, K.K.A., Jayawardena, T.U., Kim, S.-Y., Kim, H.-S., Ahn, G., Kim, J., Jeon, Y.-J., 2019. Fucoidan isolated from invasive Sargassum horneri inhibit LPS-induced inflammation via blocking NF-κB and MAPK pathways. Algal Res. 41. doi:10.1016/j.algal.2019.101561

Sfecci, E., Lacour, T., Amade, P., Mehiri, M., 2016. Polycyclic guanidine alkaloids from *Poecilosclerida* marine sponges. Mar. Drugs. 1660-3397. 14 (4), 77. doi:10.3390/md14040077.

Shang, Q., Shan, X., Cai, C., Hao, J., Li, G., Yu, G., 2016. Dietary fucoidan modulates the gut microbiota in mice by increasing the abundance of: *Lactobacillus* and Ruminococcaceae. Food Funct. 2042-650X. 7 (7), 3224–3232. Available from: http://pubs.rsc.org/en/journals/journal/fo. doi:10.1039/c6fo00309e.

Shukla, A., Chen, C., Jellusova, J., Leung, C.R., Kao, E., Bhat, N., Lin, W.W., Apgar, J.R., Rickert, R.C., 2019. Self-reactive B cells in the GALT are actively curtailed to prevent gut inflammation. JCI Insight. 2379-3708. 4 (16). Available from: https://insight.jci.org/articles/view/130621/pdf. doi:10.1172/jci.insight.130621.

Siddiqui, M.T., Cresci, G.A.M., 2021. The immunomodulatory functions of butyrate. J. Inflamm. Res. 1178-7031. 14, 6025–6041. Available from: http://www.dovepress.com/journal-of-inflammation-research-journal. doi:10.2147/JIR.S300989.

Singh, S., Kumar, S., Singh, A.K., Varshney, M., Roy, S., 2025. Exploring marine alkaloids: a natural approach to cancer treatment. Curr. Pharm. Biotechnol. 1873-4316. 26 (1), 63–79. doi:10.2174/0113892010316791240611093022.

Sun, J., Feng, Y., Wang, Y., Q, J.I., Cai, G., Shi, L., Wang, Y., Huang, Y., Zhang, J., Li, Q., 2019. α-hederin induces autophagic cell death in colorectal cancer cells through reactive oxygen species dependent AMPK/mTOR signaling pathway activation. Int. J. Oncol. 54 (5), 1601–1612. doi:10.3892/ijo.2019.4757.

Tammam, M.A., Aly, O., Pereira, F., Mahdy, A., El-Demerdash, A., 2024. Unveiling the potential of marine-derived diterpenes from the order Alcyonacea as promising anti-obesity agents. Curr. Res. Biotechnol. 2590-2628. 7. Available from: https://www.sciencedirect.com/science/journal/25902628. doi:10.1016/j.crbiot.2024.100175.

Tojjari, A., Choucair, K., Sadeghipour, A., Saeed, A., Saeed, A., 2023. Anti-inflammatory and immune properties of polyunsaturated fatty acids (PUFAs) and their impact on colorectal cancer (CRC) prevention and treatment. Cancers. 2072-6694. 15 (17), 4294. doi:10.3390/cancers15174294.

Ullah, A., Rahman, L., Yazdani, M.B., Irfan, M., Khan, W.S., Rehman, A., (2021).

Veríssimo, A.C.S., Pacheco, M., Silva, A.M.S., Pinto, D.C.G.A., 2021. Secondary metabolites from marine sources with potential use as leads for anticancer applications. Molecules. 1420-3049. 26 (14). Available from: https://www.mdpi.com/1420-3049/26/14/4292/pdf. doi:10.3390/molecules26144292.

Vladkova, T., Georgieva, N., Staneva, A., Gospodinova, D., 2022. Recent progress in antioxidant active substances from marine biota. Antioxidants. 2076-3921. 11 (3), 439. doi:10.3390/antiox11030439.

Wall, R., Ross, R.P., Fitzgerald, G.F., Stanton, C., 2010. Fatty acids from fish: the anti-inflammatory potential of long-chain omega-3 fatty acids. Nutr. Rev. 1753-4887. 68 (5), 280–289. doi:10.1111/j.1753-4887.2010.00287.x.

Wan, M.L.Y., Ling, K.H., El-Nezami, H., Wang, M.F., 2019. Influence of functional food components on gut health. Crit. Rev. Food Sci. Nutr. 1549-7852. 59 (12), 1927–1936. Available from: https://www.tandfonline.com/loi/bfsn20. doi:10.1080/10408398.2018.1433629.

Xu, F., Yang, F., Qiu, Y., Wang, C., Zou, Q., Wang, L., Li, X., Jin, M., Liu, K., Zhang, S., Zhang, Y., Li, B., 2024a. The alleviative effect of C-phycocyanin peptides against TNBS-induced inflammatory bowel disease in zebrafish via the MAPK/Nrf2 signaling pathways. Fish Shellfish Immunol. 1095-9947. 145. Available from: https://www.sciencedirect.com/science/journal/10504648. doi:10.1016/j.fsi.2023.109351.

Xu, J., Peng, W.R., Zhang, D., Sun, H.X., Li, L., Sun, F., Gu, Z.C., Lin, H.W., 2024b. Marine sponge-derived alkaloid ameliorates DSS-induced IBD via inhibiting IL-6 expression through modulating JAK2-STAT3-SOCS3 pathway. Int. Immunopharmacol. 1878-1705. 129. Available from: https://www.sciencedirect.com/science/journal/15675769. doi:10.1016/j.intimp.2024.111576.

Yan, D., Li, Y., Liu, Y., Li, N., Zhang, X., Yan, C., 2021. Antimicrobial properties of Chitosan and Chitosan derivatives in the treatment of enteric infections. Molecules. 1420-3049. 26 (23), 7136. doi:10.3390/molecules26237136.

Yan, H., Ajuwon, K.M., 2017. Butyrate modifies intestinal barrier function in IPEC-J2 cells through a selective upregulation of tight junction proteins and activation of the Akt signaling pathway. PLoS One. 1932-6203. 12 (6). Available from: http://journals.plos.org/plosone/article/file?id=10.1371/journal.pone.0179586&type=printable. doi:10.1371/journal.pone.0179586.

Yang, J.Y., Lim, S.Y., 2021. Fucoidans and bowel health. Mar. Drugs. 1660-3397. 19 (8). Available from: https://www.mdpi.com/1660-3397/19/8/436/pdf. doi:10.3390/md19080436.

Yu, J., Wang, X., Du, P., Shi, H., 2024. The therapeutic potential and application of marine alkaloids in treating breast cancer. Front. Mar. Sci. 2296-7745. 11. doi:10.3389/fmars.2024.1440928.

Zhao, H., Wu, L., Yan, G., Chen, Y., Zhou, M., Wu, Y., Li, Y., 2021. Inflammation and tumor progression: signaling pathways and targeted intervention. Signal Transduct. Target. Ther. 2059-3635. 6 (1). doi:10.1038/s41392-021-00658-5.

CHAPTER

18

Impact of seaweed on body weight management

Kirinde Gedara Isuru Sandanuwan Kirindage and Arachchige Maheshika Kumari Jayasinghe

Department of Food Technology and Nutrition, Chonnam National University, Republic of Korea

18.1 Introduction

According to the World Health Organization (WHO) in 2024, the global prevalence of obesity has reached alarming levels (World Health Organization, 2024a). The abnormal or excessive body fat accumulation is identified as obesity and overweight. For adults, being overweight is typically defined as having a body mass index (BMI) above 25, while obesity is classified as a BMI over 30. According to the WHO growth reference charts, having a BMI-for-age one standard deviation above the median is defined as overweight, while having a BMI-for-age two standard deviations above is considered obesity in children aged 5–19 years. For children under 5 years old, overweight is determined as having a weight-for-height measurement greater than two standard deviations above the WHO Child Growth Standards median (World Health Organization, 2024b). Shockingly, one out of every eight individuals worldwide was dealing with obesity, representing a significant uptick from previous years in 2022. As reported in 2022, 2.5 billion adults were categorized as overweight, with 890 million falling into the obese category. The statistics are equally concerning for children, with 37 million under the age of 5 years and over 390 million aged 5–19 years identified as overweight, including 160 million with obesity (World Health Organization, 2024a). A global prevalence of obesity is an escalating public health hurdle, urging the need for effective and sustainable weight management approaches. However, no single treatment or drug can effectively control these conditions due to their different etiological and pathological origins. Understanding the link between inactive routines, eating patterns, and the rising obesity rates highlights the urgency to explore new intervention methods (Kumanyika et al., 2008).

There are several key considerations in managing body weight. These include practicing portion control, maintaining a high nutrient density, choosing a balanced diet, consuming

Harvesting the Sea
https://doi.org/10.1016/B978-0-443-29190-6.00022-2

fiber-rich foods, limiting processed foods, monitoring sugar and fat intake, staying hydrated, practicing mindful eating, and engaging in regular physical activity (Dent et al., 2020). Hence, one promising area of study is investigating the potential role of functional foods in managing body weight, which has become a key focus of scientific research (Sandner et al., 2020). Opting for nutrient-dense options that provide essential vitamins, minerals, and nutrients without an excess of calories is important when choosing foods for weight management. In addition, incorporating plenty of fiber-rich foods into the diet can help promote feelings of fullness, support digestion, and regulate blood sugar levels.

Reported evidence suggests that marine organisms generate a variety of novel bioactive natural compounds such as polysaccharides, glycoproteins, polyunsaturated fatty acids (PUFAs), vital minerals, vitamins, carotenoids, polyphenols, enzymes, soluble and nonsoluble dietary fibers, mycosporine-like amino acids, bioactive proteins, and peptides (Pradhan et al., 2022; Charoensiddhi et al., 2017). Seaweed, also known as marine macroalgae, has played important roles in culinary traditions worldwide, valued not just for its distinct flavors but also for its potential health benefits (Figueroa et al., 2023). Around 600 species have been reported worldwide as dietary seaweed (Salido et al., 2024). Their historical use in traditional cuisines highlights their cultural significance and prompts further investigation into their modern-day role as functional foods (Blikra et al., 2021). Recent scientific research has revealed a wide range of potential health advantages associated with consuming seaweed (Murai et al., 2021; Lomartire et al., 2021). Developing seaweed-based food supplements holds significant potential in body weight management for several reasons. Seaweed is low in calories but rich in essential nutrients such as vitamins, minerals, and antioxidants (Shannon and Abu-Ghannam, 2019). Incorporating seaweed into food supplements can help individuals achieve a sense of fullness while providing necessary nutrients, potentially aiding in weight management (Trigo et al., 2023). The integration of seaweed into the discourse on weight management presents a promising avenue for scientific inquiry and translational applications. Hence, this chapter elaborates on the existing knowledge, elucidates gaps in understanding, and thoroughly examines the potential of seaweed as an agent in the pursuit of healthier body weights.

18.2 Nutritional composition of seaweeds and their effects on weight management

Seaweed possesses a remarkable array of nutritive values that make it a promising ingredient in functional food development for body weight management (Fig. 18.1). Dietary fibers in seaweed offer significant potential for promoting satiety and reducing calorie intake by slowing down digestion and promoting a feeling of fullness (Trigo et al., 2023; Wendin and Undeland, 2020). In addition, seaweeds are low in calories and fat while being packed with essential vitamins and minerals crucial for overall health (Michalak and Chojnacka, 2018). The high protein content found in certain seaweed species offers a valuable source of plant-based protein, which can help support muscle mass and metabolic function during weight management (Raja et al., 2022). Incorporating seaweeds into functional foods such as soups, salads, snacks, and beverages provides an opportunity to enhance their nutritional profile

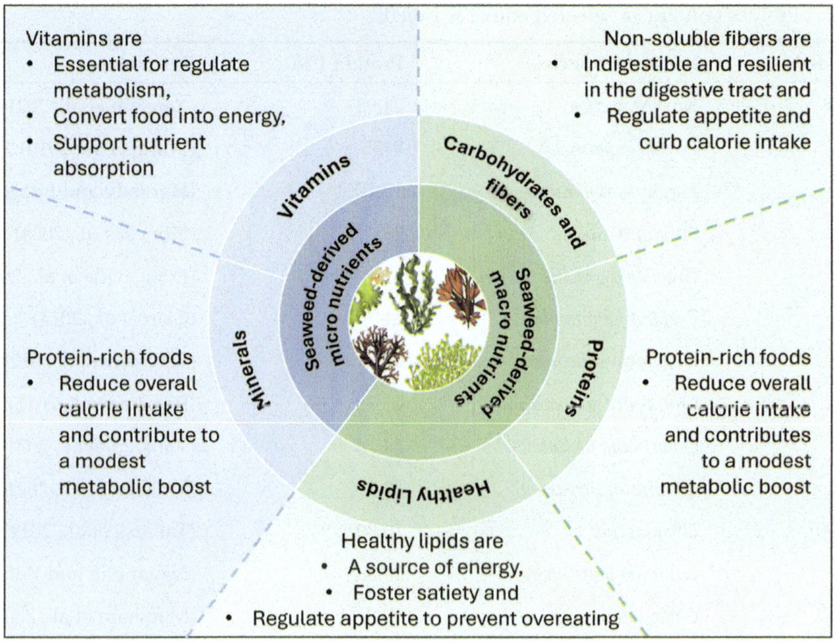

FIG. 18.1 Seaweed-derived macro- and micronutrients and their functions in body weight management.

while offering potential benefits for appetite control, metabolism regulation, and overall body weight management.

18.2.1 Macronutrients

18.2.1.1 Protein content

Protein-rich foods reduce overall calorie intake, which is important in maintaining a healthy body weight. In addition, protein contributes to a modest metabolic boost, which supports muscle maintenance. Seaweeds are vital protein sources that significantly enhance the macronutrient profile of marine organisms and have an exceptional amino acid content similar to conventional protein sources. Certain types of seaweed have been shown to contain up to 47% protein, which is comparable to typical protein sources, including meat, eggs, soybeans, and milk (Naseri et al., 2020). There are several types of proteins or forms of proteins, such as peptides, enzymes, glycoproteins, lectins, mycosporine-like amino acids, and phycobiliproteins, in seaweeds (Pliego-Cortés et al., 2020). Therefore, utilizing seaweed for protein production has several benefits compared to traditional high-protein sources (Van Krimpen et al., 2013). The protein content of seaweeds differs according to species, environmental factors, and growth stage (Thiviya et al., 2022). Generally, red algae (8%–47% of dry weight) and green algae (9%–33% of dry weight) contain a higher level of protein content than brown algae (4%–24% of dry weight). As reported, Table 18.1 indicates the protein content of selected edible seaweeds on a dry-weight basis.

TABLE 18.1 Protein content of selected edible seaweeds.

Type of seaweed	Seaweed Species	Protein (%)	Reference
Red seaweeds	*Savoiea arctica*	24.95	(Yanshin et al., 2021)
	Palmaria palmata	8–35	(Wang et al., 2010)
	Porphyra umbilicalis	15–37	(Harnedy and Fitzgerald, 2011)
	Porphyra dioica	29	(Nunes et al., 2024)
	Gracilaria gracilis	31–45	(Francavilla et al., 2015)
	Porphyra columbina	24	(Cian et al., 2014)
	Gracilaria salicornia	9.58	(Tabarsa et al., 2012)
	Osmundea pinnatifida	20.64	(Patarra et al., 2011)
	Odonthalia dentata	14.78	(Yanshin et al., 2021)
	Ceramium virgatum	32.25	(Yanshin et al., 2021)
Green seaweeds	*Ulva lactuca*	10.69	(Tabarsa et al., 2012)
	Caulerpa lentillifera	19.38	(Nagappan and Vairappan, 2014)
	Cladophora rupestris	29.8	(Marsham et al., 2007)
	Caulerpa racemosa	17.36	(Nagappan and Vairappan, 2014)
	Ulva rotundata	10	(Dumay and Morançais, 2016)
Brown seaweeds	*Saccharina japonica*	12	(Dickson et al., 2018)
	Fucus vesiculosus	12.9	(Lorenzo et al., 2017)
	Undaria pinnatifida	11–24	(Fleurence et al., 2018)
	Hizikia fusiforme	9.9–18.41	(Meinita et al., 2021)
	Fucus vesiculosus	12.9	(Lorenzo et al., 2017)
	Alaria esculenta	9–20	(Pereira, 2011)
	Laminaria digitata	8–15	(Marsham et al., 2007)
	Ascophyllum nodosum	8.2	(Tibbetts et al., 2016)
	Saccharina latissima	8.1	(Tibbetts et al., 2016)

18.2.1.2 Carbohydrates and fiber

Seaweed contains a range of carbohydrates in the form of non-soluble fibers. Non-soluble fibers are pivotal for body weight management due to their distinctive impact on digestion. These fibers are indigestible and resilient in the digestive tract and offer numerous advantages conducive to weight control. Non-soluble fibers help regulate appetite and curb calorie intake by imparting a sense of fullness and satiety. Moreover, their ability to promote regular bowel movements ensures efficient digestion and waste elimination, indirectly supporting weight management by maintaining digestive health. Fiber also aids in stabilizing blood sugar levels, mitigating spikes that can trigger cravings for high-calorie foods. Furthermore, they hinder the absorption of certain nutrients, like fats, leading to a reduction in overall calorie absorption.

As one of the main macronutrients found in marine macroalgae, carbohydrates can range in total content from 5% to 75% (w/w, Dry Weight basis) based on the species, age, time of year, and the site of harvesting (Hentati et al., 2020a). Seaweeds have a broad range of carbohydrates, including sugars, polysaccharides, and dietary fiber, with varying structural and physicochemical properties, along with interesting functional characteristics (Hentati et al., 2020a; Peñalver et al., 2020). Polysaccharides from marine seaweeds are found in sulfated and nonsulfated forms, and the type of polysaccharides depends on the macroalgae species (Gullón et al., 2020). As reported, brown seaweeds are characterized by the presence of fucoidans, laminarins, and alginic acids (or alginates); red seaweeds contain agars, carrageenans, xylans, sulfated galactan, floridean starch, and porphyran, while green seaweeds are rich in ulvans (Kraan, 2012; Hentati et al., 2018; 2020a). These polysaccharides have unique chemical structures and functional properties, making them valuable in various industrial applications, including food, pharmaceuticals, and cosmetics. Table 18.2 shows the carbohydrate content in different seaweeds on a dry-weight basis.

The components that constitute dietary fiber include cellulose, hemicellulose, pectic substances, gums, resistant starch, and inulin. These compounds can also be associated with lignin and/or other noncarbohydrate elements such as polyphenols, waxes, saponins, cutins,

TABLE 18.2 Carbohydrate content of different seaweeds.

Type of seaweed	Seaweed species	Carbohydrate %	Reference
Red seaweeds	*Gracilaria cervicornis*	63.10	(Marinho-Soriano et al., 2006)
	Kappaphycus alvarezii	27.40	(Fayaz et al., 2005)
	Ahnfeltia plicata	59.10	(Parjikolaei et al., 2016)
	Osmundea pinnatifida	17.61	(Paiva et al., 2014)
	Jania adhaerens	34.54	(Hentati et al., 2019)
Green seaweeds	*Ulva reticulata*	55.77	(Ratana-arporn and Chirapart, 2006)
	Ulva rigida	42.60	(Taboada et al., 2010)
	Caulerpa lentillifera	38.66	(Matanjun et al., 2009)
	Caulerpa taxifolia	23.86	(Kokilam and Vasuki, 2014)
Brown seaweeds	*Saccorhiza polyschides*	45.60	(Rodrigues et al., 2015)
	Sargassum naozhouense	47.43	(Peng et al., 2013)
	Ecklonia radiata	66.90	(Smith et al., 2010)
	Durvillaea antarctica	70.90	(Ortiz et al., 2006)
	Hormosira banksii	62.90	(Smith et al., 2010)
	Cystoseira compressa	39.11	(Hentati et al., 2019)

and resistant proteins (Peñalver et al., 2020). Dietary fibers can be classified into two main types, namely, soluble and insoluble fibers. The ability of soluble fibers to form a gel-like substance helps to control blood sugar, decrease cholesterol, and encourage the growth of beneficial gut bacteria. Insoluble fibers do not form gels when in contact with water, aiding in regularity and alleviating constipation. A balanced diet that includes a sufficient amount of dietary fiber is necessary to support general health, minimize the risk of chronic diseases, and maintain good digestive function (MacArtain et al., 2007). Seaweeds can range in their dietary fiber content from 36% to 60% based on dry mass, which is comparable to or even higher than that of food derived from terrestrial plants (Peñalver et al., 2020). The amount of soluble fiber in seaweeds varies based on the species and growth environment, ranging from approximately 55%–70% of the total dietary fiber content (Rajapakse and Kim, 2011). As reported, Table 18.3 indicates the dietary fiber content in different seaweeds.

18.2.1.3 *Healthy lipids*

Healthy fats, or lipids, are indispensable for body weight management, offering a multitude of benefits over the misconception of fats as solely detrimental to weight gain and obesity. Specifically, healthy lipids are a source of energy, fostering satiety and regulating appetite to prevent overeating. Moreover, healthy fats facilitate the absorption of fat-soluble vitamins, which are crucial for metabolic function. Furthermore, they play a pivotal role in hormone production and in supporting metabolic balance. The anti-inflammatory properties of some fats help mitigate obesity-related complications. A balanced intake of healthy lipids contributes to energy balance, helping individuals achieve and maintain a healthy weight and metabolic health.

TABLE 18.3 Dietary fiber content of different seaweeds (g/100 g).

Type of seaweed	Seaweed species	Soluble fiber	Insoluble fiber	Reference
Red seaweeds	*Porphyra/Pyropia* spp.	17.9	16.8	(Jiménez-Escrig and Sánchez-Muniz, 2000)
	Garateloupia turuturu	48.1 ± 1.0	12.3 ± 1.2	(Denis et al., 2010)
	Chondrus crispus	22.25 ± 0.99	12.04 ± 2.89	(Rupérez and Saura-Calixto, 2001)
Green seaweeds	*Ulva* spp.	21.9 ± 0.9	18.7 ± 2.1	(Peña-Rodríguez et al., 2011)
	Caulerpa lentillifera	17.21 ± 0.87	15.78 ± 1.20	(Matanjun et al., 2009)
Brown seaweeds	*Sargassum fusiforme*	32.9	16.3	(Jiménez-Escrig and Sánchez-Muniz, 2000)
	Undaria pinnatifida	30.0	5.3	(Jiménez-Escrig and Sánchez-Muniz, 2000)
	Himanthalia elongata	23.63 ± 0.48	13.51 ± 0.45	(Gómez-Ordóñez et al., 2010)
	Saccharina latissima	17.12 ± 0.84	13.11 ± 0.56	(Gómez-Ordóñez et al., 2010)

Lipids found in seaweeds play a pivotal role, offering a diverse range of health benefits, despite their relatively low content in seaweeds, typically ranging from 1% to 5% on a dry basis (Peñalver et al., 2020). Seaweed synthesizes higher amounts of PUFAs, with a significant proportion of omega-3 (ω-3) fatty acids constituting the main component. Certain seaweeds, mostly brown and some red algae species, produce these ω-3 fatty acids in response to environmental factors such as water temperature and nutrient availability. Among them, eicosapentaenoic acid (EPA) and docosahexaenoic acid (DHA) are the two important fatty acids of marine origin belonging to ω-3 fatty acids. EPA is the main PUFA found in most seaweeds, and its amount may exceed 30% of the total fatty acid content (Rajapakse and Kim, 2011). Generally, red seaweed contains high amounts of EPA, palmitic acid, oleic acid, and ω-6 fatty acids, including arachidonic acid. In addition, brown seaweed has significant levels of oleic acid, linoleic acid, and α-linolenic acid, with a minor amount of EPA. Conversely, large concentrations of DHA, oleic acid, palmitic acid, and α-linolenic acid are found in green algae (Fleurence et al., 1994; Kumari et al., 2010). Red seaweeds and brown seaweeds provide a balanced profile of ω-3 and ω-6 fatty acids (Mabeau and Fleurence, 1993; Sánchez-Machado et al., 2004). Thus, a balanced diet that includes seaweed may benefit from its high concentration of essential fatty acids.

18.2.2 Micronutrients

18.2.2.1 Vitamins

Vitamins are indispensable micronutrients important in body weight management, playing diverse roles in metabolic processes. Vitamins are essential for regulating metabolism, converting food into energy, and supporting nutrient absorption, all of which influence body composition. Some vitamins, such as B-complex vitamins, facilitate energy metabolism and appetite regulation. Fat-soluble vitamins, like A, D, E, and K, are vital for nutrient absorption and utilization. Moreover, vitamins contribute to muscle function, stress management, and energy production, all of which are integral to maintaining a healthy weight.

Marine seaweed is an excellent source of vitamins with many health benefits. Vitamin content in seaweeds depends on several factors, such as their habitat, species, salinity, availability of light, seawater temperature, and season (Peñalver et al., 2020; Škrovánková, 2011). Seaweeds contain a variety of vitamins, including water-soluble vitamins consisting of the B complex and C, as well as lipid-soluble vitamins such as A and E. Seaweeds are also an excellent source of vitamin E, which enhances immunity and protects cells from oxidative damage (Blaner et al., 2021). Moreover, some green and brown seaweeds contain high amounts of vitamin C, averaging between 500 and 3000 mg/kg of dry matter. Green seaweed, *Enteromorpha* sp., and *Ulva lactuca,* showed vitamin C levels of 150 and 460 mg/kg of wet matter, respectively (Mišurcová, 2011). As reported, red seaweeds, *Porphyra umbilicalis, Porphyra nereocystis,* and *Porphyra perforata* indicated higher vitamin C contents (Peñalver et al., 2020; Mišurcová, 2011). Furthermore, vitamin C is present in large amounts in brown seaweeds, *Himanthalia elongata, Crassiphycus changii,* and *Ecklonia arborea* (Hernández-Carmona et al., 2009; Norziah and Ching, 2000; Ferraces-Casais et al., 2012). As an antioxidant, vitamin C supports collagen formation, improves immunity, and promotes the healing of wounds. In addition, the B vitamin complex found in seaweeds includes thiamine (B_1), riboflavin (B_2), niacin (B_3), pyridoxine (B_6), folate (B_9), and cobalamin (B_{12}), all of which are necessary for the neurological system, energy metabolism, and the synthesis of red blood cells (Ferraces-Casais et al., 2012).

18.2.2.2 Minerals

Seaweed is well known for being a rich source of minerals due to its ability to absorb inorganic elements from the surrounding environment. The mineral content of seaweeds varies according to a variety of endogenous and external causes, and it is correlated with the amount of minerals present in the saltwater or growing medium (Mišurcová, 2011; Mišurcová et al., 2011). Algae are recognized for having a higher mineral content than terrestrial plants, resulting from the rich mineral components found in the maritime environment where they grow (Peñalver et al., 2020). Consequently, seaweed is a prospective source of vital minerals for functional foods, food supplements, and nutraceuticals, and it can significantly increase daily mineral intake. Seaweeds are rich in macronutrients, including sodium, calcium, magnesium, potassium, chlorine, sulfur, iodine, and phosphorus, whereas micronutrients include iron, zinc, copper, selenium, manganese, nickel, and cobalt (Muñoz and Díaz, 2020). Iodine is mostly found in seaweed, with brown algae having the highest concentration (1500–8000 ppm), whereas red and green algae often have lower amounts (Rajapakse and Kim, 2011).

18.3 Potential seaweed-derived bioactive compounds and their effects on weight management

18.3.1 Alginate

Alginates are anionic polysaccharides formed by alginic acid reacting with sodium, calcium, and magnesium (Tao et al., 2024). Alginic acid is the free acid form of alginate, mainly found in the intercellular matrix and cell walls of brown marine algae, having the chemical formula (C6H8O6)n (Tao et al., 2024; Fernando et al., 2022b; Raus et al., 2021). Alginates consist of two main monosaccharide units: β-(1 → 4) linked d-mannuronic acid (referred to as the M unit) and α-(1 → 4) linked l-guluronic acid (known as the G unit), typically having average molecular weights ranging from 48 to 186 kDa (Ching et al., 2017; Hasnain et al., 2020). As reported, alginate comprises hydroxyl and carboxyl functional groups. When aqueous alginate solutions come into contact with either acid or calcium ions, they undergo gel formation (Fernando et al., 2022a).

The acidic environment in the stomach allows alginate to be administered as a solution and then form a gel in the stomach due to its unique pH-responsive gelation property (Foster et al., 2012; Fan et al., 2024). In contrast to alginate, other viscous polysaccharides need to be administered in gel form to remain a gel in the stomach (Tønnesen and Karlsen, 2002). From the consumers' point of view, foods with high viscosity or gel strength are often unpalatable due to taste concerns. The use of liquid forms of alginate minimizes that problem. Gels in the stomach are associated with increased feelings of fullness and satiety

(Wanders et al., 2014). Alginate, with its distinctive molecular structure, can be administered in a low-viscosity form and spontaneously gel in the stomach, enhancing feelings of fullness after consumption. Research conducted with healthy adult males has demonstrated that adding alginate to milk-based liquid meals or chocolate milk replacements increased participants' sensation of fullness, decreased premeal glycemia, insulinemia, and appetite compared to controls (Khoury et al., 2014). Interestingly, drinks containing sodium alginate mitigated post-meal plasma glucose and insulin elevation (Khoury et al., 2014). Christine and colleagues conducted a study involving 29 overweight or obese adult females to investigate the potential of calcium-gelled alginate–pectin beverages in reducing satiety and food intake (Pelkman et al., 2007). The findings suggest that these beverages have the potential to minimize satiety and food intake. Furthermore, the consumption of alginate with high gel strength extends the duration of postmeal hunger (Cassidy et al., 2018). As reported, alginate remains in the stomach for up to 6–8 h in gel form (Sarkar et al., 2020). Upon ingestion, alginate-based beverages significantly attenuated the postprandial glycemic response and slowed down the gastric emptying rate (Williams et al., 2004; Khoury et al., 2015; Jensen et al., 2012). Alginate intake reduces nutrient absorption and influences the feeding behavior of individuals (Dettmar et al., 2011; Hoad et al., 2004).

Blood glucose levels play a critical role in body weight management due to their direct impact on metabolism and energy balance. When blood glucose levels are elevated, such as after consuming high-carbohydrate meals, the body releases insulin to facilitate glucose uptake into cells for energy production. However, excess glucose not utilized for immediate energy needs is stored as glycogen in the liver and muscles or converted into fat for long-term storage (Remesar and Alemany, 2020). This process can contribute to weight gain over time if energy intake consistently exceeds expenditure (Barnes et al., 2020). Alginate led to lower postmeal peak glucose concentrations and reduced total glucose uptake for a reasonable duration (Giuntini et al., 2022). Scientific evidence has proven that sodium alginates could reduce weight gain, adiposity index, and improve glucose homeostasis in the high-fat diet-fed mice experimental model (Wang et al., 2018, 2020). A randomized clinical trial conducted by Kato and the team found that incorporating calcium alginates into noodles effectively suppressed postprandial blood glucose concentration and glucose absorption (Kato et al., 2018). Simultaneously, blood calcium concentration increased without affecting other parameter values (Kato et al., 2018).

One study reported that supplementation with alginate-rich, iodine-reduced kelp powder significantly decreases body fat composition in overweight subjects with a $BMI \geq 25$ and $<30\ kg/m^2$ (Aoe et al., 2021). In a nutshell, the incorporation of alginate into the diet contributes to better blood sugar control and reduced cravings for sugary foods, which are beneficial for weight management. Above all, studies showed that alginates have potential applications in weight management and satiety effects for humans based on dietary interventions. However, the viscosity of alginate may impede the absorption of beneficial dietary components like β-carotene and minerals such as calcium, iron, chromium, and cobalt (Qin et al., 2018). In addition, alginate indicates inhibitory potential on various digestive enzymes, which contributes to its slimming effects (Chater et al., 2015). However, this property might pose risks, especially in vulnerable populations like the elderly, pregnant women, and infants, when included in their diets.

18.3.2 Fucoidans

A sulfated polysaccharide, commonly known as fucoidan, extracted from various species of brown seaweed, has garnered significant attention in recent years due to its potential health benefits, such as immunomodulating, radioprotective, antiinflammatory, and antioxidant properties (Jayasinghe et al., 2022, 2023; Kirindage et al., 2023; Zvyagintseva et al., 2021). Previous studies have mentioned that the consumption of fucoidan is nontoxic for humans as a food supplement (Citkowska et al., 2019). Although sulfation and acetylation patterns, structure, molecular weight, and composition of the fucoidan are affected by seaweed characteristics, geographic, seasonal, and environmental factors, as well as collecting, reproductive status, and extraction conditions, ensuring the purity of preparation, it has a common structure that gives specific characteristics to fucoidan (Kirindage et al., 2023). Mainly, fucose in the fucoidans forms the core monomeric module, which makes fucoidan a polymer. The linkage between the monomeric units is either α-(1→2) or α-(1→3) (Zvyagintseva et al., 1999). Galactose, mannose, xylose, and glucuronic acid are other sugar residues commonly found in fucoidans. Among its diverse range of bioactive properties, fucoidan has been investigated for its role in body weight management, particularly as an anti-obesity and antidiabetic agent.

Fucoidan has been reported as an adipogenesis inhibitor. Adipogenesis refers to the formation of fat cells, while lipogenesis involves the synthesis of fats. By controlling these processes, individuals can regulate their body fat accumulation. One study showcased how fucoidan from *Sargassum thunbergii* can combat obesity by reducing lipid accumulation and suppressing adipogenic and lipogenic proteins in 3T3-L1 cells. These results were further confirmed in mouse experiments. The study proposes that fucoidan holds promise in alleviating hyperlipidemia and dyslipidemia associated with obesity (Lee et al., 2024). Another study conducted by Shao-Hua Wang et al. found that fucoidan extracted from *Sargassum siliquosum* indicates antilipogenesis activities with a 28.9% reduction in lipid synthesis in hepatic cells (Wang et al., 2021). As reported, fucoidan extracted from the sea cucumber *Thelenota ananas* can prevent obesity by regulating the proliferation and differentiation of adipocytes. Moreover, results indicate that fucoidan downregulates the transcriptional factors such as SREBP-1c, PPARγ, and C/EBPα in adipocytes (Liu et al., 2021).

Another important factor in body weight management is improving insulin sensitivity due to its central role in regulating glucose metabolism and lipid storage (Li et al., 2022b). When insulin sensitivity is impaired, excess glucose remains in the bloodstream, leading to elevated insulin levels and subsequent fat storage. This contributes to weight gain and increases the risk of developing obesity and related complications. Improving insulin sensitivity promotes the utilization of glucose for energy rather than its conversion into fat, facilitating weight loss and overall metabolic health. A recent study revealed intriguing results regarding the effects of *Sargassum fusiforme* fucoidan on blood glucose levels and insulin resistance. The research demonstrated significant reductions in fasting blood glucose and insulin resistance index, coupled with enhanced glucose tolerance. These findings indicate that fucoidan has the potential to alleviate insulin resistance induced by a high-fat diet in mice. The mechanism behind this action involves the activation of the Nrf2 pathway and the modification of gut microbiota (Zhang et al., 2020). In addition, another study found that *Sargassum fusiforme* fucoidan increased tauroursodeoxycholic acid levels and inhibited the farnesoid X receptor signaling pathway to reduce colon-derived biosynthesis of ceramide, thereby improving insulin resistance in diet-induced obese mice (Zhang et al., 2021). Moreover, a study conducted with low molecular weight fucoidan indicated a remarkable reduction in serum glucose, triglyceride, cholesterol, and low-density lipoprotein levels, and gradually reduced body weight by reducing lipid parameters in db/db mice (Jeong et al., 2013).

Regulation of lipid metabolism by fucoidans has been reported by many researchers as a strategy for body weight management. Lipids are essential for various physiological functions, such as energy production, cell membrane structure, and hormone synthesis. However, an imbalance in lipid metabolism can lead to adverse effects, such as obesity and metabolic disorders like diabetes and cardiovascular disease. Effective body weight management involves controlling lipid metabolism through a range of methods, such as dietary interventions, exercise, and pharmacological approaches (Mehta et al., 2021). Shan Li et al. conducted a study on fucoidans extracted from the sea cucumber *Pearsonothuria graeffei* to demonstrate their potential for preventing metabolic syndromes. The results indicated that fucoidan intervention significantly decreased weight gain, intestinal lipid metabolism, and fat accumulation in high-fat diet-fed mice. Moreover, fucoidan improved the serum lipid profile by reducing serum concentrations of total cholesterol, triglycerides, high-density lipoprotein cholesterol, low-density lipoprotein cholesterol, apolipoprotein B48, and apolipoprotein B100 (Li et al., 2022b).

Studies suggest that fucoidan may exert appetite-suppressing effects by influencing satiety hormones and neurotransmitters involved in appetite regulation. It helps to reduce calorie intake. In addition, fucoidan has the capacity to pass along the gastrointestinal tract without being metabolized due to its fibrous quality, and its viscosity can slow down digestion (MacArtain et al., 2007). A clinical trial conducted to investigate the incorporation of fucoidan-rich *Laminaria digitata* and *Undaria pinnatifida* into diets in healthy adults provided evidence for lowering the postprandial glucose and insulin response. Moreover, it increased the postprandial feeling of satiety and fullness in healthy subjects of both sexes. The results from subjective appetite sensation assessments using visual analog scales indicated a decreased appetite sensation after intake of the above meals (Zaharudin et al., 2021). By promoting a feeling of fullness and reducing food intake, fucoidan may help to control calorie consumption and prevent overeating.

As described before, body weight gain is highly affected by blood sugar levels. Type 2 diabetes mellitus is characterized by insulin resistance and impaired glucose regulation, often associated with obesity and metabolic dysfunction. Fucoidan has emerged as a potential therapeutic agent for managing diabetes due to its ability to target multiple pathways involved in glucose metabolism. Research conducted by Daub et al. identified fucoidan refined from *Ecklonia maxima* as a mixed-type inhibitor of α-glucosidase, which shows potential for managing type 2 diabetes (Daub et al., 2020). MicroRNAs have emerged as promising therapeutic targets with significant roles in regulating insulin resistance. Fucoidan, known for its potential as an α-amylase inhibitor, presents a promising avenue for managing type 2 diabetes mellitus. Scientific investigations suggest that fucoidan exhibits a crucial antidiabetic effect by modulating miRNA expression, underscoring its potential therapeutic value in diabetes management (Zhao et al., 2020). Many other studies have reported that fucoidan is a protective, nontoxic compound in alleviating the symptoms and complications of type 2 diabetes mellitus (Danilova et al., 2022).

18.3.3 Fucoxanthin

Fucoxanthin is a characteristic carotenoid found in edible brown seaweed (Tsukui et al., 2009). Fucoxanthin stands out among carotenoids due to its distinctive structure, featuring an uncommon allenic bond, nine conjugated double bonds, a 5,6-monoepoxide, as well as several oxygenic functional groups, such as hydroxyl, epoxy, carbonyl, and carboxyl groups (Aman et al., 2005).

One of the previous studies explored how fucoxanthin affects obesity in mice fed a high-fat diet. The findings revealed that fucoxanthin led to notable reductions in body weight and the weights of visceral fat pads. In addition, diets containing fucoxanthin caused significant changes in adipocyte sizes and mRNA expression levels of enzymes involved in both lipogenesis and fatty acid β-oxidation. Notably, fucoxanthin downregulated the activities of several lipogenic enzymes in epididymal adipose tissue, coinciding with a decrease in fatty acid β-oxidation activity (Woo et al., 2009). The activity of β-oxidation plays a role in lipogenesis. β-oxidation enzymes break down fatty acids into acetyl-CoA for energy production. These enzymes work in four steps and include acyl-CoA dehydrogenase, enoyl-CoA hydratase, 3-hydroxyacyl-CoA dehydrogenase, and thiolase. They release acetyl-CoA, NADH, and FADH2 (Adeva-Andany et al., 2019). Acetyl-CoA can be used for energy or stored as triglycerides, indirectly affecting lipogenesis by regulating its precursor availability (Wallace and Metallo, 2020). Xiaojie Hu et al. have pointed out the effects of combined fucoxanthin and conjugated linoleic acid on high-fat diet-induced obese rats. Interestingly, sample treatment reduced serum concentrations of triacylglycerol, glucose, and leptin. The mechanism was explained through β-oxidation of fatty acids and triacylglycerol hydrolysis (Hu et al., 2012). Another study reported that fucoxanthin-rich ethanol extract obtained from seaweed affects the plasma and hepatic lipid profile, fecal lipids, and body fat mass, and alters hepatic cholesterol metabolism, fatty acid synthesis, and lipid absorption in high-fat-fed C57BL/6J mice (Jeon et al., 2010).

The uncoupling protein-1 (UCP-1), primarily found in brown adipose tissue, plays a pivotal role in body weight management by regulating thermogenesis, the process of heat production (Gonzalez-Barroso et al., 2000). Activation of UCP-1 boosts metabolic rate and energy expenditure, promoting fat burning. Ultimately, food supplements or ingredients that have the potential to activate UCP-1 help regulate body temperature and contribute to overall energy balance and weight management. Koo et al. investigated the potential antiobesity effects of a standardized extract from *Phaeodactylum tricornutum*, which includes fucoxanthin. They examined its impact on lipid accumulation in 3T3-L1 adipocytes and in obese mice induced by a high-fat diet. Their findings revealed significant reductions in both body weight gain and adipose tissue weight. This effect was attributed to the activation of UCP-1 and the inhibition of peroxisome proliferator-activated receptor γ and CCAAT/enhancer-binding protein α. In addition, treatment with the extract notably decreased fat mass in mice subjected to a high-fat diet (Koo et al., 2019). Another study explored the potential anti-obesity and antidiabetic properties of lipids rich in fucoxanthin extracted from *Undaria pinnatifida* in mice with diet-induced obesity. The study revealed that administering the sample significantly curbed both body weight and white adipose tissue gain in the mice. Furthermore, the mice fed a diet enriched with fucoxanthin-rich *U. pinnatifida* lipids showed normalization of hyperglycemia, hyperinsulinemia, and hyperleptinemia. In addition, the results indicate that incorporating fucoxanthin-rich *U. pinnatifida* lipids into the diet may help mitigate disruptions in lipid metabolism and insulin resistance (Maeda et al., 2009). The potential of fucoxanthin and its metabolites, such as fucoxanthinol, in attenuating inflammatory changes in the interaction between adipocytes and macrophages has been reported in previous studies. The findings suggested that fucoxanthin, found in edible algae, holds promise as a food ingredient for managing insulin resistance associated with obesity and for preventing metabolic syndrome (Maeda et al., 2015). A double-blind, placebo-controlled study conducted in another research investigation showcased notable decreases in relative body weight, BMI, and visceral fat area among participants receiving fucoxanthin. The group administered 1 mg/day of fucoxanthin exhibited significantly lower values in total fat mass, subcutaneous fat area, waist circumference, and right thigh circumference. The study suggests that fucoxanthin may have the potential to ameliorate moderate overweight conditions in both men and women (Hitoe and Shimoda, 2017).

Kh. Muradian et al. summarized the effects of fucoxanthin supplementation on adiposity and lipid metabolism in rodent models of genetically or diet-induced obesity as follows. Fourteen studies evidenced that fucoxanthin supplementation attenuates body weight gain, while a considerable number of studies mentioned that High-Density Lipoprotein cholesterol is increased in the plasma lipid profile. However, changes in total cholesterol in the plasma lipid profile were not significant. Hormones involved in body weight regulation, such as adiponectin and leptin, were increased in plasma, while insulin levels decreased. Moreover, adipogenic transcriptional factors and lipogenesis (expression and/or activity of lipogenic enzymes) were downregulated, and adipogenesis was suppressed by the fucoxanthin supplementation. Lipolysis (expression and/or activity of lipolytic enzymes) and UCPs were upregulated by the fucoxanthin (Muradian et al., 2015). Furthermore, Kazuno Miyashita and Masashi Hosokawa conducted a review on the potential of fucoxanthin in treating obesity and related disorders. They proposed that further research is necessary to elucidate the molecular mechanisms behind the ability of fucoxanthin to increase energy expenditure. Nonetheless, fucoxanthin holds promise as a valuable asset in managing obesity and its associated conditions (Miyashita and Hosokawa, 2017).

18.4 Seaweeds and gut health

Seaweed offers a plethora of advantages for the gastrointestinal tract due to the presence of various bioactive compounds such as dietary fibers, polyphenols, polysaccharides, and unique pigments like fucoxanthin. First, the high fiber content present in seaweeds, including soluble fibers such as alginate and carrageenan, acts as prebiotics, serving as a food source for beneficial gut bacteria (Charoensiddhi et al., 2020). This promotes the growth of beneficial bacteria such as Bifidobacteria and Lactobacilli while inhibiting the proliferation of harmful pathogens, thereby enhancing gut microbial diversity and balance. In addition, the polysaccharides found in seaweeds possess anti-inflammatory and antioxidant properties, which can help alleviate gastrointestinal inflammation and oxidative stress, contributing to overall gut health (Besednova et al., 2020; Michalak et al., 2022). Moreover, seaweed-derived polyphenols have been shown to modulate gut microbiota composition and improve gut barrier function, thus reducing the risk of intestinal permeability and associated disorders such as leaky gut syndrome (Zang et al., 2023). Furthermore, compounds like fucoxanthin exhibit potential antiobesity effects by promoting lipid metabolism and reducing adiposity, indirectly benefiting gut health by mitigating obesity-related gut dysbiosis and inflammation (Hu et al., 2012; Jeon et al., 2010; Maeda et al., 2015; Muradian et al., 2015; Miyashita and Hosokawa, 2017). Overall, incorporating seaweed into the diet may offer a multifaceted approach to improving gut health, making it a promising dietary aid for promoting gastrointestinal well-being and preventing associated disorders.

18.5 Safety considerations and potential risks of seaweed consumption

While seaweed offers various health benefits, it is important to consider safety considerations and potential risks associated with its consumption. Seaweed can accumulate heavy metals and contaminants from the marine environment, such as mercury, arsenic, and cadmium (Khandaker et al., 2021; Corrias et al., 2020). Prolonged consumption of seaweed harvested from polluted waters may lead to heavy metal toxicity, posing health risks, particularly for vulnerable populations like pregnant women and children. Moreover, excessive consumption of certain seaweed species, such as *Laminaria* spp. and *Undaria* spp., known for their elevated iodine content, can lead to thyroid dysfunction or worsen preexisting thyroid conditions (Bertinato, 2021; González et al., 2020). Furthermore, seaweed, particularly if harvested from polluted waters, may carry bacteria, viruses, or parasites, posing a risk of foodborne illnesses (Elekwachi and Tlusty, 2023). In addition, certain individuals may experience allergic reactions to seaweed proteins or compounds, resulting in symptoms like skin rashes, itching, or gastrointestinal discomfort (Hefft and Fornaciari, 2023). Excessive consumption of seaweed, particularly that high in soluble fibers like alginate, may lead to gastrointestinal issues such as bloating, gas, or diarrhea. Therefore, it is essential to source seaweed from reputable suppliers, consume it in moderation, and be aware of any potential allergic reactions or sensitivities. Consulting with a healthcare professional before incorporating seaweed into the diet, especially for individuals with preexisting health conditions or concerns, is advisable to mitigate potential risks and ensure safe consumption.

References

Adeva-Andany, M.M., Carneiro-Freire, N., Seco-Filgueira, M., Fernández-Fernández, C., Mouriño-Bayolo, D., 2019. Mitochondrial β-oxidation of saturated fatty acids in humans. Mitochondrion. 18728278. 46, 73–90. Available from: https://www.journals.elsevier.com/mitochondrion. doi:10.1016/j.mito.2018.02.009.

Aman, R., Schieber, A., Carle, R., 2005. Effects of heating and illumination on Trans-cis isomerization and degradation of β-carotene and lutein in isolated spinach chloroplasts. J. Agric. Food Chem. 00218561. 53 (24), 9512–9518. doi:10.1021/jf050926w.

Aoe, S., Yamanaka, C., Ohtoshi, H., Nakamura, F., Fujiwara, S., 2021. Effects of daily kelp (Laminaria japonica) intake on body composition, serum lipid levels, and thyroid hormone levels in healthy japanese adults: a randomized, double-blind study. Mar. Drugs. 1660-3397. 19 (7), 352. doi:10.3390/md19070352.

Barnes, R.A., Wong, T., Ross, G.P., Griffiths, M.M., Smart, C.E., Collins, C.E., MacDonald-Wicks, L., Flack, J.R., 2020. Excessive weight gain before and during gestational diabetes mellitus management: what is the impact? Diabetes

Care. 19355548. 43 (1), 74–81. Available from: https://care.diabetesjournals.org/content/43/1/74. doi:10.2337/dc19-0800.

Bertinato, J., 2021. Iodine nutrition: disorders, monitoring and policies. Adv. Food Nutr. Res. 96, 365–415. Available from http://www.elsevier.com/wps/find/bookdescription.cws_home/704624/description#description. doi:10.1016/bs.afnr.2021.01.004.

Besednova, N.N., Zaporozhets, T.S., Kuznetsova, T.A., Makarenkova, I.D., Kryzhanovsky, S.P., Fedyanina, L.N., Ermakova, S.P., 2020. Extracts and marine algae polysaccharides in therapy and prevention of inflammatory diseases of the intestine. Federation Mar. Drugs. 16603397. 18 (6). Available from: https://www.mdpi.com/1660-3397/18/6/289/pdf. doi:10.3390/md18060289.

Blaner, W.S., Shmarakov, I.O., Traber, M.G., 2021. Vitamin A and vitamin e: will the real antioxidant please stand up? Annu. Rev. Nutr. 15454312. 41, 105–131. Available from: http://www.annualreviews.org/journal/nutr. doi:10.1146/annurev-nutr-082018-124228.

Blikra, M.J., Altintzoglou, T., Løvdal, T., Rognså, G., Skipnes, D., Skåra, T., Sivertsvik, M., Fernández, E.N., 2021. Seaweed products for the future: using current tools to develop a sustainable food industry. Trends Food Sci. Technol. 09242244. 118, 765–776. Available from: http://www.elsevier.com/wps/find/journaldescription.cws_home/601278/description#description. doi:10.1016/j.tifs.2021.11.002.

Cassidy, Y.M., McSorley, E.M., Allsopp, P.J., 2018. Effect of soluble dietary fibre on postprandial blood glucose response and its potential as a functional food ingredient. J. Funct. Foods. 17564646. 46, 423–439. Available from: http://www.elsevier.com/wps/find/journaldescription.cws_home/717426/description#description. doi:10.1016/j.jff.2018.05.019.

Charoensiddhi, S., Abraham, R.E., Su, P., Zhang, W., 2020. Seaweed and seaweed-derived metabolites as prebiotics. Adv. Food Nutr. Res. 91, 97–156. Available from http://www.elsevier.com/wps/find/bookdescription.cws_home/704624/description#description. doi:10.1016/bs.afnr.2019.10.001.

Charoensiddhi, S., Conlon, M.A., Franco, C.M.M., Zhang, W., 2017. The development of seaweed-derived bioactive compounds for use as prebiotics and nutraceuticals using enzyme technologies. Trends Food Sci. Technol. 09242244. 70, 20–33. Available from: http://www.elsevier.com/wps/find/journaldescription.cws_home/601278/description#description. doi:10.1016/j.tifs.2017.10.002.

Chater, P.I., Wilcox, M.D., Brownlee, I.A., Pearson, J.P., 2015. Alginate as a protease inhibitor in vitro and in a model gut system; Selective inhibition of pepsin but not trypsin. Carbohydr. Polym. 01448617. 131, 142–151. Available from: http://www.elsevier.com/wps/find/journaldescription.cws_home/405871/description#description. doi:10.1016/j.carbpol.2015.05.062.

Ching, S.H., Bansal, N., Bhandari, B., 2017. Alginate gel particles–A review of production techniques and physical properties. Crit. Rev. Food Sci. Nutr. 15497852. 57 (6), 1133–1152. Available from: www.tandf.co.uk/journals/titles/10408398.asp.. doi:10.1080/10408398.2014.965773.

Cian, R.E., Fajardo, M.A., Alaiz, M., Vioque, J., González, R.J., Drago, S.R., 2014. Chemical composition, nutritional and antioxidant properties of the red edible seaweed Porphyra columbina. Int. J. Food Sci. Nutr. 14653478. 65 (3), 299–305. Available from: www.tandf.co.uk/journals/titles/09637486.asp.. doi:10.3109/09637486.2013.854746.

Citkowska, A., Szekalska, M., Winnicka, K., 2019. Possibilities of fucoidan utilization in the development of pharmaceutical dosage forms. Mar. Drugs. 1660-3397. 17 (8), 458. doi:10.3390/md17080458.

Corrias, F., Atzei, A., Addis, P., Secci, M., Russo, M., Angioni, A., 2020. Integrated environmental evaluation of heavy metals and metalloids bioaccumulation in invertebrates and seaweeds from different marine coastal areas of sardinia, Mediterranean sea. Environ. Pollut. 02697491. 266, 115048. doi:10.1016/j.envpol.2020.115048.

Danilova, I.G., Aboushanab, S.A., Sokolova, K.V., Ravishankar, G.A., Rao, A.R., Kovaleva, E.G., 2022. Anti-diabetic properties of Fucoidan from different fucus species. Rao, A.R., Ravishankar, G.A., Sustainable Global Resources of Seaweeds Volume 2: Food, Pharmaceutical and Health Applications. Springer International Publishing, 579–595. Available from: https://link.springer.com/book/10.1007/978-3-030-92174-3. doi:10.1007/978-3-030-92174-3_31.

Daub, C.D., Mabate, B., Malgas, S., Pletschke, B.I., 2020. Fucoidan from Ecklonia maxima is a powerful inhibitor of the diabetes-related enzyme, α-glucosidase. Int. J. Biol. Macromol. 18790003. 151, 412–420. Available from: www.elsevier.com/locate/ijbiomac.. doi:10.1016/j.ijbiomac.2020.02.161.

Denis, C., Morançais, M., Li, M., Deniaud, E., Gaudin, P., Wielgosz-Collin, G., Barnathan, G., Jaouen, P., Fleurence, J., 2010. Study of the chemical composition of edible red macroalgae Grateloupia turuturu from Brittany (France). Food Chem. 03088146. 119 (3), 913–917. doi:10.1016/j.foodchem.2009.07.047.

Dent, R., McPherson, R., Harper, M.-E., 2020. Factors affecting weight loss variability in obesity. Metabolism. 00260495. 113, 154388. doi:10.1016/j.metabol.2020.154388.

Dettmar, P.W., Strugala, V., Richardson, J.C., 2011. The key role alginates play in health. Food Hydrocoll. 0268005X. 25 (2), 263–266. Available from: https://www.journals.elsevier.com/food-hydrocolloids. doi:10.1016/j.foodhyd.2009.09.009.

Dickson, R., Ryu, J.H., Liu, J.J., 2018. Optimal plant design for integrated biorefinery producing bioethanol and protein from Saccharina japonica: a superstructure-based approach. Energy. 03605442. 164, 1257–1270. Available from: www.elsevier.com/inca/publications/store/4/8/3/. doi:10.1016/j.energy.2018.09.007.

Dumay, J., Morançais, M., 2016. Proteins and Pigments. Fleurence, B., Levine, I., Seaweed in Health and Disease Prevention. Elsevier Inc., 275–318. Available from: http://www.sciencedirect.com/science/book/9780128027721. doi:10.1016/B978-0-12-802772-1.00009-9.

Elekwachi, D.H., Tlusty, M.F., 2023. Review of the Literature on the Seaweed Industry and Food Safety Issues. Sea Grant Law Policy J. 12, 89.

Fan, Y., Hong, R., Sun, X., Luo, Q., Wei, H., Chen, Y., Zhang, Z., Zhou, X., Wan, J., 2024. Gastric acid-responsive deformable sodium alginate/Bletilla striata polysaccharide in situ gel for the protection and treatment of alcohol-induced peptic ulcers. Int. J. Biol. Macromol. 01418130. 258, 128815. doi:10.1016/j.ijbiomac.2023.128815.

Fayaz, M., Namitha, K.K., Murthy, K.N.C., Swamy, M.M., Sarada, R., Khanam, S., Subbarao, P.V., Ravishankar, G.A., 2005. Chemical composition, iron bioavailability, and antioxidant activity of Kappaphycus alvarezzi (Doty). J. Agric. Food Chem. 00218561. 53 (3), 792–797. doi:10.1021/jf0493627.

Fernando, I.P.S., Kirindage, K.G.I.S., Jayasinghe, A.M.K., Han, E.J., Lee, C.M., Heo, S.J., Ahn, G., 2022a. Alginate nanocapsules by water-in-oil emulsification and external gelation for drug delivery to fine dust stimulated keratinocytes. Int. J. Biol. Macromol. 18790003. 218, 102–114. Available from: www.elsevier.com/locate/ijbiomac.. doi:10.1016/j.ijbiomac.2022.07.097.

Fernando, I.P.S., Kirindage, K.G.I.S., Jeon, H.N., Han, E.J., Jayasinghe, A.M.K., Ahn, G., 2022b. Preparation of microspheres by alginate purified from Sargassum horneri and study of pH-responsive behavior and drug release. Int. J. Biol. Macromol. 18790003. 202, 681–690. Available from: www.elsevier.com/locate/ijbiomac. doi:10.1016/j.ijbiomac.2022.01.171.

Ferraces-Casais, P., Lage-Yusty, M.A., de Quirós, A.R.-B., López-Hernández, J., 2012. Evaluation of bioactive compounds in fresh edible seaweeds. Food Anal. Methods. 1936-9751. 5 (4), 828–834. doi:10.1007/s12161-011-9321-2.

Figueroa, V., Farfán, M., Aguilera, J.M., 2023. Seaweeds as novel foods and source of culinary flavors. Food Rev. Int. 8755-9129. 39 (1), 1–26. doi:10.1080/87559129.2021.1892749.

Fleurence, J., Gutbier, G., Mabeau, S., Leray, C., 1994. Fatty acids from 11 marine macroalgae of the French Brittany coast. J. Appl. Phycol. 15735176. 6 (5-6), 527–532. doi:10.1007/BF02182406.

Fleurence, J., Morançais, M., Dumay, J., 2018. Seaweed proteins. In: Yada, R.Y., Ed.. Proteins in Food Processing (2nd ed). Elsevier (Woodhead Publishing), pp. 245–262. doi:10.1016/b978-0-08-100722-8.00010-3.

Foster, K.A., Morgen, M., Murri, B., Yates, I., Fancher, R.M., Ehrmann, J., Gudmundsson, O.S., Hageman, M.J., 2012. Utility of in situ sodium alginate/karaya gum gels to facilitate gastric retention in rodents. Int. J. Pharm. 18733476. 434 (1-2), 406–412. doi:10.1016/j.ijpharm.2012.06.009.

Francavilla, M., Manara, P., Kamaterou, P., Monteleone, M., Zabaniotou, A., 2015. Cascade approach of red macroalgae Gracilaria gracilis sustainable valorization by extraction of phycobiliproteins and pyrolysis of residue. Bioresour. Technol. 09608524. 184, 305–313. doi:10.1016/j.biortech.2014.10.147.

Giuntini, E.B., Sardá, F.A.H., de Menezes, E.W., 2022. The effects of soluble dietary fibers on glycemic response: an overview and futures perspectives. Foods. 23048158. 11 (23). Available from: www.mdpi.com/journal/foods.. doi:10.3390/foods11233934.

Gómez-Ordóñez, E., Jiménez-Escrig, A., Rupérez, P., 2010. Dietary fibre and physicochemical properties of several edible seaweeds from the northwestern Spanish coast. Food Res. Int. 09639969. 43 (9), 2289–2294. doi:10.1016/j.foodres.2010.08.005.

Gonzalez-Barroso, M.D.M., Ricquier, D., Cassard-Doulcier, A.-.M., 2000. The human uncoupling protein-1 gene (UCP1): present status and perspectives in obesity research. Obes. Rev. 1467-7881. 1 (2), 61–72. doi:10.1046/j.1467-789x.2000.00009.x.

González, A., Paz, S., Rubio, C., Gutiérrez, Á.J., Hardisson, A., 2020. Human Exposure to Iodine from the Consumption of Edible Seaweeds. Biol. Trace Elem. Res. 15590720. 197 (2), 361–366. Available from: http://www.springer.com/humana+press/journal/12011. doi:10.1007/s12011-019-01996-w.

Gullón, B., Gagaoua, M., Barba, F.J., Gullón, P., Zhang, W., Lorenzo, J.M., 2020. Seaweeds as promising resource of bioactive compounds: overview of novel extraction strategies and design of tailored meat products. Trends Food Sci. Technol. 09242244. 100, 1–18. Available from: http://www.elsevier.com/wps/find/journaldescription.cws_home/601278/description#description. doi:10.1016/j.tifs.2020.03.039.

Harnedy, P.A., Fitzgerald, R.J., 2011. Bioactive proteins, peptides, and amino acids from macroalgae. J. Phycol. 15298817. 47 (2), 218–232. doi:10.1111/j.1529-8817.2011.00969.x.

Hasnain, M.S., Jameel, E., Mohanta, B., Dhara, A.K., Alkahtani, S., Nayak, A.K., 2020. Alginates: sources, structure, and properties. Nayak, A.K., Hasnain, M.S. Eds., Alginates in Drug Delivery. Elsevier, 1–17. Available from: https://www.sciencedirect.com/book/9780128176405/alginates-in-drug-delivery. doi:10.1016/B978-0-12-817640-5.00001-7.

Hefft, D.I., Fornaciari, G., 2023. Safety and allergenicity of seaweeds. Fornaciari, G., Hefft, D.I. Eds., Applications of Seaweeds in Food and Nutrition. Elsevier, 195–203. Available from: https://www.sciencedirect.com/book/9780323918039. doi:10.1016/B978-0-323-91803-9.00013-5.

Hentati, F., Barkallah, M., Atitallah, A.B., Dammak, M., Louati, I., Pierre, G., Fendri, I., Attia, H., Michaud, P., Abdelkafi, S., 2019. Quality characteristics and functional and antioxidant capacities of algae-fortified fish burgers prepared from common barbel (Barbus barbus). Biomed. Res. Int. 23146141. 2019. Available from: http://www.hindawi.com/journals/biomed/. doi:10.1155/2019/2907542.

Hentati, F., Delattre, C., Ursu, A.V., Desbrières, J., Cerf, D.L., Gardarin, C., Abdelkafi, S., Michaud, P., Pierre, G., 2018. Structural characterization and antioxidant activity of water-soluble polysaccharides from the Tunisian brown seaweed Cystoseira compressa. Carbohydr. Polym. 01448617. 198, 589–600. Available from: http://www.elsevier.com/wps/find/journaldescription.cws_home/405871/description#description. doi:10.1016/j.carbpol.2018.06.098.

Hentati, F., Tounsi, L., Djomdi, D., Pierre, G., Delattre, C., Ursu, A.V., Fendri, I., Abdelkafi, S., Michaud, P., 2020a. Bioactive polysaccharides from seaweeds. Molecules. 14203049. 25 (14). Available from: https://www.mdpi.com/1420-3049/25/14/3152/pdf. doi:10.3390/molecules25143152.

Hernández-Carmona, G., Carrillo-Domínguez, S., Arvizu-Higuera, D.L., Rodríguez-Montesinos, Y.E., Murillo-Álvarez, J.I., Muñoz-Ochoa, M., Castillo-Domínguez, R.M., 2009. Monthly variation in the chemical composition of Eisenia arborea J.E. Areschoug. J. Appl. Phycol. 09218971. 21 (5), 607–616. doi:10.1007/s10811-009-9454-5.

Hitoe, S., Shimoda, H., 2017. Seaweed fucoxanthin supplementation improves obesity parameters in mildly obese Japanese subjects. Funct. Foods Health Dis. 21603855. 7 (4), 246–262. Available from: https://ffhdj.com/index.php/ffhd/article/view/333. doi:10.31989/ffhd.v7i4.333.

Hoad, C.L., Rayment, P., Spiller, R.C., Marciani, L., Alonso, B.D.C., Traynor, C., Mela, D.J., Peters, H.P.F., Gowland, P.A., 2004. In vivo imaging of intragastric gelation and its effect on satiety in humans. J. Nutr. 00223166. 134 (9), 2293–2300. Available from: http://jn.nutrition.org. doi:10.1093/jn/134.9.2293.

Hu, X., Li, Y., Li, C., Fu, Y., Cai, F., Chen, Q., Li, D., 2012. Combination of fucoxanthin and conjugated linoleic acid attenuates body weight gain and improves lipid metabolism in high-fat diet-induced obese rats. Arch. Biochem. Biophys. 10960384. 519 (1), 59–65. doi:10.1016/j.abb.2012.01.011.

Jayasinghe, A.M.K., Kirindage, K.G.I.S., Fernando, I.P.S., Han, E.J., Oh, G.W., Jung, W.K., Ahn, G., 2022. Fucoidan isolated from sargassum confusum suppresses inflammatory responses and oxidative stress in TNF-α/IFN-γ-stimulated HaCaT keratinocytes by activating Nrf2/HO-1 signaling pathway. Mar. Drugs. 16603397. 20 (2). Available from: https://www.mdpi.com/1660-3397/20/2/117/pdf. doi:10.3390/md20020117.

Jayasinghe, A.M.K., Kirindage, K.G.I.S., Fernando, I.P.S., Kim, K.N., Oh, J.Y., Ahn, G., 2023. The anti-inflammatory effect of low molecular weight fucoidan from sargassum siliquastrum in lipopolysaccharide-stimulated RAW 264.7 macrophages via Inhibiting NF-κB/MAPK signaling pathways. Mar. Drugs. 16603397. 21 (6). Available from: http://www.mdpi.com/journal/marinedrugs. doi:10.3390/md21060347.

Jensen, M.G., Knudsen, J.C., Viereck, N., Kristensen, M., Astrup, A., 2012. Functionality of alginate based supplements for application in human appetite regulation. Food Chem. 03088146. 132 (2), 823–829. doi:10.1016/j.foodchem.2011.11.042.

Jeon, S.M., Kim, H.J., Woo, M.N., Lee, M.K., Shin, Y.C., Park, Y.B., Choi, M.S., 2010. Fucoxanthin-rich seaweed extract suppresses body weight gain and improves lipid metabolism in high-fat-fed C57BL/6J mice. Biotechnol. J. 18607314. 5(9), 961–969. Available from: http://onlinelibrary.wiley.com/doi/10.1002/biot.201000215/pdf. doi:10.1002/biot.201000215

Jeong, Y.T., Kim, Y.D., Jung, Y.M., Park, D.C., Lee, D.S., Ku, S.K., Li, X., Lu, Y., Chao, G.H., Kim, K.J., Lee, J.Y., Baek, M.C., Kang, W., Hwang, S.L., Chang, H.W., 2013. Low molecular weight fucoidan improves endoplasmic reticulum stress-reduced insulin sensitivity through ampactivated protein kinase activation in l6 myotubes and restores lipid homeostasis in a mouse model of type 2 diabetess. Mol. Pharmacol. 15210111. 84(1), 147–157. Available from: http://molpharm.aspetjournals.org/content/84/1/147.full.pdf+html. doi: 10.1124/mol.113.085100

Jiménez-Escrig, A., Sánchez-Muniz, F.J., 2000. Dietary fibre from edible seaweeds: chemical structure, physicochemical properties and effects on cholesterol metabolism. Nutr. Res. 02715317. 20 (4), 585–598. doi:10.1016/s0271-5317(00)00149-4.

Kato, T., Idota, Y., Shiragami, K., Koike, M., Nishibori, F., Tomokane, M., Seki, T., Itabashi, K., Hakoda, K., Takahashi, H., Yano, K., Kobayashi, T., Obara, N., Ogihara, T., 2018. Randomized, double-blind, crossover clinical trial of the effect of calcium alginate in noodles on postprandial blood glucose level. Biol. Pharm. Bull. 13475215. 41 (9), 1367–1371. Available from: https://www.jstage.jst.go.jp/article/bpb/41/9/41_b18-00156/_pdf/-char/en. doi:10.1248/bpb.b18-00156.

Khandaker, M.U., Chijioke, N.O., Heffny, N.A.B., Bradley, D.A., Alsubaie, A., Sulieman, A., Faruque, M.R.I., Sayyed, M., Al-Mugren, K., 2021. Elevated concentrations of metal (loids) in seaweed and the concomitant exposure to humans. Foods 10(2), 381.

Khoury, D.E., Goff, H.D., Anderson, G.H., 2015. The role of alginates in regulation of food intake and glycemia: a gastroenterological perspective. Crit. Rev. Food Sci. Nutr. 15497852. 55 (10), 1406–1424. Available from: www.tandf.co.uk/journals/titles/10408398.asp. doi:10.1080/10408398.2012.700654.

Khoury, D.E., Goff, H.D., Berengut, S., Kubant, R., Anderson, G.H., 2014. Effect of sodium alginate addition to chocolate milk on glycemia, insulin, appetite and food intake in healthy adult men. Eur. J. Clin. Nutr. 0954-3007. 68 (5), 613–618. doi:10.1038/ejcn.2014.53.

Kirindage, K.G.I.S., Jayasinghe, A.M.K., Cho, N., Cho, S.H., Yoo, H.M., Fernando, I.P.S., Ahn, G., 2023. Fine-dust -induced skin inflammation: low-molecular-weight fucoidan protects keratinocytes and underlying fibroblasts in an integrated culture model. Mar. Drugs. 16603397. 21 (1). Available from: http://www.mdpi.com/journal/marinedrugs. doi:10.3390/md21010012.

Kokilam, G., Vasuki, S., 2014. Biochemical and phytochemical analysis on Ulva fasciata and Caulerpa taxifolia. Int. J. Pharm. Pharm. Sci. Res. 4 (1), 7–11.

Koo, S.Y., Hwang, J.H., Yang, S.H., Um, J.I., Hong, K.W., Kang, K., Pan, C.H., Hwang, K.T., Kim, S.M., 2019. Anti-obesity effect of standardized extract of microalga phaeodactylum tricornutum containing fucoxanthin. Mar. Drugs. 16603397. 17 (5). Available from: https://www.mdpi.com/1660-3397/17/5/311/pdf. doi:10.3390/md17050311.

Kraan, S., 2012. Algal Polysaccharides, Novel Applications and Outlook. In: Chang, C.-F.. (Ed..), Carbohydrates – Comprehensive Studies on Glycobiology and Glycotechnology, InTech. doi: 10.5772/51572

Kumanyika, S.K., Obarzanek, E., Stettler, N., Bell, R., Field, A.E., Fortmann, S.P., Franklin, B.A., Gillman, M.W., Lewis, C.E., Poston, W.C., Stevens, J., Hong, Y., 2008. Population-based prevention of obesity: the need for comprehensive promotion of healthful eating, physical activity, and energy balance: a scientific statement from American Heart Association Council on Epidemiology and Prevention, Interdisciplinary Committee for prevention (formerly the expert panel on population and prevention science). Circulation. 15244539. 118 (4), 428–464. doi:10.1161/CIRCULATIONAHA.108.189702.

Kumari, P., Kumar, M., Gupta, V., Reddy, C.R.K., Jha, B., 2010. Tropical marine macroalgae as potential sources of nutritionally important PUFAs. Food Chem. 03088146. 120 (3), 749–757. doi:10.1016/j.foodchem.2009.11.006.

Lee, H.G., Jayawardhana, H.H.A.C.K., Yang, F., Nagahawaththa, D.P., Liyanage, N.M., Song, K.M., Choi, Y.S., Lee, S. H., Jeon, Y.J., Kang, M.C., 2024. Anti-obesity effects of fucoidan from Sargassum thunbergii in adipocytes and high fat diet induced obese mice through inhibiting adipogenic specifi c transcription factor. Food Sci. Hum. Wellness. 22134530. 13 (3), 1608–1616. Available from: https://www.sciopen.com/article/10.26599/FSHW.2022.9250136. doi:10.26599/FSHW.2022.9250136.

Li, S., Li, M., Guo, R., Zhao, T., Gao, X., Li, K., Guo, X., Li, J., Li, D., 2022b. Fucoidans from Pearsonothuria graeffei prevent obesity by regulating intestinal lipid metabolism and inflammation related signalling pathways. Food Funct. 2042650X. 13 (23), 12234–12245. Available from: http://pubs.rsc.org/en/journals/journal/fo. doi:10.1039/d2fo02588d.

Liu, Y., Wang, N., Tian, Y., Chang, Y., Wang, J., 2021. Fucoidans from Thelenota ananas with 182.4 kDa exhibited optimal anti-adipogenic activities by modulating the Wnt/β-Catenin pathway. J. Ocean Univ. China. 19935021. 20 (4), 921–930. Available from: http://www.springerlink.com/content/1672-5182. doi:10.1007/s11802-021-4681-8.

Lomartire, S., Marques, J.C., Gonçalves, A.M.M., 2021. An overview to the health benefits of seaweeds consumption. Mar. Drugs. 16603397. 19 (6). Available from: https://www.mdpi.com/1660-3397/19/6/341. doi:10.3390/md19060341.

Lorenzo, J.M., Agregán, R., Munekata, P.E.S., Franco, D., Carballo, J., Şahin, S., Lacomba, R., Barba, F.J., 2017. Proximate composition and nutritional value of three macroalgae: ascophyllum nodosum, Fucus vesiculosus and Bifurcaria bifurcata. Mar. Drugs. 16603397. 15 (11). Available from: http://www.mdpi.com/1660-3397/15/11/360/pdf. doi:10.3390/md15110360.

Mabeau, S., Fleurence, J., 1993. Seaweed in food products: biochemical and nutritional aspects. Trends Food Sci. Technol. 09242244. 4 (4), 103–107. doi:10.1016/0924-2244(93)90091-N.

MacArtain, P., Gill, C.I.R., Brooks, M., Campbell, R., Rowland, I.R., 2007. Nutritional value of edible seaweeds. Nutr. Rev. 00296643. 65 (12), 535–543. Available from: http://nutritionreviews.oxfordjournals.org/. doi:10.1301/nr.2007.dec.535-543.

Maeda, H., Hosokawa, M., Sashima, T., Murakami-Funayama, K., Miyashita, K., 2009. Anti-obesity and anti-diabetic effects of fucoxanthin on diet-induced obesity conditions in a murine model. Mol. Med. Rep. 17913004. 2 (6), 897–902. Available from: http://www.spandidos-publications.com/serveFile/mmr_2_6_897_PDF.pdf?type=article&article_id=mmr_2_6_897&item=PDF. doi:10.3892/mmr_00000189.

Maeda, H., Kanno, S., Kodate, M., Hosokawa, M., Miyashita, K., 2015. Fucoxanthinol, metabolite of fucoxanthin, improves obesity-induced inflammation in adipocyte cells. Mar. Drugs. 16603397. 13 (8), 4799–4813. Available from: http://www.mdpi.com/1660-3397/13/8/4799/pdf. doi:10.3390/md13084799.

Marinho-Soriano, E., Fonseca, P.C., Carneiro, M.A.A., Moreira, W.S.C., 2006. Seasonal variation in the chemical composition of two tropical seaweeds. Bioresour. Technol. 09608524. 97 (18), 2402–2406. doi:10.1016/j.biortech.2005.10.014.

Marsham, S., Scott, G.W., Tobin, M.L., 2007. Comparison of nutritive chemistry of a range of temperate seaweeds. Food Chem. 03088146. 100 (4), 1331–1336. doi:10.1016/j.foodchem.2005.11.029.

Matanjun, P., Mohamed, S., Mustapha, N.M., Muhammad, K., 2009. Nutrient content of tropical edible seaweeds, Eucheuma cottonii, Caulerpa lentillifera and Sargassum polycystum. J. Appl. Phycol. 09218971. 21 (1), 75–80. doi:10.1007/s10811-008-9326-4.

Mehta, M., Istfan, N.W., Apovian, C.M., 2021. Obesity: overview of weight management. Endocr. Pract. 19342403. 27 (6), 626–635. Available from: https://www.sciencedirect.com/journal/endocrine-practice. doi:10.1016/j.eprac.2021.04.001.

Meinita, M.D.N., Harwanto, D., Sohn, J.H., Kim, J.S., Choi, J.S., 2021. Hizikia fusiformis: pharmacological and nutritional properties. Foods. 23048158. 10 (7). Available from: https://www.mdpi.com/2304-8158/10/7/1660/pdf. doi:10.3390/foods10071660.

Michalak, I., Chojnacka, K., 2018. Seaweeds as a component of the human diet. In: Chojnacka, K., Ed.. Algae Biomass: Characteristics and Applications. Springer, pp. 57–71. doi:10.1007/978-3-319-74703-3_6.

Michalak, I., Tiwari, R., Dhawan, M., Alagawany, M., Farag, M.R., Sharun, K., Emran, T.B., Dhama, K., 2022. Antioxidant effects of seaweeds and their active compounds on animal health and production–a review. Vet. Q. 18755941. 42 (1), 48–67. Available from: http://www.tandf.co.uk/journals/TVEQ. doi:10.1080/01652176.2022.2061744.

Mišurcová, L., 2011. Chemical composition of seaweeds. Kim, S.-K., Handbook of Marine Macroalgae: Biotechnology and Applied Phycology. John Wiley and Sons, 171–192. Available from: http://onlinelibrary.wiley.com/book/10.1002/9781119977087. doi:10.1002/9781119977087.ch7.

Mišurcová, L., Machů, L., Orsavová, J., 2011. Seaweed minerals as nutraceuticals. Adv. Food Nutr. Res. 64, 371–390. Available from http://www.elsevier.com/wps/find/bookdescription.cws_home/704624/description#description. doi:10.1016/B978-0-12-387669-0.00029-6.

Miyashita, K., Hosokawa, M., 2017. Fucoxanthin in the management of obesity and its related disorders. J. Funct. Foods. 17564646. 36, 195–202. Available from: http://www.elsevier.com/wps/find/journaldescription.cws_home/717426/description#description. doi:10.1016/j.jff.2017.07.009.

Muñoz, I.L., Díaz, N.F., 2020. Minerals in edible seaweed: health benefits and food safety issues. Crit. Rev. Food Sci. Nutr. 15497852. 62 (6), 1592–1607. Available from: https://www.tandfonline.com/loi/bfsn20. doi:10.1080/10408398.2020.1844637.

Muradian, K., Vaiserman, A., Min, K.J., Fraifeld, V.E., 2015. Fucoxanthin and lipid metabolism: a minireview. Nutr. Metab. Cardiovasc. Dis. 15903729. 25 (10), 891–897. Available from: http://www.elsevier.com/wps/find/journaldescription.cws_home/704955/description#description. doi:10.1016/j.numecd.2015.05.010.

Murai, U., Yamagishi, K., Kishida, R., Iso, H., 2021. Impact of seaweed intake on health. Eur. J. Clin. Nutr. 14765640. 75 (6), 877–889. Available from: http://www.nature.com/ejcn/index.html. doi:10.1038/s41430-020-00739-8.

Nagappan, T., Vairappan, C.S., 2014. Nutritional and bioactive properties of three edible species of green algae, genus Caulerpa (Caulerpaceae). J. Appl. Phycol. 15735176. 26 (2), 1019–1027. Available from: www.wkap.nl/journalhome.htm/0921-8971. doi:10.1007/s10811-013-0147-8.

Naseri, A., Marinho, G.S., Holdt, S.L., Bartela, J.M., Jacobsen, C., 2020. Enzyme-assisted extraction and characterization of protein from red seaweed Palmaria palmata. Algal Res. 22119264. 47. Available from: http://www.sciencedirect.com/science/journal/aip/22119264. doi:10.1016/j.algal.2020.101849.

Norziah, M.H., Ching, C.Y., 2000. Nutritional composition of edible seaweed Gracilaria changgi. Food Chem. 03088146. 68 (1), 69–76. doi:10.1016/S0308-8146(99)00161-2.

Nunes, R., Ferreira-Santos, P., Moreira, C., Teixeira, J.A., Rocha, C.M.R., 2024. Tuning the extraction methodology targeting protein-enriched fractions from red algae. Future Foods. 26668335. 9, 100335. doi:10.1016/j.fufo.2024.100335.

Ortiz, J., Romero, N., Robert, P., Araya, J., Lopez-Hernández, J., Bozzo, C., Navarrete, E., Osorio, A., Rios, A., 2006. Dietary fiber, amino acid, fatty acid and tocopherol contents of the edible seaweeds Ulva lactuca and Durvillaea Antarctica. Food Chem. 03088146. 99 (1), 98–104. doi:10.1016/j.foodchem.2005.07.027.

Paiva, L., Lima, E., Patarra, R.F., Neto, A.I., Baptista, J., 2014. Edible Azorean macroalgae as source of rich nutrients with impact on human health. Food Chem. 18737072. 164, 128–135. Available from: www.elsevier.com/locate/foodchem.. doi:10.1016/j.foodchem.2014.04.119.

Parjikolaei, B.R., Bruhn, A., Eybye, K.L., Mørk Larsen, M., Rasmussen, M.B., Christensen, K.V., Fretté, X.C., 2016. Valuable biomolecules from nine North Atlantic red macroalgae: amino acids, fatty acids, carotenoids, minerals and metals. Nat. Res. 2158-706X. 7 (4), 157–183. doi:10.4236/nr.2016.74016.

Patarra, R.F., Paiva, L., Neto, A.I., Lima, E., Baptista, J., 2011. Nutritional value of selected macroalgae. J. Appl. Phycol. 09218971. 23 (2), 205–208. doi:10.1007/s10811-010-9556-0.

Pelkman, C.L., Navia, J.L., Miller, A.E., Pohle, R.J., 2007. Novel calcium-gelled, alginate-pectin beverage reduced energy intake in nondieting overweight and obese women: interactions with dietary restraint status. Am. J. Clin. Nutr. 00029165. 86 (6), 1595–1602. Available from: http://www.ajcn.org/cgi/reprint/86/6/1595. doi:10.1093/ajcn/86.5.1595.

Peña-Rodríguez, A., Mawhinney, T.P., Ricque-Marie, D., Cruz-Suárez, L.E., 2011. Chemical composition of cultivated seaweed Ulva clathrata (Roth) C. Agardh. Food Chem. 18737072. 129 (2), 491–498. doi:10.1016/j.foodchem.2011.04.104.

Peñalver, R., Lorenzo, J.M., Ros, G., Amarowicz, R., Pateiro, M., Nieto, G., 2020. Seaweeds as a functional ingredient for a healthy diet. Mar. Drugs. 16603397. 18 (6). Available from: https://www.mdpi.com/1660-3397/18/6/301/pdf. doi:10.3390/md18060301.

Peng, Y., Xie, E., Zheng, K., Fredimoses, M., Yang, X., Zhou, X., Wang, Y., Yang, B., Lin, X., Liu, J., Liu, Y., 2013. Nutritional and chemical composition and antiviral activity of cultivated seaweed sargassum naozhouense Tseng et Lu. Mar. Drugs. 16603397. 11 (1), 20–32. Available from: http://www.mdpi.com/1660-3397/11/1/20/pdf. doi:10.3390/md11010020.

Pereira, L., 2011. A review of the nutrient composition of selected edible seaweeds. In: Seckbach, J. (Ed.), Seaweed: Ecology, Nutrient Composition and Medicinal Uses. Nova Science Publishers, Inc., pp. 15–49. Available from https://www.novapublishers.com/catalog/product_info.php?products_id=27922.

Pliego-Cortés, H., Wijesekara, I., Lang, M., Bourgougnon, N., Bedoux, G., 2020. Current knowledge and challenges in extraction, characterization and bioactivity of seaweed protein and seaweed-derived proteins. Adv. Bot. Res. 00652296. 95, 289–326. Available from: http://www.elsevier.com/wps/find/bookdescription.cws_home/675801/description#description. doi:10.1016/bs.abr.2019.11.008.

Pradhan, B., Bhuyan, P.P., Patra, S., Nayak, R., Behera, P.K., Behera, C., Behera, A.K., Ki, J.S., Jena, M., 2022. Beneficial effects of seaweeds and seaweed-derived bioactive compounds: current evidence and future prospective. Biocatal. Agric. Biotechnol. 18788181. 39. Available from: http://www.journals.elsevier.com/biocatalysis-and-agricultural-biotechnology/. doi:10.1016/j.bcab.2021.102242.

Qin, Y., Jiang, J., Zhao, L., Zhang, J., Wang, F., 2018. Applications of Alginate as a Functional Food Ingredient. Kasapis, S.M., Norton, T., McCleary, J.B.J. Eds., Biopolymers for Food Design. Elsevier Inc., 409–429. Available from: http://www.sciencedirect.com/science/book/9780128114490. doi:10.1016/B978-0-12-811449-0.00013-X.

Raja, K., Kadirvel, V., Subramaniyan, T., 2022. Seaweeds, an aquatic plant-based protein for sustainable nutrition - A review. Future Foods. 26668335. 5 (100142). doi:10.1016/j.fufo.2022.100142.

Rajapakse, N., Kim, S.K., 2011. Nutritional and digestive health benefits of seaweed. Adv. Food Nutr. Res. 64, 17–28. Available from: http://www.elsevier.com/wps/find/bookdescription.cws_home/704624/description#description. doi:10.1016/B978-0-12-387669-0.00002-8.

Ratana-arporn, P., Chirapart, A., 2006. Nutritional evaluation of tropical green seaweeds Caulerpa lentillifera and Ulva reticulata. Kasetsart J. Nat. Sci. 00755192. 40, 75–83.

Raus, R.A., Nawawi, W.M.F.W., Nasaruddin, R.R., 2021. Alginate and alginate composites for biomedical applications. Asian J. Pharm. Sci. 2221285X. 16 (3), 280–306. Available from: http://www.journals.elsevier.com/asian-journal-of-pharmaceutical-sciences/recent-articles/. doi:10.1016/j.ajps.2020.10.001.

Remesar, X., Alemany, M., 2020. Dietary energy partition: the central role of glucose. Int. J. Mol. Sci. 1422-0067. 21 (20), 7729. doi:10.3390/ijms21207729.

Rodrigues, D., Freitas, A.C., Pereira, L., Rocha-Santos, T.A.P., Vasconcelos, M.W., Roriz, M., Rodríguez-Alcalá, L.M., Gomes, A.M.P., Duarte, A.C., 2015. Chemical composition of red, brown and green macroalgae from Buarcos bay in Central West Coast of Portugal. Food Chem. 18737072. 183, 197–207. Available from: www.elsevier.com/locate/foodchem.. doi:10.1016/j.foodchem.2015.03.057.

Rupérez, P., Saura-Calixto, F., 2001. Dietary fibre and physicochemical properties of edible Spanish seaweeds. Eur. Food Res. Technol. 14382385. 212 (3), 349–354. Available from: http://link.springer.de/link/service/journals/00217/index.htm. doi:10.1007/s002170000264.

Salido, M., Soto, M., Seoane, S., 2024. Seaweed: nutritional and gastronomic perspective. A review. Algal Res. 22119264. 77 (103357). doi:10.1016/j.algal.2023.103357.

Sánchez-Machado, D.I., López-Cervantes, J., López-Hernández, J., Paseiro-Losada, P., 2004. Fatty acids, total lipid, protein and ash contents of processed edible seaweeds. Food Chem. 03088146. 85 (3), 439–444. doi:10.1016/j.foodchem.2003.08.001.

Sandner, G., König, A., Wallner, M., Weghuber, J., 2020. Functional foods - dietary or herbal products on obesity: application of selected bioactive compounds to target lipid metabolism. Curr. Opin. Food Sci. 22147993. 34, 9–20. Available from: http://www.journals.elsevier.com/current-opinion-in-food-science/. doi:10.1016/j.cofs.2020.09.011.

Sarkar, S., Das, D., Dutta, P., Kalita, J., Wann, S.B., Manna, P., 2020. Chitosan: a promising therapeutic agent and effective drug delivery system in managing diabetes mellitus. Carbohydr. Polym. 01448617. 247, 116594. doi:10.1016/j.carbpol.2020.116594.

Shannon, E., Abu-Ghannam, N., 2019. Seaweeds as nutraceuticals for health and nutrition. Phycologia. 23302968. 58 (5), 563–577. Available from: https://www.tandfonline.com/action/showAxaArticles?journalCode=uphy20. doi:10.1080/00318884.2019.1640533.

Škrovánková, S., 2011. Seaweed vitamins as nutraceuticals. Adv. Food Nutr. Res. 64, 357–369. Available from http://www.elsevier.com/wps/find/bookdescription.cws_home/704624/description#description. doi:10.1016/B978-0-12-387669-0.00028-4.

Smith, J.L., Summers, G., Wong, R., 2010. Nutrient and heavy metal content of edible seaweeds in New Zealand. N. Z. J. Crop Hortic. Sci. 0114-0671. 38 (1), 19–28. doi:10.1080/01140671003619290.

Tabarsa, M., Rezaei, M., Ramezanpour, Z., Waaland, J.R., 2012. Chemical compositions of the marine algae Gracilaria salicornia (Rhodophyta) and Ulva lactuca (Chlorophyta) as a potential food source. J. Sci. Food Agric. 10970010. 92 (12), 2500–2506. doi:10.1002/jsfa.5659.

Taboada, C., Millán, R., Míguez, I., 2010. Composition, nutritional aspects and effect on serum parameters of marine algae Ulva rigida. J. Sci. Food Agric. 10970010. 90 (3), 445–449. Available from: http://www3.interscience.wiley.com/cgi-bin/fulltext/123193344/PDFSTART. doi:10.1002/jsfa.3836Spain.

Tao, L., Shi, C., Zi, Y., Zhang, H., Wang, X., Zhong, J., 2024. A review on the chemical modification of alginates for food research: chemical nature, modification methods, product types, and application. Food Hydrocoll. 0268005X. 147 (109338). doi:10.1016/j.foodhyd.2023.109338.

Thiviya, P., Gamage, A., Gama-Arachchige, N.S., Merah, O., Madhujith, T., 2022. Seaweeds as a source of functional proteins. Phycology. 2673-9410. 2 (2), 216–243. doi:10.3390/phycology2020012.

Tibbetts, S.M., Milley, J.E., Lall, S.P., 2016. Nutritional quality of some wild and cultivated seaweeds: nutrient composition, total phenolic content and in vitro digestibility. J. Appl. Phycol. 15735176. 28 (6), 3575–3585. Available from: www.wkap.nl/journalhome.htm/0921-8971.. doi:10.1007/s10811-016-0863-y.

Tønnesen, H.H., Karlsen, J., 2002. Alginate in drug delivery systems. Drug Dev. Ind. Pharm. 28 (6), 621–630. doi:10.1081/DDC-120003853.

Trigo, J.P., Palmnäs-Bédard, M., Juanola, M.V.L., Undeland, I., 2023. Effects of whole seaweed consumption on humans: current evidence from randomized-controlled intervention trials, knowledge gaps, and limitations. Front. Nutr. 2296861X. 10. Available from: journal.frontiersin.org/journal/nutrition doi:10.3389/fnut.2023.1226168.

Tsukui, T., Baba, N., Hosokawa, M., Sashima, T., Miyashita, K., 2009. Enhancement of hepatic docosahexaenoic acid and arachidonic acid contents in C57BL/6J mice by dietary fucoxanthin. Fish. Sci. 14442906. 75 (1), 261–263. doi:10.1007/s12562-008-0018-4.

Van Krimpen, M., Bikker, P., Van der Meer, I., Van der Peet-schwering, C.M.C., Vereijken, J.M., Cultivation, processing and nutritional aspects for pigs and poultry of European protein sources as alternatives for imported soybean products. (2013).

Wallace, M., Metallo, C.M., 2020. Tracing insights into de novo lipogenesis in liver and adipose tissues. Semin. Cell Dev. Biol. 10963634. 108, 65–71. Available from: http://www.elsevier.com/inca/publications/store/6/2/2/9/4/4/index.htt. doi:10.1016/j.semcdb.2020.02.012.

Wanders, A.J., Mars, M., Borgonjen-van den Berg, K.J., de Graaf, C., Feskens, E.J.M., 2014. Satiety and energy intake after single and repeated exposure to gel-forming dietary fiber: post-ingestive effects. Int. J. Obes. 0307-0565. 38 (6), 794–800. doi:10.1038/ijo.2013.176.

Wang, S.H., Huang, C.Y., Chen, C.Y., Chang, C.C., Huang, C.Y., Dong, C.D., Chang, J.S., 2021. Isolation and purification of brown algae fucoidan from Sargassum siliquosum and the analysis of anti-lipogenesis activity. Biochem. Eng. J. 1873295X. 165. Available from: www.elsevier.com/locate/bej.. doi:10.1016/j.bej.2020.107798.

Wang, T., Jónsdóttir, R., Kristinsson, H.G., Hreggvidsson, G.O., Óskar Jónsson, J., Thorkelsson, G., Ólafsdóttir, G., 2010. Enzyme-enhanced extraction of antioxidant ingredients from red algae Palmaria palmata. LWT. 00236438. 43 (9), 1387–1393. doi:10.1016/j.lwt.2010.05.010.

Wang, X., Liu, F., Gao, Y., Xue, C.H., Li, R.W., Tang, Q.J., 2018. Transcriptome analysis revealed anti-obesity effects of the Sodium Alginate in high-fat diet -induced obese mice. Int. J. Biol. Macromol. 18790003. 115, 861–870. Available from: www.elsevier.com/locate/ijbiomac.. doi:10.1016/j.ijbiomac.2018.04.042.

Wang, Y., Li, L., Ye, C., Yuan, J., Qin, S., 2020. Alginate oligosaccharide improves lipid metabolism and inflammation by modulating gut microbiota in high-fat diet fed mice. Appl. Microbiol. Biotechnol. 14320614. 104 (8), 3541–3554. Available from: https://link.springer.de/link/service/journals/00253/index.htm. doi:10.1007/s00253-020-10449-7.

Wendin, K., Undeland, I., 2020. Seaweed as food–Attitudes and preferences among Swedish consumers. A pilot study. Int. J. Gastronomy Food Sci. 22, 100265.

Williams, J.A., Lai, C.S., Corwin, H., Ma, Y., Maki, K.C., Garleb, K.A., Wolf, B.W., 2004. Inclusion of guar gum and alginate into a crispy bar improves postprandial glycemia in human. J. Nutr. 00223166. 134 (4), 886–889. Available from: http://jn.nutrition.org. doi:10.1093/jn/134.4.886.

Woo, M.N., Jeon, S.M., Shin, Y.C., Lee, M.K., Kang, M.A., Choi, M.S., 2009. Anti-obese property of fucoxanthin is partly mediated by altering lipid-regulating enzymes and uncoupling proteins of visceral adipose tissue in mice. Mol. Nutr. Food Res. 16134133. 53(12), 1603–1611. Available from: http://www3.interscience.wiley.com/cgi-bin/fulltext/122653728/PDFSTART. doi: 10.1002/mnfr.200900079

World Health Organization, (2024a). Obesity and overweight. https://www.who.int/news-room/fact-sheets/detail/obesity-and-overweight

World Health Organization, (2024b). Obesity and overweight. https://www.who.int/westernpacific/health-topics/obesity#tab=tab_1

Yanshin, N., Kushnareva, A., Lemesheva, V., Birkemeyer, C., Tarakhovskaya, E., 2021. Chemical composition and potential practical application of 15 red algal species from the white sea coast (the Arctic Ocean). Molecules. 1420-3049. 26 (9), 2489. doi:10.3390/molecules26092489.

Zaharudin, N., Tullin, M., Pekmez, C.T., Sloth, J.J., Rasmussen, R.R., Dragsted, L.O., 2021. Effects of brown seaweeds on postprandial glucose, insulin and appetite in humans – a randomized, 3-way, blinded, cross-over meal study. Clin. Nutr. 15321983. 40 (3), 830–838. Available from: http://www.elsevier-international.com/journals/clnu/. doi:10.1016/j.clnu.2020.08.027.

Zang, L., Baharlooeian, M., Terasawa, M., Shimada, Y., Nishimura, N., 2023. Beneficial effects of seaweed-derived components on metabolic syndrome via gut microbiota modulation. Front. Nutr. 2296861X. 10. Available from: journal.frontiersin.org/journal/nutrition doi:10.3389/fnut.2023.1173225.

Zhang, Y., Liu, J., Mao, G., Zuo, J., Li, S., Yang, Y., Thring, R.W., Wu, M., Tong, H., 2021. Sargassum fusiforme fucoidan alleviates diet-induced insulin resistance by inhibiting colon-derived ceramide biosynthesis. Food Funct. 2042650X. 12 (18), 8440–8453. Available from: http://pubs.rsc.org/en/journals/journal/fo. doi:10.1039/d1fo01272j.

Zhang, Y., Zuo, J., Yan, L., Cheng, Y., Li, Q., Wu, S., Chen, L., Thring, R.W., Yang, Y., Gao, Y., Wu, M., Tong, H., 2020. Sargassum fusiforme fucoidan alleviates high-fat diet-induced obesity and insulin resistance associated with the improvement of hepatic oxidative stress and gut microbiota profile. J. Agric. Food Chem. 15205118. 68 (39), 10626–10638. Available from: http://pubs.acs.org/journal/jafcau. doi:10.1021/acs.jafc.0c02555.

Zhao, C., Lai, S., Wu, D., Liu, D., Zou, X., Ismail, A., El-Seedi, H., Arroo, R.R.J., Xiao, J., 2020. miRNAs as regulators of antidiabetic effects of fucoidans. eFood. 26663066. 1 (1), 2–11. Available from: onlinelibrary.wiley.com/journal/26663066 . doi:10.2991/efood.k.190822.001.

Zvyagintseva, T.N., Shevchenko, N.M., Popivnich, I.B., Isakov, V.V., Scobun, A.S., Sundukova, E.V., Elyakova, L.A., 1999. A new procedure for the separation of water-soluble polysaccharides from brown seaweeds. Carbohydr. Res. 322 (1-2), 32–39. doi:10.1016/S0008-6215(99)00206-2.

Zvyagintseva, T.N., Usoltseva, R.V., Shevchenko, N.M., Surits, V.V., Imbs, T.I., Malyarenko, O.S., Besednova, N.N., Ivanushko, L.A., Ermakova, S.P., 2021. Structural diversity of fucoidans and their radioprotective effect. Federation Carbohydr. Polym. 01448617. 273. Available from: http://www.elsevier.com/wps/find/journaldescription.cws_home/405871/description#description. doi:10.1016/j.carbpol.2021.118551.

PART VI

Emerging prospects of functional foods from marine bioresources

CHAPTER

19

Future outlook of functional foods derived from marine bioresources

J.A.A.C. Wijesinghe and K.H.I.N.M. Herath

Department of Biosystems Engineering, Faculty of Agriculture and Plantation Management, Wayamba University of Sri Lanka, Makandura, Gonawila, Sri Lanka

19.1 Introduction

Foods that offer health benefits beyond their nutritional value are considered functional foods. These foods either decrease the impact of diseases or increase the physical status of the body. These functional foods have garnered considerable attention over the last few decades, shifting the focus beyond traditional aspects like body composition and energy provision to encompass broader health benefits and food wellness (Milner, 2000). In addition, foods with added, removed, or reduced components are also considered functional foods that exert health benefits. Foods are also considered functional when many ingredients are added, changed, or substituted to alter their health benefits (Roberfroid, 2000). The functional food market is expanding and is expected to hit 91 billion USD in the near future (Essa et al., 2023).

All natural foods contain ingredients that exert functional properties. Natural plant- and animal-based sources containing phytochemicals influence physiological functions, warranting focus on their effective utilization in promoting health. These phytochemicals are not nutritional components but bioactive components that are able to prevent noncommunicable diseases (Yiwen et al., 2024). Noncommunicable diseases are widely reported worldwide, which may be caused by aging or due to unhealthy lifestyles. Functional food provides numerous health benefits, such as enhancing the immune system, lowering the likelihood of cardiovascular disease (CVD), osteoporosis, obesity, and certain cancers, and contributing to improved memory and overall physical health (Essa et al., 2023).

Numerous reservoirs can be discovered in terrestrial and marine habitats, including sources of useful substances. Compared to the aquatic environment, the terrestrial ecosystem has been studied far more extensively as a source of bioactive substances (Lordan et al., 2011). However, a wide range of sources in marine habitats, including marine microorganisms, plants, and sponges, contain various bioactive substances. Marine bioresources remain an

Harvesting the Sea
https://doi.org/10.1016/B978-0-443-29190-6.00010-6

untapped area of functional food that can be applied to different food processing and fortification.

The marine environment offers an abundance of bioactive compounds with distinct characteristics and significant potential for biotechnological applications. Many of these compounds can serve the food industry as natural preservatives, colorants, stabilizers, gelling agents, and more. Some exhibit health-promoting properties, making them ideal for functional foods, nutraceuticals, dietary supplements, and prebiotics. Advancing our understanding and harnessing the potential of marine resources to create value-added food products requires a multidisciplinary approach.

19.2 Utilization of marine bioresources in food

19.2.1 Marine bioresources

A wide distribution of organisms, including fish, shellfish, seaweed, microalgae, and other marine flora and fauna, is observed in marine environments (Rotter et al., 2024). As the global population continues to grow and the demand for sustainable food sources increases, marine bioresources have become an essential asset for the food industry.

Seaweeds, photosynthetic organisms, produce abundant biologically active compounds, even those not observed in other organisms. Their ability to adapt to extreme conditions, such as fluctuations in salinity, temperature, and nutrients, allows them to synthesize more biologically active secondary metabolites (Plaza et al., 2008). Seaweeds are a rich source of bioactive components, and most are used as food due to their health benefits (Sanjeewa et al., 2024). Seaweeds are low in calories and contain bioactive components such as dietary fiber, proteins, lipids, polysaccharides, pigments, amino acids, phenolic compounds, minerals, and vitamins (Fig. 19.1) (Choudhary et al., 2021).

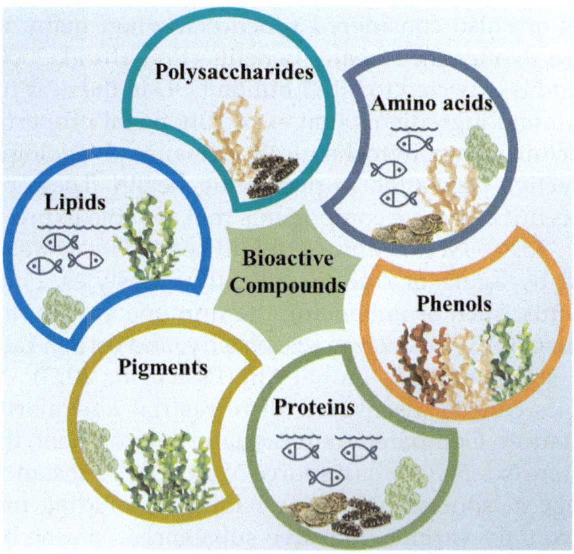

FIG. 19.1 Marine bioactive compounds.

Fish and shellfish are staples and are the most important functional foods, containing antioxidants, calcium, fatty acids, and proteins (Ribeiro et al., 2019; Tan et al., 2023).

Marine microorganisms have gained significant attention as a rich source of biologically active compounds. They exhibit diverse physiological adaptations to the marine environment (Dewapriya and Kim, 2014). Marine actinomycetes are gram-positive bacteria that thrive at a moderate pH and are associated with sponges, corals, and vertebrates. Their adaptation to diverse marine environments results in the production of unique bioactive compounds in marine microorganisms (Dewapriya and Kim, 2014).

19.2.2 Marine by-products

Among the fish and shellfish caught globally, only 50%–60% are utilized for human consumption, while the rest are discarded or barely utilized. Marine by-products are also rich sources of vitamins, minerals, lipids, and proteins. The production of fish oil is a significant method of utilizing fish by-products. According to the Food and Agriculture Organization (FAO), by 2022, by-products accounted for 34% of global fishmeal production and 53% of global fish oil production (Mkadem and Kaanane, 2024). Shellfish species like crabs, lobsters, and shrimp contain inedible parts that are rich in bioactive compounds such as chitin and chitosan, minerals, n-3 fatty acids, phospholipids, carotenoids, amino acids, and proteins, which could be utilized in the food industry through an extraction process (Mkadem and Kaanane, 2024).

19.2.3 Functional properties of marine-based polysaccharides

Polysaccharides are the major macromolecules in seaweed, comprising >80% of its weight. Polysaccharides function as energy reserves and structural molecules in marine organisms. Seaweeds are rich in sulfated polysaccharides, and fucoidan, laminarin, and alginate are the most prominent polysaccharides found in seaweed (Choudhary et al., 2021). Polysaccharides have several applications in the food industry, such as stabilizers, emulsifiers, thickeners, and food additives (Himashree and Animesh Singh Sengar, 2022). Seaweed polysaccharides have shown promising health benefits for development into functional products, including antiinflammatory, anticancer, antioxidant, antiviral, and antiallergy properties (Table 19.1).

Dietary fibers are polysaccharides that can be used as texture modifiers in the food industry due to their viscosity and gelling properties (Jiménez-Escrig and Sánchez-Muniz, 2000). Seaweed contains 25%–75% of its dry weight in dietary fiber, 50%–80% of which is soluble fiber. Seaweed-derived dietary fiber exhibits beneficial properties, including antiinflammation, antioxidant, anticoagulation, and immunomodulation (Praveen et al., 2019), (Sanjeewa et al., 2018). Dietary fiber intake has been inversely linked to the risk of type II diabetes and CVD (Huang et al., 2022).

Sulfated polysaccharides are a characteristic group of dietary fiber that is rarely observed in terrestrial plants (Huang et al., 2022). Sea cucumbers and microalgae also contain sulfated polysaccharides in lesser abundance compared to seaweed (Huang et al., 2022).

TABLE 19.1 Polysaccharides from seaweed.

Compound	Seaweed	Bioactive properties	References
Sulfated fucans	*Polysiphonia lanosa, Furcellaria lumbricalis, Palmaria palmata, Ascophyllum nodosum, Fucus vesiculosus, Ulva lactuca*	Antiviral activity against influenza A/PR/8/34 (H1N1) virus	(Jiao et al., 2012)
Sulfated polysaccharide	*Aulerpa curpressoides*	Antiinflammatory	(Rodrigues et al., 2012)
	Sargassum horneri	Antiinflammation	(Sanjeewa et al., 2017)
	Fucus vesiculosus	Antiinflammation Antimicrobial Antioxidant	(Rossi et al., 2024)
Crude polysaccharide	*Halimeda tuna*	Immunomodulatory	(Kraiem et al., 2024)
		Antiinflammatory	
		Antihemolytic Antioxidant	
Carrageenan	*Kappaphycus alvarezii*	Anticancer Antidiabetic	(Suganya et al., 2016)
		Antimicrobial	(Khan et al., 2022)
Alginate	*Sargassum wightii*	Antioxidant	(Deepak et al., 2024)
		Antiinflammation	(Kokova et al., 2023; Fu et al., 2022)
		Antiallergic	(Boming et al., 2020)
Fucoidan	*Fucus vesiculosus*	Antiinflammation	(Pozharitskaya et al., 2020)
	Undaria pinnatifida	Anticancer	(Boo et al., 2013)
	Hizikia fusiforme	Antiphotoaging	(Wang et al., 2020)
		Antimelanogenesis	
	Ecklonia maxima	Antifine dust effect	(Nagahawatta et al., 2022)
	Costaria costata	Antiinflammation	(Zhang and Lee, 2024)
Laminarin	*Laminaria hyperborea*	Antioxidant	(Rajauria et al., 2021)
	Laminaria japonica	Immune activation	(Karuppusamy et al., 2022)

19.2.4 Functional properties of pigments derived from marine organisms

There is an increasing demand for natural colorants in the food industry since artificial colorants exert detrimental effects on health with long-term usage. These natural pigments, or bio-pigments, are prominent alternatives to artificial colorants in the food industry. Even though great attention is given to terrestrial plants for extracting biopigments, the marine environment is home to nearly 80% of all known life forms, making it a significant and underutilized source of biopigments with many industrial applications (Moovendhan, 2024; Buckner, 2002). Marine organisms are well known for being abundant sources of natural pigments, including chlorophylls, carotenoids, and phycobiliproteins (Manivasagan et al., 2018). These natural pigments are nontoxic, are rich in nutritional value, and have health benefits. Bioactive components like flavonoids, polysaccharides, and fatty acids improve the food's gelling, emulsification, and thickening properties. Microalgae are one of the main sources of commercially available, naturally derived pigments, and they are becoming more valuable in the worldwide food pigment market.

Chlorophylls are greenish, lipid-soluble nanoparticles with a porphyrin ring found in all algae, higher plants, and cyanobacteria. Chlorophylls a, b, c, d, e, and f are the most prominent chlorophyll pigments observed in marine bioresources. Chlorophyll further serves as an antioxidant, in addition to being a colorant (Chen et al., 2023). Chlorophyll a is the most abundant photosynthetic pigment in cyanobacteria and red microalgae (Hachicha et al., 2022).

Carotenoids are abundant pigments in autotrophic marine resources such as bacteria, fungi, and algae. Apart from that, a few marine heterotrophic organisms also contain carotenoids by accumulation via food chains or through modifications that occur in metabolic reactions (Chen et al., 2023; Galasso et al., 2017). Among the carotenoids from plant sources, more than one-third are identified in seaweeds and microalgae (Chen et al., 2023). Carotenoids typically make up about 0.1% of the dry weight of algae. However, certain species, like *Dunaliella salina*, can accumulate up to 14% of their dry weight when cultivated under high salinity, intense light, and adequate nutrient conditions (Prasanna et al., 2007). Carotenoids found in marine bioresources are categorized into two groups: xanthophylls and carotenes (Chen et al., 2023; Hadjkacem et al., 2024). Fucoxanthin is a xanthophyll carotenoid widely found in microalgae and brown seaweed. Fucoxanthin extracted from brown algae exerts functional properties such as antiinflammatory, antiobesity, antiviral, antidiabetic, and antifibrogenic effects (Jung et al., 2012; Kang et al., 2024; Kim et al., 2019; Nguyen et al., 2014). Carotenes include compounds such as β-carotene, α-carotene, and lycopene (Galasso et al., 2017). The xanthophyll carotenoid astaxanthin is approved by the United States Food and Drug Administration as a food colorant in animal and fish feed and as a dietary supplement (Zhang and Wang, 2015). Marine astaxanthin is prominently identified in marine algae and crustaceans and exerts anticancer and antioxidative effects (Zhang and Wang, 2015). Lutein extracted from the microalgae *Tetraselmis* sp. exerts biological activities such as antioxidant, antiinflammatory, and antiviral effects, promising its utilization in product development (Kim et al., 2023). Lutein purified from the green alga *Chlorella ellipsoidea* can mitigate type-2 diabetes (Qi and Kim, 2018). Carotenoids from marine bioresources and their functional properties are indicated in Table 19.2.

Phycobiliproteins are highly concentrated in red seaweed, accounting for about 50% of their total proteins, and are the only water-soluble pigment in red seaweed (Dumay et al., 2014). The extraction and purification of phycoerythrin can be both costly and time-consuming, presenting a current challenge to streamline the process and minimize expenses and duration (Cotas et al., 2020). However, phycobiliproteins are used in the food industry as a pigment.

TABLE 19.2 Carotenoids extracted from marine bioresources and their functional properties.

Pigment of carotenoids	Species	Bioactivities	References
Fucoxanthin (a xanthophyll)	*Sargassum siliquastrum*	Antiviral	(Kang et al., 2024)
	Sargassum siliquastrum	Antiinflammation	(Nguyen et al., 2014; Heo et al., 2012)
	Eisenia bicyclisUndaria pinnatifida	Antidiabetic	(Jung et al., 2012)
	Halopteris scoparia	Antidiabetic, Protects from gastrointestinal infections	(Hadjkacem et al., 2024)
	Conticribra weissflogii	Antiinflammation	(Jingqian et al., 2019)
Astaxanthin (a xanthophyll)	*Hydrolisates of shrimp Parapenaeus longirostris*	Antioxidant	(Messina et al., 2021)
Lutein (a xanthophyll)	*Tetraselmis* sp.	Antioxidant,Antiviral, Antiinflammation	(Kim et al., 2023)
	Chlorella ellipsoidea	Antidiabetic	(Qi and Kim, 2018)
β-Carotene (a carotene)	*Halopteris scoparia*	Antidiabetic, Protects from gastrointestinal infections	(Hadjkacem et al., 2024)

19.2.5 Functional properties of lipids identified from marine organisms

Omega-3 (ω3) PUFAs, namely, eicosapentaenoic acid (EPA) and docosahexaenoic acid (DHA), are vital components of human nutrition, offering numerous health benefits. Marine fish and shellfish are the primary sources of PUFAs. PUFAs are found in higher quantities in the flesh of oily fish and in smaller quantities in lean fish, such as cod. According to Calder (2015), a lean fish meal (cod) contains 0.2–0.3 g of marine ω3 fatty acids, whereas an oily fish meal (salmon or mackerel) contains 1.5–3.0 g, which is 10 times greater than that of lean fish (Calder, 2015). Microalgae also produce lipids, which must be extracted and isolated by separating the single cells from the culture media (Barta et al., 2021). Marine invertebrates also produce PUFAs, and it has been identified that marine bacteria containing ω3-PUFAs survive in the bivalve *Chlamys islandica* and Astarte sp., as well as the amphipod *Gammarus wilkitzkii* (Monroig et al., 2013). Marine microalgae or seaweed also contain considerable amounts of PUFAs, even though their lipid content is lower (<4 g/100 g dry weight) (Kumari et al., 2010). This highlights the potential utilization of seaweed in producing low-fat food.

Marine PUFAs are utilized as supplements in the community. They also exhibit a cancer-fighting effect and have the potential to mitigate cancer complications (Babuskin et al., 2014). Song et al. (2017) revealed that high consumption of marine PUFAs decreases the risk of colorectal cancer-specific mortality after diagnosis (Song et al., 2017). Fish oil attenuates the development and severity of arthritis by reducing the activation of immune cells and the production of inflammatory eicosanoids derived from this activation (Miles and Calder,

2012). DHA and EPA exert antioxidant potential against the reactive oxygen species produced in seawater (Okuyama et al., 2008).

19.2.6 Functional properties of phenolic compounds reported from marine organisms

Polyphenols are a diverse group of bioactive compounds widely present in various marine sources, such as microalgae, macroalgae (seaweeds), cyanobacteria, corals, sponges, mollusks, and marine microbes (Ravichandran et al., 2024). Extracts from seaweed or marine algae enriched with phenolic compounds have been patented and marketed. Seaweed contains a large amount of polyphenolic compounds, which help neutralize free radicals. Free radicals play a significant role in the antioxidant activity of seaweed, particularly brown seaweed, due to its high content of bioactive compounds like polyphenols and fucoidans that scavenge free radicals effectively (Gomez-Zavaglia et al., 2019). Currently, the levels of extractable polyphenols from algae are relatively lower compared to other phytochemicals. Polyphenolic compounds extracted from seaweed exhibit antioxidant, antiinflammatory, antiallergic, antiviral, antidiabetic, and therapeutic potential against CVDs (Herath et al., 2018, 2021; Lee and Jeon, 2013; Besednova et al., 2021; Nagahawatta et al., 2024; Sanjeewa et al., 2020).

19.2.7 Functional properties of vitamins and minerals identified from marine organisms

Vitamins are organic compounds essential to our body but are not synthesized and must be obtained through the diet. Seaweed is rich in water-soluble and fat-soluble vitamins (Jingxiang et al., 2023). Among the seaweeds, *Ulva* sp. is a source of vitamins A, B1, and B2, widely used to control and treat vitamin deficiencies. Green seaweed and red seaweed are rich in vitamin C (ascorbic acid), ranging from 500 mg/kg to 3000 mg/kg (dry weight). *Ulva* sp. (a green seaweed) has been identified as being significantly rich in vitamin C (Nida Khan and Mamat, 2024). Sodium, potassium, and calcium are the most abundant minerals identified in seaweed (Jingxiang et al., 2023). The mineral content in seaweed is higher than that of terrestrial plants. Furthermore, *Ulva* sp. is a rich source of calcium, iron, and magnesium (Nida Khan and Mamat, 2024).

19.2.8 Future of marine bioresource-based functional food

Advancements in biotechnology and extraction methods are expected to significantly enhance the sustainable and efficient utilization of marine resources and ensure the preservation of their valuable bioactive compounds. Emerging technologies, such as encapsulation and nanotechnology, will play a key role in increasing the bioavailability and stability of functional ingredients derived from marine sources. In addition, the integration of artificial intelligence and data analytics is set to revolutionize the discovery of novel bioactive compounds within marine ecosystems, streamlining the process of functional food development.

Sustainability will remain a top priority, with initiatives focused on repurposing by-products from marine industries, including fish waste and seaweed residues, to minimize environmental impact. Growing consumer awareness regarding the health advantages of marine-derived nutrients, coupled with the global trend toward personalized nutrition, is expected to drive innovation in customized functional food solutions. The future of marine bioresource-based functional foods is shaped by a merging of health-focused innovations and environmental sustainability.

References

Babuskin, S., Krishnan, K.R., Babu, P.A.S., Sivarajan, M., Sukumar, M., 2014. Functional foods enriched with marine microalga Nannochloropsis oculata as a source of ω-3 fatty acids. Food Technol. Biotechnol. 13342606. 52 (3), 292–299. Available from http://www.ftb.com.hr/images/pdfarticles/2014/July-September/ftb_52-3_292-299.pdf.

Barta, D.G., Coman, V., Vodnar, D.C., 2021. Microalgae as sources of omega-3 polyunsaturated fatty acids: biotechnological aspects. Algal Res. 22119264. 58. Available from: http://www.sciencedirect.com/science/journal/aip/22119264. doi:10.1016/j.algal.2021.102410.

Besednova, N.N., Andryukov, B.G., Zaporozhets, T.S., Kryzhanovsky, S.P., Fedyanina, L.N., Kuznetsova, T.A., Zvyagintseva, T.N., Shchelkanov, M.Y., 2021. Antiviral effects of polyphenols from marine algae. Federation Biomed. 22279059. 9 (2), 1–23. Available from: https://www.mdpi.com/2227-9059/9/2/200/pdf. doi:10.3390/biomedicines9020200.

Boming, Y., Decheng, B., Yao, L., Tong, L., Liang, G., Hong, X., Xiuting, L., Hui, L., Zhangli, H., Xu, X., 2020. The inhibitory activity of alginate against allergic reactions in an ovalbumin-induced mouse model. Food Funct. 2042-6496. 11 (3), 2704–2713. doi:10.1039/d0fo00170h.

Boo, H.J., Hong, J.Y., Kim, S.C., Kang, J.I., Kim, M.K., Kim, E.J., Hyun, J.W., Koh, Y.S., Yoo, E.S., Kwon, J.M., Kang, H. K., 2013. The anticancer effect of fucoidan in PC-3 prostate cancer cells. Mar. Drugs. 16603397. 11 (8), 2982–2999. Available from: http://www.mdpi.com/1660-3397/11/8/2982/pdf. doi:10.3390/md11082982.

Buckner, A.J.I.S.T., 2002. Life-saving prod from coral reefs. Issues Sci. Technol. 18, 39–44.

Calder, P.C., 2015. Marine omega-3 fatty acids and inflammatory processes: effects, mechanisms and clinical relevance. Biochim. Biophys. Acta Mol. Cell Biol. Lipids. 18792618. 1851 (4), 469–484. Available from: https://www.elsevier.com/locate/bbalip. doi:10.1016/j.bbalip.2014.08.010.

Chen, Z., Weihao, W., Wen, Y., Zhang, L., Yanglin, W., Farid, M.S., El-Seedi, H.R., Capanoglu, E., Zhao, C., 2023. Recent advances of natural pigments from algae. Food Production, Processing and Nutrition. 2661-8974. 5 (1). doi:10.1186/s43014-023-00155-y.

Choudhary, B., Chauhan, O.P., Mishra, A., 2021. Edible seaweeds: a potential novel source of bioactive metabolites and nutraceuticals with human health benefits. Front. Mar. Sci. 22967745. 8. Available from: https://www.frontiersin.org/journals/marine-science#. doi:10.3389/fmars.2021.740054.

Cotas, J., Leandro, A., Pacheco, D., Gonçalves, A.M.M., Pereira, L., 2020. A comprehensive review of the nutraceutical and therapeutic applications of red seaweeds (Rhodophyta). Life. 2075-1729. 10 (3), 19. doi:10.3390/life10030019.

Deepak, P., Perumal, P., Balakrishnan, R., Balasubramanian, B., Velmurugan, P., 2024. Assessment of antioxidant and digestive enzyme inhibition by phyco-molecules isolated from marine brown alga Sargassum wightii. Food Humanity. 29498244. 2, 100226. doi:10.1016/j.foohum.2024.100226.

Dewapriya, P., Kim, S.K., 2014. Marine microorganisms: an emerging avenue in modern nutraceuticals and functional foods. Food Res. Int. 09639969. 56, 115–125. doi:10.1016/j.foodres.2013.12.022.

Dumay, J., Morançais, M., Munier, M., Guillard, C.L., Fleurence, J., 2014. Phycoerythrins. In: Sea Plants, vol. 71. Elsevier, pp. 321–343. doi:10.1016/b978-0-12-408062-1.00011-1.

Essa, M.M., Bishir, M., Bhat, A., Chidambaram, S.B., Al-Balushi, B., Hamdan, H., Govindarajan, N., Freidland, R.P., Qoronfleh, M.W., 2023a. Functional foods and their impact on health. J. Food Sci. Technol. 09758402. 60 (3), 820–834. Available from: https://www.springer.com/journal/13197. doi:10.1007/s13197-021-05193-3.

Fu, Y.H., Tao, X.Y., Yang, D., Li, X., Zhou, D.Y., Dai, Y.L., Jeon, Y.J., 2022. Anti-adhesive properties of calcium alginate from sargassum fusiforme against particulate matter-induced inflammation. Curr. Issues Mol. Biol. 14673045. 44 (2), 626–639. Available from: https://www.mdpi.com/1467-3045/44/2/43/pdf. doi:10.3390/cimb44020043.

Galasso, C., Corinaldesi, C., Sansone, C., 2017. Carotenoids from marine organisms: biological functions and industrial applications. Antioxidants. 20763921. 6 (4). Available from: http://www.mdpi.com/2076-3921/6/4/96/pdf. doi:10.3390/antiox6040096.

Gomez-Zavaglia, A., Lage, M.A.P., Jimenez-Lopez, C., Mejuto, J.C., Simal-Gandara, J., 2019. The potential of seaweeds as a source of functional ingredients of prebiotic and antioxidant value. Antioxidants. 2076-3921. 8 (9), 406. doi:10.3390/antiox8090406.

Hachicha, R., Elleuch, F., Hlima, H.B., Dubessay, P., de Baynast, H., Delattre, C., Pierre, G., Hachicha, R., Abdelkafi, S., Michaud, P., Fendri, I., 2022. Biomolecules from Microalgae and Cyanobacteria: applications and Market Survey. Appl. Sci. 2076-3417. 12 (4), 1924. doi:10.3390/app12041924.

Hadjkacem, F., Elleuch, J., Pierre, G., Fendri, I., Michaud, P., Abdelkafi, S., 2024. Production and purification of fucoxanthins and β-carotenes from Halopteris scoparia and their effects on digestive enzymes and harmful bacteria. Environ. Technol. 0959-3330. 45 (15), 2923–2934. doi:10.1080/09593330.2023.2195562.

Heo, S.J., Yoon, W.J., Kim, K.N., Oh, C., Choi, Y.U., Yoon, K.T., Kang, D.H., Qian, Z.J., Choi, I.W., Jung, W.K., 2012. Anti-inflammatory effect of fucoxanthin derivatives isolated from Sargassum siliquastrum in lipopolysaccharide-stimulated RAW 264.7 macrophage. Food Chem. Toxicol. 18736351. 50 (9), 3336–3342. doi:10.1016/j.fct.2012.06.025.

Herath, K.H.I.N.M., Cho, J., Kim, A., Eom, T.K., Kim, J.-S., Kim, J.-B., Doh, Y.H., Jee, Y., 2018. Phenolic acid and flavonoid-rich fraction of Sasa quelpaertensis Nakai leaves prevent alcohol induced fatty liver through AMPK activation. J. Ethnopharmacol. 03788741. 224, 335–348. doi:10.1016/j.jep.2018.06.008.

Herath, K.H.I.N.M., Kim, H.J., Lee, J.H., Jun Geon, J., Hak-Sun, Y., Jeon, Y.-J., Kim, H.J., Jee, Y., 2021. Sargassum horneri (Turner) C. Agardh containing polyphenols attenuates particulate matter-induced inflammatory response by blocking TLR-mediated MYD88-dependent MAPK signaling pathway in MLE-12 cells. J. Ethnopharmacol. 03788741. 265, 113340. doi:10.1016/j.jep.2020.113340.

Himashree, P., Animesh Singh Sengar, C.K.S., 2022. Food thickening agents: sources, chemistry, properties and applications - A review. Int. J. Gastron. Food Sci. 1878-450X. 27, 100468. Available from https://www.sciencedirect.com/science/article/pii/S1878450X22000038. doi:10.1016/j.ijgfs.2022.100468.

Huang, W., Tan, H., Nie, S., 2022. Beneficial effects of seaweed-derived dietary fiber: highlights of the sulfated polysaccharides. Food Chem. 03088146. 373 (Pt B), 131608. doi:10.1016/j.foodchem.2021.131608.

Jiao, G., Guangli, Y., Wang, W., Zhao, X., Zhang, J., Ewart, S.H., 2012. Properties of polysaccharides in several seaweeds from Atlantic Canada and their potential anti-influenza viral activities. J. Ocean Univ. China. 1672-5182. 11 (2), 205–212. doi:10.1007/s11802-012-1906-x.

Jiménez-Escrig, A., Sánchez-Muniz, F.J., 2000. Dietary fibre from edible seaweeds: chemical structure, physicochemical properties and effects on cholesterol metabolism. Nutr. Res. 02715317. 20 (4), 585–598. doi:10.1016/s0271-5317(00)00149-4.

Jingqian, S., Guo, K., Huang, M., Liu, Y., Zhang, J., Sun, L., Daliang, L., Pang, K.-L., Wang, G., Chen, L., Liu, Z., Chen, Y., Chen, Q., Huang, L., 2019. Fucoxanthin, a marine xanthophyll isolated from conticribra weissflogii nd-8: preventive anti-inflammatory effect in a mouse model of sepsis. Front. Pharmacol. 1663-9812. 10 (JULY). doi:10.3389/fphar.2019.00906.

Jingxiang, X., Liao, W., Liu, Y., Guo, Y., Jiang, S., Zhao, C., 2023. An overview on the nutritional and bioactive components of green seaweeds. Food Prod. Process. Nutr. 2661-8974. 5 (1). doi:10.1186/s43014-023-00132-5.

Jung, H.A., Islam, M.N., Lee, C.M., Jeong, H.O., Chung, H.Y., Woo, H.C., Choi, J.S., 2012. Promising antidiabetic potential of fucoxanthin isolated from the edible brown algae Eisenia bicyclis and Undaria pinnatifida. Fish. Sci. 14442906. 78 (6), 1321–1329. Available from: http://www.springer.com/life+sci/zoology/journal/12562?detailsPage=contactPublishing. doi:10.1007/s12562-012-0552-y.

Kang, N., Kim, E.-A., Park, A., Heo, S.-Y., Heo, J.-H., Heo, S.-J., 2024. Antiviral potential of fucoxanthin, an edible carotenoid purified from sargassum siliquastrum, against zika virus. Mar. Drugs. 1660-3397. 22 (6), 247. doi:10.3390/md22060247.

Karuppusamy, S., Rajauria, G., Fitzpatrick, S., Lyons, H., Mcmahon, H., Curtin, J., Tiwari, B.K., O'Donnell, C., 2022. Biological Properties and Health-Promoting Functions of Laminarin: A Comprehensive Review of Preclinical and Clinical Studies. Mar. Drugs. 20 (12), 772. doi:10.3390/md20120772.

Khan, M.S., S, R., S, H., 2022. Synthesis and characterization of Kappaphycus alvarezii derived silver nanoparticles and determination of antibacterial activity. Mater. Chem. Phys. 02540584. 282. Available from: http://www.journals.elsevier.com/materials-chemistry-and-physics. doi:10.1016/j.matchemphys.2022.125985.

Kim, E.-A., Kang, N., Heo, S.-Y., Jae-Young, O., Lee, S.-H., Cha, S.-H., Kim, W.-K., Heo, S.-J., 2023. Antioxidant, antiviral, and anti-inflammatory activities of lutein-enriched extract of tetraselmis species. Mar. Drugs. 1660-3397. 21 (7), 369. doi:10.3390/md21070369.

Kim, M.B., Bae, M., Hu, S., Kang, H., Park, Y.K., Lee, J.Y., 2019. Fucoxanthin exerts anti-fibrogenic effects in hepatic stellate cells. Biochem. Biophys. Res. Commun. 10902104. 513 (3), 657–662. Available from: http://www.sciencedirect.com/science/journal/0006291X. doi:10.1016/j.bbrc.2019.04.052.

Kokova, V., Lukova, P., Baldzhieva, A., Katsarov, P., Delattre, C., Molinié, R., Petit, E., Elboutachfaiti, R., Murdjeva, M., 2023. Elisaveta apostolova, extraction, structural characterization, and in vivo anti-inflammatory effect of alginate from cystoseira crinita (Desf.) Borry harvested in the Bulgarian Black Sea. Mar. Drugs. 1660-3397. 21 (4), 245. doi:10.3390/md21040245.

Kraiem, M., Hamouda, S.B., Eleroui, M., Ajala, M., Feki, A., Dghim, A., Boujhoud, Z., Bouhamed, M., Badraoui, R., Pujo, J.M., Essafi-Benkhadir, K., Kallel, H., Amara, I.B., 2024. Anti-Inflammatory and immunomodulatory properties of a crude polysaccharide derived from green seaweed halimeda tuna: computational and experimental evidences. Mar. Drugs. 1660-3397. 22 (2), 85. doi:10.3390/md22020085.

Kumari, P., Kumar, M., Vishal Gupta, C.R.K.R., Jha, B., 2010. Tropical marine macroalgae as potential sources of nutritionally important PUFAs. Food Chem. 03088146. 120 (3), 749–757. doi:10.1016/j.foodchem.2009.11.006.

Lee, S.H., Jeon, Y.J., 2013. Anti-diabetic effects of brown algae derived phlorotannins, marine polyphenols through diverse mechanisms. Fitoterapia. 18736971. 86 (1), 129–136. doi:10.1016/j.fitote.2013.02.013.

Lordan, S., Ross, R.P., Stanton, C., 2011. Marine bioactives as functional food ingredients: potential to reduce the incidence of chronic diseases. Mar. Drugs. 16603397. 9 (6), 1056–1100. Available from: http://www.mdpi.com/1660-3397/9/6/1056/pdf. doi:10.3390/md9061056.

Manivasagan, P., Bharathiraja, S., Moorthy, M.S., Mondal, S., Seo, H., Lee, K.D., Junghwan, O., 2018. Marine natural pigments as potential sources for therapeutic applications. Crit. Rev. Biotechnol. 0738-8551. 38 (5), 745–761. doi:10.1080/07388551.2017.1398713.

Messina, C.M., Manuguerra, S., Arena, R., Renda, G., Ficano, G., Randazzo, M., Fricano, S., Sadok, S., Santulli, A., 2021. In vitro bioactivity of astaxanthin and peptides from hydrolisates of shrimp (Parapenaeus longirostris) by-products: from the extraction process to biological effect evaluation, as pilot actions for the strategy "from waste to profit". Mar. Drugs. 1660-3397. 19 (4), 216. doi:10.3390/md19040216.

Miles, E.A., Calder, P.C., 2012. Influence of marine n-3 polyunsaturated fatty acids on immune function and a systematic review of their effects on clinical outcomes in rheumatoid arthritis. Br. J. Nutr. 14752662. 107 (2), S171. doi:10.1017/S0007114512001560.

Milner, J.A., 2000. Functional foods: the US perspective. Am. J. Clin. Nutr. 00029165. 71 (6), 1654S. doi:10.1093/ajcn/71.6.1654s.

Mkadem, H., Kaanane, A., 2024. A comprehensive review on marine by-products use for the recovery of value-added products. Curr. Opin. Green Sustain. Chem. 24522236. 50, 100972. doi:10.1016/j.cogsc.2024.100972.

Monroig, O., Tocher, D.R., Navarro, J.C., 2013. Biosynthesis of polyunsaturated fatty acids in marine invertebrates: recent advances in molecular mechanisms. Mar. Drugs. 16603397. 11 (10), 3998–4018. Available from: http://www.mdpi.com/1660-3397/11/10/3998/pdf. doi:10.3390/md11103998.

Moovendhan, M., 2024. Marine algal pigments: exploring biomedical applications and their role in natural products research. Nat. Prod. Res. 1478-6419. 1–2. doi:10.1080/14786419.2024.2383275.

Nagahawatta, D.P., Liyanage, N.M., Jayawardena, T.U., Jeon, Y.-J., 2024. Marine polyphenols in cardiovascular health: unraveling structure–activity relationships, mechanisms, and therapeutic implications. Int. J. Mol. Sci. 1422-0067. 25 (15), 8419. doi:10.3390/ijms25158419.

Nagahawatta, D.P., Liyanage, N.M., Jayawardhana, H.H.A.C.K., Lee, H.-G., Jayawardena, T., Jeon, Y.-J., 2022. Anti-fine dust effect of fucoidan extracted from ecklonia maxima leaves in macrophages via inhibiting inflammatory signaling pathways. Mar. Drugs. 1660-3397. 20 (7), 413. doi:10.3390/md20070413.

Nguyen, V.T., Qian, Z.J., Lee, B., Heo, S.J., Kim, K.N., Jeon, Y.J., Park, W.S., Choi, I.W., Jang, C.H., Ko, S.C., Park, S.J., Kim, Y.T., Kim, G.H., Lee, D.S., Yim, M.J., Je, J.Y., Jung, W.K., 2014. Fucoxanthin derivatives from Sargassum siliquastrum inhibit matrix metalloproteinases by suppressing NF-kB and MAPKs in human fibrosarcoma cells. Algae. 20930860. 29 (4), 355–366. Available from: http://koreascience.or.kr/journal/JournalIssueList.jsp?kojic=JORHBK&volno=29&issno=4&cnov_no=&year_GB=2014&vn_cd=v29n4. doi:10.4490/algae.2014.29.4.355.

Nida Khan, K.S., Mamat, R., 2024. Eco-friendly nutrient from ocean: exploring Ulva seaweed potential as a sustainable food source. J. Agric. Food Res. 26661543. 17, 101239. doi:10.1016/j.jafr.2024.101239.

Okuyama, H., Orikasa, Y., Nishida, T., 2008. Significance of antioxidative functions of eicosapentaenoic and docosahexaenoic acids in marine microorganisms. Appl. Environ. Microbiol. 00992240. 74 (3), 570–574. doi:10.1128/AEM.02256-07.

Plaza, M., Cifuentes, A., Ibáñez, E., 2008. In the search of new functional food ingredients from algae. Trends Food Sci. Technol. 09242244. 19 (1), 31–39. doi:10.1016/j.tifs.2007.07.012.

Pozharitskaya, O.N., Obluchinskaya, E.D., Shikov, A.N., 2020. Mechanisms of bioactivities of fucoidan from the brown seaweed fucus vesiculosus L. Of the barents sea. Federation Mar. Drugs. 16603397. 18 (5). Available from: https://www.mdpi.com/1660-3397/18/5/275. doi:10.3390/md18050275.

Prasanna, R., Sood, A., Suresh, A., Nayak, S., Kaushik, B., 2007. Potentials and applications of algal pigments in biology and industry. Acta Bot. Hung. 0236-6495. 49 (1-2), 131–156. doi:10.1556/abot.49.2007.1-2.14.

Praveen, M.A., Parvathy, K.R.K., Balasubramanian, P., Jayabalan, R., 2019. An overview of extraction and purification techniques of seaweed dietary fibers for immunomodulation on gut microbiota. Trends Food Sci. Technol. 09242244. 92, 46–64. Available from: http://www.elsevier.com/wps/find/journaldescription.cws_home/601278/description#description. doi:10.1016/j.tifs.2019.08.011.

Qi, J., Kim, S.M., 2018. α-glucosidase inhibitory activities of lutein and zeaxanthin purified from green alga Chlorella ellipsoidea. J. Ocean Univ. China. 19935021. 17 (4), 983–989. Available from: http://www.springerlink.com/content/1672-5182. doi:10.1007/s11802-018-3465-2.

Rajauria, G., Ravindran, R., Garcia-Vaquero, M., Rai, D.K., Sweeney, T., O'Doherty, J., 2021. Molecular characteristics and antioxidant activity of laminarin extracted from the seaweed species Laminaria hyperborea, using hydrothermal-assisted extraction and a multi-step purification procedure. Food Hydrocoll. 0268005X. 112, 106332. doi:10.1016/j.foodhyd.2020.106332.

Ravichandran, J., Babu, L., Jayalakshmi, K.B., Manivel, M., Ramudu, K.N., Ponesakki, G., Stephen, N.M., 2024. Phenolic compounds from marine organisms and their biofunctional properties. Science and Engineering of Polyphenols: Fundamentals and Industrial Scale Applications, pp. 452–478. doi:10.1002/9781394203932.

Ribeiro, A.R., Altintzoglou, T., Mendes, J., Nunes, M.L., Dinis, M.T., Dias, J., 2019. Farmed fish as a functional food: perception of fish fortification and the influence of origin – insights from Portugal. Aquaculture. 00448486. 501, 22–31. doi:10.1016/j.aquaculture.2018.11.002.

Roberfroid, M.B., 2000. Concepts and strategy of functional food science: the European perspective. Am. J. Clin. Nutr. 00029165. 71 (6), 1660S. doi:10.1093/ajcn/71.6.1660s.

Rodrigues, J.A.G., Vanderlei, E.D.S.O., Silva, L.M.C.M., de Araújo, I.W.F., de Queiroz, I.N.L., de Paula, G.A., Abreu, T.M., Ribeiro, N.A., Bezerra, M.M., Chaves, H.V., Lima, V., Jorge, R.J.B., Monteiro, H.S.A., Leite, E.L., Benevides, N.M.B., 2012. Antinociceptive and anti-inflammatory activities of a sulfated polysaccharide isolated from the green seaweed Caulerpa cupressoides. Pharmacol. Rep. 17341140. 64 (2), 282–292. doi:10.1016/s1734-1140(12)70766-1.

Rossi, L., Canala, B., Fifi, A.P., Frazzini, S., 2024. In vitro evaluation of functional properties of extracts of Fucus vesiculosus obtained with different conventional solvents. Algal Res. 22119264. 84, 103787. doi:10.1016/j.algal.2024.103787.

Rotter, A., Varamogianni-Mamatsi, D., Pobirk, A.Z., Matjaž, M.G., Cueto, M., Díaz-Marrero, A.R., Jónsdóttir, R., Sveinsdóttir, K., Catalá, T.S., Romano, G., Guler, B.A., Atak, E., Zrimec, M.B., Bosch, D., Deniz, I., Gaudêncio, S.P., Grigalionyte-Bembič, E., Klun, K., Zidar, L., Rius, A.C., Baebler, Š., Lukić Bilela, L., Rinkevich, B., Mandalakis, M., 2024. Marine cosmetics and the blue bioeconomy: from sourcing to success stories. iScience. 25890042. 27 (12), 111339. doi:10.1016/j.isci.2024.111339.

Sanjeewa, K.K.A., Fernando, I.P.S., Kim, E.A., Ahn, G., Jee, Y., Jeon, Y.J., 2017. Anti-inflammatory activity of a sulfated polysaccharide isolated from an enzymatic digest of brown seaweed Sargassum horneri in RAW

264.7 cells. Nutr. Res. Pract. 20056168. 11 (1), 3–10. Available from: https://www.ncbi.nlm.nih.gov/pmc/articles/PMC5300944/pdf/nrp-11-3.pdf. doi:10.4162/nrp.2017.11.1.3.

Sanjeewa, K.K.A., Jayawardena, T.U., Herath, K.H.I.N.M., Jeon, Y.J., 2024. The role of seaweeds in blue bioeconomy. In: The Role of Seaweeds in Blue Bioeconomy. Bentham Science Publishers, 1–353. Available from: https://benthambooks.com/book/9789815223644/. doi:10.2174/97898152236441240101.

Sanjeewa, K.K.A., Jayawardena, T.U., Kim, S.Y., Lee, H.G., Je, J.G., Jee, Y., Jeon, Y.J., 2020. Sargassum horneri (Turner) inhibit urban particulate matter-induced inflammation in MH-S lung macrophages via blocking TLRs mediated NF-κB and MAPK activation. J. Ethnopharmacol. 18727573. 249. Available from: https://www.elsevier.com/locate/jethpharm. doi:10.1016/j.jep.2019.112363.

Sanjeewa, K.K.A., Kang, N., Ahn, G., Jee, Y., Kim, Y.T., Jeon, Y.J., 2018. Bioactive potentials of sulfated polysaccharides isolated from brown seaweed Sargassum spp in related to human health applications: a review. Food Hydrocoll. 0268005X. 81, 200–208. Available from: https://www.journals.elsevier.com/food-hydrocolloids. doi:10.1016/j.foodhyd.2018.02.040.

Song, M., Zhang, X., Meyerhardt, J.A., Giovannucci, E.L., Ogino, S., Fuchs, C.S., Chan, A.T., 2017. Marine ω-3 polyunsaturated fatty acid intake and survival after colorectal cancer diagnosis. Gut 66, 1790–1796. doi:10.1136/gutjnl-2016-311990.

Suganya, A.M., Sanjivkumar, M., Chandran, M.N., Palavesam, A., Immanuel, G., 2016. Pharmacological importance of sulphated polysaccharide carrageenan from red seaweed Kappaphycus alvarezii in comparison with commercial carrageenan. Biomed. Pharmacother. 19506007. 84, 1300–1312. Available from: https://www.elsevier.com/locate/biomedpharm. doi:10.1016/j.biopha.2016.10.067.

Tan, K., Lim, L., Peng, Y., Cheong, K.-L., 2023. Effects of food processing on the lipid nutritional quality of commercially important fish and shellfish. Food Chem. X 25901575 (20), 101034. doi:10.1016/j.fochx.2023.101034.

Wang, L., Jae-Young, O., Kim, Y.-S., Lee, H.-G., Lee, J.-S., Jeon, Y.-J., 2020. Anti-photoaging and anti-melanogenesis effects of fucoidan isolated from hizikia fusiforme and its underlying mechanisms. Mar. Drugs. 1660-3397. 18 (8), 427. doi:10.3390/md18080427.

Yiwen, Z., Lubin, N., Ruifang, Z., Yan, C., Shifan, H., Ağagündüz, D., 2024. Phytochemicals in chronic noncommunicable diseases: a bibliometric study. J. Food Biochem. 0145-8884. 2024 (1). doi:10.1155/2024/5294512.

Zhang, L., Wang, H., 2015. Multiple mechanisms of anti-cancer effects exerted by astaxanthin. Mar. Drugs. 16603397. 13 (7), 4310–4330. Available from: http://www.mdpi.com/1660-3397/13/7/4310/pdf. doi:10.3390/md13074310.

Zhang, W., Lee, P.C.W., Jin, J.O., 2024. Anti-inflammatory effect of fucoidan from costaria costata inhibited lipopolysaccharide-induced inflammation in mice. Mar. Drugs. 1660-3397. 22 (9), 401. doi:10.3390/md22090401.

CHAPTER

20

Marine macroalgae as new source of bioactive compounds with potential applications in metabolic dysfunction-associated fatty liver disease

Maitane González-Arceo[a,*], *Laura Arellano-García*[a,b,*], *Saioa Gómez-Zorita*[a,b,c], *and María P. Portillo*[a,b,c]

[a]Nutrition and Obesity Group, Department of Pharmacy and Food Science, Faculty of Pharmacy and Lucio Lascaray Research Centre, University of the Basque Country [UPV/EHU], Vitoria-Gasteiz, Spain

[b]CIBERobn Physiopathology of Obesity and Nutrition, National Institute of Health Carlos III, Madrid, Spain

[c]BIOARABA Health Research Institute, Vitoria-Gasteiz, Spain

20.1 Introduction

Metabolic-associated fatty liver disease (MAFLD) is the most prevalent hepatic disease worldwide. In fact, it represents a serious public health problem that is estimated to affect around 30% of adults globally, with prevalence varying significantly depending on the region, age, or sex (Fouad et al., 2024; Teng et al., 2023). MAFLD is characterized by excessive fat accumulation in the liver (hepatocytes exceeding 5% of liver weight or 5% of hepatocytes containing lipid vacuoles) in individuals who consume little or no alcohol (<20 g ethanol/day in women and <30 g ethanol/day in men). This disease encompasses stages of varying severity, ranging from relatively benign steatosis, where there is only excessive fat

* These authors contributed equally to this work.

Harvesting the Sea
https://doi.org/10.1016/B978-0-443-29190-6.00024-6

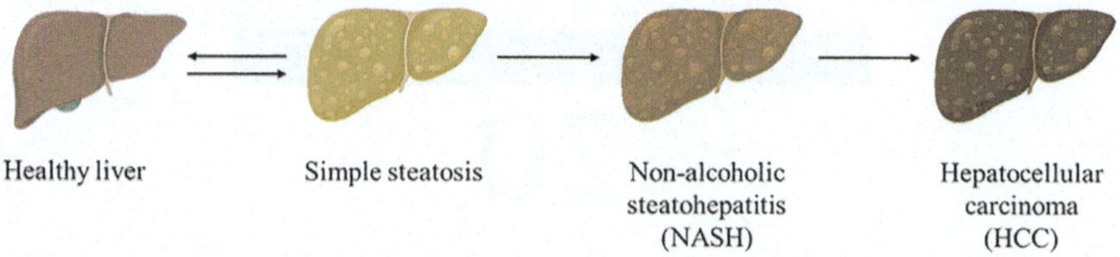

FIG. 20.1 Schematic representation of MAFLD progression. HCC: hepatocellular carcinoma; NASH: nonalcoholic steatohepatitis.

accumulation in the hepatocytes, to more harmful stages, including steatohepatitis, fibrosis, cirrhosis, and hepatocellular carcinoma (Fig. 20.1). Although the accumulation of lipids is not harmful in itself, this condition increases the risk of developing steatohepatitis, a more severe form of the disease characterized by steatosis accompanied by inflammation and hepatocyte injury or ballooning. Most patients with MAFLD do not develop steatohepatitis; in fact, only 10%–20% of affected individuals progress to this stage (Grander et al., 2023). This progression is not linear and can include periods of stability and regression (Younossi et al., 2019). However, the development of steatohepatitis is associated with increased mortality due to higher rates of liver cancer- and cardiovascular-related mortality (Pourteymour et al., 2023). Long-term tissue damage leads to fibrosis and, subsequently, to cirrhosis, which can drive the development of hepatocellular carcinoma (Manne et al., 2018).

The increasing prevalence of MAFLD is closely linked to the global increase in obesity, insulin resistance, type 2 diabetes, and metabolic syndrome, making it the hepatic manifestation of a set of systemic metabolic disorders. The prevalence rates among patients with severe obesity amount to 90% and 75% in patients with type 2 diabetes mellitus (Grander et al., 2023). Thus the term MAFLD has been recently adopted as a better replacement for the former nonalcoholic fatty liver disease (Chan et al., 2022). This new nomenclature highlights metabolic dysfunction as a key indicator of fatty liver-associated risks (Gofton et al., 2023).

Lifestyle modifications, including a healthy diet and physical activity, are considered the first-line treatment for MAFLD; however, the long-term effectiveness of this approach is often limited by poor patient adherence. While many clinical trials are currently underway, there is no specifically approved pharmacological treatment for MAFLD. Thus the scientific community is actively exploring novel molecules and alternative treatments to complement these limited therapeutic strategies.

The search for new therapies to prevent and/or treat MAFLD relies heavily on the use of experimental models that mimic the human disease. These models are essential for identifying compounds that can effectively halt or even reverse its progression. In this context, a variety of in vitro and in vivo models can be used. There are several in vitro models of hepatic steatosis available, such as primary cells generally induced to develop MAFLD with fatty acids and immortalized cells. However, these models do not mimic the pathogenesis of the disease. For that reason, years ago, coculture models and 3D multicellular organ culture systems began to be used. Despite these advances, certain aspects of MAFLD pathology still cannot be fully replicated using in vitro cell culture models. Regarding animal models, mainly rodents (mice and rats) are used. These animals can develop steatosis, mainly induced by different diets or genetic modifications, which better represent what occurs in humans.

Nevertheless, these animal models do not completely replicate the characteristics of MAFLD in humans. Thus it is ultimately necessary to test the effects of bioactive compounds in humans with fatty liver (Wu et al., 2023b).

Algae are a large and diverse group of photosynthetic organisms that can be found in a wide range of aquatic environments. They are generally categorized into two main groups: microalgae and macroalgae. Microalgae are microscopic, single-celled organisms, while macroalgae, commonly known as seaweeds, are multicellular organisms primarily found in marine environments. The classification of algae is a complex and evolving field, but macroalgae are most often grouped by their pigmentation into brown seaweeds (*Phaeophyta*), red seaweeds (*Rhodophyta*), and green seaweeds (*Chlorophyta*) (Biris-Dorhoi et al., 2020).

Throughout history, algae have served as a valuable food source, particularly in East Asian countries such as Japan and Korea. In these regions, the consumption of algae is deeply integrated into their culture due to their significant nutritional and health benefits. In contrast, the historical consumption of algae in Europe has been less prominent. However, the upward trends in the European algae sector indicate a growing interest in their use as food and as a functional ingredient (Mendes et al., 2022).

Nowadays, algae are gaining attention because they are considered good candidates as a sustainable alternative to traditional crops to address the increase in population and the corresponding rise in food demand. In fact, in terms of sustainability, algae production does not require as much arable land as traditional crops and can grow with minimal nutrients (Mendes et al., 2022). Beyond this, algae are also attracting more attention due to their high-value nutritional composition and health-promoting properties (Wu et al., 2023a).

From a nutritional perspective, seaweeds are recognized as a highly valuable food source. They provide essential macronutrients, including proteins, carbohydrates, and lipids, such as polyunsaturated fatty acids (PUFAs), and micronutrients (vitamins and minerals). Beyond their nutritional composition, seaweeds are also a rich source of secondary bioactive metabolites. These are compounds that, while not essential for the organism's immediate survival, confer a wide range of health benefits. Some of these compounds include pigments and polyphenols, which possess antioxidant properties capable of neutralizing reactive oxygen species and mitigating oxidative stress. They also exhibit antiinflammatory effects, modulating immune responses and reducing inflammation. Beyond these properties, seaweeds are being investigated for their antimicrobial and antiviral activities (Biris-Dorhoi et al., 2020; Zhou et al., 2022). Of particular relevance, many of these compounds are able to modulate lipid metabolism (Łagowska et al., 2025), making them promising candidates for the development of nutritional interventions aimed at improving hepatic and metabolic health. In fact, seaweed consumption has been linked to improvements in liver-related parameters in patients with MAFLD (Adnan and Pramaningtyas, 2022). However, since human studies on this topic remain scarce, further research is needed to definitively confirm these effects, as well as to determine the mechanisms involved.

While macroalgae hold therapeutic potential, their use in nutritional interventions must be carefully evaluated for potential associated risks, especially in certain population groups. Primarily, this risk is associated with the bioaccumulation of heavy metals, such as arsenic, lead, and mercury, from surrounding areas. Even if some algae do not surpass the legal limits regarding heavy metals, frequent and excessive consumption can lead to harmful effects associated with the bioaccumulation of these compounds in the human organism (Mendes et al., 2022). In addition, the high iodine content in some algae can lead to thyroid dysfunction; thus special attention must be paid to pregnant women, children, and individuals with thyroid dysfunction, heart disease, or renal insufficiency. Although food allergies are uncommon, they can be a concern (Mendes et al., 2022; Raja et al., 2021). Consequently, these considerations must be taken into account to ensure their effective use in nutritional interventions. However, clinical studies indicate that seaweed compounds are generally safe and well tolerated when consumed in small or moderate quantities.

20.2 Components of seaweed with a potential effect on MAFLD

Seaweeds contain various components, such as proteins, polyphenols, peptides, fatty acids, pigments, and carbohydrates, among others, that can improve MAFLD by modulating lipid metabolism or enhancing hepatic function. In this book chapter section, we will focus on polyphenols, peptides, fatty acids, pigments, and carbohydrates because they are the main representative bioactive components with the potential to prevent or treat this condition.

20.2.1 Polyphenols in seaweed

Polyphenols are secondary metabolites that exert protective effects against various stimuli. They represent a defense mechanism of plants and seaweeds against UV radiation and infections, among other factors. Moreover, they are the main group of phytochemicals found in the human diet. Human studies provide evidence that dietary polyphenols play a crucial role in human health. In this context, seaweeds are a natural source of polyphenols such as phenolic acids (e.g., gallic acid, vanillic acid) or flavonoids (e.g., catechins, kaempferol), among others (Huamán-Castilla et al., 2024; Ouahabi et al., 2024; Bouzenad et al., 2024). In addition, they contain polyphenols that are exclusive to seaweed, such as bromophenols and phlorotannins (Peñalver et al., 2020; Shannon et al., 2021).

Total polyphenol content is higher in brown seaweed (dry mass) than in green or red ones. In brown seaweed, it can rise up to 20% of the dry weight, and in the case of red and green algae, its content represents from 1% to 5% of the dry weight, or even less (Peñalver et al., 2020; Shannon et al., 2021). Bromophenols are most commonly found in red and green seaweeds, with lower content in brown genera, and consist of one or several benzene rings, a varying degree of bromine and hydroxyl substituents. By contrast, phlorotannins, consisting of phloroglucinol monomeric units (1,3,5-trihydroxybenzene), are present only in brown seaweeds (Shannon et al., 2021). Phlorotannins can be divided into six groups: fucols (phenyl association), phlorethols and fuhalols (ether association), fucophloroethols (phenyl and ether associations), eckols (dibenzodioxin association), and halogenated phlorotannins (with bromine, chlorine, or iodine) (Chen et al., 2021).

Different beneficial effects on cancer, diabetes, and obesity, among other diseases, have been attributed to these polyphenols. However, it should be mentioned that each polyphenol has a unique structure and, thus, probably a different bioactivity.

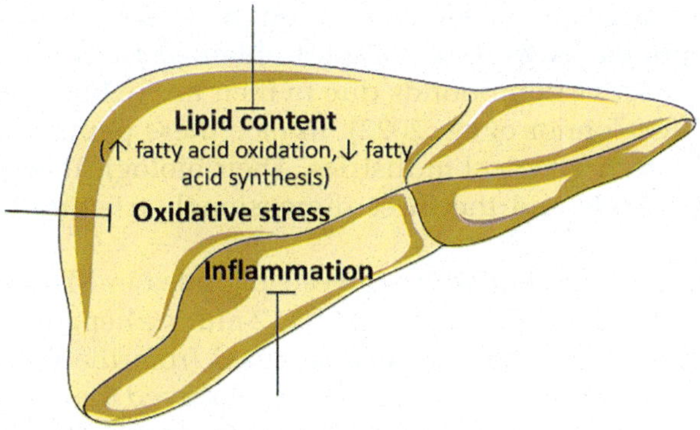

FIG. 20.2 Effect of polyphenols on MAFLD.

Regarding the effect of polyphenol extracts from brown seaweed on MAFLD (Fig. 20.2), it has been observed that a polyphenol extract from the brown alga *Ecklonia cava,* used in a model of mice showing obesity induced by high-fat feeding, reduces hepatic lipid content. This effect is mediated by a decrease in fatty acid synthesis (lipogenesis and de novo lipogenesis) and an increase in fatty acid oxidation. Moreover, it is also able to reduce hepatic inflammation and oxidative stress (Hyeyoon et al., 2015; Park et al., 2015). The phlorotannin diphlorethohydroxycarmalol is able to reduce palmitate-induced lipid accumulation in HepG2 cell line (Cha et al., 2020) and in the livers of male mice fed a high-fat diet, at least in part by reducing de novo lipogenesis (Ding et al., 2019). Dieckol, another phlorotannin, reduces hepatic lipid accumulation associated with a decrease in food intake in male mice fed a high-fat diet (Byun et al., 2021). Finally, concerning clinical trials, in women with MAFLD, it has been observed that xanthigen reduces hepatic fat content and improves liver function tests (Abidov et al., 2010).

As far as we know, there are no studies carried out with phenolic extracts from red or green seaweed, nor with bromophenols. However, bromophenols have certain antidiabetic, antiobesity, and antioxidant properties that are related to MAFLD (Dong et al., 2020).

Although all the studies show beneficial effects on MAFLD, the number of studies is still scarce, especially in the case of clinical trials. Thus further research is needed to confirm the efficacy of seaweed polyphenols in patients with MAFLD. Moreover, it would be of interest to investigate whether the combination of these polyphenols with pharmacological treatments and/or dietary interventions could enhance their effectiveness, making them useful as adjuvant therapy. Similarly, given the generally low bioavailability of polyphenols, it would be important to study this aspect and, if necessary, explore strategies to improve it, thereby increasing their potential efficacy.

20.2.2 Peptides in seaweed

The structural characteristics and biological functions of seaweed proteins are still not well documented. Peptides from seaweed are generally composed of approximately 3–40 amino acids and are mainly generated from the enzymatic hydrolysis of their proteins, either naturally in the digestive system or through controlled technological processes. Naturally present peptides, as well as peptides obtained from protein hydrolysates, are a valuable source of bioactive compounds due to their antioxidant, antiinflammatory, and anticancer properties (O'Connor et al., 2022). In fact, these peptides are currently used in pharmaceutical and biotechnological industries. Their biological activity is determined by the sequence of amino acids and the three-dimensional structure that they adopt in the organism.

Little is known about the potential of seaweed peptides in MAFLD. In this regard, our research group carried out an experiment in AML-12 murine hepatocytes to study the effects of two extracts rich in proteins and peptides obtained from the red alga *Gracilaria vermiculophylla* in the prevention of lipid accumulation. We observed that these extracts were able to reduce triglyceride accumulation when incubated simultaneously with palmitic acid. This delipidating effect may be due, at least in part, to an increase in lipid oxidation in the mitochondria and probably to a decrease in fatty acid uptake into the hepatocytes (González-Arceo et al., 2023).

In a study carried out in vivo, the effect of *Spirulina platensis* protein hydrolysate was investigated in mice fed a high-fat diet. Although the extract was not able to modify lipid weight (its composition was not measured), it modulated genes related to fatty acid oxidation and triglyceride synthesis (Zhao et al., 2019).

20.2.3 Fatty acids in seaweed

Despite the fatty acid composition of several seaweed species having been analyzed, a large number of species remain unexplored. The lipid content of marine macroalgae typically ranges from 1% to 5% of its dry weight and includes essential n-3 PUFAs, such as eicosapentaenoic acid (EPA) and docosahexaenoic acid (DHA), as well as n-6 PUFAs like arachidonic acid. Seaweeds also contain saturated fatty acids, such as palmitic acid, although these are less interesting for health effects and are less abundant compared to monounsaturated fatty acids like oleic acid. n-3 PUFAs are linked to antiinflammatory and antioxidant effects that help lower the risk of cardiovascular diseases, among other beneficial effects. In fact, different metaanalyses have revealed that n-3 PUFAs exert beneficial effects in the treatment of MAFLD in both children and adults (Guo et al., 2018; Dionysopoulos et al., 2023).

Regarding the effect of algae lipids on MAFLD, there is only limited information. The only manuscript found is a study carried out in obese *db/db* mice that analyzed the effect of *Pyropia yezoensis* red seaweed lipids on the development of hepatic steatosis. The major fatty acids found were EPA (62%), which are bound to polar lipids (easier digestion, increased bioavailability, and uptake into tissues compared to other lipids), and palmitic acid (25%). These compounds were able to reduce liver weight and its lipid content, thus reducing steatosis. Moreover, they also exerted hepatoprotective effects (increased serum adiponectin) and reduced hepatic injury (decreased serum transaminases). The delipidating effect observed may be attributed, at least in part, to an inhibitory effect on de novo lipogenesis (Gotoh et al., 2018).

20.2.4 Pigments in seaweed

Seaweed's natural pigments are responsible for their different colors. In terms of pigment composition, seaweeds can be classified into three main groups according to the specific pigments that give them their color: chlorophylls, such as chlorophyll a or b (*Chlorophyceae* or green algae); carotenoids, such as fucoxanthin (*Phaeophyceae* or brown algae); and phycobiliproteins, such as phycocyanin (*Rhodophyceae* or red algae).

All seaweeds contain chlorophylls, which are greenish, lipid-soluble natural pigments characterized by a porphyrin ring. In seaweeds, four types of chlorophylls have been identified, the most prominent of which is chlorophyll *a*. Chlorophylls b and c (c_1 and c_2) have also been found. Carotenoids, including carotenes, xanthophylls or oxycarotenoids, and lutein, are mainly found in brown seaweeds. Phycobiliproteins can be classified into four main groups: phycoerythrins, phycocyanins, phycoerythrocyanins, and allophycocyanins (Gomes et al., 2022).

Several studies have reported various bioactivities of pigments, including antioxidant, anticancer, antiangiogenic, antiinflammatory, and antiobesity effects, among others. Nevertheless, although multiple activities and applications have been demonstrated, their effect on MAFLD remains insufficiently explored. To the best of our knowledge, no studies have investigated the effects of chlorophylls and phycobiliproteins from macroalgae on fatty liver disease.

In contrast, there are numerous studies on the effects of macroalgal carotenoids on this pathology. In this context, a lipid-lowering effect in the liver has been reported for fucoxanthin, the most studied carotenoid found in brown algae. In FL83B murine hepatocytes, when this pigment was coincubated with oleic acid, it was able to reduce lipid accumulation by modulating fatty acid oxidation and reducing de novo lipogenesis. It was also able to reduce oxidative stress in these cells (Chang et al., 2018). In human HepG2 cultured cells, when it was coincubated with palmitic acid, it reduced lipid accumulation by increasing fatty acid oxidation and reducing de novo lipogenesis (Yang et al., 2024). Similarly, in another study, the authors also observed the delipidating effect in the same cell line when fucoxanthin was coincubated with oleic acid and palmitic acid (Bermont et al., 2025). In healthy human liver cells, fucoxanthin also reduced lipid accumulation when it was coincubated with oleic acid and palmitic acid at the same time and reduced alanine aminotransferase (ALT) and aspartate aminotransferase (AST) levels, which are indicators of cell injury. Moreover, it also reduced oxidative stress and inflammation in the hepatocytes. The mechanism of action was the same as described in the other cultured cells (Jiena et al., 2022).

In *ob/ob* mice (diet supplemented with 0.2% fucoxanthin for 24 weeks), fucoxanthin reduced hepatic fat accumulation by the same mechanism described in the in vitro studies (Yang et al., 2024). In a mouse model of diet-induced nonalcoholic steatohepatitis (a choline-deficient, L-amino acid-defined, high-fat diet), it also reduced hepatic lipid accumulation, oxidative stress, inflammation, and fibrosis when it was supplemented at 0.2% in the diet for 4 weeks (Takatani et al., 2020).

Regarding clinical studies, in a study carried out in healthy adults, it was observed that fucoxanthin (FX) administered for 24 weeks (5 mg FX for 12 weeks followed by 2.5 mg FX for 12 weeks) in the diet did not modify the hepatorenal index, which is used to evaluate the severity of fat accumulation in the liver. However, a trend in the reduction of glucose and insulin levels in the group fed the pigment, when compared with the placebo group, was observed. This study found that it was well tolerated, and no adverse effects were reported (Bermont et al., 2025). Nevertheless, it is important to mention that the lack of effects observed may be due, at least in part, to the fact that the subjects were healthy. It is possible that the effects would not be the same in patients with fatty liver disease, who would represent the target population for this treatment.

20.2.5 Carbohydrates in seaweed

Carbohydrates are among the most abundant constituents of seaweed, where they perform structural, storage, and protective functions (Singh et al., 2024). Although the carbohydrate composition of marine macroalgae varies with species, season, environmental conditions, and location, these macromolecules represent 30%–60% of the total biomass of seaweed (Singh et al., 2024). Polysaccharides, which are the major form of carbohydrates in these organisms, are high-molecular-weight polymers composed of monosaccharides linked by glycosidic bonds (Stiger-Pouvreau et al., 2016). There is a vast diversity of polysaccharides, depending on their monosaccharide composition, branching patterns, degree of sulfation, molecular size, and other factors (Stiger-Pouvreau et al., 2016). This variability is responsible not only for their physicochemical properties, but also for the biological activities of these complex molecules. Indeed, the available evidence suggests that marine macroalgae-derived polysaccharides exert antiobesity, antidiabetic, antioxidant, anticancer, antiinflammatory, and anticoagulant effects (Yang et al., 2022). Nonetheless, each macroalgae taxon has a distinct polysaccharide composition and, thus, different bioactive properties. In this regard, brown algae are rich in alginates, fucoidans, and laminarin; red algae mainly contain agar, carrageenans, xylans, and floridean starch; and green algae are characterized by ulvans and cellulose (Singh et al., 2024).

As for the effects of algae carbohydrates on MAFLD, an in vitro study showed that treating FaO rat hepatoma cells with a fucoidan-rich extract of the brown algae *Cytoseira compressa* at a dose of 50 μg/mL reduced intracellular triglyceride content by 35% (El Rashed et al., 2021). In addition, the authors confirmed that this extract diminished the expression of genes related to the regulation of de novo lipogenesis. Similar results have been reported by Park et al. (2016), who observed that treating HepG2 cells with doses ranging from 30 to 100 μg/mL of a fucoidan extract obtained from the brown algae *Fucus vesiculosus* significantly reduced the gene expression of de novo lipogenesis markers.

As evidenced by studies using rodent models of MAFLD, algae-derived polysaccharides have consistently demonstrated beneficial effects in the prevention and treatment of hepatic steatosis (Chu et al., 2022; Kawauchi et al., 2019; Ren et al., 2019, 2017, 2018a,2018b; Liu et al., 2021; Zhang et al., 2021; Heeba and Morsy, 2015; Sharma and Baskaran, 2021). Regarding studies analyzing their potential for the prevention of MAFLD, Chu et al. (2022) reported that C57BL/6J mice (7 weeks old) fed a high-fat diet (60% of kcal from fat) and supplemented with different doses of an arabinogalactan sulfate-rich extract obtained from *Chaetomorpha linum* (50 or 150 mg/kg) for 10 weeks experienced reduced body weight. Moreover, the higher dose of this green seaweed extract was shown to reduce liver lipid accumulation, as evidenced by histological and oil-red staining analyses. Similar results were reported in another study involving 6-week-old mice fed a methionine- and choline-deficient diet supplemented with 5% sodium-alginate polysaccharide for 8 weeks. This study observed reductions in steatosis and inflammation levels, as well as in the NAS score (MAFLD activity score), which represents the sum of scores for steatosis, lobular inflammation, and ballooning in the treated animals (Kawauchi et al., 2019). Another study reported that supplementing 5-week-old mice fed a high-fat diet with 2.5% or 5% w/w of a polysaccharide extract obtained from the brown macroalgae *Laminaria japonica,* rich in alginate and fucoidan, for 8 weeks significantly reduced serum and hepatic triglycerides levels. While both doses reduced ALT serum levels, only the higher dose reduced AST levels. To understand the mechanisms underlying the beneficial effects of L. *japonica* extract on hepatic lipid deposition, the authors analyzed related markers and found decreased gene expression of de novo lipogenesis markers in animals treated with the higher dose (Zhang et al., 2021). Regarding the effect of seaweed polysaccharide extracts on the treatment of liver steatosis, Sharma and Baskaran (2021)) reported that supplementing 29-week-old diet-induced obese mice with 2% of a fucoidan and laminaran-rich polysaccharide extract obtained from *Padina tetrastromatica* for 5 weeks significantly reduced body and liver weight. Although no differences were observed in serum triglyceride content, a reduction in hepatic lipid accumulation was noted in animals treated with the extract. These changes were accompanied by a reduction in liver oxidative stress markers and an improvement in antioxidant defense (Sharma and Baskaran, 2021).

Concerning the beneficial effects of algae polysaccharides in rat MAFLD models, Ren et al. (2017, 2018a, 2018b) reported that supplementing female Sprague-Dawley rats (8 weeks old) with either 200 or 400 mg/kg of sulfated-ester polysaccharides extracted from *Enteromorpha prolifera* for 5 weeks significantly reduced the liver index, as well as liver fat accumulation (Ren et al., 2017). Moreover, these authors confirmed that this green algae extract reduced serum transaminase levels (ALT and AST) and de novo lipogenesis (Ren et al., 2018b). Indeed, Ren et al. (2018a) suggested that the beneficial effects exerted by *E. prolifera* polysaccharide-rich extracts could derive from their ability to increase the endogenous production of hydrogen sulfide (H_2S), which has been reported to play a significant role in the regulation of hepatic lipogenesis and β-oxidation [58,59]. In addition, supplementing male Wistar rats (6 weeks old) fed a high-fat diet with 100 mg/kg of fucoidan fractions of *Saccharina sculpera* for 8 weeks evidenced the potential antioxidant effects of these bioactive compounds (Ren et al., 2019). Moreover, the authors reported that *S. sculpera* polysaccharides reduced hepatic lipid accumulation, mainly by decreasing de novo lipogenesis and boosting fatty-acid oxidation. As for evidence regarding the effects of polysaccharides obtained from red seaweed, Liu et al. (2021) treated male Wistar rats (6 weeks old) fed a high-fat diet with 5% *Gelidium amansii* polysaccharide-rich extracts for 8 weeks and reported a lowering effect on liver weight. This reduction was accompanied by a diminished content of hepatic and serum triglycerides, as well as serum transaminase levels. The mechanisms of action are similar to those reported by other authors, as *G. amansii* was shown to reduce hepatic lipogenesis. Interestingly, a reduced food intake without changes in feed efficiency ratio and higher glucagon-like peptide 1 levels were observed in animals treated with the macroalgae extract, suggesting a potential role of this seaweed in appetite regulation (Liu et al., 2021). Moreover, the authors reported that the observed increase in excreted triglycerides and bile acid content could be derived from the higher fiber content of macroalgae and its effect on the regulation of bile acid synthesis. Apart from studies analyzing the preventive potential of algae polysaccharides in MAFLD models, their effects on steatosis treatment have also been addressed. In this regard, adult male Wistar rats fed a high-fat diet for 12 weeks and treated thrice per week in the last 4 weeks of the experimental period with 100 mg/kg of fucoidan extract obtained from brown algae showed decreased liver index, serum transaminase and triglycerides levels, and lower steatosis, inflammation, and fibrosis scores. These changes were accompanied by a reduction in key markers related to oxidative stress and inflammation in the treated group (Heeba and Morsy, 2015).

As for the evidence from clinical trials, to our knowledge, no studies have been found that were performed on humans treated with macroalgae polysaccharide extracts. Thus further research is warranted on this topic to confirm the beneficial effects observed in the in vitro and in vivo reports.

References

Abidov, M., Ramazanov, Z., Seifulla, R., Grachev, S., 2010. The effects of Xanthigen ™ in the weight management of obese premenopausal women with non-alcoholic fatty liver disease and normal liver fat. Diabetes Obes. Metab. 1462-8902. 12 (1), 72–81. doi:10.1111/j.1463-1326.2009.01132.x.

Adnan, M.L., Pramaningtyas, M.D., 2022. The effects of seaweed supplementation consumption for improvement of liver injury in patients with non-alcoholic fatty liver disease: a systematic review. Clin. Exp. Hepatol. 24498238. 8 (3), 171–177. Available from: https://www.termedia.pl/doi/10.5114/ceh.2022.118275. doi:10.5114/ceh.2022.118275.

Bermont, A., Cohen, D.L., Richter, V., Hadar, R., Tam, J., Wandel, A., Brawer, S., Grundman, O., Shirin, H., 2025. FucoVital® inhibits fatty acid accumulation in liver cells in vitro and reduces triglycerides levels in patients with NAFLD. PharmaNutrition. 22134344. 31, 100429. doi:10.1016/j.phanu.2025.100429.

Biris-Dorhoi, E.-S., Michiu, D., Pop, C.R., Rotar, A.M., Tofana, M., Pop, O.L., Socaci, S.A., Farcas, A.C., 2020. Macroalgae—A Sustainable Source of Chemical Compounds with Biological Activities. Nutrients. 2072-6643. 12 (10), 3085. doi:10.3390/nu12103085.

Bouzenad, N., Ammouchi, N., Chaib, N., Messaoudi, M., Bousabaa, W., Bensouici, C., Sawicka, B., Atanassova, M., Ahmad, S.F., Zahnit, W., 2024. Exploring bioactive components and assessing antioxidant and antibacterial activities in five seaweed extracts from the northeastern coast of Algeria. Mar. Drugs. 1660-3397. 22 (6), 273. doi:10.3390/md22060273.

Byun, K.-A., Seyeon, O., Son, M., Park, C.-H., Son, K.H., Byun, K., 2021. Dieckol decreases caloric intake and attenuates nonalcoholic fatty liver disease and hepatic lymphatic vessel dysfunction in high-fat-diet-fed mice. Mar. Drugs. 1660-3397. 19 (9), 495. doi:10.3390/md19090495.

Cha, S.-H., Hwang, Y., Heo, S.-J., Jun, H.-S., 2020. Diphlorethohydroxycarmalol attenuates palmitate-induced hepatic lipogenesis and inflammation. Mar. Drugs. 1660-3397. 18 (9), 475. doi:10.3390/md18090475.

Chan, K.E., Koh, T.J.L., Tang, A.S.P., Quek, J., Yong, J.N., Tay, P., Tan, D.J.H., Lim, W.H., Lin, S.Y., Huang, D., Chan, M., Khoo, C.M., Chew, N.W.S., Kaewdech, A., Chamroonkul, N., Dan, Y.Y., Noureddin, M., Muthiah, M., Eslam, M., Ng, C.H., 2022. Global prevalence and clinical characteristics of metabolic-associated fatty liver disease: a meta-analysis and systematic review of 10 739 607 individuals. J. Clin. Endocrinol. Metab. 19457197. 107 (9), 2691–2700. Available from: https://academic.oup.com/jcem. doi:10.1210/clinem/dgac321.

Chang, Y.H., Chen, Y.L., Huang, W.C., Liou, C.J., 2018. Fucoxanthin attenuates fatty acid-induced lipid accumulation in FL83B hepatocytes through regulated Sirt1/AMPK signaling pathway. Biochem. Biophys. Res. Commun. 10902104. 495 (1), 197–203. Available from: http://www.sciencedirect.com/science/journal/0006291X. doi:10.1016/j.bbrc.2017.11.022.

Chen, L., Liu, R., Xin, H., Pei, S., Duo, L., 2021. Effects of brown seaweed polyphenols, a class of phlorotannins, on metabolic disorders via regulation of fat function. Food Funct. 2042-6496. 12 (6), 2378–2388. doi:10.1039/d0fo02886j.

Chu, X., Zhou, Y., Zhang, S., Liu, S., Guoyun, L., Xin, Y., 2022. Chaetomorpha linum polysaccharides alleviate NAFLD in mice by enhancing the PPARα/CPT-1/MCAD signaling. Lipids Health Dis. 1476-511X. 21 (1). doi:10.1186/s12944-022-01730-x.

Ding, Y., Wang, L., SeungTae, I., Hwang, O., Kim, H.-S., Kang, M.-C., Lee, S.-H., 2019. Anti-obesity effect of diphlorethohydroxycarmalol isolated from brown alga ishige okamurae in high-fat diet-induced obese mice. Mar. Drugs. 1660-3397. 17 (11), 637. doi:10.3390/md17110637.

Dionysopoulos, G., Kalopitas, G., Vadarlis, A., Bakaloudi, D.R., Gkiourtzis, N., Karanika, E., Tsekitsidi, E., Chourdakis, M., 2023. Can omega-3 fatty acids be beneficial in pediatric NAFLD? A systematic review and meta-analysis. Crit. Rev. Food Sci. Nutr. 1040-8398. 63 (27), 8545–8553. doi:10.1080/10408398.2022.2062589.

Dong, H., Dong, S., Hansen, P.E., Stagos, D., Lin, X., Liu, M., 2020. Progress of bromophenols in marine algae from 2011 to 2020: structure, bioactivities, and applications. Mar. Drugs. 1660-3397. 18 (8), 411. doi:10.3390/md18080411.

Fouad, Y., Alboraie, M., Gamal Shiha, 2024. Epidemiology and diagnosis of metabolic dysfunction-associated fatty liver disease. Hepatol. Int. 1936-0533. 18 (S2), 827–833. doi:10.1007/s12072-024-10704-3.

Gofton, C., Upendran, Y., Zheng, M.-H., George, J., 2023. MAFLD: how is it different from NAFLD? Clin. Mol. Hepatol. 2287-2728. 29 (Suppl), S17. doi:10.3350/cmh.2022.0367.

Gomes, L., Monteiro, P., Cotas, J., Gonçalves, A.M.M., Fernandes, C., Gonçalves, T., Pereira, L., 2022. Seaweeds' pigments and phenolic compounds with antimicrobial potential. Biomol. Concepts. 1868503X. 13 (1), 89–102. Available from: http://www.degruyter.com/view/j/bmc.2013.4.issue-1/issue-files/bmc.2013.4.issue-1.xml. doi:10.1515/bmc-2022-0003.

González-Arceo, M., Trepiana, J., Aguirre, L., Ibarruri, J., Martínez-Sanz, M., Cebrián, M., Recio, I., Portillo, M.P., Gó mez-Zorita, S., 2023. Anti-steatotic effects of chlorella vulgaris, nannochloropsis gaditana and gracilaria vermiculophylla algae extracts in AML-12 hepatocytes. Nutrients. 2072-6643. 15 (8), 1960. doi:10.3390/nu15081960.

Gotoh, N., Nagao, K., Ishida, H., Nakamitsu, K., Yoshinaga, K., Nagai, T., Beppu, F., Yoshinaga-Kiriake, A., Watanabe, H., Yanagita, T., 2018. Metabolism of natural highly unsaturated fatty acid, tetracosahexaenoic acid (24:6n-3), in C57BL/KsJ-*db*/*db* mice. J. Oleo Sci. 1345-8957. 67 (12), 1597–1607. doi:10.5650/jos.ess18167.

Grander, C., Grabherr, F., Tilg, H., 2023. Non-alcoholic fatty liver disease: pathophysiological concepts and treatment options. Cardiovasc. Res. 0008-6363. 119 (9), 1787–1798. doi:10.1093/cvr/cvad095.

Guo, X.F., Yang, B., Tang, J., Li, D., 2018. Fatty acid and non-alcoholic fatty liver disease: meta-analyses of case-control and randomized controlled trials. Clin. Nutr. 15321983. 37 (1), 113–122. Available from: http://www.elsevier-international.com/journals/clnu/. doi:10.1016/j.clnu.2017.01.003.

Heeba, G.H., Morsy, M.A., 2015. Fucoidan ameliorates steatohepatitis and insulin resistance by suppressing oxidative stress and inflammatory cytokines in experimental non-alcoholic fatty liver disease. Environ. Toxicol. Pharmacol. 18727077. 40 (3), 907–914. Available from: www.elsevier.com/locate/etap.. doi:10.1016/j.etap.2015.10.003.

Huamán-Castilla, N.L., Allcca-Alca, E.E., Nina, F.H., León-Calvo, N.C., Vilca, F.Z., Chura, Y.L.V., 2024. Pressurized liquid extraction of antioxidant and α-amylase-inhibitory compounds from red seaweed using water–ethanol mixtures. Molecules. 14203049. 29 (21). Available from: http://www.mdpi.com/journal/molecules. doi:10.3390/molecules29215018.

Hyeyoon, E., Jeon, Y.-J., Lee, M., Lim, Y., 2015. Brown alga ecklonia cava polyphenol extract ameliorates hepatic lipogenesis, oxidative stress, and inflammation by activation of ampk and sirt1 in high-fat diet-induced obese mice. J. Agric. Food Chem. 0021-8561. 63 (1), 349–359. doi:10.1021/jf502830b.

Jiena, Y., Zheng, J., Tian, X., Baogui, X., Yuan, F., Wang, B., Yang, Z., Huang, F., 2022. Fucoxanthin attenuates free fatty acid-induced nonalcoholic fatty liver disease by regulating lipid metabolism/oxidative stress/inflammation via the AMPK/Nrf2/TLR4 signaling pathway. Mar. Drugs. 1660-3397. 20 (4), 225. doi:10.3390/md20040225.

Kawauchi, S., Horibe, S., Sasaki, N., Tanahashi, T., Mizuno, S., Hamaguchi, T., Rikitake, Y., 2019. Inhibitory effects of sodium alginate on hepatic steatosis in mice induced by a methionine- and choline-deficient diet. Mar. Drugs. 1660-3397. 17 (2), 104. doi:10.3390/md17020104.

Łagowska, K., Jurgoński, A., Mori, M., Yamori, Y., Murakami, S., Ito, T., Toda, T., Pieczyńska-Zając, J.M., Bajerska, J., 2025. Effects of dietary seaweed on obesity-related metabolic status: a systematic review and meta-analysis of randomized controlled trials. Nutr. Rev. 0029-6643. 83 (2), e116. doi:10.1093/nutrit/nuae042.

Liu, S.H., Hung, H.I., Yang, T.H., Pan, C.L., Chiang, M.T., 2021. Polysaccharide-rich red algae (Gelidium amansii) hot-water extracts ameliorate the altered plasma cholesterol and hepatic lipid homeostasis in high-fat diet-fed rats. J. Food Drug Anal. 10219498. 29 (1), 46–56. Available from: https://www.jfda-online.com/cgi/viewcontent.cgi?article=1181&context=journal. doi:10.38212/2224-6614.1181.

Manne, V., Handa, P., Kowdley, K.V., 2018. Pathophysiology of nonalcoholic fatty liver disease/nonalcoholic steatohepatitis. Clin. Liver Dis. 15578224. 22 (1), 23–37. Available from: http://www.elsevier.com/inca/publications/store/6/2/3/2/8/8/index.htt. doi:10.1016/j.cld.2017.08.007.

Mendes, M., Navalho, S., Ferreira, A., Paulino, C., Figueiredo, D., Silva, D., Gao, F., Gama, F., Bombo, G., Jacinto, R., Aveiro, S., Schulze, P., Gonçalves, A.T., Pereira, H., Gouveia, L., Patarra, R., Abreu, M.H., Silva, J., Navalho, J., Varela, J., Speranza, L., 2022. Algae as food in Europe: an overview of species diversity and their application. Foods. 2304-8158. 11 (13), 1871. doi:10.3390/foods11131871.

O'Connor, J., Garcia-Vaquero, M., Meaney, S., Tiwari, B.K., 2022. Bioactive peptides from algae: traditional and novel generation strategies, structure-function relationships, and bioinformatics as predictive tools for bioactivity. Mar. Drugs. 1660-3397. 20 (5), 317. doi:10.3390/md20050317.

Ouahabi, S., Daoudi, N.E., Loukili, E.H., Asmae, H., Merzouki, M., Bnouham, M., Challioui, A., Hammouti, B., Fauconnier, M.-L., Rhazi, L., Gotor, A.A., Depeint, F., Ramdani, M., 2024. Investigation into the phytochemical composition, antioxidant properties, and in-vitro anti-diabetic efficacy of ulva lactuca extracts. Mar. Drugs. 1660-3397. 22 (6), 240. doi:10.3390/md22060240.

Park, E.Y., Choi, H., Yoon, J.Y., Lee, I.Y., Seo, Y., Moon, H.S., Hwang, J.H., Jun, H.S., 2015. Polyphenol-rich fraction of Ecklonia cava improves nonalcoholic fatty liver disease in high fat diet-fed mice. Mar. Drugs. 16603397. 13 (11), 6866–6883. Available from: http://www.mdpi.com/1660-3397/13/11/6866/pdf. doi:10.3390/md13116866.

Park, J., Yeom, M., Hahm, D.H., 2016. Fucoidan improves serum lipid levels and atherosclerosis through hepatic SREBP-2-mediated regulation. J. Pharmacol. Sci. 13478648. 131 (2), 84–92. Available from: http://www.jstage.jst.go.jp/browse/. doi:10.1016/j.jphs.2016.03.007.

Peñalver, R., Lorenzo, J.M., Ros, G., Amarowicz, R., Pateiro, M., Nieto, G., 2020. Seaweeds as a functional ingredient for a healthy diet. Mar. Drugs. 1660-3397. 18 (6), 301. doi:10.3390/md18060301.

Pourteymour, S., Drevon, C.A., Dalen, K.T., Norheim, F.A., 2023. Mechanisms behind NAFLD: a system genetics perspective. Curr. Atheroscler. Rep. 15346242. 25 (11), 869–878. Available from: https://www.springer.com/journal/11883. doi:10.1007/s11883-023-01158-3.

Raja, R., Hemaiswarya, S., Arunkumar, K., Carvalho, I.S., 2021. Algae for food: cultivation, processing and nutritional benefits algae for food: cultivation, processing and nutritional benefits. CRC Press, pp. 1–278. Available from https://www.taylorfrancis.com/books/9781000457100. doi:10.1201/9781003165941.

Rashed, Z.E., Lupidi, G., Grasselli, E., Canesi, L., Khalifeh, H., Demori, I., 2021. Antioxidant and antisteatotic activities of fucoidan fractions from marine and terrestrial sources. Molecules. 1420-3049. 26 (15), 4467. doi:10.3390/molecules26154467.

Ren, D., Wang, Q., Yang, Y., Yue, H., Song, Y., Yunhai, H., Liu, S., Long, W., 2019. Hypolipidemic effects of fucoidan fractions from Saccharina sculpera (Laminariales, Phaeophyceae). Int. J. Biol. Macromol. 01418130. 140, 188–195. doi:10.1016/j.ijbiomac.2019.08.002.

Ren, R., Gong, J., Zhao, Y., Zhuang, X., Yin, Y., 2017. Wenting Lin, Sulfated polysaccharides from Enteromorpha prolifera suppress SREBP-2 and HMG-CoA reductase expression and attenuate non-alcoholic fatty liver disease induced by a high-fat diet. Food Funct. 2042-6496. 8 (5), 1899–1904. doi:10.1039/c7fo00103g.

Ren, R., Gong, J., Zhao, Y., Zhuang, X., Yin, Y., Huang, F., Lin, W., 2018b. Sulfated polysaccharide from Enteromorpha prolifera suppresses SREBP-1c and ACC expression to lower serum triglycerides in high-fat diet-induced hyperlipidaemic rats. J. Funct. Foods. 17564646. 40, 722–728. doi:10.1016/j.jff.2017.12.010.

Ren, R., Yang, Z., Zhao, A., Huang, Y., Lin, S., Gong, J., Chen, J., Zhu, P., Huang, F., Lin, W., 2018a. Sulfated polysaccharide from Enteromorpha prolifera increases hydrogen sulfide production and attenuates non-alcoholic fatty liver disease in high-fat diet rats. Food Funct. 2042-6496. 9 (8), 4376–4383. doi:10.1039/c8fo00518d.

Shannon, E., Conlon, M., Hayes, M., 2021. Seaweed components as potential modulators of the gut microbiota. Mar. Drugs. 1660-3397. 19 (7), 358. doi:10.3390/md19070358.

Sharma, P.P., Baskaran, V., 2021. Polysaccharide (laminaran and fucoidan), fucoxanthin and lipids as functional components from brown algae (Padina tetrastromatica) modulates adipogenesis and thermogenesis in diet-induced obesity in C57BL6 mice. Algal Res. 54, 102187.

Singh, A., Pal, B., Singh, K.S., 2024. Carbohydrate and pigment composition of macroalgae in a kelp-dominated Arctic fjord. Reg. Stud. Mar. Sci. 23524855. 77, 103644. doi:10.1016/j.rsma.2024.103644.

Stiger-Pouvreau, V., Bourgougnon, N., Deslandes, E., 2016. Carbohydrates from seaweeds. Nabavi, S.M., Ed., Seaweed in Health and Disease Prevention. Elsevier Inc., 223–274. Available from: http://www.sciencedirect.com/science/book/9780128027721. 10.1016/B978-0-12-802772-1.00008-7.

Takatani, N., Kono, Y., Beppu, F., Okamatsu-Ogura, Y., Yamano, Y., Miyashita, K., Hosokawa, M., 2020. Fucoxanthin inhibits hepatic oxidative stress, inflammation, and fibrosis in diet-induced nonalcoholic steatohepatitis model mice. Biochem. Biophys. Res. Commun. 0006291X. 528 (2), 305–310. doi:10.1016/j.bbrc.2020.05.050.

Teng, M.L.P., Cheng Han, N., Huang, D.Q., Chan, K.E., Tan, D.J.H., Lim, W.H., Yang, J.D., Tan, E., Muthiah, M.D., 2023. Global incidence and prevalence of nonalcoholic fatty liver disease. Clin. Mol. Hepatol. 2287-2728. 29 (Suppl), S32. doi:10.3350/cmh.2022.0365.

Wu, J.Y., Tso, R., Teo, H.S., Haldar, S., 2023a. The utility of algae as sources of high value nutritional ingredients, particularly for alternative/complementary proteins to improve human health. Front. Nutr. 2296861X. 10. Available from: journal.frontiersin.org/journal/nutrition doi:10.3389/fnut.2023.1277343.

Wu, Y., Jin, X., Zhang, Y., Liu, J., Mingjiang, W., Tong, H., 2023b. Bioactive compounds from brown algae alleviate nonalcoholic fatty liver disease: an extensive review. J. Agric. Food Chem. 0021-8561. 71 (4), 1771–1787. doi:10.1021/acs.jafc.2c06578.

Yang, S., Jinhai, L., Yan, L., Wu, Y., Zhang, L., Boyang, L., Tong, H., Lin, X., 2024. Molecular mechanisms of fucoxanthin in alleviating lipid deposition in metabolic associated fatty liver disease. J. Agric. Food Chem. 0021-8561. 72 (18), 10391–10405. doi:10.1021/acs.jafc.4c00590.

Yang, Z., Wang, H., Liu, N., Zhao, K., Sheng, Y., Pang, H., Shao, K., Zhang, M., Shangyong, L., Ningning, H., 2022. Algal polysaccharides and derivatives as potential therapeutics for obesity and related metabolic diseases. Food Funct. 2042-6496. 13 (22), 11387–11409. doi:10.1039/d2fo02185d.

Younossi, Z.M., Marchesini, G., Pinto-Cortez, H., Petta, S., 2019. Epidemiology of nonalcoholic fatty liver disease and nonalcoholic steatohepatitis: implications for liver transplantation. Transplantation. 00411337. 103 (1), 22–27. Available from: journals.lww.com/transplantjournal/pages/default.aspx. doi:10.1097/TP.0000000000002484.

Zhang, Y., Yang, L., Zhao, N., Hong, Z., Cai, B., Qingqing, L., Yang, T., Shi, L., Jianlin, H., 2021. Soluble polysaccharide derived from laminaria japonica attenuates obesity-related nonalcoholic fatty liver disease associated with gut microbiota regulation. Mar. Drugs. 1660-3397. 19 (12), 699. doi:10.3390/md19120699.

Zhao, B., Cui, Y., Fan, X., Ping, Q., Liu, C., Zhou, X., Zhang, X., Brown, L., 2019. Anti-obesity effects of Spirulina platensis protein hydrolysate by modulating brain-liver axis in high-fat diet fed mice. PLoS One. 1932-6203. 14 (6), e0218543. doi:10.1371/journal.pone.0218543.

Zhou, L., Kunning, L., Duan, X., Hill, D., Barrow, C., Dunshea, F., Martin, G., Suleria, H., 2022. Bioactive compounds in microalgae and their potential health benefits. Food Biosci. 22124292. 49, 101932. doi:10.1016/j.fbio.2022.101932.

Index

Note: Page numbers followed by "*f*" indicate figures, "*t*" indicate tables.

P

R

S

Printed and bound by CPI Group (UK) Ltd, Croydon, CR0 4YY

14/09/2026

02216409-0018